# Human Genetics

## Concepts and Applications

Ricki Lewis
*The University at Albany*
*CareNet Medical Group,*
*Schenectady, New York*

**Fifth Edition**

Boston Burr Ridge, IL Dubuque, IA Madison, WI New York
San Francisco St. Louis Bangkok Bogotá Caracas Kuala Lumpur
Lisbon London Madrid Mexico City Milan Montreal New Delhi
Santiago Seoul Singapore Sydney Taipei Toronto

**McGraw-Hill Higher Education**

*A Division of The* **McGraw-Hill** *Companies*

HUMAN GENETICS: CONCEPTS AND APPLICATIONS
FIFTH EDITION

Published by McGraw-Hill, a business unit of The McGraw-Hill Companies, Inc., 1221 Avenue of the Americas, New York, NY 10020. 

Some ancillaries, including electronic and print components, may not be available to customers outside the United States.

 This book is printed on recycled, acid-free paper containing 10% postconsumer waste.

International 1 2 3 4 5 6 7 8 9 0 VNH/VNH 0 9 8 7 6 5 4 3 2
Domestic 1 2 3 4 5 6 7 8 9 0 VNH/VNH 0 9 8 7 6 5 4 3 2

ISBN 0–07–246276–0 (hardcover)
ISBN 0–07–246268–X (softcover)
ISBN 0–07–119849–0 (ISE)

Publisher: *Martin J. Lange*
Senior sponsoring editor: *Patrick E. Reidy*
Senior developmental editor: *Deborah Allen*
Senior development manager: *Kristine Tibbetts*
Marketing manager: *Tami Petsche*
Senior project manager: *Joyce M. Berendes*
Production supervisor: *Sherry L. Kane*
Coordinator of freelance design: *Michelle D. Whitaker*
Cover designer: *Maureen McCutcheon*
Cover images: *Child's Hand Against Adult Hand: Werner Bokelberg/Getty Images, Inc.; DNA Strand: Photo Researchers, Inc.*
Lead photo research coordinator: *Carrie K. Burger*
Photo research: *Toni Michaels/PhotoFind LLC*
Supplement producer: *Brenda A. Ernzen*
Media project manager: *Jodi K. Banowetz*
Media technology associate producer: *Janna Martin*
Compositor: *Precision Graphics*
Typeface: *10/12 Minion*
Printer: *Von Hoffman Press, Inc.*

The credits section for this book begins on page C-1 and is considered an extension of the copyright page.

**Library of Congress Cataloging-in-Publication Data**

Lewis, Ricki.
Human genetics : concepts and applications / Ricki Lewis. — 5th ed.
p. cm.
Includes index.
ISBN 0–07–246268–X (softcover : alk. paper) — ISBN 0–07–119849–0 (ISE alk. paper)
1. Human genetics. I. Title.

QH431 .L41855 2003
599.93′5—dc21 2002016627
CIP

INTERNATIONAL EDITION ISBN 0–07–119849–0


www.mhhe.com

# About the Author

Ricki Lewis has built a multifaceted career around communicating the excitement of life science, especially genetics and biotechnology. She earned her Ph.D. in genetics in 1980 from Indiana University, working with homeotic mutations in *Drosophila melanogaster.*

Ricki is an author of *Life,* an introductory biology text; *Human Genetics: Concepts and Applications;* co-author of two human anatomy and physiology textbooks; and author of *Discovery: Windows on the Life Sciences,* an essay collection about research and the nature of scientific investigation. As a Contributing Editor to *The Scientist,* a magazine read by scientists worldwide, she writes frequently on the latest research and news in biotechnology. Since 1980, Ricki has published more than 3,000 articles in a variety of magazines, including a cover story on DNA fingerprinting in *Discover* and book reviews for *The New York Times.* Ricki participates in Science Forum, a monthly call-in science program on public radio, and is a frequent invited speaker. She is an adjunct professor at Miami University and the University at Albany, where she has taught a variety of life science courses, and also taught at Empire State College and several community colleges. She brought science experiments to grade school classrooms for three years as part of a traveling science museum, for which she obtained a Howard Hughes Medical Institute grant. Ricki has been a genetic counselor for a large private medical practice in Schenectady, NY, since 1984, where she helps people make decisions about reproductive choices.

Ricki lives in upstate New York with chemist husband Larry, three daughters, and various cats and guinea pigs.

*dedicated to Shirley Epstein Aaronson, who encouraged an inquisitive child to become a scientist*

# Brief Contents

# List of Boxes

## Readings



## In Their Own Words



## Bioethics: Choices for the Future

# Contents

# Part Two



## Chapter 4



## Chapter 5



## Chapter 6

## Chapter 7

### Multifactorial Traits 137



## Chapter 8

### The Genetics of Behavior 155



# Part Three



## Chapter 9

### DNA Structure and Replication 171



## Chapter 10

### Gene Action and Expression 189



## Chapter 11

### Gene Mutation 211

## Chapter 20



## Chapter 21



## Chapter 22

# Visual Preview

The next few pages show you the tools found throughout the text to provide a clear framework for learning the fundamental concepts of human genetics.

## Bioethics: Choices for the Future

Discussions of difficult issues illuminate the complexities of applying genetic principles to everyday life.

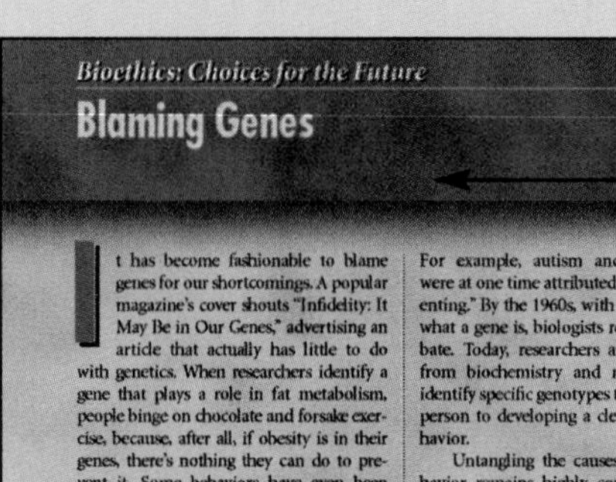

*Bioethics: Choices for the Future*

**Blaming Genes**

It has become fashionable to blame genes for our shortcomings. A popular magazine's cover shouts "Infidelity: It May Be in Our Genes," advertising an article that actually has little to do with genetics. When researchers identify a gene that plays a role in fat metabolism, people binge on chocolate and forsake exercise, because, after all, if obesity is in their genes, there's nothing they can do to prevent it. Some behaviors have even been blamed on a poorly-defined gene for "thrill-seeking" (figure 1).

Behavioral genetics has a checkered past. Early in the 20th century, it was part of eugenics— the idea that humans can improve a population's collection of genes, or gene pool. The horrific experiments and exterminations the Nazis performed in the

For example, autism and
were at one time attributed
enting." By the 1960s, with
what a gene is, biologists re
bate. Today, researchers a
from biochemistry and n
identify specific genotypes t
person to developing a de
havior.

Untangling the causes
havior remains highly co
scientific conference to ex
pects of violence was cancel
psychiatrist objected that "b
ics is the same old stuff in
another way for a violent,
say people's problems are
because they carry 'bad' ge
searchers on the trail of p

## Chapter Opener

An outline of major topics with an introductory narrative prepares you for what you will learn in the chapter.

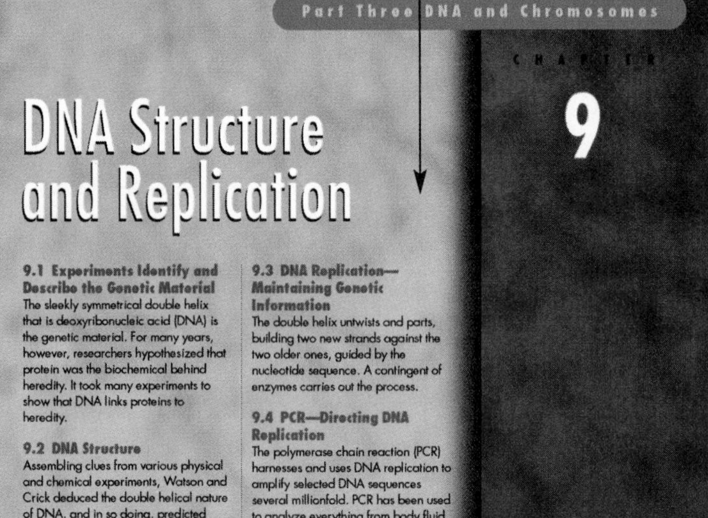

Part Three DNA and Chromosomes

CHAPTER 9

**DNA Structure and Replication**

**9.1 Experiments Identify and Describe the Genetic Material**
The sleekly symmetrical double helix that is deoxyribonucleic acid (DNA) is the genetic material. For many years, however, researchers hypothesized that protein was the biochemical behind heredity. It took many experiments to show that DNA links proteins to heredity.

**9.2 DNA Structure**
Assembling clues from various physical and chemical experiments, Watson and Crick deduced the double helical nature of DNA, and in so doing, predicted

**9.3 DNA Replication—Maintaining Genetic Information**
The double helix untwists and parts, building two new strands against the two older ones, guided by the nucleotide sequence. A contingent of enzymes carries out the process.

**9.4 PCR—Directing DNA Replication**
The polymerase chain reaction (PCR) harnesses and uses DNA replication to amplify selected DNA sequences several millionfold. PCR has been used to analyze everything from body fluid

## In-Chapter Study Aids

In addition to numerous tables and figures, you will find
**Key Terms** printed in bold type and included in a glossary at the end of the text.
**Technology Timelines** that trace the developments and discoveries leading to today's technologies.
**CD ROM Icons** that identify topics supported by *Genetics: From Genes to Genomes.*

bacteria divide, they yield many copies, or clones, of the foreign DNA and produce many copies of the protein the foreign DNA specifies. In the 1980s, researchers began to apply recombinant DNA technology to multicellular organisms, producing transgenic plants and animals. Researchers add foreign DNA at the one-cell stage (a gamete or fertilized ovum). The transgenic organism that develops from the original altered cell carries the genetic change in every cell. Yet another biotechnology, called **gene targeting,** adds precision to transgenic technology. Gene targeting "knocks out" or "knocks in" the gene of interest at a particular chromosomal locus, where it trades places with an existing gene.

The ability to combine genes from different types of organisms has raised legal questions—is a recombinant or transgenic organism an invention, deserving of patent protection? By definition, to earn a patent an invention must be new, useful, and not obvious (see Technology Timeline on page 000).

Patent law has had to evolve in parallel to the rise of biotechnology. Early on, DNA sequences could be patented. In the mid-1990s, however, when the National Institutes of Health and biotech companies began seeking patent protection for thousands of pieces of protein-encoding DNA sequences, called expressed sequences tags (ESTs), the government's Patent and Trademark Office began to tighten the requirement for utility. Today, a DNA sequence alone is not patentable. It must be useful as a tool for research or as a novel and improved diagnostic test.

Despite the increasing stringency of patent requirements, problems still arise. A biotechnology company in the United States, for example, holds a patent on the BRCA1 gene that includes any diagnostic tests based on the gene sequence. That company's tests, however, do not cover all mutations in the gene. A French physician working with a

**Technology TIMELINE**

Patenting Life and Genes

| | |
|---|---|
| 1790 | U.S. patent act is enacted. An invention must be new, useful, and no |
| | earn a patent. |
| 1873 | Louis Pasteur is awarded first patent on a life form for yeast used in processes. |
| 1930 | New plant variants can be patented. |
| 1980 | First patent is awarded on a genetically engineered organism, a bact four plasmids (DNA rings) that enable it to metabolize components The plasmids are naturally occurring, but do not all occur naturally ulated bacteria. |
| 1988 | First patent is awarded for a transgenic organism, a mouse that ma man protein in its milk. Harvard University granted patent for "On transgenic for cancer. |
| 1992 | Biotechnology company is awarded a broad patent covering all for cotton. Groups concerned that this will limit the rights of subsisten test the patent several times. |
| 1996–1999 | Companies patent partial gene sequences and certain disease-causi basis for developing specific medical tests. |
| 2000 | With gene and genome discoveries pouring into the Patent and Tra requirements for showing utility of a DNA sequence are tightened. |

is redundancy. For the same gene, it is possible to patent:

- Genomic DNA (the protein-encoding sequence as well as noncoding regions)
- expressed sequence tags
- cDNA (only the protein-encoding part of a gene)
- mutations
- SNPs

A researcher or company wanting to develop a tool or test based on a protein might infringe upon five different patents, based on essentially the same information. Now,

sulting recombinant DN
led to transgenic technol
geting. This chapter consi

**18.2 Recombina Technology**

In February 1975, 140 m
convened at Asilomar, a s
center on California's M
to discuss the safety and
new type of experiment.
found a simple way to co
two species and were co
periments requiring the

## Focus on Concepts

**Numbered Headings** identify each major topic and are directly related to the chapter introduction and the chapter summary.
**Key Concepts** are summarized at the end of each major section.

remain to be seen—as is the case for any form of agriculture.

**KEY CONCEPTS**

Economic impacts of GM foods are difficult to predict. These products may displace existing ones, or may not be equitably distributed. Ecological effects can be modeled in greenhouses and field tests, but GM organisms can escape their containments. Industrial control of many aspects of agricultural biotechnology has contributed to the negative image.

**20.5 The Impact of Genomics**

If GM organisms can survive their negative image, genomics will provide researchers with many more traits to work with. However, plant genomics lags behind similar efforts in animals and microorganisms—only the model experimental plant

ated with an international co
the same goal and has made
tion freely available to researc
Investigators need not h
genome in hand, however,
many-gene approach of genom
another popular crop, the
National Science Foundatio
"potato functional genomic
which a nonprofit orga
Institute for Genomic Researc
fers DNA microarrays that h
pressed sequence tags (ESTs)
pieces of protein-encoding ge
spond to cDNAs reverse tra
the mRNAs present in a partic
Different potato DNA micro
spond to the different tiss
shoot, stem or tuber—just as
microarrays represent differe
pression in nerve or muscle, s
Traditional breeding of
been tricky, because the leaves
alkaloid compounds that must
edible varieties. Cultivars (c
eties) represent many years
maximize taste and texture,

This text is unparalleled in its practicality and sense of reality. You will read true stories based on the author's own experience as a scientist, genetic counselor, and journalist. She regularly interviews not only leading researchers but also people who suffer from genetic disorders.

## "In Their Own Words"

Personal interviews with real people provide a different view from the standard textbook descriptions, or essays written by researchers.

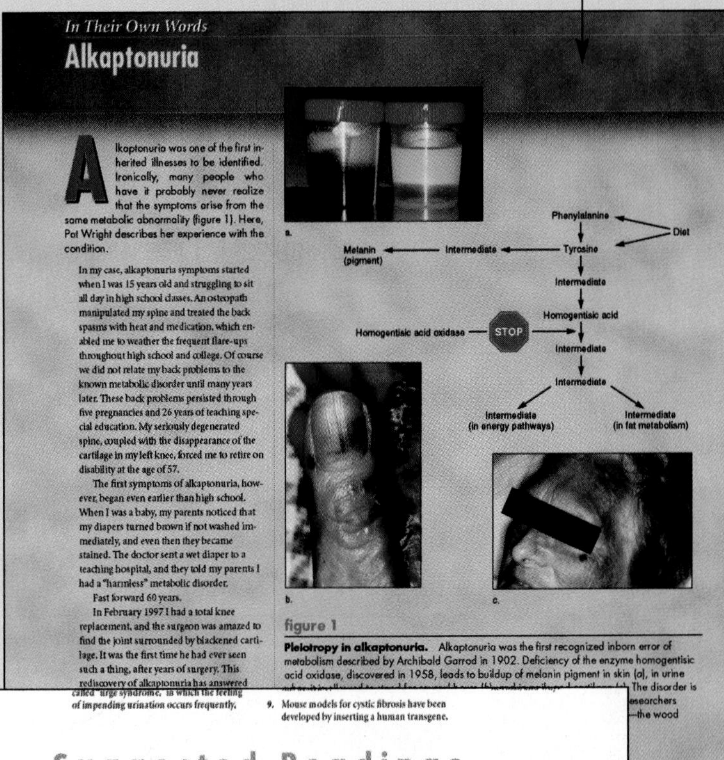

*In Their Own Words*

**Alkaptonuria**

Alkaptonuria was one of the first inherited illnesses to be identified. Ironically, many people who have it probably never realize that the symptoms arise from the same metabolic abnormality (figure 1). Here, Pat Wright describes her experience with the condition.

In my case, alkaptonuria symptoms started when I was 15 years old and struggling to sit all day in high school classes. An osteopath manipulated my spine and treated the back spasms with heat and medication, which enabled me to weather the frequent flare-ups throughout high school and college. Of course we did not relate my back problems to the known metabolic disorder until many years later. These back problems persisted through five pregnancies and 26 years of teaching special education. My seriously degenerated spine, coupled with the disappearance of the cartilage in my left knee, forced me to retire on disability at the age of 57.

The first symptoms of alkaptonuria, however, began even earlier than high school. When I was a baby, my parents noticed that my diapers turned brown if not washed immediately, and even then they became stained. The doctor sent a wet diaper to a teaching hospital, and they told my parents I had a "harmless" metabolic disorder.

Fast forward 60 years.

In February 1997 I had a total knee replacement, and the surgeon was amazed to find the joint surrounded by blackened cartilage. It was the first time he had ever seen such a thing, after years of surgery. This rediscovery of alkaptonuria has answered ...



**figure 1**

**Pleiotropy in alkaptonuria.** Alkaptonuria was the first recognized inborn error of metabolism described by Archibald Garrod in 1902. Deficiency of the enzyme homogentisic acid oxidase, discovered in 1958, leads to buildup of melanin pigment in skin (a), in urine ... The disorder is ... researchers ... the wood

... for a hip ... 15, 2001, I had a ... and I ... py maintenance ... ity and pain. ... go! The left hip is ...

Pat Wright

## End-of-Chapter Study Aids

**Chapter Summary** is presented in list format, organized by major topic.
**Review Questions** reinforce major concepts.
**Applied Questions** allow you to use genetics to solve real-life problems.
**Suggested Readings** cites the articles that were the sources of chapter information.
**On the Web** lists links that immerse you in the modern world of human genetics without ever leaving your computer. Includes **OMIM** references.

altered. The organism develops, including the change in each cell and passing it to the next generation. Heterozygotes for the transgene are then bred to yield homozygotes.

8. DNA is introduced into cells through **liposomes, electroporation, microinjection,** and **particle bombardment.**

10. **Gene targeting** uses the natural attraction of a DNA sequence for its complementary sequence, called **homologous recombination,** to swap one gene for another. It is more precise than transgenic technology, which inserts a foreign gene but does not direct it to a specific chromosomal site.

... to yield homozygotes.

12. Knockouts have the gene of interest inactivated. Knockins replace one gene with another allele with altered function.

13. Knockout mice with inactivated genes can model human disease. Sometimes, knockout mice reveal that a gene product is not vital to survival.

### Review Questions

1. Define each of the following terms:
   a. biotechnology
   b. recombinant DNA technology
   c. transgenic technology
   d. gene targeting
   e. homologous recombination
2. Describe the roles of each of the following tools in a biotechnology:
   a. restriction enzymes
   b. embryonic stem cells
   c. cloning vectors
3. How do researchers use antibiotics to select cells containing recombinant DNA?
4. List the components of an experiment to produce recombinant human insulin in *E. coli* cells.
5. Why would recombinant DNA technology be impossible if the genetic code was not universal?
6. Why must manipulations to create a transgenic organism take place at the single-cell stage?
7. Describe three ways to insert foreign DNA into cells.
8. Why isn't transgenic technology as precise as gene targeting?
9. How does Mendel's law of segregation for a monohybrid cross apply to carrying out transgenesis and gene targeting experiments?

### Applied Questions

1. Researchers have engineered a promoter that stimulates the expression of a particular gene in the nectary, or nectar-making organ of a flowering plant. By attaching the promoter to a gene of interest, they can produce the desired protein in the nectar, which bees collect and concentrate into honey. Then, the drug is extracted from the honey. What information is required to ensure that this is a safe new way to manufacture drugs?
2. Genetic engineering can creatively combine parts of organisms. From the following three lists, devise an experiment to produce a particular protein (choose one item from each list), and suggest what ...

| Organism | Biological Fluid | Protein Product |
|---|---|---|
| pig | milk | human beta globin chains |
| cow | semen | human collagen |
| goat | silk | human EPO |
| chicken | egg white | human tPA |
| aspen tree | sap | human interferon |
| silkworm | blood plasma | jellyfish GFP |
| rabbit | honey | human clotting factor |
| mouse | saliva | alpha-1- ... |

... years it was obtained from the hooves and hides of cows collected from slaughterhouses. Human collagen can be manufactured in transgenic mice. Describe the advantages of the mouse system for obtaining collagen.

4. How might cloning be used to speed transgenesis?
5. Tobacco plants given a transgene from bacteria enables them to dismantle certain buried explosives and remove these organic pollutants from soil. What information is necessary to determine whether growing such plants is safe?
6. A human oncogene called *ras* is inserted ...

... called "urge syndrome," in which the feeling of impending urination occurs frequently.

9. Mouse models for cystic fibrosis have been developed by inserting a human transgene.

### Suggested Readings


Bobrow, Martin, and Sandy Thomas. February 15, 2001. Patents in a genetic age. *Nature* 409:763–64. The Patent and Trademark Office can hardly keep up with single-gene applications. What will happen in this new age of genomics?

Cibelli, J. B., et al. May 22, 1998. "Cloned transgenic calves produced from nonquiescent fetal fibroblasts." *Science* 280:1256–58. Cloning can speed transgenesis.

Fox, Jeffrey L. July 2001. Fake biotech drugs raise concerns. *Nature Biotechnology* 19:603. Drug counterfeiting hasn't hurt anyone yet, but is potentially very dangerous.

Golovan, Serguei P., et al. August 2001. Pigs expressing salivary phytase produce low-phosphorus manure. *Nature Biotechnology* 19:741–42.

Lewis, Ricki. April 3, 2000. Clinton, Blair stoke debate on gene data. *The Scientist* 14:1. The public is very concerned about patenting genes.

Lewis, Ricki. November 8, 1999. Semen pharming. *The Scientist* 13:31. Boar semen may be a rich source of biopharmaceuticals.

Lewis, Ricki. October 26, 1998. How well do mice model humans? *The Scientist* 12:1. Many patient support groups for inherited diseases sponsor development of transgenic or knockout/knockin mice corresponding to the condition.

Marshall, Eliot. August 22, 1997. A bitter battle over insulin gene. *Science*, vol. 277. A legal dispute over experiments conducted during the early days of recombinant DNA technology continues.

Russo, Eugene. April 3, 2000. Reconsidering Asilomar. *The Scientist* 14:15. On Asilomar's 25th anniversary, those ... were there agree that it couldn't ha... again, due to the influences of the ... community and consumer activism.

Sagar, Ambuj, et al. January 2000. The t... the commoners: Biotechnology an... publics. *Nature Biotechnology* 18:2. Biotechnology affects politics and economics, and vice versa.

Schnieke, A. E., et al. December 19, 1997. Human factor IX transgenic sheep produced by transfer of nuclei fro... transfected fetal fibroblasts. *Science* 278:2130–33. Genetically engineer... supply human clotting factors.


### On the Web

Be sure to check out the additional resources on our website at www.mhhe.com/lewisgenetics5.
On the web for this chapter, you will find additional study questions, vocabulary review, useful links to case studies, tutorials, popular press coverage, and much more. To investigate specific topics mentioned in this chapter, also try the links below:

U.S. Patent and Trademark Office www.uspto.gov/

Food and Drug Administration www.fda.gov/

Genentech (recombinant DNA-derived drugs) www.gene.com/Medicine/index.html

FDA-approved recombinant DNA-derived drugs www.accessexcellence.org/AB/BA/The_Biopharmaceuticals.html

"Genetically engineered foods: Safety issues associated with antibiotic resistance genes," by A. Salyers. www.healthsci.tufts.edu/apua/salyersreport.htm

The Jackson Laboratory www.jax.org

Online Mendelian Inheritance in Man www.ncbi.nlm.nih.gov/entrez/query.fcgi?db=OMIM

alpha-anti trypsin (AAT) deficiency 107400
benign erythrocytosis 263400
BRCA1 113705
factor VIII deficiency (hemophilia ...) 306700
growth hormone deficiency 13925...
Huntington disease 143100
insulin dependent diabetes mellitu...
neurofibromatosis type I 162200
sickle cell disease 603903

# Preface

## Introduction

Very few events in human history can be said, in retrospect, to divide time. September 11, 2001, is one such date.

I was revising this edition on that bright and clear Tuesday morning, looking forward to penning an upbeat preface celebrating the human genome annotation proceeding in various laboratories. It was not to be. Now as I write this, the largest such lab is instead applying the high-throughput DNA sequencing that it used to sequence the human genome to analyzing thousands of bits of teeth and bones that arrive daily in evidence bags. Somber lab workers are extracting the mitochondrial DNA that persists after the genetic material of softer tissues is obliterated by fire and crushing pressure. Earlier, closer to that date that divided time, DNA fingerprinters at another biotech company probed softer samples shipped from the wreckage, along with cheekbrush samples bearing DNA from relatives, and bits of skin and hair left clinging to toothbrushes and hairbrushes and clothing on a day that everyone thought would be like any other. It was an astonishing and horrifying contrast to the depiction of DNA fingerprinting in the first chapter of the fourth edition of this book—tracing the ancestry of wine grapes.

Times have changed.

With DNA sequencing subverted to a purpose that no one could have predicted, revising a textbook didn't, at first, seem very important anymore. But in the weeks that followed September 11, as the belated recognition and response to bioterrorism exposed a frighteningly pervasive lack of knowledge of basic biology among our leaders, the importance of the average citizen's understanding of what genes are and what they do emerged. At the same time, new questions arose. Should researchers continue to publish new genome sequences? Suddenly, those wondrous reports of unexpected gene discoveries mined from microbial genomes held the seeds of potential weaponry.

Times have changed.

Before September 11, politicians hotly debated stem cells, renegade scientists touted their human cloning efforts, and environmentalists donned butterfly suits and destroyed crops to protest the perceived threat of corn genetically altered to escape the jaws of caterpillars. Gene therapy struggled to regain its footing in the wake of a tragic death in 1999, while a spectacularly successful new cancer drug, based on genetic research, hit the market. With time, interest in these areas will return, and maybe we will even begin to care again about the ancestry of wine grapes. *Human Genetics: Concepts and Applications,* fifth edition will guide the reader in understanding genetics and genomics and applying it to daily life. That has not changed.

## What's New and Exciting About This Edition

### Focus on Genomics—Of SNPs, Chips, and More

While Mendel's laws, the DNA double helix, protein synthesis and population dynamics will always form the foundation of genetics, the gradual shift to a genomic view opens many new research doors, and introduces new ways of thinking about ourselves. Completion of the human genome draft sequence has catapulted human genetics from the one-gene-at-a-time approach of the last half of the last century to a more multifactorial view. Genes and the environment interact to mold who we are. It is a little like jumping from listening to individual instruments to experiencing a symphony created by an entire orchestra.

The fourth edition of *Human Genetics: Concepts and Applications* introduced genomics; in the fifth edition, the impact of this new view of genes is so pervasive that it is integrated into many chapters, rather than saved for a final chapter. Rather than bludgeon the reader with details, acronyms and jargon, the approach to genomics is in context—association studies in chapter 7, human genome annotation in chapter 10, filling in chromosome details in chapter 12, and glimpses into human evolution in chapter 15. Immunity is presented in chapter 16 from the point of view of the pathogen, courtesy of genomes. Because of the integration of the genomic view throughout the text, the final chapter is free to tell the story of how this view came to be—and where it will go.

## New Chapter on Behavior

The evolution of genetic thought, from a Mendelian paradigm to a much broader consideration of genes against a backdrop of environmental influences, is perhaps nowhere more evident than in the study of human behavior. With each edition, coverage of behavior has expanded until, like a cell accumulating cytoplasm, a division was in order. The resulting binary fission of the fourth edition's chapter 7—Multifactorial and Behavioral Traits—naturally yielded a chapter on methods and basic concepts, and another on specific interesting behaviors.

Chapter 7 in this fifth edition, Multifactorial Traits, retains the classical adoption/twin/empiric risk approaches, and introduces association studies, which are critical in analyzing the traits and disorders described in depth in chapter 8, The Genetics of Behavior.

The topics for chapter 8 came from two general sources—my curiosity, and information from several human genome conferences held since 2000. The chapter opens with a focus on new types of evidence about the role of genes in behavior, then applies these new tools to dissect the genetic underpinnings of:

- Eating disorders
- Sleep
- Intelligence
- Drug addiction
- Mood disorders (depression and bipolar disorder)
- Schizophrenia

The chapter is entirely new, with many compelling examples from the biomedical literature and interviews with researchers.

## Fabulous New Art

Long-time users of *Human Genetics: Concepts and Applications* will note at a glance that all of the art is new. Vibrant new colors and closer attention to clarity of concepts ease the learning experience and make studying this complex subject less intimidating. Some of the figures are also available as Active Art, which enables the learner to manipulate portions of the illustration to review the steps to a process. Entirely new illustrations include:

| | |
|---|---|
| 7.11 | Association studies are correlations of SNP profiles |
| 8.6 | How alcohol alters gene expression in the brain |
| 10.18 | One prion, multiple conformations |
| 10.19 | Proteomics meets medicine |
| 10.20 | Exon shuffling expands gene number |
| 10.21 | Genome economy occurs in several ways |
| 11.12 | Myotonic dystrophies—novel mutation mechanism |
| 12.4 | Subtelomeres |
| 15.8 | A human HOX mutation causes synpolydactyly |
| 15.11 | Probing the molecules of extinct organisms |
| 16.19 | M cells set up immunity in the digestive tract |
| 19.1,2,3 | Three gene therapies |
| 20.9 | The global GM foods picture |
| 22.4 | Two routes to the human genome sequence |
| 22.9 | Genome sequencing, from start to finish |
| 22.10 | Comparative genomics |

Several new photos put faces on genetic diseases.

## Tables Tell the Tale

A student reviewing for an impending exam should be able to get the gist of a chapter in 10 minutes by examining the tables—if the tables are appropriately chosen and presented, as they are in this book. Table 8.5, for example, reviews every behavioral trait or disorder discussed in this new chapter, in the order of the subsections.

Most tables summarize and organize facts, easing studying. A few tables add information (table 12.1 Five Autosomes, table 14.1 Founder Populations; table 16.8 Sequenced Genomes of Human Pathogens), and some provide perspective (table 1.1 Effects of Genes on Health). Chapter 10, Gene Action and Expression, a top candidate for "toughest chapter," illustrates how the tables tell the tale:

| | |
|---|---|
| Table 10.1 | How RNA and DNA Differ |
| Table 10.2 | Major Types of RNA |
| Table 10.3 | Deciphering RNA Codons and the Amino Acids They Specify |
| Table 10.4 | The Genetic Code |
| Table 10.5 | The Non-protein Encoding Parts of the Genome |

The final table in chapter 10 is new, a summary of answers to the question, certain to be posed by students and instructors alike, "If less than 2 percent of the genome encodes protein, what does the rest of it do?" This is a table that will obviously evolve with each edition as we learn more.

## New "In Their Own Words" and Bioethics Boxes

"*In Their Own Words*" essays are written by individuals who experience inherited disease, as patients, family members, or researchers. New essays in the fifth edition introduce:

- Patricia Wright, who only recently discovered that she has had signs and symptoms of alkaptonuria all her life. (chapter 5)
- Francis Barany, a microbiologist who nearly burned his leg off searching for heat-loving bacteria with useful enzymes in a Yellowstone Park hot springs. (chapter 9)
- Toby Rodman, an immunologist and octogenarian who discovered a new source of antibodies that may protect against HIV infection. (chapter 16)

They join from past editions Don Miller, the first recipient of gene therapy for hemophilia; Stefan Schwartz, who has Klinefelter disease, and Kathy Naylor, whose little girl died of cri-du-chat syndrome.

*Bioethics: Choices for the Future* essays continue their look at controversies that arise from genetic technology. These essays explore population databases (chapter 1), cloning and stem cell research (chapter 3), sex reassignment (chapter 6), xenotransplants (chapter 16), Canavan disease as a test of fair use of genetic tests (chapter 19) and GM foods (chapter 20). Bioethical issues weave throughout the narrative as well. New section 21.4, for example, examines the dilemma of what to do with *in vitro* fertilized "spares."

## Significant Changes in Content

The two obvious changes in content are the addition of a chapter devoted to behavior, and a substantial new section in chapter 10, "The Human Genome Sequence Reveals Unexpected Complexity." This section is essentially a summary of the mid-February 2001 issues of *Science* and *Nature*, which covered the annotation of the draft human genome sequence, aka "the golden path." The rest of the chapter has been rewritten to embrace the new genome information as well.

Favorite examples and stories have been retained, and new ones added, many gleaned from my articles in *The Scientist.* They include:

- A breast cancer DNA "chip" that predicts which drugs will work on which women (chapter 1)
- Greatly expanded coverage of stem cells (chapters 2 and 3)
- Relationship between Mendel's second law and DNA microarrays (chapter 4)
- Clearer coverage of mitochondrial genes (chapter 5)
- Moved and expanded coverage of DNA repair (chapter 11)
- Updates on chromosome structure with new coverage of centromeres and subtelomeres (chapter 12)
- Applications of DNA fingerprinting to events of 9-11-01 (chapter 13)
- New coverage of genetic basis of resistance to AIDS drugs (chapter 14)
- New section on genome distinctions between humans and chimps (chapter 15)
- Genome information applied to immunity, with new sections on crowd diseases, bioweapons, and pathogen genomes (chapter 16)
- Genetic modification of pig excrement to reduce pollution (chapter 18)
- Gene therapy for Canavan disease (chapter 19)
- Impact of genomics on agricultural biotechnology (chapter 20)
- History of the human genome project (chapter 22)

## Supplements

As a full service publisher of quality educational products, McGraw-Hill does much more than just sell textbooks to your students. We create and publish an extensive array of print, video, and digital supplements to support instruction on your campus. Orders of new (versus used) textbooks help us to defray the cost of developing such supplements, which is substantial. Please consult your local McGraw-Hill representative to learn about the availability of the supplements that accompany *Human Genetics: Concepts and Applications.*

### For the Student

**Online Learning Center** Get online at www.mhhe.com/lewisgenetics5

Explore this dynamic site designed to help you get ahead and stay ahead in your study of human genetics. Some of the activities you will find on the website include:

Self-quizzes to help you master material in each chapter
Flash cards to ease learning of new vocabulary
Case Studies to practice application of your knowledge of human genetics
Links to resource articles, popular press coverage, and support groups

**Genetics: From Genes to Genomes CD-ROM** This easy-to-use CD covers the most challenging concepts in the course and makes them more understandable through presentation of full-color animations and interactive exercises. Icons in the text indicate related topics on the CD.

***Case Workbook in Human Genetics,* third edition by Ricki Lewis, ISBN 0-07-246274-4** This workbook is specifically designed to support the concepts presented in *Human Genetics* through real cases adapted from recent scientific and medical journals, with citations included. With cases now specifically related to each chapter in the book, the workbook provides practice for constructing and interpreting pedigrees; applying Mendel's laws; reviewing the relationships of DNA, RNA, and proteins; analyzing the effects of mutations; evaluating phenomena that distort Mendelian ratios; designing gene therapies; and applying new genomic approaches to understanding inherited disease. An **Answer Key** is available for the instructor.

### For the Instructor

**Online Learning Center** Find complete teaching materials online at www.mhhe.com/lewisgenetics5 including:

A complete *Instructor's Manual,* prepared by Cran Lucas of Louisiana State University, is available online. Download the complete document or use it as a chapter resource as you prepare lectures or exams. Features of the manual include:

Chapter outlines and overviews
Chapter-by-chapter resource guide to use of visual supplements
Answers to questions in the text
Additional questions and answers for each chapter
Internet resources and activities

**Downloadable Art** is provided for each chapter in jpeg format for use in class presentations or handouts. In this edition, every piece of art from the text is provided as well as every table, and a number of photographs.

Instructors will also find a link to *Pageout: The Course Website Development Center* to create your own course website. Pageout's powerful features help create a customized, professionally designed website, yet it is incredibly easy to use. There is no need to know any coding. Save time and valuable resources by typing your course information into the easy-to-follow templates.

**Test Item File** Multiple choice questions and answers that may be used in testing are provided for each chapter. Prepared by Cran Lucas of Louisiana State University, this resource covers the important concepts in each chapter and provides a variety of levels of testing. The file is available through PageOut and is also available on a cross-platform CD to adopters of the text.

**Overhead Transparencies** A set of 100 full-color transparencies showing key illustrations from the text is available for adopters. Additional images are available for downloading from the text website.

**Digital Content Manager** New to this edition is an instructor's CD containing a powerful visual package for preparing your lectures in human genetics. On this CD, you will find:

**All Text Art** in a format compatible with presentation or word processing software
**Powerpoint Presentations** covering each chapter of the text
**New Active Art!** Build images from simple to complex to suit your lecture style.


## Acknowledgments

Many heartfelt thanks to Deborah Allen for guiding yet another edition of this, my favorite book, and to Joyce Berendes and Carol Kromminga and the superb artists at Precision Graphics for making this book possible. Many thanks also to my wonderful family, cats, guinea pigs, and Speedy the relocated tortoise.


## Reviewers

Many improvements in this edition are a direct result of the suggestions from reviewers and diarists who provided feedback for this edition and previous editions of *Human Genetics: Concepts and Applications.* To each of them, a sincere thanks. We also thank the students in Ruth Sporer's Human Genetics class at the University of Pennsylvania for their review of the fourth edition, Ivan E. Leigh of West Chester, Pennsylvania for his careful review of the manuscript from the perspective of a mature student, and Clifton Poodry, Director of Minority Opportunities in Research Division of NIH, for his advice about handling issues of diversity and difference with sensitivity throughout the book.

### Reviewers for This Edition

**Michael Appleman**
*University of Southern California*
**Ruth Chesnut**
*Eastern Illinois University*
**Meredith Hamilton**
*Oklahoma State University*
**Martha Haviland**
*Rutgers University*
**Trace Jordan**
*New York University*
**A. Jake Lusis**
*University of California at Los Angeles*
**Charlotte K. Omoto**
*Washington State University*
**Bernard Possidente**
*Skidmore College*
**Ruth Sporer**
*University of Pennsylvania*
**John Sternick**
*Mansfield University*
**Dan Wells**
*University of Houston*

*We also thank these instructors for their thoughtful feedback on the Fourth Edition.*

**Sidney L. Beck**
*DePaul University*
**Hugo Boschmann**
*Hesston College*
**Hessel Bouma III**
*Calvin College*
**David Fan**
*University of Minnesota*
**Russ Feirer**
*St. Norbert College*
**Rosemary Ford**
*Washington College*
**Gail E. Gasparich**
*Towson University*
**Werner Heim**
*The Colorado College*
**Tasneem F. Khaleel**
*Montana State University–Billings*
**Marion Klaus**
*Sheridan College–Wyoming*
**Ann Hofmann**
*Madisonville Community College*
**Thomas P. Lehman**
*Morgan Community College*
**Tyre J. Proffer**
*Kent State University*
**Shyamal K. Majumdar**
*Lafayette College*
**James J. McGivern**
*Gannon University*
**Philip Meneely**
*Haverford College*
**Karen E. Messley**
*Rock Valley College*
**Nawin C. Mishra**
*University of South Carolina*
**Grant G. Mitman**
*Montana Tech of The University of Montana*
**Venkata Moorthy**
*Northwestern Oklahoma State University*
**Tim Otter**
*Albertson College of Idaho*
**Oluwatoyin O. Osunsanya**
*Muskingum College*
**Joan M. Redd**
*Walla Walla College*
**Dorothy Resh**
*University of St. Francis*
**Nick Roster**
*Eastern Wyoming College*
**Lisa M. Sardinia**
*Pacific University*
**Brian W. Schwartz**
*Columbus State University*
**Jeanine Seguin**
*Keuka College*
**Keith L. Sternes**
*Sul Ross State University*
**Edwin M. Wong**
*Western Connecticut State University*

### Reviewers for Previous Editions

**Michael Abruzzo**
*California State University at Chico*
**Mary K. Bacon**
*Ferris State University*
**Susan Bard**
*Howard Community College*
**Sandra Bobick**
*Community College of Allegheny County*
**Robert E. Braun**
*University of Washington*
**James A. Brenneman**
*University of Evansville*
**Virginia Carson**
*Chapman University*
**Mary Curtis, M.D.**
*University of Arkansas at Little Rock*

**Mary Beth Curtis**
*Tulane University*
**Ann Marie DiLorenzo**
*Montclair State College*
**Frank C. Dukepoo**
*Northern Arizona University*
**Robert Ebert**
*Palomar College*
**Larry Eckroat**
*Pennsylvania State University at Erie*
**Jack Fabian**
*Keene State College*
**David Fromson**
*California State University–Fullerton*
**Elizabeth Gardner**
*Pine Manor College*
**Michael A. Gates**
*Cleveland State University*
**Donald C. Giersch**
*Triton College*
**Miriam Golomb**
*University of Missouri–Columbia*
**Meredith Hamilton**
*Oklahoma State University*
**Greg Hampikian**
*Clayton College and State University*
**George A. Hudock**
*Indiana University*
**Neil Jensen**
*Weber State College*
**William J. Keppler**
*Florida International University*
**Valerie Kish**
*University of Richmond*
**Arthur L. Koch**
*Indiana University*
**Richard Landesman**
*University of Vermont*
**Mira Lessick**
*Rush University*
**Cran Lucas**
*Louisiana State University at Shreveport*
**Jay R. Marston**
*Lane Community College*
**Joshua Marvit**
*Penn State University*
**James J. McGivern**
*Gannon University*
**Denise McKenney**
*University of Texas of the Permian Basin*
**Wendell H. McKenzie**
*North Carolina State University*
**Mary Rengo Murnik**
*Ferris State University*
**Michael E. Myszewski**
*Drake University*
**Donald J. Nash**
*Colorado State University*
**Charlotte K. Omoto**
*Washington State University*
**David L. Parker**
*Northern Virginia Community College—Alexandria Campus*
**Jack Parker**
*Southern Illinois University at Carbondale*
**Michael James Patrick**
*Seton Hill College*
**Bernard Possidente**
*Skidmore College*
**Albert Robinson**
*SUNY at Potsdam*
**Peter A. Rosenbaum**
*SUNY–Oswego*
**Peter Russel**
*Chaffey College*
**Polly Schulz**
*Portland Community College*
**Georgia Floyd Smith**
*Arizona State University*
**Jolynn Smith**
*Southern Illinois University at Carbondale*
**Anthea Stavroulakis**
*Kingsborough Community College*
**Margaret R. Wallace**
*University of Florida*
**Robert Wiggers**
*Stephen F. Austin State University*
**Roberta B. Williams**
*University of Nevada–Las Vegas*
**H. Glenn Wolfe**
*University of Kansas*
**Virginia Wolfenberger**
*Texas Chiropractic College*
**Janet C. Woodward**
*St. Cloud State University*
**Connie Zilles**
*West Valley College*

CHAPTER

1

# Overview of Genetics

## 1.1 A Look Ahead

Testing for inherited diseases and susceptibilities will become standard practice as health care becomes increasingly individualized. Tests that detect specific variations in genetic material will enable physicians to select treatments that a person can tolerate and that are the most likely to be effective.

## 1.2 From Genes to Genomes

DNA sequences that constitute genes carry information that tells cells how to manufacture specific proteins. A gene's effects are evident at the cell, tissue, organ, and organ system levels. Traits with large inherited components can be traced and predicted in families. Genetic change at the population level underlies evolution. Comparing genomes reveals that humans have much in common with other species.

## 1.3 Genes Do Not Usually Function Alone

In the twentieth century, genetics dealt almost entirely with single-gene traits and disorders. Today it is becoming clear that multiple genes and the environment mold most traits.

## 1.4 Geneticists Use Statistics to Represent Risks

Risk is an estimate of the likelihood that a particular individual will develop a particular trait. It may be absolute for an individual, or relative based on comparison to other people.

## 1.5 Applications of Genetics

Genetics impacts our lives in diverse ways. Genetic tests can establish identities and diagnose disease. Genetic manipulations can provide new agricultural variants.

Genetics is the study of inherited variation and traits. Sometimes people confuse genetics with genealogy, which considers relationships but not traits. With the advent of gene-based tests that can predict future disease symptoms, some have even compared genetics to fortune telling! But genetics is neither genealogy nor fortune telling—it is a life science. Although genetics is often associated with disease, our genes provide a great variety of characteristics that create much of our individuality, from our hair and eye color, to the shapes of our body parts, to our talents and personality traits.

**Genes** are the units of heredity, the sets of biochemical instructions that tell cells, the basic units of life, how to manufacture certain proteins. These proteins ultimately underlie specific traits; a missing protein blood-clotting factor, for example, causes the inherited disease hemophilia. A gene is composed of the molecule **deoxyribonucleic acid,** more familiarly known as **DNA.** Some traits are determined nearly entirely by genes; most traits, however, have considerable environmental components. The complete set of genetic information characteristic of an organism, including protein-encoding genes and other DNA sequences, constitutes a **genome.**

Genetics is unlike other life sciences in how directly and intimately it affects our lives. It obviously impacts our health, because we inherit certain diseases and disease susceptibilities. But principles of genetics also touch history, politics, economics, sociology, and psychology, and they force us to wrestle with concepts of benefit and risk, even tapping our deepest feelings about right and wrong. A field of study called **bioethics** was founded in the 1970s to address many of the personal issues that arise in applying medical technology. Bioethicists have more recently addressed concerns that new genetic knowledge raises, issues such as privacy, confidentiality, and discrimination.

An even newer field is **genomics,** which considers many genes at a time. The genomic approach is broader than the emphasis on single-gene traits that pervaded genetics in the twentieth century. It also enables us to compare ourselves to other species—the similarities can be astonishing and quite humbling!

New technology has made genomics possible. Researchers began deciphering genomes in 1995, starting with a common bacterium. Some three dozen genome projects later, by 2000, a private company and an international consortium of researchers added *Homo sapiens* to the list, with completion of a "first draft" sequence of the human genome. The genomes of more than 100 species have been sequenced.

It will take much of the new century to understand our genetic selves. Following is a glimpse of how two young people might encounter genomics in the not-too-distant future. All of the tests mentioned already exist.

## 1.1 A Look Ahead

The year is 2005. Human genomics has not yet progressed to the point that newborns undergo whole-genome screens—that is still too expensive—but individuals can take selected gene tests tailored to their health histories. Such tests can detect gene variants that are associated with increased risk of developing a particular condition. Young people sometimes take such tests—if they want to—when there are ways to prevent, delay, or control symptoms. Consider two 19-year-old college roommates who choose to undergo this type of genetic testing.

Mackenzie requests three panels of tests, based on what she knows about her family background. An older brother and her father smoke cigarettes and are prone to alcoholism, and her father's mother, also a smoker, died of lung cancer. Two relatives on her mother's side had colon cancer. Mackenzie also has older relatives on both sides who have Alzheimer disease. She asks for tests to detect genes that predispose her to developing addictions, certain cancers, and inherited forms of Alzheimer disease.

Laurel, Mackenzie's roommate, requests a different set of tests, based on her family history. She has always had frequent bouts of bronchitis that often progress to pneumonia, so she requests a test for cystic fibrosis (CF). Usually a devastating illness, CF has milder forms whose symptoms are increased susceptibility to respiratory infections. These cases often go unrecognized as CF, as Laurel knows from reading a journal article for a biology class last year. Because her sister and mother also get bronchitis often, she suspects mild CF in the family.

Laurel requests tests for type II (non-insulin-dependent) diabetes mellitus, because several of her relatives developed this condition as adults. She knows that medication can control the abnormal blood glucose level, but that dietary and exercise plans are essential, too. If she knows she is at high risk of developing the condition, she'll adopt these habits right away. However, Laurel refuses a test for inherited susceptibility to Alzheimer disease, even though a grandfather died of it. She does not want to know if this currently untreatable condition is likely to lie in her future. Because past blood tests revealed elevated cholesterol, Laurel seeks information about her risk of developing traits associated with heart and blood vessel (cardiovascular) disease.

Each student proceeds through the steps outlined in figure 1.1. The first step is to register a complete family history. Next, each young woman swishes a cotton swab on the inside of her cheek to obtain cells, which are then sent to a laboratory for analysis. There, DNA is extracted and cut into pieces, then tagged with molecules that fluoresce under certain types of light. The students' genetic material is then applied to "DNA chips," which are small pieces of glass or nylon to which particular sequences of DNA have been attached. Because the genes on the chip are aligned in fixed positions, this device is also called a DNA microarray.

A typical DNA microarray bears hundreds or even thousands of DNA pieces. One of Mackenzie's DNA chips bears genes that regulate her circadian (daily) rhythms and encode the receptor proteins on nerve cells that bind neurotransmitters. If Mackenzie encounters addictive substances or activities in the future, having certain variants of these genes may increase her risk of developing addictive behaviors. Another DNA chip screens for gene variants that greatly increase risk for lung cancer, and a third DNA chip detects genes associated with colon cancer. Her fourth DNA chip is smaller, bearing the genes that correspond to four types of inherited Alzheimer disease.

Laurel's chips are personalized to suit her family background and specific requests. The microarray panel for CF is straightforward—it holds 400 DNA sequences corresponding to variants of the CF gene known to be associated with the milder symptoms that appear in Laurel's family. The microarray for diabetes bears gene variants that reflect how Laurel's body

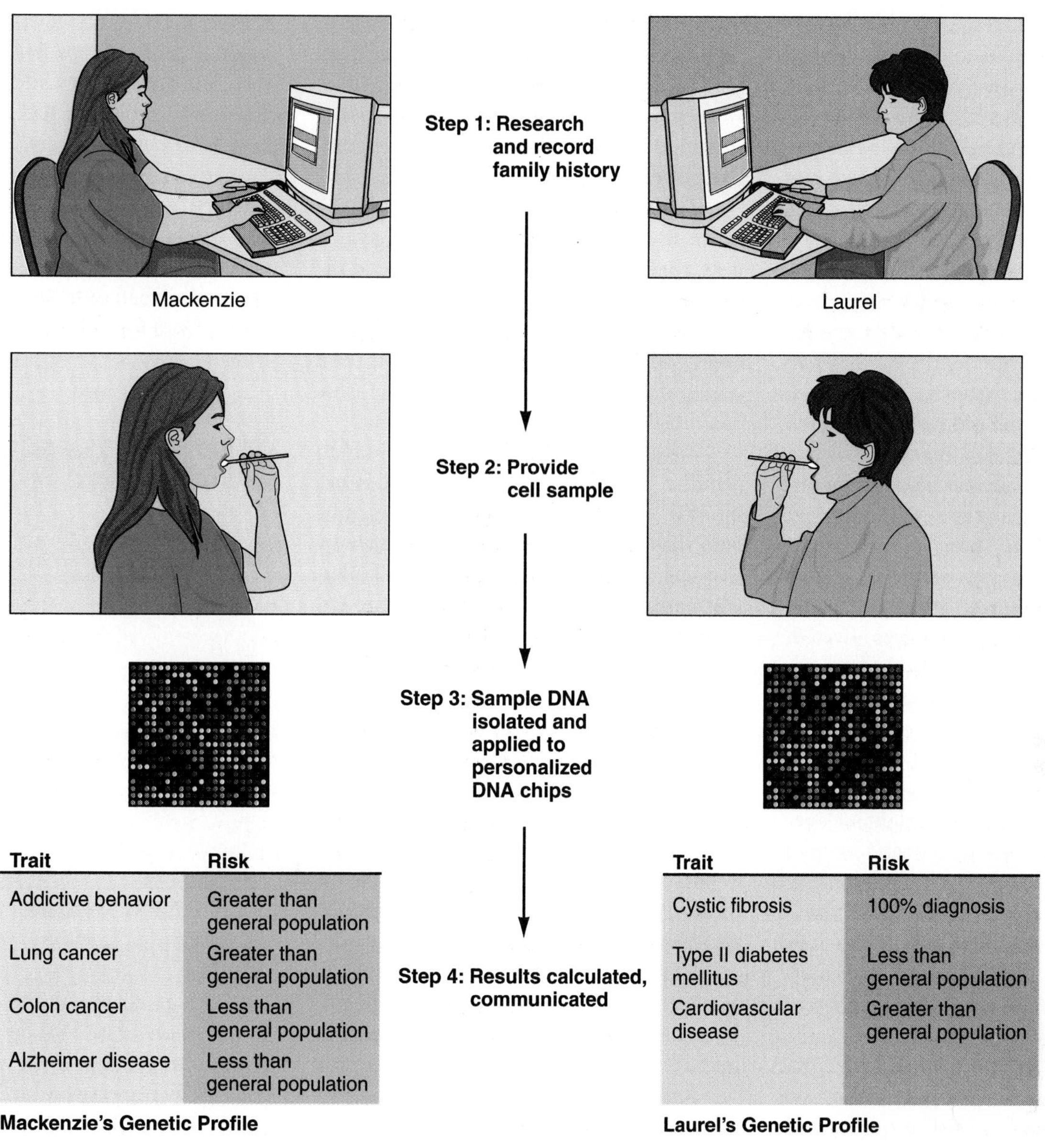

Step 1: Research and record family history
Mackenzie
Laurel
Step 2: Provide cell sample
Step 3: Sample DNA isolated and applied to personalized DNA chips
Step 4: Results calculated, communicated

| Trait | Risk |
|---|---|
| Addictive behavior | Greater than general population |
| Lung cancer | Greater than general population |
| Colon cancer | Less than general population |
| Alzheimer disease | Less than general population |

**Mackenzie's Genetic Profile**

| Trait | Risk |
|---|---|
| Cystic fibrosis | 100% diagnosis |
| Type II diabetes mellitus | Less than general population |
| Cardiovascular disease | Greater than general population |

**Laurel's Genetic Profile**

**figure 1.1**

**Genetic testing.** Tests like these will soon become a standard part of health care.

handles glucose transport and uptake into cells. The DNA microarray for cardiovascular disease is the largest and most diverse. It includes thousands of genes whose protein products help to determine and control blood pressure, blood clotting, and the synthesis, transport, and metabolism of cholesterol and other lipids.

A few days later, the test results are in, and a very important part of the process occurs—meeting a **genetic counselor,** who explains the findings. Mackenzie learns that she has inherited several gene variants that predispose her to addictive behaviors and to developing lung cancer—a dangerous combination. But she does not have genes that increase her risk for inherited forms of colon cancer or Alzheimer disease. Mackenzie is relieved. She knows to avoid alcohol and especially smoking, but is reassured that her risks of inherited colon cancer and Alzheimer disease are no greater than they are for the general population—in fact, they are somewhat less.

Laurel finds out that she indeed has a mild form of cystic fibrosis. The microarray also indicates which types of infections she is most susceptible to, and which antibiotics will most effectively treat her frequent episodes of bronchitis and pneumonia. She might even be a candidate for gene therapy—periodically inhaling a preparation containing the normal version of the CF-causing gene engineered into a "disabled" virus that would otherwise cause a respiratory infection. The diabetes test panel reveals a risk that is lower than that for the general population. Laurel also learns she has several gene variants that raise her blood cholesterol level. By following a diet low in fat and high in fiber, exercising regularly, and frequently checking her cholesterol levels, Laurel can help keep her heart and blood vessels healthy. On the basis of the cardiovascular disease microarray panel, her physician can also tell which cholesterol-lowering drug she will respond to best, should lifestyle changes be insufficient to counter her inherited tendency to accumulate cholesterol and other lipids in the bloodstream.

The DNA microarray tests that Mackenzie and Laurel undergo will become part of their medical records, and tests will be added as their interests and health status change. For example, shortly before each young woman tries to become pregnant, she and her partner will take prenatal DNA microarray panels that detect whether or not they are carriers for any of several hundred illnesses, tailored to their family backgrounds and ethnic groups. Carriers can pass an inherited illness to their offspring even when they are not themselves affected. If Laurel, Mackenzie or their partners carry inherited conditions, DNA microarray tests can determine whether their offspring inherit the illness.

Impending parenthood isn't the only reason Laurel and Mackenzie might seek genetic testing again. If either young woman suspects she may have cancer, for example, DNA microarrays called expression panels can determine which genes are turned on or off in affected cells compared to nonaffected cells of the same type. Such information can identify cancer cells very early, when treatment is more likely to work. These devices also provide information on how quickly the disease will progress, and how tumor cells and the individual's immune system are likely to respond to particular drugs. A DNA microarray can reveal that a particular drug will produce intolerable side effects before the patient has to experience that toxicity.

The first DNA microarray to analyze cancer, the "lymphochip," was developed before completion of the human genome project. It identifies cancer-causing and associated genes in white blood cells. A different DNA microarray test, for breast cancer, is used on samples of breast tissue to track the course of disease and assess treatment. The "chip" was featured on a cover of *Nature* magazine with the headline, "portrait of a breast cancer." In one experiment, DNA microarray tests were performed on tumor cells of 20 women with advanced breast cancer before and after a 3-month regimen of chemotherapy. The gene pattern returned to normal only in the three women who ultimately responded to the treatment, demonstrating the test's predictive power.

Though Laurel and Mackenzie will gain much useful information from the genetic tests, their health records will be kept confidential. Laws prevent employers and insurers from discriminating against anyone based on genetic information. This is a practical matter—everyone has some gene variants that are associated with disease.

With completion of the human genome project, the medical world is exploding with new information. One company has already invented a five-inch by five-inch wafer that houses up to 400 DNA microarrays, each the size of a dime and containing up to 400,000 DNA pieces.

New health care professionals are being trained in genetics and the new field of genomics; older health care workers are also learning how to integrate new genetic knowledge into medical practice. Another change is in the breadth of genetics. In the past, physicians typically encountered genetics only as rare disorders caused by single genes, such as cystic fibrosis, sickle cell disease, and muscular dystrophy, or chromosome disorders, such as trisomy 21 Down syndrome. Today, medical science is beginning to recognize the role that genes play in many types of conditions (table 1.1).

A study of the prevalence of genetic disorders among 4,224 children admitted to Rainbow Babies and Children's Hospital in Cleveland in 1996 revealed that genes contribute much more to disease than many medical professionals had thought. Nearly three-quarters of the children, admitted for a variety of problems, had an underlying genetic disorder or susceptibility. Specifically, 35 percent had clearly genetic conditions (the first two entries in table 1.1); 36.5 percent had an underlying condition with a genetic predisposition, such as asthma, cancer, or type 1 diabetes mellitus; and the rest were hospitalized for an injury or had no underlying disease.

**Key Concepts**

Genetics investigates inherited traits and their variations. Genes, composed of DNA, are the units of inheritance, and they specify particular proteins. Not all DNA encodes protein. A genome is the complete genetic instructions for an organism. Human genome information will personalize medicine and predict future illness.

## 1.2 From Genes to Genomes

Genetics is all about the transmission of information at several levels (figure 1.2). At the molecular level, DNA comprises genes, which are part of chromosomes. Each of our trillions of cells contains two sets of chromosomes, each set a copy of the genome. Cells interact and aggregate into

**table 1.1**

**Effects of Genes on Health**

| Type of Disorder or Association | Example | Chapter |
|---|---|---|
| Single gene (Mendelian) | Cystic fibrosis | 4, 5 |
| Chromosomal disorder | Down syndrome | 12 |
| Complex (multifactorial) disorder | Diabetes mellitus | 3, 7, 8 |
| Cancer (somatic mutation) | Breast cancer | 17 |
| Single nucleotide polymorphisms (SNPs) | Associated with various conditions in different populations | 7, 8, 22 |

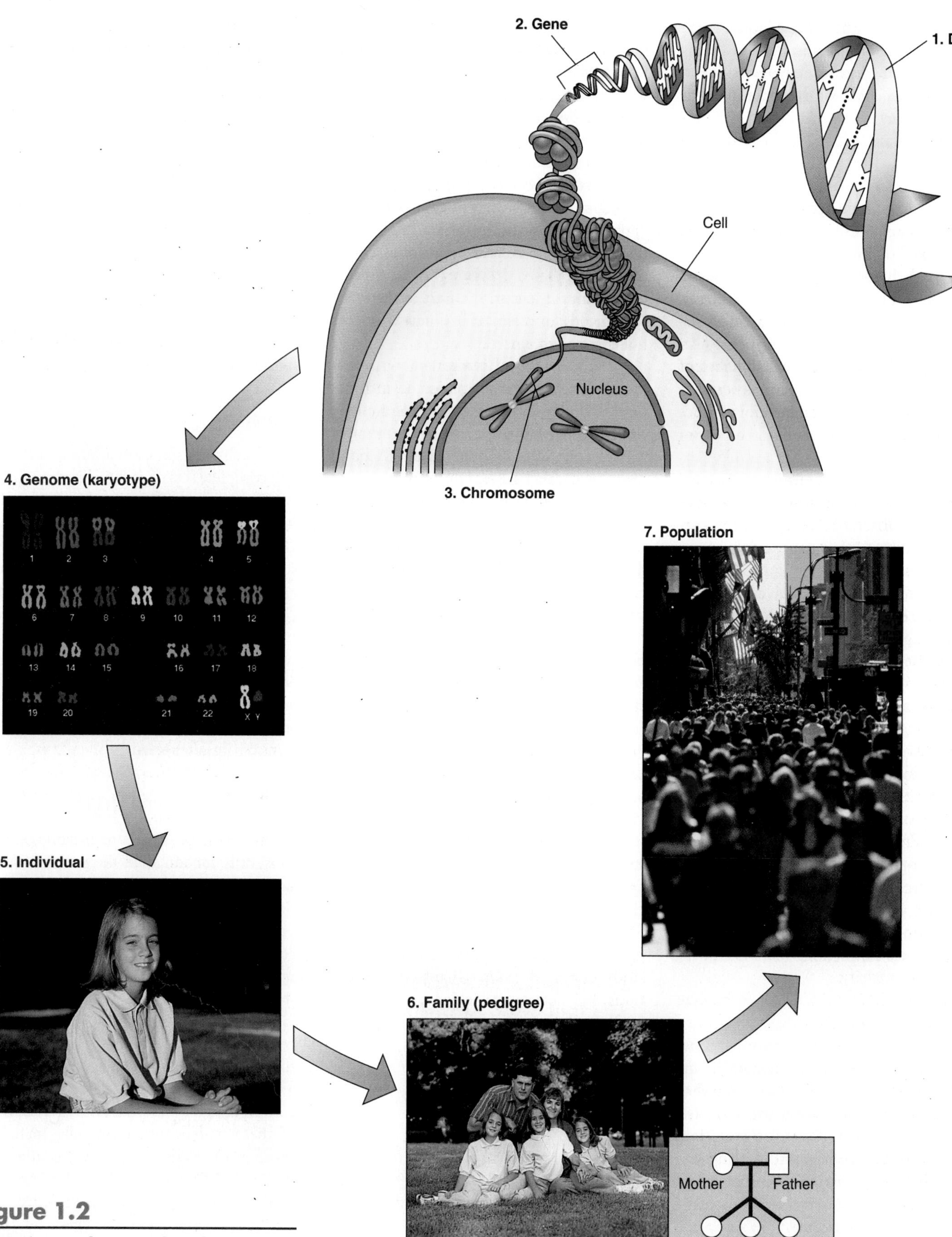


**figure 1.2**

**Genetics can be considered at several levels.**

tissues, which in turn combine to form organs and organ systems. At the family level, inherited disease may be evident. Finally, genetic changes in populations underlie evolution.

## DNA

Genes consist of sequences of four types of DNA building blocks—adenine, guanine, cytosine, and thymine, abbreviated A, G, C, and T. Each base bonds to a sugar and a phosphate group to form a unit called a nucleotide. DNA bases are also called nitrogenous (nitrogen-containing) bases. In genes, DNA bases provide an alphabet of sorts. Each three DNA bases in a row specifies the code for a particular amino acid, and amino acids are the building blocks of proteins.

An intermediate language also encoded in nitrogenous bases is contained in **ribonucleic acid (RNA).** One type of RNA carries a copy of a DNA sequence and presents it to other parts of the cell. In this way, the information encoded in DNA can be used to produce RNA molecules, which are then used to manufacture protein. DNA remains in the nucleus to be passed on when a cell divides. Only about 1.5 percent of the DNA in the human genome encodes protein. Researchers have not yet discovered the function of much of the rest, but they are learning more as they analyze genome information. Similarly, not all functions of RNA are understood. The definition of "gene" has changed over the past half century to embrace new knowledge. It might be most accurate, in light of all that remains to be learned from human genome information, to define a gene as a sequence of DNA that has a known function.

## Gene

Individual genes come in variants that differ from each other by small changes in the DNA base sequence. The variants of a gene are called **alleles,** and these changes in DNA sequence arise by a process called **mutation.** Some mutations are harmful, causing disease; others provide variation, such as freckled skin; and some mutations may actually be helpful. In some people, for example, a rare mutation renders their cells unable to bind HIV, making them resistant to HIV infection. This genetic variant would probably have remained unknown had AIDS not arisen. Many mutations have no visible effect at all because they do not change the encoded protein in a way that affects its function, just as a minor spelling error does not destroy the meaning of a sentence.

Parts of the DNA sequence can vary among individuals, yet not change external appearance or health. A variant in sequence that is present in at least 1 percent of a population is called a **polymorphism.** A polymorphism can occur in a part of the DNA that encodes protein, or in a part that does not encode protein.

"Polymorphism" is a general term that literally means "many forms." It includes disease-causing variants. The terminology can be somewhat confusing. A mutation is actually a type of polymorphism. A polymorphism can be helpful, harmful, or, in most instances, have no effect at all (that we know of). The term polymorphism has been part of the language of genetics for decades, but has recently begun to attract a great deal of attention from other fields, such as information technology and medicine. This is because of the realization that polymorphisms can be used in DNA microarray panels to predict risks of developing specific medical conditions.

Researchers have identified more than 3 million **single nucleotide polymorphisms** (SNPs, pronounced "snips"). SNPs are single base sites that differ among individuals. The human genome may include up to 20 million SNPs, or 1 in every 1,250 or so DNA nucleotides, although they are not evenly distributed. DNA microarrays include both disease-causing mutations and SNPs that merely mark places where people differ. A technique called an association study examines DNA variants in populations and detects particular combinations of SNPs that are found almost exclusively among people with a particular disorder, but not otherwise.

## Chromosome

Genes are part of larger structures called **chromosomes,** which also include proteins that the DNA wraps around. A human cell has 23 pairs of chromosomes. Twenty-two pairs are **autosomes,** or chromosomes that do not differ between the sexes. The autosomes are numbered from 1 to 22, with 1 being the largest. The other two chromosomes, the X and the Y, are **sex chromosomes.** The Y chromosome bears genes that determine maleness. In humans, lacking a Y makes one a female.

Missing even small portions of a chromosome has a devastating effect on health, because many genes are deleted. To detect chromosome abnormalities, geneticists use charts called **karyotypes** that order the chromosome pairs from largest to smallest. The chromosomes are stained with dyes or fluorescent chemicals that create different patterns to highlight abnormalities (see figure 1.2).

## Genome

The 46 chromosomes in a human cell hold two complete sets of genetic information, or two copies of each chromosome type. The human genome probably contains from 28,000 to 34,000 protein-encoding genes, scattered among three billion DNA bases among each set of 23 chromosomes. (Higher estimates may count repeated genes more than once.) Two entire genomes are tucked into each of a person's many, many cells. As noted geneticist Hermann J. Muller wrote in 1947, "In a sense we contain ourselves, wrapped up within ourselves, trillions of times repeated."

## Cells, Tissues, and Organs

A human body consists of trillions of cells. Most cells contain all of the genetic instructions, but cells differ in appearance and function by using only some of their genes, in a process called **differentiation.** Specialized cells with related functions aggregate and interact to form tissues, which in turn form the organs and organ systems of the individual. Organs also include less specialized cells, called **stem cells,** that retain the ability to differentiate further, should the need arise—perhaps when an injury requires that certain cells be replaced. Some repositories of these replenishing stem cells, including those in the brain, have only recently been discovered. Others, such as the bone marrow cells that continually replenish the blood, are better known. A new field called regenerative medicine uses stem cells to replace degenerating cells that cause conditions such as Parkinson disease and Huntington disease.

## Individual

Two terms distinguish between the alleles that are *present* in an individual and the alleles that are *expressed.* The **genotype** refers to the underlying instructions (alleles present), and the **phenotype** is the visible trait, biochemical change, or effect on health (alleles expressed). Alleles are further distinguished by how many copies it takes to affect the phenotype. A **dominant** allele produces an effect when present in just one copy (on one chromosome), whereas a **recessive** allele must be present on both chromosomes to be expressed. (Alleles on the Y chromosome are an exception; recessive alleles on the X chromosome in males are expressed because there is no second X chromosome to block expression.)

## Family

Individuals are genetically connected into families. Traditionally, the study of traits in families has been called transmission genetics or Mendelian genetics. Molecular genetics, which considers DNA, RNA, and proteins, often begins with transmission genetics, when an interesting trait or illness in a family comes to a researcher's attention. Charts called pedigrees are used to represent the members of a family and to indicate which individuals have particular inherited traits. Figure 1.2 shows a pedigree, but an unusual one—a family with identical triplets.

## Population

Above the family level of genetic organization is the population. In a strict biological sense, a population is a group of interbreeding individuals. In a genetic sense, a population is a large collection of alleles, distinguished by the frequency of particular alleles. People from Sweden, for example, would have a greater frequency of alleles that specify light hair and skin than people from a population in Ethiopia who tend to have dark hair and skin. The fact that groups of people look different and may suffer from different health problems reflects the frequencies of their distinctive sets of alleles. All the alleles in a population constitute the **gene pool.** (An individual does not have a gene pool.)

Population genetics is very important in applications such as health care and forensics. It is also the very basis of evolution. In fact, evolution is technically defined as "changing allele frequencies in populations," as the chapters in part 4 describe. These small-scale genetic changes foster the more obvious species distinctions most often associated with evolution.

## Evolution

Geneticists have known for decades that comparing DNA sequences for individual genes, or the amino acid sequences of the proteins that the genes encode, can reveal how closely related different types of organisms are. The underlying assumption is that the more similar the sequences are, the more recently two species diverged from a shared ancestor. Figure 15.7 shows such analysis for cytochrome C, a protein essential for extracting energy from nutrients.

Genomewide studies are even more startling than comparing single genes. Humans, for example, share more than 98 percent of the DNA sequence with chimpanzees. Our genomes differ more in the organization of genes and in the number of copies of genes than in the overall sequence. Still, learning the functions of the human-specific genes may explain the anatomical differences between us and them. Our kinship with other species extends much farther back in time than to chimpanzees, who are in a sense our evolutionary first cousins. Humans also share many DNA sequences with pufferfish, fruit flies, mice, and even bacteria. At the level of genetic instructions for building a body, we are not very different from other organisms.

Comparisons of person to person at the genome level reveal more sameness—we are incredibly like one another. DNA sequence similarity among humans exceeds 99.9 percent. Studies of polymorphisms among different modern ethnic groups reveal that modern humans arose and came out of Africa and haven't changed very much since. The gene pools of all groups are subsets of the modern African gene pool. Genome analyses also confirm what biologists have maintained for many years—that race is a social concept, not a biological one. "Race" is actually defined by fewer than 0.01 percent of our genes. Put another way, two members of different races may in fact have more genes in common than two members of the same race. Very few, if any, gene variants are unique to any one racial or ethnic group. Imagine if we defined race by a different small set of genes, such as the ability to taste bitter substances!

Table 1.2 defines some of the terms used in this section.

**Key Concepts**

Genetics can be considered at different levels: DNA, genes, chromosomes, genomes, individuals, families, and populations. • A gene can exist in more than one form, or allele. • Comparing genomes among species reveals evolutionary relatedness.

## 1.3 Genes Do Not Usually Function Alone

For much of its short history, the field of genetics dealt almost exclusively with the thousands of traits and illnesses that are clearly determined by single genes. These **Mendelian traits** are named for Gregor Mendel, who derived the laws of gene transmission by studying single-gene traits in peas (the topic of chapter 4). A compendium called "Mendelian Inheritance in Man" has, for decades, listed and described all known single-gene traits and disorders in humans. The computerized version, "Online Mendelian Inheritance in Man," is today a terrific resource. "OMIM" numbers are listed at the end of each chapter for disorders mentioned in the narrative. Sequencing the human genome, however, has revealed redundant entries in lists of single-gene disorders, whose actual number may be as low as 1,100. For some genes, OMIM lists different allele combinations as distinct disorders, such as different types of anemia that result from mutations in the same gene.

Genetics is far more complicated than a one-gene-one-disease paradigm. Most genes do not function alone, but are influenced by the actions of other genes, and sometimes by factors in the environment as well. Traits

**table 1.2**

**A Mini-Glossary of Genetic Terms**

| Term | Definition |
|---|---|
| Allele | An alternate form of a gene; a gene variant. |
| Autosome | A chromosome not normally involved in determining sex. |
| Chromosome | A structure, consisting of DNA and protein, that carries the genes. |
| DNA | Deoxyribonucleic acid; the molecule whose building block sequence encodes the information that a cell uses to construct a particular protein. |
| Dominant | An allele that exerts a noticeable effect when present in just one copy. |
| Gene | A sequence of DNA that has a known function, such as encoding protein or controlling gene expression. |
| Gene pool | All of the genes in a population. |
| Genome | A complete set of genetic instructions in a cell, including DNA that encodes protein as well as other DNA. |
| Genomics | The new field of investigating how genes interact, and comparing genomes. |
| Genotype | The allele combination in an individual. |
| Karyotype | A size-order display of chromosomes. |
| Mendelian trait | A trait that is completely determined by a single gene. |
| Multifactorial trait | A trait that is determined by one or more genes and by the environment. Also called a complex trait. |
| Mutation | A change in a gene that affects the individual's health, appearance, or biochemistry. |
| Pedigree | A diagram used to follow inheritance of a trait in a family. |
| Phenotype | The observable expression of an allele combination. |
| Polymorphism | A site in a genome that varies in 1 percent or more of a population. |
| Recessive | An allele that exerts a noticeable effect only when present in two copies. |
| RNA | Ribonucleic acid; the chemical that enables a cell to synthesize proteins using the information in DNA sequences. |
| Sex chromosome | A chromosome that carries genes whose presence or absence determines sex. |

**table 1.3**

**Mendelian or Multifactorial Genetic Disorders**

| Mendelian Disorders | Multifactorial Disorders |
|---|---|
| Achondroplasia | Breast cancer |
| Cystic fibrosis | Bipolar affective disorder |
| Duchenne muscular dystrophy | Cleft palate |
| Hemochromatosis | Dyslexia |
| Hemophilia | Diabetes mellitus |
| Huntington disease | Hypertension |
| Neurofibromatosis | Migraine |
| Osteogenesis imperfecta | Neural tube defects |
| Sickle cell disease | Schizophrenia |
| Tay-Sachs disease | Seizure disorders |

with several determinants are called **multifactorial,** or complex, traits. (The term *complex traits* has different meanings in a scientific and a popular sense, so this book uses the more precise term *multifactorial.*) Table 1.3 lists some Mendelian and multifactorial conditions, and figure 1.3 gives an example of each. Confusing matters even further is the fact that some illnesses occur in different forms—some inherited, some not, some Mendelian, some multifactorial. Usually the inherited forms are rarer, as is the case for Alzheimer disease, breast cancer, and Parkinson disease.

Researchers can develop treatments based on the easier-to-study inherited form of an illness, which can then be used to treat more common, multifactorial forms. For example, the statin drugs that millions of people take to lower cholesterol were developed from work on children with familial hypercholesterolemia, which affects one in a million individuals (see figure 5.2).

Knowing whether a trait or illness is inherited in a Mendelian or multifactorial manner is important for predicting recurrence risk. The probability that a Mendelian trait will occur in another family member is simple to calculate using the laws that Mendel derived. In contrast, predicting the recurrence of a multifactorial trait is difficult because several contributing factors are at play. Inherited breast cancer illustrates how the fact that genes rarely act alone can complicate calculation of risk.

a.

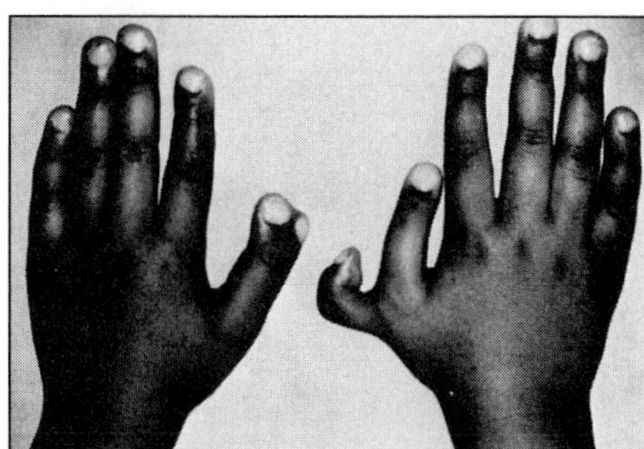

b.

**figure 1.3**

**Mendelian versus multifactorial traits.** (*a*) Hair color is multifactorial, controlled by at least three genes plus environmental factors, such as the bleaching effects of sun exposure. (*b*) Polydactyly—extra fingers and/or toes—is a Mendelian trait, determined by a single gene.

Mutations in a gene called BRCA1 cause fewer than 5 percent of all cases of breast cancer. But studies of the disease incidence in different populations have yielded confusing results. In Jewish families of eastern European descent (Ashkenazim) with many affected members, inheriting the most common BRCA1 mutation means an 86 percent chance of developing the disease over a lifetime. But women from other ethnic groups who inherit this allele may have only a 45 percent chance of developing breast cancer, because they have different alleles of other genes with which BRCA1 interacts than do the eastern European families.

Environmental factors may also affect the gene's expression. For example, exposure to pesticides that mimic the effects of estrogen may be an environmental contributor to breast cancer. It can be difficult to tease apart genetic and environmental contributions to disease. BRCA1 breast cancer, for example, is especially prevalent among women who live in Long Island, New York. This population includes both many Ashkenazim, and widespread exposure to pesticides.

Increasingly, predictions of inherited disease are considered in terms of "modified genetic risk," which takes into account single genes as well as environmental and family background information. A modified genetic risk is necessary to predict BRCA1 breast cancer occurrence in a family.

The fact that the environment modifies the actions of genes counters the idea that an inherited trait is unchangeable, which is termed **genetic determinism.** The idea that "we are our genes" can be very dangerous. In terms of predictive testing for inherited disease, effects of the environment require that results be presented as risks rather than foregone conclusions. That is, a person might be told that she has a 45 percent chance of developing BRCA1 breast cancer, not "you will get breast cancer."

Genetic determinism as part of social policy can be particularly harmful. In the past, for example, the assumption that one ethnic group is genetically less intelligent than another led to lowered expectations and fewer educational opportunities for those perceived to be biologically inferior. Environment, in fact, has a huge impact on intellectual development. The bioethics essay in chapter 8 considers genetic determinism further.

**Key Concepts**

Inherited traits are determined by one gene (Mendelian) or specified by one or more genes and the environment (multifactorial). Even the expression of single genes is affected to some extent by actions of other genes. Genetic determinism is the idea that an inherited trait cannot be modified.

## 1.4 Geneticists Use Statistics to Represent Risks

Predicting the inheritance of traits in individuals is not a precise science, largely because of the many influences on gene function and the uncertainties of analyzing several factors. Genetic counselors calculate risks for clients who want to know the chance that a family member will inherit a particular disease—or has inherited it, but does not yet exhibit the symptoms.

In general, risk assessment estimates the degree to which a particular event or situation represents a danger to a population. In genetics, that event is the likelihood of inheriting a particular gene or gene combination. The genetic counselor can infer that information from a detailed family history, or from the results of tests that identify a gene variant or a protein that is absent or abnormal.

Risks can be expressed as absolute or relative figures. **Absolute risk** is the probability that an individual will develop a particular condition. **Relative risk** is the likelihood that an individual from a particular population will develop a condition in comparison to individuals in another group, which is usually the general population. Relative risk is a ratio of the probability in one group compared to another. In genetics, relative risks might be calculated by evaluating any situation that might elevate the risk of developing a particular condition, such as one's ethnic group, age, or exposure to a certain danger. The threatening situation is called a **risk factor.** For example, chromosome abnormalities are more common in the offspring of older mothers. Pregnant women who undergo testing for Down syndrome caused by an extra chromosome 21 are compared by age to the general population of pregnant women to derive the relative risk that they are carrying a fetus that has the syndrome. The risk factor is age.

Determining a relative risk may seem unnecessary, because an absolute risk applies to an individual. However, relative risks help to identify patients who are most likely to have the conditions for which absolute risks can be calculated. Health care providers use relative risk estimates to identify individuals who are most likely to benefit from particular medical tests. A problem that genetic counselors face in assessing risk, however, is that statistics tend to lose their meaning in a one-on-one situation. To a couple learning that their fetus has Down syndrome, the fact that the relative risk was low based on population statistics pertaining to their age group is immaterial.

Mathematically, absolute and relative risk are represented in different ways. Odds and percentages are used to depict absolute risk. For example, Mackenzie's absolute risk of developing inherited Alzheimer disease over her lifetime is 4 in 100 (the odds) or 4 percent. Determining her relative risk requires knowing the risk to the general population. If that risk is 10 in 100, then Mackenzie's relative risk is 4 percent divided by 10 percent, or 0.4. A relative risk of less than 1 indicates the chance of developing a particular illness is less than that of the general population; a value greater than 1 indicates risk above that of the general population. For example, Mackenzie's 0.4 relative risk means she has 40 percent as much risk of inheriting Alzheimer disease as the average person in the general population; a relative risk of 8.4, by contrast, indicates a greater-than-8-fold risk compared to an individual in the general population. Determining the risks for Alzheimer disease is actually much more complicated than is depicted in this hypothetical case. Several genes are involved, the percentage of inherited cases isn't known, and prevalence is highly associated with age. Elevated risk is linked to having more than one affected relative and an early age of onset. But Alzheimer disease is a very common illness—about 40 percent of people over age 85 have the condition.

Environment probably plays a role in causing Alzheimer disease too. One study of several hundred nuns is investigating nongenetic contributing factors to Alzheimer disease. So far, the study has shown that nuns who expressed complex thinking in writings early in life had a lower risk of developing Alzheimer disease than nuns with more simplistic literary styles. However, the meaning of such an association, if any, is unclear.

Risk estimates can change depending upon how the groups under comparison are defined. For a couple who has a child with an extra chromosome, such as a child with Down syndrome, the risk of this happening again is 1 in 100, a figure derived from looking at many families who have at least one such child. Therefore, the next time the couple has a child, two risk estimates are possible for Down syndrome—1 in 100, based on the fact that they already have an affected child, and the risk associated with the woman's age. The genetic counselor presents the highest risk, to prepare the family for a worst-case scenario. Consider a 23-year-old and a 42-year-old woman who have each had one child with the extra chromosome of Down syndrome (figure 1.4). Each faces a recurrence risk of 1 in 100 based on medical history, but the two women have different age-associated risks—the 23-year-old's is 1 in 500, but the 42-year-old's is 1 in 63. The counselor provides the 1 in 100 figure to the younger woman, but the age-associated 1 in 63 figure to the older woman.

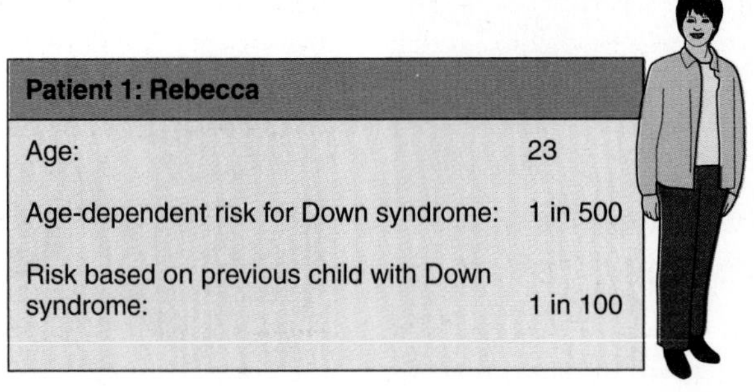


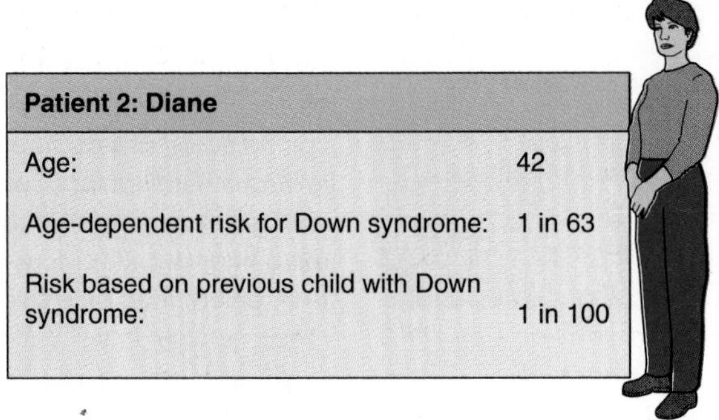


**figure 1.4**

**Relative risk.** Risk may differ depending on how the population group is defined.

Geneticists derive risk figures in several ways. **Empiric risk** comes from population-level observations, such as the 1 in 100 risk of having a second child with an extra chromosome. Another type of risk estimate derives from Mendel's laws. A child whose parents are both carriers of the Mendelian disorder sickle cell disease, for example, faces a 1 in 4, or 25 percent, chance of inheriting the disease. This child also has a 1 in 2, or 50 percent, chance of being a carrier, like the parents. The risk is the same for each offspring. It is a common error to conclude that if two carrier parents have a child with an inherited disorder, the next three children are guaranteed to be healthy. This isn't so, because each conception is an independent event.

**KEY CONCEPTS**

Risk is an estimate of the likelihood that a particular individual will develop a particular condition. Absolute risk is the probability that an individual will develop a certain condition. Relative risk is based on the person's population group compared to another population group.

## 1.5 Applications of Genetics

Barely a day goes by without some mention of genetics in the news. This wasn't true just a few years ago. Genetics is impacting a variety of areas in our everyday lives. Following are looks at some of the topics that are discussed more fully in subsequent chapters.

## Establishing Identity—From Forensics to Rewriting History

Comparing DNA sequences among individuals can establish, or rule out, that the DNA came from the same person, from blood relatives, or from unrelated people. Such DNA typing or fingerprinting has many applications.

Until September, 2001, the media reported on DNA fingerprinting sporadically, and usually in the context of plane crashes or high profile crimes. The same technology became critical in identifying those killed at the World Trade Center. At two biotechnology companies, researchers compared DNA sequences in bone and teeth collected from the scene to hair and skin samples from hairbrushes, toothbrushes, and clothing of missing people, as well as to DNA from relatives.

In more conventional forensic applications, a DNA match for rare sequences between a tissue sample left at a crime scene and a sample from a suspect is strong evidence that the accused person was at the crime scene (or that someone cleverly planted evidence of that person's presence). It has become almost routine for DNA typing to exonerate prisoners, some who had been awaiting execution. DNA typing can add objectivity to a case skewed by human subjectivity, when combined with other types of evidence. Consider what happened to Ronald Jones.

In 1985, at age 34, Jones confessed under police pressure to raping and stabbing to death a young Illinois mother of three. The woman, he claimed, was a prostitute. Jones soon recanted the confession, but he was prosecuted and convicted, possibly because he fit a stereotype of someone capable of committing this crime—he has an IQ of 80, and at the time he was homeless, he was an alcoholic, and he begged on the streets. Results of a DNA test performed in 1989, when the technique was not well developed, were "inconclusive." Jones continued to proclaim his innocence. A team of lawyers believed him, and in 1995, after DNA typing had overturned several dozen convictions, they requested that another DNA test be performed on sperm samples saved from the victim. The DNA test revealed that the man who raped and murdered the young woman was *not* Ronald Jones.

Illinois has been a trendsetter in DNA typing. In 1996, DNA tests exonerated the Ford Heights Four, men convicted of a gang rape and double murder who had spent eighteen years in prison, two of them on death row. In 1999, the men received compensation of $36 million. A journalism class at Northwestern University initiated the investigation that gained the men freedom. The case led to new laws granting death row inmates new DNA tests if their convictions could have arisen from mistaken identity.

DNA evidence can shed light on historical mysteries, too. Consider the offspring of Thomas Jefferson's slave, Sally Hemings. In 1802, Jefferson had been accused of fathering her eldest son, but DNA analysis eventually ruled that out. In 1998, DNA testing compared DNA sequences on the Y chromosomes of descendants of several males important to the case. Y chromosomes were analyzed because they are passed only from father to son.

The results were clear. Jefferson's male descendants had very distinctive Y chromosome DNA sequences, as did the descendants of Hemings' youngest son. Technically, DNA results can disprove paternity, but not prove it—they just provide evidence of an extremely high probability that a man could have fathered a particular child. A brother of Thomas Jefferson would have had such similar DNA that he could not have been excluded as a possible father of Sally Hemings' youngest son.

DNA analysis of bone cells from a child buried in a Roman cemetery in the year 450 A.D. revealed sequences known to come from the parasite that causes malaria. The genetic evidence is consistent with other signs of malaria, such as unusually porous bones and literary references to an epidemic contributing to the fall of the Roman Empire.

History taken even farther back overlaps with Biblical times, and DNA typing can clarify these ancient relationships, too. For example, comparing Y chromosomes reveals that a small group of Jewish people, the cohanim, share distinctive DNA sequences. The cohanim are known as priests and have a special status in the religion. By considering the number of DNA differences between cohanim and other Jewish people, how long it takes DNA to mutate, and the average human generation time (25 years), researchers extrapolated that the cohanim can trace their Y chromosomes to an origin about 2,100 to 3,250 years ago. This is consistent with the time of Moses. According to religious documents, Moses' brother Aaron was the first priest. Interestingly, the Jewish priest DNA signature also appears among the Lemba, a population of South Africans with black skin. Researchers thought to look at them for the telltale gene variants because their customs suggest a Jewish origin—they do not eat pork (or hippopotamus), they circumcise their newborn sons, and they celebrate a weekly day of rest. This story therefore involves genetics, religion, history, and anthropology.

DNA fingerprinting is also used in agriculture. Researchers from France and the United States collected leaves and DNA fingerprints for 300 varieties of wine grapes. The goal was to identify the two parental types that gave rise to the sixteen major types of wine. The researchers already knew that one parent was the bluish-purple Pinot grape, but the DNA analysis revealed that the second parent was a variety of white grape called Gouais blanc (figure 1.5). This surprised wine authorities, because Gouais blanc grapes are so unpopular that they haven't been grown in France or the United States for many years and were actually banned during the Middle Ages. Identifying this second parent provided very valuable information for vintners—if they maintain both parental stocks, they can preserve the gene pool from which the sixteen major wines derive. The finding also confirmed a long-held belief that Pinot and Chardonnay wine grapes are related.

## Health Care

Inherited illnesses caused by a single gene differ from other illnesses in several ways (table 1.4). First, the recurrence risk of such disorders can be predicted by the laws of inheritance, discussed in chapter 4. In contrast, an infectious disease requires that a pathogen be passed from one person to another—a far less predictable circumstance.

A second key difference between inherited illnesses and most other types of medical conditions is that in some situations, an inherited illness can be predicted before

a.

b.

**figure 1.5**

**Surprising wine origins.** (*a*) Gouais blanc and (*b*) Pinot (noir) grapes gave rise to nineteen modern popular wines, including Chardonnay.

**table 1.4**

**How Genetic Diseases Differ from Other Diseases**

1. Can predict recurrence risk in other family members.
2. Presymptomatic testing is possible.
3. Different populations may have different characteristic frequencies.
4. Correction of the underlying genetic abnormality may be possible.

symptoms appear. This is because the genes causing the problem are present in every cell from conception, even though they are not expressed in every cell. Cystic fibrosis, for example, affects the respiratory system and the pancreas, but cells taken from the inside of the cheek or from blood can reveal a mutation. Such genetic information can be considered along with symptoms in refining a diagnosis. Bioethicists debate the value of predicting an untreatable inherited condition years before symptoms arise. Huntington disease, for example, causes personality changes and worsening uncontrollable physical movements, usually beginning at around age 40. Most physicians advise against presymptomatic testing of people under 18 years of age. But older young adults might seek such testing in order to help make decisions about whether to have children and risk passing on the disease-causing gene. The fact that an inherited illness can be passed by a healthy individual raises questions about reproductive choices.

A third aspect of genetic disease is that, because of the structure of human populations, certain inherited disorders are much more common in some populations than others. For economic reasons, it is sensible to offer costly genetic screening tests only to populations in which the detectable gene variant is fairly common. Jewish people of eastern European descent, for example, develop about a dozen genetic diseases at much higher frequencies than other populations, and some companies offer tests that screen for all of these diseases at once. "Jewish disease screens" and other tests targeted to specific population groups are not meant to discriminate, but simply to recognize a biological fact.

Genetic disease also differs from others in that it can sometimes be treated by **gene therapy.** Gene therapy replaces a malfunctioning gene in the affected parts of the body, in effect correcting the gene's faulty instructions. "In Their Own Words" on page 13 is the first of a recurring feature in which people describe their experiences with an inherited illness. In this entry, Don Miller, the first recipient of gene therapy to treat hemophilia, describes his life with this disorder that impairs the ability of the blood to clot. Sadly, not all gene therapy attempts are as successful as Don Miller's is so far. In September 1999, an 18-year-old died from gene therapy. His story is told in chapter 19.

Some people who know they can transmit an inherited illness elect not to have children, or to use donated sperm or ova to avoid passing the condition to their offspring. A technique called preimplantation genetic diagnosis screens eight-celled embryos in a laboratory dish, which allows couples to choose those free of the mutation to complete development in the uterus. Another alternative is to have a fetus tested to determine whether the mutant allele has been inherited. In cases of devastating illnesses, this information may prompt the parents to terminate the pregnancy. Or, the same information may enable parents to prepare for the birth of a disabled or ill child.

## Agriculture

The field of genetics arose from agriculture. Traditional agriculture is the controlled breeding of plants and animals to select new combinations of inherited traits in livestock, fruits, and vegetables that are useful to us. Yet traditional agriculture is imprecise, in that it shuffles many genes—and therefore traits—at a time. The application of DNA-based techniques—biotechnology—enables researchers to manipulate one gene at a time, adding control and precision to agriculture. Biotechnology also enables researchers to create organisms that harbor genes that they would not naturally have.

Foods and other products altered by the introduction of genes from other types of organisms, or whose own gene expression is enhanced or suppressed, are termed "genetically modified," or GM. More specifically, an organism with genes from another species is termed **transgenic.** A GM transgenic "golden" rice, for example, manufactures beta-carotene (a precursor of vitamin

*In Their Own Words*

## Living with Hemophilia

Don Miller was born in 1949 and is semiretired from running the math library at the University of Pittsburgh. Today he has a sheep farm. On June 1, 1999, he was the first hemophilia patient to receive a disabled virus that delivered a functional gene for clotting factor VIII to his bloodstream. Within weeks he began to experience results. Miller is one of the first of a new breed of patients—people helped by gene therapy. Here he describes his life with hemophilia.

The hemophilia was discovered when I was circumcised, and I almost bled to death, but the doctors weren't really sure until I was about eighteen months old. No one where I was born was familiar with it.

When I was three, I fell out of my crib and I was black and blue from my waist to the top of my head. The only treatment then was whole blood replacement. So I learned not to play sports. A minor sprain would take a week or two to heal. One time I fell at my grandmother's house and had a 1-inch-long cut on the back of my leg. It took five weeks to stop bleeding, just leaking real slowly. I didn't need whole blood replacement, but if I moved a little the wrong way, it would open and bleed again.

I had transfusions as seldom as I could. The doctors always tried not to infuse me until it was necessary. Of course there was no AIDS then, but there were problems with transmitting hepatitis through blood transfusions, and other blood-borne diseases. All that whole blood can kill you from kidney failure. When I was nine or ten I went to the hospital for intestinal polyps. I was operated on and they told me I'd have a 10 percent chance of pulling through. I met other kids there with hemophilia who died from kidney failure due to the amount of fluid from all the transfusions. Once a year I went to the hospital for blood tests. Some years I went more often than that. Most of the time I would just lay there and bleed. My joints don't work from all the bleeding.

By the time I got married at age 20, treatment had progressed to gamma globulin from plasma. By then I was receiving gamma globulin from donated plasma and small volumes of cryoprecipitate, which is the factor VIII clotting protein that my body cannot produce pooled from many donors. We decided not to have children because that would end the hemophilia in the family.

I'm one of the oldest patients at the Pittsburgh Hemophilia Center. I was HIV negative, and over age 25, which is what they want. By that age a lot of people with hemophilia are HIV positive, because they lived through the time period when we had no choice but to use pooled cryoprecipitate. I took so little cryoprecipitate that I wasn't exposed to very much. And, I had the time. The gene therapy protocol involves showing up three times a week.

The treatment is three infusions, one a day for three days, on an outpatient basis. So far there have been no side effects. Once the gene therapy is perfected, it will be a three-day treatment. A dosage study will follow this one, which is just for safety. Animal studies showed it's best given over three days. I go in once a week to be sure there is no adverse reaction. They hope it will be a one-time treatment. The virus will lodge in the liver and keep replicating.

In the eight weeks before the infusion, I used eight doses of factor. In the 14 weeks since then, I've used three. Incidents that used to require treatment no longer do. As long as I don't let myself feel stressed, I don't have spontaneous bleeding. I've had two nosebleeds that stopped within minutes without treatment, with only a trace of blood on the handkerchief, as opposed to hours of dripping.

I'm somewhat more active, but fifty years of wear and tear won't be healed by this gene therapy. Two of the treatments I required started from overdoing activity, so now I'm trying to find the middle ground.

A) and stores twice as much iron as unaltered rice. Once these traits are bred into a commercial strain of rice, the new crop may help prevent vitamin A and iron deficiencies in malnourished people living in developing nations (figure 1.6*a*). The new rice variant's valuable traits result from the introduction of genes from petunia and bacteria, as well as boosted expression of one of the plant's own genes.

Another GM crop is "*bt* corn." A gene from the bacterium *Bacillus thuringiensis* (hence, *bt*) encodes a protein that kills certain insect larvae, including the European corn borer, which devastates corn crops. Organic farmers have applied the bacterium's natural protein as a pesticide for decades, but *bt* corn can make its own. The GM crop provides increased yield and lessens reliance on synthetic chemical pesticides.

GM animals are used to produce pharmaceuticals, usually by receiving genes from other species that they express in their milk. For example, when sheep are given the human gene that encodes the factor VIII clotting factor absent in people with hemophilia, they secrete the protein in their milk. This provides a much purer and safer preparation than the pooled blood extracts that once transmitted infections.

Many people object to genetic manipulation, particularly in Europe, where more than 75 percent of the population opposes experimenting with, growing, and marketing GM foods. Protesters have destroyed GM crops, and sometimes unaltered plants,

a.

b.

**figure 1.6**

**Genetic modification.** Modern techniques hold enormous promise, though they engender fear in some. (*a*) Beta-carotene, a precursor of vitamin A, turns this rice yellow. The genetically modified grain also has twice as much iron as nonmodified plants. (*b*) This killer tomato cartoon illustrates the fears some people have of genetic manipulation.

Seymore Chwast, The Pushpin Group, Inc.

too, because the plants look normal. In Seattle, members of a radical environmental activist group mistakenly destroyed 100 very rare trees growing in a laboratory—only 300 exist in the wild. The laboratory trees were not genetically modified.

Reasons cited to boycott GM foods vary. Some are practical—such as the possibility of having an allergic reaction to one plant-based food because it produces a protein normally found in another type of plant. Without food labeling, a consumer would not know that one plant has a protein that it normally would not produce. Another concern is that field tests may not adequately predict the effects of altered plants on ecosystems. For example, *bt* corn has been found growing in places where it was not planted.

Some of the changes made possible through genetic modification are profound. For example, a single gene exchange can create a salmon that grows to twice its normal size, or enable a fish that normally lives in temperate waters to survive in cold water thanks to an antifreeze gene from another type of fish. What effects will these animals have on ecosystems?

Some objections to the genetic modification of plants and animals, however, arise from lack of knowledge about genetics (figure 1.6*b*). A public opinion poll in the United Kingdom discovered that a major reason citizens claim they wish to avoid GM foods is that they do not want to eat DNA! One British geneticist wryly observed that the average meal provides some 150,000 kilometers (about 93,000 miles) of DNA.

Ironically, the British public had been eating GM foods for years before the current concern arose. A very popular "vegetarian cheese" was manufactured using an enzyme made in GM yeast; the enzyme was once extracted from calf stomachs. Researchers inserted the cow gene into the yeast genome, which greatly eased production and collection of the needed enzyme. Similarly, a tomato paste made from tomatoes with a gene added to delay ripening vastly outsold regular tomatoes in England—because it was cheaper! Similarly, a poll of U.S. citizens found that a large majority would not knowingly eat GM foods. Most of the participants were shocked when the interviewer informed them that they had been eating these foods for years.

## A Word on Genetic Equity

DNA microarray tests, gene therapy, new drugs based on genetic information—all are or will be expensive and not widely available for years. In a poor African nation where 2 out of 5 children have AIDS and many others die from other infectious diseases, these biotechnologies are likely to remain in the realm of science fiction for quite some time. It was ironic that scientific journals announced the sequencing of the human genome the same week that the cover of a news magazine featured the AIDS crisis in Africa. It was starkly obvious that while people in economically and politically stable nations may contemplate the coming of genome-based individualized health care, others in less-fortunate situations just try to survive from day to day.

Human genetic information can, however, ultimately benefit everyone. Consider drug development. Today, there are fewer than 500 types of drugs. Genome informa-

*Bioethics: Choices for the Future*

# National Genetic Databases—Is GATTACA Coming?

In several parts of the world, projects are underway to record genetic and other types of health information on citizens. The plans vary in how people participate, but raise similar underlying questions and concerns: How will the information be used? Who will have access to it? How can people benefit from the project?

*Brave New World* is a novel published in 1932 that depicts a society where the government controls reproduction. When the idea for population genetic databases was initially discussed in the mid 1990s in several nations, reaction was largely negative, with people citing this novel as being prophetic. Then in 1998 came the film *GATTACA,* in which a government of a very repressive society records the genome sequence of every citizen. A cell from a stray eyelash gives away the main character's true identity.

The first attempt to establish a population genetic database occurred in Iceland, where a company, deCODE Genetics, received government permission in 1998 to collect existing health and genealogy records, to be supplemented with DNA sequence data. Some Icelandic families trace back more than 1,000 years, and have family trees etched in blood on old leather. The government has promised the people health benefits if the collected information leads to new treatments. Participation in the Icelandic database is presumed. That is, citizens have to file a special form to "opt out" of the project, and this is not as simple as just mailing a letter. Many geneticists from nations where consent must be both voluntary and informed objected to this practice. A large part of the population is participating in the project, because it is administratively difficult not to.

The story is different in Estonia, where more than 90 percent of the 1.4 million population favors a gene pool project. The Estonian Genebank Foundation runs the program, with a for-profit and a nonprofit organization sharing control. The people must volunteer. The project has access to patient registries for cancer, Parkinson disease, diabetes mellitus, and osteoporosis. When patients show up for appointments, they learn about the project and are given the option to fill out a lengthy questionnaire on their health history and donate a blood sample for DNA analysis. The plan is to search for single nucleotide polymorphism (SNP) patterns that are associated with these and other multifactorial disorders, then develop new diagnostic tests and possibly treatments based on the information. Because these diseases are common in many western nations, what is learned from the Estonians may apply to many others. A pilot project consisting of 10,000 individuals is under way to learn if the approach works.

Many population-based studies of gene variants are in the planning stages. Bioethicists have suggested several strategies to ensure that individuals can only benefit from such projects. Some suggestions to assure fair use of genetic information include:

- Preserving choice in seeking genetic tests
- Protecting the privacy of individuals by legally restricting access to genome information
- Tailoring tests to those genes that are most relevant to an individual
- Refusing to screen for trivial traits in offspring, such as eye or hair color, or traits that have a large and controllable environmental component, such as intelligence
- Educating the public to make informed decisions concerning genetic information, including evaluating the risks and benefits of medical tests, judging the accuracy of forensic data, or eating genetically modified foods

If these goals are met, human genome information will reveal the workings of the human body at the molecular level, and add an unprecedented precision and personalization to health care.

tion from humans and also from the various pathogens and parasites that cause illness will reveal new drug targets. Medical organizations all over the world are discussing how nations can share new diagnostic tests and therapeutics that arise from genome information. Over the past few years, genetics has evolved from a basic life science and an obscure medical specialty, to a multifaceted discipline that will impact us all.

## Key Concepts

Genetics has applications in diverse areas. Matching DNA sequences can clarify relationships, which is useful in forensics, establishing paternity, and understanding certain historical events. • Inherited disease differs from other disorders in its predictability; the possibility of presymptomatic detection; characteristic frequencies in different populations; and the potential of gene therapy to correct underlying abnormalities. • Agriculture, both traditional and biotechnological, applies genetic principles. • Information from the human genome project has tremendous potential but must be carefully managed.

# Summary

## 1.1 A Look Ahead

1. Genes are the instructions to manufacture proteins, which determine inherited traits.
2. A **genome** is a complete set of genetic information. A cell contains two genomes of DNA.
3. Before whole genome screens become available, people will choose specific gene tests, based on family history, to detect or even predict preventable or treatable conditions. DNA microarrays detect many genes at once.
4. Genes contribute to rare as well as common disorders.

## 1.2 From Genes to Genomes

5. **Genes** are sequences of **DNA** that encode both the amino acid sequences of proteins and the RNA molecules that carry out protein synthesis. **RNA** carries the gene sequence information so that it can be utilized, while the DNA is transmitted when the cell divides. Much of the genome does not encode protein.
6. Variants of a gene arise by **mutation.** Variants of the same gene are **alleles.** They may differ slightly from one another, but they encode the same product. A **polymorphism** is a general term for a particular site or sequence of DNA that varies in 1 percent or more of a population. The **phenotype** is the gene's expression. An allele combination constitutes the **genotype.** Alleles may be **dominant** (exerting an effect in a single copy) or **recessive** (requiring two copies for expression).
7. **Chromosomes** consist of DNA and protein. The 22 types of **autosomes** do not include genes that specify sex. The X and Y **sex chromosomes** bear genes that determine sex.
8. The human genome is about 3 billion DNA bases. Cells **differentiate** by expressing subsets of genes. **Stem cells** retain the ability to divide without differentiating.
9. Pedigrees are diagrams that are used to study traits in families.
10. Genetic populations are defined by their collections of alleles, termed the **gene pool.**
11. Genome comparisons among species reveal evolutionary relationships.

## 1.3 Genes Do Not Usually Function Alone

12. Single genes determine **Mendelian traits.**
13. **Multifactorial traits** reflect the influence of one or more genes and the environment. Recurrence of a Mendelian trait is based on Mendel's laws; predicting recurrence of a multifactorial trait is more difficult.
14. Genetic determinism is the idea that expression of an inherited trait cannot be changed.

## 1.4 Geneticists Use Statistics to Represent Risks

15. Risk assessment estimates the probability of inheriting a particular gene. **Absolute risk,** expressed as odds or a percentage, is the probability that an individual will develop a particular trait or illness over his or her lifetime.
16. **Relative risk** is a ratio that estimates how likely a person is to develop a particular phenotype compared to another group, usually the general population.
17. Risk estimates are **empiric,** based on Mendel's laws, or modified to account for environmental influences.

## 1.5 Applications of Genetics

18. DNA typing can exclude an individual from being biologically related to someone else or from having committed a crime.
19. Inherited diseases are distinctive in that recurrence risks are predictable, and a causative mutation may be detected before symptoms arise. Some inherited disorders are more common among certain population groups. **Gene therapy** attempts to correct certain genetic disorders.
20. A **transgenic** organism harbors a gene or genes from a different species.

# Review Questions

1. Place the following terms in size order, from largest to smallest, based on the structures or concepts that they represent:
   a. chromosome
   b. gene pool
   c. gene
   d. DNA
   e. genome
2. How is genomics changing the emphasis of the field of human genetics?
3. Distinguish between:
   a. an autosome and a sex chromosome
   b. genotype and phenotype
   c. DNA and RNA
   d. recessive and dominant traits
   e. absolute and relative risks
   f. pedigrees and karyotypes
4. List three ways that inherited disease differs from other types of illnesses.
5. Cystic fibrosis is a Mendelian trait; height is a multifactorial trait. How do the causes of these characteristics differ?
6. How does an empiric risk estimate differ from a risk estimate for a Mendelian disorder?
7. Why is it inaccurate to assume that all mutations are harmful?

# Applied Questions

1. Breast cancer caused by the BRCA1 gene affects 1 in 800 women in the general U.S. population. Among Jewish people of eastern European descent, it affects 2 in 100. What is the relative risk for this form of breast cancer among eastern European Jewish women in the United States?
2. In the general U.S. population, 12 in 10,000 people inherit a form of amyloidosis, in which protein deposits destroy organs. Among a group of 1,000 individuals of an isolated religious sect living in Pennsylvania, 56 have the disorder. What is the relative risk that a person in this Pennsylvania community will develop the condition?
3. A man undergoes a DNA chip test for several genes that predispose to developing prostate cancer. He learns that overall, his relative risk is 1.5, compared to the risk in the general population. He is overjoyed, and tells his wife that his risk of developing prostate cancer is only 1.5 percent. She says that he is incorrect, that his risk is 50 percent greater than that of the average individual in the general population. Who is correct?
4. Crohn disease is an inflammation of the small intestine that causes painful cramps. About 15 percent of the half million people in the United States who have the condition have inherited a mutation that increases their susceptibility. A person who inherits the mutation has about a 25 percent chance of developing Crohn disease. Cite two reasons why the risk of developing Crohn disease for a person who inherits the mutation is not 100 percent.
5. Benjamin undergoes a genetic screening test and receives the following relative risks:

   | | |
   |---|---|
   | – addictive behaviors | 0.6 |
   | – coronary artery disease | 2.3 |
   | – kidney cancer | 1.4 |
   | – lung cancer | 5.8 |
   | – diabetes | 0.3 |
   | – depression | 1.2 |

   Which conditions is he more likely to develop than someone in the general population, and which conditions is he less likely to develop?
6. The Larsons have a child who has inherited cystic fibrosis. Their physician tells them that if they have other children, each faces a 1 in 4 chance of also inheriting the illness. The Larsons tell their friends, the Espositos, of their visit with the doctor. Mr. and Mrs. Esposito are expecting a child, so they ask their physician to predict whether he or she will one day develop multiple sclerosis—Mr. Esposito is just beginning to show symptoms. They are surprised to learn that, unlike the situation for cystic fibrosis, recurrence risk for multiple sclerosis cannot be easily predicted. Why not?
7. A woman picked up scabies (genital crab lice) during a rape. The mites puncture human skin and drink blood. Describe how a forensic entomologist (an insect expert) might use DNA fingerprinting on material from mites collected from the victim and three suspects to identify the rapist. (This is a real case.)
8. What precautions should be taken to ensure that GM foods are safe?
9. Burlington Northern Santa Fe Railroad asked its workers for a blood sample, and then supposedly tested for a gene variant that predisposes a person for carpal tunnel syndrome, a disorder of the wrists that is caused by repetitive motions. The company threatened to fire a worker who refused to be tested; the worker sued the company. The Equal Employment Opportunity Commission ruled in the worker's favor, agreeing that the company's action violated the Americans with Disabilities Act.
   a. Do you agree with the company or the worker? What additional information would be helpful in taking sides?
   b. How is the company's genetic testing not based on sound science?
   c. How can tests such as those described for the two students at the beginning of this chapter be instituted in a way that does not violate a person's right to privacy, as the worker in the railroad case contended?

# Suggested Readings


Cohen, Sharon. August 15, 1999. Science and investigations free death row inmates. *The Associated Press.* A fascinating account of cases where DNA testing made a difference.

Emery, John, and Susan Hayflick. April 28, 2001. The challenge of integrating genetic medicine into primary care. *British Medical Journal* 322:1027–30. Genetic information is entering medical practice.

Foster, Eugene A., et al. November 5, 1998. Jefferson fathered slave's last child. *Nature* 396:27–28. DNA testing can rewrite history.

Holon, Tom. February 19, 2001. Gene pool expeditions. *The Scientist* 15(4):1. A look at population genetic database projects in Estonia and the island of Tonga.

Lewis, Ricki. February 12, 2002. Race and the clinic: good science or political correctness? *The Scientist* 16(4):14. Race may not be a biological concept, but differences in gene frequencies among people of different skin colors may be clinically significant.

Lewis, Ricki. September 3, 2001. Where the bugs are: forensic entomology. *The Scientist* 15(17):10. DNA fingerprinting of insects provides helpful clues to solving crimes.

Lewis, Ricki. July 24, 2000. Keeping up: Genetics to genomics in four editions. *The Scientist* 14:46. A look at the evolution of this textbook.

Lewis, Ricki. July 19, 1999. Iceland's public supports database, but scientists object. *The Scientist* 13:1. Geneticists attempted to prevent genetic technology from interfering with individual freedom and the right to privacy.

Lewis, Ricki. October 13, 1997. Genetic testing for cancer presents complex challenge. *The Scientist* 11:1. Testing for BRCA1 is a genetic counseling nightmare.

Lewis, Ricki, and Barry A. Palevitz. October 11, 1999. Science vs PR: GM crops face heat of debate. *The Scientist* 13:1. Consumers have objections to GM crops that are not always based on scientific facts.

Paabo, Svante. February 16, 2001. The human genome and our view of ourselves. *Science* 291:1219. Knowing the sequence of the human genome provides new ways of looking at ourselves.

Singer, Peter A. and Abdallah S. Daar. October 5, 2001. Harnessing genomics and biotechnology to improve global health equity. *Science* 294:87–89. The New African Initiative is an effort to ensure that all cultures have access to biotechnology.

Skorecki, Karl, et al. January 2, 1997. Y chromosomes of Jewish priests. *Nature* 385:32. Tracing Y chromosomes confirms history.

Verhovek, Sam Howe, and Carol Kaesuk Yoon. May 23, 2001. Foes of genetic engineering are suspects in northwest fires. *The New York Times,* p. F1. Radical opponents of GM organisms sometimes target the wrong experiments.

Wade, Nicholas. May 9, 1999. Group in Africa has Jewish roots, DNA indicates. *The New York Times,* p. F1. DNA and customs reveal that the Lemba are Jewish.

Wilford, John Noble. February 20, 2001. DNA shows malaria helped topple Rome. *The New York Times,* p. F1. DNA clues in the bones of a three-year-old from 1,500 years ago suggest that an epidemic of malaria may have contributed to the fall of the Roman Empire.

Ye, Xudong, et al. January 14, 2000. Engineering the provitamin A (beta carotene) biosynthetic pathway into (carotenoid-free) rice endosperm. *Science* 287:303–5. "Golden rice" may prevent human malnutrition.

# On the Web

Check out the resources on our website at
**www.mhhe.com/lewisgenetics5**

On the web for this chapter you will find additional study questions, vocabulary review, useful links to case studies, tutorials, popular press coverage, and much more. To investigate specific topics mentioned in this chapter, also try the links below:

Alliance of Genetic Support Groups
**www.geneticalliance.org/**

Alzheimer's Association **www.alz.org**

American Cancer Society **www.cancer.org**

Cystic Fibrosis Foundation **www.cff.org**

Genetic Interest Group **www.gig.org.uk**

Glossary of Genetic Terms
**www.nhgri.nih.gov/DIR/VIP/Glossary**

Mannvernd (Icelandic scientists against population genetic database)
**www.mannvernd.is/english/index.html**

Online Mendelian Inheritance in Man
**www.ncbi.nlm.nih.gov/entrez/query.fcgi?db=OMIM**
Alzheimer disease 104300, 104311, 104760, 600759
BRCA1 breast cancer 113705
cystic fibrosis 219700
familial hypercholesterolemia 143890
hemophilia A 306700
sickle cell disease 603903

National Coalition for Health Professional Education in Genetics **www.nchpeg.org**

National Hemophilia Foundation
**www.hemophilia.org**

National Human Genome Research Institute
**www.nhgri.nih.gov**

National Society of Genetic Counselors
**www.nsgc.org**

Northwestern University Law School Center for Wrongful Convictions
**www.cjr.org/year/99/3/innocent/index.asp**

The Scientist **www.the-scientist.com**
(to read articles by the author)

CHAPTER

2

# Cells

### 2.1 The Components of Cells

Inherited characteristics can ultimately be explained at the cellular level. The genetic headquarters, the nucleus, oversees the coordinated functions of specialized organelles, the cell membrane, and the cytoskeleton.

### 2.2 Cell Division and Death

As a human grows, develops, and heals, cells form and die. Both cell division and cell death are highly regulated, stepwise events under genetic control.

### 2.3 Cell-Cell Interactions

Cells must communicate with each other. They do so by receiving and responding to signals, and by physically contacting one another. Signal transduction and cellular adhesion are genetically controlled processes.

### 2.4 Stem Cells and Cell Specialization

Cells specialize as subsets of genes are turned on and off. Pockets of stem cells within organs retain the potential to produce cells that differentiate, making growth, repair, and some regeneration possible.

### 2.5 Viruses and Prions—Not Cells, But Infectious

All living organisms consist of cells. Viruses and prions are not cells, but they can cause infections. A virus is a nucleic acid in a protein coat. A prion is an infectious protein.

The activities and abnormalities of cells underlie inherited traits, quirks, and illnesses. The muscles of a boy with muscular dystrophy weaken because they lack a protein that normally supports the cells' shape during forceful contractions. A child with cystic fibrosis chokes on sticky mucus because the cells lining her respiratory tract produce a malfunctional protein that in its normal form would prevent too much water from leaving the secretions. The red blood cells of a person with sickle cell disease contain an abnormal form of hemoglobin that aggregates into a gel-like mass when the oxygen level is low. The mass bends the red cells into sickle shapes, and they wedge within the tiniest vessels, cutting off the blood supply to vital organs (figure 2.1).

Understanding what goes wrong in certain cells when a disease occurs suggests ways to treat the condition—we learn what must be repaired or replaced. Understanding cell function also reveals how a healthy body works, and how it develops from one cell to trillions. Our bodies include many variations on the cellular theme, with such specialized cell types as bone and blood, nerve and muscle, and even variations of those. Cells interact. They send, receive, and respond to information. Some aggregate with others of like function, forming tissues, which in turn interact to form organs and organ systems. Other cells move about the body. Cell numbers are important, too—they are critical to development, growth, and healing. These processes reflect a precise balance between cell division and cell death.

## 2.1 The Components of Cells

All cells share certain features that enable them to perform the basic life functions of reproduction, growth, response to stimuli, and energy use. Body cells also have specialized features, such as the contractile proteins in a muscle cell, and the hemoglobin that fills red blood cells. The more than 260 specialized or differentiated cell types in a human body arise because the cells express different genes.

Figure 2.1 illustrates three cell types in humans, and figure 2.22 shows some others. The human body's cells fall into four broad categories: epithelium (lining cells), muscle, nerve, and connective tissues (blood, bone, cartilage, adipose, and others).

Other multicellular organisms, including other animals, fungi, and plants, also have differentiated cells. Some single-celled organisms, such as the familiar paramecium and ameba, have very distinctive cells that are as complex as our own. Most of the planet, however, is occupied by simpler single-celled organisms that are nonetheless successful life forms, because they occupied earth long before we did, and are still abundant today.

Biologists recognize three broad varieties of cells that define three major "domains" of life: the Archaea, the Bacteria, and

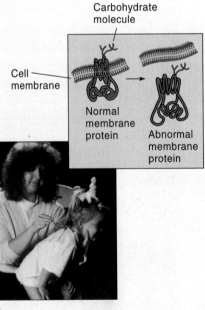

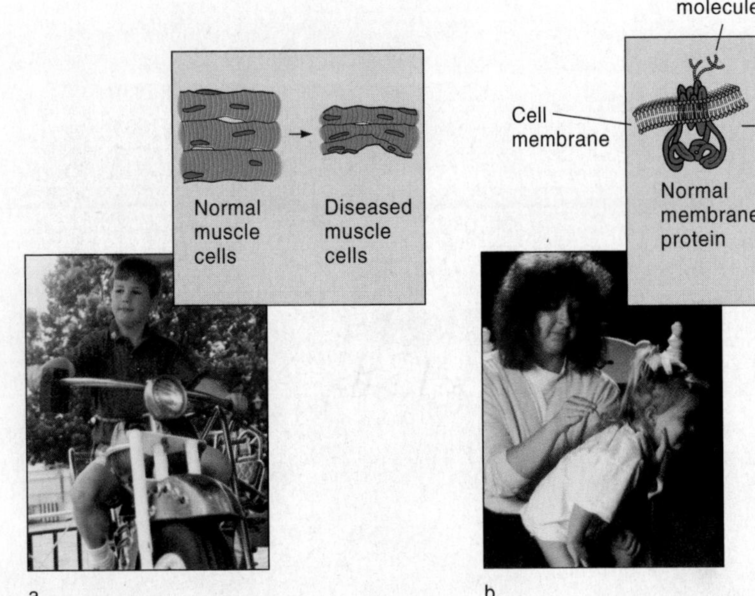


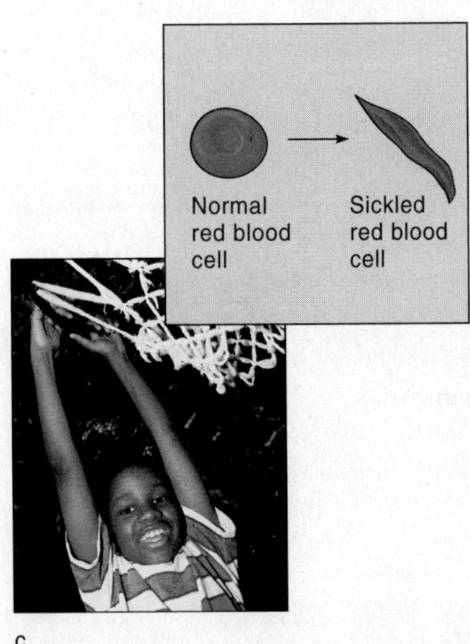


**figure 2.1**

**Genetic disease at the whole-person and cellular levels.** (*a*) This young man has Duchenne muscular dystrophy. The condition has not yet severely limited his activities, but he shows an early sign of the illness—overdeveloped calf muscles that result from his inability to rise from a sitting position the usual way. Lack of the protein dystrophin causes his skeletal muscle cells to collapse when they contract. (*b*) A parent gives this child "postural drainage" therapy twice a day to shake free the sticky mucus that clogs her lungs due to cystic fibrosis. The cells lining her respiratory passages lack a cell membrane protein that controls the entry and exit of salts. (*c*) Seye Arise was born with sickle cell disease, enduring the pain of blocked circulation and several strokes. At age four, he received a bone marrow transplant from his brother Moyo. Today he is fine! The new bone marrow produced red blood cells with a healthy doughnut shape—not the sickle shape his genes dictated.

the Eukarya. A **domain** is a designation that is broader than the familiar kingdom.

The Archaea and Bacteria are both single-celled, but they differ in the sequences of many of their genetic molecules and in the types of molecules in their membranes. Because these differences are much less obvious to us than those between, for example, a tree and a bird, biologists only recently recognized the Archaea and Bacteria as separate domains. In the past, they were lumped together as **prokaryotes,** in recognition of the fact that they both lack a **nucleus,** the structure that contains the genetic material in the cells of other types of organisms. Instead of being contained in a nucleus, the DNA of Archaea and Bacteria is complexed with protein in an area called the nucleoid. Many biologists still distinguish organisms by absence of a nucleus (prokaryotes) or presence of a nucleus (eukaryotes).

The third domain of life, the Eukarya or **eukaryotes,** includes single-celled organisms that have nuclei, as well as all multicellular organisms. The cells of all three domains contain globular structures of RNA and protein called **ribosomes.** Ribosomes provide structural and enzymatic support for protein synthesis.

## Chemical Constituents of Cells

Cells are composed of molecules. The chemicals of life (biochemicals) tend to form large macromolecules. The macromolecules that make up and fuel cells include **carbohydrates** (sugars and starches), **lipids** (fats and oils), **proteins,** and **nucleic acids.** Cells require vitamins and minerals in much smaller amounts, but they are also essential to health.

Carbohydrates provide energy and contribute to cell structure. Lipids form the basis of several types of hormones, provide insulation, and store energy. Proteins have many diverse functions in the human body. They participate in blood clotting, nerve transmission, and muscle contraction and form the bulk of the body's connective tissue. **Enzymes** are proteins that are especially important because they speed, or catalyze, biochemical reactions so that they occur swiftly enough to sustain life.

Most important to the study of genetics are the nucleic acids deoxyribonucleic acid (DNA) and ribonucleic acid (RNA). DNA and RNA form a living language that translates information from past generations into specific collections of proteins that give a cell its individual characteristics. The set of proteins that a cell can manufacture is called its **proteome.**

Macromolecules often combine to form larger structures within cells. For example, the membranes that surround cells and compartmentalize their interiors consist of double layers (bilayers) of lipids embedded with carbohydrates and proteins.

Life is based on the chemical principles that govern all matter. Genetics is based on a highly organized subset of the chemical reactions of life. Reading 2.1 describes some drastic effects that result from abnormalities in the major classes of biochemicals.

## Organelles

A eukaryotic cell holds a thousand times the volume of a bacterial or archaeal cell (figure 2.2). In order to carry out the activities of life in such a large cell, structures called **organelles** divide the labor, partitioning off certain areas or serving a specific function. Saclike organelles sequester biochemicals that might harm other cellular constituents. Some organelles consist of membranes studded with enzymes arranged in a particular physical order that parallels their sequential participation in the chemical reactions that produce a particular molecule. In general, organelles keep related biochemicals and structures close enough to one another to interact efficiently. This eliminates the need to maintain a high concentration of a particular biochemical throughout the cell. Organelles enable a cell to retain as well as use its genetic instructions; acquire energy; secrete substances; and dismantle debris. The coordinated functioning of the organelles in a eukaryotic cell is much like the organization of departments in a department store (figure 2.3).

The most prominent organelle, the nucleus, is enclosed in a layer called the nuclear envelope. Portals called nuclear pores are rings of proteins that allow certain biochemicals to exit or enter the nucleus

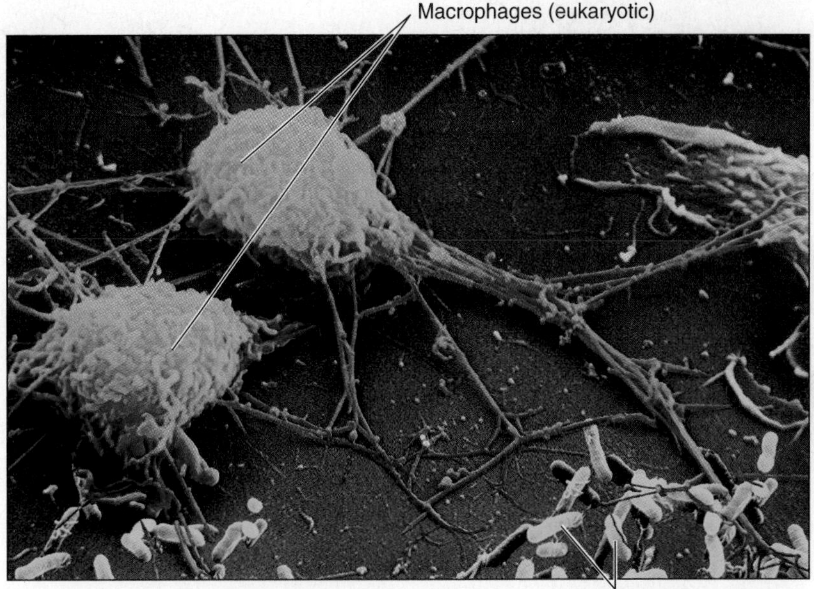


**figure 2.2**

**Eukaryotic and prokaryotic cells.** A human cell is eukaryotic and much more complex than a bacterial cell, while an archaean cell looks much like a bacterial cell. Here, human macrophages capture bacteria. Note how much larger the human cells are than the bacterial cells.

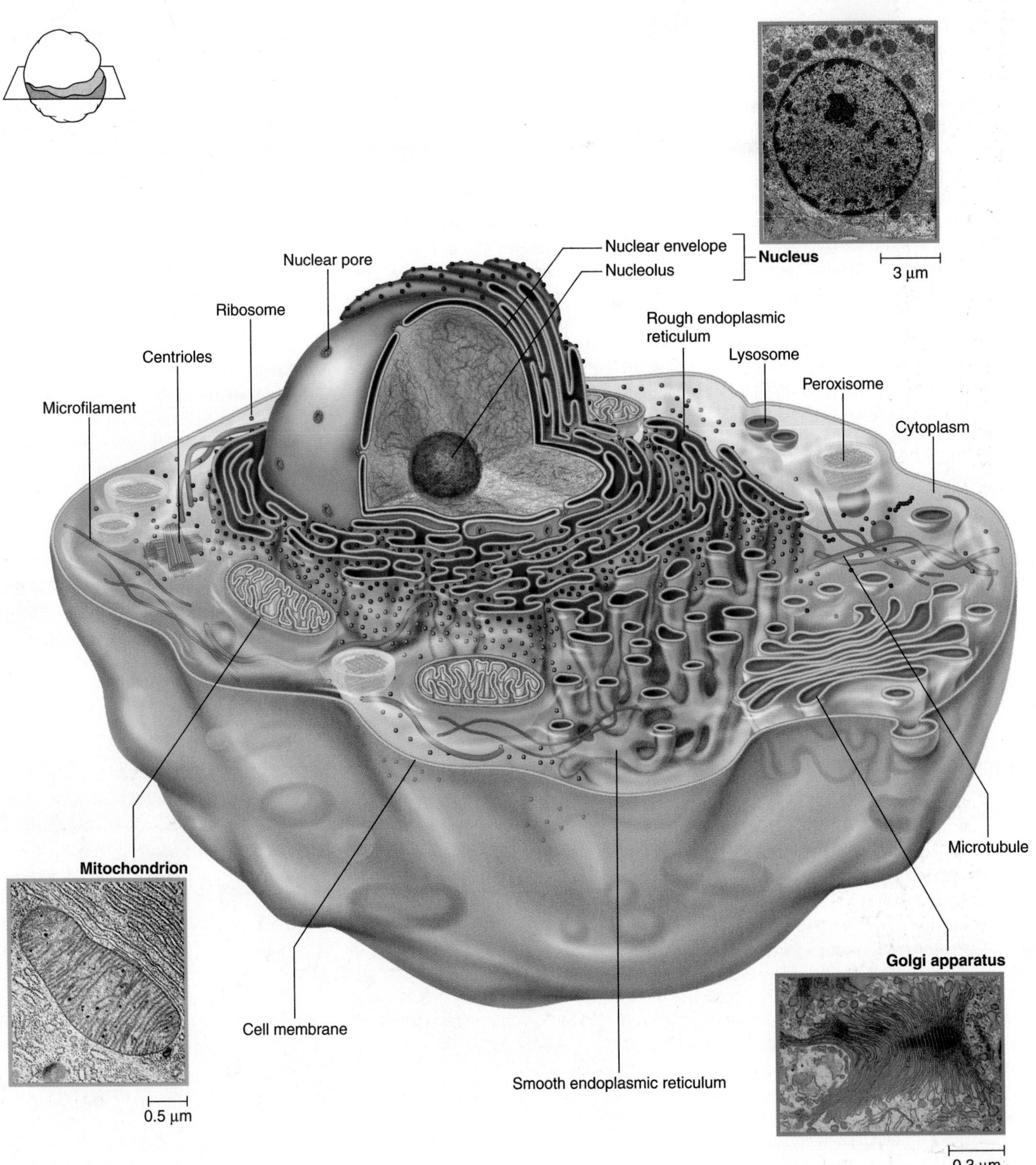


## figure 2.3

**Generalized animal cell.** Organelles provide specialized functions for the cell.

*Reading 2.1*

# Inherited Illness at the Molecular Level

Enzymes are proteins that catalyze (speed) chemical reactions. Therefore, enzymes control a cell's production of all types of macromolecules. When the gene that encodes an enzyme mutates, the result can be too much or too little of the product of the specific biochemical reaction that the enzyme catalyzes. Following are examples of inherited diseases that reflect imbalance or abnormality of particular molecules.

## Carbohydrate

The new parents grew frustrated as they tried to feed their baby, who yowled and pulled up her chubby legs in pain a few hours after each formula feeding. Finally, a doctor identified the problem—the baby lacked the enzyme lactase, which enables the digestive system to break down milk sugar, which is the carbohydrate lactose. Bacteria multiplied in the undigested lactose in the child's intestines, producing gas, cramps, and bloating. Switching to a soybean-based, lactose-free infant formula helped. She had inherited *lactose intolerance.*

## Lipid

A sudden sharp pain began in the man's arm and spread to his chest—the first sign of a heart attack. At age 36, he was younger than most people who suffer heart attacks, but he had inherited a gene that halved the number of protein receptors for cholesterol on his liver cells. Because cholesterol could not enter the liver cells efficiently, it built up in his arteries, constricting blood flow in his heart and eventually causing a mild heart attack. A fatty diet had accelerated his *familial hypercholesterolemia,* an inherited form of heart disease.

## Protein

The first sign that the newborn was ill was also the most innocuous—his urine smelled like maple syrup. Tim slept most of the time, and he vomited so often that he hardly grew. A blood test revealed that Tim had inherited *maple syrup urine disease.* He could not digest three types of amino acids (protein building blocks), so these amino acids accumulated in his bloodstream. A diet very low in these amino acids has helped Tim, but this treatment is new and his future uncertain.

## Nucleic Acids

From birth, Michael's wet diapers contained orange, sandlike particles, but otherwise he seemed healthy. By six months of age, though, urination was obviously painful. A physician also noted that Michael's writhing movements were involuntary rather than normal attempts to crawl.

When the doctor inspected the orange particles in Michael's diaper, she suspected *Lesch-Nyhan syndrome,* a disorder caused by extremely low levels of an enzyme called HGPRT. A blood test confirmed the diagnosis. The near absence of the enzyme blocked Michael's body from recycling two of the four types of DNA building blocks, instead converting them into uric acid, which forms crystals in urine.

Other symptoms that lay in Michael's future were not as easy to understand—severe mental retardation and seizures. Most inexplicable would be the aggressive and self-destructive behavior that is a hallmark of Lesch-Nyhan syndrome. By age three or so, Michael would respond to stress by uncontrollably biting his fingers, lips, and shoulders. He would probably die before the age of 30 of kidney failure or infection.

## Vitamins

Vitamins enable the body to use the carbohydrates, lipids, and proteins we eat. Julie inherited *biotinidase deficiency,* which greatly slows the rate at which her body can use the vitamin biotin.

If Julie hadn't been diagnosed in a state-sponsored newborn screening program and started on biotin supplements shortly after birth, her future would have been grave. By early childhood, she would have shown a variety of biotin-deficiency symptoms, including mental retardation, seizures, skin rash, and loss of hearing, vision, and hair. Her slow growth, caused by her body's inability to extract energy from nutrients, would have eventually proved lethal.

## Minerals

Ingrid is in her thirties, but she lives in the geriatric ward of a state mental hospital, unable to talk or walk. Although her grin and drooling make her appear mentally deficient, Ingrid is alert and communicates using a computer. In 1980, she was a vivacious, normal high-school senior. Then symptoms of *Wilson disease* began to appear, as her weakened liver could no longer control the excess copper her digestive tract absorbed from food.

The initial symptoms of Ingrid's inherited copper poisoning were stomachaches, headaches, and an inflamed liver (hepatitis). By 1983, very odd changes began—slurred speech; loss of balance; a gravelly, low-pitched voice; and altered handwriting. Ingrid received many false diagnoses, including schizophrenia, multiple sclerosis, and Parkinson disease, before a psychiatrist noted the greenish rings around her irises (caused by copper buildup) and diagnosed Wilson disease. Only then did Ingrid receive helpful treatment. A drug, penicillamine, enabled her to excrete the excess copper in her urine, which turned the color of bright new pennies. Although Ingrid's symptoms did not improve, the treatment halted the course of the illness. Without the drug, she would have soon died.

(figure 2.4). Within the nucleus, an area that appears darkened under a microscope, the nucleolus ("little nucleus"), is the site of ribosome production. The nucleus is filled with DNA that is complexed with many proteins to form chromosomes. Other proteins form fibers that give the nucleus a roughly spherical shape. RNA is abundant too, as are the enzymes and protein factors required to synthesize RNA from DNA. The material in the nucleus, minus these contents, is called nucleoplasm.

The remainder of the cell—that is, everything but the nucleus, organelles, and cell membrane—is the **cytoplasm.** Other cellular components include stored proteins, carbohydrates, and lipids; pigment molecules; and various other small chemicals.

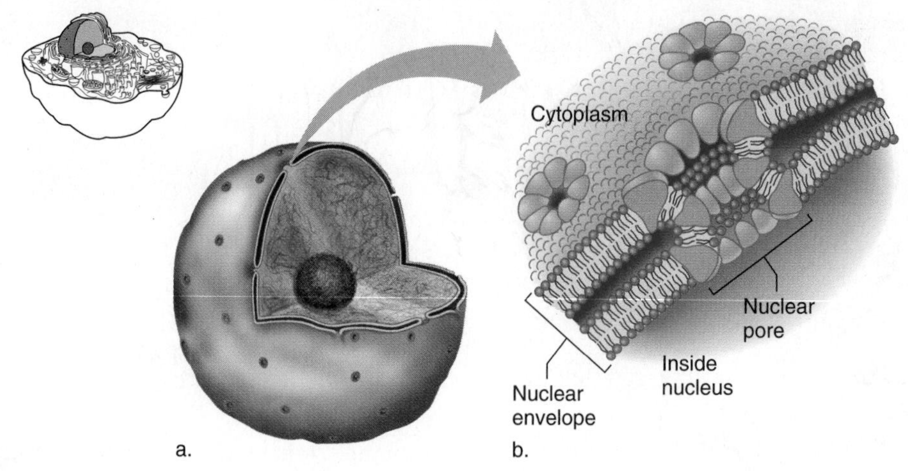


**figure 2.4**

**The nucleus.** (*a*) The largest structure within a typical eukaryotic cell, the nucleus is surrounded by two membrane layers, which make up the nuclear envelope (*b*). Pores through the envelope allow specific molecules to move in and out of the nucleus.

### *Secretion—The Eukaryotic Production Line*

Organelles interact to coordinate basic life functions and sculpt the characteristics of specialized cell types. The activities of several types of organelles may be coordinated to provide a complex function such as secretion.

Secretion begins when the body sends a biochemical message to a cell to begin producing a particular substance. For example, an infant suckling a mother's breast causes the release of hormones from her brain that signal cells in her breast to begin producing milk (figure 2.5). In response, information in certain genes is copied into molecules of **messenger RNA** (mRNA), which then exit the nucleus (see step 2 in figure 2.5). In the cytoplasm, the messenger RNAs, with the help of ribosomes and another type of RNA called **transfer RNA,** direct the manufacture of milk proteins.

Most protein synthesis occurs on a maze of interconnected membranous tubules and sacs that winds from the nuclear envelope to the cell membrane. This membrane labyrinth is the **endoplasmic reticulum** (ER). The portion of the ER nearest the nucleus, which is flattened and studded with ribosomes, is called the rough ER because it appears fuzzy when viewed under an electron microscope. Messenger RNA attaches to the ribosomes on the rough ER. Amino acids from the cytoplasm are then linked, following the instructions in the mRNA's sequence, to form particular proteins that will either exit the cell or become part of membranes

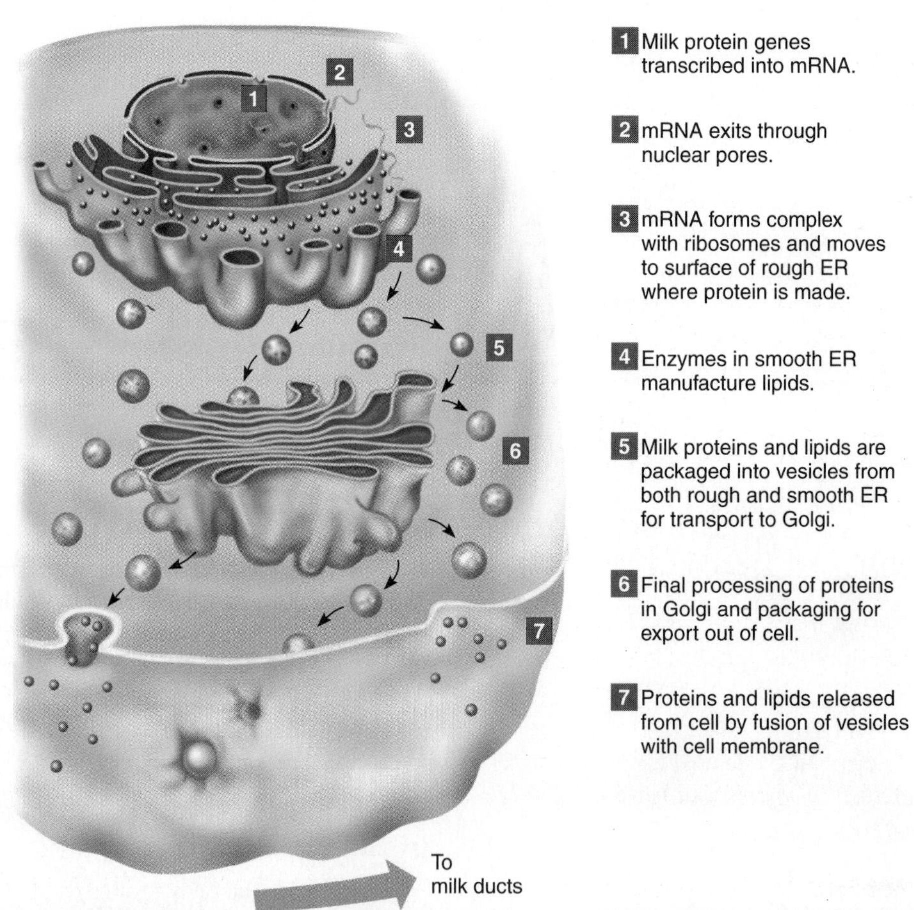

1
2
3
4
5
6
7
To milk ducts
1 Milk protein genes transcribed into mRNA.
2 mRNA exits through nuclear pores.
3 mRNA forms complex with ribosomes and moves to surface of rough ER where protein is made.
4 Enzymes in smooth ER manufacture lipids.
5 Milk proteins and lipids are packaged into vesicles from both rough and smooth ER for transport to Golgi.
6 Final processing of proteins in Golgi and packaging for export out of cell.
7 Proteins and lipids released from cell by fusion of vesicles with cell membrane.


**figure 2.5**

**Secretion.** Milk production and secretion illustrate organelle functions and interactions in a cell from a mammary gland: (*1*) through (*7*) indicate the order in which organelles participate in this process.

(step 3, figure 2.5). Proteins are also synthesized on ribosomes not associated with the ER. These proteins remain in the cytoplasm. (Recent experiments on mammalian cells growing in culture suggest that some proteins may be synthesized in the nucleus.)

The ER acts as a quality control center for the cell. Its chemical environment enables the protein that the cell is manufacturing to start folding into the three-dimensional shape necessary for its specific function. Misfolded proteins are pulled out of the ER and degraded, much as an obviously defective toy might be pulled from an assembly line at a toy factory and discarded.

As the rough ER winds out toward the cell membrane, the ribosomes become fewer, and the diameters of the tubules widen, forming a section called the smooth ER. Here, lipids are made and added to the proteins arriving from the rough ER (step 4, figure 2.5). The lipids and proteins travel until the tubules of the smooth ER eventually narrow and end. Then they exit in membrane-bound, saclike organelles called **vesicles** that pinch off from the tubular endings of the membrane (step 5, figure 2.5).

A loaded vesicle takes its contents to the next stop in the secretory production line, the **Golgi apparatus.** This processing center is a stack of flat, membrane-enclosed sacs. Here, sugars are synthesized and linked to form starches or attach to proteins to form **glycoproteins** or to lipids to form **glycolipids.** Proteins finish folding in the Golgi apparatus (step 6). The components of complex secretions, such as milk, are temporarily stored here. Droplets then bud off the Golgi apparatus in vesicles that move outward to the cell membrane, fleetingly becoming part of the membrane until they are released (secreted) to the cell's exterior (step 7, figure 2.5). Some substances, such as lipids, retain a layer of surrounding membrane when they leave the cell.

## *Cellular Digestion—Lysosomes and Peroxisomes*

Eukaryotic cells break down molecules and other structures as well as produce them. Organelles called **lysosomes** are membrane-bounded sacs that contain enzymes that dismantle captured bacterial remnants, worn-out organelles, and other debris (figure 2.6). Lysosomal enzymes also break down some digested nutrients into forms that the cell can use. Lysosomes fuse with vesicles carrying debris from the outside or from within the cell, and the lysosomal enzymes then degrade the contents. A lysosome loaded with such "garbage" moves toward the cell membrane and fuses with it, dumping its contents to the outside. The word *lysosome* means "body that lyses;" *lyse* means "to cut."

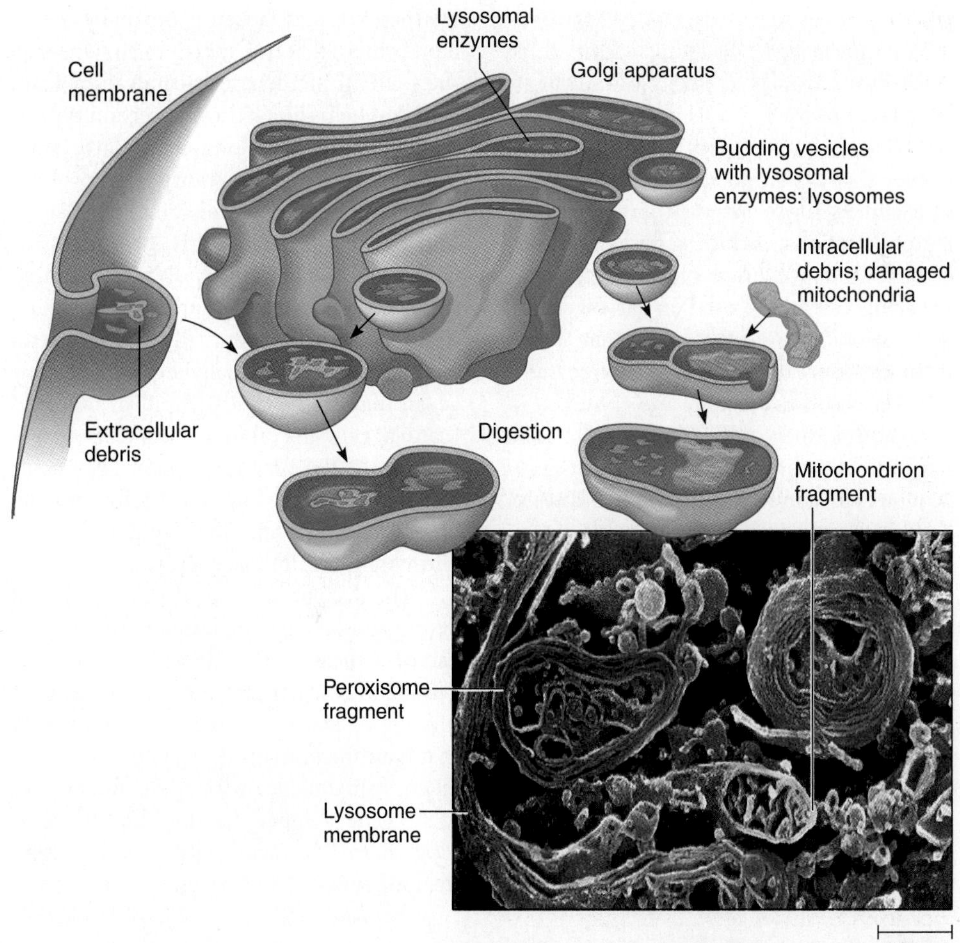


## figure 2.6

**Lysosomes.** Lysosomes fuse with vesicles or damaged organelles, activating the enzymes within to recycle the molecules for use by the cell. Lysosomal enzymes also dismantle bacterial remnants. These enzymes require a very acidic environment to function.

Lysosomal enzymes originate on the ER. These enzymes require a very acidic environment, and the organelle maintains this environment without harming other cellular constituents.

Cells differ in the number of lysosomes they contain. Certain white blood cells and macrophages (see figure 2.2) are the body's scavengers, moving about and engulfing bacteria. They are loaded with lysosomes. Liver cells require many lysosomes to break down cholesterol and toxins.

All lysosomes contain more than 40 types of digestive enzymes, which must be in correct balance to maintain health. Absence or malfunction of just one type of enzyme causes a **lysosomal storage disease.** In these inherited disorders, the molecule that the missing or abnormal enzyme normally degrades accumulates. The lysosome swells, crowding organelles and interfering with the cell's functions. In Tay-Sachs disease, for example, deficiency of an enzyme that normally breaks down lipid in the cells that surround nerve cells buries the nervous system in lipid. An affected infant begins to lose skills at about six months of age, then gradually loses sight, hearing, and the ability to move, typically dying within three years. Even before birth, the lysosomes of affected cells swell.

**Peroxisomes** are sacs with outer membranes that are studded with several types of enzymes. These enzymes perform a variety of functions, including breaking down certain lipids and rare biochemicals, synthesizing bile acids used in fat digestion, and detoxifying compounds that result from exposure to toxic oxygen-free radicals. Peroxisomes are large and abundant in liver and kidney cells (figure 2.7). (The Suggested Readings in chapter 12 lists a recent article about a peroxisomal disease, Zellweger syndrome.)

Absence or malfunction of a single enzyme in a lysosome or peroxisome results in a general type of disorder called an inborn error of metabolism. The nature of the accumulating or missing substance determines specific symptoms.

The 1992 film *Lorenzo's Oil* recounted the true story of a child with an inborn error of metabolism caused by an absent peroxisomal enzyme. Six-year-old Lorenzo Odone had adrenoleukodystrophy (ALD). His peroxisomes lacked a normally abundant protein that transports an enzyme into the peroxisome, where it catalyzes a reaction that helps break down a certain type of lipid called a very-long-chain fatty acid. Without the enzyme transporter protein, the cells of the brain and spinal cord accumulate the fatty acid. Early symptoms include low blood sugar, skin darkening, muscle weakness, and heartbeat irregularities. The patient eventually loses control over the limbs and usually dies within a few years. Ingesting a type of lipid in rapeseed (canola) oil—the oil in the film title—slows buildup of the very-long-chain fatty acids for a few years, but eventually impairs blood clotting and other vital functions. Ultimately, the illness progresses.

The disappointment over the failure of "Lorenzo's oil" may be lessened by the new use of a drug to activate a different gene. This gene's protein product can replace the missing or abnormal one in ALD. In mice that have the human ALD gene, and in cells taken from children with ALD, the replacement gene stopped the buildup of very-long-chain fatty acids, and also increased the number of peroxisomes.

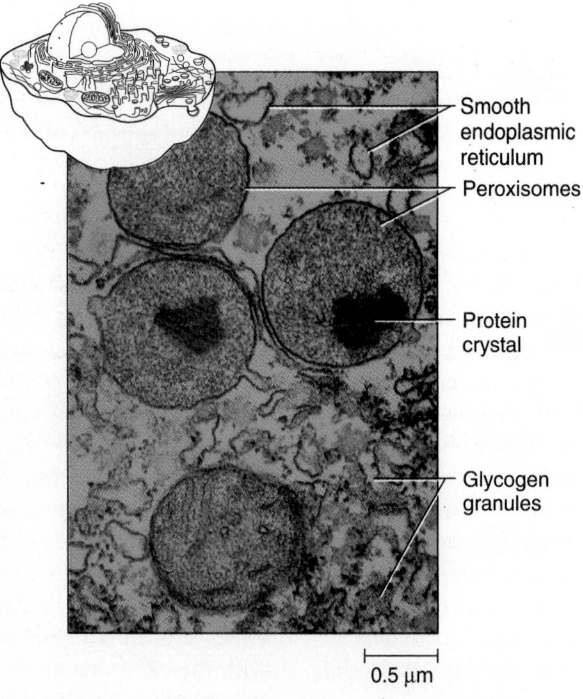


**figure 2.7**

**Peroxisomes.** The high concentration of enzymes within peroxisomes results in crystallization of the proteins, giving peroxisomes a characteristic appearance. Peroxisomes are abundant in liver cells, where they assist in detoxification processes.

### *Energy Production—Mitochondria*

The activities of secretion, as well as the many chemical reactions taking place in the cytoplasm, require enormous and continual energy. Organelles called **mitochondria** provide energy by breaking down the products of digestion (nutrients).

A mitochondrion has an outer membrane similar to those in the ER and Golgi apparatus and an inner membrane that is folded into structures called cristae (figure 2.8). These folds hold enzymes that catalyze the biochemical reactions that release energy from the chemical bonds of nutrient molecules. The bonds that hold together a molecule called adenosine triphosphate (ATP) capture this energy. ATP is, therefore, a cellular energy currency.

The number of mitochondria in a cell varies from a few hundred to tens of thousands, depending upon the cell's activity level. A typical liver cell, for example, has about 1,700 mitochondria, but a muscle cell, with its very high energy requirements, has many more.

Mitochondria are especially interesting to geneticists because, like the nucleus, they contain DNA, although a very small amount. Another unusual characteristic of mitochondria is that they are almost always inherited from the mother only, because mitochondria are in the middle regions of sperm cells but usually not in the head regions that enter eggs. Mitochondria that do enter with a sperm are usually destroyed in the very early embryo. A class of inherited diseases whose symptoms result from abnormal mitochondria are always passed from mother to offspring. These illnesses usually produce extreme muscle weakness, because muscle activity requires many mitochondria. Chapter 5 discusses mitochondrial inheritance. Evolutionary biologists study mitochondrial genes to trace the beginnings of humankind, discussed in chapter 15.

Table 2.1 summarizes the structures and functions of organelles.

## The Cell Membrane

Just as the character of a community is molded by the people who enter and leave it, the special characteristics of different cell types are shaped in part by the substances that enter and leave. The **cell membrane**

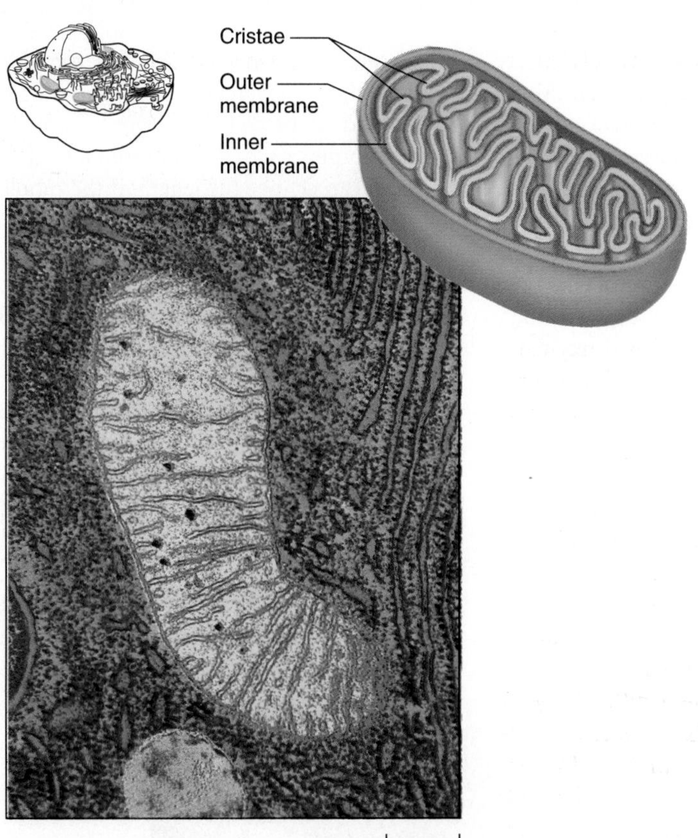


**figure 2.8**

**A mitochondrion.** Cristae, infoldings of the inner membrane, increase the available surface area containing enzymes for energy reactions in this mitochondrion.

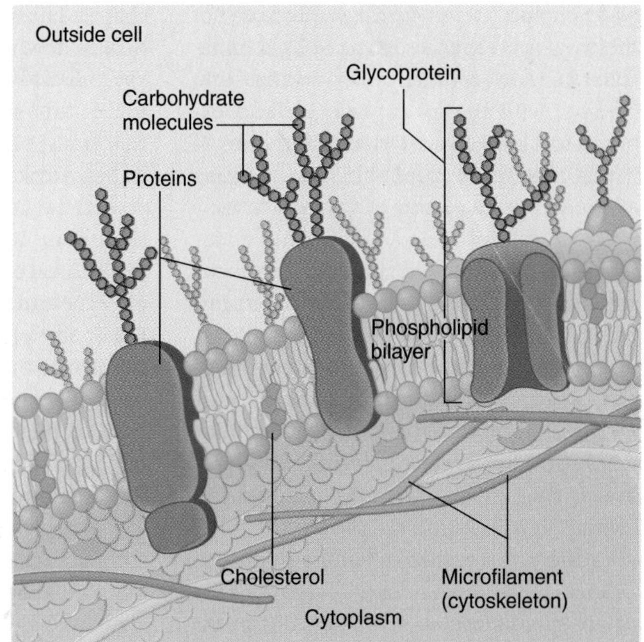


**figure 2.9**

**Anatomy of a cell membrane.** In a cell membrane, mobile proteins are embedded throughout a phospholipid bilayer, producing a somewhat fluid structure. An underlying mesh of protein fibers supports the cell membrane. Jutting from the membrane's outer face are carbohydrate molecules linked to proteins (glycoproteins) and lipids (glycolipids).

**table 2.1**

**Structures and Functions of Organelles**

| Organelle | Structure | Function |
|---|---|---|
| Endoplasmic reticulum | Membrane network; rough ER has ribosomes, smooth ER does not | Site of protein synthesis and folding; lipid synthesis |
| Golgi apparatus | Stacks of membrane-enclosed sacs | Site where sugars are made and linked into starches, or joined to lipids or proteins; proteins finish folding; secretions stored |
| Lysosome | Sac containing digestive enzymes | Degrades debris, recycles cell contents |
| Mitochondrion | Two membranes; inner membrane enzyme-studded | Releases energy from nutrients |
| Nucleus | Porous sac containing DNA | Separates DNA from rest of cell |
| Peroxisome | Sac containing enzymes | Catalyzes several reactions |
| Ribosome | Two associated globular subunits of RNA and protein | Scaffold and catalyst for protein synthesis |
| Vesicle | Membrane-bounded sac | Temporarily stores or transports substances |

controls this process. It forms a selective barrier that completely surrounds the cell and monitors the movements of molecules in and out of the cell. The chemicals that comprise the cell membrane and how they associate with each other determine which substances can enter or leave. Similar membranes form the outer boundaries of several organelles, and some organelles consist entirely of membranes.

A biological membrane is built of a double layer (bilayer) of molecules called **phospholipids** (figure 2.9). A phospholipid is a lipid (fat) molecule with attached phosphate groups ($PO_4$, a phosphorus atom bonded to four oxygen atoms). The ability of phospholipid molecules to organize themselves into sheetlike structures makes membrane formation possible. Phospholipids do this because their ends have opposite reactions to water. The phosphate end of a phospholipid is attracted to water, and thus is hydrophilic (water-loving); the other end, which consists of two chains of fatty acids, moves away from water, and is therefore hydrophobic (water-fearing). Because of these water preferences, phospholipid molecules in water spontaneously arrange into bilayers, with the hydrophilic surfaces exposed to the watery exterior and interior of the cell, and the hydrophobic surfaces facing each other on the inside of the bilayer, away from the water.

The phospholipid bilayer forms the structural backbone of a biological membrane. Embedded in the bilayer are proteins, some traversing the entire bilayer, others poking out from either or both faces. Other molecules can attach to these membrane proteins, forming **glycoproteins** and **glycolipids.** The proteins, glycoproteins, and glycolipids that jut from a cell membrane create the surface topographies that are so important in a cell's interactions with other cells.

Many molecules that extend from the cell membrane serve as **receptors.** These are structures that have indentations or other shapes that fit and hold molecules outside the cell. The molecule that binds to the receptor, called the **ligand,** sets into motion a cascade of chemical reactions that carries out a particular cellular activity. This process of communication from outside to inside the cell is termed **signal transduction.** Other membrane proteins enable a cell to stick to other cells in a process called **cellular adhesion.** Signal transduction and cellular adhesion are discussed in greater detail in section 2.3. The surfaces of your own cells indicate that they are part of your body, and also that they have differentiated in a particular way.

The phospholipid bilayer is oily and many of the proteins move within it like ships on a sea. The inner hydrophobic region of the phospholipid bilayer blocks entry and exit to most substances that dissolve in water. However, certain molecules can cross the membrane through proteins that form passageways, or when they are escorted by a "carrier" protein. Some membrane proteins form channels for ions, which are atoms or molecules that bear an electrical charge. Reading 2.2 describes how faulty ion channels can cause disease.

## The Cytoskeleton

The cytoskeleton is a meshwork of tiny protein rods and tubules that molds the distinctive structures of cells, positioning organelles and providing three-dimensional shapes. The proteins of the cytoskeleton are broken down and built up as a cell performs specific activities. Some cytoskeletal elements function as rails, forming conduits that transport cellular contents; other parts of the cytoskeleton, called motor molecules, power the movement of organelles along these rails by converting chemical energy to mechanical energy.

The cytoskeleton includes three major types of elements—**microtubules, microfilaments,** and **intermediate filaments** (figure 2.10). They are distinguished by protein type, diameter, and how they aggregate into larger structures. Other proteins connect these components to each other, creating the meshwork that provides the cell's strength and ability to resist forces, which maintains shape.

Long, hollow microtubules provide many cellular movements. A microtubule is composed of pairs (dimers) of a protein, called tubulin, assembled into a hollow tube. The cell can change the length of the tubule by adding or removing tubulin molecules.

Cells contain both formed microtubules and individual tubulin molecules. When the call requires microtubules to carry out a specific function—dividing, for example—the free tubulin dimers self-assemble into more tubules. After the cell divides, some of the microtubules fall apart into individual tubulin dimers. This replenishes the cell's supply of building blocks. Cells are in a perpetual state of flux, building up and breaking down microtubules. Some drugs used to treat cancer affect the microtubules that pull a cell's duplicated

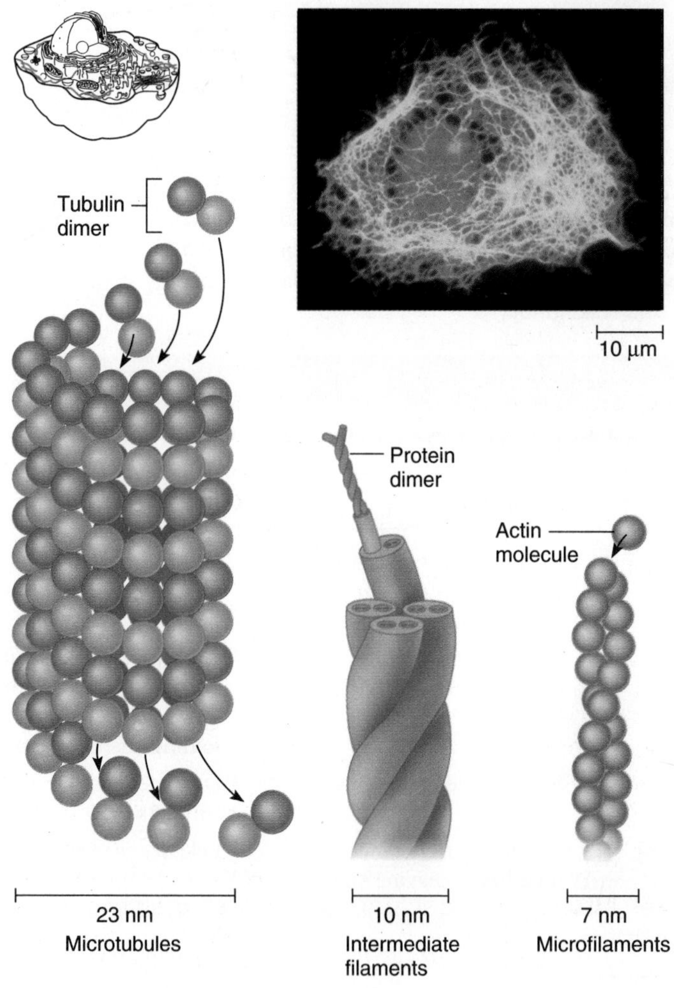


**figure 2.10**

**The cytoskeleton is made of protein rods and tubules.** The three major components of the cytoskeleton are microtubules, intermediate filaments, and microfilaments. Through special staining, the cytoskeleton in this cell glows yellow under the microscope. (nm stands for nanometer, which is a billionth of a meter.)

Reading 2.2

## Inherited Diseases Caused by Faulty Ion Channels

What do collapsing horses, irregular heartbeats in teenagers, and cystic fibrosis have in common? All result from abnormal ion channels in cell membranes.

Ion channels are protein-lined tunnels in the phospholipid bilayer of a biological membrane. These passageways permit electrical signals to pass in and out of membranes in the form of ions (charged particles).

Ion channels are specific for calcium ($Ca^{+2}$), sodium ($Na^+$), potassium ($K^+$), or chloride ($Cl^-$). A cell membrane may have a few thousand ion channels specific for each of these ions. Ten million ions can pass through an ion channel in one second! The following disorders result from abnormal ion channels.

### Hyperkalemic Periodic Paralysis and Sodium Channels

The quarter horse was originally bred in the 1600s to run the quarter mile, but one of the four very fast stallions used to establish much of today's population of 3 million animals inherited *hyperkalemic periodic paralysis* (HPP). The horse, otherwise a champion, collapsed from sudden attacks of weakness and paralysis.

HPP results from abnormal sodium channels in the cell membranes of muscle cells. But the trigger for the temporary paralysis is another ion: potassium. A rising blood potassium level, which may follow intense exercise, slightly alters a muscle cell membrane's electrical charge. Normally, this slight change would have no effect, but in horses with HPP, sodium channels open too widely, allowing too much sodium into the cell. The muscle cell cannot respond to nervous stimulation for awhile, and the racehorse falls.

People can inherit HPP, too. In one family, several members collapsed suddenly after eating bananas! These fruits are very high in potassium, which caused the symptoms.

### Long-QT Syndrome and Potassium Channels

Four children in a Norwegian family were born deaf, and three of them died at ages four, five, and nine. All of the children had inherited from unaffected carrier parents *long-QT syndrome associated with deafness.* ("QT" refers to part of a normal heart rhythm on an EKG.) They had abnormal potassium channels in the heart muscle and in the inner ear. In the heart, the malfunctioning channels disrupted electrical activity, causing a fatal disturbance to the heart rhythm. In the inner ear, the abnormal channels caused an increase in the extracellular concentration of potassium ions, impairing hearing.

### Cystic Fibrosis and Chloride Channels

A 17th century English saying, "A child that is salty to taste will die shortly after birth," described the consequence of abnormal chloride channels in CF. The chloride channel is called CFTR, for cystic fibrosis transductance regulator. In most cases of CF, CFTR protein remains in the cytoplasm, unable to reach the cell membrane, where it would function (see figure 2.1*b*). CF is inherited from carrier parents. The major symptoms of difficulty breathing, frequent severe respiratory infections, and a clogged pancreas that disrupts digestion all result from buildup of extremely thick mucous secretions.

Abnormal chloride channels in cells lining the lung passageways and ducts of the pancreas cause the symptoms of CF. The primary defect in the chloride channels also causes sodium channels to malfunction. The result: salt trapped inside cells draws moisture in and thickens surrounding mucus.

---

chromosomes apart, either by preventing tubulin from assembling into microtubules, or by preventing microtubules from breaking down into free tubulin dimers. In each case, cell division stops.

Microtubules also form structures called cilia that move or enable cells to move. Cilia are hairlike structures that move in a coordinated fashion, producing a wavelike motion. An individual cilium is constructed of nine microtubule pairs that surround a central, separated pair. A type of motor protein called dynein connects the outer microtubule pairs and also links them to the central pair. Dynein supplies the energy to slide adjacent microtubules against each other. This movement bends the cilium. Coordinated movement of these cellular extensions sets up a wave that moves the cell or propels substances along its surface. Cilia beat particles up and out of respiratory tubules, and move egg cells in the female reproductive tract. Reading 2.3 describes a condition caused by abnormal dynein that alters the positions of certain organs in the body.

Another component of the cytoskeleton is the microfilament, which is a long, thin rod composed of the protein actin. In contrast to microtubules, microfilaments are not hollow and are narrower. Microfilaments provide strength for cells to survive stretching and compressive forces. They also help to anchor one cell to another and provide many other functions within the cell through proteins that interact with actin. When any of these proteins is absent or abnormal, a genetic disease results.

Intermediate filaments are so named because their diameters are intermediate between those of the other cytoskeletal elements. Unlike microtubules and microfilaments, which consist of a single protein, intermediate filaments are made of different proteins in different specialized cell types. However, all intermediate filaments share a common overall organization of dimers entwined into nested coiled rods. Intermediate

# Reading 2.3

# A Heart in the Wrong Place

In the original *Star Trek* television series, Dr. McCoy often complained when examining Mr. Spock that Vulcan organs weren't where they were supposed to be, based on human anatomy. The good doctor would have had a hard time examining humans with a condition called *situs inversus,* in which certain normally asymmetrically located organs develop on the wrong side of the body.

In a normal human body, certain organs lie either on the right or the left of the body's midline. The heart, stomach, and spleen are on the left, and the liver is on the right. The right lung has three sections, or lobes; the left lung has two. Other organs twist and turn in either a right or left direction. All these organs originate in the center of an initially symmetrical embryo, and then the embryo turns, and the organs migrate to their final locations.

Misplaced body parts are a symptom of *Kartagener syndrome,* in which the heart, spleen, or stomach may be on the right, both lungs may have the same number of lobes, the small intestine may twist the wrong way, or the liver may span the center of the body (figure 1). Many people with this syndrome die in childhood from heart abnormalities. Kartagener syndrome was first described in 1936 by a Swiss internist caring for a family with several members who had strange symptoms—chronic cough, sinus pain, poor sense of smell, male infertility, and misplaced organs—usually a heart on the right.

Many years later, researchers identified another anomaly in patients with Kartagener syndrome, which explained how the heart winds up on the wrong side of the chest. All affected individuals lack dynein, the protein that enables microtubules to slide past one another and generate motion. Without dynein, cilia cannot wave. In the upper respiratory tract, immobile cilia allow debris and mucus to accumulate, causing the cough, clogged sinuses, and poor sense of smell. Lack of dynein also paralyzes sperm tails, producing male infertility. But how could dynein deficiency explain a heart that develops on the right instead of the left?

One hypothesis is based on the fact that dynein helps establish the spindle, the structure that determines the orientation of dividing cells in the embryo with respect to each other. The dynein defect may, early on, set cells on a developmental pathway that diverts normal migration of the heart from the embryo's midline to the left.

Another explanation for how organs end up in the wrong locations comes from mice genetically altered to lack a gene whose protein product is required for assembling cilia. About 50 percent of the mice have reversed organs. All of the animals either lack cilia that are normally present on cells of the early embryo, called node cells, or the cilia cannot move. Node cells are the sites where the first differences between right and left arise, and they set the pattern for organ placement. Normally, the cilia rotate counterclockwise, which moves fluids to the left. This movement creates a gradient (changing concentration across an area) of molecules, called morphogens, that control development. The differing concentrations of specific morphogens in different parts of the body may send signals that guide the development of organs. Without cilia that can move, organs wind up on the left or the right at random. This is why only 50 percent of the genetically altered mice, and presumably not all people who inherit Kartagener syndrome, have organs in the wrong place. Just by chance, some of them develop normally.

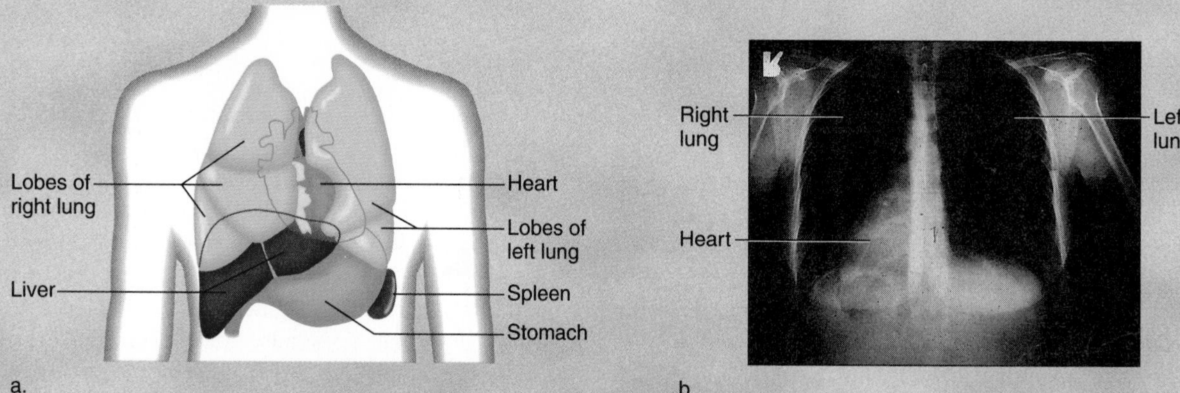


**figure 1**

**Situs inversus.** The drawing on the left (*a*) shows the normal position of the heart, lungs, liver, spleen, and stomach. In situs inversus, certain organs form on the wrong side of the body (*b*). Note the location of the heart.

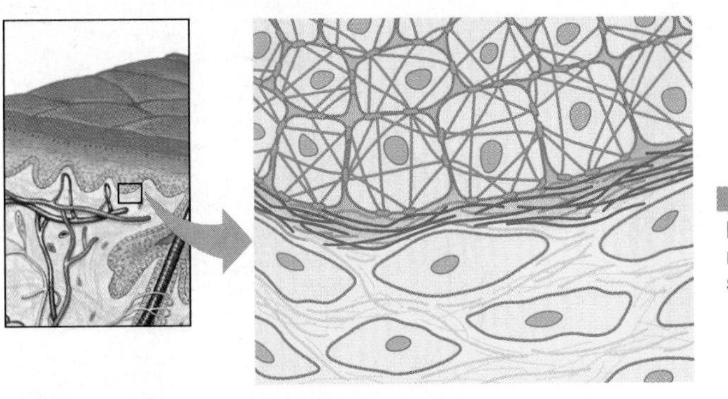

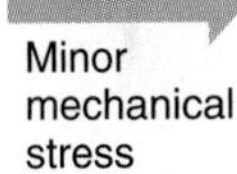


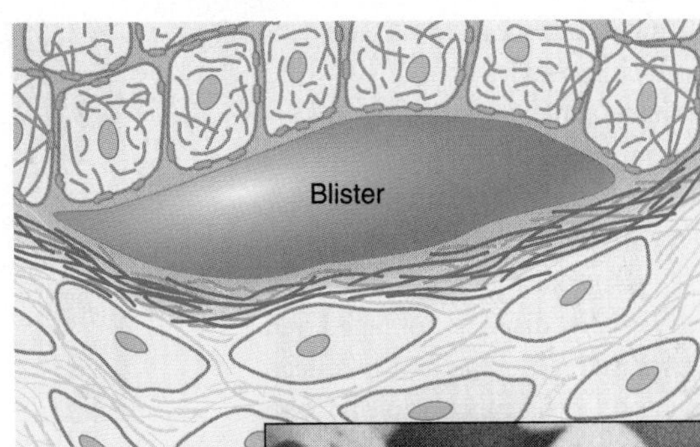


a.

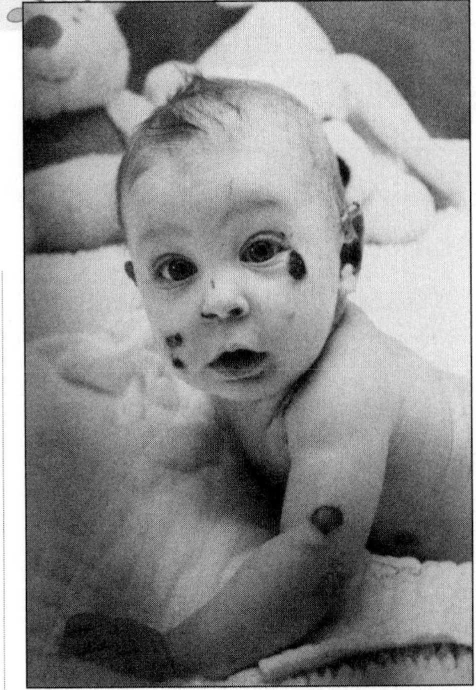

b.

**figure 2.11**

**Intermediate filaments in skin.** Keratin intermediate filaments internally support cells in the basal (bottom) layer of the epidermis (*a*). Abnormal intermediate filaments in the skin cause epidermolysis bullosa, a disease characterized by the ease with which skin blisters (*b*).

filaments are scarce in many cell types, but are very abundant in skin.

Intermediate filaments in actively dividing skin cells in the bottommost layer of the epidermis form a strong inner framework which firmly attaches the cells to each other and to the underlying tissue. These cellular attachments are crucial to the skin's barrier function. In a group of inherited conditions called epidermolysis bullosa, intermediate filaments are abnormal, which causes the skin to blister easily as tissue layers separate (figure 2.11).

Disruption in the structures of cytoskeletal proteins, or in how they interact, can be devastating. Consider hereditary spherocytosis, which disturbs the interface between the cell membrane and the cytoskeleton in red blood cells.

The doughnut shape of normal red blood cells enables them to squeeze through the narrowest blood vessels. Rods of a protein called spectrin form a meshwork beneath the cell membrane, strengthening the cell, and proteins called ankyrins attach the spectrin rods to the cell membrane (figure 2.12). Spectrin also attaches to the microfilaments and microtubules of the cytoskeleton. Spectrin molecules are like steel girders, and ankyrins are like nuts

a.

Extracellular matrix
(outside of cell)
Glycoprotein
Carbohydrate
molecules
Phospholipid
bilayer
Ankyrin
Interior
face of cell
membrane
Spectrin
Cytoplasm

b.

**figure 2.12**

**The red blood cell membrane.** (*a*) The cytoskeleton that supports the cell membrane of a red blood cell enables it to withstand the great turbulent force of circulation. (*b*) In the cell membrane, proteins called ankyrins bind molecules of spectrin from the cytoskeleton to the interior face. On its other end, ankyrin binds proteins that help ferry molecules across the cell membrane. In hereditary spherocytosis, abnormal ankyrin causes the cell membrane to collapse—a problem for a cell whose function depends upon its shape.

and bolts. If either ankyrins or spectrin is absent, the cell collapses.

In hereditary spherocytosis, the ankyrins are abnormal, parts of the red blood cell membrane disintegrate, and the cell balloons out. The bloated cells obstruct narrow blood vessels—especially in the spleen, the organ that normally disposes of aged red blood cells. Anemia develops as the spleen destroys red blood cells more rapidly than the bone marrow can replace them, producing great fatigue and weakness. Removing the spleen can treat the condition.

**KEY CONCEPTS**

Cells are the units of life. They consist mostly of carbohydrates, lipids, proteins, and nucleic acids. • Organelles subdivide specific cell functions. They include the nucleus, the endoplasmic reticulum (ER), Golgi apparatus, mitochondria, lysosomes, and peroxisomes. • The cell membrane is a flexible, selective phospholipid bilayer with embedded proteins. • The cytoskeleton is an inner structural framework made of protein rods, tubules, connectors, and motor molecules.

## 2.2 Cell Division and Death

The cell numbers in a human body must be in balance to develop normally and maintain health. The process of mitotic cell division, or **mitosis,** provides new cells by forming two cells from one. Mitosis occurs in **somatic cells** (all cells but the sperm and eggs). Although it seems counterintuitive, some cells must die as a body forms, just as a sculptor must take away some clay to shape the desired object. A foot, for example, might start out as a webbed triangle of tissue, with digits carved from it as certain cells die. This type of cell death, which is a normal part of development, is called **apoptosis.** It is a precise, genetically programmed sequence of events, as is mitosis (figure 2.13).

### The Cell Cycle

Many cell divisions transform a single fertilized egg into a many-trillion-celled person. A series of events called the **cell cycle** describes when a cell is dividing or not dividing.

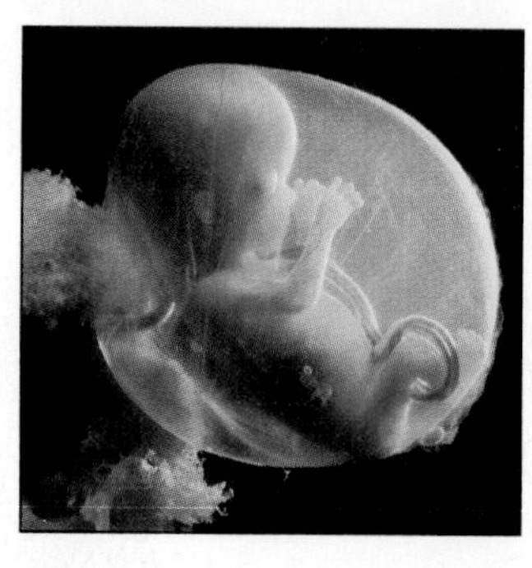

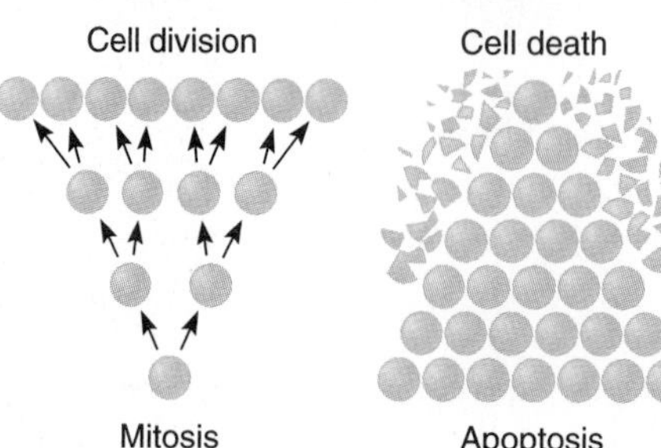


**figure 2.13**

**Mitosis and apoptosis mold a body.** Biological structures in animal bodies enlarge, allowing organisms to grow, as opposing processes regulate cell number. Cell numbers increase from mitosis and decrease from apoptosis.

Cell cycle rate varies in different tissues at different times. A cell lining the small intestine's inner wall may divide throughout life; a cell in the brain may never divide; a cell in the deepest skin layer of a 90-year-old may divide more if the person lives long enough. Frequent mitosis enables the embryo and fetus to grow rapidly. By birth, the mitotic rate slows dramatically. Later, mitosis must maintain the numbers and positions of specialized cells in tissues and organs.

The cell cycle is a continual process, but we divide it into stages based on what we see. The two major stages are **interphase** (not dividing) and mitosis (dividing) (figure 2.14). In mitosis, a cell's replicated chromosomes are distributed into two daughter cells. This maintains the set of 23 chromosome pairs characteristic of a human somatic cell. Another form of cell division, meiosis, produces sperm or eggs. These cells contain half the usual amount of genetic material, or 23 single chromosomes. Chapter 3 discusses meiosis.

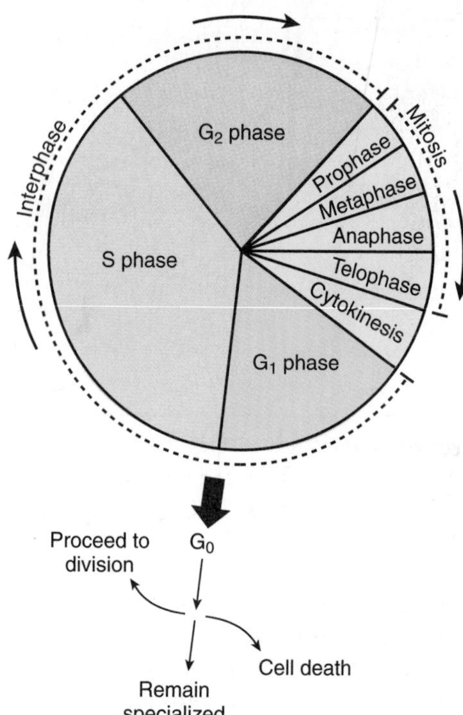


**figure 2.14**

**The cell cycle.** The cell cycle is divided into interphase, when cellular components are replicated to prepare for division, and mitosis, when the cell splits in two, distributing its contents into two daughter cells. Interphase is divided into two gap phases ($G_1$ and $G_2$), when the cell duplicates specific molecules and structures, and a synthesis phase (S), when it replicates the genetic material. Mitosis is divided into four stages, plus cytokinesis, which is when the cells physically separate. Another stage, $G_0$, is a "time-out" when a cell "decides" which course of action to follow.

#### *Interphase—A Time of Great Activity*

Interphase is a very active time. The cell continues the basic biochemical functions of life and also replicates its DNA and other subcellular structures for distribution to daughter cells.

Interphase is divided into two **gap (G) phases** and one **synthesis (S) phase.** A cell can exit the cell cycle at $G_1$ to enter $G_0$, a quiescent phase. A cell in $G_0$ can maintain its specialized characteristics, but it does not replicate its DNA or divide. It is a cellular "time out." During the first gap phase ($G_1$), the cell resumes synthesis of proteins, lipids, and carbohydrates following mitosis. These molecules will surround the two new cells that form from the original one. $G_1$ is the period of the cell cycle that varies the most in duration among different cell types. Slowly dividing cells, such as those in the liver, may exit at $G_1$ and enter $G_0$, where they remain for years. In contrast, the rapidly dividing cells in bone marrow speed through $G_1$ in 16 to 24 hours. Early cells of the embryo may skip $G_1$ entirely.

During the next period of interphase, S phase, the cell replicates its entire genome, so that each chromosome consists of two copies joined at an area called the **centromere.** In most human cells, S phase takes 8 to 10 hours. Many proteins are also synthesized during this phase, including those that form the mitotic **spindle** structure that will pull the chromosomes apart. Microtubules form structures called **centrioles** near the nucleus. Centriole microtubules are oriented at right angles to each other, forming paired oblong structures that organize other microtubules into the spindle.

The second gap phase, $G_2$, occurs after the DNA has been replicated but before mitosis begins. A cell in this phase synthesizes more proteins. Membranes are assembled from molecules made during $G_1$ and stored as small, empty vesicles beneath the cell membrane. These vesicles will be used to enclose the two daughter cells.

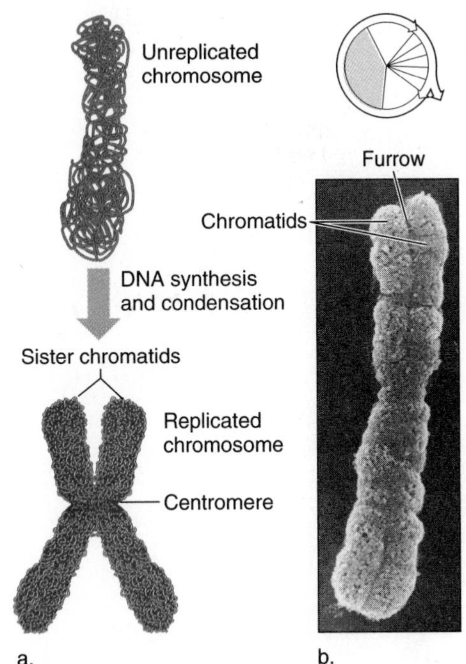


**figure 2.15**

**Replicated and unreplicated chromosomes.** Chromosomes are replicated during S phase, before mitosis begins. Two genetically identical chromatids of a replicated chromosome join at the centromere (*a*). In the photograph (*b*), a human chromosome is in the midst of forming two chromatids. A longitudinal furrow extends from the chromosome tips inward.

## *Mitosis—The Cell Divides*

As mitosis begins, the chromosomes are replicated and condensed enough to be visible, when stained, under a microscope. The two long strands of identical chromosomal material of a replicated chromosome are called **chromatids** (figure 2.15). They are joined at their centromeres. At a certain point during mitosis, a replicated chromosome's two centromeres part, allowing each chromatid pair to separate into two individual chromosomes.

During **prophase,** the first stage of mitosis, DNA coils tightly, shortening and thickening the chromosomes, which enables them to more easily separate (figure 2.16). Microtubules assemble to form the spindle from tubulin building blocks in the cytoplasm. Toward the end of prophase, the nuclear membrane breaks down. The nucleolus is no longer visible.

**Metaphase** follows prophase. Chromosomes attach to the spindle at their centromeres and align along the center of the cell, which is called the equator. When the centromeres part, each daughter cell receives one chromatid from each replicated chromosome. Metaphase chromosomes are under great tension, but they appear motionless because they are pulled with equal force on both sides, like a tug-of-war rope pulled taut.

Next, during **anaphase,** the cell membrane indents at the center, where the metaphase chromosomes line up. A band of microfilaments forms on the inside face of the cell membrane, and it constricts the cell down the middle. Then the centromeres part, which relieves the tension and releases one chromatid from each pair to move to opposite ends of the cell—like a tug-of-war rope breaking in the middle and the participants falling into two groups. Microtubule movements stretch the dividing cell. During the very brief anaphase stage, a cell temporarily contains twice the normal number of chromosomes because each chromatid becomes an independently moving chromosome, but the cell has not yet physically divided.

In **telophase,** the final stage of mitosis, the cell looks like a dumbbell with a set of chromosomes at each end. The spindle falls apart, and nucleoli and the membranes around the nuclei re-form at each end of the elongated cell. Division of the genetic material is now complete. Next, during a process called **cytokinesis,** organelles and macromolecules are distributed between the two daughter cells. Finally, the microfilament band contracts like a drawstring, separating the newly formed cells.

## *Control of the Cell Cycle*

When and where a somatic cell divides is crucial to health, and regulation of mitosis is a daunting task. Quadrillions of mitoses occur in a lifetime, and these cell divisions do not occur at random. Too little mitosis, and an injury may go unrepaired; too much, and an abnormal growth forms.

Groups of interacting proteins function at times called **checkpoints** to ensure that chromosomes are faithfully replicated and apportioned into daughter cells (figure 2.17). A "DNA damage checkpoint," for example, temporarily pauses the cell cycle while special proteins repair damaged DNA. The cell thus gains time to recover from an injury. An "apoptosis checkpoint" turns on as mitosis begins. During this checkpoint, proteins called survivins override signals telling the cell to die, keeping it in mitosis rather than apoptosis. Later during mitosis, the "spindle assembly checkpoint" oversees construction of the spindle and the binding of chromosomes to it.

Cells obey an internal "clock" that tells them how many times to divide. Mammalian cells grown (cultured) in a dish divide about 40 to 60 times. A connective tissue cell from a fetus, for example, divides on average about 50 times. But a similar cell from an adult divides only 14 to 29 times. The number of divisions left declines with age.

How can a cell "know" how many divisions it has undergone and how many remain? The answer lies in the chromosome

## figure 2.16

**Mitosis in a human cell.** During interphase (*a*), chromosomes are not yet condensed, and hence not usually visible. (*b*) In prophase, chromosomes are condensed and visible when stained. The spindle assembles, centrioles appear at opposite poles of the cell, and the nuclear membrane breaks down. (*c*) During metaphase, chromosomes align along the spindle. (*d*) In anaphase, the centromeres part and the chromatids separate. (*e*) In telophase, the spindle disassembles and the nuclear membrane re-forms. In a separate process, cytokinesis, the cytoplasm and other cellular structures distribute and pinch off into two daughter cells.

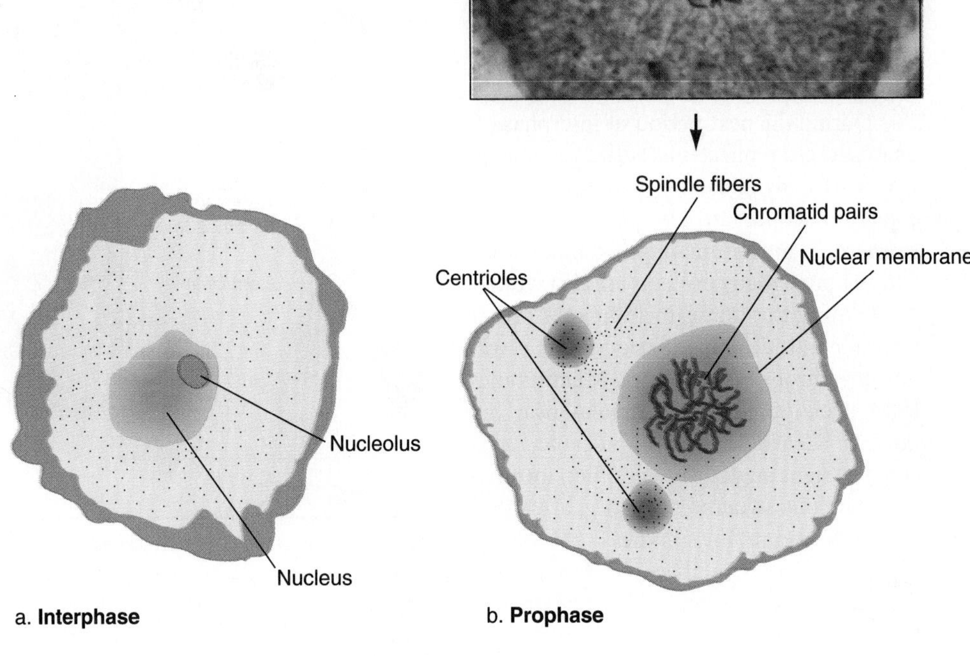


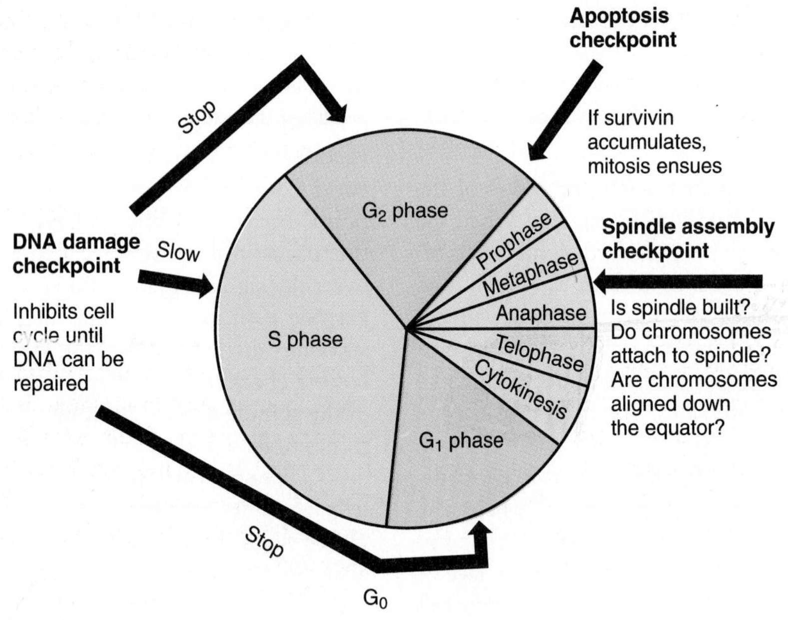


## figure 2.17

**Cell cycle checkpoints.** Checkpoints ensure that events occur in the correct sequence. Many types of cancer result from deranged checkpoints.

tips, called **telomeres** (figure 2.18). Telomeres function like a cellular fuse that burns down as pieces are lost from the very ends. Telomeres have hundreds to thousands of repeats of a specific six-nucleotide DNA sequence. At each mitosis, the telomeres lose 50 to 200 of these nucleotides, gradually shortening the chromosome like a fuse. After about 50 divisions, a critical amount of telomere DNA is lost, which signals mitosis to stop. The cell may remain alive but not divide again, or it may die. An enzyme called telomerase keeps chromosome tips long in eggs and sperm, in cancer cells, and in a few types of normal cells (such as bone marrow cells) that must supply many new cells. However, most cells do not produce telomerase, and their chromosomes gradually shrink. Telomerase includes a six-base RNA sequence that functions as a model, or template, used to add DNA nucleotides to telomeres. Figure 17.1 depicts the action at telomeres.

Outside factors also affect a cell's mitotic clock. Crowding can slow or halt mito-

Centrioles

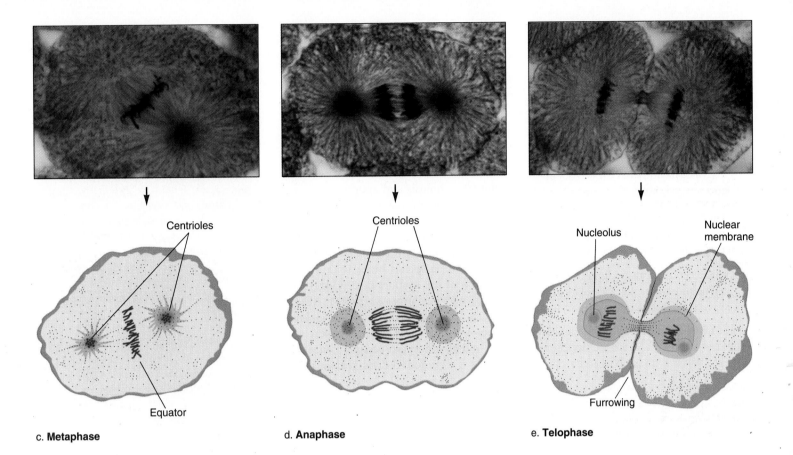

Equator

c. **Metaphase**

Centrioles

d. **Anaphase**

Nucleolus

Nuclear membrane

Furrowing

e. **Telophase**

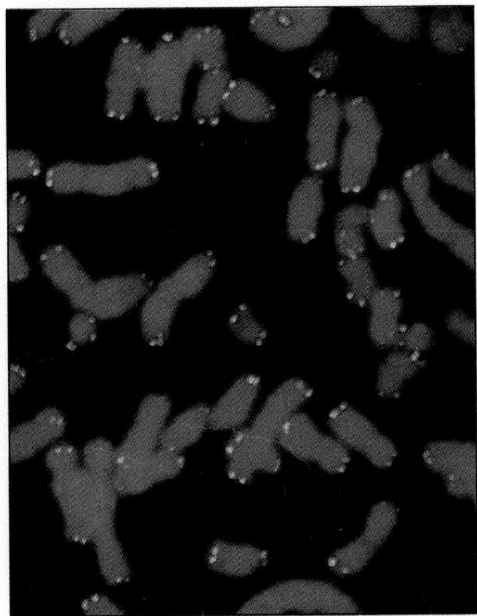

**figure 2.18**

**Telomeres.** Fluorescent tags mark the telomeres in this human cell.

sis. Normal cells growing in culture stop dividing when they form a one-cell-thick layer lining the container. If the layer tears, the cells that border the tear grow and divide to fill in the gap, but stop dividing once it is filled. Perhaps a similar mechanism in the body limits mitosis.

Chemical signals control the cell cycle from outside as well as from inside the cell. Hormones and growth factors are biochemicals from outside the cell that influence mitotic rate. A hormone is a substance synthesized in a gland and transported in the bloodstream to another part of the body, where it exerts a specific effect. Hormones secreted in the brain, for example, signal the cells lining a woman's uterus to build up each month by mitosis in preparation for possible pregnancy. Growth factors act more locally. Epidermal growth factor, for example, stimulates cell division beneath a scab.

Two types of proteins, cyclins and kinases, interact inside cells to activate the genes whose products carry out mitosis. The two types of proteins form pairs. Levels of cyclins fluctuate regularly throughout the cell cycle, while kinase levels stay the same. A certain number of cyclin-kinase pairs turn on the genes that trigger mitosis. Then, as mitosis begins, enzymes degrade cyclin. The cycle starts again as cyclin begins to build up during the next interphase.

## Apoptosis

Apoptosis rapidly and neatly dismantles a cell into neat, membrane-bounded pieces that a phagocyte (a cell that engulfs and destroys another) can mop up. It is a little like taking the contents of a messy room and packaging them into garbage bags—then disposing of it all.

Like mitosis, apoptosis is a continuous process that occurs in a series of steps. It begins when a "death receptor" on the doomed cell's membrane receives a signal to die. Within seconds, enzymes called caspases are activated inside the cell, stimulating each other and snipping apart various cell components.

These killer enzymes take several actions at once. They

- destroy the cytoskeletal threads that support the nucleus so that it collapses, causing the genetic material within to condense.
- demolish the enzymes that replicate and repair DNA.
- activate enzymes that chew DNA up into similarly sized small pieces.
- tear apart the rest of the cytoskeleton.
- destroy the cell's ability to adhere to other cells.
- send a certain phospholipid from the cell membrane's inner face to its outer surface, where it attracts phagocytes.

From the outside, a cell in the throes of apoptosis has a characteristic appearance (figure 2.19). It rounds up as contacts with other cells are cut off, and the cell membrane undulates and forms bulges called blebs. The nucleus bursts, releasing DNA pieces that align in a way that resembles ladders when dyed and displayed in an electrical field. Then the cell shatters. Almost instantly, pieces of membrane encapsulate the cell fragments, which prevents inflammation. Within an hour, the cell is gone.

From the embryo onward through development, mitosis and apoptosis are synchronized, so that tissue neither overgrows nor shrinks. In this way, a child's liver remains much the same shape as she grows into adulthood. During early development, mitosis and apoptosis orchestrate the ebb and flow of cell number as new structures form. Later, these processes protect. Mitosis fills in new skin to heal a scraped knee; apoptosis peels away sunburnt skin cells that might otherwise become cancerous. Cancer is a profound derangement of the balance between cell division and death, with mitosis too frequent or occurring too many times, or apoptosis too infrequent.

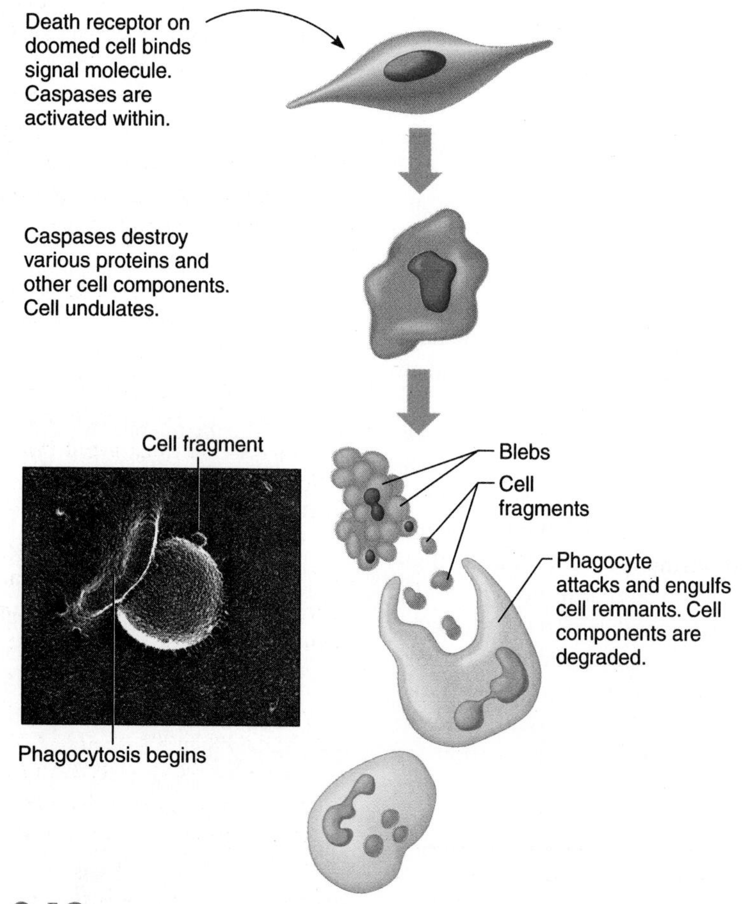
Death receptor on doomed cell binds signal molecule. Caspases are activated within.
Caspases destroy various proteins and other cell components. Cell undulates.
Cell fragment
Blebs
Cell fragments
Phagocyte attacks and engulfs cell remnants. Cell components are degraded.
Phagocytosis begins

**figure 2.19**

**Death of a cell.** A cell undergoing apoptosis loses its characteristic shape, forms blebs, and finally falls apart. Caspases destroy the cell's insides. Phagocytes digest the remains.

### Key Concepts

Mitosis and apoptosis regulate cell numbers during development, growth, and healing. The cell cycle includes interphase and mitosis. Interphase includes two gap (G) phases and a synthesis (S) phase that prepares the cell for mitosis. During S phase, DNA is replicated. Proteins, carbohydrates, and lipids are synthesized during $G_1$ and more proteins are synthesized in $G_2$. During the mitotic stages of prophase, metaphase, anaphase, and telophase, replicated chromosomes condense, align, split, and distribute into daughter cells. The cell cycle is controlled by checkpoints, telomeres, hormones, and growth factors from outside the cell, and cyclins and kinases from within. • During apoptosis, cells receive a death signal, activate caspases, and break apart in an orderly fashion.

## 2.3 Cell-Cell Interactions

Precisely coordinated biochemical steps orchestrate the cell-cell interactions that make multicellular life possible. Defects in cell communication and interaction cause certain inherited illnesses. We look now at two broad types of interactions among cells.

### Signal Transduction 

Organelles, cell membranes, and cytoskeletons are dynamic structures that communicate with each other and with the environment outside the cell. A key player in this constant biological communication is the cell membrane, where signal transduction takes place. *Transduce* means to change one form of something (such as energy or information) into another.

In signal transduction, molecules on the cell membrane assess, transmit, and amplify incoming messages to the cell's interior. Various types of stimuli are thus transduced into the biochemical language of the cell. Some signals must reach receptors for the cell to function normally; others, such as a signal to divide when cell division is not warranted, must be ignored.

Signal transduction is carried out by the interaction between cytoplasmic proteins and proteins embedded in the cell membrane that extend from one or both faces. The transduction process is a complex series of chemical interactions that begins at the cell surface. In the first step, a receptor directly binds an incoming stimulus, called the **first messenger** (figure 2.20). The responding receptor contorts in a way that touches a nearby protein called a regulator. Next, the regulator protein activates a nearby enzyme, which catalyzes (speeds) a specific chemical reaction. The product of this reaction is called the **second messenger.** The second messenger lies at the crux of the entire process; it elicits the cell's response, typically by activating certain enzymes. A single stimulus can trigger the production of many second messenger molecules, and therefore signal transduction amplifies incoming information. Because cascades of proteins carry out signal transduction, it is a genetically controlled process.

Defects in signal transduction underlie many inherited disorders. In neurofibromatosis type I (NF1), for example, tumors (usually benign) grow in nervous tissue under the skin and in parts of the nervous system. At the cellular level, NF1 occurs when cells fail to block transmission of a growth factor signal that triggers cell division. Affected cells misinterpret the signal and divide when inappropriate. Several new cancer drugs work by plugging up receptors for growth factors on cancer cells.

## Cellular Adhesion

Cellular adhesion is a precise sequence of interactions among the proteins that join cells. Inflammation—the painful, red swelling at a site of injury or infection—involves cellular adhesion. Inflammation occurs when white blood cells (leukocytes) move in the blood vessels to the endangered body part, where they squeeze between cells of the blood vessel walls to reach the site of injury or infection. (The specific process is called "leukocyte trafficking.") **Cellular adhesion molecules,** or CAMs, help guide white blood cells to the injured area. Because CAMs are proteins, cellular adhesion is genetically controlled.

Three types of CAMs act in sequence during the inflammatory response (figure 2.21). First, **selectins** attach to the white blood cells, and slow them to a roll by also binding to carbohydrates on the capillary wall. (This is a little like putting out your arms to slow a ride down a slide.) Next, clotting blood, bacteria, or decaying tissues release chemical attractants that signal white blood cells to stop. The chemical attractants activate CAMs called **integrins,** which latch onto the white blood cells, and CAMs called **adhesion receptor proteins,** which extend from the capillary wall at the injury site. The integrins and adhesion receptor proteins then guide the white blood cells between the tilelike lining cells to the other side—the injury site.

What happens if the signals that direct white blood cells to injury sites fail? A young woman named Brooke Blanton knows the answer all too well. Her first symptom was teething sores that did not heal. These and other small wounds never accumulated the pus (bacteria, cellular debris, and white blood cells) that indicates that the body is fighting infection. Brooke has a type of disorder called a leukocyte-adhesion deficiency. Her body lacks the CAMs that enable white blood cells to stick to blood vessel walls, and as a result, her blood cells zip right past wounds. Brooke must avoid injury and infection, and she receives anti-infective treatments for even the slightest wound.

More common disorders may also reflect abnormal cell adhesion. Lack of cell adhesion eases the journey of cancer cells from one part of the body to another. Arthritis may occur when the wrong adhesion molecules rein in white blood cells, inflaming a joint where no injury exists.

Cell adhesion is critical to many other functions. CAMs guide cells surrounding an embryo to grow toward maternal cells and form the placenta, the supportive organ linking a pregnant woman to the fetus. Sequences of CAMs also help establish connections among nerve cells that underlie learning and memory.

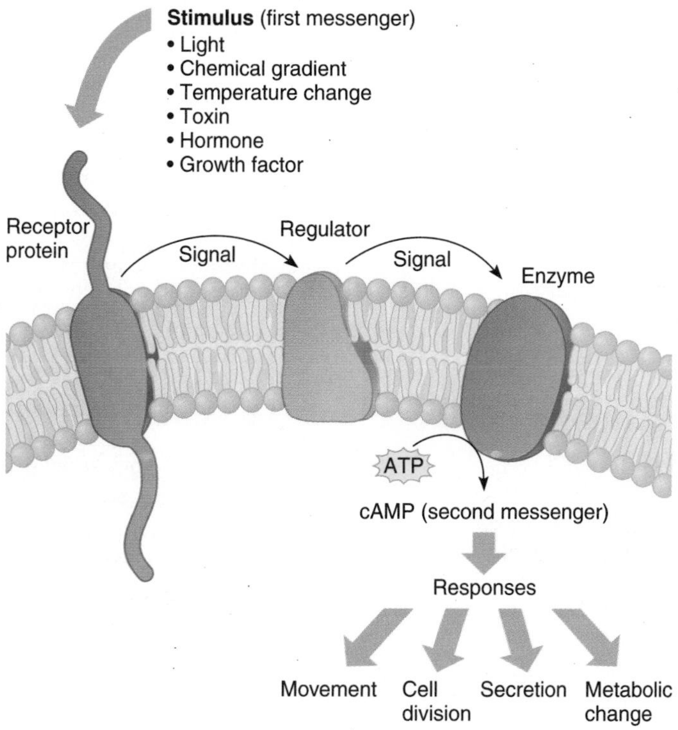

**figure 2.20**

**Signal transduction.** A first messenger binds a receptor, triggering a cascade of biochemical activity at the cell's surface. An enzyme catalyzes a reaction inside the cell that circularizes ATP to cyclic AMP, the second messenger. cAMP stimulates various responses, such as cell division, metabolic changes, and muscle contraction. The splitting of ATP also provides energy.

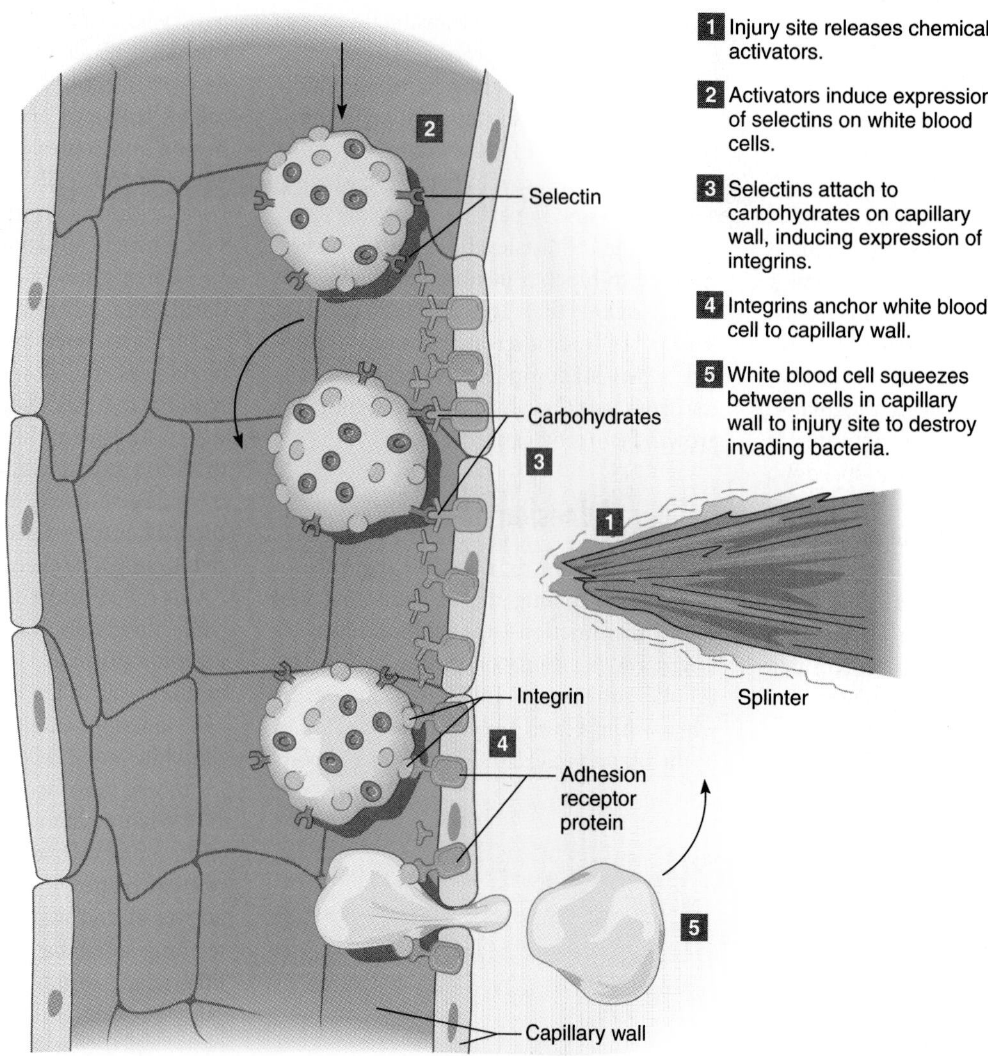
1 Injury site releases chemical activators.
2 Activators induce expression of selectins on white blood cells.
3 Selectins attach to carbohydrates on capillary wall, inducing expression of integrins.
4 Integrins anchor white blood cell to capillary wall.
5 White blood cell squeezes between cells in capillary wall to injury site to destroy invading bacteria.
Selectin
Carbohydrates
Integrin
Adhesion receptor protein
Splinter
Capillary wall

**figure 2.21**

**Cellular adhesion aids in defenses.** Cellular adhesion molecules (CAMs), including selectins, integrins, and adhesion receptor proteins, direct white blood cells to injury sites.

Researchers are using understanding of cell adhesion to develop new drugs. For example, a compound that blocks integrins is being tested against the inflammation associated with multiple sclerosis and Crohn disease.

**KEY CONCEPTS**

In signal transduction, cell surface receptors receive information from first messengers (stimuli) and pass them to second messengers, which then trigger a cellular response. • Cellular adhesion molecules (CAMs) guide white blood cells to injury sites using a sequence of cell-protein interactions.

## 2.4 Stem Cells and Cell Specialization

Cells build bodies, but to do so, they must aggregate in specific combinations and patterns to form tissues, organs, and organ systems. Genes control this process throughout development and life.

Sperm fertilizes egg, and the genetic material from two individuals mingles in a new nucleus. Hours later, that initial cell divides, becoming two, and then later, it does so again, producing a total of four cells, and so on. The cells form a ball, and then the ball hollows out. Up until this point, one cell is pretty much like another, both in appearance and in position. Then changes begin. A few of the cells collect on the inner face of the ball and spread out to form a sheet, and then the sheet folds, forming two layers, then a third in the middle. From these three layers arise the body's tissues.

Cell differentiation begins as different genes turn on or off. A cell destined to be part of the heart muscle, for example, produces huge amounts of proteins that can contract. By about the fourth week after sperm meets egg, those cells will beat in unison, forming the start of the heart. Another cell might express genes enabling it to produce neurotransmitters, eventually becoming a nerve cell.

Not all cells specialize. Some cells in nearly every tissue, called stem cells, are less differentiated than others (figure 2.22). A stem cell divides to yield a more specialized

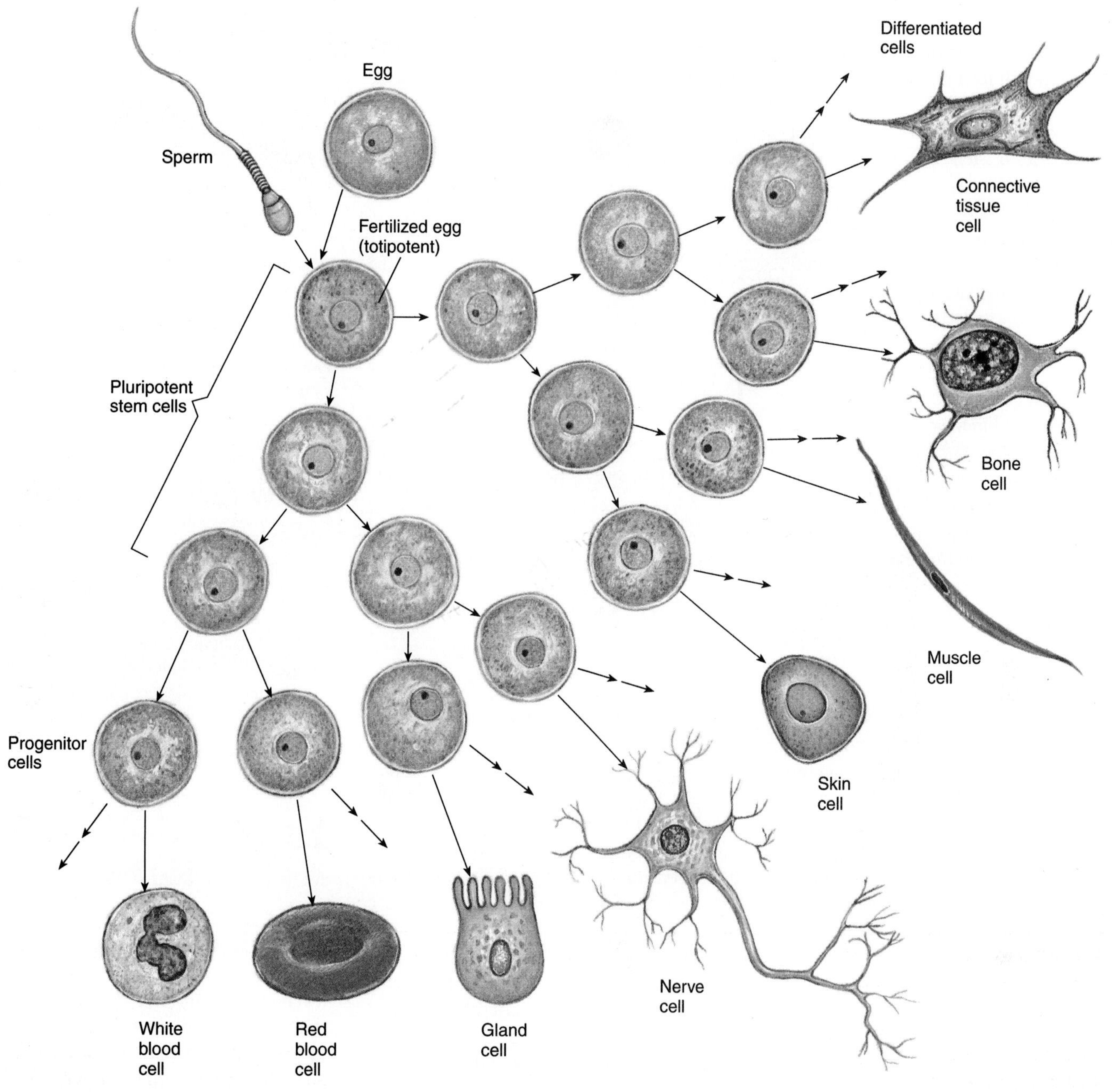


**figure 2.22**

**Cell differentiation.** The trillions of cells in an adult human derive from the original fertilized egg cell by mitosis and differential gene expression. As different genes turn on or off in different cells, the characteristics of specific cell types emerge. Only the fertilized ovum is totipotent, its daughter cells capable of becoming anything. Stem cells with many, but not all, possibilities for specialization as they divide are called pluripotent. Cells that are themselves unspecialized but give rise to differentiated cells when they divide are termed progenitor cells. (Relative cell sizes are not to scale.)

daughter cell, as well as another stem cell. Stem cells are present in the embryo and the fetus, and also after birth, when they serve as repositories for growth and cell replacement to heal injuries.

Cell specialization is often described in terms of potential. A fertilized ovum, as the first cell of a new organism, is the only type of cell that is **totipotent**—literally capable of yielding daughter cells that can become any cell type. Early in development, cells are **pluripotent,** which means that they yield daughter cells that can differentiate into many, but not all, cell varieties. Pluripotent stem cells do not show signs of specialization themselves, but their developmental future, or fate, is more restricted than that of the fertilized ovum. Further in development, pluripotent stem cells give rise to the even more restricted **progenitor cells.** Such cells are committed to follow a particular developmental pathway, leading toward a specific cell type. A pluripotent stem cell, for example, might produce daughter cells that differentiate into any of several connective tissues, including blood. A progenitor cell in bone marrow, in contrast, produces daughter cells that differentiate only into immature red blood cells, which in turn specialize further to become red blood cells.

Biologists have long known about certain repositories of stem cells that persist in adults, yet are discovering new ones all the time. Well-studied stem cells include those in the skin, small intestine, and bone marrow, all tissues whose cells are replaced very often. One type, mesenchymal stem cells, are tucked into bone marrow and other places, but they can migrate to the site of an injury. Here, responding to signals, they differentiate into blood, muscle, cartilage, bone, fat cells, and the outer layers of blood vessels—whatever is needed to repair the damage. Other types of stem cells have been more recently discovered, such as those in the brain, heart, and muscles. By watching the fates of single cells, researchers have shown that the combination of nutrients, hormones, and growth factors in the surroundings determines which sets of genes the cell will express—that is, how the cell will differentiate.

Stem cell discoveries can reveal surprising relationships among cell types. Consider hair and skin. In 1990, researchers hypothesized that pluripotent stem cells in the upper portions of hair follicles could regenerate skin, because burn patients regrow skin only if some hair follicles remain. The new skin seemed to grow out of the hair follicles. When 10 years later researchers could label DNA, they followed the fates of the suspected cells, and saw the bulge just above the hair root produce both hair and surrounding skin. This shared cellular ancestry had not been thought possible because hair and skin are structurally very different.

Researchers are applying stem cell biology in a new field called regenerative medicine that uses stem cells to replace or rejuvenate abnormal or damaged tissue. Experiments in nonhuman animals demonstrate several approaches. For example, stem cells that give rise to cardiac muscle can mend damaged hearts, and stem cells that differentiate into cartilage can be used to repair injuries in this hard-to-heal tissue.

One form of regenerative medicine already in practice is storing blood-forming (hematopoietic) stem cells from umbilical cords. This resource is used to treat blood or immune system disorders that the child may develop (table 2.2), but it is also increasingly used to provide replacement cells

**table 2.2**

**Uses of Umbilical Cord Stem Cells**

| |
|---|
| **Anemias** |
| Aplastic anemia |
| Fanconi anemia |
| Sickle cell disease |
| Thalassemia |
| **Cancers** |
| Leukemias |
| Neuroblastoma |
| Non-Hodgkins lymphoma |
| **Clotting disorders** |
| **Inborn errors of metabolism** |
| **Inherited immune deficiencies** |

for anyone who might need them, such as people who need bone marrow transplants. The cells are particularly valuable because they are much less likely to provoke an immune system rejection response than are more mature cells. Some younger siblings have been conceived to provide older sisters or brothers with life-saving stem cells (see figure 21.1).

In another application of stem cell technology, people being treated for cancer can store hematopoietic stem cells separated from their own blood or bone marrow before undergoing intensive radiation treatment and chemotherapy. When their blood counts fall in response to the therapy, they can receive infusions of their own stem cells to restore their blood supplies. This practice enables them to tolerate higher doses of the treatments, which are more likely to be effective against the cancer.

Stem cell research and technology raise controversy when the source is embryos or fetuses, because they are destroyed. Some nations restrict the sources of stem cells for research for this reason. However, umbilical cord stem cells, cells growing in laboratory dishes, and cells from the patient or another person, can be obtained without harming anyone. Bioethics: Choices for the Future in chapter 3 returns to this topic.

Once governments decide on an ethically acceptable source of stem cells, the first applications of regenerative medicine will use single cell types. Future applications will generate tissues that consist of more than one cell type, and ultimately, organs. But first biologists must learn how cells aggregate and interact to form tissues, then discover how those tissues build organs—the essence of development. Somewhere encoded in the genome are the instructions not only for differentiation, but for guiding cells to recognize each other and to interact to build larger structures.

**Key Concepts**

Stem cells retain the potential to differentiate, which means they can enable a tissue to grow or to repair itself. Stem cells may be pluripotent or progenitor cells; only a fertilized ovum is totipotent. Regenerative medicine uses stem cells to replace or rejuvenate injured or diseased tissue.

## 2.5 Viruses and Prions—Not Cells, But Infectious

More complex than chemicals, yet not nearly as organized as cells, viruses and the even simpler prions straddle the boundary between the nonliving and the living. Although not technically alive, these entities still have powerful effects on health. Viruses cause infectious diseases, not inherited ones. Prion diseases are usually infectious, and not inherited.

### A Virus—Nucleic Acid in a Protein Coat

A **virus** is a single or double strand of a nucleic acid (DNA or RNA) wrapped in a protein coat, and in some types, also in an outer envelope. A virus can reproduce only if it can enter a host cell and use its energy resources and protein synthetic machinery.

A virus may have only a few protein-encoding genes, but many copies of the same protein can assemble to form an intricate covering, like the panes of glass in a greenhouse. Ebola virus, for example, is an extremely simple, but deadly virus. In contrast, the virus that causes smallpox has more than 100 different types of proteins. HIV also has a complex structure (figure 2.23).

Human chromosomes harbor viral DNA sequences that are vestiges of past infections, perhaps in distant ancestors. Many DNA viruses reproduce by inserting DNA into the host cell's genetic material. In contrast, it takes several steps for an RNA virus to insert DNA into a human chromosome, because it must first be copied from the RNA. A viral enzyme called reverse transcriptase does this. The DNA that represents the RNA virus then inserts into the host cell's chromosome. (In some viruses the invading DNA is replicated separate from a host chromosome.) Certain RNA viruses are called "retro" because they function opposite the usual direction of genetic information—from DNA to RNA to protein. HIV is a retrovirus.

Once viral DNA integrates into the host's DNA, it can either remain and replicate along with the host's DNA without causing harm, or it can take over and kill the cell. Viral genes direct the host cell to replicate viral DNA and then use it to manufacture viral proteins at the expense of the cell's normal activities. The cell fills with viral DNA and protein, which assemble into new viruses. The cell bursts, releasing many new viruses into the body.

Viruses infect all types of organisms, and are diverse. They were discovered in tobacco plants, but also infect microorganisms, fungi, and, of course, animals.

## A Prion—One Protein That Takes Multiple Forms

Another type of infectious agent simpler than a cell is a **prion,** which stands for "proteinaceous infectious particle." (It is pronounced "pree-on.") A prion is a glycoprotein that naturally occurs in the brain, but it can exist in more than one three-dimensional shape, or conformation. When it assumes one particular shape, it begins a chain reaction that converts other prions into the infective form (figure 2.24). This rare form causes a type of disease called a transmissible spongiform encephalopathy in 85 types of mammals, including humans. The name of the disease reflects the damage it causes—the brain is shot full of holes and ends up resembling a sponge. Nerve cells die, and star-shaped supportive cells overgrow.

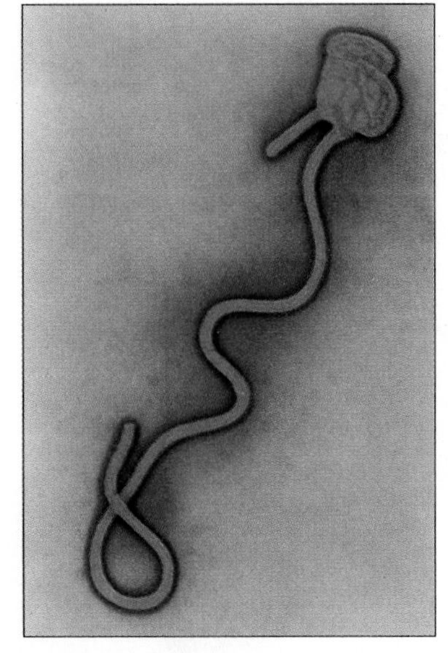


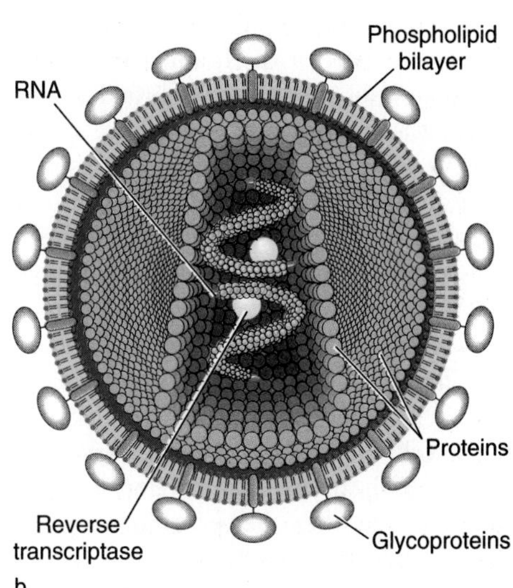


**figure 2.23**

**Virus structure.** Viruses are nucleic acids in protein coats. (*a*) Ebola virus is a single strand of RNA and just seven proteins. People become infected when they touch the body fluids of those who have died of the infection. Symptoms progress rapidly, from headache and fever to vomiting of blood, to tearing apart of the internal organs. (*b*) The human immunodeficiency virus (HIV), which causes AIDS, consists of RNA surrounded by several protein layers. Once inside a human cell, the virus uses reverse transcriptase to synthesize a DNA copy of its RNA. The virus then inserts this copy into the host DNA.

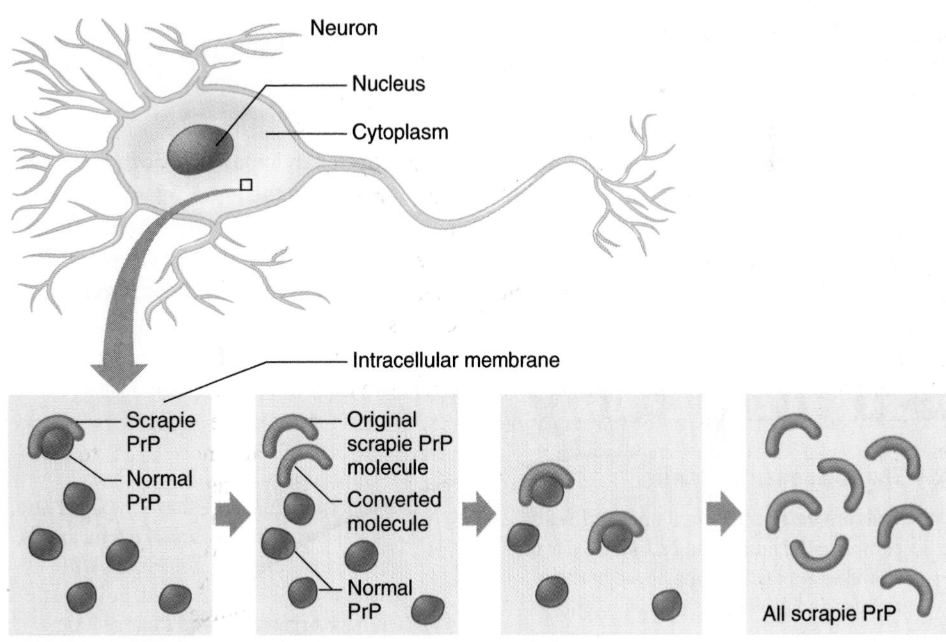


**figure 2.24**

**Prions change shape.** A prion disease may begin when a single scrapie PrP contacts a normal PrP and causes it to switch into the abnormal conformation. As the change spreads, disease occurs, usually as accumulated abnormal prion protein clogs brain tissue.

Prion diseases cause extreme weight loss and poor coordination, with other symptoms, such as dementia or relentless insomnia, reflecting the part of the brain that is eaten away. These diseases are typically fatal within 18 months once symptoms appear.

About 10 percent of people who suffer from prion diseases inherit the condition because they have mutations in the gene that encodes prion protein, called PrP. Most cases, however, are acquired. A person is somehow exposed to prions in the abnormal shape, and this triggers the conversion of their own normal prions. Only a handful of these illnesses are known. They were first discovered in sheep, which develop a disease called scrapie when they eat prion-infected brains from other sheep.

A dramatic example of a prion disease in humans was kuru, which affected the Foré people in a remote mountainous area of New Guinea (figure 2.25). In the Foré language, *kuru* means to tremble. The disease began with wobbling legs, quickly followed by trembling hands and fingers. Gradually, the entire body became wracked with uncontrollable shaking. Speech slurred and faded, thinking slowed, and after several months, the person could no longer walk or eat. Death typically came within a year.

The fact that only women and young children developed kuru at first suggested that the disease might be inherited, but D. Carleton Gajdusek, the physician who has spent much of his lifetime studying the Foré, learned that the preparation of human brain for a cannibalism ritual probably passed on the abnormal prions. When the people discarded cannibalism in the 1970s, the disease gradually vanished. Gajdusek vividly described the Foré preparation of human brains at a time when he thought the cause was viral:

**figure 2.25**

**Kuru.** Kuru is a prion disease that affected the Foré people of New Guinea until they gave up the cannibalism ritual that spread an abnormal form of prion protein.

> **Children participated in both the butchery and the handling of cooked meat, rubbing their soiled hands in their armpits or hair, and elsewhere on their bodies. They rarely or never washed. Infection with the kuru virus was most probably through the cuts and abrasions of the skin or from nose picking, eye rubbing, or mucosal injury.**

Although kuru vanished, other prion diseases surfaced. In the 1970s and 1980s, several people acquired the similar Creutzfeldt-Jakob disease (CJD). This time, the route of transmission was either through corneal transplants, in which abnormal prions entered the brain through the optic nerve, or from human growth hormone taken from cadavers and used to treat short stature in children. The most familiar prion disease is probably "mad cow disease" and the variant CJD that it has caused in more than 120 people in the United Kingdom since 1995. The abnormal prions were likely acquired when the people ate infected beef.

Researchers have studied the PrP gene in great detail, and discovered that several specific polymorphisms (variants) that affect different sites in the protein interact in ways that make some people resistant to prion diseases, yet others highly susceptible. These mutations are discussed further in chapter 11. The function of normal prion protein isn't known, but it resides in the cell membranes of brain neurons.

### KEY CONCEPTS

Viruses and prions are simpler than cells, but are infectious. A virus is a nucleic acid in a protein coat that takes over a cell's protein-synthesizing machinery. • A prion is an infectious protein. Susceptibility to prion disorders is inherited.

# Summary

### 2.1 The Components of Cells

1. Cells are the fundamental units of life and comprise the human body. Inherited traits and illnesses can be understood at the cellular and molecular levels.
2. All cells share certain features, but they are also specialized because they express different subsets of genes. Cells consist primarily of water and several types of **macromolecules: carbohydrates, lipids, proteins,** and **nucleic acids.**
3. The three domains of life—Archaea, Bacteria, and Eukarya—have characteristic cells. Those of Archaea and Bacteria are simple, small, and lack organelles, with genetic material in a nucleoid. **Eukaryotic** cells have **organelles,** and their genetic material is contained in a **nucleus.**
4. Organelles sequester related biochemical reactions, improving efficiency of life functions and protecting the cell. Along with organelles, the cell consists of **cytoplasm** and other chemicals.
5. The nucleus contains DNA and a **nucleolus,** which is a site of ribosome synthesis. **Ribosomes** provide scaffolds for protein synthesis; they exist free in the cytoplasm or complexed with the **rough ER.**
6. In secretion, the rough ER is the site of protein synthesis and folding, the **smooth ER** is the site of lipid synthesis, transport, and packaging, and the **Golgi apparatus** packages secretions into vesicles, which exit through the cell membrane. Enzymes in **mitochondria** extract energy from

nutrients. **Lysosomes** contain enzymes that dismantle debris, and **peroxisomes** house enzymes that perform a variety of functions.

7. The **cell membrane** is a protein-studded **phospholipid bilayer.** It controls which substances exit and enter the cell, and how the cell interacts with other cells.
8. The **cytoskeleton** is a protein framework of hollow **microtubules,** made of tubulin, and solid **microfilaments,** which consist of actin. **Intermediate filaments** are made of more than one protein type and are abundant in skin. The cytoskeleton and the cell membrane distinguish different types of cells.

## 2.2 Cell Division and Death

9. Coordination of cell division (**mitosis**) and cell death (**apoptosis**) maintains cell numbers, enabling structures to enlarge during growth and development but preventing abnormal growth.
10. The **cell cycle** describes whether a cell is dividing (mitosis) or not (**interphase**). Interphase consists of two gap phases when proteins and lipids are produced, and a synthesis phase when DNA is replicated.
11. Mitosis proceeds in four stages. In **prophase,** replicated chromosomes consisting of two **chromatids** condense, the **spindle** assembles, the nuclear membrane breaks down, and the nucleolus is no longer visible. In **metaphase,** replicated chromosomes align along the center of the cell. In **anaphase,** the **centromeres** part, equally dividing the now unreplicated chromosomes into two daughter cells. In **telophase,** the new cells separate. **Cytokinesis** apportions other cellular components.
12. Internal and external factors control the cell cycle. **Checkpoints** are times when proteins regulate the cell cycle. **Telomere** size counts mitoses. Crowding, hormones, and growth factors signal cells from the outside; interactions of cyclins and kinases trigger mitosis from inside.
13. Apoptosis, or programmed cell death, is a normal part of development. It begins when a death receptor on the cell membrane receives a death signal, then activates caspases that tear apart the cell in an orderly fashion. Membrane surrounds the pieces, preventing inflammation.

## 2.3 Cell-Cell Interactions

14. In **signal transduction,** a stimulus activates a cascade of action among membrane proteins, culminating in production of a **second messenger** that turns on enzymes that provide the response.
15. **Cellular adhesion molecules** enable cells to interact. In leukocyte trafficking, selectins slow the movement of white blood cells, and integrins and adhesion receptor proteins guide the cell through a capillary wall to an injury site.

## 2.4 Stem Cells and Cell Specialization

16. Cell differentiation reflects the expression of subsets of the genome.
17. Stem cells retain the ability to divide and produce daughter cells that specialize in particular ways.
18. **Totipotent** stem cells (fertilized ova) can become anything. **Pluripotent** stem cells can differentiate as any of a variety of cell types. **Progenitor cells** can specialize as any of a restricted number of cell types.

## 2.5 Viruses and Prions—Not Cells, But Infectious

19. **Viruses** consist of single- or double-stranded DNA or RNA wrapped in protein and sometimes in an outer envelope.
20. A virus enters a host cell and replicates its genetic material, then uses the host's protein-synthesizing machinery to mass produce itself.
21. A **prion** is a protein that exists in normal and abnormal forms. Presence of an abnormal form triggers the conversion of the normal form, resulting in spongiform encephalopathies.

# Review Questions

1. List the steps to the following processes:
   a. signal transduction
   b. cellular adhesion
   c. the cell cycle
   d. apoptosis
   e. mitosis
   f. secretion
2. Explain the functions of the following proteins:
   a. tubulin and actin
   b. caspases
   c. cyclins and kinases
   d. checkpoint proteins
   e. cellular adhesion molecules
3. List four types of controls on cell cycle rate.
4. How can all of a person's cells contain exactly the same genetic material, yet diverge into bone cells, nerve cells, muscle cells, and connective tissue cells?
5. Distinguish between
   a. a bacterial cell and a eukaryotic cell.
   b. interphase and mitosis.
   c. mitosis and apoptosis.
   d. rough ER and smooth ER.
   e. microtubules and microfilaments.
6. What functions do each of the following organelles perform?
   a. mitochondria
   b. lysosomes
   c. peroxisomes
   d. smooth ER
   e. rough ER
   f. Golgi apparatus
   g. nucleus
7. What advantage does compartmentalization provide to a large and complex cell?
8. What role does the cell membrane play in signal transduction?
9. Distinguish among the terms "totipotent," "pluripotent," and "progenitor cell."
10. What are the differences among cells, viruses, and prions?

# Applied Questions

1. How might abnormalities in each of the following contribute to cancer?
   a. cellular adhesion
   b. signal transduction
   c. balance between mitosis and apoptosis
   d. cell cycle control
   e. telomerase activity
2. Why do many inherited conditions result from defective enzymes?
3. In neuronal ceroid lipofuscinosis, the nervous system degenerates from birth. The child experiences seizures, loss of vision, and lack of coordination, dying in early childhood. At the molecular level, the child lacks an enzyme that normally breaks down certain proteins, causing them to accumulate and destroying the nervous system. Name two organelles that are involved in this illness.
4. How do stem cells maintain their populations within tissues that consist of mostly differentiated cells?
5. Explain why too frequent or too infrequent mitosis, and too frequent or too infrequent apoptosis, can endanger health.
6. Why wouldn't a cell in an embryo likely be in phase $G_0$?
7. A defect in which organelle would cause fatigue?
8. Human embryonic stem cells are pluripotent, and for this reason, they are valuable in regenerative medicine research. However, they are obtained from embryos or fetuses (humans before birth). Suggest an alternative source of stem cells for research.
9. Describe three ways that drugs can be used to treat cancer, based on disrupting microtubule function, telomere length, and signal transduction.
10. How can signal transduction, the cell membrane, and the cytoskeleton function together?
11. What abnormality at the cellular or molecular level lies behind each of the following disorders?
    a. cystic fibrosis
    b. adrenoleukodystrophy
    c. hereditary spherocytosis
    d. neurofibromatosis type 1
    e. leukocyte adhesion deficiency
    f. kuru
    g. syndactyly

# Suggested Readings


Inwald, D., et al. January 2001. Adhesion molecule deficiencies. *Journal of Clinical Pathology* 54:1. The discovery of cellular adhesion molecules explained disorders in which wounds do not heal.

Lewis, Ricki. March 6, 2000. A paradigm shift in stem cell research? *The Scientist* 14:1. Stem cells may be more widespread, and more plastic, than anyone imagined.

Lewis, Ricki. April 12, 1999. Human mesenchymal stem cells differentiate in the lab. *The Scientist* 13:1. Fetal cells will not be necessary for regenerative medicine.

Lewis, Ricki. December 1998. Telomere tales. *BioScience* vol. 48. Long of interest to cell biologists, telomeres are now impacting on medicine.

Mazzarello, Paolo, and Marina Bentivoglio. April 9, 1998. The centenarian Golgi apparatus. *Nature* vol. 392. The Golgi apparatus was discovered a century ago.

Nichols, William C., and David Ginsburg. June 1999. From the ER to the Golgi: Insights from the study of combined factors V and VIII deficiency. *American Journal of Human Genetics* 64:1493–98. Secretion is a highly selective event.

Pollack, Andrew W. December 18, 2001. Scientists seek ways to rebuild the body, bypassing the embryos. *The New York Times,* F1. Using stem cells from adults would remove the controversy from stem cell research.

Raff, Martin. November 12, 1998. Cell suicide for beginners. *Nature* vol. 396. A review of the steps of apoptosis.

Renehan, Andrew G., et al. June 21, 2001. What is apoptosis and why is it important? *British Medical Journal* 322:1536.

Taylor, Gina, et al. August 18, 2000. Involvement of follicular stem cells in forming not only the follicles but also the epidermis. *Cell* 102:451–561. The descendants of the same stem cell need not resemble each other, as researchers learned recently for hair and skin.

Wade, Nicholas. December 18, 2001. In tiny cells, glimpses of body's master plan. *The New York Times,* F1. A primer on stem cells.

Wanders, Ronald J. A. November 1998. A happier sequel to Lorenzo's Oil? *Nature Medicine* 4:1245–46. Sometimes dietary intervention can alleviate symptoms of inborn errors of metabolism.

# On the Web

Check out the resources on our website at

**www.mhhe.com/lewisgenetics5**

On the web for this chapter you will find additional study questions, vocabulary review, useful links to case studies, tutorials, popular press coverage, and much more. To investigate specific topics mentioned in this chapter, also try the links below:

Ion Channel Diseases **www.neuro.wustl.edu/neuromuscular/mother/chan.html**

Maple Syrup Urine Disease Family Support Group **www.msud-support.org**

The New York Times **nytimes.com/science**

Online Mendelian Inheritance in Man
**www.ncbi.nlm.nih.gov/entrez/query.fcgi?db=OMIM**
adrenoleukodystrophy 300100
biotinidase deficiency 253260
familial hypercholesterolemia 143890
hereditary spherocytosis 182900
HGPRT deficiency (Lesch-Nyhan syndrome) 30800
Kartagener syndrome 244400
leukocyte adhesion deficiency 116920
long-QT syndrome 192500, 600919, 152427
maple syrup urine disease 248600
neurofibromatosis type 1 16220
prion protein 176640
Wilson disease 277900
Tay-Sachs disease 272800

National Neurofibromatosis Foundation **www.nf.org**

National Tay-Sachs and Allied Diseases Association **www.ntsad.org**

Periodic Paralysis Association **www.periodicparalysis.org**

Sudden Arrhythmia Death Syndromes Foundation (Long QT Syndrome) **www.sads.org**

Umbilical Cord Cell Banking **www.cordblood.com**

United Leukodystrophy Foundation **www.ulf.org**

CHAPTER

# 3

# Development

### 3.1 The Reproductive System

Males and females have paired gonads that house reproductive cells, and networks of tubes and associated glands that nurture the development of sperm and oocytes.

### 3.2 Meiosis

Meiosis is a form of cell division that halves the two genomes of a somatic cell to produce haploid gametes. It ensures that the chromosome number remains constant from generation to generation, and it also recombines the genetic contributions of each parent, creating great genetic variability.

### 3.3 Gamete Maturation

A sperm and an oocyte look nothing alike, yet each houses a haploid package of genetic material. A sperm is specialized to deliver its package; an oocyte accumulates reserves to support early development.

### 3.4 Prenatal Development

Nearly all human prenatal development occurs during the first eight weeks, the period of the embryo. During the remaining months of gestation, structures grow and specialize.

### 3.5 Birth Defects

Malfunctioning genes or environmental insults can derail development, with devastating results. The nature of a birth defect depends upon which structures were forming at the time of the abnormal gene action or environmental intervention.

### 3.6 Maturation and Aging

Development hardly ceases with birth—decades of growth and elaboration of structures continue. Single-gene and multifactorial conditions may be expressed at different times, or speed aging-associated changes. A region of chromosome 4 contains genes involved in determining life span.

Genes orchestrate our physiology from shortly after conception through adulthood. As a result, disorders caused by the malfunction of single genes, or genetic predispositions, affect people of all ages. Certain single gene mutations act before birth, causing broken bones, dwarfism, or even cancer. Many other mutant genes exert their effects during childhood, and it may take parents months or even years to realize that there is a health problem. Duchenne muscular dystrophy (figure 2.1*a*), for example, usually begins as clumsiness in early childhood. Inherited forms of heart disease and breast cancer often appear in early or middle adulthood, which is sooner than multifactorial forms of these conditions typically begin. Pattern baldness is an inherited trait that may not become obvious until well into adulthood.

This chapter explores the stages of the human life cycle, which is the developmental backdrop against which genes function.

## 3.1 The Reproductive System

The formation of a new individual begins with the **sperm** from a male and the ovum (more precisely called an **oocyte**) from a female. Sperm and oocytes are called **gametes,** or sex cells. They provide a mechanism for forming a new individual and mixing genetic contributions from past generations. As a result, each person has a unique combination of inherited traits (except for identical multiples).

Sperm and oocytes are produced in the reproductive system, which is organized similarly in the male and female. Each system has paired structures, called **gonads,** where the sperm and oocytes are manufactured; tubules to transport these cells; and hormones and secretions that control the process.

### The Male

Sperm cells develop within a 125-meter-long network of **seminiferous tubules,** which are packed into paired, oval organs called **testes** (sometimes called testicles) (figure 3.1). The testes are the male gonads. They lie outside the abdomen within a sac called the scrotum. Lying outside the abdominal cavity allows the testes to maintain a lower temperature than the rest of the body, which is necessary for sperm to develop properly. Leading from each testis is a tightly coiled tube, the **epididymis,** in which sperm cells mature and are stored; each epididymis continues into another tube, the **vas deferens.** Each vas deferens bends behind the bladder to join the **urethra,** the tube that carries both sperm and urine out through the penis.

Along the sperm's path, three glands produce secretions. The vasa deferentia pass through the **prostate gland,** which produces a thin, milky, alkaline fluid that activates the sperm to swim. Opening into the vas deferens is a duct from the **seminal vesicles,** which secrete fructose (a sugar that supplies sperm with energy), plus hormonelike prostaglandins, which may stimulate contractions in the female that help sperm and oocyte meet. The **bulbourethral glands,** each about the size of a pea, join the urethra where it passes through the body wall. They secrete an alkaline mucus that coats the urethra before sperm are released. All of these secretions combine to form the **seminal fluid** that carries sperm.

During sexual arousal, the penis becomes erect so that it can penetrate and deposit sperm in the female reproductive tract. At the peak of sexual stimulation, a pleasurable sensation called **orgasm** occurs, accompanied by rhythmic muscular contractions that eject the sperm from each vas deferens through the urethra and out the penis. The discharge of sperm from the penis, called ejaculation, delivers about 200 to 600 million sperm cells.

### The Female

The female sex cells develop within paired organs in the abdomen called **ovaries** (figure 3.2), which are the female gonads. Within each ovary of a newborn female are about a million immature oocytes. Each individual oocyte is surrounded by nourishing **follicle cells,** and each ovary houses oocytes in different stages of development. After puberty, once a month, one ovary releases the most mature oocyte. Beating cilia sweep the mature oocyte into the fingerlike projections of one of two

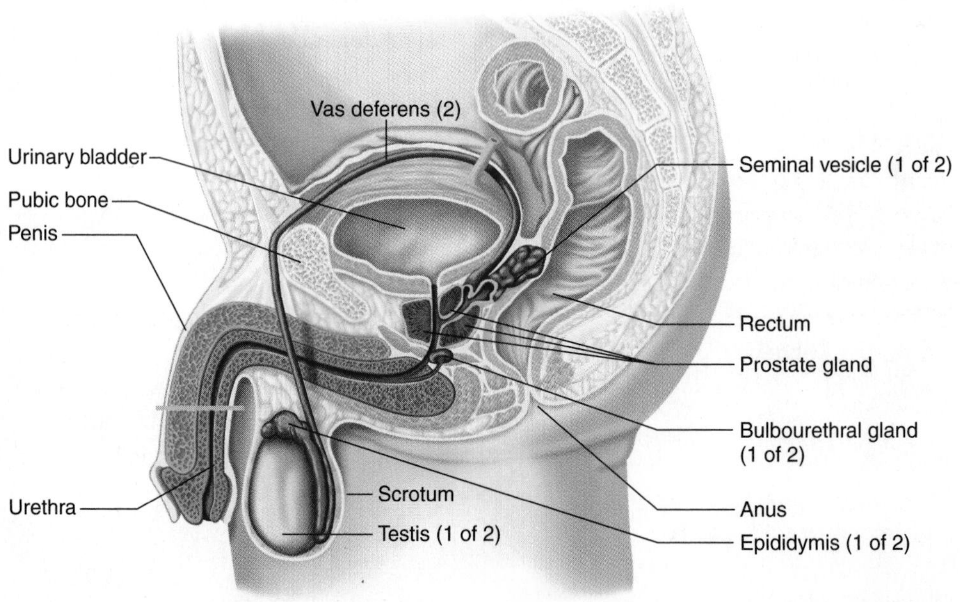


**figure 3.1**

**The human male reproductive system.** Sperm cells are manufactured within the seminiferous tubules, which tightly wind within the testes, which descend into the scrotum. The prostate gland, seminal vesicles, and bulbourethral glands add secretions to the sperm cells to form seminal fluid. Sperm mature and are stored in the epididymis and exit through the vas deferens. The paired vasa deferentia join in the urethra, through which seminal fluid exits the body.

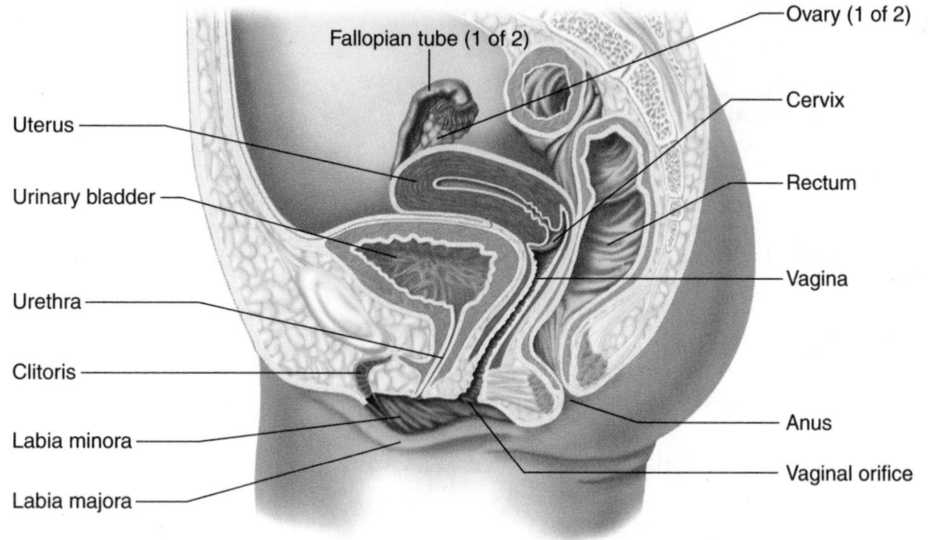


**figure 3.2**

**The human female reproductive system.** Oocytes are packed into the paired ovaries. Once a month after puberty, one oocyte is released from an ovary and is drawn into a nearby fallopian tube. If a sperm fertilizes the oocyte in the fallopian tube, the fertilized ovum continues into the uterus, where for nine months it develops into a new individual. If the oocyte is not fertilized, the body expels it, along with the built-up uterine lining.

**fallopian tubes.** The tube carries the oocyte into a muscular saclike organ. This is the **uterus,** or womb.

The released oocyte may encounter a sperm, usually in a fallopian tube. If the sperm enters the oocyte so that the genetic material of the two cells merges into a new nucleus, the result is a **fertilized ovum.** This cell undergoes a series of rapid cell divisions while moving through the tube and within days nestles into the lining of the uterus. Here, if all goes well, it will continue to develop. If fertilization does not occur, the oocyte, along with much of the uterine lining, is shed as the menstrual flow. Hormones coordinate the monthly menstrual cycle.

The lower end of the uterus narrows and leads to the **cervix,** which opens into the tubelike vagina that exits from the body. The vaginal opening is protected on the outside by two pairs of fleshy folds. At the upper juncture of both pairs is a two-centimeter-long structure called the clitoris, which is anatomically similar to the penis. Rubbing the clitoris triggers female orgasm. Hormones control the cycle of oocyte maturation and the preparation of the uterus to nurture a fertilized ovum.

**Key Concepts**

Reproductive systems have paired gonads that house reproductive cells and tubes. Sperm develop in the seminiferous tubules, mature and collect in each epididymis, enter the vasa deferentia, and move through the urethra in the penis. The prostate gland adds an alkaline fluid, seminal vesicles add fructose and prostaglandins, and bulbourethral glands secrete mucus to form seminal fluid. • In the female, ovaries contain oocytes. Each month, one oocyte is released from an ovary and enters a fallopian tube, which leads to the uterus. If the oocyte is fertilized, it begins a series of rapid cell divisions and nestles into the uterine lining and develops. Otherwise, it exits the body with the menstrual flow. Hormones control the monthly cycle of oocyte development.

## 3.2 Meiosis

Gametes form from special cells, called germline cells, in a type of cell division called **meiosis** that halves the chromosome number. A further process, maturation, sculpts the distinctive characteristics of sperm and oocyte. The organelle-packed oocyte has 90,000 times the volume of the sperm. Unlike other cells in the human body, gametes contain 23 different chromosomes—half the usual amount of genetic material, but a complete genome. Somatic (nonsex) cells contain 23 pairs or 46 chromosomes. The chromosome pairs are called **homologous pairs,** or homologs for short. Homologs have the same genes in the same order but may carry different alleles, or forms of the same gene. Gametes are **haploid** ($1n$), which means that they have only one of each type of chromosome and therefore one copy of the human genome. Somatic cells are **diploid** ($2n$), signifying that they have two copies of the genome.

Halving the number of chromosomes during gamete formation makes sense. If the sperm and oocyte each contained 46 chromosomes, the fertilized ovum would contain twice the normal number of chromosomes, or 92. Such a genetically overloaded cell, called a polyploid, usually does not develop. About one in a million newborns is polyploid, but these infants have abnormalities in all organ systems and live no more than a few days. However, studies on spontaneously aborted embryos indicate that about 1 percent of conceptions have three chromosome sets instead of the normal two, indicating that most of these individuals do not survive to be born.

Meiosis not only produces cells with only one copy of the genome, but it also mixes up combinations of traits. For example, a person might produce one gamete containing alleles encoding green eyes and freckles, yet another encoding brown eyes and no freckles. Meiosis explains why siblings differ from each other genetically and from their parents.

In a much broader sense, meiosis, as the mechanism of sexual reproduction, provides genotypic diversity, which can help a population to survive a challenging environment. A population of sexually reproducing organisms is made up of individuals with different genotypes and phenotypes. In contrast, a population of asexually reproducing organisms consists of identical individuals. Should a new threat arise, such as an infectious disease

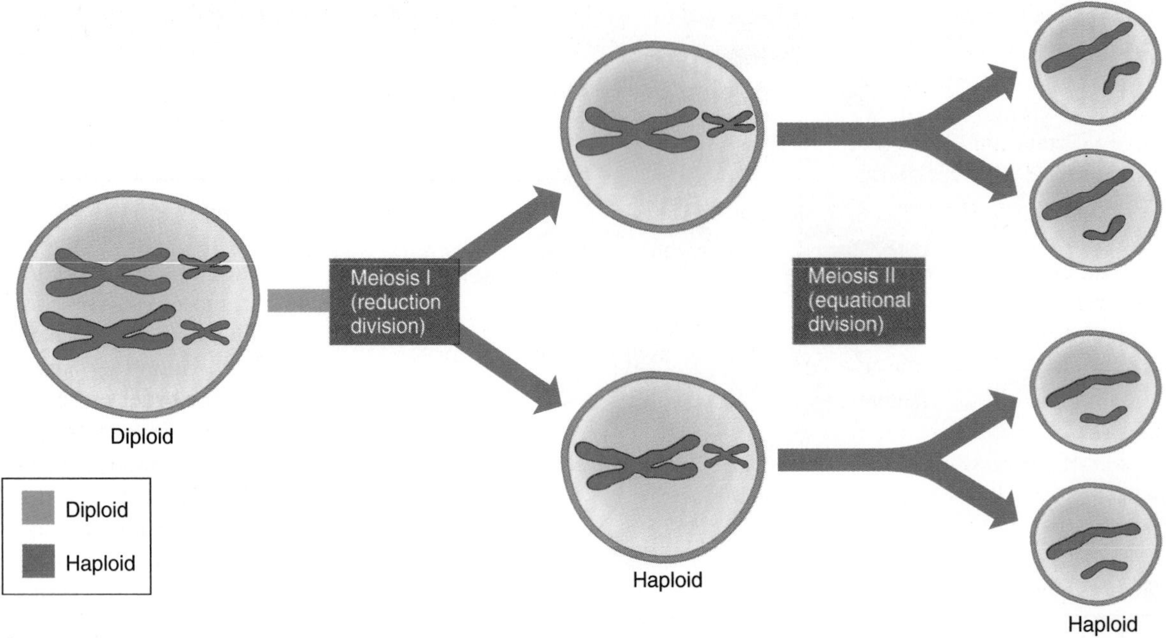


## figure 3.3

**Overview of meiosis.** Meiosis is a form of cell division in which certain cells are set aside to give rise to haploid gametes. The first meiotic division reduces the number of chromosomes to 23, all in the replicated form. In the second meiotic division the cells essentially undergo mitosis. The result of the two divisions of meiosis is four haploid cells. Homologous pairs of chromosomes are indicated by size, and parental origin of chromosomes by color.

**MEIOSIS I**

Nucleus

Spindle fibers

Nuclear envelope

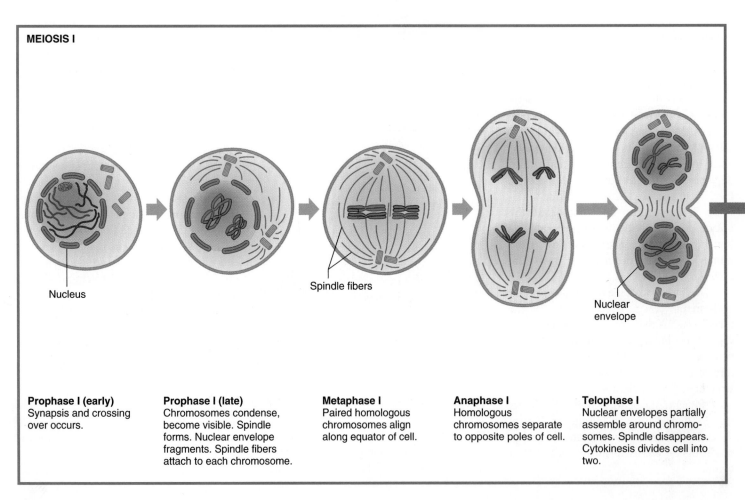

**Prophase I (early)** Synapsis and crossing over occurs.

**Prophase I (late)** Chromosomes condense, become visible. Spindle forms. Nuclear envelope fragments. Spindle fibers attach to each chromosome.

**Metaphase I** Paired homologous chromosomes align along equator of cell.

**Anaphase I** Homologous chromosomes separate to opposite poles of cell.

**Telophase I** Nuclear envelopes partially assemble around chromosomes. Spindle disappears. Cytokinesis divides cell into two.

## figure 3.4

**Meiosis.**

that kills only individuals with a certain genotype, then the entire asexual population could be wiped out. In a sexually reproducing population, by contrast, individuals that inherited a certain combination of genes might survive. This differential survival of certain genotypes is the basis of evolution, discussed in chapter 15. Some microorganisms that can reproduce asexually or sexually revert to the sexual route when the environment changes.

Meiosis entails two divisions of the genetic material. The first division is called **reduction division** (or meiosis I) because it reduces the number of chromosomes from 46 to 23. The second division, called the **equational division** (or meiosis II), produces four cells from the two cells formed in the first division. Figure 3.3 shows an overview of the process and figure 3.4 depicts the major events of each stage.

As in mitosis, meiosis occurs after an interphase period when DNA is replicated (table 3.1). For each chromosome pair in the cell undergoing meiosis, one homolog comes from the person's mother, and one from the father. The two colors in the illustrations represent the contributions of the two parents.

After interphase, prophase I (so called because it is the prophase of meiosis I) begins as the replicated chromosomes condense and become visible when stained. A spindle forms. Toward the middle of prophase I, the homologs line up next to one another, gene by gene, in an event called **synapsis.** A mixture of RNA and protein

**table 3.1**

**Comparison of Mitosis and Meiosis**

| Mitosis | Meiosis |
|---|---|
| One division | Two divisions |
| Two daughter cells per cycle | Four daughter cells per cycle |
| Daughter cells genetically identical | Daughter cells genetically different |
| Chromosome number of daughter cells same as that of parent cell ($2n$) | Chromosome number of daughter cells half that of parent cell ($1n$) |
| Occurs in somatic cells | Occurs in germline cells |
| Occurs throughout life cycle | In humans, completes after sexual maturity |
| Used for growth, repair, and asexual reproduction | Used for sexual reproduction, in which new gene combinations arise |

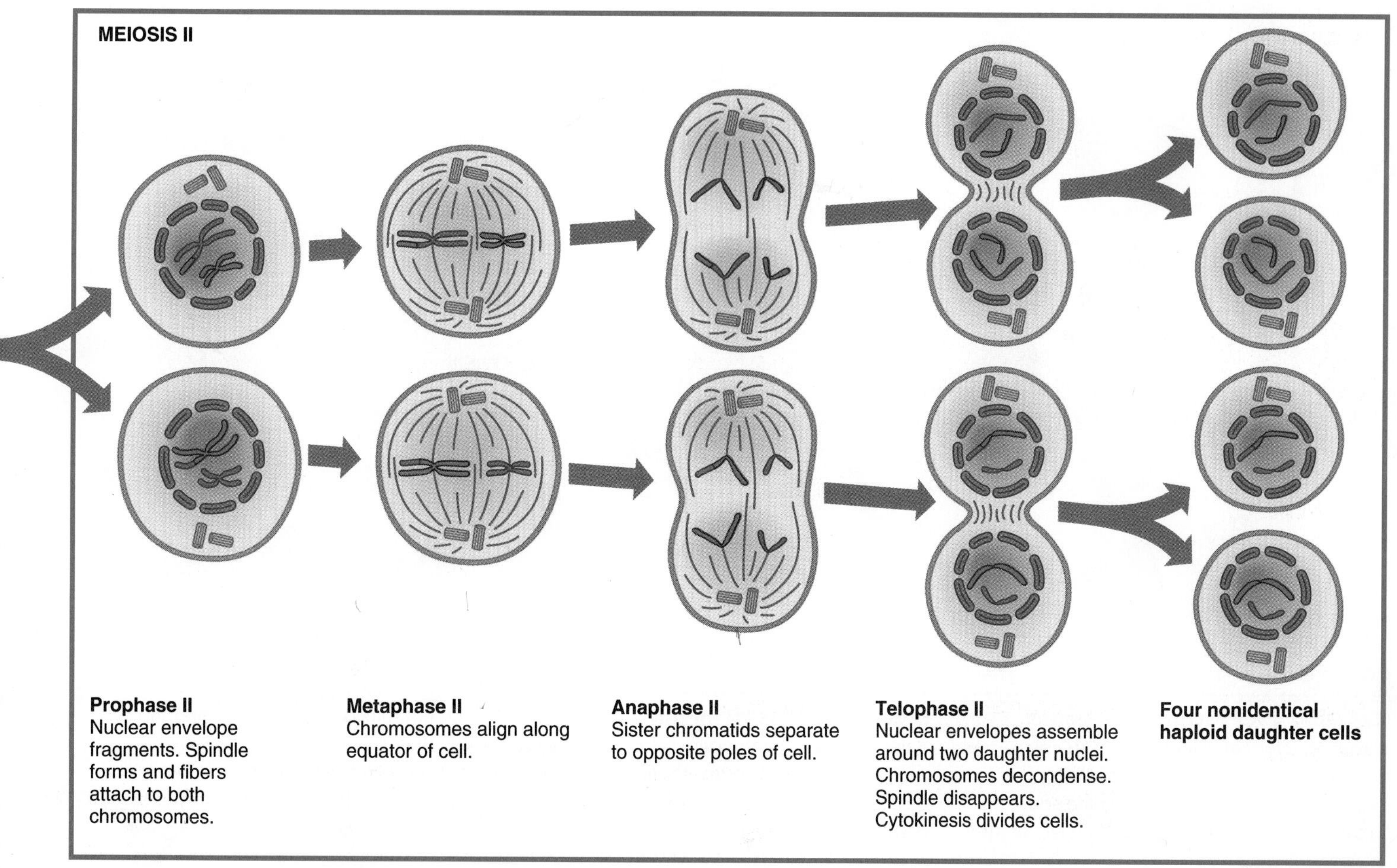
MEIOSIS II
Prophase II
Nuclear envelope fragments. Spindle forms and fibers attach to both chromosomes.
Metaphase II
Chromosomes align along equator of cell.
Anaphase II
Sister chromatids separate to opposite poles of cell.
Telophase II
Nuclear envelopes assemble around two daughter nuclei. Chromosomes decondense. Spindle disappears. Cytokinesis divides cells.
Four nonidentical haploid daughter cells

holds the chromosome pairs together. At this time, the homologs exchange parts in a process called **crossing over** (figure 3.5). All four chromatids that comprise each homologous chromosome pair at this time are pressed together as exchanges occur. The four-chromatid arrangement is called a tetrad. After crossing over, each homolog contains genes from each parent. (Prior to this, all of the genes on a homolog were derived from one parent.) New gene combinations arise from crossing over when the parents carry different alleles. Toward the end of prophase I, the synapsed chromosomes separate but remain attached at a few points along their lengths.

To understand how crossing over mixes trait combinations, consider a simplified example. Suppose that homologs carry genes for hair color, eye color, and finger length. One of the chromosomes carries alleles for blond hair, blue eyes, and short fingers. Its homolog carries alleles for black hair, brown eyes, and long fingers. After crossing over, one of the chromosomes might bear alleles for blond hair, brown eyes, and long fingers, and the other might bear alleles for black hair, blue eyes, and short fingers. The daughter cells that result from meiosis carry a mix of the parent cell traits.

Meiosis continues in metaphase I, when the homologs align down the center of the cell. Each member of a homolog pair attaches to a spindle fiber at opposite poles. The pattern in which the chromosomes align during metaphase I is important in generating genetic diversity. For each homolog pair, the pole the maternally or paternally derived member goes to is random. The situation is analogous to the number of different ways that 23 boys and 23 girls could line up in boy-girl pairs. The greater the number of chromosomes, the greater the genetic diversity generated at this stage.

For two pairs of homologs, four ($2^2$) different metaphase configurations are possible. For three pairs of homologs, eight ($2^3$) different combinations can occur. Our 23 chromosome pairs can line up in 8,388,608 ($2^{23}$) different ways. This random arrangement of the members of homolog pairs in metaphase is called **independent assortment** (figure 3.6). It accounts for a basic law of inheritance discussed in the next chapter.

Homologs separate in anaphase I and finish moving to opposite poles by telophase I. Unlike in mitosis, the centromeres of each homolog in meiosis I remain together. During a second interphase, chromosomes unfold into very thin threads.

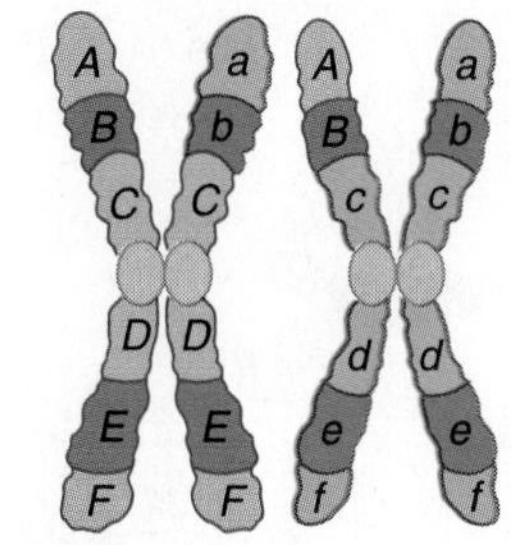


## figure 3.5

**Crossing over recombines genes.** Crossing over helps to generate genetic diversity by mixing up parental traits. The capital and lowercase forms of the same letter represent different forms (alleles) of the same gene.

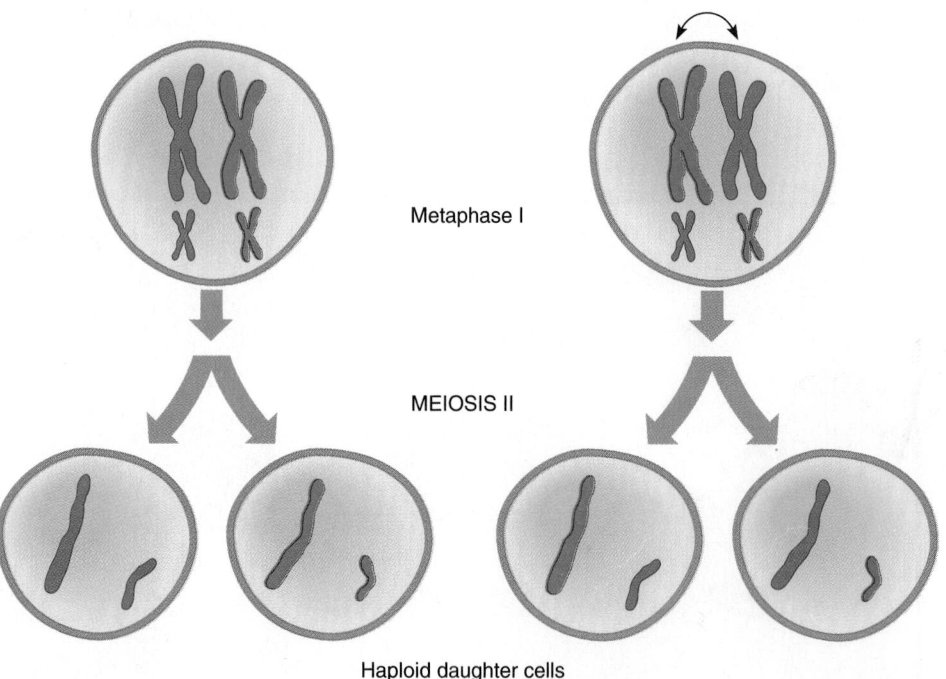


## figure 3.6

**Independent assortment.** The pattern in which homologs align during metaphase I determines the combination of maternally and paternally derived chromosomes in the daughter cells. Two pairs of chromosomes can align in two different ways to produce four different possibilities in the daughter cells. The potential variability generated by meiosis skyrockets when one considers all 23 chromosome pairs and the effects of crossing over.

Proteins are manufactured, but the genetic material is not replicated a second time. It is the single DNA replication, followed by the double division of meiosis, that halves the chromosome number.

Prophase II marks the start of the second meiotic division. The chromosomes are again condensed and visible. In metaphase II, the replicated chromosomes align down the center of the cell. In anaphase II, the centromeres part, and the newly formed chromosomes, each now in the unreplicated form, move to opposite poles. In telophase II, nuclear envelopes form around the four nuclei, which then separate into individual cells. The net result of meiosis is four haploid cells, each carrying a new assortment of genes and chromosomes that represent a single copy of the genome.

Meiosis generates astounding genetic variety. Any one of a person's more than eight million possible combinations of chromosomes can meet with any one of the more than eight million combinations of a partner, raising potential variability to more than 70 trillion ($8,388,608^2$) genetically unique individuals! Crossing over contributes even more genetic variability.

### Key Concepts

The haploid sperm and oocyte are derived from diploid germline cells by meiosis and maturation. Meiosis maintains the chromosome number over generations and mixes gene combinations. In the first meiotic (or reduction) division, the number of chromosomes is halved. In the second meiotic (or equational) division, each of two cells from the first meiotic division divides again, yielding four cells from the original one. Chromosome number is halved because the DNA replicates once, but the cell divides twice. Crossing over and independent assortment generate further genotypic diversity by creating new combinations of alleles.

## 3.3 Gamete Maturation

Meiosis occurs in both sexes, but the sperm and oocyte look very different. Although each type of gamete is haploid, different distributions of other cell components create their distinctions. The cells of the maturing male and female proceed through similar stages, but with sex-specific terminology and different timetables. A male begins manufacturing sperm at puberty and continues throughout life, whereas a female begins meiosis when she is a fetus. Meiosis in the female is completed only if a sperm fertilizes the oocyte.

### Sperm Development

**Spermatogenesis,** the formation of sperm cells, begins in a diploid cell called a **spermatogonium** (figure 3.7). This cell divides mitotically, yielding two daughter cells: one daughter cell continues to specialize into a mature sperm, and the other remains unspecialized as a stem cell.

Bridges of cytoplasm join several spermatogonia, and their daughter cells enter meiosis together. As they mature, these spermatogonia accumulate cytoplasm and replicate their DNA, becoming **primary spermatocytes.**

During reduction division (meiosis I), each primary spermatocyte divides, forming two equal-sized haploid cells called **secondary spermatocytes.** In meiosis II, each secondary spermatocyte divides to

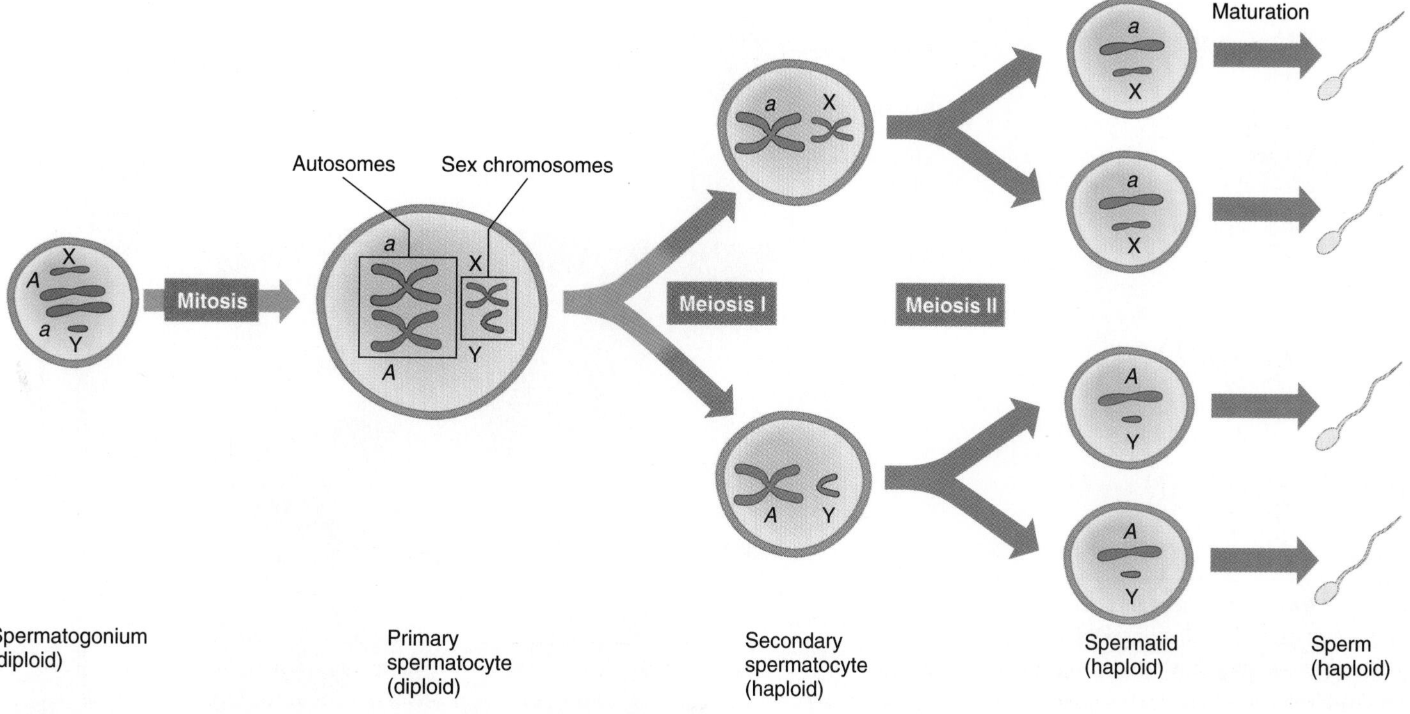


**figure 3.7**

**Sperm formation (spermatogenesis).** Primary spermatocytes have the normal diploid number of 23 chromosome pairs. The large pair of chromosomes represents autosomes (nonsex chromosomes). The X and Y chromosomes are sex chromosomes.

yield two equal-sized **spermatids.** Each spermatid then develops the characteristic sperm tail, or flagellum. The base of the tail has many mitochondria, which produce ATP molecules that propel sperm inside the female reproductive tract. After spermatid differentiation, some of the cytoplasm connecting the cells falls away, leaving mature, tadpole-shaped **spermatozoa,** or sperm. Figure 3.8 presents an anatomical view showing where the stages of spermatogenesis occur within the seminiferous tubules.

A sperm, which is a mere 0.006 centimeters (0.0023 inch) long, must travel about 18 centimeters (7 inches) to reach an oocyte. Each sperm cell consists of a tail, body or midpiece, and head region (figure 3.9). A membrane-covered area on the front end, the **acrosome,** contains enzymes that help the cell penetrate the protective layers around the oocyte. Within the bulbous sperm head, DNA is wrapped around proteins. The sperm's DNA at this time is genetically inactive. A male manufactures trillions of sperm in his lifetime. Although many of these will come close to an oocyte, very few will actually touch one.

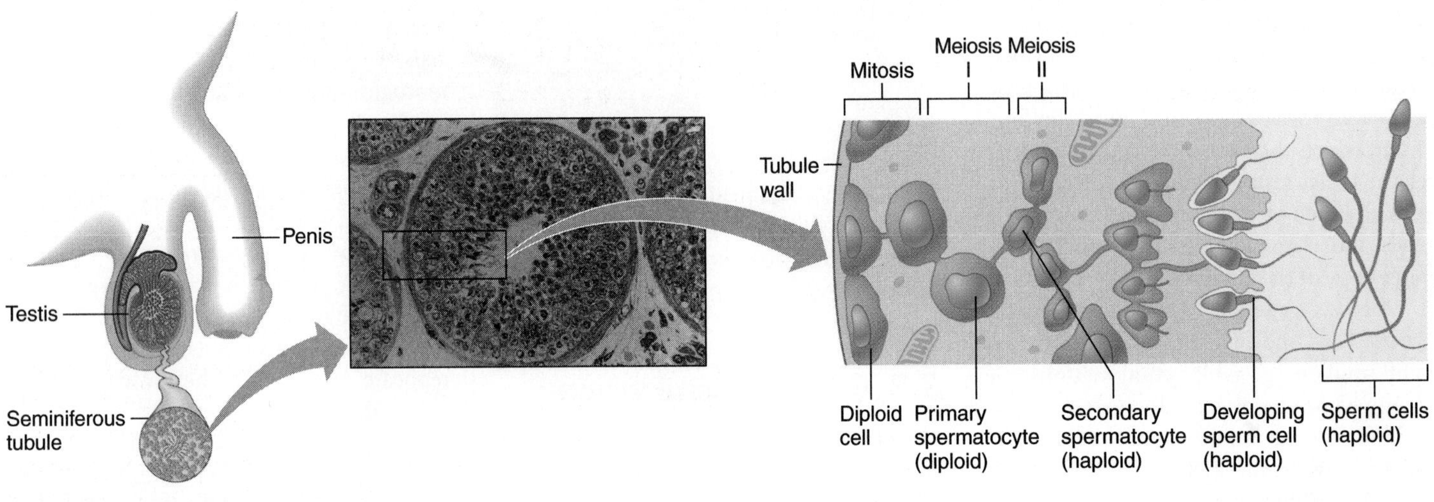


## figure 3.8

**Meiosis produces sperm cells.** Diploid cells divide through mitosis in the linings of the seminiferous tubules. Some of the daughter cells then undergo meiosis, producing haploid spermatocytes, which differentiate into mature sperm cells.

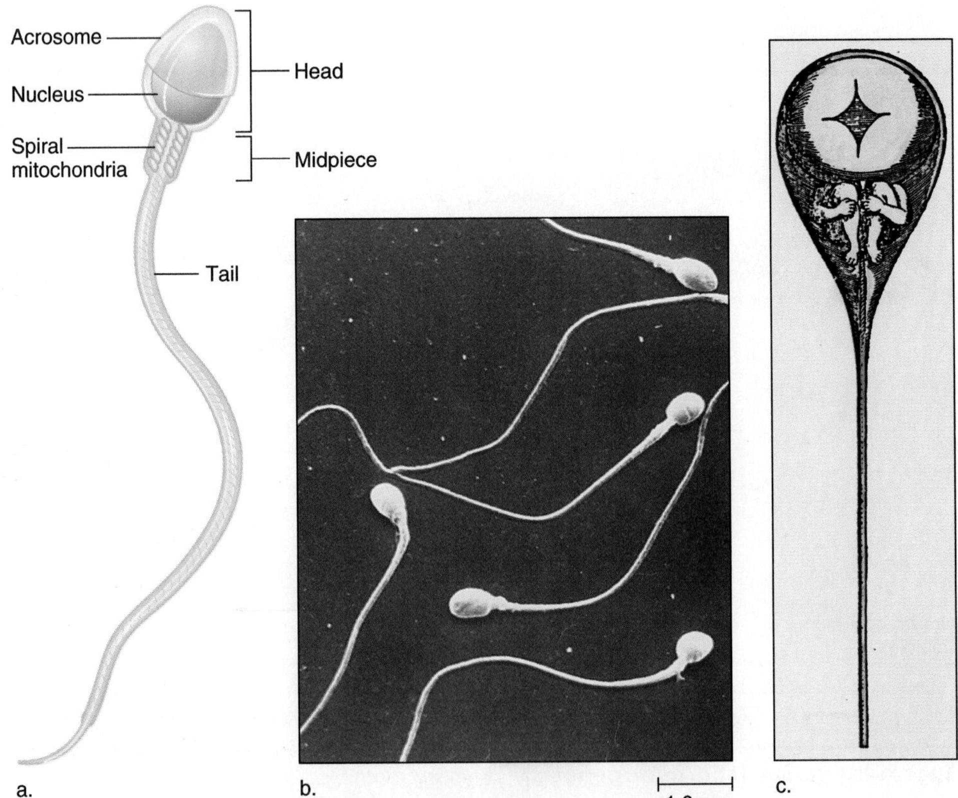


## figure 3.9

**Sperm.** (*a*) A sperm contains distinct regions that assist in delivering DNA to an oocyte. (*b*) Scanning electron micrograph of human sperm cells. (*c*) This 1694 illustration by Dutch histologist Niklass Hartsoeker presents a once-popular hypothesis that a sperm carries a preformed human called a homunculus.

Meiosis in the male has built-in protections that help prevent sperm from causing birth defects. Spermatogonia that are exposed to toxins tend to be so damaged that they never mature into sperm. More mature sperm cells exposed to toxins are often so damaged that they cannot swim. However, seminal fluid can carry certain drugs along with sperm cells. These can affect a fetus by harming the uterus or by entering the woman's circulation and passing to the placenta, the organ connecting a pregnant woman to the fetus. Cocaine can affect a fetus by attaching to thousands of binding sites on sperm without harming the cells or impeding their movements. Therefore, sperm can ferry cocaine to an oocyte, potentially affecting the embryo that may develop.

## Oocyte Development

Meiosis in the female, called **oogenesis** (egg making), begins, as does spermatogenesis, with a diploid cell. This cell is called an **oogonium.** Unlike the male cells, oogonia are not attached, but each is surrounded by follicle cells. Each oogonium grows, accumulates cytoplasm, and replicates its DNA, becoming a **primary oocyte.** The ensuing meiotic division in oogenesis, unlike that in spermatogenesis, produces cells of different sizes.

In meiosis I, the primary oocyte divides into two cells: a small cell with very little cytoplasm, called a **polar body,** and a much larger cell called a **secondary oocyte** (figure 3.10). Each cell is haploid. In meiosis II, the tiny polar body may divide to yield two polar bodies of equal size, or it may simply decompose. The secondary oocyte, however, divides unequally in meiosis II to produce another small polar body and the mature egg cell, or ovum, which contains a large volume of cytoplasm. Figure 3.11 summarizes meiosis in

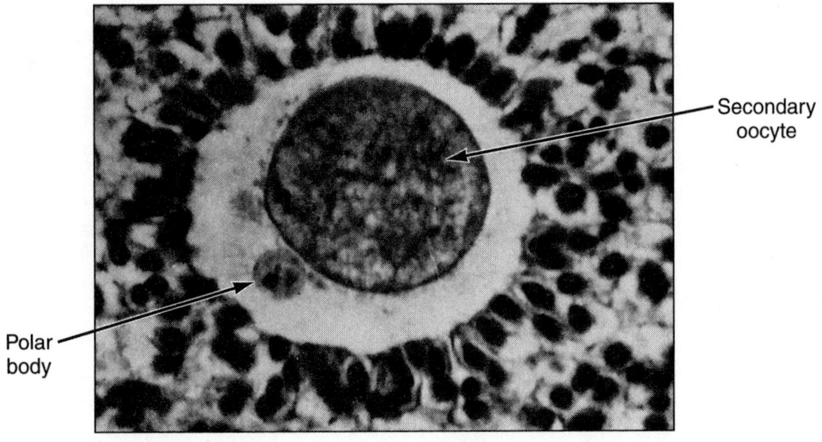


**figure 3.10**

**Meiosis in a female produces a secondary oocyte and a polar body.** The unequal division enables the cell destined to become a fertilized ovum to accumulate the bulk of the cytoplasm and organelles from the primary oocyte, but with only one genome's worth of DNA. (×700)

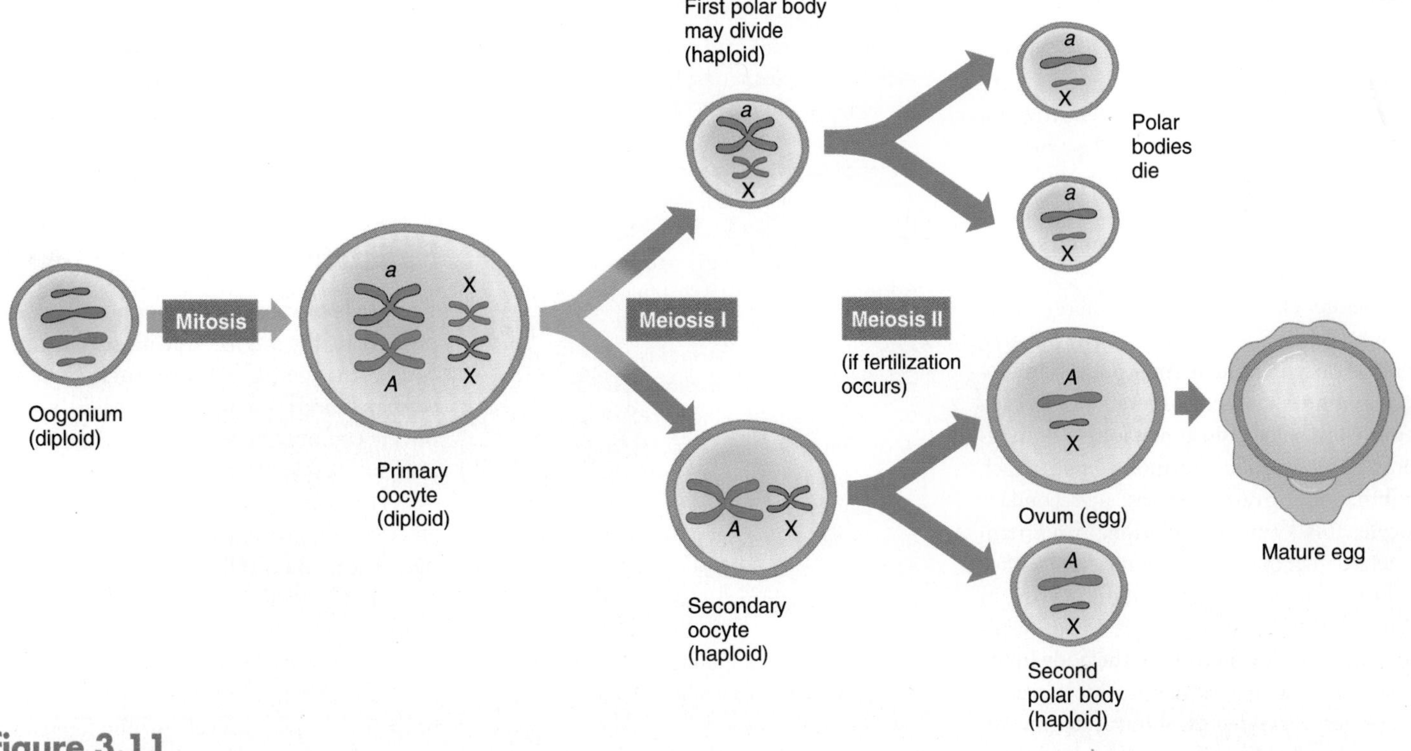


**figure 3.11**

**Ovum formation (oogenesis).** Primary oocytes have the diploid number of 23 chromosome pairs. Meiosis in females is uneven, concentrating most of the cytoplasm into one large cell, an oocyte (or egg). The other products of meiosis, called polar bodies, contain the other three sets of chromosomes and are normally discarded.

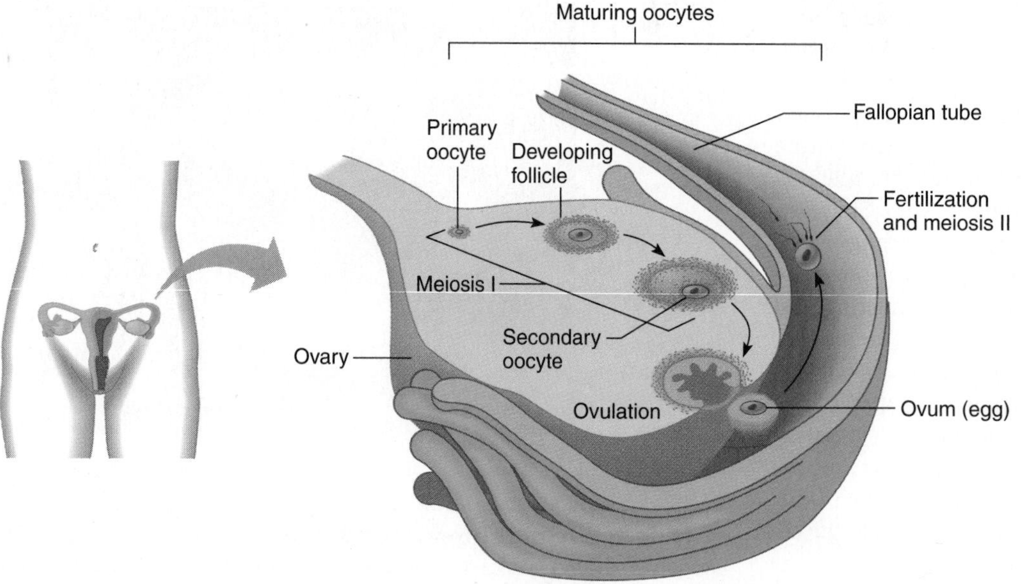


**figure 3.12**

**The making of oocytes.** Oocytes develop within the ovary in protective follicles. An ovary contains many oocytes in various stages of maturation. After puberty, each month the most mature oocyte in one ovary bursts out, an event called ovulation.

the female, and figure 3.12 provides an anatomical view of the process.

Most of the cytoplasm among the four meiotic products in the female concentrates in only one cell, the ovum. The woman's body absorbs the polar bodies, and they normally play no further role in development. Rarely, a sperm fertilizes a polar body. The woman's hormones respond as if she is pregnant, but a clump of cells that is not an embryo grows for a few weeks, and then leaves the woman's body. This event is a type of miscarriage called a "blighted ovum." Reading 3.1 describes a very unusual child whose beginnings trace back to a fertilized polar body.

Before birth, a female's million or so oocytes arrest in prophase I. By puberty, 400,000 oocytes remain. After puberty, meiosis I continues in one or several oocytes each month, but halts again at metaphase II. In response to specific hormonal cues each month, one ovary releases a secondary oocyte; this event is **ovulation.** If a sperm penetrates the oocyte membrane, then female meiosis completes, and a fertilized ovum forms. If the secondary oocyte is not fertilized, it degenerates and leaves the body in the menstrual flow, meiosis never completed.

A female ovulates about 400 oocytes between puberty and menopause. Sperm cells are likely to enter only a few of these oocytes. Only 1 in 3 of the oocytes that do meet and merge with a sperm cell will continue to grow, divide, and specialize to eventually form a new individual.

**Key Concepts**

Spermatogonia divide mitotically, yielding one stem cell and one cell that accumulates cytoplasm and becomes a primary spermatocyte. In meiosis I, each primary spermatocyte halves its genetic material to form two secondary spermatocytes. In meiosis II, each secondary spermatocyte divides, yielding two equal-sized spermatids attached by bridges of cytoplasm. Maturing spermatids separate and shed some cytoplasm. A mature sperm has a tail, body, and head, with an enzyme-containing acrosome covering the head. • An oogonium accumulates cytoplasm and replicates its chromosomes, becoming a primary oocyte. In meiosis I, the primary oocyte divides, forming a small polar body and a large, haploid secondary oocyte. In meiosis II, the secondary oocyte divides, yielding another small polar body and a mature haploid ovum. Oocytes arrest at prophase I until puberty, when one or several oocytes complete the first meiotic division during ovulation. The second meiotic division completes at fertilization.

## 3.4 Prenatal Development

A prenatal human is considered an **embryo** for the first eight weeks. During this time, rudiments of all body parts form. The embryo in the first week is considered to be in a "preimplantation" stage because it has not yet settled into the uterine lining. Prenatal development after the eighth week is the fetal period, when structures grow and specialize. The human organism between the start of the ninth week and birth is a **fetus.**

### Fertilization

Hundreds of millions of sperm cells are deposited in the vagina during sexual intercourse. A sperm cell can survive in the woman's body for up to six days, but the oocyte can only be fertilized in the 12 to 24 hours after ovulation.

The woman's body helps sperm reach an oocyte. A process in the female called **capacitation** chemically activates sperm, and the oocyte secretes a chemical that attracts sperm. Sperm are also assisted by contractions of the female's muscles, by their moving tails, and by upwardly moving mucus propelled by cilia on cells of the female reproductive tract. Still, only 200 or so sperm near the oocyte.

A sperm first contacts a covering of follicle cells, called the corona radiata, that guards a

*Reading 3.1*

# The Strange Case of F.D.: A Human Partial Parthenogenote

In some animal species, reproduction can occur without a mate, when an oocyte doubles its DNA. Reproduction in a sexually reproducing organism, but from one parent, is called parthenogenesis, which is Greek for "virgin birth." Animals as diverse as salamanders, lizards, snakes, turkeys, roundworms, flatworms, aphids, and various single-celled pond dwellers do it. The most familiar examples are certain male bees, wasps, and ants, which develop from unfertilized eggs.

Mammalian parthenogenotes are less successful. A mouse embryo derived from an activated oocyte shrivels up and ceases developing. Humans fare even worse. If an oocyte somehow doubles its genetic material and starts to develop, it doesn't form an embryo at all, but a tumorlike mass called a teratoma. If a sperm doubles its DNA and divides, it forms a different type of abnormal growth. But biologists thought it might be possible for a mammal to be a partial parthenogenote—that is, part of the body is derived from an unfertilized maternal cell. The birth of a most unusual child in 1991, whom the medical journals called "F.D.," showed that this can indeed happen.

Shortly after F.D.'s birth, his parents noticed that his head was lopsided, with the left side skewed and drooping. Photographs of the boy from one side look completely normal, yet from the other side, he has a jutting chin, sunken eye, and oddly clenched jaw. Not obvious are a cleft palate and other abnormal throat structures. F.D. has been relatively healthy.

Because individuals who have chromosome abnormalities often have unusual facial features, doctors took a blood sample to examine the child's chromosomes. To their surprise, they found two X chromosomes in the white blood cells, indicating a female. But F.D. looked like a boy!

The doctors continued to examine the sex chromosomes in different types of cells in F.D. They found some to be XX and some to be XY. It was clear that F.D.'s body consists of two cell populations, one XX and one XY.

Abnormal meiosis in the mother can explain the unusual conception of F.D. One polar body evidently replicated its DNA to become an XX cell, and a second polar body was fertilized by a Y-bearing sperm. The two cells remained attached, and developed into F.D., such that half the body is chromosomally male, and the other half female (figure 1). Fortunately, F.D. looks and feels male, and can one day become a father because his reproductive system descended from the XY cell.

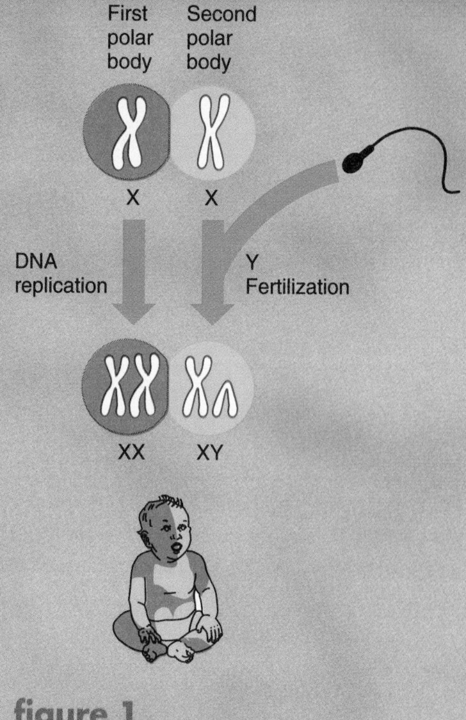


**figure 1**

**A most unusual child.** "F.D." arose from a polar body that replicated its DNA but did not divide, and a second polar body, that was fertilized. The two initial cells remained together, producing a child with two cell populations. He is a partial parthenogenote.

secondary oocyte. The sperm's acrosome then bursts, releasing enzymes that bore through a protective layer of glycoprotein, called the zona pellucida, beneath the corona radiata. Fertilization, or conception, begins when the outer membranes of the sperm and secondary oocyte meet (figure 3.13). The encounter is dramatic. A wave of electricity spreads physical and chemical changes across the entire oocyte surface—changes that keep other sperm out. More than one sperm can enter an oocyte, but the resulting cell typically has too much genetic material to develop normally.

Usually only the sperm's head enters the oocyte. Within 12 hours of the sperm's penetration, the nuclear membrane of the ovum disappears, and the two sets of chromosomes, called **pronuclei,** approach one another. Within each pronucleus, DNA replicates. Fertilization completes when the two genetic packages meet, forming the genetic instructions for a new individual. The fertilized ovum is called a zygote. Bioethics: Choices for the Future at the end of this chapter describes cloning, another way to begin development of an organism.

## Early Events—Cleavage and Implantation

About a day after fertilization, the zygote divides by mitosis, beginning a period of frequent cell division called **cleavage** (figure 3.14). The resulting early cells are called **blastomeres.** When the blastomeres form a solid ball of 16 or more cells, the embryo is called a **morula** (Latin for "mulberry," which it resembles).

During cleavage, organelles and molecules from the secondary oocyte's cytoplasm

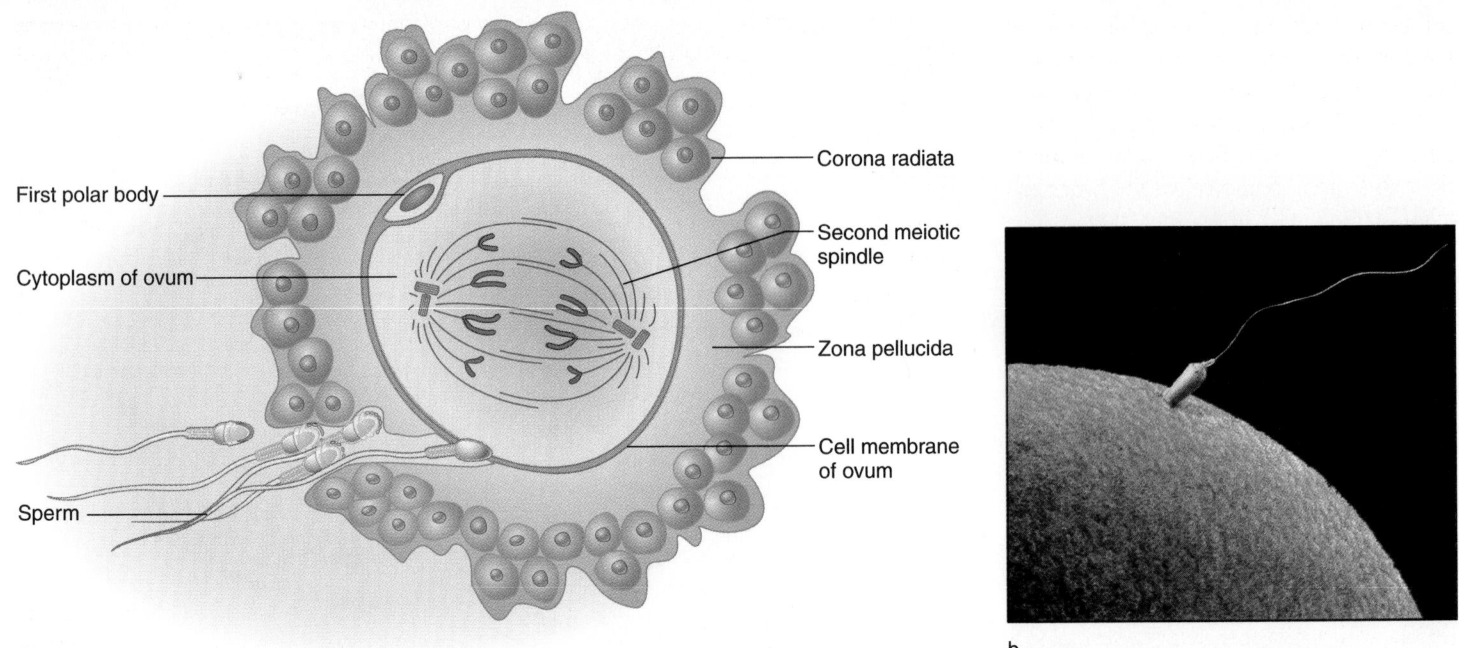


## figure 3.13

**Fertilization.** (*a*) Fertilization by a sperm cell induces the oocyte (arrested in metaphase II) to complete meiosis. Before fertilization occurs, the sperm's acrosome bursts, spilling forth enzymes that help the sperm's nucleus enter the oocyte. (*b*) A series of chemical reactions ensues that helps to ensure that only one sperm nucleus enters an oocyte.

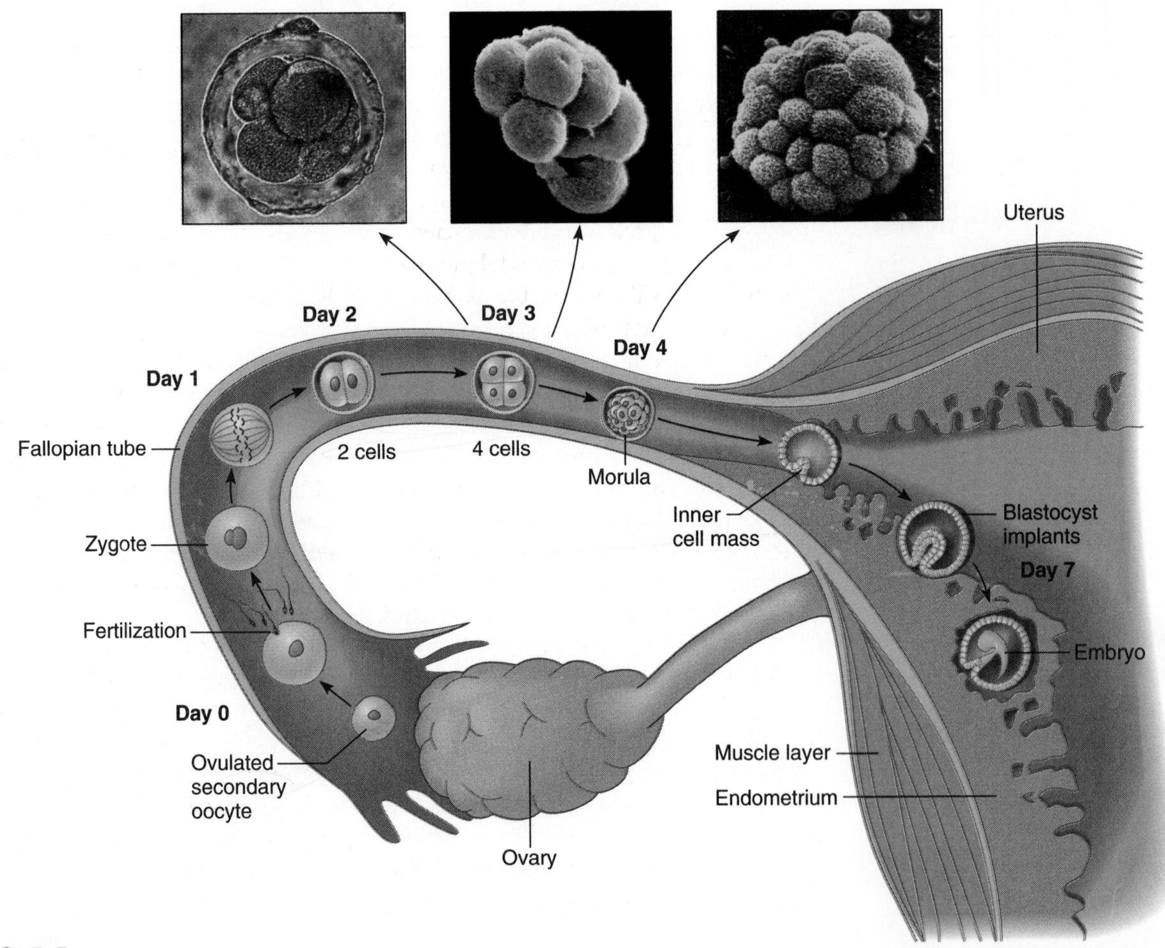


## figure 3.14

**From ovulation to implantation.** The zygote forms in the fallopian tube when a sperm nucleus fuses with the nucleus of an oocyte. The first divisions ensue while the zygote moves toward the uterus. By day 7, it begins to implant in the uterine lining.

still control cellular activities, but some of the embryo's genes begin to function. The ball of cells hollows out, and its center fills with fluid. It is now a **blastocyst,** the "cyst" referring to the fluid-filled center. Some of the cells form a clump called the **inner cell mass.** This is the first event that distinguishes cells from each other in terms of their relative positions, other than the inside and outside of the morula. The cells of the inner cell mass will continue developing to form the embryo.

A week after conception, the blastocyst begins to nestle into the rich lining of the woman's uterus. This event, called implantation, takes about a week. As it starts, the outermost cells of the embryo, called the **trophoblast,** secrete the "pregnancy hormone," **human chorionic gonadotropin** (hCG), which prevents menstruation. hCG detected in a woman's urine or blood is one sign of pregnancy.

## The Embryo Forms

During the second week of prenatal development, a space called the **amniotic cavity** forms between the inner cell mass and the outer cells anchored to the uterine lining. Then the inner cell mass flattens into a two-layered disc. The layer nearest the amniotic cavity is the **ectoderm** (Greek for "outside skin"). The inner layer, closer to the blastocyst cavity, is the **endoderm** (Greek for "inside skin"). Shortly after, a third layer, the **mesoderm** ("middle skin"), forms in the middle. This three-layered structure is called the primordial embryo, or the **gastrula.**

Once these three layers, called **primary germ layers,** form, the fates of many cells become determined, which means that they are destined to develop as a specific cell type. Each layer gives rise to certain structures (figure 3.15). Cells in the ectoderm become skin, nervous tissue, or parts of certain glands. Endoderm cells form parts of the liver and pancreas and the linings of many organs. The middle layer of the embryo, the mesoderm, forms many structures, including muscle, connective tissues, the reproductive organs, and the kidneys. However, certain stem cells in each of the germ layers retain the ability to produce daughter cells that specialize as cells from other germ layers. Table 3.2 summarizes the stages of early prenatal development.

**table 3.2**

**Stages and Events of Early Human Prenatal Development**

| Stage | Time Period | Principal Events |
|---|---|---|
| Fertilized ovum | 12–24 hours following ovulation | Oocyte fertilized; zygote has 23 pairs of chromosomes and is genetically distinct |
| Cleavage | 30 hours to third day | Mitosis increases cell number |
| Morula | Third to fourth day | Solid ball of cells |
| Blastocyst | Fifth day through second week | Hollowed ball forms trophoblast (outside) and inner cell mass, which implants and flattens to form embryonic disc |
| Gastrula | End of second week | Primary germ layers form |

## Supportive Structures

As the embryo develops, structures form that support and protect it. These include chorionic villi, the placenta, the yolk sac, the allantois, the umbilical cord, and the amniotic sac.

By the third week after conception, fingerlike projections called **chorionic villi** extend from the area of the embryonic disc close to the uterine wall, dipping into pools of the woman's blood. Her blood system and that of the embryo are separate, but nutrients and oxygen diffuse across the chorionic villi from her circulation to the embryo, and wastes leave the embryo's circulation and enter the woman's circulation to be excreted.

By 10 weeks, the placenta is fully formed. This structure will connect the woman to the fetus for the rest of the pregnancy. The placenta secretes hormones that maintain pregnancy and alter the woman's metabolism to shuttle nutrients to the fetus.

Other structures nurture the developing embryo. The **yolk sac** manufactures blood cells, as does the **allantois,** a membrane surrounding the embryo that gives rise to the umbilical blood vessels. The umbilical cord forms around these vessels and attaches to the center of the placenta. Toward the end of the embryonic period, the yolk sac shrinks, and the amniotic sac swells with fluid that cushions the embryo and maintains a constant temperature and pressure. The amniotic fluid contains fetal urine and cells.

Two of the supportive structures that develop during pregnancy are the basis of prenatal tests, discussed in chapter 12. In amniocentesis, a sample of amniotic fluid is taken after the fourteenth week of pregnancy, and fetal cells in the fluid are examined for biochemical and chromosome anomalies. Chorionic villus sampling examines chromosomes from cells snipped off the chorionic villi at 10 weeks. Because the villi cells and the embryo's cells come from the same fertilized ovum, an abnormal chromosome detected in villi cells should also be present in the embryo.

## On the Matter of Multiples

Twins and other multiples arise during the earliest stages of development.

Twins are either fraternal or identical. Fraternal, or **dizygotic** (DZ), twins result when two sperm fertilize two oocytes. This can happen if ovulation occurs in two ovaries in the same month, or if two oocytes leave the same ovary, and are both fertilized. DZ twins are no more alike than any two siblings, although they share a very early environment, in the uterus. The tendency to have DZ twins can run in families in which certain women may ovulate two oocytes a month.

Identical, or **monozygotic** (MZ), twins descend from a single fertilized ovum, and therefore are genetically identical. They are, in essence, natural clones. Three types of MZ twins form, depending upon when the fertilized ovum or very early embryo splits (figure 3.16). This difference in timing determines which supportive structures the twins share. About a third of all MZ twins

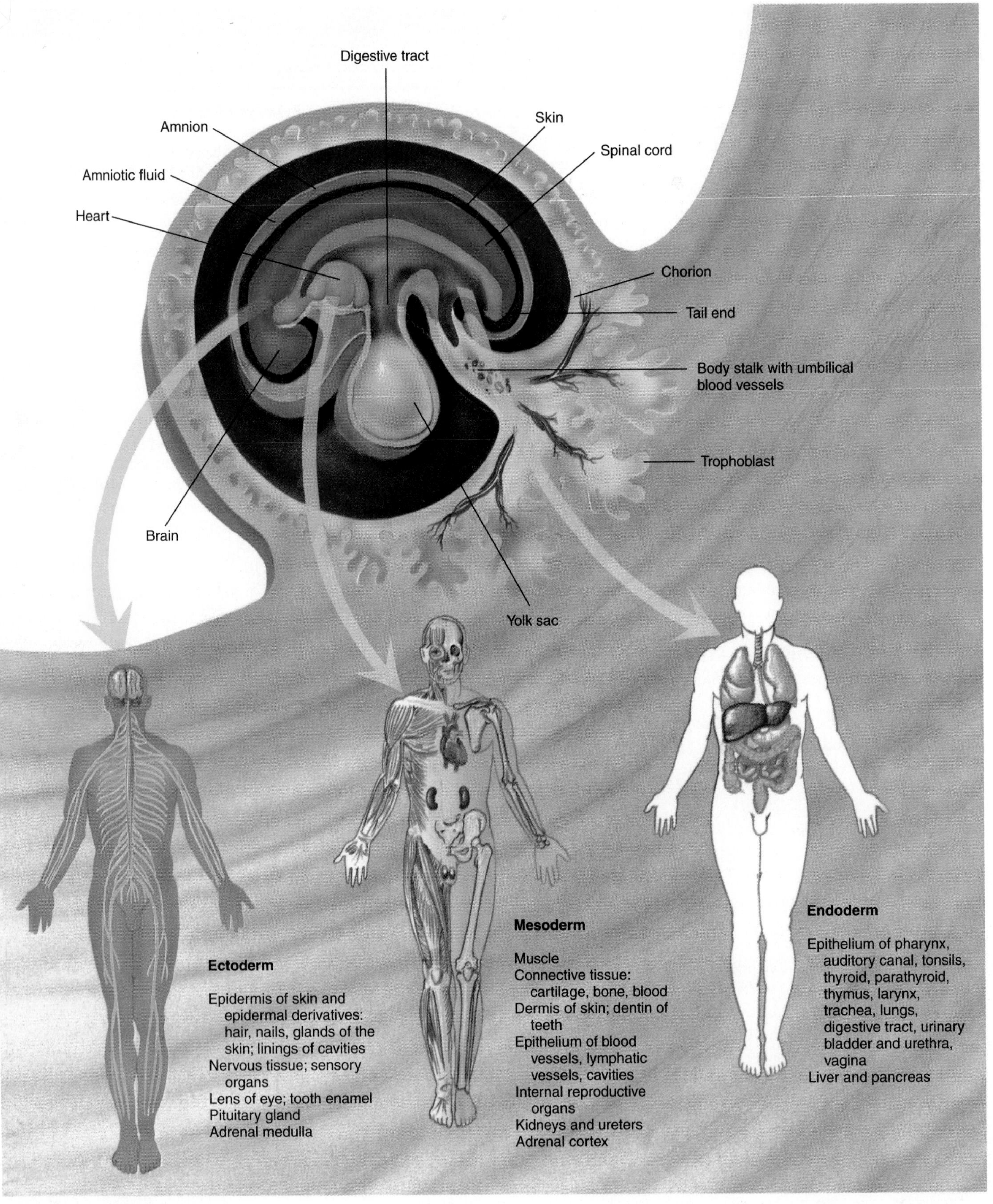


**figure 3.15**

**The primordial embryo.** When the three basic layers of the embryo form, many cells become "determined" to follow a specific developmental pathway. However, each layer retains stem cells as the organism develops, and these may be capable of producing daughter cells that can specialize as any of many cell types, some not even associated with the layer of origin.

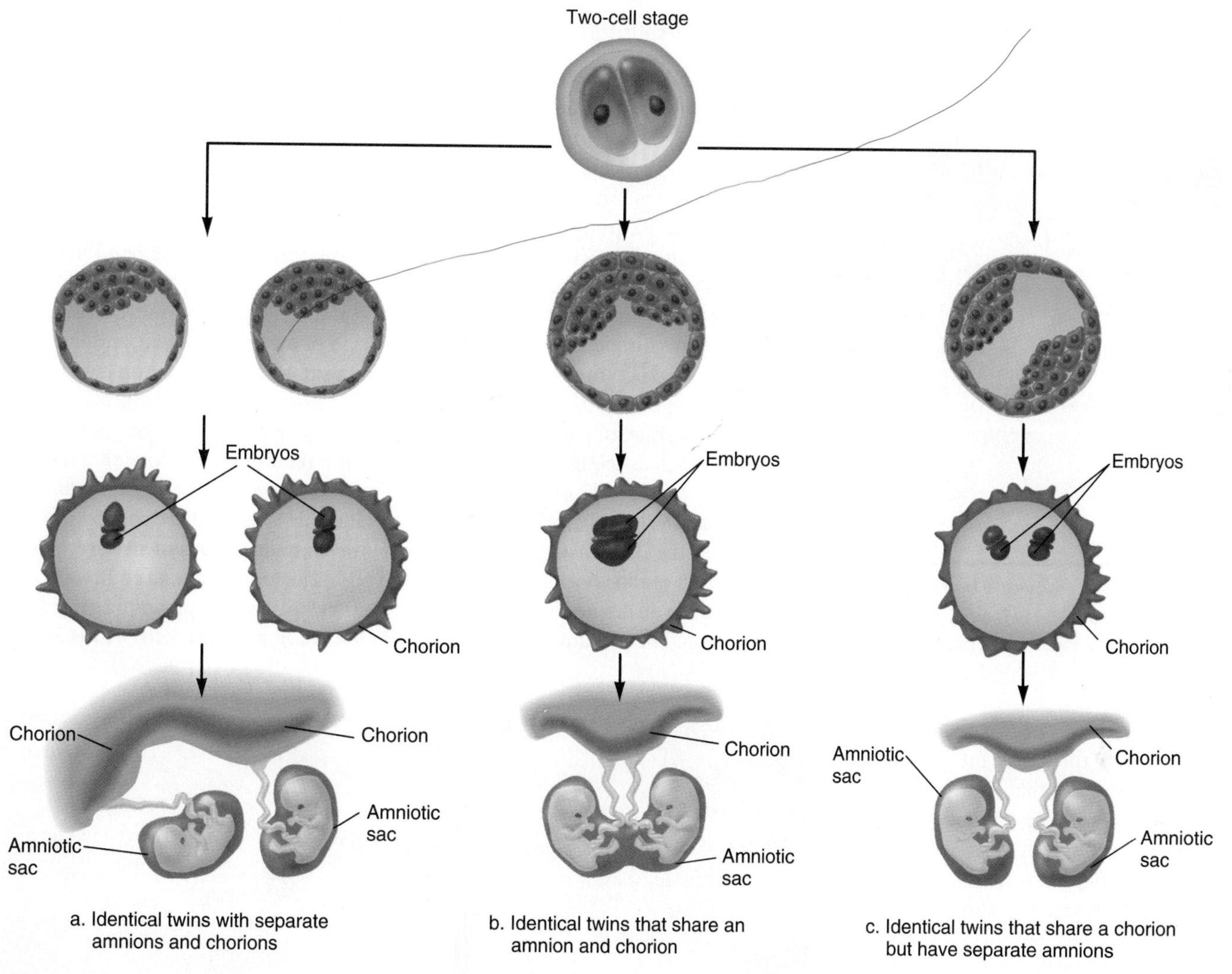


**figure 3.16**

**Facts about twins.** Identical twins originate at three points in development. (*a*) In about one-third of identical twins, separation of cells into two groups occurs before the trophoblast forms on day 5. These twins have separate chorions and amnions. (*b*) About 1 percent of identical twins share a single amnion and chorion, because the tissue splits into two groups after these structures have already formed. (*c*) In about two-thirds of identical twins, the split occurs after day 5 but before day 9. These twins share a chorion but have separate amnions. Fraternal twins result from two sperm fertilizing two secondary oocytes. The twins develop their own amniotic sacs, yolk sacs, allantois, placentae, and umbilical cords.

have completely separate chorions and amnions, and about two-thirds share a chorion but have separate amnions. Slightly less than 1 percent of MZ twins share both amnion and chorion. (The amnion is the sac that contains fluid that surrounds the fetus. The chorion develops into the placenta.) The significance of these differences, if any, is that they determine whether MZ twins develop in slightly different uterine environments. For example, if one chorion develops more attachment sites to the maternal circulation, one twin may receive more nutrients and grow to a slightly higher birth weight.

Rarely, an embryo divides into twins after the point at which the two groups of cells can develop as two individuals, resulting in conjoined or "Siamese" twins. The latter name comes from Chang and Eng, who were born in Thailand, then called Siam, in 1811. They were joined by a band of tissue from the navel to the breastbone, and could easily have been separated today. Chang and Eng lived for 63 years, attached, and each married.

In the case of Abigail and Brittany Hensel, shown in figure 3.17, the separation occurred after the ninth day of development, but before the fourteenth day. Biologists determined this because the girls' shared organs contain representatives of ectoderm, mesoderm, and endoderm; that is, when the lump of cells divided incompletely, the three primary germ layers had not yet completely sorted themselves out. The Hensel girls formed from an extremely rare event called incomplete twinning. Each twin has her own neck, head, heart, stomach, and gallbladder. Each has one leg and one arm, and a third arm between their heads was surgically removed. Each girl also has her own nervous system! The twins share a large liver, a single bloodstream, and all organs below the navel. They have three lungs and three kidneys. Because Abby and Britty were

**figure 3.17**

**Conjoined twins.** Abby and Britty Hensel are the result of incomplete twinning during the first 2 weeks of prenatal development.

strong and healthy, doctors suggested surgery to separate them. But their parents, aware from other cases that only one child would likely survive a separation, chose to let their daughters be.

Higher multiples are usually not identical. The triplets in figure 1.2 are rare, indeed. Emily, Ellen, and Katie Nolan probably arose from a fertilized ovum that divided and separated, and then one of those cells divided again.

In North America, twins occur in about 1 in 81 pregnancies, which means that 1 in 40 of us is a twin. However, not all twins survive to be born. One study of twins detected early in pregnancy showed that up to 70 percent of the eventual births are of a single child. This is called the "vanishing twin" phenomenon.

MZ twins occur in 3 to 4 pregnancies per 1,000 births worldwide. The mother's age, health, and number of other children have nothing to do with MZ twins—they just happen. Not so for DZ twins. They occur at different rates in different populations, ranging from 2 to 7 per 1,000 births in Asians, to 45 to 50 per 1,000 among black Africans. DZ twins are more likely to be conceived by women who are heavier, taller, between the ages of 35 and 39, and already mothers. Chapter 7 discusses how geneticists study twins to tease apart the influences of heredity and the environment.

## The Embryo Develops

As the days and weeks proceed, different rates of cell division in different parts of the embryo fold the forming tissues into intricate patterns. In a process called embryonic induction, the specialization of one group of cells causes adjacent groups of cells to specialize. Gradually, these changes mold the three primary germ layers into organs and organ systems. **Organogenesis** is the transformation of the simple three layers of the embryo into distinct organs. During the weeks of organogenesis, the developing embryo is particularly sensitive to environmental influences such as chemicals and viruses.

During the third week of prenatal development, a band called the **primitive streak** appears along the back of the embryo. The primitive streak gradually elongates to form an axis that other structures organize around as they develop. The primitive streak eventually gives rise to connective tissue precursor cells and the **notochord,** a structure that forms the basic framework of the skeleton. The notochord induces overlying ectoderm to specialize into a hollow **neural tube,** which develops into the brain and spinal cord (central nervous system). Many nations designate day 14 of prenatal development and primitive streak formation as the point beyond which they ban research on the human embryo. The reason is that the primitive streak is the first sign of a nervous system, and this is also the time at which implantation is completed.

Appearance of the neural tube marks the beginning of organ development. Shortly after, a reddish bulge containing the heart appears. The heart begins to beat around day 18, and this is easily detectable by day 22. Soon the central nervous system starts to form.

The fourth week of embryonic existence is one of spectacularly rapid growth and differentiation (figure 3.18). Arms and legs be-

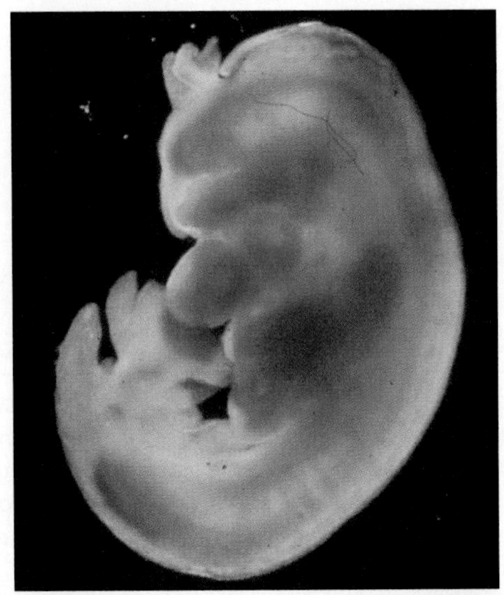

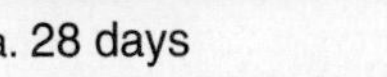

a. 28 days 4–6 mm

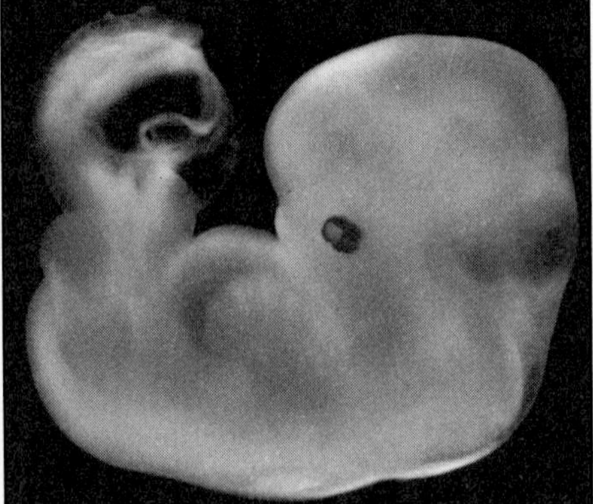

b. 42 days 12–15 mm

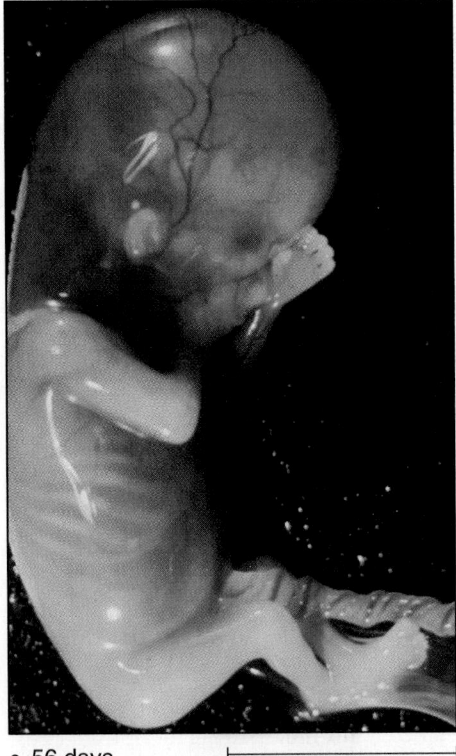

c. 56 days 23–32 mm

**figure 3.18**

**Human embryos.** Embryos are shown at (*a*) 28 days, (*b*) 42 days, and (*c*) 56 days.

gin to extend from small buds on the torso. Blood cells form and fill primitive blood vessels. Immature lungs and kidneys appear.

If the neural tube does not close normally at about day 28, a neural tube defect results, leaving open an area of the spine through which parts of the brain or spinal cord protrude (see Reading 15.1). If this happens, a substance from the fetus's liver called alpha fetoprotein (AFP) leaks at an abnormally rapid rate into the pregnant woman's circulation. A maternal blood test at the fifteenth week of pregnancy measures levels of AFP. If it is elevated, further tests measure AFP in the amniotic fluid, and ultrasound is used to visualize a defect. (Ultrasound scanning bounces sound waves off the fetus, creating an image.)

By the fifth and sixth weeks, the embryo's head appears to be too large for the rest of its body. Limbs end in platelike structures with tiny ridges, and gradually apoptosis sculpts the fingers and toes. The eyes are open, but they do not yet have lids or irises. By the seventh and eighth weeks, a skeleton composed of cartilage forms. The embryo is now about the length and weight of a paper clip. At eight weeks of gestation, the prenatal human has rudiments of all of the structures that will be present at birth. It is now a fetus.

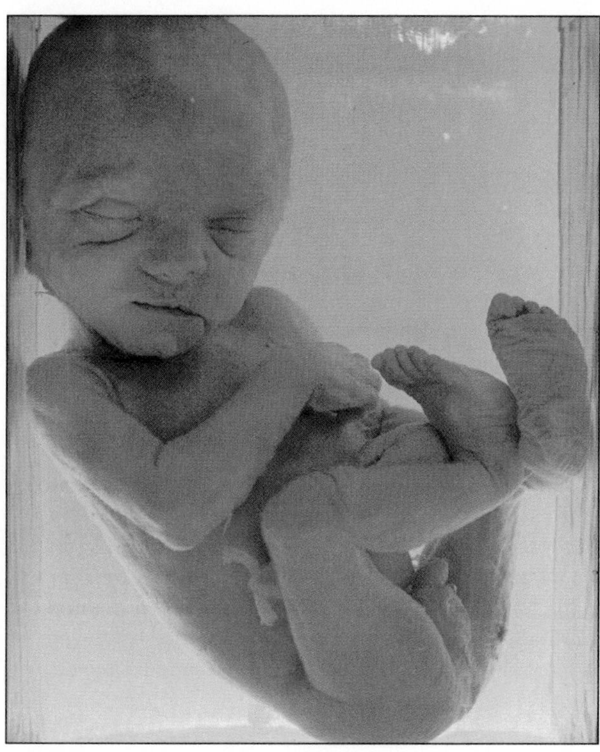

**figure 3.19**

**A fetus at 24 weeks.** This is the stage after which a fetus can survive outside of the uterus.

## The Fetus

During the fetal period, body proportions approach those of a newborn (figure 3.19). Initially, the ears lie low, and the eyes are widely spaced. Bone begins to replace the softer cartilage. As nerve and muscle functions become coordinated, the fetus moves.

Sex is determined at conception, when a sperm bearing an X or Y chromosome meets an oocyte, which always carries an X chromosome. An individual with two X chromosomes is a female, and one with an X and a Y is a male. A gene on the Y chromosome, called SRY (for "sex-determining region of the Y"), determines maleness. Differences between the sexes do not appear until week 6, after the SRY gene is activated in males. Male hormones then stimulate male reproductive organs and glands to differentiate from existing, indifferent structures. In a female, the indifferent structures of the early embryo develop as female organs and glands. Differences may begin to be noticeable on ultrasound scans by 12 to 15 weeks. Sexuality is discussed further in chapter 6.

By week 12, the fetus sucks its thumb, kicks, makes fists and faces, and has the beginnings of baby teeth. It breathes amniotic fluid in and out, and urinates and defecates into it. The first trimester (three months) of pregnancy ends.

By the fourth month, the fetus has hair, eyebrows, lashes, nipples, and nails. By 18 weeks, the vocal cords have formed, but the fetus makes no sounds because it doesn't breathe air. By the end of the fifth month, the fetus curls into a head-to-knees position. It weighs about 454 grams (1 pound). During the sixth month, the skin appears wrinkled because there isn't much fat beneath it (figure 3.19). The skin turns pink as capillaries fill with blood. By the end of the second trimester, the woman feels distinct kicks and jabs and may even detect a fetal hiccup. The fetus is now about 23 centimeters (9 inches) long.

In the final trimester, fetal brain cells rapidly form networks as organs elaborate and grow. A layer of fat forms beneath the skin. The digestive and respiratory systems mature last, which is why infants born prematurely often have difficulty digesting milk and breathing. Approximately 266 days after a single sperm burrowed its way into an oocyte, a baby is ready to be born.

The birth of a live, healthy baby is against the odds. Of every 100 secondary oocytes exposed to sperm, 84 are fertilized. Of these 84, 69 implant in the uterus, 42 survive 1 week or longer, 37 survive 6 weeks or longer, and only 31 are born alive. Of those fertilized ova that do not survive, about half have chromosomal abnormalities that are too severe for development to proceed.

Fetal tissues may be useful medically because they have unique healing properties. Because they are not yet completely specialized, fetal tissues implanted into an adult may not evoke an immune response, as transplanted adult tissues do. Furthermore, many fetal cells have a greater capacity to divide than their adult counterparts. Fetal tissue may be useful to replace tissue in neurodegenerative disorders such as Parkinson disease and Alzheimer disease, and to repair injured spinal cords. However, such uses of fetal tissue are highly controversial because the tissue cannot come from naturally aborted material, which often has abnormal chromosomes. Experiments using fetal cell implants have had mixed results, because researchers cannot yet adequately control the growth of the implants—which can cause symptoms when the brain is the target. Chapter 16 discusses implants and transplants further.

## Key Concepts

Following sexual intercourse, sperm are capacitated and drawn to the secondary oocyte. Acrosomal enzymes assist the sperm's penetration of the oocyte, and chemical and electrical changes in the oocyte's surface block additional sperm entry. The two sets of chromosomes meet, forming a zygote. • Cleavage cell divisions form a morula and then a blastocyst. The outer layer of cells invades and implants in the uterine lining. The inner cell mass develops into the embryo. Certain blastocyst cells secrete hCG. • Germ layers form in the second week. Cells in a specific germ layer later become part of particular organ systems as a result of differential gene expression. • During week 3, chorionic villi extend toward the maternal circulation, and the placenta begins to form. Nutrients and oxygen enter the embryo, and wastes pass from the embryo into the maternal circulation. The yolk sac and allantois manufacture blood cells, the umbilical cord forms, and the amniotic sac expands with fluid. • Monozygotic twins arise from a single fertilized ovum and may share supportive structures. Dizygotic twins arise from two fertilized ova. • During week 3, the primitive streak appears, followed rapidly by the central nervous system, heart, notochord, neural tube, limbs, digits, facial features, and other organ rudiments. By week 8, all of the organs that will be present in the newborn have begun to develop. • During the fetal period, structures grow, specialize, and begin to interact. • Bone replaces cartilage in the skeleton, body growth catches up with the head, and sex organs become more distinct. In the final trimester, the fetus moves and grows rapidly, and fat fills out the skin.

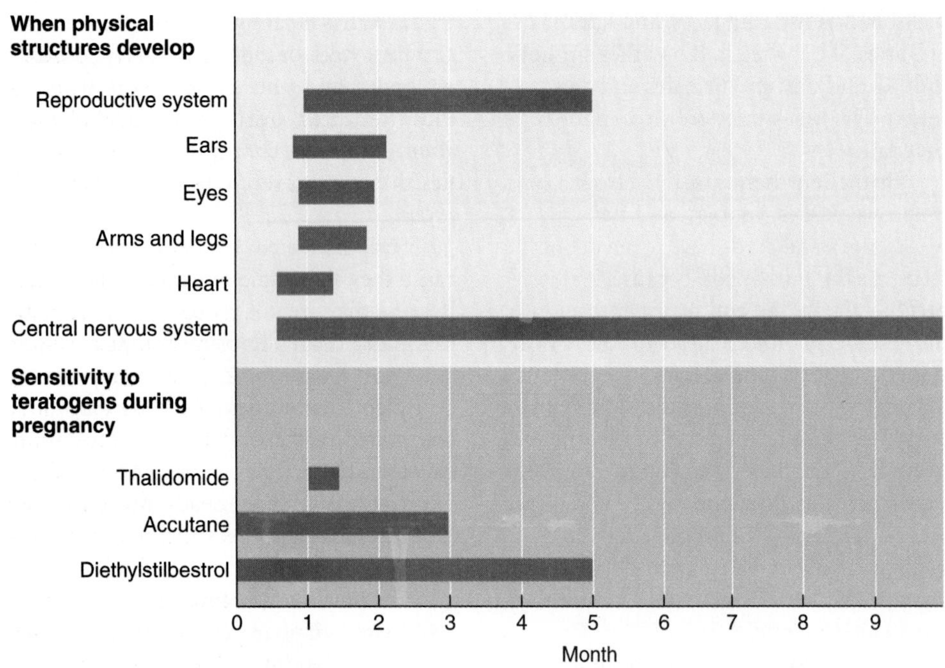


**figure 3.20**

**Sensitive periods of development.** The nature of a birth defect resulting from drug exposure depends upon which structures were developing at the time of exposure. The time when a particular structure is vulnerable is called the critical period. Accutane is an acne medication. Diethylstilbestrol (DES) was used in the 1950s to prevent miscarriage. Thalidomide was used to prevent morning sickness.

## 3.5 Birth Defects

When genetic abnormalities or toxic exposures affect an embryo or fetus, developmental errors occur, resulting in birth defects. Only a genetically caused birth defect can be passed to future generations. Although development can be derailed in many ways, about 97 percent of newborns appear healthy at birth.

### The Critical Period

The specific nature of a birth defect usually depends on which structures are developing when the damage occurs. The time when genetic abnormalities, toxic substances, or viruses can alter a specific structure is its **critical period** (figure 3.20). Some body parts, such as fingers and toes, are sensitive for short periods of time. In contrast, the brain is sensitive throughout prenatal development, and connections between nerve cells continue to change throughout life. Because of the brain's continuous critical period, many birth defect syndromes include mental retardation.

About two-thirds of all birth defects arise from a disruption during the embryonic period. More subtle defects that become noticeable only after infancy, such as learning disabilities, are often caused by interventions during the fetal period. A disruption in the first trimester might cause mental retardation; in the seventh month of pregnancy, it might cause difficulty in learning to read.

Some birth defects can be attributed to an abnormal gene that acts at a specific point in prenatal development. In a rare inherited condition called phocomelia, for example, an abnormal gene halts limb development from the third to the fifth week of the embryonic period, causing the infant to be born with "flippers" in place of arms and legs. The risk that a genetically caused birth defect will affect a particular family member can be calculated.

Many birth defects are caused not by genes but by toxic substances the pregnant woman encounters. These environmentally caused problems will not affect another family member unless the exposure occurs again. Chemicals or other agents that cause birth defects are called **teratogens** (Greek for "monster-causing"). While it is best to avoid teratogens while pregnant, some women may need to remain on a potentially teratogenic drug to maintain their own health.

## Teratogens

Most drugs are not teratogens. Table 3.3 lists some that are.

### *Thalidomide*

The idea that the placenta protects the embryo and fetus from harmful substances was tragically disproven between 1957 and 1961, when 10,000 children were born in Europe with what seemed, at first, to be phocomelia. Because doctors realized that this genetic disorder is very rare, they began to look for another cause. They soon discovered that the mothers had all taken a mild tranquilizer, thalidomide, early in pregnancy, during the time an embryo's limbs form. Thalidomide lessened the nausea of morning sickness. The "thalidomide babies" were born with incomplete or missing legs and arms.

The United States was spared from the thalidomide disaster because an astute government physician noted the drug's adverse effects on laboratory monkeys. Still, several "thalidomide babies" were born in South America in 1994, where pregnant women were given the drug. In spite of its teratogenic effects, thalidomide is still a valuable drug—it is used to treat leprosy and certain blood and bone marrow cancers.

### *Cocaine*

Cocaine is very dangerous to the unborn. It can cause spontaneous abortion by inducing a stroke in the fetus. Cocaine-exposed infants who do survive are more distracted and unable to concentrate on their surroundings than unexposed infants. Other health and behavioral problems arise as these children grow. A problem in evaluating the prenatal effects of cocaine is that affected children are often exposed to other environmental influences that could also account for their symptoms.

### *Cigarettes*

Chemicals in cigarette smoke stress a fetus. Carbon monoxide crosses the placenta and prevents the fetus's hemoglobin molecules from adequately binding oxygen. Other chemicals in smoke prevent nutrients from reaching the fetus. Smoke-exposed placentas lack important growth factors, causing poor growth before and after birth. Cigarette smoking during pregnancy increases the risk of spontaneous abortion, stillbirth, prematurity, and low birth weight.

### *Alcohol*

A pregnant woman who has just one or two drinks a day, or perhaps a large amount at a single crucial time in prenatal development, risks fetal alcohol syndrome (FAS) in her unborn child. Because even small amounts of alcohol can harm nerve cells, doctors advise women to entirely avoid alcohol when pregnant. In the future, tests for variants of genes that encode proteins that regulate alcohol metabolism will be able to predict which women and fetuses are at elevated risk for developing fetal alcohol syndrome.

A child with FAS has a characteristic small head, and a flat face and nose (figure 3.21). Growth is slow before and after birth. Intellectual impairment ranges from minor learning disabilities to mental retardation. Teens and young adults who have FAS are short and have small heads. More than 80 percent of them retain the facial characteristics of a young child with FAS.

The long-term mental effects of prenatal alcohol exposure are more severe than the physical vestiges. Many adults with FAS function at early grade-school level. They often lack social and communication skills and find it difficult to understand the consequences of actions, form friendships, take initiative, and interpret social cues.

Aristotle noticed problems in children of alcoholic mothers more than 23 centuries

**table 3.3**

**Teratogenic Drugs**

| Drug | Medical Use | Risk to Fetus |
|---|---|---|
| Alkylating agents | Cancer chemotherapy | Growth retardation |
| Aminopterin, methotrexate | Cancer chemotherapy | Skeletal and brain malformations |
| Coumadin derivatives | Seizure disorders | Tiny nose<br>Hearing loss<br>Bone defects<br>Blindness |
| Diphenylhydantoin (Dilantin) | Seizures | Cleft lip, palate<br>Heart defects<br>Small head |
| Diethylstilbestrol | Repeat miscarriage | Vaginal cancer, vaginal adenosis<br>Small penis |
| Isotretinoin (Accutane) | Severe acne | Cleft palate<br>Heart defects<br>Abnormal thymus<br>Eye defects<br>Brain malformation |
| Lithium | Bipolar disorder | Heart and blood vessel defects |
| Penicillamine | Rheumatoid arthritis | Connective tissue abnormalities |
| Progesterone in birth control pills | Contraception | Heart and blood vessel defects<br>Masculinization of female structures |
| Tetracycline | Antibiotic | Stained teeth |
| Thalidomide | Morning sickness | Limb defects |

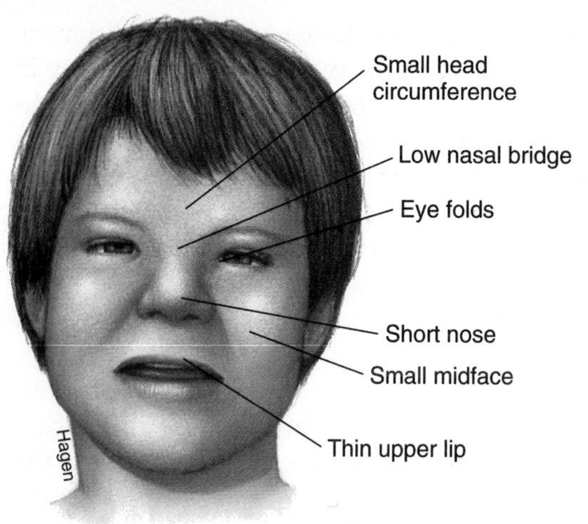

a.
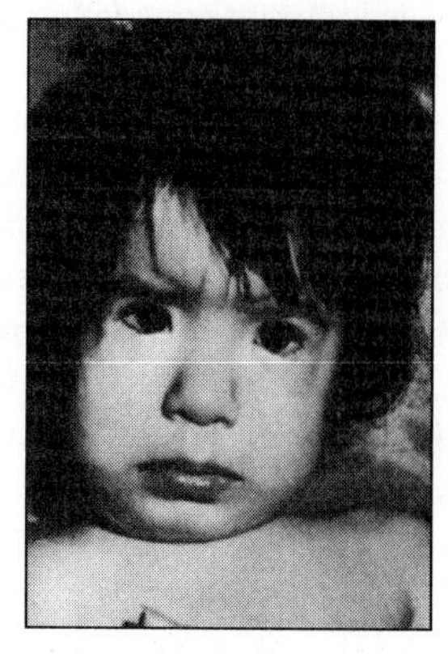
b.
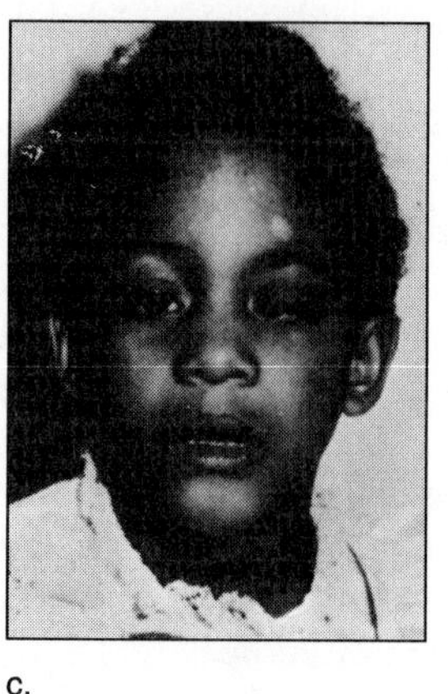
c.
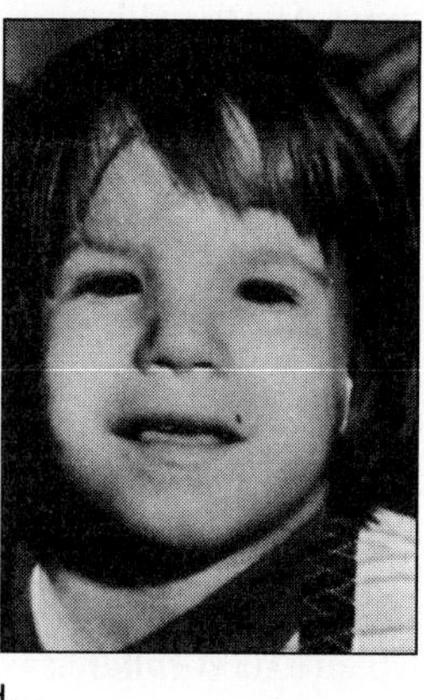
d.

**figure 3.21**

**Fetal alcohol syndrome.** Some children whose mothers drank alcohol during pregnancy have characteristic flat faces (*a*) that are strikingly similar in children of different races (*b, c,* and *d*).

ago. In the United States today, 1 to 3 of every 1,000 infants has the syndrome, adding up to 2,000 to 12,000 affected children born each year. Many more children have milder "alcohol-related effects." A fetus of a woman with active alcoholism has a 30 to 45 percent chance of harm from prenatal alcohol exposure.

### *Nutrients*

Certain nutrients ingested in large amounts, particularly vitamins, act as drugs in the human body. The acne medicine isotretinoin (Accutane) is a vitamin A derivative that causes spontaneous abortions and defects of the heart, nervous system, and face in exposed embryos. The tragic effects of this drug were first noted nine months after dermatologists began prescribing it to young women in the early 1980s. Another vitamin A-based drug used to treat psoriasis, as well as excesses of vitamin A itself, also cause birth defects. Some forms of vitamin A are stored in body fat for up to three years.

Excessive exposure to vitamin C can also harm a fetus. The fetus becomes accustomed to the large amounts the woman takes; after birth, when the vitamin exposure suddenly ceases, the baby may develop symptoms of vitamin C deficiency (scurvy). Such a baby bruises easily and is prone to infection.

Malnutrition also threatens the fetus. A woman must consume extra calories while she is pregnant or breastfeeding. Obstetrical records of pregnant women before, during, and after World War II link inadequate nutrition in early pregnancy to an increase in the incidence of spontaneous abortion. The aborted fetuses had very little brain tissue. Poor nutrition later in pregnancy affects the development of the placenta and can cause low birth weight, short stature, tooth decay, delayed sexual development, and learning disabilities. Some effects of prenatal malnutrition may not become apparent for years, a phenomenon discussed in the next section.

### *Occupational Hazards*

Some people encounter teratogens in the workplace. Researchers note increased rates of spontaneous abortion and children born with birth defects among women who work with textile dyes, lead, certain photographic chemicals, semiconductor materials, mercury, and cadmium. Men whose jobs expose them to sustained heat, such as smelter workers, glass manufacturers, and bakers, may produce sperm that can fertilize an oocyte and then cause spontaneous abortion or a birth defect. A virus or a toxic chemical carried in semen may also cause a birth defect.

### *Viral Infection*

Viruses are small enough to cross the placenta and reach a fetus. Some viruses that cause mild symptoms in an adult, such as the virus that causes chicken pox, may devastate a fetus. Men can transmit infections to an embryo or fetus during sexual intercourse.

HIV can reach a fetus through the placenta or infect a newborn via blood contact during birth. Fifteen to 30 percent of infants born to HIV-positive women are HIV positive themselves. The risk of transmission is significantly reduced if the woman takes anti-HIV drugs while pregnant. All fetuses of HIV-infected women are at higher risk for low birth weight, prematurity, and stillbirth if the woman's health is failing.

Australian physicians first noted the teratogenic effects of the rubella virus that causes German measles in 1941. In the United States, rubella did not gain public attention until the early 1960s, when an epidemic of the usually mild illness caused 20,000 birth defects and 30,000 stillbirths. Women who contract the virus during the first trimester of pregnancy run a high risk of bearing children with cataracts, deafness, and heart defects. Rubella's effects on fetuses exposed during the second or third trimesters of pregnancy include learning disabilities, speech and hearing problems, and type I diabetes mellitus.

The incidence of these problems, called "congenital rubella syndrome," has dropped markedly thanks to widespread vaccination. However, the syndrome resurfaces in unvaccinated populations. A resurgence in 1991 was attributed to a cluster of unvaccinated Amish women in rural Pennsylvania. In that isolated group, 14 of every 1,000 newborns had congenital rubella syndrome, compared to the incidence in the general U.S. population of 0.006 per 1,000. Another resurgence, in Arkansas in 1999, was traced to unvaccinated Mexican immigrants.

Herpes simplex virus can harm a fetus or newborn whose immune system is not yet completely functional. Forty percent of babies exposed to active vaginal herpes lesions become infected, and half of these infants die. Of those infants who are infected but survive, 25 percent sustain severe nervous system damage, and another 25 percent have widespread skin sores. A woman who has sores at the time of delivery can have a surgical delivery to protect the child.

Pregnant women are routinely checked for hepatitis B infection, which in adults causes liver inflammation, great fatigue, and other symptoms. Each year in the United States, 22,000 infants are infected with this virus during birth. These babies are healthy, but are at high risk for developing serious liver problems as adults. By identifying infected women, a vaccine can be given to their newborns, which can help prevent complications.

**Key Concepts**

The critical period is the time during prenatal development when a structure is sensitive to damage from a faulty gene or environmental insult. Most birth defects develop during the embryonic period and are more severe than problems that arise later. • Teratogens, agents that cause birth defects, include drugs, cigarettes, certain nutrients, malnutrition, occupational hazards, and viral infections.

## 3.6 Maturation and Aging

Aging begins at conception. Later on, as we age, the life spans of cells are reflected in the waxing and waning of biological structures and functions. Although some aspects of our anatomy and physiology peak very early—such as the number of brain cells or hearing acuity, which do so in childhood—age 30 seems to be a turning point for decline. Some researchers estimate that, after this age, the human body becomes functionally less efficient by about 0.8 percent each year.

Many diseases that begin in adulthood, or are associated with aging, have genetic components. Often these disorders are multifactorial, because it takes many years for environmental exposures to alter gene expression in ways that noticeably affect health. Following is a closer look at some of the ways that genes affect health throughout life.

### Adult-Onset Inherited Disorders

Human prenatal development can be viewed as a highly regulated program of genetic switches that are turned on at specific places and times. Environmental factors can affect how certain of these genes are expressed before birth, creating risks that are realized much later. Specifically, adaptations that enable a fetus to grow despite near-starvation become risk factors for certain common illnesses of adulthood, such as coronary artery disease, stroke, hypertension, and type II diabetes mellitus. A fetus that does not receive adequate nutrition has "intrauterine growth retardation" (IUGR). Such an individual is born on time, but is very small. Premature infants, in contrast, are small but are born early, and are not predisposed to conditions resulting from IUGR.

More than 100 studies clearly correlate low birth weight due to IUGR with increased incidence of the classic cardiovascular disease risk factors, and of the diseases themselves. Much of the data come from war records—enough time has elapsed to study effects of prenatal malnutrition as people age. A study of nearly 15,000 people born in Sweden from 1915 to 1929 correlates IUGR to heightened cardiovascular disease risk after age 65. Similarly, an analysis of individuals who were fetuses during a seven-month famine in the Netherlands in 1943 indicates a high rate of diabetes today. Experiments on sheep and rat fetuses that were intentionally starved support these historical findings.

How can poor nutrition before birth reverberate as disease many decades later? Researchers hypothesize that in order to survive, the starving fetus redirects its circulation to protect vital organs such as the brain. At the same time, the body's muscle mass and hormone production change in ways that conserve energy. Growth-retarded babies have too little muscle tissue, and since muscle is the primary site of action of insulin, glucose metabolism is altered. Thinness at birth, and the accelerated weight gain in childhood that often occurs to compensate, sets the stage for coronary heart disease and type II diabetes. In addition, abnormal levels of stress hormones, stiffer arteries, and too few kidney tubules can contribute to hypertension later in life.

Genetic disease may begin at any time (table 3.4). In general, conditions that affect children are recessive. Figure 3.22*a* illustrates a fetus who already has the broken bones characteristic of osteogenesis imperfecta. Dominantly inherited conditions more often start to affect health in early to middle adulthood. This is the case for polycystic kidney disease. Cysts that may have been present in the kidneys during one's twenties begin causing bloody urine, high blood pressure, and abdominal pain as one enters the thirties. Similarly, hundreds of benign polyps, symptomatic of familial polyposis of the colon (discussed in chapter 17), may coat the inside of the large intestine of a 20-year-old, but they do not cause bloody stools until the fourth decade, when some of them may be cancerous. The joint destruction of osteoarthritis may begin in one's thirties but not become painful for another 10 or 20 years. The personality changes, unsteady gait, and diminishing mental faculties of Huntington disease typically begin near age 40.

Five to 10 percent of Alzheimer disease cases are inherited, and first produce symptoms in the forties and fifties. Mutations in four genes are known to directly cause familial Alzheimer disease, and mutation in a fifth gene may be associated with increased risk of developing the condition. Chapter 11 discusses mutations that cause Alzheimer disease.

German neurologist Alois Alzheimer first identified the condition in 1907 as affecting people in mid-adulthood. It became known as "presenile dementia." However, the memory loss and inability to reason are so common—affecting 5 percent of U.S. citizens over age 65 and 40 percent of those over 85—that for many years physicians regarded

## table 3.4

**Onset of Genetic Disorders Along the Timeline of a Human Life**

| Prenatal period | Birth | 10 years | 20 years | 30 years | 40 years | 50 years | 60 years | 70 years |
|---|---|---|---|---|---|---|---|---|
| Osteogenesis imperfecta | Adrenoleukodystrophy | Familial hypertrophic cardiomyopathy | Multiple endocrine neoplasia | Hemochromatosis | Gout | Fatal familial insomnia | | |
| Pituitary dwarfism | Chronic granulomatous disease | Wilson disease | Marfan syndrome | Breast cancer | Huntington disease | Alzheimer disease | | |
| Lissencephaly | von Willebrand disease | | | Polycystic kidney disease | Pattern baldness | Porphyria | | |
| Wilms' tumor | Xeroderma pigmentosum | | | | | Amyotrophic lateral sclerosis | | |
| Polydactyly | Diabetes insipidus | | | | | | | |
| | Color blindness | | | | | | | |
| | Familial hypercholesterolemia | | | | | | | |
| | Albinism | | | | | | | |
| | Duchenne muscular dystrophy | | | | | | | |
| | Menkes disease | | | | | | | |
| | Sickle cell disease | | | | | | | |
| | Rickets | | | | | | | |
| | Cystic fibrosis | | | | | | | |
| | Hemophilia | | | | | | | |
| | Tay-Sachs disease | | | | | | | |
| | Phenylketonuria | | | | | | | |
| | Progeria | | | | | | | |

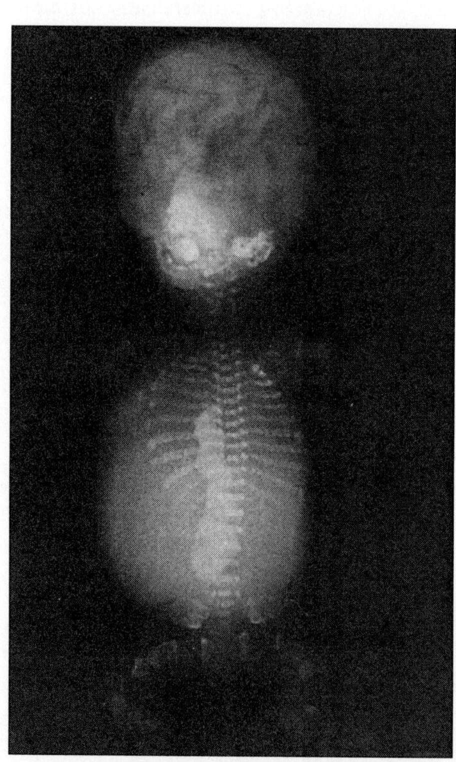

a.

b.

**figure 3.22**

**Genes act at various stages of development and life.** (*a*) Osteogenesis imperfecta breaks bones, even before birth. This fetus has broken leg bones. (*b*) At the funeral of former president Richard M. Nixon in April 1994, all was not right with former president Ronald Reagan. He was forgetful and responded inappropriately to questions. Six months later he penned a moving letter confirming that he had Alzheimer disease. He wrote, "My fellow Americans, I have recently been told that I am one of the millions of Americans who will be afflicted with Alzheimer's disease." By 1997, Reagan no longer knew the names of his closest relatives. By 1999, he didn't remember anyone, and by 2001 he no longer recalled being president. Because of the late onset of symptoms, Ronald Reagan's Alzheimer disease is probably not due to the malfunction of a single gene, but is multifactorial.

these symptoms as part of normal aging. Today, Alzheimer disease is considered a disorder. It strikes four million people in the United States annually, although most cases are not inherited. The noninherited cases usually begin later in life.

Alzheimer disease starts gradually. Mental function declines steadily for 3 to 10 years after the first symptoms appear. Confused and forgetful, Alzheimer patients often wander away from family and friends. Finally, the patient cannot perform basic functions such as speaking or eating and usually must be cared for in a hospital or nursing home. A patient's inability to recognize loved ones can be heartbreaking (figure 3.22*b*). Death is often due to infection.

On autopsy, the brains of Alzheimer disease patients are found to contain deposits of a protein called beta-amyloid in learning and memory centers. Alzheimer brains also contain structures called neurofibrillary tangles, which consist of a protein called tau. Tau binds to and disrupts microtubules in nerve cell branches, destroying the shape of the cell.

## Accelerated Aging Disorders

Genes control aging both passively (as structures break down) and actively (by initiating new activities). A class of inherited diseases that accelerate the aging timetable vividly illustrates the role genes play in aging.

The most severe aging disorders are the **progerias.** In Hutchinson-Gilford syndrome, one form of progeria, a child appears normal at birth but slows in growth by the first birthday. Within just a few years, the child ages with shocking rapidity, acquiring wrinkles, baldness, and the facial features characteristic of advanced age (figure 3.23). The body ages on the inside as well, as arteries clog with fatty deposits. The child usually dies of a heart attack or a stroke by age 12, although some patients live into their twenties. Only a few dozen cases of this syndrome have ever been reported.

An adult form of progeria called Werner syndrome becomes apparent before age 20, causing death before age 50 from diseases associated with aging. People with Werner syndrome develop, as young adults, atherosclerosis, diabetes mellitus, hair graying and loss, osteoporosis, cataracts, and wrinkled skin. Curiously, they do not develop Alzheimer disease or hypertension.

**figure 3.23**

**Progeria.** The Luciano brothers inherited progeria and appear much older than their years.

Not surprisingly, the cells of progeria patients show aging-related changes. Recall that normal cells growing in culture divide about 50 times before dying. Cells from progeria patients die in culture after only 10 to 30 divisions. Understanding how and why progeria cells race through the aging process may help us to understand genetic control of normal aging.

## Is Longevity Inherited?

Aging reflects genetic activity plus a lifetime of environmental influences. Families with many very aged members have a fortuitous collection of genes plus shared environmental influences, such as good nutrition, excellent health care, devoted relatives, and other advantages. A genome-level approach to identifying causes of longevity has identified a region of chromosome 4 that houses gene variants that are associated with long life. Researchers compared the genomes of 137 sets of siblings who lived beyond 91 years, searching for genome regions that all of them share. The implicated region of chromosome 4 may house 100 to 500 genes. Perhaps researchers will be able to narrow that number down to a few genes that influence longevity.

It is difficult to tease apart inborn from environmental influences on life span. One approach compares adopted individuals to both their biological and adoptive parents. In one study, Danish adoptees with one biological parent who died of natural causes before age 50 were more than twice as likely to die before age 50 themselves as were adoptees whose biological parents lived beyond this age. This suggests an inherited component to longevity. Interestingly, adoptees whose natural parents died early due to infection were more than five times as likely to also die early of infection, perhaps because of inherited immune system deficiencies. Age at death of the adoptive parents had no influence on that of the adopted individuals. Chapter 7 explores the "nature versus nurture" phenomenon more closely.

### Key Concepts

Genes affect health throughout life. Most single-gene disorders are recessive and strike early in life. Some single-gene disorders have an adult onset. Starvation before birth can set the stage for later disease by affecting gene expression in certain ways. • The progerias are single-gene disorders that speed aging. • Families with many aged members can probably thank their genes as well as the environment. Adoption studies reveal the effects of genes versus environmental influences. Chromosome 4 houses genes that affect longevity.

# Considering Cloning and Stem Cell Technology

Fictional scientists have cloned Nazis, politicians, and dinosaurs, but in actuality, the closest that cloning has come to humans is a sheep named Dolly, a mouse named Cumulina, a few farm animals, and a six-celled human embryo at a Massachusetts biotechnology company. Cloning a mammal from the nucleus of a somatic cell is extraordinarily difficult, and even when successful, most clones die in infancy. However, most public objection to cloning is based on ethics, rather than technical difficulty. Anti-cloning arguments range from objection to destroying embryos and fetuses, to the unnaturalness of the process, to the idea of individuality.

Cloning is the creation of a genetic replica of an individual. The technique transfers a nucleus from a somatic cell into a cell whose nucleus has been removed (figure 1), and then a new individual develops from the manipulated cell. There are two goals of cloning. Reproductive cloning seeks to create a baby using the nucleus from the cell of a particular individual who will then, supposedly, be duplicated. Creating a "copy" of a child killed in an accident by transferring a nucleus from one of her cells into an enucleated oocyte is reproductive cloning. In contrast, the goal of therapeutic cloning is to use very early embryos as sources of stem cells. Specifically, a nucleus from a person's somatic cell is transferred to an enucleated oocyte. The embryo develops until the inner cell mass forms, and these cells are then used to establish cultures of stem cells that are genetically identical to the donor of the somatic cell nucleus. The idea is that the stem cells match cells of the person, so the immune system will not reject them. Therapeutic cloning holds much promise to treat spinal cord injuries as well as neurodegenerative disorders such as Parkinson disease and Alzheimer disease, both prevalent in our aging population.

The bioethical issues that reproductive and therapeutic cloning raise are complex. Some people argue that therapeutic cloning violates rights of early-stage embryos. Proponents of the research argue that not to allow this research violates the rights of people who might benefit from stem cell therapy. They suggest that "leftover" fertilized ova currently deep frozen in many reproductive health clinics could be used in therapeutic cloning (chapter 21 discusses this further). As the debate rages, research in some nations has been put on hold, as politicians and the public argue the pros and cons of the research.

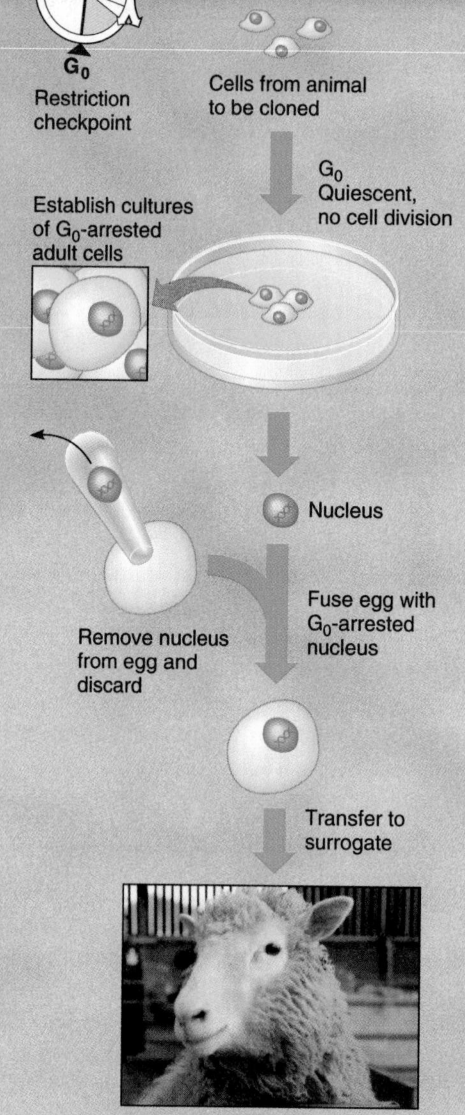

**figure 1**

**Cloning mammals requires cell cycle synchronization.** At first, researchers thought that using a nucleus that was in $G_0$ was necessary to clone a mammal, but today cloning is accomplished using donor nuclei from a variety of stages. However, it is important that the cell that donates the nucleus and the recipient cell be in the same stage of the cell cycle.

The U.S. House of Representatives voted overwhelmingly on July 31, 2001 to pass legislation that would outlaw creating or selling "any embryo produced by human cloning," for reproductive or therapeutic purposes. The ruling is more extreme than restricting government funding of such research, because this allowed researchers to use private funds. Instead, scientists caught defying the regulation would be treated as criminals, facing fines of up to $1 million or a decade in prison. Researchers and patient groups protested vigorously to the inclusion of therapeutic cloning in the ban, but attempts to exempt this application were overruled. Fortunately, cells from adults may be useful in stem cell therapies, one day making therapeutic cloning unnecessary. Since most biologists object to reproductive cloning, that was not as big an issue.

Several consequences may stem from restricting funding for cloning research. Investigators may leave countries that enforce restrictions. Some researchers have already left the U.S. for the U.K., which has a history of supporting basic research in human developmental and reproductive biology. Scientists who claim to be already cloning humans are not doing so in the U.S.

Besides the ethical debate over cloning, biology suggests that premises behind the attempt may be flawed, for a clone is not an exact replica of an individual. Many of the distinctions between an individual and a clone arise from epigenetic phenomena—that is, effects that do not change genes themselves, but alter their expression. There are other distinctions too: (Parentheses indicate chapters that discuss these subjects further.)

- Telomeres of chromosomes in the donor nucleus are shorter than those in the recipient cell (chapter 2).
- In normal development, for some genes, one copy is turned off, depending upon which parent transmits it.

That is, some genes must be inherited from either the father or the mother to be active, a phenomenon called genomic imprinting. Genes in a donor nucleus do not pass through a germline before they go back to the beginning of development, and thus are not imprinted. Effects of lack of imprinting in clones aren't known (chapter 5).

- DNA from the donor cell had years to accumulate mutations. Such a somatic mutation might not be noticeable if it occurs in one of millions of somatic cells, but it could be devastating if that somatic cell nucleus is used to program development of an entire new individual. Donor DNA introducing mutations from the previous life may contribute to the very low success rates of animal cloning experiments (chapter 11).
- At a certain time in early prenatal development in all female mammals, one X chromosome is inactivated. Whether the inactivated X chromosome is from the mother or the father occurs at random in each cell, creating an overall mosaic pattern of expression for genes on the X chromosome. The pattern of X inactivation of a female clone would most likely not match that of her nucleus donor (chapter 6).
- Mitochondria contain DNA. A clone's mitochondria descend from the recipient cell, not the donor cell.

The environment is another powerful factor in why a clone isn't really a clone. Figure 2 shows one effect of the environment on gene expression. Although the calves in the figure were cloned from identical nuclei, they have slightly different coat color patterns. When the calves were embryos, cells destined to produce pigment moved about in a unique way in each calf, producing different color patterns. In humans, experience, nutrition, stress, exposure to infectious diseases, and other environmental influences mold who we are as well as do our genes. Identical twins, although they have the same DNA sequence (except for somatic mutations), are not exact replicas of each other. Similarly, cloning a deceased child would probably disappoint parents seeking to recapture their lost loved one.

A compelling argument against reproductive cloning that embraces both ethics and biology is that it would be cruel to create a child who would most likely suffer. Cloning rarely works, according to results on other types of animals. In the 1980s, a company called Granada Genetics cloned cattle from fetal cells, but the newborns were huge and required a great deal of care, a problem repeatedly seen in other cloned animals since then. There is no reason to assume that a newborn human clone will fare any better than have other species.

**figure 2**

**Cloned calves.** Clones are not identical in phenotype. The color patterns on these cloned calves differ because of cell movements in the embryo.

Why does cloning so often fail? The reasons may lie in the fact that, as one researcher puts it, "The whole natural order [meaning meiosis] is broken." Recall that meiosis in the female completes at fertilization. In cloning, a diploid nucleus is plunked into oocyte cytoplasm, where signals direct it to do what a female secondary oocyte tends to do—shed half of itself as a polar body. If the out-of-place donor nucleus does this, the new cell is haploid and will not develop. Another possible problem occurs if the donor nucleus replicates its DNA. A genetic overload results, again preventing development.

If human cloning ever becomes reality, we will probably learn that we are not merely the products of our genes. This idea is the essence of the ethical objection to cloning, that we are dissecting and defining our very individuality, reducing it to a biochemistry so supposedly simple that we can duplicate it (figure 3). We probably can't.

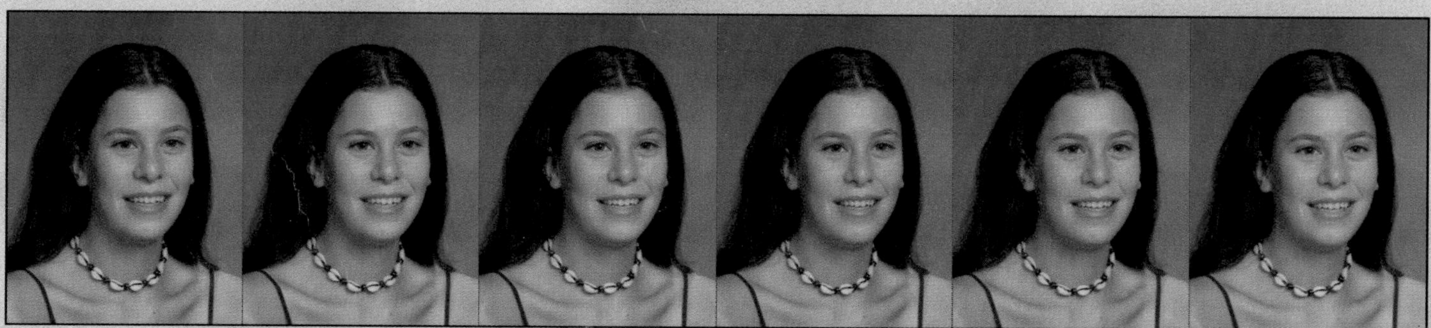

**figure 3**

**Cloning creates genetic replicas.** The author has cloned her daughter Sarah (not really).

# Summary

## 3.1 The Reproductive System

1. The male and female reproductive systems include paired **gonads** and networks of tubes in which **sperm** and **oocytes** are manufactured.
2. Male **gametes** originate in **seminiferous tubules** within the paired **testes.** They then pass through the **epididymis** and **vasa deferentia,** where they mature before exiting the body through the **urethra** during sexual intercourse. The **prostate gland,** the **seminal vesicles,** and the **bulbourethral glands** add secretions.
3. Female gametes originate in the **ovaries.** Each month after puberty, one ovary releases an oocyte into a **fallopian tube.** The oocyte then moves to the **uterus.**

## 3.2 Meiosis

4. **Meiosis** reduces the chromosome number to one genome in gametes, which maintains the chromosome number from generation to generation. Meiosis ensures genetic variability by partitioning different combinations of genes into gametes as a result of **crossing over** and **independent assortment** of chromosomes.
5. Meiosis I, **reduction division,** halves the number of chromosomes. Meiosis II, **equational division,** produces four cells from the two that result from meiosis I, without another DNA replication.
6. Crossing over occurs during prophase I. It mixes up paternally and maternally derived genes on homologous chromosome pairs.
7. Chromosomes segregate and independently assort in metaphase I, which determines the distribution of genes from each parent in the gamete.

## 3.3 Gamete Maturation

8. Maturation completes gamete manufacture. **Spermatogenesis** begins with **spermatogonia,** which accumulate cytoplasm and replicate their DNA to become **primary spermatocytes.** After meiosis I, the cells become haploid **secondary spermatocytes.** In meiosis II, the secondary spermatocytes divide to each yield two **spermatids,** which then differentiate into **spermatozoa.**
9. In **oogenesis,** some oogonia grow and replicate their DNA, becoming **primary oocytes.** In meiosis I, the primary oocyte divides to yield one large **secondary oocyte** and a much smaller **polar body.** In meiosis II, the secondary oocyte divides to yield the large ovum and another small polar body. Female meiosis is completed at fertilization.

## 3.4 Prenatal Development

10. In the female, sperm are **capacitated** and drawn chemically and physically towards a secondary oocyte. One sperm burrows through the oocyte's protective layers with **acrosomal** enzymes. Fertilization occurs when the sperm and oocyte fuse and their genetic material combines in one nucleus, forming the **zygote.** Electrochemical changes in the egg surface block further sperm entry. Cleavage begins and a 16-celled **morula** forms. Between days 3 and 6, the morula arrives at the uterus and hollows, forming a **blastocyst** made up of **blastomeres.** The **trophoblast** and **inner cell mass** form. Around day 6 or 7, the blastocyst implants and trophoblast cells secrete **hCG,** which prevents menstruation.
11. During the second week, the **amniotic cavity** forms as the inner cell mass flattens. **Ectoderm** and **endoderm** form, and then **mesoderm** appears, establishing the **primary germ layers.** Cells in each germ layer begin to develop into specific organs. During the third week, the **placenta, yolk sac, allantois,** and umbilical cord begin to form as the amniotic cavity swells with fluid. **Monozygotic** twins result from splitting of one fertilized ovum. **Dizygotic** twins result from two fertilized ova. Organs form throughout the embryonic period. Structures gradually appear, including the **primitive streak,** the **notochord** and **neural tube,** arm and leg buds, the heart, facial features, and the skeleton.
12. The **fetal** period begins after the eighth week. Organ rudiments laid down in the embryo grow and specialize. The developing organism moves and reacts, and gradually, its body proportions resemble those of a baby. In the last trimester, the brain develops rapidly, and fat is deposited beneath the skin. The digestive and respiratory systems mature last.

## 3.5 Birth Defects

13. Birth defects can result from a malfunctioning gene or an environmental intervention.
14. A substance that causes birth defects is a **teratogen.** Environmentally caused birth defects are not transmitted to future generations.
15. The time when a structure is sensitive to damage from an abnormal gene or environmental intervention is its critical period.

## 3.6 Maturation and Aging

16. Genes cause or predispose us to illness throughout life. Single-gene disorders that strike early tend to be recessive, whereas adult-onset single-gene conditions are often dominant.
17. Malnutrition before birth can alter gene expression in ways that increase risk of type II diabetes mellitus and cardiovascular disease much later in life.
18. The **progerias** are single-gene disorders that increase the rate of aging-associated changes.
19. Long life is due to genetics and environmental influences.

# Review Questions

1. How many sets of human chromosomes are present in each of the following cell types?
   a. an oogonium
   b. a primary spermatocyte
   c. a spermatid
   d. a cell from either sex during anaphase of meiosis I
   e. a cell from either sex during anaphase of meiosis II
   f. a secondary oocyte
   g. a polar body derived from a primary oocyte
2. List the structures and functions of the male and female reproductive systems.

3. A dog has 39 pairs of chromosomes. Considering only independent assortment of chromosomes, how many genetically different puppies are possible when two dogs mate? Is this number an underestimate or overestimate of the actual total? Why?
4. How does meiosis differ from mitosis?
5. What do oogenesis and spermatogenesis have in common, and how do they differ?
6. How does gamete maturation differ in the male and female?
7. Describe the events of fertilization.
8. Exposure to teratogens tends to produce more severe health effects in an embryo than in a fetus. Why?
9. List four teratogens, and explain how they disrupt prenatal development.
10. Cite two pieces of evidence that genes control aging.
11. Describe how F.D. (Reading 3.1) and a clone (Bioethics Box) began development in unusual ways.

# Applied Questions

1. Based on your knowledge of human prenatal development, at what stage do you think it is ethical to ban experimentation? Cite reasons for your answer. (The options of banning the research altogether, or of allowing research at any stage, are as valid an option as pinpointing a particular stage.)
2. Two groups of researchers reported culturing of human embryonic stem cells in 1998. These cells can theoretically develop into any cell type, and will be valuable in regenerative medicine. However, there were objections to the research because the cells came from prenatal humans—specifically, one research group obtained them from the inner cell mass, and the other from a seven-week embryo. Compare the structures that are present at these two stages of human prenatal development.
3. Some Vietnam War veterans who were exposed to the herbicide Agent Orange claim that their children—born years after the exposure—have birth defects caused by dioxin, a contaminant in the herbicide. What types of cells in these men would the chemical have to affect in order to cause birth defects years later?
4. In about 1 in 200 pregnancies, a sperm fertilizes a polar body instead of an oocyte. A mass of tissue that is not an embryo develops. Why can't a polar body support development of an entire embryo?
5. Should a woman be held legally responsible if she drinks alcohol, smokes, or abuses drugs during pregnancy and it harms her child? Should liability apply to all substances that can harm a fetus, or only to those that are illegal?
6. What difficulties might be encountered in studying the inheritance of longevity?
7. On an episode of television's "The X-Files," a man can use his own body to regenerate any damaged body part. When FBI agent Dr. Dana Scully examines his cells, she finds they are all cancerous, and they all have long telomeres. How might this man have arisen?
8. What types of evidence have led researchers to hypothesize that a poor prenatal environment can raise the risk for certain adult illnesses? How are genes part of this picture?

# Suggested Readings


Cibelli, Jose B. et al. January 2002. The first human cloned embryo. *Scientific American* 286(1):44–51. It was only a few cells, but caused a flutter of headlines.

Green, Ronald M. 2001. *The Human Embryo Debates: Bioethics in the Vortex of Controversy*. New York: Oxford University Press. Experimenting with human embryos, to learn about development or provide new treatments, is highly controversial.

Haslet, Chris, and John Savill. June 21, 2001. Why is apoptosis important to clinicians? *British Medical Journal* 322:1499–1500. Understanding apoptosis is leading to development of new drugs.

Hayflick, Leonard. January 27, 2000. New approaches to old age. *Nature* 403:365. We live longer than our genes dictate.

Holden, Constance. August 10, 2001. Would cloning ban affect stem cells? *Science* 293:1025. Legislation may block promising research.

Lewis, R. October 30, 2000. New light on fetal origins of adult disease. *The Scientist* 14:1. Starvation during prenatal development can alter gene expression in ways that raise risks of developing certain multifactorial disorders in adulthood.

Mayor, Susan. June 30, 2001. Ban on human reproductive cloning demanded. *British Medical Journal* 322:1566. A spiritual group's claim that it is cloning a human prompted an international outcry.

Puca, Annibale, et al. August 28, 2001. A genome-wide scan for linkage to human exceptional longevity identifies a locus on chromosome 4. *Proceedings of the National Academy of Sciences* 98:10505–10508. Comparing long-lived siblings revealed part of chromosome 4 that includes genes involved in life span.

Short, R.V. February 17, 2000. Where do babies come from? *Nature* 403:705. William Harvey puzzled over the beginning of life.

Smith, Bradley R. March 1999. Visualizing human embryos. *Scientific American* 280:76–83. Spectacular images of human embryos.

Wilmut, Ian. 2000. *The Second Creation: The Age of Biological Control by the Scientists That Cloned Dolly.* London: Headline. How Dolly came to be, by those who did it.

The December 18, 2001 issue of *The New York Times,* section F, has several articles and spectacular illustrations and photos of human prenatal development.

# On the Web

Check out the resources on our website at

**www.mhhe.com/lewisgenetics5**

On the web for this chapter you will find additional study questions, vocabulary review, useful links to case studies, tutorials, popular press coverage, and much more. To investigate specific topics mentioned in this chapter, also try the links below:

Alzheimer's Association **www.alz.org**

Association of Birth Defect Children **www.birthdefects.org**

Charcot-Marie-Tooth Disease Foundation **www.charcot-marie-tooth.org**

Huntington Disease Society of America **www.hdsa.mgh.harvard.edu**

March of Dimes **www.modimes.org**

Motherisk Program **www.motherisk.org**

Multi-Dimensional Human Embryo Project **www.embryo.soad.umich.edu/index.html**

National Bioethics Advisory Commission **http://bioethics_gov/pubs.html**

National Organization on Fetal Alcohol Syndrome **www.nofas.org**

The New York Times **nytimes.com/science**

Online Mendelian Inheritance in Man **www.ncbi.nlm.nih.gov/entrez/query.fcgi?db=OMIM**
Alzheimer disease 104300, 104310, 104311, 600759
Charcot-Marie-Tooth disease 118200, 118220, 214400
Huntington disease 143100
osteogenesis imperfecta 166210, 166200, 166260

Organization of Teratology Information Services **www.orpheus.ucsd.edu/ctis/**

Osteogenesis Imperfecta Foundation **www.oif.org**

Scientific American Human Cloning **www.sciam.com/explorations/2001/112401ezzell/**

Spermatology home page **www.numbut.murdoch.edu.au/spermatology/spermhp.html**

CHAPTER

4

# Mendelian Inheritance

## 4.1 Following the Inheritance of One Gene—Segregation

Farmers, gardeners, and scientists have long noticed that in diverse organisms—from trees and flowers, to sheep and peas—some traits seem to disappear in one generation, then reappear in a future generation. Gregor Mendel was the first to methodically conduct experiments from which he inferred how traits are transmitted—without any knowledge of the structures that might do this. The two concepts he distilled from experiments with pea plants explain trait transmission in any species with two sets of chromosomes, including our own.

## 4.2 Single-gene Inheritance in Humans

Single-gene traits and illnesses in humans are rare. Applying Mendel's laws enables us to predict the chances of inheriting particular traits. Autosomal dominant traits can affect both sexes and do not skip generations. Autosomal recessive traits can affect both sexes and can skip generations. Whether an allele is dominant or recessive depends upon how it affects the abundance or activity of the gene's encoded protein product.

## 4.3 Following the Inheritance of Two Genes—Independent Assortment

Mendel's second set of experiments followed two genes with two variants on two different chromosomes. He was ahead of his time—today, geneticists are increasingly examining the effects and interactions of multiple genes.

## 4.4 Pedigree Analysis

A pedigree diagram that is used to follow trait transmission may seem simple and even archaic in this age of sequencing genomes. But human genetics is ultimately about families—and the study of families begins with pedigrees.

**figure 4.1**

**Inherited similarities.** Facial similarities are not always as obvious as those between rocker Steve Tyler of Aerosmith and his actress daughter, Liv.

Inherited similarities can be startling. When Aerosmith singer Steve Tyler first met his daughter, Liv, seven years old at the time, he knew with one glance that she was his child. Even today, father and daughter have strikingly similar facial features, unobscured by their different sexes and ages (figure 4.1).

## 4.1 Following the Inheritance of One Gene—Segregation

Awareness of heredity appears as early as an ancient Jewish law that excuses a boy from circumcision if relatives bled to death from the ritual. Nineteenth-century biologists thought that body parts controlled trait transmission and gave the units of inheritance, today called genes, such colorful names as pangens, idioblasts, bioblasts, gemules, or just characters. An investigator who used the term *elementen* made the most lasting impression on what would become the science of genetics. His name was Gregor Mendel.

### Mendel the Man

Mendel spent his early childhood in a small village in what is now the Czech Republic, near the Polish border. His father was a farmer, and his mother was the daughter of a gardener, so Mendel learned early how to tend fruit trees. He did so well in school that at age 10 he left home to attend a special school for bright students, supporting himself by tutoring. After a few years at a preparatory school, Mendel became a priest at the Augustinian monastery of St. Thomas in Brno. But this was not a typical monastery—the priests were also teachers, and they did research in natural science. It was the perfect place for Mendel.

The young man learned how to artificially pollinate crop plants to control their breeding. He wanted to teach natural history, but had difficulty passing the necessary exams, a victim of test anxiety. At age 29, he was such an effective substitute teacher that he was sent to earn a college degree. At the University of Vienna, courses in the sciences and statistics fueled his long-held interest in plant breeding. Then Mendel began to think about experiments to address a compelling question that had confounded other plant breeders—why did certain traits disappear in one generation, only to reappear in the next? To solve this puzzle, Mendel bred hybrids and applied the statistics he had learned in college.

From 1857 to 1863, Mendel crossed and cataloged some 24,034 plants, through several generations. He deduced that consistent ratios of traits in the offspring indicated that the plants transmitted distinct units, or "elementen." He derived two hypotheses to explain how inherited traits are transmitted. Mendel described his work to the Brno Medical Society in 1865 and published it in the organization's journal the next year. The remarkably clear paper discusses plant hybridization, the reappearance of traits in the third generation, and the joys of working with peas, plus Mendel's data.

Mendel's hypotheses became laws because they apply as much to pea plant traits as they do to human families with inherited disease. We can best appreciate Mendel's inferences by realizing what he did not know. He deduced how chromosomes transmit genes to the next generation without knowing what either of these structures are!

Finally, his treatise was published in English in 1901. Then three botanists (Hugo DeVries, Karl Franz Joseph Erich Correns, and Seysenegg Tschermak) rediscovered the laws of inheritance independently and eventually found Mendel's paper. All credited Mendel, who came to be regarded as the "father of genetics." As is the way of science, many others repeated and confirmed Mendel's work. However, researchers in this new field of genetics also identified situations that seem to disrupt Mendelian ratios, the subject of chapter 5.

In the 20th century, researchers discovered the molecular basis of the pea plant traits that Mendel studied. The "short" and "tall" plants reflect expression of a gene that enables a plant to produce the hormone gibberellin, which elongates the stem. One tiny change to the DNA, and a short plant results. Likewise, "round" and "wrinkled" peas arise from the *R* gene, whose encoded protein connects sugars into branching polysaccharides. Seeds with a mutant gene cannot attach the sugars, water exits the cells, and the peas wrinkle.

Today genetics, DNA, and even genomes are familiar. However, the study of heredity on an organismal level, as Mendel did, is still intriguing. New parents don't usually ask about their newborn's DNA sequences, but note such obvious traits as sex, hair color, and facial features. When we select mates we don't usually consider DNA sequences, but more visible manifestations of our genes. But we can appreciate Gregor Mendel's genius in light of what we know about DNA.

### Mendel's Experiments

The beauty of studying heredity is that different types of organisms can be used to demonstrate underlying principles common to all species. However, some organisms are easier to work with than others. Peas were a good choice for probing trait transmission because they are easy to grow, develop quickly, and have many traits that take one of two easily distinguishable forms. It was this last quality of "differentiating characters" that particularly interested Mendel. Figure 4.2 illustrates the seven traits that he followed.

Mendel's first experiments dealt with single traits that have two expressions, such as "short" and "tall." He set up all combinations of possible artificial pollinations, manipulating fertilizations to cross tall with tall, short with short, and tall with short, which produces hybrids. Mendel noted that short plants crossed to other short plants were "true-breeding," always producing short plants.

The crosses of tall plants to each other were more confusing. Tall plants were only sometimes true-breeding. Some tall plants, when crossed with short plants, produced only tall plants, but certain other tall plants crossed with each other yielded surprising results—about one-quarter of the plants in the next generation were short. It appeared as if in some tall plants, tallness could mask shortness (figure 4.3). The trait that masks

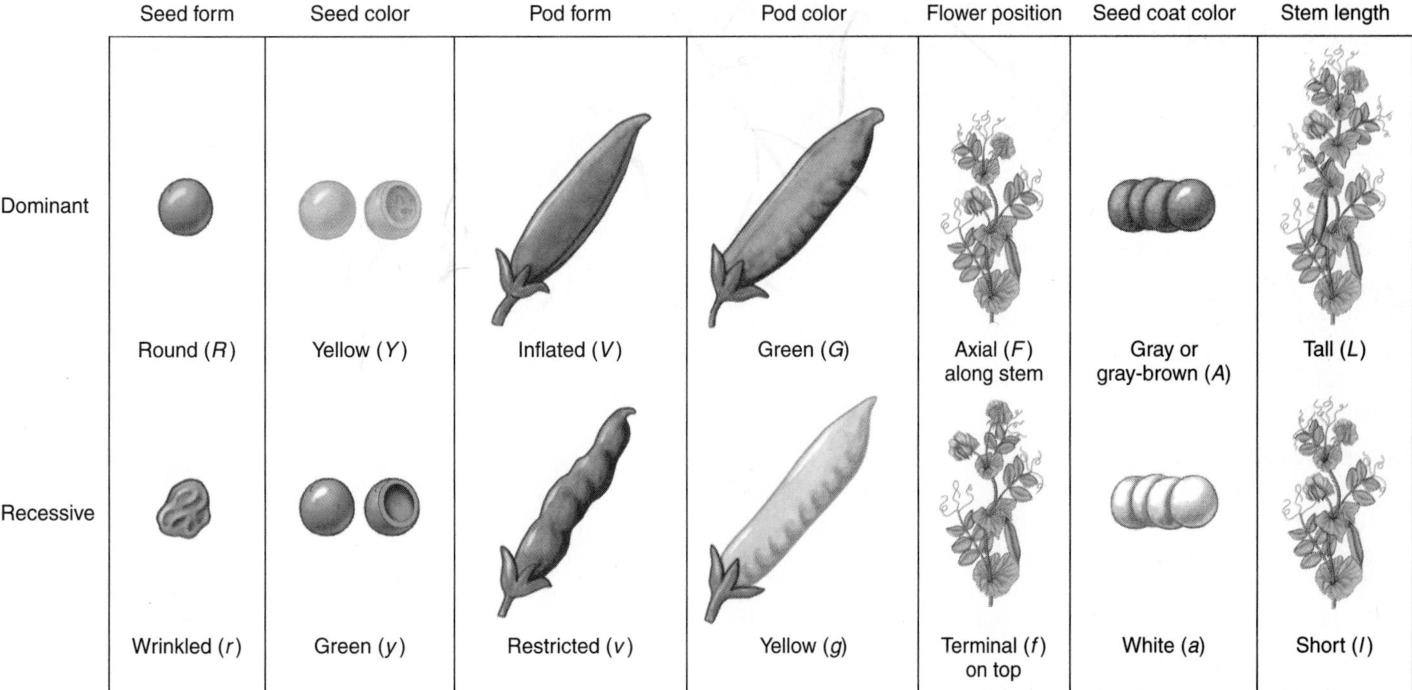


**figure 4.2**

**Traits Mendel studied.** Gregor Mendel studied the transmission of seven traits in the pea plant. Each trait has two easily distinguished expressions, or phenotypes.

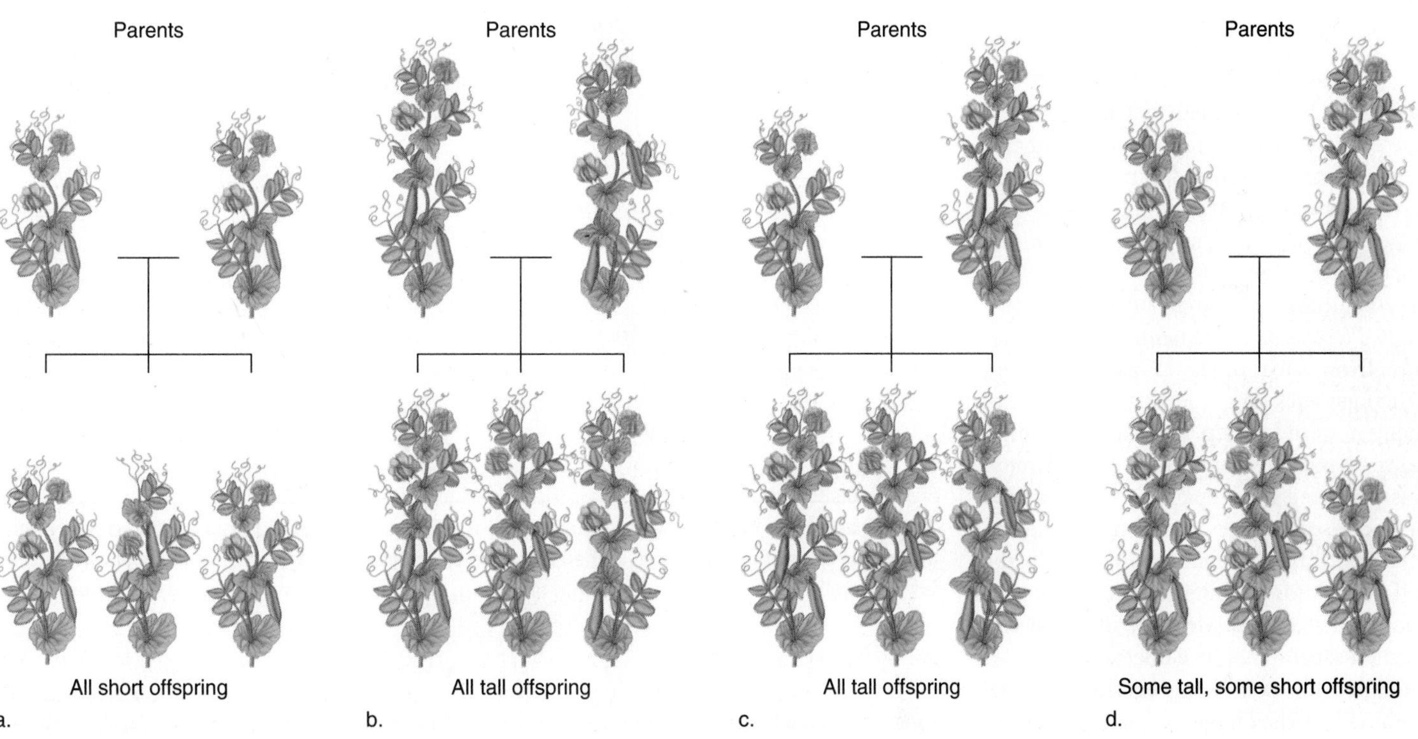


**figure 4.3**

**Mendel crossed short and tall pea plants.** (*a*) When Mendel crossed short pea plants with short pea plants, all of the progeny were short. (*b*) Some tall plants crossed to tall plants yielded only tall plants. (*c*) Certain tall plants crossed with short plants produced all tall plants. (*d*) Other tall plants crossed with short plants produced some tall plants and some short plants.

the other is said to be **dominant;** the masked trait is **recessive.** Wrote Mendel,

> In the case of each of the seven crosses the hybrid character resembles that of one of the parental forms so closely that the other either escapes observation completely or cannot be detected with certainty. . . . The expression "recessive" has been chosen because the characters thereby designated withdraw or entirely disappear in the hybrids, but nevertheless reappear unchanged in their progeny.

Mendel conducted up to 70 hybrid crosses for each of the seven traits.

To further investigate these non-true-breeding tall plants, Mendel set aside the short plants that had arisen from crossing the hybrid tall plants, and crossed the remaining tall offspring to themselves. Because one trait is followed and the parents are hybrids, this is called a **monohybrid cross.** He saw the ratio of one-quarter short to three-quarters tall. Two-thirds of those tall plants were non-true-breeding, and the remaining third were true-breeding (figure 4.4). He wrote "of those forms which possess the dominant character in the first generation, two-thirds have the hybrid character, while one-third remains consistent with the dominant character."

In these initial experiments with one trait, Mendel confirmed that hybrids hide one expression of a trait, which reappears when hybrids are crossed. But Mendel went farther, trying to explain how this happened. He suggested that gametes distribute "elementen,"—what we now call genes—because these cells physically link generations. Paired sets of elementen would separate from each other as gametes form. When gametes join at fertilization, the elementen would group into new combinations. Mendel reasoned that each element was packaged in a separate gamete, and if opposite-sex gametes combine at random, then he could mathematically explain the different ratios of traits produced from his pea plant crossings. Mendel's idea that elementen separate in the gametes would later be called the **law of segregation.** Mendel eventually turned his energies to monastery administration when other scientists did not recognize the importance of his work.

When Mendel's ratios were demonstrated again and again in several species in the early 1900s, at the same time that chromosomes were being described for the first time, it quickly became apparent that elementen and chromosomes had much in common. Both paired elementen and pairs of chromosomes separate at each generation and pass—one from each parent—to offspring. Both elementen and chromosomes are inherited in random combinations. Chromosomes provided a physical mechanism for Mendel's hypotheses. In 1909, Mendel's elementen were renamed genes (Greek for "give birth to"), but for the

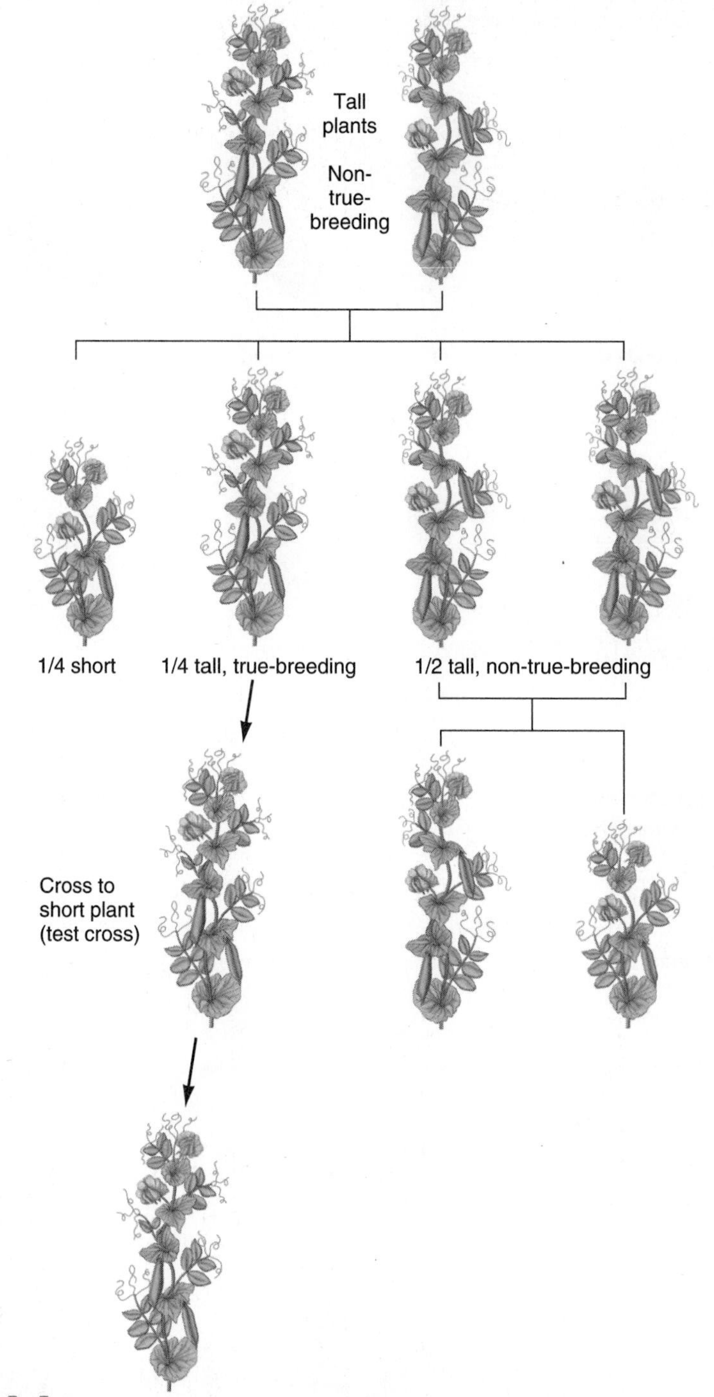


**figure 4.4**

**A monohybrid cross.** When Mendel crossed tall plants that did not breed true with each other, one-quarter of the plants in the next generation were short, and three-quarters were tall. Of these tall plants, one-third were true-breeding, and the other two-thirds were not true-breeding (hybrid).

next several years, the units of inheritance remained a mystery. By the 1940s, scientists began investigating the gene's chemical basis. We pick up the historical trail at this point in chapter 9.

## Terms and Tools to Follow Segregating Genes

Figures 4.3 and 4.4 depict the law of segregation at the whole-organism level—pea plants. Since the law reflects events of meiosis, we can also describe it at more microscopic levels—that is, in terms of chromosomes and genes.

Because a gene is a long sequence of DNA, it can vary in many ways. An individual with two identical alleles for a gene is **homozygous** for that gene. An individual with two different alleles is **heterozygous.** Mendel's "non-true-breeding" plants were heterozygous, also called "hybrid."

When a gene has two alleles, it is common to symbolize the dominant allele with a capital letter and the recessive with the corresponding small letter. If both alleles are recessive, the individual is homozygous recessive. Two small letters, such as *tt* for short plants, symbolize this. An individual with two dominant alleles is homozygous dominant. Two capital letters, such as *TT* for tall pea plants, represent homozygous dominance. Another possible allele combination is one dominant and one recessive allele—*Tt* for non-true-breeding tall pea plants, or heterozygotes.

An organism's appearance does not always reveal its alleles. Both a *TT* and a *Tt* pea plant are tall, but the first is a homozygote and the second a heterozygote. The **genotype** describes the organism's alleles, and the **phenotype** describes the outward expression of an allele combination. A pea plant with a tall phenotype may have genotype *TT* or *Tt*. A **wild type** phenotype is the most common expression of a particular allele combination in a population. A **mutant** phenotype is a variant of a gene's expression that arises when the gene undergoes a change, or **mutation.**

When analyzing genetic crosses, the first generation is the parental generation, or $P_1$; the second generation is the first filial generation, or $F_1$; the next generation is the second filial generation, or $F_2$, and so on. If you considered your grandparents the $P_1$ generation, your parents would be the $F_1$ generation, and you and your siblings are the $F_2$ generation.

Mendel's observations on the inheritance of single genes reflect the events of meiosis. When a gamete is produced, the two copies of a particular gene separate, as the homologs that carry them do. In a plant of genotype *Tt,* for example, gametes carrying either *T* or *t* form in equal numbers during anaphase I. When gametes meet to start the next generation, they combine at random. That is, a *t*-bearing oocyte is neither more or less attractive to a sperm than is a *T*-bearing oocyte. These two factors—equal allele distribution into gametes and random combinations of gametes—underlie Mendel's law of segregation (figure 4.5).

Mendel's crosses of short and tall plants make more sense in terms of meiosis. He crossed short plants (*tt*) with true-breeding tall plants (*TT*). The resulting seeds grew into $F_1$ plants that were all tall (genotype *Tt*). Next, he crossed the $F_1$ plants, a monohybrid cross. The three possible genotypic outcomes of such a cross are *TT, tt,* and *Tt.* A *TT* individual results when a *T* sperm fertilizes a *T* oocyte; a *tt* plant results when a *t*

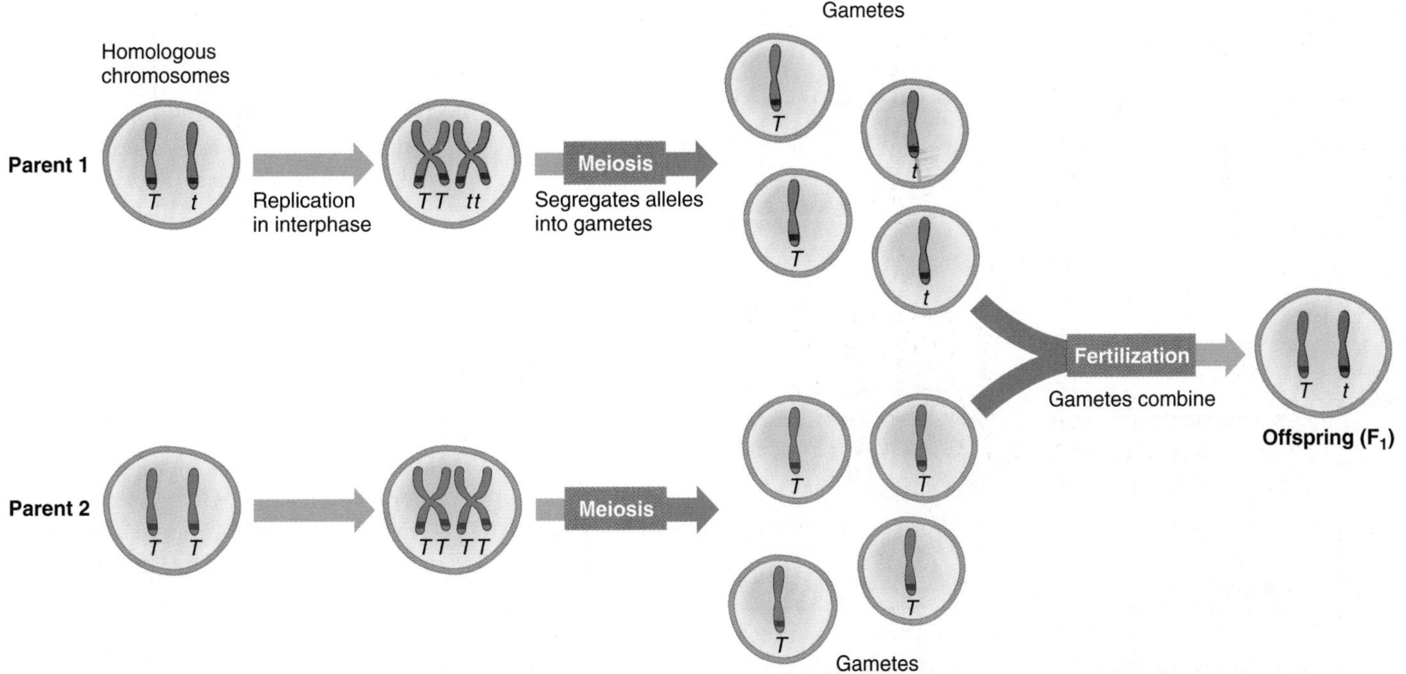


**figure 4.5**

**Mendel's first law—gene segregation.** During meiosis, homologous pairs of chromosomes (and the genes that compose them) separate from one another and are packaged into separate gametes. At fertilization, gametes combine at random to form the individuals of a new generation. Green and blue denote different parental origins of the chromosomes. In this example, offspring of genotype *TT* are also generated.

oocyte meets a *t* sperm; and a *Tt* individual results when either a sperm fertilizes a *T* oocyte, or a *T* sperm fertilizes a *t* oocyte.

Because two of the four possible gamete combinations produce a heterozygote, and each of the others produces a homozygote, the genotypic ratio expected of a monohybrid cross is 1 *TT*: 2 *Tt*: 1 *tt*. The corresponding phenotypic ratio is three tall plants to one short plant, a 3:1 ratio. Mendel saw these results for all seven traits that he studied, although as table 4.1 shows, the ratios were not exact. Today we use a diagram called a **Punnett square** to derive these ratios (figure 4.6). Experiments yield numbers of offspring that approximate these ratios.

Mendel distinguished the two genotypes resulting in tall progeny—*TT* from *Tt*—with additional crosses. He bred tall plants of unknown genotype with short (*tt*) plants. If a tall plant crossed with a *tt* plant produced both tall and short progeny, Mendel knew it was genotype *Tt*; if it produced only tall plants, he knew it must be *TT*. Crossing an individual of unknown genotype with a homozygous recessive individual is called a test cross. The homozygous recessive is the only genotype that can be identified by the phenotype—that is, a short plant is only *tt*. The homozygous recessive is therefore a "known" that can reveal the unknown genotype of another individual when the two are crossed.

**table 4.1**

**Mendel's Law of Segregation**

| Experiment | Total | Dominant | Recessive | Ratio |
|---|---|---|---|---|
| 1. Seed form | 7,324 | 5,474 | 1,850 | 2.96:1 |
| 2. Seed color | 8,023 | 6,022 | 2,001 | 3.01:1 |
| 3. Seed coat color | 929 | 705 | 224 | 3.15:1 |
| 4. Pod form | 1,181 | 882 | 299 | 2.95:1 |
| 5. Pod color | 580 | 428 | 152 | 2.82:1 |
| 6. Flower position | 858 | 651 | 207 | 3.14:1 |
| 7. Stem length | 1,064 | 787 | 277 | 2.84:1 |
| | | | | Average = 2.98:1 |

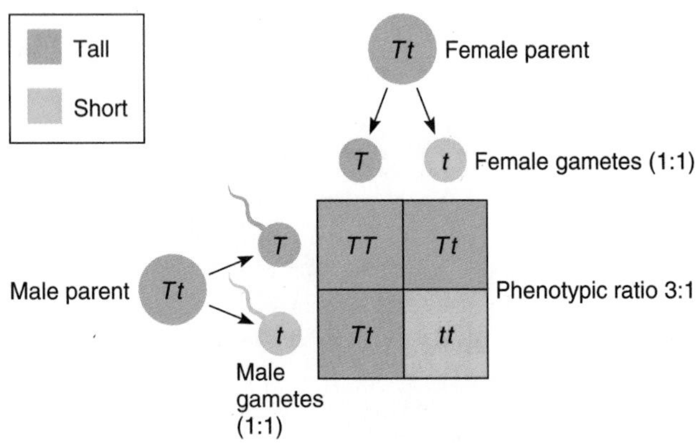


**figure 4.6**

**A Punnett square.** A diagram of gametes and how they can combine in a cross between two particular individuals is helpful in following the transmission of traits. The different types of female gametes are listed along the top of the square, and male gametes are listed on the left-hand side. Each compartment within the square contains the genotype that results when gametes that correspond to that compartment join. The Punnett square here describes Mendel's monohybrid cross of two tall pea plants. Among the progeny, tall plants outnumber short plants 3:1. Can you determine the genotypic ratio?

**KEY CONCEPTS**

From observing crosses in which two tall pea plants produced short offspring, along with other crosses, Mendel deduced that "characters" for height segregate during meiosis, then combine at random with those from the opposite gamete at fertilization. • A homozygote has two identical alleles, and a heterozygote has two different alleles. The allele expressed in a heterozygote is dominant; the allele not expressed is recessive. A monohybrid cross yields a genotypic ratio of 1:2:1 and a phenotypic ratio of 3:1. Punnett squares help calculate expected genotypic and phenotypic ratios among progeny. A test cross uses a homozygous recessive individual to reveal an unknown genotype.

## 4.2 Single-gene Inheritance in Humans

Mendel's first law addresses traits determined by single genes, as demonstrated in pea plants. Transmission of single genes in humans is called Mendelian, unifactorial, or single-gene inheritance.

Even the most familiar Mendelian disorders, such as sickle cell disease and Duchenne muscular dystrophy, are rare, compared to infectious diseases, cancer, and multifactorial disorders. Mendelian conditions typically affect 1 in 10,000 or fewer individuals. Table 4.2 lists some Mendelian disorders, and Reading 4.1 considers some interesting inherited quirks described in *Online Mendelian Inheritance in Man.*

### Modes of Inheritance

**Modes of inheritance** are rules that explain the common patterns that inherited characteristics follow as they are passed through families. Knowing the mode of inheritance makes it possible to calculate the probability that a particular couple will have a child who inherits a particular condition.

Mendel derived his laws by studying traits carried on autosomes (non-sex chromosomes). The way those laws affect the modes of inheritance depend on whether a trait is transmitted on an autosome or a sex

**table 4.2**

**Some Mendelian Disorders in Humans**

| Disorder | Symptoms |
|---|---|
| **Autosomal Recessive** | |
| Ataxia telangiectasis | Facial rash, poor muscular coordination, involuntary eye movements, high risk for cancer, sinus and lung infections |
| Cystic fibrosis | Lung infections and congestion, poor fat digestion, male infertility, poor weight gain, salty sweat |
| Familial hypertrophic cardiomyopathy | Overgrowth of heart muscle causes sudden death in young adults |
| Gaucher disease | Swollen liver and spleen, anemia, internal bleeding, poor balance |
| Hemochromatosis | Body retains iron; high risk of infection, liver damage, excess skin pigmentation, heart and pancreas damage |
| Maple syrup urine disease | Lethargy, vomiting, irritability, mental retardation, coma, and death in infancy |
| Phenylketonuria | Mental retardation, fair skin |
| Sickle cell disease | Joint pain, spleen damage, high risk of infection |
| Tay-Sachs disease | Nervous system degeneration |
| **Autosomal Dominant** | |
| Achondroplasia | Dwarfism with short limbs, normal size head and trunk |
| Familial hypercholesterolemia | Very high serum cholesterol, heart disease |
| Huntington disease | Progressive uncontrollable movements and personality changes, beginning in middle age |
| Lactose intolerance | Inability to digest lactose causes cramps after eating this sugar |
| Marfan syndrome | Long limbs, sunken chest, lens dislocation, spindly fingers, weakened aorta |
| Myotonic dystrophy | Progressive muscle wasting |
| Neurofibromatosis (I) | Brown skin marks, benign tumors beneath skin |
| Polycystic kidney disease | Cysts in kidneys, bloody urine, high blood pressure, abdominal pain |
| Polydactyly | Extra fingers and/or toes |
| Porphyria variegata | Red urine, fever, abdominal pain, headache, coma, death |

chromosome, and whether an allele is recessive or dominant. Autosomal dominant and autosomal recessive are the two modes of inheritance directly derived from Mendel's laws. An extension of the laws to the X chromosome is discussed in chapter 5. (Y linked inheritance is very rare because the Y chromosome has very few genes.)

### *Autosomal Dominant Inheritance*

In autosomal dominant inheritance, a trait can appear in either sex because an autosome carries the gene. If a child has the trait, at least one parent must also have it. Autosomal dominant traits do not skip generations. If no offspring inherit the trait in one generation, its transmission stops because the offspring can pass on only the recessive form of the gene. Figure 4.7 uses a Punnett square to predict the genotypes and

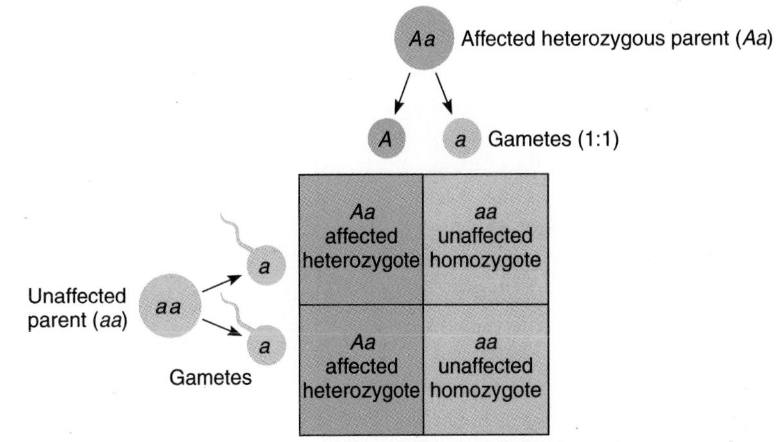


**figure 4.7**

**Autosomal dominant inheritance.** When one parent is affected and the other is not, each offspring has a 50 percent probability of inheriting the mutant allele and the condition.

Reading 4.1

# It's All in the Genes

Do you have uncombable hair, misshapen toes or teeth, or a pigmented tongue tip? Are you unable to smell a squashed skunk, or do you sneeze repeatedly in bright sunlight? Do you lack teeth, eyebrows, eyelashes, nasal bones, thumbnails, or fingerprints? If so, your unusual trait may be one of thousands described in *Online Mendelian Inheritance in Man* ("OMIM," at www.ncbi.nlm.nih.gov/entrez/query.fcgi?db=OMIM). A team at Johns Hopkins University led by renowned geneticist Victor McKusick updates OMIM daily. Entering a disease name retrieves family histories, clinical descriptions, the mode of inheritance, and molecular information on the causative gene and protein products. Woven in amidst the medical terminology and genetic jargon are the stories behind some fascinating inherited traits.

Genes control whether hair is blond, brown, or black, has red highlights, and is straight, curly, or kinky. Widow's peaks, cowlicks, a whorl in the eyebrow, and white forelocks run in families; so do hairs with triangular cross sections. Some people have multicolored hairs, like cats; others have hair in odd places, such as on the elbows, nose tip, knuckles, palms, or soles. Teeth can be missing or extra, protuberant or fused, present at birth, shovel shaped, or "snowcapped." A person can have a grooved tongue, duckbill lips, flared ears, egg-shaped pupils, three rows of eyelashes, spotted nails, or "broad thumbs and great toes." Extra breasts are known in humans and guinea pigs, and one family's claim to genetic fame is a double nail on the littlest toe.

Unusual genetic variants can affect metabolism, producing either disease or harmless, yet noticeable, effects. Members of some families experience "urinary excretion of odoriferous component of asparagus" or "urinary excretion of beet pigment," producing a strange odor or dark pink urine stream after consuming the offending vegetable. In blue diaper syndrome, an infant's urine turns blue on contact with air, thanks to an inherited inability to break down an amino acid.

One bizarre inherited illness is the jumping Frenchmen of Maine syndrome. This exaggerated startle reflex was first noted among French-Canadian lumberjacks from the Moosehead Lake area of Maine, whose ancestors were from the Beauce region of Quebec. Physicians first reported the condition at a medical conference in 1878. Geneticists videotaped the startle response in 1980, and the condition continues to appear in genetics journals. OMIM offers a most vivid description:

> **If given a short, sudden, quick command, the affected person would respond with the appropriate action, often echoing the words of command. . . . For example, if one of them was abruptly asked to strike another, he would do so without hesitation, even if it was his mother and he had an ax in his hand.**

The jumping Frenchmen of Maine syndrome may be an extreme variant of the more common Tourette syndrome, which causes tics and other uncontrollable movements. Figure 1 illustrates some other genetic variants.

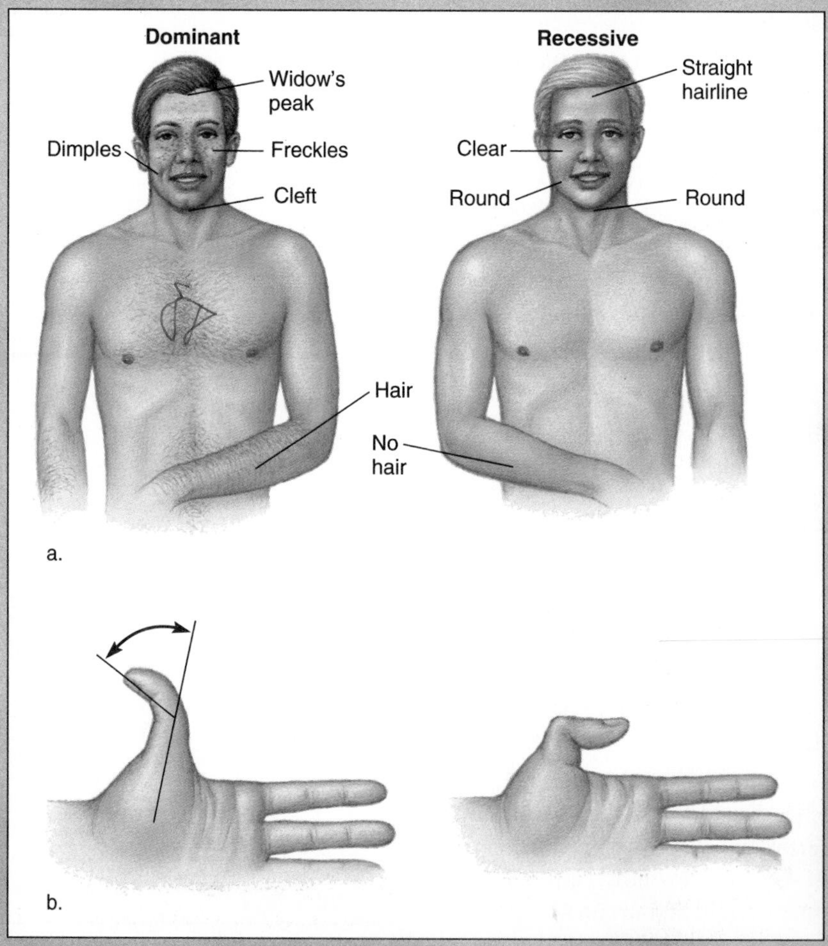


**figure 1**

**Inheritance of some common traits.** (*a*) Freckles, dimples, hairy arms, widow's peak, and a cleft chin are examples of dominant traits. (*b*) The ability to bend the thumb backward or forward is inherited.

phenotypes of offspring of a mother who has an autosomal dominant trait and a father who does not.

A medical student named James Poush discovered an autosomal dominant trait in his own family while in high school. He published the first full report on distal symphalangism. James had never thought that his stiff fingers and toes with their tiny nails were odd; others in his family had them, too. But when he studied genetics, he realized that his quirk might be a Mendelian trait. James identified 27 affected individuals among 156 relatives, and concluded that the trait is autosomal dominant. Of 63 relatives with an affected parent, 27 (43 percent) expressed the phenotype—this was close to the 50 percent expected for autosomal dominant inheritance. Table 4.3 lists the criteria for an autosomal dominant trait.

### *Autosomal Recessive Inheritance*

An autosomal recessive trait can appear in either sex. Affected individuals have a homozygous recessive genotype, whereas in heterozygotes—also called carriers—the wild type allele masks expression of the mutant allele. Consider the Deford family. Sportswriter Frank Deford and his wife, Carol, had their daughter Alex in 1972. She died at age eight of cystic fibrosis. Frank Deford wrote about his feelings at the time of diagnosis in a book, *Alex, the Life of a Child:*

> **I went to the encyclopedia and read about this cystic fibrosis. To me, at that point, it was one of those vague diseases you hear about now and then, sounding like some kind of clinging vine or a guy who kicks field goals with the side of his foot for the Kansas City Chiefs. . . . One out of every 20 whites carries the defective gene, as I do, as Carol does, as perhaps 10 million other Americans do—a population about the size of Illinois or Ohio. . . . If Carol and I had been lucky, if our second-born had not been cursed with cystic fibrosis, then we probably never would have had another child, and our bad genes would merely have been passed on, blissfully unknown to us.**

Frank Deford described the very essence of recessive inheritance: skipped generations in the expression of a trait. Because the human generation time is so long—about 30 years—we can usually trace a trait or illness for only two or three generations. For this reason, the Defords did not know of any other affected relatives.

Alex had inherited a mutant allele on chromosome 7 from each of her parents. The Defords were unaffected by the illness because they each also had a dominant allele that encodes enough of a functional protein for health. They are carriers (figure 4.8).

Mendel's first law can be used to calculate the probability that an individual will have either of two phenotypes. The probabilities of each possible genotype are added. For example, the chance that a child whose parents are both carriers of cystic fibrosis will *not* have the condition is the sum of the probability that she has inherited two normal alleles (1/4) plus the chance that she herself is a heterozygote (1/2), or 3/4. Note that this also equals one minus the probability that she is a homozygous recessive who has the condition.

The ratios that Mendel's first law predicts for autosomal recessive inheritance apply to each offspring anew, just as a tossed coin has a 50 percent chance of coming up heads with each throw, no matter how many heads have already been thrown. Misunderstanding of this concept leads to a common problem in genetic counseling. Many people conclude that if they have already had a

**table 4.3**

**Criteria for an Autosomal Dominant Trait**

1. Males and females can be affected. Can have male-to-male transmission.
2. Males and females transmit the trait with equal frequency.
3. Successive generations are affected.
4. Transmission stops if a generation arises in which no one is affected.

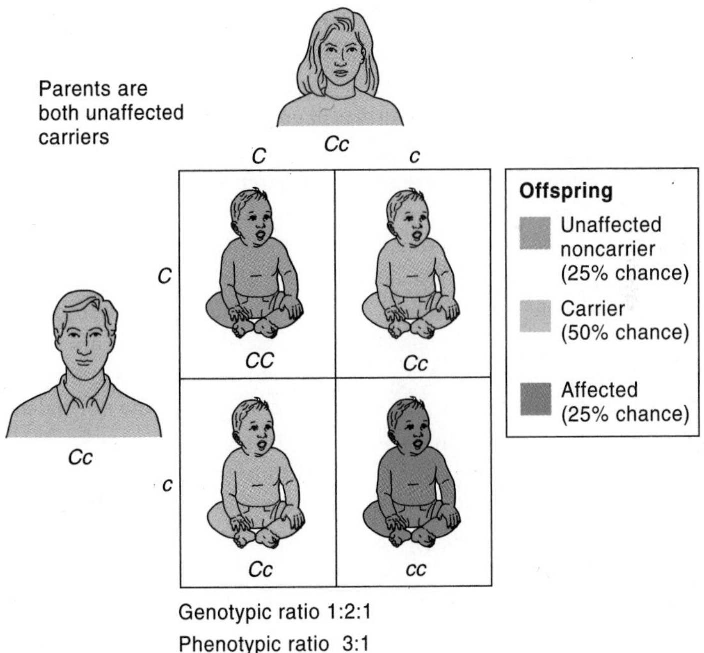


**figure 4.8**

**Autosomal recessive inheritance.** A 1:2:1 genotypic ratio results from a monohybrid cross, whether in peas or people. When parents are carriers for the same autosomal recessive trait or disorder, such as cystic fibrosis (CF), each child faces a 25 percent risk of inheriting the condition (a sperm carrying a CF allele fertilizing an oocyte carrying a CF allele); a 50 percent chance of being a carrier like the parents (a CF-carrying sperm with a wild type oocyte, and vice versa); and a 25 percent chance of inheriting two wild type alleles. Here, the colors denote different genotypes.

child affected by an autosomal recessive illness, then their next three children are guaranteed to escape it. This isn't true. Each child faces the same 25 percent risk of inheriting the condition.

Most autosomal recessive conditions occur unexpectedly in families. However, blood relatives who have children together have a much higher risk of having a child with an autosomal recessive condition. Marriage between relatives produces consanguinity, which means "shared blood"—an accurate description, although it is figurative—genes are not passed in blood. People who are related can trace their families back to a common ancestor. A man and woman who are unrelated have eight different grandparents. But first cousins have only six different grandparents, because they share a pair through their parents, who are siblings (see figure 4.14*c*). Consanguinity increases the likelihood of disease because the parents may have inherited the same mutant recessive allele from the same grandparent. That is, there is a greater chance of two relatives inheriting the same disease-causing recessive alleles than two unrelated people having the same alleles by chance.

The nature of the phenotype should be considered when evaluating transmission of Mendelian traits. For example, each adult sibling of a person who is a known carrier of Tay-Sachs disease has a two-thirds chance of being a carrier. The probability is two-thirds, and not one-half, because there are only three genotypic possibilities for an adult—homozygous for the normal allele, or a carrier who inherits the mutant allele from either the mother or father. A homozygous recessive individual for Tay-Sachs disease would never have survived childhood. We will return to this concept in chapter 5.

Geneticists who study human traits and illnesses can hardly set up crosses as Mendel did, but they can pool information from families whose members have the same trait or illness, based on symptoms, biochemical tests, or genetic tests. Consider a simplified example of 50 couples in which both partners are carriers of sickle cell disease. If 100 children are born, about 25 of them would be expected to have sickle cell disease. Of the remaining 75, theoretically 50 would be carriers like their parents, and the remaining 25 would have two nonmutant alleles. Table 4.4 lists criteria for an autosomal recessive trait.

**table 4.4**

**Criteria for an Autosomal Recessive Trait**

1. Males and females are affected.
2. Affected males and females can transmit the trait, unless it causes death before reproductive age.
3. The trait can skip generations.
4. Parents of affected individual are heterozygous or also have the trait.

## On the Meaning of Dominance and Recessiveness

Determining whether an allele is dominant or recessive is critical in medical genetics because it helps predict which individuals are at high risk of inheriting a particular condition. Dominance and recessiveness reflect the characteristics or abundance of the protein products.

Mendel based his definitions of dominance and recessiveness on what he could see—one allele masked the other. Today we can often add a cellular or molecular explanation for whether an allele is dominant or recessive. Consider inborn errors of metabolism, which are caused by the absence of an enzyme, such as the lysosomal storage diseases described in chapter 2. These disorders tend to be recessive because cells of a carrier make half the normal amount of the enzyme, but this is sufficient to maintain health. The one normal allele, therefore, compensates for the mutant one, to which it is dominant. The situation is similar in pea plants, in which short stem length is an inborn error in the sense that it results from deficiency of an enzyme that activates a growth hormone; the *Tt* plants evidently produce enough hormone to attain the same height as *TT* plants.

A recessive trait is sometimes called a "loss of function" because the recessive allele usually causes the loss of normal protein production and function. In contrast, some dominantly inherited disorders result from action of an abnormal protein that interferes with the function of the normal protein. Huntington disease is an example of such a "gain of function" disorder. The mutant allele, which is dominant, encodes an abnormally elongated protein that prevents the normal protein from functioning in the brain. Researchers determined that Huntington disease represents a gain of function because individuals who are missing one copy of the gene do not have the illness.

Recessive disorders tend to be more severe, and produce symptoms at much earlier ages, than do dominant disorders (figure 3.22). Disease-causing recessive alleles can remain, and even flourish, in populations because heterozygotes carry them without becoming ill, and pass them to future generations. In contrast, if a dominant mutation arises that causes severe illness early in life, people who have the allele are too ill or do not live long enough to reproduce, and the allele eventually grows rare in the population unless replaced by mutation. Dominant disorders whose symptoms do not appear until adulthood, or that do not drastically disrupt health, tend to remain in a population because they do not affect health until after a person has reproduced. Therefore, the dominant conditions that remain tend to be those that first cause symptoms in middle adulthood.

**Key Concepts**

A Mendelian trait is caused by a single gene. Modes of inheritance reveal whether a Mendelian trait is dominant or recessive and whether the gene that controls it is carried on an autosome or a sex chromosome. Autosomal dominant traits do not skip generations and can affect both sexes; autosomal recessive traits can skip generations and can affect both sexes. Rare autosomal recessive disorders sometimes recur in families when blood relatives have children together. Mendel's first law, which can predict the probability that a child will inherit a Mendelian trait, applies anew to each child. • At the biochemical level, dominance refers to the ability of a protein encoded by one allele to compensate for a missing or abnormal protein encoded by another allele.

## 4.3 Following the Inheritance of Two Genes—Independent Assortment

The law of segregation follows the inheritance of two alleles for a single gene. In a second set of experiments, Mendel examined the inheritance of two different traits, each attributable to a gene with two different alleles. The second law states that for two genes on different chromosomes, the inheritance of one does not influence the chance of inheriting the other. The two genes thus "independently assort" because they are packaged into gametes at random, as figure 4.9 shows. Two genes that are far apart on the same chromosome also appear to independently assort, because so many crossovers occur between them that the effect is as if they are carried on separate chromosomes (see figure 3.5).

Mendel looked at seed shape, which was either round or wrinkled (determined by the *R* gene), and seed color, which was either yellow or green (determined by the *Y* gene). When he crossed true-breeding plants that had round, yellow seeds to true-breeding plants that had wrinkled, green seeds, all the progeny had round, yellow seeds. These offspring were double heterozygotes, or dihybrids, of genotype *RrYy*. From their appearance, Mendel deduced that round is dominant to wrinkled, and yellow to green.

Next, he bred the dihybrid plants to each other in a **dihybrid cross,** so named because two individuals heterozygous for two genes are crossed. Mendel found four types of seeds in the next, third generation: 315 plants with round, yellow seeds; 108 plants with round, green seeds; 101 plants with wrinkled, yellow seeds; and 32 plants with wrinkled, green seeds. These classes occurred in a ratio of 9:3:3:1.

Mendel then took each plant from the third generation and crossed it to plants with wrinkled, green seeds (genotype *rryy*). These test crosses established whether each plant in the third generation was true-breeding for both genes (genotypes *RRYY* or *rryy*), true-breeding for one gene but heterozygous for the other (genotypes *RRYy*, *RrYY*, *rrYy*, or *Rryy*), or heterozygous for both genes (genotype *RrYy*). Mendel could explain the 9:3:3:1 proportion of progeny classes only if one gene does not influence transmission of the other. Each parent

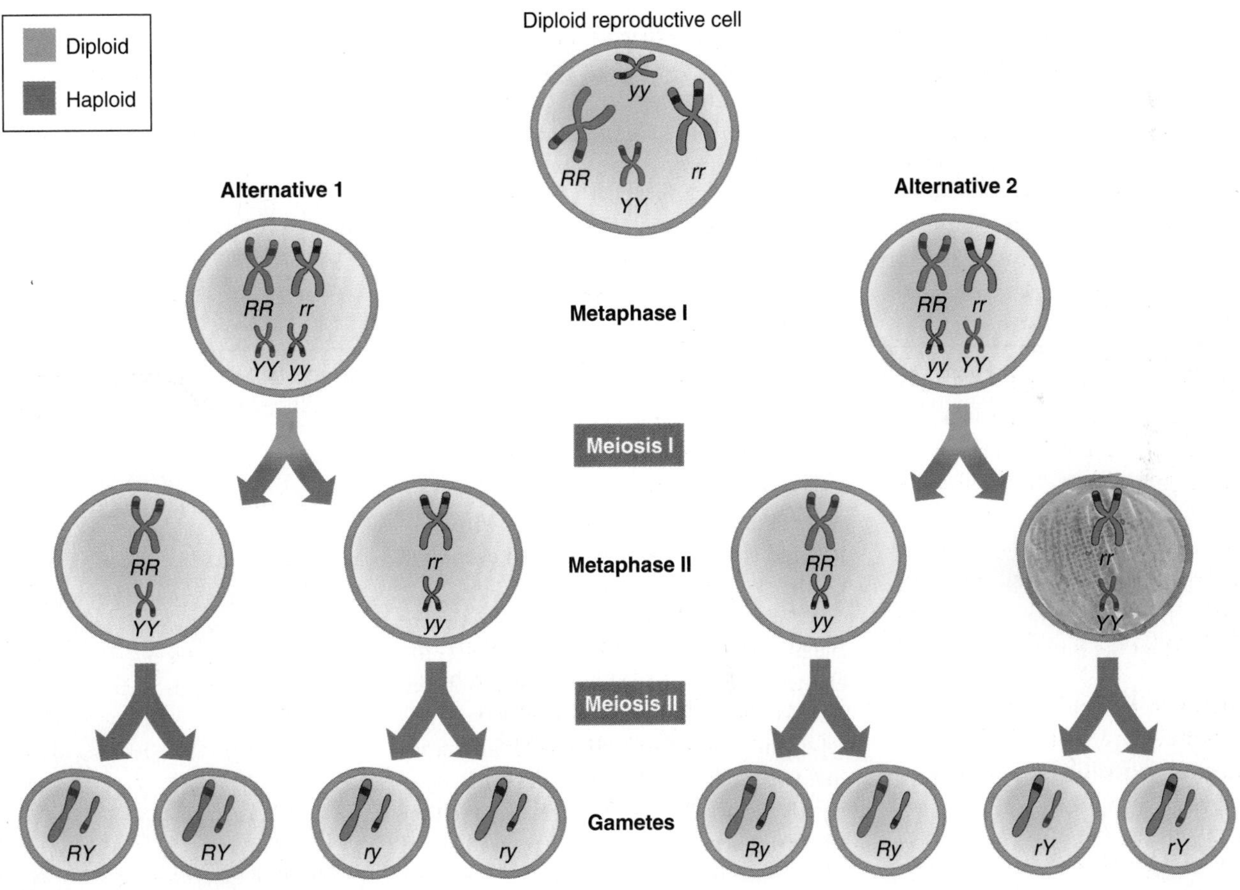


**figure 4.9**

**Mendel's second law—independent assortment.** The independent assortment of genes carried on different chromosomes results from the random alignment of chromosome pairs during metaphase of meiosis I. An individual of genotype *RrYy*, for example, manufactures four types of gametes, containing the dominant alleles of both genes (*RY*), the recessive alleles of both genes (*ry*), and a dominant allele of one with a recessive allele of the other (*Ry* or *rY*). The allele combination depends upon which chromosomes are packaged together into the same gamete—and this happens at random.

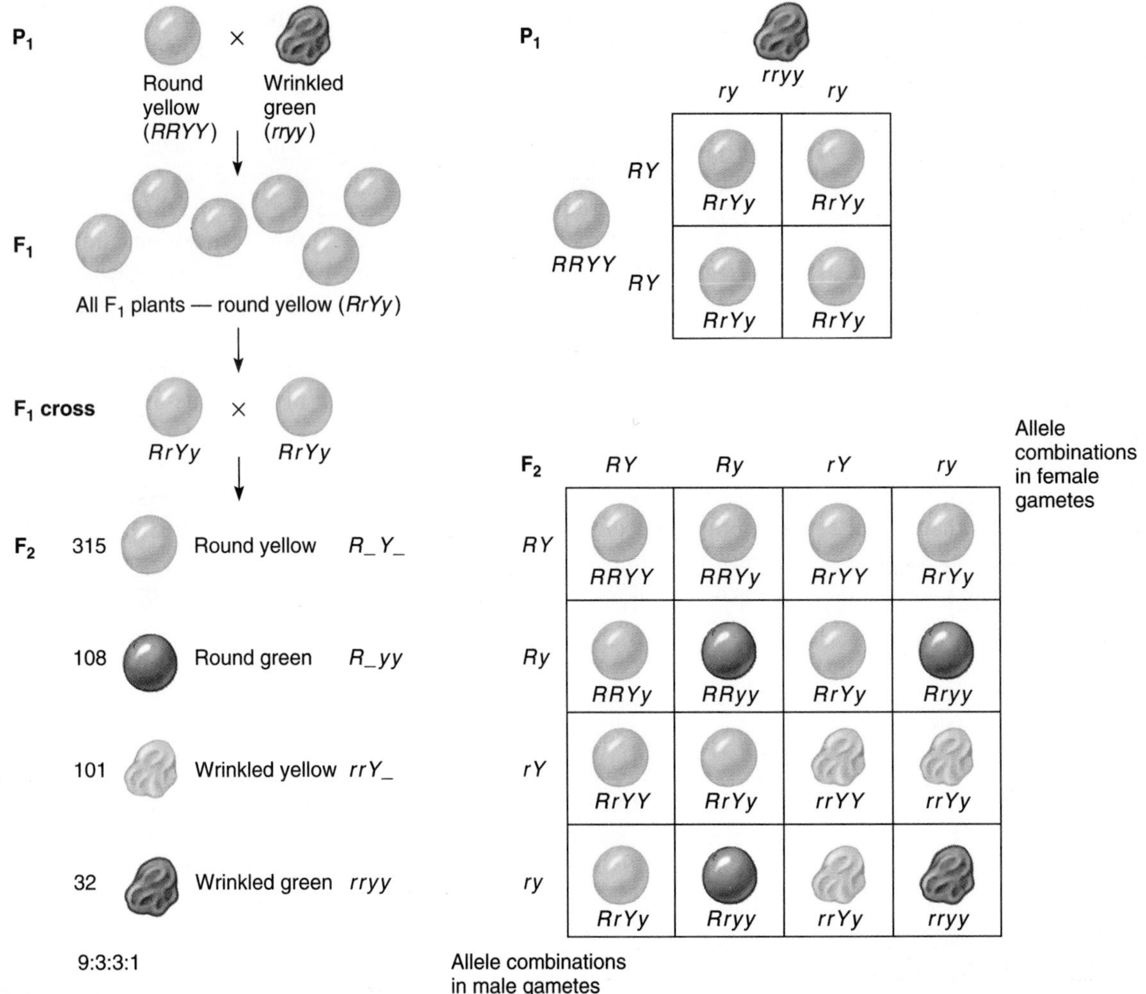


**figure 4.10**

**Plotting a dihybrid cross.** A Punnett square can represent the random combinations of gametes produced by dihybrid individuals. An underline in a genotype (in the $F_2$ generation) indicates that either a dominant or recessive allele is possible. The numbers in the $F_2$ generation are Mendel's experimental data.

would produce equal numbers of four different types of gametes: *RY, Ry, rY,* and *ry.* Note that each of these combinations has one gene for each trait. A Punnett square for this cross shows that the four types of seeds:

1. round, yellow (RRYY, RrYY, RRYy, and RrYy)
2. round, green (*RRyy* and *Rryy*)
3. wrinkled, yellow (*rrYY* and *rrYy*) and
4. wrinkled, green (rryy)

are present in the ratio 9:3:3:1, just as Mendel found (figure 4.10).

A Punnett square for three genes has 64 boxes; for four genes, 256 boxes. An easier way to predict genotypes and phenotypes in crosses involving more than one gene is to use the mathematical laws of probability on which Punnett squares are based. Probability predicts the likelihood that an event will occur.

An application of probability theory called the product rule can predict the chance that parents with known genotypes can produce offspring of a particular genotype. The product rule states that the chance that two independent events will both occur equals the product of the chance that either event will occur. Consider the probability of obtaining a plant with wrinkled, green peas (genotype *rryy*) from dihybrid (*RrYy*) parents. Do the reasoning for one gene at a time, then multiply the results (figure 4.11).

A Punnett square for *Rr* crossed to *Rr* shows that the probability of *Rr* plants producing *rr* progeny is 25 percent, or 1/4. Similarly, the chance of two *Yy* plants pro-

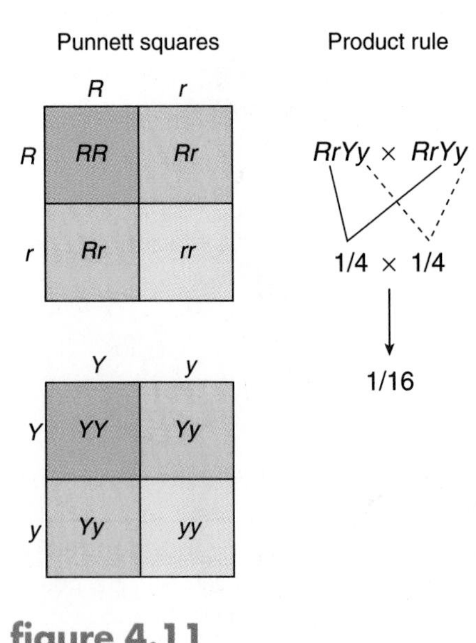


**figure 4.11**

**The product rule.**

ducing a *yy* plant is 1/4. Therefore, the chance of dihybrid parents (*RrYy*) producing homozygous recessive (*rryy*) offspring is 1/4 multiplied by 1/4, or 1/16. Now consult the 16-box Punnett square for Mendel's dihybrid cross again (figure 4.10). Only one of the 16 boxes is *rryy*, just as the product rule predicts. Figure 4.12 depicts how probability and Punnett squares can be used to predict offspring genotypes and phenotypes for three traits simultaneously in humans.

Until recently, Mendel's second law has not been nearly as useful in medical genetics as the first law, because not enough genes were known to follow the transmission of two or more traits at a time. But human genome information and DNA microarray technology are changing that practice. A DNA microarray used to analyze breast cancer, for example, screens for thousands of different genes and their variants. The increasingly computational nature of genetics in the 21st century has produced an entirely new field, called bioinformatics. So, in that sense, genetics is continuing the theme of mathematical analysis that Gregor Mendel began more than a century ago.

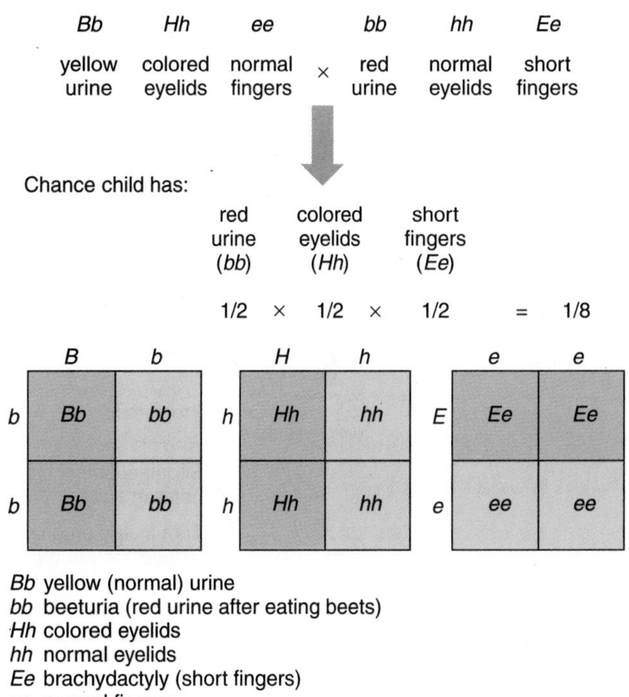


**figure 4.12**

**Using probability to track three traits.** A man with normal urine, colored eyelids, and normal fingers wants to have children with a woman who has red urine after she eats beets, normal eyelids, and short fingers. The chance that a child of theirs will have red urine after eating beets, colored eyelids, and short fingers is 1/8.

### Key Concepts

Mendel's second law, the law of independent assortment, considers genes transmitted on different chromosomes. In a dihybrid cross of heterozygotes for seed color and shape, Mendel saw a phenotypic ratio of 9:3:3:1. He concluded that transmission of one gene does not influence that of another. Meiotic events explain independent assortment. Punnett squares and probability can be used to follow independent assortment. Knowing the human genome sequence has made possible the analysis of more than one gene at a time.

## 4.4 Pedigree Analysis

In this era of genome sequencing, families are still often the starting point of genetic investigation—and families still represent the context in which most people think of genetics. For researchers, families are tools, and the bigger the family the better—the more children in a generation, the easier it is to discern modes of inheritance. Geneticists use charts called **pedigrees** to display family relationships and to follow which relatives have specific phenotypes and, sometimes, genotypes. A human pedigree serves the same purpose as one for purebred dogs or cats or thoroughbred horses—it helps keep track of relationships and traits.

A pedigree is built of shapes connected by lines. Vertical lines represent generations; horizontal lines that connect two shapes at their centers depict parents; shapes connected by vertical lines joined horizontally on one line above them represent siblings. Squares indicate males; circles, females; and diamonds, individuals of unspecified sex. Figure 4.13 shows these and other commonly used pedigree symbols. Colored shapes indicate individuals who express the trait under study, and half-filled shapes represent known carriers. A genetic counselor will often sketch out a pedigree while interviewing a client, then use a computer program and add test results that might indicate genotypes to fill in the pedigree and finalize a report.

### Pedigrees Then and Now

The earliest pedigrees were genealogical, indicating family relationships but not traits. Figure 4.14 shows such a pedigree for a highly inbred part of the ancient Egyptian royal family. The term *pedigree* arose in the 15th century, from the French *pie de grue*, which means "crane's foot." Pedigrees at that time, typically depicting large families, showed parents linked by curved lines to their offspring. The overall diagram often resembled a bird's foot.

An extensive family tree of several European royal families indicates which members had the clotting disorder hemophilia (see figure 6.8). This pedigree is believed to be one of the first to trace an inherited illness. The mutant gene probably originated in Queen Victoria of England in the 19th century. In 1845, a genealogist named Pliny Earle constructed a pedigree of a family with color blindness, using

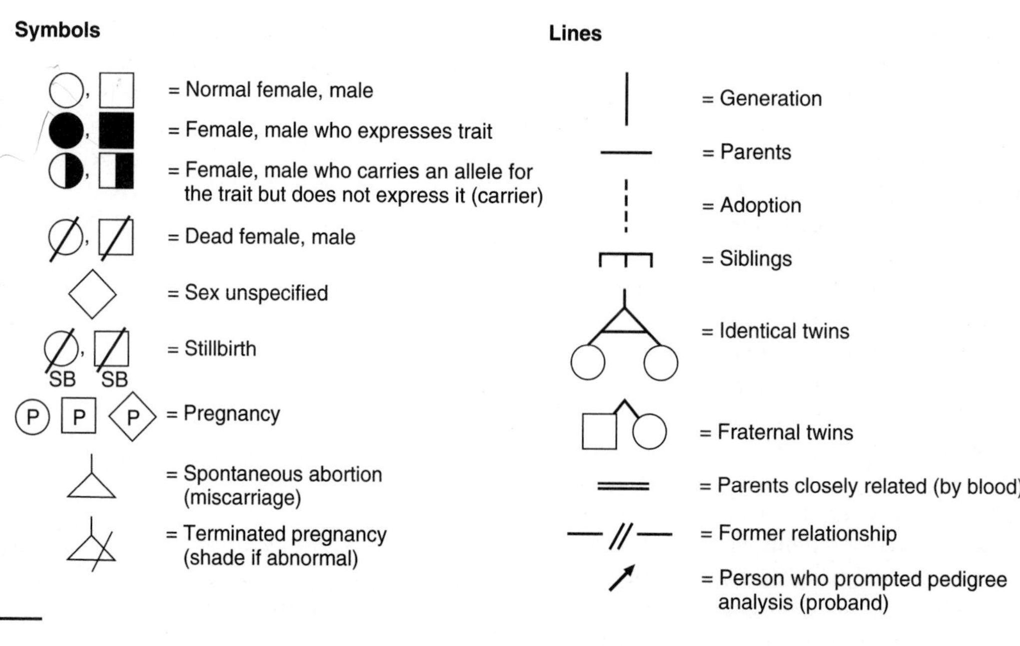


## figure 4.13

**Pedigree components.** Symbols representing individuals are connected to form pedigree charts, which display the inheritance patterns of particular traits.



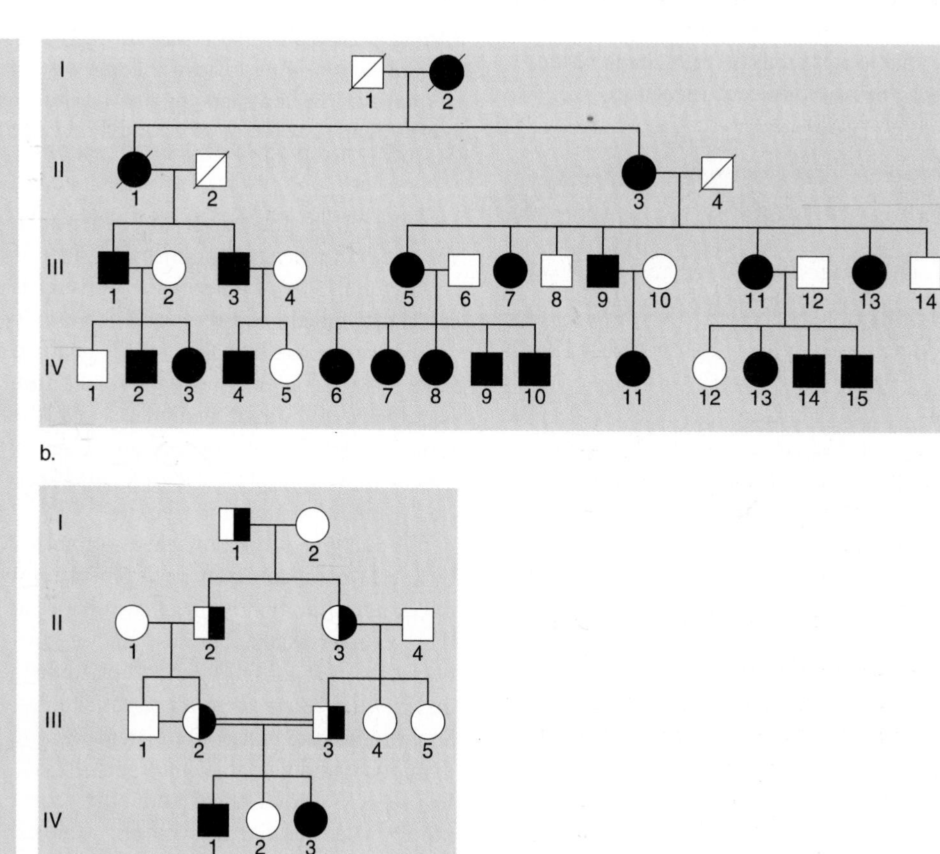


## figure 4.14

**Some unusual pedigrees.** *(a)* A partial pedigree of Egypt's Ptolemy dynasty shows only genealogy, not traits. It appears almost ladderlike because of the extensive inbreeding. From 323 B.C. to Cleopatra's death in 30 B.C., the family experienced five brother-sister pairings, plus an uncle-niece relationship. Cleopatra married her brother, Ptolemy XIII, when he was 10 years old! These marriage patterns were an attempt to preserve the royal blood. *(b)* In contrast to the Egyptian pedigree, a family with polydactyly (extra fingers and toes) extends laterally, because there are many children. *(c)* The most common form of consanguinity is marriage of first cousins. They share grandparents, and therefore risk passing on the same recessive alleles to offspring.

musical notation—half notes for unaffected females, quarter notes for color blind females, and filled-in and squared-off notes to represent the many color-blind males. In the early 20th century, pedigrees took on a negative note when eugenicists attempted to use the diagrams to show that traits such as criminality, feeblemindedness, and promiscuity were the consequence of faulty genes.

Today, pedigrees are important both for helping families identify the risk of transmitting an inherited illness and as starting points for gene searches. In some instances, however, genealogy remains an important motivation for developing pedigrees. Groups of people who have kept meticulous family records can be invaluable in helping researchers follow the inheritance of particular genes. In the United States, the Mormons and the Amish have aided genetic studies, and the national genetic databases discussed in Bioethics: Choices for the Future in chapter 1 rely on family pedigrees and health records.

Very large pedigrees are helpful in gene hunts because they provide researchers with information on many individuals with a particular disorder. The researchers can then search these peoples' DNA to identify a particular sequence they have all inherited but that is not found in healthy family members. Discovery of the gene that causes Huntington disease, for example, took researchers to a remote village in Venezuela to study an enormous family whose pedigree looked more like wallpaper than a family tree. The gene was eventually traced to a sailor believed to have introduced the mutant gene in the 19th century.

## Pedigrees Display Mendel's Laws

A person familiar with Mendel's laws can often tell a mode of inheritance just by looking carefully at a pedigree. Consider an autosomal recessive trait, albinism. The homozygous recessive individual lacks an enzyme necessary to manufacture the pigment melanin and, as a result, has very pale hair and skin. Recall that an autosomal recessive trait can affect both sexes and can (but doesn't necessarily) skip generations. Figure 4.15 shows a pedigree for albinism. If a condition is known to be inherited as an autosomal recessive trait, carrier status can be inferred for individuals who have affected (homozygous recessive) children. In figure 4.15, because individuals III-1 and III-3 are affected, individuals II-2 and II-3 must be carriers. One partner from each pair of grandparents must also be a carrier, which can be determined using a carrier test or inferred from family history.

An autosomal dominant trait does not skip generations and can affect both sexes. Transmission stops whenever an individual does not inherit the causative allele. A typical pedigree for an autosomal dominant trait has some squares and circles filled in to indicate affected individuals in each generation. Figure 4.16 is a pedigree for distal symphalangism.

Sometimes a pedigree may be inconclusive, which means that either autosomal recessive or autosomal dominant inheritance can explain the pattern of filled-in

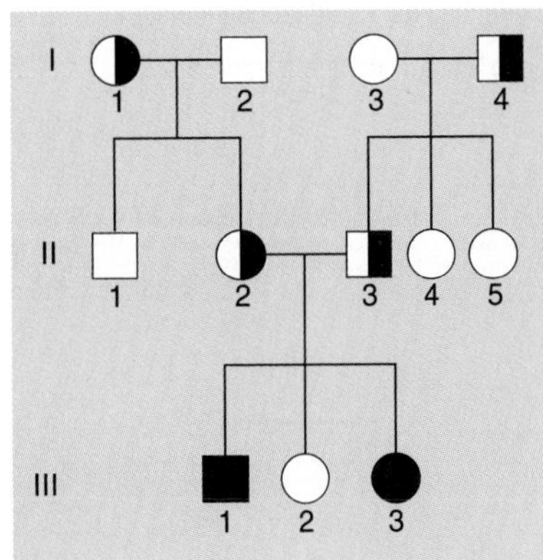

**figure 4.15**

**A pedigree for an autosomal recessive trait.** Albinism affects males and females, and can skip generations, as it does here in the first and second generations. Albinism occurs because the homozygous recessive individual lacks an enzyme needed to produce melanin, which colors the eyes, skin, and hair.

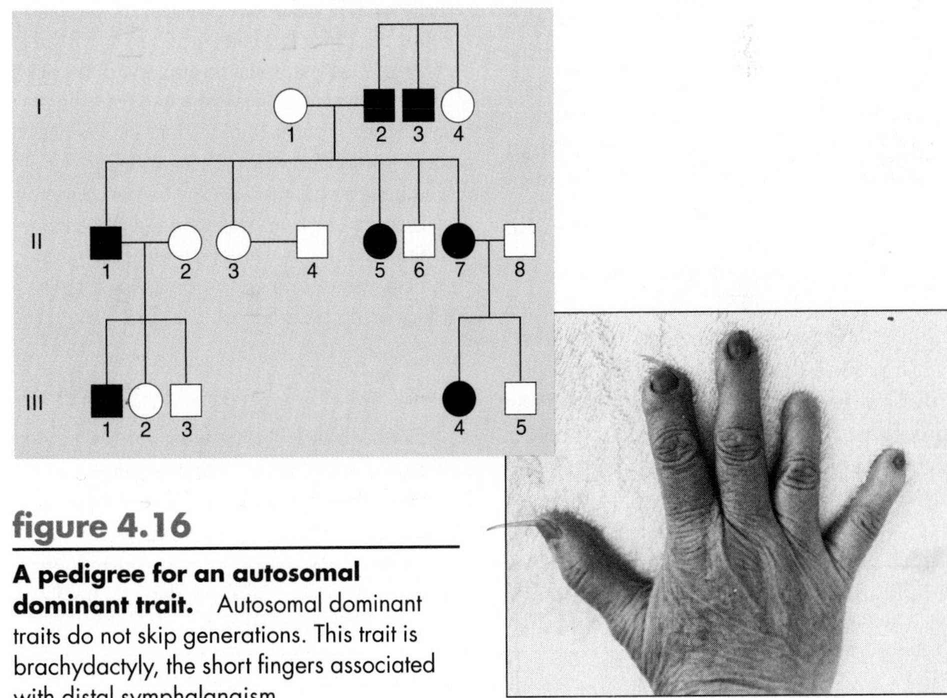


**figure 4.16**

**A pedigree for an autosomal dominant trait.** Autosomal dominant traits do not skip generations. This trait is brachydactyly, the short fingers associated with distal symphalangism.

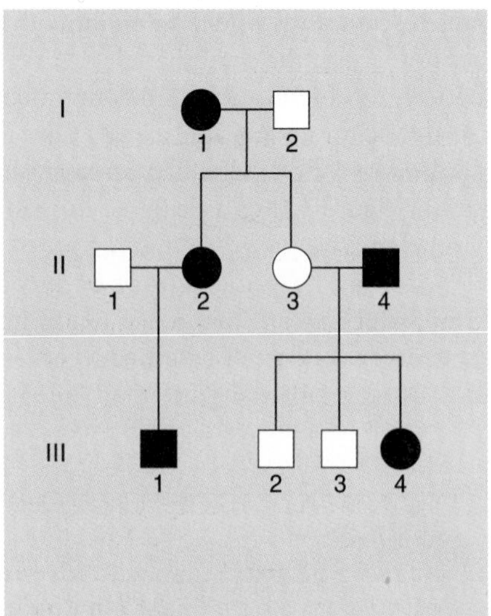


## figure 4.17

**An inconclusive pedigree.** This pedigree could account for an autosomal dominant trait or an autosomal recessive trait that does not prevent affected individuals from having children.

symbols. Figure 4.17 shows one such pedigree, for a type of hair loss called alopecia. According to the pedigree, this trait can be passed in an autosomal dominant mode because it affects males and females and is present in every generation. However, the pedigree can also depict autosomal recessive inheritance if the individuals represented by unfilled symbols are carriers. Inconclusive pedigrees tend to arise when families are small and the trait is not severe enough to keep heterozygotes from having children. Further family information or biochemical tests to detect carriers can sometimes clarify an inconclusive pedigree.

Often genetic counselors are asked to predict the probability, or risk of recurrence, of a condition in a particular individual. This requires thinking through Mendel's laws for more than one generation. Pedigrees and Punnett squares can be helpful in doing these calculations. Consider the family depicted in figure 4.18.

Michael Stewart has sickle cell disease, which is inherited as an autosomal recessive condition. This means that his unaffected parents, Kate and Brad, must each be heterozygotes (carriers). Michael's sister, Ellen, also healthy, is expecting her first child.

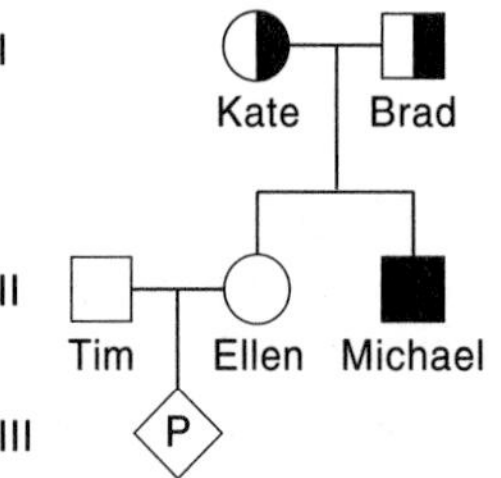


a. Ellen's brother, Michael, has sickle cell disease.

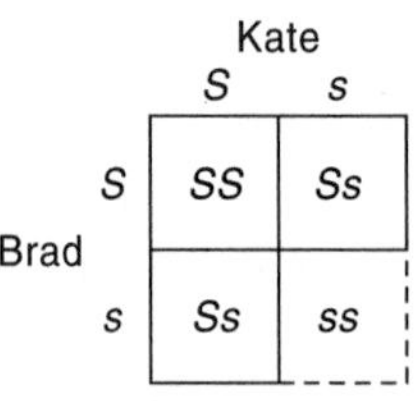


b. Probability that Ellen is a carrier: $^2/_3$

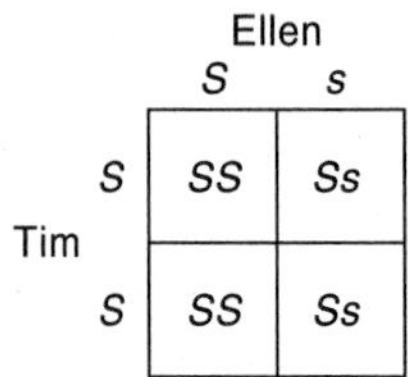


c. If Ellen is a carrier, chance that fetus is a carrier: $^1/_2$

Total probability = $^2/_3 \times ^1/_2 = ^1/_3$

## figure 4.18

**Making predictions.** Ellen's brother, Michael, has sickle cell disease, as depicted in this pedigree (*a*). Ellen wonders what the chance is that her fetus has inherited the sickle cell allele from her. First, she must calculate the chance that she is a carrier. The Punnett square in (*b*) shows that this risk is 2 in 3. (She must be genotype *SS* or *Ss* but cannot be *ss* because she does not have the disease.) The risk that the fetus is a carrier, assuming that the father is not a carrier, is half Ellen's risk of being a carrier, or 1 in 3 (*c*).

Ellen's husband, Tim, has no family history of sickle cell disease. Ellen wants to know the risk that her child will inherit the mutant allele from her and be a carrier.

Ellen's request really contains two questions. First, what is the risk that she herself is a carrier? Because Ellen is actually the product of a monohybrid cross, and we know that she is not homozygous recessive, she has a 2 in 3 chance of being a carrier, as the Punnett square indicates.

If Ellen is a carrier, then the next question is, what are the chances that she will pass the mutant allele to an offspring? That chance is 1 in 2, because she has two copies of the gene, and according to Mendel's first law, only one goes into each gamete.

To calculate the overall risk to Ellen's child, we can apply the product rule and multiply the probability that Ellen is a carrier by the chance that if she is, she will pass the mutant allele on. If we assume Tim is not a carrier, Ellen's chance of giving birth to a child who carries the mutant allele is therefore 2/3 times 1/2, which equals 2/6, or 1/3. Ellen thus has a 1 in 3 chance of giving birth to a child who is a carrier for sickle cell disease.

Pedigrees can be difficult to construct and interpret for several reasons. People sometimes hesitate to supply information because they are embarrassed by symptoms affecting behavior or mental stability. Tracing family relationships can be complicated by adoption, children born out of wedlock, serial relationships, blended families, and assisted reproductive technologies such as surrogate mothers and artificial insemination by donor (chapter 21). Moreover, many people cannot trace their families back more than three or four generations, and so lack sufficient evidence to reveal a mode of inheritance. Still, the pedigree is perhaps the most classic genetic tool, and it remains a powerful way to see, at a glance, how a trait passes from generation to generation—just what Gregor Mendel studied in peas.

### Key Concepts

Pedigrees are charts that depict family relationships and the transmission of inherited traits. Squares represent males, and circles, females; horizontal lines indicate parents, vertical lines show generations, and elevated horizontal lines depict siblings. Symbols for heterozygotes are half-shaded, and symbols for individuals who express the trait under study are completely shaded. • Pedigrees can reveal mode of inheritance. Along with Punnett squares, they are tools that apply Mendel's first law to predict recurrence risks of inherited disorders or traits.

# Summary

## 4.1 Following the Inheritance of One Gene—Segregation

1. Gregor Mendel described the two basic laws of inheritance using pea plant crosses. The laws derive from the actions of chromosomes during meiosis, and apply to all diploid organisms.
2. Mendel used a statistical approach to investigate why some traits seem to disappear in the hybrid generation. The **law of segregation** states that alleles of a gene are distributed into separate gametes during meiosis. Mendel demonstrated this using seven traits in pea plants.
3. A diploid individual with two identical alleles of a gene is **homozygous.** A **heterozygote** has two different alleles of a gene. A gene may have many alleles.
4. A **dominant** allele masks the expression of a recessive allele. An individual may be homozygous dominant, homozygous **recessive,** or heterozygous.
5. Mendel repeatedly found that when he crossed two true-breeding types, then bred the resulting hybrids to each other, the two variants of the trait appeared in a 3:1 phenotypic ratio. Crossing these progeny further revealed a genotypic ratio of 1:2:1.
6. A **Punnett square** is a chart used to follow the transmission of alleles. It is based on probability.

## 4.2 Single gene Inheritance in Humans

7. Traits or disorders caused by single genes are called Mendelian or unifactorial traits.
8. **Modes of inheritance** enable geneticists to predict phenotypes. In **autosomal dominant** inheritance, males and females may be affected, and the trait does not skip generations. Inheritance of an **autosomal recessive** trait may affect either males or females and may skip generations. Autosomal recessive conditions are more likely to occur in families with **consanguinity.** Recessive disorders tend to be more severe, and cause symptoms earlier than dominant disorders.
9. Dominance and recessiveness reflect how alleles affect the abundance or activity of the protein product.

## 4.3 Following the Inheritance of Two Genes—Independent Assortment

10. Mendel's second law, the law of **independent assortment,** follows the transmission of two or more genes on different chromosomes. It states that a random assortment of maternally and paternally derived chromosomes during meiosis results in gametes that have different combinations of these genes.
11. The chance that two independent genetic events will both occur is equal to the product of the probabilities that each event will occur on its own. This principle, called the product rule, is useful in calculating the risk that certain individuals will inherit a particular genotype and in following the inheritance of two genes on different chromosomes.

## 4.4 Pedigree Analysis

12. A **pedigree** is a chart that depicts family relationships and patterns of inheritance for particular traits. A pedigree can be inconclusive.

# Review Questions

1. How does meiosis explain Mendel's laws of segregation and independent assortment?
2. How was Mendel able to derive the two laws of inheritance without knowing about chromosomes?
3. Distinguish between
   a. autosomal recessive and autosomal dominant inheritance.
   b. Mendel's first and second laws.
   c. a homozygote and a heterozygote.
   d. a monohybrid and a dihybrid cross.
   e. a Punnett square and a pedigree.
4. Why would Mendel's results for the dihybrid cross have been different if the genes for the traits he followed were located near each other on the same chromosome?
5. Why are extremely rare autosomal recessive disorders more likely to appear in families in which blood relatives have children together?
6. How does the pedigree of the ancient Egyptian royal family in figure 4.14*a* differ from a pedigree a genetic counselor might use today?
7. People who have Huntington disease inherit one mutant and one normal allele. How would a person who is homozygous dominant for the condition arise?
8. Which figures in this chapter depict a single-gene cross at the organismal level? At the chromosomal level? At the gene level?
9. What is the probability that two individuals with an autosomal recessive trait, such as albinism, will have a child with the same genotype and phenotype as they do?

# Applied Questions

1. On the soap opera "The Young and the Restless," several individuals suffer from a rapid aging syndrome in which a young child is sent off to boarding school and returns three months later an angry teenager. In the Newman family, siblings Nicholas and Victoria aged from ages six and eight to sixteen and eighteen within a few months. Their parents, Victor and Nikki, are not affected; in fact, they never seem to age at all.
   a. What is the mode of inheritance of the rapid aging disorder affecting Nicholas and Victoria?
   b. How do you know what the mode of inheritance is?
   c. Draw a pedigree to depict this portion of the Newman family.
2. Achondroplasia is a common form of hereditary dwarfism that causes very short

limbs, stubby hands, and an enlarged forehead. Below are four pedigrees depicting families with this specific type of dwarfism: What is the most likely mode of inheritance? Cite a reason for your answer.

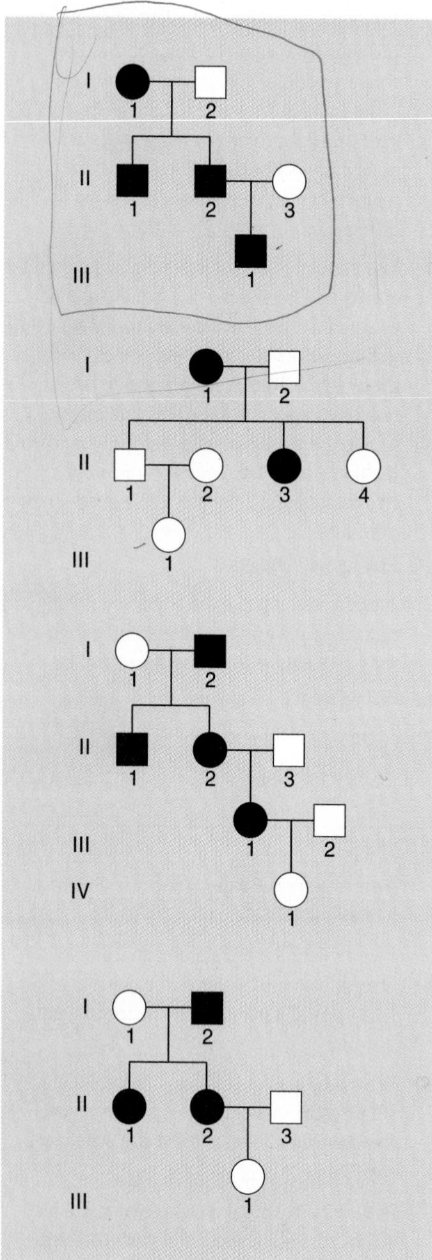

3. Draw a pedigree to depict the following family:

   One couple has a son and daughter with normal skin pigmentation. Another couple has one son and two daughters with normal skin pigmentation. The daughter from the first couple has three children with the son of the second couple. Their son and one daughter have albinism; their other daughter has normal skin pigmentation.

4. Chands syndrome is an autosomal recessive condition characterized by very curly hair, underdeveloped nails, and abnormally shaped eyelids. In the following pedigree, which individuals must be carriers?

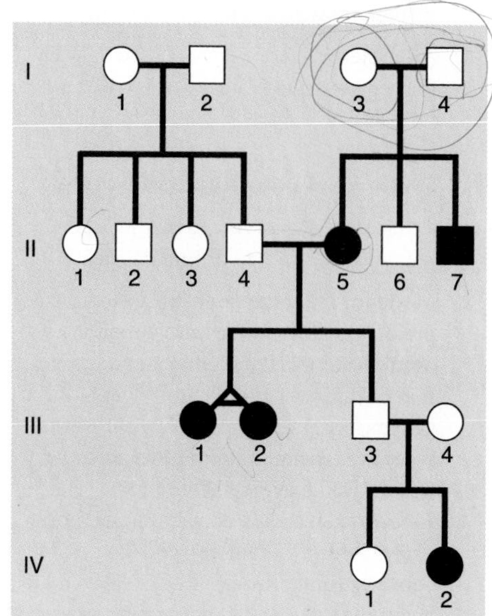

5. Caleb has a double row of eyelashes, which he inherited from his mother as a dominant trait. His maternal grandfather is the only other relative to have the trait. Veronica, a woman with normal eyelashes, falls madly in love with Caleb, and they marry. Their first child, Polly, has normal eyelashes. Now Veronica is pregnant again and hopes they will have a child who has double eyelashes. What chance does a child of Veronica and Caleb have of inheriting double eyelashes? Draw a pedigree of this family.

6. Congenital insensitivity to pain with anhidrosis is an extremely rare autosomal recessive condition that causes fever, inability to sweat (anhidrosis), mental retardation, inability to feel pain, and self-mutilating behavior. Researchers compared the following three families with this condition:

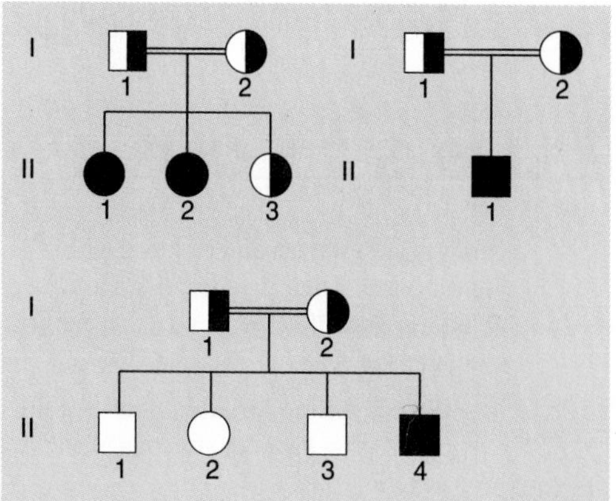

What do these families have in common that might explain the appearance of this rare illness?

7. The child in figure 4.12 who has red urine after eating beets, colored eyelids, and short fingers, is of genotype *bbHhEe*. The genes for these traits are on different chromosomes. If he has children with a woman who is a trihybrid for each of these genes, what are the genotypic and phenotypic ratios for their offspring?

8. In the pedigree below, individual III-1 died at age two of Tay-Sachs disease, an autosomal recessive disorder. Which other family members must be carriers and which could be?

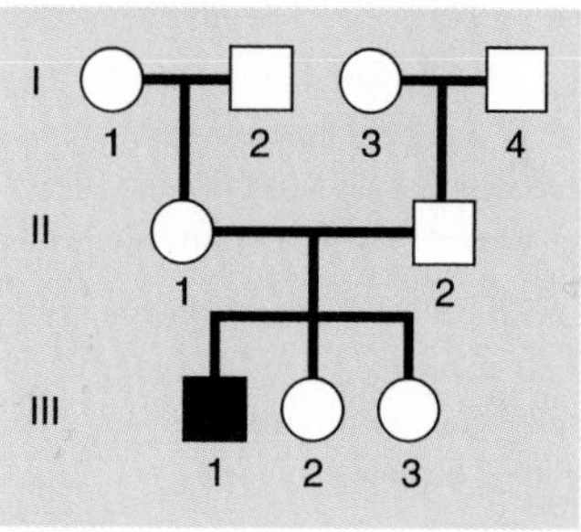

9. The Jackson Laboratory in Bar Harbor, Maine, has supplied genetically identical mice to researchers for decades. These mice are obtained by repeatedly crossing siblings to generate strains that are homozygous for all genes. Cloning would generate individuals that are identical, but may be heterozygous for some genes (see Bioethics: Choices for the Future in chapter 3). Why would the Jackson lab mice mated to each other yield offspring genetically identical to themselves, while cloned mice might not (assuming you could generate male and female cloned mice, which would require two sets of experiments)?

10. Recall Mackenzie, the young woman from chapter 1 who underwent genetic testing. The tests revealed that she has a 1 in 10 chance of developing lung cancer, and a 2 in 1,000 chance of developing colon cancer. What is the probability that she will develop both cancers?

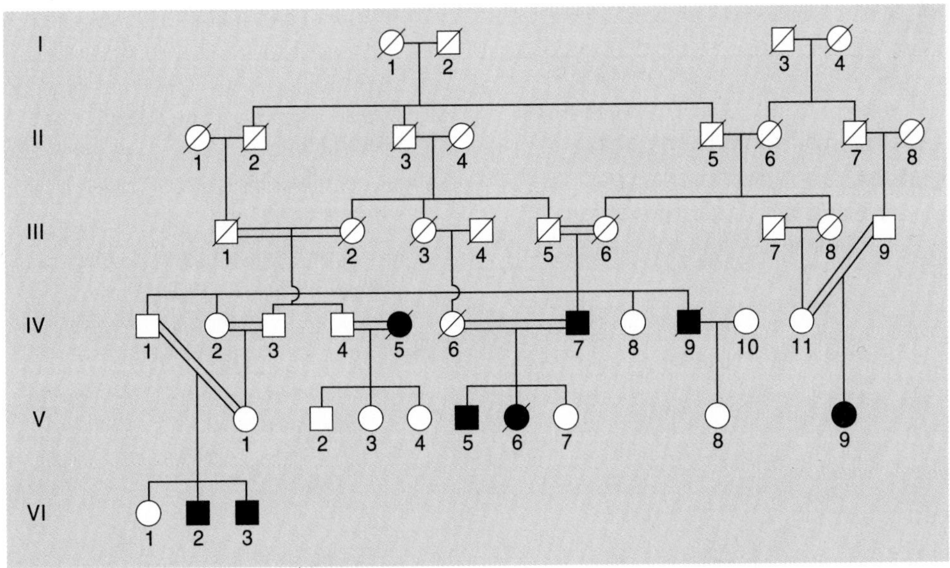


11. Sclerosteosis causes overgrowth of the skull and jaws that produces a characteristic face, gigantism, facial paralysis, and hearing loss. The overgrowth of skull bones can cause severe headaches and even sudden death. In the above pedigree for a family afflicted by sclerosteosis:
    a. What is the relationship between the individuals who are connected by slanted double lines?
    b. Sclerosteosis is an autosomal recessive condition. Which individuals in the pedigree must be carriers?

12. According to this pedigree from the soap opera "All My Children," is Charlie the product of a consanguineous relationship? The trait being tracked is freckles.

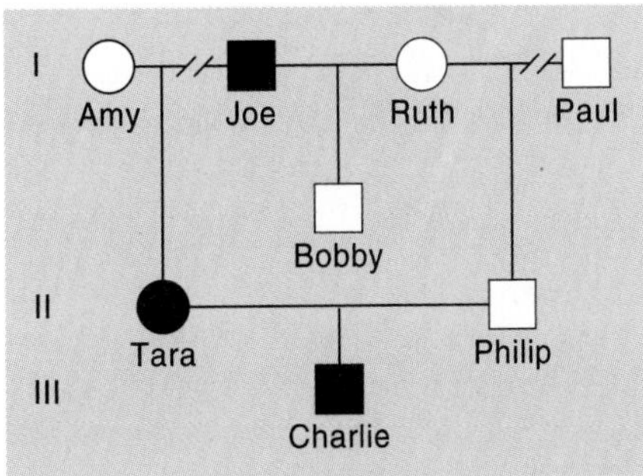


13. On "General Hospital," six-year-old Maxi suffered from Kawasaki syndrome, an inflammation of the heart. She desperately needed a transplant, and received one from BJ, who died in a bus accident. Maxi and BJ had the same unusual blood type, which is inherited. According to the pedigree below, how are Maxi and BJ related?

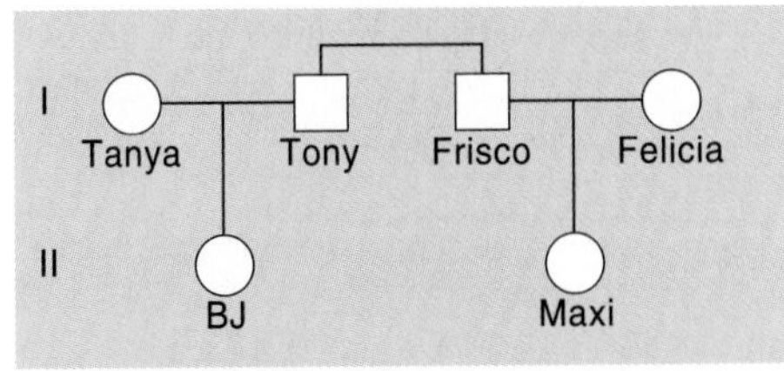


14. A man has a blood test for Tay-Sachs disease and learns that he is a carrier. His body produces half the normal amount of the enzyme hexoseaminidase. Why doesn't he have symptoms? (This disease causes degeneration of the nervous system in early childhood.)

# Suggested Readings

Bhattacharyya, Madan K., et al. January 12, 1990. The wrinkled-seed character of pea described by Mendel is caused by a transposon-like insertion in a gene encoding starch-branching enzyme. *Cell* 60:115–27. A description of one of the traits that Mendel studied, at the molecular level.

Gustafsson, A. February 1979. Linnaeus' Peloria: The history of a monster. *Theoretical Applied Genetics* 54:241–48. Charles Darwin demonstrated Mendel's first law using mutant flowers, but he didn't realize it.

Henig, Robin Marantz. 2000. *The Monk in the Garden: The Lost and Found Genius of Gregor Mendel, the Father of Genetics.* Boston: Houghton Mifflin Co. A complete biography of the "father of genetics."

Lester, Diane R., et al. August 1997. Mendel's stem length gene (Le) encodes a gibberellin 3β-hydroxylase. *Plant Cell* 9:1435–43. The molecular and cellular basis of Mendel's short and tall traits involves regulation of a growth hormone.

Lewis, Ricki. December 1994. The evolution of a classic genetic tool. *BioScience* 44:5–8. Pedigrees may include molecular information.

Mendel, Gregor. March 1866. Experiments in plant hybridization. *Journal of the Royal Horticultural Society,* pp. 3–47. Mendel's original paper is surprisingly understandable.

Orel, Vitězslav. 1996. *Gregor Mendel: The First Geneticist.* New York: Oxford University Press. A detailed account of Gregor Mendel and his work.

# On the Web

Check out the resources on our website at

**www.mhhe.com/lewisgenetics5**

On the web for this chapter, you will find additional study questions, vocabulary review, useful links to case studies, tutorials, popular press coverage, and much more. To investigate specific topics mentioned in this chapter, also try the links below:

Online Mendelian Inheritance in Man **www.ncbi.nlm.nih.gov/entrez/query.fcgi?db=OMIM**

MendelWeb **ftp.netspace.org/MendelWeb/Mendel.htm**

Multimedia Mendelian Genetics Primer **www.vector.cshl.org/**

National Organization for Rare Disorders **www.nord-rdb.com/~orphan**

CHAPTER

5

# Extensions and Exceptions to Mendel's Laws

### 5.1 When Gene Expression Appears to Alter Mendelian Ratios

Phenotypic ratios of offspring are rarely as predictable and clear-cut as Mendel's experiments indicated. Allele characteristics and interactions create many variations on phenotypic themes.

### 5.2 Maternal Inheritance and Mitochondrial Genes

Cells house mitochondria, and mitochondria house their own tiny chromosomes, each complete with 37 genes, but with different alleles possible in different mitochondria. Traits encoded by mitochondrial genes pass only from mothers to offspring. Disorders resulting from mitochondrial gene mutations tend to affect muscles and energy level.

### 5.3 Linkage

The pairs of traits that Mendel studied in peas to derive the laws of segregation and independent assortment travel on different chromosomes. Linked genes travel on the same chromosome. Whether or not genes are on the same chromosome determines the proportions of progeny classes. Genetic linkage is the basis of gene mapping, which locates genes on chromosomes.

The transmission of inherited traits is not always as straightforward as Mendel's pea experiments indicated. Variations in gene expression and in allele and gene interactions can alter expected phenotypic ratios. In some situations, Mendel's laws do not apply, such as when genes are in mitochondria or are "linked" on the same chromosome. This chapter examines extensions and exceptions to Mendel's laws.

## 5.1 When Gene Expression Appears to Alter Mendelian Ratios

Mendel's crosses yielded offspring that were easily distinguished from each other. A pea is either yellow or green, round or wrinkled; a plant is either tall or short. For some characteristics, though, offspring classes do not occur in the proportions that Punnett squares or probabilities predict. In other cases, transmission patterns of a visible trait are not consistent with the mode of inheritance. In these instances, Mendel's laws operate, and the underlying genotypic ratios persist, but either the nature of the phenotype or influences from other genes or the environment alter phenotypic ratios. Following are several circumstances that appear to contradict Mendel's laws—although the laws actually still apply.

### Lethal Allele Combinations

"Lethal" means deadly, so any genotype (allele combination) that causes the death of an individual is literally lethal. In a population and evolutionary sense, though, a lethal genotype has a more specific meaning—it causes death before the individual can reproduce, which prevents passage of his or her genes to the next generation. Huntington disease, which is ultimately fatal and begins in middle age, is lethal to the individual, but not lethal in a population sense because it does not typically cause symptoms or death until after a person has had children.

In organisms used in experiments, such as fruit flies, pea plants, or mice, lethal allele combinations remove an expected progeny class following a specific cross. For example, a cross of two heterozygotes in which the homozygous recessive progeny die as embryos would leave only heterozygotes and homozygous dominant individuals as survivors.

In humans, early-acting lethal alleles cause many spontaneous abortions (technically called "miscarriages" if they occur after the embryonic period). When a man and woman each carry a recessive lethal allele for the same gene, each pregnancy has a 25 percent chance of spontaneously aborting—this represents the homozygous recessive class. Sometimes a double dose of a dominant allele is lethal. This is the case for Mexican hairless dogs. Inheriting one dominant allele confers the coveted hairlessness trait, but inheriting two dominant alleles is lethal to the unlucky embryo. Breeders cross hairless to hairy ("powderpuff") dogs, rather than hairless to hairless, to avoid losing the lethal homozygous dominant class—a quarter of the pups, as figure 5.1 indicates.

### Multiple Alleles

A person has two alleles for any autosomal gene (one allele on each homolog), but a gene can exist in more than two allelic forms in a population because it can mutate in many ways. Different allele combinations can produce variations in phenotype.

Correlations between genotypes and phenotypes would enable physicians to pre-

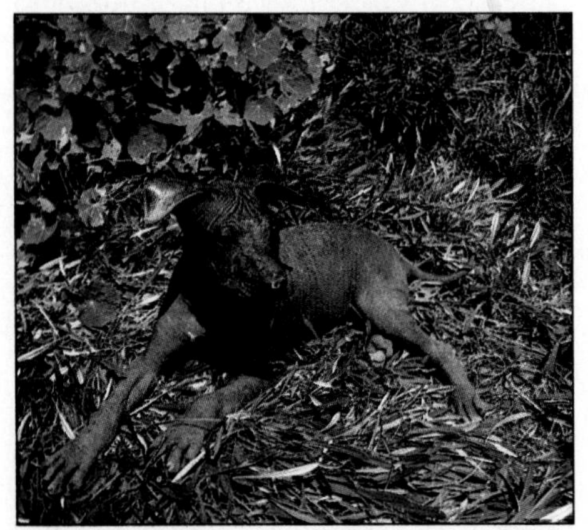

a.

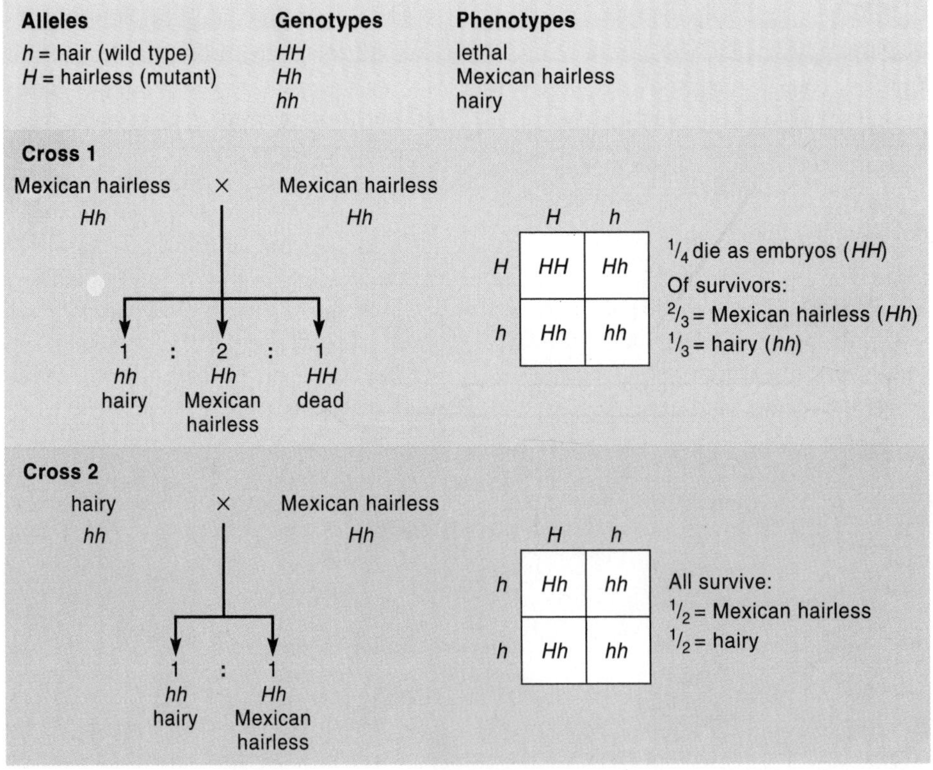


b.

**figure 5.1**

**Lethal alleles.** (*a*) This Mexican hairless dog has inherited a dominant allele that makes it hairless. Inheriting two such dominant alleles is lethal during embryonic development. (*b*) Breeders cross Mexican hairless dogs to hairy ("powderpuff") dogs to avoid dead embryos and stillbirths that represent the *HH* genotypic class.

dict the course of a particular illness following a genetic test at birth, or even earlier. Such information would also be useful for determining the prognosis for adult-onset genetic disorders. Unfortunately, many attempts to correlate phenotype to genotype have encountered unexpected complexity, usually because other genes exert effects on a disease-causing gene's expression.

One disorder for which genotype does predict phenotype has a cumbersome name—very-long-chain acyl-CoA dehydrogenase deficiency. This inborn error of metabolism causes deficiency of an enzyme produced in the mitochondria that helps provide energy to the heart and skeletal (voluntary) muscles. If a person's two alleles are so altered that no enzyme is made, symptoms of heart failure and an enlarged liver begin in infancy, causing death in early childhood. But several other alleles permit cells to manufacture some enzyme, lessening the severity of the symptoms. Combinations of these other alleles thus cause a milder childhood form of the illness, and an even milder adult form. Knowing whether their newborn has the severe or the mild form tells parents what to expect.

Phenylketonuria (PKU) is another inborn error of metabolism where phenotype/genotype correlations are useful in planning care. When the implicated enzyme is completely absent, the individual is profoundly mentally retarded, the result of the amino acid phenylalanine that the enzyme normally breaks down building up in brain cells. However, eating a special diet extremely low in phenylalanine from birth to at least eight years of age, and possibly much longer, can allow normal brain development. The more than 300 mutant alleles known for this gene combine to form four basic phenotypes: classic PKU with profound mental retardation; moderate PKU; mild PKU; and simply excreting excess of the amino acid in the urine. Knowing a child's allele combination at birth can give parents an idea of how strict the dietary regimen need be, and how long it must continue.

The existence of multiple alleles—several hundreds of them—has greatly complicated carrier testing for cystic fibrosis (CF), as well as the ability to predict phenotypes from genotypes. When the CF gene was discovered in 1989, researchers identified one mutant allele, called Δ*F508*, that causes about 70 percent of cases in many populations (see figure 2.1*b*). Within months, researchers were adding more alleles to the list, and finding that not all allele combinations cause the exact same symptoms.

People homozygous for Δ*F508* have a classic combination of severe symptoms, including frequent serious respiratory infections, very sticky mucus in the lungs, and poor weight gain due to insufficient pancreatic function. Soon after Δ*F508* was described, an allele was discovered that causes only an increased susceptibility to bronchitis and pneumonia; then another was found that causes only absence of the vas deferens! It was beginning to look as if CF is actually many different diseases, with all variations caused by a nonfunctional cell membrane protein that alters salt transport in and out of certain cells. Researchers eventually found that certain allele combinations are associated with different degrees of pancreatic function, but the lung symptoms do not correlate well to any particular genotype other than the Δ*F508* homozygote. Mendel's short and tall pea plants seem much simpler! Chapter 11, on mutation, offers more examples of genotype/phenotype correlations.

## Different Dominance Relationships

In complete dominance, one allele is expressed, while the other isn't. Alternatively, some genes show **incomplete dominance,** in which the heterozygous phenotype is intermediate between that of either homozygote. (Technically, this is a lack of dominance.)

In a sense, enzyme deficiencies in which a threshold level is necessary for health illustrate both complete and incomplete dominance—depending upon how one evaluates the phenotype. For example, on a whole-body level, Tay-Sachs disease displays complete dominance because the heterozygote (carrier) is as healthy as a homozygous dominant individual. However, if phenotype is based on enzyme level, then the heterozygote is indeed intermediate between the homozygous dominant (full enzyme level) and homozygous recessive (no enzyme).

A more obvious incompletely dominant trait in humans is hair curliness. The homozygous dominant condition is straight hair, and the homozygous recessive phenotype is curly hair. The heterozygote has wavy hair.

A classic example of incomplete dominance occurs in the snapdragon plant. A red-flowered plant of genotype *RR* crossed to a white-flowered *rr* plant can give rise to a *Rr* plant—which has pink flowers. This intermediate color is presumably due to an intermediate amount of pigment.

Familial hypercholesterolemia (FH) is another example of incomplete dominance in humans. A person with two disease-causing alleles lacks receptors on liver cells that take up cholesterol from the bloodstream. A person with one disease-causing allele has half the normal number of receptors. Someone with two normal alleles has the normal number of receptors. Figure 5.2 shows how measurement of plasma cholesterol reflects these three genotypes. The phenotypes parallel the number of receptors—those with two mutant alleles die as children of heart attacks, those with one mutant allele may suffer heart attacks in young adulthood, and those with two wild type alleles do not develop this inherited form of heart disease.

Different alleles that are both expressed in a heterozygote are **codominant.** The ABO blood group is based on the expression of codominant alleles. Blood types are determined by the patterns of cell surface molecules on red blood cells. Most of these molecules, called antigens, are proteins embedded in the cell membrane with attached sugars that poke out from the cell surface. People who belong to blood group A have an allele that encodes an enzyme that adds a certain final piece to a certain sugar. The final sugar is called antigen A. In people with blood group B, the allele and its encoded enzyme are slightly different, which adds a different piece to the sugar, producing antigen B. Blood group O reflects yet a third allele of this gene. It is missing just one DNA building block, but this change is sufficient to drastically change the encoded enzyme in a way that robs the sugar chain of its final piece (figure 5.3).

ABO blood types in the past have been described as variants of a gene called "*I*," although OMIM now abbreviates the designations. Table 5.1 shows the old and the new terminology. The older "*I*" system makes the codominance easier to understand. The three alleles are $I^A$, $I^B$, and *i*. People with blood type A have antigen A on the surfaces of their red blood cells, and may be of genotype $I^AI^A$ or $I^Ai$. People with blood type B

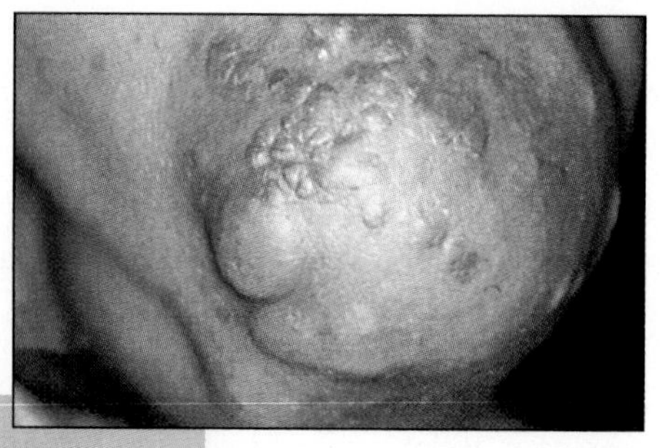

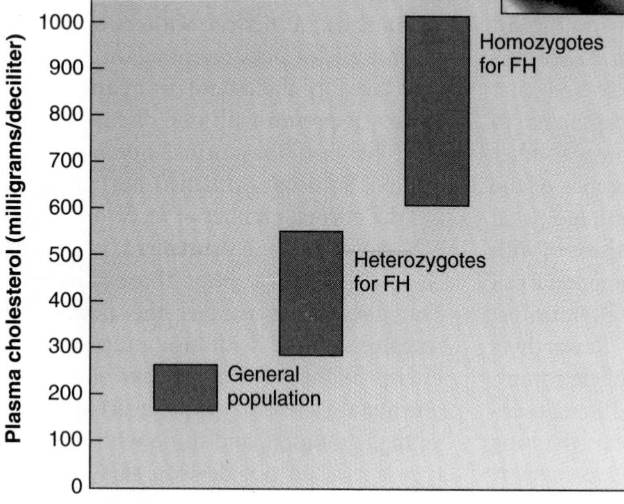


**figure 5.2**

**Incomplete dominance.** A heterozygote for familial hypercholesterolemia has approximately half the normal number of cell surface receptors in the liver for LDL cholesterol. A homozygous dominant individual, who has the severe form of FH, has liver cells that lack the receptors. The photograph shows cholesterol deposits on the elbow of an affected young man.

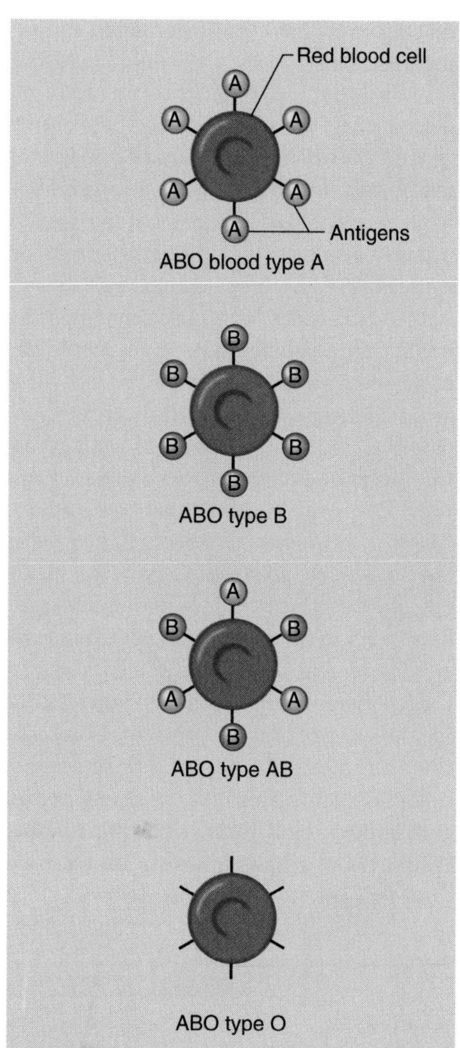


**figure 5.3**

**ABO blood types illustrate codominance.** ABO blood types are based on antigens on red blood cell surfaces. The size of the A and B antigens is greatly exaggerated.

**table 5.1**

**The ABO Blood Group**

| Genotypes | Phenotypes: Antigens on Surface | Phenotypes: ABO Blood Type |
|---|---|---|
| $I^AI^A$ | A | Type A |
| $I^Ai$ | A | |
| $I^BI^B$ | B | Type B |
| $I^Bi$ | B | |
| $I^AI^B$ | AB | Type AB |
| $ii$ | None | Type O |

have antigen B on their red blood cell surfaces, and may be of genotype $I^BI^B$ or $I^Bi$. People with the rare blood type AB have both antigens A and B on their cell surfaces, and are genotype $I^AI^B$. People with blood type O have neither antigen, and are genotype *ii*.

Television program plots, particularly soap operas, often misuse ABO blood type terminology, assuming that a child's ABO type must match that of a parent. This is not true, because a person with type A or B blood can be heterozygous. A person who is genotype $I^Ai$ and a person who is $I^Bi$ can have offspring together of any ABO genotype or phenotype. Figure 5.4 illustrates how this can happen.

## Epistasis—When One Gene Affects Expression of Another

Mendel's laws can appear to not operate when one gene masks or otherwise affects the expression of a different gene, a phenomenon called **epistasis.** (Do not confuse this with dominance relationships between alleles of the *same* gene.) The Bombay phenotype, for example, is a result of two interacting genes: the *I* and *H* genes. The relationship of these two genes affects the expression of the ABO blood type.

The normal *H* allele encodes an enzyme that inserts a sugar molecule, called antigen H, onto a particular glycoprotein on the surface of an immature red blood cell. The recessive *h* allele produces an inactive form of the enzyme that cannot insert the sugar. (The *H* gene's product is fucosyltransferase 1, and in OMIM the *H* gene is called FUT1.) The *A* and *B* antigens are attached to the *H* antigen. As long as at least one *H* allele is present, the ABO genotype

| Type B \ Type A | $I^A$ | $I^A$ |
|---|---|---|
| $I^B$ | $I^AI^B$ AB | $I^AI^B$ AB |
| $I^B$ | $I^AI^B$ AB | $I^AI^B$ AB |

| Type B \ Type A | $I^A$ | $i$ |
|---|---|---|
| $I^B$ | $I^AI^B$ AB | $I^Bi$ B |
| $I^B$ | $I^AI^B$ AB | $I^Bi$ B |

| Type B \ Type A | $I^A$ | $I^A$ |
|---|---|---|
| $I^B$ | $I^AI^B$ AB | $I^AI^B$ AB |
| $i$ | $I^Ai$ A | $I^Ai$ A |

| Type B \ Type A | $I^A$ | $i$ |
|---|---|---|
| $I^B$ | $I^AI^B$ AB | $I^Bi$ B |
| $i$ | $I^Ai$ A | $ii$ O |

## figure 5.4

**Codominance.** Even though the $I^A$ and $I^B$ alleles of the $I$ gene are codominant, they still follow Mendel's law of segregation. These Punnett squares follow the genotypes that could result by crossing a person with type A blood with a person with type B blood.

dictates the ABO blood type. However, in a person with genotype *hh*, there is no *H* antigen to bind to the *A* and *B* antigens, and they fall away. The person has blood type O based on phenotype (a blood test), but may have any ABO genotype.

Individuals with the *hh* genotype are very rare with one notable exception—residents of Reunion Island, which is in the Indian Ocean east of Madagascar. Apparently the settlers of this isolated island included at least one *hh* or *Hh* individual, and, with time and large families and some consanguinity, the allele spread in the population. Figure 5.5 illustrates one of the rare cases where an *hh* genotype obscured the ABO genotype, from the soap opera "General Hospital."

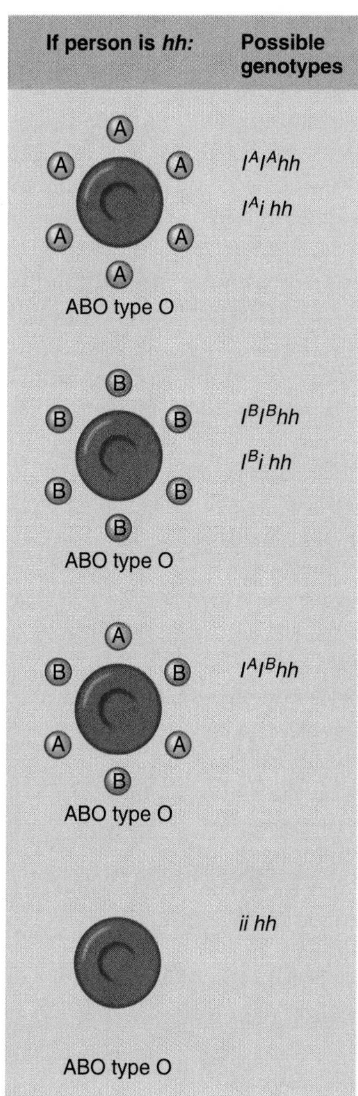


## figure 5.5

**Epistasis masks gene expression.** Once upon a time, Monica had a baby. Is the father Alan or Rick? Monica's blood type is A, Alan's is AB, and Rick's is O. The child's blood type is O. Considering only the ABO blood type, it looked as if Rick was the father, and Monica is $I^Ai$. But, after a long story, the baby was found to have the Bombay phenotype, with genotype *hh,* and Alan's family to have several inexplicable blood types. Monica and Alan turned out to be *Hh*, but Rick is *HH*. The baby—Alan Jr. or "AJ"—had a type O phenotype, but his genotype could be any of the ABO two-allele combinations. He can't be Rick's son, or he would not be genotype *hh*. The figure shows how a person who is *hh* can have any ABO genotype.

## Penetrance and Expressivity

The same allele combination can produce different degrees of a phenotype in different individuals, even siblings. Different expressions of the same genotype reflect the fact that a gene does not act alone. Nutrition, toxic exposures, other illnesses, and actions of other genes may influence expression of a gene.

Cystic fibrosis illustrates how other genes can modify a gene's expression. Consider two individuals who are homozygous for the *ΔF508* allele, the most severe genotype for this illness. One is much sicker than the other, possibly because she also inherited genes predisposing her to develop asthma and respiratory allergies. Understanding CF has proven much more difficult than anyone expected, both because of the large number of alleles, and because of possible influences from other genes. One research group is studying the interaction between CF and another very common inherited disease, hemochromatosis. This "iron overload" disease deposits iron in several organs, including the pancreas. Being a heterozygote for hemochromatosis could impair pancreas function in a way that worsens CF symptoms.

Many Mendelian traits and illnesses do have distinctive phenotypes, despite all of these influences. The terms penetrance and expressivity are used to describe degrees of gene expression. Penetrance refers to the all-or-none expression of a genotype; expressivity refers to severity or extent.

Most disease-causing allele combinations are **completely penetrant,** which means that everyone who inherits the combination has some symptoms. A genotype is **incompletely penetrant** if some individuals do not express the phenotype, or have no symptoms. Polydactyly, having extra fingers or toes, is incompletely penetrant (see figure 1.3). Some people who inherit the dominant allele have more than five digits on a hand or foot, yet others who are known to have the allele (because they have an affected parent and child) have the normal number of fingers and toes. The penetrance of a gene is described numerically. If 80 of 100 people who have inherited the dominant polydactyly allele have extra digits, the genotype is 80 percent penetrant.

A phenotype is **variably expressive** if the symptoms vary in intensity in different people. One person with polydactyly might have an extra digit on both hands and a foot; another might have two extra digits on both hands and both feet; a third person might have just one extra fingertip.

It is hard to imagine how other genes or the environment can influence the numbers of fingers or toes. In the case of familial hypercholesterolemia, variable expressivity more clearly reflects influences of other genes and the environment (see figure 5.2). Heterozygotes tend to develop heart disease due to high serum cholesterol in middle adulthood. The age of onset of symptoms depends on diet and exercise habits, use of cholesterol-lowering drugs, and on other genes. About 60 percent of such individuals die prematurely due to heart disease—meaning that 40 percent can slow the progression of the illness.

## Pleiotropy—One Gene, Many Effects

A Mendelian disorder with many symptoms, or a gene that controls several functions or has more than one effect, is termed **pleiotropic.** Such conditions can be difficult to trace through families because people with different subsets of symptoms may appear to have different disorders. This was the case for porphyria variegata, an autosomal dominant, pleiotropic, inborn error of metabolism. The disease affected several members of the royal families of Europe.

King George III ruled England during the American Revolution (figure 5.6). At age 50, he first experienced abdominal pain and constipation, followed by weak limbs, fever, a fast pulse, hoarseness, and dark red urine. Next, nervous system symptoms began, including insomnia, headaches, visual problems, restlessness, delirium, convulsions, and stupor. His confused and racing thoughts, combined with actions such as ripping off his wig and running about

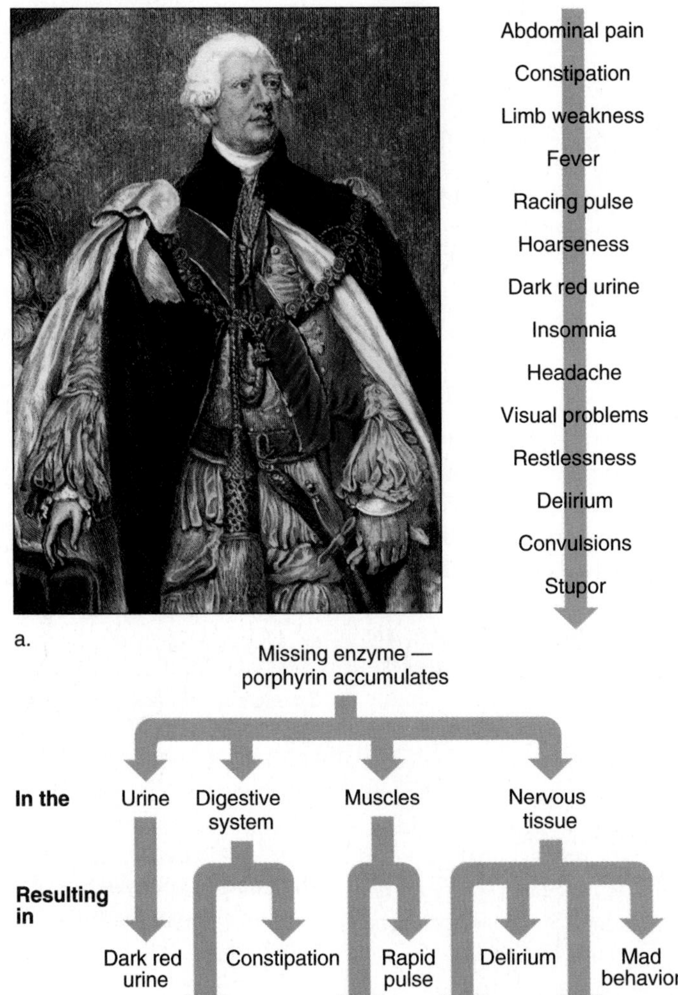


**figure 5.6**

**Pleiotropy.** King George III (*a*) suffered from the autosomal dominant disorder porphyria variegata—and so did several other family members. Because of pleiotropy, the family's varied illnesses and quirks appeared to be different, unrelated disorders. In King George, symptoms appeared every few years in a particular order (*b*).

# *In Their Own Words*

# Alkaptonuria

Alkaptonuria was one of the first inherited illnesses to be identified. Ironically, many people who have it probably never realize that the symptoms arise from the same metabolic abnormality (figure 1). Here, Pat Wright describes her experience with the condition.

In my case, alkaptonuria symptoms started when I was 15 years old and struggling to sit all day in high school classes. An osteopath manipulated my spine and treated the back spasms with heat and medication, which enabled me to weather the frequent flare-ups throughout high school and college. Of course we did not relate my back problems to the known metabolic disorder until many years later. These back problems persisted through five pregnancies and 26 years of teaching special education. My seriously degenerated spine, coupled with the disappearance of the cartilage in my left knee, forced me to retire on disability at the age of 57.

The first symptoms of alkaptonuria, however, began even earlier than high school. When I was a baby, my parents noticed that my diapers turned brown if not washed immediately, and even then they became stained. The doctor sent a wet diaper to a teaching hospital, and they told my parents I had a "harmless" metabolic disorder.

Fast forward 60 years.

In February 1997 I had a total knee replacement, and the surgeon was amazed to find the joint surrounded by blackened cartilage. It was the first time he had ever seen such a thing, after years of surgery. This rediscovery of alkaptonuria has answered many questions that arose over the years. It explains the dark blue-gray ears, the degenerated spine, knees, and shoulder joints. It may also explain the steadily increasing aortic stenosis and hearing loss, and perhaps the kidney stones and gallstones.

I was told I had degenerative disc disease, fibromyalgia, osteoarthritis, or degenerative joint disease. Over the years I have tried every known treatment for these conditions, with little or no success. I do recommend a regular exercise routine. I attended arthritis water aerobics classes and worked with a personal trainer for three years before and after the knee replacement, to improve my strength, range of motion and mobility. I think this was vital, because, though I could not do much about the condition of the joints, I could improve my muscle tone, strength, and endurance.

In January of 2000, I had my aortic valve replaced so that I might withstand further joint surgery. The following April I began aquatic physical therapy to both rehabilitate the heart and ready my body for a hip replacement in July. On June 15, 2001, I had a total left shoulder replacement and I returned to an aquatic therapy maintenance program to help with mobility and pain. Three more major joints to go! The left hip is complaining the loudest.

Pat Wright

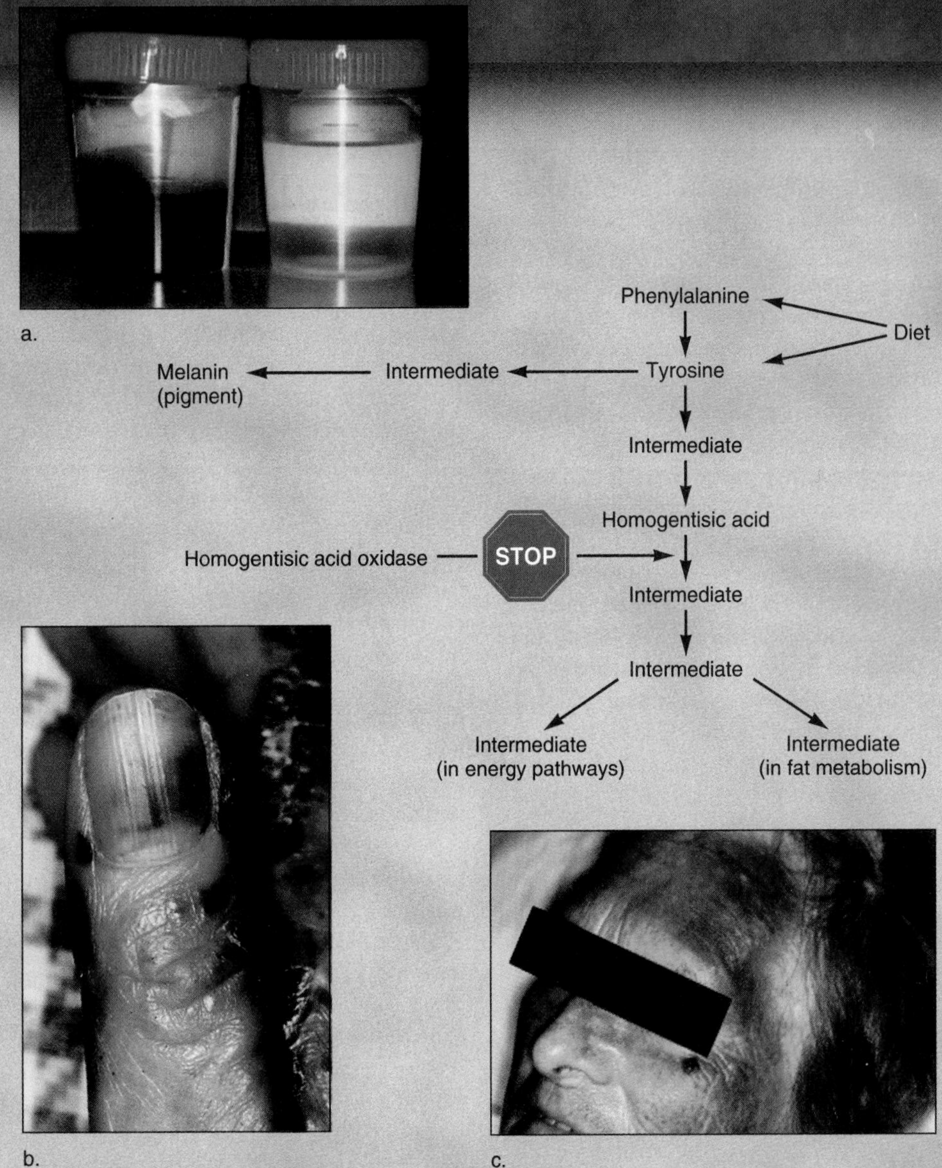


**figure 1**

**Pleiotropy in alkaptonuria.** Alkaptonuria was the first recognized inborn error of metabolism described by Archibald Garrod in 1902. Deficiency of the enzyme homogentisic acid oxidase, discovered in 1958, leads to buildup of melanin pigment in skin (*a*), in urine when it is allowed to stand for several hours (*b*), and in nails and cartilage (*c*). The disorder is inherited as an autosomal recessive trait, and the gene is on chromosome 3. Researchers discovered a form of alkaptonuria in lab mice whose cages needed changing—the wood shavings, soaked in urine and left too long, turned blue-black!

naked while at the peak of a fever, convinced court observers that the king was mad. Just as Parliament was debating his ability to rule, he mysteriously recovered.

But George's plight was far from over. He relapsed 13 years later, then again three years after that. Always the symptoms appeared in the same order, beginning with abdominal pain, fever, and weakness, and progressing to nervous system symptoms. Finally, an attack in 1811 placed George in a prolonged stupor, and the Prince of Wales dethroned him. George III lived for several more years, experiencing further episodes.

In George III's time, doctors were permitted to do very little to the royal body, and they simply made their diagnoses based on what the king told them. Twentieth-century researchers found that porphyria variegata caused George's red urine. Because of the absence of an enzyme, a part of the blood pigment hemoglobin called a porphyrin ring is routed into the urine instead of being broken down and metabolized by cells. Porphyrin builds up and attacks the nervous system, causing many of the symptoms. Examination of physicians' reports on George's relatives—easy to obtain for a royal family—showed that several of them had symptoms as well. Before anyone realized that porphyria variegata is pleiotropic, the royal disorder appeared to be several different illnesses. Today, porphyria variegata remains rare, and people with it are often misdiagnosed as having a seizure disorder. Unfortunately, some seizure medications and anesthetics worsen the symptoms.

Pleiotropy occurs when a single protein affects different parts of the body or participates in more than one type of biochemical reaction. Consider Marfan syndrome, an autosomal dominant defect in an elastic connective tissue protein called fibrillin. The protein is abundant in the lens of the eye, in the aorta (the largest artery in the body, leading from the heart), and in the bones of the limbs, fingers, and ribs. Once researchers knew this, the Marfan syndrome symptoms of lens dislocation, long limbs, spindly fingers, and a caved-in chest made sense. The most serious symptom is a life-threatening weakening in the aortic wall, sometimes causing the vessel to suddenly burst. However, if the weakening is detected early, a synthetic graft can be inserted to replace the weakened section of artery wall.

Pleiotropy can be confusing even to medical doctors. *The New England Journal of Medicine* ran a contest to see if readers could identify the cause of three seemingly unrelated symptoms, illustrated in the figure in the "In Their Own Words" box. These symptoms are characteristic of the inborn error of metabolism alkaptonuria.

## Phenocopies—When It's Not in the Genes

An environmentally caused trait that appears to be inherited is called a **phenocopy.** Such a trait can either produce symptoms that resemble a Mendelian disorder's symptoms or mimic inheritance patterns by occurring in certain relatives. For example, the limb birth defect caused by the drug thalidomide, discussed in chapter 3, is a phenocopy of the inherited illness phocomelia. Physicians realized that an environmental disaster had occurred when they began seeing many children born with what looked like the very rare phocomelia. A birth defect caused by exposure to a teratogen was a more likely explanation than a sudden increase in incidence of a rare inherited disease.

An infection can be a phenocopy. Children who have AIDS may have parents who also have the disease, but these children acquired AIDS by viral infection, not by inheriting a gene. Similarly, infection or injury of the pancreas can damage insulin metabolism in a way that mimics diabetes, which is a multifactorial disorder. A phenocopy caused by a highly contagious infection can seem to be inherited if it affects more than one family member.

True phenocopies are rare. However, common symptoms may seem to resemble those of an inherited condition until medical tests rule heredity out. For example, an underweight child who has frequent colds may show some signs of cystic fibrosis, but may instead suffer from malnutrition. A negative test for several CF alleles would alert a physician to look for another cause.

## Genetic Heterogeneity—More than One Way to Inherit a Trait

Different genes can produce the same phenotype, a phenomenon called **genetic heterogeneity.** This redundancy of function can make it appear that Mendel's laws are not operating. For example, 132 forms of deafness are transmitted as autosomal recessive traits. If a man who is heterozygous for a particular type of deafness gene on one chromosome has a child with a woman who is heterozygous for another type of deafness gene on a different chromosome, then that child faces only the same risk as anyone in the general population of inheriting either form of deafness. He or she *doesn't* face the 25 percent risk that Mendel's law predicts for a monohybrid cross because the parents are heterozygous for *different* genes whose encoded proteins affect hearing. Cleft palate and albinism are other traits that are genetically heterogeneic—that is, different genes can cause them.

Genetic heterogeneity can occur when genes encode different enzymes that participate in the same biochemical pathway. For example, 11 biochemical reactions lead to blood clot formation. Clotting disorders may result from abnormalities in the genes that specify any of these enzymes, leading to several types of bleeding disorders.

Similarly, the various forms of retinitis pigmentosa (RP) reflect mutations in different genes. RP is a gradual degeneration of the rod cells that provide black and white vision. One form of the disease affects rhodopsin, the visual pigment that captures light energy. Other forms of the disease target the enzymes necessary to manufacture rhodopsin. Yet others affect proteins in the membranes of nerve cells that send visual information to the brain for interpretation. At least 14 different genes are known to cause RP, and the disease can have an autosomal dominant, autosomal recessive, or an X-linked (on the X chromosome) mode of inheritance.

Table 5.2 summarizes several of the phenomena that appear to alter Mendelian inheritance.

Gregor Mendel derived the two laws of inheritance working with traits conferred by genes located on different chromosomes in the nucleus. When genes do not conform to these conditions, however, the associated traits may not appear in Mendelian ratios. The remainder of this chapter considers two types of gene transmission that do not fulfill the requirements for Mendelian inheritance.

**table 5.2**

**Factors That Alter Mendelian Phenotypic Ratios**

| Phenomenon | Effect on Phenotype | Example |
|---|---|---|
| Lethal alleles | A phenotypic class dies very early in development | Spontaneous abortion |
| Multiple alleles | Produces many variants or degrees of a phenotype | Cystic fibrosis |
| Incomplete dominance | A heterozygote's phenotype is intermediate between those of two homozygotes | Familial hypercholesterolemia |
| Codominance | A heterozygote's phenotype is distinct from and not intermediate between those of the two homozygotes | ABO blood types |
| Epistasis | One gene masks or otherwise affects another's phenotype | Bombay phenotype |
| Penetrance | Some individuals with a particular genotype do not have the associated phenotype | Polydactyly |
| Expressivity | A genotype is associated with a phenotype of varying intensity | Polydactyly |
| Pleiotropy | The phenotype includes many symptoms, with different subsets in different individuals | Porphyria variegata |
| Phenocopy | An environmentally caused condition has symptoms and a recurrence pattern similar to those of a known inherited trait | Infection |
| Genetic heterogeneity | Different genotypes are associated with the same phenotype | Hearing impairment |

**KEY CONCEPTS**

A number of factors can appear to disrupt Mendelian ratios. A lethal allele combination is never seen as a progeny class. • Different allele combinations may produce different phenotypes. In incomplete dominance, the heterozygote phenotype is intermediate between those of the homozygotes, and in codominance, two different alleles are each expressed. • In epistasis, one gene influences expression of another. • Genotypes vary in penetrance (percent of individuals affected) and expressivity (intensity of symptoms) of the phenotype.

A gene with more than one expression is pleiotropic. • A trait caused by the environment but resembling a known genetic trait or occurring in certain family members is a phenocopy. • Genetic heterogeneity occurs when different genes cause the same phenotype.

## 5.2 Maternal Inheritance and Mitochondrial Genes

The basis of the law of segregation is that both parents contribute genes equally to offspring. This is not the case for genes in mitochondria. Recall from chapter 2 that mitochondria are the organelles that house the biochemical reactions that provide cellular energy. Mitochondria in human cells contain several copies of a "mini-chromosome" that carries just 37 genes. Sometimes it is called the twenty-fifth chromosome.

The inheritance patterns and mutation rates for mitochondrial genes differ from those for genes in the nucleus. Mitochondrial genes are maternally inherited. They are passed only from an individual's mother because sperm almost never contribute mitochondria when they fertilize an oocyte. In the rare instances where mitochondria from sperm enter an oocyte, they are usually selectively destroyed early in development. Pedigrees that follow mitochondrial genes show a woman passing the trait to all her children, while a male cannot pass the trait to any of his children. The pedigree in figure 5.7 illustrates such a maternal inheritance pattern indicating transmission of a mutation in a mitochondrial gene.

Unlike DNA in the nucleus, mitochondrial DNA does not cross over. Mitochondrial DNA also mutates faster than nuclear DNA because it lacks DNA repair enzymes (discussed in chapter 9), and the mitochondrion is the site of the energy

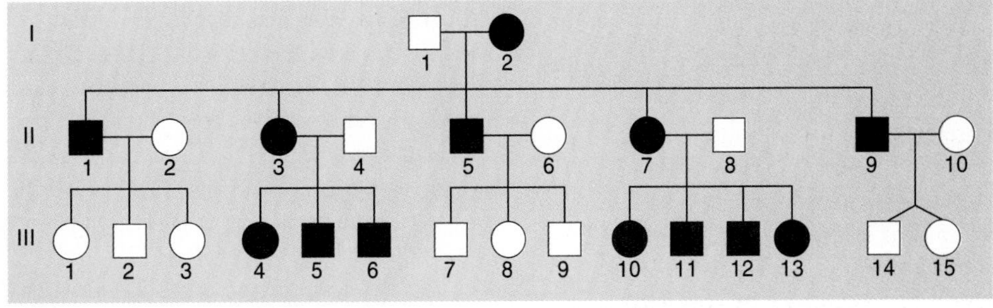


**figure 5.7**

**Inheritance of mitochondrial genes.** Mothers pass mitochondrial genes to all offspring. Fathers do not transmit mitochondrial genes, because sperm do not contribute mitochondria to fertilized ova.

reactions that produce oxygen free radicals that damage DNA. Also unlike nuclear DNA, mitochondrial DNA is not wrapped in histone proteins, nor are genes "interrupted" by DNA sequences that do not encode protein called introns. Finally, inheritance of mitochondrial genes differs from inheritance of nuclear genes simply because a cell has one nucleus but many mitochondria—and each mitochondrion harbors several copies of its "chromosome" (figure 5.8 and table 5.3). Mitochondria with different alleles for the same gene can reside in the same cell.

**table 5.3**

| Features of Mitochondrial DNA |
|---|
| No crossing over |
| No DNA repair |
| Maternal inheritance |
| Many copies per mitochondrion and per cell |
| High exposure to oxygen free radicals |
| No histones |
| No introns |

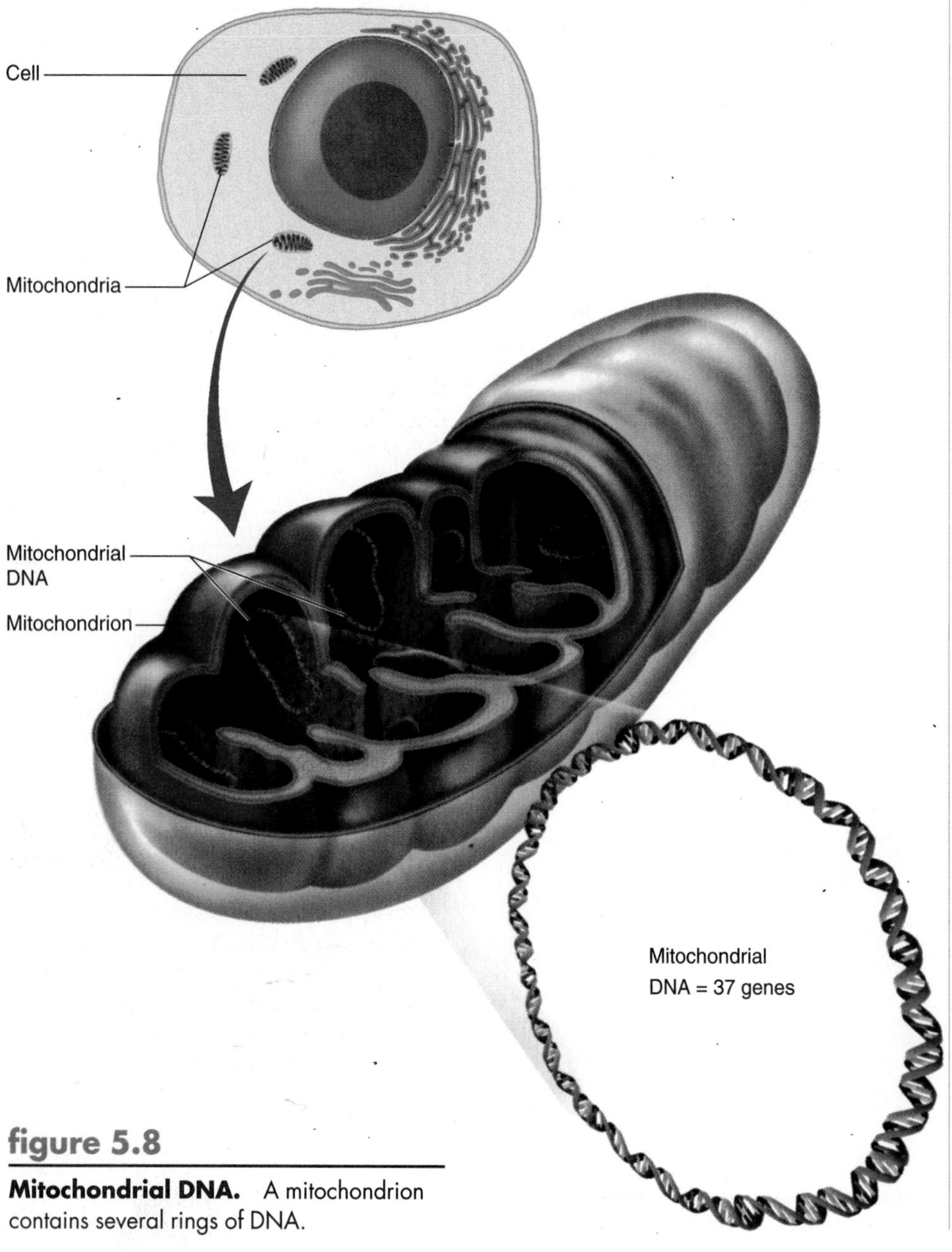


**figure 5.8**

**Mitochondrial DNA.** A mitochondrion contains several rings of DNA.

## Mitochondrial Disorders

Mitochondrial genes encode proteins that participate in protein synthesis and energy production. Twenty-four of the 37 genes encode RNA molecules (22 transfer RNAs and two ribosomal RNAs) that help assemble proteins. The other 13 mitochondrial genes encode proteins that function in cellular respiration, the biochemical reactions that use energy from digested nutrients to produce ATP, the biological energy molecule.

In diseases resulting from mutations in mitochondrial genes, symptoms arise from tissues whose cells have many mitochondria, such as skeletal muscle. It isn't surprising that a major symptom is often great fatigue. Inherited illnesses called mitochondrial myopathies, for example, produce weak and flaccid muscles and intolerance to exercise. Skeletal muscle fibers appear red and ragged, their abundant abnormal mitochondria visible beneath the cell membrane.

A defect in an energy-related gene can produce symptoms other than fatigue. This is the case for Leber's hereditary optic neuropathy (LHON), which impairs vision. First described in 1871, with its maternal transmission noted, LHON was not associated with a mitochondrial mutation that impairs cellular energy reactions until 1988. Symptoms of LHON usually begin in early adulthood with a loss of central vision. Eyesight worsens and color vision vanishes as the central portion of the optic nerve degenerates.

A mutation in a mitochondrial gene that encodes a tRNA or rRNA can be devastating because it impairs the cell's ability to manufacture proteins. Consider what happened to Linda Schneider, a once active and articulate dental hygienist and travel agent. In her forties, Linda gradually began to slow down at work. She heard a buzzing in her ears and developed difficulty talking and walking. Then her memory would fade in and out, she would become lost easily in familiar places, and her conversation would not make sense. Her condition worsened, and she developed diabetes, seizures, and pneumonia and became deaf and demented. After many false diagnoses, which included stroke, Alzheimer disease, and a prion disorder, she was finally found to have a mitochondrial illness called MELAS, which stands for "mitochondrial myopathy encephalopathy lactic acidosis syndrome." Linda died. Her son and daughter will likely develop the condition because of the transmission pattern of a mitochondrial mutation.

A controversial technique called ooplasmic transfer attempts to enable a couple to avoid transmitting a mitochondrial disorder. In the procedure, mitochondria from a healthy woman's oocyte are injected into the oocyte of a woman who is infertile. Then, the bolstered oocyte is fertilized in a laboratory dish by the infertile woman's partner's sperm, and the zygote implanted in her uterus. So far, 30 children have been born from the technique, apparently free of mitochondrial disease. Headlines sensationalized the fact that technically the children each have three parents and were born as a result of genetic manipulation.

## Heteroplasmy Complicates Mitochondrial Inheritance

The fact that a cell contains many mitochondria makes possible a condition called **heteroplasmy,** in which a particular mutation may be present in some mitochondrial chromosomes, but not others (figure 5.9). At each cell division, the mitochondria are distributed at random into daughter cells. Because of the laws of probability, over time, the chromosomes within a mitochondrion tend to be all wild type or all mutant for any particular gene.

Heteroplasmy has several consequences for the inheritance of mitochondrial phenotypes. Expressivity may vary widely among siblings, depending upon how many mutation-bearing mitochondria were in the oocyte that became each brother or sister. Severity of symptoms is also influenced by which tissues have cells whose mitochondria bear the mutation. This is the case for a family with Leigh syndrome, which affects the enzyme that directly produces ATP. Two boys died of the severe form of the disorder because the brain regions that control movement rapidly degenerated. Another child was blind and had central nervous system degeneration. Several relatives, however, suffered only mild impairment of their peripheral vision. The more severely affected family members had more brain cells that received the mutation-bearing mitochondria.

The most severe mitochondrial illnesses are heteroplasmic. This is because homoplasmy—all mitochondria bearing the mutant gene—too severely impairs protein synthesis or energy production for embryonic development to complete. Often, severe heteroplasmic mitochondrial disorders do not produce symptoms until adulthood because it takes many cell divisions, and therefore much time, for a cell to receive enough mitochondria bearing mutant genes to cause symptoms. LHON usually does not affect vision until adulthood for this reason.

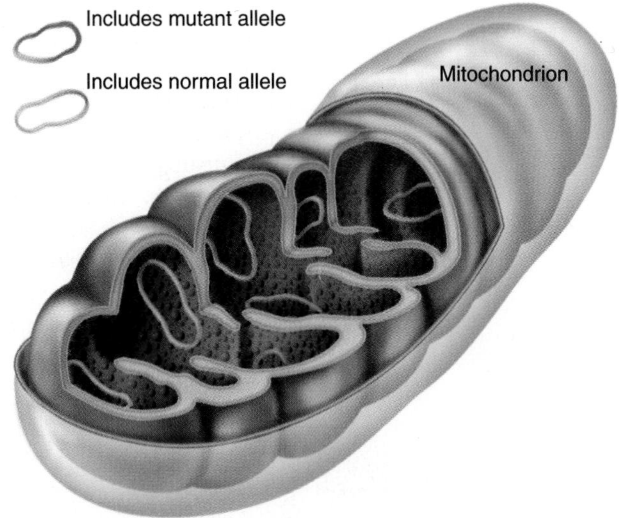


**figure 5.9**

**Heteroplasmy.** Mitochondrial "chromosomes" can have different alleles.

## Mitochondrial DNA Studies Clarify the Past

Interest in mitochondrial DNA extends beyond the medical. Mitochondrial DNA provides a powerful forensic tool used to link suspects to crimes, identify war dead, and support or challenge historical records. The technology, for example, identified the son of Marie Antoinette and Louis XVI, who supposedly died in prison at age 10. In 1845, the boy was given a royal burial, but some people thought he was an imposter. The boy's heart had been stolen at the autopsy, and through a series of bizarre events, wound up, dried out, in the possession of the royal family. Recently, researchers compared mitochondrial DNA sequences from cells in the boy's heart to corresponding sequences in heart and hair cells from Marie Antoinette (her decapitated body identified by her fancy underwear), two of her sisters, and the still-living Queen Anne of Romania and her brother. The genetic evidence showed that the unfortunate boy was indeed the prince, Louis XVII.

Mitochondrial DNA comparisons have solved other historical mysteries (see Reading 9.1). More recently, comparison of mitochondrial DNA sequences was the only type of DNA fingerprinting that could provide information from hard remains, such as teeth and bone bits, that survived the crushing inferno of the World Trade Center site in the weeks after September 11, 2001.

### Key Concepts

Mitochondrial genes are maternally inherited and mutate rapidly. A cell contains many mitochondria, which have multiple copies of the mitochondrial genome. • Mitochondrial genes encode RNAs that function generally in protein synthesis, or proteins that are part of energy metabolism. • In heteroplasmy, cells contain mitochondria that have different alleles of a particular gene.

## 5.3 Linkage

Most of the traits that Mendel studied in pea plants were conferred by genes on different chromosomes. (Some were actually at opposite ends of the same chromosome.) When genes are located close to each other on the same chromosome, they usually do not separate during meiosis. Instead, they are packaged into the same gametes (figure 5.10). **Linkage** refers to the transmission of genes on the same chromosome. Linked genes do not assort independently and do not result in the predicted Mendelian ratios for crosses tracking two or more genes. Understanding linkage has been critical in identifying disease-causing genes.

### Linkage Was Discovered in Pea Plants

The unexpected ratios indicating linkage were first observed by William Bateson and R. C. Punnett in the early 1900s, again in pea plants. Bateson and Punnett crossed true-breeding plants with purple flowers and long pollen grains (genotype *PPLL*) to true-breeding plants with red flowers and round pollen rains (genotype *ppll*). The plants in the next generation, of genotype *PpLl*, were then crossed among each other. But this dihybrid cross did not yield the expected 9:3:3:1 phenotypic ratio that Mendel's second law predicts (figure 5.11).

Bateson and Punnett noticed that two types of third generation peas—those with the parental phenotypes *P_L_* and *ppll*—were more abundant than predicted, while the other two progeny classes—*ppL_* and *P_ll*—were less common (the blank indicates that the allele can be dominant or re-

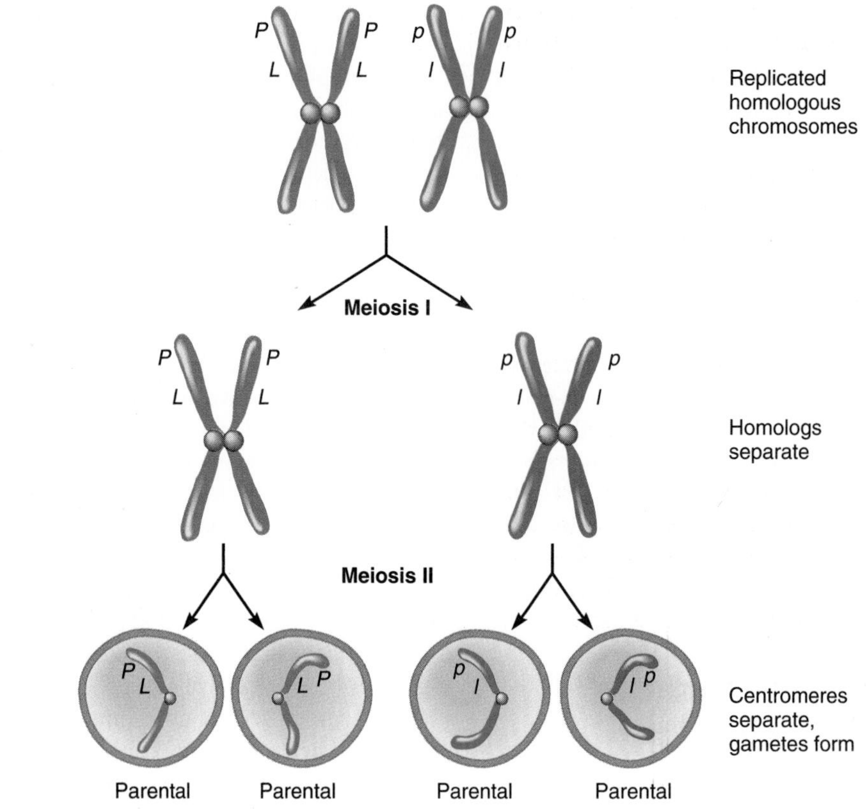


**figure 5.10**

**Inheritance of linked genes.** Genes that are linked closely to one another on the same chromosome are usually inherited together when that chromosome is packaged into a gamete.

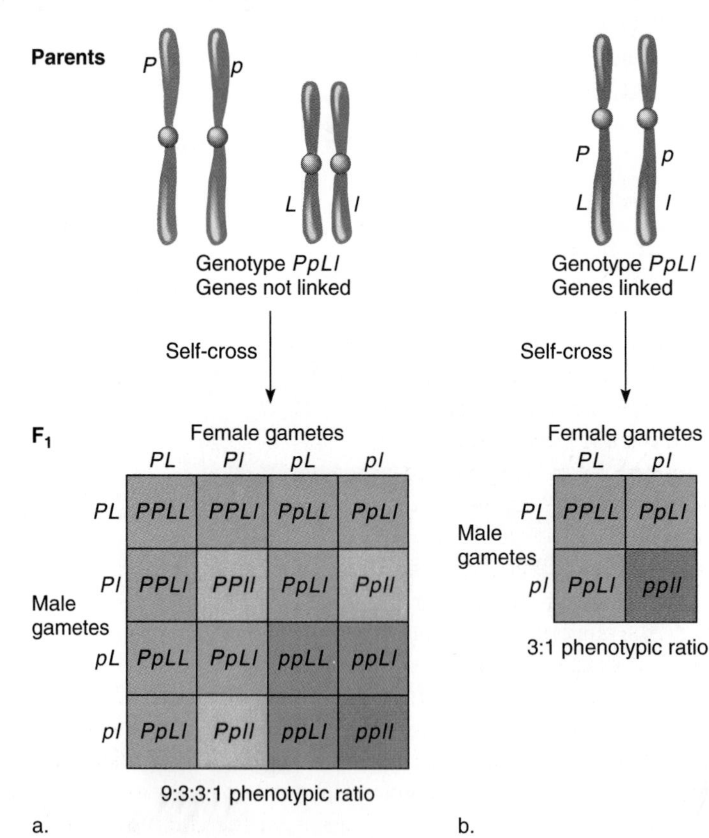


**figure 5.11**

**Expected results of a dihybrid cross.** (*a*) When genes are not linked, they assort independently. The gametes then represent all possible allele combinations. The expected phenotypic ratio of a dihybrid cross would be 9:3:3:1. (*b*) If genes are linked on the same chromosome, only two allele combinations are expected in the gametes. The phenotypic ratio would be 3:1, the same as for a monohybrid cross.

cessive). The more prevalent parental allele combinations, Bateson and Punnett hypothesized, could reflect genes that are transmitted on the same chromosome and that therefore do not separate during meiosis (see figure 5.10). The two less-common offspring classes could also be explained by a meiotic event—crossing over. Recall that this is an exchange between homologs that mixes up maternal and paternal gene combinations without disturbing the sequence of genes on the chromosome (figure 5.12).

Progeny that exhibit this mixing of maternal and paternal alleles on a single chromosome are called **recombinant.** *Parental* and *recombinant* are relative terms. Had the parents in Bateson and Punnett's crosses been of genotypes *ppL_* and *P_ll,* then *P_L_* and *ppll* would be recombinant rather than parental classes.

Two other terms are used to describe the configurations of linked genes in dihybrids. Consider a pea plant with genotype *PpLl.* These alleles can be arranged on the chromosomes in either of two ways. If the two dominant alleles are on one chromosome and the two recessive alleles are transmitted on the other, the genes are in "coupling." In the opposite configuration the genes are in "repulsion" (figure 5.13). Whether alleles in a dihybrid are in coupling or repulsion is important in distinguishing recombinant from parental progeny classes in specific crosses.

## Linkage Maps 

As Bateson and Punnett were learning about linkage from pea crosses, geneticist Thomas Hunt Morgan and his coworkers at Columbia University were observing the results of crosses in the fruit fly *Drosophila melanogaster.* These researchers compared progeny class sizes to assess whether various combinations of two traits were linked. They soon realized that the pairs of traits fell into four groups. Within each group, crossed dihybrids did not produce offspring according to the proportions Mendel's second law predicts. Not coincidentally, the number of these linkage groups—four—is exactly the number of chromosome pairs in the fly. The traits fell into four groups based on progeny class proportions because the genes controlling traits that are inherited together are transmitted on the same chromosome.

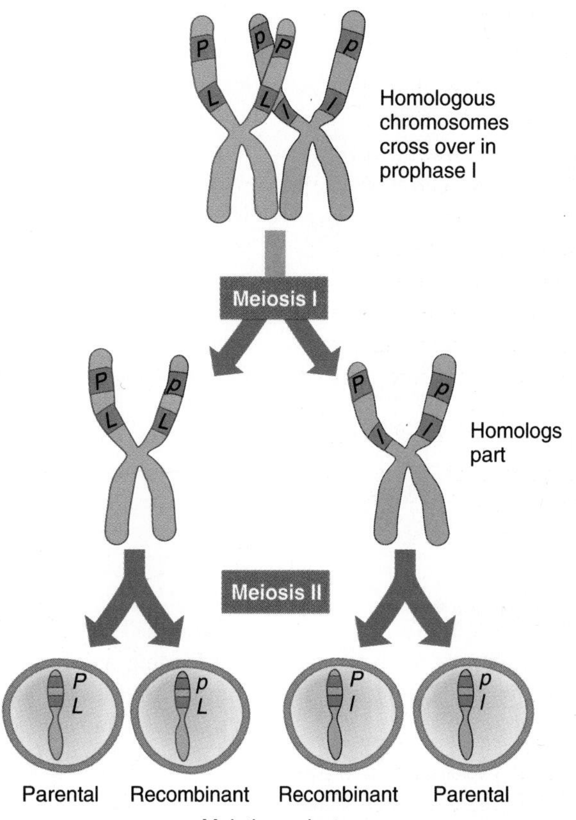


**figure 5.12**

**Crossing over disrupts linkage.** The linkage between two genes can be interrupted if the chromosome they are located on crosses over with its homolog at a point between the two genes. Crossing over packages recombinant arrangements of the genes into gametes.

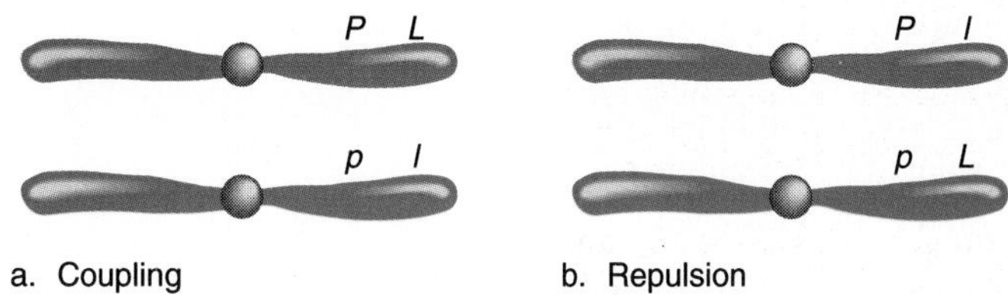


**figure 5.13**

**Allele configuration is important.** Parental chromosomes can be distinguished from recombinant chromosomes only if the allele configuration of the two genes is known—they are either in coupling (*a*) or in repulsion (*b*).

Morgan wondered why the size of the recombinant classes varied depending upon which genes were studied. Might the differences reflect the physical relationship of the genes on the chromosome? Exploration of this idea fell to an undergraduate, Alfred Sturtevant. In 1911, Sturtevant developed a theory and technique that would profoundly affect the fledgling genetics of his day and medical genetics today. He proposed that the

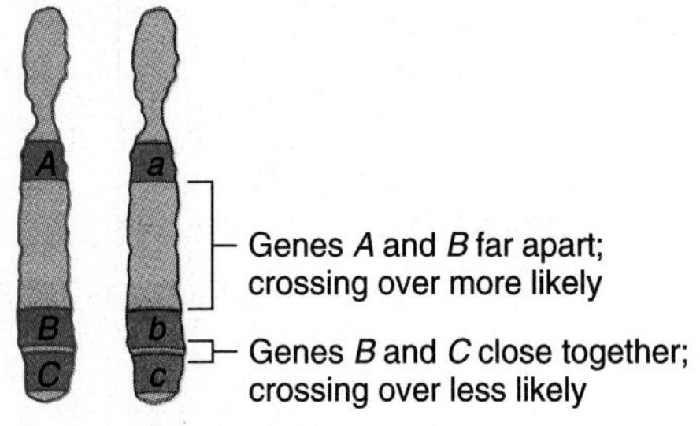


**figure 5.14**

**Breaking linkage.** Crossing over is more likely to occur between the widely spaced linked genes *A* and *B* or between *A* and *C* than between the more closely spaced linked genes *B* and *C*, because there is more room for an exchange to occur.

farther apart two genes are on a chromosome, the more likely they are to cross over simply because more physical distance separates them (figure 5.14).

The correlation between crossover frequency and the distance between genes is used to construct **linkage maps,** which are diagrams that show the order of genes on chromosomes and the relative distances between them. The distance is represented using "map units" called centimorgans (cm), where 1 cm equals 1 percent recombination (see figure 22.3). The frequency of a crossover between any two linked genes is inferred from the proportion of offspring that are recombinant. Genes at opposite ends of the same chromosome often cross over, generating a large recombinant class, but a crossover would rarely separate genes lying very close on the chromosome.

As the 20th century progressed, geneticists in Columbia University's "fly room" mapped several genes on all four chromosomes of the insect, and in other labs many genes were assigned to the human X chromosome. Localizing genes on the X chromosome was easier than doing so on the autosomes, because in human males, with their single X chromosome, recessive alleles on the X are expressed, a point we will return to in the next chapter.

By 1950, geneticists began to contemplate the daunting task of mapping genes on the 22 human autosomes. To start, a gene must be matched to its chromosome. Gene mapping began with the association of a particular chromosome abnormality with a physical trait. Matching phenotypes to chromosomal variants, a field called **cytogenetics,** is the subject of chapter 12.

In 1968, researchers assigned the first human gene to an autosome. R. P. Donohue was observing chromosomes in his own white blood cells when he noticed a dark area consistently located near the centromere of one member of his largest chromosome pair (chromosome 1). He then examined chromosomes from several family members for the dark area, noting also whether each family member had a blood type called Duffy. (Recall that blood types are inherited and refer to the patterns of proteins on red blood cell surfaces.) Donohue found that the Duffy blood type was linked to the chromosome variant. That is, he could predict a relative's Duffy blood type by whether or not the chromosome had the telltale dark area.

Finding a chromosomal variation and using it to detect linkage to another gene is a valuable but rare clue. More often, researchers must rely on the sorts of experiments Sturtevant conducted on his flies—calculating percent recombination (crossovers) between two genes whose locations on the chromosome are known. However, because humans do not have hundreds of offspring, as fruit flies do, obtaining sufficient data to establish linkage relationships requires observing the same traits in many families and pooling the information.

## Examples of Linked Genes in Humans

As an idealized example of determining the degree of linkage by percent recombination, consider the traits of Rh blood type and a form of anemia called elliptocytosis (figure 5.15). An $Rh^+$ phenotype corresponds to genotypes *RR* or *Rr*. (This is a simplification.) The anemia corresponds to genotypes *EE* or *Ee*.

Suppose that in 100 one-child families, one parent is Rh negative with no anemia (*rree*), and the other parent is Rh positive with anemia (*RrEe*), and the *R* and *E* (or *r* and *e*) alleles are in coupling. Of the 100 offspring, 96 have parental genotypes (*re/re* or *RE/re*) and four individuals are recombinants for these two genes (*Re/re* or *rE/re*). Percent recombination is therefore 4 percent, and the two linked genes are 4 cm apart.

Consider another pair of linked genes in humans. Nail-patella syndrome is an autosomal dominant trait that causes absent or underdeveloped fingernails and toenails, and painful arthritis, especially in the knee and elbow joints. It was identified in 1897, and affects only 1 in 50,000 people. The gene is located 10 map units from the *I* gene that determines the ABO blood type, on chromosome 9. Geneticists determined the map distance by pooling information from many families. The information can be used

| | Parent 1 | | Parent 2 | |
|---|---|---|---|---|
| Phenotype | $Rh^-$, no anemia | | $Rh^+$, anemia | |
| Genotype | *rree* | | *RrEe* | |
| Allele Configuration | $\frac{r\ e}{r\ e}$ | | $\frac{R\ E}{r\ e}$ | |
| Gametes: | sperm | frequency | oocytes | Progeny: |
| Parental | *re* | 48% | *RE* | $Rh^+$, anemia |
| | | 48% | *re* | $Rh^-$, no anemia |
| Recombinants | *re* | 2% | *Re* | $Rh^+$, no anemia |
| | | 2% | *rE* | $Rh^-$, anemia |

**figure 5.15**

**Tracing the inheritance of linked genes.** If we know the allele configurations of the parental generation, we can calculate the parental and recombinant class frequencies by pooling family data. Note that in this case the gamete frequencies correspond to the progeny class frequencies because the sperm are all the same genotype for these two genes, even if crossing over occurs.

to predict genotypes and phenotypes in offspring, as in the following example.

Greg and Susan each have nail-patella syndrome. Greg has type A blood, and Susan has type B blood. They want to know what the chance is that a child of theirs would inherit normal nails and knees and type O blood. Fortunately, information was available on Greg and Susan's parents, which enabled a genetic counselor to deduce their allele configurations (figure 5.16).

Greg's mother has nail-patella syndrome and type A blood. His father has normal nails and type O blood. Therefore, Greg must have inherited the dominant nail-patella syndrome allele ($N$) and the $I^A$ allele from his mother, on the same chromosome. We know this because Greg has type A blood and his father has type O blood—he couldn't have gotten the $I^A$ allele from his father. Greg's other chromosome 9 must carry the alleles $n$ and $i$. His alleles are therefore in coupling.

Susan's mother has nail-patella syndrome and type O blood, and so Susan inherited $N$ and $i$ on the same chromosome. Because her father has normal nails and type B blood, her homolog bears alleles $n$ and $I^B$. Her alleles are in repulsion.

Determining the probability that a child of theirs would have normal nails and knees and type O blood is actually the easiest question the couple could ask. The only way this genotype can arise from theirs is if an *ni* sperm (which occurs with a frequency of 45 percent, based on pooled data) fertilizes an *ni* oocyte (which occurs 5 percent of the time). The result—according to the product rule—is a 2.25 percent chance of producing a child with the *nnii* genotype.

Calculating other genotypes for their offspring is more complicated, because more combinations of sperm and oocytes could account for them. For example, a child with nail-patella syndrome and type AB blood could arise from all combinations that include $I^A$ and $I^B$ as well as at least one $N$ allele (assuming that the *NN* genotype has the same phenotype as the *Nn* genotype).

A linkage map begins to emerge when percent recombination is known between all possible pairs of three linked genes. Consider genes *x*, *y*, and *z*. If the percent recombination between *x* and *y* is 10, between *x* and *z* is 4, and between *z* and *y* is 6, then the order of the genes on the chromosome is *x*-*z*-*y* (figure 5.17).

|  | Greg |  | Susan |
|---|---|---|---|
| Phenotype | nail-patella syndrome, type A blood |  | nail-patella syndrome, type B blood |
| Genotype | $NnI^A\_$ |  | $NnI^B\_$ |
| Allele configuration | $\frac{N\ I^A}{n\ i}$ |  | $\frac{N\ i}{n\ I^B}$ |
| Gametes: | sperm | frequency | oocytes |
| Parental | $N\ I^A$ | 45% | $N\ i$ |
|  | $n\ i$ | 45% | $n\ I^B$ |
| Recombinants | $N\ i$ | 5% | $N\ I^B$ |
|  | $n\ I^A$ | 5% | $n\ i$ |

$N$ = nail-patella syndrome
$n$ = normal

**figure 5.16**

**Inheritance of nail-patella syndrome.** Greg inherited the $N$ and $I^A$ alleles from his mother; that is why the alleles are on the same chromosome. His $n$ and $i$ alleles must therefore be on the homolog. Susan inherited alleles $N$ and $i$ from her mother, and $n$ and $I^B$ from her father. Population-based probabilities are used to calculate the likelihood of phenotypes in the offspring of this couple.

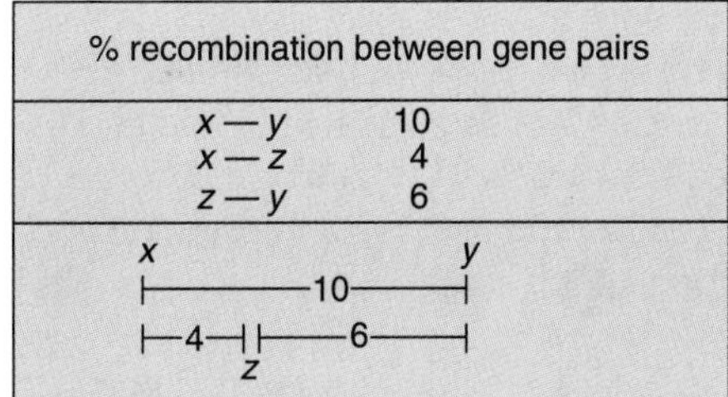


**figure 5.17**

**Recombination mapping.** If we know the percent recombination between all possible pairs of three genes, we can determine their relative positions on the chromosome.

Knowing the percent recombination between linked genes was useful in ordering them on genetic maps in a crude sense. The sequencing of the human genome has revealed an unexpected complexity. That is, crossing over is by no means equally likely to occur throughout the genome. Comparisons of DNA sequences among individuals in different populations reveals that some sequences are nearly always inherited together, more often than would be predicted from their frequency in the population. This nonrandom association between DNA sequences is called **linkage disequilibrium** (LD). The human genome seems to consist of many blocks of "LD" interspersed with areas where crossing over is prevalent. Understanding LD can reveal how populations interacted in historical and evolutionary time, and is being used to predict disease, discussed in chapter 7.

## The Evolution of Gene Mapping

Linkage mapping has had an interesting history. In the first half of the 20th century, gene maps for nonhuman organisms, such as fruit flies, were constructed based on recombination frequencies between pairs of visible traits. Beginning in the 1950s, linkage data on such traits in humans began to accumulate. At first, it was mostly a few visible or measurable traits linked to blood types or proteins that are part of blood plasma (the liquid portion of the blood). These plasma proteins come in variant forms that can be detected by how quickly they migrate in an electrical field, using a technique called electrophoresis.

In 1980 came a great stride in linkage mapping. Researchers began using nuances in the DNA sequence at sites near genes of interest as landmarks. These "markers" do not necessarily encode a protein that causes a phenotype—they might be differences that alter where a DNA cutting enzyme cuts, or differing numbers of short repeated sequences of DNA with no obvious function, or single nucleotide polymorphisms (SNPs).

Computers tally how often genes and markers are inherited together. Gene mappers express the "tightness" of linkage between a marker and the gene of interest as a **LOD score,** which stands for "logarithm of the odds." A LOD score indicates the likelihood that particular crossover frequency data indicate linkage.

A LOD score of three or greater signifies linkage. It means that the observed data are 1,000 ($10^3$) times more likely to have occurred if the two DNA sequences (a disease-causing allele and its marker) are linked than if they reside on different chromosomes and just happen to often be inherited together by chance. It is somewhat like deciding whether two coins tossed together 1,000 times always come up both heads or both tails by chance, or because they are attached together side by side in that position, as linked genes are. If the coins land with the same side up in all 1,000 trials, it indicates they are very likely linked together.

Even though a marker is not the gene of interest, it makes possible a molecular level diagnosis—even before symptoms appear. A marker test requires DNA samples from several family members to follow linkage of the marker to the disease-causing gene. Using markers revolutionized and greatly sped gene mapping, making the genome project possible.

Today linkage information is often depicted as a list of molecular markers used to distinguish parts of chromosomes among the members of a family, often placed beneath the traditional symbols of a pedigree. Such a panel of markers, called a **haplotype,** is a set of DNA sequences that are inherited together on the same chromosome due to linkage disequilibrium. Haplotypes can look complicated, because markers are often given names that have meaning only to their discoverers. They read like license plates, bearing labels such as D9S1604. The haplotypes in the pedigree in figure 5.18, for a family with cystic fibrosis are simplified. Each number beneath a symbol represents a "license plate" haplotype. Haplotypes make it possible to track which parent transmits which genes and chromosomes to offspring. In figure 5.18, knowing the haplotype of individual II-2 reveals which chromosome in parent I-1 contributes the mutant allele. Because Mr. II-2 received haplotype 3233 from his affected mother, his other haplotype, 2222, comes from his father. Since Mr. II-2 is affected and his father is not, the father must be a heterozygote, and 2222 the haplotype that is linked to the mutant CFTR allele.

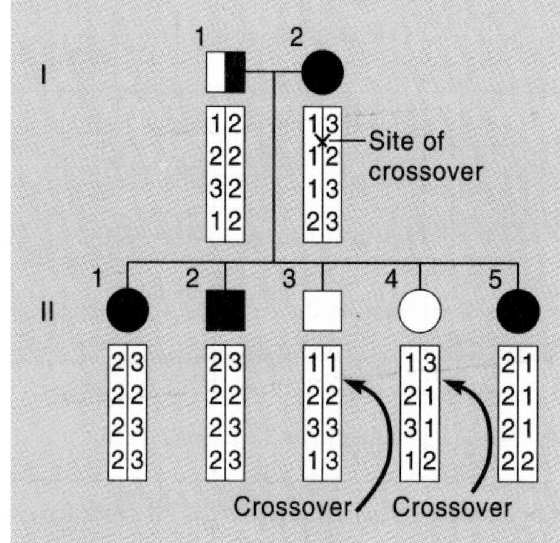


**figure 5.18**

**Haplotypes.** The numbers in bars beneath pedigree symbols enable researchers to track specific chromosome segments with markers. Disruptions of a marker sequence indicate the sites of crossovers.

It has been interesting and exciting to watch gene mapping evolve, from the initial crude associations between blood types and chromosomal quirks, to today's sophisticated SNP maps with their millions of signposts along the genome. Throughout the 1990s, each October, *Science* magazine published a human genome map. The number of identified genes steadily grew as chromosome depictions became ever more packed with information. From that information, during your lifetime, will spring a revolution in health care and in how we understand ourselves.

### Key Concepts

Genes on the same chromosome are linked, and they are inherited in different patterns than the unlinked genes Mendel studied. Crosses involving linked genes produce a large parental class and a small recombinant class (caused by crossing over). • The farther apart two genes are on a chromosome, the more likely they are to recombine. Linkage maps are based on this relationship between crossover frequency and distance between genes on the same chromosome. Linkage disequilibrium is a linkage combination that is stronger than that predicted by gene frequencies in a population.

Cytogenetics can be used to associate a phenotype to a chromosomal aberration. • Linkage maps reflect the percent recombination between linked genes. LOD scores describe the tightness of linkage and thereby the proximity of a gene to a marker. Haplotypes indicate linked DNA sequences.

# Summary

### 5.1 When Gene Expression Appears to Alter Mendelian Ratios

1. Homozygosity for lethal recessive alleles stops development before birth, eliminating an offspring class.
2. A gene can have multiple alleles because its sequence can be altered in many ways. Different allele combinations produce different variations of the phenotype.
3. Heterozygotes of **incompletely dominant** alleles have phenotypes intermediate between those associated with the two homozygotes. **Codominant** alleles are both expressed.
4. In **epistasis,** one gene affects the expression of another.

5. An **incompletely penetrant** genotype is not expressed in all individuals who inherit it. Phenotypes that vary in intensity among individuals are **variably expressive.**
6. **Pleiotropic** genes have several expressions.
7. A **phenocopy** is a characteristic that appears to be inherited but is environmental.
8. In **genetic heterogeneity,** two or more genes specify the same phenotype.

### 5.2 Maternal Inheritance and Mitochondrial Genes

9. Only females transmit mitochondrial genes; males can inherit such a trait but cannot pass it on.
10. Mitochondrial genes do not cross over, do not repair DNA, and lack introns.
11. The 37 mitochondrial genes encode tRNA, rRNA, or proteins involved in energy reactions.
12. Many mitochondrial disorders are **heteroplasmic,** with mitochondria in a cell harboring different alleles.

### 5.3 Linkage

13. Genes on the same chromosome are **linked** and, unlike genes that independently assort, produce a large number of individuals with parental genotypes and a small number of individuals with **recombinant** genotypes.
14. **Linkage maps** are developed from studies of linked genes. Researchers can examine a group of known linked DNA sequences (a **haplotype**) to follow the inheritance of certain chromosomes.
15. Knowing whether linked alleles are in coupling or repulsion, and using crossover frequencies determined by pooling data, one can predict the probabilities of certain genotypes appearing in progeny.
16. Genetic linkage maps assign distances to linked genes based on crossover frequencies.

## Review Questions

1. Explain how each of the following phenomena can disrupt Mendelian phenotypic ratios.
   a. lethal alleles
   b. multiple alleles
   c. incomplete dominance
   d. codominance
   e. epistasis
   f. incomplete penetrance
   g. variable expressivity
   h. pleiotropy
   i. a phenocopy
   j. genetic heterogeneity
2. How does the relationship between dominant and recessive alleles differ from epistasis?
3. Why can transmission of an autosomal dominant trait with incomplete penetrance look like autosomal recessive inheritance?
4. How does inheritance of ABO blood type exhibit both complete dominance and codominance?
5. Distal symphalangism (see section 4.2) is variably expressive and incompletely penetrant. What does this mean for people who might inherit this condition?
6. Describe why inheritance of mitochondrial DNA and linkage are exceptions to Mendel's laws.
7. What is the physical basis of the epistasis that causes the Bombay phenotype?
8. How does a pedigree for a maternally inherited trait differ from one for an autosomal dominant trait?
9. What might be a confounding factor in attempting to correlate different genotypes with different expressions of a Mendelian illness?
10. If researchers could study pairs of human genes as easily as they can study pairs of genes in fruit flies, how many linkage groups would they detect?

## Applied Questions

1. For each of the diseases described in situations *a* through *i*, indicate which of the following phenomena (A–H) is at work. A disorder may result from more than one of these causes.
   A. lethal alleles
   B. multiple alleles
   C. epistasis
   D. incomplete penetrance
   E. variable expressivity
   F. pleiotropy
   G. a phenocopy
   H. genetic heterogeneity

   a. A woman has severe neurofibromatosis type I. She has brown spots on her skin and several large tumors beneath her skin. A gene test shows that her son has inherited the disease-causing autosomal dominant allele, but he has no symptoms.
   b. A man and woman have six children. They also had two stillbirths—fetuses that died shortly before birth.
   c. Most children with cystic fibrosis have frequent lung infections and digestive difficulties. Some people have mild cases, with onset of minor respiratory problems in adulthood. Some men have cystic fibrosis, but their only symptom is infertility.
   d. In Labrador retrievers, the *B* allele confers black coat color and the *b* allele brown coat color. The *E* gene controls expression of the *B* gene. If a dog inherits the *E* allele, the coat is golden no matter what the *B* genotype is. A dog of genotype *ee* expresses the *B* (black) phenotype.
   e. Two parents are heterozygous for genes that cause albinism, but each gene specifies a different enzyme in the biochemical pathway that leads

to skin pigment synthesis. Their children thus do not face a 25 percent risk of having albinism.

f. Alagille syndrome, in its most severe form, prevents the formation of ducts in the gallbladder, causing liver damage. Affected children also usually have heart murmurs, unusual faces, a line in the eye, and butterfly-shaped vertebrae. Such children often have one seemingly healthy parent who, when examined, proves to also have a heart murmur, unusual face, and butterfly vertebrae.

g. Two young children in a family have terribly decayed teeth. Their parents think it is genetic, but the true cause is a babysitter who puts them to sleep with juice bottles.

h. A woman develops dark patches on her face. Her family physician suspects that she may have alkaptonuria, an inherited deficiency of the enzyme homogentisic acid oxidase that produces darkened skin, a stiff spine, darkened ear tips, and urine that turns black when it contacts the air. However, a dermatologist the woman is referred to discovers that she has been using a facial cream containing hydroquinone, which is known to cause dark skin patches in dark-skinned people.

i. An apparently healthy 24-year-old basketball player dies suddenly during a game when her aorta, the largest artery, ruptures. A younger brother is nearsighted and has long and thin fingers, and an older sister is extremely tall, with long arms and legs. An examination reveals that the older sister, too, has a weakened aorta. All of these siblings have Marfan syndrome, although they are affected to different degrees.

**2.** If many family studies for a particular autosomal recessive condition reveal fewer affected individuals than Mendel's law predicts, the explanation may be either incomplete penetrance or lethal alleles. How might you use haplotypes to determine which of these two possibilities is the causative factor?

**3.** A man who has type O blood has a child with a woman who has type A blood. The woman's mother has AB blood, and her father, type O. What is the probability that the child has each of the following blood types?

a. type O

b. type A

c. type B

d. type AB

**4.** Two people who are heterozygous for familial hypercholesterolemia decide to have children, but they are concerned that a child might inherit the severe form of the illness. What is the probability that this will happen?

**5.** A young woman taking a genetics class learns about ABO blood types. She knows that she is type O but also knows that one of her parents is type AB and the other is type B. Adultery or adoption are not possible, so the woman wonders how her blood type could have arisen. Suggest how type AB and A parents could produce a child whose blood type is O.

**6.** Enzymes are used in blood banks to remove the *A* and *B* antigens from blood types A and B. This makes the blood type O.

a. Does this alter the phenotype or the genotype?

b. Removing the *A* and *B* antigens from red blood cells is a phenocopy of what genetic phenomenon?

**7.** Friedreich's ataxia, which impairs the ability to feel and move the limbs, usually begins in early adulthood. The molecular basis of the disease is impairment of ATP production in mitochondria, but the mutant gene is in the nucleus of the cells. Would this disorder be inherited in a Mendelian fashion? Explain your answer.

**8.** What is the chance that Greg and Susan, the couple with nail-patella syndrome, could have a child with normal nails and type AB blood?

**9.** A gene called secretor is located one map unit from the *H* gene that confers the Bombay phenotype on chromosome 19. Secretor is dominant, and a person of either genotype *SeSe* or *Sese* secretes the ABO and *H* blood type antigens (cell surface proteins) in saliva and other body fluids. This secretion, which the person is unaware of, is the phenotype. A man has the Bombay phenotype and is not a secretor. A woman does not have the Bombay phenotype and is a secretor. She is a dihybrid whose alleles are in coupling. What is the chance that a child of theirs will have the same genotype as the father?

**10.** A Martian creature called a gazook has 17 chromosome pairs. On the largest chromosome are genes for three traits—round or square eyeballs (*R* or *r*); a hairy or smooth tail (*H* or *h*); and nine or 11 toes (*T* or *t*). Round eyeballs, hairy tail, and nine toes are dominant to square eyeballs, smooth tail, and 11 toes. A trihybrid male has offspring with a female who has square eyeballs, a smooth tail, and 11 toes on each of her three feet. She gives birth to 100 little gazooks, who have the following phenotypes:

- 40 have round eyeballs, a hairy tail, and nine toes
- 40 have square eyeballs, a smooth tail, and 11 toes
- six have round eyeballs, a hairy tail, and 11 toes
- six have square eyeballs, a smooth tail, and nine toes
- four have round eyeballs, a smooth tail, and 11 toes
- four have square eyeballs, a hairy tail, and nine toes

a. Draw the allele configurations of the parents.

b. Identify the parental and recombinant progeny classes.

c. What is the crossover frequency between the *R* and *T* genes?

## Suggested Readings


Chinnery, Patrick F., et al. March 1999. Clinical mitochondrial genetics. *Journal of Medical Genetics* 36:425–36. Mitochondrial disorders are often misdiagnosed.

Ferriman, Annabel. May 12, 2001. First cases of human germline genetic modification announced. *British Medical Journal* 322:1144. People objected to ooplasmic transfer because the resulting children have three genetic parents.

Goldstein, David B. October 2001. Islands of linkage disequilibrium. 29:109–11. The human genome consists of modular blocks of genes that travel together.

Risch, Neil. August 1999. A genomic screen of autism: Evidence for a multilocus etiology. *American Journal of Human Genetics* 65:493–507. Many disorders can be caused by any of several genes.

Sijbrands, Eric J. G., et al. April 28, 2001. Mortality over two centuries in large pedigree with familial hypercholesterolemia: Family tree mortality study. *British Medical Journal* 322:1019–23. Variable expressivity in FH reflects diet and exercise habits, and effects of other genes.

Smeitink, Jan, and Lambert van den Heuvel. June 1999. Human mitochondrial complex I in health and disease. *American Journal of Human Genetics* 64:1505–9. Many inherited diseases affecting muscles stem from mutations in mitochondrial genes.


## On the Web

Check out the resources on our website at

**www.mhhe.com/lewisgenetics5**

On the web for this chapter, you will find additional study questions, vocabulary review, useful links to case studies, tutorials, popular press coverage, and much more. To investigate specific topics mentioned in this chapter, also try the links below:

Alkaptonuria Hotline
**www.goodnet.com/nee72478/enable/hotline.htm**

American Porphyria Foundation
**www.enterprise.net/apf**

Cystic Fibrosis Foundation **www.cff.org**

International Mitochondrial Disease Network
**www.imdn.org**

Marfan Disease Organization **www.marfan.org**

Nail-Patella Syndrome Worldwide Inc.
**www.nailpatella.org**

NIH Consensus Development Program, PKU
**consensus.nih.gov**

Online Mendelian Inheritance in Man
**www.ncbi.nlm.nih.gov/entrez/query.fcgi?db=OMIM**
cystic fibrosis 219700
familial hypercholesterolemia 143890
FUT1 deficiency (Bombay phenotype) 211100
Leber's hereditary optic neuropathy 535000
MELAS 540000
nail-patella syndrome 161200
phenylketonuria 261600
phocomelia 269000
polydactyly 174200
porphyria variegata 176020
retinitis pigmentosa 600105, 268025, 312600
very long chain acyl-CoA dehydrogenase deficiency 201475

United Mitochondrial Diseases Foundation
**www.umdf.org**

Rr hh Tt

CHAPTER

6

# Matters of Sex

## 6.1 Sexual Development

Whether we develop as female or male is determined when two X chromosomes, or an X and a Y chromosome, come together at fertilization. Our sexual selves depend further upon whether one key gene is present and active six weeks after fertilization. The hormone-guided events of development, and possibly our feelings and experiences, also help to mold our sexual identities.

## 6.2 Traits Inherited on Sex Chromosomes

The tiny Y chromosome passes few traits, while the gene-packed X carries many. Understanding Mendel's laws and the mechanism of sex chromosome distribution at fertilization enables us to follow traits caused by the actions of proteins encoded by genes on the sex chromosomes.

## 6.3 X Inactivation Equalizes the Sexes

To make the number of genes on the two X chromosomes of a female equivalent to that of the single X of males, one X in each cell of a female is inactivated early in prenatal development. The female is a mosaic of gene expression for this chromosome. X inactivation has noticeable effects when a female is heterozygous for certain genes.

## 6.4 Gender Effects on Phenotype

Some genes—autosomal as well as X- or Y-linked—are expressed in one sex but not the other, or may be inherited as a dominant trait in one but a recessive in the other. In genomic imprinting, gene expression depends upon which parent transmits a particular gene.

Whether we are male or female is enormously important in our lives, affecting our relationships, how we think and act, and how others perceive us. Sex is ultimately a genetic phenomenon, but is also layered with psychological and sociological components.

Maleness or femaleness is determined at conception, when he inherits an X and a Y chromosome, or she inherits two X chromosomes. Another level of sexual identity comes from the control that hormones exert over the development of reproductive structures. Finally, both biological factors and social cues influence sexual feelings, including the strong sense of whether we are male or female.

## 6.1 Sexual Development

In prenatal humans, the sexes look alike until the ninth week of prenatal development. During the fifth week, all embryos develop two unspecialized gonads, which are organs that will develop as either testes or ovaries. Each such "indifferent" gonad forms near two ducts that give it two developmental options. If one set of tubes, called the Müllerian ducts, continue to develop, they eventually form the sexual structures characteristic of a female. If the other set, the Wolffian ducts, persist, male sexual structures form. The choice of either one of these two developmental pathways occurs during the sixth week, and it depends upon the sex chromosome constitution. If a gene on the Y chromosome called *SRY* (for "sex-determining region of the Y") is activated, hormones steer development along a male route. In the absence of *SRY* activation, a female develops (figure 6.1). Femaleness was long considered to be a "default" option in development, but sex determination is more accurately described as a choice imposed on ambiguous precursor structures.

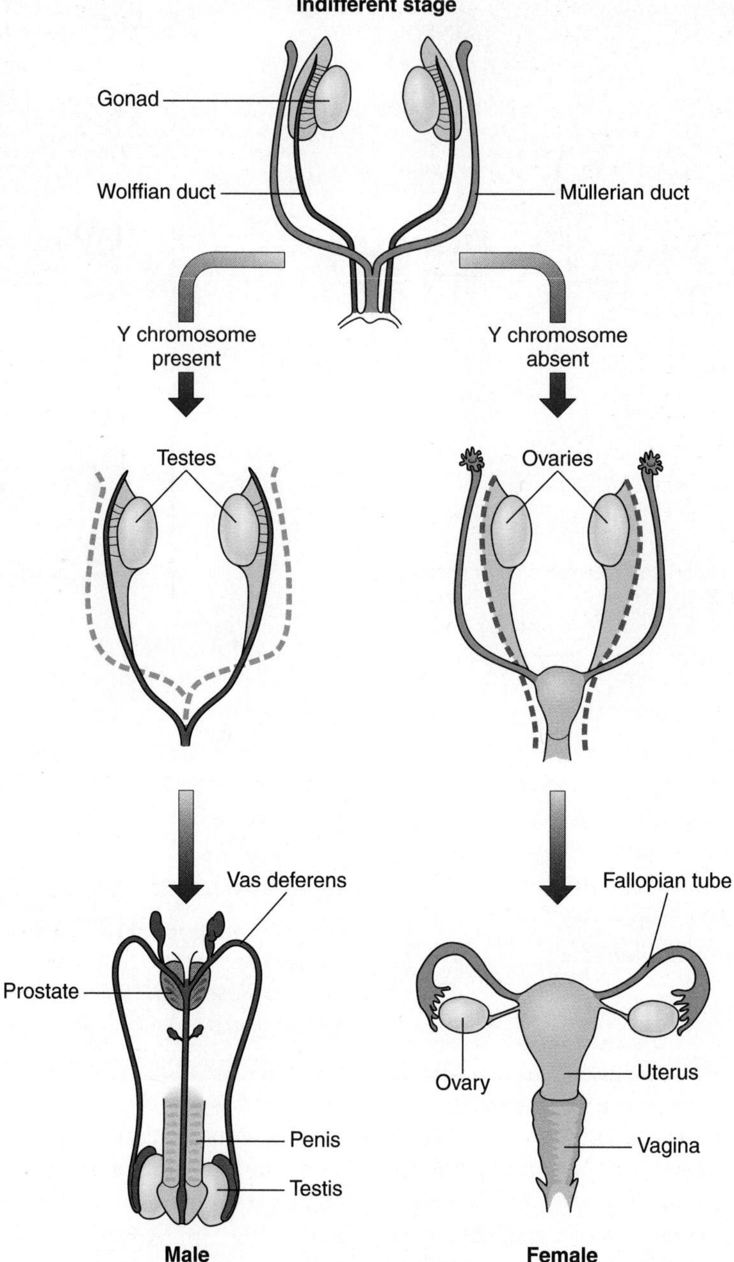


**figure 6.1**

**Male or female?** The duct system in the early human embryo may develop into male reproductive organs *or* female reproductive organs.

### Sex Chromosomes

Recall that males and females have equal numbers of autosomes, but males have one X chromosome and one Y chromosome, and females have two X chromosomes (figure 6.2). The sex with two different sex chromosomes is called the **heterogametic sex,** and the other, with two of the same sex chromosomes, is the **homogametic sex.** These general terms are necessary in a broader biological sense, because not all species have homogametic females and heterogametic males, as mammals do. In birds, for example, it is opposite—males are ZZ and females are ZW.

The X chromosome in humans contains more than a thousand identified genes. The much smaller Y chromosome has 85 protein-encoding genes, about half of which have been identified. In meiosis in a male, the X and Y chromosomes act as if they are a pair of homologs. We introduce the Y chromosome here, then consider the X in section 6.2.

Linkage maps for the Y chromosome are very sparse, compared to those for the other chromosomes, for several reasons: it does not have a homolog with which to cross over; it has many repeated DNA se-

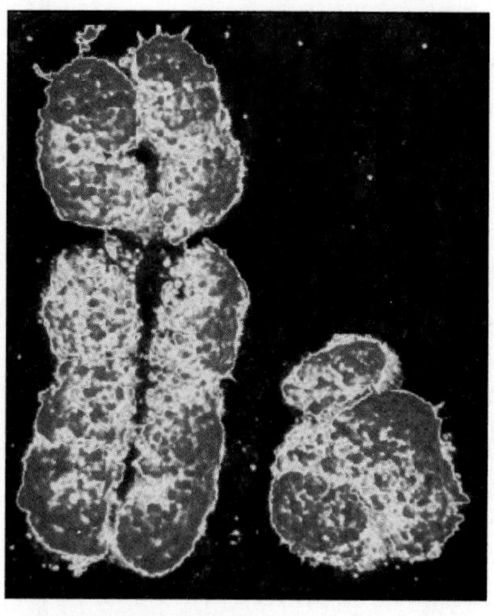


## figure 6.2

**The X and Y chromosomes.** In humans, females are the homogametic sex (XX) and males are the heterogametic sex (XY).

quences; and it has very few genes. Recall from chapter 5 that crossing over provides the information used to create genetic maps. For the Y chromosome, researchers have had to infer the functions and locations of genes by examining men who are missing parts of the chromosome and determining how they differ from normal. Since the 1970s, for example, infertility researchers have associated several types of male fertility problems with extremely small deletions of Y chromosome material. Today, information from the human genome project is filling in our knowledge of genes on the Y chromosome.

Figure 6.3 shows a partial map of the human Y chromosome. The chromosome has a short arm and a long arm. The genes are considered as three broad groups based on similarity to genes on the X chromosome.

The first group consists of genes located at both tips of the Y chromosome. These areas are called **pseudoautosomal regions,** termed PAR1 and PAR2. They comprise only 5 percent of the chromosome. The pseudoautosomal genes are so-called because they have counterparts on the X chromosome, and can even cross over with them. These genes encode a variety of proteins that function in both sexes, participating in or controlling such activities as bone growth, signal transduction, the synthesis of hormones and receptors, and energy metabolism.

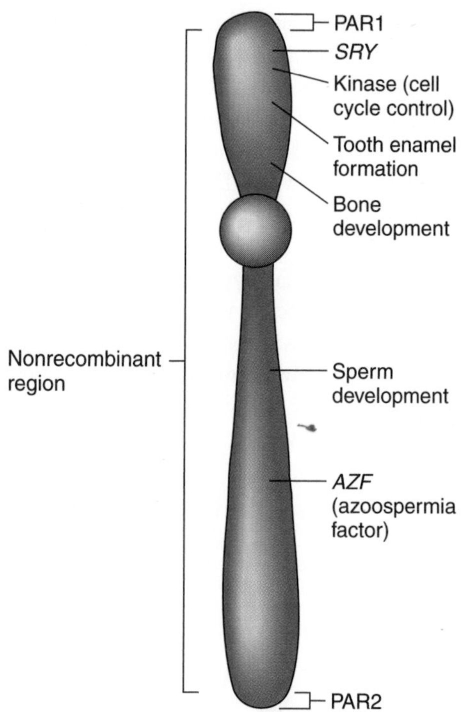


## figure 6.3

**Anatomy of the Y chromosome.** The Y chromosome has two pseudoautosomal regions whose genes match genes on the X chromosome, and a large central area that does not recombine, comprising about 95 percent of the chromosome. A few of the genes are indicated. *SRY* determines sex. *AZF* encodes a protein essential to producing sperm; mutations in it cause infertility. "PAR" stands for pseudoautosomal region.

The second functional group of Y chromosome genes includes the **X-Y homologs.** These genes are very similar in DNA sequence to certain genes on the X chromosome, but they are not identical, as are the pseudoautosomal genes. X-Y homologs are expressed in nearly all tissues, including those found only in males. Existence of the pseudoautosomal genes and the X-Y homologs may be evidence that the Y chromosome is a stripped-down version of an ancestral, X-like chromosome.

The third functional group of genes includes those unique to the Y chromosome. The *SRY* gene is part of this group, and it is distinguished by being present in only one copy. The other genes occur in multiple copies. They encode proteins that participate in cell cycle control and in the regulation of gene expression, enzymes, and receptors for immune system biochemicals.

The tiny Y chromosome was first visualized with the use of a microscope in 1923, and researchers soon recognized its association with maleness. For many years, they sought to identify the gene or genes that determine sex. Important clues came from two very interesting types of people—men who have two X chromosomes (XX male syndrome), and women who have one X and one Y chromosome (XY female syndrome). Both of these are the reverse of the normal situation. A close look at the composition of these people's sex chromosomes revealed that the XX males actually had a small piece of a Y chromosome, and the XY females lacked a small part of the Y chromosome. The part of the Y chromosome present in the XX males was the same part that was missing in the XY females. This critical area accounted for half a percent of the Y chromosome, about 300,000 DNA bases. Finally, in 1990, two groups of researchers isolated and identified the *SRY* gene in this implicated area.

## The Phenotype Forms

The *SRY* gene encodes a type of protein called a **transcription factor,** which controls the expression of other genes. The SRY transcription factor stimulates male development by sending signals to the indifferent gonads. In response, **sustentacular cells** in the developing testis secrete **anti-Müllerian** hormone, which causes degeneration of and prevents further development of female structures (uterus, fallopian tubes, and upper vagina). At the same time, **interstitial cells** in the testis secrete testosterone, which stimulates development of the epididymides, vas deferentia, seminal vesicles, and ejaculatory ducts. Some testosterone is also converted to **dihydrotestosterone (DHT),** which directs the development of the urethra, prostate gland, penis, and scrotum.

Because so many steps contribute to male prenatal sexual development, genetic abnormalities can intervene at several different points. The result may be an XY individual with a block in the gene- and hormone-controlled elaboration of male structures. A chromosomal he is a phenotypic she. For example, in androgen insensitivity syndrome, caused by a mutation in a gene on the X chromosome, absence of receptors for testosterone stops cells in early reproductive structures from receiving the signal to develop as male.

In a group of disorders called male pseudohermaphroditism (figure 6.4), testes are usually present (indicating that the *SRY* gene is functioning) and anti-Müllerian hormone is produced, so the female set of tubes degenerates. However, a block in testosterone synthesis prevents the fetus from developing male structures. The child appears to be a girl. But at puberty, the adrenal glands, which sit atop the kidneys, begin to produce testosterone (as they normally do in a male). This leads to masculinization: The voice deepens, and muscles build up into a masculine physique; breasts do not develop, nor does menstruation occur. The clitoris may enlarge so greatly under the adrenal testosterone surge that it looks like a penis. Individuals with a form of this condition in the Dominican Republic are called *guevedoces,* which means "penis at age 12." In a more common condition, congenital adrenal hyperplasia due to 21-hydroxylase deficiency, an enzyme block leads to excess testosterone production. The mutant gene is on chromosome 6. The condition causes overgrowth of the clitoris or penis, so much so that a girl may appear to be a boy.

A hermaphrodite is an individual with both male and female sexual structures. The word comes from the Greek god of war, Hermes, and the goddess of love, Aphrodite. "Pseudohermaphroditism" refers to the presence of both types of structures, but at different stages of life.

Prenatal tests that detect chromosomal sex have changed the way that pseudohermaphroditism is diagnosed. Before these tests were available (and today in affected individuals who have not undergone the tests), the condition was detected only after puberty, when masculinization occurred in a person who

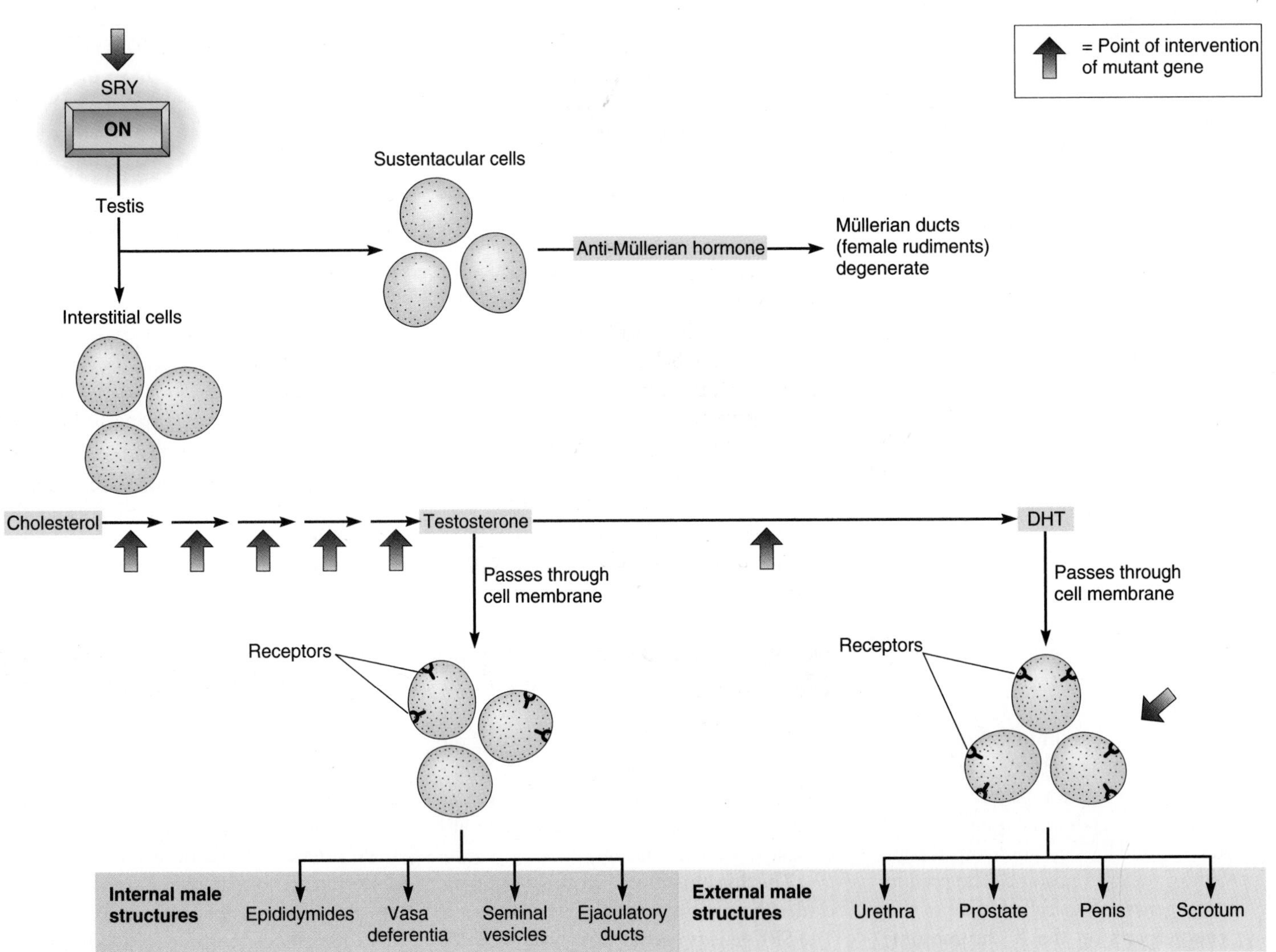


**figure 6.4**

**A chromosomal "he" develops as a phenotypic "she."** Male pseudohermaphroditism results from disruptions in differentiation and development of reproductive structures. Mutations can do this at several points. Before puberty, the person appears female, although the sex chromosomes are XY. At puberty, the adrenal glands produce testosterone, which belatedly turns on some of the genes that control development of male structures and secondary sex characteristics such as a deepening voice, hair growth, and muscle growth.

looked female. Today, pseudohermaphroditism is indicated when a prenatal chromosome check reveals an X and a Y chromosome, but the newborn is a phenotypic girl.

## Gender Identity—Is Homosexuality Inherited?

No one really knows why we have feelings of belonging to one gender or the other, but these feelings are intense. On p. 132, Bioethics: Choices for the Future describes people whose gender identity persists even after surgery alters their phenotype in a way that contradicts their sex chromosome constitution.

In homosexuality, a person's phenotype and genotype are consistent, but physical attraction is toward members of the same sex. Homosexuality is seen in all cultures so far examined and has been observed for thousands of years.

Evidence is accumulating that homosexuality is at least partially inherited. Earlier studies cite the feelings that homosexual individuals have as young children, well before they know of the existence or the meaning of the term.

In humans, twin studies suggest a genetic influence on homosexuality. A 1991 study found that identical twins are more likely to both be homosexual than are both members of fraternal same-sex twin pairs. Specifically, in 52 percent of identical twin pairs in which one or both were homosexual, both brothers were homosexual, but this was true for only 22 percent of fraternal twin pairs. Also, two brain areas are of different sizes in homosexual versus heterosexual men.

In 1993, National Cancer Institute researcher Dean Hamer brought the tools of molecular biology to bear on the age-old question of whether homosexuality is inherited. Hamer traced the inheritance of five DNA marker sequences on the X chromosome in 40 pairs of homosexual brothers. Although these DNA sequences are highly variable in the general population, they were identical in 33 of the sibling pairs. Hamer interpreted the finding to mean that genes causing or predisposing a person to homosexuality reside on the X chromosome. However, the work only identified DNA sequences that are more common among homosexual twins—it did not identify a causative gene.

Hamer's report continues to cause controversy today. One research group confirmed and extended the work, finding that when two brothers are homosexual and have another brother who is heterosexual, the heterosexual brother does not share the X chromosome markers. This study also did not find the X chromosome markers between pairs of lesbian sisters. Several research groups have refuted Hamer's findings. But a gene controlling homosexuality need not reside on a sex chromosome, which is where Hamer looked. Ongoing studies are searching among the autosomes for such genes.

In yet another approach to understand the biological basis of homosexuality, researchers have genetically manipulated male fruit flies to display what looks like homosexual behavior. A mutant allele of an eye color gene, called "*white,*" causes the flies to have white eyes when expressed in cells of the eye only. Wild type eye color is red. Researchers altered male fly embryos so that the resulting adult insects expressed the *white* gene in every cell. The altered male flies displayed what appears to be mating behavior with each other (figure 6.5), presumably as a result of the altered gene expression.

The ability to genetically induce homosexual behavior suggests genetic control. The biochemical basis of the phenotype makes sense; the *white* gene's product, an enzyme that controls eye color, enables cells to use the amino acid tryptophan, which is required to manufacture the hormone serotonin. When all the fly's cells express the mutant *white* gene, instead of just eye cells, serotonin levels in the brain drop, and this may cause the unusual behavior. In other animals, lowered brain serotonin is associated with homosexual behavior.

Table 6.1 summarizes the several components of sexual identity.

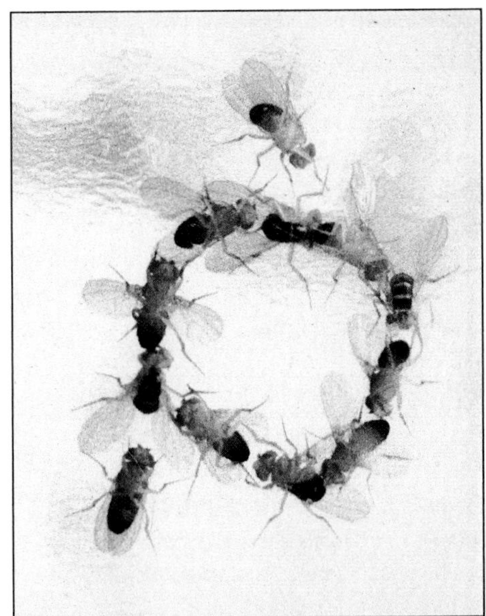

**figure 6.5**

**Is homosexuality inherited?** The ability to genetically alter male fruit flies, causing them to display mating behavior toward each other, adds to evidence that homosexuality is at least partially an inherited trait.

**table 6.1**

**Sexual Identity**

| Level | Events | Timing |
|---|---|---|
| Chromosomal/ genetic | XY = male<br>XX = female | Fertilization |
| Gonadal sex | Undifferentiated structure becomes testis or ovary | 9–16 weeks after fertilization |
| Phenotypic sex | Development of external and internal reproductive structures continues as male or female in response to hormones | 8 weeks after fertilization, puberty |
| Gender identity | Strong feelings of being male or female develop | From childhood, possibly earlier |

## Key Concepts

The human female is homogametic, with two X chromosomes, and the male is heterogametic, with one X and one Y chromosome. The Y chromosome has about 85 genes, and includes two small pseudoautosomal regions and a large area that does not recombine. • Activation of the *SRY* gene on the Y chromosome causes the undifferentiated gonad to develop into a testis. Then, sustentacular cells in the testis secrete anti-Müllerian hormone, which stops development of female structures. Interstitial cells in the testis secrete testosterone, which stimulates development of the epididymides, vasa deferentia, seminal vesicles, and ejaculatory ducts. Testosterone is also converted to DHT, which directs development of the urethra, prostate gland, penis, and scrotum. • Genes may contribute to homosexuality.

## 6.2 Traits Inherited on Sex Chromosomes

Genes carried on the Y chromosome are said to be **Y-linked,** and those on the X chromosome are **X-linked.**

Y-linked traits are rare, because the chromosome has few genes, and many have counterparts on the X chromosome. These traits are passed from male to male, because a female does not have a Y chromosome. No other Y-linked traits besides infertility (which obviously can't be passed on) are yet clearly defined, although certain gene products are known. Claims that "hairy ears" is a Y-linked trait did not hold up—it turned out that families hid their affected female members!

Genes on the X chromosome have different patterns of expression in females and males, because a female has two X chromosomes, and a male just one. In females, X-linked traits are passed just like autosomal traits—that is, two copies are required for expression of a recessive allele, and one copy for a dominant allele. In males, however, a single copy of an X-linked allele results in expression of the trait or illness, because there is no copy of the gene on a second X chromosome to mask the other's effect. A man inherits an X-linked trait only from his mother, because he gets his Y chromosome from his father. The human male is considered to be **hemizygous** for X-linked traits, because he has only one set of X-linked genes.

Understanding how sex chromosomes are inherited is important in predicting phenotypes and genotypes in offspring. A male inherits his Y chromosome from his father and his X chromosome from his mother (figure 6.6). A female inherits one X chromosome from each parent. If a mother is heterozygous for a particular X-linked gene, her son or daughter has a 50 percent chance of inheriting either allele from her. X-linked traits are always passed on the X chromosome from mother to son or from either parent to daughter, but there is no direct male-to-male transmission of X-linked traits.

### X-Linked Recessive Inheritance

An X-linked recessive trait is expressed in females if it is present in two copies. A common situation is for an X-linked trait to pass from a heterozygous mother to an affected son. Table 6.2 summarizes the transmission of an X-linked recessive trait.

If an X-linked condition is not lethal, a man may be healthy enough to transmit it to offspring. Consider the small family depicted in figure 6.7, an actual case. A middle-aged man who had rough, brown, scaly skin did not realize his condition was inherited until his daughter had a son. By a year of age, the boy's skin resembled his grandfather's. In the condition, called ichthyosis, an enzyme deficiency blocks removal of cholesterol

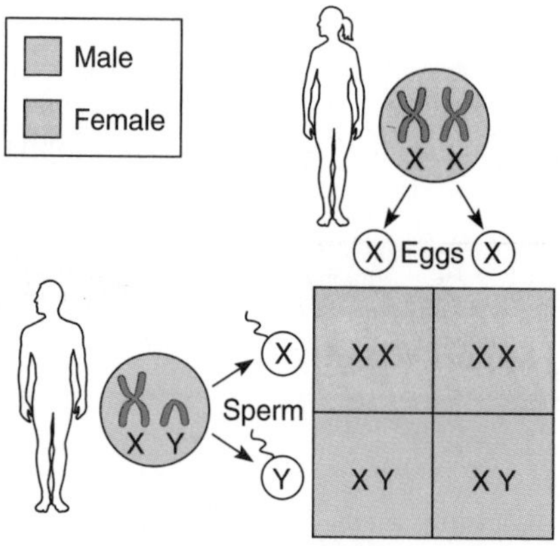

**table 6.2**

**Criteria for an X-Linked Recessive Trait**

1. Always expressed in the male.
2. Expressed in a female homozygote but not in a heterozygote.
3. Passed from heterozygote or homozygote mother to affected son.
4. Affected female has an affected father and a mother who is affected or a heterozygote.

**figure 6.6**

**Sex determination in humans.** An oocyte has a single X chromosome. A sperm cell has either an X chromosome or a Y chromosome. If a Y-bearing sperm cell with a functional *SRY* gene fertilizes an oocyte, the zygote is a male (XY). If an X-bearing sperm cell fertilizes an oocyte, then the zygote is a female (XX).

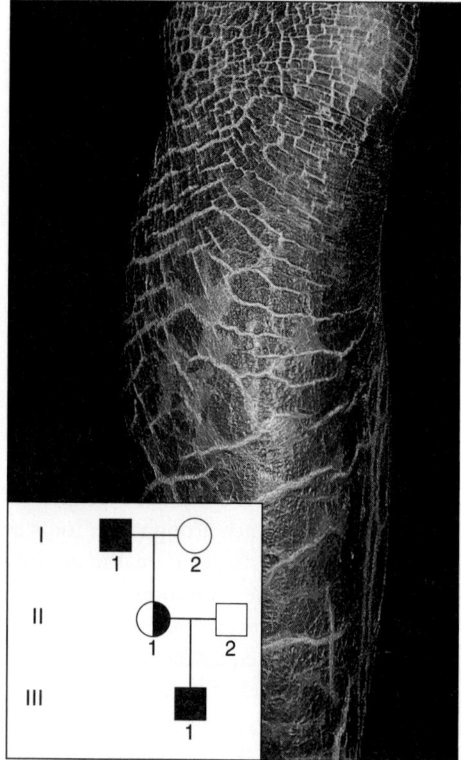


**figure 6.7**

**An X-linked recessive trait.** Ichthyosis is transmitted as an X-linked recessive trait. A grandfather and grandson were affected in the family that the pedigree depicts.

from skin cells. As a result, the upper skin layer cannot peel off as it normally does, causing a brown, scaly appearance. A test of the daughter's skin cells revealed that she produced half the normal amount of the enzyme, indicating that she is a carrier.

Color blindness is another X-linked recessive trait that does not hamper the ability of a man to have children. About 8 percent of males of European ancestry are color blind, as are 4 percent of males of African descent. Only 0.4 percent of females in both groups are color blind. Reading 6.1 takes a closer look at this interesting trait.

Figure 6.8 shows part of a very extensive pedigree for another X-linked recessive trait, hemophilia A. Note the combination of pedigree symbols and a Punnett

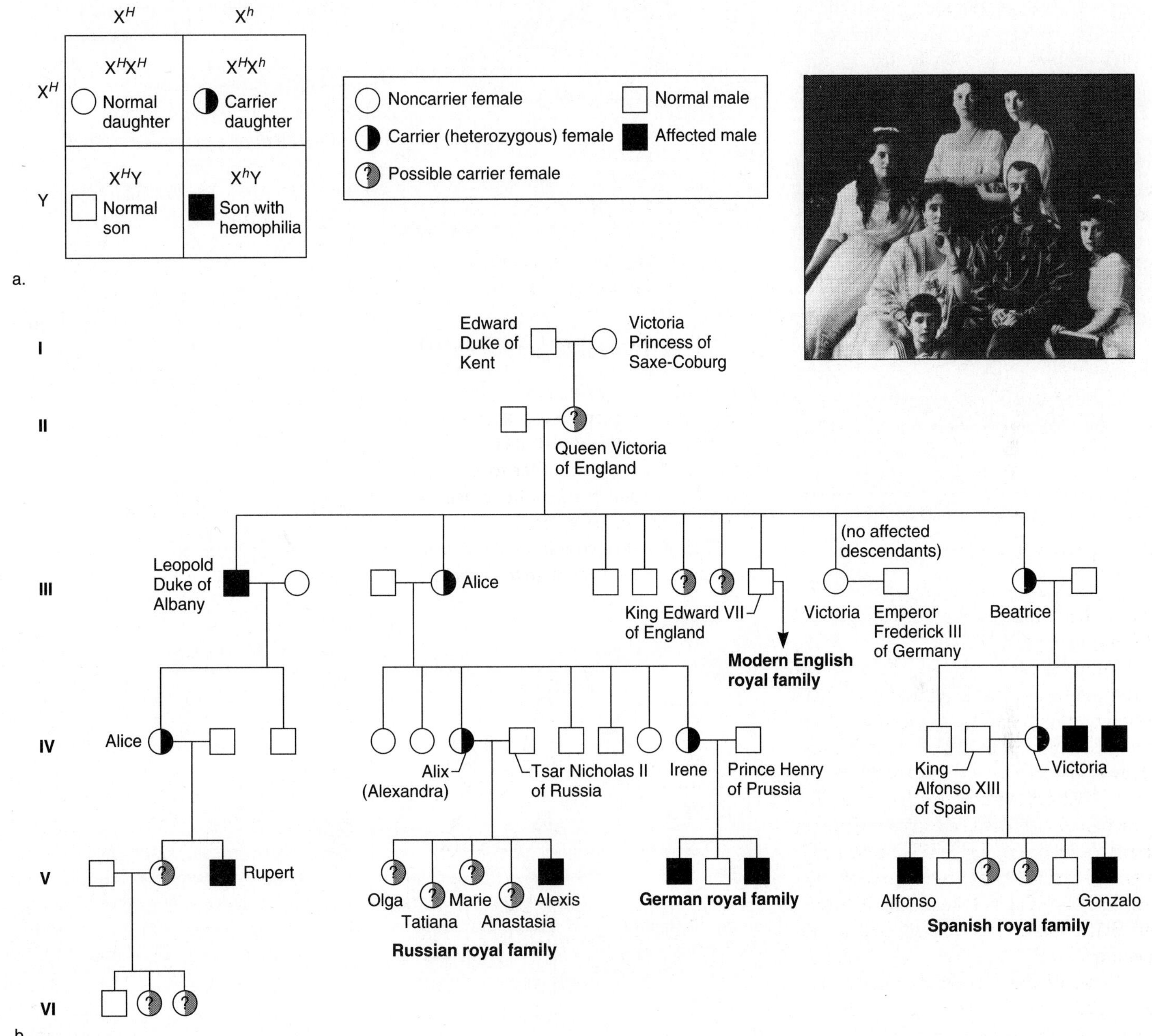


**figure 6.8**

**Hemophilia A.** (*a*) This X-linked recessive disease usually passes from a heterozygous woman (designated $X^HX^h$, where $X^h$ is the hemophilia-causing allele) to heterozygous daughters or hemizygous sons. (*b*) The disorder has appeared in the royal families of England, Germany, Spain, and the former Soviet Union. The mutant allele apparently arose in Queen Victoria, who was either a carrier or produced oocytes in which the gene mutated. She passed the alleles to Alice and Beatrice, who were carriers, and to Leopold, who had a case so mild that he fathered children. In the fourth generation, Alexandra was a carrier and married Nicholas II, Tsar of Russia. Alexandra's sister Irene married Prince Henry of Prussia, passing the allele to the German royal family, and Beatrice's descendants passed it to the Spanish royal family. This figure depicts only part of the enormous pedigree. The modern royal family in England does not carry hemophilia.

Reading 6.1

# Of Preserved Eyeballs and Duplicated Genes—Color Blindness

John Dalton, a famous English chemist, saw things differently than most people. In a 1794 lecture, he described his visual world. Sealing wax that appeared red to other people was as green as a leaf to Dalton and his brother. Pink wildflowers were blue, and Dalton perceived the cranesbill plant as "sky blue" in daylight, but "very near yellow, but with a tincture of red," in candlelight. He concluded, "that part of the image which others call red, appears to me little more than a shade, or defect of light." The Dalton brothers had the X-linked recessive trait of color blindness.

Curious about the cause of his color blindness, Dalton asked his personal physician, Joseph Ransome, to dissect his eyes after he died. Ransome snipped off the back of one eye, removing the retina, where the cone cells that provide color vision are nestled among the more abundant rod cells that impart black-and-white vision. Because Ransome could see red and green normally when he peered through the back of his friend's eyeball, he concluded that it was not an abnormal filter in front of the eye that altered color vision.

Fortunately, Ransome stored the eyes in dry air, where they remained relatively undamaged. In 1994, Dalton's eyes underwent DNA analysis at London's Institute of Ophthalmology. The research showed that Dalton's remaining retina lacked one of three types of pigments, called photopigments, that enable cone cells to capture certain wavelengths of light.

## Color Vision Basics

Cone cells are of three types, defined by the presence of any of three types of photopigments. An object appears colored because it reflects certain wavelengths of light, and each cone type captures a particular range of wavelengths. The brain then interprets the incoming information as a visual perception, much as an artist mixes the three primary colors to create many hues and shadings. Color vision results from the brain's interpretation of information from three types of input cells.

Each photopigment consists of a vitamin A-derived portion called retinal and a protein portion called an opsin. The presence of retinal in photopigments explains why eating carrots, rich in vitamin A, pro-

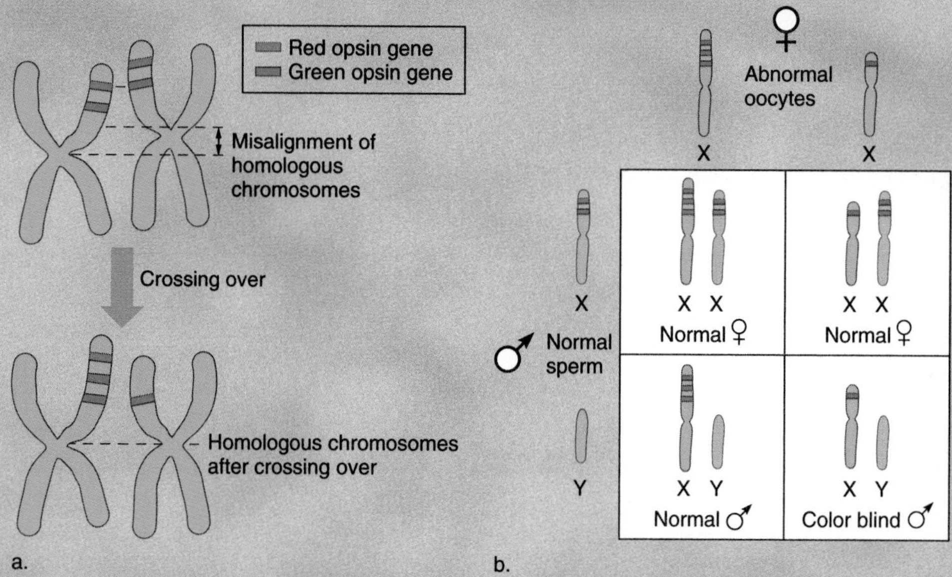


**figure 1**

**How color blindness arises.** (*a*) The sequence similarities among the opsin genes responsible for color vision may cause chromosome misalignment during meiosis in the female. Offspring may inherit too many, or too few, opsin genes. A son inheriting an X chromosome missing an opsin gene would be color blind. A daughter, unless her father is color blind, would be a carrier. (*b*) A missing gene causes X-linked color blindness.

square to trace transmission of the trait. Dominant and recessive alleles are indicated by superscripts to the X and Y chromosomes.

In Their Own Words in chapter 1 vividly describes one man's experience with this illness. In the royal families of England, Germany, Spain, and Russia, the mutant allele arose in one of Queen Victoria's X chromosomes; it was either a new mutation, or she inherited it. In either case, she passed it on through carrier daughters and one mildly affected son.

The transmission pattern of hemophilia A is consistent with the criteria for an X-linked recessive trait listed in table 6.2. A daughter can inherit an X-linked recessive disorder or trait if her father is affected and her mother is a carrier, because the daughter inherits one affected X chromosome from each parent. Without a biochemical test, though, an unaffected woman would not know she is a carrier of an X-linked recessive trait unless she has an affected son. A motes good vision. The presence of opsins—because they are controlled by genes—explains why color blindness is inherited. The three types of opsins correspond to short, middle, and long wavelengths of light. Mutations in opsin genes cause the three types of color blindness.

A gene on chromosome 7 encodes shortwave opsins, and mutations in it produce the rare autosomal "blue" form of color blindness. Dalton had deuteranopia (green color blindness), which means his eyes lacked the middle-wavelength opsin. In the third type, protanopia (red color blindness), long-wavelength opsin is absent. Deuteranopia and protanopia are X-linked.

## Molecular Analysis

John Dalton wasn't the only person interested in color blindness to work on his own tissue. Johns Hopkins University researcher Jeremy Nathans does so today—and, obviously, not posthumously. Nor is he color blind.

Nathans examined the color vision genes in his own cells. First, he used a cow version of a protein called rhodopsin that provides black-and-white vision to identify the human counterpart of this gene. Hypothesizing that the nucleotide sequence in rhodopsin would be similar to that in the three opsin genes, and therefore able to bind to them, Nathans used the human rhodopsin gene as a "probe" to search his own DNA for genes with similar sequences. He found three. One was on chromosome 7, the other two on the X chromosome.

Nathans's opsin genes were not entirely normal, however, and this gave him a big clue as to how the trait arises and why it is so common. On his X chromosome, Nathans has one red opsin gene and two green genes, instead of the normal one of each. Because the red and green genes have similar sequences, Nathans reasoned, they can misalign during meiosis in the female (figure 1). The resulting oocytes would then have either two or none of one opsin gene type. An oocyte lacking either a red or a green opsin gene would, when fertilized by a Y-bearing sperm, give rise to a color blind male.

People who are color blind must get along in a multicolored world. To help them overcome the disadvantage of not seeing important color differences, researchers have developed computer algorithms that convert colored video pictures into shades they can see. Figure 2 shows one of the tests typically used to determine whether someone is color blind. Absence of one opsin type prevents affected individuals from seeing a different color in certain of the circles in such a drawing. As a result, their brains cannot perceive a particular embedded pattern that other people can see.

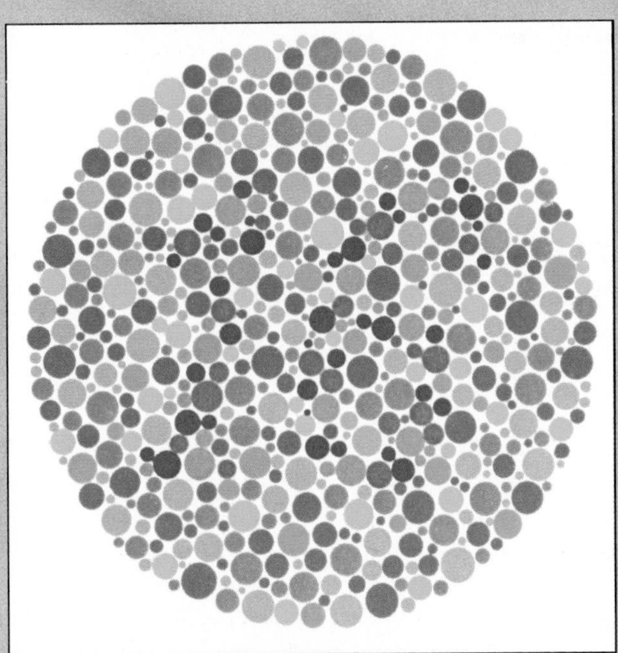

**figure 2**

Males with red-green color blindness cannot see the number 16 within this pattern of circles, as a person with normal color vision can.

Reproduced from *Ishihara's Tests for Colour Blindness,* published by Kanehara & Co. Ltd, Tokyo, Japan. Tests for color blindness cannot be conducted with this material. For accurate testing, the original plates should be used.

genetic counselor can estimate a potential carrier's risk by using probabilities derived from Mendel's laws, plus knowledge of X-linked inheritance patterns.

Consider a woman whose brother has hemophilia A. Both her parents are healthy, but her mother is a carrier because her brother is affected. The woman's chance of being a carrier is 1/2 (or 50 percent), which is the chance that she has inherited the X chromosome bearing the hemophilia allele from her mother. The chance of the woman conceiving a son is 1/2, and of that son inheriting hemophilia is 1/2. Using the product rule, the risk that she will have a son with hemophilia, out of all the possible children she can conceive, is $1/2 \times 1/2 \times 1/2$, or 1/8.

Table 6.3 lists several X-linked disorders. Most genes on the X chromosome actually are not related to sex determination, and are necessary for normal development or physiology in both sexes.

## table 6.3

**Some Disease-related Genes on the Human X Chromosome***

| Condition | Dominance (D or r) | Description |
|---|---|---|
| **Eye** | | |
| Green color blindness | r | Abnormal green cone pigments in retina |
| Megalocornea | r | Enlarged cornea |
| Norrie disease | r | Abnormal growth of retina, eye degeneration |
| Red color blindness | r | Abnormal red cone pigments in retina |
| Retinitis pigmentosa | r | Constriction of visual field, night blindness, clumps of pigment in eye |
| Retinoschisis | r | Retina degenerates and splits |
| **Inborn Errors of Metabolism** | | |
| Agammaglobulinemia | r | Lack of certain antibodies |
| Chronic granulomatous disease | r | Skin and lung infections, enlarged liver and spleen |
| Diabetes insipidus | r | Copious urination |
| Fabry disease | r | Abdominal pain, skin lesions, kidney failure |
| G6PD deficiency and favism | r | Hemolytic anemia after eating fava beans |
| Hemophilia A | r | Absent clotting factor VIII |
| Hemophilia B | r | Absent clotting factor IX |
| Hypophosphatemia | D + r | Vitamin D-resistant rickets |
| Hunter syndrome | r | Deformed face, dwarfism, deafness, mental retardation, heart defects, enlarged liver and spleen |
| Ornithine transcarbamylase deficiency | r | Mental deterioration, ammonia accumulation in blood |
| Severe combined immune deficiency | r | Lack of immune system cells |
| Wiskott-Aldrich syndrome | r | Bloody diarrhea, infections, rash, too few platelets |
| **Nerves and Muscles** | | |
| Lesch-Nyhan syndrome | r | Mental retardation, self-mutilation, urinary stones, spastic cerebral palsy |
| Menkes disease | r | Kinky hair, abnormal copper transport, brain atrophy |
| Muscular dystrophy, Becker and Duchenne forms | r | Progressive muscle weakness |
| **Other** | | |
| Amelogenesis imperfecta | D | Abnormal tooth enamel |
| Alport syndrome | r | Deafness, inflamed kidney tubules |
| Anhidrotic ectodermal dysplasia | r | Absence of teeth, hair, and sweat glands |
| Congenital generalized hypertrichosis | D | Dense hair growth over large areas of the body |
| Ichthyosis | r | Rough, scaly skin on scalp, ears, neck, abdomen, and legs |
| Incontinentia pigmenti | D | Swirls of skin color, hair loss, seizures, abnormal teeth |
| Rett syndrome | D | Mental retardation, neurodegeneration |

*Some of these conditions may be inherited through genes on the autosomes as well.

## X-Linked Dominant Inheritance

Most X-linked conditions and traits are recessive, as table 6.3 indicates. However, a few are dominant. Again, gene expression differs between the sexes (table 6.4).

A female who inherits a dominant X-linked allele has the associated trait or illness, but a male who inherits the allele is usually more severely affected because he has no other allele to offset it. The children of a normal man and a woman with a dominant, disease-causing gene on the X-chromosome face the risks summarized in figure 6.9.

An example of an X-linked dominant condition is incontinentia pigmenti, another "inborn error" that Archibald Garrod described in 1906 (see In Their Own Words in chapter 5). The name incontinentia pigmenti derives from the major sign in females with the disorder—swirls of pigment in the skin that arise when melanin penetrates the deeper skin layers. A newborn girl with incontinentia pigmenti has yellow, pus-filled vesicles on her limbs that come and go over the first few weeks. Then the lesions become warty and are eventually replaced by brown splotches that may remain for life, although they fade with time. Other symptoms include patches of hair loss; visual problems due to abnormal and scarred blood vessels in the retina; missing, peg-shaped, or underdeveloped teeth, and seizures. Rarely, mental retardation, paralysis, and developmental delay are part of the clinical picture.

Males with the condition are so severely affected that they do not survive to be born. This is why women with the disorder have a high rate of miscarriage, about 25 percent.

**table 6.4**

**Criteria for an X-Linked Dominant Trait**

1. Expressed in female in one copy.
2. Expressed much more severely in male.
3. High rates of miscarriage, due to early lethality in males.

The gene that causes incontinentia pigmenti is called *NEMO,* which stands for NF-kB essential modulator. The gene product activates NF-kB, which in turn activates various genes that carry out the immune response and apoptosis. Most affected individuals are missing a major part of the causative gene, and this deletion somehow affects differentiation of tissues that derive from ectoderm, such as skin, hair, nails, eyes, and the brain. Genetic tests that detect the deletion can confirm a diagnosis. Before such tests became available, children with incontinentia pigmenti were sometimes incorrectly diagnosed with severe skin infections, and then treated inappropriately with antibiotics. The genetic test can also identify women who have very mild cases, with the only symptoms abnormal teeth and bald patches.

In another X-linked dominant condition, congenital generalized hypertrichosis (CGH), a person has many extra hair follicles, and hence denser and more abundant hair. Unlike hirsutism, which is caused by a hormonal abnormality that makes a woman grow hair in places where it is usually more pronounced in males, CGH

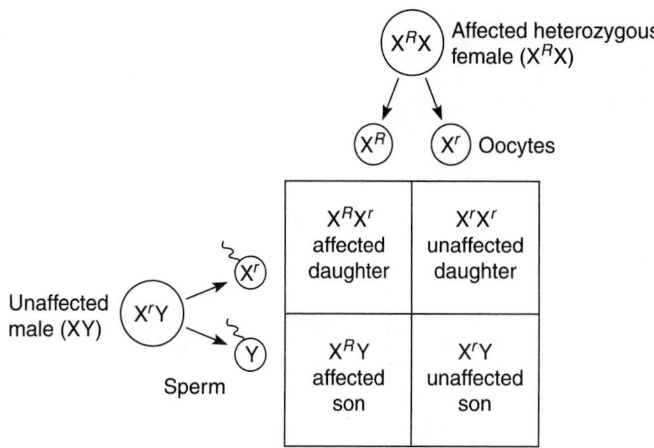


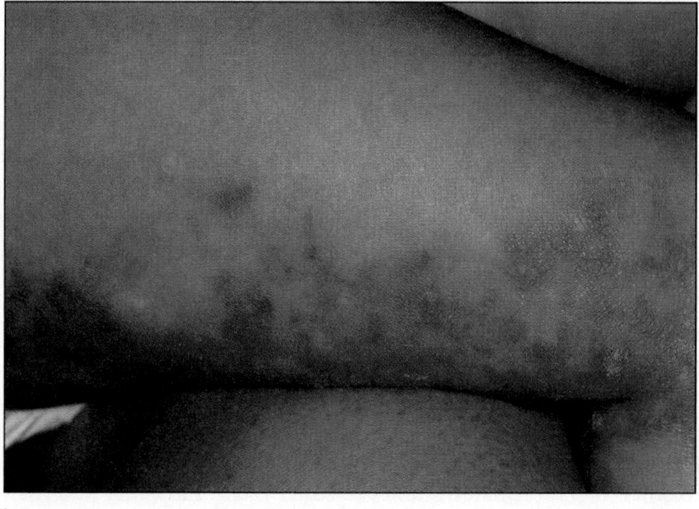


**figure 6.9**

**X-linked dominant inheritance.** *(a)* A woman who is a carrier of an X-linked dominant trait passes it to sons with a probability of 1 in 2, and to a carrier daughter with the same chance. Males are generally more severely affected than females. *(b)* Note the characteristic patchy pigmentation of this leg of a girl who has incontinentia pigmenti.

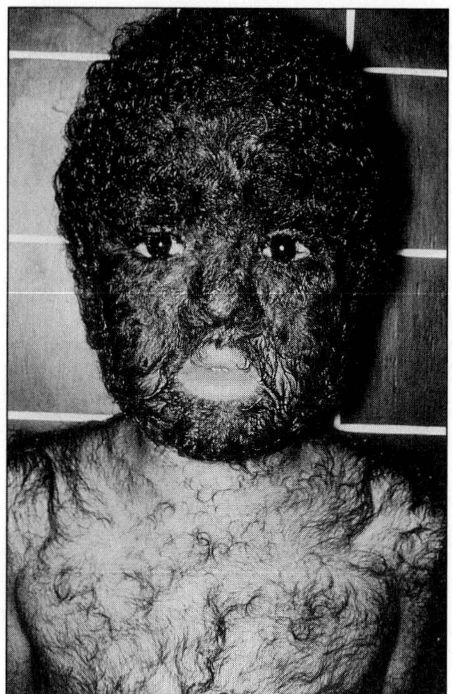

a.

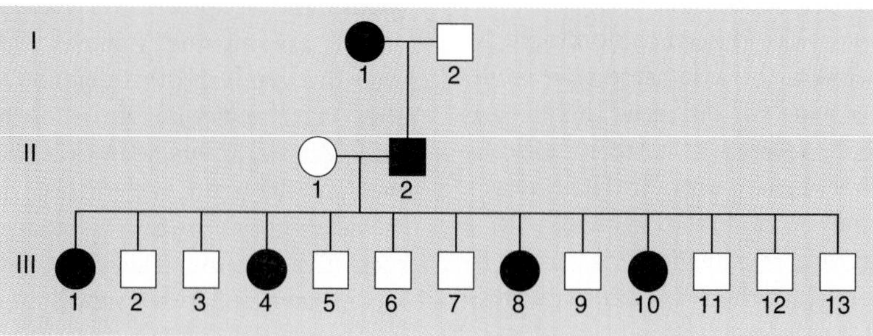


b.

**figure 6.10**

**An X-linked dominant condition.** (*a*) This six-year-old child has congenital generalized hypertrichosis (CGH). (*b*) In this partial pedigree of a large Mexican family with CGH, the affected male in the second generation has passed the condition to all of his daughters and none of his sons. This is because he transmits his X chromosome only to females.

causes excess facial and upper body hair that covers extensive areas of skin (figure 6.10*a*). The hair growth is milder and patchier in females because of hormonal differences and the mitigating presence of a second X chromosome.

Researchers studied a large Mexican family that had 19 relatives with CGH. The pattern of inheritance was distinctive for X-linked dominant inheritance, which is quite rare. In one portion of the pedigree, depicted in figure 6.10*b*, an affected man passed the trait to all four of his daughters, but to none of his nine sons. Because sons inherit the X chromosome from their mother, and only the Y from their father, they could not have inherited CGH from their affected father.

The mutant gene that causes CGH is atavistic, which means that it controls a trait also present in ancestral species. A version of the gene is probably present in chimpanzees and other hairy primates. Sometime in our distant past, the functional form of the gene must have mutated in a way that enables humans to grow dense hair only on their heads and in areas dictated by sex hormones. Chapter 15 discusses a particular mutation that may explain how humans lost body hair.

**Key Concepts**

Y-linked traits are passed on the Y chromosome, and X-linked traits on the X. Because a male is hemizygous, he expresses the genes on his X chromosome, whereas a female expresses recessive alleles on the X chromosome only if she is homozygous. X-linked recessive traits pass from carrier mothers to sons with a probability of 50 percent. • X-linked dominant conditions are expressed in both males and females but are more severe in males.

## 6.3 X Inactivation Equalizes the Sexes

Females have two alleles for every gene on the X chromosome, whereas males have only one. In mammals, a mechanism called **X inactivation** balances this inequality. Early in the embryonic development of a female, most of the genes on one X chromosome in each cell are inactivated. Which X chromosome is turned off in each cell—the one inherited from the mother or the one from the father—is random. As a result, a female mammal expresses the X chromosome genes inherited from her father in some cells and those from her mother in others (figure 6.11).

By studying rare human females who have lost a small part of one X chromosome, researchers identified a specific region, the **X inactivation center,** that shuts off much of the chromosome. A few genes on the chromosome, however, remain active. (Genes in the PAR and some others escape inactivation.) The inactivation process is under the control of a gene called *XIST.* The *XIST* gene encodes an RNA that binds to a specific site on the same (inactivated) X chromosome. From this point, the X chromosome is inactivated. Experiments in mice demonstrate the central role of the *XIST* gene in silencing the X chromosome. If *XIST* RNA is applied to an autosome, that chromosome is inactivated.

Once an X chromosome is inactivated in one cell, all its daughter cells have the same inactivated X chromosome. Because the inactivation occurs early in development, the adult female has patches of tissue that are phenotypically different in their expression of X-linked genes. But now that each cell in her body has only one active X

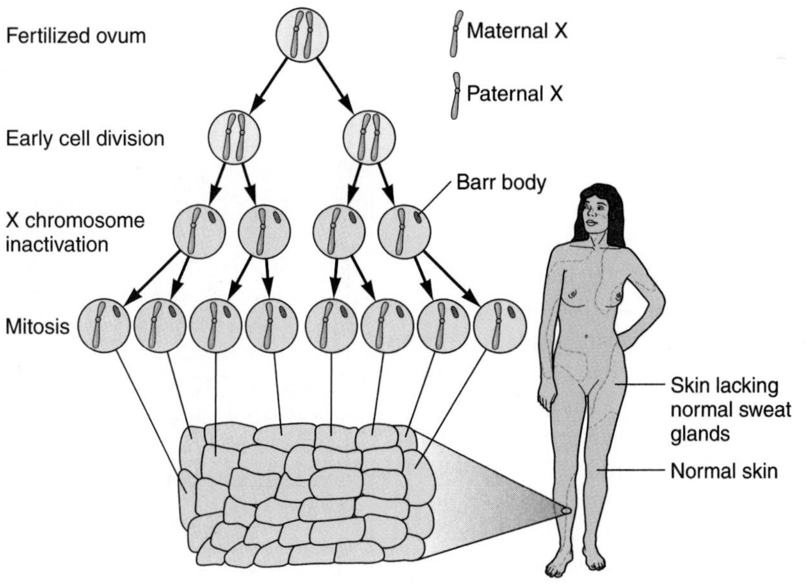


**figure 6.11**

**X inactivation.** A female is a mosaic for expression of genes on the X chromosome because of the random inactivation of either the maternal or paternal X in each cell early in prenatal development. In anhidrotic ectodermal dysplasia, a woman has patches of skin that lack sweat glands and hair.

chromosome, she is chromosomally equivalent to the male.

X inactivation can alter the phenotype (gene expression), but not the genotype. It is not permanent, because the inactivation is reversed in germline cells destined to become oocytes. Therefore, a fertilized ovum does not have an inactivated X chromosome. We can observe X inactivation at the cellular level because the turned-off X chromosome absorbs a stain much faster than the active X. The basis of this differential staining is that inactivated DNA has chemical methyl groups ($CH_3$) on it that prevent it from being transcribed into RNA, and also enable it to absorb stain.

The nucleus of a female cell in interphase has one dark-staining X chromosome called a **Barr body.** This structure is named after Murray Barr, a Canadian researcher who noticed these dark bodies in 1949 in the nerve cells of female cats. A normal human male cell has no Barr body because his one X chromosome remains active (figure 6.12).

In 1961, English geneticist Mary Lyon proposed that the Barr body is the inactivated X chromosome and that the turning off occurs early in development. She reasoned that for homozygous X-linked genotypes, X inactivation would have no effect. No matter which X chromosome is turned off, the same allele is left to be expressed. For heterozygotes, however, X inactivation leads to expression of one allele or the other. Usually this doesn't affect health, because enough cells express the functional gene product. However, some traits reveal striking evidence of X inactivation. For example, the swirls of skin color in incontinentia pigmenti patients reflect patterns of X inactivation in cells in the skin layers. Where the normal allele for melanin pigment is shut off, pale swirls develop. Where pigment is produced, brown swirls result.

The mosaic nature of the female due to X inactivation is also seen in the expression of the X-linked recessive condition anhidrotic ectodermal dysplasia. In heterozygous females, patches of skin in which the normal allele is inactivated lack sweat glands, whereas patches descended from cells in which the mutant allele is inactivated do have sweat glands (see figure 6.11).

Sometimes a female who is heterozygous for an X-linked recessive gene expresses the associated condition because the tissues

a.

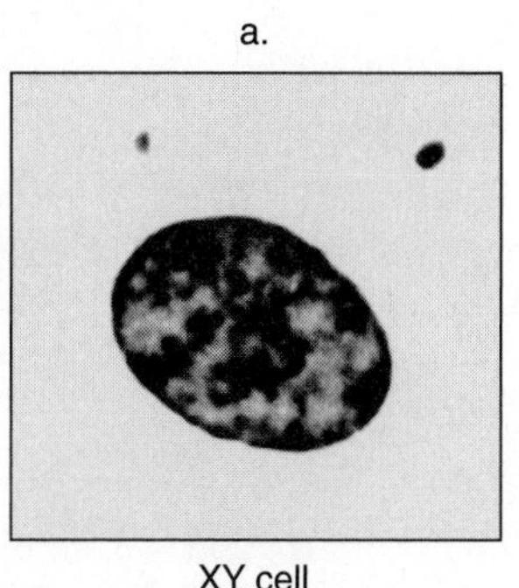

XY cell
No Barr bodies

b.

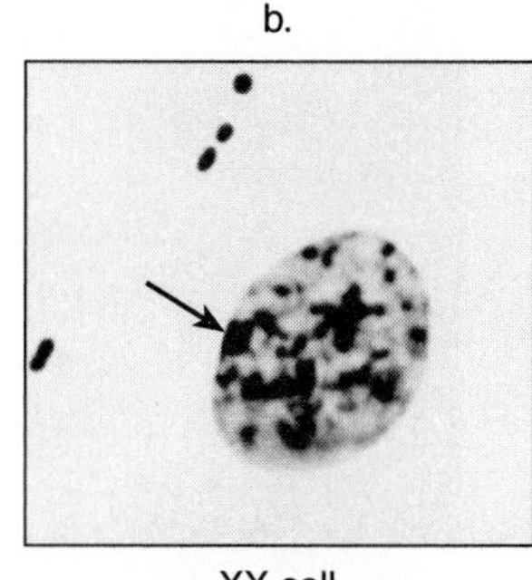

XX cell
One Barr body

c.

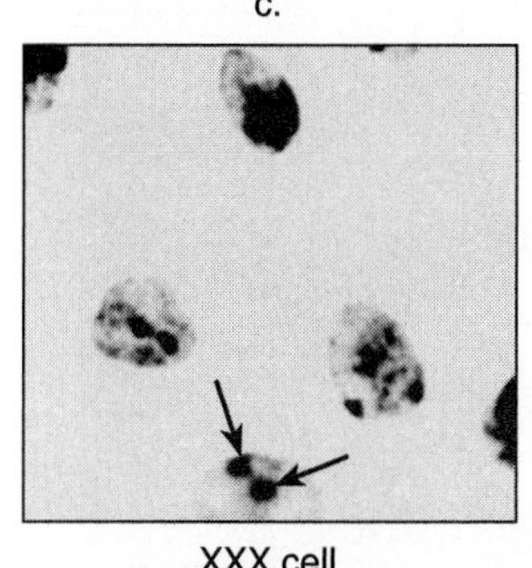

XXX cell
Two Barr bodies

d.

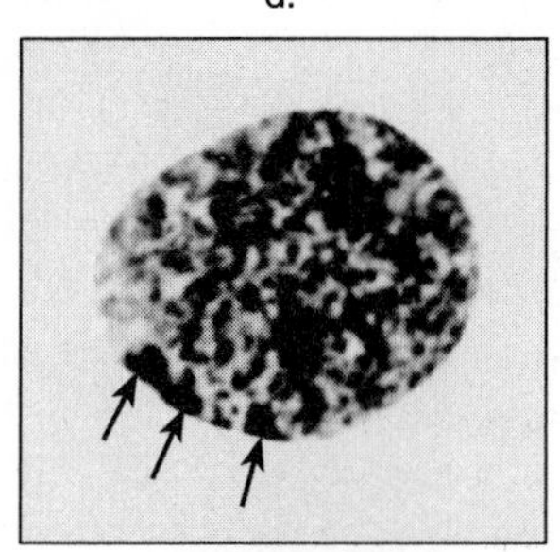

XXXX cell
Three Barr bodies

**figure 6.12**

**Barr body.** One X chromosome is inactivated in each cell of a female mammal. The turned-off X chromosome absorbs a stain faster than the active X chromosome, forming a dark spot called a Barr body. A normal male cell has no Barr body (*a*), and a normal female cell has one (*b*). Individual (*c*) has two Barr bodies and three X chromosomes. She is normal in appearance, behavior, and intellect, but has a lower IQ than her siblings. Rarely, a female has two extra X chromosomes, as in (*d*).

that the illness affects happen to have the normal version of the allele inactivated. This can happen in a carrier of hemophilia A. If the X chromosome carrying the normal allele for the clotting factor is turned off in many immature blood platelet cells, then the woman's blood will take longer than normal to clot—causing mild hemophilia. A carrier of an X-linked trait who expresses the phenotype is called a **manifesting heterozygote.** In incontinentia pigmenti, X-inactivation is skewed in the opposite direction: The mutation-bearing X chromosome is preferentially silenced. Perhaps females in whom the normal X chromosome is silenced in most cells are so severely affected that they do not survive to be born.

A striking and familiar example of X inactivation is the coat colors of tortoiseshell and calico cats. An X-linked gene confers brownish-black (dominant) or yellowish-orange (recessive) color. A female cat heterozygous for this gene has patches of either color, forming a tortoiseshell pattern that reflects different cells expressing either of the two different alleles (figure 6.13). The earlier the X inactivation, the larger the patches, because more cell divisions can occur afterward. White patches may occur due to epistasis by an autosomal gene that shuts off pigment synthesis. A cat with patches against such a white background is a calico. Can you see why tortoiseshell and calico cats are nearly always female? The only way a male can have these coat color patterns is if he inherits an extra X chromosome.

X inactivation can be used to detect carriers of some X-linked disorders. This is the case for Lesch-Nyhan syndrome, in which an affected boy has cerebral palsy, bites his or her fingers and lips to the point of mutilation, is mentally retarded, and passes painful urinary stones. Mutation results in defective or absent HGPRT, an enzyme. A woman who carries Lesch-Nyhan syndrome can be detected when hair bulbs from widely separated parts of her head are tested for HGPRT. (Hair is used for the test because it is accessible and produces the enzyme.) If some hairs contain HGPRT but others do not, she is a carrier. The hair cells that lack the enzyme have turned off the X chromosome that carries the normal allele; the hair cells that manufacture the normal enzyme have turned off the X chromosome that carries the disease-causing allele. The woman is healthy because her brain has enough HGPRT, but each son has a 50 percent chance of inheriting the disease.

**KEY CONCEPTS**

In female mammals, X inactivation compensates for differences between males and females in the numbers of gene copies on the X chromosome. Early in development, one X chromosome in each cell of the female is turned off. The effects of X inactivation can be noticeable for heterozygous alleles expressed in certain tissues.

## 6.4 Gender Effects on Phenotype

An X-linked recessive trait generally is more prevalent in males than females, as color blindness strikingly illustrates. Other situations, however, can affect gene expression in the sexes differently.

a.

b.

**figure 6.13**

**Visualizing X inactivation.** X inactivation is obvious in tortoiseshell (*a*) and calico (*b*) cats. X inactivation is rarely observable in humans because most cells do not remain together during development, as a cat's skin cells do.

## Sex-limited Traits

A **sex-limited trait** affects a structure or function of the body that is present in only males or only females. Such a gene may be X-linked or autosomal.

Understanding sex-limited inheritance is important in animal breeding. In cattle, for example, milk yield and horn development are traits that affect only one sex each, but the genes controlling them can be transmitted by either parent. In humans, beard growth and breast size are sex-limited traits. A woman does not grow a beard because she does not manufacture the hormones required for facial hair growth. She can, however, pass to her sons the genes specifying heavy beard growth.

An inherited medical condition that arises during pregnancy is by definition a sex-limited trait, since males do not become pregnant. Preeclampsia may be one such trait. This is a sudden increase in blood pressure that occurs in a pregnant woman as the birth nears. It is fatal to 50,000 women worldwide each year. For many years, obstetricians routinely asked their patients if their mothers had preeclampsia, because it has a tendency to occur in women whose mothers were affected. However, in 1998, a study of 1.7 million pregnancies in Norway revealed that if a man's first wife had preeclampsia, his second wife had double the relative risk of developing the condition too. The Norway study led to the hypothesis that a male can transmit a tendency to develop preeclampsia.

Another study in 2001 on 298 men and 237 women in Utah supports the hypothesis that preeclampsia risk is sex limited. This investigation found that women whose mothers-in-law had experienced preeclampsia when pregnant with the womens' husbands were of approximately twice the relative risk of developing the condition themselves. One hypothesis to explain the results of the two studies is that a gene from the male affects the placenta in a way that elevates the pregnant woman's blood pressure. Genes that may confer the increased risk include those whose protein products participate in blood clotting, blood glucose control, and blood pressure. An alternative explanation might be that an infection transmitted by the male causes preeclampsia. Further investigations are warranted.

## Sex-influenced Traits

In a **sex-influenced trait,** an allele is dominant in one sex but recessive in the other. Again, such a gene may be X-linked or autosomal. The difference in expression can be caused by hormonal differences between the sexes. For example, an autosomal gene for hair growth pattern has two alleles, one that produces hair all over the head and another that causes pattern baldness (figure 6.14). The baldness allele is dominant in males but recessive in females, which is why more men than women are bald. A heterozygous male is bald, but a heterozygous female is not. A bald woman is homozygous recessive. Even a bald woman tends to have some wisps of hair, whereas an affected male may be completely hairless on the top of his head.

a.

b.

c.

d.

**figure 6.14**

**Pattern baldness.** This sex-influenced trait was seen in the Adams family. John Adams (1735–1826) (*a*) was the second president of the United States. He was the father of John Quincy Adams (1767–1848) (*b*), the sixth president. John Quincy was the father of Charles Francis Adams (1807–1886) (*c*), a diplomat and the father of historian Henry Adams (1838–1918) (*d*).

## Genomic Imprinting

The ever-meticulous Gregor Mendel set up his pea plant crosses in two ways, with the female conferring a particular allele in some crosses, and the male in others. He noted no differences in gene expression dependent on parental sex, at least for the handful of traits he studied. The same is not true for mammals, including humans.

Up to 1 percent of our genes exhibit **genomic imprinting,** which is a difference in the expression of a gene or chromosomal region depending upon whether it is inherited from the father or the mother. It is also known as a "parent-of-origin" effect. The physical basis of genomic imprinting is that a gene from one parent is "silenced." The precise mechanisms of silencing are not well understood. Blocking of gene expression may result from the binding of methyl groups ($CH_3$) to certain chromosomal regions; a localized contortion of DNA and its surrounding proteins (chromatin); or proteins that bind to chromatin.

When silenced DNA is replicated during mitosis, the pattern of blocked gene expression is placed, or imprinted, exactly on the new DNA, covering the same genes it did in the parental DNA. In this way, the "imprint" of inactivation is perpetuated as if each such gene "remembers" which parent it arose from. For a particular imprinted gene, it is the copy from the father or the mother that is always silenced, even in different individuals. Imprinting is a type of **epigenetic** intervention, which is a layer of meaning that is stamped upon a gene without changing its DNA sequence.

Although imprints are passed from cell to daughter cell within an individual, they are removed and stamped anew onto genes with each generation (figure 6.15). The methyl groups fall away during meiosis. When a woman's primary oocytes undergo meiosis, the methyl groups come off, so that the secondary oocyte bears no imprint. Once a sperm joins the oocyte, the imprint of the new individual is set, corresponding to its sex. The same genome that bore a female imprint when it was part of a primary oocyte acquires a male imprint when it becomes part of a male embryo. Similarly, the imprint on a primary spermatocyte is removed and reset at fertilization. In this way, women can have sons and men can have daughters without passing on their sex-specific parental imprints.

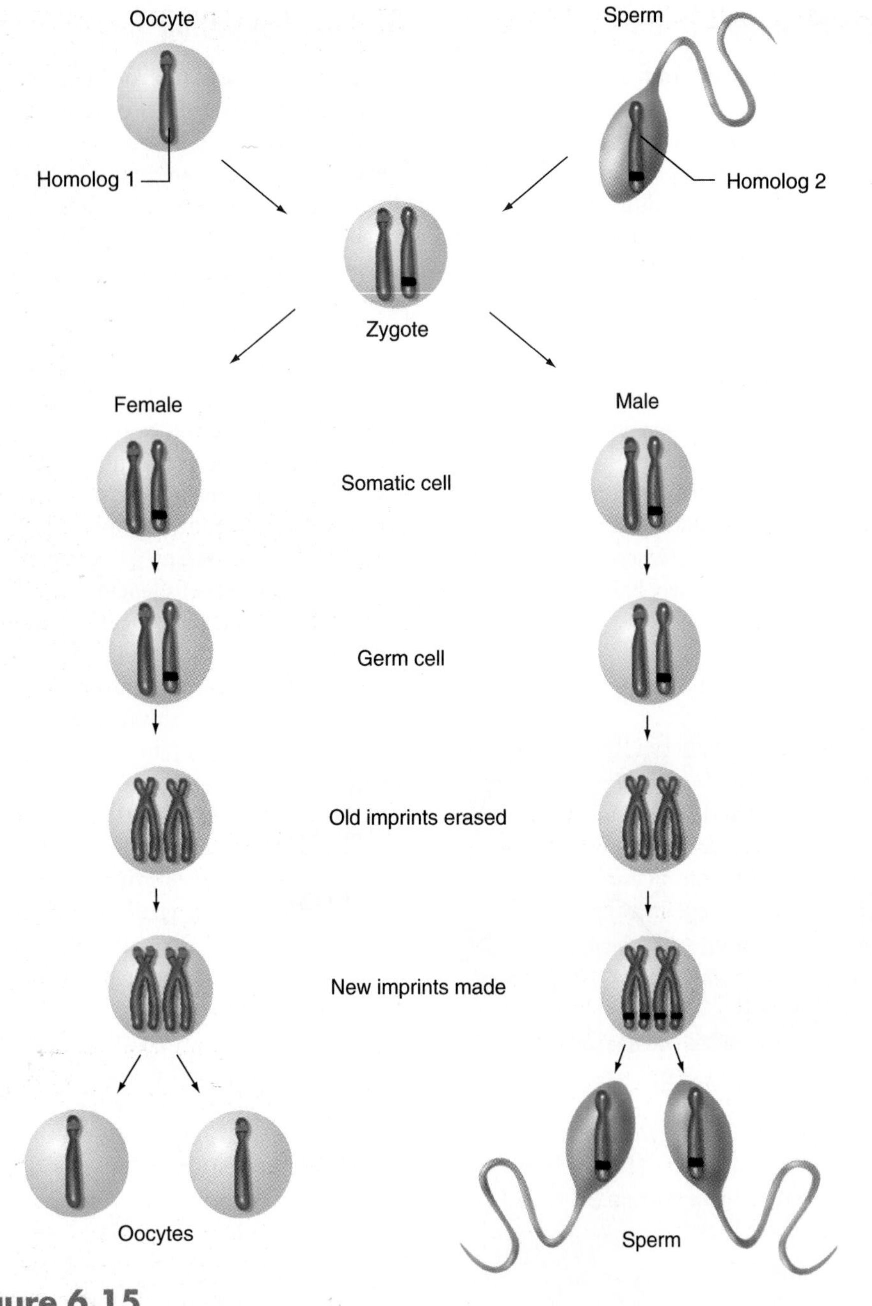


**figure 6.15**

**Genomic imprinting.** Imprints are erased during meiosis, then reinstituted according to the sex of the new individual.

While raising many questions, genomic imprinting also explains certain observations. For decades, geneticists had noted that in some disorders, symptoms start at a younger age, and progress more rapidly, when they are inherited from the father than the mother, or vice versa. Huntington disease, for example, typically begins near age 40, with uncontrollable movements and personality changes. Most cases that begin much earlier and are more severe, however, are inherited from the father. Puzzled geneticists attributed this trend to a statistical quirk such as reporting bias or small sample sizes in studies. But it has biological meaning—the cases *are* more severe when the mutant allele comes from the father, perhaps because the mother's normal allele is imprinted into silence.

Imprinting was discovered in the early 1980s, through experiments on early mouse embryos and examination of certain rare pregnancy problems in humans. To investigate the roles of the genomes passed from the male and the female to the fertilized ovum in mice, researchers created cells that contained two male pronuclei or two female pronuclei. The results were bizarre. When the fertilized ovum contained two male genomes, a normal placenta developed, but the embryo was tiny and quickly ceased to develop. A zygote with two female pronuclei, on the other hand, developed into an embryo, but the placenta was grossly abnormal. It looked like the male genome controlled placenta development, and the female genome, embryo development.

The mouse results were consistent with abnormalities of human development (table 6.5). In the rare cases where two sperm fertilize an oocyte, resulting in an individual with two male genomes and one female genome, an abnormal growth forms that is mostly placental tissue. In the even rarer situation where two sperm enter an oocyte that lacks a female pronucleus, a growth called a hydatidiform mole forms, which, again, is mostly placenta. (This control of the male genome on placental development might explain how a man can pass on a tendency to develop preeclampsia, which affects the placenta.) When an oocyte contains two female genomes and is fertilized by a single sperm, the embryo is normal, but the placenta is malformed. If a fertilized ovum contains only two female genomes but no male genome, a mass of random differentiated tissue, called a teratoma, grows. An embryo does not form.

Clearly, it takes two opposite-sex parents to produce a healthy embryo and placenta. More subtly, these types of abnormalities indicate that early in development, genes from a female parent direct different activities than do genes from a male parent. The requirement for a male or female contribution to a zygote may explain why cloned mammals are almost always unhealthy.

The first imprinted gene was discovered in mice in 1991. A mouse embryo manufactures insulin-like growth factor only if the normal allele comes from the father and the mother's version of the gene is "silenced." As is true for X inactivation, the effects of genomic imprinting are obvious only when an individual is heterozygous for the imprinted gene and the normal copy of the gene is silenced.

## table 6.5

**For Parental Genomes, Sex Matters**

| Genomes in Fertilized Ovum | Effect |
|---|---|
| **Mice** | |
| 2 male pronuclei, no female pronucleus | Normal placenta, tiny embryo |
| 2 female pronuclei, no male pronucleus | Normal embryo, malformed placenta |
| **Humans** | |
| 2 male pronuclei, 1 female pronucleus | Mostly placental tissue |
| 2 male pronuclei, no female pronucleus | Hydatidiform mole (mostly placental tissue) |
| 2 female pronuclei, 1 male pronucleus | Normal embryo, malformed placenta |
| 2 female pronuclei, no male pronucleus | Teratoma (mass of differentiated tissue) |

When geneticists learned about genomic imprinting, they went back to pedigrees tracing incompletely penetrant traits. In these families, a particular individual is known to have inherited a genotype associated with a particular phenotype—such as a person whose parent and child have polydactyly—but has no signs of the trait him- or herself. An imprinted gene that silences the mutant allele could explain these cases: The predicted genotype is present, but not expressed. The effects of genomic imprinting can also be severe. This occurs when a person lacks one copy of a particular gene, a type of mutation called a deletion. If the normal copy of the gene is silenced, the person has no other allele to provide the gene's normal function.

A striking example of genomic imprinting involves two different syndromes that arise from small deletions in the same region of chromosome 15 (figure 6.16 and table 6.6). A child with Prader-Willi syndrome is obese, has small hands and feet, eats uncontrollably, and does not mature sexually. The other condition, Angelman syndrome, was once called "happy puppet syndrome" because of its symptoms—frequent laughter, extended tongue, large jaw, poor muscle coordination that gives a "floppy" appearance, and convulsions that make the arms flap. He or she is severely mentally retarded. In many cases of Prader-Willi syndrome, only the mother's chromosome 15 region is expressed; in Angelman syndrome, the father's gene (or genes) is expressed. It appears as if

## table 6.6

**Prader-Willi Versus Angelman Syndromes**

| | Prader-Willi Syndrome | Angelman Syndrome |
|---|---|---|
| Incidence | 1/15,000 | 1/20,000 |
| Symptoms | Obesity<br>Small hands and feet<br>Developmental delay | Severe mental retardation<br>Seizures<br>"Happy puppet" appearance |
| Cause | Usually deletion of paternal chromosome 15 segment, rarely uniparental disomy of maternal genes | 70 percent have deletion of maternal chromosome 15 segment, 30 percent have uniparental disomy of parental genes |

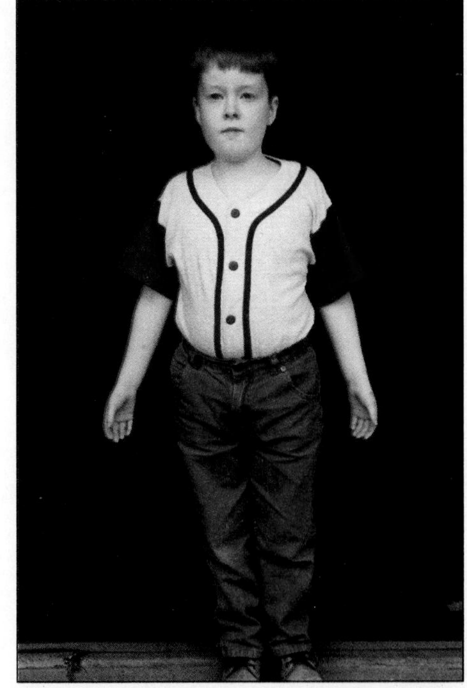

a.

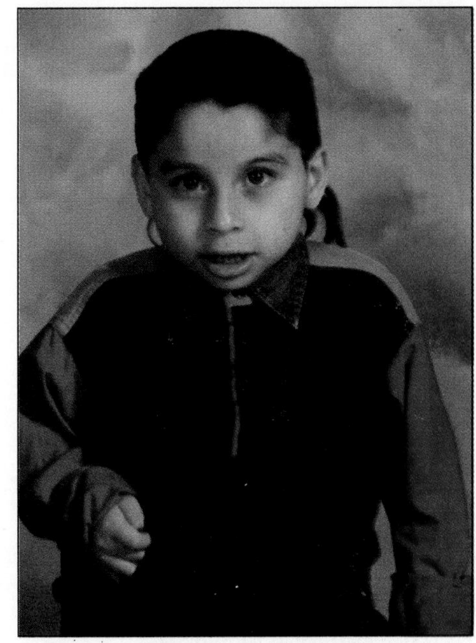

b.

**figure 6.16**

**Prader-Willi and Angelman syndromes.** Two syndromes result from missing genetic material in the same chromosomal region. (*a*) Tyler has Prader-Willi syndrome. Note his small hands and feet. (*b*) Angelman syndrome also causes mental retardation, but the other symptoms differ from those of Prader-Willi syndrome. Note the distinctive features of this child's face, described originally as a "happy puppet."

# Sex Reassignment: Making a Biological "He" into a Social "She"

Bruce Reimer was born in 1965, an identical twin. At age eight months, most of Bruce's penis was accidentally burned off during a botched circumcision. After much agonizing, and on the advice of physicians and psychologists, the parents decided that it would be best for Bruce to "reassign" his gender as female. And so at 22 months of age, corrective surgery created Brenda from Bruce. In numerous articles and books, the prominent psychologist in charge hailed the transformation as a resounding success. The case came to serve as a precedent for early surgical intervention for hundreds, maybe thousands, of other children born with "ambiguous genitalia" or structures characteristic of both sexes. All of these conditions are termed "intersexes." About 1 in 2,000 newborns are intersexes, with a few others, the result of a surgical accident. A rare form of congenital intersex with an effect similar to that of a surgical accident is called micropenis, and it affects 1 in 50,000 newborns.

## Gender Identity Can't Be Changed

Evidence is emerging that sex reassignment, although well-intentioned, is perhaps not the best choice of treatment for some intersexual individuals. Reassessment of sex reassignment began when the Reimer case came to public attention. Reality for young Brenda was far different from the published descriptions.

Always uneasy in her dress-clad body, Brenda suffered relentless ridicule and confusion, because it was always obvious to her and others that she was more than a "tomboy"—she was a boy, despite her surgically altered appearance. Comparison to twin Brian made matters even worse. When confronted with the truth at age 14, "Brenda" threatened suicide unless allowed to live as the correct gender. And so at age 14, she became David Reimer. He eventually married, adopted his stepchildren, and is today a grandfather. He told his story to journalist John Colapinto, who published riveting accounts in *Rolling Stone* magazine and in a book. At about the same time, 1997, a groundbreaking paper by Keith Sigmundson, David's psychiatrist in his hometown of Winnipeg, and Milton Diamond of the University of Hawaii, supported David's contention that gender identity is due more to nature than nurture.

A study published in 2000 added more evidence. William Reiner of Johns Hopkins University followed the fates of 14 children born with a form of intersex called cloacal exstrophy. "These kids have a 'pelvic field' defect that is probably a problem in genetic timing in the embryo, a control gene that is turned off too soon, or on too late. A number of anomalies are associated with the problem. A boy has no penis, but normal testicles. In a girl, there is no clitoris but usually a vagina. She is raised as a girl," he explains. The individuals in the study were XY, and had normal testicles and hormone levels, but no penis. Twelve of them were reassigned as female—and all behaved as boys throughout childhood. Six of them declared themselves male sometime between the ages of 5 and 12 years, as David Reimer had at age 14. The two children who were not surgically converted into female are normal males who lack penises, something that surgery later in life may be able to correct. Since then, other studies have confirmed Reiner's findings.

## The Surgical Yardstick

In the past, physicians based the decision to remove a small or damaged penis and reassign sex as female on a yardstick, of sorts. If a newborn's stretched organ exceeded an inch, he was deemed a he. If the protrusion was under three-eighths of an inch, she was deemed a she. Organs that fell in between were shortened into a clitoris during the first week of life, and girlhood officially began. Further plastic surgeries and hormone treatments during puberty completed the superficial transformation, with external female tissue sculpted from scrotal tissue. The reverse, creating a penis, is much more difficult and was therefore usually delayed several months. These surgeries usually destroy fertility and sexual sensation.

Easier to surgically treat are babies with congenital adrenal hyperplasia (CAH) due to 21-hydroxylase deficiency, the most common cause of intersex. The individual is XX, but overproduces masculinizing hormones (androgens). The result is a girl with a clitoris so large that it looks like a small penis. Thirty years ago, surgeons would cut away most of the extra tissue in a baby, and create a vagina from skin flaps. Recently, with the discovery that these females needed a second surgery in adolescence anyway, treatment is being postponed, giving these young women the chance to take part in decisions affecting their bodies.

Delaying surgery until a person can decide for his or herself, as is the case for CAH, may be the best approach for intersex individuals. Sex reassignment surgery is a bioethical issue that involves paternalism, confidentiality, the doctor-patient relationship, and the promise of physicians to "do no harm." Sums up Alice Dreger of Michigan State University, who has researched and written extensively on intersexuality, "Gender identity is very complicated, and it looks from the evidence like the various components interact and matter in different ways for different individuals. That's why unconsenting children and adults should never be subjected to cosmetic, medically unnecessary surgeries designed to alter their sexual tissue. We cannot predict what parts they may want later."

the differences between these two disorders derive from the gender of the parent whose gene(s) is expressed, because the same gene or genes are implicated.

The function of genomic imprinting isn't known, but researchers hypothesize that because many imprinted genes take part in early development, it may be a way to finely regulate the amounts of key proteins in the embryo. The fact that some genes lose their imprints after birth supports this idea of early importance. It also appears that imprinted genes occur in clusters, and that clusters are under the control of other regions of DNA called imprinting centers. Perhaps one gene in a cluster is essential for early development, and the others become imprinted simply because they are nearby.

The region of chromosome 15 implicated in Prader-Willi and Angelman syndromes is unstable. The genes that are deleted in the two syndromes are normally bracketed by regions where the DNA sequence is highly repeated. Such repeats can lead to chromosome instability because during meiosis, repeats located at opposite ends of the cluster on the two homologs can pair, which loops out the genes in between them. A germ cell bearing a deleted region results, although the parent in whom the cell originated has a normal chromosome 15. Symptoms of Prader-Willi arise because several paternal genes that are not normally imprinted (that is, are normally active) are missing. In Angelman syndrome, a normally active single maternal gene is deleted. It is apparently the location of imprinted genes in an unstable chromosomal region that sets the stage for these two conditions to arise.

Some cases of Prader-Willi syndrome have been traced to a tiny deletion of a controlling imprinting center. The deletion abolishes the normal erasing of imprints in germ cells. So, if a male has such a deletion, when he manufactures sperm cells, the copy of chromosome 15 he inherited from his mother does not acquire his paternal imprint. If that sperm fertilizes an oocyte, which has the maternal imprint, the resulting zygote has two maternally active chromosome 15s, and the offspring has Prader-Willi syndrome.

Researchers have so far identified genomic imprinting in 31 human genes. It has been implicated in several disorders, including rare forms of diabetes mellitus, autism, and tumor syndromes. Clues to a condition that indicate it is associated with genomic imprinting include increased severity when it is inherited from one parent seen in many families, and also a phenomenon called uniparental disomy. This term literally means "two bodies from one parent," and refers to an offspring who inherits both copies of a gene from one parent. Chapter 12 discusses uniparental disomy further. If a disease is known to result from uniparental disomy, as are Prader-Willi and Angelman syndromes, then the causative genes are imprinted.

Despite attempts to equalize treatment of the sexes in terms of sociology, the sexes are clearly different, genetically speaking. It's interesting that some manifestations of sexual distinctions are long known and well understood, such as the transmission of hemophilia in the royal families of Europe, and the fact that calico cats are always female. Yet at the same time, we have barely begun to comprehend the mechanism and function of genomic imprinting. Perhaps information from the human genome project will further enlighten us about the meaning of being male or female.

### Key Concepts

A sex-limited trait affects body parts or functions present in only one gender. • A sex-influenced allele is dominant in one sex but recessive in the other. • In genomic imprinting, the phenotype differs depending on whether a gene is inherited from the mother or the father. Methyl groups may bind to DNA and temporarily suppress gene expression in a pattern determined by the individual's sex.

## Summary

### 6.1 Sexual Development

1. Sexual identity includes sex chromosome makeup; gonadal specialization; phenotype (reproductive structures); and gender identity.
2. In humans, the male is the **heterogametic sex,** with an X and a Y chromosome. The female, with two X chromosomes, is the **homogametic sex.** The ***SRY*** gene on the Y chromosome determines sex.
3. The human Y chromosome includes a few hundred genes, with two **pseudoautosomal regions** and a large region that does not recombine. Y-linked genes may correspond to X-linked genes, be similar to them, or be unique to the Y chromosome.
4. If the SRY gene is expressed, undifferentiated gonads develop as testes. If SRY is not expressed, the gonads develop as ovaries.
5. Starting about eight weeks after fertilization, **sustentacular cells** in the testes **secrete anti-Müllerian hormone,** which prevents development of female structures, and **interstitial cells** produce **testosterone,** which triggers development of the epididymides, vasa deferentia, seminal vesicles, and ejaculatory ducts.
6. Testosterone converted to DHT controls development of the urethra, prostate gland, penis, and scrotum. If SRY doesn't turn on, the Müllerian ducts continue to develop into female reproductive structures.
7. Homosexuality is probably influenced by genes. Evidence for an inherited component to homosexuality is accumulating.

### 6.2 Traits Inherited on Sex Chromosomes

8. Y-linked traits are rare. They are passed from fathers to sons only.
9. Males are **hemizygous** for genes on the X chromosome and express such genes because they do not have another allele on a homolog. An X-linked trait passes from mother to son because he inherits his X chromosome from his mother and his Y chromosome from his father.
10. An X-linked allele may be dominant or recessive. X-linked dominant traits are usually more devastating to males than females.

### 6.3 X Inactivation Equalizes the Sexes

11. **X inactivation** shuts off one X chromosome in each cell in female mammals, making them mosaics for heterozygous genes on the X chromosome. This phenomenon evens out the dosages of genes on the sex chromosomes between the sexes.
12. A female who expresses the phenotype corresponding to an X-linked gene she carries is a **manifesting heterozygote.**

### 6.4 Gender Effects on Phenotype

13. **Sex-limited traits** may be autosomal or sex-linked, but they only affect one sex because of anatomical or hormonal gender differences.
14. A **sex-influenced gene** is dominant in one sex but recessive in the other.
15. In **genomic imprinting,** the phenotype corresponding to a particular genotype differs depending on whether the parent who passes the gene is female or male. Imprints are erased during meiosis and reassigned based on the sex of a new individual. Methyl groups that temporarily suppress gene expression are the physical basis of genomic imprinting.

# Review Questions

1. How is sex expressed at the chromosomal, gonadal, phenotypic, and gender identity levels?
2. How do genes in the pseudoautosomal region of the Y chromosome differ from X-Y homologs?
3. What are the phenotypes of the following individuals?
   a. a person with a mutation in the *SRY* gene, rendering it nonfunctional
   b. a normal XX individual
   c. an XY individual with a block in testosterone synthesis
4. List the cell types and hormones that participate in development of male reproductive structures.
5. List the events that must take place for a fetus to develop as a female.
6. Cite evidence that may point to a hereditary component to homosexuality.
7. Why would it be extremely unlikely to see a woman who is homozygous dominant for an X-linked dominant disease?
8. Why are male calico cats very rare?
9. How might X inactivation cause patchy hairiness on women who have congenital generalized hypertrichosis (CGH), even though the disease-causing allele is dominant?
10. How does X inactivation even out the "doses" of X-linked genes between the sexes?
11. Traits that appear more frequently in one sex than the other may be caused by genes that are inherited in an X-linked, sex-limited, or sex-influenced fashion. How might you distinguish among these possibilities?
12. Cite evidence that genetic contributions from both parents are necessary for normal prenatal development.

# Applied Questions

1. In Hunter syndrome, lack of the enzyme iduronate sulfate sulfatase leads to buildup of carbohydrates called mucopolysaccharides. In severe cases, this may cause the liver, spleen, and heart to swell. In mild cases, deafness may be the only symptom. A child with this syndrome is also deaf and has unusual facial features. Hunter syndrome is inherited as an X-linked recessive condition. Intellect is usually unimpaired and life span can be normal. A man who has mild Hunter syndrome has a child with a woman who is a carrier.
   a. What is the probability that if the child is a boy, it would inherit Hunter syndrome?
   b. What is the chance that if the child is a girl, she would inherit Hunter syndrome?
   c. What is the chance that a girl would be a carrier?
   d. How might a carrier of this condition experience symptoms?
2. Coffin-Lowry syndrome is a rare disorder. Symptoms include short, tapered fingers; abnormal finger and toe bones; puffy hands; soft, elastic skin; curved fingernails; facial anomalies; and sometimes hearing loss and heart problems. Geneticists believe it is X-linked recessive, but girls are affected to a much lesser degree than are boys. Suggest two explanations for why girls tend to have milder cases.
3. Amelogenesis imperfecta is an X-linked dominant condition that affects tooth enamel. Affected males have extremely thin enamel layers all over each tooth. Female carriers, however, have grooved teeth from the uneven deposition of enamel. Explain the difference in phenotype between the sexes for this condition.
4. If Alfonso in figure 6.8 were to have children with distant relative Olga, what is the chance that a son of theirs would inherit hemophilia?
5. If the daughter in figure 6.7 had had a normal amount of the enzyme, what other pattern of inheritance might explain the pedigree?
6. Insulin-like growth factor II is an imprinted gene. The maternal allele is inactivated. This gene encodes a protein that enables an embryo to gain weight. How does the action of this gene support David Haig's theory of genomic imprinting?
7. Reginald has mild hemophilia A that he can control by taking a clotting factor. He marries Lydia, whom he met at the hospital where he and Lydia's brother, Marvin, receive their treatment. Lydia and Marvin's mother and father, Emma and Clyde, do not have hemophilia. What is the probability that Reginald and Lydia's son will inherit hemophilia A?
8. Harold works in a fish market, but the odor does not bother him because he has anosmia, an X-linked recessive lack of sense of smell. Harold's wife, Shirley, has a normal sense of smell. Harold's sister, Maude, also has a normal sense of smell, as does her husband, Phil, and daughter, Marsha, but their identical twin boys, Alvin and Simon, cannot detect odors. Harold

and Maude's parents, Edgar and Florence, can smell normally. Draw a pedigree for this family, indicating people who must be carriers of the anosmia gene.

**9.** Metacarpal 4–5 fusion is an X-linked recessive condition in which certain finger bones are fused. It occurs in many members of the Flabudgett family, depicted in the pedigree below:

a. Why are three females affected, considering that this is an X-linked condition?

b. What is the risk that individual III-1 will have an affected son?

c. What is the risk that individual III-5 will have an affected son?

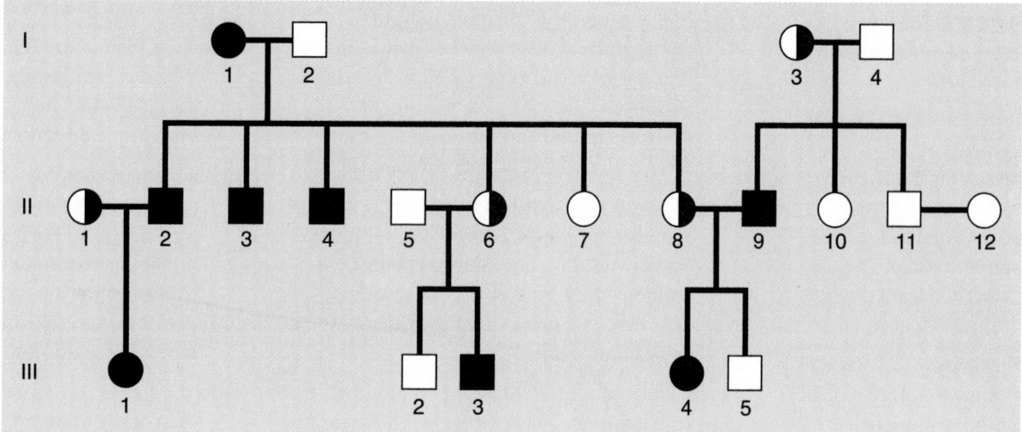


**10.** Herbert is 58-years-old and bald. His wife, Sheri, also has pattern baldness. What is the risk that their son, Frank, will lose his hair?

**11.** The Addams family knows of three relatives who had kinky hair disease, an X-linked recessive disorder in which a child does not grow, experiences brain degeneration, and dies by the age of two. Affected children have peculiar white stubby hair, from which the disorder takes its name.

Wanda Addams is hesitant about having children because her two sisters have each had sons who died from kinky hair disease. Her mother had a brother who died of the condition, too. The pedigree for the family is shown at the right.

a. Fill in the symbols for the family members who must be carriers of kinky hair disease.

b. What is the chance that Wanda is a carrier?

c. If Wanda is a carrier, what is the chance that a son of hers would inherit kinky hair disease?

d. Why don't any women in the family have kinky hair disease?

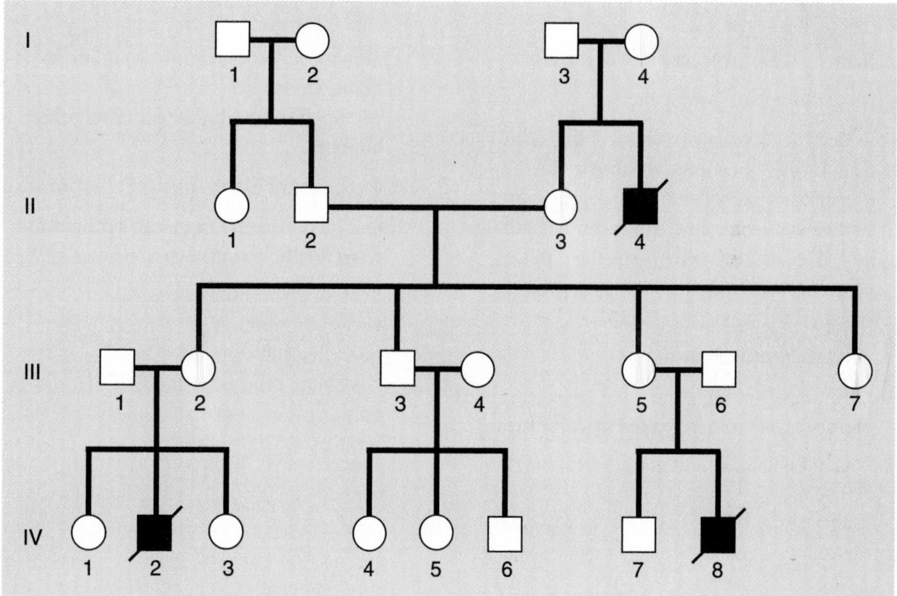


Colapinto, J. *As Nature Made Him.* 2000. New

# Suggested Readings


York: HarperCollins. The true story of David Reimer, whose sex reassignment surgery was a tragic failure.

Crowe, Mark A., and William D. James. November 17, 1993. X-linked ichthyosis. *The Journal of the American Medical Association.* A grandfather and grandson each had unusual scaly skin.

Esplin, M. Sean, et al. March 22, 2001. Paternal and maternal components of the predisposition to preeclampsia. *The New England Journal of Medicine* 344, no. 12:867–72. The tendency to develop high blood pressure toward the end of pregnancy may be inherited from the father.

Hall, Brian K. June 1995. Atavisms and atavistic mutations. *Nature Genetics,* vol. 10. Is CGH a "throwback" mutation to our hairier forebears?

Hamer, D. July 16, 1993. A linkage between DNA markers on the X-chromosome and male sexual orientation. *Science,* vol. 261. The original controversial report on a possible "gay gene."

Hanel, M. L., and R. Wevrick. March 2001. The role of genomic imprinting in human developmental disorders: Lessons from Prader-Willi syndrome. *Clinical Genetics* 59:156–64. Two completely different disorders result from expression of the same maternal versus paternal genes.

Hunt, David M., et al. February 17, 1995. The chemistry of John Dalton's color blindness. *Science,* vol. 267. The famous chemist requested that his eyeballs be examined after his death to localize the cause of his color blindness.

Lewis, Ricki. July 10, 2000. Re-evaluating sex reassignment. *The Scientist* 14:14. Research supports anecdotal evidence that sex reassignment is often unsuccessful.

Lyon, Mary F. July 1998. X-chromosome inactivation spreads itself: Effects in autosomes. *The American Journal of Human Genetics* 63:17–19. The discoverer of X inactivation maintains that we still have more to learn about this process.

Makrinov, Eleni. June 2001. TTYZ: A multicopy Y-linked gene family. *Genome Research* 11, no. 6:935–45. The Y chromosome has few protein-encoding genes, but many repeats.

Smahi, Asmae, et al. May 25, 2000. Genomic rearrangement in *NEMO* impairs NF-kB activation and is a cause of incontinentia pigmenti. *Nature* 405:466–72. Malfunction of a transcription factor causes symptoms in ectodermal derivatives in this X-linked dominant condition.

Tilford, Charles A. February 15, 2001. A physical map of the human Y chromosome. *Nature* 409:943–45. The Y chromosome carries very few protein-encoding genes.


# On the Web

Check out the resources on our website at

**www.mhhe.com/lewisgenetics5**

On the web for this chapter, you will find additional study questions, vocabulary review, useful links to case studies, tutorials, popular press coverage, and much more. To investigate specific topics mentioned in this chapter, also try the links below:

Angelman Syndrome Foundation
**www.chem.ucsd.edu/nsf**

Color Blindness
**members.aol.com/nocolorvsn/color.htm**

The Human Genome: A guide to online information resources.
**www.ncbi.nlm.nih.gov/genome/guide/human/**

Intersex Society of North America **www.isna.org**

National Incontinentia Pigmenti Foundation
**www.medhelp.org/www/nipf.htm**

Online Mendelian Inheritance in Man
**www.ncbi.nlm.nih.gov/entrez/query.fcgi?db=OMIM**
Angelman syndrome 105830
anhidrotic ectodermal dysplasia 129490, 224900, 305100
color blindness 303700, 303800, 303900
congenital generalized hypertrichosis 307150
incontinentia pigmenti 308300
Lesch-Nyhan syndrome (HGPRT deficiency) 308000
Prader-Willi syndrome 176270
pseudohermaphroditism 264300
SRY 480000
XX male syndrome 278850
XX female syndrome 306100

CHAPTER

# 7

# Multifactorial Traits

## 7.1 Genes and the Environment Mold Most Traits

Environmental influences mold many human traits. Fingerprint pattern, skin color, disease susceptibilities, and even intelligence and behavior are molded by both genes and the environment.

## 7.2 Methods Used to Investigate Multifactorial Traits

Separating genetic from environmental influences on phenotype is enormously difficult. Observations on how common a trait is in a population, combined with theoretical predictions of the percentage of genes that certain relatives should share, enable us to calculate heritability. This is an estimate of genetic contribution to variations among individuals in a trait. Twins are important in studying multifactorial traits.

## 7.3 Some Multifactorial Traits

Cardiovascular health and body weight are two common characteristics that are influenced by the interactions of specific predisposing genes and environmental factors.

A woman who is a prolific writer has a daughter who becomes a successful novelist. An overweight man and woman have obese children. A man whose father suffers from alcoholism has the same problem. Are these characteristics—writing talent, obesity, and alcoholism—inherited or imitated? Or are they a combination of nature (genetics) and nurture (the environment)?

Most of the traits and medical conditions mentioned so far in this book are single-gene characteristics, inherited according to Mendel's laws, or linked on the same chromosome. Many single gene disorders are very rare, each affecting one in hundreds or even thousands of individuals. Using Mendel's laws, geneticists can predict the probability that certain family members will inherit single gene conditions. Most more common traits and diseases, though, can seem to "run in families" with no apparent pattern, or they occur sporadically, with just one case in a family. Genes rarely act completely alone. Even single gene disorders are modified by environmental factors or other genes. This chapter discusses the nature of non-Mendelian characteristics, and the tools used to study them. Chapter 8 focuses on the most difficult traits to assess for their inherited components—behaviors.

## 7.1 Genes and the Environment Mold Most Traits

On the first page of the first chapter of *On the Origin of Species,* Charles Darwin noted that two factors are responsible for variation among organisms—"the nature of the organism and the nature of the conditions." Darwin's thoughts were a 19th-century musing on "heredity versus the environment." Though this phrase might seem to indicate otherwise, genes and the environment are not adversaries. They are two forces that interact, and they do so in ways that mold many of our characteristics.

A trait can be described as either Mendelian or **polygenic.** A single gene is responsible for a Mendelian trait. A polygenic trait, as its name implies, reflects the activities of more than one gene, and the effect of these multiple inputs is often additive, although not necessarily equal. Both Mendelian and polygenic traits can also be **multifactorial,** which means they are influenced by the environment (multifactorial traits are also called complex traits). Pure polygenic traits—those not influenced by the environment—are very rare. Multifactorial traits include common characteristics such as height and skin color, illnesses, and behavioral conditions and tendencies. Behavioral traits are not inherently different from other types of traits; they simply involve the functioning of the brain, rather than another organ. Figure 7.1 depicts the comparative contributions of genes and the environment to several disorders and events.

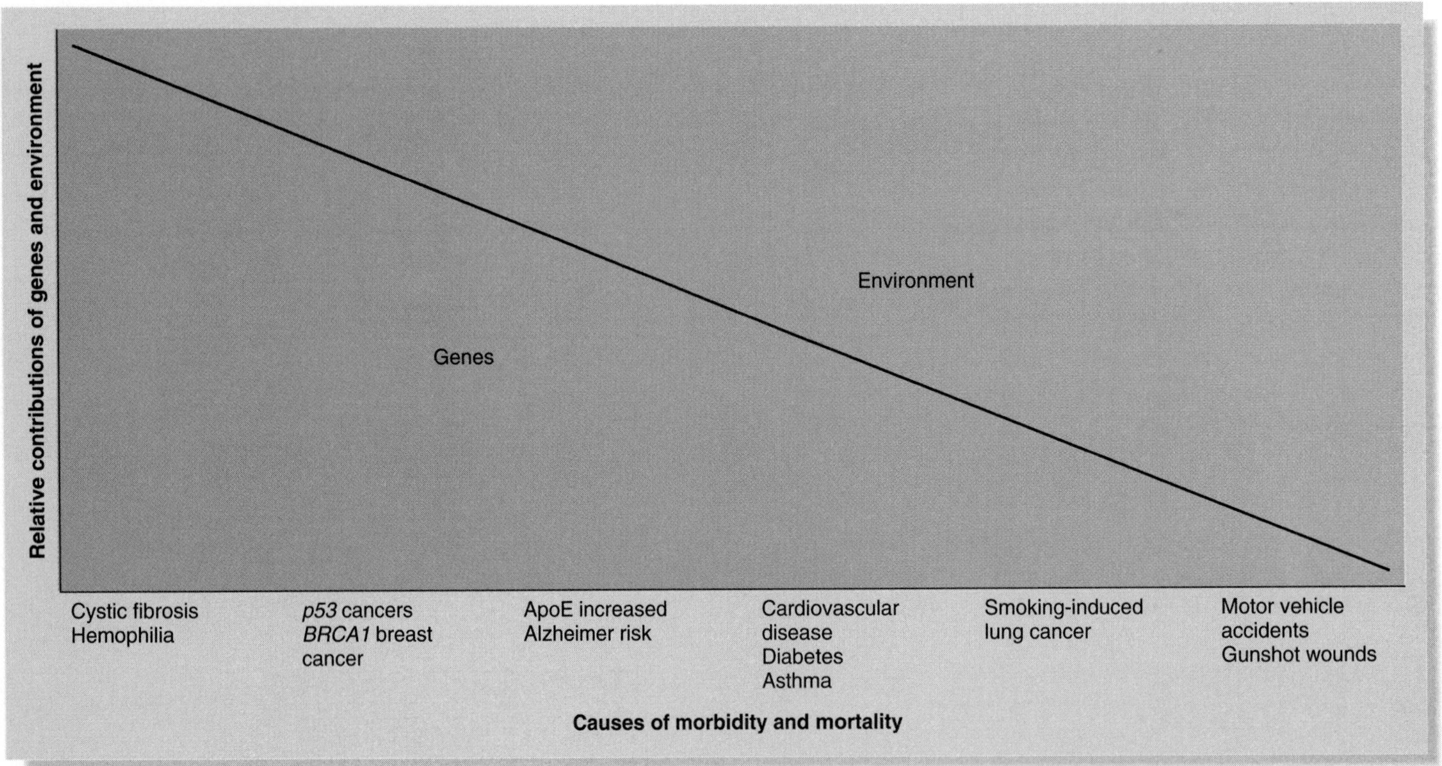


**figure 7.1**

**Genes and environment.** The comparative contributions of genes and the environment in causing traits and disorders form a continuum. In this schematic representation, the examples toward the left are caused predominantly by genes, and those to the right, more by environmental effects. The classic multifactorial conditions are in the middle—these are polygenic and influenced by the environment.

In contrast to a single gene disorder, a complex multifactorial condition may be caused by the additive contributions of several genes, each of which confers susceptibility. For example, we know that multiple sclerosis (MS) has a genetic component, because siblings of an affected individual are 25 times as likely to develop MS as siblings of people who do not have MS. One model of MS origin is that five susceptibility genes have alleles, each of which increases the risk of developing the condition. Those risks add up, and, in the presence of an appropriate (and unknown) environmental trigger, the disease begins.

## Polygenic Traits Are Continuously Varying

For a polygenic trait, the combined action of many genes often produces a continuum of the phenotype, which is called a continuously varying or quantitative trait. The parts of chromosomes that contribute to polygenic traits are therefore called quantitative trait loci, or QTLs. A multifactorial trait is continuously varying only if it is also polygenic. That is, it is the genetic component of the trait that contributes the continuing variation of the phenotype. The individual genes that confer a polygenic trait follow Mendel's laws (if they are unlinked), but together they do not produce Mendelian ratios. They all contribute to the phenotype, without showing dominance or recessiveness with respect to each other. For example, the multiple genes that regulate height and skin color result in continuously varying traits that exhibit a range of possible phenotypes. Mendelian traits are instead discrete or qualitative, often providing an all-or-none phenotype such as "affected" versus "normal."

A polygenic trait is variable in populations, such as the many nuances of hair color, body weight, and cholesterol levels. Some genes contribute more to a polygenic trait than others, and within genes, certain alleles have differing impacts that depend upon exactly how they influence or contribute to a trait, as well as by how common they are in a particular population. For example, a mutation in the gene that encodes the receptor that takes low-density lipoproteins (LDL cholesterol) into cells drastically raises a person's blood serum cholesterol level. But because fewer than 1 percent of the individuals in most populations have this mutation, it contributes very little to the variation in cholesterol level seen at the population level.

Although the expression of a polygenic trait is continuous, we can categorize individuals into classes and calculate the frequencies of the classes. When we do this and plot the frequency for each phenotype class, a bell-shaped curve results. This curve indicating continuous variation of a polygenic trait is strikingly similar for any trait. Even when different numbers of genes affect the trait, the curve is the same shape, as is evident in the following examples.

## Fingerprint Patterns, Height, and Eye Color

The skin on the fingertips folds into patterns of raised skin called dermal ridges that in turn align to form loops, whorls, and arches. A technique called dermatoglyphics ("skin writing") compares the number of ridges that comprise these patterns to identify and distinguish individuals (figure 7.2). Dermatoglyphics is part of genetics, because certain disorders (such as Down syndrome) are characterized by unusual ridge patterns, and of course it is also part of forensics, used for fingerprint analysis. Fingerprint pattern is a multifactorial trait.

The number of ridges in a fingerprint pattern is largely determined by genes, but also responds to the environment. During weeks 6 through 13 of prenatal development, the ridge pattern can be altered as the fetus touches the finger and toe pads to the wall of the amniotic sac. This early environmental effect explains why the fingerprints of identical twins, who share all genes, are not exactly alike.

We can quantify a fingerprint with a measurement called a total ridge count, which tallies the number of ridges comprising a whorl, loop, or arch part of the pattern for each finger. The average total ridge count in a male is 145, and in a female, 126. Plotting total ridge count reveals the bell curve characteristic of a continuously varying trait.

The effect of the environment on height is more obvious than that on fingerprint pattern—people who do not have enough to eat do not reach their genetic potential for height. Students lined up according to height vividly reveal the effects of genes and the environment on this continuously varying trait.

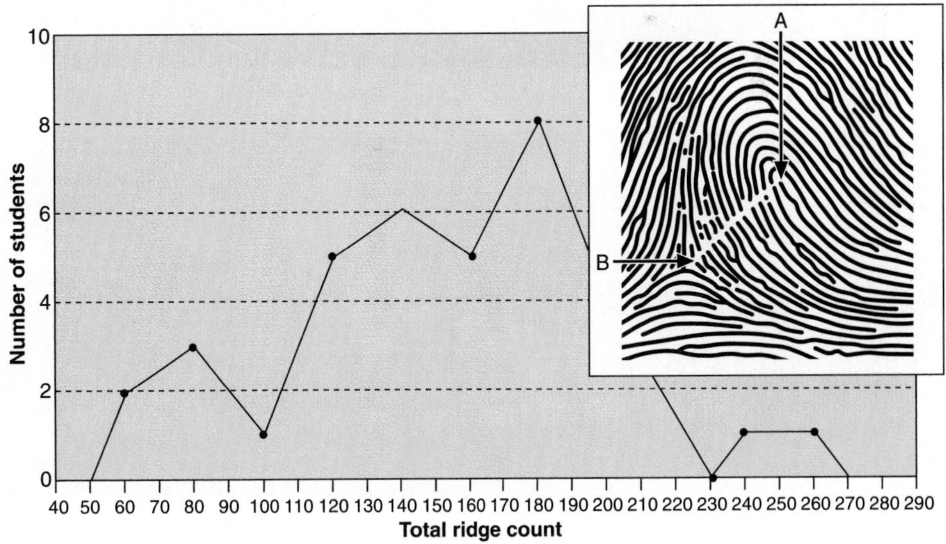


**figure 7.2**

**Anatomy of a fingerprint.** Total ridge counts of a number of individuals, plotted on a bar graph, form an approximate bell-shaped curve. This signals a multifactorial trait. The number of ridges between landmark points A and B on this loop pattern is 12. Total ridge count includes the number of ridges on all fingers.

The top part of figure 7.3 depicts students from 1920, and the bottom part, students from 1997. The similarity of the bell curve reflects the inherited component of the trait. But also note that the tallest people in the old photograph are 5′9″, whereas the tallest people in the more recent photograph are 6′5″. The difference is attributed to such factors as improved diet and better overall health.

We usually do not know exactly how many genes contribute to multifactorial traits that are also polygenic. However, geneticists can suggest models for different expressions of a trait based on a certain number of genes acting in an additive manner. Figure 7.4 shows this approach for eye color, as figure 7.5 does for skin color.

The colored part of the eye, the iris, is colored by the pigment melanin, which cells called melanocytes produce. Blue eyes have just enough melanin to make the color opaque. People with dark blue or green, brown, or black eyes make increasingly more melanin in the iris. Unlike melanin in skin melanocytes, the pigment in the eye tends to stay in the cell that produces it. If there is such a thing as a purely polygenic trait—that is, one caused by genes but not the environment—then eye color might be a candidate.

One model of how different eye colors arise considers two genes with two alleles each, that interact additively to produce five eye colors—light blue, deep blue or green, light brown, medium brown, and dark brown/black. (It seems that manufacturers of mascara follow this two-gene scheme.) If each allele contributes a certain amount of pigment, then the greater the number of such alleles, the darker the eye color. If eye color is controlled by two genes, *A* and *B*, each of which comes in two allelic forms—*A* and *a* and *B* and *b*—then the lightest color would be genotype *aabb;* the darkest, *AABB.* The bell curve arises because there are more ways to inherit light brown eyes, the midrange color, with any two contributing dominant alleles, than there are ways to inherit the other colors. Eye color may actually be more complex than a two-gene explanation.

## A Closer Look at Skin Color

Skin color is also due to melanin production. Melanocytes in skin have long extensions that snake between the other, tile-like skin cells, distributing pigment granules through the skin layers. Some melanin exits the melanocytes and enters the hardened, or keratinized, cells in the skin's upper layers. Here the melanin breaks into pieces, and as the skin cells are pushed up towards the skin's surface, the melanin bits provide color. The pigment protects against DNA damage from ultraviolet radiation, and exposure to the sun increases melanin synthe-

a.

b.

**figure 7.3**

**The inheritance of height.** Previous editions of this (and other) textbooks have used the photograph in (*a*) to illustrate the continuously varying nature of height. In the photo, taken around 1920, 175 cadets at the Connecticut Agricultural College lined up by height. In 1997, Professor Linda Strausbaugh asked her genetics students at the school, today the University of Connecticut at Storrs, to recreate the scene (*b*). They did, and confirmed the continuously varying nature of human height. But they also elegantly demonstrated how height has increased during the 20th century. Improved nutrition has definitely played a role in expressing genetic potential for height. The tallest people in the old photograph (*a*) are 5′9″ tall, whereas the tallest people in the more recent photograph (*b*) are 6′5″ tall.

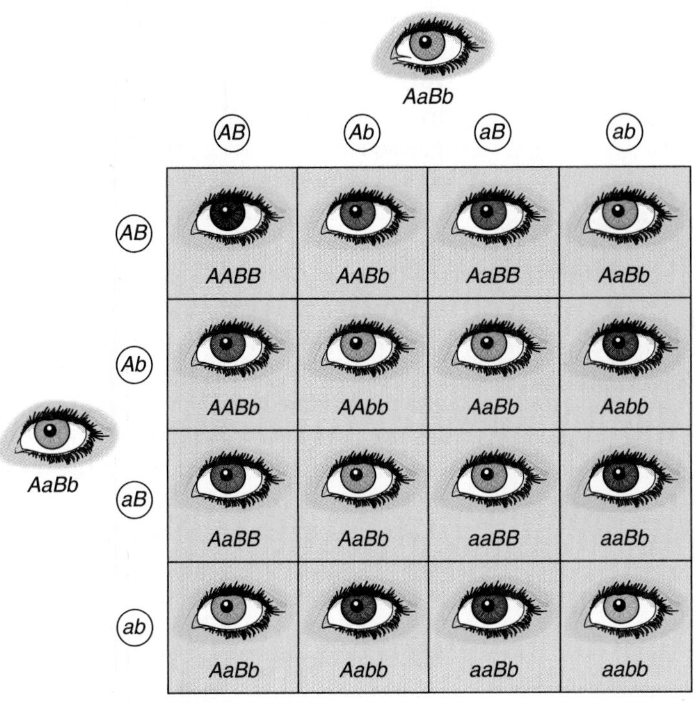


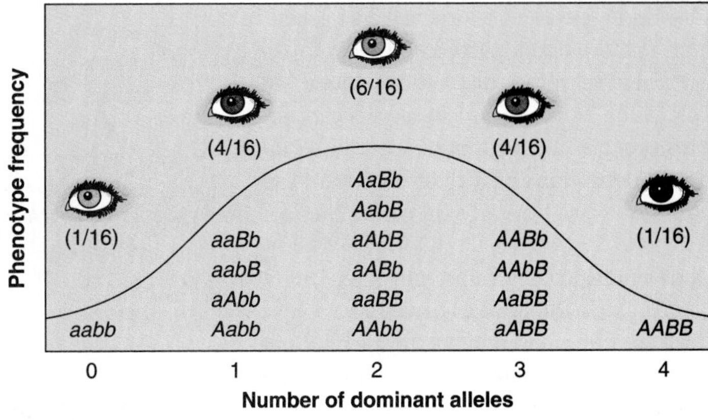


## figure 7.4

**Variations in eye color.** (*a*) A model of two genes, with two alleles each, can explain existence of five eye colors in humans. (*b*) The frequency distribution of eye colors forms the characteristic bell-shaped curve for a polygenic trait.

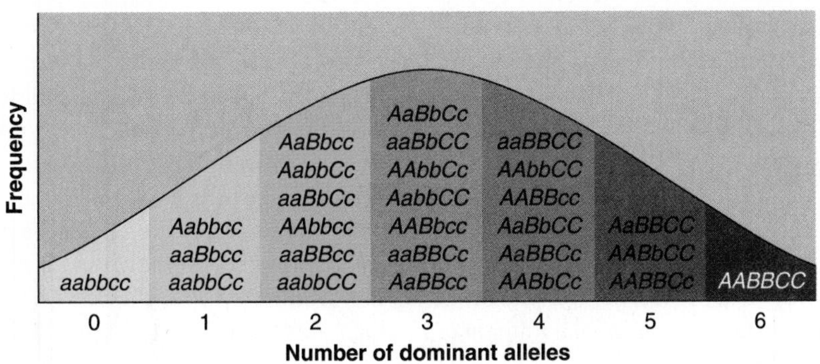


## figure 7.5

**Variations in skin color.** A model of three genes, with two alleles each, can explain some of the hues of human skin. In actuality, this trait likely involves many more than three genes.

sis. Although people come in a wide variety of hues, we all have about the same number of melanocytes per unit area of skin. Differences in skin color arise from the number and distribution of melanin pieces in the skin cells in the uppermost layers. People with albinism cannot manufacture melanin (see figure 4.15).

The definition of race based on skin color is more a social construct than a biological concept. From a genetic perspective, races are groups within species that are distinguished by different allele frequencies. Although we tend to classify people by skin color because it is an obvious visible way to distinguish individuals, skin color is not a reliable indicator of heritage. Golfer Tiger Woods, for example, is African American, Caucasian, Asian, and Native American. He illustrates that skin color alone hardly indicates a person's ethnic and genetic background (figure 7.6). The case of Thomas Jefferson and his dark-skinned descendants described in chapter 1 also illustrates how the

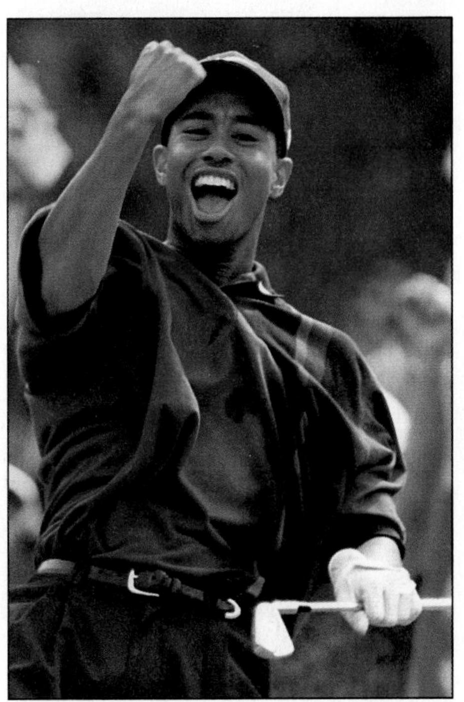

## figure 7.6

**Skin color is only one way humans differ from each other.** Golfer Tiger Woods objected to being called black; he is actually part African American, Caucasian, Asian, and Native American.

traditional definition of race considers only one trait—the distribution of melanin. When many genes are examined, two people with black skin may be less alike than either is to another person with white skin. On a population level, sub-Saharan Africans and Australian aborigines have dark skin, but they are very dissimilar in other inherited characteristics. Their dark skins may reflect the same adaptation (persistence of a valuable genetic trait) to life in a sunny, tropical climate.

Even as racial distinctions based on skin color continue to cause social problems, people are beginning to rethink accepted definitions of race, if only to recognize more variations of skin color. The U.S. Census expanded racial classifications to include multiracial groups, but this led to confusion. In 2000, nearly 7 million people checked more than one box for race. About 800,000 people claimed to be both black and white, presumably because they have one parent of each race. An editorial on the census results in a prominent medical journal concluded that this self-identification in more than one racial group "underscores the heterogeneity of the U.S. population and the futility of using race as a biologic marker."

The American College of Physicians advises its members not to indicate race on medical records, because it does not provide any valuable medical information. In fact, recent studies show that screening for variants of genes that control response to a particular drug is a better predictor of efficacy than prescribing a drug based on consideration of skin color. This study was published shortly after two others that showed that of two drugs used to treat heart failure, one worked equally well in blacks and whites, and the other was more effective in whites. Some physicians criticized the implication to not give blacks the second drug, because some people with darker skin might indeed be best helped with that drug. Human genome information, perhaps in the form of SNP patterns, will help physicians select the drugs that will treat individual patients, based on genotypes that directly relate to a drug's mechanism of action. Overall, 93 percent of inherited traits that vary are no more common in people of one skin color than any other.

Fingerprint pattern, height, eye color, and skin color are "normal" polygenic, multifactorial traits. Illnesses, too, may result from the interplay of a gene or genes with environmental influences. Reading 7.1 presents three interesting examples of environmental influences on illness.

**KEY CONCEPTS**

Polygenic traits are determined by more than one gene and vary continuously in expression. Multifactorial traits are determined by a combination of a gene or genes and the environment. • A bell curve describes the distribution of phenotypic classes of a polygenic trait, such as fingerprint pattern, height, eye color, and skin color.

## 7.2 Methods Used to Investigate Multifactorial Traits

It is much more challenging to predict recurrence risks for polygenic traits and disorders than it is for Mendelian traits. Geneticists evaluate the input of genes, using information from population and family studies.

### Empiric Risk

Using Mendel's laws, it is possible to predict the risk that a single gene trait will recur in a family if one knows the mode of inheritance—such as autosomal dominant or recessive. To predict the risk that a multifactorial trait will recur, geneticists use **empiric risks,** which are predictions of recurrence based on the trait's incidence in a specific population. Incidence is the rate at which a certain event occurs, such as the number of new cases of a particular disorder diagnosed per year in a population of known size.

Empiric risk is not a calculation, but an observation, a population statistic. The population might be broad, such as an ethnic group or residents of a geographical area, or genetically more well-defined, such as families that have a particular disease. In general, empiric risk for an individual increases with the severity of the disorder, the number of affected family members, and how closely related the person is to affected individuals.

Empiric risk may be used to predict the likelihood that a neural tube defect (NTD) will recur. In the United States, the overall population's risk of carrying a fetus with an NTD is about 1 in 1,000 (0.1 percent). For people of English, Irish, or Scottish ancestry, the risk is about 3 in 1,000. However, if a sibling has an NTD, no matter what the ethnic group, the risk of recurrence increases to 3 percent, and if two siblings are affected, the risk to a third child is even greater. By determining whether a fetus has any siblings with NTDs, a genetic counselor can predict the risk to that fetus, using the known empiric risk.

If a trait has an inherited component, then it makes sense that the closer the relationship between two individuals, one of whom has the trait, the greater the probability that the second individual has the trait, too. Studies of empiric risk support this logic. Table 7.1 summarizes the empiric risk

**table 7.1**

**Empiric Risk of Recurrence for Cleft Lip**

| Relationship to Affected Person | Empiric Risk of Recurrence |
|---|---|
| Identical twin | 40.0% |
| Sibling | 4.1% |
| Child | 3.5% |
| Niece/nephew | 0.8% |
| First cousin | 0.3% |
| General population risk (no affected relatives) | 0.1% |

*Reading 7.1*

# Disentangling Genetic from Environmental Effects

Sometimes it is difficult to determine whether differences among individuals for a trait are caused by genes, the environment, or both. The following examples illustrate how important it can be to understand effects on gene expression. Even a Mendelian (single gene) disorder can be influenced by the actions of other genes or environmental factors.

## Cystic Fibrosis

When researchers realized that cystic fibrosis (CF) was caused by several allele combinations that produce varying degrees of symptom severity, they attempted to see if genotypes correlated to specific phenotypes. Such information would inform people of how sick they would become. Unfortunately, researchers could not establish a direct correlation. Apparently, other genes influence the expression of the cystic fibrosis alleles.

CF held another surprise. In many cases, an environmental input influences the course of an inherited illness; in CF, it was the other way around. The thick mucus that builds up along airway linings provides a very attractive environment for bacteria, most notably a highly transmissible and quick-killing species called *Burkholderia cepacia.* Bacterial infection is familiar to people with CF, but in the past, they have usually been infected by *Pseudomonas aeruginosa.* Unlike a *Pseudomonas* infection, which can be present, on and off, for two decades before it kills, *B. cepacia* can do so in weeks. *B. cepacia* have appendages called cable pili that enable them to anchor tenaciously to the mucus-covered cells lining the airway. They are resistant to most antibiotic drugs.

Epidemics of *B. cepacia* were first reported a few years ago from Toronto and Edinburgh, under tragic circumstances. Because the infection is transferred so easily from person to person, it swept through summer camps for children with CF. Patients and their families were not prepared for such deadly bacterial infections. The camps and other support services are vitally important to affected families, but suddenly any patient with a *B. cepacia* diagnosis was isolated to avoid spread of the infection to others with CF.

Genetics has helped these patients a little. But rather than determining genotypes for the underlying CF, researchers are typing the DNA of the bacteria. Only patients with the more virulent strains of *B. cepacia* need be isolated. The task now is to correlate all of the information: Which CF genotypes attract which genetic variants of which bacteria?

## Type I Diabetes Mellitus

Type I or juvenile diabetes mellitus runs in families, but in no particular pattern of recurrence. The reason for the unpredictability may be environmental. In this inborn error in glucose (sugar) metabolism, the immune system attacks the pancreas. As a result, the pancreas does not produce the insulin required to route blood glucose into cells for use.

When studying the pancreases of young people who died suddenly just weeks after being diagnosed with diabetes, researchers observed severe infection of the pancreas plus an unusually strong immune response to the infection. They concluded that certain individuals may inherit a susceptibility to that type of infection or a strong immune response to it, but not develop diabetes unless such an infection occurs.

## Neural Tube Defects

Sometimes what appears to be an environmental influence is actually genetic. This may be the case for neural tube defects (NTDs), which are openings in the brain or spinal cord that occur at the end of the first month of prenatal development.

An NTD is an example of a "threshold" birth defect, in which no particular gene is implicated, but the effects of several genes and environmental influences sum, expressing the phenotype after a certain point is reached. Such a condition has a high population frequency compared to a Mendelian trait (1/1,000 for NTDs), and recurrence risk in an affected family is low (2 to 5 percent for NTDs).

Researchers have long attributed NTDs to a combination of genes and the environment, based on two observations.

1. A woman who has one affected child is at increased risk of having another affected child (implicating heredity).
2. Recurrence risk diminishes by 70 percent if a woman takes folic acid supplements shortly before and during pregnancy (implicating a vitamin deficiency).

An inherited enzyme deficiency may explain why vitamin supplementation prevents some NTDs. A large group of Irish women whose children have NTDs were found to be deficient in both folic acid and vitamin $B_{12}$. Only one biochemical reaction in the human body requires both of these vitamins; an abnormality in the enzyme that catalyzes that reaction could cause the double vitamin deficiency. Researchers are now trying to determine how disruption of this reaction causes an NTD. Meanwhile, pregnant women routinely take folic acid supplements to reduce the incidence of these birth defects. In the future, genetic tests will be able to identify women whose embryos are most likely to benefit from folic acid supplementation.

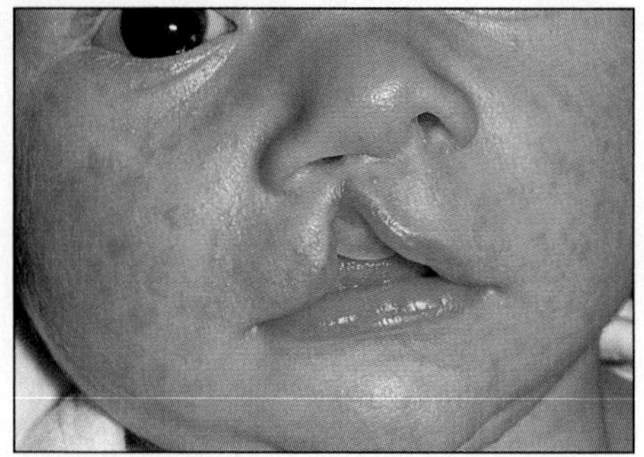

**figure 7.7**

**Cleft lip.** Cleft lip is more likely to occur in a person who has an affected relative.

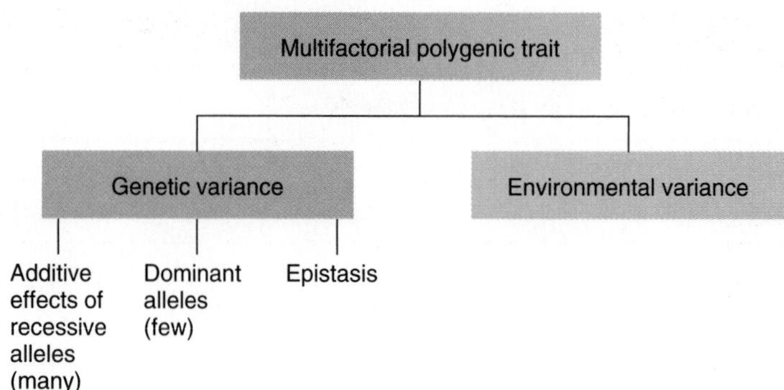


**figure 7.8**

**Heritability estimates the genetic contribution to a trait.** Observed variance in a polygenic, multifactorial trait or illness reflects genetic and environmental contributions. Genetic variants are mostly determined by additive effects of recessive alleles of different genes, but can also be influenced by effects of a few dominant alleles and by epistasis (interactions between alleles of different genes).

for cleft lip (figure 7.7) among groups of identical twins, siblings, parent-child pairs, nieces and nephews to uncles and aunts, cousins, and the general population.

Because empiric risk is based solely on observation, we can use it to derive risks of recurrence for disorders with poorly understood transmission patterns. For example, certain multifactorial disorders affect one sex more often than the other. Pyloric stenosis is an overgrowth of muscle at the juncture between the stomach and the small intestine. It is five times more common among males than females. The condition must be corrected surgically shortly after birth, or the newborn will be unable to digest foods. Empiric data show that the risk of recurrence for the brother of an affected brother is 3.8 percent, but for the brother of an affected sister, the risk is 9.2 percent.

## Heritability—The Genetic Contribution to a Multifactorial Trait

As Charles Darwin observed, some of the variation of a trait is due to heredity, and some to environmental influences. A measurement called **heritability,** designated H, estimates the percentage of the phenotypic variation for a particular trait that is due to genes in a certain population at a certain time. Figure 7.8 outlines the factors that contribute to observed variation in a trait. Heritability equals 1.0 for a trait that is completely the result of gene action, and 0 if it is entirely caused by an environmental influence. Most traits lie in between. For example, height has a heritability of 0.8, and body mass index, which is a measure of weight taking height into account, has a heritability of 0.55. Table 7.2 lists some traits and their heritabilities.

Heritability changes as the environment changes. For example, the heritability of skin color would be higher in the winter months, when sun exposure is less likely to increase melanin synthesis.

Heritability can be estimated in several ways using statistical methods. One way is to compare the actual proportion of pairs of people related in a certain manner who share a particular trait, to the expected proportion of pairs that would share it if it were inherited in a Mendelian fashion. The expected proportion is derived by knowing the blood relationships of the individuals, and using a measurement called the **correlation coefficient,** which is the proportion of genes that two people related in a certain way share (table 7.3). (It is also called the coefficient of relatedness.) A parent and child, for example, share 50 percent of their genes, because of the mechanism of meiosis. Siblings share on average 50 percent of their genes, because they have a 50 percent chance of inheriting each allele for a gene from each parent. (The designations of primary (1°), secondary (2°), and tertiary (3°) relatives are useful in genetic counseling when empiric risks are consulted.)

**table 7.2**

**Heritabilities for Some Human Traits**

| Trait | Heritability |
|---|---|
| Clubfoot | 0.8 |
| Height | 0.8 |
| Blood pressure | 0.6 |
| Body mass index | 0.5 |
| Verbal aptitude | 0.7 |
| Mathematical aptitude | 0.3 |
| Spelling aptitude | 0.5 |
| Total fingerprint ridge count | 0.9 |
| Intelligence | 0.5–0.8 |
| Total serum cholesterol | 0.6 |

If the heritability of a trait is very high, then of a group of 100 sibling pairs, nearly 50 would be expected to have it, because siblings share on average 50 percent of their genes. Height is a trait for which heritability reflects the environmental influence of nutrition. Of 100 sibling pairs in a population, for example, 40 are the same number of inches tall. Heritability for height among this group of sibling pairs is .40/.50, or 80 percent, which is the observed phenotypic variation divided by the expected phenotypic variation if environment had no influence.

Genetic variance for a polygenic trait is mostly due to the additive effects of recessive alleles of different genes. For some traits, a few dominant alleles can greatly influence phenotype, but because they are

**table 7.3**

**Correlation Coefficients for Pairs of Relatives**

| Relationship | Degree of Relationship | Percent Shared Genes (Correlation Coefficients) |
|---|---|---|
| Sibling to sibling | 1° | 50% (1/2) |
| Parent to child | 1° | 50% (1/2) |
| Uncle/aunt to niece/nephew | 2° | 25% (1/4) |
| Grandparent to grandchild | 2° | 25% (1/4) |
| First cousin to first cousin | 3° | 12 1/2% (1/8) |

rare, they do not contribute greatly to heritability. Heart disease caused by a faulty LDL (low density lipoprotein cholesterol) receptor is an example of a condition caused by a rare dominant allele, but that is also influenced by many other genes. Epistasis (interaction between alleles of different genes) can also influence heritability. Some geneticists calculate a "narrow" heritability that considers only additive recessive effects, and a "broad" heritability that also considers the effects of rare dominant alleles and epistasis. For LDL cholesterol level, for example, the narrow heritability is 0.36, but the broad heritability is 0.96, reflecting the fact that a rare dominant allele can have a large impact on LDL level.

Multifactorial inheritance has many applications in agriculture, where a breeder needs to know whether such traits as birth weight, milk yield, length of wool fiber, and egg hatchability are determined largely by heredity or the environment. The breeder can control the environmental input by adjusting the conditions under which animals are raised, and the inherited input by setting up matings between particular individuals. Studying multifactorial traits in humans is difficult, because information must be obtained from many families. Two special types of people, however, can help geneticists to tease apart the genetic and environmental components of multifactorial traits—adopted individuals and twins.

## Adopted Individuals

A person adopted by people who are not blood relatives shares environmental influences, but typically not many genes, with his or her adoptive family. Conversely, adopted individuals share genes, but not the exact environment, with their biological parents. Therefore, biologists assume that similarities between adopted people and adoptive parents reflect mostly environmental influences, whereas similarities between adoptees and their biological parents reflect mostly genetic influences. Information on both sets of parents can reveal how heredity and the environment each contribute to the development of a trait.

Many early adoption studies used the Danish Adoption Register, a compendium of all adopted Danish children and their families from 1924 to 1947. One study examined correlations between causes of death among biological and adoptive parents and adopted children. If a biological parent died of infection before age 50, the adopted child was five times more likely to die of infection at a young age than a similar person in the general population. This may be because inherited variants in immune system genes increase susceptibility to certain infections. In support of this hypothesis, the risk that an adopted individual would die young from infection did not correlate with adoptive parents' death from infection before age 50.

Although researchers concluded that length of life is mostly determined by heredity, they did find evidence of environmental influences. For example, if adoptive parents died before age 50 of cardiovascular disease, their adopted children were three times as likely to die of heart and blood vessel disease as a person in the general population. What environmental factor might account for this correlation?

## Twins

Studies that use twins to separate the genetic from the environmental contribution to a phenotype provide more meaningful information than studying adopted individuals. In fact, twin studies have largely replaced adoption methods.

Using twins to study genetic influence on traits dates to 1924, when German dermatologist Hermann Siemens compared school transcripts of identical versus fraternal twins. Noticing that grades and teachers' comments were much more alike for identical twins than for fraternal twins, he proposed that genes contribute to intelligence.

A trait that occurs more frequently in both members of identical (MZ) twin pairs than in both members of fraternal (DZ) twin pairs is at least partly controlled by heredity. Geneticists calculate the **concordance** of a trait as the percentage of pairs in which both twins express the trait.

In one study, 142 MZ twin pairs and 142 DZ twin pairs took a "distorted tunes test," in which 26 familiar songs were played, each with at least one note altered. A person was considered to be "tune deaf" if he or she failed to detect three or more of the mistakes. Concordance for "tune deafness" was 0.67 for MZ twins, but only 0.44 for DZ twins, indicating a considerable inherited component in the ability to perceive musical pitch accurately. Figure 7.9 compares twin types for a variety of hard-to-measure traits. Figure 3.16 shows how DZ and MZ twins arise.

Diseases caused by single genes that are 100 percent penetrant, whether dominant or recessive, are 100 percent concordant in MZ twins. If one twin has the disease, so does the other. However, among DZ twins, concordance generally is 50 percent for a dominant trait and 25 percent for a recessive trait. These are the Mendelian values that apply to any two siblings. For a polygenic trait with little environmental input, concordance values for MZ twins are significantly greater than for DZ twins. A trait molded mostly by the environment exhibits similar concordance values for both types of twins.

An ongoing investigation called the Twins Early Development Study shows how concordance values indicate the degree to which heredity contributes to a trait. Headed by Robert Plomin of the Institute of

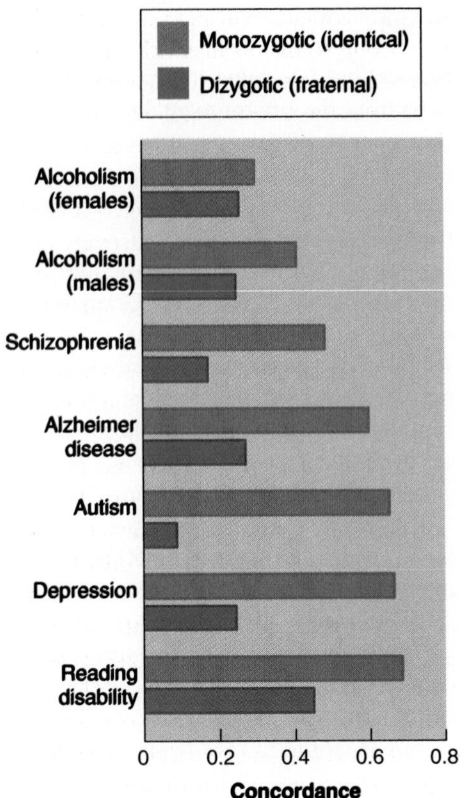


**figure 7.9**

**Twin studies.** A trait more often present in both members of MZ twin pairs than in both members of DZ twin pairs is presumed to have a significant inherited component.

Source: Robert Plomin, et al., "The Genetic Basis of Complex Human Behaviors," *Science* 17 June 1994, vol. 264, pp. 1733–39. Copyright 1994 American Association for the Advancement of Science.

Psychiatry in London, this project is following 7,756 pairs of twins born in England and Wales in 1994. One experiment looked at 2-year-olds whose language skills place them in the lowest 5 percent of children that age. With the parents' help, researchers recorded the number of words in the vocabularies of 1,044 pairs of identical twins, 1,006 pairs of same-sex fraternal twins, and 989 pairs of opposite-sex twins. The results clearly indicated a large genetic influence for children lagging behind in language skills—the concordance for the identical twins was 81 percent, but it was 42 percent for the fraternal twins. Put another way, if an identical twin fell into the lowest 5 percent of 2-year-olds for language acquisition, the chance that her identical twin would, too, was 81 percent. But if a fraternal twin was in this category, the chance that her twin would also be was only 42 percent. When similar assessments were done on twin pairs of all abilities, the differences between the concordance values was not nearly as great. This indicates that the environment plays a larger role in most children's adeptness at learning vocabulary than it does for the 5 percent who struggle to do so.

Comparing twin types has a limitation—the technique assumes that both types of twins share similar experiences. In fact, identical twins are often closer than fraternal twins. This discrepancy between how close the two types of twins are led to misleading results in twin studies conducted in the 1940s. One study concluded that tuberculosis is inherited because concordance among identical twins was higher than among fraternal twins. Actually, part of the reason for the difference in concordance was that the infectious disease was more readily passed between identical twins because their parents kept them in close physical contact.

A more informative way to use twins to assess the genetic component of a multifactorial trait is to study identical twins who were separated at birth, then raised in very different environments. Hermann Siemens suggested this in 1924, but much of the work using this "twins reared apart" approach has taken place at the University of Minnesota. Here, since 1979, hundreds of sets of identical/fraternal twins and triplets who were separated at birth have visited the laboratories of Thomas Bouchard. For a week or more, the twins and triplets undergo tests that measure physical and behavioral traits, including 24 different blood types, handedness, direction of hair growth, fingerprint pattern, height, weight, functioning of all organ systems, intelligence, allergies, and dental patterns. Researchers videotape facial expressions and body movements in different circumstances and probe participants' fears, vocational interests, and superstitions.

Twins and triplets separated at birth provide natural experiments for distinguishing nature from nurture. Many of their common traits can be attributed to genetics, especially if their environments have been very different (figure 7.10). By con-

*Separated at birth, the Mallifert twins meet accidentally.*

**figure 7.10**

trast, their differences tend to reflect differences in their upbringing, since their genes are identical (MZ twins and triplets) or similar (DZ twins and triplets).

The researchers have found that identical twins and triplets separated at birth and reunited later are remarkably similar, even when they grow up in very different adoptive families. Idiosyncrasies are particularly striking. For example, twins who met for the first time when they were in their thirties responded identically to questions; each paused for 30 seconds, rotated a gold necklace she was wearing three times, and then answered the question. Coincidence, or genetics?

The "twins reared apart" approach is not a perfectly controlled way to separate nature from nurture. Identical twins and other multiples share an environment in the uterus and possibly in early infancy that may affect later development. Siblings, whether adoptive or biological, do not always share identical home environments. Differences in sex, general health, school and peer experiences, temperament, and personality affect each individual's perception of such environmental influences as parental affection and discipline.

Adoption studies, likewise, are not perfectly controlled experiments. Adoption agencies often search for adoptive families with ethnic, socioeconomic, or religious backgrounds similar to those of the biological parents. Thus, even when different families adopt and raise separated twins, their environments might not be as different as they might be for two unrelated adoptees. However, twins and triplets reared apart are still providing intriguing insights into the number of body movements, psychological quirks, interests, and other personality traits that seem to be rooted in our genes.

## Association Studies

Empiric risk, heritability, and adoptee and twin studies are traditional ways of estimating the degree to which genes contribute to the variability of a trait or illness. Until recently, geneticists had to be satisfied with these limited measures of inheritance. With genomics, however, powerful tools are emerging that can help to identify the specific genes that contribute directly to pathogenesis, or confer susceptibility to, disease.

Identifying the single genes behind Mendelian traits and disorders has largely relied on linkage analysis, discussed in chapters 5 and 22. Researchers compiled data on families with more than one affected member, determining whether a section of chromosome (a marker) was inherited along with a disease-causing gene by inferring whether alleles of the two genes were in coupling or repulsion (see figure 5.13). Such linkage analysis in humans is extremely difficult to do because the rarity of single gene disorders makes it hard to find enough families and individuals to compare. Because detecting linkage to one gene is so difficult, using the approach to identify the several genes that contribute to a polygenic trait is even more daunting. Fortunately, another, related method is coming to the forefront of genetic research and can more powerfully detect the several genes that contribute to a polygenic trait—**SNP mapping.** Because SNP mapping tracks large populations rather than families, it is somewhat easier to find participants, although the sheer volume of data can seem overwhelming—at least to a human. (Computers can handle it!)

Recall from chapter 1 that a SNP, or single nucleotide polymorphism, is a site within a DNA sequence that varies in at least 1 percent of a population. The human genome has about 12 to 16 million SNPs among the 3 billion bases. Researchers have identified more than 3 million SNPs. The number comes from an analysis of linkage disequilibrium (LD), which is the tendency for certain SNPs to be inherited together. Linkage disequilibrium generally occurs over areas that are 5,000 to 50,000 bases long. Three million SNPs, spread out over the genome, will enable researchers to pair specific SNPs with specific genes that cause or contribute to a particular disease or trait, because the distance between the SNPs will be less than the average length of a DNA sequence that is in linkage disequilibrium. Several SNPs that are transmitted together constitute a haplotype, which is short for "haploid genotype."

SNPs are useful in **association studies,** in which researchers compare SNP patterns between a group of individuals who have a particular disorder and a group that does not. An association study uses a case-control design, which means that each individual in one group is matched to an individual in the other group so that as many characteristics as possible are shared, such as age, sex, activity level, and environmental exposures. In this way, SNP differences can be contrasted to presence or absence of the particular medical condition. Variables are limited. When many SNPs are considered, many susceptibility genes can be tracked, and patterns may emerge that can then be used to predict the course of the illness.

It will take a great deal of research to test whether SNP patterns are actually meaningful. For example, if a certain SNP pattern in a population is associated with breast cancer, then the next step might be for researchers to see if these tumors are different, at a cell and tissue level, from other cases of breast cancer. Software can compare SNP and histological data to see if the SNP pattern accurately reflects a physically distinct subtype of the disease. Then, eventually, a simple SNP scan might replace or augment microscopic analysis of the tumor. Figure 7.11 shows how SNP patterns are generated and compared in association studies.

SNP association studies, which are well under way in many biotech and pharmaceutical companies as well as academic laboratories, promise many practical benefits. Obtaining SNP maps is a much faster way to identify DNA sequence differences among individuals than sequencing everyone's genome. The correlations between SNP patterns and elevated disease risks may be able to guide medical care, including the ability to make more meaningful prognoses, and to predict an individual's likely response to a particular drug. Identifying the SNPs that travel with genes of interest will enable researchers to then search through human genome data to identify nearby genes whose protein products suggest that they could be implicated in the pathogenesis of the disorder. This was the case, for example, in Crohn disease, in which the intestines become severely inflamed in response to the presence of certain bacteria.

Two different research groups identified a gene that increases risk for developing Crohn disease. One group consulted older linkage data that indicated a causative gene on chromosome 16 in some families. These researchers considered all the genes identified so far in that region, and pursued one, called NOD2, that encodes a protein that takes part

## figure 7.11

**Association studies.** Association studies are correlations that use SNP profiles. By considering many individuals and many SNPs, researchers correlate SNP patterns with genes that affect health. Such correlations identify parts of chromosomes that are associated with increased risk of developing the particular condition. (*a*) This illustration is a schematic view of how frequently SNPs occur—about one per 1,200 bases, although they are not evenly distributed. (*b*) An individual SNP may be correlated to one gene that contributes to a particular phenotype. For polygenic traits, SNP patterns on several chromosomes might be considered.

```
ATGCTCGAGCCTAATAGCGTGACTGATCTGACTGACTGACTGATCT
CTATCGGCGATGCATCGCGTATTACTGCGCCCGGGATAGCGTATGC
GATCGATCTCGAGAGGTTCTCTAGACTGATCGATATGCGTAGCGTG
TGAAGGGTTGTCTCGCGGCCCACTCTGACTGATCGCGTTGTGTACG
ACTGTCGAACTCTGAGGTGGCCCATGTTTACGGTCTGAATCTGATC
GCTCTTCTGAGTCTGACTGTCAGTCTAGAACCGTGTCAACTGTGACT
CCTGTAAAGGGTCTGGTCCCTGACTCGGTCTCAACTGCTGCCTCTCT
TCAACGCATATGCCGTAGTCGTGTGACAATCTTCGGTGTGTCTAACT
GACTCTCGGACCTGTCATCGTGTCGACTGCTGTCACTGTCACTTCGC
GGTTAAAGGGTCGTCCCGTGTCACATGCGTATTGCGATTGAACCCG
TCGTATTCCCGTGTATCGTACGTCGTAGCTGGTGATCGTACACTGTG
TGTGACCGCGTGACTGCTGTACGATCGCGTCACTGACGTCGACTCG
ACCCTCGACGCGTCTCGATCGCTCACTCGACTATAAATGCCGTGTC
ATCCTCTGAGGCTCAGGTCGTAGCTGACGCCTGTCCACTCGATCGC
TCCTGAGTGACTGTGACTGTTCCACACGGGTGTCATCGTCTACACTG
TGACGTCGAAGCTCTGACATCGACTCTACGCGGGGTGGTGCAAATC
TTAACGATCGACTAGGTAGGGTTATACGCGTAGTCGACTGACTGAC
TGACTGACTGACTGACTGTGACTGTGTCTACTGATCTCGATCTGCTG
ACTGTCACTCTGAACGTCTGACTACTCTATGGTATTTCCGTGTATGT
GATATGAATGTTCCACAACTGTGTCACATGATCTGTGACTGTATGC
ACAACCTTGTGGGAGTCTCACTGACCTGATAGTCCCCTGAGTCGTA
CGATCGTGACTGAGTGTTGTGTACTGTACGTACGTAACTGTCTGAG
AAAGTCTTTCGCGCCCGGGCGTACTGACGATCGTACGTGTGTCCCA
GTAACGTAGTCGAGTGACGATCGATCTCGAGAGGTTCTCTAGACTG
ATCGATATGCGTAGCGTGTGAAGGGTTGTCTCGCGGCCCACTCTGA
CTGATCGCGTTGTGTACGACGATCGCTCCTGAGTGACTGTGACTGTT
CCACACGGGTGTCATCGTCTAGATGCATCGCGTATTACTGCGCCCG
GGATAGCGTATGCGATCGATCTCGAGAGGTTCTCTAGACTGATCGA
TATGCGTAGCGTGTGAAGGGTTGTCTCGCGGCCCACTCTGACTGAT
CGCGTTGTGTACGACTGTCGATCGTATCGAGGGACTCTCAGTGACG
CCTCACTGGCCGGCCGGCTGATGCATCGTCGCGATATCGGTCGATC
TGTACGCTACAAACGTGTTTGTATGTGACGCGTTAGTCGATTCGCGG
TTAAAGGGTCGTCCCGTGTCACATGCGTATTGCGATTGACTGTATG
CACAACCTTGTGGGAGTCTCACTGACCTGATAGTCTAACTGTCTGA
GAAAGTACTGTGTGTGACCGCGTGACTGCTGTACGATCGCGTCACT
GACGTCGACTCGCCGTGTATGTGATATGAATGTTCCACAACTGTGT
CACATGATCTGTGACTTTCGCGCCCGGGCGTACTGACG ATCGTACG
TGTGTCCCAGTAACGTAGTCGAGTGACGATCGATCTCGAGAGGTTC
TCTAGACTGATCGATATGCGTAGCGTGTGAAGGGTTGTCTCGCGGC
GCGATCGCGTGTGTGTGAAACGCGATCTGCGTGTAG TCTGAACGTC
TGACTACTCTATGGTATTTCCGTGTATGTGATATGAATGTTCCACAA
CTGTGTCACACGATCTCGAGAGGTTCTCTAGACTGA TCGATATGCG
TAGCGTGTGAAGGGTTGTCTCGCGGCCCACTCTGACTGATCGCGTT
GTGTACGACTGTCGATCGTATCGAGGGACTCTCAGTGACGCCTCAC
TGGCCGGCCGGGCCGTGTCATCCTCTGAGGCTCAGGTCGTAGCTGA
CGCCTGTCCACTCGATCGCTCCTGAGTGACTGTGACTGTTCCACACG
GGTGTCATCGTTCTGACTGATCGCGTTGTGTACGACGATCGCTCCTG
AGTGACTGTGACTGTTCCACACGGGTGTCATCGTCTAGATGCATCG
CGTATTACTGCGTAGTCGAGTGACGATCGATCTCGAGAGGTTCTCT
AGACTGATCGAATCTTAACGATCGACTAGGTAGGGTTATACGCGTA
GTCGACTGACTGACTGACTGACTGACTGACTGTGACTGTGTCTACT
GATCTCGATCTGCCTAAACGTCGTACTGATACACATTCCCAACTAG
```

a.

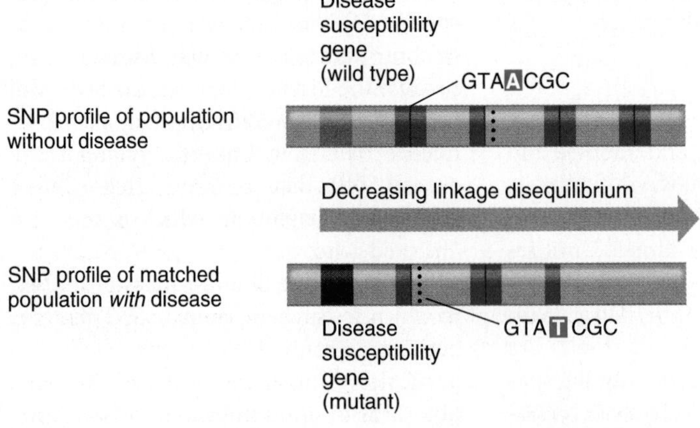


b.

in the inflammatory response. Then they examined DNA from patients and controls and found that patients were more likely to have mutations in this gene than the healthy comparisons. The second research group looked at 11 SNPs in 235 families with affected members. They found a certain pattern that was more likely than chance to be present in the affected individuals, and found that these people also had mutations in the NOD2 gene.

The Crohn disease research reveals a limitation of association studies—they are just correlations, not proofs of causation. It is rare to find a DNA sequence that contributes to a polygenic trait that is found exclusively in affected individuals. The NOD2 allele that elevates risk of developing Crohn disease, for example, is found in 5 percent of the general population, but in 15 percent of individuals with the condition. Clearly, more than one gene contributes to Crohn disease. Further studies will expand the number of SNPs and search for associations on other chromosomes.

The more complex a SNP association study, the more individuals are required to achieve statistical significance. Consider an investigation that examines 20 genes, each with 4 SNPs, that contribute to development of a particular polygenic disease. A screen would look for 80 data points per individual. With so many possible genotypes (3,160), it's clear that many thousands of individuals would have to be examined to note any correlations between the SNP pattern and disease.

Table 7.4 reviews the measures of multifactorial traits.

## table 7.4

**Terms Used in Evaluating Multifactorial Traits**

**Association Study** Detecting correlation between SNP (or other marker) patterns and increased risk of developing a particular medical condition.

**Empiric Risk** The risk of recurrence of a trait or illness based on known incidence in a particular population.

**Heritability** The percentage of phenotypic variation for a trait that is attributable to genetic differences. It equals the ratio of the observed phenotypic variation to the expected phenotypic variation for a population of individuals who are related in a particular way.

**Correlation Coefficient** The proportion of genes that two people related in a particular way share. Used to calculate heritability.

**Concordance** The percentage of twin pairs in which both twins express a trait.

Association studies that correlate SNP patterns to increased disease risk are replacing linkage studies based on families. Association studies are suited for polygenic disorders because they can track correlations to several genes.

### Key Concepts

Empiric risk applies population incidence data to predict risk of recurrence for a multifactorial trait or disorder. • Heritability measures the genetic contribution to a multifactorial trait; it is specific to a particular population at a particular time. A correlation coefficient is the proportion of genes that individuals related in a certain way are expected to share, and is used to calculate heritability. • Researchers compare traits in adopted individuals to those in their adoptive and biological parents to assess the genetic contribution to a trait. • Concordance is the percentage of twin pairs in which both express a trait. For a trait largely determined by genes, concordance is higher for MZ than DZ twins.

## 7.3 Some Multifactorial Traits

Multifactorial traits include such common conditions as heart and blood vessel (cardiovascular) disease and obesity, as well as harder-to-define traits such as intelligence and aspects of personality, mood, and behavior.

### Heart Health

Arthur Ashe was a professional tennis player who suffered a severe heart attack in his early thirties. He was in top physical shape and followed a low-fat diet, but an inherited tendency to deposit lipids (fats) on the insides of his coronary arteries led to a heart attack. He eventually died of AIDS, which he acquired from a blood transfusion during heart surgery.

In contrast to Arthur Ashe was the case of an 88-year-old man reported in a medical journal. He ate 25 eggs a day, yet had a very healthy heart and low serum cholesterol level. The lucky egg eater didn't have a sky-high cholesterol level or plaque-clogged coronary arteries because his particular metabolism, orchestrated by genes, could handle the large load of dietary lipids. The vastly different health status of Arthur Ashe and the elderly egg lover demonstrates the powerful influence of genes—Ashe followed all preventative measures and suffered a heart attack; the 88-year-old ate a diet oozing with cholesterol and enjoyed good cardiovascular health.

Genes control how well the body handles levels of lipids in the blood; how readily the blood clots; blood pressure; and the stickiness of white blood cells to the walls of blood vessels, a property controlled by cellular adhesion factors (see figure 2.21). Fats are insoluble in the fluid environment of the blood, but when bound to proteins to form large molecules called lipoproteins, fats can travel in the circulation. Several genes encode the protein parts of lipoproteins, called apolipoproteins. Some types of lipoproteins ferry lipids in the blood to tissues, where they are utilized, and other types of lipoproteins take lipids to the liver, where they are dismantled or converted to biochemicals that the body can excrete more easily. One allele of a gene that encodes apolipoprotein E, called E4, increases the risk of a heart attack threefold in people who smoke. This is clear evidence that genes and environmental factors can interact in ways that cause illness.

Maintaining a healthy cardiovascular system requires a balance: cells require sufficient lipid levels inside but cannot allow too much accumulation on the outside. Several dozen genes control lipid levels in the blood and tissues by specifying other proteins. These include enzymes that process lipids; that transport lipids; and proteins that function as receptors that admit lipids into cells.

Much of what we know about genetic control of cardiovascular health comes from studying rare inherited conditions. Identifying and understanding a genetic cause of a disease can help the larger number of people who suffer from noninherited forms of the illness. For example, the cholesterol-lowering drugs that millions of people take today, called statins, grew out of research on people with familial hypercholesterolemia, described in figure 5.2. In another example, inborn errors of metabolism that affect several types of liver enzymes led to the discovery that high blood levels of an amino acid, homocysteine, elevate the risk of developing arteriosclerosis. This is the hardening and narrowing of arteries that precede cholesterol deposition. Getting enough dietary folic acid can keep homocysteine levels within a healthy range.

As many as 20 to 50 genes regulate blood pressure. One gene encodes angiotensinogen, a protein that is elevated in the blood of people with hypertension. This protein controls blood vessel tone and fluid balance in the body. Certain alleles are found much more frequently among people with hypertension than chance would explain. Even though environmental factors, such as emotional stress, can raise blood pressure, knowing who is genetically susceptible to dangerously high blood pressure can alert doctors to monitor high-risk individuals.

An enzyme, lipoprotein lipase, specified by a gene on chromosome 8, is important in lipid metabolism. It lines the walls of the smallest blood vessels, where it breaks down fat packets released from the small intestine and liver. Lipoprotein lipase is activated by high-density lipoproteins (HDLs), and it breaks down low-density lipoproteins (LDLs). High HDL levels and low LDL levels are associated with a healthy cardiovascular system.

A group of autosomal recessive inborn errors of metabolism called type I hyperlipoproteinemias cause a deficiency of lipoprotein lipase, which in turn causes triglycerides (a type of fat) to build to dangerously high levels in the blood. Lipoprotein lipase also regulates fat cell size; fat cells usually contribute to obesity by enlarging, rather than by dividing to form more fat cells.

The fluidity of the blood is also critical to cardiovascular health. Overly active clotting factors, overly sticky white blood cells, or disregulation of homocysteine can induce formation of clots that block blood flow, usually in blood vessels in the heart or in the legs.

Genetic research into cardiovascular function is leading the way in the analysis of multifactorial disease. One pharmaceutical company offers a genetic test panel that detects either of two alleles at 35 sites within 15 genes (figure 7.12). A patient's DNA is applied to a nylon strip containing the test genes, and a dark line forms where the sample matches the test, revealing which genes predisposing to cardiovascular disease are present. DNA microarrays will test for many more characteristics that affect the heart and blood vessels.

The premise behind "multilocus genotype information" is that people have composite genetic risks that are based on the small contributions of several genes—the essence of polygenic inheritance. Computer analysis of the multigene tests account for environmental factors, such as those outlined in table 7.5. The test panels can also be used to predict how individuals will respond to certain drugs, such as those that lower cholesterol level or blood pressure. Many companies are now conducting association tests to establish disease-specific SNP profiles.

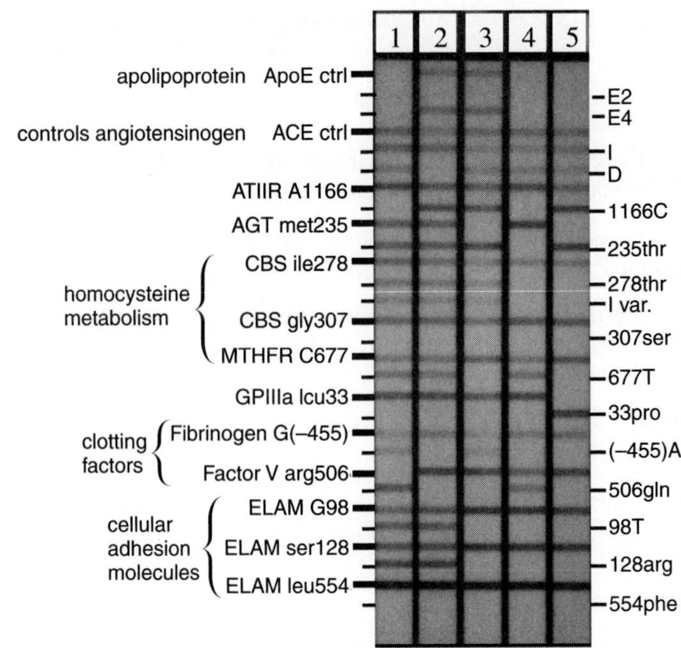


**figure 7.12**

**Assessing a complex genotype for cardiovascular disease risk.** Each strip represents an individual whose DNA is being probed for specific alleles that predispose to development of disease. The notations on the left indicate the gene, and the notations on the right indicate rare alleles. Apolipoproteins transport cholesterol. Angiotensinogen controls blood pressure. Homocysteine, in excess, may cause arteriosclerosis, which may trigger other disorders. Abnormal clotting factors or cellular adhesion molecules may lead to blockage of blood vessels.

**table 7.5**

**Risk Factors for Cardiovascular Disease**

| Uncontrollable | Controllable or Treatable |
|---|---|
| Age | Fatty diet |
| Male sex | Hypertension |
| Genes | Smoking |
| Lipid metabolism | High serum cholesterol |
| Apolipoproteins | Low serum HDL |
| Lipoprotein lipase | High serum LDL |
| Blood clotting | Stress |
| Fibrinogen | Insufficient exercise |
| Clotting factors | Obesity |
| Homocysteine metabolism | Diabetes |
| Leukocyte adhesion | |

## Body Weight

One half of all adults in the United States are obese—defined as 20 percent or more above ideal weight. Obesity raises the risk of developing hypertension, diabetes, stroke, gallstones, sleep apnea, some cancers, and psychological problems. The reason that losing weight can be difficult may reside in the genes—at least 17 genes interact to control body weight, and it is likely that subtle abnormalities or variants in many genes add up to create the tendency to gain. Still, the environment plays a significant role in body weight. Recall that heritability for body mass index, a measure of weight that accounts for height, is 0.55. Research approaches designed to tease out the genetic contributions to body weight range from the molecular level, to comparing body weights, to population studies that highlight environmental inputs.

### *Leptin and Associated Proteins*

In 1994, Jeffrey Friedman at Rockefeller University discovered a gene that encodes a protein hormone called **leptin** in mice and in humans. Normally, eating stimulates fat cells (adipocytes) to secrete the hormone. Leptin travels in the bloodstream to the brain's hypothalamus, where it binds to receptors on cells and signals them to suppress appetite and increase metabolism to digest the food already eaten. The leptin gene is one of several so-called thrifty genes that probably conserved energy in times past when periodic famine was a way of life. Low levels of leptin signal starvation, which triggers hunger and lowers metabolic rate.

When Friedman gave mice extra leptin, they ate less and lost weight. Headlines proclaimed the new magic weight loss elixir but research soon revealed that very few people could blame their obesity on low leptin levels. Figure 7.13 depicts two such rare cases—cousins who are extremely obese due to lack of leptin. Consanguinity (relatives marrying relatives) in the pedigree accounts for the transmission of the condition. Treating one of the cousins with leptin has had excellent results—he no longer overeats, has lost weight, and has shown signs of puberty. A more common situation explaining obesity, although still rare, is defective leptin receptors on cells in the hypothalamus. The receptors do not recognize the hormone signal to cease eating.

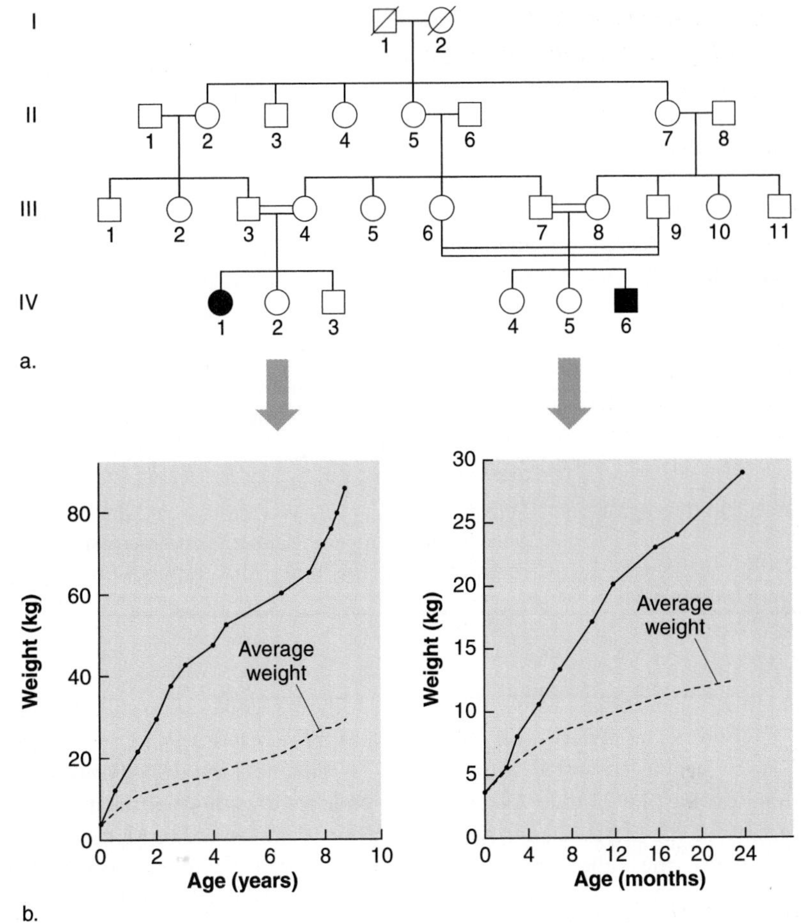


**figure 7.13**

**Leptin deficiency is rare.** The two affected cousins depicted in the pedigree (individuals IV-1 and IV-6) inherited the same recessive allele from shared ancestors (*a*). Their obesity is extreme (*b*).

Genes whose protein products are involved in other aspects of leptin function may explain more cases of obesity. Some proteins escort leptin from the circulation to the brain. A pair of specific proteins have opposite functions, one acting as an accelerator of weight gain and the other as a brake. **Neuropeptide Y** is produced in the hypothalamus in response to low leptin levels. Sensing starvation, it increases appetite. The "brake," the **melanocortin-4 receptor,** is activated when weight is gained, and it suppresses appetite. Some of these proteins may be involved in eating disorders, discussed in the next chapter. Drug companies are exploring all of these leptin-related proteins in the never-ending search for drugs that can control body weight (table 7.6).

### *Environmental Influences on Obesity*

Many studies on adopted individuals and twins suggest that obesity has a heritability of 75 percent. Although the environment contributes only about 25 percent to variability in body weight, that component can be striking when eating and exercise habits change suddenly in a population. Natural experiments illustrate this phenomenon.

On the tiny island of Naura, in Western Samoa, the residents' lifestyles changed drastically when they found a market for the tons of bird droppings on their island as commercial fertilizer. The influx of money translated into inactivity and a high-fat diet, replacing an agricultural lifestyle and a diet of fish and vegetables. Within just a generation, two-thirds of the population had become obese, and a third suffered from diabetes.

**table 7.6**

**Sites of Possible Genetic Control of Body Weight Related to Leptin**

| Protein | Function |
|---|---|
| Leptin | Stimulates cells in hypothalamus to decrease appetite and metabolize nutrients. |
| Leptin transporter | Enables leptin to cross from bloodstream into brain. |
| Leptin receptor | Binds leptin on surfaces of hypothalamus cells, triggering hormone's effects. |
| Neuropeptide Y | Produced in hypothalamus when leptin levels are low and individual loses weight, stimulating appetite and lowering rate of energy use. An appetite "accelerator" that responds to starvation. |
| Melanocortin-4 receptor | Activated when leptin levels are high and the individual gains weight, dampening appetite and increasing rate of energy use. Appetite "brake" that responds to weight gain. |

The Pima Indians offer another example of environmental effects on body weight. These people separated into two populations during the Middle Ages, one group settling in the Sierra Madre mountains of Mexico, the other in southern Arizona. By the 1970s, the Arizona Indians no longer farmed nor ate a low-fat diet, now consuming 40 percent of their calories from fat. With this drastic change in lifestyle, they developed the highest prevalence of obesity of any population on earth. (Prevalence is the total number of individuals with a certain condition in a particular population at a given time.) Half of the Arizona group had diabetes by age 35, weighing, on average, 57 pounds (26 kilograms) more than their southern relatives, who still eat a low fat diet and are very active.

The Pima Indians demonstrate that the tendency to gain weight is not sealed in the genes dealt at conception, but instead is more a tendency to gain weight if the environment provides fatty foods. Study of an aboriginal group called the Nunavut Inuit brings the research full circle, applying a molecular analysis to a specific population.

The Inuit, also known as Canadian Eskimos, account for 35 percent of the 52,000 people who live in the Northwest Territories. Like the western Samoans and Pima Indians, lifestyle changes have recently increased their average body weight. Researchers zeroed in on an inherited predisposing factor to their weight gain—a variant of a protein that is part of the signal transduction pathway depicted in figure 2.20. This particular protein enables adipocytes to swell with fat. A mutation in this gene packs too much fat into fat cells everywhere, causing an overall roundness to the physique. Researchers typed 213 healthy Inuits for this gene, finding that those with the mutation had the highest weight, waist and hip measurements, and other measures of obesity.

Interactions and contributions of genes and the environment provide some of the greatest challenges in studying human genetics. Why does one heavy smoker develop lung cancer, but another does not? Why can one person consistently overeat and never gain weight, while another does so easily? Because we exist in an environment, no gene functions in a vacuum. Subtle interactions of nature and nurture profoundly affect our lives and make us all—even identical twins—unique individuals.

**Key Concepts**

Genes that affect lipid metabolism, blood clotting, leukocyte adhesion, and blood pressure influence cardiovascular health. • Genes that encode leptin, the leptin receptor, and proteins that transmit leptin's signals affect body weight. Studies on adopted individuals and twins indicate a heritability of 75 percent for obesity. Populations that suddenly became sedentary and switched to a fatty diet reflect environmental influences on body weight.

# Summary

### 7.1 Genes and the Environment Mold Most Traits

1. **Multifactorial traits** are attributable to both the environment and genes. A **polygenic trait** is determined by more than one gene and varies continuously in its expression. The frequency distribution of phenotypes for a polygenic trait is a bell curve.

### 7.2 Methods Used to Investigate Multifactorial Traits

2. **Empiric risk** measures the likelihood that a multifactorial trait will recur based on its prevalence in a population. The risk rises as genetic closeness to an affected individual increases, as the severity of the phenotype increases, and as the number of affected relatives rises.
3. **Heritability** estimates the proportion of variation in a multifactorial trait that is attributable to genetics. It describes a trait in a particular population at a particular time. Heritability is estimated by comparing the actual incidence of a shared trait among people related in a certain way to the expected incidence (**correlation coefficient**). In some cases rare dominant alleles can contribute to heritability.
4. Characteristics shared by adopted people and their biological parents are mostly inherited, whereas similarities between adopted people and their adoptive parents reflect environmental influences.
5. **Concordance** measures the frequency of expression of a trait in both members of MZ or DZ twin pairs. The more influence genes exert over a trait, the higher the concordance value.
6. **Association studies** correlate SNP patterns to increased risk of developing a particular disorder.

### 7.3 Some Multifactorial Traits

7. Genes that control lipid metabolism and blood clotting contribute to cardiovascular health.
8. **Leptin,** its receptor, its transporter, **neuropeptide Y,** and the **melanocortin-4 receptor** are proteins that affect body weight. Fat cells secrete leptin in response to starvation, and the protein acts in the hypothalamus. Populations that switch to a fatty diet and a less-active lifestyle reveal the effects of the environment on weight.

# Review Questions

1. What is the difference between a Mendelian multifactorial trait and a polygenic multifactorial trait?
2. Which has a greater heritability—eye color or height? State a reason for your answer.
3. How can skin color have a different heritability at different times of the year, or in genetically similar populations that live in different areas?
4. How can the environment influence the course of cystic fibrosis, which is a Mendelian (single gene) disorder?
5. Describe the type of information provided by
   a. empiric calculation
   b. twin studies
   c. adoption studies
   d. association studies
6. Why does SNP mapping require enormous amounts of data?
7. Name three types of proteins that affect cardiovascular functioning, and three that affect body weight.
8. In a large, diverse population, why are medium brown skin colors more common than very white or very black skin?

# Applied Questions

1. Why is the recurrence risk for a single gene disorder so much higher than for a polygenic disorder?
2. State two ways that the study of a multifactorial trait involves statistical analysis.
3. According to table 7.2, would taking a preparatory course better help a student improve the verbal part of the SAT exam, or the math part? Provide a reason for your answer.
4. One way to calculate heritability is to double the difference between the concordance values for MZ versus DZ twins. For multiple sclerosis, concordance for MZ twins is 30 percent, and for DZ twins, 3 percent. What is the heritability? What does the heritability suggest about the relative contributions of genes and the environment in causing MS?
5. Why is a case-control experimental design valuable in an association study?
6. How can an association study be clinically valuable, if it does not lead immediately to identification of the disease-causing gene?
7. Studies among Caucasians in the United States revealed the following heritabilities for traits associated with cardiovascular health:

| | |
|---|---|
| HDL cholesterol level | 0.63 |
| triglyceride level | 0.37 |
| diastolic blood pressure | 0.21 |
| lipoprotein A activity | 0.77 |
| body mass index | 0.55 |

   List the traits in order, from most affected by genes to least.
8. In a given population at a particular time, a researcher examines 200 parent-child pairs for bushy eyebrows. In 50 pairs, both parent and child have the trait. Another researcher examines 100 sibling pairs for the trait of selfishness, as assessed by the parents. In 10 of the sibling pairs, the parents rate both children as selfish.
   a. What is the heritability of bushy eyebrows for this population?
   b. What is the heritability of selfishness?
   c. What are some problems with calculating the heritability for selfishness?
9. A G protein mutation studied in Canadian Inuits "accounted for between 1.6 percent and 3.3 percent of the total variation of the obesity-related trait." Does this statement, taken from the research paper reporting the results of experiments, mean that the mutation is the sole determinant of the obesity, or that there are other influences?
10. Would a treatment for obesity seek to increase production of neuropeptide Y and decrease the production of the melanocortin-4 receptor, or the opposite?

# Suggested Readings


Graves, Joseph L. Jr. *The Emperor's New Clothes: Biological Theories of Race at the Millennium.* 2001. Piscataway, NJ: Rutgers University Press. An examination of definitions of race, and the consequences of seeing people as black or white.

Elson, Charles O. February 21, 2002. Genes, microbes, and T cells—new therapeutic targets in Crohn's disease. *The New England Journal of Medicine.* 346(8):614–616. Gene analysis reveals the basis of susceptibility to this painful disorder.

Harris, Muriel J. May 2001. Why are the genes that cause risk of human neural tube defects so hard to find? *Teratology* 63:165–66. Because few families have more than one member with an NTD, linkage studies are lacking.

Holden, Constance. March 9, 2001. Study suggests pitch perception is inherited. *Science* 291:1879. Twin studies indicate that the ability to detect a wrong note in a tune is largely inherited.

Humphries, Steve E., et al. July 14, 2001. Apolipoprotein E3 and coronary heart disease in middle-aged men who smoke: A prospective study. *The Lancet* 358:115–19. Males who smoke and have a certain apolipoprotein gene variant are at triple the risk of suffering a heart attack.

Lewis, Ricki. February 18, 2002. Race and the clinic: Good science? *The Scientist* 16(3):16. Genotypes are better predictors of drug response than is consideration of skin color.

Lewis, Ricki. January 24, 2000. Homing in on homocysteine. *The Scientist* 14:1. Several inborn errors of metabolism can elevate plasma homocysteine, which raises the risk of cardiovascular disease.

Lewis, Ricki. July 20, 1998. Unraveling leptin pathways identifies new drug targets. *The Scientist* 12:1. Leptin, its receptor, and other proteins may inspire development of new weight-control drugs.

Mathew, Christopher. April 28, 2001. Postgenomic technologies: Hunting the genes for common disorders. *British Medical Journal* 322:1031–34. The focus of genetics is shifting from rare to common disorders.

McLeod, Howard L. November, 2001. Pharmacogenetics: More than skin deep. *Nature Medicine* 29:247–48. SNP patterns are much better predictors of drug efficacy than skin color.

Perola, Markus, et al. July 2001. Quantitative trait locus analysis of BMI and of stature, by combined analysis of genome scans of 5 Finnish study groups. *The American Journal of Human Genetics* 69:117–23. These researchers analyzed BMI and height measurements obtained in past studies to evaluate heritability.

Pritchard, Jonathan. July 2001. Are rare variants responsible for susceptibility to complex diseases? *The American Journal of Human Genetics* 69:124–37. Not all genes contribute equally to multifactorial traits.

Pritchard, Jonathan, and Molly Przeworski. July 2001. Linkage disequilibrium in humans: Models and data. *The American Journal of Human Genetics* 69:1–14. Constructing SNP maps relies on understanding linkage disequilibrium.

Todd, John A. May 31, 2001. Tackling common disease. *Nature* 411:537–39. Two different approaches led to the same predisposition gene for Crohn disease.

Wood, Alastair J. J. May 3, 2001. Racial differences in the response to drugs—pointers in genetic differences. *The New England Journal of Medicine* 344(18):1393–95. Skin color is not an adequate predictor of drug response.

# On the Web

Check out the resources on our website at

**www.mhhe.com/lewisgenetics5**

On the web for this chapter, you will find additional study questions, vocabulary review, useful links to case studies, tutorials, popular press coverage, and much more. to investigate specific topics mentioned in this chapter, also try the links below:

American Heart Association
**www.americanheart.org**

Cleft Palate Foundation **www.cleft.com**

Online Mendelian Inheritance in Man
**www.ncbi.nlm.nih.gov/entrez/query.fcgi?db=OMIM**
blue eye color 227240
brown eye color 227220
cystic fibrosis 219700
hyperlipoproteinemia 238600
leptin deficiency 164160
leptin receptor 601001
obesity 601665
type I diabetes mellitus 222100

CHAPTER

# 8

# The Genetics of Behavior

### 8.1 Genes Contribute to Most Behavioral Traits
Several genes and environmental factors contribute to most behavioral traits and disorders. Behavioral genetics is evolving from a descriptive to a more mechanistic science as researchers associate patterns of gene expression in specific brain regions with particular behaviors.

### 8.2 Eating Disorders
Anorexia and bulimia are common in affluent nations. Heritability is high, yet the behavior may be a response to powerful societal pressures. Candidate genes control appetite or regulate the neurotransmitters dopamine or serotonin. Psychological factors are harder to study.

### 8.3 Sleep
We spend much of our lives asleep. Sleep habits are largely inherited. A Utah family with many members who have a highly unusual sleep-wake cycle led researchers to discover the first "clock" gene in humans.

### 8.4 Intelligence
Investigating the genetic components to intelligence has been controversial. Although heritability is high, environmental influences are profound, particularly early in life. Psychologists use a measure called "general intelligence," whereas geneticists search for genes whose protein products can explain individual variations in intelligence.

### 8.5 Drug Addiction
Genes whose protein products affect the brain's limbic system influence individual variations in susceptibility to addiction. Candidate genes affect neurotransmission and signal transduction.

### 8.6 Mood Disorders
Deficits of the neurotransmitters serotonin or norepinephrine cause major depressive disorder. Less common are the mood swings of bipolar disorder, whose genetic roots are many and difficult to isolate.

### 8.7 Schizophrenia
A condition like no other, schizophrenia devastates the ability to think and perceive clearly. Candidate genes and environmental associations are many.

In the summer of 2001, a tennis promoter offered star athletes Steffi Graf and Andre Agassi, expecting a baby, $10 million if they would promise that their child would play a tennis match in the year 2017 against the offspring of another tennis great. *People* magazine called the child "the most DNA-advantaged prodigy in tennis" and quoted Agassi as saying, "I've got genetics on my side" when asked if Junior would win the match (figure 8.1). Once Jaden was born, speculation about his future athletic abilities continued.

The idea that children-of-athletes will grow up to be athletes illustrates the popular but flawed idea of genetic determinism, that a gene dictates every imaginable trait, even the ability to whack a ball over a net. Jaden Agassi may indeed have inherited fortuitous muscle anatomy, quick reflexes, athletic grace, and a competitive spirit from his parents. He will also experience powerful environmental cues, and, if his father's comment is meant literally, will probably have a racquet in his hands before kindergarten. Still, whether or not he chooses to follow in his parents' footsteps—or how he will react if his athletic prowess does not match people's expectations—depends upon an unpredictable combination of many factors, both genetic and environmental.

This chapter explores how researchers are disentangling the genetic and environmental threads that contribute to several familiar behaviors and disorders.

**figure 8.1**

**A sports gene?** Will the son of Steffi Graf and Andre Agassi inherit their tennis ability or his pattern baldness—or both? Genetic determinism is often an oversimplification of how genes work.

## 8.1 Genes Contribute to Most Behavioral Traits

Behavioral traits include abilities, feelings, moods, personality, intelligence, and how a person communicates, copes with rage and handles stress. Disorders with behavioral symptoms are wide-ranging and include phobias, anxiety, dementia, psychosis, addiction, and mood alteration. Very few medical conditions with behavioral components can be traced to a single gene—Huntington disease, with its characteristic anger, is a rare example. Most behavioral disorders fit the classic complex disease profile: they affect more than 1 in 1,000 individuals and are caused by several genes and the environment—that is, they are common, polygenic and multifactorial.

Until recently, geneticists and social scientists studying behavior were limited to such tools as empiric risk estimates and adoptee and twin studies, discussed in chapter 7. These approaches clearly indicate that nearly all behaviors have inherited influences. Two powerful new approaches to understanding the biological basis of behavioral traits are:

1. association studies that correlate genetic markers such as SNP (single nucleotide polymorphism) patterns with particular symptoms
2. analysis of mutations in specific candidate genes that are present exclusively in individuals with the behavior

Behavioral genetics is, by definition, a study of nervous system variation and function. Genes control the synthesis, levels, and distribution of neurotransmitters, which are the chemical messengers that connect nerve cells (neurons) into pathways. Signal transduction is also a key part of the function of the nervous system (see figure 2.20). Therefore, candidate genes for the inherited components of a variety of mood disorders and mental illnesses—as well as of normal variations in temperament and personality—affect neurotransmission and signal transduction.

Typically, traditional methods identify a large inherited component to a behavior, and then studies focus on identifying and describing candidate genes. Consider attention deficit hyperactivity disorder (ADHD).

Siblings of an affected child have a three to five times higher risk of developing the disorder than children without affected siblings. Adopted children with ADHD are more likely to have a biological parent who also had the disorder than adopted children who do not have ADHD. Twin studies indicate a heritability of 0.80. (Recall that heritability estimates the proportion of phenotypic variation due to genetics.) Linkage analysis on families with more than one affected member implicate the neurotransmitter dopamine—specifically, one gene that encodes a "transporter" protein that shuttles dopamine between neurons, and another, the dopamine D(4) receptor gene, whose encoded protein binds dopamine on the membrane of the postsynaptic (receiving) neuron. Attempts to develop new drugs to treat ADHD, then, will focus on dopamine.

Deciphering genetic components of most behavioral disorders is not as straightforward as ADHD analysis appears to be so far. Investigating the causes of autism illustrates the difficulty of reconciling empiric, adoptee, and twin data with molecular methods.

Autism is a disorder of communication—the individual does not speak or interact with others and is comfortable only with restricted or repetitive behaviors. Asperger syndrome is a related condition that does not impair language ability and may be a mild form of autism. Siblings of children with autism are 45 to 150 times more likely to develop either condition than is the general population. Put more absolutely, general population incidence of autism is only 10 to 12 per 10,000 (<0.1 percent), but to a sibling, risk of recurrence is 2 to 4 percent. Twin studies also indicate high heritability. In various studies, concordance for MZ twins ranges from 60 to 90 percent, and for DZ twins it is 0 to 20 percent. Yet the search for causative genes so far has yielded many candidates with weak linkage, rather than a few compelling candidate genes. Four whole genome scans, using many markers and hundreds of families that have more than one affected member, have implicated possible risk-raising genes on 14 different chromosomes! The genetics of autism is complex—perhaps it is actually several different disorders that have similar symptoms.

Investigating the genetics of behavior is more difficult, for several reasons, than understanding a disorder in which an abnormal protein clearly disrupts physiology in a particular way. Many behavioral disorders share symptoms, which can complicate diagnosis. However, many symptoms also fall within the range of normal behavior. It is difficult to say, for example, whether the many people who have post-traumatic stress disorder after being caught in the terrifying events of September 11, 2001, are behaving in an expected or unusual way. Studies that rely on self-reporting of symptoms may be highly subjective. A person can also, unintentionally, copy someone's unusual behavior, because he or she does not realize it is unusual. Such sources of confusion would not occur with a strictly physical condition such as sickle cell disease. However, being too quick to assign a genetic cause to a behavior can be dangerous, too, as Bioethics: Choices for the Future, "Blaming Genes" (p. 167) discusses.

Each part of this chapter begins with the traditional approaches of evaluating siblings, adoptees, and twins, and concludes with a look at the latest efforts to identify the genes that underlie certain behaviors. The eventual impact of human genome information on the study of behavior promises to be great. Wrote one psychologist, "Ultimately, the human genome sequence will revolutionize psychology and psychiatry." Specifically, a methodological search for human behavioral genes will lead to better understanding of the causes of disease and the neurobiological bases of individual differences. More practically, understanding the roles of genes in behavior will make it possible to subtype mental disorders so that effective, individualized therapies can be identified and implemented early in an illness. Table 8.1 lists some of the more common behavioral disorders.

### Key Concepts

Most behavioral traits and disorders are fairly common, polygenic and multifactorial. Traditional methods to estimate the genetic contribution to a behavior include empiric risk estimates and adoptee and twin studies. Association studies and candidate gene analyses are now extending these data. Behavioral disorders are difficult to study because symptoms overlap and behaviors can be imitated.

### table 8.1

**Prevalence of Behavioral Disorders in the U.S. Population**

| Condition | Prevalence (%) |
|---|---|
| Alzheimer disease | 4.0 |
| Anxiety | 8.0 |
| Phobias | 2.5 |
| Post-traumatic stress disorder | 1.8 |
| Generalized anxiety disorder | 1.5 |
| Obsessive compulsive disorder | 1.2 |
| Panic disorder | 1.0 |
| Attention deficit hyperactivity disorder | 2.0 |
| Autism | 0.1 |
| Drug addiction | 4.0 |
| Eating disorders | 3.0 |
| Mood disorders | 7.0 |
| Major depression | 6.0 |
| Bipolar disorder | 1.0 |
| Schizophrenia | 1.3 |

Source: Psychiatric Genomics Inc., Gaithersbury, MD. The information was collated from the Surgeon General's Report on Mental Health 1999.

## 8.2 Eating Disorders

When gymnast Christy Henrich was buried on a Friday morning in July 1994, she weighed 61 pounds and was 22 years old (figure 8.2). Three weeks earlier, she had weighed an unbelievable 47 pounds. Christy suffered from anorexia nervosa, a psychological disorder that is fairly common among professional athletes. The person perceives herself or himself as obese, even when obviously not, and starves intentionally. Christy's decline began in 1988, when a judge at a gymnastics competition told her that at 90 pounds, she was too heavy to make the U.S. Olympic team. From then on, her life consisted of starving, exercising, and taking laxatives to hasten weight loss.

For economically advantaged females in the United States, the lifetime risk of developing anorexia nervosa is 0.5 percent. Anorexia has the highest risk of death of any psychiatric disorder—15 to 21 percent. The same population group has a lifetime risk of 2.5 percent of developing another eating disorder, bulimia. A person with bulimia eats huge amounts but exercises and vomits to maintain weight.

**figure 8.2**

**Eating disorders.** World-class gymnast Christy Henrich died of complications of the self-starvation eating disorder anorexia nervosa in July, 1994. In this photo, taken 11 months before her death, she weighed under 60 pounds. Concern over weight gain propelled her down the path of this deadly nutritional illness.

About 10 percent of people with eating disorders are male. One survey of 8-year-old boys revealed that more than a third of them had attempted to lose weight. In an eating disorder called muscle dysmorphia, or, more commonly, "bigorexia," boys and young men take amino acid food supplements to bulk themselves up into what they consider to be more masculine physiques. Just as the person with anorexia looks in a mirror and sees herself as too large, a person with muscle dysmorphia sees himself as too small.

Because for many years eating disorders were associated almost exclusively with females, most available risk estimates exclude males. Twin studies repeatedly reveal a considerable genetic component to eating disorders. One typical investigation, for example, identified concordance for anorexia nervosa in 9 of 16 MZ twin pairs, but only 1 of 14 DZ pairs. The Danish twin study, which reports on thousands of twin pairs, also reveals a distinct role of genes in predisposing people to develop eating disorders. Heritability calculations range from 0.5 to 0.8, ranking eating disorders with diabetes mellitus and schizophrenia in terms of contribution of genes.

Studies of eating disorders that recur in families without twins are more difficult to interpret. It's hard to determine whether a 10-year-old girl who observes her older sister starve herself and copies the behavior does so because she has inherited genes that predispose her to develop an eating disorder, or simply because she idolizes her sister and wants to be like her.

The next step after demonstrating a role of genetics in eating disorders is to find the types of genes that could be involved. One website for eating disorders states that "high-risk genes" include those that control the traits of perfectionism, orderliness, low tolerance for new situations, maturity fears, low self-esteem, and overall anxiety. A young woman's particular combination of these traits presumably determines how she responds to powerful societal messages to be thin. While psychologists might measure these hard-to-define types of traits, geneticists look more toward particular proteins whose physiological functions suggest that the genes could, if abnormal, affect eating behavior. The genes that encode proteins that control appetite are candidates. Table 7.6 lists them—genes for leptin and its transporter and receptor, neuropeptide Y, and the melanocortin-4 receptor. Genes that regulate the neurotransmitters dopamine and serotonin may also contribute to the risk of developing an eating disorder. It will be interesting to learn which genes affect body image, and how they do so.

Complementing studies on known genes are whole genome scans that use clinical data to associate SNP patterns (see chapters 1 and 7) with eating disorders. One biotechnology company is cataloging 60 SNP sites among the genomes of 2,000 individuals representing 600 families where more than one member has anorexia nervosa. The researchers are searching for a SNP pattern—if there is one—that appears disproportionately among individuals who have anorexia. Once the SNP maps highlight chromosome regions that seem to mark predisposition to developing anorexia, researchers will look for genes in the regions whose protein products might affect appetite.

**KEY CONCEPTS**

Eating disorders are common. Twin and heritability studies indicate a high genetic contribution. Genes whose products control appetite or regulate the neurotransmitters dopamine and serotonin may cause or raise the risk of developing an eating disorder.

## 8.3 Sleep

Sleep has been called "a vital behavior of unknown function," and, indeed, without sleep, animals die. Similarly, the existence of "fatal familial insomnia" indicates the necessity of sleep. We spend a third of our lives in this mysterious state.

Genes influence sleep characteristics. When asked about sleep duration, schedule, quality, nap habits, and whether they are "night owls" or "morning people," MZ twins report significantly more in common than do DZ twins, even MZ twins separated at birth. Twin studies of brain wave patterns through four of the five stages of sleep confirm a hereditary influence. The fifth stage, REM sleep, is associated with dreaming and therefore may more closely mirror the input of experiences rather than genes.

Studies of a disorder called "narcolepsy with cataplexy" further illustrate the influence of genes on sleep behavior. Narcolepsy produces daytime sleepiness and the tendency to fall asleep very rapidly, several times a day. Cataplexy is short episodes of muscle weakness, during which the jaw sags, head drops, knees buckle, and the person falls to the ground. This often occurs during a bout of laughter—which can be quite disturbing both to the affected individual and bystanders.

Narcolepsy with cataplexy affects only 0.02 to 0.06 percent of the general populations of North America and Europe. Yet it is much more common within some families. A person with an affected first-degree relative has a 1 to 2 percent chance of developing the condition, which usually begins between the ages of 15 and 25. Concordance for narcolepsy with cataplexy among MZ twins is 25 to 31 percent.

Dogs may help us understand the complex genetics of narcolepsy. In canines, narcolepsy is a fully penetrant, autosomal recessive Mendelian trait, caused by an allele called *canarc-1*. Researchers are searching the corresponding region of the human genome for a possible narcolepsy gene. (We can't tell if dogs also have cataplexy, because they do not laugh!)

A multigenerational Utah family with many members who have an unusual sleep pattern has enabled researchers to identify the first "biological clock" gene in humans that controls sleep. The subjects have familial advanced sleep phase syndrome (FASPS) and the effect is striking—they promptly fall asleep at 7:30 each night and awaken with a jolt each morning at 4:30. The family is a geneticist's dream—a distinctive behavioral phenotype, many affected individuals, and a clear mode of inheritance (autosomal dominant) (figure 8.3).

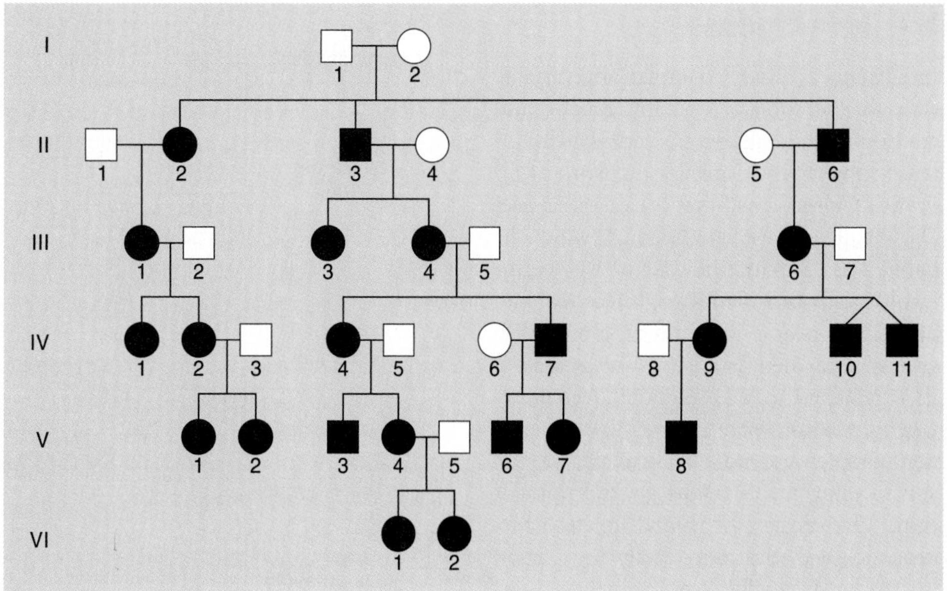


**figure 8.3**

**Inheritance of a disrupted sleep-wake cycle.** This partial pedigree depicts a large family with familial advanced sleep phase syndrome. The condition is genetically heterogeneic—that is, different families have causative mutations in different genes. In this family from Utah, the condition is autosomal dominant.

Analyzing the Utah family followed the standard approach to gene identification that chapter 7 described. A whole genome scan for short repeated DNA sequences revealed a variant area at the tip of the long arm of chromosome 2 that is found exclusively in the affected family members. Within that defined area is a gene, called *period,* that has a counterpart in golden hamsters and fruit flies that causes the same disrupted sleep-wake cycle phenotype. The humans with the condition have a single DNA base substitution in the gene. This mutation prevents the encoded protein from binding a phosphate chemical group, which it must do to pass on the signal that synchronizes the sleep-wake cycle with daily sunrise and sunset.

Despite the clear connection between sleep behavior and a specific gene in the Utah family, that gene is just one influence on this behavior. In other families with FASPS, linkage analysis did not point to this gene, meaning that the condition is genetically heterogeneic. Environmental influence is great too. Daily rhythms such as the sleep-wake cycle are set by cells that form a "circadian pacemaker" in a part of the brain called the suprachiasmatic nuclei. Genes are expressed in these cells in response to light or dark in the environment. Other environmental effects on sleeping and waking are more subtle. Knowing that the hour is late may trigger an "I should go to sleep" or "yikes, I have to get up for class" response. Boredom may dictate bedtime. Culture also affects the times that we retire and rise. In European countries, for example, dinner and bedtime are usually later than in the United States. In many New York City offices, the workday starts at 9 A.M.; on many farms, the day begins hours earlier. Understanding how the *period* gene and others control the sleep-wake cycle may lead to new treatments for jet lag, insomnia, and the form of advanced sleep phase syndrome that is common among older individuals.

**Key Concepts**

Empiric risk studies on cataplexy and twin studies on sleep habits indicate a high heritability for sleep characteristics. A Utah family with a very unusual sleep-wake cycle led researchers to identify the first "clock" gene in humans. It has counterparts in several types of animals.

## 8.4 Intelligence

Intelligence is a vastly complex and variable trait that is subject to many genetic and environmental influences, and also to intense subjectivity. Consider the work of Sir Francis Galton, a half first cousin of Charles Darwin. He investigated genius, which he defined as "a man endowed with superior faculties," by first identifying successful and prominent people in Victorian-era English society, and then assessing success among their relatives. In his 1869 book, *Hereditary Genius,* Galton wrote that relatives of eminent people were more likely to also be successful than people in the general population. The closer the blood relationship, he concluded, the more likely the person was to be successful. This, he believed, established a hereditary basis for intelligence as he measured it.

Definitions of intelligence have varied greatly. In general, intelligence refers to the ability to reason, learn, remember, connect ideas, deduce, and create. The first intelligence tests, developed in the late 19th century, assessed sensory perception and reaction times to various stimuli. In 1904, Alfred Binet at the Sorbonne developed a test with verbal, numerical, and pictorial questions. Its purpose was to predict the success of developmentally handicapped youngsters in school. The test was subsequently modified at Stanford University to represent white, middle-class Americans. An average score on this "intelligence quotient," or IQ test, was 100, with two-thirds of all people scoring between 85 and 115 in a bell curve or normal distribution (figure 8.4). An IQ between 50 and 70 is considered mild mental retardation, and below 50, severe mental retardation.

Over the years, IQ has proven to be a fairly accurate predictor of success in school and work. However, low IQ also correlates with many societal situations, such as poverty, high divorce rate, failure to complete high school, incarceration (males), and having a child out of wedlock (females). In 1994, a book called *The Bell Curve* asserted that because certain minorities are overrepresented in these groups, they must be of genetically inferior intelligence and that is why they are prone to suffering social ills. It was a controversial thesis, to put it mildly.

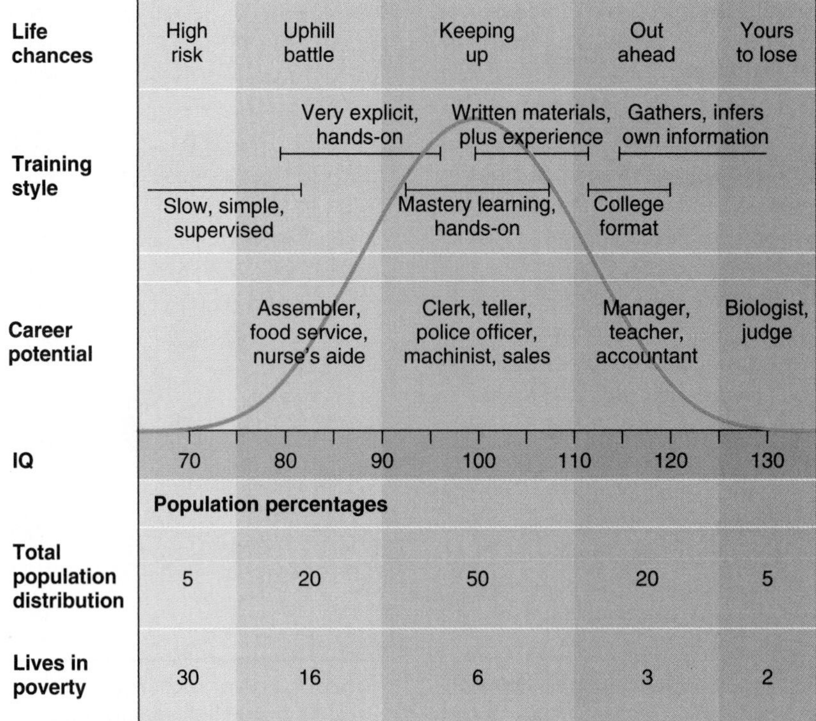


**figure 8.4**

**Success and IQ.** IQ scores predict success in school and the workplace in U.S. society. Correlations to other measures of success are controversial because they do not necessarily reflect heredity.

The IQ test is actually a battery of short exams that measure verbal fluency, mathematical reasoning, memory, and spatial visualization skills. Because people tend to earn similar scores in all these areas, psychologists hypothesized that a general or global intelligence ability, called "g," must underlie the four basic skills that IQ encompasses. Statistical analysis indeed reveals one factor that accounts for general intelligence. In contrast, similar analysis of personality revealed five contributing factors. The g value is the part of IQ that accounts for differences between individuals based on a generalized intelligence, rather than on enhanced opportunities, such as attending classes to boost test-taking skills.

Environment does not seem to play too great a role in IQ differences. Evidence includes the observation that IQ scores of adoptees, with time, are closer to those of their biological parents than to the adoptive parents. Heritability studies also reveal a declining environmental impact with age (table 8.2). This makes sense. As a person ages, he or she has more control over the environment, so genetic contributions to intelligence become more prominent.

A genetic explanation for intelligence differences has long been realized because nearly all syndromes that result from abnormal chromosomes include some degree of mental retardation. Down syndrome and fragile X syndrome (see figure 12.1 and Reading 11.1) are two of the more common chromosomal causes of mental retardation.

**table 8.2**

**Heritability of Intelligence Changes Over Time**

| Age Group | Heritability |
|---|---|
| Preschoolers | 0.4 |
| Adolescents | 0.6 |
| Adults | 0.8 |

Down syndrome is usually caused by an extra chromosome, and fragile X syndrome by an expanding gene on the X chromosome. The human genome sequence revealed that mutations in genes located next to the tips of chromosomes—the subtelomeric regions—also account for many cases of mental retardation.

The search for single genes that contribute to intelligence differences focuses on neurons and how they interact. In one study, a certain SNP pattern in a gene encoding neural cellular adhesion molecule (N-CAM) correlated with high IQ. Perhaps this gene variant facilitates certain neural connections that enhance learning ability.

Identifying the N-CAM is an example of the candidate gene approach—the function of the gene's product was known and could be related to intelligence. Whole genome scans are being done to locate other genes whose protein products affect formation of neural connections in ways that enhance intelligence. Preliminary studies singled out chromosome 4 as harboring intelligence-related genes. Further investigation identified three candidate genes on this chromosome by comparing 147 markers in one group of children with very high IQ scores to controls who had average scores.

Usually in scientific research, negative evidence is not exciting, but when investigating polygenic traits, it is important to rule out roles for certain genes. For example, association studies showed that the dopamine D(2) receptor, which affects several behaviors, and the apolipoprotein E4 allele (see section 7.3), apparently do not contribute to intelligence differences. It will be fascinating in the years to come to learn which genes control variations in intelligence—and which do not.

**Key Concepts**

Intelligence is the use of mental skills to complete complex tasks or solve problems. IQ assesses verbal fluency, mathematical reasoning, memory, and spatial visualization ability. The "g" value measures a general intelligence factor that better represents the inherited contribution to IQ. Environment has less of an influence on intelligence as a person ages. A neural adhesion molecule is one of many candidate genes for intelligence. An association study identified part of chromosome 4 as containing a gene or genes important in intelligence.

## 8.5 Drug Addiction

Drug addiction is the compulsive behavior of seeking and taking a drug despite knowing its possible adverse effects. Drug addiction has two identifying characteristics, tolerance and dependence. Tolerance is the need to take more of the drug to achieve the same effects as time goes on. Dependence is the onset of withdrawal symptoms when a person ceases taking the drug. Both tolerance and dependence contribute to the biological and psychological components of craving the drug. The behavior associated with drug addiction can be extremely difficult to break.

Drug addiction produces stable, rather than temporary, brain changes, because the craving and high risk of relapse remain even after a person has abstained for years. Heritability is 0.4 to 0.6, with a two- to three-fold increase in risk among adopted individuals who have one affected biological parent. Twin studies also indicate an inherited component to drug addiction.

Brain imaging techniques have localized the "seat" of drug addiction in the brain by highlighting the cell surface receptors that bind neurotransmitters. The brain changes that contribute to addiction occur in a group of functionally related structures called the limbic system (figure 8.5). These structures

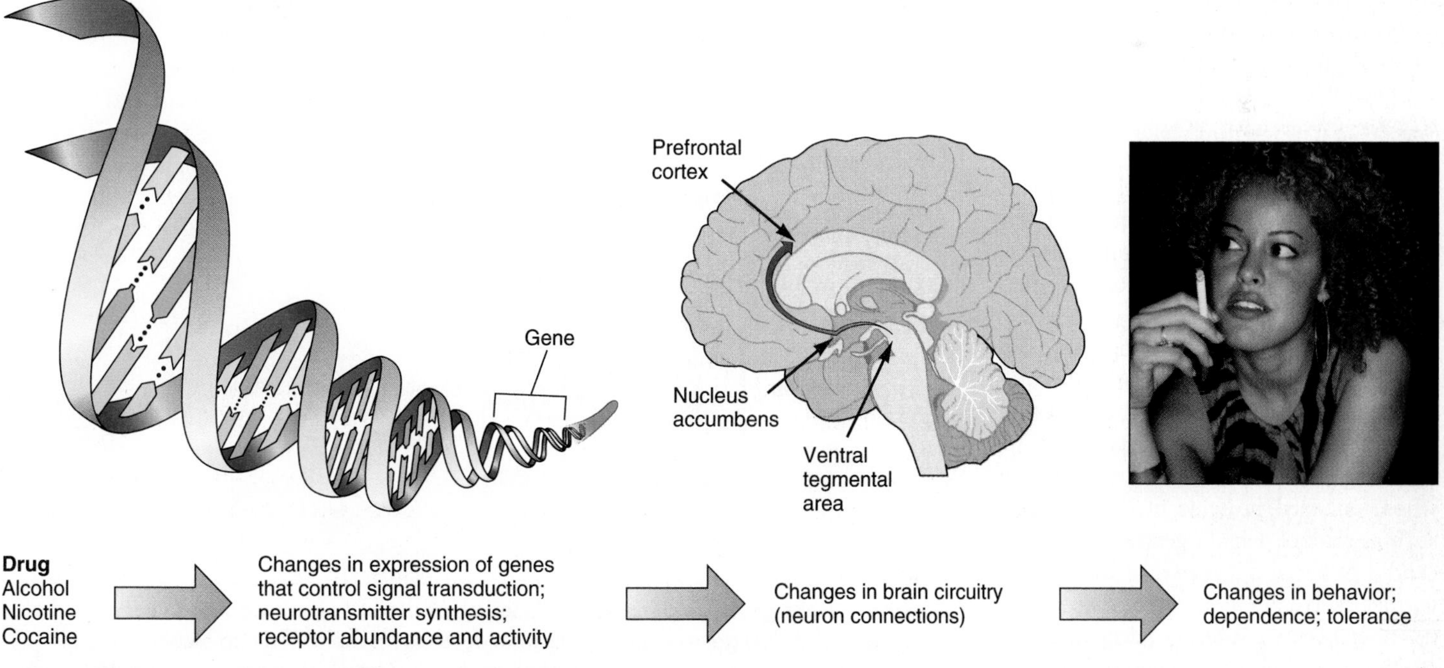


**figure 8.5**

**The events of addiction.** Addiction is manifest at several levels: molecular, neuron-neuron interactions, and behavioral responses.

are the nucleus accumbens, the prefrontal cortex, and the ventral tegmental area. The effects of cocaine seem to be largely confined to the nucleus accumbens, whereas alcohol affects the prefrontal cortex.

Although the specific genes and proteins that are implicated in addiction to different substances may vary, several general routes of interference in brain function are at play. Proteins involved in drug addiction are those that:

- are part of the biosynthetic pathways of neurotransmitters, such as enzymes
- form reuptake transporters, which remove excess neurotransmitter from the space between two neurons (the synapse)
- form receptors on the receiving cell (postsynaptic neuron) that are activated or inactivated when specific neurotransmitters bind
- are part of the signal transduction pathway in the postsynaptic neuron

An interesting aside is the fact that drugs of abuse are often plant-derived chemicals, such as cocaine, opium, and tetrahydrocannabinol (THC), the active ingredient in marijuana. These substances bind to receptors that are part of human neurons, which indicates that our bodies have their own versions of these substances. The human equivalents of the opiates are the endorphins and enkephalins, discovered in the 1970s. More recently described is anandamide, a biochemical that binds the same receptors as does THC. The endorphins and enkephalins relieve pain. Anandamide modulates how brain cells respond to stimulation by binding to neurotransmitter receptors on presynaptic (sending) neurons.

DNA microarray technology has revolutionized how researchers study the biology of addiction by revealing the expression of many genes at a time. In the past, research efforts tended to focus on individual genes whose encoded proteins fit the picture—such as alcohol dehydrogenase, an obvious candidate because it is part of the pathway to metabolize ethanol. Another gene that has often been the subject of drug addiction studies is an allele of the gene that encodes the dopamine D(2) receptor. People who are homozygous for the A1 allele of the D(2) dopamine receptor gene are overrepresented among people with alcoholism—45 percent compared to 14.5 percent among people who do not have alcoholism. However, this association is too imprecise to be very useful in identifying those at high risk for alcoholism. Also, the A1/A1 genotype is more prevalent among individuals with other behavioral conditions, including addictions to opioids, cocaine, and nicotine; autism; obesity; attention deficit hyperactivity disorder; and Tourette syndrome.

DNA microarray tests scan thousands of genes at a time before and after exposure to a particular drug, and, typically, 1 to 5 percent of the genes are expressed. Consider a comparison of gene expression profiles in human brain cells, 10 samples from people who were known to have been alcoholics, and 10 from people who had died from other causes (figure 8.6). Of 4,000 genes screened, researchers discovered that 160 varied in expression by at least 40 percent between the two types of brains. In addition to indicating genes involved in signal transduction and neurotransmitter activity, the study also highlighted genes that function in the cell cycle and particularly in apoptosis, and others that help a cell survive oxidative damage. Perhaps the genes that affect neurotransmitter function and signal transduction underlie the tolerance and dependency of addiction in general, and the other categories of genes that are activated reflect the body's response to the specific toxic ef-

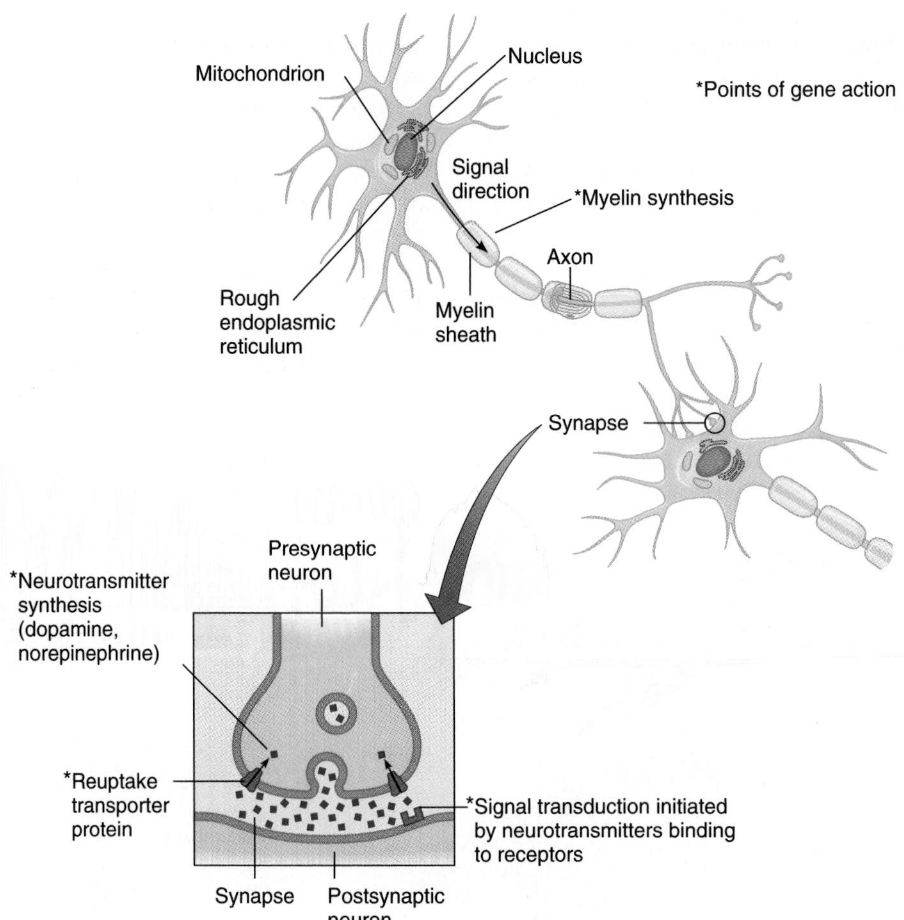


## figure 8.6

**Alcohol alters gene expression.** Researchers used DNA microarrays to compare the expression of 4,000 genes in brain cells from deceased people who had alcoholism and deceased people who had not had alcoholism. Another set of microarray experiments compared expression of 6,000 genes in mouse brain neurons growing in culture, with or without exposure to alcohol. The studies identified dozens of genes whose expression increases or decreases in the presence of alcohol.

fects of alcohol. The DNA microarray tests also revealed a subset of genes whose actions are unique to brain cells—genes that control the synthesis of myelin, which is the white, fatty substance that ensheaths neurons and enables them to communicate rapidly with one another. Impairment of myelin synthesis with chronic alcohol use is consistent with imaging studies that show a shrinkage of the brain's white matter that accompanies alcoholism.

**Key Concepts**

Drug addiction is dependency on a drug despite knowing it is harmful. Structures in the limbic system are the most directly involved in drug addiction. A candidate gene for addiction encodes the dopamine D(2) receptor. DNA microarray studies reveal many genes whose protein products affect neurotransmission and signal transduction, as well as genes that control myelin deposition on neurons.

## 8.6 Mood Disorders

Mood disorders are especially difficult to separate into genetic and environmental components because, although considered to be illnesses, they may also appear to be extremes of normal behavior. For example, a person who has previously been happy but inexplicably becomes lethargic, sad, and no longer enjoys activities that once gave pleasure, may receive a diagnosis of **major depressive disorder** (MDD) (also called clinical depression). A person with the exact same symptoms, but who can trace the onset to the death of a loved one, may be the victim of extended grief and loss, not clinical depression. Context is important.

The two most prevalent mood disorders are major depressive disorder and **bipolar affective disorder** (also called bipolar disorder or manic-depression). MDD affects 6 percent of the U.S. population at any given time, and occurs in more women than men. Lifetime risk of MDD for the general population is 5 to 10 percent. Often depression is chronic, interspersed with acute episodes that may be provoked by stress. Fifteen percent of people hospitalized for severe, recurrent depression ultimately end their lives.

Bipolar disorder is much rarer than MDD, affecting 1 percent of the population and with a general population lifetime risk of 0.5 to 1.0 percent. In the condition, weeks or months of depression alternate with periods of mania. In a classic period of mania, the person is hyperactive and restless, and may experience a flow of ideas and excitement that is so fast that the words come out in a rush. Ideas may be fantastic, and behavior reckless. For example, a person who is normally quiet and frugal might, when manic, suddenly make large monetary donations and spend lavishly—very out-of-character behavior. In one subtype of bipolar disorder, the "up" times are termed hypomania, and they seem more a temporary reprieve from the doldrums than the starkly aberrant behavior of a full manic period. It's easy to see how even clinicians can confuse MDD with bipolar disorder and hypomania.

At the root of depression, and possibly of bipolar disorder too, is deficiency of the neurotransmitter serotonin, which affects mood, emotion, appetite, and sleep. Levels of norepinephrine, another type of neurotransmitter, are important as well. Although researchers do not know how these neurotransmitter levels fall, the success of certain antidepressant drugs that target either or both neurotransmitters implicates them in the disease process. Millions of people take selective serotonin reuptake inhibitors (SSRIs) to prevent presynaptic neurons from admitting serotonin from the synapse. This leaves more of the neurotransmitter available to stimulate the postsynaptic cell (figure 8.7), which apparently offsets the neurotransmitter deficit. SSRIs include Prozac, Paxil, and Zoloft. Older antidepressants called tricyclics target norepinephrine, and newer drugs affect both serotonin and norepinephrine levels.

Researchers do not know how the distribution of serotonin deficiency in the brain of a depressed person differs from normal. One study of the brains of 220 people who had died while clinically depressed

Nondepressed individual

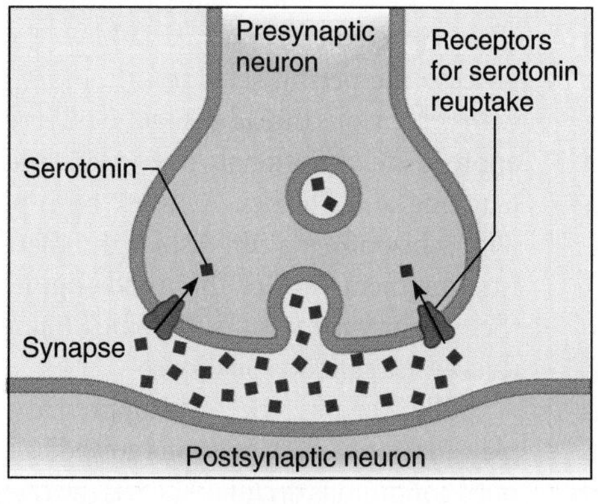


Depressed individual, untreated

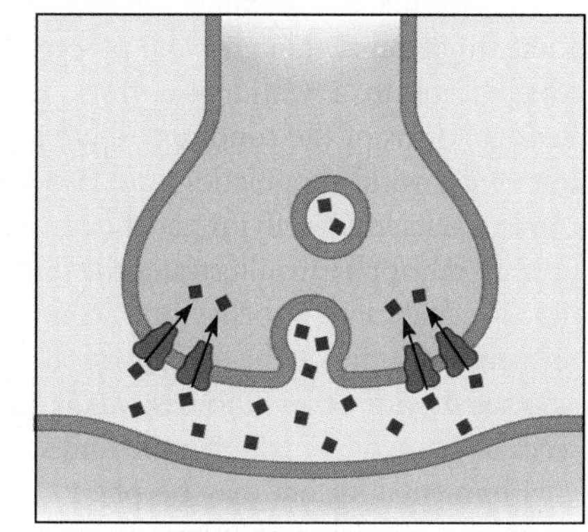

Depressed individual, treated with SSRI

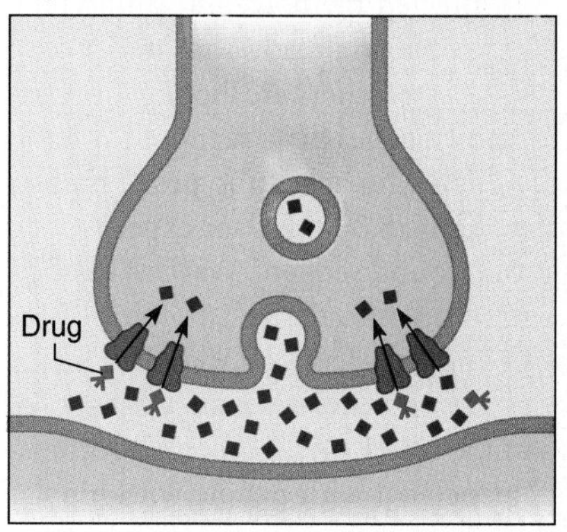


**figure 8.7**

**Anatomy of an antidepressant.** Selective serotonin reuptake inhibitors (SSRIs) are antidepressant drugs that act as their name states—they block the reuptake of serotonin, making more of the neurotransmitter available in the spaces between neurons (synapses). This restores a neurotransmitter deficit that presumably causes the symptoms. Overactive or overabundant reuptake receptors can cause the deficit. The precise mechanism of SSRIs is not well understood.

revealed a generalized decrease in serotonin activity in many brain regions, but a concentrated area of poor activity in the prefrontal cortex of those individuals who had been suicidal. The genetic connection to the declining neurotransmitter levels that appears to set the stage for depression is that the serotonin transporter is a protein.

Assigning specific genes or even chromosomal regions to bipolar disorder has lagged behind such efforts for depression. Over the past 30 years, linkage studies in large families, or association studies in very isolated populations such as the Amish, have indicated potential genes that predispose to bipolar disorder on chromosomes 4, 10, 18, 22, and the mitochondrial DNA (presumably because in some families only the females seem to pass on the trait). For depression, evidence points strongly toward malfunction of the serotonin transporter coupled with a precipitating environmental trigger. For bipolar disorder, evidence is insufficient to distinguish among a polygenic model of many genes each with an additive effect; roles for several genes, each with a large and independent effect (genetic heterogeneity); or epistasis, where different genes interact and control each other's expression.

It may be just a matter of time before the genes that contribute to bipolar disorder are identified from human genome project data and their interactions worked out. The National Institute of Mental Health has formed a consortium of nine U.S. centers to decipher the causes of bipolar disorder. The project has enrolled 500 families with more than one affected member, and genotyping of 1,200 people is already well under way. Meanwhile, researchers are looking for variants of known genes that are more common in individuals who have subtypes of bipolar disorder. Figure 8.8 shows one experimental design that found sequence variants associated with the serotonin transporter in people with bipolar disorder who had many episodes of psychosis, and variants of monoamine oxidase A, another neurotransmitter, associated with people with bipolar disorder who were also suicidal. Another candidate gene for bipolar disorder encodes one of the six types of receptors for serotonin. In addition, abundant evidence rules out certain neurotransmitters as having a role in bipolar disorder.

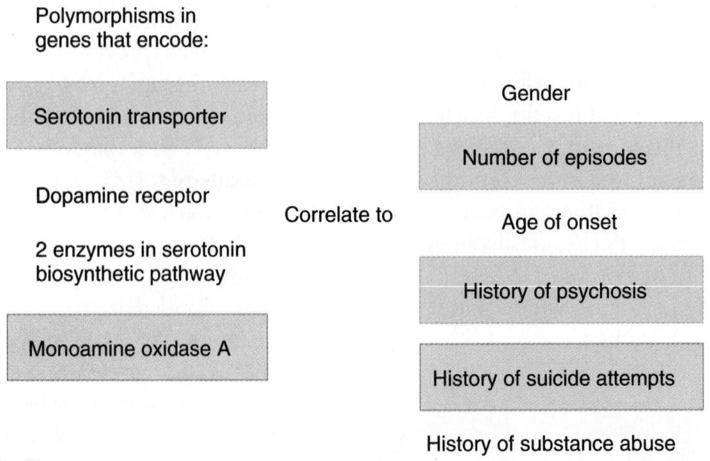

**figure 8.8**

**Bipolar affective disorder may have subtypes.** The fact that groups of patients with similar symptoms have polymorphisms in the same candidate genes is evidence that the condition is genetically heterogeneic—that is, it may be several disorders with similar and overlapping symptoms.

**KEY CONCEPTS**

Major depressive disorder is common compared to bipolar disorder, and is likely caused by deficits of serotonin, norepinephrine, or both. Bipolar disorder is associated with several chromosomal sites, and its genetic roots are difficult to isolate. Many candidate genes for MDD and bipolar disorder are currently under investigation.

## 8.7 Schizophrenia

**Schizophrenia** is a debilitating loss of the ability to organize thoughts and perceptions, which leads to a withdrawal from reality. Various forms of the condition affect 1 percent of the world's population, and 10 percent commit suicide. Identifying genetic contributions to schizophrenia illustrates the difficulties in analyzing a behavioral condition. Specifically, some of the symptoms are also associated with other illnesses; many genes seem to cause or contribute to it; and several environmental factors may be phenocopies. Yet the words of one sufferer reflect the idea that there is a genetic component. Wrote Irish journalist Harry Cavendish, in the medical journal *The Lancet,* "Schizophrenia is not easy to develop, but if it is going to happen to you it does."

The first signs of schizophrenia often are unrecognized. They are cognitive, affecting thinking. In late childhood or early adolescence, a person might suddenly have trouble paying attention in school, and learning may become difficult as memory falters and information processing skills lag. Symptoms of psychosis begin between ages 17 and 27 for males and 20 and 37 for females. The psychosis includes delusions and hallucinations—sometimes heard, sometimes seen—that can make everyday events terrifying. A person with schizophrenia may hear a voice giving instructions. What others perceive as irrational fears, such as being followed by monsters, are very real to the person with schizophrenia. Meanwhile, cognitive skills continue to decline. Speech reflects the garbled thought process; the person skips from topic to topic with no obvious thread of logic, or displays inappropriate emotional responses, such as laughing at sad news. Artwork by a person with schizophrenia can display the characteristic fragmentation of the mind (figure 8.9). (Schizophrenia means "split mind," but it does not cause a split personality.)

The course of schizophrenia often plateaus (evens out), or occurs episodically. It is not a continuous decline, which is the case for a dementia. Schizophrenia has been frequently misdiagnosed since its description a century ago. Because it causes a variety of symptoms that vary in expression, people with schizophrenia are sometimes mistakenly diagnosed

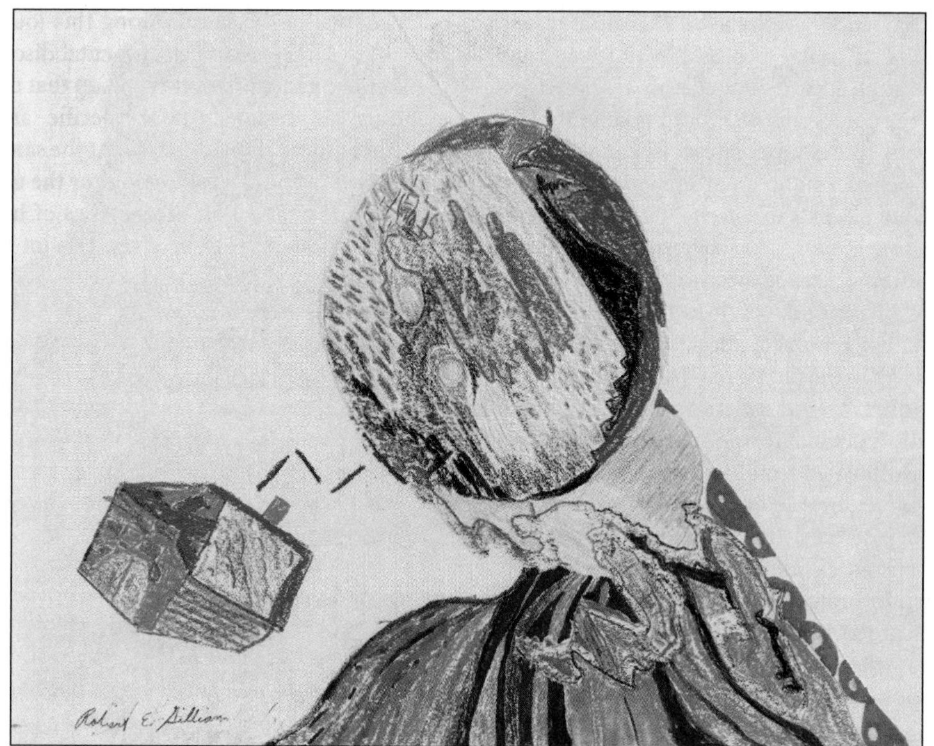

**figure 8.9**

**Schizophrenia alters thinking.** People with schizophrenia sometimes communicate the disarray of their thoughts by drawing characteristically disjointed pictures.

with depression or bipolar affective disorder. However, schizophrenia primarily affects thinking; these other conditions mostly affect mood. It is a very distinctive mental illness.

A heritability of 0.8 and empiric risk values indicate a strong role for genes in causing schizophrenia (table 8.3). Because most of the symptoms are behavioral, however, it is possible to develop some of them—such as disordered thinking—from living with people who have schizophrenia. Although concordance is high, a person who has an identical twin with schizophrenia has a 54 percent chance of *not* developing schizophrenia. Therefore, the condition has a significant environmental component, too.

One environmental hypothesis for the cause of schizophrenia is that influenza in a woman during the second trimester of pregnancy predisposes offspring to developing schizophrenia. People with schizophrenia are more likely to have been born in the spring than is the general population, with the second trimester spanning the winter months. During this time, the influenza virus can cross the placenta and alter brain cells. Table 8.4 lists other environmental factors that have been associated with schizophrenia.

Many studies indicate that several susceptibility genes cause or contribute to the development of schizophrenia. In fact, researchers have identified possible susceptibility genes on 14 different chromosomes! Early investigations focused on affected individuals who also had visible chromosome abnormalities, such as two Chinese brothers who had a duplication of part of chromosome 5. Assuming that the mental illness and the extra chromosomal material were related, researchers looked to this chromosome in other affected families, and identified five families in Iceland and two in England that had similar mutations.

The chromosome 5 association with schizophrenia, however, did not hold up for other families, indicating genetic heterogeneity—that is, the same symptoms can arise from mutations in different genes. One study examined an isolated population in northeast Finland where the lifetime risk of schizophrenia is more than three times that in most other populations studied. Meticulous church records enabled researchers to trace the disease in these extended and

**table 8.3**

**Risk of Recurrence for Schizophrenia within Families**

| Relationship of Affected Person | Risk of Recurrence |
|---|---|
| MZ twins, offspring of affected person | 46.0% |
| Sibling, DZ twins | 15% |
| Child of one affected parent | ≈13.0% |
| Child of two affected parents | 40.0% |
| Nice/nephew, half-siblings | 3.0–5.0% |
| First cousin | 2.0% |
| General population risk (no affected relatives) | 0.8% |

**table 8.4**

| Environmental Risk Factors for Schizophrenia |
|---|
| Maternal malnutrition |
| Infection by Borna virus |
| Fetal oxygen deprivation |
| Obstetric or birth complication |
| Psychoactive drug use (phencyclidine) |
| Traumatic brain injury |
| Herpes infection at time of birth |

related families to one couple among 40 founding families who lived in the area in 1650, when the records began. Researchers performed genome-wide scans and detected sites on four chromosomes—1, 4, 9, and 11—where unusual alleles are present only in people with schizophrenia. The northeast Finnish population was ideal for this type of study because their isolation and inbreeding means that many of their genes are identical. Genetic heterogeneity is less likely to be a complicating factor than in other populations. In isolated and inbred populations, unusual DNA sequences, such as those associated with schizophrenia, are easier to detect than in more diverse populations.

The largest schizophrenia study to date is based on the world's largest known affected family. They live in northern Sweden, and geneticists have family records for 3,400 individuals extending back 12 generations to the 17th century. Researchers conducted a whole genome scan using 371 markers on a branch of the family that consists of 210 people—43 with schizophrenia. The result was identification of a part of chromosome 6 that is exclusively coinherited with schizophrenia in this family.

The search for candidate genes for schizophrenia centers on two neurotransmitters, dopamine and glutamate. Brain scans reveal that in schizophrenia, dopamine levels are out of balance—too high in certain areas correlate to psychosis, and too low levels elsewhere correlate to cognitive decline. Perhaps an abnormal dopamine transporter causes the distinctive symptoms. Use of a street drug, PCP (phencyclidine), suggested the possible link between schizophrenia and altered glutamate activity because PCP blocks a type of glutamate receptor. Both dopamine and glutamate act in the same parts of the cerebral cortex. It is possible that what physicians diagnose as schizophrenia is actually several disorders that range from Mendelian disorders to cases caused entirely by environmental factors.

Genes have long been associated with physical traits, such as eye and hair color, height, and family facial resemblances. Behavioral traits have been much more difficult to describe, categorize, and attribute to genetic and/or environmental influences. Table 8.5 lists the heritabilities and candidate genes for the behavioral traits and conditions discussed in this chapter.

The influx of information from the human genome project will provide many pieces to the puzzle of what makes a human a human, and an individual an individual—personality, mood, feelings, and how we think. Assembling those puzzle pieces will require additional information from neuroscience. Identifying which neurotransmitters and which transporter and receptor proteins are implicated in particular disorders is only a start. We must also learn how these genetic instructions are imposed upon and influence the pattern of neuronal connections in the brain. A[illegible] ney of discovery, subtypes of m[illegible] ders will be genetically subtyped[illegible] agnosis will become more sp[illegible] treatment more individualized. [illegible] time, we must be careful to moni[illegible] of new information on the geneti[illegible] havior to understand ourselves, yet [illegible] discriminate.

## Key Concepts

Schizophrenia affects thinking and causes delusions and hallucinations, usually beginning in young adulthood. Studies have implicated several chromosomal regions as well as several possible environmental causes. Dopamine and glutamate are the neurotransmitters that appear to be altered in schizophrenia.

**table 8.5**

**Review of Behavioral Traits and Disorders**

| Condition | Heritability | Candidate Genes |
|---|---|---|
| ADHD | 0.80 | Dopamine transporter |
| | | Dopamine D(4) receptor (DRD4) |
| Eating disorders | 0.50–0.80 | Leptin |
| | | Leptin transporter |
| | | Leptin receptor |
| | | Neuropeptide Y |
| | | Melanocortin-4 |
| Intelligence | 0.80 | Neural cellular adhesion molecule (NCAM) |
| Addiction | 0.40–0.60 | Dopamine D(2) receptor (DRD2) |
| | | Myelin synthesis |
| Depression | 0.40–0.54 | Serotonin synthesis, transporter, receptor |
| | | Norepinephrine synthesis, transporter, receptor |
| Bipolar disorder | 0.80 | Serotonin transporter, receptor |
| | | Monoamine oxidase A control |
| Schizophrenia | 0.80 | Dopamine synthesis, transporter, receptor |
| | | Glutamate synthesis, transporter, receptor |

# Blaming Genes

It has become fashionable to blame genes for our shortcomings. A popular magazine's cover shouts "Infidelity: It May Be in Our Genes," advertising an article that actually has little to do with genetics. When researchers identify a gene that plays a role in fat metabolism, people binge on chocolate and forsake exercise, because, after all, if obesity is in their genes, there's nothing they can do to prevent it. Some behaviors have even been blamed on a poorly-defined gene for "thrill-seeking" (figure 1).

Behavioral genetics has a checkered past. Early in the 20th century, it was part of eugenics—the idea that humans can improve a population's collection of genes, or gene pool. The horrific experiments and exterminations the Nazis performed in the name of eugenics, however, turned many geneticists away from studying the biology of behavior. Social scientists then dominated the field and attributed many behavioral disorders to environmental influences. For example, autism and schizophrenia were at one time attributed to "adverse parenting." By the 1960s, with a clearer idea of what a gene is, biologists reentered the debate. Today, researchers apply knowledge from biochemistry and neurobiology to identify specific genotypes that predispose a person to developing a clearly defined behavior.

Untangling the causes of human behavior remains highly controversial. One scientific conference to explore genetic aspects of violence was cancelled after a noted psychiatrist objected that "behavioral genetics is the same old stuff in new clothes. It's another way for a violent, racist society to say people's problems are their own fault, because they carry 'bad' genes." Genetic researchers on the trail of physical explanations for behaviors counter that their work can help uncover ways to alter or prevent dangerous behaviors. Attempts to hold meetings that discuss the genetics of violence still elicit public protests.

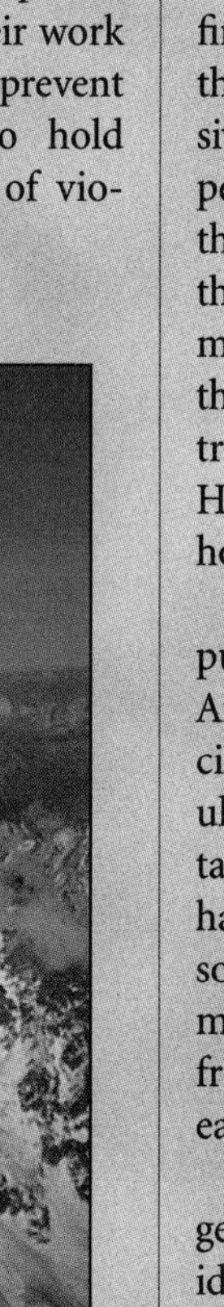

**figure 1**

**A thrill-seeking gene?** These air surfers were dropped from a helicopter over a mountain. Does a gene variant make them seek thrills?

Even in the rare instances when a behavior is associated with a particular DNA polymorphism or even linked to a specific mutation, environmental influences remain important. Consider a 1993 study of a Dutch family that had "a syndrome of borderline mental retardation and abnormal behavior." Family members had committed arson, attempted rape, and shown exhibitionism. Researchers found a mutation in a gene that made biological sense. Alteration of a single DNA base in the X-linked gene encoding an enzyme called monoamine oxidase A (MAOA) rendered the enzyme nonfunctional. This enzyme normally catalyzes reactions that metabolize the neurotransmitters dopamine, serotonin, and norepinephrine, and it is therefore important in conducting nerve messages. Studies since 1993 have confirmed that some combinations of alleles of the MAOA gene correlate with highly aggressive behavior, and others with calmer temperaments. The direct effect of mutations in the MAOA gene still isn't known. Perhaps the inherited enzyme deficiency causes slight mental impairment, and this interferes with the person's ability to cope with certain frustrating situations, resulting in violence. Hence, the argument returns once again to how genes interact with the environment.

The study on the Dutch family was publicized and applied to other situations. An attorney tried to use the "MAOA deficiency defense" to free a client from a scheduled execution for committing murder. A talk-show host suggested that people who had inherited the "mean gene" be sterilized so they couldn't pass on the tendency. This may have been meant as a joke, but it is frighteningly close to the eugenics practiced early in the last century.

With the information from the human genome project, researchers will continue to identify gene variant combinations that, given specific environmental triggers, predispose an individual to certain antisocial behaviors. It will become increasingly important to determine how to effectively use this information in a way that will help people.

# Summary

## 8.1 Genes Contribute to Most Behavioral Traits

1. Most behavioral traits and conditions are multifactorial and are more common than most single gene disorders.
2. Behavioral geneticists today are supplementing traditional ways to assess heritability with linkage analyses, association studies, and human genome sequence data to identify candidate genes and follow their expression in the nervous system.
3. Candidate genes for behavioral traits and disorders affect neurotransmission and signal transduction.
4. Analyzing behaviors is difficult because symptoms of different syndromes overlap, study participants can provide biased information, and behaviors can be imitated.

## 8.2 Eating Disorders

5. Eating disorders affect both sexes and are prevalent in the United States and Canada. Twin studies indicate high heritability.
6. Candidate genes for eating disorders include those whose protein products control appetite and the neurotransmitters dopamine and serotonin.

## 8.3 Sleep

7. Twin studies and the existence of single gene disorders that affect the sleep-wake cycle reveal a large inherited component to sleep behavior.
8. A large family with familial advanced sleep phase syndrome enabled researchers to identify the first "clock" gene in humans. The *period* gene enables a person to respond to day and night environmental cues. It was already known in several other species.

## 8.4 Intelligence

9. Intelligence is difficult to define and to measure. IQ testing predicts success in school or work, but is being replaced by the **general intelligence (g)** value. This is a global ability that underlies population variance in IQ test performance.
10. The fact that heritability for intelligence increases with age suggests that environmental factors are more important early in life.
11. The fact that many chromosomal disorders affect intelligence suggests high heritability. An N-CAM is a candidate gene.

## 8.5 Drug Addiction

12. Defining characteristics of drug addiction are tolerance and dependence. Addiction produces stable brain changes, yet heritability is not as high as for some other behavioral conditions.
13. A long-known candidate gene for drug addiction is the dopamine D(2) receptor. DNA microarray tests on gene expression in brains of people with alcoholism help to identify genes involved in neurotransmission, signal transduction, cell cycle control, apoptosis, surviving oxidative damage, and myelination of neurons.

## 8.6 Mood Disorders

14. **Major depressive disorder** is relatively common and associated with deficits of serotonin and/or norepinephrine.
15. **Bipolar affective disorder,** which consists of depressive periods interspersed with times of mania or hypomania, is much rarer. Linkage and association studies implicate several chromosomal sites as housing genes that raise the risk of developing this disorder.

## 8.7 Schizophrenia

16. **Schizophrenia** greatly disrupts ability to think and perceive the world. Onset is typically in early adulthood, and the course is episodic or steady but not degenerative.
17. Empiric risk estimates and heritability indicate a large genetic component, yet there are certain environmental associations too.
18. Many chromosomes house candidate genes for schizophrenia. Dopamine and glutamate are the implicated neurotransmitters.

# Review Questions

1. In general, what types of gene products are responsible for variations in behaviors?
2. How does the type of information about a behavioral disorder that is revealed in empiric risk, adoptee, and twin studies differ from the type of information that linkage or genome-wide scans provide?
3. Why is the genetics of ADHD easier to analyze than that of autism?
4. Which behaviors are traced to altered activities in the following regions of the brain?
   a. suprachiasmatic nuclei
   b. nucleus accumbens
   c. prefrontal cortex
5. Why is identifying a candidate gene only a first step in understanding how behavior arises and varies among individuals?
6. Name a candidate gene for the following traits or disorders:
   a. intelligence
   b. cocaine addiction
   c. alcoholism
   d. bipolar affective disorder
   e. schizophrenia
7. Describe three factors that can complicate the investigation of a behavioral trait.
8. What is the difference in effect on phenotype of a behavior that is polygenic compared to one that is genetically heterogeneic?
9. Why does the heritability of intelligence decline with age?

# Applied Questions

1. Are experiments that show no association between a candidate gene and a behavioral trait valuable? Why or why not?
2. Serotonin levels are implicated in eating disorders, major depressive disorder, and bipolar disorder.
   a. How can an abnormality in one type of neurotransmitter contribute to different disorders?
   b. What is another neurotransmitter that is implicated in more than one behavioral disorder?
3. How has DNA microarray technology changed the study of the genetics of behavior?
4. What might be the advantages and disadvantages of a SNP profile or other genotyping test done at birth that indicates whether or not a person is at high risk for developing
   a. a drug addiction.
   b. an eating disorder.
   c. depression.
5. The U.S. government prohibits recreational use of cocaine and opiates, which are physically addictive drugs, but not of alcohol and nicotine (cigarettes), which are also physically addictive. Do you think that the legal status of any of these drugs should be changed, and if so, how and why? What measures, if any, should the government use in deciding which drugs to outlaw?
6. Do you think that having a genotype known to predispose someone to aggressive or violent behavior should be a valid legal defense? Cite a reason for your answer.
7. In some association studies of depression and bipolar disorder, correlations to specific alleles are only evident when participants are considered in subgroups based on symptoms. What might be a biological basis for this finding?
8. Many older individuals experience advanced sleep phase syndrome. Even though this condition is probably a normal part of aging, how might research on the Utah family with an inherited form of the condition help researchers develop a drug to help the elderly sleep through the night and awaken later in the morning?
9. What is an alternate explanation for the idea that genetic differences account for the overrepresentation of minority groups among people with low IQ scores in the United States?
10. Fifty years ago, people who were so unhappy that they could barely get out of bed in the morning and carry out daily activities were either ignored, or, if symptoms became worse, hospitalized for a "nervous breakdown." Today, depression is identified and diagnosed routinely, and people take medication for it that very often works. Similarly, 50 years ago, a child who could not sit still for very long in a classroom was considered a "behavior problem." Today, the child is often diagnosed with attention deficit hyperactivity disorder, and prescribed a medication that is often effective. Do you think that people with such disorders were better off back then, or now? State reasons for your answer.
11. Wolfram syndrome is a rare autosomal recessive disorder that causes severe diabetes, impaired vision, and neurological problems. Examinations of hospital records and self-reports reveal that blood relatives of Wolfram syndrome patients have an eightfold risk over the general population of developing serious psychiatric disorders such as depression, violent behavior, and suicidal tendencies. Can you suggest further experiments and studies to test the hypothesis that these mental manifestations are a less severe expression of Wolfram syndrome?
12. A study of 2,685 twin pairs showed that female MZ twins are six times as likely as female DZ twins to both have alcoholism. Does this finding suggest a large genetic or environmental component to alcoholism?

# Suggested Readings


Bradbury, Jane. May 19, 2001. Teasing out the genetics of bipolar disorder. *Lancet* 257:1596. Many genes lead to the mood swings of bipolar disorder.

Bulik, C. M., et al. June 2000. An empirical study of the classification of eating disorders. *American Journal of Psychiatry* 157:886–95. Biotech companies seek the genes that predispose people to develop eating disorders—but it will be hard to sift through the powerful environmental influences.

Cavendish, Harry. December 2001. Saved by Ezmerelda. *The Lancet,* suppl., 560. This journalist describes what it's like to live with schizophrenia.

Clayton, J. D., et al. February 15, 2001. Keeping time with the human genome. *Nature* 409:829–31. Several genes control the sleep-wake cycle.

Du, L., et al. February 2000. Association of polymorphism of serotonin 2A receptor gene with suicidal ideation in major depressive disorder. *American Journal of Medical Genetics* 96:56–60. Depression may be caused by too few receptors for serotonin.

Gottfredson, Linda S. Winter 1998. The general intelligence factor. *Scientific American* 9, no. 4:24–31. IQ is being replaced with the "g" value, which may more accurately reflect genetic influences on intelligence.

Hamer, Dean, and Peter Copeland. 1988. *Living with Our Genes.* New York: Doubleday. A highly readable account of behavioral genetics.

Lesch, Peter K. December 2001. Weird world inside the brain. *The Lancet,* suppl., 559. Schizophrenia is actually several disorders.

Lewis, Ricki. June 25, 2001. Focusing on endocannabinoid. *The Scientist* 15, no. 13:14. Marijuana exerts its effects because our brain cells have receptors for it.

Lindholm, Eva, et al. July 2001. A schizophrenia-susceptibility locus at 6q25, in one of the world's largest reported pedigrees. *The American Journal of Human Genetics* 69:96–105. Different large families with many cases of schizophrenia suggest that several genes can cause the condition.

McGuffin, Peter, et al. February 16, 2001. Toward behavioral genomics. *Nature* 291:1232–33. Part of the human genome project is to identify genes that affect behavior.

Nestler, Eric J. November 2000. Genes and addiction. *Nature Genetics* 26:277–80. Drug-seeking behavior reflects aberrant brain function, which reflects the actions of certain gene variants.

Plomin, R., and I. Craig. April 2001. Genetics, environment, and cognitive abilities: Review and work in progress towards a genome scan for quantitative trait locus associations using DNA pooling. *British Journal of Psychiatry,* Suppl 2001:S41–S48. Researchers are developing shortcuts to searching through genome information for the genes that influence behavior and personality.

Rayl, A. J. S. February 5, 2001. Microarrays on the mind. *The Scientist* 15, no. 3:16. DNA microarrays reveal changes in gene expression that accompany drinking alcohol.

Thaker, Gunvant K., and William T. Carpenter, Jr. June 2001. Advances in schizophrenia. *Nature Medicine* 7, no. 6:667–71. Brain imaging studies combined with gene expression profiles should reveal the origins of schizophrenia.

Toh, Kong L., et al. February 9, 2001. An hPer2 phosphorylation site mutation in familial advanced sleep phase syndrome. *Science* 291:1040–43. A family with strange sleeping habits led researchers to the discovery of the first "clock" gene in humans.

# On the Web

Check out the resources on our website at

**www.mhhe.com/lewisgenetics5**

On the web for this chapter, you will find additional study questions, vocabulary review, useful links to case studies, tutorials, popular press coverage, and much more. To investigate specific topics mentioned in this chapter, also try the links below:

Autism Genetic Resource Exchange **www.agre.org**

Depression and Related Affective Disorders Association **www.med.jhu.edu/drada**

National Institute for Drug Abuse **www.nida.nih.gov**

Online Mendelian Inheritance in Man **www.ncbi.nlm.nih.gov/entrez/query.fcgi?db=OMIM**

alcoholism 103780
attention deficit hyperactivity disorder 143465
autism and Asperger syndrome 209850
bipolar affective disorder 125480
dopamine D(2) receptor 126450
dopamine D(4) receptor 126452
familial advanced sleep phase syndrome 604348
fatal familial insomnia 600072
leptin receptor 601001
major depressive (affective) disorder 125480
monoamine oxidase 309850
narcolepsy 161450
schizophrenia 181500, 603342
tobacco addiction 188890

CHAPTER

9

# DNA Structure and Replication

### 9.1 Experiments Identify and Describe the Genetic Material
The sleekly symmetrical double helix that is deoxyribonucleic acid (DNA) is the genetic material. For many years, however, researchers hypothesized that protein was the biochemical behind heredity. It took many experiments to show that DNA links proteins to heredity.

### 9.2 DNA Structure
Assembling clues from various physical and chemical experiments, Watson and Crick deduced the double helical nature of DNA, and in so doing, predicted how the molecule replicates. Some DNA nucleotide sequences encode information that tells a cell how to synthesize a particular protein.

### 9.3 DNA Replication—Maintaining Genetic Information
The double helix untwists and parts, building two new strands against the two older ones, guided by the nucleotide sequence. A contingent of enzymes carries out the process.

### 9.4 PCR—Directing DNA Replication
The polymerase chain reaction (PCR) harnesses and uses DNA replication to amplify selected DNA sequences several millionfold. PCR has been used to analyze everything from body fluid samples to extinct animals to moose meat turned into hamburger.

DNA is perhaps the most interesting of molecules, its twofold function a direct consequence of its structure. Encoded in some of the sequences of DNA building blocks is information that cells use to synthesize proteins. Yet at the same time, DNA can copy itself, ensuring that the information embedded in its structure is passed from one cell generation to the next. As Francis Crick, codiscoverer of the structure of DNA, pointed out, "A genetic material must carry out two jobs: duplicate itself and control the development of the rest of the cell in a specific way."

Recognizing and understanding DNA's vital role in life was a long time coming. This chapter explores the experiments that led to the identification and description of the genetic material, as well as the details of DNA structure and replication (figure 9.1). Chapter 10 examines how a cell accesses the protein-encoding information encoded in its DNA, and explores possible functions of the rest of the genome.

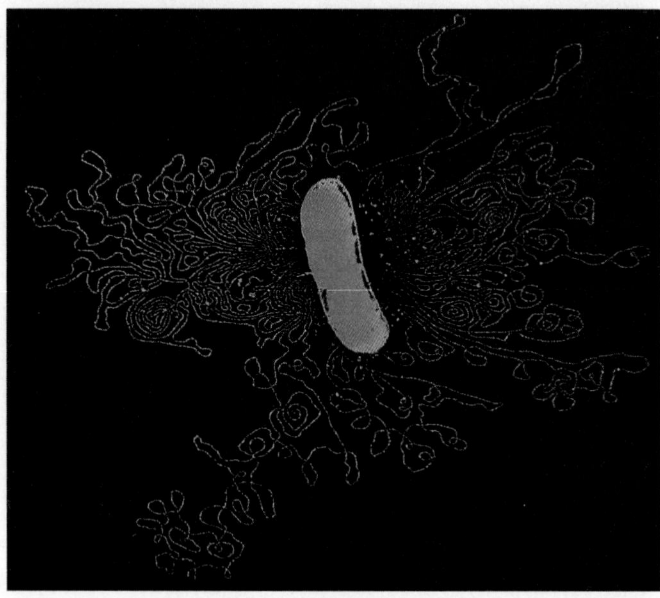

**figure 9.1**

**DNA is highly packaged.** DNA bursts forth from this treated bacterial cell, illustrating how DNA is tightly wound into a single cell.

## 9.1 Experiments Identify and Describe the Genetic Material

DNA was first described in the mid–18th century, when Swiss physician and biochemist Friedrich Miescher isolated nuclei from white blood cells in pus on soiled bandages. In the nuclei, he discovered an unusual acidic substance containing nitrogen and phosphorus. He and others found it in cells from a variety of sources. Because the material resided in cell nuclei, Miescher called it **nuclein** in an 1871 paper; subsequently, it was called a nucleic acid. But few people appreciated the importance of Miescher's discovery, because at the time, the study of heredity focused on the association between inherited disease and protein.

English physician Archibald Garrod, in 1902, was the first to link inheritance in humans and protein. He noted that people who had certain "inborn errors of metabolism" lacked certain enzymes. One of the first inborn errors that he described was alkaptonuria, the subject of In Their Own Words in chapter 5. Other researchers added evidence of a link between heredity and enzymes from other species—fruit flies with unusual eye colors, and bread molds with nutritional deficiencies. Both organisms had absent or abnormal specific enzymes. As researchers wondered what, precisely, was the connection between enzymes and heredity, they returned to Miescher's discovery of nucleic acids.

### DNA Is the Hereditary Molecule

In 1928, English microbiologist Frederick Griffith took the first step in identifying DNA as the genetic material. Griffith noticed that mice with a certain type of pneumonia harbored one of two types of *Diplococcus pneumoniae* bacteria. Type R bacteria are rough in texture. Type S bacteria are smooth, because they are enclosed in a polysaccharide capsule. Mice injected with type R bacteria did not develop pneumonia, but mice injected with type S did. The polysaccharide coat seemed to be necessary for infection.

When type S bacteria were heated—which killed them but left their DNA intact—they no longer could cause pneumonia in mice. However, when Griffith injected mice with a mixture of type R bacteria plus heat-killed type S bacteria—neither of which, alone, was deadly to the mice—the mice died of pneumonia (figure 9.2). Their bodies contained live type S bacteria, encased in polysaccharide. What was happening?

The answer came in 1944. U.S. physicians Oswald Avery, Colin MacLeod, and Maclyn McCarty hypothesized that something in the heat-killed type S bacteria "transformed" the normally harmless type R strain into a killer. Could a nucleic acid be the "transforming principle?" They suspected it might be, because treating type R bacteria with an enzyme that dismantles protein (a protease) did not prevent the transformation of a nonkilling to a killing strain, but treating it with an enzyme that dismantles DNA only (deoxyribonuclease) disrupted the transformation. Avery, MacLeod, and McCarty then confirmed that DNA transformed the bacteria. They isolated DNA from heat-killed type S bacteria and injected it along with type R bacteria into mice (figure 9.3). The mice died, and their bodies contained active type S bacteria. The conclusion: the DNA passed from type S bacteria to type R, enabling it to manufacture the smooth coat necessary for infection.

### DNA Is the Hereditary Molecule—and Protein Is Not

In 1953, U.S. microbiologists Alfred Hershey and Martha Chase confirmed that DNA is the genetic material. They used *E. coli* bacteria

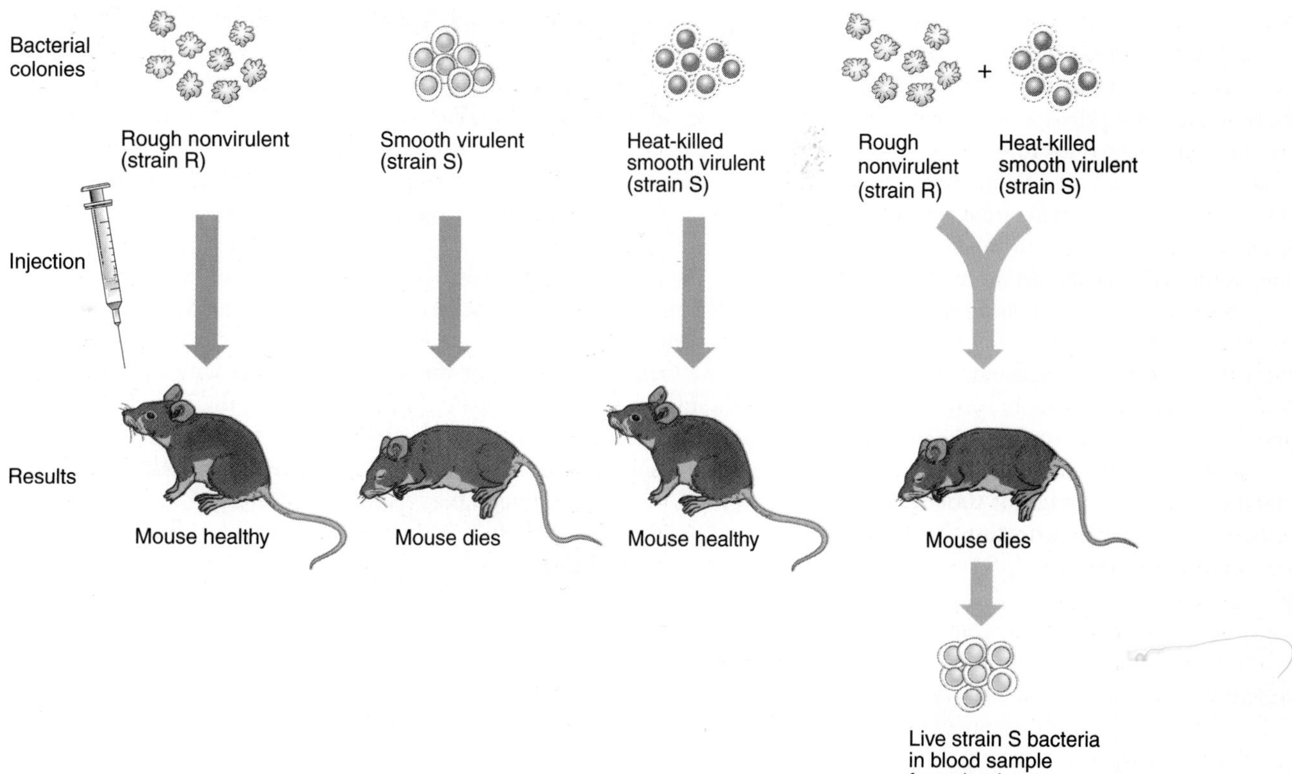


**figure 9.2**

**Discovery of a "transforming principle."** Griffith's experiments showed that a molecule in a lethal strain of bacteria can transform nonkilling bacteria into killers.

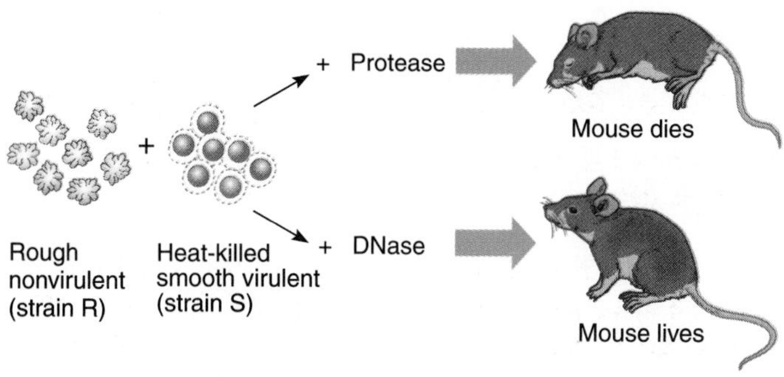


**figure 9.3**

**DNA is the "transforming principle."** Avery, MacLeod, and McCarty identified Griffith's transforming principle as DNA. By adding enzymes that either destroy proteins (protease) or DNA (DNase) to the types of solutions that Griffith used in his experiments, they demonstrated that DNA transforms bacteria—and that protein does not.

infected with a virus that consisted of a protein "head" surrounding DNA. Viruses infect bacterial cells by injecting their DNA inside them. The viral protein coats remain outside the bacterial cells.

Hershey and Chase showed that viruses grown with radioactive sulfur became radioactive, and the protein coats emitted the radioactivity. When they repeated the experiment with radioactive phosphorus, the viral DNA emitted radioactivity. This showed that sulfur is found in protein but not in nucleic acid, and that phosphorus is found in nucleic acid but not in protein. DNA is the only phosphorus-containing molecule in the virus.

Next, the researchers "labeled" two batches of virus by growing one in a medium containing radioactive sulfur (designated $^{35}S$) and the other in a medium containing radioactive phosphorus (designated $^{32}P$). The viruses grown on sulfur had their protein marked but not their DNA, because protein incorporates sulfur but DNA does not. Conversely, the viruses grown on labeled phosphorus had their DNA marked but not their protein, because this element is found in DNA but not protein. (Recall that Miescher noted the presence of phosphorus in DNA from soiled bandages.)

After allowing several minutes for the virus particles to bind to the bacteria and inject their DNA into them, Hershey and Chase agitated each mixture in a blender, shaking free the empty virus protein coats. The contents of each blender were collected in test tubes, then centrifuged (spun at high speed). This settled the bacteria at the bottom of each tube because virus coats drift down more slowly than bacteria.

At the end of the procedure, Hershey and Chase examined fractions containing the virus coats from the top of each test tube and the infected bacteria that had settled to the bottom of each test tube (figure 9.4). In the tube containing viruses labeled with sulfur, the virus coats were radioactive, but the virus-infected bacteria, containing viral DNA, were not. In the other tube, where the virus had incorporated radioactive phosphorus, the virus coats carried no radioactive label, but the infected bacteria were radioactive. This meant that the part of the virus that could enter bacteria and direct them to mass produce more virus was the part that had incorporated the phosphorus label—the DNA. The genetic material, therefore, was DNA, and not protein. It is DNA that transmits information from generation to generation.

## Deciphering the Structure of DNA

In 1909, Russian-American biochemist Phoebus Levene identified the 5-carbon sugar **ribose** as part of some nucleic acids, and in 1929, he discovered a new, similar sugar in other nucleic acids—**deoxyribose.** By identifying these two sugars, Levene had revealed a major chemical distinction between RNA and DNA. RNA contains ribose, and DNA contains deoxyribose.

Levene then discovered that the three parts of a nucleic acid—a sugar, a nitrogen-containing base, and a phosphorus-containing component—are present in equal proportions. He deduced that a nucleic acid building block must contain one of each component. Furthermore, although the sugar and phosphate portions were always the same, the nitrogen-containing bases were of four types. For several years, scientists erroneously thought that these nitrogen-containing bases are present in equal amounts. But if this were the case, DNA could not encode as much information as it could if there were no restrictions on the relative numbers of each base type. Imagine how much less useful the English language would be if all the letters in a document had to occur with equal frequency.

In the early 1950s, two lines of experimental evidence converged to provide the direct clues that finally revealed DNA's structure. Chemical analysis by Austrian-American biochemist Erwin Chargaff showed that DNA in several species con-

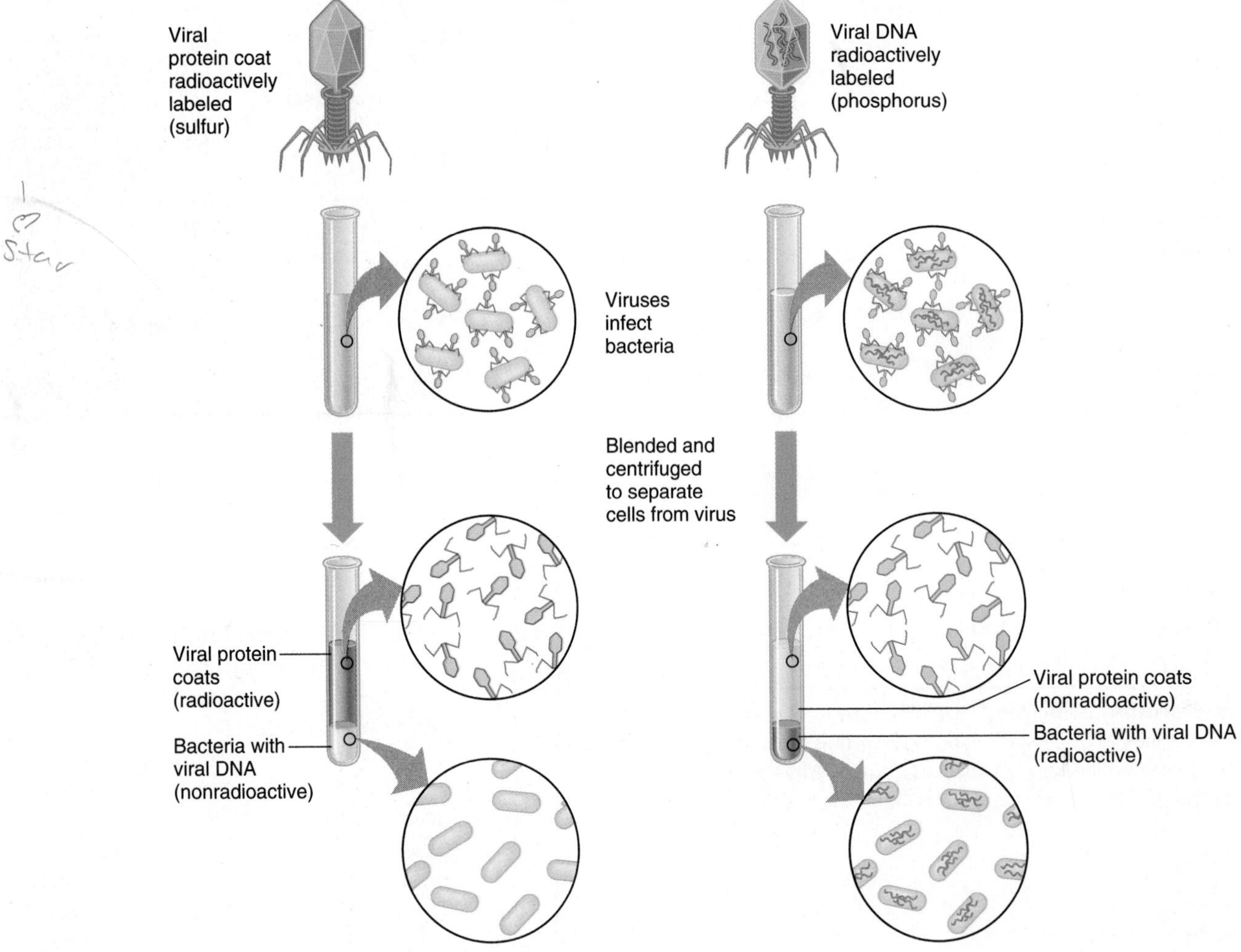


**figure 9.4**

**DNA is the hereditary material; protein is not.** Hershey and Chase used different radioactive isotopes to distinguish the viral protein coat from the genetic material (DNA). These "blender experiments" showed that DNA is what the virus transfers to the bacterium. DNA is the genetic material.

tains equal amounts of the bases **adenine** and **thymine** and equal amounts of the bases **guanine** and **cytosine.** These bases are abbreviated A, T, G, and C, respectively. Next, English physicist Maurice Wilkins and English chemist Rosalind Franklin bombarded DNA with X rays using a technique called X-ray diffraction, then observed the pattern in which the X rays were deflected (figure 9.5*a*). This pattern revealed a regularly repeating structure of building blocks.

In 1953, American biochemist James Watson and English physicist Francis Crick worked together in England to build a replica of the DNA molecule using ball-and-stick models, using the clues provided by others' experiments (figure 9.5*b*). Their model included equal amounts of guanine and cytosine and equal amounts of adenine and thymine, and it satisfied the symmetry shown in the X-ray diffraction pattern. The result of their insight: the now-familiar double helix.

Rosalind Franklin was a young woman when she contributed the X-ray diffraction data that would prove so pivotal to the elucidation of the structure of DNA. She did not share in the Nobel Prize, however, because she died from cancer before it was awarded. It is likely that she developed the cancer from her years of working with radioactive chemicals. Franklin is one of the unsung heroines of biology.

Table 9.1 summarizes the experiments that led to discovering the structure of DNA.

## Key Concepts

DNA contains the information the cell requires to synthesize protein. DNA replicates. Miescher first isolated DNA in 1869, naming it nuclein. Garrod first linked heredity to enzymes. • DNA was chemically characterized in the 1940s. Griffith identified a substance capable of transmitting infectiousness, which Avery, MacLeod, and McCarty showed was DNA. • Hershey and Chase confirmed that DNA, and not protein, is the genetic material. • Using Chargaff's discovery that the number of adenine bases equals the number of thymines, and the number of guanines equals the number of cytosines, along with Franklin's discovery that DNA is regular and symmetrical in structure, Watson and Crick deciphered the structure of DNA.

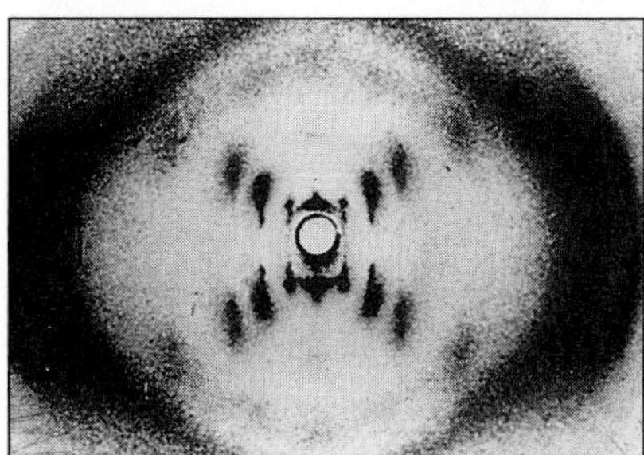

a.

b.

**figure 9.5**

**Deciphering DNA structure.** (*a*) The X-ray diffraction pattern of DNA, obtained by Rosalind Franklin, was crucial to Watson and Crick's elucidation of the molecule's structure. The "X" in the center indicates a helix, and the darkened regions reveal symmetrically organized subunits. (*b*) James Watson (*left*) and Francis Crick.

**table 9.1**

**The Road to the Double Helix**

| Investigator | Contribution | Timeline |
|---|---|---|
| Friedrich Miescher | Isolated nuclein in white blood cell nuclei | 1869 |
| Frederick Griffith | Transferred killing ability between strains of bacteria | 1928 |
| Oswald Avery, Colin MacLeod, and Maclyn McCarty | Discovered that DNA transmits killing ability in bacteria | 1940s |
| Alfred Hershey and Martha Chase | Determined that the part of a virus that infects and replicates is its nucleic acid and not its protein | 1950 |
| Phoebus Levene, Erwin Chargaff, Maurice Wilkins, and Rosalind Franklin | Discovered DNA components, proportions, and positions | 1909–early 1950s |
| James Watson and Francis Crick | Elucidated DNA's three-dimensional structure | 1953 |

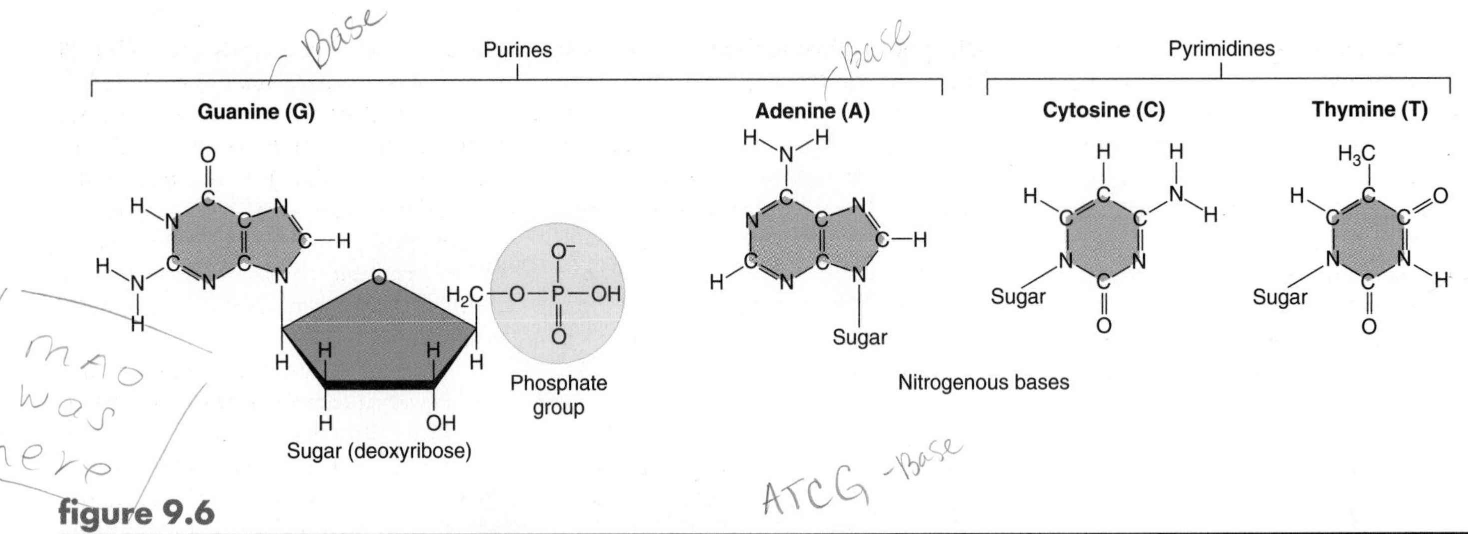


**figure 9.6**

**DNA bases are the informational parts of nucleotides.** Each nucleotide of a nucleic acid consists of a 5-carbon sugar, a phosphate group, and an organic, nitrogenous base. The DNA bases adenine and guanine are purines, each composed of a six-member organic ring plus a five-member ring. Cytosine and thymine are pyrimidines, each built of a single six-member ring.

## 9.2 DNA Structure

A **gene** is a long section of a DNA molecule whose sequence of building blocks specifies the sequence of amino acids in a particular protein. The activity of the protein is responsible for the phenotype associated with the gene. The fact that different building blocks combine to form nucleic acids enables them to carry information, as the letters of an alphabet combine to form words. In contrast, a complex carbohydrate is a string of identical sugars. Like an alphabet consisting of only one letter, it does not encode information in a sequence. (Complex carbohydrates can, however, impart information in their branch patterns and number of building blocks.)

The nature of an inherited trait reflects the function of a protein. Inherited traits are diverse because proteins have diverse functions. Pea color, plant height, and the chemical reactions of metabolism are all the consequence of enzyme activity. Proteins such as collagen and elastin provide structural support in connective tissues, and actin and myosin form muscle. Hemoglobin transports oxygen, and antibodies protect against infection. Malfunctioning or inactive proteins, which reflect genetic defects, can devastate health, as we have seen for many hereditary disorders. Most of the amino acids that are assembled into proteins ultimately come from the diet; the body synthesizes the others.

Biological proteins are extremely diverse in three-dimensional shape and function, but they all are specified by DNA. The structure of DNA is easiest to understand beginning with the smallest components. A single building block of DNA is a **nucleotide.** It consists of one deoxyribose sugar, one phosphate group (which is a phosphorus atom bonded to four oxygen atoms), and one nitrogenous base. Figure 9.6 shows the components of a nucleotide and the chemical structures of the four types of bases. Adenine (A) and guanine (G) are **purines,** which have a two-ring structure. Cytosine (C) and thymine (T) are **pyrimidines,** which have a single-ring structure. Reading 9.1 on page 181 explains how clues in DNA sequences solved a mystery of history.

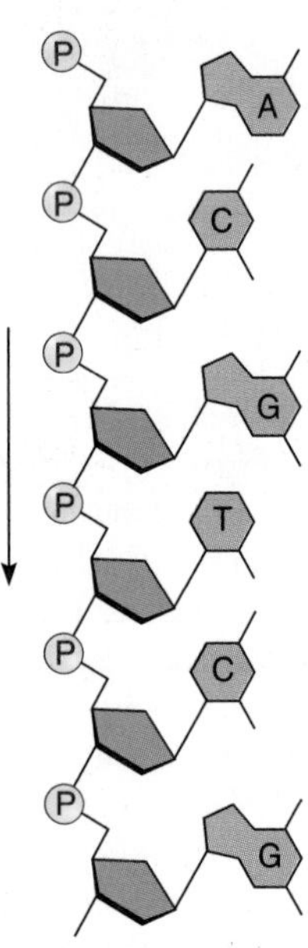


**figure 9.7**

**A chain, a double chain, then a helix.** A single DNA strand consists of a chain of nucleotides that forms when the deoxyribose sugars (green) and phosphates (yellow) bond to create a sugar-phosphate backbone. The bases A, C, G, and T are indicated in blue.

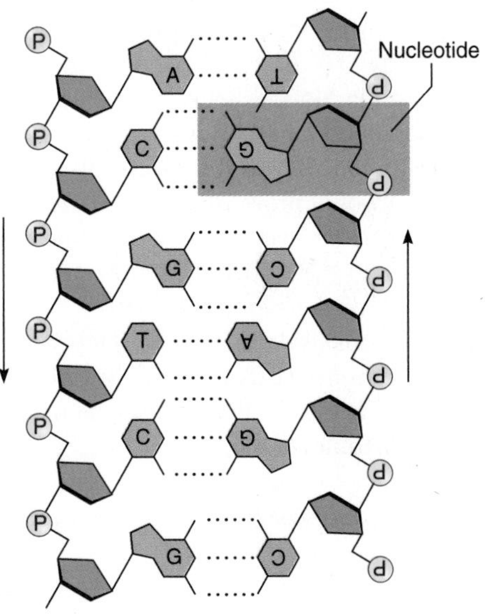


**figure 9.8**

**DNA consists of two chains of nucleotides.** The nitrogenous bases of one strand are held to the nitrogenous bases of the second strand by hydrogen bonds (dotted lines). Note that the sugars point in opposite directions—that is, the strands are antiparallel.

Nucleotides join into long chains when chemical bonds form between the deoxyribose sugars and the phosphates, which creates a continuous **sugar-phosphate backbone** (figure 9.7). Two such chains of nucleotides align head-to-toe, as figure 9.8 depicts. M. C. Escher's drawing of hands in figure 9.9*c* resembles the spatial relationship of the two strands of the DNA double helix.

The opposing orientation of the two aligned nucleotide chains is a property of DNA called **antiparallelism.** It derives from the structure of the sugar-phosphate backbone. Antiparallelism becomes evident when the carbons of the sugars are assigned numbers to indicate their positions in the molecule. The carbons are numbered from 1 to 5, starting with the first carbon moving clockwise from the oxygen in each sugar in figure 9.9*a*. One chain runs from the 5 carbon (top of the figure) to the 3 carbon, but the chain aligned with it runs from the 3 to the 5 carbon. These ends are called "5 prime" and "3 prime," abbreviated 5′ and 3′.

The symmetrical double helix of DNA forms when nucleotides containing A pair with those containing T, and nucleotides containing G pair with those carrying C. Because purines have two rings and pyrimidines one, the consistent pairing of a purine with a pyrimidine ensures that the double helix has the same width throughout. These specific purine-pyrimidine couples are called **complementary base pairs.** Chemical attractions called hydrogen bonds hold the base pairs together. As figure 9.10 shows, two hydrogen bonds join A and T, and three hydrogen bonds join G and C.

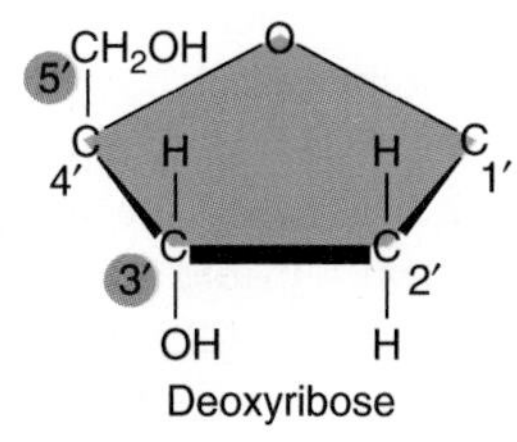


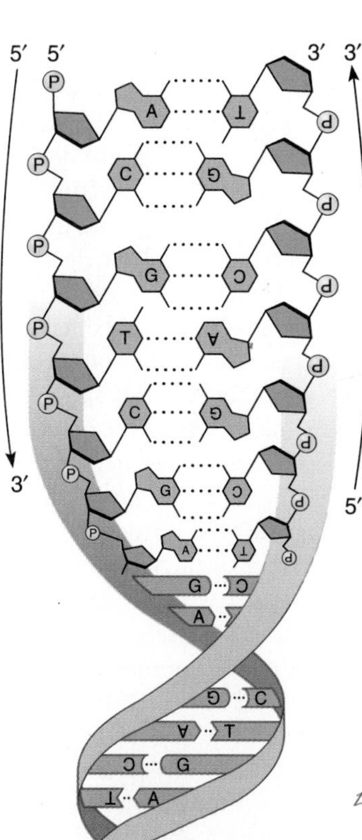


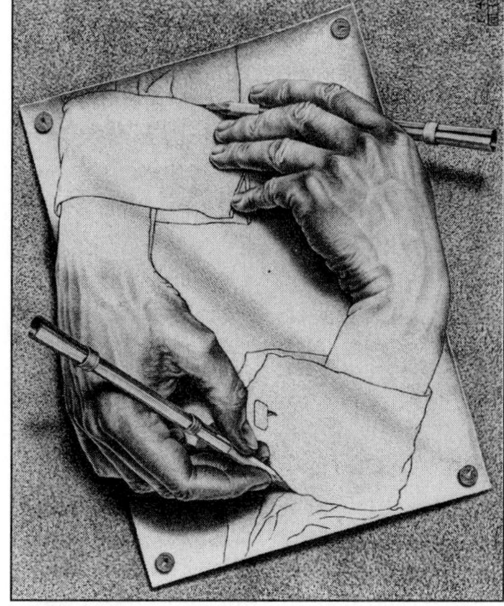


**figure 9.9**

**DNA is directional.** (*a*) The antiparallel nature of the DNA double helix becomes apparent when the carbons in the sugar are numbered. (*b*) One half of a double helix runs in a 5′ to 3′ direction, and the other half runs in a 3′ to 5′ direction. (*c*) Artist M. C. Escher captured the essence of antiparallelism in his depiction of hands.

Thymine (T)

Adenine (A)

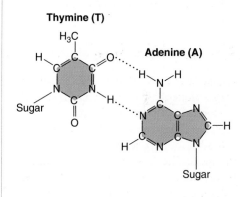

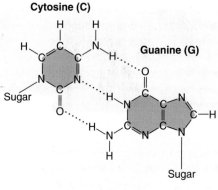

Cytosine (C)

Guanine (G)

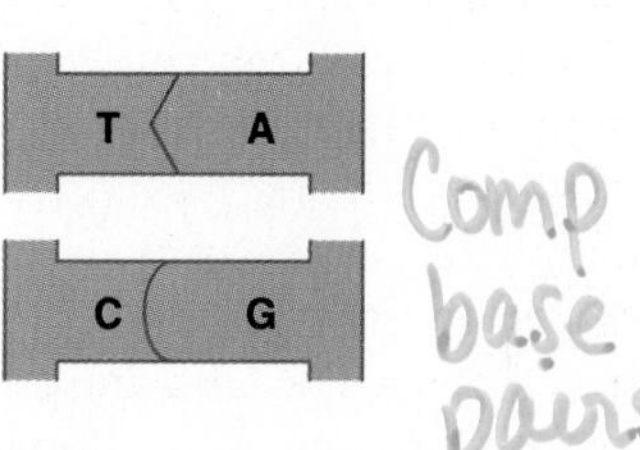


**figure 9.10**

**DNA base pairs.** The key to the constant width of the DNA double helix is the pairing of purines with pyrimidines. Specifically, two hydrogen bonds join adenine and thymine, while three hydrogen bonds connect cytosine and guanine.

Complementary base pairing is crucial to the utilization of genetic information, and to many biotechnologies. Finally, the DNA assumes the double helix form when the antiparallel, base-paired strands twist about one another in a regular fashion.

DNA molecules are incredibly long. A single molecule several inches long must fold up to fit inside a cell that is only 1 millionth of an inch across—and actually within the nucleus of that cell. Various types of proteins assist in compressing the DNA without damaging or tangling it. Scaffold proteins form frameworks that guide DNA strands. Then, the DNA coils around proteins called **histones,** forming a "beads-on-a-string"-like structure. The bead part is called a **nucleosome.** It is a little like taking a very long, thin piece of thread, and wrapping parts of it around small spools or your fingers, to keep it from unraveling and tangling. In this way, the DNA wraps at several levels, until it is compacted into a chromosome (figure 9.11). It packs so tightly during mitosis that the chromosomes become condensed enough to be visible when stained. The DNA can unwind locally when it replicates or when it is being copied into RNA, which is the intermediate step in synthesizing protein.

A chromosome consists of about a third DNA, a third histone proteins, and a third other types of DNA binding proteins. A small amount of RNA may also be associated with a chromosome. Altogether, the chromosome material is called **chromatin,** which means "colored material."

### Key Concepts

The DNA double helix is a ladder-like structure, its backbone comprised of alternating deoxyribose and phosphate groups and its rungs formed by complementary pairs of A-T and G-C bases. A and G are purines; T and C are pyrimidines. The DNA double helix is antiparallel, its strands running in an opposite head-to-toe manner. DNA winds tightly about histone proteins, forming nucleosomes, which in turn wind into a tighter structure to form chromatin.

## 9.3 DNA Replication—Maintaining Genetic Information

As soon as Watson and Crick deciphered the structure of DNA, its mechanism for replication became obvious to them.

### Replication Is Semiconservative

Watson and Crick had a flair for the dramatic, ending their report on the structure of DNA with the tantalizing statement, "It has not escaped our notice that the specific pairing we have postulated immediately suggests a possible copying mechanism for the genetic material." They envisioned the immense molecule unwinding, exposing unpaired bases that would attract their mates, and neatly knitting two double helices from one. This route to replication is called **semiconservative,** because each new DNA molecule conserves half of the original double helix. However, double-stranded DNA cannot be considered two single helices because when DNA is single-stranded, it does not form a helix.

Not everyone thought DNA replication was so straightforward. One critic was Max Delbrück, another of the founders of molecular biology. When Watson shared his and Crick's results with Delbrück before publication of their paper, Delbrück wrote back that "for a DNA molecule of molecular weight 3,000,000 there would be about 500 turns around each other. These would have to be untwiddled to separate the strands." Although "untwiddled" is hardly a scientific term, the problem at hand was clear—separating the DNA strands was like having to keep two pieces of thread the length of a football field from tangling.

Because separating the DNA strands seemed so difficult, some researchers suggested that the molecule replicated some other way. Gunther Stent, another of the two dozen or so molecular biologists of the time, named three possible mechanisms of DNA replication. Replication might be **conservative,** with one double helix specifying creation of a second double helix. Or replication might be **dispersive,** with a double helix shattering into pieces that would then join with newly synthesized DNA pieces to form two molecules. Figure 9.12 distinguishes among these three hypothesized mechanisms of DNA replication.

The problem of how the genetic material replicates was discussed even before Watson and Crick unveiled the structure of DNA in 1953. In 1941, English geneticist J. B. S. Haldane wrote, "How can one distinguish between model and copy? Perhaps you could use heavy nitrogen atoms in the food supplied to your cell, hoping that the 'copy' genes would contain it while the models did not."

Following publication of Watson and Crick's paper, Delbrück suggested experiments that might reveal how DNA replicates that were conceptually similar to Haldane's idea. If newly synthesized DNA from a virus could incorporate radioactive phosphorus, then the radiation could be detected by its ability to expose photographic film, and the new DNA thus revealed. Delbrück did the experiments, but by 1956, his results weren't clear enough to distinguish among the semiconservative, conservative, or dispersive models. But just over a year later, the very experiments that Haldane suggested answered the question of how DNA replicates, although the two researchers who did them later claimed to be unaware of Haldane's suggestion of 1941. The series of experiments are often hailed as one of the best illustrations of scientific inquiry, because they not only supported one hypothesis, but disproved the other two.

A young researcher named Matthew Meselson labeled newly synthesized DNA with a heavy form of nitrogen ($^{15}N$), which could then be distinguished from older DNA synthesized with the more common lighter form, $^{14}N$. Meselson and coworker Franklin Stahl then examined DNA replication in bacteria. The idea was that DNA that incorporated the heavy nitrogen could be separated from DNA that incorporated the normal lighter nitrogen by its greater density. DNA in which one half of the double helix was light and one half heavy would be of intermediate density. In their **density shift experiments,** Meselson and Stahl grew cells on media with the different forms of nitrogen, which would enable them to trace replication through generations of DNA. After growing cells, they broke them open, extracted DNA, and spun the DNA in a centrifuge. The heavier DNA sank to the bottom of the centrifuge tube, the light DNA

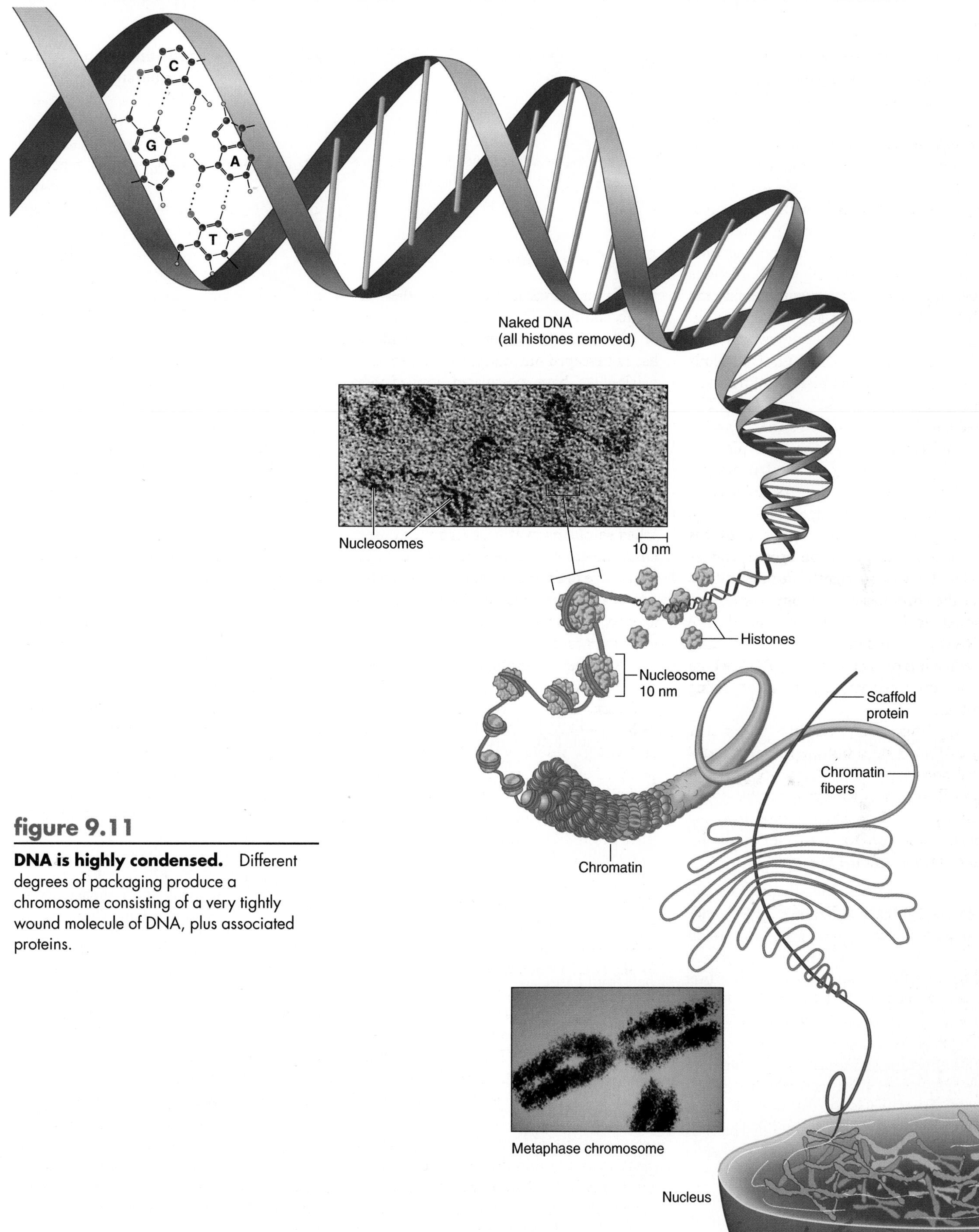


## figure 9.11

**DNA is highly condensed.** Different degrees of packaging produce a chromosome consisting of a very tightly wound molecule of DNA, plus associated proteins.

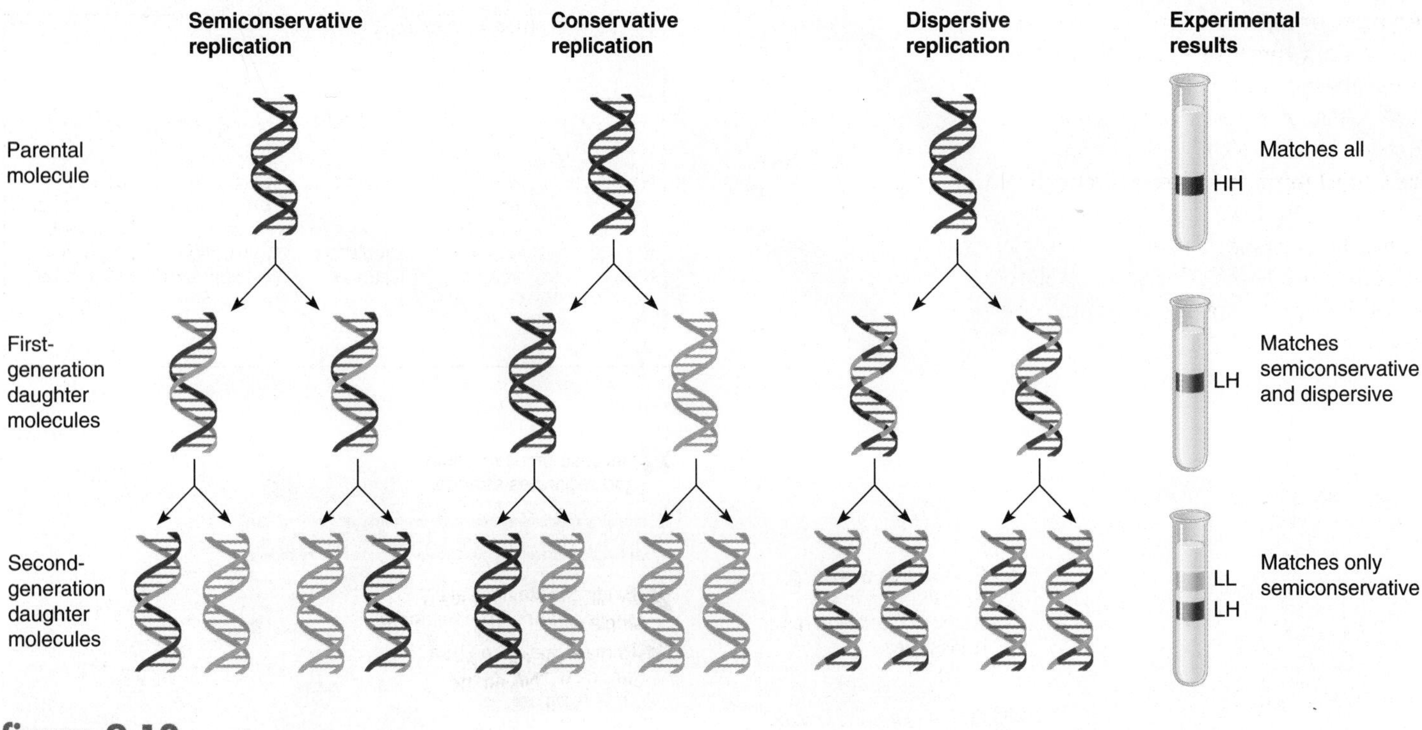


**figure 9.12**

**DNA replication is semiconservative.** Density shift experiments distinguished the three hypothesized mechanisms of DNA replication. DNA molecules containing light nitrogen are designated "LL" and those with heavy nitrogen, "HH." Molecules containing both isotopes are designated "LH."

rose to the top, and the heavy-light double helices settled in the middle area of the tube. The "shift" referred to the fact that bacterial cultures were shifted from a medium with heavy nitrogen to one with light nitrogen.

Meselson and Stahl grew *E. coli* on media containing $^{15}N$ for several generations. As a result, the bacteria had completely heavy DNA. The researchers knew this because only "heavy-heavy" molecules appeared in the tube. They then shifted the bacteria to media containing $^{14}N$, allowing enough time for the bacteria to divide only once (about 30 minutes). The proportions of heavy and light nitrogen in the tubes over the next two replications would reveal the type of mechanism at play.

When Meselson and Stahl collected the DNA this time and centrifuged it, the double helices were all of intermediate density, occupying a region in the middle of the tube, indicating that they contained half $^{14}N$ and half $^{15}N$. This pattern was consistent with semiconservative DNA replication—but it was also consistent with a dispersive mechanism. In contrast, the result of conservative replication would have been one band of material in the tube completely labeled with $^{15}N$, corresponding to one double helix, and another totally "light" band containing $^{14}N$ only, corresponding to the other double helix. This did not happen.

Meselson and Stahl definitively distinguished among the three possible routes to DNA replication, supporting the semiconservative mode and disproving the others, by extending the experiment one more generation. For the semiconservative mechanism to hold up, each hybrid (half $^{14}N$ and half $^{15}N$) double helix present after the first generation following the shift to $^{14}N$ medium would part and assemble a new half from bases labeled only with $^{14}N$. This would produce two double helices with one $^{15}N$ (heavy) and one $^{14}N$ (light) chain, plus two double helices containing only $^{14}N$. The tube would have one heavy-light band and one light-light band. This is indeed what Meselson and Stahl saw.

The conservative mechanism would have yielded two bands in the tube in the third generation, indicating three completely light double helices for every completely heavy one. The third generation for the dispersive model would have been a single large band, somewhat higher than the second generation band because additional $^{14}N$ would have been randomly incorporated.

Other researchers demonstrated the semiconservative mode of DNA replication using other ways to label newly synthesized DNA. For example, in other species, introducing a radioactive label into growth medium and then removing it and monitoring cell division reveals chromosomes in their replicated form, with just one chromatid displaying the radioactivity by exposing photographic film. These experiments extended Meselson and Stahl's results by demonstrating semiconservative DNA replication in the cells of more complex organisms and at the whole-chromosome level.

## Steps and Participants in DNA Replication

After experiments demonstrated the semiconservative nature of DNA replication, the next challenge was to decipher the steps of the process.

When a long length of DNA replicates, it must break, unwind, build a new nucleotide chain, and mend. A contingent of enzymes carries out the process, shown as an overview in figure 9.13. Enzymes called **helicases** unwind and hold apart replicating DNA so that other enzymes can guide the assembly of a new DNA strand. A helicase looks like a bagel through which replicating DNA is threaded. One DNA strand shoots through the bagel's hole, at the rate of 300 nucleotides per second, while the other strand is moved out of the way. Helicases can also repair errors in replicated DNA.

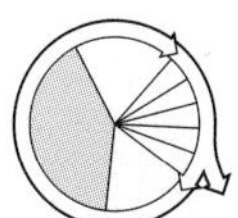

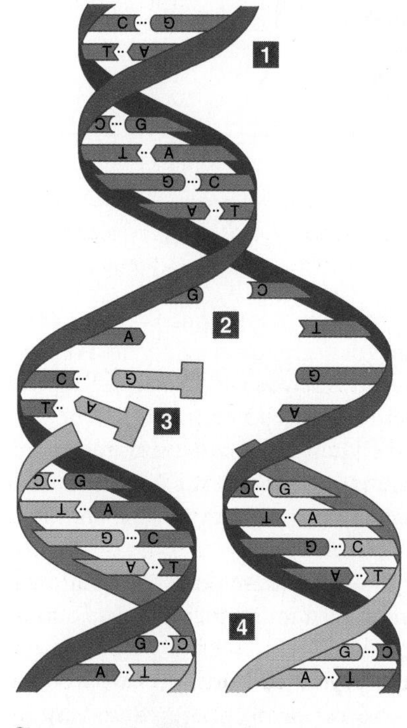

1 Parent DNA molecule; two complementary strands of base-paired nucleotides.

2 Parental strands unwind and separate at several points along the DNA molecule, forming replication forks.

3 Each parental strand provides a template that attracts and binds complementary bases, A with T and G with C.

4 Sugar-phosphate backbone of daughter strands closed. Each new DNA molecule consists of one parental and one daughter strand, as a result of semiconservative replication.

**figure 9.13**

**Overview of DNA replication.**

**Enzymes in DNA replication**

**Helicase** unwinds parental double helix

**Binding proteins** stabilize separate strands

**Primase** adds short primer to template strand

**DNA polymerase** binds nucleotides to form new strands

**Ligase** joins Okazaki fragments and seals other nicks in sugar-phosphate backbone

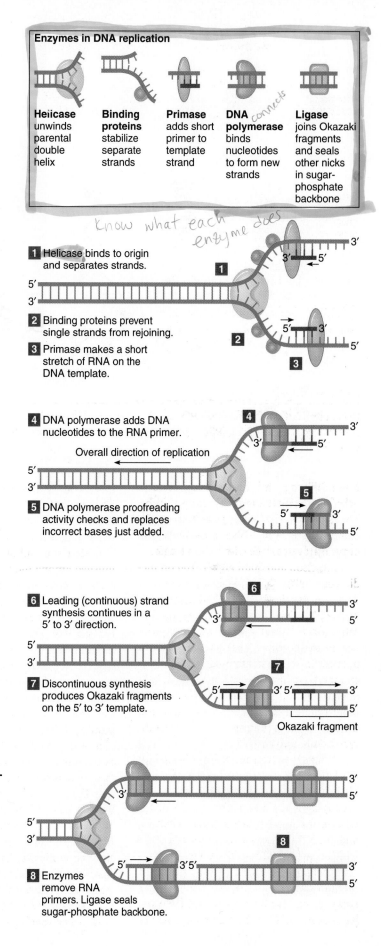

1 Helicase binds to origin and separates strands.

2 Binding proteins prevent single strands from rejoining.

3 Primase makes a short stretch of RNA on the DNA template.

4 DNA polymerase adds DNA nucleotides to the RNA primer.

Overall direction of replication

5 DNA polymerase proofreading activity checks and replaces incorrect bases just added.

6 Leading (continuous) strand synthesis continues in a 5′ to 3′ direction.

7 Discontinuous synthesis produces Okazaki fragments on the 5′ to 3′ template.

Okazaki fragment

8 Enzymes remove RNA primers. Ligase seals sugar-phosphate backbone.

**figure 9.14**

**DNA replication takes many steps.** Because DNA can only be replicated in a 5′ to 3′ direction, the process occurs continuously complementary to one strand (called the leading strand) and discontinuously, as Okazaki fragments form against the other strand of parental DNA (called the lagging strand). Various enzymes replicate DNA. First, a helicase opens the initiation site. Then primase adds a short RNA primer, which is later eaten away by an exonuclease and replaced with DNA. DNA polymerase extends the new strands and proofreads the base sequences, and ligase seals the sugar-phosphate backbone and joins Okazaki fragments.

## Reading 9.1

# DNA Makes History

One night in July 1918, Tsar Nicholas II of Russia and his family met gruesome deaths at the hands of Bolsheviks in a Ural mountain town called Ekaterinburg (figure 1). Captors led the tsar, tsarina, three of their daughters, the family physician, and three servants to a cellar and shot them, bayoneting those who did not immediately die. The executioners then stripped the bodies and loaded them onto a truck, planning to hurl them down a mine shaft. But the truck broke down, and the killers instead placed the bodies in a shallow grave, then damaged them with sulfuric acid to mask their identities.

In another July—many years later, in 1991—two Russian amateur historians found the grave. Because they were aware that the royal family had spent its last night in Ekaterinburg, they alerted the government that they might have unearthed the long-sought bodies of the Romanov family. An official forensic examination soon determined that the skeletons represented nine individuals. The sizes of the skeletons indicated that three were children, and the porcelain, platinum, and gold in some of the teeth suggested royalty. Unfortunately, the acid had so destroyed the facial bones that some conventional forensic tests were not feasible. But one type of evidence survived—DNA. Thanks to DNA amplification made possible by the polymerase chain reaction (PCR—see section 9.4), researchers obtained enough genetic material to solve the mystery.

British researchers eagerly examined DNA from cells in the skeletal remains. DNA sequences specific to the Y chromosome enabled the investigators to distinguish males from females. Then the genetic material of mitochondria, inherited from mothers only, established one woman as the mother of the children.

But a mother, her children, and companions were not necessarily a royal family. The researchers had to connect the skeletons to known royalty. To do so, they again turned to DNA. However, an inherited quirk proved, at first, to be quite confusing.

The challenge in proving that the male remains with fancy dental work were once Tsar Nicholas II centered around nucleotide position 16169 of a mitochondrial gene whose sequence is highly variable among individuals. About 70 percent of the bone cells examined from the remains had cytosine (C) at this position, and the remainder had thymine (T). Skeptics at first suspected contamination or a laboratory error, but when the odd result was repeated, researchers realized that this historical case had revealed a genetic phenomenon called heteroplasmy. The bone cells apparently harbored two populations of mitochondria, one type with C at this position, the other with T.

The DNA of a living blood relative of the tsar, Countess Xenia Cheremeteff-Sfiri, had only T at nucleotide site 16169. Xenia is the great-granddaughter of Tsar Nicholas II's sister. However, mitochondrial DNA from Xenia and the murdered man matched at every other site. DNA of another living relative, the Duke of Fife, the great-grandson of Nicholas's maternal aunt, matched Xenia at the famed 16169 site. A closer relative, Nicholas's nephew Tikhon Kulikovsky, refused to lend his DNA, citing anger at the British for not assisting the tsar's family during the Bolshevik revolution.

But the story wasn't over. It would take an event in yet another July, in 1994, to clarify matters.

Attention turned to Nicholas's brother, Grand Duke of Russia Georgij Romanov. In 1899, Georgij had died at age 28 of tuberculosis. His body was exhumed in July 1994, and researchers sequenced the troublesome mitochondrial gene in bone cells from his leg. They found a match! Georgij's mitochondrial DNA had the same polymorphic site as the man murdered in Siberia, who was, therefore, Tsar Nicholas II. The researchers calculated the probability that the remains are truly those of the tsar, rather than resembling Georgij's unusual DNA sequence by chance, as 130 million to 1. The murdered Russian royal family can finally rest in peace, thanks to DNA analysis.

**figure 1**

**DNA fingerprinting sheds light on history.** DNA analysis identified the remains of the murdered members of the Romanov family—and an interesting genetic phenomenon.

A human chromosome replicates at hundreds of points along its length, and then the individual pieces are joined. A point at which DNA replicates resembles a fork in a road, and the locally opened portion of a DNA double helix is called a **replication fork.** Many replication forks are necessary to copy so much DNA.

DNA replication begins when a helicase breaks the hydrogen bonds that connect a base pair (figure 9.14). This first step occurs at a region called an origin of replication site. Another enzyme, called **primase,** then attracts complementary RNA nucleotides to build a short piece of RNA, called an **RNA primer,** at the start of each segment of DNA to be replicated. The RNA primer is required because the major replication enzyme, **DNA polymerase** (DNAP), can only add bases to an existing strand, and adds new bases one at a time, starting at the RNA primer. Next, the RNA primer attracts DNA polymerase, which brings in DNA nucleotides complementary to the exposed bases on the parental strand, which serves as a mold, or template. (A polymerase is an enzyme that builds a polymer, which is a chain of chemical building blocks.) The new DNA strand grows as hydrogen bonds form between the complementary bases. Special binding proteins keep the two strands apart.

DNA polymerase also "proofreads" as it goes, excising mismatched bases and inserting correct ones. At the same time, another enzyme, a type of **exonuclease,** removes the RNA primer and replaces it with the correct DNA bases. Finally, enzymes called **ligases** seal the sugar-phosphate backbone. Ligase comes from a Latin word meaning "to tie." Table 9.2 lists the enzymes that replicate DNA.

DNA polymerase works directionally, adding new nucleotides to the exposed 3′ end of the sugar in the growing strand. Replication proceeds in a 5′ to 3′ direction, because this is the only chemical configuration in which DNAP can add bases. How can the growing fork proceed in one direction, when 5′ to 3′ elongation requires movement in both directions? The answer is that on at least one strand, replication is discontinuous. That is, it is accomplished in small pieces from the inner part of the fork outward in a pattern similar to backstitching. The pieces are then ligated to build the new strand. These smaller pieces, up to 150 nucleotides long, are called Okazaki fragments, after their discoverer (see figure 9.14).

### Key Concepts

Meselson and Stahl, and others, grew cells on media containing labeled DNA precursors. They then followed the distribution of the label to show that DNA replication is semiconservative, not conservative or dispersive.

Enzymes orchestrate DNA replication. DNA replication occurs simultaneously at several points on each chromosome, and the pieces are joined. At each initiation site, primase directs synthesis of a short RNA primer. DNA eventually replaces this RNA primer. DNA polymerase adds complementary DNA bases to the RNA primer, building a new half of a helix against each template. Finally, ligase joins the sugar-phosphate backbone. DNA is synthesized in a 5′ to 3′ direction, discontinuously on one strand.

**table 9.2**

**Enzymes that Replicate DNA**

| Enzyme | Function |
|---|---|
| Helicase | Unwinds and holds apart DNA section to be replicated |
| Primase | Begins new DNA molecule with RNA primer |
| DNA polymerase | Brings in new DNA nucleotides to existing strand |
| Exonuclease | Removes RNA primer |
| Ligase | Seals sugar-phosphate backbone |

## 9.4 PCR—Directing DNA Replication

Every time a cell divides, it replicates all of its DNA. A technology called the **polymerase chain reaction** (PCR) uses DNA polymerase to rapidly produce millions of copies of a specific DNA sequence of interest. PCR is perhaps the most successful technology to ever emerge from life science—it is rare to open a general biology journal today and not see mention of it in research reports.

PCR is useful in any situation where a small amount of DNA or RNA would provide information if it were mass produced. Applications span a variety of fields. In forensics, PCR is used routinely to establish blood relationships, to identify remains, and to help convict criminals or exonerate the falsely accused. When used to amplify the nucleic acids of microorganisms, viruses, and other parasites, PCR is important in agriculture, veterinary medicine, environmental science, and human health care. For example, PCR can detect HIV in a human blood sample long before antibodies to the virus are detectable, an ability with the potential to help stem the spread of the infection by identifying HIV sooner. In genetics, PCR is both a crucial laboratory tool to identify genes as well as the basis of many tests to detect inherited disease, sometimes even years before symptoms arise. Table 9.3 lists some uses of PCR.

The polymerase chain reaction was born in the mind of Kary Mullis on a moonlit night in northern California in 1983. As he drove up and down the hills, Mullis was thinking about the incredible precision and power of DNA replication and, quite suddenly, a way to tap into that power popped into his mind. He excitedly explained his idea to his girlfriend and then went home to think it through further. "It was difficult for me to sleep with deoxyribonuclear bombs exploding in my brain," he wrote much later, after PCR had revolutionized the life sciences.

The idea behind PCR was so stunningly straightforward that Mullis had trouble convincing his superiors at Cetus Corporation that he was really onto something. He spent the next year using the technique to amplify a well-studied gene so he could prove that his brainstorm was not just a flight of fancy. One by one, other researchers glimpsed Mullis's vision of that starry night. After convincing

## table 9.3

### Uses of PCR

PCR has been used to amplify:

- DNA to diagnose an inherited disease in a cremated man from skin cells left in his electric shaver.
- Short repeated sequences of DNA extracted from bits of human tissue from the site of the World Trade Center in the days following September 11, 2001, for use in identifying victims.
- A bit of DNA in a preserved quagga (a relative of the zebra) and a marsupial wolf, which are recently extinct animals.
- Genes from microorganisms that cannot be cultured for study.
- Mitochondrial DNA from various modern human populations, indicating that *Homo sapiens* originated in Africa, supporting fossil evidence.
- Similar genes from several species, revealing evolutionary relationships.
- DNA from the brain of a 7,000-year-old human mummy.
- Genetic material from saliva, hair, skin, and excrement of organisms that we cannot catch to study. The prevalence of a rare DNA sequence among all bird droppings from a certain species in an area can be extrapolated to estimate population size.
- DNA in the digestive tracts of carnivores, to reveal food web interactions.
- DNA in deteriorated roadkills and carcasses washed ashore, to identify locally threatened species.
- DNA in products illegally made from endangered species, such as powdered rhinoceros horn, sold as an aphrodisiac.
- DNA sequences in animals indicating that they carry the bacteria that cause Lyme disease, providing clues to how the disease is transmitted.
- DNA from genetically altered bacteria that are released in field tests, to follow their dispersion.
- DNA from one cell of an 8-celled human embryo to detect a genotype that causes cystic fibrosis.
- DNA from poached moose meat in hamburger.
- DNA from human remains in Jesse James's grave, to make a positive identification.
- DNA from the guts of genital crab lice on a rape victim, which matched DNA of the suspect.
- DNA in artificial knee joints, indicating infection.
- DNA from dried semen on a blue dress belonging to a White House intern, which helped to identify the person with whom the intern had a sexual encounter.
- DNA from cat hair used to link a murder suspect to a crime. Cat hair on the victim matched that of the suspect's pet, Snowball.

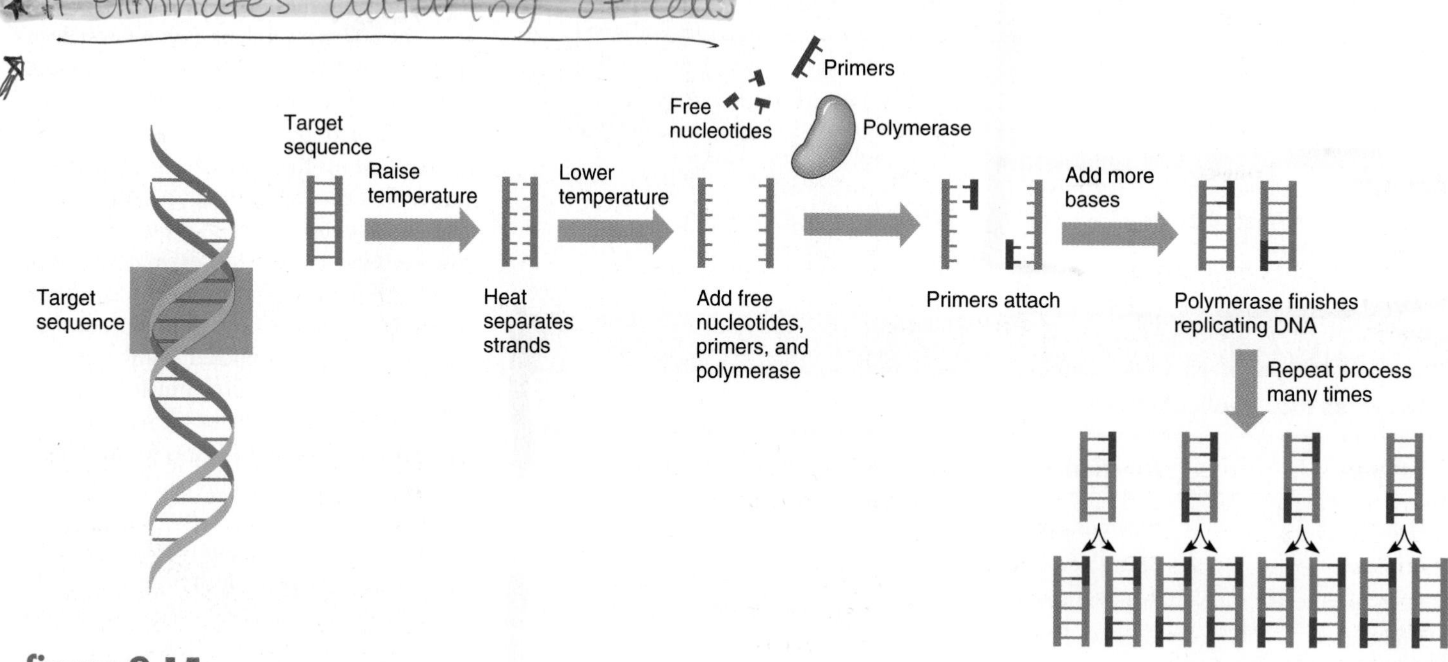


**figure 9.15**

**Amplifying a specific DNA sequence.** In the polymerase chain reaction, specific primers are used to bracket a DNA sequence of interest, along with a thermostable DNA polymerase and plenty of free nucleotides. The reaction rapidly builds up millions of copies of the target sequence.

*In Their Own Words*

# In Search of Heat-Loving Microorganisms

"I still have nightmares about being boiled alive," recalls Francis Barany, a microbiologist at Weill Medical College in New York City. Barany fell into a pool of boiling mud in Yellowstone National Park in 1991, while searching for bacteria.

The steaming water and mud of hot springs today are found only near volcanoes and certain lakes, where they are topped with colored mats of algae that are home to a few types of thermophilic ("heat loving") microorganisms. Francis Barany was looking for bacteria that make chemicals that do not fall apart under high-temperature conditions, as many chemicals do. He had developed a new method to mass-produce specific pieces of DNA, which would form the basis of tests to diagnose cancer and genetic and infectious diseases. Because the technique required heating the DNA, briefly and repeatedly, so that strands could form against old strands, he needed a DNA-copying enzyme that could tolerate heat. Since hot springs microbes reproduce readily, their DNA replication enzymes must be able to withstand high temperatures, Barany reasoned. He wasn't the first inquisitive microbiologist to look among Yellowstone's thousands of hot springs.

In 1965, Indiana University microbiologist Thomas Brock had discovered a microbe at Yellowstone by dipping glass slides into hot springs, and collecting the heat-loving bacteria that accumulated. It was one of the first observations that life could exist at high temperatures. A quarter century later, Barany ventured into the hot springs. Here is his story:

**It was evening, and we had been collecting so well that we thought we might as well finish, and the next day relax. We ended the day at Artist's Paint Pot, a hot springs and mud hole where the water had dried out, leaving hot mud. Hot springs are just beautiful, deep blue in the interior, the surface yellow-brown, where the bacteria are. It was constantly steaming, with bubbling water.**

**While collecting my thirteenth sample, the ground simply gave way, collapsed under me, and my left leg sank into the hot mud. It was 70° C; that's 158° F! It blistered the skin right off. I jumped out, and hopped onto land, screaming. Someone poured cold water on my leg, ripped my jeans off. I remember screaming that I had a permit to collect, and to save that thirteenth sample!**

Barany indeed found some novel enzymes in that thirteenth sample. A few months later, he was still hobbling about, taking pain medication, and telling everyone about his accident. But he felt healed enough to marry the woman who pulled him out of the hot springs, and today they have two little girls. He used the enzyme to develop a DNA amplification technique, called the ligase detection reaction, which he and others are currently using in SNP analyses of people with certain types of cancer. The technique is also being used in a Jewish genome project.

Today, discovering "extremophiles," as organisms that live in what to us are hostile conditions are called, for their hardy "extremozymes," a term for enzymes that function under extremes of temperature, pressure, acidity or salinity, is big business. One biotechnology company, for example, offers enzymes from life in Arctic pools, Indonesian volcanoes, Costa Rican jungles, and the rock beneath the Savannah River in South Carolina.

Extremozymes have varied uses. A detergent that removes lint and fuzz from clothing comes from an organism living in a salty mountain lagoon. Another extremozyme breaks up and bleaches wood pulp, and is used to make paper.

Products from microbes living in Antarctic sea ice are used in food processing done at low temperatures to retard spoilage, and in cold-wash laundry detergents. We use extremozymes to stonewash jeans, convict criminals using DNA fingerprinting techniques, and to diagnose infections and produce the antibiotic drugs that treat them. Extremozymes are also often safer than the industrial chemicals that they replace, which may require toxic substances to produce.

If the biological source of an extremozyme is difficult to grow, culture, or raise, researchers can isolate the gene and transfer it to a species that is easier to work with. Expression of a gene from one species in another is recombinant DNA technology, the subject of chapter 18.

his colleagues at Cetus, Mullis published a landmark 1985 paper and filed patent applications, launching the field of gene amplification. Mullis received only a $10,000 bonus from Cetus for his invention, which the company sold to another for $300 million. Mullis did, however, win a Nobel Prize.

PCR rapidly replicates a selected sequence of DNA in a test tube (figure 9.15). The requirements include:

1. Knowing parts of a target DNA sequence to be amplified.
2. Two types of lab-made, single-stranded, short pieces of DNA called primers. These are complementary in sequence to opposite ends of the target sequence.
3. A large supply of the four types of DNA nucleotide building blocks.
4. Taq1, a DNA polymerase produced by *Thermus aquaticus*, a microbe that inhabits hot springs. This enzyme is adapted to its host's hot surroundings and makes PCR easy because it does not fall apart when DNA is heated. (In Their Own Words describes one researcher's painful discovery of such a heat-tolerant enzyme.)

In the first step of PCR, heat is used to separate the two strands of the target DNA. Next, the temperature is lowered, and the two short DNA primers and Taq1 DNA polymerase are added. The primers bind by complementary base pairing to the separated target strands. In the third step, more DNA nucleotide bases are added. The DNA polymerase adds bases to the primers and builds a sequence complementary to the target sequence. The newly synthesized strands then act as templates in the next round of replication, which is initiated immediately by raising the temperature. All of this is done in an automated device that controls the key temperature changes.

The pieces of DNA accumulate geometrically. The number of amplified pieces of DNA equals $2^n$, where n equals the number of temperature cycles. After just 20 cycles, 1 million copies of the original sequence have accumulated in the test tube.

PCR's greatest strength is that it works on crude samples of rare, old, and minute sequences, as Reading 9.1 describes. PCR's greatest weakness, ironically, is its exquisite sensitivity. A blood sample submitted for diagnosis of an infection contaminated by leftover DNA from a previous test, or a stray eyelash dropped from the person running the reaction, can yield a false result, although this rarely happens today. The technique is also limited because a user must know the sequence to be amplified.

Five weeks after the publication of Watson and Crick's critical paper describing the structure of DNA, they published another detailing how a sequence of DNA can spell out a sequence of amino acids. That, and more, is the subject of the next chapter.

**KEY CONCEPTS**

PCR is a technique that induces rapid DNA replication of a small, selected part of an organism's genome. It has many uses in forensics, agriculture, and medicine.

# Summary

### 9.1 Experiments Identify and Describe the Genetic Material

1. DNA encodes information that the cell uses to synthesize protein. DNA can also replicate, so that its information is passed on.
2. Many experimenters described DNA and showed it to be the hereditary material. Miescher identified DNA in white blood cell nuclei. Garrod conceptually connected heredity to symptoms caused by enzyme abnormalities. Griffith identified a substance that transmits infectiousness in pneumonia-causing bacteria; Avery, MacLeod, and McCarty discovered that the transforming principle is DNA; and Hershey and Chase confirmed that the genetic material is DNA and not protein.
3. Levene described the three components of a DNA building block and found that they appear in DNA in equal amounts. Chargaff discovered that the amount of **adenine** (A) equals the amount of **thymine** (T), and the amount of **guanine** (G) equals that of **cytosine** (C). Watson and Crick put all these clues together to propose DNA's double helix structure.

### 9.2 DNA Structure

4. A **nucleotide** is a DNA building block. It consists of a deoxyribose, a phosphate, and a **nitrogenous base.**
5. The rungs of the DNA double helix consist of hydrogen-bonded complementary base pairs (A with T, and C with G). The rails are chains of alternating sugars and phosphates that run **antiparallel** to each other. DNA is highly coiled.

### 9.3 DNA Replication—Maintaining Genetic Information

6. Meselson and Stahl demonstrated the **semiconservative** nature of DNA replication with **density shift experiments.**
7. During replication, the DNA unwinds locally at several initiation points. **Replication forks** form as the hydrogen bonds break between an initial base pair. **Primase** builds a short RNA primer, which DNA eventually replaces. Next, **DNA polymerase** fills in DNA bases, and **ligase** seals the sugar-phosphate backbone.
8. Replication proceeds in a 5′ to 3′ direction, so the process must be discontinuous in short stretches on one strand.

### 9.4 PCR—Directing DNA Replication

9. Gene amplification techniques, such as **PCR,** utilize the power and specificity of DNA replication enzymes to selectively amplify certain sequences.
10. In PCR, primers corresponding to a gene of interest direct the polymerization of supplied nucleotides to construct many copies of the gene.

# Review Questions

1. The function of DNA is to specify and regulate the cell's synthesis of protein. If a cell contains all the genetic material required to carry out protein synthesis, why must its DNA be replicated?
2. Match the experiment described on the left to the concept it illustrates on the right:

| | |
|---|---|
| 1. Density shift experiments | a. DNA is the hereditary material |
| 2. Discovery of an acidic substance that includes nitrogen and phosphorus on dirty bandages | b. Complementary base pairing is part of DNA structure and maintains a symmetrical double helix |
| 3. "Blender experiments" that showed that the part of a virus that infects bacteria contains phosphorus, but not sulfur | c. Identification of nuclein |
| 4. Determination that DNA contains equal amounts of guanine and cytosine, and of adenine and thymine | d. DNA, not protein, is the hereditary material |
| 5. Discovery that bacteria can transfer a "factor" that transforms a harmless strain into a lethal one | e. DNA replication is semiconservative, and not conservative or dispersive |

3. What part of the DNA molecule encodes information?
4. Explain how DNA is a directional molecule in a chemical sense.
5. Place the following proteins in the order in which they begin to function in DNA replication.

   ligase　　primase
   exonuclease　　helicases
   DNA polymerase

6. Write the sequence of a strand of DNA replicated from each of the following base sequences:

   a. T C G A G A A T C T C G A T T
   b. C C G T A T A G C C G G T A C
   c. A T C G G A T C G C T A C T G

7. Place in increasing size order:

   nucleosome
   histone protein
   chromatin

8. Define:

   a. antiparallelism
   b. semiconservative replication
   c. complementary base pairing

9. Describe two experiments that supported one hypothesis while also disproving another.
10. List the steps in DNA replication.

# Applied Questions

1. In Bloom syndrome, ligase malfunctions. As a result, replication forks move too slowly. Why does this happen?
2. DNA contains the information that a cell uses to synthesize a particular protein. How do proteins assist in DNA replication?
3. To diagnose a rare form of encephalitis (brain inflammation), a researcher needs a million copies of a viral gene. She decides to use the polymerase chain reaction on a sample of the patient's cerebrospinal fluid, which bathes his infected brain. If one cycle of PCR takes two minutes, how long will it take the researcher to obtain her millionfold amplification?
4. Why would a DNA structure in which each base type could form hydrogen bonds with any of the other three base types not produce a molecule that could be replicated easily?
5. A person with deficient or abnormal ligase may have an increased cancer risk and chromosomes that cannot heal breaks. The person is, nevertheless, alive. Why are there no people who lack DNA polymerase?
6. Until recently, HIV infection was diagnosed by detecting antibodies in a person's blood or documenting a decline in the number of the type of white blood cell that HIV initially infects. Why is PCR detection more sensitive?
7. Which do you think was the more far-reaching accomplishment, determining the structure of DNA, or sequencing the human genome? State a reason for your answer.

# Suggested Readings


Firneisz, Gabor, et al. July 7, 2001. Postcremation diagnosis from an electric shaver. *The Lancet* 358:34. Five years after he was cremated, a man received a diagnosis of the genetic disease that killed him, thanks to stray skin cells recovered from his razor.

Holmes, Frederic Lawrence. 2001. *Meselson, Stahl, and the Replication of DNA: A History of the Most Beautiful Experiment in Biology.* New Haven: Yale University Press. The story of the discovery of "untwiddling" DNA.

Kilesnikov, Lev L., et al. February 15, 2001. Anatomical appraisal of the skulls and teeth associated with the family of Tsar Nicolay Romanov. *Anatomical Record,* 265:15–32. Forensic analysis of skeletal remains supports DNA evidence that the royal Romanovs were brutally murdered.

Lederberg, Joshua. February 1994. Honoring Avery, MacLeod, and McCarty: The team that transformed genetics. *Genetics* 136:423–27. Classical experiments identified DNA as the genetic material.

Piper, Anne. April 1998. Light on a dark lady. *Trends in Biological Sciences* 23:151–54. A tribute to Rosalind Franklin by her best friend.

Watson, James D. 1968. *The Double Helix.* New York: New American Library. An exciting, personal account of the discovery of DNA structure.

Watson, James D., and F. H. C. Crick. April 25, 1953. Molecular structure of nucleic acids: A structure for deoxyribose nucleic acid. *Nature* 171, no. 4356:737–8. The original paper describing the structure of DNA.


# On the Web

Check out the resources on our website at

**www.mhhe.com/lewisgenetics5**

On the web for this chapter, you will find additional study questions, vocabulary review, useful links to case studies, tutorials, popular press coverage, and much more. To investigate specific topics mentioned in this chapter, also try the links below:

The Polymerase Chain Reaction Jump Station
**www.horizonpress.com/pcr/**

Possible models for DNA replication
**www.accessexcellence.com/AB/GG/possible.html**

Sites with information on DNA replication
**http://dir.niehs.nih-gov/dirlmg/repl.html**

CHAPTER

10

# Gene Action and Expression

**10.1 Transcription—The Link Between Gene and Protein**
The proteins that a particular type of cell manufactures constitute its proteome. The information in the nucleotide base sequence of a protein-encoding gene is transcribed into mRNA, which is then translated into the amino acid sequence of a protein.

**10.2 Translating a Protein**
The genetic code is the correspondence between RNA base triplets and particular amino acids. It is universal—the same messenger RNA codons specify the same amino acids in humans, hippos, herbs, and bacteria. Ribosomes provide structural support and enzyme activity for transfer RNA molecules to align the designated amino acids against a messenger RNA. The amino acids link, building proteins.

**10.3 The Human Genome Sequence Reveals Unexpected Complexity**
The human genome sequence reveals that the classic one gene-one protein paradigm is an oversimplification of the function of DNA. Only 1.5 percent of the genome encodes protein, yet those 31,000 or so genes specify more than 100,000 different proteins.

DNA replication preserves genetic information by endowing each new cell with a complete set of operating instructions. A cell uses some of the information to manufacture proteins. To do this, first the process of **transcription** copies a particular part of the DNA sequence of a chromosome into an RNA molecule that is complementary to one strand of the DNA double helix. Then the process of **translation** uses the information copied into RNA to manufacture a specific protein by aligning and joining the specified amino acids. The overall events of transcription and translation are referred to as **gene expression.**

Watson and Crick, shortly after publishing their structure of DNA in 1953, described the relationship between nucleic acids and proteins as a directional flow of information called the "central dogma" (figure 10.1). Francis Crick explained in 1957, "The specificity of a piece of nucleic acid is expressed solely by the sequence of its bases, and this sequence is a code for the amino acid sequence of a particular protein." But understanding the central dogma was only a beginning. Today, nearly half a century later, researchers all over the world are detailing the patterns of gene expression in the many types of cells that build a human body, as well as in diseased cells. The central dogma explained how a gene encodes a protein; it did not explain how a cell "knows" which genes to express. What, for example, directs a bone cell to transcribe the genes that control the synthesis of collagen, and not to transcribe those that specify muscle proteins? How does a stem cell in bone marrow "know" when to divide and send daughter cells on pathways to differentiate as white blood cells, red blood cells, or platelets—and how does the balance of blood cell production veer from normal when a person has leukemia? Genes do more than encode proteins. They also control each other's functioning, in sometimes complex hierarchies.

Knowing the human genome sequence has revealed new complexities and seeming contradictions. The neat one gene-one protein picture painted by the work that followed Watson and Crick's description of DNA is correct, but a gross oversimplification. Our 31,000 or so genes actually encode between 100,000 to 200,000 proteins. Yet only a small part of our genome encodes protein. Even if the few researchers who hypothesize that the human genome contains 60,000 or more genes are correct, that is still a very small portion of the genome. This chapter presents the classical elucidation of the **genetic code**—the correspondence between gene and protein—and concludes with a consideration of the new questions posed by human genome sequence information.

**figure 10.1**

**DNA to RNA to protein.** The central dogma of biology states that information stored in DNA is copied to RNA (transcription), which is used to assemble proteins (translation). DNA replication perpetuates genetic information. This figure repeats within the chapter, with the part under discussion highlighted.

## 10.1 Transcription—The Link Between Gene and Protein

RNA is of crucial importance in the flow of genetic information, serving as a bridge between gene and protein. Cells replicate their DNA only during S phase of the cell cycle. In contrast, transcription and translation occur continuously during the cell cycle, to supply the proteins essential for life as well as those that give a cell its specialized characteristics.

RNA and DNA share an intimate relationship, as figure 10.2 depicts. RNA is synthesized against one side of the double helix, called the **template strand,** with the assistance of an enzyme called **RNA polymerase.** The other side of the DNA double helix is the **coding strand.** RNA comes in three major types, and several less abundant types, distinguished by their functions (although we do not yet know all of them). RNA also differs in a few ways from DNA. We begin our look at transcription—the prelude to translation—by considering RNA.

### RNA Structure and Types

RNA and DNA are both nucleic acids, consisting of sequences of nitrogen-containing bases joined by sugar-phosphate backbones. However, RNA is usually single-stranded, whereas DNA is double-stranded. Also, RNA has the pyrimidine base **uracil** in place of DNA's thymine. As their names imply, RNA nucleotides include the sugar ribose, rather than the deoxyribose that is part of DNA. Functionally, DNA stores genetic information, whereas RNA actively utilizes that information to enable the cell to synthesize a particular protein. Table 10.1 and

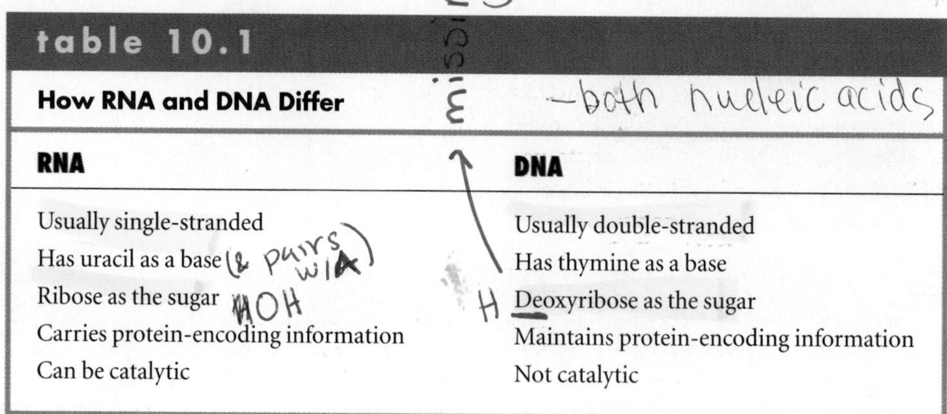

**table 10.1**

**How RNA and DNA Differ**

| RNA | DNA |
| --- | --- |
| Usually single-stranded | Usually double-stranded |
| Has uracil as a base | Has thymine as a base |
| Ribose as the sugar | Deoxyribose as the sugar |
| Carries protein-encoding information | Maintains protein-encoding information |
| Can be catalytic | Not catalytic |

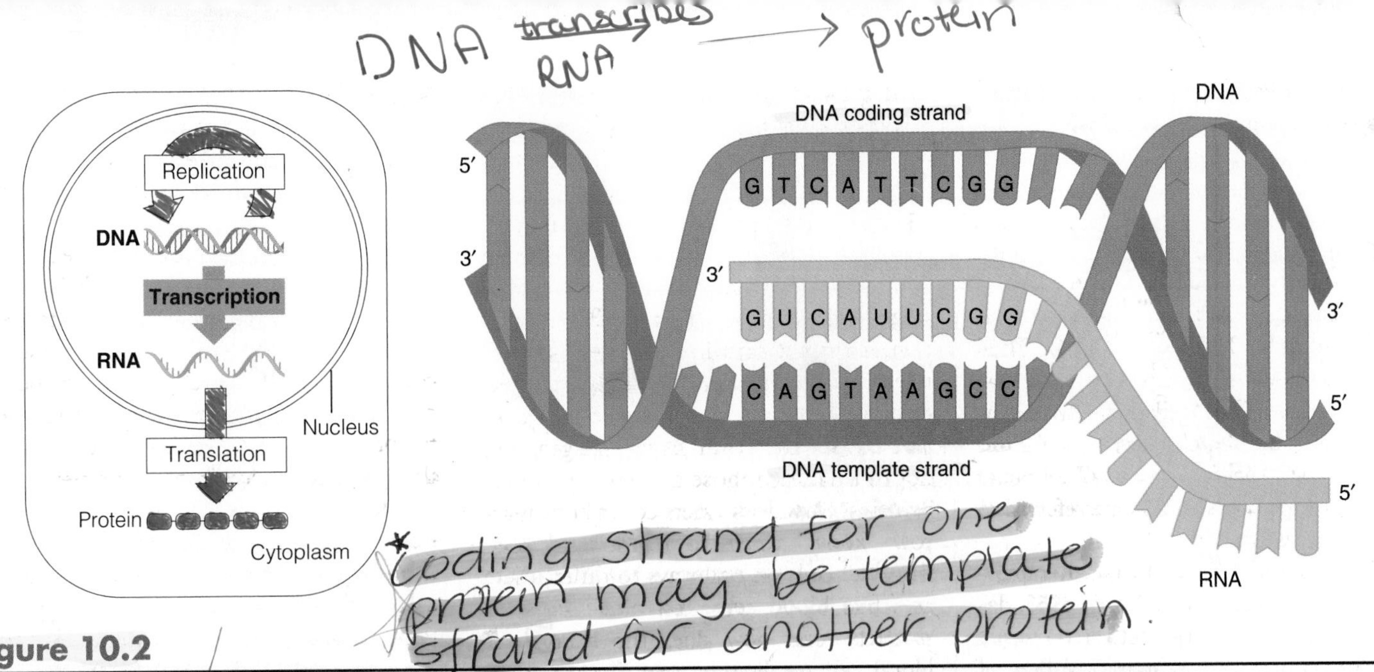


**figure 10.2**

**The relationship among RNA, the DNA template strand, and the DNA coding strand.** The RNA sequence is complementary to that of the DNA template strand and therefore is the same sequence as the DNA coding strand, with uracil (U) in place of thymine (T).

figure 10.3 summarize the differences between RNA and DNA.

As RNA is synthesized along DNA, it folds into three-dimensional shapes, or **conformations,** that are determined by complementary base pairing within the same RNA molecule. These shapes are very important for RNA's functioning. The three major types of RNA are messenger RNA, ribosomal RNA, and transfer RNA (table 10.2).

**Messenger RNA** (mRNA) carries the information that specifies a particular protein product. Each three mRNA bases in a row form a genetic code word, or **codon,** that specifies a certain amino acid. Because genes vary in length, so do mRNA molecules. Most mRNAs are 500 to 2,000 bases long. At any one time, different cell types have different mRNA molecules, present in differing amounts, that reflect their functions. A muscle cell has many mRNAs that specify the abundant contractile proteins actin and myosin, whereas a skin cell contains many mRNAs corresponding to the gene that encodes the scaly protein keratin. Geneticists use DNA microarrays to identify the types and amounts of mRNAs by converting the mRNAs of a cell into DNA called complementary or cDNA.

The information encoded in an mRNA sequence cannot be utilized without the participation of two other major classes of RNA. **Ribosomal RNA** (rRNA)

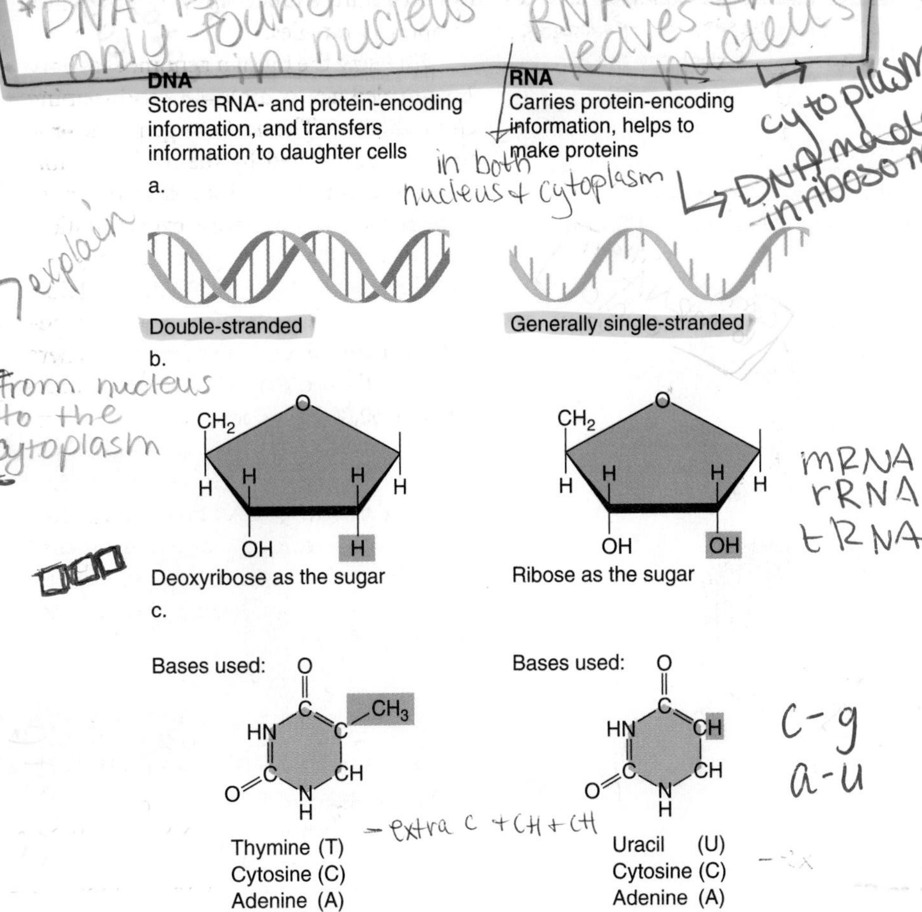


**figure 10.3**

**DNA and RNA differences.** (*a*) DNA is double-stranded; RNA is usually single-stranded (*b*). DNA nucleotides include deoxyribose, whereas RNA nucleotides have ribose (*c*). Finally, DNA nucleotides include the pyrimidine thymine, whereas RNA has uracil (*d*).

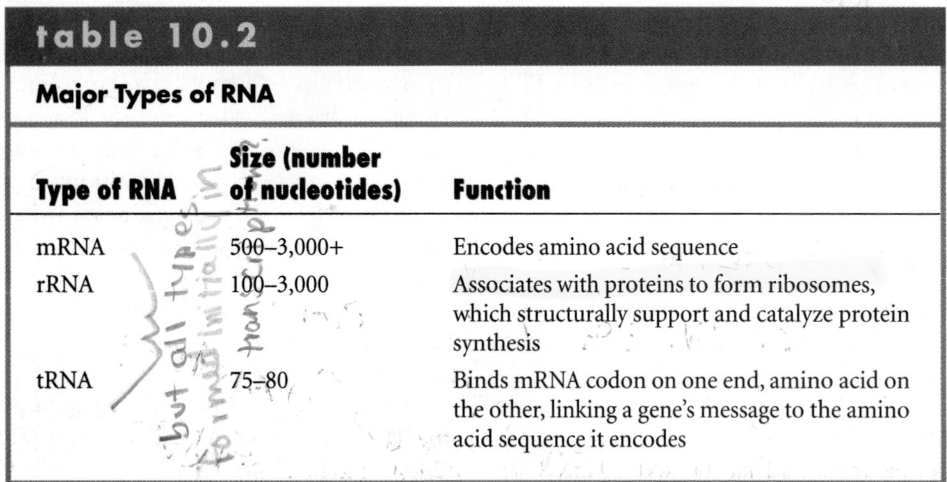

**table 10.2**

**Major Types of RNA**

| Type of RNA | Size (number of nucleotides) | Function |
|---|---|---|
| mRNA | 500–3,000+ | Encodes amino acid sequence |
| rRNA | 100–3,000 | Associates with proteins to form ribosomes, which structurally support and catalyze protein synthesis |
| tRNA | 75–80 | Binds mRNA codon on one end, amino acid on the other, linking a gene's message to the amino acid sequence it encodes |

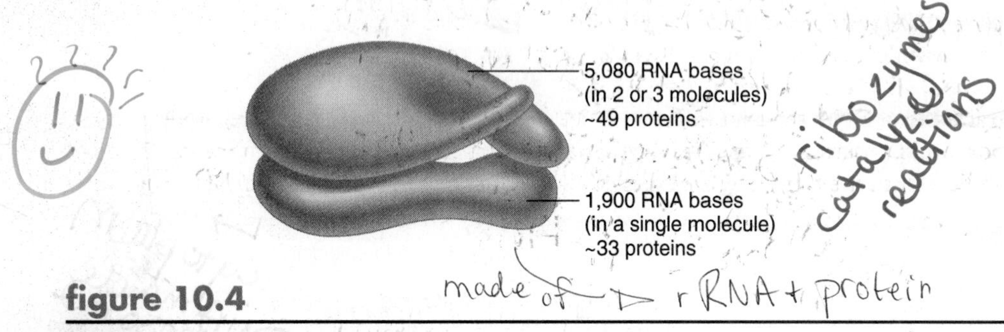


**figure 10.4**

**The ribosome.** A ribosome from a eukaryotic cell, shown here, has two subunits, containing 82 proteins and 4 rRNA molecules altogether.

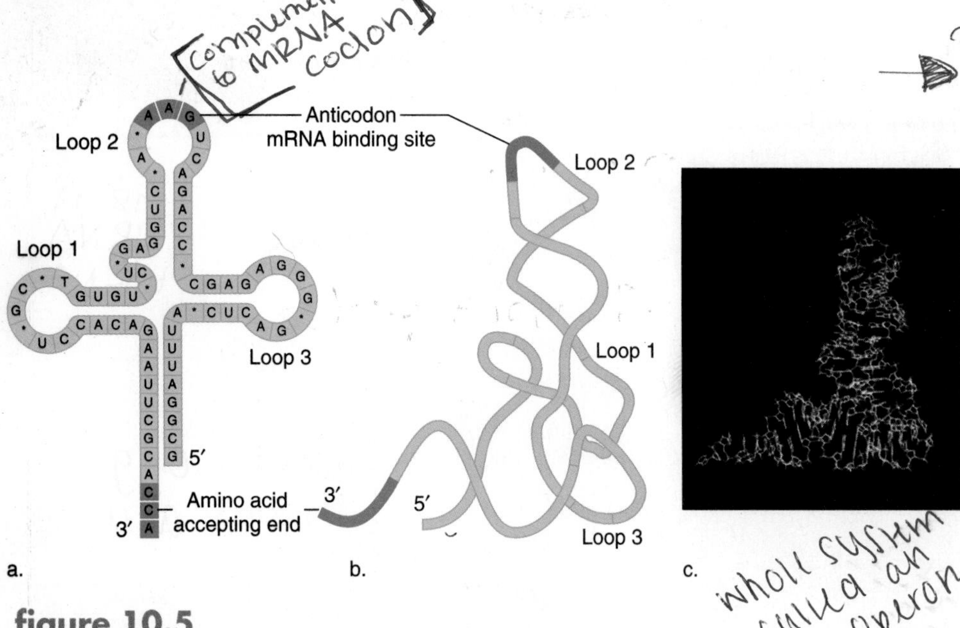


**figure 10.5**

**Transfer RNA.** (*a*) Certain nucleotide bases within a tRNA molecule hydrogen bond with each other to give the molecule a "cloverleaf" conformation that can be represented in two dimensions. The darker bases at the top form the anticodon, the sequence that binds a complementary mRNA codon. Each tRNA terminates with the sequence CCA, where a particular amino acid covalently bonds. Three-dimensional representations of a tRNA (*b*) and (*c*) depict the loops that interact with the ribosome to give tRNA its functions in translation.

molecules range from 100 to nearly 3,000 nucleotides long. This type of RNA associates with certain proteins to form a ribosome. Recall from chapter 2 that a ribosome is a structural support for protein synthesis (figure 10.4). A ribosome has two subunits that are separate in the cytoplasm but join at the initiation of protein synthesis. The larger ribosomal subunit has three types of rRNA molecules, and the small subunit has one. Ribosomal RNA, however, is much more than a structural support. Certain rRNAs catalyze the formation of bonds between amino acids. Such an RNA with enzymatic function is called a **ribozyme.** Other rRNAs help to align the ribosome and mRNA.

The third major type of RNA molecule is **transfer RNA** (tRNA). These molecules are "connectors" that bind mRNA codons at one end and specific amino acids at the other. A tRNA molecule is only 75 to 80 nucleotides long. Some of its bases weakly bond with each other, folding the tRNA into loops that form a characteristic cloverleaf shape (figure 10.5). One loop of the tRNA has three bases in a row that form the **anticodon,** which is complementary to an mRNA codon. The end of the tRNA opposite the anticodon strongly bonds to a specific amino acid. A tRNA with a particular anticodon sequence always carries the same amino acid. (There are 20 types of amino acids in organisms.) For example, a tRNA with the anticodon sequence GAA always picks up the amino acid phenylalanine. Special enzymes attach amino acids to tRNAs that bear the appropriate anticodons.

## Transcription Factors

Study of the control of gene expression began in 1961, when French biologists François Jacob and Jacques Monod described the remarkable ability of *E. coli* to produce the enzymes to metabolize the sugar lactose only when lactose is present in the cell's surroundings. What "tells" a simple bacterial cell to transcribe those products it needs—at exactly the right time?

Jacob and Monod discovered and described how a modified form of lactose turned on the genes whose encoded proteins break down the sugar. Jacob and Monod named the set of genes that are coordinately controlled an operon. Wrote Jacob and

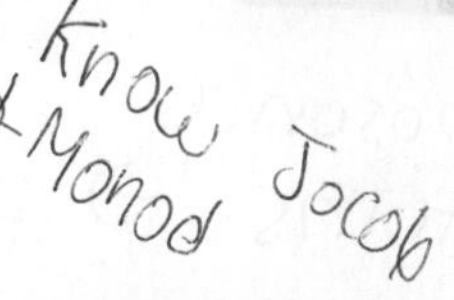

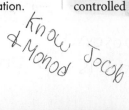

Monod in 1961, "The genome contains not only a series of blueprints, but a coordinated program of protein synthesis and means of controlling its execution." Operons were originally described in several types of bacteria, but the genome sequence of the roundworm *Caenorhabditis elegans* revealed that nearly a quarter of its genes are organized into operon-like groups, too.

In bacteria, operons act like switches, turning gene transcription on or off. In multicellular eukaryotes like ourselves, genetic control is more complex because different cell types express different subsets of genes. To manage such complexity, groups of proteins called **transcription factors** come together, forming an apparatus that binds DNA at certain sequences and initiates transcription at specific sites on a chromosome. The transcription factors, activated by signals from outside the cell, set the stage for transcription to begin by forming a pocket for RNA polymerase—the enzyme that actually builds an RNA chain.

Several types of transcription factors are required to transcribe a gene. Because transcription factors are proteins, they too are gene-encoded. The DNA sequences that transcription factors bind may be located near the genes they control, or as far as 40,000 bases away. DNA may form loops so that the genes encoding proteins that act together come near each other for transcription. Proteins in the nucleus may help bring certain genes and their associated transcription factors in close proximity, much as books on a specialized topic might be grouped together in a library for easier access.

About 2,000 transcription factors are known, and defects in them cause some diseases. For example, Huntington disease results from a defect in a gene whose encoded protein, huntington, is a transcription factor. Part of its normal function is to turn on transcription of a gene that encodes brain-derived neurotrophic factor (BDNF). Neurons in a part of the brain called the striatum require BDNF to stay alive. With abnormal huntington, not enough BDNF interacts with the striatal neurons, and they die. It may take many years to produce the uncontrollable movements and other changes characteristic of the disorder.

Many transcription factors have regions in common, called motifs, that fold into similar three-dimensional shapes, or conformations. These motifs generally enable the transcription factor to bind DNA. They have very colorful names, such as "helix-turn-helix," "zinc fingers," and "leucine zippers," that reflect their distinctive shapes.

## Steps of Transcription

How do transcription factors and RNA polymerase "know" where to bind to DNA to begin transcribing a specific gene? Transcription factors and RNA polymerase are attracted to a **promoter,** which is a special sequence that signals the start of the gene. Figure 10.6 shows one order in which transcription factors bind, to set up a site to receive RNA polymerase, called a preinitiation complex. The first transcription factor to bind, called a TATA binding protein, is attracted to a DNA sequence called a TATA box, which consists of the base sequence TATA surrounded by long stretches of G and C. Once the first transcription factor binds, it attracts others in groups and finally RNA polymerase joins the complex, binding just in front of the start of the gene sequence. The coming together of these components constitutes transcription initiation.

Complementary base pairing underlies transcription, just as it does DNA replication. In the next stage, transcription elongation, enzymes unwind the DNA double helix, and RNA nucleotides bond with exposed complementary bases on the DNA template strand (figure 10.2). RNA polymerase adds the RNA nucleotides in the sequence the DNA specifies, moving along the DNA strand in a 3′ to 5′ direction, synthesizing the RNA molecule in a 5′ to 3′ direction. A terminator sequence in the DNA indicates where the gene's RNA-encoding region ends. This is transcription termination.

For a particular gene, RNA is transcribed using only one strand of the DNA double helix as the template. The other

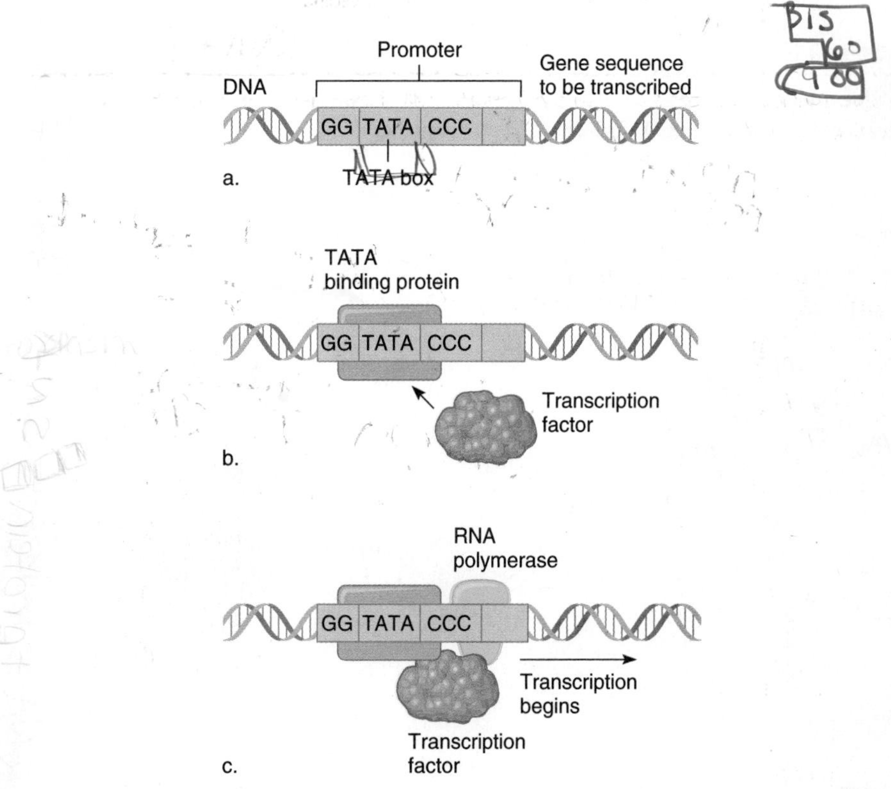


**figure 10.6**

**Setting the stage for transcription to begin.** (*a*) The promoter region of a gene has specific sequences recognized by proteins that initiate transcription. (*b*) A binding protein recognizes the TATA region and binds to the DNA. This allows other transcription factors to bind. (*c*) The presence of the necessary transcription factors allows RNA polymerase to bind and begin making RNA.

DNA strand that isn't transcribed is called the coding strand because its sequence is identical to that of the RNA, except with thymine (T) in place of uracil (U). Several RNAs may be transcribed from the same DNA template strand simultaneously (figure 10.7). Since RNA is relatively short-lived, a cell must constantly transcribe certain genes to maintain supplies of essential proteins. However, different genes on the same chromosome may be transcribed from different halves of the double helix.

To determine the sequence of RNA bases transcribed from a gene, write the RNA bases that are complementary to the template DNA strand, using uracil opposite adenine. For example, if a DNA template strand has the sequence

CCTAGCTAC

then it is transcribed into RNA with the sequence

GGAUCGAUG

The coding DNA sequence is:

GGATCGATG

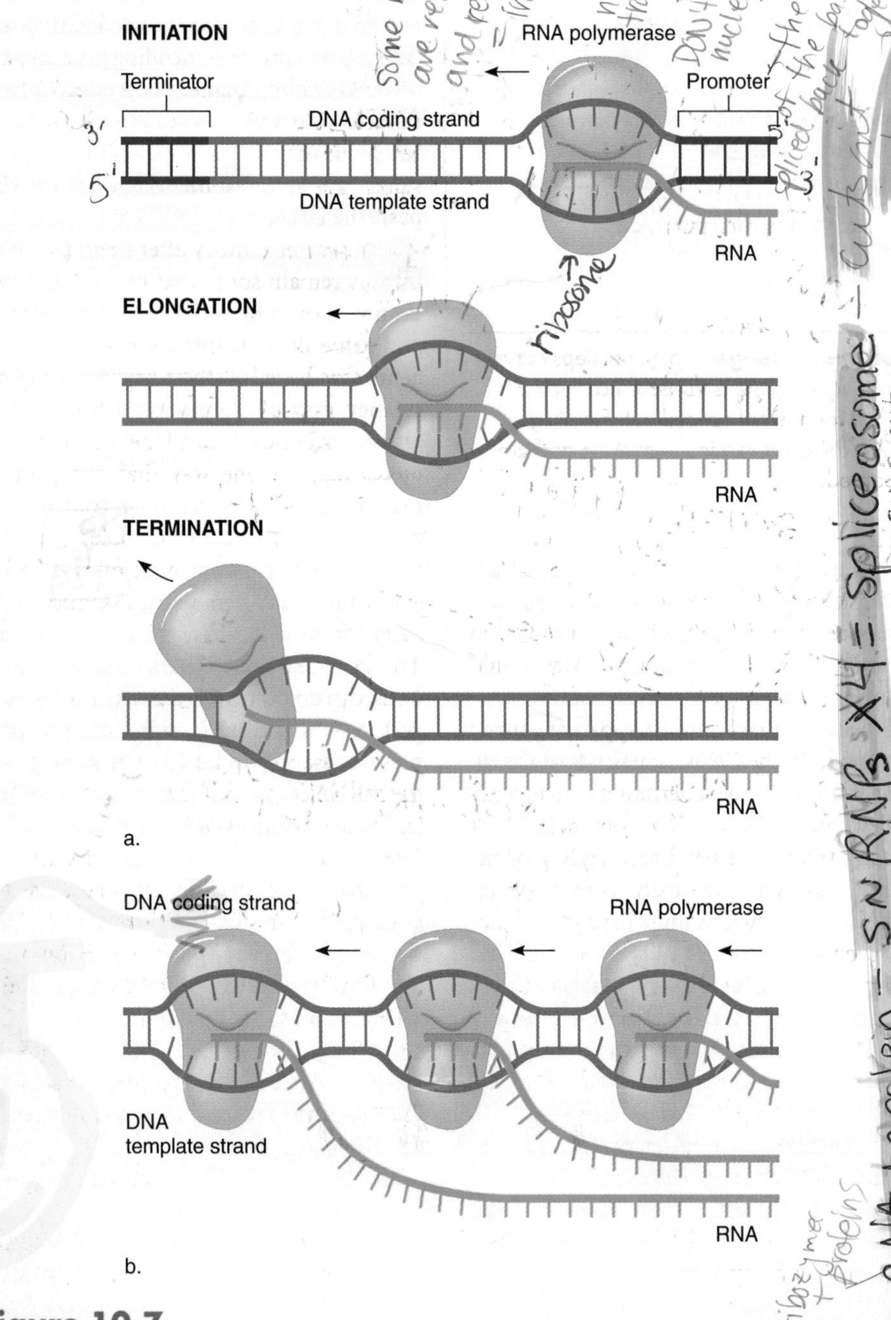


**figure 10.7**

**Transcription of RNA from DNA.** (*a*) Transcription occurs in three stages: initiation, elongation, and termination. Initiation is the control point that determines which genes are transcribed and when RNA nucleotides are added during elongation, and a terminator sequence in the gene signals the end of transcription. (*b*) Many identical copies of RNA are simultaneously transcribed, with one RNA polymerase starting after another.

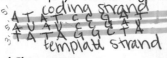

## RNA Processing

In bacteria and archaea, RNA is translated into protein as soon as it is transcribed from DNA because a nucleus does not physically separate the two processes. In cells of eukaryotes, mRNA must first exit the nucleus to enter the cytoplasm, where protein synthesis occurs. RNA is altered before it participates in protein synthesis in these more complex cells. (However, recent evidence indicates that some protein synthesis does occur in the nucleus.)

After mRNA is transcribed, a short sequence of modified nucleotides, called a cap, is added to the 5′ end of the molecule. At the 3′ end, a special polymerase adds 200 adenines, forming a "poly A tail." The cap and poly A tail may mark which mRNAs should exit the nucleus.

In addition to these modifications, not all of an mRNA is translated into an amino acid sequence in eukaryotic cells. Parts of mRNAs called **introns** (short for "intervening sequences") are transcribed but are later removed. The ends of the remaining molecule are spliced together before the mRNA is translated.

The mRNA prior to intron removal is called pre-mRNA. Introns are excised by small RNA molecules that are ribozymes, which associate with proteins to form small nuclear ribonucleoproteins (snRNPs), or "snurps." Four snurps form a structure called a spliceosome that cuts introns out and knits exons together to form the mature mRNA that exits the nucleus. The parts of mRNA that are translated are called **exons** (figure 10.8).

Introns range in size from 65 to 10,000 or more bases; the average intron is 3,365 bases. The average exon, in contrast, is 145

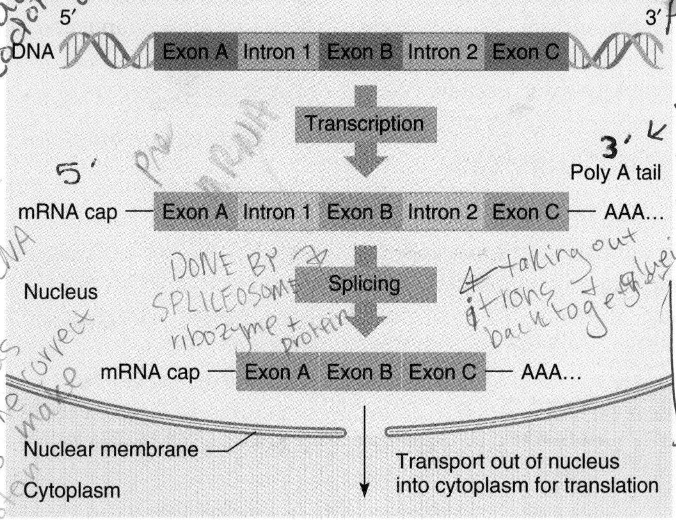

**figure 10.8**

**Messenger RNA processing—the maturing of the message.** Several steps carve the mature mRNA. First, a large region of DNA containing the gene is transcribed. Then a modified nucleotide cap and poly A tail are added, and introns are spliced out. Finally, the mature mRNA is transported out of the nucleus. Some mRNAs remain in the nucleus and are translated there—something that was only recently learned.

bases long. Many genes are riddled with introns—the human collagen gene, for example, contains 50. The gene whose absence causes Duchenne muscular dystrophy is especially interesting in its intron/exon organization. The gene is 2,500,000 bases, but its corresponding mRNA sequence is only 14,000 bases. The gene contains 80 introns. The number, size, and organization of introns vary from gene to gene. In many genes, introns take up more space than exons. We know from the human genome sequence that the coding portion of the average human gene is 1,340 bases, whereas the average total size of a gene is 27,000 bases.

The discovery of introns in 1977 arose out of then-new DNA sequencing technology. Certain gene sequences, when compared to the protein sequences that they encode, were found to have extra sections. The experiment that led to the detection of introns examined the 600-base rabbit beta globin gene, which encodes a 146 amino acid chain. Because three DNA bases encode one amino acid, a gene of only 438 bases would suffice to specify 146 amino acids ($3 \times 146 = 438$), but that is not the case. Since then, introns have been found in many genes, almost exclusively in eukaryotes.

The existence of introns surprised geneticists, who likened gene structure to a sentence in which all of the information contributes to the meaning. Why would protein-encoding genes consist of meaningful pieces scattered amidst apparent genetic gibberish? In the 1980s, geneticists confident that introns were aberrations arrogantly called them "junk DNA," Francis Crick among them. The persistent finding of introns, however, eventually convinced researchers that introns must have some function, or they would not have been retained through evolution. Said one speaker at a recent genomics conference, "Anyone who still thinks that introns have no function, please volunteer to have them removed, so we can see what they do." He had no takers.

The human genome project has revealed that the pervasiveness and size of our introns distinguishes our genome from those of our closest relatives. The human genome has many introns, with an average size 10 times that of an intron in the fruit fly or roundworm. The exon/intron organization that is a hallmark of many human genes can be compared to the signal/noise in a message. Our genome has more "noise" than do others so far sequenced, and we do not know why. Introns have complicated the "annotation" phase of the human genome project, which locates the protein-encoding parts. Finding the exons in a gene is a little like a word search puzzle, in which a square of letters harbors hidden words. Computer programs can hunt among strings of A, T, G, and C—raw DNA sequence—for the telltale signs of a protein-encoding gene, and then distinguish the exons from the introns. For example, a short sequence that indicates the start of a protein-encoding gene is called an open reading frame. Another clue is that dinucleotide repeats, such as CGCGCG, often flank introns, forming splice sites that signal the spliceosome where to cut and paste the DNA.

A quarter century after their discovery, introns remain somewhat of a mystery. We do not know why some genes have introns and some do not. Introns may be ancient genes that have lost their original function, or they may be remnants of the DNA of viruses that once infected the cell. Genes-in-pieces may be one way that our genome maximizes its informational content. For example, introns may enable exons to combine in different ways, even from different genes, much as a person can assemble many outfits from a few basic pieces of clothing. The fact that some disease-causing mutations disrupt intron/exon splice sites suggests that this cutting and pasting of gene parts is essential to health. For some genes, the mRNA is cut to different sizes in different tissues, which is called alternate splicing. The trimming of genes may explain how cells that make up different tissues use the same protein in slightly different ways. This is the case for apolipoprotein B (apo B). Recall from chapter 7 that apolipoproteins transport fats. In the small intestine, the mRNA that encodes apo B is short, and the protein binds and carries dietary fat. In the liver, however, the mRNA is not shortened, and the longer protein transports fats manufactured in the liver, which do not come from food.

Once introns have been spliced out, enzymes check the remaining RNA molecule for accuracy, much as enzymes proofread newly replicated DNA. Messenger RNAs that are too short or too long may be stopped from exiting the nucleus. A proofreading mechanism also monitors tRNAs, ensuring that the correct conformation takes shape.

## KEY CONCEPTS

RNA differs from DNA in that it is single-stranded, contains uracil instead of thymine and ribose instead of deoxyribose, and has different functions. Messenger RNA transmits information to build proteins. Each three mRNA bases in a row forms a codon that specifies a particular amino acid. Ribosomal RNA and proteins form ribosomes, which physically support protein synthesis and help catalyze bonding between amino acids. Transfer RNAs connect particular mRNA codons to particular amino acids.
• Bacterial operons are simple gene control systems. In more complex organisms, transcription factors control gene expression. • Transcription proceeds as RNA polymerase inserts complementary RNA bases opposite the template strand of DNA. • Messenger RNA gains a modified nucleotide cap and a poly A tail. Introns are transcribed and cut out, and exons are reattached. Introns are common, numerous, and large in human genes. Certain genes are transcribed into different-sized RNAs in different cell types.

## 10.2 Translating a Protein

Transcription copies the information encoded in a DNA base sequence into the complementary language of RNA. The next step is translating mRNA into the specified sequence of amino acids. Particular mRNA codons (three bases in a row) correspond to particular amino acids (figure 10.9). This correspondence between the chemical languages of mRNA and protein is the genetic code.

Francis Crick hypothesized that an "adaptor" would enable the RNA message to attract and link amino acids into proteins. He wrote in an unpublished paper in early 1955, "In its simplest form there would be 20 different kinds of adaptor molecule, one for each amino acid, and 20 different enzymes to join the amino acids to their adaptors." He was describing tRNAs, but the solution was more complex than originally thought. In the 1960s, many researchers deciphered the genetic code, determining which mRNA codons correspond to which amino acids. Marshall Nirenberg led the effort, for which he won the Nobel Prize.

The news media, on announcing the sequencing of the human genome in June 2000, widely reported that the "human genetic code" had been cracked. This was not the case. The genetic code is not unique to humans, and it was cracked decades ago. The code is the correspondence between nucleic acid triplet and amino acid, not the sequence itself.

## Deciphering the Genetic Code

The researchers who deciphered the genetic code used a combination of logic and experiments. More recently, annotation of the human genome sequence has confirmed and extended the earlier work, revealing new nuances in genetic coding. Figure 10.9 summarizes the relationship of DNA, RNA, and protein. It is helpful in understanding how the genetic code works to ask the questions that researchers asked in the 1960s.

### *Question 1—How Many RNA Bases Specify One Amino Acid?*

Because the number of different protein building blocks (20) exceeds the number of different mRNA building blocks (4), each codon must contain more than one mRNA base. In other words, if a codon consisted of only one mRNA base, then codons could specify only four different amino acids, with one corresponding to each of the four bases: A, C, G, and U (figure 10.10). If each codon

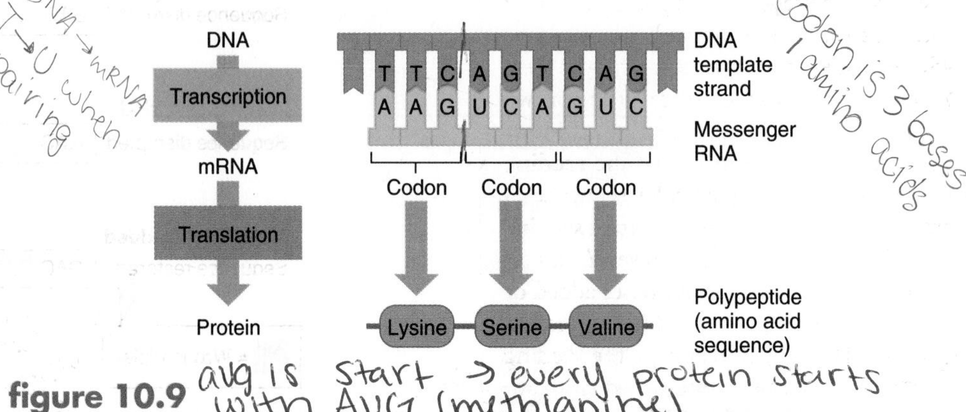


**figure 10.9**

**From DNA to RNA to protein.** Messenger RNA is transcribed from a locally unwound portion of DNA. In translation, transfer RNA matches up mRNA codons with amino acids.

| Size of a genetic code word (codon)<br>To encode a protein alphabet of 20 amino acids: | Logic |
|---|---|
| mRNA genetic code of 1 letter<br>U<br>C<br>A<br>G | 4 combinations:<br>**not sufficient** |
| mRNA genetic code of 2 letters<br>UU CC AA GG UC CU UA<br>AU UG GU CA AC CG GC<br>AG GA | 16 combinations:<br>**not sufficient** |
| mRNA genetic code of 3 letters<br>UUU UUC UUA UUG UCU UCC UCA UCG<br>UAU UAC UAA UAG UGU UGC UGA UGG<br>CUU CUC CUA CUG CCU CCC CCA CCG<br>CAU CAC CAA CAG CGU CGC CGA CGG<br>AUU AUC AUA AUG ACU ACC ACA ACG<br>AAU AAC AAA AAG AGU AGC AGA AGG<br>GUU GUC GUA GUG GCU GCC GCA GCG<br>GAU GAC GAA GAG GGU GGC GGA GGG | 64 combinations:<br>**sufficient** |

**figure 10.10**

**Codon size.** An exercise in logic reveals the triplet nature of the genetic code.

consisted of two bases, then 16 ($4^2$) different amino acids could be specified, one corresponding to each of the 16 possible orders of two RNA bases. This still is inadequate to encode the 20 amino acids found in organisms. If a codon consisted of three bases, then the genetic code could specify as many as 64 ($4^3$) different amino acids. Because 20 different amino acids require at least 20 different codons, the minimum number of bases in a codon is three.

Francis Crick and his coworkers conducted experiments on a type of virus called T4 that confirmed the triplet nature of the genetic code. They exposed the viruses to chemicals that add or remove one, two, or three bases, and examined a viral gene whose sequence and protein product were known. Altering the sequence by one or two bases produced a different amino acid sequence. The change disrupted the **reading frame,** which is the particular sequence of amino acids encoded from a certain starting point in a DNA sequence. However, adding or deleting three contiguous bases added or deleted only one amino acid in the protein product. This did not disrupt the reading frame. The rest of the amino acid sequence was retained. The code, the researchers deduced, is triplet (figure 10.11). To confirm the triplet nature of the genetic code, other experiments showed that if a base is added at one point in the gene, and a base deleted at another point, then the reading frame is disrupted only between these sites, resulting in a protein with a stretch of the wrong amino acids. In yet other experiments, if one base was inserted but two removed, the reading frame never returned to the specified normal amino acid sequence.

## *Question 2—Does a DNA Sequence Contain Information in an Overlapping Manner?*

Consider the hypothetical mRNA sequence AUCAGUCUA. If the genetic code is triplet and a DNA sequence accessed in a nonoverlapping manner (that is, each three bases in a row form a codon, but any one base is part of only one codon), then this sequence contains only three codons: AUC, AGU, and CUA. If the DNA sequence is overlapping, the sequence contains seven codons: AUC, UCA, CAG, AGU, GUC, UCU, and CUA.

An overlapping DNA sequence seems economical in that it packs maximal information into a limited number of bases (figure 10.12). However, an overlapping sequence constrains protein structure because certain amino acids must always be followed by certain others. For example, the amino acid the first codon specifies, AUC, would always be followed by an amino acid whose codon begins with UC. Experiments that sequence proteins show that no specific type of amino acid always follows another. Any amino acid can follow any other amino acid in a protein's sequence. Therefore, the protein-encoding DNA sequence is not overlapping. There are a few exceptions,

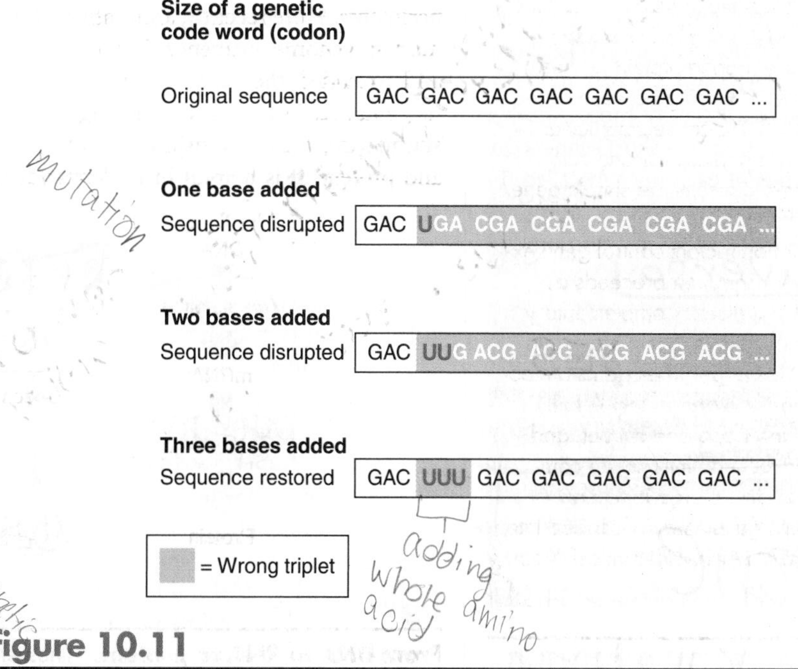


**figure 10.11**

**Three at a time.** Adding or deleting one or two nucleotides to a DNA sequence disrupts the encoded amino acid sequence. However, adding or deleting three bases does not disrupt the reading frame. Therefore, the code is triplet. This is a simplified representation of the Crick experiment.

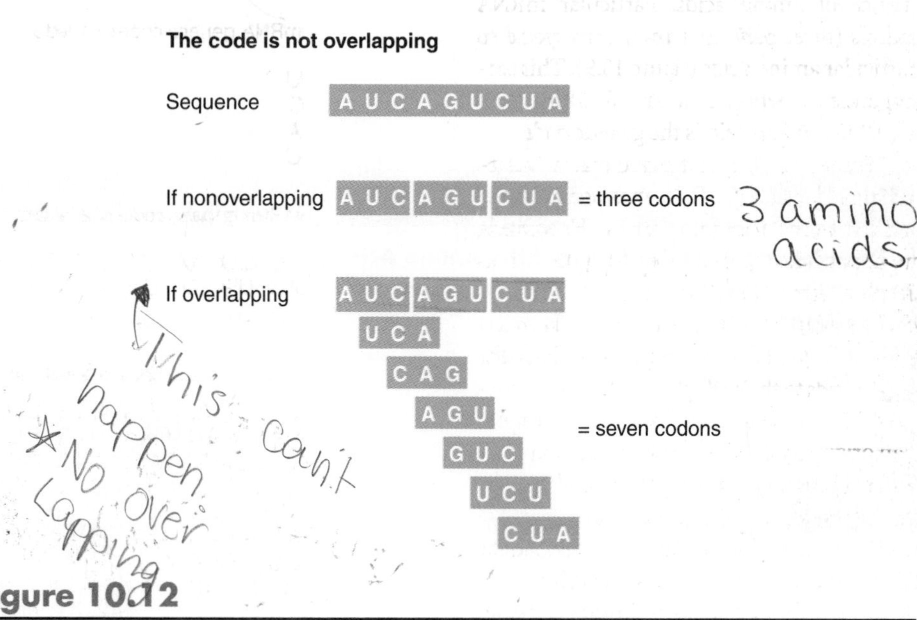


**figure 10.12**

**The genetic code does not overlap.** An overlapping genetic code may seem economical, but it is restrictive, dictating that certain amino acids must follow others in a protein's sequence. This does not happen; therefore, the genetic code is nonoverlapping.

particularly in viruses, where the same DNA sequence can be read from different starting points.

### Question 3—Can mRNA Codons Signal Anything Other Than Amino Acids?

Chemical analysis eventually showed that the genetic code contains directions for starting and stopping translation. The codon AUG signals "start," and the codons UGA, UAA, and UAG each signify "stop." Another form of "punctuation" is a short sequence of bases at the start of each mRNA, called the leader sequence, that enables the mRNA to hydrogen bond with rRNA in a ribosome.

### Question 4—Do All Species Use the Same Genetic Code?

All species use the same mRNA codons to specify the same amino acids, despite the popular idea of a "human" genetic code. This universality of the genetic code is part of the abundant evidence that all life on earth evolved from a common ancestor. No other mechanism as efficient at directing cellular activities has emerged and persisted. The only known exceptions to the universality of the genetic code are a few codons in the mitochondria of certain single-celled organisms. The ability of mRNA from one species to be translated in a cell of another species has made recombinant DNA technology possible, in which bacteria manufacture proteins normally made in the human body. Chapter 18 explains how such proteins are used as drugs.

### Question 5—Which Codons Specify Which Amino Acids?

In 1961, Marshall Nirenberg and his coworkers at the National Institute of Health began deciphering which codons specify which amino acids, using a precise and logical series of experiments. First they synthesized mRNA molecules in the laboratory. Then they added them to test tubes that contained all the chemicals and structures needed for translation, which they had extracted from *E. coli* cells. Which amino acid would each synthetic RNA specify?

The first synthetic mRNA tested had the sequence UUUUUU. . . . In the test tube, this was translated into a peptide consisting entirely of one amino acid type: phenylalanine. Thus was revealed the first entry in the genetic code dictionary: the codon UUU specifies the amino acid phenylalanine. The number of phenylalanines always equaled one-third the number of mRNA bases, confirming that the genetic code is triplet and nonoverlapping. The next three experiments revealed that AAA codes for the amino acid lysine, GGG for glycine, and CCC for proline.

Next, the researchers synthesized chains of alternating bases. Synthetic mRNA of the sequence AUAUAU . . . introduced codons AUA and UAU. When translated, the mRNA yielded an amino acid sequence of alternating isoleucines and tyrosines. But was AUA the code for isoleucine and UAU for tyrosine, or vice versa? Another experiment answered the question.

An mRNA of sequence UUUAUAUUUAUA encoded alternating phenylalanine and isoleucine. Because the first experiment showed that UUU codes for phenylalanine, the researchers deduced that AUA must code for isoleucine. If AUA codes for isoleucine they reasoned, looking back at the previous experiment, then UAU must code for tyrosine. Table 10.3 summarizes some of these experiments.

By the end of the 1960s, researchers had deciphered the entire genetic code (table 10.4). Sixty of the possible 64 codons were found to specify particular amino acids, while the others indicate "stop" or "start." This means that some amino acids are specified by more than one codon. For example, both UUU and UUC encode phenylalanine. Different codons that specify the same amino acid are called synonymous codons, just as synonyms are words with the same meaning. The genetic code is said to be **degenerate** because each amino acid is not uniquely specified. Synonymous codons often differ from one another by the base in the third position. The corresponding base of a tRNA's anticodon is called the "wobble" position because it can bind to more than one type of base in synonymous codons. The degeneracy of the genetic code provides protection against mutation, because changes in the DNA that cause the substitution of a synonymous codon would not affect the protein's amino acid sequence.

**table 10.3**

**Deciphering RNA Codons and the Amino Acids They Specify**

| Synthetic RNA | Encoded Amino Acid Chain | Puzzle Piece |
|---|---|---|
| UUUUUUUUUUUUUUUUUU | Phe-Phe-Phe-Phe-Phe-Phe | UUU = Phe |
| AAAAAAAAAAAAAAAAAA | Lys-Lys-Lys-Lys-Lys-Lys | AAA = Lys |
| GGGGGGGGGGGGGGGGGG | Gly-Gly-Gly-Gly-Gly-Gly | GGG = Gly |
| CCCCCCCCCCCCCCCCCC | Pro-Pro-Pro-Pro-Pro-Pro | CCC = Pro |
| AUAUAUAUAUAUAUAUAU | Ile-Tyr-Ile-Tyr-Ile-Tyr | AUA = Ile or Tyr |
| | | UAU = Ile or Tyr |
| UUUAUAUUUAUAUUUAUA | Phe-Ile-Phe-Ile-Phe-Ile | AUA = Ile |
| | | UAU = Tyr |

**table 10.4**

**The Genetic Code**

Second Letter

| First Letter | U | | C | | A | | G | | Third Letter |
|---|---|---|---|---|---|---|---|---|---|
| U | UUU | Phenylalanine (Phe) | UCU | | UAU | Tyrosine (Tyr) | UGU | Cysteine (Cys) | U |
| | UUC | | UCC | Serine (Ser) | UAC | | UGC | | C |
| | UUA | Leucine (Leu) | UCA | | UAA | "stop" | UGA | "stop" | A |
| | UUG | | UCG | | UAG | "stop" | UGG | Tryptophan (Trp) | G |
| C | CUU | | CCU | | CAU | Histidine (His) | CGU | | U |
| | CUC | Leucine (Leu) | CCC | Proline (Pro) | CAC | | CGC | Arginine (Arg) | C |
| | CUA | | CCA | | CAA | Glutamine (Gln) | CGA | | A |
| | CUG | | CCG | | CAG | | CGG | | G |
| A | AUU | | ACU | | AAU | Asparagine (Asn) | AGU | Serine (Ser) | U |
| | AUC | Isoleucine (Ile) | ACC | Threonine (Thr) | AAC | | AGC | | C |
| | AUA | | ACA | | AAA | Lysine (Lys) | AGA | Arginine (Arg) | A |
| | AUG | Methionine (Met) and "start" | ACG | | AAG | | AGG | | G |
| G | GUU | | GCU | | GAU | Aspartic acid (Asp) | GGU | | U |
| | GUC | Valine (Val) | GCC | Alanine (Ala) | GAC | | GGC | Glycine (Gly) | C |
| | GUA | | GCA | | GAA | Glutamic acid (Glu) | GGA | | A |
| | GUG | | GCG | | GAG | | GGG | | G |

The human genome project picked up where the genetic code experiments of the 1960s left off by identifying the DNA sequences that are transcribed into tRNAs. That is, 61 different tRNAs could theoretically exist, one for each codon that specifies an amino acid (the 64 triplets minus 3 stop codons). However, only 49 different genes were found to encode tRNAs. This is because the same type of tRNA can detect synonymous codons that differ only in whether the wobble (third) position is U or C. The same type of tRNA, for example, binds to both UUU and UUC codons, which specify the amino acid phenylalanine. Synonymous codons ending in A or G use different tRNAs.

The monumental task of deciphering the genetic code was, in an intellectual sense, even more important than today's sequencing of genomes, for it revealed the "rules" that essentially govern life at the cellular level. As with genome projects, many research groups contributed to the effort of solving the genetic code problem. Because genetics was still a very young science, the code breakers came largely from the ranks of chemistry, physics, and math.

Some of the more exuberant personalities organized an "RNA tie club" and inducted a new member whenever someone added a new piece to the puzzle of the genetic code, anointing him (there were no prominent hers) with a tie and tie pin emblazoned with the structure of the specified amino acid (figure 10.13). By the end of the 1960s, researchers had deciphered the entire genetic code.

## Building a Protein

Protein synthesis requires mRNA, tRNA molecules carrying amino acids, ribosomes, energy-storing molecules such as adenosine triphosphate (ATP), and various protein factors. These pieces come together at the

**figure 10.13**

**The RNA tie club.** In 1953, physicist-turned-biologist George Gamow started the RNA tie club, to "solve the riddle of RNA structure and to understand the way it builds proteins." The club would have 20 members, one for each amino acid. Each honored member received a tie and tie pin labeled with the name of the particular amino acid he had worked on. Francis Crick (upper left) was tyrosine; James Watson (lower right) was proline.

beginning of translation in a stage called **translation initiation** (figure 10.14).

First, the mRNA leader sequence hydrogen bonds with a short sequence of rRNA in a small ribosomal subunit. The first mRNA codon to specify an amino acid is always AUG, which attracts an initiator tRNA that carries the amino acid methionine (abbreviated *met*). This methionine signifies the start of a polypeptide. The small ribosomal subunit, the mRNA bonded to it, and the initiator tRNA with its attached methionine form the **initiation complex.**

To start the next stage of translation, called **elongation,** a large ribosomal subunit attaches to the initiation complex. The codon adjacent to the initiation codon (AUG), which is GGA in figure 10.15*a*, then bonds to its complementary anticodon, which is part of a free tRNA that carries the amino acid glycine. The two amino acids (*met* and *gly* in the example), which are still attached to their tRNAs, align.

The part of the ribosome that holds the mRNA and tRNAs together can be described as having two sites. The positions of the sites on the ribosome remain the same with respect to each other as translation proceeds, but they cover different parts of the mRNA as the ribosome moves. The **P site** holds the growing amino acid chain, and the **A site** right next to it holds the next amino acid to be added to the chain. In figure 10.15, when the protein-to-be consists of only the first two amino acids, *met* occupies the P site and *gly* the A site.

With the help of rRNA that functions as a ribozyme, the amino acids link by an attachment called a peptide bond. Then the first tRNA is released. It will pick up another amino acid and be used again. The ribosome and its attached mRNA are now bound to a single tRNA, with two amino acids extending from it at the P site. This is the start of a polypeptide.

Next, the ribosome moves down the mRNA by one codon. The region of the mRNA that was at the A site is thus now at the P site. A third tRNA enters, carrying its amino acid (*cys* in figure 10.15*b*). This third amino acid aligns with the other two and forms a peptide bond to the second amino acid in the growing chain, now extending from the P site. The tRNA attached to the second amino acid is released and recycled. The polypeptide continues to build, one amino acid at a time. Each piece is brought in by a tRNA whose anticodon corresponds to a consecutive mRNA codon as the ribosome moves down the mRNA (figure 10.15*c*).

Elongation halts when one of the mRNA "stop" codons (UGA, UAG, or UAA) is reached, because no tRNA molecules correspond to these codons. The last tRNA leaves the ribosome, the ribosomal subunits separate from each other and are recycled, and the new polypeptide is released.

Protein synthesis is economical. A cell can produce large amounts of a particular protein from just one or two copies of a gene. A plasma cell in the immune system, for

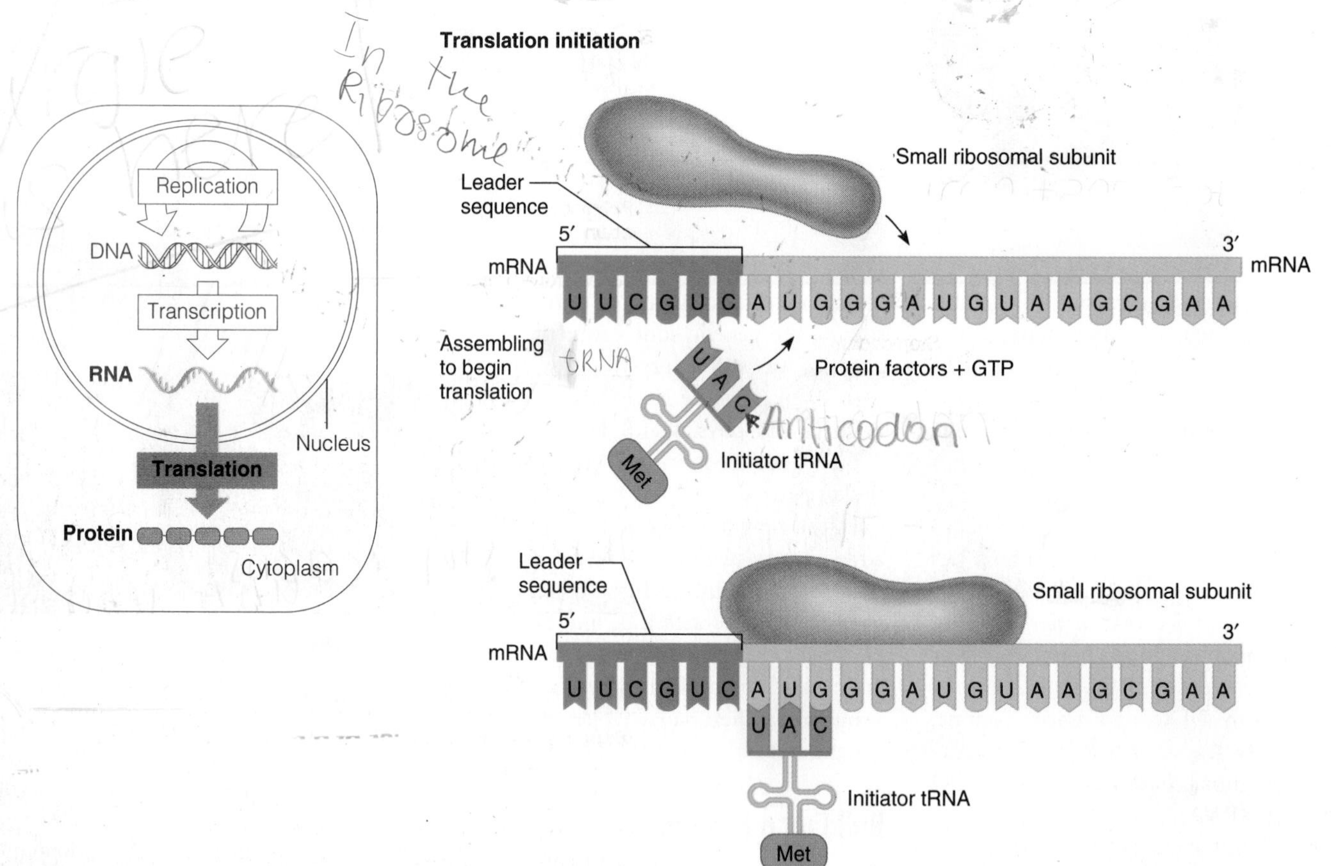


**figure 10.14**

**Translation begins.** Initiation of translation brings together a small ribosomal subunit, mRNA, and an initiator tRNA, and aligns them in the proper orientation to begin translation.

**TRANSLATION ELONGATION**

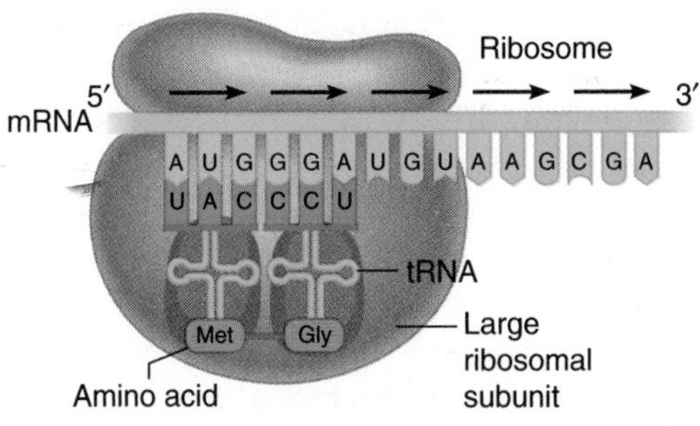


a.

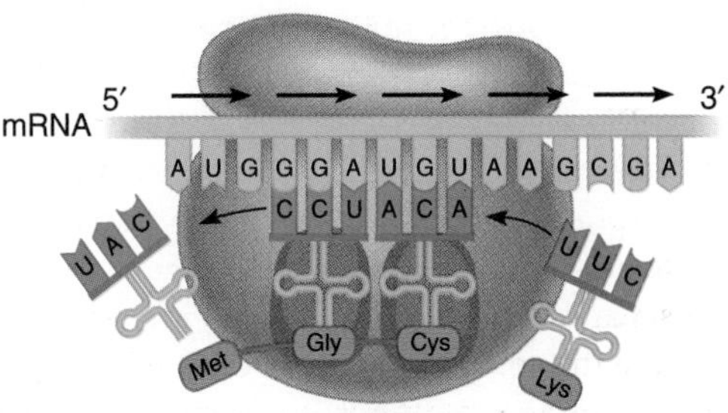


b.

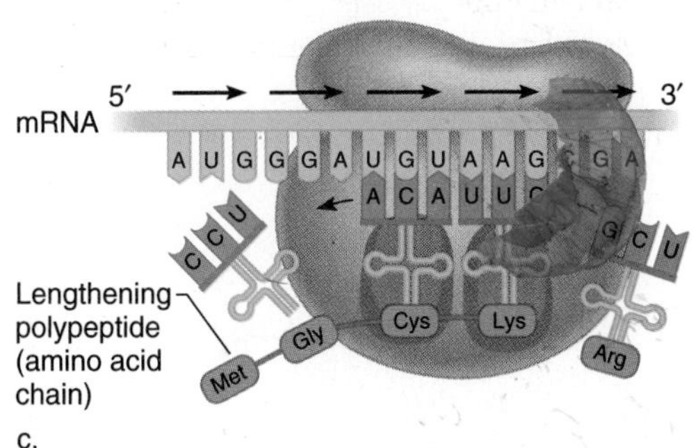


c.

**TRANSLATION TERMINATION**

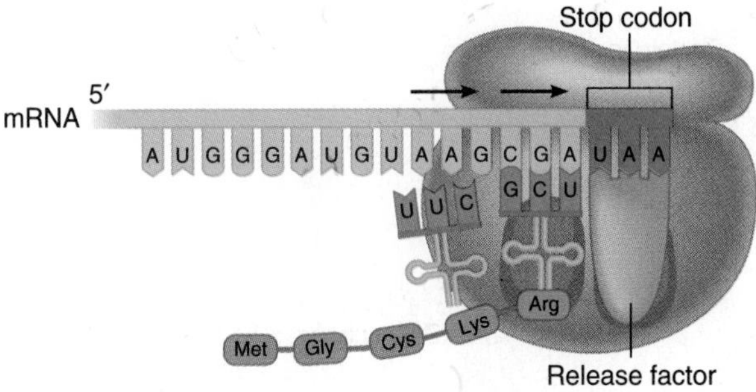


d. Ribosome reaches stop codon

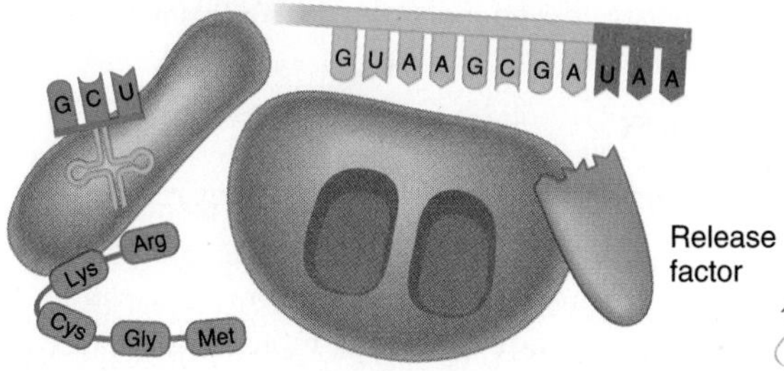


e. Once stop codon is reached, elements disassemble

## figure 10.15

**Building a polypeptide.** A large ribosomal subunit binds to the initiation complex, and a tRNA bearing a second amino acid (glycine, in this example) forms hydrogen bonds between its anticodon and the mRNA's second codon (*a*). The methionine brought in by the first tRNA forms a peptide bond with the amino acid brought in by the second tRNA, and a third tRNA arrives, in this example carrying the amino acid cysteine (*b*). A fourth amino acid is linked to the growing polypeptide chain (*c*), and the process continues until a termination codon is reached. (*d*) A protein release factor binds to the stop codon, releasing the completed protein from the tRNA and (*e*) freeing all of the components of the translation machine.

example, manufactures 2,000 identical antibody molecules per second. To mass-produce proteins on this scale, RNA, ribosomes, enzymes, and other proteins must be continually recycled. Transcription always produces multiple copies of a particular mRNA, and each mRNA may be bound to dozens of ribosomes, as figure 10.16 shows. As soon as one ribosome has moved far enough along the mRNA, another ribosome will attach. In this way, many copies of the encoded protein will be made from the same mRNA.

## Protein Folding

As a protein is synthesized, it folds into a three-dimensional shape (conformation) that helps determine its function. This folding occurs because of attractions and repulsions between the protein's atoms. In addition,

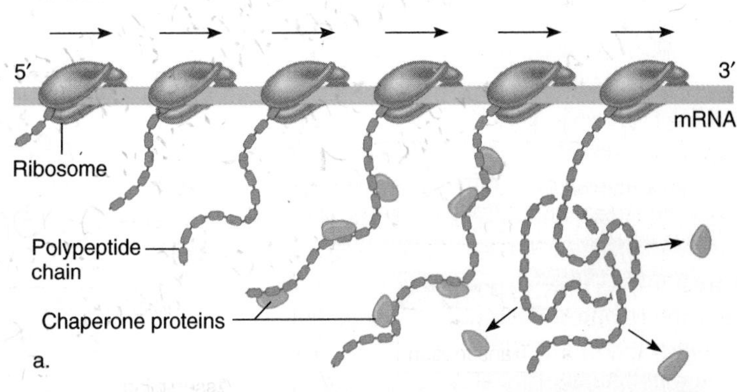


a.

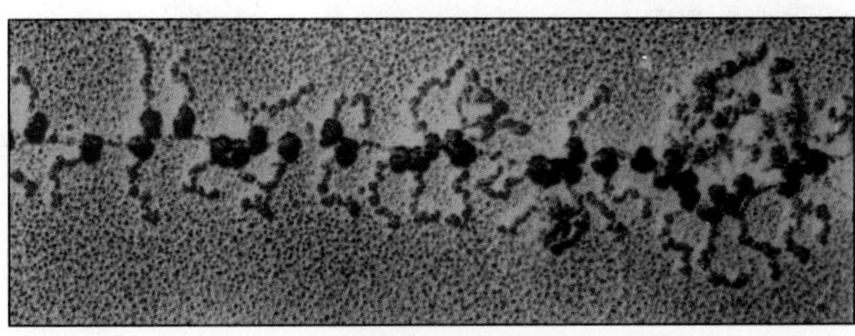

b.

## figure 10.16

**Making multiple copies of a protein.** Several ribosomes can translate the same protein from a single mRNA at the same time. (*a*) The ribosomes have different-sized polypeptides dangling from them—the closer a ribosome is to the end of a gene, the longer its polypeptide. Chaperone proteins help fold the polypeptide into its characteristic conformation. (*b*) In the micrograph, the ribosomes on the left have just begun translation and the polypeptides are short. Further along in translation, the polypeptides are longer. The chaperones are not visible.

thousands of water molecules surround a growing chain of amino acids, and, because some amino acids are attracted to water and some repelled by it, the water contorts the protein's shape. Sulfur atoms also affect overall conformation by bridging the two types of amino acids that contain them.

The conformation of a protein may be described at several levels. Figure 10.17 shows the four levels for hemoglobin, which carries oxygen in the blood. The amino acid sequence of a polypeptide chain determines its **primary (1°) structure.** Chemical attractions between amino acids that are close together in the 1° structure fold the polypeptide chain into its **secondary (2°) structure,** which may take the distinctive forms of loops, coils, barrels, helices, or sheets. Secondary structures wind into larger **tertiary (3°) structures** as more widely separated amino acids attract or repel in response to water molecules. Finally, proteins consisting of more than one polypeptide form a **quaternary (4°) structure.** Hemoglobin has four polypeptide chains. The liver protein ferritin has 20 identical polypeptides of 200 amino acids each. In contrast, the muscle protein myoglobin is a single polypeptide chain.

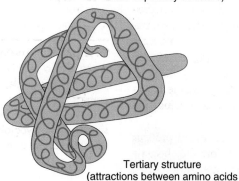

For many years, biochemists thought that protein folding was straightforward; the amino acid sequence dictated specific attractions and repulsions between parts of a protein, contorting it into its final form as it emerged from the ribosome. But these attractions and repulsions are not sufficient to fold the polypeptide into the highly specific form essential to its function. A protein apparently needs help to fold correctly.

An amino acid chain may start to fold as it emerges from the ribosome. Localized regions of shape form, and possibly break apart and form again, as translation proceeds. Experiments that isolate proteins as they are synthesized show that other proteins oversee the process of proper folding. These accessory proteins include enzymes that foster chemical bonds and chaperone proteins, which stabilize partially folded regions that are important to the molecule's final form.

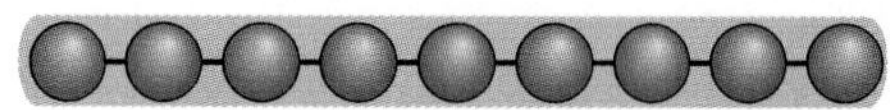


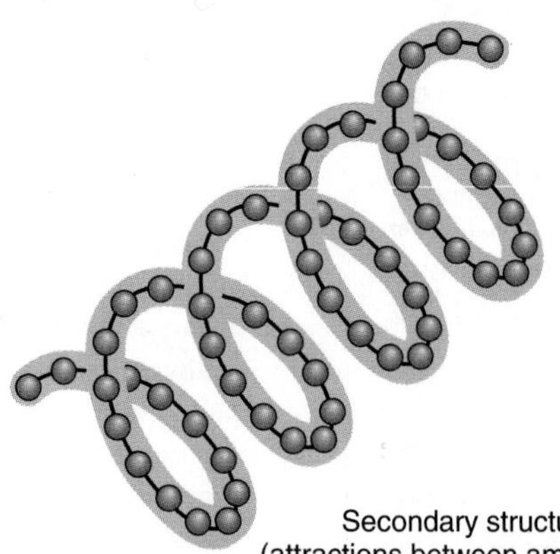

**figure 10.17**

**Protein conformation.** The hemoglobin molecule consists of four polypeptides, called globins. Each globin surrounds a smaller organic compound that holds an iron atom. Chapter 11 revisits the hemoglobin molecule.

Just as repair enzymes check newly replicated DNA for errors and RNAs are proofread, proteins scrutinize a folding protein to detect and dismantle incorrectly folded regions. Errors in protein folding can cause illness. Some mutations that cause cystic fibrosis, for example, prevent the encoded protein from assuming its final form and anchoring in the cell membrane, where it normally controls the flow of chloride ions. One type of Alzheimer disease is associated with a protein called amyloid that forms an abnormal, gummy mass instead of remaining as distinct molecules, because of improper folding.

Some members of a class of inherited disorders called triplet repeats tack extra glutamines onto particular proteins. The extra amino acids alter the ability of the protein to fold into its characteristic conformation. Most of the triplet repeat disorders—so-called because extra DNA triplets encode the extra amino acids—affect the brain. They are discussed further in chapter 11.

Yet another type of disorder that arises from an abnormal protein conformation is the spongiform encephalopathies, such as "mad cow disease" and similar conditions in sheep, humans, and several other types of mammals. Recall from chapter 2 that these disorders are caused by abnormal aggregation of proteins called prions. Unlike other proteins that misfold to cause disease, the normal and abnormal forms of prion protein have the same primary structure, but they are capable of folding into at least eight three-dimensional shapes (figure 10.18). Just as researchers assumed that genes were continuous, so too did they assume incorrectly that a protein can fold into just one conformation.

In addition to folding, certain proteins must be altered further before they become functional. Sometimes enzymes must shorten a polypeptide chain for it to become active. Insulin, which is 51 amino acids long, for example, is initially translated as the polypeptide proinsulin, which is 80 amino acids long. Some proteins must have sugars attached for them to become functional.

The linguistic nature of the flow of genetic information makes it ideal for computer analysis. The view of DNA sequences as a language emerged in the 1960s, as experiments revealed the linear relationship between nucleic acid sequences in genes and amino acid sequences in proteins. Yet the "rules" by which DNA sequences specify protein shapes are still not well understood, even as we routinely decipher the sequences of entire genomes.

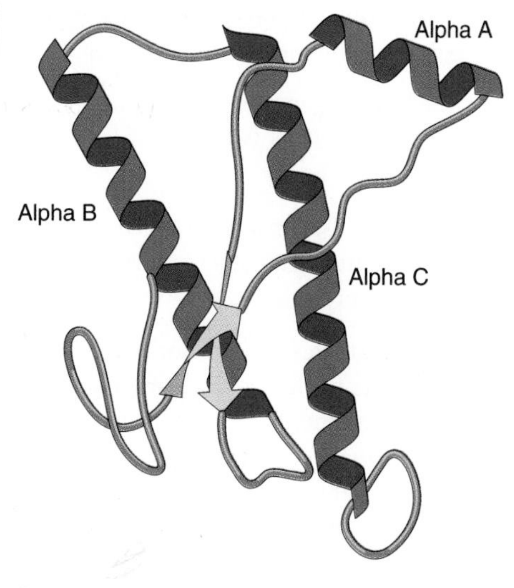

a. Cellular prion protein (noninfectious)

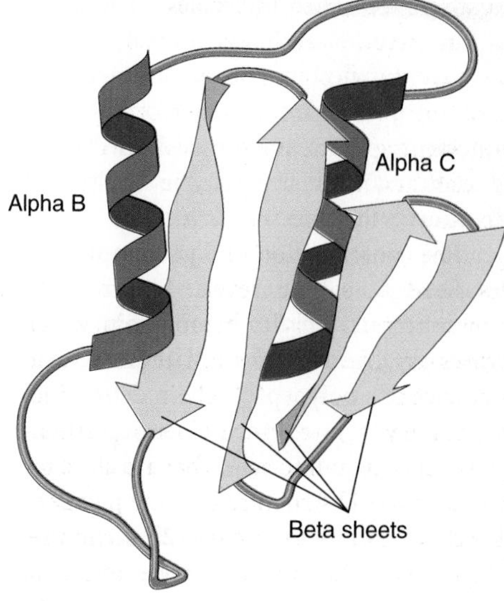

b. Scrapie prion protein (infectious)

**figure 10.18**

**One protein, multiple conformations.** Biochemists once thought that the primary structure of a protein dictated one conformation. This is not true. The cellular form of prion protein, for example, does not cause disease (*a*). The scrapie form is infectious—it converts the cellular form to more of itself (*b*). Infectious prions cause scrapie in sheep, bovine spongiform encephalopathy in cows, and variant Creutzfeldt-Jakob disease in humans. Recent work showing that myoglobin also assumes different forms suggests that there is much that we do not know about protein conformation.

### Key Concepts

The genetic code is triplet, nonoverlapping, continuous, universal, and degenerate. • As translation begins, mRNA, tRNA with bound amino acids, ribosomes, energy molecules, and protein factors assemble. The mRNA binds to rRNA in the small subunit of a ribosome, and the first codon attracts a tRNA bearing methionine. Next, as the chain elongates, the large ribosomal subunit attaches and the appropriate anticodon parts of tRNAs bind to successive codons in the mRNA. As the amino acids attached to the aligned tRNA molecules form peptide bonds, a polypeptide grows. The ribosome moves down the mRNA to the region that holds the amino acid chain, called the P site, and the area where a new tRNA binds, called the A site. When the ribosome reaches a "stop" codon, protein synthesis ceases. RNA, ribosomes, enzymes, and key proteins are recycled. • Protein folding begins as translation proceeds, with enzymes and chaperone proteins assisting. A protein can fold in more than one way.

## 10.3 The Human Genome Sequence Reveals Unexpected Complexity

For nearly half a century after Watson and Crick deduced the structure of DNA, a view of the genome as a set number of genes that specify an equal number of proteins ruled. Even the finding that many genes are split into coding (exon) and noncoding (intron) regions did little, at first, to shake the one gene–one protein way of thinking. All that has changed with the sequencing of the human genome.

Until intense genome sequencing efforts began in the 1990s, most researchers focused on mapping, identifying, and discovering the functions of individual genes. With the genome sequence in hand, researchers can now estimate the number of protein-encoding genes and categorize proteins by function. **Proteomics** is the study of the entire collection of proteins produced in a particular cell. Charts such as figure 10.19 are being compiled for all cell types, at all stages of development, and under different conditions, to catalog the functioning of the human body at the molecular and cellular levels, in health and disease. But proteomics is just the start of genome analysis. Learning the protein-producing capabilities of cells seems simple compared to other information revealed in the genome project. Consider these facts:

- The human genome consists of approximately 3.2 billion DNA base pairs.
- Only about 1.5 percent of the human genome sequence encodes protein.
- The human genome includes approximately 31,000 protein-encoding genes.
- Human cells, considered together, can produce from 100,000 to 200,000 different proteins.

These statistics just do not add up in a one gene–one protein paradigm. Two central questions emerge, and they are the focus of this final section of the chapter: (1) How can 31,000 genes encode 100,000 to 200,000 proteins? Even if the number of protein-encoding genes exceeds 31,000, they are still fewer than the number of proteins. (2) What does the other 98.5 percent of the human genome—the part that doesn't encode protein—do?

## Genome Economy: Reconciling Gene and Protein Number

The discovery of introns in 1977 first planted the idea that a number of genes could specify a larger number of proteins by mixing and matching gene parts. Research since then has revealed that about a third of the protein-encoding portion of the

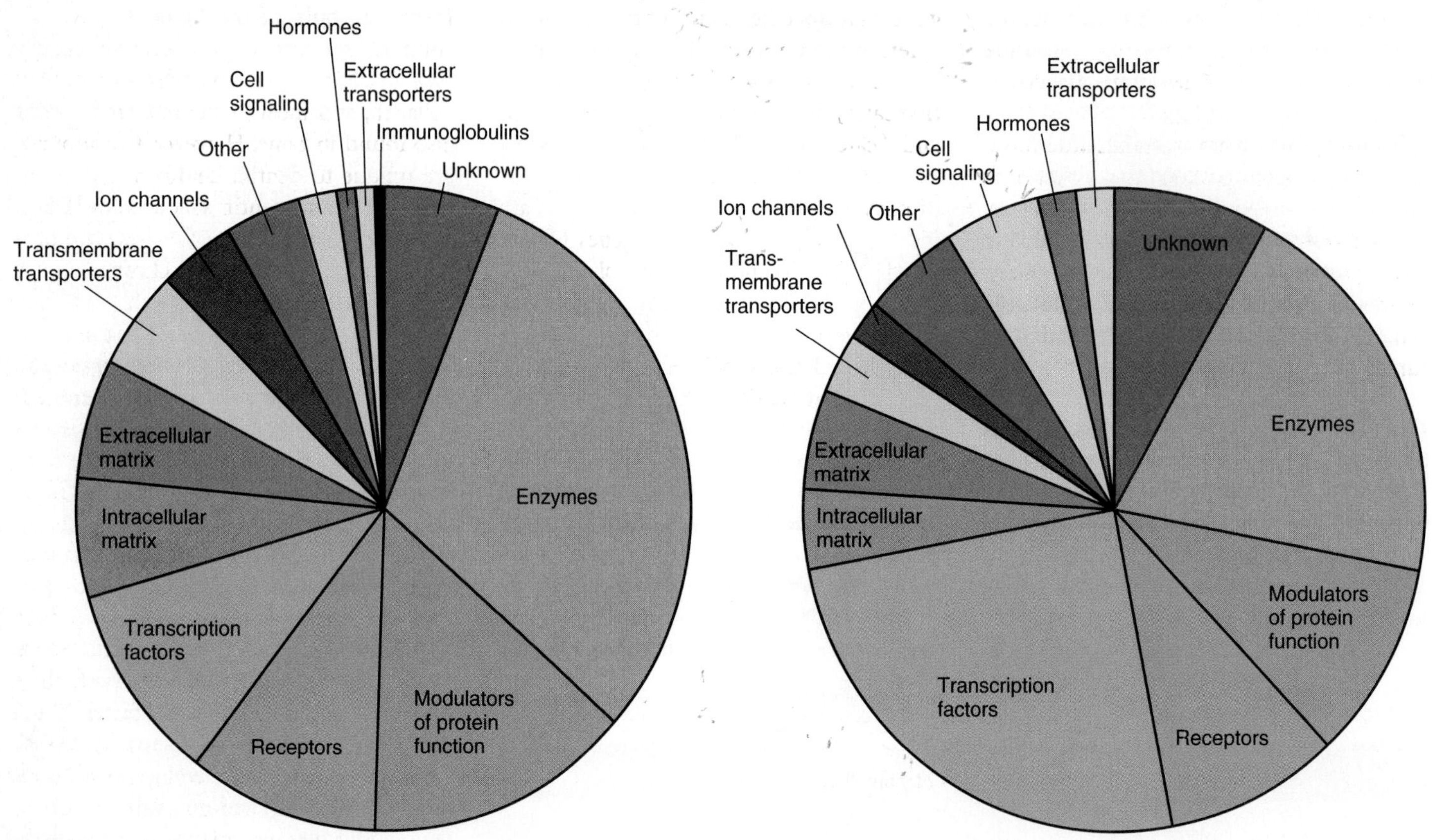


a. Distribution of health-related proteins from conception through old age.

b. Distribution of health-related proteins from conception to birth.

**figure 10.19**

**Proteomics meets medicine.** One way to analyze the effects of genes is to categorize them by the functions of their protein products, and then to chart the relative abundance of each class at different stages of development or life, and in sickness and in health. The pie chart in (*a*) considers 14 categories of proteins that when abnormal or missing cause disease, and their relative abundance from conception through advanced age. The pie chart in (*b*) displays the same protein categories for the prenatal period, from conception to birth. Note, for example, that transcription factor genes are more highly expressed in the embryo and fetus, presumably because of the extensive cell differentiation that is a hallmark of this period. The relative expression of genes that encode enzymes is slightly less in the prenatal period than at other times because before birth, some metabolic needs are met by the pregnant woman. These depictions represent just one of the many new ways of looking at gene action.

genome—at least 10,000 genes—mix and match exons, each of which encodes a segment of a protein called a domain. For example, the DNA that encodes a blood-clotting protein called tissue plasminogen activator (t-PA) includes sequences from genes that encode three other proteins (plasminogen, epidermal growth factor, and fibronectin)—that's four proteins from three genes. This process of combining exons is called **exon shuffling.** Figure 10.20 illustrates schematically how two genes can give rise to seven proteins. Sequencing of extensive regions of chromosomes confirms the disparity between gene and protein number that researchers first discovered in considering the exon/intron structures of genes one at a time. For example, in a large part of chromosome 22, the first autosome to be sequenced, 245 genes are associated with 642 mRNA transcripts.

Introns may seem wasteful, little more than vast stretches of DNA bases that outnumber and outsize exons. But researchers are discovering that a DNA sequence that is an intron in one context may encode protein in another. Consider prostate specific antigen (PSA), a protein found on certain cell surfaces that is overproduced in some cases of prostate cancer (figure 10.21*a*). The gene for PSA has five exons and four introns, but it also encodes a second, different protein, called PSA-linked molecule (PSA-LM). Both genes have the same beginning DNA sequence, but the remainder of the PSA-LM gene is part of the fourth intron of the PSA sequence! The proteins seem to have antagonistic functions. That is, when the level of one is high, the other is low. Future blood tests to detect elevated risk of prostate cancer will likely consider levels of both proteins.

In another situation where introns may account for the overabundance of proteins compared to genes, a DNA sequence that is an intron in one gene's template strand may encode protein on the coding strand. That is, what is the template strand for one gene may be the coding strand for the other. This is the case for the gene that specifies neurofibromin, which when mutant causes neurofibromatosis. (This is an autosomal dominant condition that causes benign tumors beneath the skin and "café au lait" spots on the skin.) Encoded within an intron of the neurofibromin gene, but on the coding strand, are instructions for three other genes (figure 10.21*b*). It's even possible for a single RNA to be patched together from instructions on both strands, an apparently rare occurrence called trans-splicing. A gene in the fruit fly that controls chromosome structure in early development is transcribed from four exons on one strand and from two exons in the opposite orientation on the other strand. A small region of overlap allows the two mRNAs to complementary base pair. Then, a kind of cut-and-paste operation links all of the transcribed exons into a single mRNA molecule. Trans-splicing has not been identified in the human genome yet, but probably exists.

Still another way that a gene can maximize its informational content is for its encoded protein to be cut to yield two products. An inherited disorder called dentinogenesis imperfecta revealed this mechanism (figure 10.21*c*). The condition causes discolored, misshapen teeth with peeling enamel, due to abnormal dentin, which is the bone-like substance beneath the enamel that forms the bulk of the tooth. Dentin is a complex mixture of extracellular matrix proteins. Ninety percent of dentin protein is collagen, and most of the rest are proteins also found in bone. However, two proteins are unique to dentin: dentin phosphoprotein (DPP) and dentin sialoprotein (DSP). The single gene that encodes these two proteins is part of an area of chromosome 4 that seems to be devoted to teeth.

It was abnormal DPP that had been associated with dentinogenesis imperfecta. However, DPP, because it is abundant, accounting for 50 percent of the noncollagen protein in dentin compared to 6 percent for DSP, may have overshadowed a contributory role for DSP as well. Both proteins are translated from a single mRNA molecule as the precursor protein dentin sialophosphoprotein (DSPP), which is then cut to yield DPP and DSP. The DPP may be much more abundant because it is longer-lived than DSP. That is, DSP is degraded faster. These two proteins remain somewhat of a mystery. Often when genes with similar functions are right next to each other on a chromosome, as they are, it is because they arise from gene duplication. The two genes are then very similar in sequence. This is not the case with the DNA sequences that encode DPP and DSP.

Exon shuffling is probably fairly common—researchers had predicted its existence since shortly after the discovery of introns. Less well known and studied are genes in introns, trans-splicing, and two proteins encoded on one gene. Still, software can search for evidence of these ways to maximize genome information.

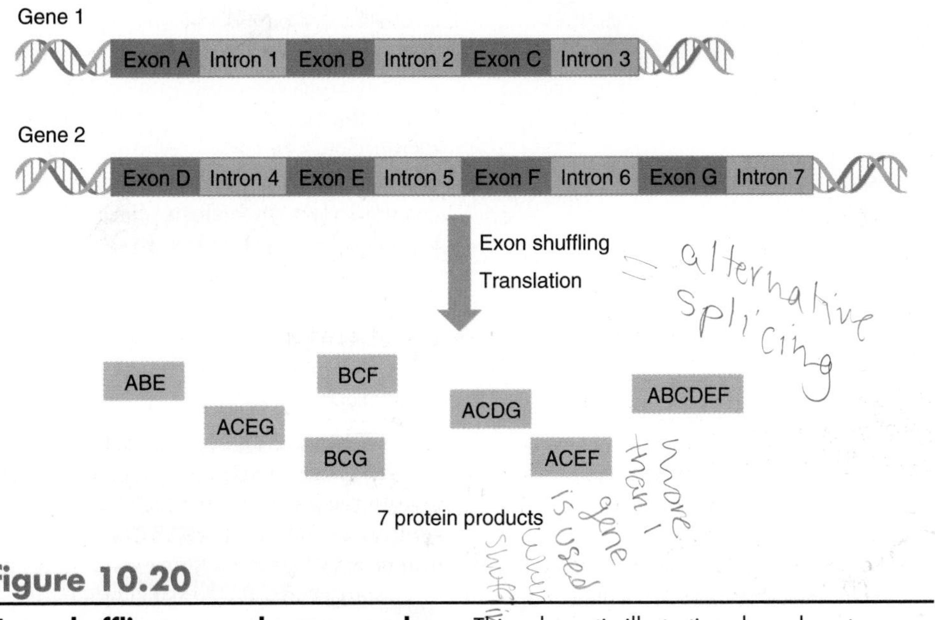


**figure 10.20**

**Exon shuffling expands gene number.** This schematic illustration shows how two genes can encode seven different proteins. Considering that many genes have many more introns than this one, it's clear that exon shuffling can generate many different proteins.

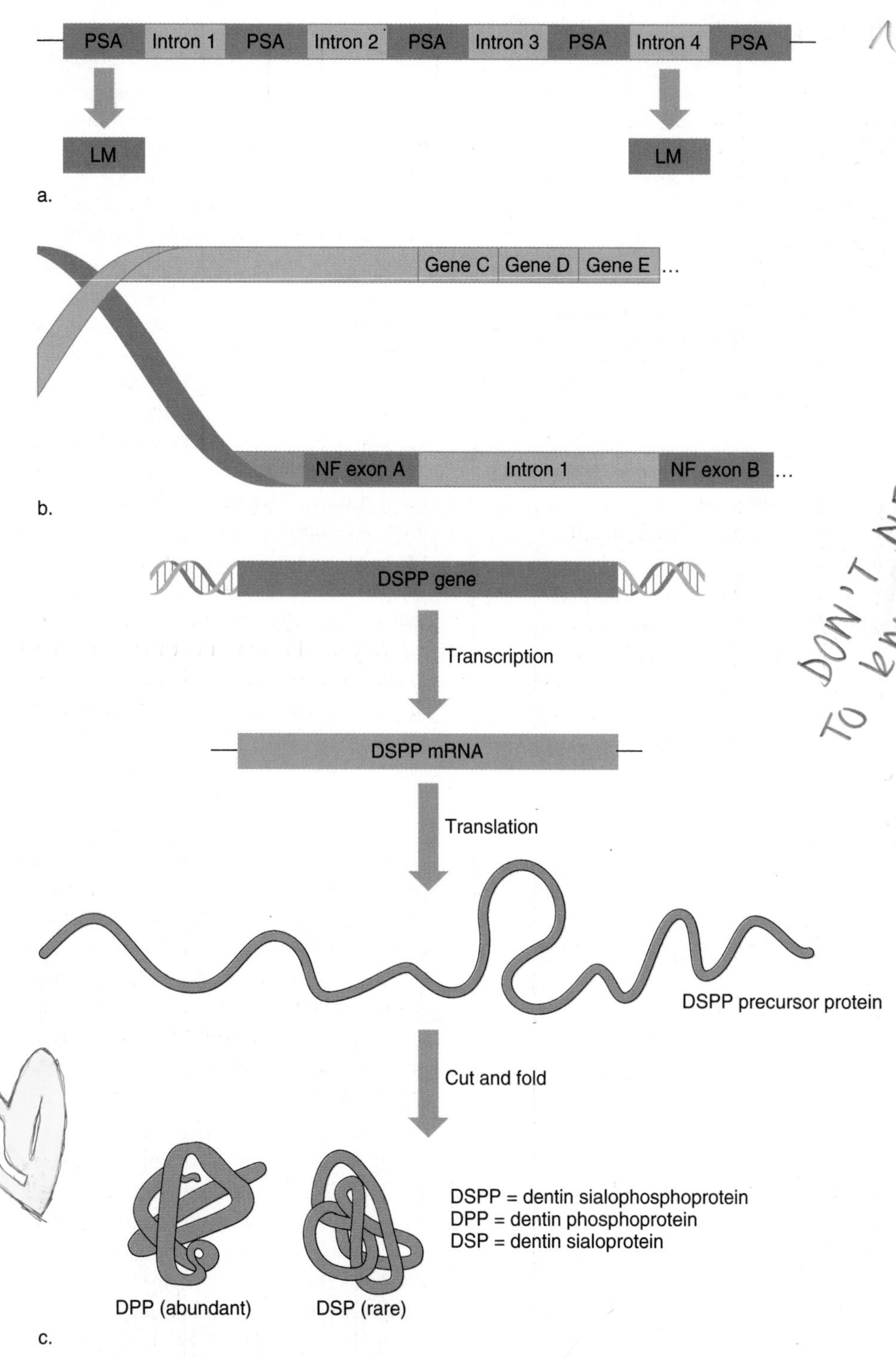


**figure 10.21**

**Genome economy occurs in several ways.** (*a*) Embedded in the PSA gene are really two protein-encoded sequences—the PSA portion consists of five exons. The PSA-LM part consists of two exons, one of which lies within an intron of PSA. (*b*) An intron of the neurofibromin gene harbors three genes on the opposite strand. (*c*) The dentin sialophosphoprotein (DSPP) gene encodes a long protein that is cleaved to yield dentin phosphoprotein (DPP) and dentin sialoprotein (DSP). The observation that DPP is more abundant than DSP indicates that their rate of degradation differs, because their rate of synthesis is presumably the same, since they come from the same transcript.

## What Does the Other 98.5 Percent of the Human Genome Do?

The second quandary revealed in the human genome sequence is the role of the "other" 98.5 percent of the DNA bases. In general, this noncoding DNA falls into four categories: (1) RNAs other than mRNA (called noncoding or ncRNAs), (2) introns, (3) promoters and other control sequences, and (4) repeated sequences. Table 10.5 summarizes some of the functions of DNA other than encoding protein.

### *Noncoding RNAs*

About a third of the human genome is transcribed into RNA types other than mRNA, the noncoding RNAs. The two best-studied ncRNAs are already familiar—tRNA and rRNA. The first draft sequence of the human genome identified 497 types of tRNA genes. The rate of transcription of a cell's tRNA genes seems to be attuned to the specific proteins that the cell manufactures—its proteome. This thrifty expression of genes is a little like the operons that enable bacteria to "sense" when enzymes are needed to dismantle certain nutrients. Human tRNA genes are dispersed among the chromosomes in clusters—25 percent of them are in a 4-million-base (4 megabase) region on chromosome 6. Altogether they account for 0.1 percent of the genome. Our 497 types of tRNA genes may seem like a lot, but frogs have thousands! This may reflect the fact that frog eggs are huge and contain may types of proteins. The 243 types of rRNA genes are clustered on six chromosomes, each cluster harboring 150 to 200 copies of a 44,000 base repeat sequence. Once transcribed from these clustered genes, the rRNAs go to the nucleolus, where they are cut to their final forms by yet another type of ncRNA called small nucleolar RNAs (snoRNAs).

Hundreds of thousands of ncRNAs are neither tRNA nor rRNA, nor snoRNAs, nor the other less-abundant types described in table 10.5. Instead, they are transcribed from DNA sequences called **pseudogenes.** A pseudogene is very similar in sequence to a particular protein-encoding gene, and it may be transcribed into RNA but it is not translated into protein. Pseudogenes may be

## table 10.5

**The Nonprotein Encoding Parts of the Human Genome**

| | Function or Characteristic |
|---|---|
| Noncoding RNA genes | |
| tRNA genes | Connect mRNA codon to amino acid |
| rRNA genes | Parts of ribosomes |
| Pseudogenes | DNA sequences that are very similar to known gene sequences and may be transcribed but are not translated |
| Small nucleolar RNAs | Process rRNA in nucleolus |
| Small nuclear RNAs | Parts of spliceosomes |
| Telomerase RNA | Part of ribonucleoprotein that adds bases to chromosome tips |
| Xist RNA | Inactivates one X chromosome in cells of females |
| Vault RNA | Part of "vault," a large ribonucleoprotein complex of unknown function |
| Introns | Parts of protein-encoding genes that are transcribed but cut out before the encoded protein is translated |
| Promoters and other control sequences | Guide enzymes that carry out DNA replication, transcription, or translation |
| Repeats | |
| Transposons | Repeats that move around the genome |
| Telomeres | Chromosome tips whose lengths control the cell cycle |
| Centromeres | Provide backdrop for proteins that form attachments for spindle fibers |
| Duplications of 10 to 300 kilobases | Unknown |
| Simple short repeats | Unknown |

remnants of genes past, once-functional variants that diverged from the normal sequence too greatly to encode a working protein. Pseudogenes are incredibly common in the human genome. For example, at least 324 pseudogenes shadow our 497 tRNA genes.

### *Repeats*

The human genome is riddled with highly repetitive sequences that appear to be gibberish, at least if we restrict the definition of genetic meaning to encoding protein. It is entirely possible that repeats represent a different type of genetic information, perhaps using a language in which meaning lies in a repeat size or number. Some types of repeats may serve a structural function of helping to hold a chromosome together.

The most abundant type of repeat is a sequence of DNA that can jump about the genome, called a transposable element, or **transposon** for short. Originally identified in corn by Barbara McClintock in the 1940s, and then in bacteria as the age of molecular biology dawned in the 1960s, transposons are now known to comprise about 45 percent of the human genome sequence. They are considered to be repeats because they are typically present in many copies. Some transposons include parts that encode enzymes that enable them to leave one chromosomal site and integrate into another. In a way, such a transposon is like a stripped-down virus or retrovirus.

Transposons are classified by size, whether they are transcribed into RNA, which enzymes they use to move, and whether they resemble bacterial transposons. For example, a class of transposons called long interspersed elements (LINEs) are 6,000 bases long and are transcribed and then trimmed to 900 bases before they reinsert into a chromosome. In contrast, short interspersed elements (SINEs) are 100 to 500 bases long and use enzymes that are encoded in LINEs. A major class of SINEs are called Alu repeats. Each Alu repeat is about 300 bases long, and a genome may contain 300,000 to 500,000 of them. Researchers still do not know what Alu repeats do, but they comprise 2 to 3 percent of the genome, and they have been increasing in number over time because they can copy themselves. Other rarer classes of repeats include those that comprise telomeres, centromeres, and rRNA gene clusters; duplications of 10,000 to 300,000 bases (10 to 300 kilobases); copies of pseudogenes; and simple repeats of one, two, or three bases. Many repeats arise from RNAs that are reverse transcribed into DNA and are then inserted into chromosomes.

Our understanding of the functions of repeats lags far behind our knowledge of the roles of the various noncoding RNA genes. Repeats may make sense in light of evolution, past and future. Pseudogenes are likely vestiges of genes that functioned in our nonhuman ancestors. Perhaps the repeats that seem to have no obvious function today can serve as raw material from which future genes may arise.

#### Key Concepts

Only a tiny proportion of the genome encodes protein, yet the number of proteins greatly outnumbers known protein-encoding genes. Ways that the genome is economical include exon shuffling, introns that encode protein, and single genes that encode a precursor protein that is cut in two. • The 98.5 percent of the human genome that does not specify protein encodes many types of RNA as well as introns, promoters, and other control sequences and repeats. We do not know the functions of some types of repeats.

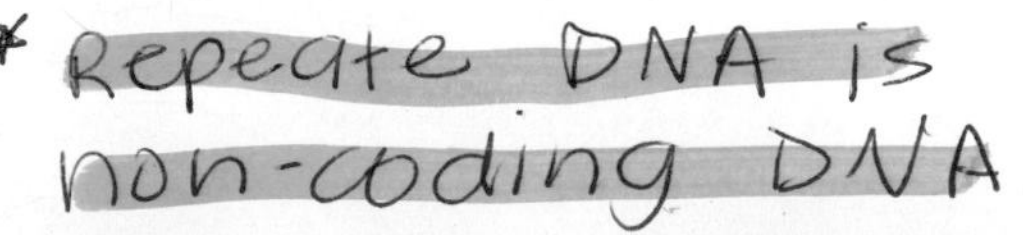
* Repeate DNA is
non-coding DNA

# Summary

### 10.1 Transcription—The Link Between Gene and Protein

1. The DNA sequence of a gene that encodes a protein is **transcribed** into RNA and **translated** into protein. The overall process is called **gene expression.** But much of the genome does not encode protein. The proteins produced in a particular cell type constitute its **proteome.**
2. RNA is transcribed from the **template strand** of DNA. The other strand is called the **coding strand.**
3. RNA is a single-stranded nucleic acid similar to DNA but containing uracil and ribose rather than thymine and deoxyribose.
4. Several types of RNA participate in protein synthesis (translation). **Messenger RNA** (mRNA) carries a protein-encoding gene's information. **Ribosomal RNA** (rRNA) associates with certain proteins to form ribosomes, which physically support protein synthesis. **Transfer RNA** (tRNA) is cloverleaf-shaped, with a three-base sequence called the **anticodon** that is complementary to mRNA on one end. It bonds to a particular amino acid at the other end.
5. **Transcription factors** regulate which genes or subsets of genes are transcribed in a particular cell type. Operons control gene expression in bacteria.
6. Transcription begins when transcription factors help **RNA polymerase** bind to a gene's starting region, or promoter. RNAP then adds RNA nucleotides to a growing chain, in a sequence complementary to the DNA template strand.
7. After a gene is transcribed, the mRNA receives a "cap" of modified nucleotides at one end and a poly-A tail at the other.
8. Many genes are in pieces. After transcription, segments called **exons** are translated into protein, but segments called **introns** are removed. Introns may outnumber and outsize exons. Researchers aren't certain what introns do, but they are probably not "junk."

### 10.2 Translating a Protein

9. Each three consecutive mRNA bases form a **codon** that specifies a particular amino acid. The **genetic code** is the correspondence between each codon and the amino acid it specifies. Of the 64 different possible codons, 61 specify amino acids and three signal stop. Because 61 codons specify the 20 amino acids, more than one type of codon may encode a single amino acid. The genetic code is nonoverlapping, triplet, universal, and degenerate.
10. In the 1960s, researchers used logic and clever experiments that used synthetic RNAs to decipher the genetic code. The sequencing of the human genome has refined that information.
11. Translation requires tRNA, ribosomes, energy-storage molecules, enzymes, and protein factors. An **initiation complex** forms when mRNA, a small ribosomal subunit, and a tRNA carrying methionine join. The amino acid chain begins to elongate when a large ribosomal subunit joins the small one. Next, a second tRNA binds by its **anticodon** to the next mRNA codon, and its amino acid bonds with the first amino acid. Transfer RNAs add amino acids, forming a polypeptide. The ribosome moves down the mRNA as the chain grows. The P site bears the amino acid chain, and the A site holds the newest tRNA. When the ribosome reaches a "stop" codon, it falls apart into its two subunits and is released. The new polypeptide breaks free.
12. After translation, some polypeptides are cleaved, and some aggregate to form larger proteins. The cell uses or secretes the protein, which must have a particular **conformation** to be active and functional.
13. A protein's **primary structure** is its amino acid sequence. Its **secondary structure** forms as amino acids close in the primary structure attract one another. **Tertiary structure** appears as more widely separated amino acids approach or repel in response to water molecules. **Quaternary structure** forms when a protein consists of more than one polypeptide. Chaperone proteins help mold conformation. Some proteins can fold into several conformations, some of which can cause disease.

### 10.3 The Human Genome Sequence Reveals Unexpected Complexity

14. Only 1.5 percent of the 3.2 billion base pairs of the human genome encode protein, yet those 31,000 or so genes specify 100,000 to 200,000 distinct proteins.
15. Several mechanisms explain how a set number of genes can encode a larger number of proteins. These include exon shuffling, use of introns, and cutting proteins translated from a single gene.
16. The 98.5 percent of the human genome that does not encode protein apparently encodes several types of RNA, control sequences, and repeats.

# Review Questions

1. Explain how complementary base pairing is responsible for
   a. the structure of the DNA double helix.
   b. DNA replication.
   c. transcription of RNA from DNA.
   d. the attachment of mRNA to a ribosome.
   e. codon/anticodon pairing.
   f. tRNA conformation.
2. A retrovirus has RNA as its genetic material. When it infects a cell, it uses enzymes to copy its RNA into DNA, which then integrates into the host cell's chromosome. Is this flow of genetic information consistent with the central dogma? Why or why not?
3. Genomics is highly dependent upon computer algorithms that search DNA sequences for indications of specialized functions. Explain the significance of detecting the following sequences:
   a. a promoter
   b. a sequence of 75 to 80 bases that folds into a cloverleaf shape
   c. a gene with a sequence very similar to that of a known protein-encoding gene, but that isn't translated into protein
   d. 200 copies of a 44,000 base long sequence

e. RNAs with poly A tails

f. a sequence that is very similar to part of a known virus that is found at several sites in a genome

4. Many antibiotic drugs work by interfering with protein synthesis in the bacteria that cause infections. Explain how each of the following antibiotic mechanisms disrupts genetic function in bacteria.

   a. Transfer RNAs misread mRNA codons, binding with the incorrect codon and bringing in the wrong amino acid.

   b. The first amino acid is released from the initiation complex before translation can begin.

   c. Transfer RNA cannot bind to the ribosome.

   d. Ribosomes cannot move.

   e. A tRNA picks up the wrong amino acid.

5. Define and distinguish between transcription and translation.

6. List the differences between RNA and DNA.

7. Where in a cell do DNA replication, transcription, and translation occur?

8. How does transcription control cell specialization?

9. How can the same mRNA codon be at an A site on a ribosome at one time, but at a P site at another?

10. Describe the events of transcription initiation.

11. List the three major types of RNA and their functions.

12. State three ways that RNA is altered after it is transcribed.

13. What are the components of a ribosome?

14. Why was the discovery of introns a surprise? of ribozymes?

15. Why would an overlapping genetic code be restrictive?

16. How are the processes of transcription and translation economical?

17. How does the shortening of proinsulin to insulin differ from the shortening of apolipoprotein B?

18. What factors determine how a protein folds into its characteristic conformation?

19. Why would two-nucleotide codons be insufficient to encode the number of amino acids in biological proteins?

20. Cite two ways that RNA helps in its own synthesis, and two ways that proteins help in their own synthesis.

21. In the 1960s, a gene was defined as a continuous sequence of DNA, located permanently at one place on a chromosome, that specifies a sequence of amino acids. State three ways that this statement is incomplete.

22. Until recently, a protein was thought to have only one conformation. Which protein provides evidence that this is incorrect?

23. How can one of the two dental proteins implicated in dentinogenesis imperfecta be much more abundant than the other, if they are both transcribed and translated from the same gene?

24. The four mRNA codons that specify the amino acid leucine are CUU, CUC, CUA, and CUG. Only three types of tRNAs recognize these four codons. How is this possible? Which two codons does a single tRNA recognize?

# Applied Questions

1. The *BRCA1* gene that, when missing several bases, causes a form of breast cancer has 24 exons and 23 introns.

   a. How many splice sites does the gene contain? (A splice site is the junction of an exon and an intron.)

   b. In a woman with *BRCA1* breast cancer, an entire exon is missing, or "skipped." How many splice sites does her affected copy of the gene have?

2. When researchers compared the number of mRNA transcripts that correspond to a part of chromosome 19 to the number of protein-encoding genes in the region, they found 1,859 transcripts and 544 genes. State three mechanisms that could account for the discrepancy.

3. List the sequences of RNA that would be transcribed from the following DNA template sequences.

   a. TTACACTTGCTTGAGAGTC

   b. ACTTGGGCTATGCTCATTA

   c. GGCTGCAATAGCCGTAGAT

   d. GGAATACGTCTAGCTAGCA

4. Given the following partial mRNA sequences, reconstruct the corresponding DNA template sequences.

   a. GCUAUCUGUCAUAAAAGAGGA

   b. GUGGCGUAUUCUUUUCCGGGUAGG

   c. GAGGGAAUUCUUUCUCAACGAAGU

   d. AGGAAAACCCCUCUUAUUAUAGAU

5. List three different mRNA sequences that could encode the following amino acid sequence:

   histidine-alanine-arginine-serine-leucine-valine-cysteine

6. Write a DNA sequence that would encode the following amino acid sequence:

   valine-tryptophan-lysine-proline-phenylalanine-threonine

7. In the film *Jurassic Park,* which is about genetically engineered dinosaurs, a cartoon character named Mr. DNA talks about the billions of genetic codes in the DNA. Why is this statement incorrect?

8. In investigating the genetic code, when the researchers examined synthetic RNA of sequence ACACACACACACACA, they found that it encoded the amino acid sequence *thr-his-thr-his-thr-his.* How did the researchers determine the codon assignments for ACA and CAC?

9. Cystic fibrosis is caused by an abnormal chloride channel protein that makes up part of the cell membrane. How might a defect in protein folding cause the cystic fibrosis phenotype at the cellular level?

10. Figure 10.19 shows the distribution of types of proteins that, when abnormal or absent from a certain cell type, cause disease. Such charts have been constructed for different stages of development—prenatal, under a year, childhood, puberty to age 50, and over age 50. Explain the observation that transcription factors account for:

    - 9 percent of proteins overall (throughout development and life)
    - 25 percent of proteins before birth
    - 7 percent of proteins from birth to one year
    - 6 percent of proteins from childhood to age 50 years
    - 5 percent of proteins for those over 50 years

11. Titin is a muscle protein named for its gargantuan size—its gene has the largest known coding sequence—80,781 DNA bases. How many amino acids long is it?

12. On the television program *The X Files,* Agent Scully discovers an extraterrestrial life form that has a triplet genetic code, but with five different bases, instead of the four of earthly inhabitants. How many different amino acids can this code specify?

13. In malignant hyperthermia, a person develops a life-threateningly high fever after taking certain types of anesthetic drugs. In one family, the mutation deletes three contiguous bases in exon 44. How many amino acids are missing from the protein?

14. A mutation in a gene called RPGR-interacting protein causes visual loss. The encoded protein is 1,259 amino acids long. What is the minimal size of this gene?

15. Parkinson disease caused rigidity, tremors, and other motor symptoms. Only 2 percent of cases are inherited, and these tend to have an early onset of symptoms. Some inherited cases result from mutations in a gene that encodes the protein parkin, which has 12 exons. Indicate whether each of the following mutations in the parkin gene would result in a smaller protein, a larger protein, or not change the size of the protein.
    a. deletion of exon 3
    b. deletion of six contiguous nucleotides in exon 1
    c. duplication of exon 5
    d. disruption of the splice site between exon 8 and intron 8
    e. deletion of intron 2

16. How can repeated sequences impart information?

# Suggested Readings


Caron, Huib, et al. February 16, 2001. The human transcriptome map: Clustering of highly expressed genes in chromosomal domains. *Science* 291:1289–92. The transcriptome map includes gene expression profiles by chromosome region for various normal and diseased cells.

Dahlberg, Albert. May 4, 2001. The ribosome in action. *Science* 292:868–69. A review of recent research reveals the structures of interacting ribosomes, tRNA, and mRNA, in bacteria.

Gilbert, Walter. February 9, 1978. Why genes in pieces? *Nature* 271:501. A classic and insightful look at the enigma of introns.

Hoagland, Mahlon. 1990. *Toward the habit of truth.* New York: W. W. Norton. The story of the RNA tie club, by a member.

Jiminez-Sanchez, Gerardo, et al. February 15, 2001. Human disease genes. *Nature* 409:853–58. Human inherited disease results from defects in proteins that fall into a few categories.

Kay, Lily E. 2001. *Who Wrote the Book of Life?* Stanford, CA: Stanford University Press. The story of how a group of mostly physicists-turned-biologists deciphered the genetic code, in the 1960s.

Lewis, Ricki. February 1996. On cracked codes, cell walls, and human fungi. *The American Biology Teacher* 58:16. A funny look at errors in genetic code usage.

Lewis, Ricki, and Barry Palevitz. June 11, 2001. Genome economy. *The Scientist* 15:19. The article that forms the basis for section 10.3 of this book shows how the genome specifies many proteins with few genes.

Pollack, Andrew. July 24, 2001. Scientists are starting to add letters to life's alphabet. *The New York Times,* p. F1. Investigators at the Scripps Research Institute in La Jolla, California, are attempting to create life forms that use a more extensive genetic code.

Prusiner, Stanley. May 17, 2001. Shattuck lecture—neurodegenerative diseases and prions. *The New England Journal of Medicine* 344:1516–20. The normal and pathogenic forms of prion protein have the same amino acids sequence, but different conformations.

Solovitch, Sara. July 2001. The citizen scientists. *Wired.* Frustrated by nonexistent or slow research, parents of sick children have been instrumental in discovering the genes behind some rare disorders.

Tupler, Rosella. February 15, 2001. Expressing the human genome. *Nature* 409:832–33. Sequencing the genome will seem easy compared to the task of tracking gene expression in the more than 260 different cell types.

Vogel, Gretchen. February 16, 2001. Why sequence the junk? *Science* 291:1184. DNA as "junk" is a value judgment often based on lack of data.


# On the Web

Check out the resources on our website at

**www.mhhe.com/lewisgenetics5**

On the web for this chapter, you will find additional study questions, vocabulary review, useful links to case studies, tutorials, popular press coverage, and much more. To investigate specific topics mentioned in this chapter, also try the links below:

Annotated Human Genomic Database
**www.DoubleTwist.com/genome**

Cold Spring Harbor Laboratory Learning Center
**vector.cshl.org/resources/resources.html**

Functional Genomics Website
**www.sciencegenomics.org**

Genome Jokes and Cartoons
**cagle.slate.msn.com/news/gene**

Genome News Network
**www.celera.com/genomics/genomic.cfm**

The Human Transcriptome Map
**http://bioinfo.amc.uva.nl/HTM/l**

National Center for Biotechnology Information Splash Page
**www.ncbi.nlm.nih.gov/genome/guide/human**

Online Mendelian Inheritance in Man
**www.ncbi.nlm.nih.gov/entrez/query.fcgi?db=OMIM**
dentinogenesis imperfecta 125490
Duchenne muscular dystrophy 310200
epidermal growth factor 131530
fibronectin 135600
Huntington disease 143100
neurofibromatosis I 162200
tissue plasminogen activator 173370

CHAPTER

11

# Gene Mutation

### 11.1 Mutations Can Alter Proteins—Three Examples

A mutation changes a gene's sequence, which may or may not disrupt the encoded protein in a way that causes a mutant phenotype. On an individual level, a mutation may cause an inherited illness; yet at the population and evolutionary levels, DNA's ability to mutate enables life to overcome environmental challenges.

### 11.2 Causes of Mutation

Mutation is a part of life. Naturally occurring errors in DNA replication may result in spontaneous mutation. Mutagens such as chemicals or radiation can cause mutation by adding, deleting, or replacing DNA bases. Induced mutation is a research tool, but it may also result from exposure to environmental agents.

### 11.3 Types of Mutations

DNA mutates in many ways. Bases may be substituted, deleted, inserted, or moved. Mutations that disrupt the reading frame—that is, the sequence of DNA triplets—tend to be the most drastic in effect.

### 11.4 The Importance of a Mutation's Position in the Gene

Because of the informational nature of gene structure, a single gene can mutate at many sites. Whether the change affects the phenotype depends upon the protein's structure and function. Some mutations exert no noticeable effect, because they do not occur in a part of the protein that is crucial to its function.

### 11.5 Factors That Lessen the Effects of Mutation

The genetic code protects against mutation, because many changes result in a codon that specifies the same or a structurally similar amino acid. Recessive inheritance protects against deleterious effects of mutation if one functioning allele provides enough of the encoded protein for health. Some mutations are expressed only in response to certain environmental triggers.

### 11.6 DNA Repair

DNA replication, like any good manufacturing process, is routinely checked for errors and damage repaired.

A **mutation** is a change in a gene's nucleotide base sequence that affects the phenotype. It is a type of polymorphism. A mutation can occur at the molecular level, substituting one DNA base for another or adding or deleting a few bases, or at the chromosome level, the subject of chapter 12. Chromosomes can exchange parts, and genetic material (transposons) can even jump from one chromosome to another. This chapter discusses mutations at the molecular level. Such a mutation can occur in the part of a gene that encodes a protein, in a sequence that controls transcription, in an intron, or at a site critical to intron removal and exon splicing.

The effects of mutation vary. A mutation can completely halt production of a protein, lower the amount of a protein synthesized, overproduce it, or impair the protein's function. A mutation may even offer protection. For example, about 1 percent of the population is homozygous for a recessive allele of a gene that encodes a cell surface protein called CCR5 (see figure 16.14). To enter a human's T cell, HIV must bind to CCR5 as well as another protein. The mutation prevents CCR5 from traveling from the cytoplasm to the cell surface. HIV cannot bind and the person cannot become infected. People lucky enough to have inherited a double dose of this mutant allele are also at a much lower risk of developing asthma and rejecting transplanted tissue, and if they develop an autoimmune disorder (such as multiple sclerosis or rheumatoid arthritis), the symptoms are unusually mild. Heterozygotes are partially protected against HIV infection—they are at considerably lower risk of being infected should they become exposed to the virus.

The term *mutation* refers to genotype—that is, a change at the DNA or chromosome level. The familiar term **mutant** refers to an unusual phenotype. Whether a mutation causes a mutant phenotype depends upon precisely how the alteration affects the gene's product or activity. A mutant phenotype usually connotes an abnormal or unusual characteristic. However, it may also mean an unusual variant that is nevertheless "normal," such as a red-haired child in a class of brunettes and blondes (figure 11.1). In an evolutionary sense, mutation has been essential to life, because it produces individuals with variant phenotypes who are better able to survive specific environmental challenges, including disease. Disease-resistant gene variants that arise by mutation tend to become more common in populations over time when they exert a protective effect because they give the people with the mutation a survival advantage. Chapter 14 further discusses the role of mutations in populations.

**figure 11.1**

**Red hair.** This child's red curls make her the proud possessor of an unusual genetic variant.

A mutation may be present in all the cells of an individual or just in some cells. In a **germline mutation** (also called a constitutional mutation), the change occurs during the DNA replication that precedes meiosis. The resulting gamete and all the cells that descend from it after fertilization have the mutation. In contrast, in a **somatic mutation,** the change happens during DNA replication before a *mitotic* cell division. All the cells that descend from the original changed cell are altered, but this might only comprise a small part of the body. Somatic mutations are responsible for certain cancers (see Reading 17.2 and figure 17.4).

## 11.1 Mutations Can Alter Proteins—Three Examples

Research on inherited disease has traditionally begun with studies of mutations. Medical geneticists tried to identify precisely how a specific mutation alters the phenotype in a way that harms health.

### The Beta Globin Gene

The first genetic illness to be understood at the molecular level was sickle cell disease. Researchers knew in the 1940s that an inherited anemia (weakness and fatigue caused by too few red blood cells) was associated with sickle-shaped red blood cells (figure 11.2). In 1949, Linus Pauling and coworkers discovered that hemoglobin from healthy people and from people with the anemia, when placed in a solution in an electrically charged field (a technique called electrophoresis), moved to different positions. Hemoglobin from parents of people with the anemia, who were carriers, showed movement to both positions.

The researchers suspected that a physical difference accounted for the different electrophoretic mobilities of normal versus sickled hemoglobin. But how could they identify the part of the protein portion of hemoglobin affected in sickle cell disease? Hemoglobin is a very large molecule. It consists of four globular-shaped polypeptide chains, each surrounding an organic molecule called heme that includes an iron atom (figure 11.3). Two of the chains are called beta chains, and two are called alpha chains.

The globin genes occur in clusters, the beta genes on chromosome 11 and the alpha chain genes on chromosome 16. The beta cluster also includes globin chains that encode protein subunits that are part of the hemoglobin molecule during embryonic or fetal development. These other polypeptide chains have differing affinities for oxygen that match the differing routes and requirements of oxygen delivery before and after birth. Eventually researchers learned that an abnormality in the beta globin polypeptide causes sickle cell disease.

Protein chemist V. M. Ingram cut normal and sickle hemoglobin with a protein-

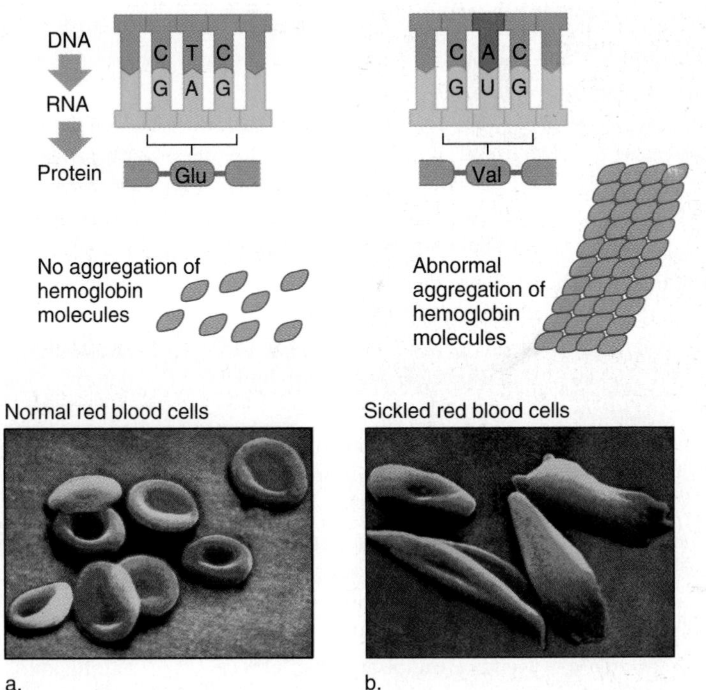


## figure 11.2

**Sickle cell disease results from a single base change.** Hemoglobin carries oxygen throughout the body. When normal (*a*), the globular molecules do not aggregate, enabling the cell to assume a rounded shape. In sickle cell disease (*b*), a single DNA base change replaces one amino acid in the protein with a different one (valine replaces glutamic acid). The result is a change in the surface of the molecule that causes hemoglobins to aggregate into long, curved rods that deform the red blood cell. Sets of 14 rods twist together.

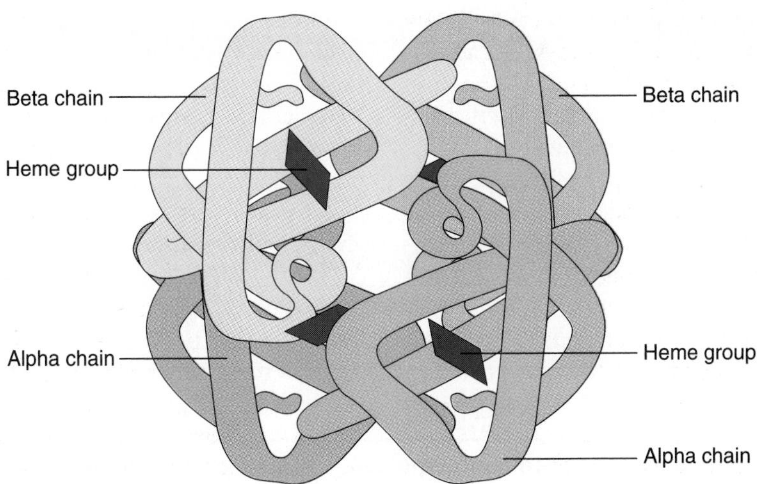


## figure 11.3

**The structure of hemoglobin.** More than 750 different mutations of the human globin genes are known. Mutations can (1) disrupt the binding of the globin chains to each other or to the iron-containing groups; (2) change a stop codon to one specifying an amino acid, elongating the beta globin chain; or (3) change an amino acid-specifying codon into a stop codon, shortening the beta chain. Many beta chain polymorphisms change a codon into one specifying the same or a similar amino acid, or occur in a part of the chain not essential to its function.

digesting enzyme, separated the pieces, stained them, and displayed them on filter paper. The patterns of fragments—known as peptide fingerprints—were different for the two types of hemoglobin. This meant, Ingram deduced, that the two molecules differ in amino acid sequence. Then he homed in on the difference. One piece of the molecule in the fingerprint, fragment four, occupied a different position in the two types of hemoglobin. Because this peptide was only 8 amino acids long, Ingram needed to decipher only that short sequence—rather than the 146 amino acids of a full beta globin sequence—to find the site of the mutation. It was a little like knowing which sentence on a page contains a typographical error.

Using then newly invented protein sequencing techniques, Ingram identified the tiny mutation responsible for sickle cell disease. It is a substitution of the amino acid valine for the glutamic acid that normally is the sixth amino acid in the beta globin polypeptide chain. At the DNA level, the change was even smaller—a CTC to a CAC, corresponding to RNA codons GAG and GUG, which was learned after researchers deciphered the genetic code. Eventually, researchers found that this mutation changes the surfaces of hemoglobin molecules, which causes them to link when in low-oxygen conditions. This action makes red blood cells rigid and fragile, and they bend into sickle shapes that cause anemia, joint pain, and organ damage when the cells lodge in narrow blood vessels, cutting off local blood supplies.

Sickle cell disease was the first inherited illness linked to a molecular abnormality, but it wasn't the first known condition that results from a mutation in the beta globin genes. In 1925, Thomas Cooley and Pearl Lee described severe anemia in Italian children, and in the decade following, others described a milder version of this "Cooley's anemia," also in Italian children. The disease was named thalassemia, from the Greek for "sea," in light of its high prevalence in the Mediterranean area. The two disorders turned out to be versions of the same illness. The severe form, sometimes called thalassemia major, results from a homozygous mutation in the beta globin gene.

The milder form, called thalassemia minor, is associated with heterozygosity for the mutation.

Once the structure of the hemoglobin molecule had been worked out, along with knowing that different globins function in the embryonic and fetal periods, the molecular basis of thalassemia became clear. The disorder that is common in the Mediterranean area is more accurately called beta thalassemia, because the symptoms result from too few beta globin chains, which are needed to build enough hemoglobin molecules to effectively deliver oxygen to tissues. Symptoms of anemia, such as fatigue and bone pain, arise during the first year of life as the child depletes fetal hemoglobin supplies, and the "adult" beta globin genes are not transcribed and translated on schedule.

As severe beta thalassemia continues, red blood cells die because of the relative excess of alpha globin chains, and the liberated iron slowly destroys the heart, liver, and endocrine glands. People with the disorder can receive periodic blood transfusions to control the anemia, but this treatment hastens iron buildup and organ damage. Using drugs called chelators that entrap the iron can extend life past early adulthood, but this treatment is very costly and not available in developing nations.

## Disorders of Orderly Collagen

Much of the human body consists of the protein collagen. It accounts for more than 60 percent of the protein in bone and cartilage and provides 50–90 percent of the dry weight of skin, ligaments, tendons, and the dentin of teeth. Collagen is in parts of the eyes and the blood vessel linings, and it separates cell types in tissues. It is also a major component of connective tissue. Mutations in the genes that encode collagen lead to a variety of medical problems (table 11.1). This is not surprising given collagen's diverse functions and locations,

Mutations in the collagen genes are particularly devastating, because the encoded protein has an extremely precise conformation that is easily disrupted, even by slight alterations that might have little effect in proteins that have other shapes (figure 11.4). Collagen is sculpted from a longer precursor molecule called procollagen, which consists of many repeats of a specific amino acid sequence (glycine, proline, and a modified proline). Three procollagen chains entwine. Two of the chains are identical and are encoded by one gene, and the other is encoded by a second gene. The electrical charges and in-

**table 11.1**

**Collagen Disorders**

| Disorder | Defect | Signs and Symptoms |
|---|---|---|
| Alport syndrome | Mutation in type IV collagen that interferes with tissue boundaries | Deafness and inflamed kidneys |
| Aortic aneurysm | Missense mutation substitutes an *arg* for *gly* in alpha 1 gene | Aorta bursts |
| Chondrodysplasia | Deletion, insertion, or missense mutation replaces *gly* with bulky amino acids | Stunted growth, deformed joints |
| Dystrophic epidermolysis bullosa | Collagen fibrils that attach epidermis to dermis break down | Skin blisters on any touch |
| Ehlers-Danlos syndrome | Missense mutations replace *gly* with bulky amino acids; deletions or missense mutations disrupt intron/exon splicing | Stretchy, easily scarred skin, lax joints |
| Osteoarthritis | Missense mutation substitutes *cys* for *arg* in alpha 1 gene | Painful joints |
| Osteogenesis imperfecta type I | Inactivation of α allele reduces collagen triple helices by 50% | Easily broken bones; blue eye whites; deafness |
| Stickler syndrome | Nonsense mutation in procollagen | Joint pain, degeneration of vitreous gel and retina |

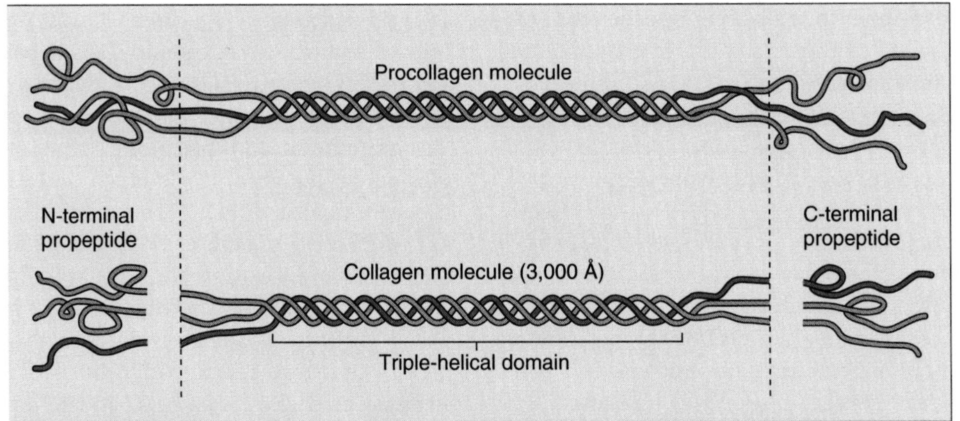


**figure 11.4**

**Collagen has a very precise conformation.** The α1 collagen gene encodes the two blue polypeptide chains, and the α2 procollagen gene encodes the third, red chain. The procollagen triple helix is shortened before it becomes functional, forming the fibrils and networks that comprise much of the human body.

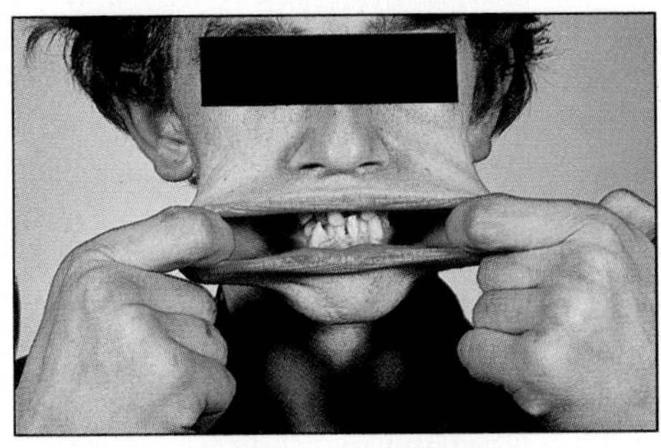

figure 11.5

**A disorder of connective tissue produces stretchy skin.** A mutation that blocks the trimming of procollagen chains to collagen causes the characteristic stretchy skin of Ehlers-Danlos syndrome type I. Researchers were able to find and study this gene by using a version from calves, which also inherit the condition.

teractions of these amino acids with water cause the procollagen chains to coil into a very regular triple helix, with space in the middle only for glycine, a very small amino acid. The ragged ends of the polypeptides are snipped off by enzymes to form mature collagen. The collagen fibrils continue to associate outside the cell, building the fibrils and networks that hold the body together.

The boy in figure 11.5 has a form of Ehlers-Danlos syndrome. A mutation prevents his procollagen chains from being cut down to size, and as a result, mature collagen molecules cannot assemble. Instead they form ribbonlike fibrils that lack the tensile strength needed to keep the skin from becoming too stretchy. Other collagen mutations include missing procollagen chains, kinks in the triple helix, failure to cut mature collagen, and defects in aggregation outside the cell.

Aortic aneurysm is a serious connective tissue disorder for which detection of the causative mutation before symptoms arise can be lifesaving. An early sign is a weakened aorta (the largest blood vessel in the body, which emerges from the heart), which can suddenly burst. Knowing that one has inherited the mutant collagen gene can signal the need for frequent ultrasound exams to detect aortic weakening early enough to treat it surgically. (Aortic aneurysm is also part of Marfan syndrome. In this case, the mutation is in a gene that encodes fibrillin, a protein that is much scarcer in connective tissue than is collagen.)

## A Mutation That Causes Early-Onset Alzheimer Disease

The story of the discovery of the gene that causes an early-onset, autosomal dominant form of Alzheimer disease began in the 1880s, when a woman named Hannah, born in Latvia, developed progressive dementia. Hannah's condition was highly unusual in that she was only in her early forties when the classic forgetfulness that heralds the disease's onset began. Apparently this form of the illness originated, in this family, in Hannah. Many of her descendants also experienced dementia—some as early as in their thirties.

In 1974, Hannah's grandson and great-grandson, both physicians, constructed an extensive pedigree tracing Alzheimer disease in their family. They circulated the pedigree among geneticists, hoping to elicit interest in identifying the family's mutation, offering their own and relatives' DNA for testing. Research teams in Mexico, the United States, and Canada began the search in 1983. By 1992, they narrowed the investigation to a portion of chromosome 14, and three years later, they pinpointed the gene. It encodes a protein called presenilin 1 that forms a receptor anchored in the membrane of a Golgi apparatus or a vesicle (figure 11.6). Normally, the protein monitors the cell's storage or use of beta amyloid, the substance that accumulates in the brains of people with Alzheimer disease. Members of families that have early-onset Alzheimer disease due to mutation in this gene have elevated levels of presenilin 1 in their bloodstreams before symptoms begin. Somehow the abnormality in presenilin disrupts amyloid production, folding, or function.

So far, researchers have identified more than 30 mutations that substitute one amino acid for another in the gene for presenilin 1, impairing its function sufficiently to cause the beta amyloid buildup that eventually causes the symptoms of Alzheimer disease. Mutations in at least four other genes can cause or increase the risk of developing Alzheimer disease. Table 11.2 offers other examples of how mutations impair health.

### Key Concepts

Mutations add, delete, or rearrange genetic material, in a germline cell or in a somatic cell. Learning exactly how a mutation alters a protein can help explain how disease arises. In sickle cell disease, a mutation causes hemoglobin to crystallize in a low-oxygen environment, bending red blood cells into sickle shapes and impairing circulation. In beta thalassemia, beta globin is absent or scarce, causing too few complete hemoglobin molecules and buildup of free alpha globin chains and iron. • Mutations in collagen genes often disrupt the protein's precise organization. • In one form of Alzheimer disease, a mutation in a receptor protein may lead to beta-amyloid buildup.

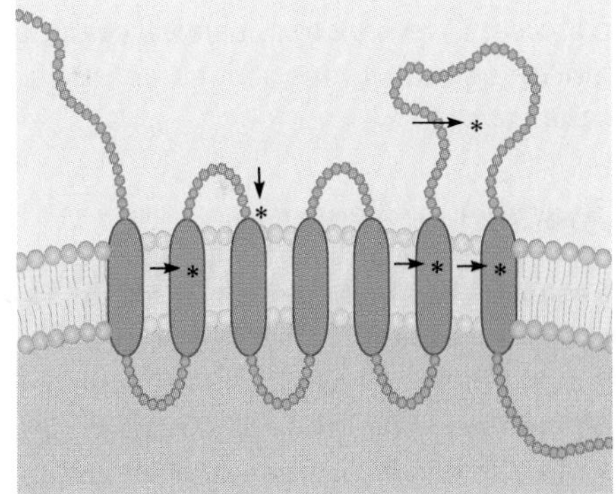

figure 11.6

**One cause of Alzheimer disease.** When geneticists searched the DNA of people with very early-onset inherited Alzheimer disease, they identified a gene on chromosome 14 whose protein product, shown here, fits the well-known pattern of a receptor anchored into a membrane at seven points. This protein resides in vesicles derived from the Golgi apparatus. When abnormal, it somehow enables amyloid proteins to accumulate outside cells. Asterisks indicate sites where mutations in the gene disrupt the protein in a way that causes symptoms.

**table 11.2**

**How Mutation Causes Disease**

| Disease | Signs and Symptoms (Phenotype) | Protein | Genetic Defect (Genotype) |
|---|---|---|---|
| Cystic fibrosis | Frequent lung infection, pancreatic insufficiency | Cystic fibrosis transmembrane regulator (CFTR) | Missing single amino acid or other defect alters conformation of chloride channels in certain epithelial cell membranes. Water enters cells, drying out secretions. |
| Duchenne muscular dystrophy | Gradual loss of muscle function | Dystrophin | Deletion in dystrophin gene eliminates this protein, which normally binds to inner face of muscle cell membranes, maintaining cellular integrity. Cells and muscles weaken. |
| Familial hypercholesterolemia | High blood cholesterol, early heart disease | LDL receptor | Deficient LDL receptors cause cholesterol to accumulate in blood. |
| Hemophilia A | Slow or absent blood clotting | Factor VIII | Absent or deficient clotting factor causes hard-to-control bleeding. |
| Huntington disease | Uncontrollable movements, personality changes | Huntingtin | Extra bases in the gene add amino acids to the protein product, altering function in a way that causes brain degeneration. |
| Marfan syndrome | Long limbs, weakened aorta, spindly fingers, sunken chest, lens dislocation | Fibrillin | Too little elastic connective tissue protein in lens and aorta. |
| Neurofibromatosis type 1 | Benign tumors of nervous tissue beneath skin | Neurofibromin | Defect in protein that normally suppresses activity of a gene that causes tumor formation. |

## 11.2 Causes of Mutation

A mutation can occur spontaneously or be induced by exposure to a chemical or radiation. An agent that causes mutation is called a **mutagen.**

### Spontaneous Mutation

A spontaneous mutation can show up as a surprise. For example, two healthy people of normal height may have a child of extremely short stature. The child has achondroplasia (a form of dwarfism) caused by an autosomal dominant mutation; therefore, each of the son's children will face a 50 percent chance of inheriting the condition. How could this happen when there are no other affected family members? If the mutation is dominant, why are the parents of normal height? The child has a genetic condition, but he did not inherit it. Instead, he originated it. The boy with achondroplasia arose from a *de novo,* or new, mutation in either his mother's oocyte or father's sperm cell. This is a spontaneous mutation—that is, it is not caused by known exposure to a mutagen. Instead, a spontaneous mutation usually originates as an error in DNA replication.

One cause of spontaneous mutation stems from the chemical tendency of free nitrogenous bases to exist in two slightly different structures, called tautomers. For extremely short times, each base is in an unstable tautomeric form. If, by chance, such an unstable base is inserted into newly forming DNA, an error will be generated when that strand replicates and is perpetuated. Figure 11.7 shows how this can happen.

#### *Spontaneous Mutation Rate*

The spontaneous mutation rate varies for different genes. The dominant gene that causes neurofibromatosis type 1 (NF1), for example, has one of the highest mutation rates known, arising in 40 to 100 of every million gametes (table 11.3). NF1 affects 1 in 3,000 births, about half of which occur in families with no prior cases. The large size of this gene may contribute to its high mutability—there are more ways for its sequence to change, just as there are more opportunities for a misspelling to occur in a long sentence than in a short one.

Based on the prevalence of certain disease-causing genes, geneticists estimate that each human gene has about a 1 in 100,000 chance of mutating. Each of us probably carries a few new spontaneously mutated genes. Mitochondrial genes mutate at a higher rate than nuclear genes because they lack DNA repair mechanisms. These are discussed in section 11.6.

Estimates of the spontaneous mutation rate for a particular gene are usually derived from observations of new, dominant conditions, such as the boy with achondroplasia. This is possible because a new dominant mutation is detectable simply by observing the phenotype. In contrast, a new recessive mutation would not be obvious until two heterozygotes produced a homozygous recessive offspring with a noticeable phenotype.

The spontaneous mutation rate for autosomal genes can be estimated using the following formula: number of *de novo* cases/2X, where X is the number of individuals examined. The denominator has a factor of 2 to account for the nonmutated homologous chromosome.

Spontaneous mutation rates in human genes are difficult to assess because our generation time is long—usually 20 to 30 years. In bacteria, a new generation arises every half hour or so, and mutation is therefore much more frequent. This ability to rapidly

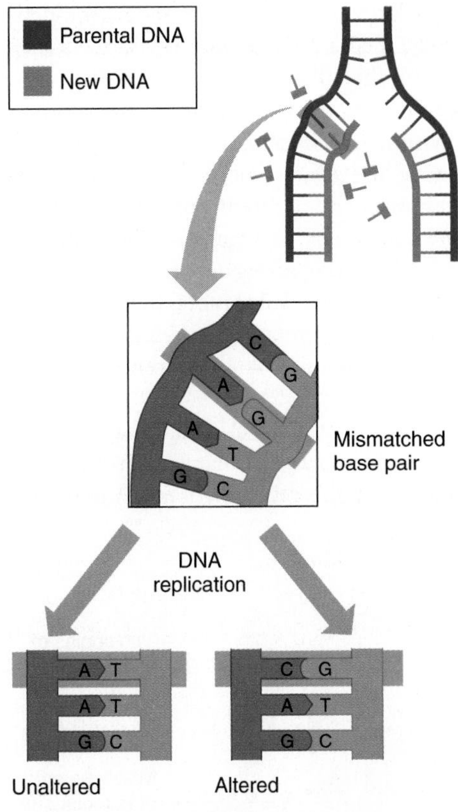


**figure 11.7**

**Spontaneous mutation.** DNA bases are very slightly chemically unstable and, for brief moments, they exist in altered forms. If a replication fork encounters a base in its unstable form, a mismatched base pair can result. After another round of replication, one of the daughter cells has a different base pair than the one in the corresponding position in the original DNA.

mutate can be harmful to human health when disease-causing bacteria become resistant to the antibiotic drugs we use to destroy them. Chapter 14 addresses this pressing health concern from an evolutionary viewpoint.

The genetic material of viruses also spontaneously mutates rapidly. This is why an influenza vaccine manufactured to fight one year's predominant strain may be ineffective by the next flu season. Genetic changes can alter the virus's surface to such an extent that the vaccine no longer recognizes the virus. A high viral spontaneous mutation rate is also one reason why it has been difficult to develop a vaccine to protect against HIV infection.

### *Mutational Hot Spots*

Mutations may occur anywhere in a gene, but in some genes they are more likely to occur in certain regions called hot spots. Sequences that are mutational hot spots are often not random. Many hot spots occur where the DNA sequence is repetitive. It is as if the molecules that guide and carry out replication become "confused" by short repeated sequences, as an editor scanning a manuscript might miss the spelling errors in the words "happpiness" and "banananana" (figure 11.8). In alkaptonuria, for example, an analysis of many different mutations in the causative gene found that 35 percent of them occurred at or near one or more CCC repeats, even though these repeats account for only 9 percent of the gene. Alkaptonuria is featured in In Their Own Words, chapter 5.

The increased incidence of mutations in repeated DNA sequences has a physical basis. Within a gene, when DNA strands locally unwind to permit replication, symmetrical or repeated sequences allow base pairing to occur between bases located on the same strand, such as a stretch of ATATAT pairing with TATATA. This action interferes with replication and repair enzymes, increasing the chance of an error. Mutations in the gene for clotting factor IX, which causes hemophilia B, for example, occur 10 to 100 times as often at any of 11 sites in the gene that have extensive direct repeats of CG (CGCGCG . . .) (figure 11.8).

Small additions and deletions of DNA are more likely to occur near sequences called palindromes (figure 11.8). These sequences read the same, in a 5′ to 3′ direction, on complementary strands. Put another way, the sequence on one strand is the reverse of the sequence on the complementary strand. Palindromes probably increase the spontaneous mutation rate by disturbing replication.

The blood disorder alpha thalassemia illustrates the confusing effect of direct (as opposed to inverted) repeats of an entire gene. A person who does not have the disorder has four genes that specify alpha

**table 11.3**

**Mutation Rates of Some Genes That Cause Inherited Disease**

| | Mutations per Million Gametes | Signs and Symptoms |
|---|---|---|
| **X-linked** | | |
| Duchenne muscular dystrophy | 40–105 | Muscle atrophy |
| Hemophilia A | 30–60 | Severe impairment of blood clotting |
| Hemophilia B | 0.5–10 | Mild impairment of blood clotting |
| **Autosomal Dominant** | | |
| Achondroplasia | 10 | Very short stature |
| Aniridia | 2.6 | Absence of iris |
| Huntington disease | <1 | Uncontrollable movements, personality changes |
| Marfan syndrome | 4–6 | Long limbs, weakened blood vessels |
| Neurofibromatosis type 1 | 40–100 | Brown skin spots, benign tumors under skin |
| Osteogenesis imperfecta | 10 | Easily broken bones |
| Polycystic kidney disease | 60–120 | Benign growths in kidneys |
| Retinoblastoma | 5–12 | Malignant tumor of retina |

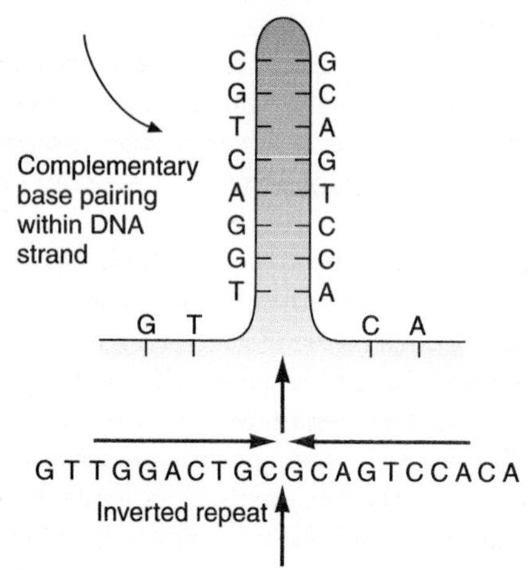


## figure 11.8

**DNA symmetry may increase the likelihood of mutation.** These examples show repetitive DNA sequences and symmetrical sequences that may "confuse" replication enzymes, causing errors.

globin chains, two next to each other on each chromosome 16. Homologs with repeated genes can misalign during meiosis when the first sequence on one chromosome lies opposite the second sequence on the homolog. If crossing over occurs, a sperm or oocyte can form that has one or three of the alpha globin genes instead of the normal two (figure 11.9). Fertilization with a normal gamete then results in a zygote with one extra or one missing alpha globin gene.

A person with only three alpha globin genes produces enough hemoglobin, and is considered to be a healthy carrier. Rarely, individuals arise with only two copies of the gene, and they are mildly anemic and tire easily. A person with a single alpha globin gene is severely anemic, and a fetus lacking alpha globin genes does not survive. Alpha thalassemia is a common single gene disorder because carriers have an advantage—they are protected against malaria. An infectious illness transmitted by mosquitoes in the tropics, malaria is the world's number one killer. The protective effect of being heterozygous for any of several inherited disorders, called balanced polymorphism, is discussed in chapter 15.

## Induced Mutations

Researchers can sometimes infer a gene's normal function by observing what happens when mutation alters it. But the spontaneous mutation rate is far too low to be a practical source of genetic variants for experiments. Instead, mutants are made. Over the years, geneticists have used many mutagens on a variety of experimental organisms to infer normal gene functions. Many collections are available. For example, a researcher can obtain mutant fruit flies from a facility at Indiana University, or mutant mice from the Jackson Laboratory in Bar Harbor, Maine.

### *Intentional Use of Mutagens*

Geneticists use chemicals or radiation to induce mutation. Chemicals called alkylating agents, for example, remove a DNA base, which is replaced with any of the four bases—three of which will be a mismatch against the template strand. Similarly, dyes called acridines can either add or remove a single DNA base. Because the genetic code is read three bases in a row, adding or deleting a single base can destroy a gene's information, altering the amino acid sequence of the encoded protein. Several other mutagenic chemicals alter base pairs, so that an A-T replaces a G-C, or vice versa, changing a gene's sequence. X rays and other forms of radiation delete a few bases or break chromosomes.

Researchers have developed several *in vitro* (in the test tube) protocols for testing the mutagenicity of a substance. The most famous of these, the Ames test, developed by Bruce Ames of the University of California, assesses how likely a substance is to harm the DNA of rapidly reproducing bacteria. One version of the test uses a strain of *Salmonella* that cannot grow when the amino acid histidine is absent from its media. If exposure to a substance enables the bacteria to grow on the deficient media, then it has undergone a mutation that allows it to do so. Another variation of the Ames test incorporates mammalian liver tissue into the media to make the results more applicable to the response of a multicellular organism. Because many mutagens are also carcinogens (cancer-causing agents), scientists often study the substances the Ames test identifies as mutagens to see if they cause cancer. Table 11.4 lists some common mutagens.

A limitation of using a mutagen is that it cannot alter a particular gene in exactly the way that a researcher might like. A technique

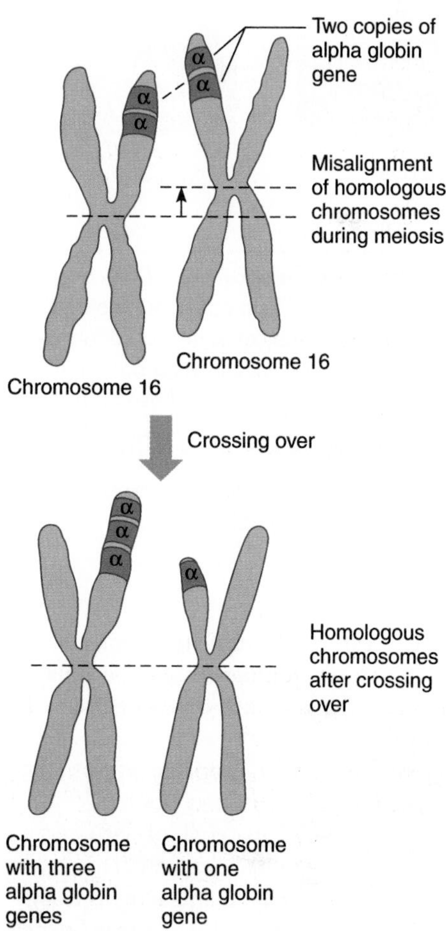


## figure 11.9

**Gene duplication and deletion.** The repeated nature of the alpha globin genes makes them prone to mutation by mispairing during meiosis. A person missing one alpha globin gene can develop anemia.

called **site-directed mutagenesis** makes this possible, using the polymerase chain reaction (see figure 9.15). The PCR primers include a specific base change, but are still similar enough in sequence to base pair with the gene of interest in the DNA sample. However, when the gene is amplified, the intentional change is replicated from it, just as an error in a manuscript is printed in every copy of a book. Site-directed mutagenesis can alter the directions for synthesis of proteins one amino acid at a time, and is faster and more precise than waiting for nature or a mutagen to produce a useful variant. It also makes possible studying lethal mutations that can theoretically exist, but never do because they are so drastic. Experimental organisms that have such mutations can be studied before they cease developing.

### *Accidental Exposures to Mutagens*

In contrast to the intentional use of mutagens in research is unintentional mutagen exposure. This occurs from contact on the job that occurred before the danger was known, from industrial accidents, from medical treatments such as chemotherapy and radiation, and from exposure to weapons that emit radiation.

An example of an environmental disaster that released mutagenic radiation was a steam explosion at a nuclear reactor in the former Soviet Union on April 25, 1986. Between 1:23 and 1:24 A.M., Reactor 4 at the Chernobyl Nuclear Power Station in Ukraine exploded, sending a great plume of radioactive isotopes into the air that spread for thousands of miles. The reactor had been undergoing a test, its safety systems temporarily disabled, when it became overloaded and rapidly flared out of control. Twenty-eight people died in the days following the explosion, of radiation exposure.

Acute radiation poisoning is not a genetic phenomenon. Evidence of a mutagenic effect has come from the observed increased rate of thyroid cancer among children who were living in nearby Belarus. Rates have risen tenfold. The thyroid glands of young people soak up iodine, which in a radioactive form bathed the area in the first days after the explosion. Cancer rates have also risen among "liquidators," the workers who cleaned up after the disaster. Analysis of evidence of radiation exposure in their teeth is being used to assess whether or not cancer correlates to the degree of exposure.

Another way researchers tracked mutation rate in the wake of the Chernobyl explosion was to compare the lengths of short DNA repeats called minisatellite sequences in children born in 1994 and in their parents, who lived in the Mogilev district of Belarus at the time of the accident and have remained there. Minisatellites are the same length within all cells of an individual. A minisatellite size in a child that does not match the size of either parent indicates that a mutation occurred in a parent's gamete. Such mutation was twice as likely to occur in exposed families than in control families living elsewhere. Minisatellite sequences mutate faster than other genes in the nucleus because they are repeats. The increased mutation rate enables researchers to chart genetic change.

**table 11.4**

**Commonly Encountered Mutagens**

| Mutagen | Source |
|---|---|
| Aflatoxin B | Fungi growing on peanuts and other foods |
| 2-amino 5-nitrophenol | Hair dye components |
| 2,4-diaminoanisole | " |
| 2,5-diaminoanisole | " |
| 2,4-diaminotoluene | " |
| p-phenylenediamine | " |
| Furylfuramide | Food additive |
| Nitrosamines | Pesticides, herbicides, cigarette smoke |
| Proflavine | Antiseptic in veterinary medicine |
| Sodium nitrite | Smoked meats |
| Tris (2,3-dibromopropyl phosphate) | Flame retardant in children's sleepwear |

**table 11.5**

**Sources of Radiation Exposure**

| Source | Percentage of Total |
|---|---|
| Natural (cosmic rays, sunlight, earth's crust) | 81% |
| Medical X rays | 11% |
| Nuclear medicine procedures | 4% |
| Consumer products | 3% |
| Other (nuclear fallout, occupational) | <1% |

## Natural Exposure to Mutagens

The simple act of being alive on planet earth exposes us to radiation that can cause mutation. Such natural environmental sources of radiation include cosmic rays, sunlight, and radioactive minerals in the earth's crust, such as radon. Contributions from medical X rays and occupational radiation hazards are comparatively minor (table 11.5). Job sites with increased radiation exposure include weapons facilities, research laboratories, health care facilities, nuclear power plants, and certain manufacturing plants.

Radiation exposure is measured in units called millirems; the average annual exposure in the northern hemisphere is 360 millirems.

Most of the potentially mutagenic radiation to which we are exposed is of the ionizing type, which means that it has sufficient energy to remove electrons from atoms. Unstable atoms that emit ionizing radiation exist naturally, and some are made by humans.

Ionizing radiation is of three major types. Alpha radiation is the least energetic and short-lived, and most of it is absorbed by the skin. Uranium and radium emit alpha radiation. Beta radiation can penetrate the body farther, and emitters include tritium (an isotope of hydrogen), carbon-14, and strontium-70. Both alpha and beta rays tend not to harm health, although each can do damage if inhaled or eaten. In contrast is the third type of ionizing radiation, gamma rays. These can penetrate all the way through the body, damaging tissues as they do so. Plutonium and cesium isotopes used in weapons emit gamma rays, and this form of radiation is intentionally used to kill cancer cells.

X rays are the major source of exposure to human-made radiation, and they are not a form of ionizing radiation. They have less energy and do not penetrate the body to the extent that gamma rays do.

The effects of radiation damage to DNA depend upon the functions of the mutated genes. Mutations in genes called oncogenes or tumor suppressors, discussed in chapter 17, can cause cancer. Radiation damage can be widespread too. Exposing cells to radiation and then culturing them causes a genome-wide destabilization, such that mutations may occur even after the cell has divided a few times. Cell culture studies have also identified a "bystander effect," in which radiation seems to harm even cells not directly exposed. Researchers have noted which cells in an experiment receive radiation, then detect chromosome breakage in cells located nearby as well as in the exposed cells. This effect is not well understood.

Chemical mutagens are in the environment too. Evaluating the risk of a specific chemical exposure causing a mutation is very difficult, largely because people vary greatly in inherited susceptibilities, and are exposed to many chemicals. The risk that exposure to a certain chemical will cause a mutation is often less than the natural variability in susceptibility within a population, making it nearly impossible to track the true source and mechanism of any mutational event. Data from the human genome project will be increasingly used to determine specific risks for specific employees who might encounter a mutagen in the workplace, but such testing raises ethical concerns. The Bioethics box in this chapter discusses testing for sensitivity to the element beryllium.

**KEY CONCEPTS**

Genes have different mutation rates. Spontaneous mutations result when unusual tautomers of bases during replication change the sequence. Spontaneous mutations occur more frequently in microorganisms and viruses because they reproduce often, and lack repair mechanisms. Mutations are more likely to happen when the nearby DNA is repetitive or unusually symmetrical. • Mutagens are chemicals or radiation that increase the likelihood of mutation. Researchers use mutagens to more quickly obtain mutants, which are studied to reveal normal gene function. Site-directed mutagenesis uses PCR primers with intentional base mismatches to engineer and amplify specific mutations. Accidental exposure to mutagens may come on the job, from nuclear accidents, medical treatments or weapons. • Natural radiation sources include cosmic rays, sunlight, and radioactive elements in the earth's crust. Gamma rays are the most damaging form of ionizing radiation.

## 11.3 Types of Mutations

Mutations are classified by exactly how they alter DNA. Table 11.6 summarizes the types of genetic changes described in this section using an analogy to an English sentence.

### Point Mutations

A **point mutation** is a change in a single DNA base. It is a **transition** if a purine replaces a purine (A to G or G to A) or a pyrimidine replaces a pyrimidine (C to T or T to C). It is a **transversion** if a purine replaces a pyrimidine or vice versa (A or G to T or C). A point mutation can have any of several consequences—or it may have no obvious effect at all on the phenotype, acting as a silent mutation.

#### *Missense Mutations*

A point mutation that changes a codon that normally specifies a particular amino acid into one that codes for a different amino acid is called a **missense mutation.** If the substituted amino acid alters the protein's conformation sufficiently or occurs at a site critical to its function, signs or symptoms of disease or an observable variant of a trait may result.

The point mutation that causes sickle cell disease (see figure 11.2) is a missense mutation. The DNA sequence CTC encodes the mRNA codon GAG, which specifies glutamic acid. In sickle cell disease, the mutation changes the DNA sequence to CAC, which encodes GUG in the mRNA, which specifies valine. This mutation changes the protein's shape, which alters its function.

A missense mutation can also affect the phenotype by decreasing the amount of a gene's product. For example, in 15 percent of people who have Becker muscular dystrophy—a milder adult form of the condition—the muscle protein dystrophin is normal, but its levels are reduced. The missense mutation causing the protein shortage is in the promoter for the dystrophin gene, which slows the transcription rate. Since cells then produce fewer mRNAs that encode dystrophin, the protein is scarce. Muscle function suffers.

Another way that missense mutations can affect proteins is to disrupt the trimming of long precursor molecules. Such a missense mutation causes the type of Ehlers-Danlos syndrome that affects the boy in figure 11.5.

A missense mutation can profoundly affect a gene's product if it alters a site where introns are removed from the mRNA. Retaining an intron adds bases to the pro-

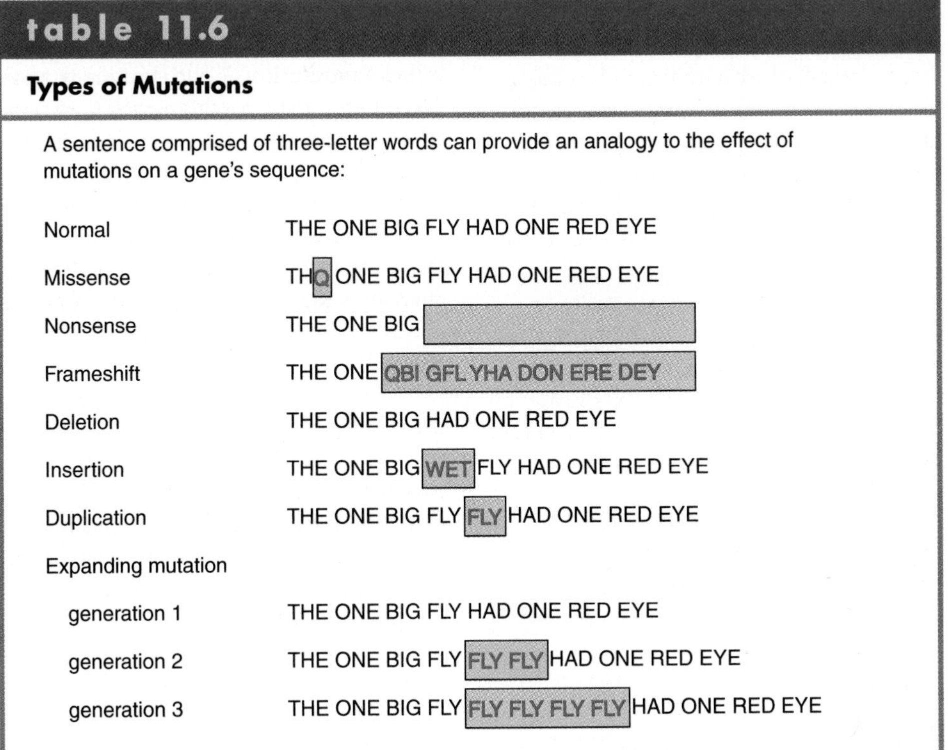

**table 11.6**

**Types of Mutations**

A sentence comprised of three-letter words can provide an analogy to the effect of mutations on a gene's sequence:

| | |
|---|---|
| Normal | THE ONE BIG FLY HAD ONE RED EYE |
| Missense | THQ ONE BIG FLY HAD ONE RED EYE |
| Nonsense | THE ONE BIG |
| Frameshift | THE ONE QBI GFL YHA DON ERE DEY |
| Deletion | THE ONE BIG HAD ONE RED EYE |
| Insertion | THE ONE BIG WET FLY HAD ONE RED EYE |
| Duplication | THE ONE BIG FLY FLY HAD ONE RED EYE |
| Expanding mutation | |
| generation 1 | THE ONE BIG FLY HAD ONE RED EYE |
| generation 2 | THE ONE BIG FLY FLY FLY HAD ONE RED EYE |
| generation 3 | THE ONE BIG FLY FLY FLY FLY FLY HAD ONE RED EYE |

tein coding portion of an mRNA. For example, in one family with severe cystic fibrosis, a missense mutation alters an intron site so that it is not removed. The encoded protein is too bulky to move to its normal position in the cell membrane, where it should enable salt to exit the cell. As a result, chloride (a component of salt) accumulates in cells and water moves in, drying and thickening the mucus outside the cells

A missense mutation need not alter the amino acid sequence to cause harm if it disrupts intron/exon splicing. For example, a missense mutation in the BRCA1 gene that causes breast cancer went undetected for a long time because it does not alter the amino acid sequence. Instead, the protein is missing several amino acids. What happens is that the missense mutation creates an intron splicing site where there should not be one, and an entire exon is "skipped" when the mRNA is translated into protein, as if it is an intron. This mutation, therefore, is a deletion (missing material), but caused by a missense mutation.

### *Nonsense Mutations*

A point mutation that changes a codon specifying an amino acid into a "stop" codon—UAA, UAG, or UGA in mRNA—is a **nonsense mutation.** A premature stop codon shortens the protein product, which can profoundly influence the phenotype. Nonsense mutations are predictable by considering which codons can mutate to a "stop" codon.

The most common cause of factor XI deficiency, a blood clotting disorder, is a nonsense mutation that changes the GAA codon specifying glutamic acid to UAA, signifying "stop." The shortened clotting factor cannot halt the profuse bleeding that occurs during surgery or from injury. In the opposite situation, when a normal stop codon mutates into a codon that specifies an amino acid, the resulting protein is longer than normal, because translation proceeds through what is normally a stop codon.

## Deletions and Insertions Can Cause Frameshifts

In genes, the number three is very important, because triplets of DNA bases specify amino acids. Adding or deleting a number of bases that is not a multiple of three devastates a gene's function because it disrupts the gene's reading frame, which refers to the nucleotide position where the DNA begins to encode protein. Such a change is called a **frameshift mutation.** The mutation that disables the CCR5 HIV receptor is a frameshift.

A **deletion mutation** removes genetic material. A deletion that removes three or a multiple of three bases will not cause a frameshift, but can still alter the phenotype. Deletions range from a single DNA nucleotide to thousands of bases to larger pieces of chromosomes. The next chapter considers large deletions. Many common inherited disorders result from deletions. About two-thirds of people with Duchenne muscular dystrophy, for example, are missing large sections of the very extensive gene that encodes dystrophin. Some cases of male infertility are caused by tiny deletions in the Y chromosome.

An **insertion mutation** adds genetic material and it, too, can offset a gene's reading frame. In one form of Gaucher disease, for example, an inserted single base prevents production of an enzyme that normally breaks down glycolipids in lysosomes. The resulting buildup of glycolipid enlarges the liver and spleen and causes easily fractured bones and neurological impairment. Gaucher disease is common among Jewish people of eastern European descent. Although most cases arise from a missense mutation, some families have the insertion mutation. Gaucher disease provides a good illustration of how different types of mutations in the same gene result in the same or a similar phenotype.

Another type of insertion mutation repeats part of a gene's sequence. The insertion is usually adjacent or close to the original sequence, like a typographical error repeating a word word. Two copies of a gene next to each other is called a **tandem duplication.** A form of Charcot-Marie-Tooth disease, which causes numb hands and feet, results from a one and a half million base long tandem duplication.

Figure 11.10 compares the effects on protein sequence of missense, nonsense, and frameshift mutations in the gene that encodes the LDL receptor, causing familial hypercholesterolemia (see figure 5.3). These three mutations exert very different effects on the protein. A missense mutation replaces one amino acid with another, bending the protein in a way that impairs its function. A nonsense mutation is much more drastic, removing a part of the protein. A frameshift mutation introduces a section of amino acids not normally found in the protein.

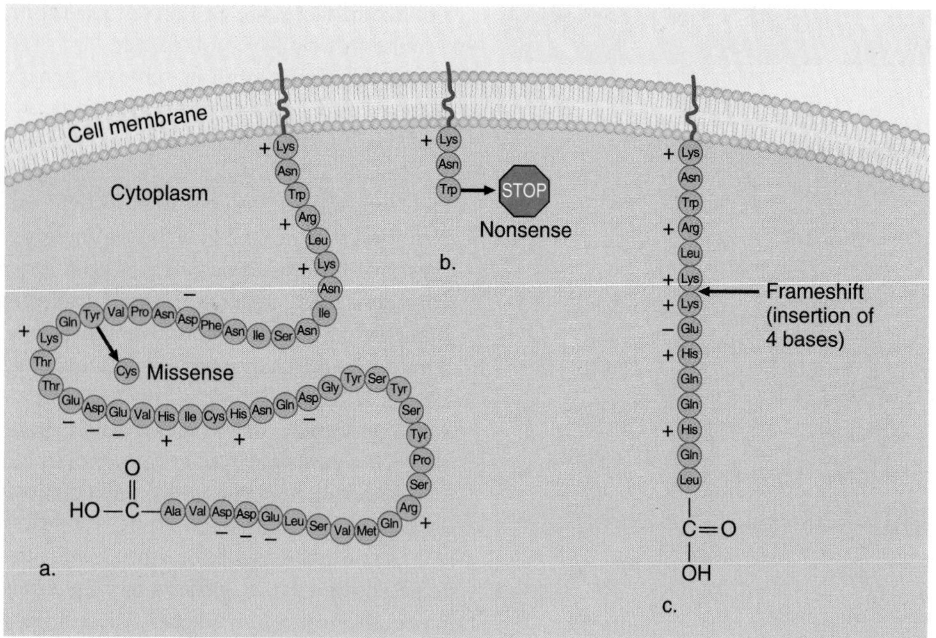


**figure 11.10**

**Different mutations in a gene can cause the same disorder.** In familial hypercholesterolemia, several types of mutations may disrupt the portion of the LDL receptor normally anchored in the cytoplasm. LDL receptor (*a*) bears a missense mutation—a substitution of a cysteine for a tyrosine. The receptor is bent enough to impair its function, causing disease. The short LDL receptor in (*b*) results from a nonsense mutation, in which a stop codon replaces a tryptophan codon. In (*c*), a 4-base insertion disrupts the gene sequence of the LDL receptor, which throws off the reading frame. A sequence of amino acids not normally in this protein continues until a stop codon occurs.

## Pseudogenes and Transposons Revisited

Recall from chapter 10 that a pseudogene is a stretch of DNA with a sequence very similar to that of another gene. A pseudogene is not translated into protein, although it may be transcribed. The pseudogene may have descended from the original gene sequence, which was duplicated when DNA strands misaligned during meiosis. When this happens, a gene and its pseudogene end up right next to each other on the chromosome. The original gene or the copy then mutated to such an extent that it was no longer functional and became a pseudogene. Its duplicate, though, lived on as the functional gene.

Although a pseudogene is not translated, its presence can interfere with the expression of the functional gene and cause a mutation. For example, some cases of Gaucher disease can result from a crossover between the working gene and its pseudogene, which has 95 percent of the same sequence located 16,000 bases away. The result is a fusion gene, which is a sequence containing part of the functional gene and part of the pseudogene. The fusion gene does not retain enough of the normal gene sequence to enable the enzyme to be synthesized. Gaucher disease results.

Chapter 10 also considered transposons, or "jumping genes." Transposons can alter gene function in several ways. They can disrupt the site they jump from, shut off transcription of the gene they jump into, or alter the reading frame of their destination if they are not a multiple of three bases. For example, a boy with X-linked hemophilia A had a transposon in his factor VIII gene—a sequence that was also in his carrier mother's genome, but on her chromosome 22. Apparently, in the oocyte, the transposon jumped into the factor VIII gene on the X chromosome, causing the boy's hemophilia.

## Expanding Repeats Lead to Protein Misfolding

Until 1992, myotonic dystrophy was a very puzzling disorder because it worsened and began at an earlier age as it passed from one generation to the next. This phenomenon is called "anticipation," and for many years was thought to be psychological. A grandfather might experience only mild weakness in his forearms, and cataracts. In the next generation, a daughter might have more noticeable arm and leg weakness, and a characteristic flat facial expression. By the third generation, children who inherit the genes might experience severe muscle impairment—worse if the affected parent was the mother.

With the ability to sequence genes, researchers found evidence that myotonic dystrophy was indeed worsening with each generation because the gene was expanding! The gene for the disorder, on chromosome 19, has an area rich in repeats of the DNA triplet CTG. A person who does not have myotonic dystrophy usually has from 5 to 37 copies of the repeat, whereas a person with the disorder has from 50 to thousands of copies (figure 11.11). Myotonic dystrophy is an example of an **expanding triplet repeat** disorder.

So far, expanding triplet repeats have been implicated in more than a dozen human inherited disorders. Usually, a repeat number of fewer than 40 copies is stably transmitted to the next generation and doesn't produce symptoms. Larger repeats are unstable, increasing in number with each generation and causing symptoms that are more severe and begin sooner. Reading 11.1 describes the first triplet repeat disorder to be discovered, fragile X syndrome.

As researchers discovered more expanding triplet repeat disorders, a pattern emerged. All of the conditions affect the brain, and most seem to involve protein misfolding. Somehow, the extra amino acids that an expanded gene can add to a protein affects the way that proteins fold and stick to each other or to other types of proteins.

The more repeats, the earlier symptoms begin and the more severe they are. For example, in Huntington disease, the extra amino acids disable the huntingtin protein, which normally stimulates transcription and translation of brain-derived neurotrophic factor (BDNF). In turn, BDNF normally acts in a part of the cerebral cortex called the striatum, where it keeps neurons alive by blocking apoptosis. Without functional huntingtin, these cells die, eventually causing symptoms. Because several transcription factors bind to huntingtin, it may have several effects.

Protein misfolding also accounts for certain other disorders that do not contain triplet repeat mutations, including certain inherited forms of Alzheimer disease, amyotrophic lateral sclerosis, and Parkinson disease. The affected proteins in triplet repeat disorders can misfold in several ways, with varied consequences. They may:

- fold incorrectly.
- fold, unfold, and refold incorrectly.
- link to each other into an intractable mass.
- be unable to bind to other proteins that they normally would contact.
- be cut too short.
- be unable to process another type of protein, causing it to form clumps.
- enter the nucleus when they normally do not, or be unable to enter when they normally do.

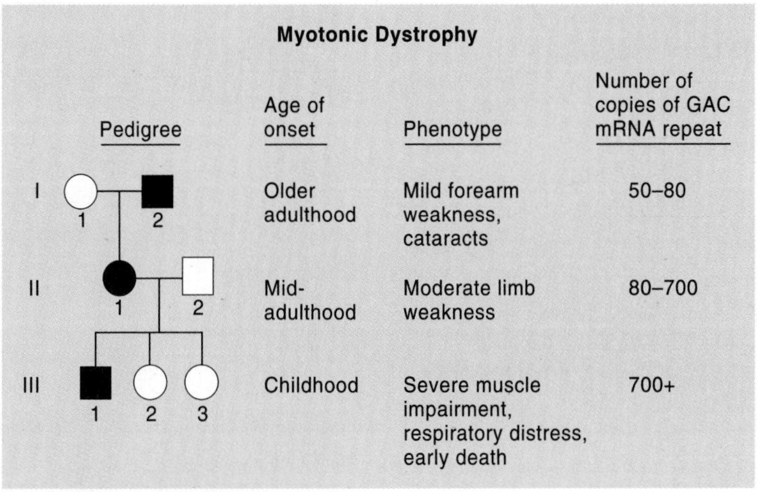

The triplet repeat disorders are described as causing a "dominant toxic gain of function." This means that they tend to be autosomal dominant (although a few are X-linked), and they usually cause something to happen that does not happen normally, rather than removing a function, such as is often associated with a recessive enzyme defect. The idea of a gain of function arose from the observation that deletions of these genes do not cause symptoms. Table 11.7 describes several triplet repeat disorders. Particularly common are the "polygln diseases" that have repeats of the mRNA codon CAG, which encodes the amino acid glutamine (*gln*).

For some triplet repeat disorders, the mutation thwarts gene expression before a protein is even manufactured. In myotonic dystrophy type I—the gene variant on chromosome 19 in which triplet repeats were

**Myotonic Dystrophy**

| Pedigree | Age of onset | Phenotype | Number of copies of GAC mRNA repeat |
|---|---|---|---|
| I (1, 2) | Older adulthood | Mild forearm weakness, cataracts | 50–80 |
| II (1, 2) | Mid-adulthood | Moderate limb weakness | 80–700 |
| III (1, 2, 3) | Childhood | Severe muscle impairment, respiratory distress, early death | 700+ |

**figure 11.11**

**Expanding genes explain anticipation.** In some disorders, symptoms that worsen from one generation to the next—a phenomenon termed *anticipation*—have a physical basis: the gene is expanding.

**table 11.7**

**Triplet Repeat Disorders**

| Disease | mRNA Repeat | Normal Number of Copies | Disease Number of Copies | Symptoms |
|---|---|---|---|---|
| Fragile X syndrome | CGG or CCG | 6–50 | 200–2,000 | Mental retardation, large testicles, long face |
| Friedreich ataxia | GAA | 6–29 | 200–900 | Loss of coordination and certain reflexes, spine curvature, knee and ankle jerks |
| Haw River syndrome | CAG | 7–25 | 49–75 | Loss of coordination, uncontrollable movements, dementia |
| Huntington disease | CAG | 10–34 | 40–121 | Personality changes, uncontrollable movements |
| Jacobsen syndrome | CGG | 11 | 100–1,000 | Poor growth, abnormal face, slow movement |
| Myotonic dystrophy type I | CTG | 5–37 | 80–1,000 | Progressive muscle weakness; heart, brain, and hormone abnormalities |
| Myotonic dystrophy type II | CCTG | <10 | >100 | Progressive muscle weakness; heart, brain, and hormone abnormalities |
| Spinal and bulbar muscular atrophy | CAG | 14–32 | 40–55 | Muscle weakness and wasting in adulthood |
| Spinocerebellar ataxia (5 types) | CAG | 4–44 | 40–130 | Loss of coordination |

*Reading 11.1*

## Fragile X Syndrome—The First of the Triplet Repeat Disorders

In the 1940s, geneticists hypothesized that a gene on the X chromosome confers mental retardation, because more affected individuals are male. It wasn't until 1969, though, that a clue emerged to the genetic basis of X-linked mental retardation. Two retarded brothers and their mother had an unusual X chromosome. The tips at one chromosome end dangled, separated from the rest of each chromatid by a thin thread (figure 1). When grown under specific culture conditions (lacking folic acid), this part of the X chromosome was very prone to breaking—hence, the name fragile X syndrome. Although fragile X syndrome was first detected at the chromosomal level, it turned out that the cause is a mutation at the DNA level—but of a type never seen before.

Fragile X syndrome is second only to Down syndrome in genetic or chromosomal causes of mental retardation. Worldwide, it affects 1 in 2,000 males, accounting for 4 to 8 percent of all males with mental retardation. One in 4,000 females is affected. They usually have milder cases because of the presence of a second, normal X chromosome.

Youngsters with fragile X syndrome do not appear atypical, but by young adulthood, certain similarities become apparent. The fragile X patient has a very long, narrow face. The ears protrude, the jaw is long, and the testicles are very large. Mental impairment and behavioral problems vary, and are related to difficulty in handling environmental stimuli. They include mental retardation, learning disabilities, repetitive speech, hyperactivity, shyness, social anxiety, a short attention span, language delays, and temper outbursts.

Fragile X syndrome is inherited in an unusual pattern. Because the fragile chromosome is the X, the associated syndrome should be transmitted as any X-linked trait is, from carrier mother to affected son. However, penetrance is quite low. One-fifth of males who inherit the chromosomal abnormality have no symptoms. However, because they pass on the affected chromosome to all their daughters—half of whom have some degree of mental impairment—they are called "transmitting males." A transmitting male's grandchildren may inherit the condition.

Researchers in the 1980s were on the right track when they proposed two states of the X chromosome region responsible for fragile X signs and symptoms—a premutation form that does not cause symptoms but does transmit the condition, and a full mutation form that usually causes mental impairment. Still, it took a molecular-level look in 1991 to begin to clarify the inheritance pattern of fragile X syndrome.

In unaffected individuals, the fragile X area contains 6 to 50 repeats of the DNA sequence CGG, as part of a gene called the fragile X mental retardation gene (FMR1). In people who have the fragile chromosome and show its effects, this region is greatly expanded to 200 to 2,000 CGG repeats. Transmitting males, as well as females with mild symptoms, often have a premutation consisting of an intermediate number of repeats—50 to 200 copies.

The FMR1 gene encodes fragile X mental retardation protein (FMRP). This protein, when abnormal, binds to and disables several different mRNA molecules whose encoded proteins are crucial for brain neuron function. The fact that a mutation in FMR1 ultimately affects several proteins explains the several signs and symptoms of fragile X syndrome.

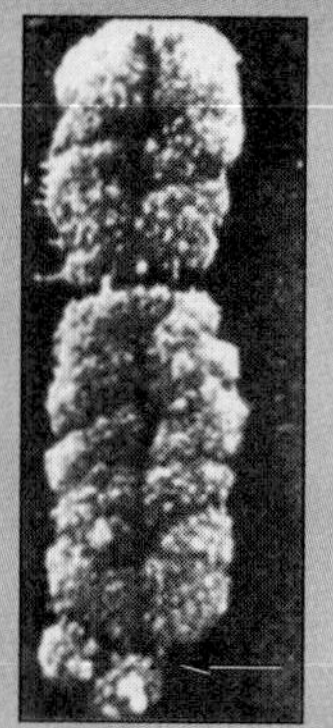

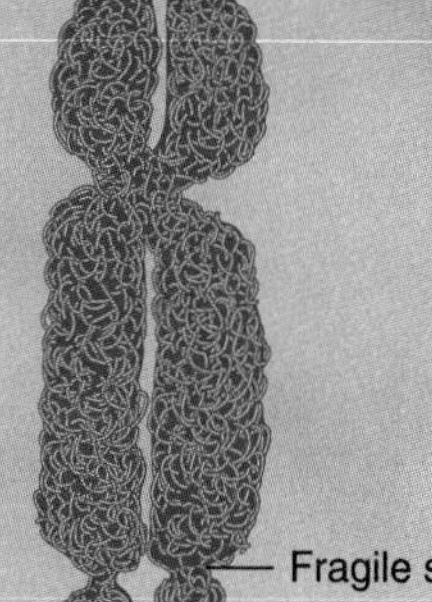


a.

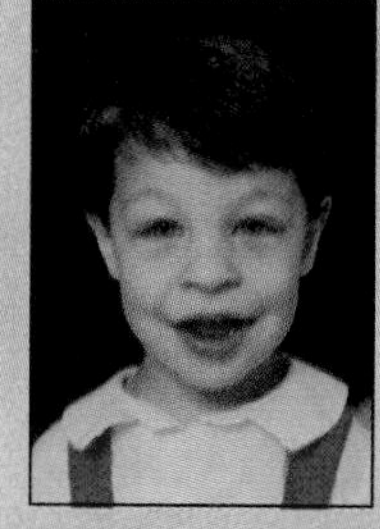

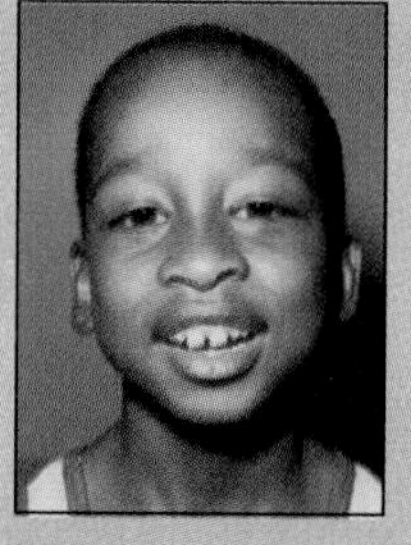

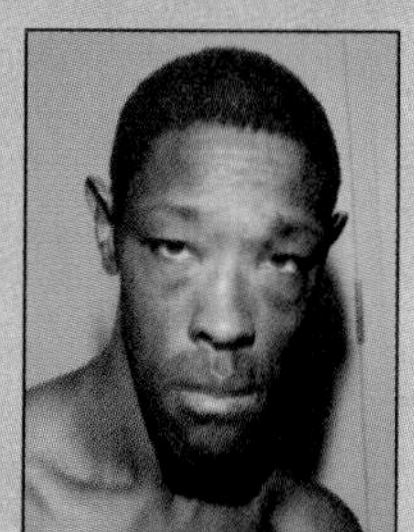

b.

**figure 1**

**Fragile X syndrome.** A fragile site on the tip of the long arm of the X chromosome (*a*) is associated with mental retardation and a characteristic long face that becomes pronounced with age (*b*).

discovered—the expansion occurs in the initial untranslated region of the gene, resulting in a huge mRNA. When genetic testing became available for myotonic dystrophy, researchers discovered a second form of the illness. Myotonic dystrophy type II is caused by an expanding quadruple repeat of (CCTG)n in a gene on chromosome 3. Affected individuals have more than 100 copies of the repeat, compared to the normal fewer than 10 copies (figure 11.12). When researchers realized that this second repeat mutation was also in a non-protein-encoding part of the gene—an intron—a mechanism of disease became apparent: the mRNA is simply too big to get out of the nucleus! In myotonic dystrophy type I, the excess material is tacked onto the front end of the gene; in type II, it appears in an intron that is not excised. Further evidence must support or refute this hypothesis, and identify this mechanism in other genes.

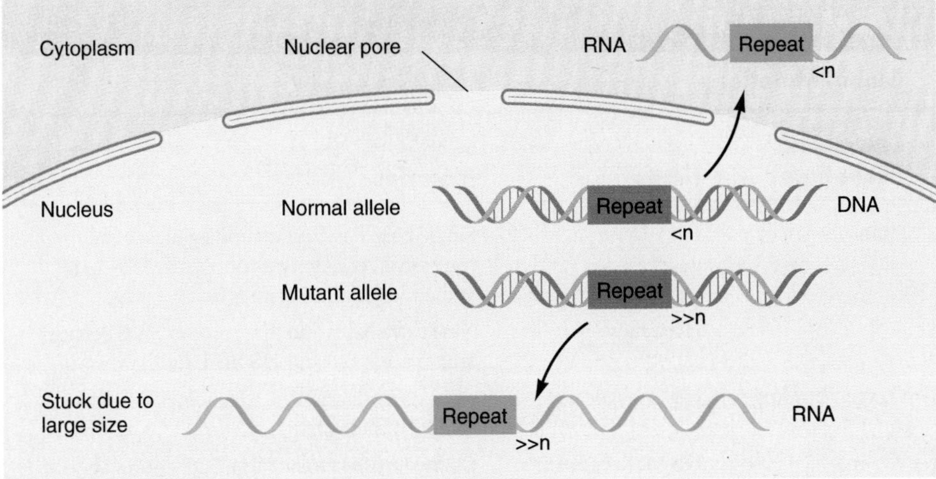


**figure 11.12**

**Triplet or quadruple repeats need not be in exons.** In both types of myotonic dystrophy, the gene is expanded in a part that does not encode protein. Part of the mechanism of disease may be that the mRNA is too large to exit the nucleus.

The lesson learned from the myotonic dystrophies is that a mutation need not affect the protein-encoding portions of genes—exons—to affect the phenotype. A more subtle and general lesson learned from the expanding triplet and now quadruple repeat disorders is that a DNA sequence is more than just one language that can be translated into another. Whether a sequence is random—CGT CGT ATG CAT CAG, for example—or highly repetitive—such as CAG CAG CAG CAG and on and on—can affect transcription, translation, or the ways that proteins interact. Several lines of evidence indicate that DNA may harbor more meaning than simple specification of amino acid sequences:

- Synthetic DNAs with random sequences migrate differently in electrical fields than do synthetic DNAs of the same length with repeated sequences.
- Transcription factors interact differently with random and repeated sequences.
- Repeats of CTG and CGG make the double helix more flexible.
- DNA with many CTG repeats winds more tightly about histone proteins to form nucleosomes than other DNA sequences do, affecting the accessibility of genes for transcription.
- DNA with many repeats can fold into a variety of structures as it is being replicated or transcribed. These include loops, triple and quadruple helices, and "slipped-strand" formations where the same repeats on homologs misalign, increasing the chance of mutation.

**Key Concepts**

A point mutation, which alters a single DNA base, can occur in a part of a gene that encodes protein, in a control region, or in a splicing junction between introns and exons. In a transversion, a purine replaces a pyrimidine, or vice versa; in a transition, a purine replaces a purine or a pyrimidine replaces a pyrimidine. A missense mutation replaces one amino acid with another. A nonsense mutation alters an amino-acid-coding codon into a "stop" codon, shortening the protein. A stop codon that changes to an amino-acid-coding codon, conversely, lengthens the protein. • Inserting or deleting bases can upset the DNA reading frame, causing a frameshift mutation or otherwise altering protein function. Tandem duplications repeat a section of a gene. Pseudogenes are nonfunctional sequences very similar to a nearby functional gene. Transposons can move, insert into genes, and cause illness. • Expanded repeats disturb brain function, possibly by adding function by causing protein misfolding. Repeats may affect exons, causing a too-large protein to be synthesized, or affect noncoding gene regions, resulting in mRNAs too large to exit the nucleus. The degree of repetition in a DNA region may affect its structure and function.

## 11.4 The Importance of a Mutation's Position in the Gene

The degree to which a mutation alters the phenotype depends to a great extent upon where in the gene the change occurs, and how the mutation affects the conformation or activity of the encoded protein. A mutation that replaces an amino acid with a very similar one would probably not affect the phenotype greatly, because it wouldn't substantially change the conformation of the protein. Even substituting a very different amino acid would not have much effect if the change is in part of the protein not crucial to its function. Yet, sickle cell disease and the disorders of collagen illustrate that even a small change in a gene can drastically alter the encoded protein's function. The effects of specific mutations are well-studied in the hemoglobin molecule, and less understood, but also fascinating, in the gene that encodes prion protein.

### Globin Variants

Because the globin gene mutations were the first to be studied in humans, and because some variants are easily detected using electrophoresis, hundreds of mutations are known. Mutations in the globin genes can cause anemia with or without sickling, or cyanosis (a blue pallor due to poor oxygen binding), or, rarely, boost the molecule's affinity for oxygen. Some globin gene variants exert no effect on the phenotype at all, and are thus termed "clinically silent." Oddly, hemoglobin S and hemoglobin C each change the sixth amino acid in the beta globin polypeptide, but in different ways. Homozygotes for hemoglobin S have sickle cell disease, yet homozygotes for hemoglobin C are healthy. Both types of homozygotes are resistant to malaria, a phenomenon discussed in chapter 14. Table 11.8 lists some hemoglobin variants with differing effects, along with the nature of the mutations that define them.

An interesting consequence of certain mutations in either the alpha or beta globin chains is hemoglobin M. Normally, the iron in hemoglobin is in the ferrous form, which means that it has two positive charges.

In hemoglobin M, the mutation stabilizes the ferric form, which has three positive charges and cannot bind oxygen. Fortunately, an enzyme converts the abnormal ferric iron to the normal ferrous form, so that the only symptom is usually cyanosis. The condition has been known for more than 200 years in a small town in Japan. Many people there have "blackmouth," referring to the cyanosis caused by the faulty hemoglobin.

## Inherited Susceptibility to Prion Disorders

We know a great deal less about the inherited prion disorders, caused by mutations in the prion protein gene, than we do about mutations in the globin genes, but here, too, certain mutations exert drastic effects on health, and others don't. Recall from chapter 2 that a prion is a protein with an unknown function in its stable form that behaves in an infectious manner in its abnormal form, converting other prion proteins to the abnormal form and spreading brain destruction. A prion disease can be inherited, or acquired, such as developing variant Creutzfeldt-Jakob disease from eating beef from a cow that had bovine spongiform encephalopathy. The prion protein has at least eight distinct conformations. The normal form of the protein has a central core made up of helices. In the disease-causing form, the helices open into a sheet (see figure 10.18). Precise genetic changes make all the difference in the form of the prion protein—and in the health of the person.

The nature of the amino acid at position 129 in the prion protein is key to developing disease. In people who inherit these disorders, amino acid 129 is either valine in all copies of the protein (genotype VV) or methionine in all copies (genotype MM). These people are homozygous for this small part of the gene. Most people, however, are heterozygotes, with valine in some prion proteins and methionine in others (genotype VM). One hypothesis is that having two different amino acids at this position enables the proteins to assemble and carry out their normal function, without damaging the brain. Further studies on the gene revealed that a mutation at a different site raises the risk even higher. Comparing the prion proteins of healthy individuals to those who inherited a prion disorder showed that normally prion protein folds so that amino acid 129 is near amino acid 178, which is aspartic acid. People who inherit prion diseases are not only homozygous for the gene at position 129, they also display another mutation that changes amino acid 178 to asparagine. Interestingly, people with two valines at 129 develop a condition called fatal familial insomnia, whereas those with two methionines develop a form of Creutzfeldt-Jakob syndrome.

Other genes can affect susceptibility to prion disorders too. So far, 80 percent of the more than 120 people to contract the infectious vCJD lack a specific allele of an immune system gene. Studies on mice suggest that several other genes may affect susceptibility to prion disorders.

Although we still have much to learn about the genetic underpinnings of the strange prion disorders, researchers are already applying the little that is known. For example, sheep and cows, which are prone, respectively, to the prion disorders scrapie and bovine spongiform encephalopathy, can be genetically altered to have valine and methionine at position 129 and aspartic acid at position 178, the genotypes that seem to prevent prion disorders. Perhaps this will be done in the future

**table 11.8**

**Globin Mutations**

| Associated Phenotype | Name | Mutation |
|---|---|---|
| Clinically silent | Hb Wayne | Single base deletion in alpha gene causes frameshift, changing amino acids 139–141 and adding amino acids |
| | Hb Grady | Nine extra bases add three amino acids between amino acids 118 and 119 of alpha chain |
| Oxygen binding | Hb Chesapeake | Change from arginine to leucine at amino acid 92 of beta chain |
| | Hb McKees Rock | Change from tyrosine to STOP codon at amino acid 145 in beta chain |
| Anemia | Hb Constant Spring | Change from STOP codon to glutamine elongates alpha chain |
| | Hb S | Change from glutamic acid to valine at amino acid 6 in beta chain causes sickling |
| | Hb Leiden | Amino acid 6 deleted from beta chain |
| Protection against malaria | Hb C | Change from glutamic acid to lysine at amino acid 6 in beta chain causes sickling |

**Key Concepts**

Whether a mutation alters the phenotype, and how it does so, depends upon where in the protein the change occurs. Mutations in the globin genes are well-studied and diverse; they may cause anemia or cyanosis or they may be silent. Hemoglobin M affects the state of the bound iron. • Mutations in two parts of the prion protein gene predispose an individual to developing a prion disorder.

## 11.5 Factors That Lessen the Effects of Mutation

Mutation is a natural consequence of DNA's ability to change, and it has been and continues to be essential for evolution. However, many factors minimize the deleterious effects of mutations on phenotypes.

Synonymous codons render many alterations in the third codon position "silent." For example, a change from RNA codon CAA to CAG does not alter the designated amino acid, glutamine, so a protein whose gene contains the change would not be altered. The genetic code has other nuances that protect against drastically altering proteins. Mutations in the second codon position sometimes replace one amino acid with another that has a similar conformation. Often, this does not disrupt the protein's form too drastically. For example, a GCC mutated to GGC replaces alanine with glycine; both are very small amino acids.

A **conditional mutation** affects the phenotype only under certain conditions. This can be protective if an individual avoids the exposures that trigger symptoms. Consider a common variant of the X-linked gene that encodes glucose 6-phosphate dehydrogenase (G6PD), an enzyme that immature red blood cells use to extract energy from glucose. One hundred million people worldwide have G6PD deficiency, which causes life-threatening hemolytic anemia, but only under rather unusual conditions—eating fava beans, inhaling pollen in Baghdad, or taking a certain antimalarial drug.

In the 5th century B.C., the Greek mathematician Pythagoras wouldn't allow his followers to consume fava beans—he had discovered that it would make some of them ill. During the second world war, several soldiers taking the antimalarial drug primaquine developed hemolytic anemia. A study began shortly after the war to investigate the effects of the drug on volunteers at the Stateville Penitentiary in Joliet, Illinois, and researchers soon identified abnormal G6PD in people who developed anemia when they took the drug.

What do fava beans, antimalarial drugs, and dozens of other triggering substances have in common? They "stress" red blood cells by exposing them to oxidants, chemicals that strip electrons from other compounds. Without the enzyme, the stress bursts the red blood cells.

Another example of a conditional mutation is trichothiodystrophy. Symptoms of brittle hair and nails and scaly skin arise in some patients only during periods of fever that persist long enough for hair, nail, and skin changes to become noticeable. Heat destabilizes the gene product, which is an enzyme that functions in base excision repair, discussed in the next section.

**Key Concepts**

Genetic code degeneracy ensures that some third-codon-position mutations do not alter the specified amino acid. Changes in the second codon position often replace an amino acid with a structurally similar one. Conditional mutations are expressed only in certain environments.

## 11.6 DNA Repair

Any manufacturing facility tests a product in several ways to see whether it has been built correctly. Mistakes in production are rectified before the item goes on the market—at least, most of the time. The same is true for a cell's manufacture of DNA.

DNA replication is incredibly accurate—only about 1 in 100 million bases is incorrectly incorporated. Repair enzymes oversee the fidelity of replication.

All organisms can repair their nuclear DNA, although some species do so more efficiently than others. Mitochondrial DNA cannot repair itself, which accounts for its higher mutation rate. The master at DNA repair is a large, reddish microbe that *The Guinness Book of World Records* named "the world's toughest bacterium." *Deinococcus radiodurans* was discovered in a can of spoiled ground meat at the Oregon Agricultural Experiment Station in Corvallis in 1956, where it had withstood the radiation used to sterilize the food. It tolerates 1,000 times the radiation level that a person can, and it even lives amidst the intense radiation of a nuclear reactor. The bacterium realigns its radiation-shattered pieces of genetic material and enzymes bring in new nucleotides and stitch together the pieces.

The discovery of DNA repair systems began with observations in the late 1940s that when fungi were exposed to ultraviolet radiation, those cultures nearest a window grew best. The DNA-damaging effect of ultraviolet radiation, and the ability of light to correct it, was soon observed in a variety of organisms.

### Types of DNA Repair

Since its beginning, the Earth has been periodically bathed in ultraviolet radiation. Volcanoes, comets, meteorites, and supernovas all depleted ozone in the atmosphere, which allowed ultraviolet wavelengths of light to reach organisms. The shorter wavelengths—UVA—are not dangerous, but the longer UVB wavelengths damage DNA by causing an extra covalent bond to form between adjacent (same strand) pyrimidines, particularly thymines (figure 11.13). The linked thymines are called thymine dimers. Their extra bonds kink the double helix sufficiently to disrupt replication and lead to possible insertion of a noncomplementary base. For example, an A might be inserted opposite a G or C instead of a T.

Early in the evolution of life, organisms that could handle UV damage enjoyed a survival advantage. Enzymes enabled them to do this, and thus DNA repair came to persist. In many modern species, three types of DNA repair mechanisms peruse the genetic material for mismatched base pairs. In the first type of DNA repair, enzymes called photolyases absorb energy from visible light and use it to detect and bind to pyrimidine dimers, then break the extra bond. This type of repair, called photoreactivation, is what enabled ultraviolet-damaged fungi to recover when exposed to sunlight. Humans do not have it.

In the early 1960s, researchers discovered a second type of DNA self-mending, called **excision repair,** in mutant *E. coli* that were extra sensitive to ultraviolet radiation. These bacteria were unable to repair ultraviolet-induced DNA damage. The enzymes that carry out excision repair cut the bond between the DNA sugar and base and snip out—or excise—the pyrimidine dimer and surrounding bases. Then, a DNA polymerase fills in the correct nucleotides, using the exposed template as a guide.

Two types of excision repair are recognized. **Nucleotide excision repair** replaces up to 30 nucleotides and removes errors that result from several types of insults, including exposure to chemical carcinogens, UVB in sunlight, and oxidative damage. The second type, called **base**

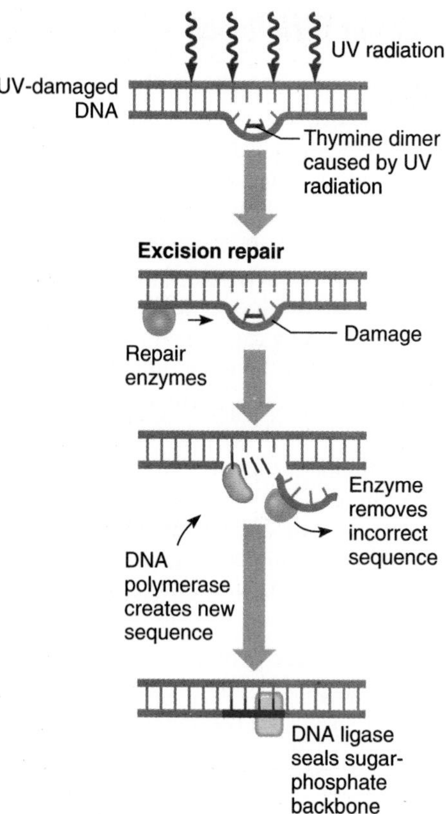


**figure 11.13**

**Excision repair.** Human DNA damaged by UV light is repaired by excision repair, in which the pyrimidine dimer and a few surrounding bases are removed and replaced.

**excision repair,** replaces one to five nucleotides at a time, but specifically corrects errors that result from oxidative damage. Oxygen free radicals are highly reactive forms of oxygen that arise during chemical reactions such as those of metabolism and transcription. Free radicals damage DNA. Genes that are very actively transcribed face greater oxidative damage from free radicals. It is base excision repair that targets this type of damage.

A third mechanism of DNA repair is called **mismatch repair.** In this type of repair, enzymes "proofread" newly replicated DNA for small loops that emerge from the double helix. The enzymes excise the mismatched base so that it can be replaced (figure 11.14). These loops indicate an area where the two strands are not precisely aligned, as they should be if complementary base pairing is occurring at every point. Such slippage and mismatching tends to occur in chromosome regions where very short DNA sequences repeat. These sequences, called microsatellites, are scattered throughout the genome. Like minisatellites, microsatellite lengths can vary from person to person, but within an individual, they are usually all the same length. Excision and mismatch repair differ in the cause of the error—ultraviolet-induced pyrimidine dimers versus replication errors—and the types of enzymes involved.

## DNA Repair Disorders

The ability to repair DNA is crucial to health. Mutations in any of the genes whose protein products take part in DNA repair can cause problems. A particular repair disorder may be genetically heterogeneic because it can be caused by mutations in any of several genes that participate in the same repair mechanism. That is, different single-gene defects can cause the same symptoms.

At the crux of the decision as to whether a cell whose DNA has been damaged can be repaired is a protein called p53. Signal transduction activates the p53 protein, stabilizing it and causing it to aggregate into complexes consisting of four proteins. These quartets bind to the DNA by recognizing four palindromic repeats (sequences that read the same in both directions) that indicate genes that slow the cell cycle. The cycle must slow so that repair can take place. If the damage is too severe to be repaired, the p53 protein quartets instead increase the rate of transcription of genes that promote apoptosis. At the same time, p53 increases transcription of yet another protein that regulates the synthesis of the p53 protein, so that it can shut off its own production.

DNA repair disorders break chromosomes and greatly increase susceptibility to certain types of cancer following exposure to ionizing radiation or chemicals that affect cell division. These conditions develop because errors in the DNA sequence accumulate and are perpetuated to a much greater extent than they are in people with functioning repair systems. Table 11.9 lists several disorders that result from faulty DNA replication or repair. We conclude this chapter with a closer look at a few of them.

### *Trichothiodystrophy*

At least five different genes can cause trichothiodystrophy. At its worst, trichothiodystrophy causes dwarfism, mental retardation, failure to develop, and childhood death, in addition to the scaly hair with low sulfur content from which the illness takes it name. Although the child may appear to be normal for a year or two, growth soon slows dramatically, and signs of premature aging begin. Hearing and vision may fail. Interestingly, the condition does not increase the risk of cancer. Symptoms reflect accumulating oxidative damage. Individuals have faulty nucleotide excision repair, base excision repair, or both (figure 11.15*a*).

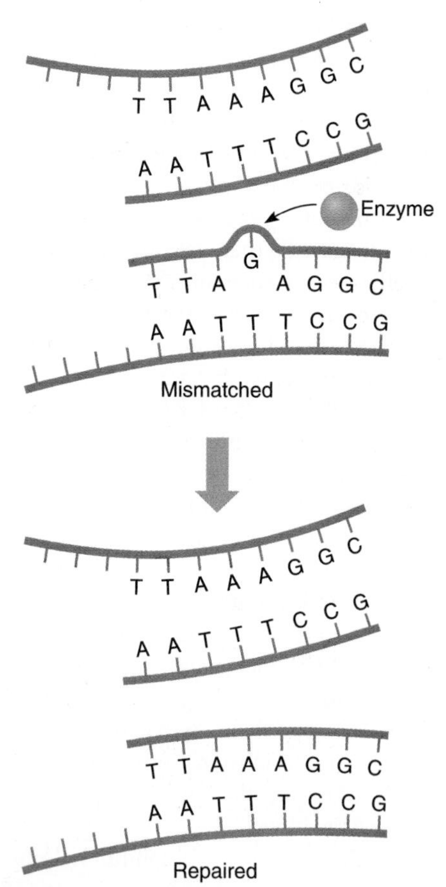


**figure 11.14**

**Mismatch repair.** In this form of DNA repair, enzymes detect loops and bulges in newly replicated DNA that indicates mispairing. The enzymes correct the error. Highly repeated sequences are more prone to this type of error.

### *Inherited Colon Cancer*

Hereditary nonpolyposis colon cancer was linked to a DNA repair defect when researchers discovered that these cancer cells exhibit different-length microsatellites within an individual. Because mismatch repair normally keeps a person's microsatellites all the same length, researchers hypothesized that people with this type of colon cancer might be experiencing a breakdown in this form of DNA repair. The causative gene is located on chromosome 2 and is remarkably similar to a corresponding mismatch repair gene in *E. coli.* Hereditary nonpolyposis colon cancer is a common repair disorder, affecting 1 in 200 people.

### *Xeroderma Pigmentosum (XP)*

A child with XP lives, intentionally, indoors in artificial light, because even the briefest exposure to sunlight causes painful blisters. Failing to cover up and use sunblock can result in skin cancer (figure 11.15*b*). More than half of all children with XP develop the cancer before they reach their teens.

XP is autosomal recessive and can reflect malfunction of any of nine enzymes that take part in either type of excision repair. It is extremely rare—only about 250 people in the world are known to have it. A family living in upstate New York runs a special summer camp for children with XP, where they turn night into day, as they have done for their own affected daughter Katie. Activities take place at night, or in special areas where the windows are covered and light comes from low-ultraviolet incandescent lightbulbs.

### *Ataxia Telangiectasis (AT)*

This multisymptom disorder is the result of a defect in a kinase that functions as a cell cycle checkpoint (see figure 2.17). Cells proceed through the cell cycle without pausing just after replication to inspect the new DNA and to repair any mispaired bases. Some cells die through apoptosis if the damage is too great to repair. As a result of the malfunctioning cell cycle, individuals who have this autosomal recessive disorder have extremely high rates of cancers, particularly of the blood. Additional symptoms include poor balance and coordination (ataxia), red marks on the face (telangiectasia), delayed sexual maturation, and high risk of contracting lung infections and developing diabetes mellitus.

**table 11.9**

**Some DNA Replication and Repair Disorders**

| Disorder | Frequency | Defect |
|---|---|---|
| Ataxia telangiectasis | 1/40,000 | Deficiency in kinase that controls the cell cycle |
| Bloom syndrome (two types) | 100 cases since 1950 | DNA ligase is inactive or heat sensitive, slowing replication |
| Fanconi anemia (several types) | As high as 1/22,000 in some populations | Deficient excision repair |
| Hereditary nonpolyposis colon cancer | 1/200 | Deficient mismatch repair |
| Werner syndrome (a progeria) | 3/1,000,000 | Deficient helicase |
| Xeroderma pigmentosum (nine types) | 1/250,000 | Deficient excision repair |
| Trichothiodystrophy (five or more types) | Fewer than 100 cases | Deficient excision repair |

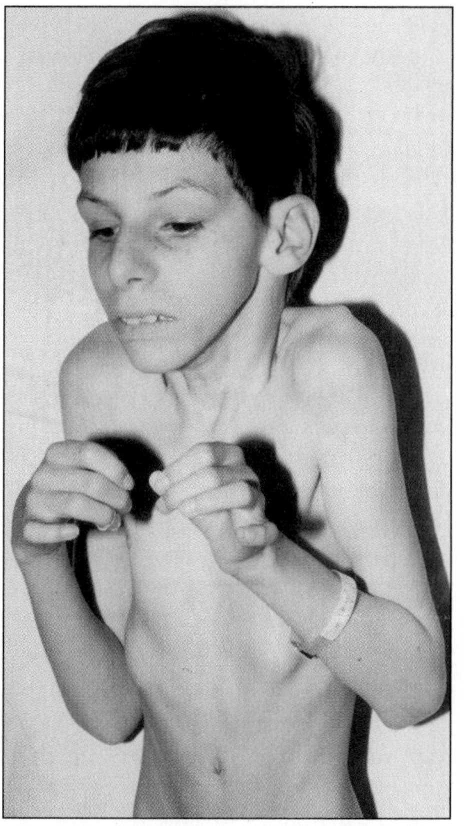

a.

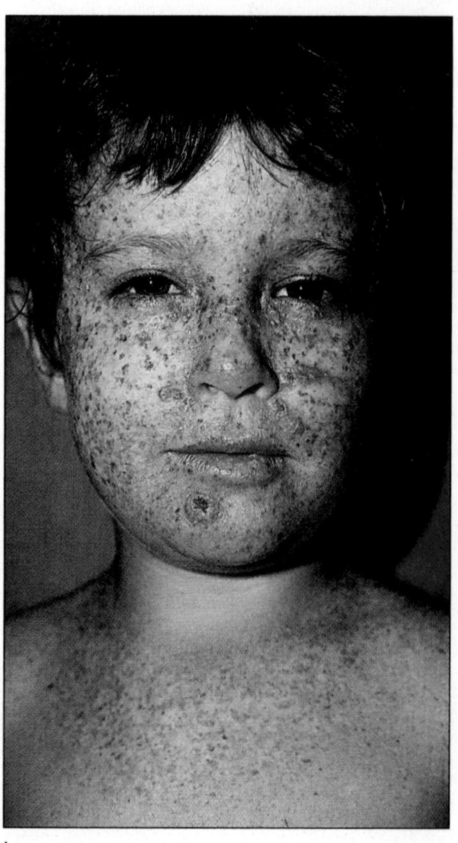

b.

**figure 11.15**

**DNA repair disorders.** (*a*) Trichothiodystrophy (a form called Cockayne syndrome) causes a child to appear aged. Excision repair fails. (*b*) The marks on this child's face are a result of sun exposure. He is highly sensitive because he has inherited xeroderma pigmentosum (XP), also an impairment of excision repair. The large lesion on his chin is a skin cancer.

# Beryllium Screening

On the surface, screening workers for a genetic variant that predisposes them to develop a possibly fatal reaction to a substance they may contact on the job may seem like a good idea. But some workers say the risks outweigh uncertain benefits. The case in point: screening for chronic beryllium disease (CBD), also called berylliosis. This condition causes the person to react to the metal beryllium, which is used in nuclear power plants, in electronics, and in manufacturing fluorescent powders. Exposed workers include those who mine beryllium, nuclear power plant employees, and support staff such as office workers who inhale beryllium dust.

## figure 1

**Beryllium screening.** Screening for beryllium sensitivity is done to protect workers, but some people regard it as an invasion of their genetic privacy. The lung tissue in the inset is granular in appearance, a sign of damage from exposure to beryllium in a sensitive individual.

A small percentage of people exposed to beryllium dust or vapor develop an immune response that damages the lungs, producing cough, shortness of breath, fatigue, loss of appetite, and weight loss. Fevers and night sweats indicate the immune system is responding to the exposure. The steroid drug prednisone can control symptoms, but isn't used until symptoms begin, which can be anywhere from a few months to 40 years after the first exposure.

The Department of Energy and some private companies have screened more than 10,000 workers exposed to beryllium on the job using a test based on an immune system response. In people who have symptoms, certain white blood cells divide in the presence of beryllium. People without symptoms who test positive go on to develop the condition about 45 percent of the time. A more precise and predictive genetic test will eventually replace the immune system test, but even this test is far from perfect. It detects homozygosity for a rare genetic variant that is part of the human leukocyte antigen complex, a group of genes that control immune system function. A person who tests positive on the genetic test has an 85 percent chance of developing CBD if exposed to beryllium (figure 1).

Experience so far with the immune system test suggests that screening for beryllium sensitivity using genetic tests will be controversial. So far, workers at the Department of Energy who test positive on the immune system test are not allowed to work near beryllium. Some of them resent this, preferring to make their own decisions concerning where they work. In addition, both the immune system and genetic tests are inconclusive—that is, they detect only a susceptibility, meaning that some people who test positive will not develop the condition. Finally, people labeled as "sensitive" fear that their test results might have negative effects on their health insurance coverage. These issues will continue to mount as DNA tests become more powerful.

The goal of CBD screening at the Department of Energy and the companies is to protect workers. Not only are sensitive individuals kept away from beryllium, but efforts have been underway for several years to minimize the dust and to make sure all beryllium workers use protective clothing and devices. Still, some people object to identifying susceptible individuals. This is yet another area where "political correctness" clashes with the reality of genetics—individuals *do* vary in their responses to many stimuli, including environmental lung irritants. To deny that is to expose certain people to potentially harmful surroundings. Genetic screening for beryllium sensitivity may set a precedent for other types of susceptibility testing that will be developed using human genome information.

These symptoms probably arise from disruption of other functions of the kinase.

AT is rare, but heterozygotes are not. They make up from 0.5 to 1.4 percent of various populations. Carriers may have mild radiation sensitivity, which causes a two- to sixfold increase in cancer risk over the general population.

We continue looking at mutation in chapter 12, at the chromosomal level.

### Key Concepts

Many genes encode enzymes that search for and correct errors in replicating DNA. A common cause of noncomplementary base insertion is an ultraviolet radiation-induced pyrimidine dimer. Photoreactivation or excision repair can unlink pyrimidine dimers. Mismatch repair corrects noncomplementary base pairs that form in newly replicated DNA. • Abnormal repair genes cause disorders that are usually associated with chromosome breaks and predisposition to cancer.

# Summary

### 11.1 Mutations Can Alter Proteins—Three Examples

1. A **mutation** is a change in a gene's nucleotide base sequence that is rare and usually causes a **mutant** phenotype.
2. A **germline mutation** originates in meiosis and affects all cells of an individual. A **somatic mutation** originates in mitosis and affects a subset of cells.
3. A mutation causes illness by disrupting the function or amount of a protein. In sickle cell disease, beta globin is misshapen; in beta thalassemia, it is absent or reduced. Mutations readily disrupt the highly symmetrical gene encoding collagen. One form of Alzheimer disease is caused by mutation in a receptor protein.

### 11.2 Causes of Mutation

4. A spontaneous mutation arises due to chemical phenomena or to an error in DNA replication. Spontaneous mutation rate is characteristic of certain genes. Spontaneous mutation is more likely to occur in repeated or symmetrical DNA sequences.
5. **Mutagens** are chemicals or forms of radiation that can induce mutation by deleting, substituting, or adding bases. An organism may be exposed to a mutagen intentionally, accidentally, or naturally.

### 11.3 Types of Mutations

6. A **point mutation** alters a single DNA base. It may be a **transition** (purine to purine or pyrimidine to pyrimidine) or a **transversion** (purine to pyrimidine or vice versa). A **missense mutation** substitutes one amino acid for another, while a **nonsense mutation** substitutes a "stop" codon for a codon that specifies an amino acid, shortening the protein product.
7. Adding or deleting genetic material may upset the reading frame or otherwise alter protein function.
8. A **pseudogene** results when a duplicate of a gene mutates. It may disrupt chromosome pairing, causing mutation.
9. **Transposons** may disrupt the functions of genes they jump into.
10. Expanding triplet repeat mutations add stretches of the same amino acid to a protein, usually one that functions in the brain. This type of mutation may add a function, often leading to a neurodegenerative disease when the number of repeats exceeds a threshold level. Expanding repeat mutations also occur in non-protein-encoding regions, resulting in mRNAs that are too large to exit the nucleus.

### 11.4 The Importance of a Mutation's Position in the Gene

11. Several types of mutations can affect a gene.
12. Mutations in the globin genes may affect the ability of the blood to transport oxygen, or have no effect.
13. Susceptibility to prion disorders requires one to inherit two mutations that affect different parts of the protein that interact as the amino acid chain folds.

### 11.5 Factors That Lessen the Effects of Mutation

14. Synonymous codons limit the effects of mutation. Changes in the second codon position often substitute a similarly-shaped amino acid, so the protein's function may not be impaired.
15. **Conditional mutations** are expressed only in response to certain environmental triggers.

### 11.6 DNA Repair

16. DNA polymerase proofreads DNA, but repair enzymes correct errors in other ways.
17. **Photoreactivation** repair uses light energy to split pyrimidine dimers that kink the DNA.
18. In **excision repair,** pyrimidine dimers are removed and the area is filled in correctly. **Nucleotide excision repair** replaces up to 30 nucleotides from various sources. **Base excision repair** fixes up to five bases that are incorrectly paired due to oxidative damage.
19. **Mismatch repair** proofreads newly replicated DNA for loops that indicate noncomplementary base pairing.
20. Mutations in repair genes lead to chromosome breakage and increased cancer risk.

# Review Questions

1. Distinguish between a germinal and a somatic mutation. Which is likely to be more severe? Which is more likely to be transmitted to offspring?
2. Why is the collagen gene particularly prone to mutation?
3. Describe how a spontaneous mutation can arise.
4. What is the physical basis of a mutational hot spot?
5. What are three different types of mutations that cause Gaucher disease?
6. Cite three ways in which the genetic code protects against the effects of mutation.
7. List four ways that DNA can mutate without affecting the phenotype.
8. What is a conditional mutation?
9. List two types of mutations that can disrupt the reading frame.
10. Why can a mutation that retains an intron's sequence and a triplet repeat mutation have a similar effect on a gene's product?
11. Cite two ways that a jumping gene can disrupt gene function.
12. Cite two reasons it takes many years to detect induction of recessive mutations in a human population.
13. What is a physical, molecular explanation for anticipation, the worsening of an inherited illness over successive generations?
14. Compare and contrast the ways that short repeats within a gene, long triplet repeats within a gene, and repeated genes can cause disease.
15. A confusing aspect of medical genetics is that in some genes, any mutation produces a variant of the same disorder, such as cystic fibrosis or Gaucher disease. For other genes, however, different mutations are associated with different diseases. Give an example from the chapter in which mutations in the same gene yield distinct medical conditions.
16. How do excision and mismatch repair differ?

# Applied Questions

1. A condition called "congenital insensitivity to pain with anhidrosis" causes loss of the ability to feel pain, inability to sweat, fever, mental retardation, and self-mutilation. The genetic cause is a mutation that replaces a glycine (*gly*) with an arginine (*arg*). List every way this change can occur.
2. Retinitis pigmentosa causes night blindness and loss of peripheral vision before age 20. A form of X-linked retinitis pigmentosa is caused by a frameshift mutation that deletes 199 amino acids. How can a simple mutation have such a drastic effect?
3. One form of Ehlers-Danlos syndrome (not the "stretchy skin" type described in the chapter) can be caused by a mutation that changes a C to a T. This change results in the formation of a "stop" codon and premature termination of procollagen. Consult the genetic code table and suggest a way that this can happen.
4. Townes-Brocks syndrome causes several unrelated problems, including extra thumbs, a closed anus, hearing loss, and malformed ears. The causative mutation occurs in a transcription factor. How can a mutation in one gene cause such varied symptoms?
5. Susceptibility to developing prion diseases entails a mutation from aspartic acid (*asp*) to asparagine (*asn*). Which nucleotide base changes make this happen?
6. Two teenage boys meet at a clinic set up to treat muscular dystrophy. The boy who is more severely affected has a two-base insertion at the start of his dystrophin gene. The other boy has the same two-base insertion but also has a third base inserted a few bases away. Explain why the second boy's illness is milder.
7. About 10 percent of cases of amyotrophic lateral sclerosis (also known as ALS and Lou Gehrig disease) are inherited. This disorder causes loss of neurological function over a five-year period. Two missense mutations cause ALS. One alters the amino acid asparagine (*asn*) to lysine (*lys*). The other changes an isoleucine (*ile*) to a threonine (*thr*). List the codons involved and describe how single-base mutations alter the amino acids they specify.
8. In one family, Tay-Sachs disease stems from a four-base insertion, which changes an amino-acid-encoding codon into a "stop" codon. What type of mutation is this?
9. Epidermolytic hyperkeratosis is an autosomal dominant condition that produces scaly skin. It can be caused by a missense mutation that substitutes a histidine (*his*) amino acid for an arginine (*arg*). Write the mRNA codons that could account for this change.
10. Fanconi anemia is an autosomal recessive condition that causes bone marrow abnormalities and an increased risk of certain cancers. It is caused by a transversion mutation that substitutes a valine (*val*) for an aspartic acid (*asp*) in the amino acid sequence. Which mRNA codons are involved?
11. Aniridia is an autosomal dominant eye condition in which the iris is absent. In one family, an 11-base insertion in the gene causes a very short protein to form. What kind of mutation must the insertion cause?
12. Two young people with skin cancer resulting from xeroderma pigmentosum meet at an event held for teenagers with cancer. However, their mutations affect different genes. They decide to marry but will not have children because they believe that each child would have a 25 percent chance of inheriting XP because it is autosomal recessive. Are they correct to be so concerned? Why or why not?
13. A biotechnology company has encapsulated DNA repair enzymes in fatty bubbles called liposomes. Why would this be a valuable addition to a suntanning lotion?

# Suggested Readings

Burkhart, James G. January 2000. Fishing for mutations. *Nature Biotechnology* 18:21. A gene transferred from zebrafish to bacteria may replace the Ames test in detecting mutagens.

Colliage, Alain, et al. August 1999. Human Ehlers-Danlos syndrome Type VIIC and bovine dermatosparaxis are caused by mutations in the procollagen 1 N-proteinase gene. *The American Journal of Human Genetics* 65:308. Using a cow gene, researchers discovered what goes wrong to cause "stretchy skin disease."

Darnell, Jennifer C., et al. November 16, 2001. Fragile X mental retardation protein targets G quartet mRNAs important for neuronal function. *Cell* 107:489–99. Fragile X syndrome results from a cascade of disrupted protein function.

Friedberg, Errol C. 1997. *Correcting the blueprint of life: An historical account of the discovery of DNA repair mechanisms.* New York: Cold Spring Harbor Laboratory Press. A look at discoveries that revealed the mechanisms of DNA repair.

Green, P. M., et al. December 1999. Mutation rates in humans. 1. Overall and sex-specific rates obtained from a population study of hemophilia B. *The American Journal of Human Genetics* 65:1572. The mutation rate for the gene that, when mutant, causes hemophilia B is two to eight mutations per million gametes.

Housman, David. May 1995. Gain of glutamines, gain of function? *Nature Genetics.* Huntington disease may add a function rather than impair one.

Ingram, V. M. 1957. Gene mutations in human hemoglobin: The chemical difference between normal and sickle cell hemoglobin. *Nature* vol. 180. The classic paper explaining the molecular basis for sickle cell disease.

Jackson, Graham S., et al. November 15, 2001. HLA-DQ7 antigen and resistance to variant CJD. *Nature* 414:269. Lack of an allele of an immune system gene may increase susceptibility to a prion disorder.

Lewis, Ricki. August 16, 2001. Understanding Huntington's disease. *The Scientist* 15:14. Huntingtin binds to several other proteins.

Modiano, David, et al. November 15, 2001. Haemoglobin C protects against clinical *Plasmodium falciparum* malaria. *Nature* 414:305–8. Two copies of hemoglobin C protects against malaria, and does not cause anemia.

Olivieri, Nancy F. July 8, 1999. The β-thalassemias. *The New England Journal of Medicine* 341:99. Beta thalassemia is a common blood disorder resulting from a mutation that severely lowers the number of beta globin chains.

Tapscott, Stephen J., and Charles A. Thornton. August 3, 2001. Reconstructing myotonic dystrophy. *Science* 293:864–67. Triplet and even quadruplet repeats can have devastating effects on gene expression, even if they are not in exons.

Withgott, Jay. July–August 2001. Evolving under UV. *Natural History.* DNA repair systems evolved to help organisms cope with environmental UV exposure.

Wood, R. D., et al. February 16, 2001. Human DNA repair genes. *Science* 292:1284–88.

# On the Web

Check out the resources on our website at

**www.mhhe.com/lewisgenetics5**

On the web for this chapter, you will find additional study questions, vocabulary review, useful links to case studies, tutorials, popular press coverage, and much more. To investigate specific topics mentioned in this chapter, also try the links below:

Coalition for Heritable Disorders of Connective Tissue **www.chdct.org**

Cooley's Anemia Foundation **www.thalassemia.org**

Ehlers-Danlos National Foundation **www.ednf.org**

FRAXA Research Foundation Inc. (Fragile X) **www.fraxa.org**

Globin gene server website **http://globin.cse.psu.edu/**

Online Mendelian Inheritance in Man **www.ncbi.nlm.nih.gov/entrez/query.fcgi?db=OMIM**

alkaptonuria 203500
alpha thalassemia 141800
Alzheimer disease 104300, 104310, 104311, 600759
ataxia telangiectasia 208900
Becker muscular dystrophy 310200
beta thalassemia 141900
Bloom syndrome 278700
Duchenne muscular dystrophy 310200
Ehlers-Danlos syndrome 130050
familial hypercholesterolemia 143890
fragile X syndrome 309550
G6PD deficiency 305900
hemoglobin M 250800
hereditary nonpolyposis colon cancer 120435
myotonic dystrophy 160900
prion protein 176640
sickle cell disease 603903
trichothiodystrophy 601675
xeroderma pigmentosum 278700

Muscular Dystrophy Association **www.mdausa.org/**

Neurofibromatosis Inc. **www.nfinc.org**

CHAPTER

12

# Chromosomes

## 12.1 Portrait of a Chromosome

Chromosomes are the bearers of genes. A chromosome includes DNA and proteins that enable it to be replicated; information in the form of protein-encoding genes; and DNA sequences that provide stability. The 24 chromosome types in a human cell are distinguished by size, shape, centromere position, and staining patterns. Chromosome charts record health information useful to individuals and families, and can provide clues to exposure to environmental pollution. Comparing chromosomes can reveal evolutionary relationships among species.

## 12.2 Visualizing Chromosomes

Chromosomes can be seen in any cell that has a nucleus. Three techniques are used to obtain fetal cells and observe their chromosomes. Techniques used to visualize chromosomes have evolved from crude stains to highly specific labeled DNA probes.

## 12.3 Abnormal Chromosome Number

Genetic health is a matter of balance. Extra sets of chromosomes, or missing or extra individual chromosomes, can devastate health. In general, additional genetic material is more tolerable than a deficit. Many embryos with abnormalities in chromosome number do not develop for very long.

## 12.4 Abnormal Chromosome Structure

Disruption in the precise sequence of events that occur during meiosis can result in chromosomes that have missing or extra genetic material or that exchange parts. An inverted gene sequence causes loops to form as chromosomes pair in meiosis, possibly leading to deletions and duplications of genetic material.

## 12.5 Uniparental Disomy—Two Genetic Contributions from One Parent

Rarely, the simultaneous occurrence of two rare events causes the inheritance of two alleles of the same gene from one parent. This can transmit an autosomal recessive disorder or disrupt control of an imprinted gene.

Genetic health is largely a matter of balance—inheriting the "correct" number of genes, usually on the "correct" number of chromosomes (46, for humans). Too much or too little genetic material, particularly among the autosomes, can cause syndromes (groups of signs and symptoms). A person with Down syndrome, for example, usually has an extra chromosome 21 and, therefore, extra copies of all the genes on that chromosome. The extra genes cause mental retardation and various medical problems, but, as figure 12.1 illustrates, people with Down syndrome can lead full and productive lives.

Abnormal numbers of genes or chromosomes are a form of mutation. Mutations range from the single-base changes described in chapter 11, to missing or extra pieces of chromosomes or entire chromosomes, to entire extra sets of chromosomes. A mutation is considered a chromosomal aberration if it is large enough to see with a light microscope using stains and/or fluorescent tags to highlight missing, extra, or moved material. The mutations described in chapter 11 and this chapter represent a continuum—they differ in scale and in our ability to detect them.

In general, excess genetic material has milder effects on health than a deficit. Still, most chromosomal abnormalities are so harmful that prenatal development ceases in the embryo. As a result, only a few—0.65 percent—of all newborns have chromosomal abnormalities that produce symptoms. An additional 0.20 percent have chromosomal rearrangements; their chromosome parts have flipped or been swapped but they do not produce symptoms unless they disrupt genes that are crucial to health.

**Cytogenetics** is the subdiscipline within genetics that links chromosome variations to specific traits, including illnesses. Data from the human genome project are adding to our knowledge of cytogenetics by identifying which genes contribute which symptoms to chromosome-related syndromes, and by comparing the gene contents of the chromosomes. For example, for decades geneticists did not understand why the most frequently seen extra autosomes in newborns are chromosomes 13, 18, and 21. The human genome sequence revealed that these chromosomes have the lowest gene densities—that is, they carry considerably fewer protein-encoding genes than the other autosomes, compared to the total amount of DNA. Therefore, extra copies of these chromosomes are tolerated sufficiently well for some individuals with them to survive to be born.

This chapter explores several ways that chromosome structure can deviate from normal and the consequences of these variations.

## 12.1 Portrait of a Chromosome

A chromosome is more than a many-million-base-long molecule of DNA. It is a structure that consists primarily of DNA and proteins that is duplicated and transmitted, via mitosis or meiosis, to the next cell generation. Cytogeneticists have long described

a.

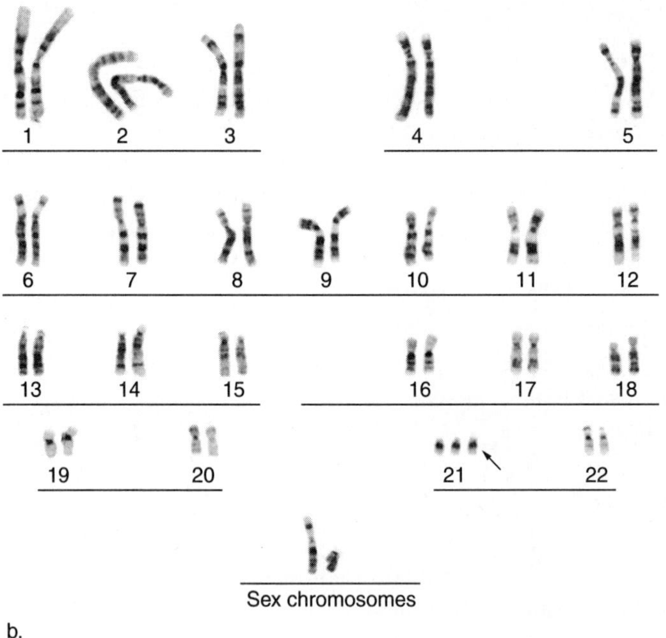

b.

**figure 12.1**

**Trisomy 21 Down syndrome.** (*a*) Wendy Weisz enjoys studying art at Cuyahoga Community College. (*b*) A karyotype (chromosome chart) for trisomy 21 Down syndrome shows the extra chromosome 21.

and distinguished the chromosome types by large-scale differences, such as size and shape. In addition, stains and dyes have traditionally been used to contrast dark **heterochromatin,** which is mostly repetitive DNA sequences, with lighter **euchromatin,** which harbors more protein-encoding genes (figure 12.2). Today, geneticists are superimposing information from the human genome sequence onto the existing views of chromosomes.

## Telomeres and Centromeres Are Essential

A chromosome must include those structures that enable it to replicate and remain intact—everything else is essentially informational cargo (the protein-encoding genes) and DNA sequences that impart stability to the overall structure. The absolutely essential parts of a chromosome, in terms of navigating cell division, are:

- telomeres
- origin of replication sites, where replication forks begin to form
- the centromere

Recall from figure 2.18 that **telomeres** are chromosome tips, each consisting of many repeats of the sequence TTAGGG that are whittled down with each mitotic cell division.

The **centromere** is the largest constriction of a chromosome, and is where spindle fibers attach. A chromosome without a centromere is no longer a chromosome—it vanishes from the cell as soon as division begins, because it cannot attach to the spindle.

In humans, many of the hundreds of thousands of DNA bases that form the centromere are repeats of a 171-base sequence called an **alpha satellite.** (In this usage, *satellite* refers to the fate of these sequences when chromosomal DNA is shattered and different pieces are allowed to settle in a density gradient. Because this part of the chromosome settles out at a density separate from the rest of the chromosome, it is called a satellite, just as the moon is a lump derived from the Earth.) The sequence of the alpha satellites might not be what is important in establishing the centromere, but rather the number of repeats is important. This idea is based on the observation that other species have alpha satellites of about the same repeat length, but with different sequences. The approximate size across species of about 170 base pairs just about equals the size of a nucleosome (see figure 9.11), the unit of chromatin that consists of DNA wrapped around octets of histone proteins. Perhaps it is the binding of histones to DNA in a certain pattern that attracts spindle fibers to the centromere.

In addition to alpha satellites, which are DNA, centromeres also include **centromere-associated proteins.** Some of these are synthesized only when mitosis is imminent. At this time, certain centromere-associated proteins form a structure called a **kinetochore,** which emanates outward from the centromere and contacts the spindle fibers. The kinetochore appears at prophase and vanishes during telophase. Another type of centromere-associated protein, the cohesins, are part of the centromere during interphase.

Centromeres are replicated toward the end of S phase, and a protein that may control this process is called centromere protein A, or CENP-A. It resembles a histone, but unlike them, CENP-A remains associated with a chromosome as it is replicated, covering about half a million DNA base pairs. When the replicated (sister) chromatids separate at anaphase, each member of the pair retains some CENP-A. The protein is therefore a "heritable centromeric molecule"—something that passes to the next cell generation that is *not* DNA. The amino acid sequence of CENP-A is nearly identical in diverse species, and it is located right in the middle of the centromere.

Researchers hypothesize that CENP-A and other centromere-associated proteins are the critical parts of centromeres, rather than the alpha satellite DNA sequences. Evidence comes from DNA sequences that function as "neocentromeres." These are sequences, found throughout the genome in noncentromeric regions, that can function as centromeres if moved, even if they lack alpha satellites.

The composition of centromeres is still not completely understood, because the techniques used to sequence the human genome, described in chapter 22, do not work on highly repeated sequences. It is known, though, that centromeres lie within vast stretches of heterochromatin, and, gradually, the terrain of the DNA comes to contain protein-encoding sequences with distance from the centromere. Outward from the centromere lie the arms of the chromosomes. Here, gene density varies greatly among the chromosome types, a finding discovered as research groups began to publish whole-chromosome sequences in 1998. Chromosome 21 is described as a "desert," harboring a million-base stretch with no protein-encoding genes at all. Chromosome 22, in contrast, is a gene jungle. These two tiniest chromosomes are at opposite ends of the gene-density spectrum. They are remarkably similar in size, but chromosome 22 contains 545 genes to chromosome 21's 225. Table 12.1 compares some basic characteristics of the first five

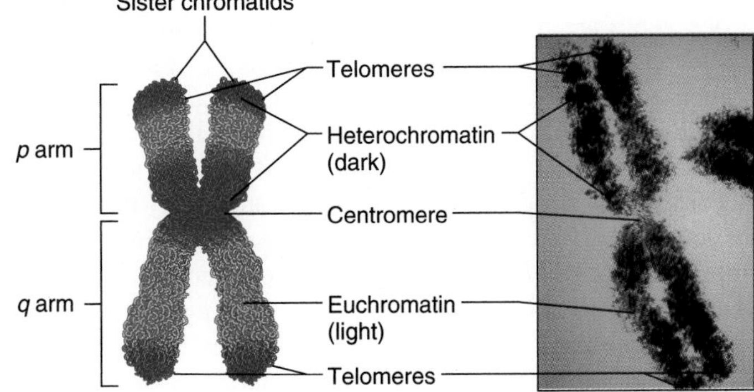


**figure 12.2**

**Portrait of a chromosome.** Tightly wound highly repetitive heterochromatin forms the centromere (the largest constriction) and the telomeres (the tips). Elsewhere, lighter-staining euchromatin includes protein-encoding genes. The centromere divides this chromosome into a short arm (*p*) and a long arm (*q*).

**table 12.1**

**Five Autosomes**

| Chromosome | Size (Megabases = Million) | % of Genome | Genes of Interest |
|---|---|---|---|
| 5 | 194.00 | 6 | Acute myelogenous leukemia<br>Basal cell carcinoma<br>Colorectal cancer<br>Dwarfism<br>Salt resistant hypertension |
| 16 | 98.00 | 3 | Adult polycystic kidney disease<br>Breast cancer<br>Crohn's disease<br>Prostate cancer |
| 19 | 60.00 | 2 | Atherosclerosis<br>Type I diabetes mellitus<br>DNA repair |
| 21 | 33.55 | 1 | Alzheimer disease<br>Amyotropic lateral sclerosis<br>Bipolar disorder susceptibility<br>Down syndrome<br>Homocystinuria<br>Usher syndrome |
| 22 | 33.46 | 1 | Cardiovascular disease<br>Chronic myelogenous leukemia<br>DiGeorge syndrome<br>Schizophrenia susceptibility |

autosomes sequenced, and figure 12.3 shows how the National Center for Biotechnology Information depicts them. Reading 12.1 describes the technique of creating chromosomes.

Between the chromosome's arms where protein-encoding genes lie amidst stretches of repeats and the telomeres are regions appropriately called **subtelomeres** (figure 12.4). Looked at from another perspective, these areas extend from 8,000 to 300,000 bases inward toward the centromere from the telomeres. Researchers had thought that these areas would be buffer zones, of sorts, leading gradually into the telomeric repeats with few, if any, protein-encoding genes. Human genome sequence information, although incomplete, has contradicted expectations by revealing several protein-encoding genes in the subtelomeres, but the transition is gradual. Areas of 50 to 250 bases, right next to the telomeres, indeed consist of 6-base repeats, many of which are very similar to TTAGGG. Then, moving inward from the 6-base buffer zone are many shorter repeats, each present in a few copies. Their function isn't known. Finally the sequence diversifies and protein-encoding genes appear.

When researchers compared subtelomeres to known gene sequences, they found more than 500 matches. About half of these identified genes are members of multigene families (groups of genes of very similar sequence next to each other) that include pseudogenes. These multigene families may be sites that reflect recent evolution. For many of the gene families in humans, our closest relatives, the apes and chimps, have only one or two genes. Such gene organization is one explanation for why our genome sequence is so very similar to that of our primate cousins—but we are clearly different animals. That is, our genomes differ more in gene copy number and chromosomal organization than in base sequence similarity. Chapter 15 compares primate genomes in greater detail.

## Karyotypes Are Chromosome Charts

Even in this age of genomics, the standard chromosome chart, or **karyotype,** remains a major clinical genetic tool. A karyotype presents chromosomes by size and by physical landmarks that appear during mitotic metaphase, when DNA coils especially tightly.

The 24 human chromosome types are numbered from largest to smallest—1 to 22—although chromosome 21 is actually the smallest. The other two chromosomes are the X and the Y. Early attempts to size-order chromosomes resulted in generalized groupings because many of the chromosomes are of similar size.

Centromere position is one distinguishing feature of chromosomes. A chromosome is **metacentric** if the centromere divides it into two arms of approximately equal length. It is **submetacentric** if the centromere establishes one long arm and one short arm, and **acrocentric** if it pinches off only a small amount of material toward one end (figure 12.5). Some species have telocentric chromosomes that have only one arm, but humans do not. The long arm of a chromosome is designated *q,* and the short arm *p,* where *p* stands for "petite."

Five human chromosomes (13, 14, 15, 21, and 22) are distinguished further by bloblike ends, called satellites, that extend from a thinner, stalklike bridge from the rest of the chromosome. (This is a different use of the word "satellite" from the centromeric repeats.) The stalklike regions are areas that do not bind stains well. The stalks carry many repeats of genes coding for ribosomal RNA and ribosomal proteins, areas called nucleolar organizing regions. They coalesce to form the nucleolus, a structure in the nucleus where ribosomal building blocks are produced and assembled.

Karyotypes are useful at several levels. When a baby is born with the distinctive facial characteristics of Down syndrome, a karyotype confirms the clinical diagnosis. Within families, karyotypes are used to identify relatives with a particular chromosomal aberration that can affect health. For example, in one family, several adult members died from a rare form of kidney

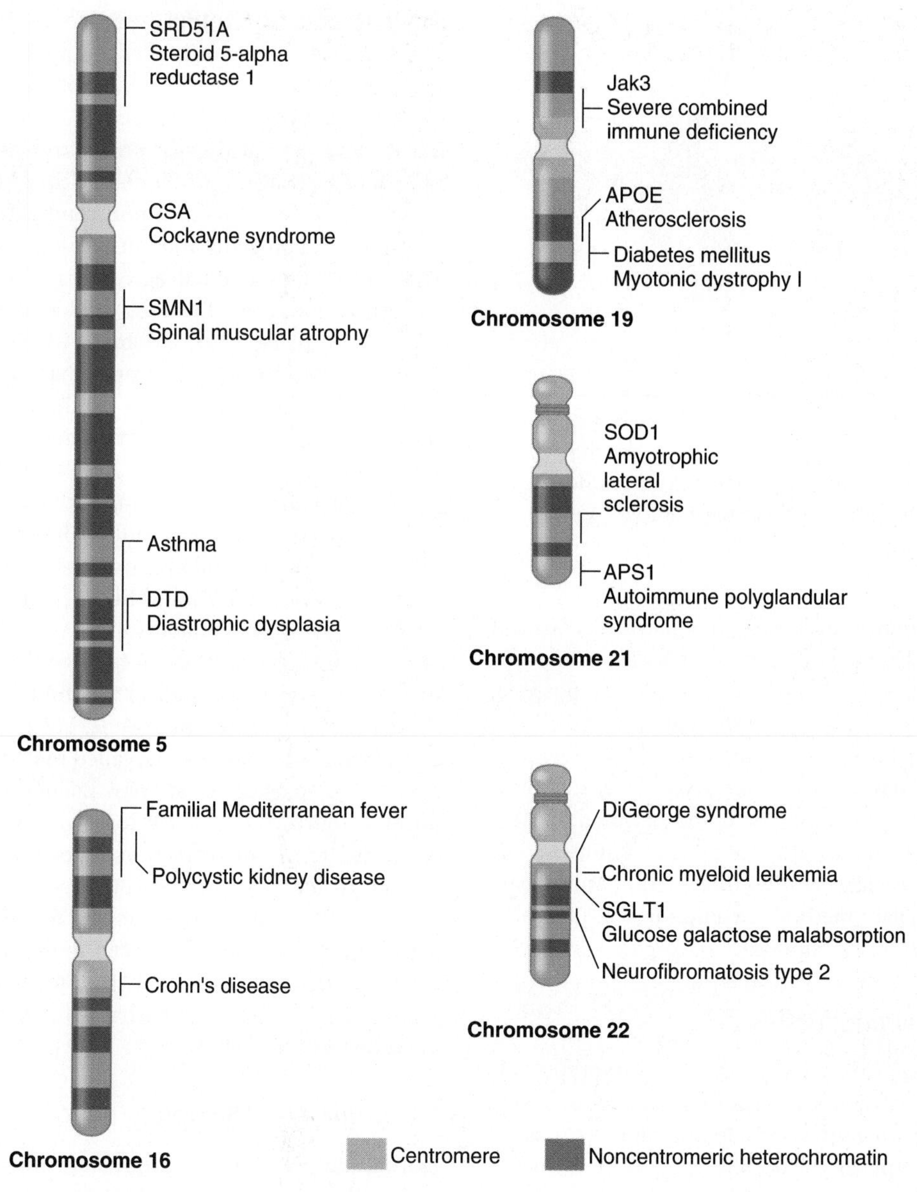


**figure 12.3**

**Disease gene maps.** The National Center for Biotechnology Information maintains depictions of human chromosomes with major health-related genes highlighted. Many of these disorders have been described in past chapters.

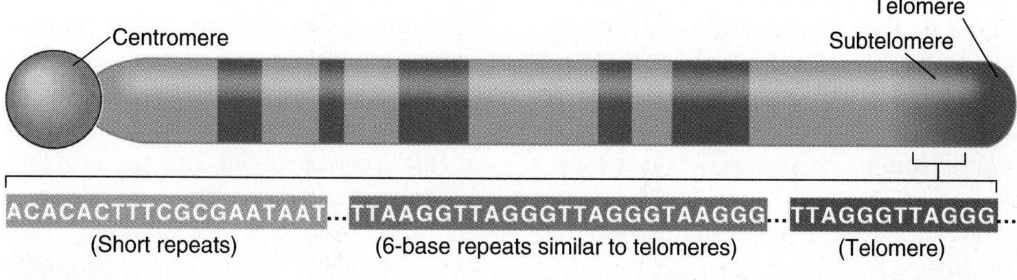


**figure 12.4**

**Subtelomeres.** The repetitive sequence of a telomere gradually diversifies toward the centromere. A subtelomere consists of from 8,000 to 300,000 bases from the telomere inward on a chromosome arm.

cancer. Because the cancer was so unusual, researchers karyotyped the affected individuals, and found that they all had an exchange called a **translocation,** between chromosome 3 and 8. When karyotypes showed that two young family members had the translocation, physicians examined and monitored their kidneys, detecting cancer very early and treating it successfully.

Karyotypes of individuals from different populations can reveal the effects of environmental toxins, if abnormalities appear only in a group exposed to a particular contaminant. Because chemicals and radiation that can cause cancer and birth defects often also break chromosomes into fragments or rings, detecting this genetic damage can alert physicians to the possibility that certain cancers will appear in the population.

Karyotypes compared between species can clarify evolutionary relationships. The more recent the divergence of two species from a common ancestor, the more closely related we presume they are, and the more alike their chromosome banding patterns should be. Our closest relative, according to karyotypes, is the pygmy chimpanzee, also known as the bonobo. The human karyotype is also remarkably similar to that of the domestic cat, and somewhat less similar to the karyotypes of mice, pigs, and cows (see figure 15.5).

## Key Concepts

Chromosomes consist of DNA, RNA, histones, and other proteins. A chromosome minimally includes telomeres, origins of replication, and centromeres, which enable the entire structure to replicate. A centromere consists of alpha satellite repeats and associated proteins, some of which form the kinetochore, to which spindle fibers attach. Centromere protein A enables the centromere to replicate. Subtelomeres contain telomerelike repeats, but also include protein-encoding multigene families. • Chromosomes are distinguishable by size, centromere location, the presence of satellites, and differential staining of heterochromatin and euchromatin. Karyotypes are size-order charts of chromosomes that provide information that is valuable at several levels.

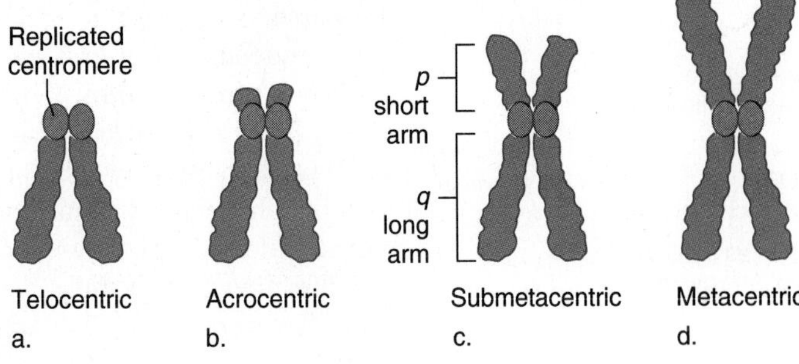


**figure 12.5**

**Centromere position is used to distinguish chromosomes.** (*a*) A telocentric chromosome has the centromere at one end. (*b*) An acrocentric chromosome has the centromere near an end. (*c*) A submetacentric chromosome's centromere creates a long arm (*q*) and a short arm (*p*). (*d*) A metacentric chromosome's centromere creates equal-sized arms.

## 12.2 Visualizing Chromosomes

Extra or missing chromosomes are easily detected by counting a number other than 46. Identifying chromosome rearrangements, such as an inverted sequence or two chromosomes exchanging parts, requires a way to distinguish among the chromosomes. A combination of stains and DNA probes applied to chromosomes allows this. A **DNA probe** is a labeled piece of DNA that binds to its complementary sequence on a particular chromosome.

### Obtaining Cells for Chromosome Study

Any cell other than a mature red blood cell (which lacks a nucleus) can be used to examine chromosomes, but some cells are easier to obtain and culture than others. For adults, white blood cells separated from a blood sample or skinlike cells collected from the inside of the cheek are usually the basis of a chromosome test. A person might require such a test if he or she has a family history of a chromosomal abnormality or seeks medical help because of infertility.

For blood-borne cancers (leukemias and lymphomas), cytogeneticists examine chromosomes from bone marrow cells, which give rise to blood cells. DNA microarray tests are replacing karyotypes in matching cancers to the most effective chemotherapies. Chromosome tests are most commonly performed on fetal cells. Couples who receive a prenatal diagnosis of a chromosome abnormality can arrange for treatment of the newborn, if this is possible; learn more about the condition and perhaps contact support groups and plan care; or terminate the pregnancy. These choices are highly individual and personal and are best made after a genetic counselor or physician provides information on the particular medical condition and available treatment options.

#### *Amniocentesis*

The first successful fetal karyotype was constructed in 1966 by a technique called **amniocentesis.** A doctor removes a small sample of fetal cells and fluids from the uterus with a needle passed through the woman's abdominal wall (figure 12.6*a*). The cells are cultured for a week to 10 days, and typically 20 cells are karyotyped. Culturing and staining chromosomes takes a week, but DNA probes can detect chromosomes in a day or two. The sampled amniotic fluid is also examined for deficient, excess, or abnormal biochemicals that could indicate particular inborn errors of metabolism. Amniocentesis takes only a minute or two and causes only a feeling of pressure. Ultrasound is used to guide the needle so it doesn't harm the fetus (figure 12.7).

Amniocentesis can detect approximately 400 of the more than 5,000 known chromosomal and biochemical problems. It is usually performed at 15 or 16 weeks gestation, when the fetus isn't yet very large but amniotic fluid is plentiful. Amniocentesis can be carried out anytime after this point.

Doctors recommend amniocentesis if the risk that the fetus has a detectable condition exceeds the risk that the procedure will cause a miscarriage, which is about 1 in 350 (table 12.2). The most common candidate for the test is a pregnant woman over age 35. This "advanced maternal age" statistically increases the risk that the fetus will have an extra or missing chromosome. Amniocentesis is also warranted if a couple has had several spontaneous abortions or children with birth defects or a known chromosome abnormality.

Another reason to seek amniocentesis is if a blood test on the pregnant woman reveals low levels of a fetal liver protein called alpha fetoprotein (AFP) and high levels of human chorionic gonadotropin (hCG). These signs may indicate a fetus with a small liver, which may reflect a condition caused by an extra chromosome, such as Down syndrome. These tests, called **maternal serum marker tests,** may include a third or fourth biochemical too. They are useful for pregnant women younger than 35 who would not routinely undergo age-related amniocentesis. Doctors use maternal serum marker tests to screen their patients to identify those who may require genetic counseling and perhaps further, more invasive testing.

#### *Chorionic Villus Sampling*

During the 10th week of pregnancy, **chorionic villus sampling** (CVS) obtains cells from the chorionic villi, the structures that develop into the placenta (figure 12.6*b*). A karyotype is prepared directly from the collected cells, rather than first culturing them, as in amniocentesis. Results are ready in days.

Because chorionic villus cells descend from the fertilized ovum, their chromosomes should be identical to those of the embryo and fetus. Occasionally, a chromosomal aberration occurs only in an embryo, or only in chorionic villi. This results in a situation called **chromosomal mosaicism**—that is, the karyotype of a villus cell differs from that of an embryo cell. Chromosomal mosaicism has great clinical consequences. If CVS indicates an aberration in villus cells that is not also in the fetus, then a couple may elect to terminate the pregnancy based on misleading information—that is, the fetus is

## Reading 12.1

# HACs—Human Artificial Chromosomes

What are the minimal building blocks necessary to form a chromosome? Cytogeneticists knew from work on other species that a chromosome consists of three basic parts:

1. Telomeres.
2. Origins of replication, estimated to occur every 50 to 350 kilobases (thousands of bases) in human chromosomes.
3. Centromeres.

All of these elements enable the entire unit to replicate during cell division and the original and replicated DNA double helices to be distributed into two cells from one.

The 24 types of human chromosomes range in size from 50 to 250 megabases (millions of bases). Researchers considered two ways to construct a chromosome—pare down an existing chromosome to see how small it can get and still hold together, or build a new chromosome from DNA pieces.

To trim an existing chromosome, researchers swapped in a piece of DNA that included telomere sequences. New telomeres formed at the insertion site, a little like prematurely ending a sentence by adding a period in the middle. This technique formed chromosomes as small as 3.5 megabases—but researchers couldn't get them out of cells for further study.

Huntington Willard and his colleagues at Case Western Reserve University tried the building-up approach. They sent separately into cultured cells telomere DNA alpha satellites and random pieces of DNA from the human genome containing origin-of-replication sites (figures 1 and 2). In the cells, some of the pieces assembled in a correct orientation to form structures that the researchers called "human artificial microchromosomes." These "HACs" are 6 to 10 megabases long. The hardy HACs withstand repeated rounds of cell division—they have the integrity of a natural chromosome.

Since then, other researchers have whittled down combinations of telomeres, neocentromeres, and other sequences of 0.7 to 1.8 megabases. Neocentromeres consist of centromere DNA sequences minus the alpha satellite repeats. The human genome contains several dozen of them, scattered among the chromosomes.

Constructing ever-smaller artificial chromosomes is revealing what it is to be a chromosome—an autonomous nucleic acid protein partnership that can replicate. More practically, artificial chromosomes may one day ferry healing genes to cells where their activity is missing or abnormal.

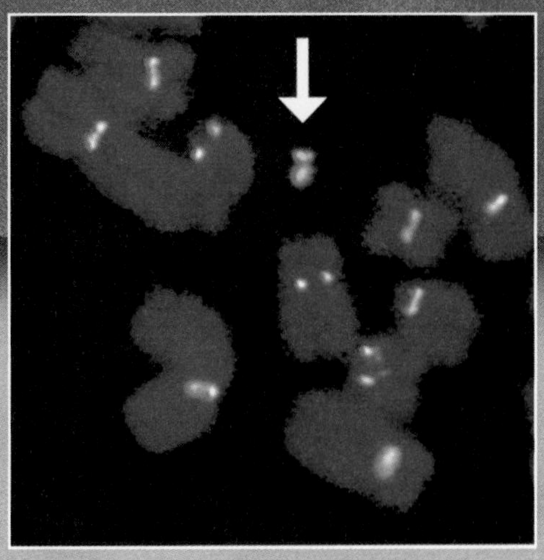

**figure 1**

**HACs.** Researchers introduced telomere sequences, centromere sequences, and other DNA pieces into cells in culture. The pieces aligned and assembled into human artificial microchromosomes.

(John Harrington, Huntington Willard, et al. *Nature Genetics* 4:345–55, 1997.)

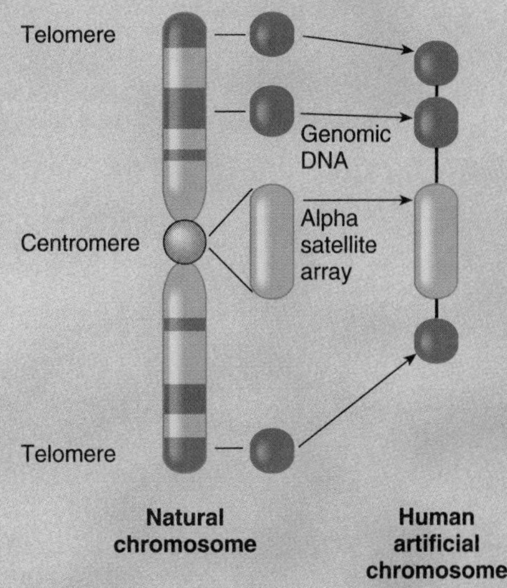


**figure 2**

**Creating a human artificial chromosome.** Human artificial chromosomes (HACs) are formed by combination of isolated telomeres, centromeric DNA from alpha satellite arrays, and genomic DNA derived from natural chromosomes.

(Modified from Willard [1998] *Curr Opin Genet Dev* 8:219–25.)

actually chromosomally normal, although CVS indicates otherwise. In the opposite situation, the results of the CVS may be normal, but the fetus has a chromosomal problem—leading to an unpleasant surprise at birth or later.

CVS is slightly less accurate than amniocentesis and in about 1 in 1,000 to 3,000 procedures, causes a fatal limb defect. Also, the sampling procedure in CVS does not include amniotic fluid, so the biochemical tests that amniocentesis allows are not possible.

Couples expecting a child are sometimes asked to choose between amniocentesis and CVS. The advantage of CVS is earlier results, but it is associated with a greater risk of spontaneous abortion. Although CVS is slightly more invasive and dangerous to the fetus, its greater risk also reflects the fact that CVS is done earlier in pregnancy. Since most spontaneous abortions occur early in pregnancy, more will follow CVS than amniocentesis. The spontaneous abortion rate after the 12th week of pregnancy is about 5 percent, and the additional risk that CVS poses is 0.8 percent. In contrast, the spontaneous abortion rate after the 14th week is 3.2 percent, and amniocentesis adds 0.3 percent to the risk.

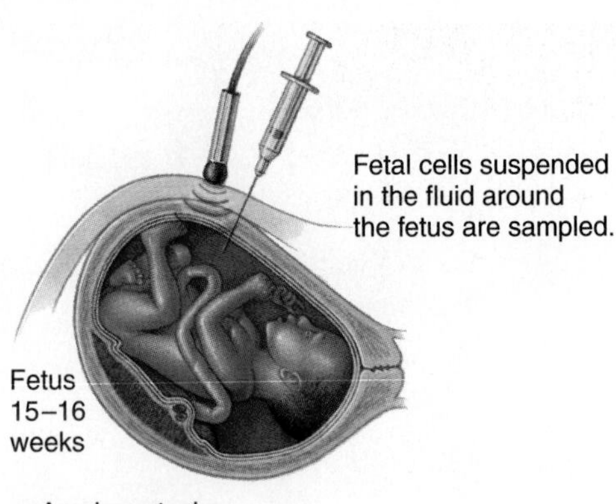

a. Amniocentesis

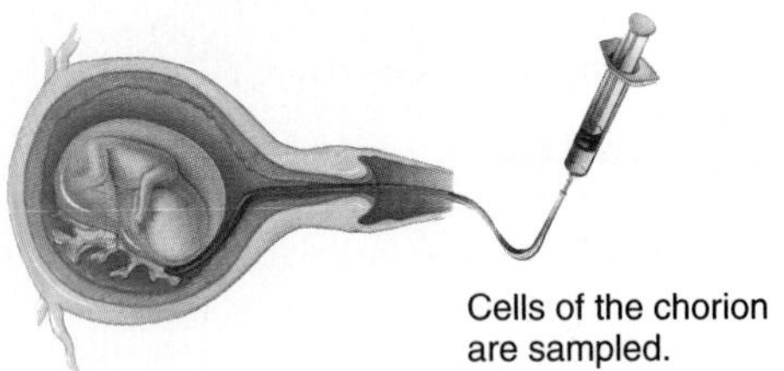

b. Chorionic villi sampling

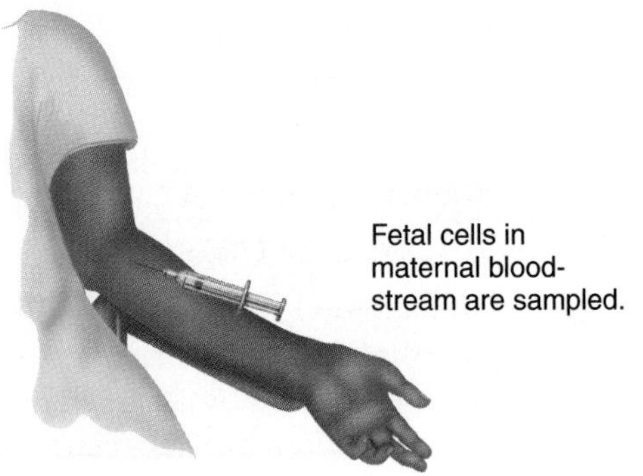

c. Fetal cell sorting

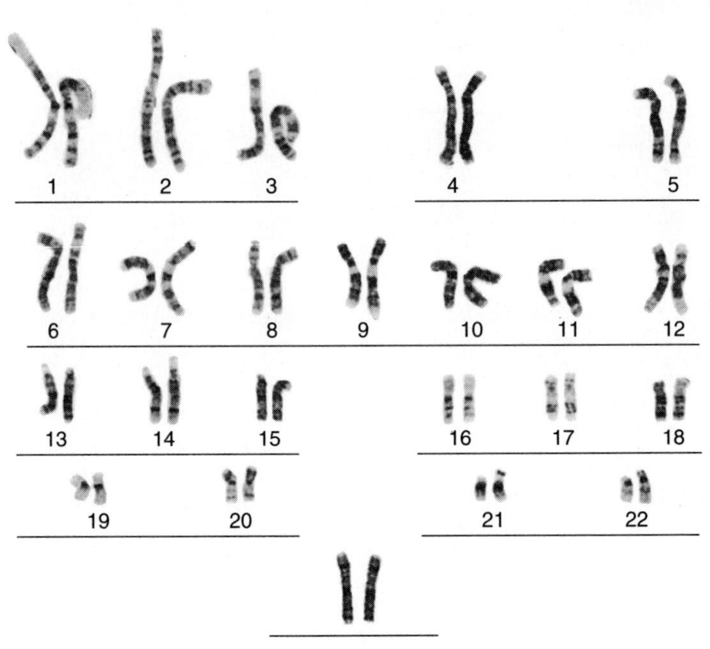

d. Fetal karyotype (normal female)

## figure 12.6

**Three ways of checking a fetus's chromosomes.** (*a*) Amniocentesis harvests fetal cells shed during development into the amniotic fluid by drawing out some of the fluid. (*b*) Chorionic villus sampling removes cells that would otherwise develop into the placenta. Since these cells came from the fertilized ovum, they have the same chromosomal constitution as the fetus. (*c*) Improved techniques at identifying and extracting specific cells allow researchers to detect fetal cells in a sample of blood from the woman. (*d*) For all three techniques, the harvested cells are allowed to reach metaphase, where chromosomes are most visible, and then broken open on a slide. The chromosomes are stained or their DNA probed, then arranged into a karyotype.

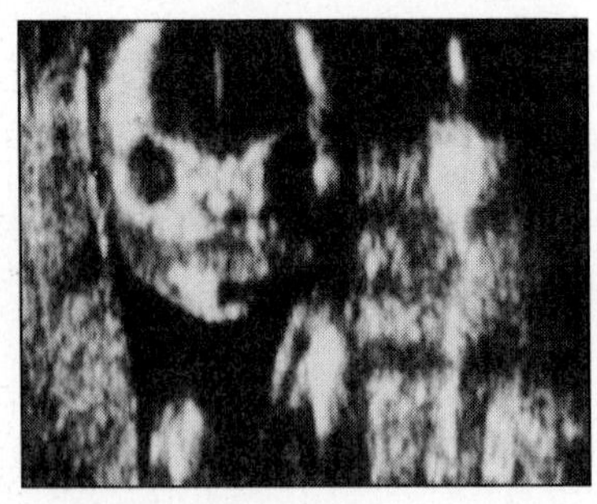

## figure 12.7

**Ultrasound.** In an ultrasound exam, sound waves are bounced off the embryo or fetus, and the pattern of deflected sound waves is converted into an image. By 13 weeks, the face can be discerned. Ultrasound can now provide images that look three-dimensional.

## table 12.2

**Amniocentesis or Chorionic Villus Sampling (CVS)?**

| Procedure | Time (weeks) | Cell Source | Route | Added Risk of Miscarriage |
|---|---|---|---|---|
| CVS | 10–12 | Chorionic villi | Vagina | 0.8% |
| Amniocentesis | 14–16 | Skin, bladder, digestive system cells in amniotic fluid | Needle in abdomen | 0.3% |

### *Fetal Cell Sorting*

**Fetal cell sorting,** a new technique that separates fetal cells from the woman's bloodstream, is safer than amniocentesis and CVS but is still experimental (figure 12.6*c*) in the United States. The technique traces its roots to 1957, when a pregnant woman died when cells from a very early embryo lodged in a major blood vessel in her lung, blocking blood flow. The fetal cells were detectable because they were from a male, and were distinguished by the telltale Y chromosome. This meant that fetal cells could enter a woman's circulation. By studying the blood of other pregnant women, researchers found that fetal cells enter the maternal circulation in up to 70 percent of pregnancies. Cells from female embryos, however, cannot be distinguished from the cells of the pregnant woman on the basis of sex chromosome analysis. But fetal cells from either sex can be distinguished from maternal cells using a device called a fluorescence-activated cell sorter. It separates fetal cells from maternal blood by identifying surface characteristics that differ from those on the woman's cells. The fetal cells are then karyotyped (figure 12.6*d*) and fetal DNA extracted and amplified for specific gene tests.

Rarely, other techniques are used to sample specific fetal tissues, such as blood, skin, liver, or muscle. These biopsy procedures are usually done by using ultrasound to guide a hollow needle through the woman's abdominal wall to reach the fetus. Such an invasive test is performed if the family has a disease affecting the particular tissue, and a DNA-based test is not available.

Instead of examining chromosomes, ultrasound can be used to identify physical features that are part of chromosomal syndromes. For example, ultrasound can detect thickened neck skin and absent or underdeveloped nasal bones that are characteristic of Down syndrome and were part of the initial description in 1866. In one study, 75 percent of fetuses with Down syndrome had these characteristics, compared to 0.5 percent of fetuses who did not have Down syndrome. Ultrasound scanning is therefore somewhat imprecise, but it is safer than obtaining chromosomes with amniocentesis or CVS. Combined with maternal serum marker testing, ultrasound scans may enable physicians to identify pregnancies in need of further testing.

## Technology TIMELINE

| | |
|---|---|
| **1923** | Theophilus Painter's chromosome sketches are published, human chromosome number thought to be 48 |
| **1951** | Method to detangle chromosomes discovered by accident |
| **1953** | Albert Levan and Joe-Hin Tjio develop "squash and stain" technique for chromosome preparation |
| **1956** | Using tissue culture cells, Levan, Tjio, and Biesele determine chromosome number to be 46 |
| **1956** | J. L. Hamerton and C. E. Ford identify 23 chromosomes in human gametes |
| **1959** | First chromosome abnormalities identified |
| **1960** | Phytohemagglutinin added to chromosome preparation protocol to separate and stimulate division in white blood cells |
| **1970s** | Several chromosome stains implemented to improve resolution of karyotypes |
| **1970s** | FISH developed |
| **1990s** | Spectral karyotyping combines FISH probes to distinguish each chromosome |

## Preparing Cells for Chromosome Observation

Microscopists have tried to describe and display human chromosomes since the late 19th century (figure 12.8 and the Technology Timeline). Then, the prevailing view held that humans had an XO sex determination

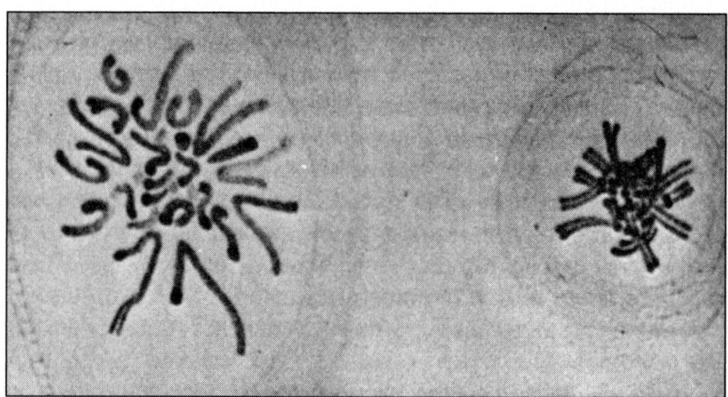

a.

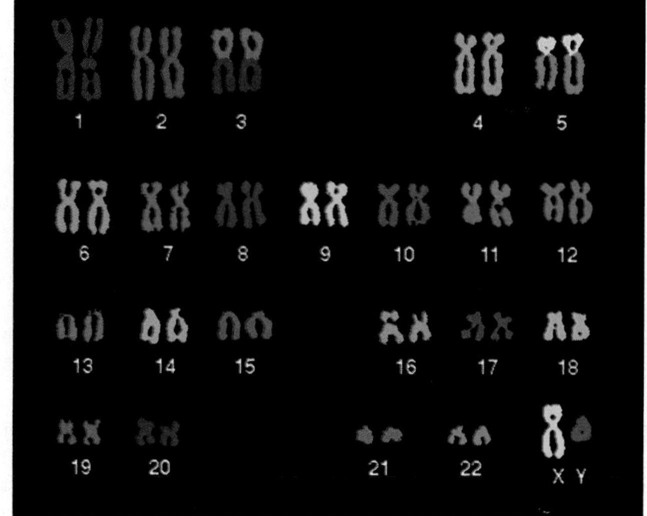


b.

**figure 12.8**

**Karyotypes old and new.** (*a*) The earliest drawings of chromosomes, by German biologist Walter Flemming, date from 1882. (*b*) This karyotype was constructed using a technique called FISH to "paint" the individual chromosomes. Figure 12.6*d* shows a normal karyotype with the chromosomes distinguished by stained bands.

system, with females having an extra chromosome (XX). Estimates of the human chromosome number ranged from 30 to 80. In 1923, Theophilus Painter published sketches of human chromosomes from three patients at a Texas state mental hospital. The patients had been castrated in an attempt to control their abusive behavior, and Painter was able to examine the tissue. He could not at first tell whether the cells had 46 or 48 chromosomes, but finally decided that he saw 48. Painter also showed that both sexes have the same chromosome number.

The difficulty in distinguishing between 46 or 48 chromosomes was physical—preparing a cell in which chromosomes do not overlap is challenging. To easily count the chromosomes, scientists had to find a way to capture them when they are the most condensed—during cell division—and also spread them apart. Since the 1950s, cytogeneticists have used colchicine, an extract of the chrysanthemum plant, to arrest cells during division.

### *Swelling, Squashing, and Untangling*

The dilemma of how to untangle the spaghettilike mass of chromosomes in a human cell was solved by accident in 1951. A technician mistakenly washed white blood cells being prepared for chromosome analysis in a salt solution that was less concentrated than the interiors of the cells. Water rushed into the cells, swelling them and separating the chromosomes.

Two years later, cell biologists Albert Levan and Joe-Hin Tjio found that when they drew cell-rich fluid into a pipette and dropped it onto a microscope slide prepared with stain, the cells burst open and freed the mass of chromosomes. Adding a glass coverslip spread the chromosomes enough that they could be counted. Another researcher, a former student of Painter named John Biesele, suggested that Levan and Tjio use cells from tissue culture, and by 1956, they finally settled the matter of how many chromosomes occupy a diploid human cell—46. In the same year, J. L. Hamerton and C. E. Ford identified the expected 23 chromosomes in human gametes. In 1960 came another advance in visualizing chromosomes—through the use of a kidney bean extract called phytohemagglutinin. Originally used to clump red blood cells to separate them from white blood cells, the substance also could stimulate division of white blood cells.

Until recently, a karyotype was constructed using a microscope to locate a cell in which the chromosomes were not touching, photographing the cell, developing a print, and cutting out the individual chromosomes and arranging them into a size-order chart. A computerized approach has largely replaced the cut-and-paste method. The device scans ruptured cells in a drop of stain and selects one in which the chromosomes are the most visible and well spread. Then image analysis software recognizes the band patterns of each stained chromosome pair, sorts the structures into a size-order chart, and prints the karyotype—in minutes. If the software recognizes an abnormal band pattern, a database pulls out identical or similar karyotypes from other patients, providing clinical information on the anomaly. Genome sequence information is also scanned. However, the expert eyes of a skilled technician are still needed to detect subtle abnormalities in chromosome structure.

### *Staining*

In the earliest karyotypes, dyes stained the chromosomes a uniform color. Chromosomes were grouped into size classes, designated A through G, in decreasing size order. In 1959, scientists described the first chromosomal abnormalities—Down syndrome (an extra chromosome 21), **Turner syndrome** (also called XO syndrome, a female with only one X chromosome), and **Klinefelter syndrome** (also called XXY syndrome, a male with an extra X chromosome). Before this, women with Turner syndrome were thought to be genetic males because they lack Barr bodies (see figure 6.13), while men with Klinefelter syndrome were thought to be genetic females because their cells have Barr bodies. The ability to visualize and distinguish the sex chromosomes, even crudely, allowed researchers to determine the causes of these conditions.

The first stains that were applied to chromosomes could highlight large deletions and duplications, but more often than not, researchers only vaguely understood the nature of a chromosomal syndrome. In 1967, a mentally retarded child with material missing from chromosome 4 would have been diagnosed as having a "B-group chromosome" disorder. Today the exact genes that are missing can be identified.

Describing smaller-scale chromosomal aberrations required better ways to distinguish among the chromosomes. In the 1970s, Swedish scientists developed more specific chromosome stains that create banding patterns unique to each chromosome. Combining stains reveals even more bands, making it easier to distinguish chromosomes. Stains are specific for AT-rich or GC-rich stretches of DNA, or for heterochromatin, which stains darkly at the centromere and telomeres.

The ability to detect missing, extra, inverted, or misplaced bands allowed researchers to link many more syndromes with specific chromosome aberrations. In the late 1970s, Jorge Yunis at the University of Minnesota improved chromosome staining further by developing a way to synchronize white blood cells in culture, arresting them in early mitosis. His approach, called high-resolution chromosome banding, revealed many more bands. Today, **fluorescence *in situ* hybridization,** or FISH, is eclipsing even high-resolution chromosome banding, enabling cytogeneticists to focus on individual genes.

### *FISHing* 

One drawback of conventional chromosome stains is that they are not specific to particular chromosomes. Rather, they generate different banding patterns among the 24 human chromosome types. FISH is much more specific because it uses DNA probes that are complementary to DNA sequences found only on one chromosome type. The probes are attached to molecules that fluoresce when illuminated, producing a flash of color precisely where the probe binds to a chromosome in a patient's sample. The technique can reveal a particular extra chromosome in a day or two.

FISH is based on a technique, developed in 1970, called *in situ* hybridization, which originally used radioactive labels rather than fluorescent ones. *In situ* hybridization took weeks to work, because it relied on exposing photographic film to reveal where DNA probes bound among the chromosomes. The danger of working with radioactivity, and the crudeness of the results, eventually prompted researchers to seek alternative ways of highlighting bound DNA probes.

FISH is used to identify specific chromosomes and to "paint" entire karyotypes, providing a different color for each chromosome. Many laboratories that perform amniocentesis or chorionic villus sampling use FISH probes specific to chromosomes 13, 18, 21, and the sex chromosomes to quickly identify the most common chromosome abnormalities. Figure 12.9 shows FISH analysis that easily identifies the extra chromosome 21 in cells from a fetus with Down syndrome.

In an application of FISH called spectral karyotyping, each chromosome is probed with several different fluorescent molecules. A computer integrates the images and creates a false color for each chromosome (figure 12.8*b*).

A new approach to prenatal chromosome analysis called quantitative PCR amplifies certain repeated sequences on chromosomes 13, 18, 21, X, and Y. The technique distinguishes paternally derived from maternally derived repeats on each homolog for these five chromosomes. An abnormal ratio of maternal to paternal repeats indicates a numerical problem, such as two copies of one parent's chromosome 21. Combined with the one chromosome 21 from the other parent, this situation would produce a fertilized ovum with three copies of chromosome 21, which causes Down syndrome. Quantitative PCR is less accurate than culturing cells or FISH to examine chromosomes, but gives results in only hours, which can greatly reduce parental anxiety.

### Chromosomal Shorthand

Geneticists abbreviate the pertinent information in a karyotype. They list chromosome number first, then sex chromosome constitution, then abnormal autosomes. Symbols describe the type of aberration, such as deletion or translocation. Numbers are listed that correspond to specific bands. A normal male is 46, XY; a normal female is 46, XX. Geneticists use this notation to describe gene locations. For example, the β-globin subunit of hemoglobin is located at 11p15.5. Table 12.3 gives some examples of chromosomal shorthand.

Chromosome information is displayed in an **ideogram,** which is a graphical representation of a karyotype (figure 12.10). Bands appear as stripes, and they are divided into numbered major regions and subregions. Specific gene loci known from mapping data are listed on the righthand side with information from the human genome sequence. Ideograms are becoming so crowded with notations indicating specific genes that they may soon become obsolete.

**table 12.3**

**Chromosomal Shorthand**

| Abbreviation | What It Means |
|---|---|
| 46,XY | Normal male |
| 46,XX | Normal female |
| 45,X | Turner syndrome (female) |
| 47,XXY | Klinefelter syndrome (male) |
| 47,XYY | Jacobs syndrome (male) |
| 46,XY del (7*q*) | A male missing part of the long arm of chromosome 7 |
| 47,XX,+21 | A female with trisomy 21 Down syndrome |
| 46,XY t (7;9)(*p*21.1; *q*34.1) | A male with a translocation between the short arm of chromosome 7 at band 21.1 and the long arm of chromosome 9 at band 34.1 |

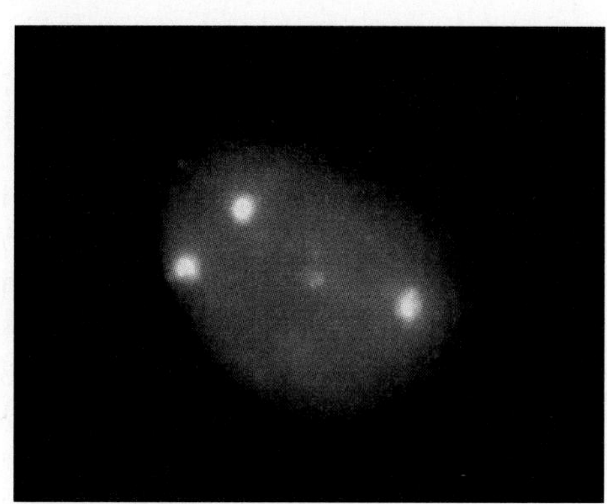

**figure 12.9**

**FISHing for genes and chromosomes.** FISH technology clearly shows three fluorescent dots that correspond to the three copies of chromosome 21 in this nucleus of a cell of an individual with Down syndrome.

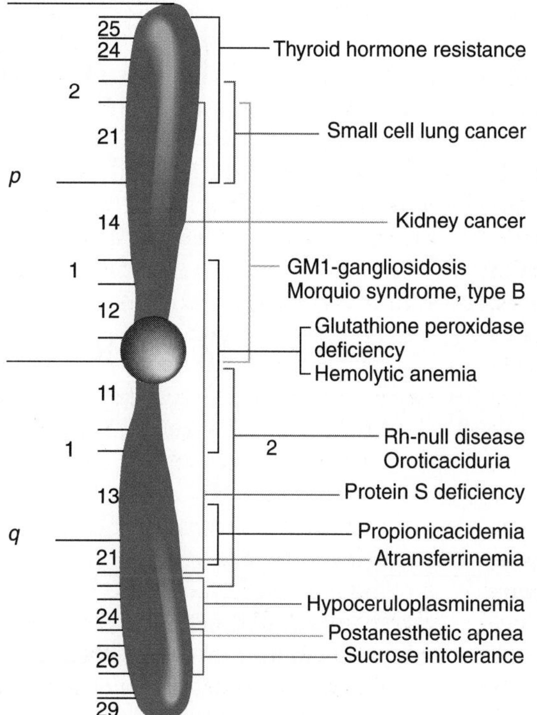


**figure 12.10**

**Ideogram.** An ideogram is a schematic chromosome map. It indicates chromosome arm (*p* or *q*), major regions delineated by banding patterns, and the loci of known genes. This is a partial map of human chromosome 3.

## Key Concepts

Karyotypes are charts that display chromosomes in size order. Chromosomes can be obtained from any cell that has a nucleus and can be cultured. Fetal karyotypes are constructed from cells obtained by amniocentesis, chorionic villus sampling, or fetal cell sorting from maternal blood. Ultrasound can detect physical signs of chromosome abnormalities.
• To detect a chromosome abnormality, cytogeneticists obtain cells; display, stain, and probe chromosomes with fluorescent molecules; and then arrange them in a karyotype. Chromosomal shorthand summarizes the number of chromosomes, sex chromosome constitution, and type of aberration. Ideograms are maps of the distinguishing features of individual chromosomes.

## 12.3 Abnormal Chromosome Number

A human karyotype is abnormal if the number of chromosomes is not 46, or if individual chromosomes have extra, missing, or rearranged genetic material. Table 12.4 summarizes the types of chromosome abnormalities in the order in which they are discussed.

Abnormal chromosomes are the most frequent cause of spontaneous abortion, accounting for at least 50 percent of them. Studies of preimplantation embryos cultured in laboratories for *in vitro* fertilization (see chapter 21) reveal that 50 to 70 percent of them have abnormal chromosomes, although this percentage is probably considerably lower in couples not seeking infertility treatment. Yet only 0.5 to 0.7 percent of newborns have abnormal chromosomes. Therefore, most embryos and fetuses with abnormal chromosomes do not develop enough to be born.

### Polyploidy

The most drastic upset in chromosome number is an entire extra set. A cell with one or more extra sets of chromosomes is **polyploid.** An individual whose cells have three copies of each chromosome is a triploid. Two-thirds of all polyploids result from fertilization of an oocyte by two sperm. The other cases of polyploidy arise from formation of a diploid gamete, such as a normal haploid sperm fertilizing a diploid oocyte. Triploids account for 15 percent of spontaneous abortions (figure 12.11). Very rarely, an infant survives for a few days, with defects in nearly all organs.

### Aneuploidy

Cells missing a single chromosome or having an extra one are termed **aneuploid,** which means "not good set." Studies on preimplantation embryos reveal that rare aneuploids can have a few missing or extra chromosomes, indicating defective meiosis in a parent. A normal chromosome number is **euploid,** which means "good set."

Most autosomal aneuploids are spontaneously aborted. Those that survive have specific syndromes, with symptoms depending upon which chromosomes are missing or extra. Mental retardation is common in an individual who survives with aneuploidy, because development of the brain is so complex and of such long duration that nearly any chromosome scale disruption includes genes whose protein products affect the brain. Sex chromosome aneuploidy usually produces less-severe symptoms.

Most children born with the wrong number of chromosomes have an extra

**table 12.4**

**Chromosome Abnormalities**

| Type of Abnormality | Definition |
|---|---|
| Polyploidy | Extra chromosome sets |
| Aneuploidy | An extra or missing chromosome |
| Monosomy | One chromosome absent |
| Trisomy | One chromosome extra |
| Deletion | Part of a chromosome missing |
| Duplication | Part of a chromosome present twice |
| Inversion | Segment of chromosome reversed |
| Translocation | Two chromosomes join long arms or exchange parts |

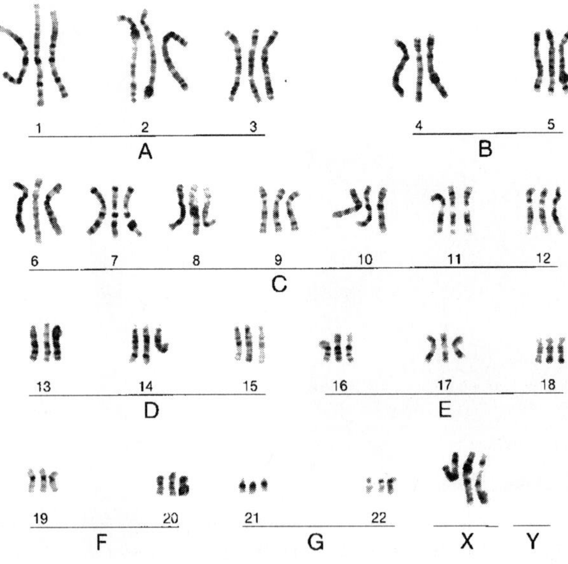


**figure 12.11**

**Polyploids in humans are lethal.** Individuals with three copies of each chromosome (triploids) account for 17 percent of all spontaneous abortions and 3 percent of stillbirths and newborn deaths.

chromosome (a **trisomy**) rather than a missing chromosome (a **monosomy**). Most monosomies are so severe that an affected embryo ceases developing. Trisomies and monosomies are named according to the chromosome involved, and the associated syndrome has traditionally been named for the investigator who first described it. Today, cytogenetic terminology is used because it is more precise. For example, Down syndrome can result from an extra chromosome 21 (a trisomy) or a translocation. The distinction is important in genetic counseling. Translocation Down syndrome, although accounting for only 4 percent of cases, has a much higher recurrence risk than the trisomy 21 form, a point we will return to later in the chapter.

The meiotic error that causes aneuploidy is called **nondisjunction.** Recall that in normal meiosis, homologs separate, and each of the resulting gametes receives only one member of each chromosome pair. In nondisjunction, a chromosome pair fails to separate at anaphase of either the first or second meiotic division. This produces a sperm or oocyte that has two copies of a particular chromosome, or none, rather than the normal one copy (figure 12.12). When such a gamete fuses with its mate at fertilization, the zygote has either 45 or 47 chromosomes, instead of the normal 46. Different trisomies are consistently caused

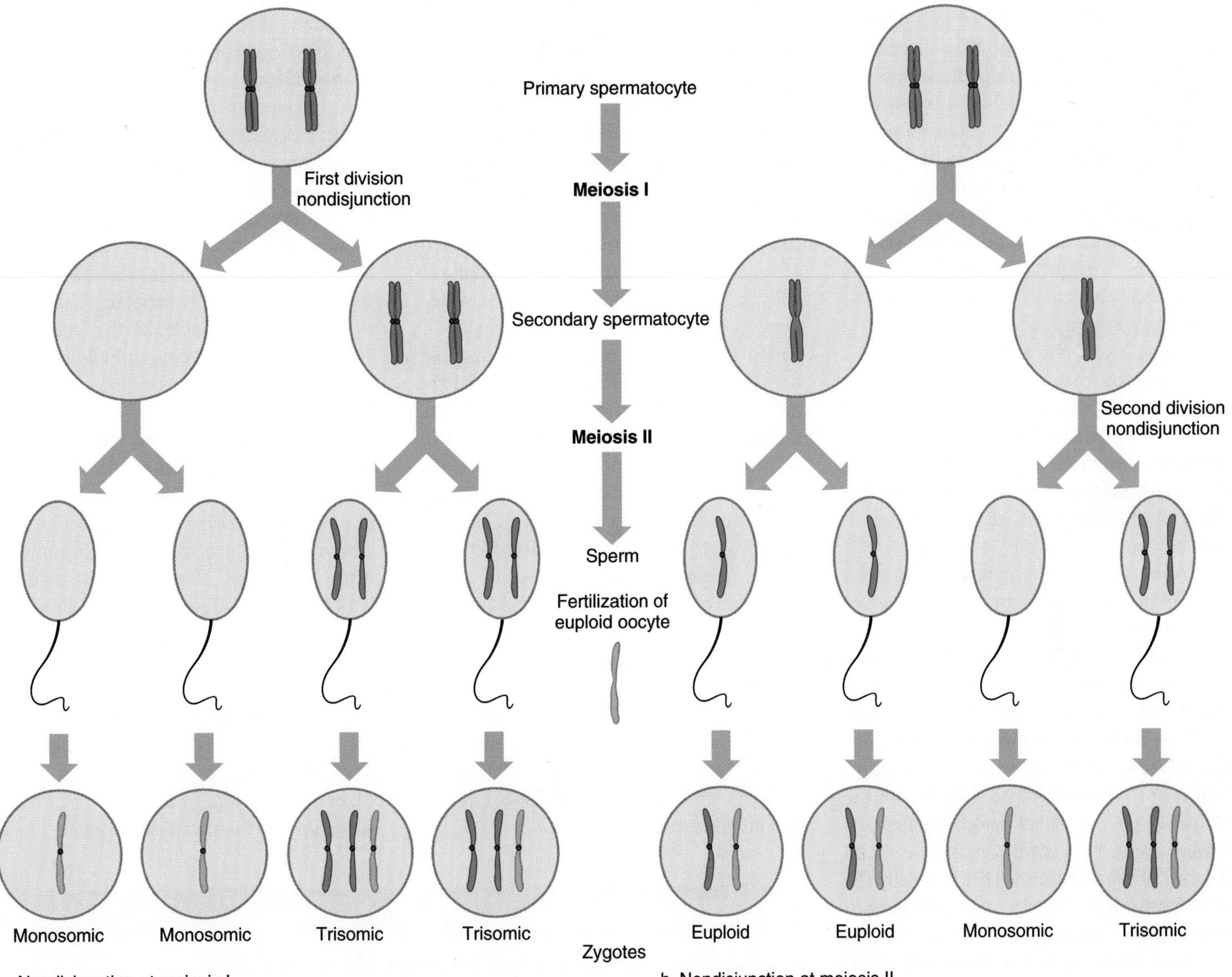


## figure 12.12

**Extra and missing chromosomes—aneuploidy.** Unequal division of chromosome pairs can occur at either the first or second meiotic division. (*a*) A single pair of chromosomes is unevenly partitioned into the two cells arising from meiosis I in a male. The result: two sperm cells have two copies of the chromosome, and two sperm cells have no copies. When a sperm cell with two copies of the chromosome fertilizes a normal oocyte, the zygote is trisomic; when a sperm cell lacking the chromosome fertilizes a normal oocyte, the zygote is monosomic. (*b*) This nondisjunction occurs at meiosis II. Because the two products of the first division are unaffected, two of the mature sperm are normal and two are aneuploid. Oocytes can undergo nondisjunction as well, leading to zygotes with extra or missing chromosomes when normal sperm cells fertilize them.

by nondisjunction in the male or female, at meiosis I or II.

A cell can have a missing or extra chromosome in 49 ways—an extra or missing copy of each of the 22 autosomes, plus the five abnormal types of sex chromosome combinations (Y, X, XXX, XXY, XYY). (Sometimes individuals have four or even five sex chromosomes.) However, only nine types of aneuploids are known in newborns. Others are seen in spontaneous abortions or fertilized ova intended for *in vitro* fertilization.

Most of the 50 percent of spontaneous abortions that result from extra or missing chromosomes are 45,X individuals (missing an X chromosome), triploids, or trisomy 16. About 9 percent of spontaneous abortions are trisomy 13, 18, or 21. Although these are the most common autosomal aneuploids seen in newborns, they are rare, affecting only 0.1 percent of all children. Put yet another way, more than 95 percent of newborns with abnormal chromosome numbers have an extra 13, 18, or 21, or an extra or missing X or Y chromosome.

Types of chromosome abnormalities seem to differ between the sexes. Abnormal oocytes mostly have extra or missing chromosomes, whereas abnormal sperm more often have structural variants, such as inversions or translocations, discussed later in the chapter.

Aneuploidy and polyploidy also arise during mitosis, producing groups of somatic cells with the extra or missing chromosome. An individual with two chromosomally distinct cell populations is a mosaic. If only a few cells are altered, health may not be affected. However, a mitotic abnormality that occurs early in development, so that many cells descend from the unusual one, can affect health. A chromosomal mosaic for a trisomy may have a mild version of the condition. This is usually the case for the 1 to 2 percent of people with Down syndrome who are mosaic. For example, a person may have the extra chromosome that causes Down syndrome in 5 of 20 sampled fetal cells. This individual would possibly not be as severely mentally impaired as a person who has the extra chromosome in every cell. The phenotype depends upon which cells have the extra chromosome. A fetus with affected cells in the brain would later show a greater mental impairment than a fetus with the affected cells mostly in the skin. Unfortunately, prenatal testing cannot reveal which cells are affected.

### *Autosomal Aneuploids*

Most autosomal aneuploids are very rarely seen in live births, due to the lethality of a large imbalance of genetic material. Following are descriptions of the most common autosomal aneuploids among liveborns, summarized in table 12.5.

**Trisomy 21 Down Syndrome** The most common autosomal aneuploid among liveborns is trisomy 21. The characteristic extra fold in the eyelids, called epicanthal folds, and flat face of a person with trisomy 21 prompted Sir John Langdon Haydon Down to coin the inaccurate term *mongoloid* when he described the syndrome in 1866. As the medical superintendent of a facility for the profoundly mentally retarded, Down noted that about 10 percent of his patients resembled people of Mongolian heritage. The resemblance is superficial and meaningless. Males and females of all ethnic groups can have Down syndrome.

Researchers suspected a link between Down syndrome and an abnormal chromosome number as long ago as 1932. In 1958, improved chromosome visualization techniques revealed 47 chromosomes in cells of a person with trisomy 21 Down syndrome. By 1959, researchers had implicated chromosome 21. In 1960, they discovered Down syndrome caused by a translocation between chromosome 21 and another chromosome, and in 1961, researchers identified mosaic Down syndrome. The affected girl had physical signs of the condition, but normal intelligence.

A person with Down syndrome is usually short and has straight, sparse hair and a tongue protruding through thick lips. The hands have an abnormal pattern of creases, the joints are loose, and poor reflexes and muscle tone give a "floppy" appearance. Developmental milestones (such as sitting, standing, and walking) come slowly, and toilet training may take several years. Intelligence varies greatly. Parents of a child with Down syndrome can help their child reach maximal potential by providing a stimulating environment.

Many people with Down syndrome have physical problems, including heart and kidney defects, and hearing and vision loss. A suppressed immune system can make influenza deadly. Digestive system blockages are common and must be surgically corrected. A child with Down syndrome is 15 times more likely to develop leukemia than a child who does not have the syndrome, but this is only a 1 percent risk. Many of the medical problems associated with Down syndrome are treatable, so that more than 70 percent of affected individuals live longer than age 30. In 1910, life expectancy was only nine years.

Persons with Down syndrome who pass age 40 often develop the black fibers and tangles of amyloid protein in their brains characteristic of Alzheimer disease, although they usually do not become severely demented. The chance of a person with trisomy 21 developing Alzheimer disease is 25 percent, compared to 6 percent for the general population. A gene on chromosome 21 causes one inherited form of Alzheimer disease. Perhaps the extra copy of the gene in trisomy 21 has a similar effect to a mutation in the gene that causes Alzheimer disease. In a person with Down syndrome, Alzheimer disease seems like an

**table 12.5**

**Comparing and Contrasting Trisomies 13, 18, and 21**

| Type of Trisomy | Incidence at Birth | Percent of Conceptions That Survive 1 Year After Birth |
|---|---|---|
| 13 (Patau) | 1/12,500–1/21,700 | <5% |
| 18 (Edward) | 1/6,000–1/10,000 | <5% |
| 21 (Down) | 1/800–1/826 | 85% |

acceleration in the forgetfulness that can accompany aging.

By looking at people who have a third copy of only part of chromosome 21, researchers are zeroing in on the genes implicated in Down syndrome. The distal third of the long arm is consistently present in people with Down syndrome who have only part of the extra chromosome 21. This region houses a few hundred genes, a few of which cause signs and symptoms of Down syndrome: a gene for an enzyme involved in aging; a leukemia-causing gene; and a gene for a lens protein whose overexpression might cause the cataracts often seen in the eyes of older people with Down syndrome.

The likelihood of giving birth to a child with Down syndrome increases dramatically with the age of the mother. The overall frequency of trisomy 21 Down syndrome is 1 in about 800 births. For women under 30, the chances are 1 in 952. But at age 35, the risk is 1 in 378, and at age 40, 1 in 106. By age 45, risk jumps to 1 in 30, and by age 48, it is 1 in 14. However, 80 percent of children with trisomy 21 are born to women under age 35. This is because younger women are more likely to become pregnant and less likely to undergo amniocentesis. About 90 percent of trisomy 21 conceptions are due to nondisjunction during meiosis I in the female. The 10 percent of cases due to the male result from nondisjunction during meiosis I or II.

The chance that trisomy 21 will recur in a family, based on empirical data (how often it actually does recur in families), is 1 percent. Genetic counselors consider this figure along with maternal age effects, presenting a worst-case scenario to the expectant couple (see figure 1.4).

The age factor in Down syndrome may reflect the fact that meiosis in the female is completed only after conception. The older a woman is, the longer her oocytes have been arrested on the brink of completing meiosis. During this time, the oocytes may have been exposed to chromosome-damaging chemicals, viruses, or radiation. A variation on this idea suggests that females have a pool of aneuploid oocytes resulting from nondisjunction. An unknown mechanism prevents these abnormal oocytes from maturing. As a woman ages, selectively releasing normal oocytes each month since puberty, the abnormal ones begin to accumulate, much as black jellybeans accumulate as people preferentially eat the colored ones.

The maternal age association with Down syndrome has been recognized since the 19th century, when physicians noticed that affected babies were often the youngest children in large families. At that time, the condition was thought to be caused by syphilis, tuberculosis, thyroid malfunction, alcoholism, or emotional trauma. In 1909, a study of 350 affected infants revealed an overrepresentation of older mothers, prompting some researchers to attribute the link to "maternal reproductive exhaustion." In 1930, another study found that the increased risk of Down syndrome correlated to maternal age, and not to the number of children in the family.

### Trisomy 18—Edward Syndrome

Trisomies 18 and 13 were described in the same research report in 1960. Only 1 in 6,000 to 10,000 newborns has trisomy 18, but as table 12.5 indicates, most affected individuals do not survive to be born. The severe symptoms of trisomy 18 explain why few affected fetuses survive and also make the syndrome relatively easy to diagnose prenatally using ultrasound—yet the symptoms are presumably milder than those associated with the majority of aneuploids, which are manifest solely as spontaneous abortions. The associated major abnormalities include heart defects, a displaced liver, growth retardation, and oddly clenched fists. After birth, additional anomalies are apparent. These include overlapping placement of fingers (figure 12.13), a narrow and flat skull, abnormally shaped and low-set ears, a small mouth and face, unusual or absent fingerprints, short large toes with fused second and third toes, and "rocker-bottom" feet.

Affected children have great physical and mental disabilities, with developmental skills stalled at the six-month level. Most cases of trisomy 18 are traced to nondisjunction in meiosis II during oocyte formation.

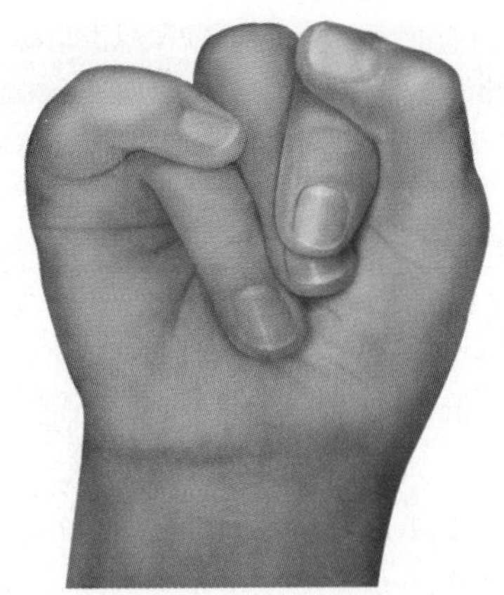

**figure 12.13**

**Trisomy (Edward syndrome).** An infant with trisomy 18 clenches its fist in a characteristic manner, with fingers overlapping.

### Trisomy 13—Patau Syndrome

Trisomy 13 is very rare, but, as is the case with trisomy 18, the number of newborns with the anomaly reflects only a small percentage of affected conceptions. Trisomy 13 has a different set of signs and symptoms than trisomy 18. Most striking, although rare, is a fusion of the developing eyes, so that a fetus has one large eyelike structure in the center of the face. More common is a small or absent eye. Major abnormalities affect the heart, kidneys, brain, face, and limbs. The nose is often malformed, and cleft lip and/or palate is present in a small head. Extra fingers and toes may occur. Appearance of a facial cleft and extra digits on an ultrasound exam are considered sufficient evidence to pursue chromosome analysis of the fetus to detect trisomy 13.

Ultrasound examinations of affected newborns reveal more extensive anomalies, including an extra spleen, abnormal liver structure, rotated intestines, and an abnormal pancreas. A few individuals have survived until adulthood, but they do not progress developmentally beyond the six-month level.

## *Sex Chromosome Aneuploids*

People with sex chromosome aneuploidy have extra or missing sex chromosomes. Table 12.6 indicates how these aneuploids can arise. Note that some conditions can result from nondisjunction in meiosis in the male or female.

**Turner Syndrome (45,X)** In 1938, at a medical conference, an endocrinologist named Henry Turner described seven young women, aged 15 to 23, who were sexually undeveloped, short, had folds of skin on the back of the neck, and had malformed elbows. (Eight years earlier, an English

**table 12.6**

**How Nondisjunction Leads to Sex Chromosome Aneuploids**

| Situation | Oocyte | Sperm | Consequence |
|---|---|---|---|
| Normal | X | Y | 46,XY normal male |
| | X | X | 46,XX normal female |
| Female nondisjunction | XX | Y | 47,XXY Klinefelter syndrome |
| | XX | X | 47,XXX triplo-X |
| | | Y | 45,Y nonviable |
| | | X | 45,X Turner syndrome |
| Male nondisjunction | X | | 45,X Turner syndrome |
| (meiosis I) | X | XY | 47,XXY Klinefelter syndrome |
| Male nondisjunction | X | XX | 47,XXX triplo-X |
| (meiosis II) | X | YY | 47,XYY Jacobs syndrome |
| | X | | 45,X Turner syndrome |

physician named Ullrich had described the syndrome in young girls and so it is called Ullrich syndrome in the U.K.) Alerted to what would become known as Turner syndrome, other physicians soon began identifying such patients in their practices. Physicians assumed that a hormonal insufficiency caused the symptoms. They were right, but there was more to the story—a chromosomal imbalance caused the hormone deficit.

In 1954, at a London hospital, P. E. Polani discovered that cells from Turner patients lacked a Barr body, the dark spot that indicates the presence of a second X chromosome. Might the lack of a sex chromosome cause the symptoms, particularly the failure to mature sexually? By 1959, karyotyping confirmed the absence of an X chromosome in cells of Turner syndrome patients. It was later learned that only 50 percent of individuals with Turner syndrome are XO. The rest have partial deletions of the X chromosome, or are mosaics, with only some cells affected.

Like the autosomal aneuploids, Turner syndrome is found more frequently among spontaneously aborted fetuses than among liveborns—99 percent of affected fetuses die before birth. The syndrome affects 1 in 2,000 female births. However, newborns with the condition are usually not seriously affected. Often they do not know they have a chromosome abnormality until they lag behind their classmates in sexual development. Two X chromosomes are necessary for normal sexual development in females.

In childhood, signs of Turner syndrome include wide-set nipples, slight webbing at the back of the neck, short stature, coarse facial features, and a low hairline at the back of the head. About half of people with Turner syndrome have impaired hearing and frequent ear infections, due to a small defect in the shape of the cochlea, which is the coiled part of the inner ear. They cannot hear certain frequencies of sound. At sexual maturity, sparse body hair develops, but the girls do not ovulate or menstruate, and they have underdeveloped breasts. The uterus is very small, but the vagina and cervix are normal. However, in the ovaries, oocytes speed through development, depleting the supply during infancy. Intelligence is normal, and these women can lead fairly normal lives if they receive hormone supplements. Using growth hormone adds up to three inches or so of height. Although women with Turner syndrome are infertile, individuals who are mosaics may have children.

Women with Turner syndrome who do become pregnant are at high risk of carrying a fetus with an abnormal number of chromosomes. In one study of 138 pregnancies, 82 produced infants, and 23 of them had birth defects, 10 of these with abnormal chromosomes. *In vitro* fertilization (IVF), discussed in chapter 21, can be used to select oocytes that have a normal chromosome number. Interestingly, Turner syndrome is the only aneuploid condition that seems unrelated to the age of the mother.

For many years, it was thought that Turner syndrome had no effects in adulthood, but this was largely because most studies of common disorders of adults did not consider chromosome status. Researchers in Edinburgh have been tracking the health of 156 women with Turner syndrome for more than 25 years as part of an abnormal karyotype register. Having Turner syndrome apparently does affect lifespan. For example, 68 percent of the 156 participants reached age 60, compared to 88 percent of the general British population. Studies that calculate relative risk for females with Turner syndrome indicate that they are more likely to develop certain disorders than the general population (figure 12.14).

The many signs and symptoms of Turner syndrome result from the loss of specific genes. For example, loss of a gonadal dysgenesis gene accounts for the ovarian failure, whereas absence of a homeobox gene causes the characteristic short stature. A gene may soon be identified for the unusual hearing defect that is part of the syndrome.

**Extra X Chromosomes** About 1 in every 1,000 females has an extra X chromosome in each of her cells, a condition called triplo-X. The only symptom seems to be tallness and menstrual irregularities. Although triplo-X females are rarely mentally retarded, they tend to be less intelligent than their siblings. The lack of symptoms associated with having extra X chromosomes reflects the protective effect of X inactivation—all but one of the X chromosomes is inactivated.

About 1 in 1,000 males has an extra X chromosome, which causes Klinefelter syndrome (XXY). Physicians first described the signs and symptoms in 1942, and geneticists identified the underlying chromosomal anomaly in 1959. Men severely affected with Klinefelter syndrome are underdeveloped sexually, with rudimentary testes and prostate glands and sparse pubic and facial hair. They have very long arms and legs, large hands and feet, and may develop breast tissue. Klinefelter syndrome is the most common genetic or chromosomal cause of male infertility, accounting for 4 to 6 percent of infertile men.

Testosterone injections during adolescence can limit limb lengthening and prompt development of secondary sexual characteristics. Boys and men with Klinefelter syndrome may be slow to learn, but they are usually not

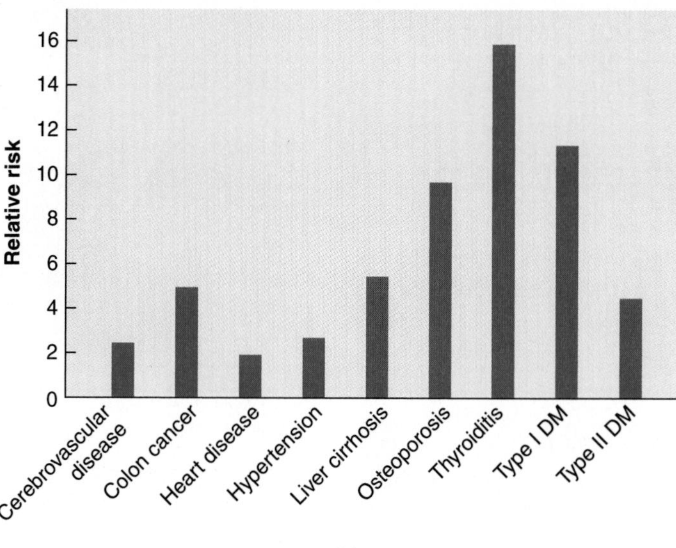


**figure 12.14**

**Relative risk of disease in Turner syndrome in adulthood.** For many years it was thought that Turner syndrome did not affect individuals in middle adulthood, because such women did not report problems. But it turned out that nobody asked! Women with Turner syndrome do not look different than anyone else, so until recently, there were no data on their higher risk of developing certain conditions.

mentally retarded unless they have more than two X chromosomes, which happens rarely.

Many textbooks include photographs of very extreme cases of Klinefelter syndrome, which may give the erroneous impression that the syndrome is always severe. Actually, many men who have the condition discover it only when they have an infertility problem. Some affected men probably never learn that they have Klinefelter syndrome. The photograph of the young man who wrote "A Personal Look at Klinefelter Syndrome" on page 252 shows that affected individuals can look quite like anyone else.

Some men with Klinefelter syndrome have fathered children, using IVF and intracytoplasmic sperm injection (ICSI), also discussed in chapter 21. Using these methods, doctors can identify and select sperm that contain only one sex chromosome and use them to fertilize oocytes. Examination of sperm from men with Klinefelter syndrome shows that many of the cells are XX or XY, instead of the normal X or Y. However, one man also produced sperm that had two chromosome 21s, which, if they fertilize an oocyte, would lead to an individual with Down syndrome. Researchers speculate that in this man, perhaps chromosome 21 was a "bystander" located near the sex chromosomes that were undergoing nondisjunction, and were distributed incorrectly too, a phenomenon called the interchromosomal effect.

**XYY Syndrome** One male in 1,000 has an extra Y chromosome. Awareness of this condition arose in 1961, when a tall, healthy, middle-aged man, known for his boisterous behavior, underwent a routine chromosome check after fathering a child with Down syndrome. The man had an extra Y chromosome. A few other cases were detected over the next several years.

In 1965, researcher Patricia Jacobs published results of a survey among 197 inmates at Carstairs, a high-security prison in Scotland. Of 12 men with unusual chromosomes, seven had an extra Y. Might their violent or aggressive behavior be linked to their extra Y chromosome? Jacobs's findings were repeated in studies in English and Swedish mental institutions. Soon after, *Newsweek* magazine ran a cover story on "congenital criminals." In 1968, defense attorneys in France and Australia pleaded their violent clients' cases on the basis of an inherited flaw, the extra Y of what became known as Jacobs syndrome. Meanwhile, the National Institute of Mental Health, in Bethesda, Maryland, held a conference on the condition, lending legitimacy to the hypothesis that an extra Y predisposes to violent behavior.

In the early 1970s, newborn screens began in hospital nurseries in England, Canada, Denmark, and Boston. XYY babies were visited by social workers and psychologists who offered "anticipatory guidance" to the parents on how to deal with their toddling future criminals. By 1974, geneticists and others halted the program, pointing out that singling out these boys on the basis of a few statistical studies was inviting self-fulfilling prophecy.

Today, we know that 96 percent of XYY males are apparently normal. The only symptoms attributable to the extra chromosome may be great height, acne, and perhaps speech and reading problems. An explanation of the continued prevalence of XYY among mental-penal institution populations may be more psychological than biological. Large body size may lead teachers, employers, parents, and others to expect more of these people, and a few of them may deal with this stress by becoming aggressive.

Jacobs syndrome can arise from nondisjunction in the male, producing a sperm with two Y chromosomes that fertilizes an X-bearing oocyte. Geneticists have never observed a sex chromosome constitution of one Y and no X. Since the Y chromosome carries little genetic material, and the gene-packed X chromosome would not be present, the absence of so many genes makes development beyond a few cell divisions in a YO embryo impossible.

**KEY CONCEPTS**

Polyploids have extra sets of chromosomes, and do not survive for long. Aneuploids have extra or missing chromosomes. Nondisjunction during meiosis causes aneuploidy. Trisomics are more likely to survive than monosomics, and sex chromosome aneuploidy is less severe than autosomal aneuploidy. Mitotic nondisjunction produces chromosomal mosaics. Down syndrome (trisomy 21) is the most common autosomal aneuploid, followed by trisomies 18 and 13. Sex chromosome aneuploids include Turner syndrome (XO), triplo-X females, Klinefelter syndrome (XXY), and XYY syndrome males.

*In Their Own Words*

# A Personal Look at Klinefelter Syndrome

I was diagnosed with Klinefelter syndrome (KS) at age 25, in February 1996. Being diagnosed has been . . . a big sigh of relief after a life of frustrations. Throughout my early childhood, teens, and even somewhat now, I was very shy, reserved, and had trouble making friends. I would fly into rages for no apparent reason. My parents knew when I was very young that there was something about me that wasn't right.

I saw many psychologists, psychiatrists, therapists, and doctors, and their only diagnosis was "learning disabilities." In the seventh grade, I was told by a psychologist that I was stupid and lazy, and I would never amount to anything. After barely graduating high school, I started out at a local community college. I received an associate degree in business administration, and never once sought special help. I transferred to a small liberal arts college to finish up my bachelor of science degree, and spent an extra year to complete a second degree. Then I started a job as a software engineer for an Internet-based company. I have been using computers for 20 years and have learned everything I needed to know on my own.

To find out my KS diagnosis, I had gone to my general physician for a physical. He noticed that my testes were smaller than they should be and sent me for blood work. The karyotype showed Klinefelter syndrome, 47,XXY. After seeing the symptoms of KS and what effects they might have, I found it described me perfectly. But, after getting over the initial shock and dealing with the denial, depression, and anger, I decided that there could be things much worse in life. I decided to take a positive approach.

There are several types of treatments for KS. I give myself a testosterone injection in the thigh once every two weeks. My learning and thought processes have become stronger, and I am much more outgoing and have become more of a leader. Granted, not all of this is due to the increased testosterone level, some of it is from a new confidence level and from maturing.

I feel that parents who are finding out prior to the birth of their son (that he will have Klinefelter syndrome) or parents of affected infants or young children are very lucky. There is so much they can do to help their child have a great life. I have had most all of the symptoms at some time in my life, and I've gotten through and done well.

Stefan Schwarz
sdschwarz13@yahoo.com

(Stefan Schwarz runs a Boston-area support group for KS.)

## 12.4 Abnormal Chromosome Structure

Structural chromosomal defects include missing, extra, or inverted genetic material within a chromosome or combined or exchanged parts of nonhomologs (translocations) (figure 12.15).

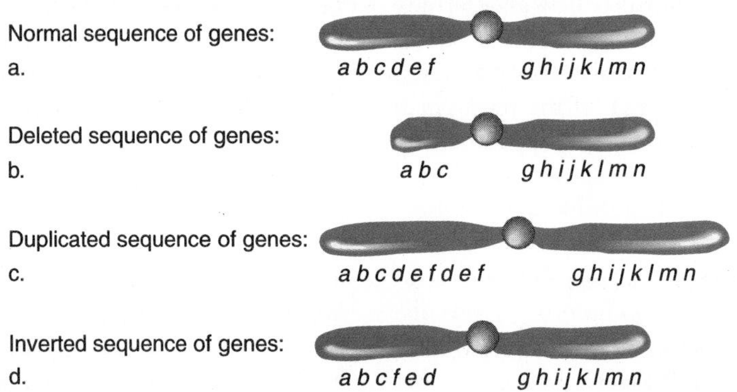


**figure 12.15**

**Chromosome abnormalities.** If a hypothetical normal gene sequence appears as shown in (*a*), then (*b*) represents a deletion, (*c*) a duplication, and (*d*) an inversion.

### Deletions and Duplications

A **deletion** is missing genetic material. Deletions range greatly in size, and the larger ones tend to have greater effects because they remove more genes. Consider cri-du-chat syndrome (French for "cat's cry"), caused by deletion of part of the short arm of chromosome 5 (also called $5p^-$ syndrome). Affected children have a high-pitched cry similar to the mewing of a cat and pinched facial features, and are mentally retarded and developmentally delayed. The chromosome region responsible for the catlike cry is distinct from the region that causes mental retardation and developmental delay, suggesting that the deletion can remove more than one gene. A cytogeneticist can determine by examining a detailed karyotype whether a child will have only the catlike cry and perhaps poor weight gain, or will have all of the signs and symptoms, which include low birth weight, poor muscle tone, a small head, and impaired language skills. In Their Own Words on page 253 describes a child who had $5p^-$ syndrome.

*In Their Own Words*

# Ashley's Message of Hope

What is it like to have a child born with cri-du-chat syndrome? How does this affect the family and its future? What kinds of assistance can the medical community offer the family?

The birth of any child raises many questions. Will she have my eyes, her dad's smile? What will she want to be when she grows up? But the biggest question for every parent is "Will she be healthy?" If complications occur during birth or if the child is born with a genetic disorder, the questions become more profound and immediate. "How did this happen?" "Where do we go from here?" "Will this happen again?"

Our daughter, Ashley Elizabeth Naylor (figure 1), was born August 12, 1988. We had a lot of mixed emotions the day of her birth, but mainly we felt fear and despair. The doctors suspected complications, which led to a cesarean section, but the exact problem was not known. Two weeks after her birth, chromosome analysis revealed cri-du-chat (cat cry) syndrome, also known as $5p^-$ syndrome because part of the short arm of one copy of chromosome 5 is missing. The prognosis was uncertain. This is a rare disorder, we were told, and little could be offered to help our daughter. The doctors used the words "profoundly retarded," which cut like a knife through our hearts and our hopes. It wasn't until a few years later that we realized how little the medical community actually knew about cri-du-chat syndrome and especially about our little girl!

Ashley defied all the standard medical labels, as well as her doctors' expectations. Her spirit and determination enabled her to walk with the aid of a walker and express herself using sign language and a communication device. With early intervention and education at United Services for the Handicapped, Ashley found the resources and additional encouragement she needed to succeed. In return, Ashley freely offered one of her best-loved and sought-after gifts—her hugs. Her bright eyes and glowing smile captured the hearts of everyone she met.

In May of 1994, Ashley's small body could no longer support the spirit that inspired so many. She passed away after a long battle with pneumonia. Her physical presence is gone, but her message remains: hope.

If you are a parent faced with similar profound questions after the birth of your child, do not assume one doctor has all the answers. Search for doctors who respect your child enough to talk to her, not just about her. Above all, find an agency or a school that can help you give your child a chance to succeed. Early education for your child and support for yourself are crucial.

If you are a student in a health field, become as knowledgeable as possible and stay current with the latest research, but most importantly, be sensitive to those who seek your help. Each word you speak is taken to heart. Information is important, but hope can make all the difference in a family's future.

Kathy Naylor

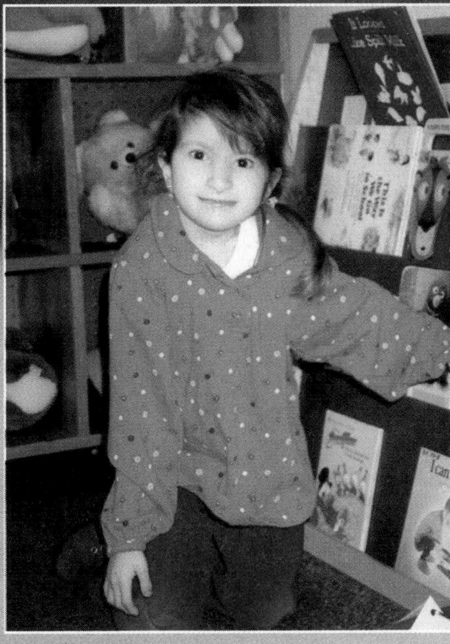

**figure 1**

Ashley Naylor brought great joy to her family and community during her short life.
Courtesy of Kathy Naylor.

A **duplication** is a region of a chromosome where genes are repeated. Duplications, like deletions, are more likely to cause symptoms if they are extensive. For example, duplications of chromosome 15 do not produce a phenotype unless they repeat several genes. Figure 12.16 shows three duplicated chromosome 15s, with increasing amounts of material repeated. Many people have the first two types of duplications and have no symptoms. However, several unrelated individuals with the third, larger duplication have seizures and are mentally retarded. A duplication elsewhere on chromosome 15 has been associated with panic attacks and other anxiety disorders.

FISH can detect tiny deletions and duplications that are smaller than the bands revealed by conventional chromosome staining. Small duplications are generally not dangerous, but some "microdeletions" have been associated with a number of syndromes. Certain microdeletions in the short arm of the Y chromosome, for example, cause male infertility. Deletions and duplications can arise from chromosome rearrangements, which include translocations, inversions, and ring chromosomes.

## Translocations

In a translocation, different (nonhomologous) chromosomes exchange or combine parts. Exposure to certain viruses, drugs, and radiation can cause translocations, but often they arise for no apparent reason.

There are two major types of translocations. In a **Robertsonian translocation,** the short arms of two different acrocentric chromosomes break, leaving sticky ends that then cause the two long arms to adhere. A new, large chromosome forms from the long arms of the two different chromosomes. Because genes on the short arms of the involved chro-

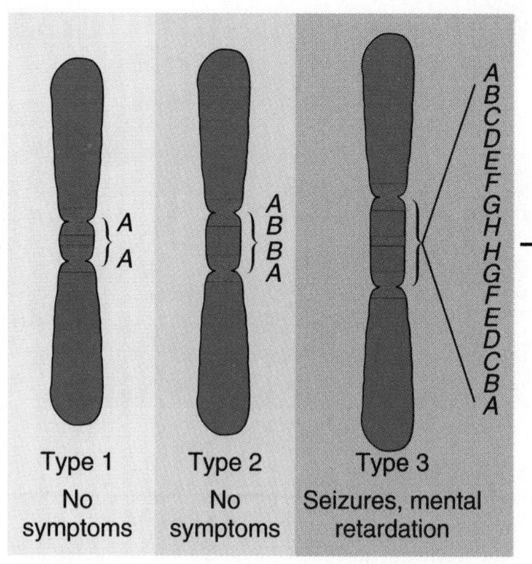


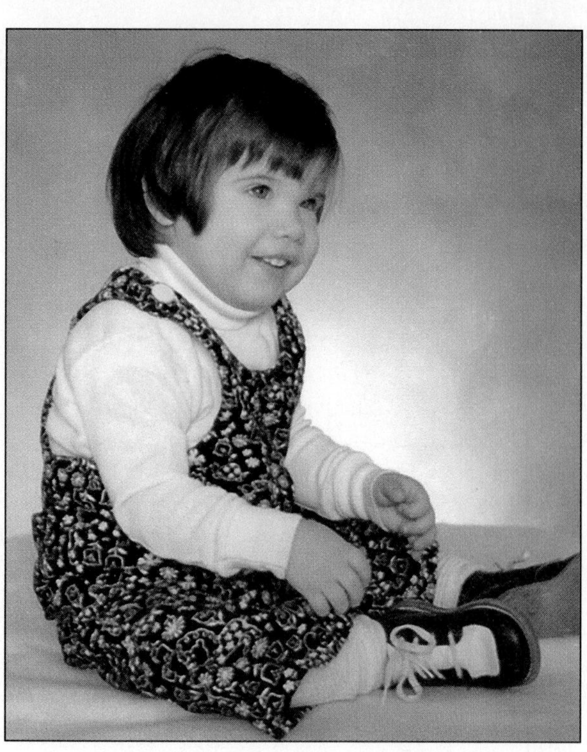


**figure 12.16**

**A duplication.** A study of duplications of parts of chromosome 15 revealed that small duplications do not affect the phenotype, but larger ones may. (*a*) The letters indicate specific DNA sequences, which serve as markers to compare chromosome regions. Note that the duplication is also inverted. (*b*) This child, who has "inv dup (15) syndrome," appears normal but has minor facial anomalies characteristic of the condition.

mosomes are repeated elsewhere, their absence in a Robertsonian translocation does not affect the phenotype. The person with the large, translocated chromosome, called a **translocation carrier,** has 45 chromosomes, but may not have symptoms if a crucial gene has not been deleted or damaged. Even so, he or she may produce unbalanced gametes—sperm or oocytes with too many or too few genes. This can lead to reproductive difficulties, such as spontaneous abortion or birth defects.

One in 20 cases of Down syndrome arises because a parent has a Robertsonian translocation between chromosome 21 and another, usually chromosome 14. The problem arises because the individual with the translocated chromosome produces some gametes that lack either of the involved chromosomes and some gametes that have extra material from one of the translocated chromosomes (figure 12.17). In such a case, each fertilized ovum has a 1 in 2 chance of ending in spontaneous abortion, and a 1 in 6 chance of developing into an individual with Down syndrome. The risk of the couple having a child with Down syndrome is theoretically 1 in 3, because the spontaneous abortions are not counted as births. However, because some Down syndrome fetuses spontaneously abort, the actual risk of a couple in this situation having a child with Down syndrome is about 15 percent. The other two outcomes—normal chromosomes and a translocation carrier like the parent—have normal phenotypes. Either a male or a female can be a translocation carrier, and the condition is not related to age.

The second most common type of Robertsonian translocation occurs between chromosomes 13 and 14, causing symptoms of Patau syndrome because of an excess of chromosome 13 material. Robertsonian translocations occur in 1 in 500 births, making them fairly common chromosomal aberrations.

In a **reciprocal translocation,** two different chromosomes exchange parts. FISH can be used to highlight the involved chromosomes (figure 12.18). If the chromosome exchange does not break any genes, then a person who has both translocated chromosomes is healthy and is also a translocation carrier. He or she has the normal amount of genetic material, but it is rearranged.

A reciprocal translocation carrier can have symptoms if one of the two breakpoints lies in a gene, disrupting its function. Figure 12.19 shows a father and son who have a reciprocal translocation between chromosomes 2 and 20 that causes a condition called Alagille syndrome. Apparently the exchange disrupts a gene on chromosome 20 that causes the condition, because families with the syndrome have deletions in this region of the chromosome. Alagille syndrome produces a characteristic face, absence of bile ducts in the liver, abnormalities of the eyes and ribs, heart defects, and severe itching. The condition is so variable that some people do not know that they have it. The father of the young man in fig-

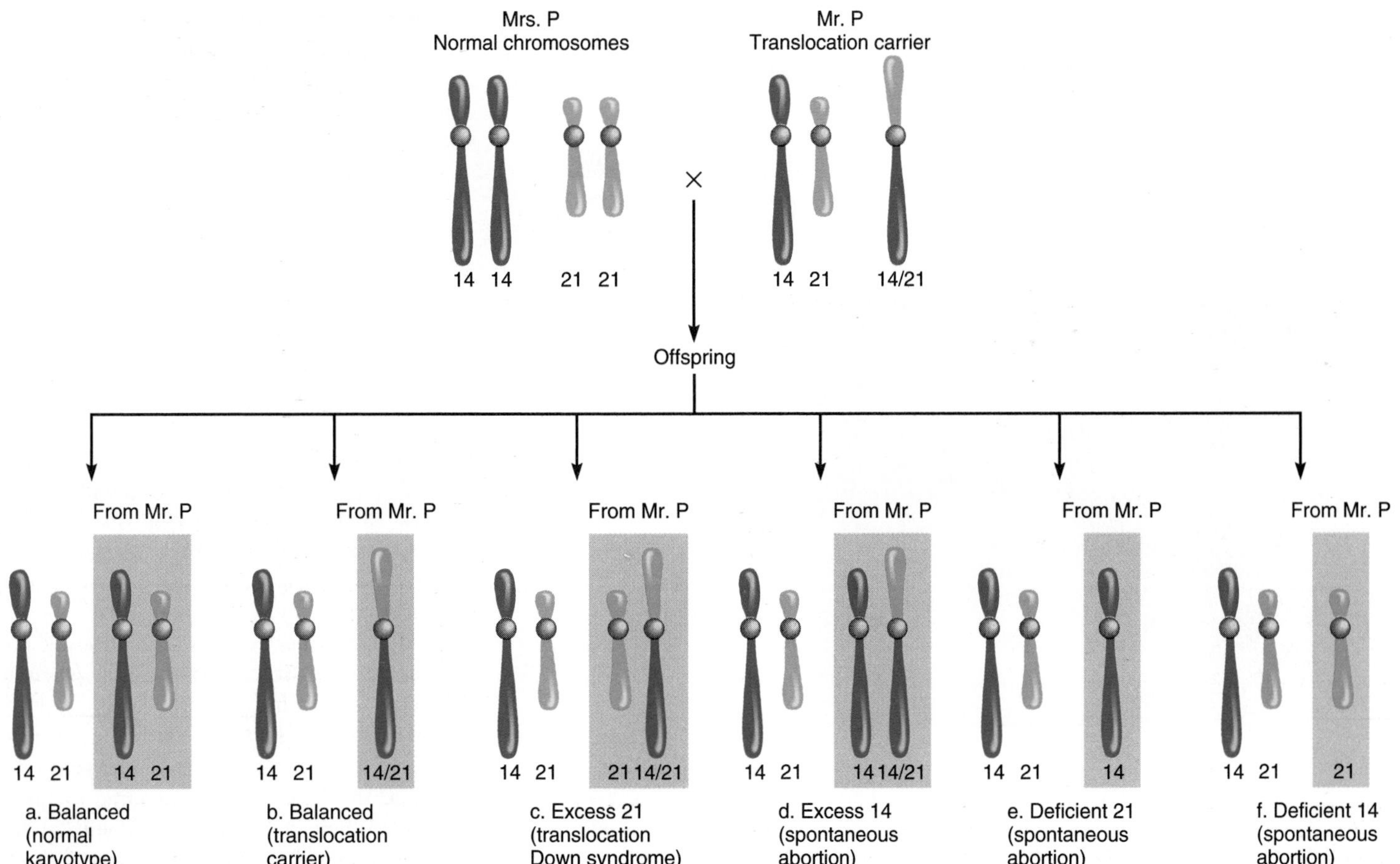


**figure 12.17**

**A Robertsonian translocation.** A Robertsonian translocation can cause spontaneous abortion or syndromes. Mr. P. is a translocation carrier. He has only 45 chromosomes because the long arm of one chromosome 14 has joined the long arm of one chromosome 21, with their centromeres joined and short arms lost. He has no symptoms. Mr. P. makes six types of sperm cells, and they determine the fate of his and Mrs. P.'s offspring. (*a*) A sperm with one normal chromosome 14 and one normal 21 yields a chromosomally normal child. (*b*) A sperm carrying the translocated chromosome produces a child who is a translocation carrier, like Mr. P. (*c*) If a sperm contains Mr. P.'s normal 21 and his translocated chromosome, the child receives too much chromosome 21 material and therefore has Down syndrome. (*d*) A sperm containing the translocated chromosome and a normal 14 leads to excess chromosomal 14 material, which is lethal in the embryo or fetus. (*e*) and (*f*) If a sperm lacks either chromosome 14 or 21, it leads to monosomies, which are lethal prenatally. (Chromosome arm lengths are not precisely accurate.)

ure 12.19 did not realize he had the syndrome until he had an affected child with a more severe case. Sometimes, a translocation arises anew in a gamete that leads to a new individual. This is called *"de novo,"* as opposed to inheriting a translocated chromosome from a parent.

A translocation carrier produces some unbalanced gametes—sperm or oocytes that have deletions or duplications of some of the genes in the translocated chromosomes. The resulting phenotype depends upon the particular genes that the chromosomal rearrangement disrupts or are extra or missing.

Information from the human genome project on particular genes and from other chromosomal abnormalities can explain how some translocations affect health. For example, a child was born with a reciprocal translocation between chromosomes 12 and 22. The distinctive symptoms of language delay, mild mental retardation, loose joints, minor facial anomalies, and a narrow, long head matched those of another chromosome problem, called 22q13.3 deletion syndrome. That condition is caused by absence of a gene (called ProSAP2) that forms scaffolds for neurons in the cerebral cortex and cerebellum. Apparently, the translocation cuts this gene, abolishing its function just as a deletion does. As a result, these parts of the brain malfunction.

A genetic counselor becomes alerted to the possibility of a translocation if a family has had multiple birth defects and spontaneous abortions. People with translocations have been very valuable to medical genetics research. Studies to identify disease-causing genes often began with people whose translocations pointed the way toward a gene of interest.

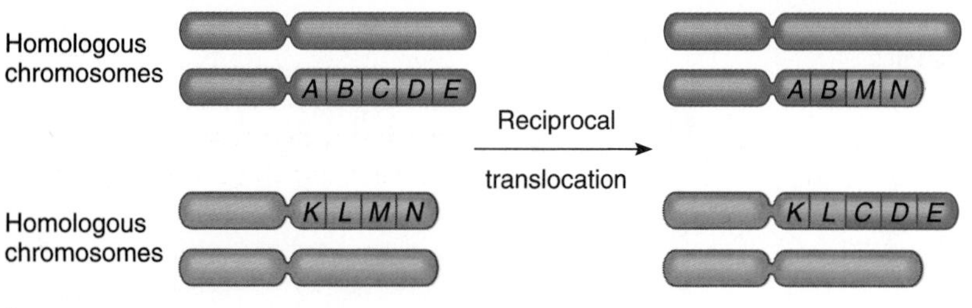


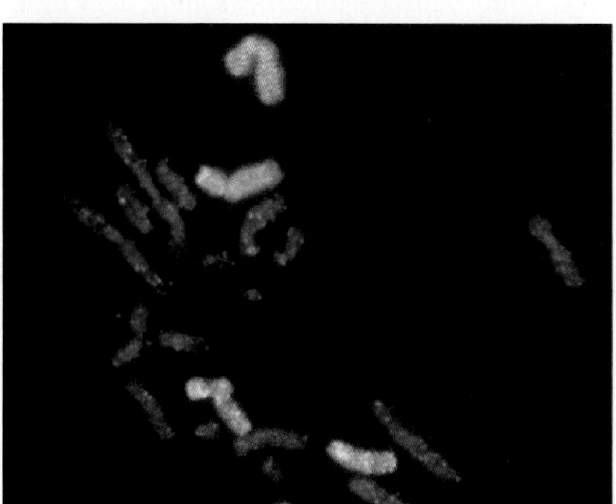


**figure 12.18**

**A reciprocal translocation.** In a reciprocal translocation, two nonhomologous chromosomes exchange parts. In (*a*), genes *C, D,* and *E* on the blue chromosome exchange positions with genes *M* and *N* on the red chromosome. (*b*) shows a reciprocal translocation that is highlighted using FISH. The pink chromosome with the dab of blue, and the blue chromosome with a small section of pink, are the translocated chromosomes.

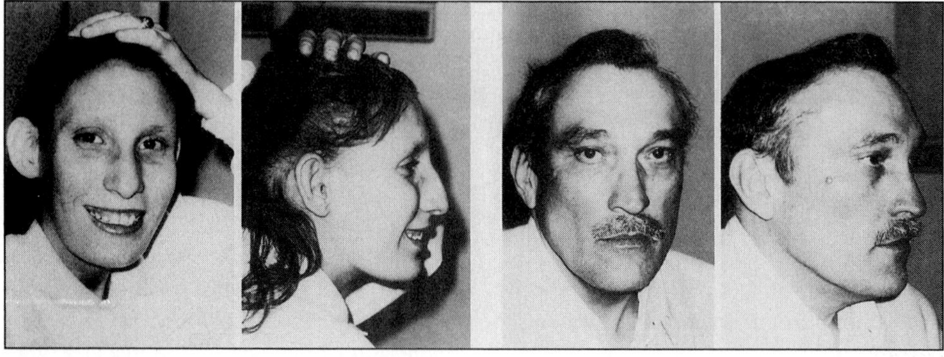

**figure 12.19**

**A translocation syndrome.** In one family with Alagille syndrome, a reciprocal translocation occurs between chromosomes 2 and 20. Distinctive facial features are part of the condition.

## Inversions 

An inverted sequence of chromosome bands indicates that part of the chromosome has flipped around. Empirical studies show that 5 to 10 percent of inversions cause health problems, probably because they disrupt important genes. Sometimes inversions are detected in fetal chromosomes, but physicians do not know whether symptoms will be associated with the problem. The parents can have their chromosomes checked. If one of them has the inversion and is healthy, then the child will most likely not have symptoms related to the inversion. If neither parent has the inversion, then the anomaly arose in a gamete, and predicting effects may depend on knowing which genes are involved.

Like a translocation carrier, an adult can be heterozygous for an inversion and be healthy, but have reproductive problems. One woman had an inversion in the long arm of chromosome 15 and had two spontaneous abortions, two stillbirths, and two children with multiple problems who died within days of birth. She did eventually give birth to a healthy child. How did the inversion cause these problems?

Inversions with such devastating effects can be traced to meiosis, when a crossover occurs between the inverted chromosome segment and the noninverted homolog. To allow the genes to align, the inverted chromosome forms a loop. When crossovers occur within the loop, in the resulting recombinant chromosomes, some areas are duplicated and some deleted. In inversions, the abnormal chromosomes result from the chromatids that crossed over.

Two types of inversions are distinguished by the position of the centromere relative to the inverted section. A **paracentric inversion** does not include the centromere (figure 12.20). A single crossover within the inverted segment gives rise to two very abnormal chromosomes. The other two chromosomes are normal. One abnormal chromosome retains both centromeres and is said to be **dicentric.** When the cell divides, the two centromeres are pulled to opposite sides of the cell, and the chromosome breaks, leaving pieces with extra or missing segments. The second type of abnormal chromosome resulting from a crossover within an inversion loop is a small piece that lacks a centromere, called an acentric fragment. When the cell divides, the fragment is lost.

A **pericentric inversion** includes the centromere within the loop. A crossover in it produces two chromosomes that have duplications and deletions, but one centromere each (figure 12.21).

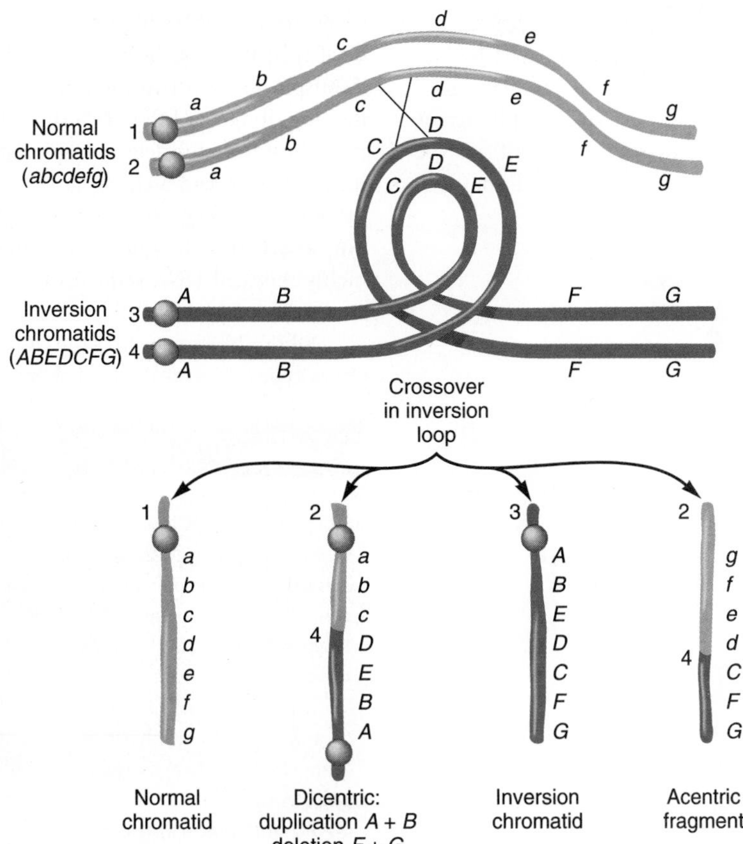


## figure 12.20

**Paracentric inversion.** A crossover between a chromosome with a paracentric inversion and its normal homolog, when in the region of the inversion, produces one normal chromatid, one inverted chromatid, one with two centromeres (called a dicentric), and one with no centromere (called an acentric fragment). The letters *a* through *g* denote genes.

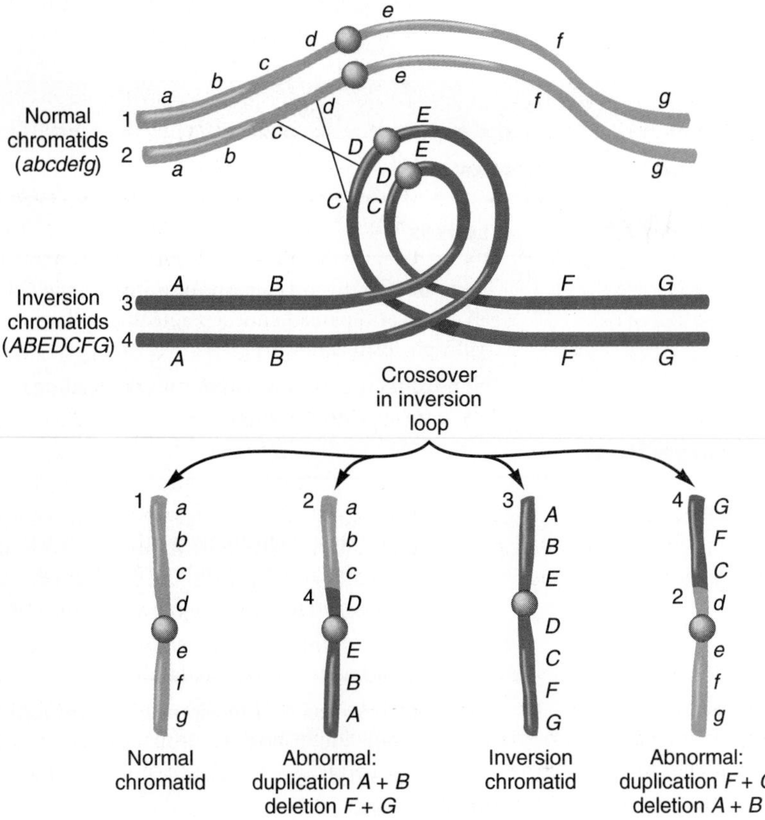


## figure 12.21

**Pericentric inversion.** A pericentric inversion in one chromosome leads to two chromatids with duplications and deletions, one normal chromatid, and one inverted chromatid if a crossover occurs.

## Isochromosomes and Ring Chromosomes

Another meiotic error that leads to unbalanced genetic material is the formation of an **isochromosome,** which is a chromosome that has identical arms. This occurs when, during division, the centromeres part in the wrong plane (figure 12.22). Isochromosomes are known for chromosomes 12 and 21 and for the long arms of the X and the Y. Some women with Turner syndrome do not have the more common monosomy XO, but an isochromosome in which the long arm of the X chromosome is duplicated but the short arm is absent.

Chromosomes shaped like rings form in 1 out of 25,000 conceptions. Ring chromosomes may arise when telomeres are lost, leaving sticky ends that close up. Exposure to radiation can form rings. They can involve any chromosome, and may occur in addition to a full diploid chromosome set, or account for one of the 46 chromosomes.

Ring chromosomes can produce symptoms when they add genetic material. For example, a small ring chromosome of DNA from chromosome 22 causes cat eye syndrome. Affected children have vertical pupils, are mentally retarded, have heart and urinary tract anomalies, and have skin growing over the anus. They have 47 chromosomes—the normal two chromosome 22s and a ring.

A study from Japan examined 15 women who have a ring X chromosome in some cells, in addition to the other two complete X chromosomes. Nine of the women were mentally retarded, and in all of them, the ring X chromosome did not include the XIST site. Recall from chapter 6 that XIST normally shuts off all but one X chromosome. In these nine women, too much expressed X chromosome material caused the mental retardation. In the other six women, the ring chromosome clearly included XIST, and so presumably was silenced.

In a very unusual case, identical twin girls each inherited two ring chromosomes. FISH analysis revealed that about two-thirds of the examined cells had one ring derived from chromosome 1 and one from chromosome 16. The remainder of the cells had either one ring or neither. Double rings had only been reported once previously—and never for twins. The twins' extra genetic material brought them many medical woes. The girls were hospitalized as newborns for several months for a variety of problems, and at two years of age, each still required supplemental oxygen, was developmentally delayed, had an undersized head, and was small. The girls' symptoms and distinctive faces led physicians to call in cytogeneticists. They identified the ring chromosomes and deduced that the twins arose from a single fertilized egg that developed the rings at the 2- or 4-cell stage, then split to yield two individuals that have the rings in most of their cells. In other cases, ring chromosomes detected on routine amniocentesis present a challenging problem in genetic counseling, because rings usually do not affect health—often, they consist of highly repeated DNA sequences that do not encode proteins.

Table 12.7 summarizes the causes of different types of chromosomal aberrations.

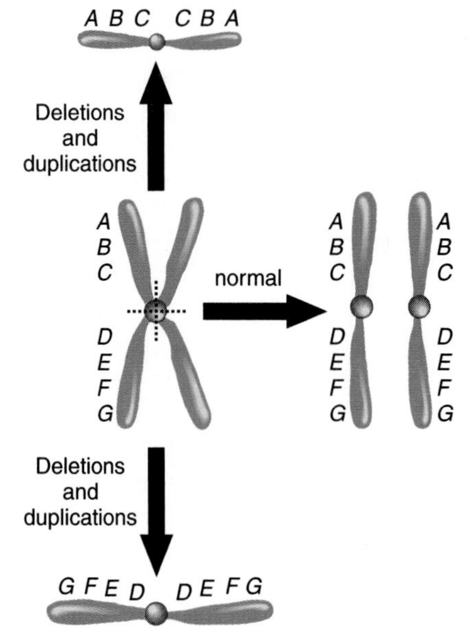


**figure 12.22**

**Isochromosomes have identical arms.** They form when chromatids divide along the wrong plane (in this depiction, horizontally rather than vertically).

### KEY CONCEPTS

Chromosome rearrangements can cause deletions and duplications. In a Robertsonian translocation, the long arms of two different acrocentric chromosomes join. In a reciprocal translocation, two chromosomes exchange parts. If a translocation leads to a deletion or duplication, or disrupts a gene, symptoms may result. Gene duplications and deletions can occur in isochromosomes and ring chromosomes, and when crossovers involve inversions. An isochromosome has two identical arms, thereby introducing duplications and deletions, and ring chromosomes can add genetic material.

**table 12.7**

**Causes of Chromosomal Aberrations**

| **Numerical Abnormalities** | **Causes** |
|---|---|
| Polyploidy | Error in cell division (meiosis or mitosis) in which all chromatid pairs do not separate in anaphase. Multiple fertilization |
| Aneuploidy | Nondisjunction (in meiosis or mitosis) leading to lost or extra chromosomes |
| **Structural Abnormalities** | |
| Deletions and duplications | Translocation<br>Crossover between a chromosome that has a pericentric inversion and its noninverted homolog |
| Translocation | Exchange between nonhomologous chromosomes |
| Inversion | Breakage and reunion of fragment in same chromosome, but with wrong orientation |
| Dicentric and acentric | Crossover between a chromosome with a paracentric inversion and its noninverted homolog |
| Ring chromosome | A chromosome loses telomeres and the ends fuse to form a circle |

## 12.5 Uniparental Disomy—Two Genetic Contributions from One Parent

If nondisjunction occurs in both of the gametes that join to become a fertilized ovum, a situation can arise in which a pair of homologs (or parts of them) comes solely from one parent, rather than one from each parent, as Mendel's law of segregation predicts. For example, if a sperm lacking a chromosome 14 fertilizes an ovum with two copies of that chromosome, it produces an individual with the normal 46 chromosomes, but one homologous pair that comes only from the female. This very rare situation of inheriting two chromosomes, or two segments of chromosomes, from one parent is called **uniparental disomy** (UPD). It means "two bodies from one parent," with "bodies" referring to chromosomes. The alternative state—the consequence of normal segregation of chromosome pairs—is termed biparental inheritance. UPD can also arise from a trisomic embryo in which some cells lose the extra chromosome, restoring diploidy, but leaving two chromosomes from one parent.

Because UPD requires the simultaneous occurrence of two very rare events—either nondisjunction of the same chromosome in gametes that join, or trisomy followed by chromosome loss, it is exceedingly rare. In addition, many cases are probably never seen, because bringing together identical homologs inherited from one parent could give the fertilized ovum a homozygous set of lethal alleles. Development would halt. Other cases of UPD may go undetected if they cause known recessive conditions and both parents are assumed to be carriers, when actually only one parent contributed to the offspring's illness. This was how UPD was discovered.

In 1988, Arthur Beaudet of the Baylor College of Medicine saw a very unusual patient with cystic fibrosis. Beaudet was comparing CF alleles in the patient to those in her parents, and he found that only the mother was a carrier—the father had two normal alleles. Didn't both parents have to be carriers for a child to inherit this autosomal recessive disorder? Beaudet did further testing and constructed haplotypes for each parent's chromosome 7, which includes the CF gene. Much to his surprise, he found that the daughter had two copies from her mother, and none from her father (figure 12.23). How did this happen?

Apparently, in the patient's mother, nondisjunction of chromosome 7 in meiosis II led to formation of an oocyte bearing two identical copies of the chromosome, instead of the usual one copy. The abnormal oocyte was then fertilized by a sperm that had also undergone nondisjunction and lacked a chromosome 7. The mother's extra genetic material compensated for the father's deficit, and an offspring developed. Unfortunately, she inherited a double dose of the mother's chromosome that carried the CF allele. In effect, inheriting two of the same chromosome from one parent shatters the protection that combining genetic material from two individuals offers, a protection that is the defining characteristic of sexual reproduction.

UPD may also cause disease if it removes the contribution of the important parent for an imprinted gene. Recall from chapter 6 that an imprinted gene is expressed if it comes from one parent, but silenced if it comes from the other (see figure 6.15). If UPD removes the parental genetic material that must be present for a critical gene to be expressed, then a mutant phenotype results. The classic example is the 20 to 30 percent of Prader-Willi syndrome and Angelman syndrome cases caused by UPD (see figure 6.16). These disorders arise from mutations in different genes that are closely linked in a region of the long arm of chromosome 15, where imprinting occurs. They both cause mental retardation and a variety of other symptoms, but are quite distinct.

In 1989, researchers found that some children with Prader-Willi syndrome have two parts of the long arm of chromosome 15 from their mothers. The disease results because the father's Prader-Willi gene must be expressed for the child to avoid the associated illness. For Angelman syndrome, the situation is reversed. Children have a double dose of their father's DNA in the same chromosomal region implicated in Prader-Willi syndrome, with no maternal contribution. The mother's gene must be present for health. UPD and imprinting are still not well understood—if you are confused, you are not alone!

People usually learn their chromosomal makeup only when something goes wrong—when they have a family history of reproductive problems, exposure to a toxin, cancer, or symptoms of a known chromosomal disorder. While researchers are busy analyzing the information from the human genome project, chromosome studies will continue to be part of medical care—beginning before birth.

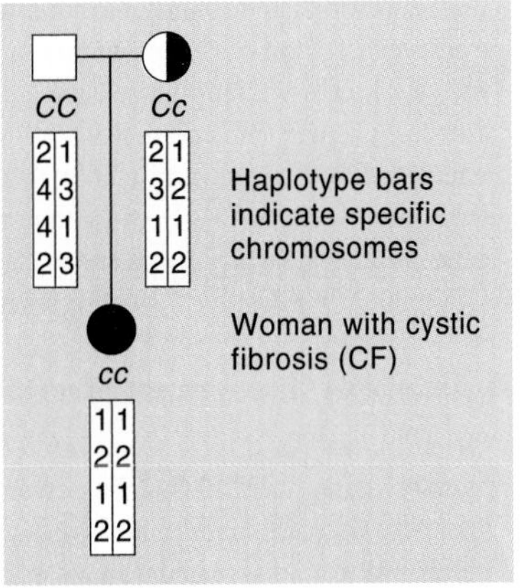


**figure 12.23**

**Uniparental disomy.** Uniparental disomy doubles part of one parent's genetic contribution. According to the law of segregation, a child with cystic fibrosis (CF) cannot be born to a person who is homozygous dominant for the wild type allele. But it can happen. In this family, the woman with CF inherited two copies of her mother's chromosome 7, and neither of her father's. Unfortunately, it was the chromosome with the disease-causing allele that she inherited in a double dose.

**Key Concepts**

Uniparental disomy (UPD) results when two chromosomes or chromosome parts are inherited from the same parent. It can arise from a trisomy and subsequent chromosome loss, or from two nondisjunction events. Uniparental disomy can cause disease if it creates a homozygous recessive condition, or if it disrupts imprinting.

# Summary

## 12.1 Portrait of a Chromosome

1. Mutation can occur at the chromosomal level. **Cytogenetics** is the study of chromosome aberrations and their effects on phenotypes.
2. **Heterochromatin** stains darkly and harbors many DNA repeats. **Euchromatin** is light staining and contains many protein-encoding genes.
3. A chromosome consists of DNA and proteins. Essential parts are the **telomeres, centromeres,** and origin of replication sites.
4. Centromeres include **alpha satellites** and **centromere-associated proteins,** some of which form **kinetochores** that contact spindle fibers. CENP-A is a protein that may control centromere duplication.
5. **Subtelomeres** have telomerelike repeats that gradually disappear, as some protein-encoding genes occur. Chromosomes vary in gene density.
6. Chromosomes are distinguishable by size, centromere position, satellites, DNA probes to specific sequences, and staining patterns.
7. A **karyotype** is a size-order chromosome chart. A **metacentric** chromosome has two fairly equal arms. A **submetacentric** chromosome has a large arm designated *q* and a short arm designated *p*. An **acrocentric** chromosome's centromere is near a tip, so that it has one long arm and one very short arm.

## 12.2 Visualizing Chromosomes

8. Chromosomes can be obtained from any cell that has a nucleus. Prenatal diagnostic techniques that obtain fetal chromosomes include **amniocentesis, chorionic villus sampling,** and **fetal cell sorting.**
9. Hand-cut karyotypes and stains to view chromosomes are giving way to computerized karyotyping and chromosome-specific **fluorescence *in situ* hybridization (FISH). Ideograms** are diagrams that display chromosome bands, FISH data, and gene loci.
10. Chromosomal shorthand indicates chromosome number, sex chromosome constitution, and the nature of the specific chromosomal abnormality.

## 12.3 Abnormal Chromosome Number

11. A **euploid** somatic human cell has 22 pairs of autosomes and a pair of sex chromosomes.
12. **Polyploid** cells have extra chromosome sets.
13. **Aneuploids** have extra or missing chromosomes. **Trisomies** (an extra chromosome) are less harmful than **monosomies** (lack of a chromosome), and sex chromosome aneuploidy is less severe than autosomal aneuploidy. **Nondisjunction** is uneven division of chromosomes in meiosis. It causes aneuploidy. Most autosomal aneuploids cease developing as embryos. The most common at birth are trisomies 21, 13, and 18, because they are gene-poor. Sex chromosome anomalies (XXY; 45, XO; XXX; XYY) are less severe.

## 12.4 Abnormal Chromosome Structure

14. Deletions and/or duplications can result from crossing over after pairing errors in synapsis. Crossing over in an inversion heterozygote can also generate deletions and duplications.
15. In a **Robertsonian translocation,** the short arms of two acrocentric chromosomes break, and the long arms join, forming an unusual, large chromosome.
16. In a **reciprocal translocation,** two nonhomologous chromosomes exchange parts. In both types of translocation, a **translocation carrier** may have an associated phenotype if the translocation disrupts a vital gene. A translocation carrier also produces a predictable percentage of unbalanced gametes, which can lead to birth defects and spontaneous abortions.
17. A heterozygote for an **inversion** may have reproductive problems if a crossover occurs between the inverted region and the noninverted homolog, generating deletions and duplications. A **paracentric inversion** does not include the centromere; a **pericentric inversion** does.
18. **Isochromosomes** repeat one chromosome arm but delete the other. They form when the centromere divides in the wrong plane during meiosis. Ring chromosomes form when telomeres are removed.

## 12.5 Uniparental Disomy—Two Genetic Contributions from One Parent

19. In **uniparental disomy,** a chromosome, or a part of one, doubly represents one parent. It can result from nondisjunction in both gametes, or from a trisomic cell that loses a chromosome.
20. Uniparental disomy causes symptoms if it creates a homozygous recessive state associated with an illness, or if it affects an imprinted gene.

# Review Questions

1. What are the essential components of a chromosome? Of a centromere?
2. How does the DNA sequence change with distance from the telomere?
3. How are centromeres and telomeres alike?
4. What happens during meiosis to produce each of the following?
   a. an aneuploid
   b. a polyploid
   c. the increased risk of trisomy 21 Down syndrome in the offspring of a woman over 40 at the time of conception
   d. recurrent spontaneous abortions to a couple in which the man has a pericentric inversion
   e. several children with Down syndrome in a family where one parent is a translocation carrier
5. A human liver has patches of cells that are octaploid—that is, they have eight sets of chromosomes. Explain how this might arise.
6. Describe an individual with each of the following chromosome constitutions. Mention the person's sex and possible phenotype.
   a. 47,XXX
   b. 45,X
   c. 47,XX, trisomy 21

7. Which chromosomal anomaly might you expect to find more frequently among the members of the National Basketball Association than in the general population? Cite a reason for your answer.
8. List three examples illustrating the idea that the extent of genetic material involved in a chromosomal aberration affects the severity of the associated phenotype.
9. List three types of chromosomal aberrations that can cause duplications and/or deletions, and explain how they do so.
10. Why would having the same inversion on both members of a homologous chromosome pair *not* lead to unbalanced gametes, as having the inversion on only one chromosome would?
11. Define or describe the following technologies:
    a. high-resolution chromosome banding
    b. FISH
    c. amniocentesis
    d. chorionic villus sampling
    e. fetal cell sorting
    f. maternal serum markers
    g. quantitative PCR
    h. human artificial chromosomes
12. What is the evidence that trisomies 13 and 18 are lethal primarily before birth? Why are they more common than trisomies 5 or 16?
13. How many chromosomes would a person have who has Klinefelter syndrome and also trisomy 21?
14. Explain why a female cannot have Klinefelter syndrome and a male cannot have Turner syndrome.
15. List three causes of Turner syndrome.

# Applied Questions

1. The following is part of a chart used to provide genetic counseling on maternal age effect on fetal chromosomes. Answer questions a–e based on this chart.

| Maternal Age | Trisomy 21 Risk | Risk for Any Aneuploid |
|---|---|---|
| 20 | 1/1,667 | 1/526 |
| 24 | 1/1,250 | 1/476 |
| 28 | 1/1,053 | 1/435 |
| 30 | 1/952 | 1/385 |
| 32 | 1/769 | 1/322 |
| 35 | 1/378 | 1/192 |
| 36 | 1/289 | 1/156 |
| 37 | 1/224 | 1/127 |
| 38 | 1/173 | 1/102 |
| 40 | 1/106 | 1/66 |
| 45 | 1/30 | 1/21 |
| 48 | 1/14 | 1/10 |

    a. If the risk at a particular obstetrical practice that amniocentesis will cause miscarriage is 1 in 250, at what age should patients in this practice undergo the test?
    b. The Willoughbys have a son who has trisomy 21 Down syndrome. The mother, Suzanne, is 24 years old and pregnant. The Martinis do not have any relatives who have Down syndrome or any other chromosomal condition. Karen Martini is pregnant, and is 32 years old. Who has the lower risk of having a child with Down syndrome, Suzanne Willoughby or Karen Martini?
    c. Why are the risks in the righthand column higher than those in the middle column?
    d. Sam and Alice Dekalb receive genetic counseling because of "advanced maternal age"—Alice is 40 years old. When amniocentesis reveals trisomy 13, the couple is shocked, explaining that they had thought the risk of a chromosomal problem was less than 1 percent. How have they misinterpreted the statistics?
    e. A 40-year-old woman wants to have children, but would like to postpone becoming pregnant until she is 45. How much will her risk of conceiving a child with trisomy 21 increase in that time?
2. Amniocentesis indicates that a fetus has the chromosomal constitution 46, XX,del(5)(p15). What does this mean? What might the child's phenotype be?
3. The medical literature includes 18 cases of children with a syndrome consisting of poor growth before birth, developmental delay, premature puberty, loose joints, a large head, short stature, and small hands. In a different syndrome, children have a small chest, ears, and facial features as well as rib and finger defects. Children with the first condition have both copies of the entire long arm of chromosome 14 from their mothers, whereas children with the second condition inherit the same chromosome part from their fathers.
    a. What is the type of chromosomal aberration responsible for these two disorders?
    b. Describe how each of the conditions might arise.
    c. Describe how these conditions might result from a deletion mutation.
4. What type of test would you use to determine whether a triploid infant resulted from a diploid oocyte fertilized by a haploid sperm, or from two sperm fertilizing one oocyte?
5. Two sets of parents who have children with Down syndrome meet at a clinic. The Phelps know that their son has trisomy 21. The Watkins have two affected children, and Mrs. Watkins has had two spontaneous abortions. Why should the Watkins probably be more concerned about future reproductive problems than the Phelps? How are the offspring of the two families different, even though they have the same symptoms?
6. For an exercise in a college genetics laboratory course, a healthy student constructs a karyotype from a cell in a drop of her blood. She finds only one chromosome 3 and one chromosome 21, plus two unusual chromosomes that do not seem to have matching partners.
    a. What type of chromosomal abnormality does she have?
    b. Why doesn't she have any symptoms?
    c. Would you expect any of her relatives to have any particular medical problems? If so, which medical conditions?
7. A fetus ceases developing in the uterus. Several of its cells are karyotyped. Approximately 75 percent of the cells are diploid, and 25 percent are tetraploid (containing four copies of each chromosome). What do you think happened? When in development did it probably occur?
8. Distinguish among Down syndrome caused by aneuploidy, mosaicism, and translocation.

9. A couple has a son diagnosed with Klinefelter syndrome. Explain how the son's chromosome constitution could have arisen from either parent.

10. DiGeorge syndrome causes abnormal parathyroid glands, disrupting blood calcium levels; heart defects; and an underdeveloped thymus gland, impairing development of the immune system. About 85 percent of patients have a microdeletion of a particular area of chromosome 22. In one family, a girl, her mother, and a maternal aunt have very mild cases of DiGeorge syndrome, and they also all have a reciprocal translocation involving chromosomes 22 and 2.

    a. How can a microdeletion and a translocation cause the same set of symptoms?
    b. Why were the cases of people with the translocation less severe than those of people with the microdeletion?
    c. What other problems might arise in the family with the translocation?

# Suggested Readings


Cicero, Simona, et al. November 17, 2001. Absence of nasal bone in fetuses with trisomy 21 at 11–14 weeks of gestation: An observational study. *The Lancet* 358:1665–67. A majority of trisomy 21 Down syndrome fetuses lack nasal bones, detectable with ultrasound.

Csonka, E., et al. December 2000. Novel generation of human satellite DNA-based artificial chromosomes in mammalian cells. *Journal of Cell Science* 113:3207–16.

Gratacos, Monica, et al. August 2001. A polymorphic genomic duplication on human chromosome 15 is a susceptibility factor for panic and phobic disorders. *Cell* 106:367–79.

Hennebicq, S. et al. June 20, 2001. Risk of trisomy 21 in offspring of patients with Klinefelter syndrome. *The Lancet* 357:2104–5. Tendency for nondisjunction for chromosomes other than the X may be part of Klinefelter syndrome.

Koch, Jørn. January 22, 2000. Neocentromeres and alpha satellite: A proposed structural code for functional human centromere DNA. *Human Molecular Genetics* 9:149. Centromeres from one species function in another, suggesting a fundamental structure.

Lewis, Ricki. June 12, 2000. Chromosome 21 reveals sparse gene content. *The Scientist* 14:1. When it comes to chromosomes, size doesn't matter—gene content and density do.

Lewis, Ricki. October 15, 2001. Mapping subtelomeres. *The Scientist* 15:20. The regions next to telomeres are more than just buffers.

Mefford, Heather C., and Barbara J. Trask. February 2002. The complex structure and dynamic evolution of human subtelomeres. *Nature Reviews: Genetics* 3(2):91–102. Subtelomeres mix repeats and protein-encoding genes.

Quaini, Federico, et al. January 3, 2002. Chimerism of the transplanted heart. *The New England Journal of Medicine* 346:5–15. FISH-highlighted Y chromosomes reveal stem cells in the heart.

Ranke, Michael R., and Paul Saenger. July 28, 2001. Turner's syndrome. *The Lancet* 358:309–14. Having an XO chromosome constitution raises the risk of some more common disorders.

Van Biema, David. July 16, 2001. When God hides his face. *Time,* pp. 62–64. The story of a Tennessee couple who lost a child to a very rare genetic disease, had a vasectomy, then conceived again. CVS diagnosed the same disorder, and the couple chose to continue the pregnancy—and buried another child.

Willard, Huntington F. May 8, 2001. Neocentromeres and human artificial chromosomes: An unnatural act. *Proceedings of the National Academy of Sciences* 98:5374–76. The human genome harbors several dozen potential centromeric sequences.


# On the Web

Check out the resources on our website at

**www.mhhe.com/lewisgenetics5**

On the web for this chapter, you will find additional study questions, vocabulary review, useful links to case studies, tutorials, popular press coverage, and much more. To investigate specific topics mentioned in this chapter, also try the links below:

5p$^-$ Society (cri du chat) **www.fivepminus.org**

Alagille Syndrome **www.alagille.org**

Chromosomes 5, 6, 19 **www.doe.gov/news/releases00/aprps/pr00104.htm**

Chromsome Deletion Outreach **members.aol.com/cdousa/cdo.htm**

Klinefelter Syndrome & Associates **www.genetic.org**

National Down Syndrome Society **www.ndss.org**

Online Mendelian Inheritance in Man **www.ncbi.nlm.nih.gov/entrez/query.fcgi?db=OMIM**
Alagille syndrome 118450
cri du chat syndrome 123450
Prader-Willi syndrome 176270
trisomy 21 190685

Prader-Willi Syndrome Association **www.pwsausa.org**

Rare Chromosome Disorder Support Group **members.aol.com/rarechromo**

Support Organization for Trisomy 18, 13 + Related Disorders **www.trisomy.org**

Turner Syndrome Society of the U.S. **www.turner-syndrome-us.org**

UniGene (chromosome charts) **www.ncbi.nlm.nih.gov/UniGene**

CHAPTER

13

# When Allele Frequencies Stay Constant

## 13.1 The Importance of Knowing Allele Frequencies

We can consider genes at the biochemical and organismal levels. At another level—the population level—allele frequencies reveal whether evolution is occurring. For most genes, it almost always is. Nonrandom mating, migration, genetic drift, mutation, and natural selection alter allele frequencies.

## 13.2 When Allele Frequencies Stay Constant—Hardy-Weinberg Equilibrium

It seems logical that dominant traits would increase in prevalence in a population, and recessive traits would decrease. In 1908, a mathematician (Hardy) and a physician (Weinberg) independently showed, using algebra, why this isn't so. They thereby founded the field of population genetics. Their equation follows unchanging allele frequencies through generations when evolution is not occurring.

## 13.3 Practical Applications of Hardy-Weinberg Equilibrium

Carrier risks for certain disorders in specific populations can be determined by applying the incidence of the homozygous recessive class to the Hardy-Weinberg algebraic equation. Conversely, allele frequencies can be used to derive genotype frequencies. The equation is modified for X-linked traits and very rare alleles.

## 13.4 DNA Fingerprinting—A Practical Test of Hardy-Weinberg Assumptions

The ability to identify and compare DNA repeats on different chromosomes, and to calculate how likely it is that certain combinations of alleles occur in a particular population, provide a powerful tool to determine the chance that two DNA samples could come from the same individual. DNA fingerprinting is an application of population genetics.

So far, we've considered the gene as an ill-defined "character" that transmits traits, and as a biochemical blueprint for building a specific protein. Genes can also be considered at the population level.

A **population** is any group of members of the same species in a given geographical area. Human populations might include the students in a class, a stadium full of people, or the residents of a community, state, or nation. **Population genetics** is a branch of genetics that considers all of the alleles in a population, which constitute the **gene pool.** The "pool" in gene pool refers to a collection of gametes. An offspring can be considered as a sample of two gametes taken from the pool. Alleles can move between populations when individuals migrate and then mate. This movement, termed **gene flow,** underlies evolution. It is at the population level that genetics reflects history, anthropology, human behavior and sociology, enabling us to trace our beginnings and to understand our present diversity. This chapter introduces the major principle of population genetics, and the next two chapters explore the impact population genetics has on evolution.

## 13.1 The Importance of Knowing Allele Frequencies

Thinking about genes at the population level begins by considering frequencies—that is, how often a particular genetic variant occurs in a particular population. Such frequencies can be calculated for alleles, genotypes, or phenotypes. For example, an allele frequency for the cystic fibrosis (CF) gene might be the number of ΔF508 alleles among the residents of San Francisco. ΔF508 is the most common allele that, when homozygous, causes the disorder. The allele frequency includes the two ΔF508 alleles in each person with cystic fibrosis, plus those carried in heterozygotes, as a proportion of all alleles for that gene in the gene pool. The genotype frequencies are the proportions of heterozygotes and the two types of homozygotes in the population. Finally, a phenotypic frequency is simply the percentage of people in the population who have CF. With genes that have multiple alleles, the situation becomes more complex.

Phenotypic frequencies are determined empirically—that is, by observing how common a condition or trait is in a population. These figures have value in genetic counseling in estimating the risk that a particular inherited disorder will occur in an individual when there is no family history of the illness. Table 13.1 shows disease incidence for phenylketonuria (PKU), an inborn error of metabolism that causes mental retardation unless the person follows a special low-protein diet from birth. Note how the frequency differs in different populations.

On a broader level, shifting allele frequencies in populations provide the small steps of genetic change, called **microevolution,** that constitute evolution. Allele frequencies can change when any of the following conditions are met:

1. Individuals of one genotype are more likely to produce offspring with each other than with those of other genotypes (*nonrandom mating*).
2. Individuals *migrate* between populations.
3. Reproductively isolated small groups form within a larger population (*genetic drift*).
4. *Mutation* introduces new alleles into a population.
5. People with a particular genotype are more likely to be able to produce viable, fertile offspring under a specific environmental condition than individuals with other genotypes (*natural selection*).

Because these factors are operating more often than not, genetic equilibrium—when allele frequencies are *not* changing—is rare. Thus, microevolution is not only possible, but also nearly unavoidable. Chapter 14 considers these factors in this order, in depth.

**table 13.1**

**Frequency of PKU in Various Populations**

| Population | Frequency of PKU |
|---|---|
| Chinese | 1/16,000 |
| Irish, Scottish, Yemenite Jews | 1/5,000 |
| Japanese | 1/119,000 |
| Swedes | 1/30,000 |
| Turks | 1/2,600 |
| United States Caucasians | 1/10,000 |

When microevolutionary changes accumulate sufficiently to keep two fertile organisms of opposite sex from successfully producing fertile offspring with one another, **macroevolution,** which is the formation of a new species, has occurred. In contrast, this chapter discusses the interesting, but unusual, situation in which allele frequencies stay constant, a condition called **Hardy-Weinberg equilibrium.**

**KEY CONCEPTS**

Population genetics is the study of allele frequencies in groups of organisms of the same species in the same geographic area. The genes in a population comprise its gene pool. Microevolution reflects changes in allele frequencies in populations. Microevolution is not occurring if allele frequencies stay constant from generation to generation, a condition called Hardy-Weinberg equilibrium. This happens only if mating is random and the population is large, with no migration, genetic drift, mutation, or natural selection.

## 13.2 When Allele Frequencies Stay Constant—Hardy-Weinberg Equilibrium

Gregor Mendel worked with phenotypes, and inferred the underlying genotypes, in crosses of individual plants. Population genetics also looks at phenotypes and genotypes, but among large numbers of individuals. Here, the underlying rules are revealed by following allele frequencies. Considering the alleles in a population—the gene pool—is more precise than assessing phenotypes, because the same phenotype can result from different genotypes. (Recall Mendel's *TT* and *Tt* tall plants.)

In 1908, a Cambridge University mathematician named Godfrey Harold Hardy (1877–1947) and Wilhelm Weinberg (1862–1937), a German physician interested in genetics, independently used algebra to explain how allele frequencies can be used to predict phenotypic and genotypic frequencies in populations of diploid, sexually reproducing organisms.

Hardy unintentionally cofounded the field of population genetics with a simple letter published in the journal *Science*—he did not consider his idea to be worthy of the more prestigious British journal *Nature.* The letter began with a curious mix of modesty and condescension:

> **I am reluctant to intrude in a discussion concerning matters of which I have no expert knowledge, and I should have expected the very simple point which I wish to make to have been familiar to biologists.**

Hardy continued to explain how mathematically inept biologists had deduced from Mendel's work that dominant traits would increase in populations, while recessive traits would become rarer. This seems to make sense, but is actually untrue, because recessive alleles are introduced by mutation or migration and maintained in heterozygotes. Hardy and Weinberg disproved the assumption that dominant traits increase while recessive traits decrease using the language of algebra.

The expression of population genetics in algebraic terms begins with the simple equation

$$p + q = 1.0$$

where $p$ represents all dominant alleles for a gene, and $q$ represents all recessive alleles. "$p + q = 1.0$" simply means that all the dominant alleles and all the recessive alleles comprise all the alleles for that gene in a population.

Next, Hardy and Weinberg described the genotypes for a gene with two alleles using the binomial expansion

$$p^2 + 2pq + q^2 = 1.0$$

In this equation, $p^2$ represents homozygous dominant individuals, $q^2$ represents homozygous recessive individuals, and $2pq$ represents heterozygotes (table 13.2). The letter $p$ designates the frequency of a dominant allele, and $q$ is the frequency of a recessive allele.

Use of the binomial expansion to describe genes in populations became known as the Hardy-Weinberg equation. It can reveal the allele frequency changes that underlie evolution. If the proportion of genotypes remains the same from generation to generation, as the equation indicates, then that gene is not evolving (changing). This situation, Hardy-Weinberg equilibrium, is an idealized state. It is possible only if the population is large, if its members mate at random, and if there is no migration, genetic drift, mutation, or natural selection taking place.

Hardy-Weinberg equilibrium is rare for protein-encoding genes that affect the phenotype, because how an organism appears and its health affect its ability to reproduce. However, Hardy-Weinberg equilibrium is seen in repeats and other DNA sequences that do not affect the phenotype, and therefore are not subject to natural selection.

To understand the concept of Hardy-Weinberg equilibrium, it helps to follow the frequency of two alleles of a particular gene from one generation to the next. This exercise demonstrates that Mendelian principles underlie population genetics calculations.

Consider an autosomal recessive trait: a middle finger shorter than the second and fourth fingers. If we know the frequencies of the dominant and recessive alleles, then we can calculate the frequencies of the genotypes and phenotypes, and then trace the trait through the next generation. Table 13.3 and figure 13.1 depict this.

The dominant allele $D$ confers normal-length fingers; the recessive allele $d$ confers a short middle finger. We can figure out the frequencies of the dominant and recessive alleles by observing the frequency of the homozygous recessive class—because this phenotype reflects only one genotype. In our example, 9 out of 100 individuals in a population have short fingers—genotype *dd*. The frequency is therefore 9/100 or 0.09. Since *dd* equals $q^2$, then $q$ equals 0.3. Since $p + q = 1.0$, knowing that $q$ (the frequency of the recessive allele) is 0.3 means that $p$ (the frequency of the dominant allele) is 0.7.

**table 13.2**

**The Hardy-Weinberg Equation**

| Algebraic Expression | What It Means |
|---|---|
| $p + q = 1.0$ (allele frequencies) | All dominant alleles plus all recessive alleles add up to all alleles for a particular gene in a population. |
| $p^2 + 2pq + q^2 = 1.0$ (genotype frequencies) | For a particular gene with 2 alleles, all homozygous dominant individuals ($p^2$) plus all heterozygotes ($2pq$) plus all homozygous recessives ($q^2$) add up to all of the individuals in the population. |

**table 13.3**

**Hardy-Weinberg Equilibrium—When Allele Frequencies Stay Constant**

| Possible Matings | | Proportion | Frequency of Offspring Genotypes | | |
|---|---|---|---|---|---|
| **Male** | **Female** | **in Population** | **DD** | **Dd** | **dd** |
| 0.49 *DD* | 0.49 *DD* | 0.2401 (*DD* × *DD*) | 0.2401 | | |
| 0.49 *DD* | 0.42 *Dd* | 0.2058 (*DD* × *Dd*) | 0.1029 | 0.1029 | |
| 0.49 *DD* | 0.09 *dd* | 0.0441 (*DD* × *dd*) | | 0.0441 | |
| 0.42 *Dd* | 0.49 *DD* | 0.2058 (*Dd* × *DD*) | 0.1029 | 0.1029 | |
| 0.42 *Dd* | 0.42 *Dd* | 0.1764 (*Dd* × *Dd*) | 0.0441 | 0.0882 | 0.0441 |
| 0.42 *Dd* | 0.09 *dd* | 0.0378 (*Dd* × *dd*) | | 0.0189 | 0.0189 |
| 0.09 *dd* | 0.49 *DD* | 0.0441 (*dd* × *DD*) | | 0.0441 | |
| 0.09 *dd* | 0.42 *Dd* | 0.0378 (*dd* × *Dd*) | | 0.0189 | 0.0189 |
| 0.09 *dd* | 0.09 *dd* | 0.0081 (*dd* × *dd*) | | | 0.0081 |
| | | Resulting offspring frequencies: | 0.49 *DD* | 0.42 *Dd* | 0.09 *dd* |

## figure 13.1

**Hardy-Weinberg equilibrium.** In Hardy-Weinberg equilibrium, allele frequencies remain constant from one generation to the next.



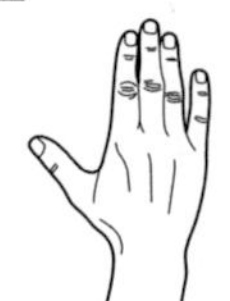

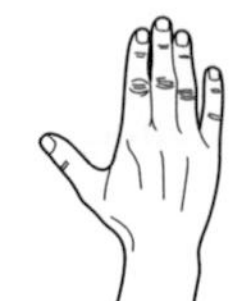



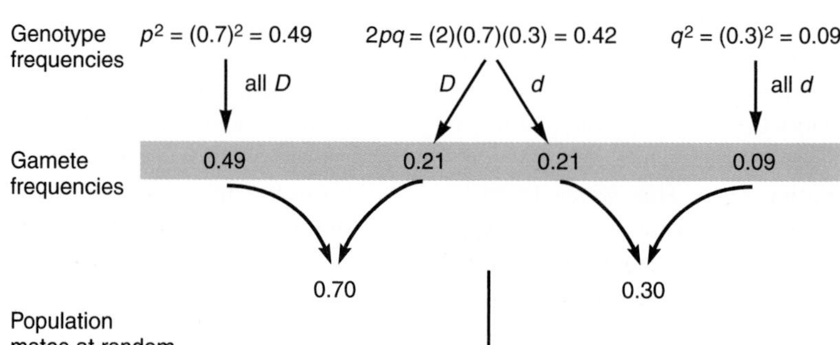



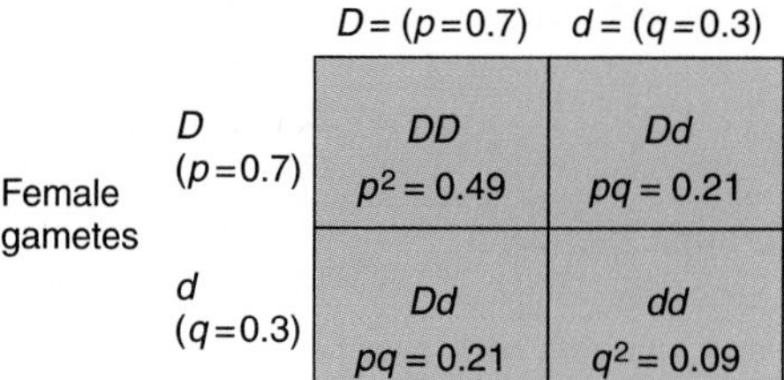



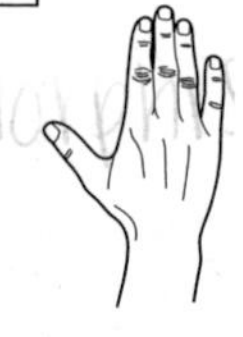

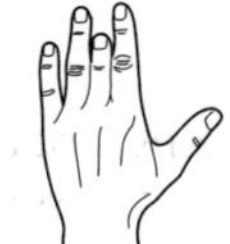



Next, we can calculate the proportions of each of the three types of genotypes that arise when gametes combine at random to form the first generation in figure 13.1:

homozygous dominant = $DD$
$= 0.7 \times 0.7 = 0.49$
= 49 percent of individuals in generation 1

homozygous recessive = $dd$
$= 0.3 \times 0.3 = 0.09$
= 9 percent of individuals in generation 1

heterozygous = $Dd + dD$
$= 2pq = (0.3)(0.7) + (0.3)(0.7) = 0.42$
= 42 percent of individuals in generation 1

The proportion of homozygous individuals is calculated simply by multiplying the allele frequency for the recessive or dominant allele by itself. The heterozygous calculation is $2pq$ because there are two ways of combining a *D* with a *d* gamete—a *D* sperm with a *d* egg, and a *d* sperm with a *D* egg.

In this population, 9 percent of the individuals have a short middle finger. Now jump ahead a few generations, and assume that people choose mates irrespective of finger length. This means that each genotype of a female (*DD, Dd,* or *dd*) is equally likely to mate with each of the three types of males (*DD, Dd,* or *dd*), and vice versa. Table 13.3 multiplies the genotype frequencies for each possible mating, which leads to offspring in the now-familiar proportions of 49 percent *DD,* 42 percent *Dd,* and 9 percent *dd.* This gene, therefore, is in Hardy-Weinberg equilibrium—the allele and genotype frequencies do not change from one generation to the next.

### Key Concepts

For any two alleles of a gene in a population, the proportion of homozygous dominants equals the square of the frequency of the dominant allele ($p^2$), and the proportion of homozygous recessives equals the square of the frequency of the recessive allele ($q^2$). The proportion of heterozygotes equals $2pq$. If we know the genotype proportions in a population, we can calculate allele frequencies. The frequency of the recessive allele equals the proportion of homozygous recessives plus one-half that of carriers, and the frequency of the dominant allele equals the proportion of homozygous dominants plus that of one-half the carriers. In Hardy-Weinberg equilibrium, allele frequencies do not change from one generation to the next.

## 13.3 Practical Applications of Hardy-Weinberg Equilibrium

The Hardy-Weinberg equation is applied to population statistics on genetic disease incidence to derive carrier risks. To determine allele frequencies for autosomal recessively inherited characteristics, we need to know the frequency of one genotype in the population. This is typically the homozygous recessive class, for the same reason that this group is used to do a test cross when tracing inheritance of a single trait—its phenotype indicates its genotype.

The known incidence of an autosomal recessive disorder can be used to help calculate the risk that a particular person is a heterozygote. Returning to the example of CF, the incidence of the disease, and therefore also of carriers, may vary greatly in different populations (table 13.4). Figure 13.2 provides another example of an illness that is common in one population, but exceedingly rare elsewhere.

Cystic fibrosis affects 1 in 2,000 Caucasian newborns. Therefore, the homozygous recessive frequency—*cc* if *c* represents the disease-causing allele—is 1/2,000, or 0.0005 in this population. This equals $q^2$. The square root of $q^2$ is about 0.022, which equals the frequency of the *c* allele. If $q$ equals 0.022, then $p$, or $1 - q$, equals 0.978. Carrier frequency is equal to $2pq$, which equals (2)(0.978)(0.022), or 0.043—about 1 in 23.

If there is no cystic fibrosis in a family, a person's risk of having an affected child, derived from population statistics, is relatively low. Consider a Caucasian couple with no family history of cystic fibrosis asking a genetic counselor to calculate the risk that they could conceive a child with this illness. The genetic counselor tells them that the chance of *each* potential parent being a carrier is about 4.3 percent, or 1 in 23. But this is only part of the picture. The chance that *both* of these people are carriers is 1/23 multiplied by 1/23—or 1 in 529—because the probability that two independent events will occur equals the product of the probability that each event will happen. However, if they *are* both carriers, each of their children would face a 1 in 4 chance of inheriting the illness, based on Mendel's first law, of gene segregation. Therefore, the risk that two unrelated Caucasian individuals with no family history of cystic fibrosis will have an affected child is $1/4 \times 1/23 \times 1/23$, or 1 in 2,116. This couple has learned their chance of producing an affected child from disease incidence statistics.

**table 13.4**

**Carrier Frequency for Cystic Fibrosis**

| Population Group | Carrier Frequency |
|---|---|
| African Americans | 1 in 66 |
| Asian Americans | 1 in 150 |
| Caucasians of European descent | 1 in 23 |
| Hispanic Americans | 1 in 46 |

For X-linked traits, different predictions of allele frequencies apply for males and females. For a female, who can be homozygous recessive, homozygous dominant, or a heterozygote, the standard Hardy-Weinberg equation of $p^2 + 2pq + q^2$ applies, as it usually would to an autosomal recessive trait. However, in males, the allele frequency is the phenotypic frequency, because a male who inherits an X-linked recessive allele exhibits it in his phenotype.

The incidence of X-linked hemophilia ($X^hY$), for example, is 1 in 10,000 male births. Therefore, $q$ (the frequency of the "h" allele) equals 0.0001. Using the formula $p + q = 1$, the frequency of the normal allele is 0.9999. The incidence of carriers ($X^hX^H$), who are all female, equals $2pq$, or (2)(0.0001)(0.9999), which equals 0.00019; this is 0.0002, or 0.02 percent, which equals about 1 in 5,000. The incidence of a female having hemophilia ($X^hX^h$) is $q^2$, or $(0.0001)^2$, or about 1 in 100 million.

In the real world of medical genetics, fairly even allele frequencies such as 0.6 and 0.4, or 0.7 and 0.3, are unusual. Mendelian disorders are usually very, very rare, and therefore, the $q$ component of the Hardy-Weinberg equation does not contribute much. Because this means that the value of $p$ approaches 1, the carrier frequency, $2pq$, is very close to just $2q$. Therefore, the carrier frequency is approximately twice the frequency of the rare, disease-causing allele.

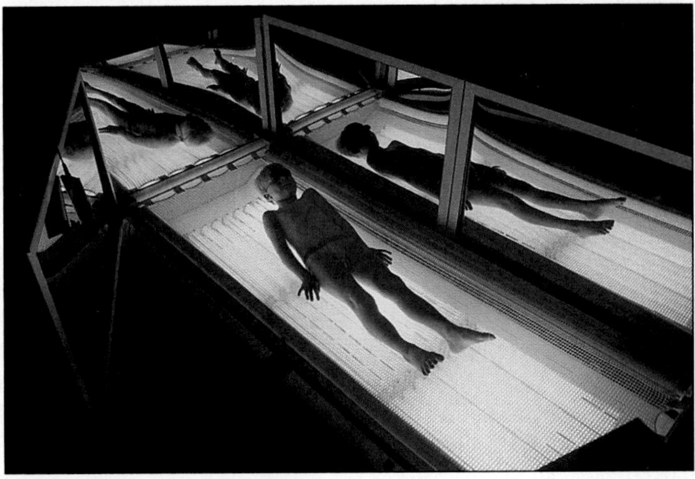

**figure 13.2**

**Disease incidence varies in populations.** Only 60 cases of Crigler-Najjar syndrome are known worldwide; many of them are in the Mennonite and Amish community of Lancaster County, Pennsylvania. Crigler-Najjar syndrome is an enzyme deficiency that causes the buildup of a substance called bilirubin, producing severe jaundice (yellowing of the skin and eyes). If untreated, the syndrome rapidly results in fatal brain damage. Fortunately, exposure to ultraviolet light breaks down excess bilirubin, but the treatment is not pleasant—children must sleep unclothed under the lights every night. The Mennonite and Amish populations have many other autosomal recessive illnesses that are extremely rare elsewhere because they descended from a few founding families and marry among themselves.

Consider Tay-Sachs disease, which occurs in 1 in 3,600 Ashkenazim (Jewish people of eastern European descent). This means that $q^2$ equals 1/3,600, or about 0.0003. The square root, $q$, equals 0.017. The frequency of the dominant allele is then 1 – 0.017, or 0.983. What is the likelihood that an Ashkenazi carries Tay-Sachs disease? It is $2pq$, or (2)(0.983)(0.017), or 0.033. Note that this is very close to double the frequency of the mutant allele, 0.017. While these examples all trace a gene with just two alleles, the Hardy-Weinberg equation can also be modified to analyze genes that have more than two alleles.

**KEY CONCEPTS**

Population calculations are applied to real-life situations by determining the value of $q^2$ (homozygous recessives) from the incidence of an inherited disease in a population, deducing $p$ and $q$, and then using the Hardy-Weinberg equation to predict the likelihood that a person is a carrier. To track the inheritance of X-linked recessive traits, we must use different values for males and females. The frequency of the recessive phenotype in males is $q$, whereas in females it is $q^2$, as in autosomal recessive inheritance. For very rare inherited disorders, the value of $p$ approaches 1, so the carrier frequency is approximately twice the frequency of the disease-causing allele ($2q$).

## 13.4 DNA Fingerprinting—A Practical Test of Hardy-Weinberg Assumptions

DNA sequences can vary from person to person, whether they code for a protein and contribute to a phenotype or not. Hardy-Weinberg equilibrium also applies to DNA sequences that do not encode protein. A DNA sequence that varies among individuals at the same site on a chromosome is a polymorphism if it occurs too frequently to be accounted for by new mutation alone. But "frequent" is still rare—it means that the variant is present in at least 1 percent of a population, giving a carrier frequency ($2q$) of 2 percent.

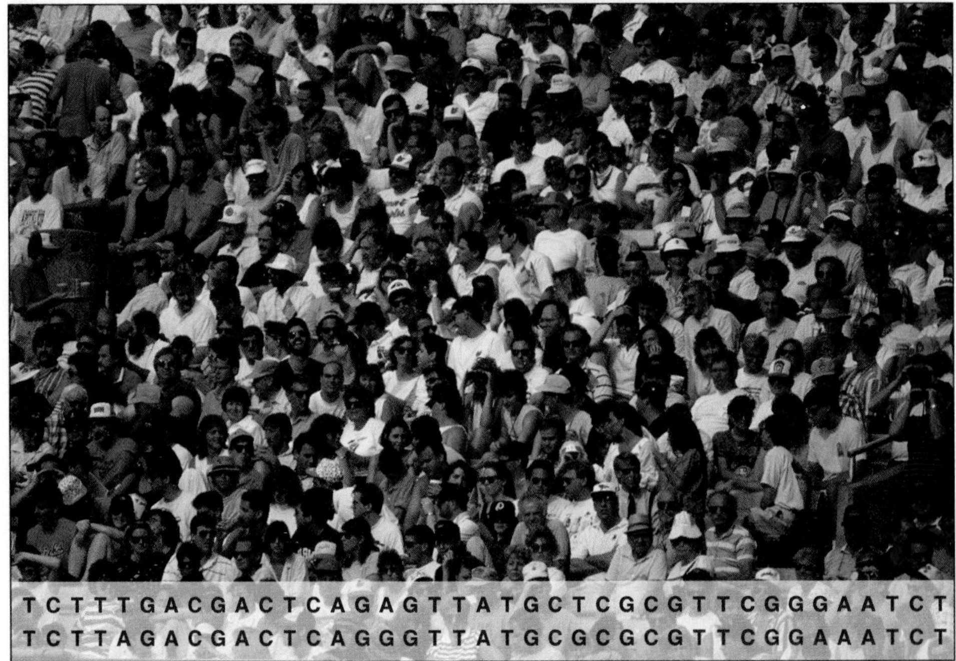


a.

| | |
|---|---|
| Individual 1 | GCATC GCATC |
| Individual 2 | GCATC GCATC GCATC GCATC |
| Individual 3 | GCATC GCATC GCATC GCATC GCATC GCATC |

b.

**figure 13.3**

**DNA fingerprinting detects genetic variation.** (*a*) SNPs are single-base differences among individuals. (*b*) Individuals also vary by the numbers of specific repeated sequences. Both SNPs and variants in repeat number can alter the sizes of fragments generated by the application of a restriction enzyme to DNA in a sample of evidence.

A polymorphism may be as small as a single nucleotide, which is the basis of SNP (single nucleotide polymorphism) research and technology (see figure 7.11). DNA fingerprinting can also be based on differences in number of repeats among individuals (figure 13.3). DNA fingerprinting calculates the probabilities that certain genetic variants—alleles, SNPs, or repeat number—occur in two places by chance. The interpretation of DNA fingerprints has many applications, as chapter 1 discusses. Although DNA fingerprinting is a molecular technique, population data are critical for interpreting the results.

Direct DNA sequencing is not necessary to spot a polymorphism or repeat number variant. **Restriction enzymes,** which cut DNA at particular short sequences, can do this. (Restriction enzymes are enzymes that naturally protect certain bacteria by cutting foreign DNA, such as from a viral infection.) If the polymorphism disrupts a restriction enzyme cutting site, then "digesting" (cutting) the DNA with the enzyme yields larger pieces than digesting the normal sequence where the cutting site is intact. Sequences that disrupt restriction enzyme cutting sites are called **restriction fragment length polymorphisms,** or RFLPs (figure 13.4). The first type of DNA fingerprinting technology combined use of RFLPs on repeats with an understanding of Hardy-Weinberg equilibrium. Such sequences are in Hardy-Weinberg equilibrium because they do not affect the phenotype, and therefore cannot influence mate choice and are not under selection.

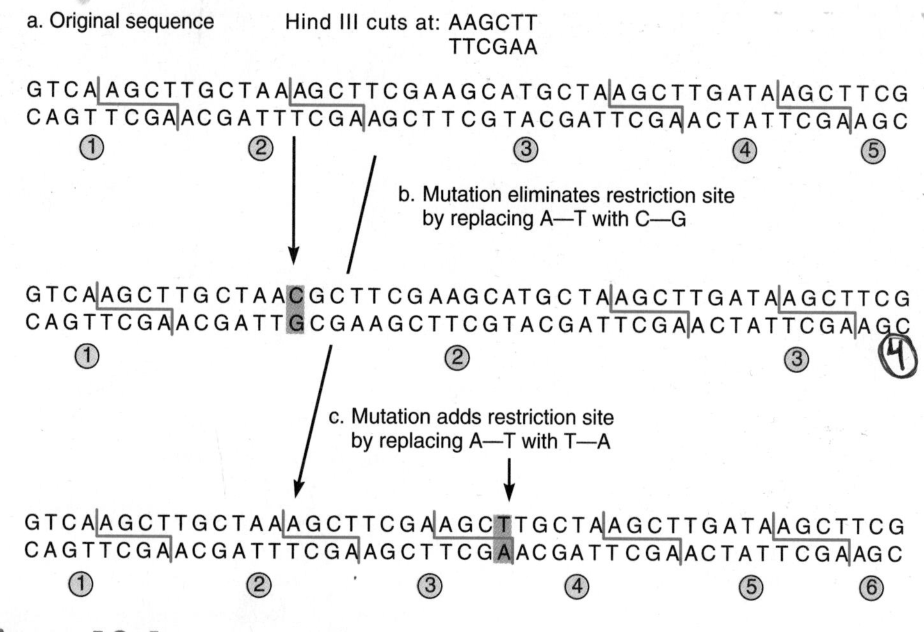


**figure 13.4**

**RFLPs reveal polymorphisms.** (*a*) Hind III is a restriction enzyme that snips DNA between the adenines (AA) of the 6-base sequence AAGCTT. Altering a base by mutation can remove (*b*) or add (*c*) a restriction site, changing the numbers and sizes of the resulting DNA fragments. Numbers indicate restriction fragments, shown in figure 13.5.

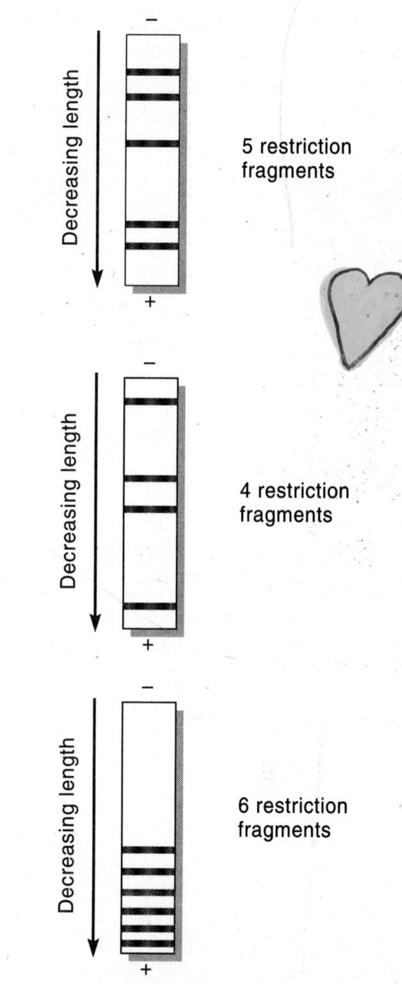


**figure 13.5**

**Electrophoresis separates DNA fragments.** Polyacrylamide gel electrophoresis separates the DNA fragments in figure 13.4 by length. The shorter fragments move closer to the positive pole.

## DNA Patterns Distinguish Individuals

DNA fingerprinting is based on a simple fact: since a 3-billion-base sequence of four different nucleotides can produce more varied combinations than there are humans, each of us (except identical twins) should have a unique DNA sequence. Still, humans are 99.9 percent identical at the DNA sequence level. However, that 0.1 percent adds up to many differences—3 million—over the 3 billion bases that make up the human genome.

Sir Alec Jeffreys at Leicester University in Great Britain developed the first type of DNA fingerprinting in the 1980s. He detected differences in the number of long repeats among individuals by cutting their DNA with restriction enzymes. Then Jeffreys measured the resulting fragments using a technique called polyacrylamide gel electrophoresis.

Polyacrylamide gel is a jellylike material that small pieces of DNA migrate through when an electrical field is applied. A positive electrode is placed at one end of the gel, and a negative electrode at the other. The DNA pieces, carrying negative charges because of their phosphate groups, move toward the positive pole. The pieces migrate according to size, the shorter pieces moving faster and thus traveling farther in a given time. The pattern that forms from the different-sized fragments, with the shorter fragments closer to the positive pole and the longer fragments farther away, creates a distinctive DNA fingerprint.

Figure 13.5 shows the electrophoresis patterns obtained for the restriction fragments described in figure 13.4. Now the meaning of "restriction fragment length polymorphisms" (RFLPs) may be clearer—they are differences in DNA fragment sizes that reflect individual DNA sequence differences that alter restriction enzyme cutting sites.

Jeffreys's first celebrated cases using DNA fingerprinting included proving that a boy was the son of a British citizen so that he could enter the country, and freeing a man held in custody for the rape of two schoolgirls. Then in 1988, Jeffreys's approach matched DNA fingerprints from suspect Tommie Lee Andrews's blood cells to sperm cells left on his victim in a notorious rape case (Reading 13.1). More recently, Jeffreys used DNA fingerprinting to demonstrate that Dolly, the Scottish sheep, is truly a clone of the six-year-old ewe that donated her nucleus.

## Population Statistics Are Used to Interpret DNA Fingerprints

The power of DNA fingerprinting is greatly enhanced by tracking repeats in several parts of the genome, just as it is easier to identify a criminal suspect with more clues to physical appearance. Each repeat used in a DNA fingerprint is considered an allele and is assigned a probability based on its observed frequency in a particular population. The sequences that are tracked must

# DNA Fingerprinting Relies on Molecular Genetics and Population Genetics

DNA fingerprinting has rapidly become a standard and powerful tool used in forensic investigations, agriculture, paternity testing, and historical investigations. Section 1.5 discussed several interesting examples. But DNA fingerprinting is a relatively new tool. In 1986, it was virtually unheard of outside of scientific circles. A dramatic rape case changed that.

Tommie Lee Andrews became the first person in the United States to be convicted of a crime on the basis of DNA evidence. Andrews thought he was being very meticulous in planning his crimes, picking his victims months before he attacked and watching them so that he knew exactly when they would be home alone. On a balmy Sunday night in May 1986, Andrews awaited Nancy Hodge, a young computer operator at Disney World, at her home in Orlando, Florida. The burly man surprised her when she was in the bathroom removing her contact lenses. He covered her face so that she could not see who was attacking her. He then raped and brutalized her repeatedly.

Andrews was very careful not to leave fingerprints, threads, hairs, or any other indication that he had ever been in Hodge's home. But he had not counted on the then-fledgling technology of DNA fingerprinting. Thanks to a clear-thinking crime victim and scientifically informed lawyers, Andrews was soon at the center of a trial—not only his trial, but one that would judge the technology that helped to convict him.

After the attack, Hodge went to the hospital, where she provided a vaginal secretion sample containing the rapist's sperm cells. Two district attorneys who had read about DNA fingerprinting sent some of the sperm to a biotechnology company that extracted DNA from the sperm cells and cut it with restriction enzymes. The sperm's DNA pieces were then mixed with labeled DNA probes that bound to complementary segments.

The same procedure of extracting, cutting, and probing the DNA was done on white blood cells from Nancy Hodge and Tommie Lee Andrews, who had been apprehended and held as a suspect in several assaults. When the radioactive DNA pieces from each sample, which were the sequences where the probes had bound, were separated and displayed according to size, the resulting pattern of bands—the DNA fingerprint—matched exactly for the sperm sample and Andrews's blood, differing from Nancy Hodge's DNA. He was the attacker (figure 1).

Because Tommie Lee Andrews is black, his allele frequencies were compared to those for a representative African American population. At his first trial in November 1987, the judge, perhaps fearful that too much technical information would overwhelm the jury, did not allow the prosecution to cite population-based statistics. Without the appropriate allele frequencies, DNA fingerprinting was reduced to a comparison of smeary lines on test papers to see whether the patterns of DNA pieces in the forensic sperm sample looked like those for Andrews's white blood cells. The probabilities determined from population-based statistics indicated that the possibility that Tommie Lee Andrews's DNA would match the evidence by chance was 1 in 10 billion. But the prosecution could not mention this.

After a mistrial was declared, the prosecution cited the precedent of using population statistics to derive databases on standard blood types. So when Andrews stood trial just three months later for the rape of a different woman, the judge permitted population analysis, and Andrews was convicted. Today, in jail, he keeps a copy of the *Discover* magazine article (written by this author) that describes his role in the first case tried using DNA fingerprinting.

The sizes of the DNA pieces in the type of DNA fingerprint used in the Andrews case vary from person to person because of differences in DNA sequence in the regions surrounding the probed genes. The discriminating power of the technology stems from the fact that there are many more ways for the 3 billion bases of the human genome to vary than there are people. However, this theoretical variation is tempered by the fact that certain gene combinations are more prevalent in some populations because of marriage, travel, and other social customs. That is, within ethnic groups, some people may be more alike genetically, and distinguishing among them might be more difficult. In one case, for example, DNA fingerprinting could not reveal whether a man or his father had committed a rape.

Today, many DNA fingerprinting analyses use much shorter DNA sequences, such as repeats of 2, 3, 4, or 5 bases. Instead of using restriction enzymes, these shorter repeats are analyzed using the polymerase chain reaction (PCR). Recall from chapter 9 that PCR uses primer sequences that flank a gene of interest and replication enzymes to rapidly mass produce a sequence. The greater the number of repeats an individual has, the larger the corresponding PCR product, because more DNA is amplified between the primers. PCR works well on degraded DNA and requires only a few cells—conditions common in biological evidence from crime scenes.

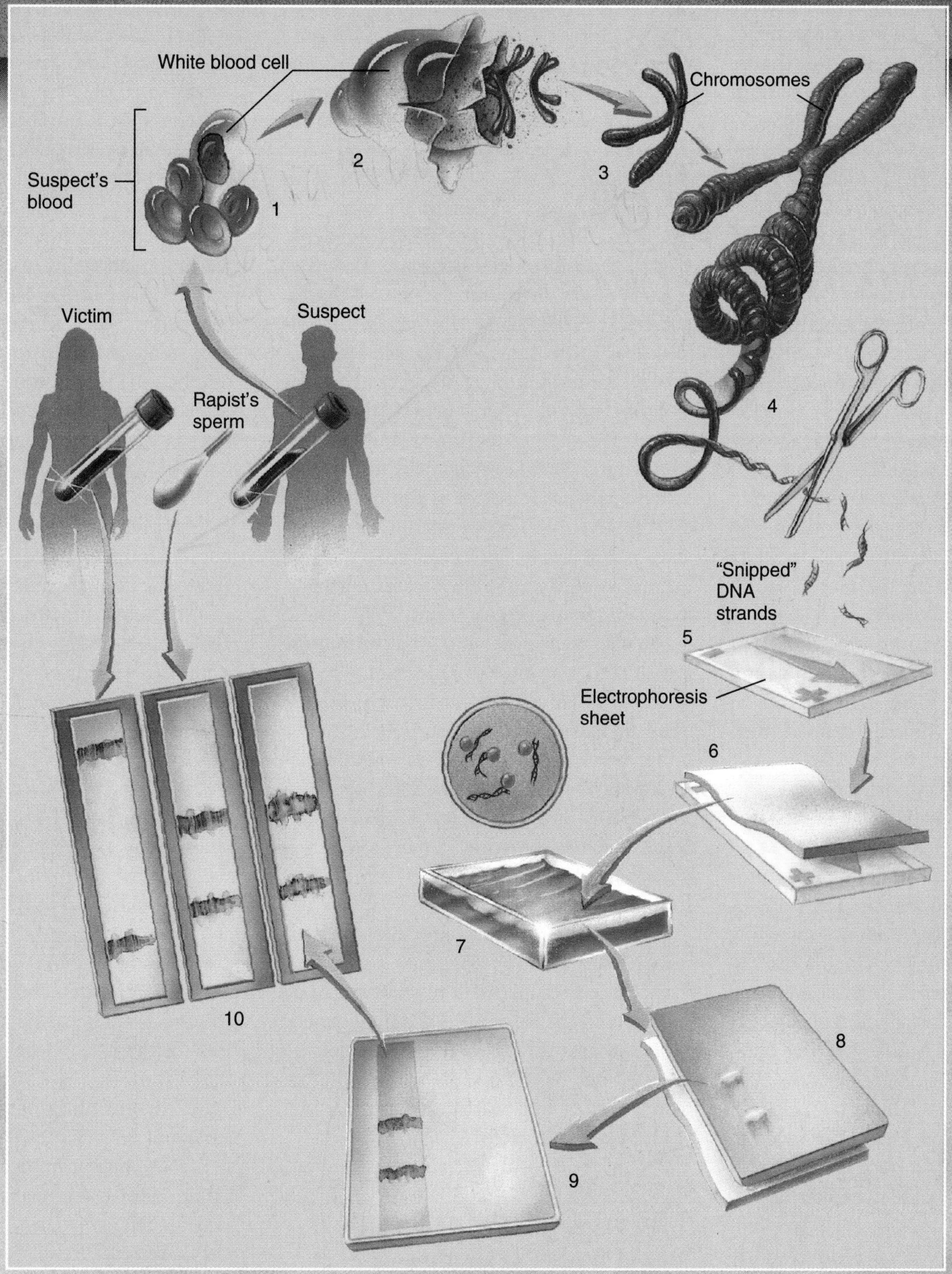


## figure 1

**DNA fingerprinting.** A blood sample (1) is collected from the suspect. White blood cells are separated and burst open (2), releasing DNA (3). Restriction enzymes snip the strands into fragments (4), and electrophoresis aligns them by size in a groove on a sheet of gel (5). The resulting pattern of DNA fragments is transferred to a nylon sheet (6). It is then exposed to radioactively tagged probes (7) that home in on the DNA areas used to establish identity. When the nylon sheet is placed against a piece of X-ray film (8) and processed, black bands appear where the probes stuck (9). This pattern of bands constitutes a DNA print (10). This print may then be compared to the victim's DNA fingerprint, the rapist's DNA fingerprint obtained from sperm cells, and other biological evidence.

come from different chromosomes and should be as close to Hardy-Weinberg equilibrium as possible.

The Hardy-Weinberg equation and the product rule are used to derive the statistics that back up a DNA fingerprint (figure 13.6). First, the DNA fingerprint pattern shows whether an individual is a homozygote or a heterozygote for each repeat, because a homozygote only has one band representing that gene. Genotype frequencies are then calculated using parts of the Hardy-Weinberg equation; that is, $p^2$ and $q^2$ denote the two homozygotes for a two-allele gene, and $2pq$ represents the heterozygote. Applying the product rule, the genotype frequencies are multiplied. The result is the probability that this particular combination of DNA sequences would occur in the population. Logic then enters the equation. If an allele combination is exceedingly rare in the population the suspect comes from, and if it is found both in the suspect's DNA and in crime scene evidence, the suspect's guilt appears to be highly likely.

The accuracy and meaning of a DNA fingerprint depend upon the population that allele frequencies are derived from. If populations are too broadly defined, then allele frequencies are typically quite low, leading to very large estimates of the likelihood that a suspect matches evidence based on chance alone.

In one oft-quoted trial, the prosecutor concluded, "The chance of the DNA fingerprint of the cells in the evidence matching blood of the defendant by chance is 1 in 738 trillion." The numbers themselves were not at fault, but some population geneticists wondered about the validity of the databases. Did they really reflect the gene pool compositions of actual human populations? By 1991, half a dozen judges had rejected DNA evidence because population geneticists had testified that the databases do not correspond directly to true allele frequencies because they greatly oversimplify human population structure. Therefore, the odds that crime scene DNA matched suspect DNA were not as reliable as originally suggested.

The first DNA fingerprinting databases used neatly shoehorned many different groups into just three seemingly homogeneous categories—people were either Caucasian, black, or Hispanic. People from Poland, Greece, or Sweden would all be considered white, and a dark-skinned person from Jamaica and one from Somalia would be lumped together as blacks. Perhaps the most incongruous of all were the Hispanics. Cubans and Puerto Ricans are part African, whereas people from Mexico and Guatemala have mostly native American gene variants. Spanish and Argentinians have neither black African nor native American genetic backgrounds. Yet these diverse peoples were considered together as a single population! Other groups were left out, such as native Americans and Asians. Ultimately, analysis of these three databases revealed significantly more homozygous recessives for certain polymorphic genes than the Hardy-Weinberg equation would predict, confirming what many had suspected—allele frequencies were not in equilibrium.

To give meaning to the allele frequencies necessary to interpret DNA fingerprints, we need more restrictive ethnic databases. A frequency of 1 in 1,000 for a particular allele in all whites may actually be much higher or lower in, for example, only Italians, because they (and many others) tend to marry among themselves. On the other hand, narrowly defined ethnic databases may be insufficient to interpret DNA fingerprints from people of mixed heritages, such as someone whose mother was Scottish/French and whose father was Greek/German.

We may need to develop mathematical models to account for real population structures. Perhaps the first step will be to understand the forces that generate genetic substructures within more broadly defined populations, which means taking into account history and human nature. Chapter 14 explores these factors.

## DNA Fingerprinting to Identify World Trade Center Victims

During the second half of September, 2001, Myriad Genetics Inc., a biotechnology company in Salt Lake City that normally provides breast cancer tests, was flooded with a very different type of medical sample—frozen DNA from people who had presumably perished in the terrorist attack on the World Trade Center on the 11th (figure 13.7). The laboratory also received cheekbrush scrapings from relatives of the missing, amassed at DNA collection centers set up throughout New York City in the days after the disaster, and tissue from the victims' toothbrushes, razors, and hairbrushes. The workers used the polymerase chain reaction to determine the numbers of copies of four-based sequences of DNA, called short tandem repeats or STRs, at 13 locations in the genome. They also determined the sex chromosome constitution. The chance that any two individuals have the same 13 markers by chance is 1 in 250 trillion, so that if the STR pattern of a sample from the crime scene matched a sample from a victim's toothbrush, identification was fairly certain. Myriad sent its results to the New York State Forensic Laboratory, where kinship analysis matched family members to victims. Meticulous records were kept along the way, so that when an individual was identified more than once, the bad news wasn't delivered to the family again.

The most difficult part of the massive DNA fingerprinting effort was right at the beginning—obtaining victim samples, because most of the bodies were incinerated. Myriad performed STR analysis on whatever pieces of soft tissue were found. Sadly, their job ended within a few weeks, but the DNA analysis continued. Bone bits, which can persist despite the ongoing fire at the site, were sent to a laboratory in Virginia where DNA was extracted, and it was sent to Celera Genomics Corp. in Rockville, Maryland, a company involved in sequencing the human genome. Here, mitochondrial DNA fingerprinting was attempted on tissue that had been too degraded to yield to the more accurate STR typing. The armed forces handled DNA fingerprinting of the victims at the Pentagon.

DNA fingerprinting can counter the uncertainties of traditional forensic techniques such as dental patterns, scars, fingerprints, and clues such as jewelry, wallets, and rolls of film found with the victim. Consider the case of Jose Guadalupe and Christopher Santora, two of the 15 firefighters lost from one engine company on September 11, 2001. Rescue workers brought a body that had been found beside a fire truck next to the destroyed towers to the Medical Examiner's office on September 13th. Other firefighters identified the remains as belonging to Guadalupe based

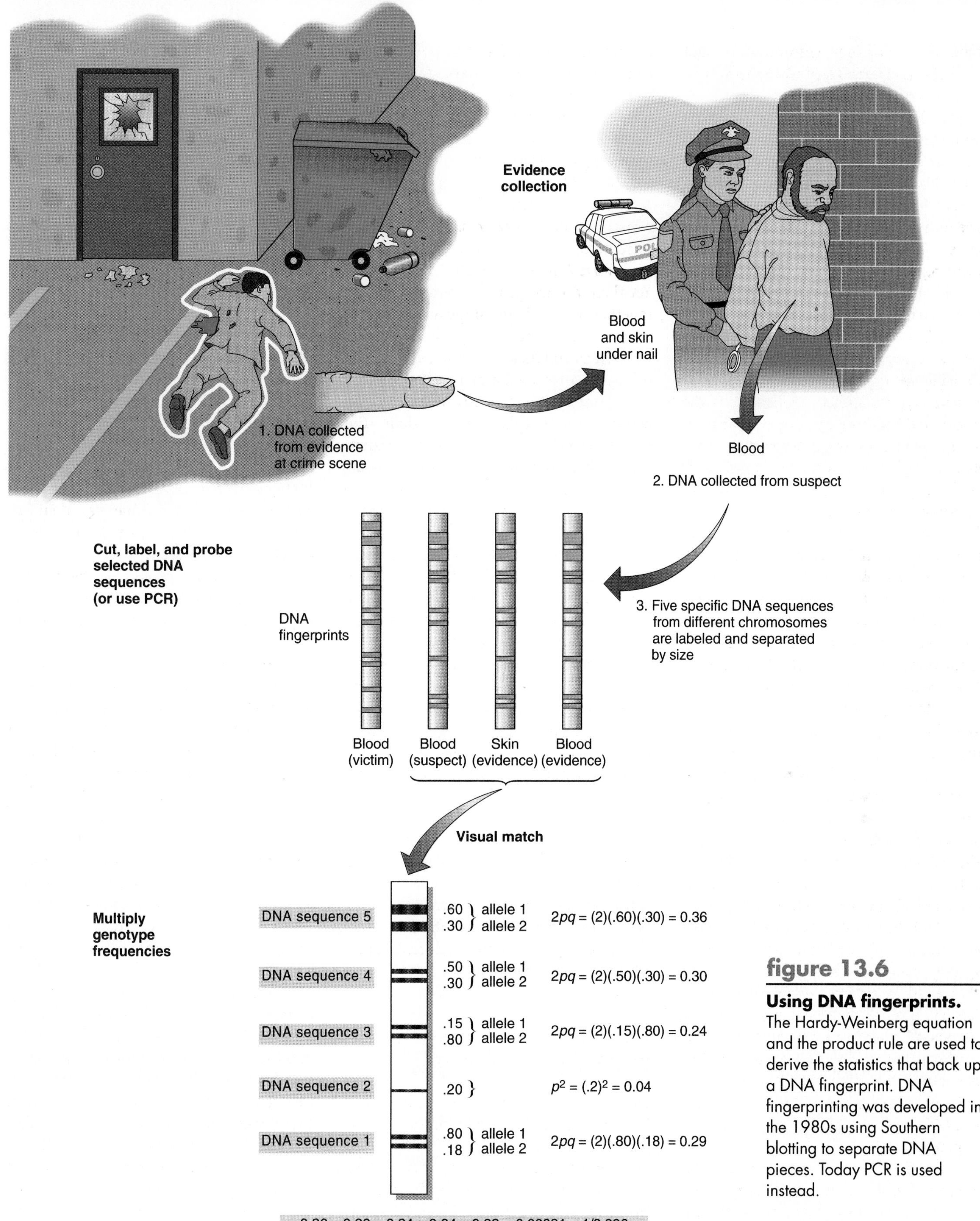


## figure 13.6

**Using DNA fingerprints.** The Hardy-Weinberg equation and the product rule are used to derive the statistics that back up a DNA fingerprint. DNA fingerprinting was developed in the 1980s using Southern blotting to separate DNA pieces. Today PCR is used instead.

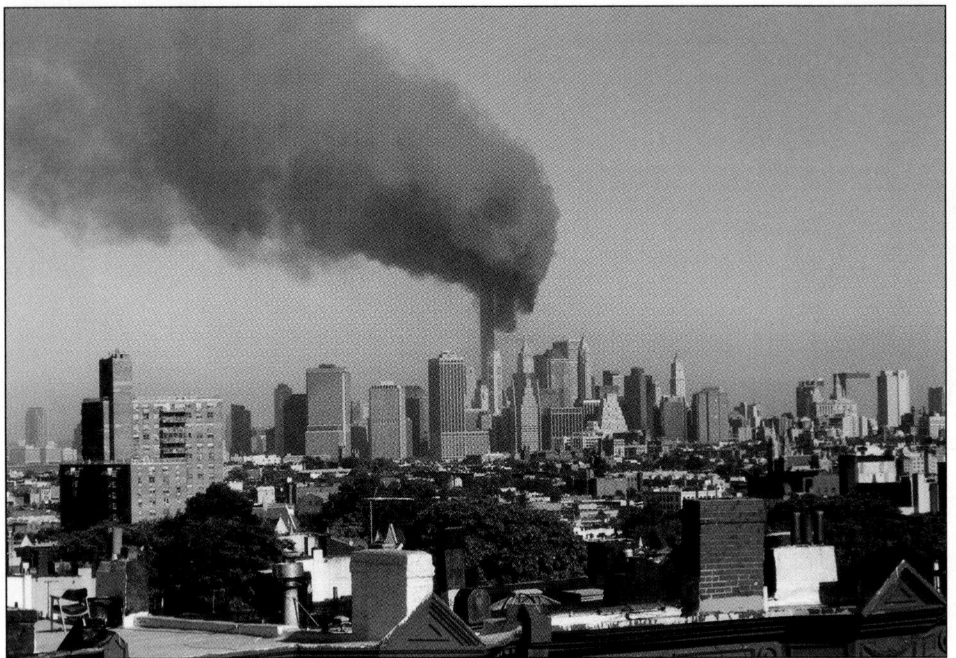

**figure 13.7**

**DNA fingerprinting and the September 11, 2001 tragedy.** It may take more than a million DNA tests to identify victims of the World Trade Center terrorist attacks.

on where the body had been found, because he had been the chauffeur of the fire truck. The body also had a gold chain that the men recognized, and X-rays revealed a birth defect in the neck bone that he was known to have had. Guadalupe was buried on October 1—but it wasn't Guadalupe who was buried. It turned out that Christopher Santora had the same necklace and the same neck condition! A DNA sample taken from the buried man's remains and from Santora's relatives matched.

The scale of the DNA fingerprinting task that followed the September 11, 2001 attack was unprecedented, and particularly difficult because there was no list of victims, as is the case in a plane crash. DNA fingerprinting of the 230 victims of a crash in 1996 took a year. The World Trade Center disaster yielded more than a million DNA samples. It was a very distressing experience for the technicians and researchers whose jobs had suddenly shifted from detecting breast cancer and sequencing genomes to helping in recovery. Said J. Craig Venter, president of Celera at the time, "I never, ever thought we would have to do DNA forensics at this level, and for this reason."

### Key Concepts

DNA fingerprints are based on SNPs or differences in the sizes of highly repeated DNA sequences among individuals, based on the idea that the human genome sequence can vary in more ways than there are people. • Population statistics are applied to determine the probability that the same pattern would occur by chance in two individuals. A limitation of the method is that databases may not adequately represent real human populations. Developing narrower ethnic databases and considering historical and social factors may make population statistics more realistic. • DNA fingerprinting of nuclear and mitochondrial DNA was done on evidence from the September 11 terrorist attacks.

# Summary

1. A **population** is a group of interbreeding members of the same species in a particular area. Their genes constitute the **gene pool.**

### 13.1 The Importance of Knowing Allele Frequencies

2. **Population genetics** considers allele, genotype, and phenotype frequencies to reveal whether or not **microevolution** is occurring. Phenotypic frequencies can be determined empirically, then used in algebraic expressions to derive other frequencies.
3. Allele frequencies change if migration, nonrandom mating, genetic drift, mutations, or natural selection operate. In **Hardy-Weinberg equilibrium,** allele frequencies are not changing.

### 13.2 When Allele Frequencies Stay Constant—Hardy-Weinberg Equilibrium

4. Hardy and Weinberg proposed using algebra to explain the constancy of allele frequencies. This would show why dominant traits do not increase and recessive traits do not decrease in populations. The Hardy-Weinberg equation is a binomial expansion used to represent genotypes in a population.
5. Hardy-Weinberg equilibrium is demonstrated by following gamete frequencies as they recombine in the next generation. In Hardy-Weinberg equilibrium, these genotypes remain constant from generation to generation if evolution is not occurring. When the equation $p^2 + 2pq + q^2$ represents a gene with one dominant and one recessive allele, $p^2$ corresponds to the frequency of homozygous dominant individuals; $2pq$ stands for heterozygotes; and $q^2$ represents the frequency of the homozygous recessive class. The frequency of the dominant allele is $p$, and of the recessive allele, $q$.

### 13.3 Practical Applications of Hardy-Weinberg Equilibrium

6. If we know either $p$ or $q$, we can calculate genotype frequencies, such as carrier risks. Often such information comes from knowing the $q^2$ class, which corresponds to the frequency of homozygous recessive individuals in a population.
7. For X-linked recessive traits, the mutant allele frequency for males equals the trait

frequency. For very rare disorders or traits, the value of $p$ approaches 1, so the carrier frequency ($2pq$) is approximately twice the frequency of the rare trait ($q$).

### 13.4 DNA Fingerprinting—A Practical Test of Hardy-Weinberg Assumptions

8. A highly variable gene variant or DNA sequence present in more than 1 percent of the population is a **polymorphism**.
9. **Restriction enzymes** or PCR are used to isolate polymorphic repeats, and polyacrylamide gel electrophoresis is used to separate and display labeled DNA fragments by size. The fragment pattern is the DNA fingerprint.
10. To interpret DNA fingerprints, allele frequencies are derived from population data, and genotype frequencies are calculated and then multiplied. The result estimates the likelihood that two individuals from a given population share the same genotype for all of the DNA sequences examined.
11. DNA fingerprinting will become a more powerful tool when we can more specifically define and analyze human populations.
12. DNA fingerprinting based on short tandem repeats and mitochondrial DNA was used to identify victims of the World Trade Center disaster on September 11, 2001.

# Review Questions

1. "We like him, he seems to have a terrific gene pool," say the parents upon meeting their daughter's latest boyfriend. Why doesn't their statement make sense?
2. What is *not* happening in a population in Hardy-Weinberg equilibrium?
3. Why is knowing the incidence of a homozygous recessive condition in a population important in deriving allele frequencies?
4. Two couples want to know their risk of conceiving a child with cystic fibrosis. In one couple, neither partner has a family history of the disease; in the other, one partner knows he is a carrier. How do their risks differ?
5. Why are short DNA repeats more likely to be in Hardy-Weinberg equilibrium than protein-encoding genes?
6. Why are specific databases necessary to interpret DNA fingerprints?
7. How is the Hardy-Weinberg equation used to predict the recurrence of X-linked recessive traits?

# Applied Questions

1. An extra row of eyelashes is an autosomal recessive trait that occurs in 900 of the 10,000 residents of an island in the south Pacific. Greta knows that she is a heterozygote for this gene, because her eyelashes are normal, but she has an affected parent. She wants to have children with a homozygous dominant man, so that the trait will not affect her offspring. What is the probability that a person with normal eyelashes in this population is a homozygote for this gene?
2. Glutaric aciduria type I causes progressive paralysis and brain damage. It is very common in the Amish of Lancaster County, Pennsylvania—0.25 percent of newborns have the disorder. Calculate the percentage of newborns that are carriers for this condition, and the percentage that do not have the disease-causing allele.
3. In a true crime that took place in Israel, a man raped a woman, first knocking her unconscious with a cement block so that she could not identify him. He was careful not to leave any hairs at the crime scene. Police tracked him down, however, because he left behind eyeglasses with unusual frames, and an optician helped locate him. The man left something else at the scene, too—a half-eaten lollipop! DNA from blood taken from the suspect did not match the victim's blood at the scene, but it did match DNA from cheek-lining cells collected from the base of the telltale lollipop. Forensic scientists obtained the DNA fingerprint shown below looking at four genes and considering their allele frequencies in the man's ethnic group in Israel.

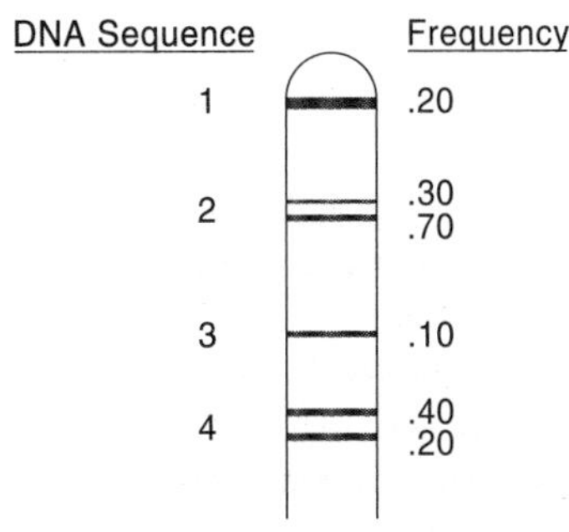


   a. Which of the tested genes has two alleles? How do you know this?
   b. What is the possibility that the suspect's DNA matches that of the lollipop rapist by chance? (Do the calculation.)
   c. The man's population group is highly inbred—many people have children with relatives. How does this information affect the accuracy or reliability of the DNA fingerprint?
   d. If a fifth gene was added to the analysis and the lollipop rapist was found to have just one allele of this gene that has a frequency of 0.45 in his population, how would it affect the probability that the man is guilty?

   (P.S.—He was so frightened by the DNA analysis that he confessed!)
4. Dystonia is a movement disorder that affects 1 in 1,000 Jewish people of eastern European descent (Ashkenazim). What is the carrier frequency in this population?
5. Factor IX deficiency is a clotting disorder affecting 1 in 190 Ashkenazim living in Israel. It affects 1 in 1,000,000 Japanese, Korean, Chinese, German, Italian, African American, English, Indian, and Arab people.
   a. What is the frequency of the mutant allele in the Israeli population?
   b. What is the frequency of the normal allele in this population?
   c. Calculate the proportion of carriers in the Israeli population.
   d. Why might the disease incidence be very high in the Israeli population but very low in others?
6. The Finnish population has a 1 percent carrier frequency for a seizure disorder

called myoclonus epilepsy. Two people who have no relatives with the illness ask a genetic counselor to calculate the risk that they will conceive an affected child, based on their belonging to this population group. What is the risk?

7. Maple syrup urine disease (MSUD) is an autosomal recessive inborn error of metabolism that causes mental and physical retardation, difficulty feeding, and a sweet odor to urine. In Costa Rica, 1 in 8,000 newborns inherits the condition. Calculate the carrier frequency of MSUD in this population.

8. The amyloidoses are a group of inborn errors of metabolism in which sticky protein builds up in certain organs. Amyloidosis caused by a mutation in the gene encoding a blood protein called transthyretin affects the heart and/or nervous system. It is autosomal recessive. In a population of 177 healthy African Americans, four proved, by blood testing, to have one mutant allele of the transthyretin gene. What is the carrier frequency in this population?

9. Ability to taste phenylthiocarbamide (PTC) is determined by the *T* gene. *TT* individuals taste a strong, bitter taste; *Tt* people experience a slightly bitter taste; *tt* individuals taste nothing.

   A fifth-grade class of 20 students tastes PTC papers, rating the experience as "very yucky" (*TT*), "I can taste it" (*Tt*), and "I can't taste it" (*tt*). For homework, the students test their parents, with these results:

   Of 6 *TT* students, 4 have 2 *TT* parents; one has one parent who is *TT* and one parent who is *Tt*. The sixth *TT* student has one parent who is *Tt* and one who is *tt*.

   Of 4 students who are *Tt*, 2 have 2 parents who are *Tt*, and 2 have one parent who is *TT* and one parent who is *tt*.

   Of the 10 students who can't taste PTC, 4 have 2 parents who also are *tt*, but 4 students have one parent who is *Tt* and one who is *tt*. The remaining 2 students have 2 *Tt* parents.

   Calculate the frequencies of the *T* and *t* alleles in the two generations. Is Hardy-Weinberg equilibrium maintained, or is this gene evolving?

## Suggested Readings


Altman, Lawrence K. September 25, 2001. Now, doctors must identify the dead. *The New York Times*, F1. It isn't doctors, but genetic researchers, who did the DNA fingerprinting of World Trade Center victims.

Crow, James F. February 15, 2001. The beanbag lives on. *Nature* 409:771. A very clear explanation of Hardy-Weinberg equilibrium.

Grady, Denise. June 29, 1999. At gene therapy's frontier, the Amish build a clinic. *The New York Times*. The gene that causes Crigler-Najjar syndrome is far from Hardy-Weinberg equilibrium.

Hoffert, Stephen P. May 11, 1998. Taking the clinic to the patients gave researcher sense of purpose. *The Scientist* 12:1. The Amish and Mennonites of Lancaster County, Pennsylvania, have high frequencies of alleles that are extremely rare elsewhere.

Kleinfeld, N. R. November 28, 2001. Error puts body of one firefighter in a grave of a firehouse colleague. *The New York Times*, F1. DNA fingerprinting adds precision to forensics.

Lewis, Ricki. June 1988. Witness for the prosecution. *Discover* vol. 9. The story of one of the first DNA fingerprinting cases.

Pollack, Andrew. Identifying the dead, 2,000 miles away. *The New York Times*, F1. Myriad Genetics Inc. never expected to be doing what it is doing.


## On the Web

Check out the resources on our website at:

**www.mhhe.com/lewisgenetics5**

On the web for this chapter, you will find additional study questions, vocabulary review, useful links to case studies, tutorials, popular press coverage, and much more. To investigate specific topics mentioned in this chapter, also try the links below:

Crigler-Najjar Syndrome
**www.crigler-najjar.com/**

DNA fingerprinting
**vector.cshl.org/resources/aboutdnafingerprinting.html**

HuGENet (population-based information on the human genome)
**www.cdc.gov/genetics/hugenet/**

Neo Gen Screening
**www.neogenscreening.com/**

Online Mendelian Inheritance in Man
**www.ncbi.nlm.nih.gov/entrez/query.fcgi?db=OMIM**
Crigler-Najjar syndrome 218800
cystic fibrosis 219700

Organic Acidemia Association Inc.
**www.oaanews.org**

CHAPTER

# 14

# Changing Allele Frequencies

### 14.1 Nonrandom Mating

Hardy-Weinberg equilibrium—unchanging allele frequencies—assumes that we all mate with equal frequency and without regard for our partners' phenotypes. Human societies do not behave this way, so the gene pool changes with each generation.

### 14.2 Migration

Despite mountains and oceans, language barriers and cultural differences, people are always on the move. Migration shakes up the gene pool, too, introducing some alleles and eliminating others.

### 14.3 Genetic Drift

Allele frequencies change when subgroups form. Such random genetic drift may take the form of a population-within-a-population, such as in an ethnic group, or when settlers leave to found a new town. In a population bottleneck, disaster strikes and numbers shrink, then rise anew as the few survivors breed—but at the expense of genetic diversity.

### 14.4 Mutation

The changeable nature of DNA continually adds a trickle of new alleles to populations.

### 14.5 Natural Selection

To a great extent, environmental conditions determine which alleles are transmitted to the next generation—and which are not.

### 14.6 Gene Genealogy

Charting allele frequencies through space and time reflects the events of history, and may even reveal evolution in action.

Historically, we seem to have gone out of our way to see that the very specific conditions necessary for Hardy-Weinberg equilibrium—unchanging allele frequencies from generation to generation—do not occur, at least for some genes. Wars and persecution selectively kill certain populations. Economic and political systems enable some groups to have more children than others. Religious restrictions and personal preferences guide our choices of mates. We travel extensively, shuttling our genes in and out of populations. Natural disasters and new infectious diseases reduce populations to a few individuals, who then rebuild the numbers, but at the expense of genetic diversity. Add to all this the forces of mutation and the reshuffling of genes that occurs in each generation, and it is clear that a gene pool is a very fluid entity. These ever-present and interacting forces of nonrandom or selective mating, migration, genetic drift, mutation, and natural selection work to differing degrees to shape populations. This chapter addresses the factors that shake up gene pools.

## 14.1 Nonrandom Mating

In the theoretical state of Hardy-Weinberg equilibrium, individuals of all genotypes are presumed equally likely to mate and to choose their partners at random. But we give great thought to selecting mates—it is hardly a random process. We choose mates based on physical appearance, ethnic background, intelligence, and shared interests. Surveys show that we marry people similar to ourselves about 80 percent of the time. Worldwide, about one-third of all marriages occur between people who were born fewer than 10 miles apart! Such nonrandom mating is a major factor in changing allele frequencies in human populations.

Another form of nonrandom mating on a population level occurs when certain individuals contribute disproportionately to the next generation. This is common in agriculture when an animal or plant with valuable characteristics is bred extensively. For example, semen from one prize bull may be used to artificially inseminate thousands of cows. Such an extreme situation can arise in a human population when a man fathers many children. A striking mutation can reveal such behavior. In the Cape population of South Africa, for example, a Chinese immigrant known as Arnold had a very rare dominant genetic disease that causes one's teeth to fall out before age 20. Arnold had seven wives. Of his 356 living descendants, 70 have the dental disorder. The frequency of this allele in the Cape population is exceptionally high, thanks to Arnold.

The high frequency of people with autosomal recessive albinism among Arizona's Hopi Indians also reflects nonrandom mating. Albinism is uncommon in the general U.S. population, but it affects 1 in 200 Hopi Indians. The reason for the trait's prevalence is cultural—men with albinism often stay back and help the women, rather than risk severe sunburn in the fields with the other men. They contribute more children to the population because they have more contact with women.

The events of history reflect patterns of nonrandom mating. When a group of people is subservient to another, genes tend to "flow" from one group to the other as the males of the ruling class mate with females of the underclass. Historical records and living records of DNA sequences on the Y chromosome, which indicate gene transmission through the male, reveal this phenomenon of directional gene flow.

Despite our preferences in mate selection, many traits do mix randomly in the next generation. This may be because we are unaware of these characteristics, they do not confer an obvious benefit, or they are not considered important in choosing a mate. In populations where AIDS is extremely rare or nonexistent, for example, the two types of mutations that render a person resistant to HIV infection are in Hardy-Weinberg equilibrium. This would change, over time, if infection arrives, because the people with these mutations would become more likely to survive to produce offspring—some of whom would perpetuate the protective mutation. Natural selection would intervene, ultimately altering allele frequencies.

Many blood types are in Hardy-Weinberg equilibrium, because we do not select life partners on the basis of blood type. Yet sometimes the opposite occurs. People with uncommon alleles meet and mate more often than might happen by chance when their families participate in a program for people with the same inherited condition, such as sickle cell disease clinics or summer camps for children with cystic fibrosis. In the reverse situation, in a very religious Jewish sect living near New York City, marriages are not permitted between carriers of the same inherited disease. This practice is a curious mix of the old and the new—arranged marriages, but use of genetic testing to identify carriers of specific disorders.

A population that practices consanguinity has very nonrandom mating. Recall from chapter 4 that a consanguineous relationship is one in which "blood" relatives have children together. On the family level, this practice increases the likelihood that harmful recessive alleles from shared ancestors will be in the same offspring, causing disease. The birth defect rate in offspring is 2.5 times normal, which is about 3 percent. On a population level, consanguinity decreases genetic diversity. The proportion of homozygotes rises as that of heterozygotes falls.

Some populations encourage marriage between cousins, resulting in an increase in the incidence of certain recessive disorders. In certain isolated middle eastern settlements, more than half of all marriages are consanguineous, with uncle-niece pairings quite common. The tools of molecular genetics can reveal these relationships. Researchers traced DNA sequences on the Y chromosome and in mitochondria among residents of an ancient, geographically isolated "micropopulation" on the island of Sardinia, near Italy. They consulted archival records dating from the village's founding by 200 settlers around 1,000 A.D. to determine familial relationships. Between 1640 and 1870, population size doubled, and it reached 1,200 by 1990. Fifty percent of the present population descends from just two paternal and four maternal lines, and 86 percent have the same X chromosome. Researchers are analyzing the medical conditions that are especially prevalent in this population, which include hypertension and a kidney disorder.

### Key Concepts

People choose mates for many reasons, and they do not contribute the same numbers of children to the next generation. This changes allele frequencies in populations. Traits lacking obvious phenotypes may be in Hardy-Weinberg equilibrium. Consanguinity in populations increases the proportion of homozygotes, at the expense of heterozygotes.

## 14.2 Migration

Large cities, with their pockets of ethnicity, defy Hardy-Weinberg equilibrium by their very existence. Waves of immigrants formed the population of New York City, for example. The original Dutch settlers of the 1600s had different alleles than those in today's metropolis, which were contributed by the English, Irish, Slavics, Africans, Hispanics, Italians, Asians, and others.

### Historical Clues

We can trace the genetic effects of migration by correlating allele frequencies in present-day populations to events in history, or by tracking how allele frequencies change from one geographical region to another, and then inferring in which directions ancient peoples traveled. Figure 14.1 depicts the great changes in frequency of the allele that causes galactokinase deficiency in several European populations. This autosomal recessive disorder causes cataracts (clouding of the lens) in infants. It is very common among a population of 800,000 gypsies, called the Vlax Roma, who live in Bulgaria. Here, it affects 1 in 1,600 to 2,500 people, and 5 percent of the people are carriers. But among all gypsies in Bulgaria as a whole, the incidence drops to 1 in 52,000. As the map in figure 14.1 shows, the disease becomes rarer with distance to the west. This pattern may have arisen when people with the allele settled in Bulgaria, with only a few individuals or families moving westward.

Allele frequencies reflect who rules whom. The frequency of ABO blood types in certain parts of the world today mirrors past Arab rule. The distribution of ABO blood types is very similar in northern Africa, the Near East, and southern Spain. These are precisely the regions where Arabs ruled until 1492. The uneven distribution of allele frequencies can also reveal when and where nomadic peoples stopped for awhile. For example, in the 18th century, European caucasians called trekboers migrated to the Cape area of South Africa. The men stayed and had children with the native women of the Nama tribe. The mixed society remained fairly isolated, leading to the distinctive allele frequencies found in the present-day people of color of the area.

Genetic analyses can account for migration by consulting historical records. This is the case for Creutzfeldt-Jakob disease (CJD). Recall from chapter 2 that this is a prion disorder that can be caused by a mutation in the prion protein gene on chromosome 20. It is rare, but more than 70 percent of affected families worldwide share the same mutation, suggesting a common origin. Researchers examined the responsible section of the chromosome in 62 affected families from 11 populations, looking at a haplotype that included repeated DNA sequences and a SNP. They identified the same haplotype in families from certain groups in Libya, Tunisia, Italy, Chile, and Spain—the exact populations that were expelled from Spain in the Middle Ages. These groups apparently took the CJD gene with them, where it persists today to cause the rare inherited form of this disease.

### Geographical and Linguistic Clues

Sometimes allele frequencies change from one neighboring population to another. Such a gradient is termed a **cline.** Changing allele frequencies usually reflect migration patterns, as immigrants introduced alleles and emigrants removed them. Clines may be gradual, reflecting unencumbered migration paths, but barriers often help cause more abrupt changes in allele frequencies. Geographical formations such as mountains and bodies of water may block migration, maintaining population differences in allele frequencies. Language differences may isolate alleles as people who cannot communicate tend not to have children together.

Allele frequencies up and down the lush strip of fertile land that hugs the Nile River illustrate the concept of clines. In one study, researchers analyzed specific DNA sequences in the mitochondrial DNA of 224 people who live on either side of the Nile, an area settled for 15,000 years. The researchers found a smooth transition of gradual change in mitochondrial DNA sequences. That is, the farther apart two individuals live along the Nile, the less alike their mitochondrial DNA. This is consistent with evidence

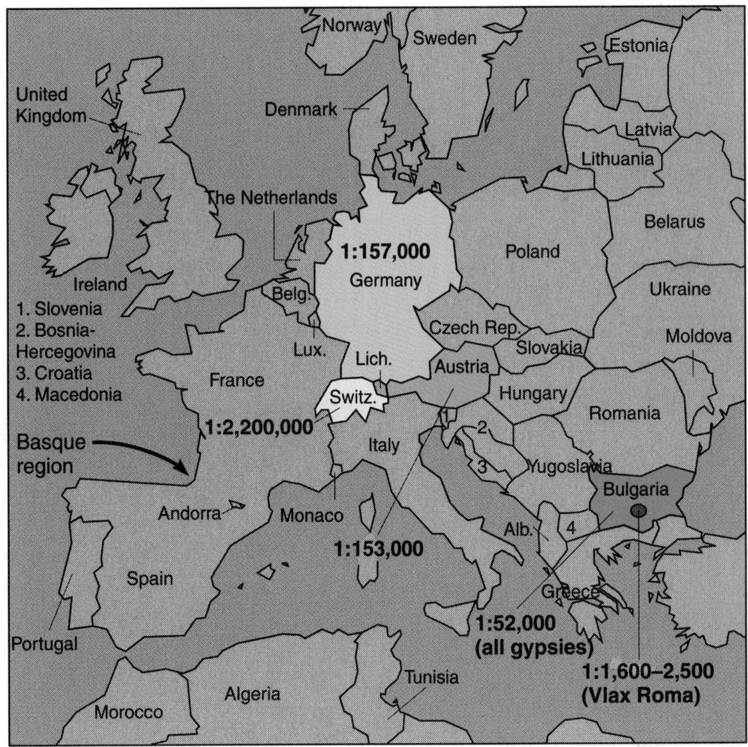


**figure 14.1**

**Galactokinase deficiency in Europe.** This autosomal recessive disorder that causes blindness varies in prevalence across Europe. It is most common among the Vlax Roma gypsies in Bulgaria. The condition becomes much rarer to the west, as indicated by the shading from dark to light green.

from mummies and historical records that the area once consisted of a series of kingdoms separated by wars and language differences. If the area had been one interacting settlement, then the DNA sequences would have been more mixed. Instead, the researchers suggest, the Nile may have served as a "genetic corridor" between Egypt and sub-Saharan Africa.

Another pattern of changing allele frequencies reflects the human dependence on communication. In one study, population geneticists correlated 20 blood types to geographically defined regions of Italy and to areas where a single dialect is spoken. They chose Italy because it is rich in family history records and linguistic variants. Six of the blood types varied more consistently with linguistically defined subregions of the country than with geographical regions. Perhaps differences in language prevent people from socializing, sequestering alleles within groups that speak the same dialect because these people marry each other.

**KEY CONCEPTS**

Migration alters allele frequencies by adding and removing alleles from populations. • Clines are gradual changes in allele frequencies between neighboring populations. Geographical barriers and language differences often create more abrupt divisions.

## 14.3 Genetic Drift

When a small group of individuals is separated from a larger population, or mates only among themselves within a larger population, allele frequencies may change as a result of chance sampling from the whole. This change in allele frequency that occurs when a small group is considered separately from the larger whole is termed **genetic drift.** It can be compared to reaching into a bag of jellybeans and, by chance, grabbing only green and yellow ones. The allele frequency changes that occur with genetic drift are random and therefore unpredictable, just as reaching into the jellybean bag a second time might yield mostly black and orange candies.

Genetic drift occurs when the population size plummets, due either to migration, to a natural disaster that isolates small pockets of a population, or to the consequences of human behavior. In a common sociological scenario, members of a small community choose to mate only among themselves, keeping genetic variants within their ethnic group. Pittsburgh, Pennsylvania, for example, is made up of many distinct neighborhoods; within these neighborhoods, the residents are more like each other genetically than they are to others in the city. New York City, too, is more a hodgepodge of groups with distinct ethnic flavors, rather than a "melting pot" of mixed heritage.

Some groups of people become isolated in several ways—geographically, linguistically, and by choice of mates. Such populations often have high incidence of several otherwise rare inherited conditions. Consider the native residents of the Basque country in the western part of the Pyrenees Mountains between France and Spain. They still speak remnants of Euskera, a language brought by the first European settlers who arrived there during the late Paleolithic period over 10,000 years ago. The Basques have unusual frequencies of ABO and Rh blood types, rare mitochondrial DNA sequences and cell surface antigen patterns, and a high incidence of a mild form of muscular dystrophy. We return to them at the end of the chapter.

### The Founder Effect

A common type of genetic drift in human populations is the **founder effect,** which occurs when small groups of people leave their homes to found new settlements. The new colony may have different allele frequencies than the original population.

Founder populations amplify certain alleles while maintaining great stretches of uniformity in other DNA sequences. This shows up as increased disease frequencies. Among a population of 18,000 who live in northeastern Finland, for example, the lifetime risk of developing schizophrenia is 3.2 percent, which is nearly triple the national average risk. This group traces its ancestry to 40 families who settled in the region at the end of the 17th century. The population has been easy to study because the Finnish church has records of births, deaths, marriages, and moves, and hospital records are available. Table 14.1 lists some other founder populations.

A powerful founder effect is seen in the French Canadian population of Quebec. Their lack of diversity in disease-causing mutations reflects a long history of isolation. Consider breast cancer caused by the BRCA1 gene. More than 500 alleles are known worldwide, yet only four are seen among French Canadians. Several inborn errors of metabolism are also more common in this group than in others. The French Canadians have what one researcher calls "optimum characteristics for gene discovery." These include many generations since founding (14), a small number of founders (about 2,500), a high rate of population expansion (74 percent increase per generation), a large present-day population (about 6 million), and minimal marriage outside the group.

The French-Canadian population exemplifies genetic drift because the people

**table 14.1**

**Founder Populations**

| Population | Number of Founders | Number of Generations | Population Size Today |
|---|---|---|---|
| Costa Rica | 4,000 | 12 | 2,500,000 |
| Finland | 500 | 80–100 | 5,000,000 |
| Hutterites | 80 | 14 | 36,000 |
| Japan | 1,000 | 80–100 | 120,000,000 |
| Iceland | 25,000 | 40 | 300,000 |
| Newfoundland | 25,000 | 16 | 500,000 |
| Quebec | 2,500 | 12–16 | 6,000,000 |
| Sardinia | 500 | 400 | 1,660,000 |

have remained mostly among themselves, within a larger population. The French founded Quebec City in 1608. Until 1660, the population grew as immigrants arrived from France, but then began to increase from births. More than 10,000 French had arrived by the time the British took over in 1759, but many of them had headed westward, taking their genes with them. Meanwhile, in Quebec, religious, language, and other cultural differences kept the French and English gene pools largely separate. And so the French Canadian population of Quebec grew from 2,000 to 4,000 founding genotypes to about 6 million individuals today.

The cultural and physical isolation in Canada created an unusual situation—a founder effect within a founder effect. In the 19th century, when agricultural lands opened up about 150 miles north of Quebec, some families migrated north. Their descendants, who remained in the remote area, form an incredibly genetically homogeneous subpopulation of founders split off from the original set of founders.

A classical example of a founder effect within a larger population is the Dunker community of Germantown, Pennsylvania. Excellent historical records combined with distinctive or measurable traits enabled geneticists to clearly track genetic drift from the larger surrounding population. The Dunkers came from Germany between 1719 and 1729, but they have lived among others since that time. Still, the frequencies of some genotypes are different among the Dunkers than among their non-Dunker neighbors, and they are also different from the frequencies seen among people living in their native German village. The Dunkers have a different distribution of blood types (table 14.2) and much higher incidence of attached earlobes, hyperextensible thumbs, hairs in the middle of their fingers, and left-handedness compared to the other two groups.

Founder effects can be studied at the phenotypic and genotypic levels. Phenotypically, a founder effect is indicated when a community of people, known from local history to have descended from a few founders, have their own collection of inherited traits and illnesses that are rare elsewhere in the world. This is striking among the Old Order Amish and Mennonites of Lancaster County, Pennsylvania. Often, worried parents would bring their ill children to medical facilities in Philadelphia, and over the years, researchers realized that these people are subject to an array of extremely rare conditions (table 14.3 and figure 14.2). For example, Victor McKusick, of *Mendelian Inheritance in Man* fame, discovered and described cartilage-hair hypoplasia. In 1965, six Amish children died at a Philadelphia hospital from chicken pox. Part of their inherited syndrome was impaired immunity, and the children could not recover from this usually mild illness.

**table 14.2**

**Genetic Drift and the Dunkers**

| | Population | | |
|---|---|---|---|
| **Blood Type** | **U.S.** | **Dunker** | **European** |
| **ABO System** | | | |
| A | 40% | 60.0% | 45% |
| B, AB | 15% | 5.0% | 15% |
| $Rh^-$ | 15% | 11.0% | 15% |
| **MN System** | | | |
| M | 30% | 44.5% | 30% |
| MN | 50% | 42.0% | 50% |
| N | 20% | 13.5% | 20% |

**table 14.3**

**Inherited Conditions Common Among the Amish and Mennonites of Lancaster County, Pennsylvania**

| Illness | Symptoms |
|---|---|
| Ataxia telangiectasia | Increased sensitivity to radiation, loss of balance and coordination, red marks on face, delayed sexual maturation, lung infections, diabetes, high risk of cancer |
| Bipolar affective disorder | Mood swings (manic depression) |
| Cartilage-hair hypoplasia (metaphyseal chondrodysplasia, McKusick type) | Dwarfism, sparse hair, anemia, poor immunity |
| Crigler-Najjar syndrome | Bilirubin buildup, jaundice, brain damage |
| Ellis-van Creveld syndrome | Dwarfism, short fingers, abnormal teeth, underdeveloped nails, polydactyly, hair "blaze" pattern |
| Glutaric aciduria type I | Paralysis, brain damage |
| Homocystinuria | Damaged blood vessels, stroke, heart attack |
| Limb-girdle muscular dystrophy | Progressive muscle weakness in limbs |
| Maple syrup urine disease | Sweet-smelling urine, sleepiness, vomiting, mental retardation |
| Metachromatic leukodystrophy | Rigid muscles, convulsions, mental deterioration |
| Morquio syndrome | Clouded corneas, abnormal skeleton and aortic valve |

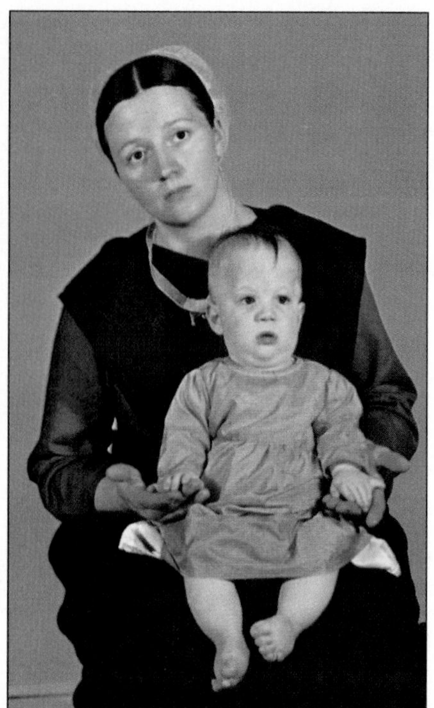

**figure 14.2**

**Ellis-van Creveld syndrome.** This Amish child has inherited Ellis-van Creveld syndrome. He has short-limbed dwarfism, extra fingers, heart disease, fused wrist bones, and had teeth at birth. The condition is autosomal recessive. The mutant allele occurs in 7 percent of the people of this community, reflecting consanguinity. Homozygous recessive individuals have this severe dwarfism, but heterozygotes have the milder condition Weyers acrodental dysostosis. These were thought to be different disorders until the gene was discovered in early 2000.

Until McKusick made the connection, other symptoms—including dwarfism, sparse hair, and anemia—were not recognized as part of a syndrome. Today, as many geneticists study inherited diseases common among the Amish and Mennonites, treatments are becoming available, from special diets to counter inborn errors of metabolism, to the "bili lights" that treat children with Crigler-Najjar syndrome (see figure 13.2), to gene therapy.

In addition to historical records, raw numbers provide evidence of a founder effect when allele frequencies in the smaller population are compared to those in the general population. The incidence of certain diseases in Lancaster County is astounding. Maple syrup urine disease, for example, affects 1 in 225,000 newborns in the United States, but 1 in 400 newborns among the Lancaster families! Similarly, a doctor new to the community involved in a gene therapy project was startled when he took a walk at night, and noted house after house where an eerie blue glow emanated from a window. Inside each house, a child with Crigler-Najjar syndrome was being treated with bili lights. In another example, in 1989, D. Holmes Morton, a research fellow at Children's Hospital in Philadelphia, discovered that several young children from Lancaster County with cerebral palsy presumably caused by oxygen deprivation at birth instead had an inborn error of metabolism called glutaric aciduria type I. Morton went from farm to farm, tracking cases against genealogical records, and found that every family that could trace its roots back to the founders had the disease! Today, 0.5 percent of newborns in this population have the condition.

Strong evidence of a founder effect is a mutation that is the same in all affected individuals in a population. The Bulgarian gypsies who have galactokinase deficiency, for example, all have a mutation that is extremely rare elsewhere. Similarly, an isolated group of Jewish people who settled in the Belmonte area of Portugal after being driven from Spain in the late 1400s has its own, personal form of retinitis pigmentosa, which causes blindness. Their gene maps to a different chromosome than the genes that cause other forms of the illness.

Very often a disease-associated allele is the same base-for-base among people in the same population, and so is the DNA surrounding the gene. This pattern indicates that a portion of a chromosome has been passed among the members of the population from the founders. For this reason, many studies that trace founder effects examine haplotypes, which indicate very tightly linked genes that reflect linkage disequilibrium.

When historical or genealological records are particularly well kept, founder effects can sometimes be traced to the very beginning. This is the case for the Afrikaner population of South Africa. The 2.5 million Afrikaners descended from a small group of Dutch, French, and German immigrants who had huge families, often with as many as 10 children. In the 19th century, some Afrikaners migrated northeast to the Transvaal Province, where they lived in isolation until the Boer War in 1902 resulted in the introduction of better transportation.

Today, 30,000 Afrikaners have porphyria variegata, an autosomal dominant deficiency of one of the enzymes required to manufacture heme, the iron-containing part of hemoglobin. Symptoms include nervous attacks, abdominal pain, very fragile and sun-sensitive skin, and a severe reaction to a particular barbiturate anesthetic. All affected people descended from one couple who came from Holland in 1688! Today's allele frequency in South Africa is far higher than that in the Netherlands because this couple had many children—who, in turn, had large families, passing on and amplifying the dominant gene.

In a similar extreme example of a founder effect, all of the cases of polydactyly (extra digits) among the Old Order Amish in Lancaster stem from one founder (see figure 14.2). Today, thanks to large families and restricted marriages, the number of cases of polydactyly among the Amish exceeds the total number in the rest of the world!

Founder effects are also evident in more common illnesses, where different populations may have different mutations in the same gene. This is the case for BRCA1 breast cancer. The disease is most prevalent among Jewish people of central and eastern European background, the Ashkenazim. Nearly all affected individuals in this population have the same point mutation. In contrast, BRCA1 breast cancer is quite rare in blacks, but it has affected families from the Ivory Coast in Africa, the Bahamas, and the southeastern United States. These families all share a 10-base deletion mutation in the BRCA1 gene, probably inherited from West Africans who were ancestors of all three modern groups. Slaves brought the disease to the United States and the Bahamas between 1619 and 1808, but some of their relatives who stayed in Africa have perpetuated the gene there as well.

## Population Bottlenecks

A **population bottleneck** occurs when many members of a group die, and only a few are left, by chance, to replenish the numbers. A bottleneck is genetically significant because the new population has only those alleles present in the small group that survived the catastrophe. An allele present in the small remnant population might become more common in the replenished population than it was in the larger group. Therefore, the new population has a much more restricted gene pool than the larger ancestral population.

Population bottlenecks sometimes occur when people (or other animals) colonize islands. An extreme example is seen among the Pingelapese people of the eastern Caroline islands in Micronesia. Between 4 and 10 percent of them are born with "Pingelapese blindness," an autosomal recessively inherited combination of color blindness, nearsightedness, and cataracts. It is also called achromatopsia. Nearly 30 percent of the Pingelapese are carriers. Elsewhere, only 1 in 20,000 to 50,000 people inherits the condition. The prevalence of this condition among the Pingelapese has been traced to a typhoon that decimated the population in 1780. Only 9 males and 10 females survived, and they founded the present-day population. The effects of this severe population bottleneck, combined with geographic and cultural isolation, resulted in the high frequency of the blindness gene.

Figure 14.3 illustrates schematically the dwindling genetic diversity that results from a population bottleneck, against a backdrop of a cheetah. Once 10,000 animals strong and widespread, today's cheetahs live in just two isolated populations of a few thousand animals in South and East Africa. The South African cheetahs are so alike genetically that even unrelated animals can accept skin grafts from each other. Researchers attribute the cheetahs' genetic uniformity to two bottlenecks—one that occurred at the end of the most recent ice age, when habitats were altered, and another that involved mass slaughter by humans in the 19th century. However, the good health of the animals today indicates that the genes that have survived enable the cheetahs to thrive in their environment.

Human-wrought disasters that kill many people can also cause population bottlenecks—perhaps even more severely, because aggression is typically directed at particular groups, while a typhoon indiscriminately kills whoever is in its path. The Chmielnicki massacre is an example of one of the many attacks aimed against the Jewish people. Overall, these acts have left a legacy of several inherited diseases that are at least 10 times more common among Jewish people than in other populations (table 14.4).

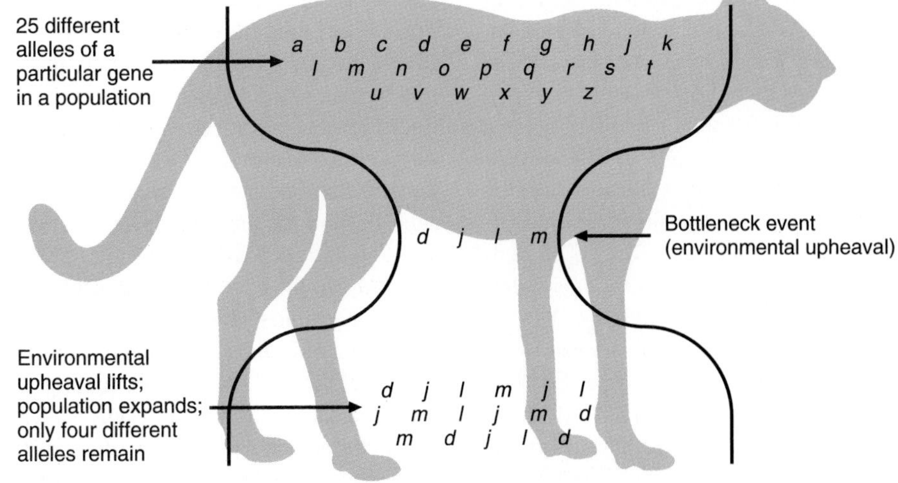


## figure 14.3

**Population bottlenecks.** A population bottleneck occurs when the size of a genetically diverse population drastically falls, remains at this level for a time, and then expands again. The new population loses some genetic diversity if different alleles are lost in the bottleneck event. Cheetahs are difficult to breed in zoos because sperm quality is poor and many newborns die—both due to lack of genetic diversity.

## table 14.4

**Autosomal Recessive Genetic Diseases Prevalent Among Ashkenazi Jewish Populations**

| Condition | Symptoms | Carrier Frequency |
|---|---|---|
| Bloom syndrome | Sun sensitivity, short stature, poor immunity, impaired fertility, increased cancer risk | 1/110 |
| Breast cancer | Malignant breast tumor caused by mutant BRCA1 or BRCA2 genes | 3/100 |
| Canavan disease | Brain degeneration, seizures, developmental delay, death by 18 months of age | 1/40 |
| Dysautonomia | No tears, cold hands and feet, skin blotching, drooling, difficulty swallowing, excess sweating | 1/32 |
| Gaucher disease | Enlarged liver and spleen, bone degeneration, nervous system impairment | 1/12 |
| Niemann-Pick disease type A | Lipid accumulation in cells, particularly in the brain; mental and physical retardation, death by age three | 1/90 |
| Tay-Sachs disease | Brain degeneration causing developmental retardation, paralysis, blindness, death by age four | 1/26 |
| Fanconi anemia type C | Deficiencies of all blood cell types, poor growth, increased cancer risk | 1/89 |

The Chmielnicki massacre began in 1648, when a Ukrainian named Bogdan Chmielnicki led a massacre against the Polish people, including peasants, nobility, and the Jewish people, in retaliation for a Polish nobleman's seizure of his possessions. By 1654, Russians, Tartars, Swedes, and others joined the Ukrainians in wave after wave of violence against the Polish people. Thousands perished, with only a few thousand Jewish people remaining.

The Jewish people have survived many massacres; after this time, like the others, their numbers grew again. From 1800 to 1939, the Jewish population in Eastern Europe swelled to several million. But the Chmielnicki massacre, like others, changed allele frequencies and contributed to the high incidence of certain inherited diseases among people of eastern European Jewish heritage.

**Key Concepts**

Genetic drift occurs when a subset of a population contains different gene frequencies than the larger population because it is a small sample. • The founder effect occurs when a few individuals leave a community to start a new settlement. The resulting population may, by chance, either lack some alleles present in the original population or have high frequencies of others. • In a population bottleneck, many members of a population die, and only a few individuals contribute genetically to the next generation.

## 14.4 Mutation

A major and continual source of genetic variation is mutation—when one allele changes into another. (Chapters 11 and 12 discussed mutation.) Genetic variability also arises from crossing over and independent assortment during meiosis, but these events recombine existing traits rather than introduce new ones. If a DNA base change occurs in a part of a gene that encodes a portion of a protein necessary for its function, then an altered trait may result. If the mutation is in a gamete, then the change can pass to future generations and therefore ultimately affect an allele's frequency in the population.

Natural selection eliminates deleterious alleles that are expressed in the phenotype and affect an individual's ability to reproduce. Yet at the same time, these alleles are maintained in heterozygotes, where they do not exert a noticeable effect, and are reintroduced by new mutation. Therefore, all populations have some alleles that would be harmful if homozygous. The collection of such deleterious alleles in a population is called its **genetic load.**

Overall, the contribution that mutation makes to counter Hardy-Weinberg equilibrium is quite small compared to the influence of migration and nonrandom mating. The spontaneous mutation rate is only about 30 bases per haploid genome in each gamete. Each of us probably has only 5 to 10 recessive lethal alleles. Fortunately, most mutations do not alter the phenotype due to the degeneracy of the genetic code and changes that do not alter protein function. Some mutations actually increase the chance of survival, such as the CCR5 mutation that blocks infection by HIV (see figure 16.14).

**Key Concepts**

Mutation alters gene frequencies by introducing new alleles. Heterozygotes and mutations maintain the frequencies of deleterious alleles in populations, even if homozygotes die.

## 14.5 Natural Selection

Environmental change can alter allele frequencies when individuals with certain phenotypes are more likely to survive and reproduce than others. This differential survival based on phenotype, and therefore genotype, is called **natural selection.** It may be negative—removal of alleles—or positive—retention of alleles.

The effect of natural selection on allele frequencies is vividly revealed by the changing virulence of certain human infectious diseases over many years. Two such illnesses are tuberculosis and AIDS.

### Tuberculosis Ups and Downs—and Ups

The spread of tuberculosis (TB) in the Plains Indians of the Qu'Appelle Valley Reservation in Saskatchewan, Canada, illustrates natural selection. When TB first appeared on the reservation in the mid-1880s, it struck swiftly and lethally, infecting many organs. Ten percent of the population died. But by 1921, TB in the Indians tended to affect only the lungs, and only 7 percent of the population died annually from it. By 1950, mortality was down to 0.2 percent.

Outbreaks of TB ran similar courses in other human populations. The disease appeared in crowded settlements where the bacteria easily spread from person to person in exhaled droplets. In the 1700s, TB raged through the cities of Europe. Immigrants brought it to the United States in the early 1800s, where it also swept the cities. Many people thought TB was hereditary until German bacteriologist Robert Koch identified the causative bacterium in 1882.

As in the Plains Indians, TB incidence and virulence fell dramatically in the cities of the industrialized world in the first half of the 20th century—before antibiotic drugs were discovered. What tamed tuberculosis?

Natural selection, operating on both the bacterial and human populations, lessened the virulence of the infection. Some people inherited resistance and passed this beneficial trait on. At the same time, the most virulent bacteria killed their hosts so quickly that the victims had no time to spread the infection. As the deadliest bacteria were selected out of the population (negative selection), and as people who inherited resistance mutations contributed disproportionately to the next generation (positive selection), TB gradually evolved from a severe, acute, systemic infection to a rare chronic lung infection. This was true until the late 1980s.

A series of recent unrelated events has created conditions just right for the resurgence of TB, but in a form that is resistant to many of the 11 drugs used to treat it. Some health officials trace the return of tuberculosis to complacency. Researchers turned to other projects when funding became scarce for this seemingly controlled disease. Patients became complacent, too. When antibiotics eased symptoms in two to three months, patients felt cured and stopped taking the drugs, even though they unknowingly continued to spread live bacteria for up to 18 months. The *Mycobacterium tuberculosis* bacteria had time to mutate, and mutant strains to flourish, eventually evolving the drug resistances that make the newest cases so difficult to treat (Reading 14.1). Tuberculosis treatment in the 1950s was actually more effective; patients were isolated for a year or longer in rest homes called sanitaria and were not released until the bacteria were gone (figure 14.4).

Today, 1 in 7 new tuberculosis cases is resistant to several drugs, and 5 percent of these patients die. People living in crowded, unsanitary conditions with poor health care are especially susceptible to drug-resistant TB. Another reservoir of new TB infection is persons infected with HIV. Tuberculosis develops so quickly in people with suppressed immunity that it can kill before physicians have determined which drugs to use—physicians have died of the infection while treating patients. The resurgence of TB should remind us never to underestimate the fact that evolution operates in all organisms—and does so unpredictably.

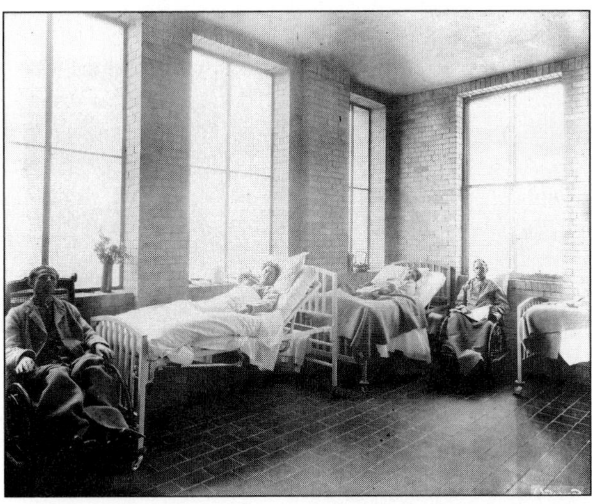

**figure 14.4**

**Tuberculosis evolves.** Earlier in the century, society controlled tuberculosis by isolating infected people in sanitaria until they could no longer pass the infection on.

## Evolving HIV

Because the RNA or DNA of viruses replicates often and is not repaired, viral mutations accumulate rapidly. Like bacteria, the viruses in a human body form a population, including naturally occurring genetic variants. In HIV infection, natural selection controls the diversity of HIV genetic variants within a human body as the disease progresses. The environmental factors that select (favor) resistant viral variants are the human immune system and the drugs we use to slow the infection.

HIV infection can be divided into three stages, both from the human and the viral perspective (figure 14.5). A person infected with HIV may experience an initial acute phase, with symptoms of fever, night sweats, rash, and swollen glands. In a second period, lasting from 2 to 15 years, health usually returns. In a third stage, immunity collapses, the virus replicates explosively, and opportunistic infections and cancer eventually cause death.

The HIV population changes and expands throughout the course of infection, even when the patient seems to remain the same for long periods of time. New mutants continuously arise, and they alter such traits as speed of replication and the patterns of molecules on the viral surface.

In the first stage of HIV infection, as the person battles acute symptoms, viral variants that can replicate very fast predominate. Soon, most of the viral particles have the same rapidly multiplying genotype. In the second stage, the immune system starts to fight back and symptoms abate, as viral replication slows and many viruses are destroyed. Now natural selection acts—those viral variants that persist reproduce and mutate, giving rise to a diverse viral population. Ironically, drugs used to treat AIDS may further select against the weakest HIV variants, inadvertently supporting the resistant variants. Gradually, the HIV population overtakes the number of immune system cells the body produces to counter it. Still, the immune system is so vast that years may pass until immunity begins to noticeably decline.

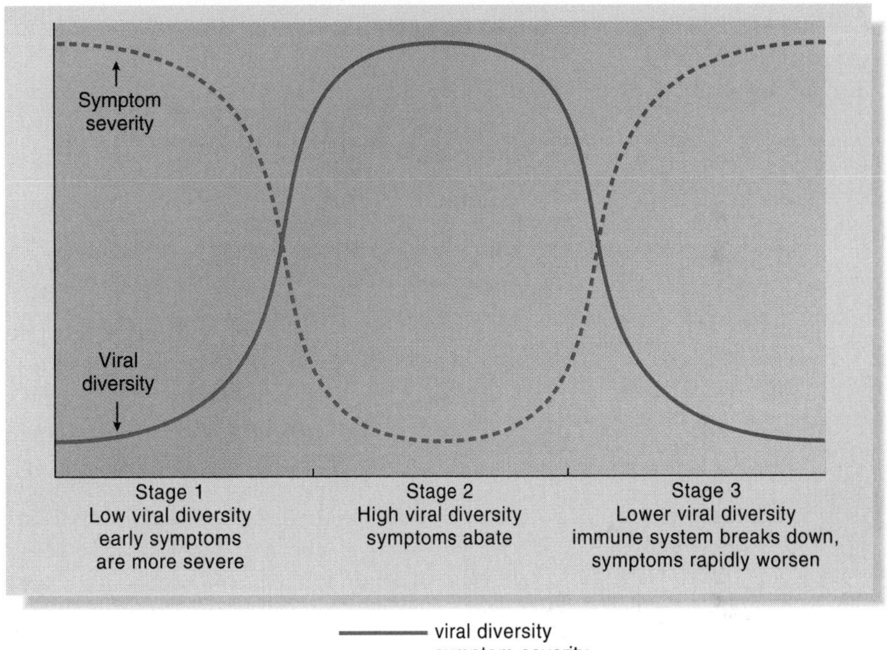


**figure 14.5**

**Natural selection of HIV.** Natural selection controls the genetic diversity of an HIV population in a person's body. Before the immune system gathers strength, and after it breaks down, HIV diversity is low. A rapidly reproducing viral strain predominates, although new mutations continually arise. During the 2- to 15-year latency period, viral variants that can evade the immune system gradually accumulate. Eventually, HIV overpowers the immune system. Evolution renders the body's fight against HIV a losing battle—unless the person is one of the lucky few to inherit a mutation that blocks HIV infection.

*Reading 14.1*

# Antibiotic Resistance—Stemming a Biological Arms Race

The sudden appearance of an infectious disease often reflects evolution. Natural selection may favor persistence of a particular pathogen under certain environmental conditions, or a bacterium or virus may be brought to a new part of the world, where host populations have not evolved resistances to it. Infectious diseases that were once prevalent are resurging, including diphtheria, tuberculosis, dengue, cholera, and yellow fever (figure 1).

Mutation also plays a role in the appearance or resurgence of an infectious disease. Bacteria and viruses can mutate to produce a new toxin, infect new species, or exist in novel places. Changes in weather patterns caused by global warming may be responsible for some recent shifts in infectious disease patterns. Unfortunately, we cannot do much to prevent mutation—it is a consequence of DNA replication. We can, however, intervene at the level of natural selection, countering environmental changes that permit disease-causing mutations to persist.

New infections appear and old ones return for these two major reasons—natural selection and mutation. The current crisis of infectious bacteria that are resistant to

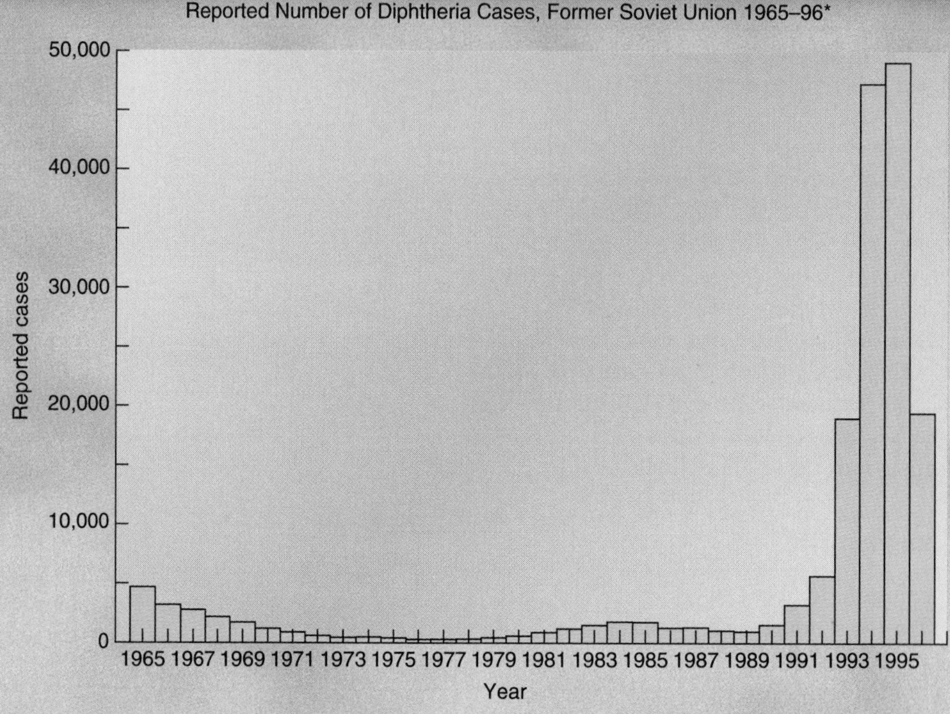


**figure 1**

**Diphtheria—an old foe returns.** In the former Soviet Union, incomplete vaccination of the population caused the reemergence of a toxic strain of diphtheria. Crowded conditions and forced migrations helped spread the infection, which the World Health Organization calls "an international public health emergency." Diphtheria causes a white coating of the throat, along with a fever and cough; it may also affect the skin, heart, kidneys, and nervous tissue.

*Source: World Health Organization, as reported in *Morbidity and Mortality Weekly Report,* March 17, 1995, p. 177, U.S. Department of Health and Human Services.

The third stage, full-blown AIDS, occurs when the number of viruses reaches a level that overwhelms the immune system. Now, with the selective pressure off, viral diversity again diminishes, and the fastest-replicating variants predominate. HIV wins. The entire scenario of HIV infection reflects the value of genetic diversity—to enable the survival of a population or species in the face of an environmental threat. When that threat—an immune attack or drugs—diminishes, one genotype may come to prevail.

Knowing that HIV diversifies early in the course of infection has yielded clinical benefits. Patients now take combinations of drugs that act in different ways to squelch several viral variants simultaneously, slowing the course of the infection. Chapter 16 considers HIV infection further.

## Balanced Polymorphism

If natural selection eliminates individuals with detrimental phenotypes from a population, then how do harmful mutant alleles persist in a gene pool? Such alleles are introduced by new mutation, and they persist in heterozygotes. Sometimes, a recessive condition remains prevalent because the heterozygote enjoys some unrelated health advantage, such as being resistant to an infectious disease or able to survive an environmental threat. This "heterozygous advantage" that maintains a recessive, disease-causing allele in a population is called **balanced polymorphism.** Recall that *polymorphism* means variant; the effect is *balanced* because the protective effect of the noninherited condition counters the negative effect of the deleterious allele, maintaining its frequency in the population. Table 14.5 lists some examples of disease-causing alleles associated with resistance to certain conditions.

### *Sickle Cell Disease and Malaria*

Recall that sickle cell disease is an autosomal recessive disorder that causes anemia, joint pain, a swollen spleen, and frequent, severe infections (see figure 2.1*c*). It illustrates antibiotics illustrates the interplay of mutation and natural selection in the spread of infectious disease.

The bacteria in a human body constitute a population of organisms, and as such they include different genetic variants. Some bacterial strains inherit the ability to survive in the presence of a particular antibiotic drug. When the bacteria infect a person, the immune system responds, causing such symptoms as inflammation and fever in response. The person goes to the doctor, and a course of antibiotics seems to help; but a month later, symptoms return. What has happened?

The drug probably killed most of the bacteria, but a few survived because they have a fortuitous (for them) mutation that enables them to withstand the antibiotic assault. Over a few weeks, as sensitive bacteria die, those few mutants reproduce, taking over the niche the antibiotic-sensitive bacteria vacated. Soon, the person has enough antibiotic-resistant bacteria to feel ill again. The next step is to try a drug that works differently—and hope that the bacteria haven't mutated around that one, too. Antibiotic resistance is, in a sense, a biological arms race.

Antibiotic drugs do not cause mutations in bacteria; they select for preexisting resistant variants. (This doesn't happen in viruses because viruses are not cells, and antibiotics have no effect on them.) Bacteria with drug-resistance mutations circumvent antibiotic actions in several ways. Penicillin kills bacteria by tearing apart their cell walls. Resistant microbes produce enzyme variants that dismantle penicillin or have altered cell walls that the drug cannot bind to. Erythromycin, streptomycin, tetracycline, and gentamicin kill bacteria by attacking their ribosomes, which are different from ribosomes in a human. Drug-resistant bacteria have altered ribosomes that the drugs cannot bind.

Bacteria acquire antibiotic resistance in several ways. Their DNA may spontaneously mutate—this is how drug-resistant tuberculosis arose. They may receive a resistance gene from another bacterium in a form of microbial sex called transformation. This is how the sexually transmitted disease gonorrhea gained resistance to penicillin. Bacteria can acquire resistance to several drugs at once by taking up a small circle of DNA, called a plasmid, from another bacterium. Not only can plasmids transmit multiple drug resistances, but they flit freely from one species of bacteria to another.

Practical steps to halt the growing threat of antibiotic-resistant bacteria include:

- Limiting antibiotic use to bacterial infections diagnosed by a physician.
- Taking the full schedule of doses so that the infection does not return in a drug-resistant form.
- Developing new antibiotic drugs. Many models for new pharmaceuticals come from natural products. Therefore, preserving natural habitats can help to preserve potential sources of new anti-infective agents.
- Improving public health measures. These include having health care workers wash their hands frequently and thoroughly, rapidly identifying and isolating patients with drug-resistant infections, improving sewage systems and water purity, and using clean needles.

**table 14.5**

**Balanced Polymorphism**

| Inherited Disease | Infectious Disease | Possible Mechanism of Heterozygote Advantage |
|---|---|---|
| Sickle cell disease | Malaria | Red blood cells inhospitable to malaria parasite |
| G6PD deficiency | Malaria | Red blood cells inhospitable to malaria parasite |
| Phenylketonuria (PKU) | Spontaneous abortion | Excess amino acid (phenylalanine) in carriers inactivates ochratoxin A, a fungal toxin that causes miscarriage |
| Tay-Sachs disease | Tuberculosis | Unknown |
| Cystic fibrosis | Diarrheal diseases | Carriers have too few chloride channels in intestinal cells, blocking cholera toxin (cholera) entry, or CFTR protein is a receptor for *Salmonella typhi* |

balanced polymorphism because carriers are resistant to malaria, an infection by the parasite *Plasmodium falciparum* that causes debilitating cycles of chills and fever. The parasite spends the first stage of its life cycle in the salivary glands of the mosquito *Anopheles gambiae.* When an infected mosquito bites a human, malaria parasites enter red blood cells, which transport them to the liver. The red blood cells burst, releasing parasites throughout the body.

In people with sickle cell disease, many of the red blood cells burst prematurely, which expels the parasites before they can cause rampant infection. The blood of a person with sickle cell disease is also thicker than normal, which may hamper the parasite's ability to infect. A carrier's blood is abnormal enough to be inhospitable to the malaria parasite.

A clue to the protective effect of sickle cell disease heterozygosity came from striking differences in the incidence of the two diseases in different parts of the world (figure 14.6). In the United States, 8 percent of African Americans are sickle cell carriers, whereas in parts of Africa, up to 45 to 50 percent are carriers. In 1949, British geneticist Anthony Allison found that the frequency of sickle cell carriers in tropical Africa was higher in regions where malaria rages all year long. Blood tests from children hospitalized with malaria found that nearly all were homozygous for the wild type sickle cell allele. The few sickle cell carriers among them had the mildest cases of malaria. Was malaria enabling the sickle cell allele to persist by felling people who did not inherit it? The fact that sickle cell disease is rarer where malaria is rare supports the idea that sickle cell heterozygosity protects against the infection.

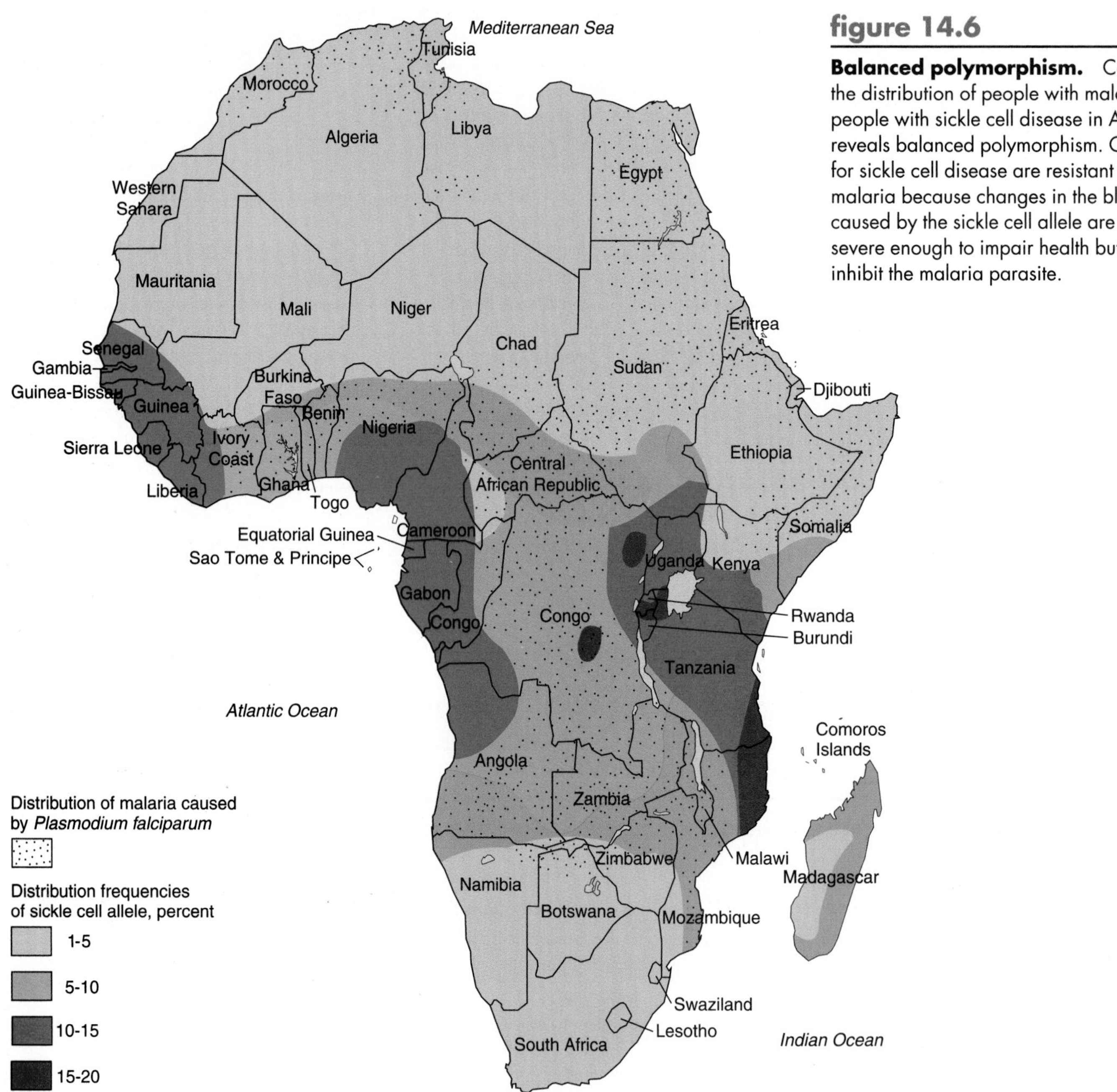


## figure 14.6

**Balanced polymorphism.** Comparing the distribution of people with malaria and people with sickle cell disease in Africa reveals balanced polymorphism. Carriers for sickle cell disease are resistant to malaria because changes in the blood caused by the sickle cell allele are not severe enough to impair health but do inhibit the malaria parasite.

The sickle cell gene may have been brought to Africa by people migrating from Southern Arabia and India, or it may have arisen by mutation directly in East Africa. However it happened, people who inherited one copy of the sickle cell allele survived or never contracted malaria. These carriers had more children and passed the protective allele to approximately half of them. Gradually, the frequency of the sickle cell allele in East Africa rose from 0.1 percent to 45 percent in 35 generations. Carriers paid the price for this genetic protection, however, whenever two of them produced a child with sickle cell disease.

### *Glucose-6-Phosphate Dehydrogenase Deficiency and Malaria*

Recall from chapter 11 that G6PD deficiency is an X-linked recessive enzyme deficiency that causes life-threatening hemolytic anemia under specific conditions, such as eating fava beans or taking certain drugs. Among African children with severe malaria, heterozygous females ($X^GX^g$) and affected (hemizygous) males ($X^gY$) for G6PD deficiency are underrepresented. This suggests that carrying or inheriting G6PD deficiency protects against malaria. Cell culture studies confirm that the parasite enters red blood cells of carriers or affected males but cannot reproduce to cause infection.

The fact that G6PD deficiency is X-linked introduces a possibility we do not see with sickle cell disease, which is autosomal recessive. Because in G6PD deficiency heterozygotes and hemizygotes (males with the disease) have an advantage, the mutant allele should eventually predominate in a malaria-exposed population as homozygotes and hemizygotes for the normal allele die of malaria. However, this doesn't happen—there are still males hemizygous and females homozygous for the normal allele. The reason again relates to natural selection.

Table 14.6 shows how natural selection acts in two directions on the two types of hemizygous males—selecting for the mutant allele because it protects against malarial infection, yet also selecting for the normal allele because it protects against an enzyme deficiency. This is the essence of balanced polymorphism.

Although the exact way that G6PD protects against malaria isn't known, studies of different mutations in different populations confirm that the beginning of a mutation's prevalence coincides with the onset of agriculture. These protective mutations are only seen where agriculture and malaria coexist—that is, the mutations do not accumulate where there is farming but not malaria. One researcher wrote that G6PD and malaria offer "a striking example of the signature of selection on the human genome."

### *PKU and Fungal Infection*

In phenylketonuria (PKU), a missing enzyme causes the amino acid phenylalanine to build up, which devastates the nervous system unless the individual follows a restrictive diet. Carriers of this autosomal recessive condition have elevated phenylalanine levels—not sufficient to cause symptoms, but high enough to be protective during pregnancy. Women who are PKU carriers have a much lower-than-average incidence of miscarriage. One explanation is that excess phenylalanine inactivates a poison, called ochratoxin A, that certain fungi produce and that is known to cause spontaneous abortion.

History provides the evidence that links PKU heterozygosity to protection against a fungal toxin. PKU is most common in Ireland and western Scotland, and many affected families living elsewhere trace their roots to this part of the world. PKU spread eastward in Europe when the Vikings brought wives and slaves back from the Celtic lands. In the moist environment of Ireland and Scotland, the fungi that produce ochratoxin A—*Aspergillis* and *Penicillium*—grow on grains. During the famines that have plagued these nations, starving people ate moldy grain. If PKU carriers were more likely to have children than noncarriers because of the protective effects of the PKU gene, over time, the disease-causing allele would increase in prevalence in the population.

### *Tay-Sachs Disease and Tuberculosis*

Being a carrier for Tay-Sachs disease may protect against tuberculosis. During World War II, TB ran rampant in eastern European Jewish settlements. Often, healthy relatives of children who had Tay-Sachs disease did not contract TB, even when repeatedly exposed. The protection against TB that Tay-Sachs disease heterozygosity apparently offered remained among the Jewish people because they were prevented from leaving the ghettos, creating a sort of enforced founder effect. The mutant Tay-Sachs allele increased in frequency as TB selectively killed those who did not carry it, and carriers had children with each other. Precisely how lowered levels of the Tay-Sachs gene product, an enzyme called hexoseaminidase A, protect against TB isn't known.

### *Cystic Fibrosis and Diarrheal Disease*

Balanced polymorphism may explain why cystic fibrosis is so common—the cellular defect that underlies CF protects against diarrheal illnesses, such as cholera and typhus.

Diarrheal disease epidemics have left their mark on many human populations. Severe diarrhea rapidly dehydrates the body and leads to shock, kidney and heart failure, and death in days. In cholera, bacteria produce a toxin that opens chloride channels in cells of the small intestine. As salt (NaCl) leaves the intestinal cells, water rushes out, producing diarrhea. Cholera opens chloride channels, releasing chloride and water. The CFTR protein does just the opposite, closing chloride channels and trapping salt and water

**table 14.6**

**G6PD Deficiency Protects Against Malaria**

| Genotypic Class | Enzyme Deficiency | Malaria Susceptibility |
|---|---|---|
| Normal male $X^GY$ | no | yes |
| G6PD male $X^gY$ | yes | no |
| Heterozygous female $X^GX^g$ | no | no |
| Homozygous female $X^GX^G$ | no | yes |
| Homozygous female $X^gX^g$ | yes | no |

in cells, which dries out mucus and other secretions. A person with CF is very unlikely to contract cholera, because the toxin cannot open the chloride channels in the small intestine cells.

Carriers of CF enjoy the mixed blessing of balanced polymorphism. They do not have enough abnormal chloride channels to cause the labored breathing and clogged pancreas of cystic fibrosis, but they do have enough of a defect to prevent the cholera toxin from taking hold. During the devastating cholera epidemics that have occurred throughout history, individuals carrying mutant CF alleles had a selective advantage, and they disproportionately transmitted those alleles to future generations.

However, because CF arose in western Europe and cholera originated in Africa, perhaps an initial increase in CF heterozygosity was a response to a different diarrheal infection. This disease may have been typhoid fever. The causative bacterium, *Salmonella typhi,* rather than a toxin, enters cells lining the small intestine—but only if functional CFTR channels are present (see figure 2.1*b*). The cells of people with severe CF manufacture CFTR proteins that never reach the cell surface, and therefore no bacteria get in. Cells of CF carriers admit some bacteria.

### *Diabetes Mellitus and Surviving Famine*

Type II (non-insulin-dependent) diabetes mellitus (NIDDM) is a gradual failure of cells to respond to insulin and take up glucose from the bloodstream. It is a complex trait—a first-degree relative has a tenfold risk of developing the condition, but studies on many populations indicate that one can inherit a susceptibility to develop NIDDM that becomes reality in the face of an unhealthy diet and lack of exercise. Arizona's Pima Indians, discussed in chapter 7, show this pronounced effect of environment on genotype.

Some geneticists suggest that NIDDM is common today because the gene or genes that predispose to it might once have been beneficial, a variation on the balanced polymorphism theme. The reasoning is that NIDDM prevents the breakdown of fat and alters the body's ability to store glucose. In times past, extra fat stores and altered glucose metabolism were insurance for surviving famine. Today, they cause weight gain and diabetes. Genetic predisposition to developing diabetes is sometimes called a "thrifty genotype."

### *P-Glycoprotein and Resistance to AIDS Drugs*

Selection is a response to environmental change. When an environment changes, a trait that has been selected for may no longer be advantageous. This is the case for a gene, MDR1, that encodes the protein portion of a glycoprotein (called P-glycoprotein) that dots the surfaces of intestinal lining cells and T lymphocytes. An allele that causes overexpression of the gene became prevalent in some populations because the gene's product enables cells to pump out poisons. But in the different circumstance of intentionally taking a poison—specifically, a drug to combat cancer—these pumps were no longer valuable. A population-based study reveals that variants in this gene may explain why drugs to fight HIV infection are less effective in Africans than in other groups.

Researchers identified a polymorphism in an exon of the MDR1 gene that correlates with overexpression, creating an allele called C. Another allele that leads to very few P-glycoproteins is called T. The resulting genotypes and phenotypes are therefore:

| | |
|---|---|
| T/T | (very few P-glycoproteins) |
| T/C | (intermediate number of P-glycoproteins) |
| C/C | (many P-glycoproteins) |

Researchers genotyped 172 individuals from West Africa, 41 African Americans, 537 Caucasians, and 50 Japanese, all healthy. The results were startling (figure 14.7). The C/C genotype is clearly overrepresented among the West Africans (83%) and the African Americans (61%), compared to Caucasians (26%) and Japanese (34%). The finding makes sense in terms of natural selection—the C/C genotype confers resistance to many bacteria and viruses that cause gastroenteritis, a major killer of children in Africa. But the same genotype ejects chemotherapeutic drugs, AIDS drugs, and antirejection transplant drugs. Using this information, physicians will be able to identify people

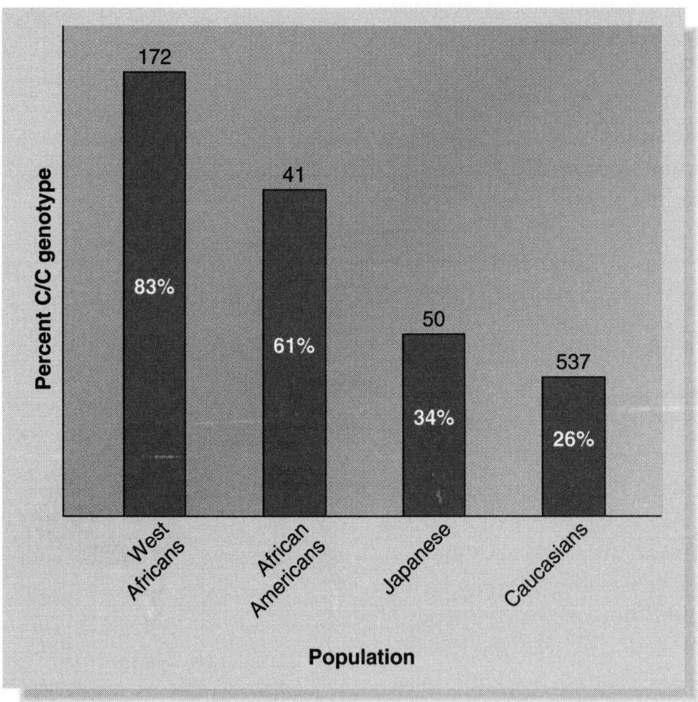


**figure 14.7**

**Resistance genotypes.** An advantageous trait can turn disadvantageous when the environment changes. The C/C genotype of the MDR1 gene enables a person's cells to pump out poisons. The genotype became prevalent in populations with African ancestry, possibly because it was protective against viral and bacterial infections of the gastrointestinal tract in the African environment. Today, the same genotype may hamper the efficacy of AIDS drugs.

with the C/C genotype, and perhaps give them higher doses of certain drugs.

The forces of nonrandom mating, migration, genetic drift, mutation, and natural selection interact in complex ways. Figure 14.8 reviews and summarizes the forces that alter gene frequencies and therefore impact on evolution. Reading 14.2 looks at intentional shrinking of gene pools—the "artificial selection" used to breed cats and dogs.

## Key Concepts

Because of natural selection, different alleles are more likely to confer a survival advantage in different environments. Cycles of infectious disease prevalence and virulence often reflect natural selection. • Balanced polymorphism is a type of natural selection in which a particular disease-causing allele is maintained or increases in a population because heterozygotes resist a certain infectious illness or environmental condition.

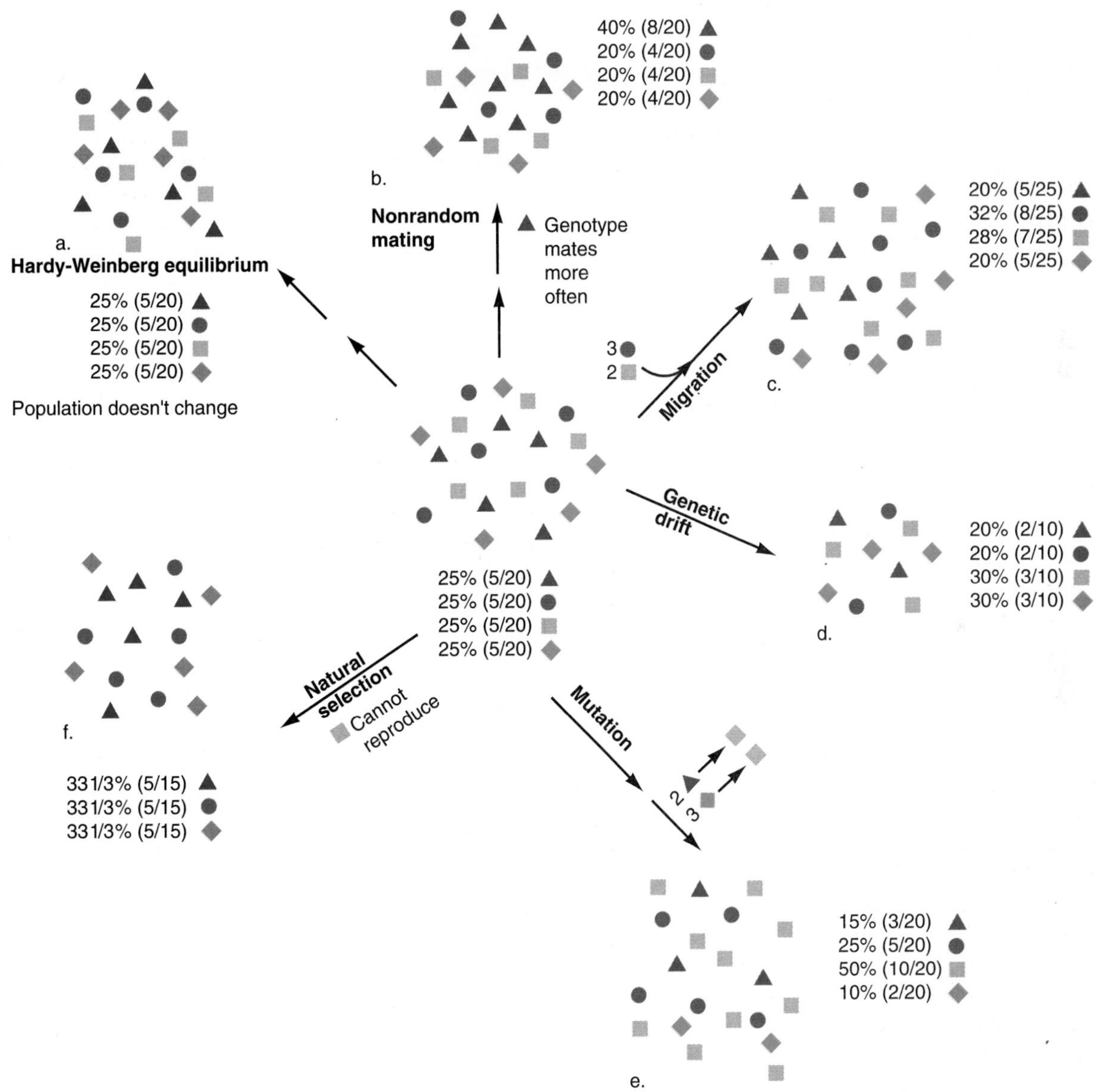


## figure 14.8

**Forces that change allele frequencies.** Several factors alter allele frequencies and thereby contribute to microevolution. The different-colored shapes represent individuals with distinctive genotypes. (*a*) In Hardy-Weinberg equilibrium, allele frequencies stay constant. (*b*) Nonrandom mating increases some allele frequencies and decreases others because individuals with certain genotypes mate more often than others. (*c*) Migration removes alleles from or adds alleles to populations. (*d*) Genetic drift samples a portion of a population, altering allele frequencies. (*e*) Mutation changes some alleles into others. (*f*) Natural selection operates when environmental conditions lower the probability that individuals of certain genotypes can reproduce.

*Reading 14.2*

# Dogs and Cats: Products of Artificial Selection

The pampered poodle may win in the show ring, but it is a poor specimen in terms of genetics and evolution. Human notions of attractiveness in pets can lead to breeds that might never have evolved naturally. Behind carefully bred traits lurk small gene pools and extensive inbreeding—all of which spell disaster to the health of many highly prized and highly priced show animals. Purebred dogs suffer from more than 300 types of inherited disorders!

The sad eyes of the basset hound make him a favorite in advertisements, but his runny eyes can be quite painful. His short legs make him prone to arthritis, his long abdomen encourages back injuries, and his characteristic floppy ears often hide ear infections. The eyeballs of the Pekingese protrude so much that a mild bump can pop them out of their sockets. The tiny jaws and massive teeth of pugdogs and bulldogs cause dental and breathing problems, as well as sinusitis, bad colds, and their notorious "dog breath." Folds of skin on their abdomens easily become infected. Larger breeds, such as the Saint Bernard, are plagued by bone problems and short life spans.

We artificially select natural oddities in cats, too. One of every 10 New England cats has six or seven toes on each paw, thanks to a multitoed ancestor in colonial Boston (figure 1). Elsewhere, these cats are rare. The sizes of the blotched tabby populations in New England, Canada, Australia, and New Zealand correlate with the time that has passed since cat-loving Britons colonized each region. The Vikings brought the orange tabby to the islands off the coast of Scotland, rural Iceland, and the Isle of Man, where these feline favorites flourished.

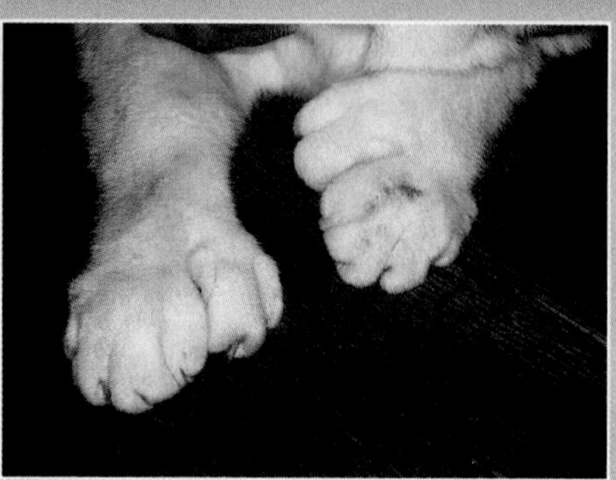

**figure 1**

Multitoed cats are common in New England but rare elsewhere.

The American curl cat's origin traces back to a stray female who wandered into the home of a cat-loving family in Lakewood, California, in 1981 and passed her unusual, curled-up ears to future generations (figure 2). The cause—a dominant gene that leads to formation of extra cartilage lining the outer ear. Cat breeders hope that the gene does not have other less lovable effects. Cats with floppy ears, for example, tend to have large feet, stubbed tails, and lazy natures.

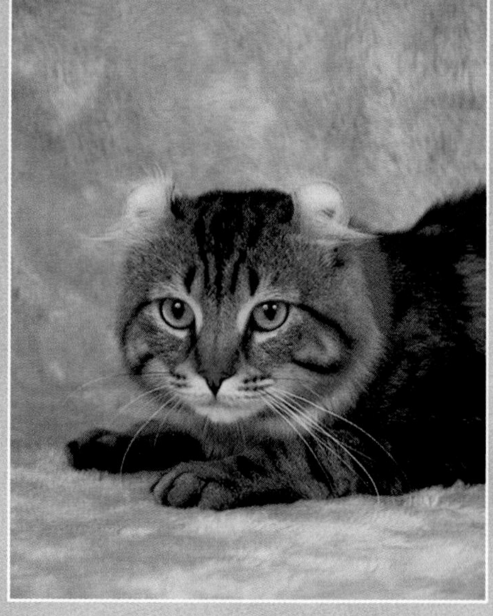

**figure 2**

An American curl cat.

## 14.6 Gene Genealogy

Identifying different mutations in a gene is useful in charting the evolution and spread of genetic variants. An assumption in deciphering gene origins is that the more prevalent an allele is, the more ancient it is, because it has had more time to spread and accumulate in a population. Correlating allele frequencies with historical, archeological, and linguistic evidence provides fascinating peeks at the growth of modern peoples.

### PKU Revisited

The diversity of PKU mutations suggests that the disease has arisen more than once. Mutations common to many groups of people probably represent more ancient mutational events, which perhaps occurred before many groups spread into disparate populations. This was the case for CJD among groups that left Spain in the Middle Ages, mentioned earlier in the chapter. In contrast, mutations found only in a small geographical region, or perhaps in a single family, are more likely to be of recent origin. They have had less time to spread. For example, Turks, Norwegians, French Canadians, and Yemenite Jews have their own PKU alleles. Analysis of the frequencies of PKU mutations in different populations, plus logic, can reveal the roles that genetic drift, mutation, and balanced polymorphism play in maintaining the mutation.

A high mutation rate cannot be the sole reason for the continued prevalence of PKU because some countries, such as Denmark, continue to have only one or two mutations. If the gene were unstable, so that it mutated frequently, all populations would have several different types of PKU mutations. This is not the case.

In some isolated populations, such as French Canadians and Yemenite Jews, migration and the founder effect have maintained certain PKU alleles. Consider the history of PKU among Yemenite Jews. In most populations, point mutations in the phenylalanine hydroxylase (PAH) gene cause PKU. Virtually all of the Yemenite Jews in Israel who have PKU instead have a 6,700-base deletion in the third exon of the PAH gene. An eclectic group of researchers—including geneticists, cell biologists, scholars in Jewish history, and pediatricians—traced the spread of this PKU mutation from North Africa to Israel.

The researchers tested for the telltale deletion in the grandparents of the 22 modern Yemenite Jewish families with PKU in Israel. By asking questions and consulting court and religious records, which this close-knit community kept meticulously, the team found that all clues pointed to San'a, the capital of Yemen.

The earliest records identify two families with PKU in San'a, and indicate that the mutation originated in one person before 1800. By 1809, religious persecution and hard economic times led nine families carrying the mutation to migrate north and settle in three towns (figure 14.9). Four of the families then moved farther northward, into four more towns. Twenty more families spread from San'a to inhabit 17 other towns. All of this migration took place from 1762 through the mid-1900s, and eventually led to Israel.

A more recent example of the effect history and politics can have on gene frequency is the influx of families with PKU into northwest Germany after the second world war, when Germans from the east moved westward. Future shifts in allele frequencies may parallel the breakdown of the former Soviet Union.

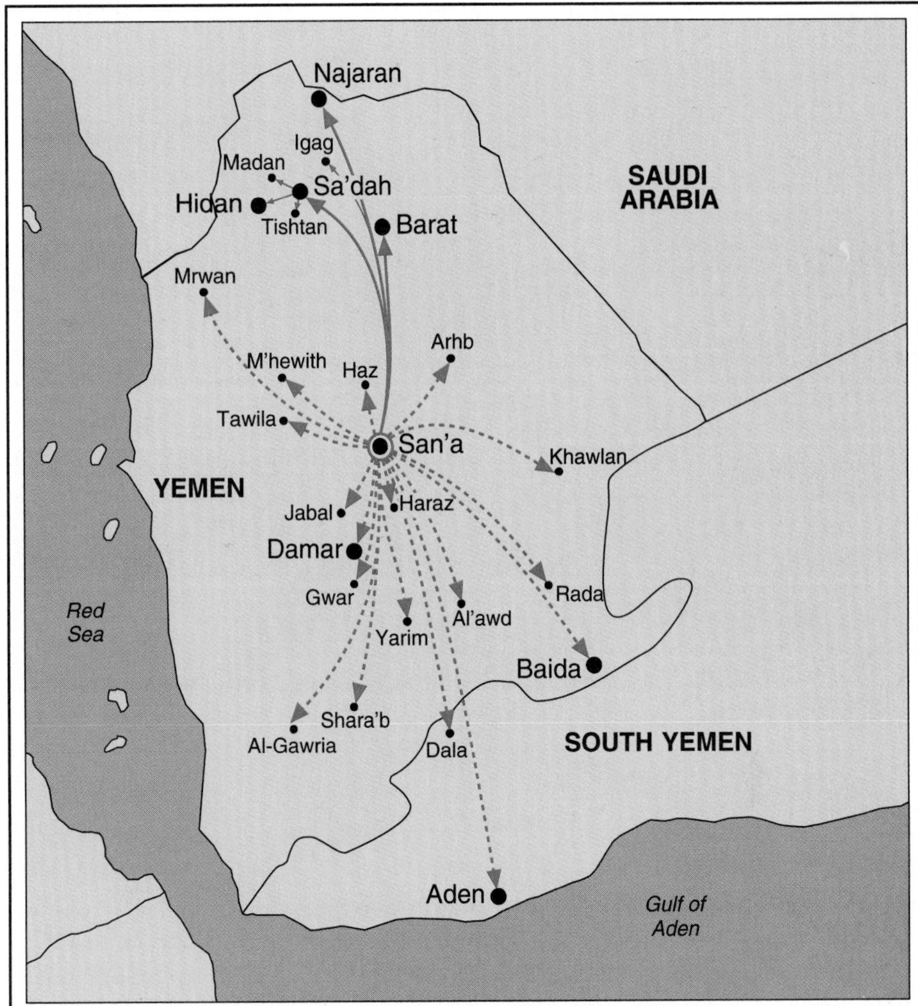


## figure 14.9

**The origin of PKU.** The exon 3 deletion in Israeli Yemenite Jews probably arose in San'a, Yemen, in the mid–18th century. The allele spread northward as families moved from San'a in 1809 (solid arrows) and subsequently spread to other regions (broken arrows).

Source: Data from Smadar Avigad et al., A single origin of phenylketonuria in Yemenite Jews, *Nature* 344:170, March 8, 1990.

## CF Revisited

Following the distributions of alleles in modern populations known to have very ancient roots offers clues to how early genetic disorders plagued humankind. For example, the CF allele ΔF508 is very prevalent among northern Europeans, yet not as common in the south (figure 14.10). This distribution might mean that early farmers migrating from the Middle East to Europe in the Neolithic period, up until about 10,000 years ago, brought the allele to Europe. At this time, people were just beginning to give up a hunter-gatherer lifestyle for semipermanent settlements, exhibiting the first activities of agriculture.

The origin, or at least the existence, of ΔF508 may go farther back, to the Paleolithic age before 10,000 years ago. People then were hunter-gatherers who occasionally lived in caves and tents. They used tools of chipped stone, followed a lunar calendar, and created magnificent cave art.

Geneticists were led back to the Paleolithic by an intriguing group of people, the Basques. Researchers studied 45 families from the Basque country (in the mountains between France and Spain)

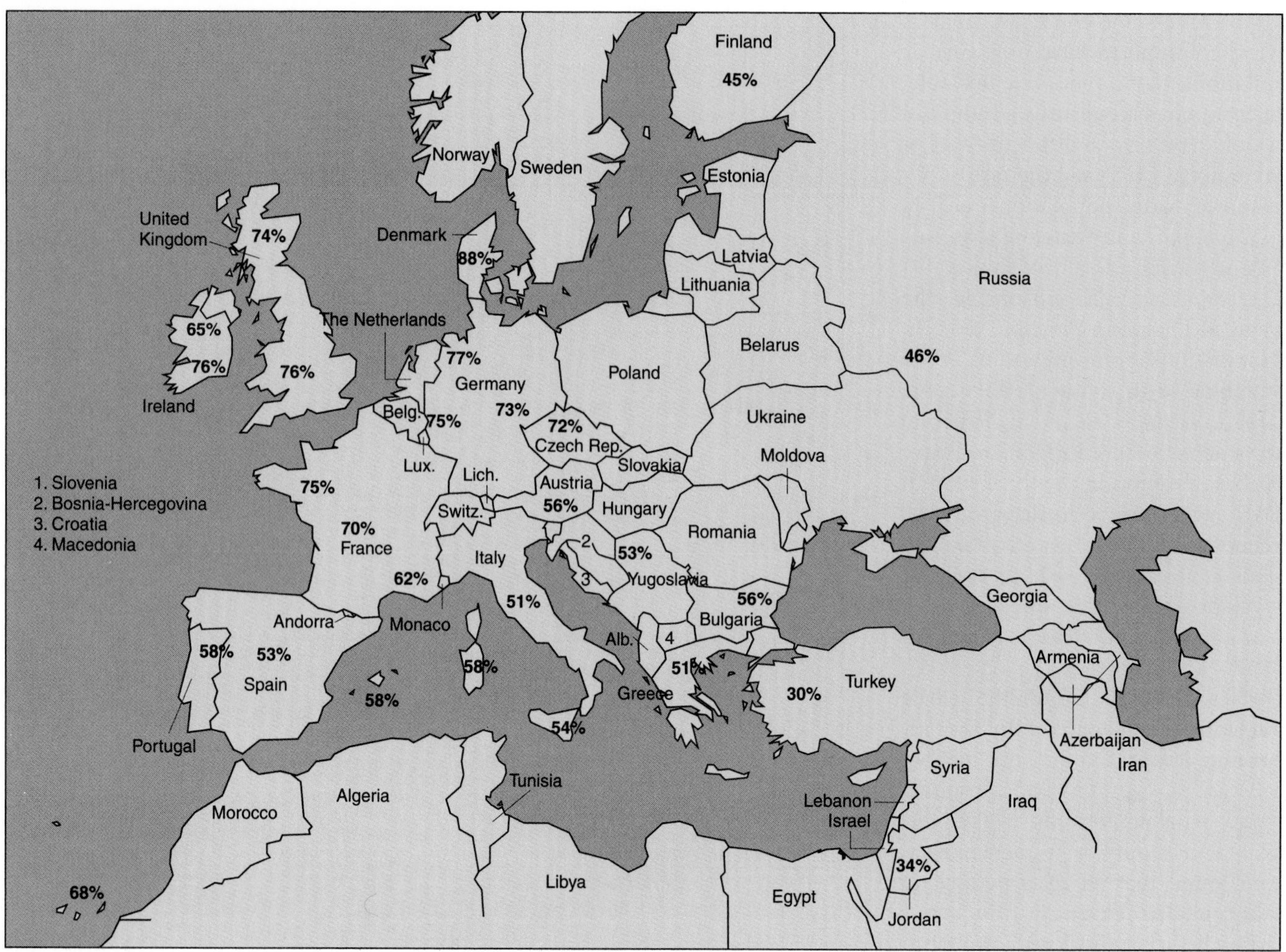


## figure 14.10

**CF allele frequencies vary.** The most common allele causing cystic fibrosis, ΔF508, occurs with vastly different frequencies in different populations. This map indicates the percentage of CF alleles that are ΔF508 in these nations.

Source: European Working Group on Cystic Fibrosis Genetics, Gradient of distribution in Europe of the major CF mutation and of its associated haplotype, *Human Genetics* 85:436–45, 1990.

with cystic fibrosis and identified affected children with four pure Basque grandparents by their distinctive double surnames. For 87 percent of the pure Basque CF patients, ΔF508 was the causative mutation—a frequency higher than for most European populations, indicating a more ancient origin. Families of "mixed Basque" background, in which Basque ancestors interbred with French or Spanish neighbors, showed ΔF508 only 58.3 percent of the time. Among the nearby Spanish population, the frequency of ΔF508 is 50 percent. Today's remaining Basque people, then, may carry a pocket of cystic fibrosis mutations from long ago.

### Key Concepts

Mutational analysis in various populations indicates that PKU originated more than once, and that genetic drift, balanced polymorphism, and perhaps mutation have affected its prevalence. Studies of allele prevalence place the origin of cystic fibrosis farther back in history than previously thought.

# Summary

### 14.1 Nonrandom Mating

1. Hardy-Weinberg equilibrium assumes that all individuals mate with the same frequency and choose mates without regard to phenotype. This rarely, if ever, happens in real human populations. We choose mates based on certain characteristics, and some individuals have many more children than others.
2. DNA sequences that do not cause a phenotype important in mate selection or reproduction may be in Hardy-Weinberg equilibrium.
3. Consanguinity increases the proportion of homozygotes in a population, which may lead to increased incidence of recessive illnesses or traits.

### 14.2 Migration

4. Changing allele frequencies from one geographic area to another constitute **clines.**
5. Clines may reflect geographical barriers or linguistic differences and may be either abrupt or gradual.
6. Human migration patterns through history explain many cline boundaries. Forces behind migration include escape from persecution and a nomadic lifestyle.

### 14.3 Genetic Drift

7. **Genetic drift** occurs when a small population separates from a larger one, and its members breed only among themselves, perpetuating allele frequencies that are not characteristic of the ancestral population.
8. Genetic drift is random and may occur within a larger group or apart from it.
9. A **founder effect** is a type of genetic drift that occurs when a few individuals found a settlement and their alleles form a new gene pool, amplifying the alleles they introduce and eliminating others.
10. A **population bottleneck** is a narrowing of genetic diversity that occurs after many members of a population die and the few survivors rebuild the gene pool.

### 14.4 Mutation

11. Mutation continually introduces new alleles into populations. It is a consequence of DNA replication errors.
12. Mutation does not have as great an influence on disrupting Hardy-Weinberg equilibrium as the other factors.
13. The **genetic load** is the collection of deleterious alleles in a population.

### 14.5 Natural Selection

14. Environmental conditions influence allele frequencies via **natural selection,** as the rise and fall of infectious disease indicates. Alleles that do not enable an individual to reproduce in a particular environment are selected against and diminish in the population, unless conditions change. Beneficial alleles are retained.
15. In **balanced polymorphism,** the frequencies of some deleterious alleles are maintained when heterozygotes have a reproductive advantage under certain conditions, such as resisting an infection.

### 14.6 Gene Genealogy

16. Frequencies of different mutations in different populations provide information on the natural history of alleles and on the relative importance of nonrandom mating, genetic drift, and natural selection in deviations from Hardy-Weinberg equilibrium.

# Review Questions

1. Give examples of how each of the following can alter gene frequencies from Hardy-Weinberg equilibrium:
   a. nonrandom mating
   b. migration
   c. a population bottleneck
   d. mutation
2. Explain the influence of natural selection on
   a. the virulence of tuberculosis.
   b. bacterial resistance to antibiotics.
   c. the changing degree of genetic diversity in an HIV population during infection.
   d. the prevalence of cystic fibrosis.
3. Why is increasing homozygosity in a population detrimental?
4. Explain how misuse of antibiotics can add to the problem of antibiotic-resistant bacteria.
5. Why might a mutant allele that causes an inherited illness when homozygous persist in a population?
6. Give an example of an inherited disease allele that protects against an infectious illness.
7. Porphyria variegata, which resulted from a founder mutation in the Afrikaner population, is inherited as an autosomal dominant with incomplete penetrance. How can this mode of inheritance complicate analysis of the condition?
8. A disease-causing allele that is very rare in most populations is unusually prevalent in a particular population. What type of information might enable you to determine whether the prevalence reflects a founder effect, balanced polymorphism, or a population bottleneck?
9. Provide two examples of how molecular evidence confirms presence of genetic uniformity.
10. Explain why table 14.2 indicates that genetic drift has occurred among the Dunkers.
11. Would a carrier test to detect the common cystic fibrosis allele ΔF508 be more accurate in France or Finland? Cite a reason for your answer.
12. What type of molecular evidence indicates a founder effect?

13. How does a founder effect differ from a population bottleneck?

14. Describe two scenarios in human populations, one of which accounts for a gradual cline, and one for an abrupt cline.

15. How do genetic drift, nonrandom mating, and natural selection interact?

16. Define:

    a. founder effect

    b. linkage disequilibrium

    c. balanced polymorphism

    d. genetic load

17. How does a knowledge of history, sociology, and anthropology help geneticists to interpret allele frequency data?

# Applied Questions

1. Use the information in chapters 13 and 14 to explain why

    a. porphyria variegata is more prevalent among Afrikaners than other South African populations.

    b. many people among the Cape population in South Africa lose their teeth before age 20.

    c. cystic fibrosis and sickle cell disease remain common Mendelian illnesses.

    d. the Pima Indians have an extremely high incidence of non-insulin-dependent diabetes mellitus.

    e. the Amish in Lancaster County and certain Pakistani groups have a high incidence of genetic diseases that are very rare elsewhere.

    f. the frequency of the allele that causes galactokinase deficiency varies across Europe.

    g. a haplotype associated with Creutzfeldt-Jakob disease is the same in populations from Spain, Chile, Libya, Italy, and Tunisia.

    h. mitochondrial DNA sequences vary gradually in populations along the Nile River valley.

    i. disease-causing BRCA1 alleles are different in Jewish people of eastern European descent and African Americans.

2. Which principles discussed in this chapter do the following science fiction film plots illustrate?

    a. In *When Worlds Collide,* the Earth is about to be destroyed. One hundred people are selected to colonize a new planet.

    b. In *The Time Machine,* set in the distant future on Earth, one group of people is forced to live on the planet's surface while another group is forced to live in caves. Over many years, they come to look and behave differently. The Morlocks that live below ground have dark skin, dark hair, and are very aggressive, whereas the Eloi that live aboveground are blond, fair-skinned, and meek.

    c. In *Children of the Damned,* all of the women in a small town are suddenly made pregnant by genetically identical beings from another planet.

    d. In *The War of the Worlds,* Martians cannot survive on Earth because they are vulnerable to infection by terrestrial microbes.

3. Treatment for PKU has been so successful that, over the past 30 years, many people who would otherwise have been profoundly mentally retarded have led normal lives and become parents. How has this treatment altered Hardy-Weinberg equilibrium for mutant alleles causing PKU?

4. Ashkenazim, French Canadians, and people who live in southwestern Louisiana all have a much higher incidence of Tay-Sachs disease than other populations, yet each of these groups has a different mutation. How is this possible?

5. Syndrome X consists of obesity, type II diabetes, hypertension, and heart disease. Researchers surveyed and sampled blood from nearly all of the 2,188 residents of the Pacific Island of Kosrae, and found that 1,709 of them are part of the same pedigree. The incidence of all of the symptoms of syndrome X is much higher in this population than for other populations. Suggest a reason for this finding, and indicate why it would be difficult to study these particular traits, even in an isolated population.

6. A single allele enables a person to digest the milk sugar lactose. People who lack this allele suffer the stomach cramps of lactose intolerance. The allele conferring "lactase persistence" has a high incidence in societies that drink a great deal of milk, including northern Europeans and nomadic tribes of Africa and the Middle East. Which of the following factors has probably most influenced the frequency of the allele that enables one to digest lactose—migration, mutation, nonrandom mating, or natural selection?

7. People with familial Mediterranean fever have an unusually low incidence of asthma. What force may help maintain this disorder in populations?

8. By which mechanisms discussed in this chapter do the following situations alter Hardy-Weinberg equilibrium?

    a. Ovalocytosis is a rare genetic abnormality that is not only symptomless, but seems to be beneficial. A protein that anchors the red blood cell membrane to the cytoplasm is abnormal, making the membrane unusually rigid. As a result, the parasites that cause malaria cannot enter the red blood cells of individuals with ovalocytosis.

b. In the mid-1700s, a multitoed male cat from England crossed the sea and settled in Boston, where he left behind quite a legacy of kittens—about half of whom also had six, seven, eight, or even nine digits on their paws. Today, in Boston and nearby regions, multitoed cats are far more common than in other parts of the United States.

c. Many slaves in the United States arrived in groups from Nigeria, which is an area in Africa with many ethnic subgroups. They landed at a few sites and settled on widely dispersed plantations. Once emancipated, former slaves in the South were free to travel and disperse.

**9.** The human population of India is divided into many castes, and the people follow strict rules about who can marry whom. Researchers from the University of Utah compared several genes among 265 Indians of different castes and 750 people from Africa, Europe, and Asia. The study found that the genes of higher Indian castes most closely resembled those of Europeans, and that the genes of the lowest castes most closely resembled those of Asians. In addition, the study found that maternally inherited genes (mitochondrial DNA) more closely resembled Asian versions of those genes, but paternally inherited genes (on the Y chromosome) more closely resembled European DNA sequences. Construct an historical scenario to account for these observations.

# Suggested Readings


Avigad, Smadar, et al. March 8, 1990. A single origin of phenylketonuria in Yemenite Jews. *Nature,* vol. 344. Correlations of allele frequencies with historical information reveal gene flow.

Diamond, Jared. February 1992. The return of cholera. *Discover.* CF heterozygosity may protect against viral illnesses.

Hurles, Matthew, et al. November 1999. Recent male-mediated gene flow over a linguistic barrier in Iberia, suggested by analysis of a Y-chromosomal DNA polymorphism. *The American Journal of Human Genetics* 65:1437. DNA sequences reveal the effects of geographic barriers on gene flow.

Krings, Matthias, et al. April 1999. mtDNA analysis of Nile River valley populations: A genetic corridor or a barrier to migration? *The American Journal of Human Genetics* 64:1166. A look at the haplotype patterns of 224 people inhabiting different towns in the Nile River valley.

Lewis, Ricki. January 7, 2002. SNPs as windows on evolution. *The Scientist* 16(1):16–18. SNP patterns reveal natural selection in action.

Lewis, Ricki, April 16, 2001. Founder populations fuel gene discovery. *The Scientist* 15(8):8. Founder populations enable geneticists to zero in on disease-causing genes.

McKusick, Victor A. March 2000. Ellis-van Creveld syndrome and the Amish. *Nature Genetics* 24:203. McKusick did some of the first genetic studies on the Amish.

May, Robert. August 1995. The rise and fall and rise of tuberculosis. *Nature Medicine,* vol. 1. Natural selection has molded the virulence of this reemerging infection.

Motulsky, Arno. February 1995. Jewish diseases and origins. *The American Journal of Human Genetics,* vol. 9. Centuries of discrimination sequestered and selected certain genetic variants in this group, creating "ethnic diseases."

Nowak, Martin A., and Andrew J. McMichael. August 1995. How HIV defeats the immune system. *Scientific American.* Genetic diversity of HIV waxes and wanes during infection in response to the selective pressures of the immune system and therapeutic drugs.

Sachs, Oliver. 1998. *The Island of the Colorblind.* New York: Random House Vintage Books. The story of Pingelapese blindness.

Schaeffeler, Elke, et al. April 4, 2001. Frequency of C3435T polymorphism of MDR1 gene in African people. *The Lancet* 358:383–84. A polymorphism that once protected against infection today blocks AIDS drugs.

Tishkoff, Sarah A., et al. July 20, 2001. Haplotype diversity and linkage disequilibrium at human G6PD: Recent origin of alleles that confer malarial resistance. *Science* 293:455–62. Rise of the protective G6PD allele parallels rise of malaria.

Vonnegut, Kurt. 1985. *Galapagos.* New York: Dell Publishing. A look at a new type of human, living a million years in the future, that evolved from a few founders.

# On the Web

Check out the resources on our website at:

**www.mhhe.com/lewisgenetics5**

On the web for this chapter, you will find additional study questions, vocabulary review, useful links to case studies, tutorials, popular press coverage, and much more. To investigate specific topics mentioned in this chapter, also try the links below:

Antimicrobial Resistance Information **sss.cdc.gov/ncidod/ar/**

Cat Fanciers Association **www.fanciers.com/**

Color Blindness **members.aol.com/nocolorvsn/color.htm**

Crigler-Najjar Syndrome **www.crigler-najjar.com/**

Kissing Cousins: The Genetic Fallout of Consanguinity **www.genesage.com/professionals/geneletter/archives/issue2/kissingcousins.html**

National Tay-Sachs and Allied Diseases Association **www.ntsad.org**

Online Mendelian Inheritance in Man **www.ncbi.nlm.nih.gov/entrez/query.fcgi?db=OMIM**

achromatopsia (color blindness) 262300
albinism 300500
BRCA1 113705
Creutzfeldt-Jakob disease 123400
Crigler-Najjar syndrome 218800
G6PD deficiency 305900
galactokinase deficiency 230200
non-insulin-dependent diabetes mellitus 125853
phenylketonuria 261600
porphyria variegata 176200
Tay-Sachs disease 272800

Tuberculosis Information **www.cdc.gov/nchstp/tb/faqs/qa.htm**

CHAPTER

# 15

# Human Origins and Evolution

### 15.1 Human Origins

We are primates, the modern descendants of a long line of animals that began with small insect eaters that lived 60 million years ago. From the apelike hominoids arose the ancestors of humans, the hominids. It is strange to think of a time when more than one type of "person" inhabited the earth—but it probably happened.

### 15.2 Molecular Evolution

Fossils provide pieces of the past, small glimpses into evolution. Chromosome patterns and the sequences of genomes, genes, and proteins can fill in many of the gaps in our knowledge of the past, and clarify species relationships. Such molecular-level evidence can also raise more questions.

### 15.3 Molecular Clocks

Applying mutation rates to molecular evidence enables researchers to construct evolutionary tree diagrams that can track time as well as relationships. Mitochondrial DNA is used to trace maternal lineages, and Y chromosome sequences to follow paternal lineages. Combined with fossil evidence and information from linguistics and anthropology, this genetic evidence is helpful in describing the origins of peoples.

### 15.4 Eugenics

Eugenics is the control of reproduction with the intent of altering the gene pool. It is common in agriculture, but deplored in many human societies. The goal of genetic testing may be misconstrued as eugenic in nature. Instead, testing aims to help individuals and families cope with difficult choices and situations.

Imagine being asked to build a story from the following elements:

1. A pumpkin that turns into a coach
2. A prince who hosts a ball
3. A poor but beautiful young woman with dainty feet who has two mean and ugly stepsisters with large feet

Chances are that unless you're familiar with the fairytale "Cinderella," you wouldn't come up with that exact story. In fact, 10 people given the same pieces of information might construct 10 very different tales.

So it is with the sparse evidence we have of our own beginnings—pieces of a puzzle in time, some out of sequence, many missing. Traditionally, paleontologists (scientists who study evidence of ancient life) have consulted the record in the earth's rocks—fossils—to glimpse the ancestors of *Homo sapiens,* our own species. Researchers assign approximate ages to fossils by observing which rock layers the fossils are located in and by extrapolating the passage of time from the ratios of certain radioactive chemicals in surrounding rock.

Other modern organisms also provide intriguing clues to the origins and relationships of species through their chromosomes and informational molecules of RNA, DNA, and proteins. Comparisons of genome organization are providing compelling clues to the past. In this chapter, we explore human origins, genetic and genomic evidence for evolution, and how we attempt to alter the evolution of our own species and others.

## 15.1 Human Origins

A species includes individual organisms that are alike enough that they can successfully produce healthy offspring. *Homo sapiens* ("the wise human") probably first appeared during the Pleistocene epoch, about 200,000 years ago. Our ancestry reaches farther back, to about 60 million years ago when rodentlike insect eaters flourished. These first primates gave rise to many new species. Their ability to grasp and to perceive depth provided the flexibility and coordination necessary to dominate the treetops.

About 30 to 40 million years ago, a monkeylike animal the size of a cat, *Aegyptopithecus,* lived in the lush tropical forests of Africa. Although the animal probably spent most of its time in the trees, fossilized remains of limb bones indicate that it could run on the ground, too. Fossils of different individuals found together indicate that they were social animals. *Aegyptopithecus* had fangs that it might have used for defense. The large canine teeth seen only in males suggest that males may have provided food for their smaller female mates. *Propliopithecus* was a monkeylike contemporary of *Aegyptopithecus.* Both animals are possible ancestors of gibbons, apes, and humans.

From 22 to 32 million years ago, Africa was home to the first **hominoids,** animals ancestral to apes and humans only. One such resident of southwestern and central Europe was called *Dryopithecus,* meaning "oak ape," because fossilized bones were found with oak leaves (figure 15.1). The way the bones fit together suggests that this animal lived in the trees but could swing and walk farther than *Aegyptopithecus.*

More abundant fossils represent the middle-Miocene apes of 11 to 16 million years ago. These apes were about the size of a human 7-year-old and had small brains and pointy snouts.

Apelike animals similar to *Dryopithecus* and the mid-Miocene apes flourished in Europe, Asia, and the Middle East during the same period. Because of the large primate

a. *Dryopithecus*

b. *Australopithecus*

c. *Homo erectus*

**figure 15.1**

**Human forerunners.** (*a*) The "oak ape" *Dryopithecus,* who lived from 22 to 32 million years ago, was more dextrous than his predecessors. (*b*) Several species of *Australopithecus* lived from 2 to slightly more than 4 million years ago. These hominids walked upright on the plains. (*c*) *Homo erectus* made tools out of bone and stone, used fire, and dwelled communally in caves from 1.6 million years ago to possibly as recently as 35,000 years ago.

population in the forest, selective pressure to venture onto the grasslands in search of food and habitat space must have been intense. Many primate species probably vanished as the protective forests shrank. Of all of the abundant middle-Miocene apes, one survived to give rise to humans and African apes. Eventually, animals ancestral to humans only, called **hominids,** arose and eventually thrived. (Some researchers use hominim instead of hominid.)

Hominoid and hominid fossils from 4 to 19 million years ago are scarce. This was the time that the stooped ape gradually became the upright ape-human. A 10-million-year-old fossilized face found in northern Greece is from an animal newly named *Ouranopithecus macedoniensis.* Its small canine teeth and thick tooth enamel suggest that it could be an immediate forerunner of hominids.

## The Australopithecines—and Others?

Four million years ago, human forebears diversified, as **bipedalism**—the ability to walk upright—opened up vast new habitats on the plains. Several species of a hominid called *Australopithecus* lived at this time, from 2 to 4 million years ago, probably following a hunter-gatherer lifestyle. Figure 15.2 depicts the probable relationships among known australopithecines, members of the genus *Homo,* and modern primates that are our closest relatives.

Australopithecines had flat skull bases, as do all modern primates except humans. They stood about 1.2 to 1.5 meters (four to five feet) tall and had brains about the size of a gorilla's, with humanlike teeth. The angle of preserved pelvic bones, plus the discovery of *Australopithecus* fossils with those of grazing animals, indicate that this ape-human had left the forest.

The most ancient species of australopithecine that we have evidence of, *Australopithecus anamensis,* lived about 4.1 million years ago. Until recently, most of our knowledge of australopithecines came from a very famous fossilized partial skeleton named "Lucy" discovered in 1974. A member of *Australopithecus afarensis,* Lucy lived about 3.6 million years ago in the grasses along a lake in the Afar region of Ethiopia (figure 15.3). Her skull was shaped more like a human's than an ape's, with a less prominent face and larger brain than her predecessors. The condition of her skeleton indicated that she died, with arthritis, at about the age of 20.

Much of what we know about the australopithecines comes from two parallel paths of footprints, preserved in volcanic ash in the Laetoli area of Tanzania. Archeologist Mary Leakey and her team discovered them by accident in 1976. The 89-foot long trail of footprints, left about 3.6 million years ago, was probably made by a large and small individual walking close together, and a third following in the steps of the larger animal in front. The shape of the prints indicates that their feet and gait were remarkably like ours.

A partial skull discovered near a lake in northern Kenya in 2001 may have belonged to a hominid that was either an australopithecine or a contemporary. The animal's tentative name, *Kenyanthropus platyops,* indicates where its remains were found and its most compelling feature—a flat face. It lived sometime between 3.2 and 3.5 million years ago in what was then grasslands bordering woods, in an area where many other species of mammals dwelled. *K. platyops* had a novel combination of traits that included small earholes similar to those of chimpanzees and *Australopithecus anamensis,* yet the teeth, small brain, and flat nose of *A. afarensis.* The huge flat face with elongated cheekbones is what compelled the paleontologists who discovered *K. platyops*—Meave and Louise Leakey and their colleagues—to place it in its own genus, calling the australopithecines a "garbage can" genus because so many fossil finds seem to end up in that category. Whatever *Kenyanthropus platyops* turns out to have been, it did not replace the australopithecines, but instead may have been a contemporary.

Another type of australopithecine, *Australopithecus africanus,* lived about 2.8 million years ago, according to scant fossil evidence, and therefore lived closer in time to members of genus *Homo* than Lucy or *Kenyanthropus platyops* did. However, because this hominid lived far from the site of the earliest known *Homo* fossils, paleontologists hypothesized that there must have been some other ape-human "missing link" that lived, intermediate in time and place, between the known australopithecines and the earliest

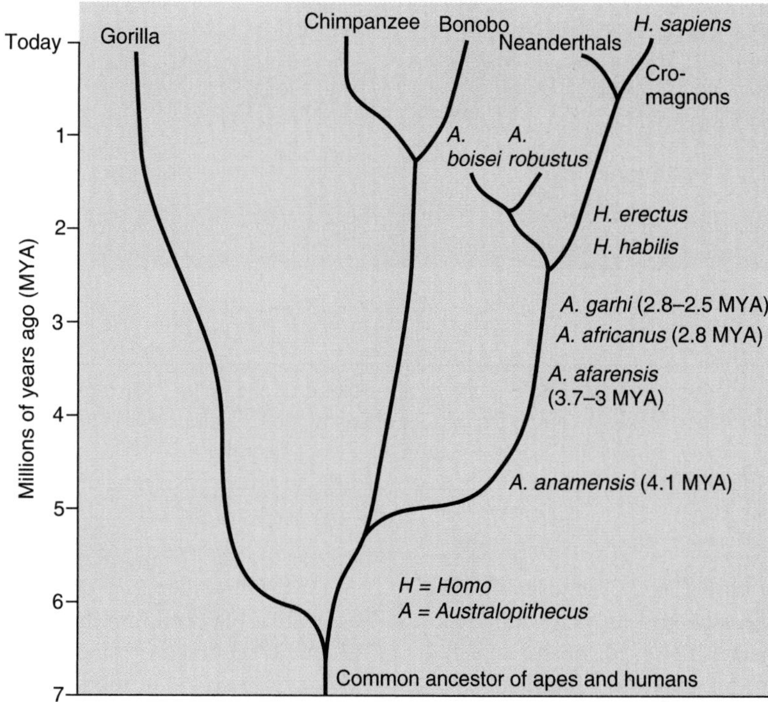


**figure 15.2**

**Evolutionary tree.** An evolutionary tree diagram indicates the relationships among primates, past and present.

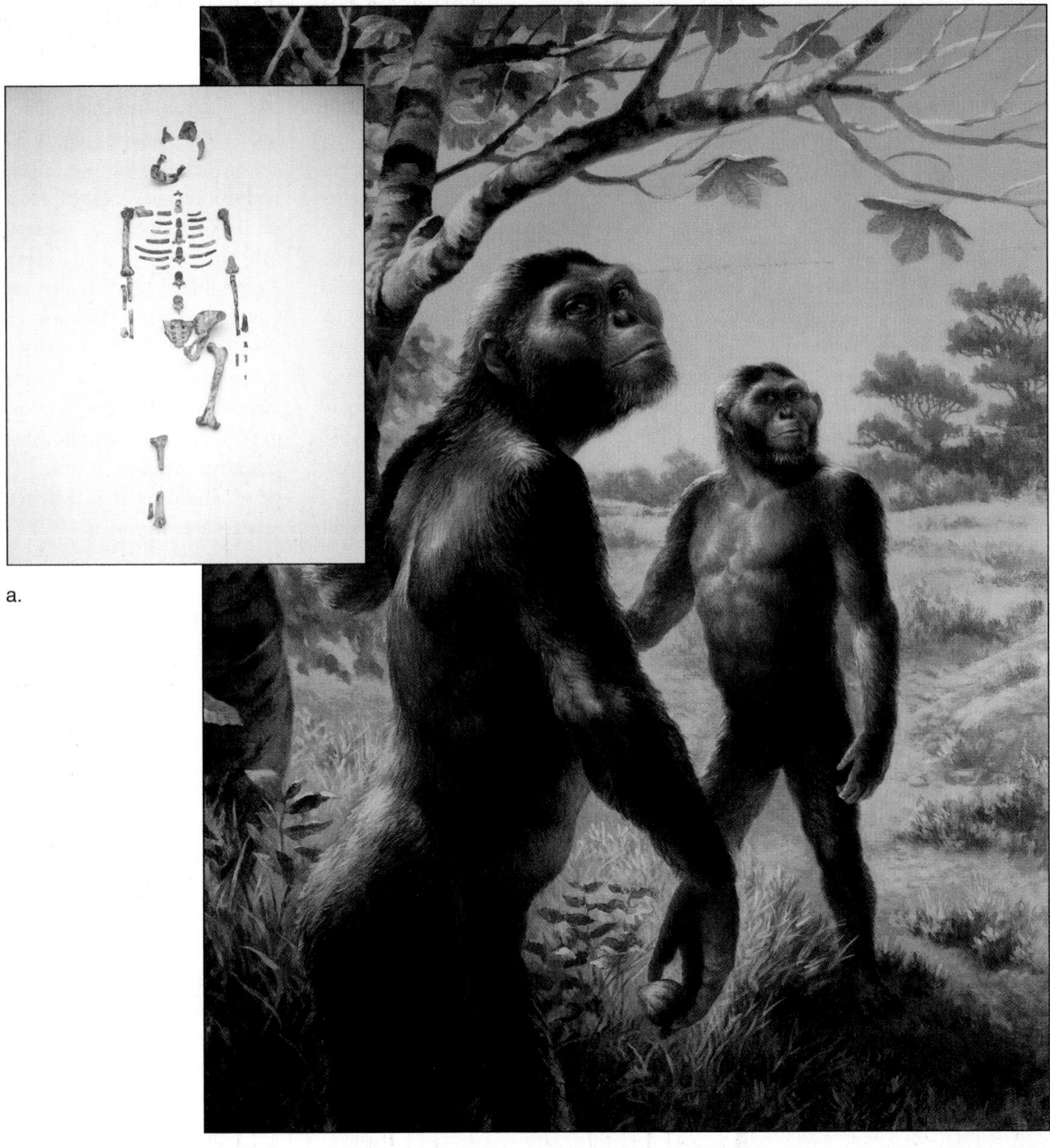

a.

b.

**figure 15.3**

**Lucy.** About 3.6 million years ago, a small-brained human ancestor walked upright in the grasses along a lake in the Afar region of Ethiopia. She skimmed the shores for crabs, turtles, and crocodile eggs to eat. Her discoverers, Donald Johanson of the Cleveland Museum of Natural History and Timothy White of the University of California at Berkeley, named her "Lucy" because they were listening to the Beatles song "Lucy in the Sky with Diamonds" when they found her. (*a*) The bones that White and Johanson discovered and (*b*) an artist's interpretation of what this animal on the road to humanity may have looked like.

members of *Homo.* Paleontologists may have found a missing link in 1999, with the discovery of evidence of three individuals from a species named *Australopithecus garhi.* Trying to understand how this human forebear lived is a little like piecing together the Cinderella story. Researchers found:

1. Remains of an antelope that had obviously been butchered. It was dismembered, and the ends of the long bones had been cleanly cut with tools, the marrow removed, meat stripped, and the tongue cleanly sliced off.
2. Limb bones from an individual who stood about 1.4 meters (about 4.5 feet) tall. The long legs were like those of a human, but the long arms were more like those of an ape.
3. A partial skull of a different individual, with a small cranium (holding a 450-cubic-centimeter brain, compared to the 1,400-or-so-cc modern human brain) and large teeth, its dimensions suggesting an apelike face on the bottom.

The *A. garhi* fossil finds are important for many reasons. The evidence is much more complete than the stray teeth and jaw bits that paleontologists have of other australopithecines. More importantly, this hominid lived in the right time and place to be, and had characteristics consistent with, the long-sought bridge between *Australopithecus* and *Homo.* Several dating techniques place its time of existence as 2.5 million years ago. The fossils were found in a desert in eastern Ethiopia, near later *Homo* fossils. The limb lengths and cranial capacity suggest a transitional form. In addition, *A. garhi* hunted.

## *Homo*

Like the Cinderella story, the tale of how *Australopithecus* became or was replaced by *Homo* is built on sparse clues. Some australopithecines were "dead ends," dying off or leading to the Neanderthals, a branch of the *Homo* family tree that eventually vanished. And of course, we do not know what we have yet to discover. The clues do suggest, however, that by 2.3 million years ago, *Australopithecus* coexisted with *Homo habilis*—a more humanlike primate who lived communally in caves and provided intensive care for its young. *Habilis* means handy, and this primate is the first to have used tools for tasks more challenging than stripping meat from bones. *H. habilis* may have descended from a group of australopithecines who ate a greater variety of foods than other ape-humans, allowing them to adapt to a wider range of habitats.

*H. habilis* coexisted with and was followed by *Homo erectus* during the Paleolithic Age (table 15.1). *H. erectus* left fossil evidence of cooperation, social organization, and tool use, including the use of fire. Fossilized teeth and jaws of *H. erectus* suggest that these primates ate meat. They were the first to have an angled skull base that permitted them to make a greater range of sounds, making speech possible. *H. erectus* fossils are widespread. They have been found in China, Java, tropical Africa, and southeast Asia, indicating that these animals could migrate farther than earlier primates. One *Homo erectus* skull was discovered recently in an antique shop in Manhattan! The astute owner took it to the nearby American Museum of Natural History, where paleontologists immediately recognized it. The skull came from Indonesia.

**table 15.1**

**Cultural Ages**

| Age | Time (years ago) | Defining Skills |
|---|---|---|
| Paleolithic | 750,000 to 15,000 | Earliest chipped tools |
| Mesolithic | 15,000 to 10,000 | Cutting tools, bows and arrows |
| Neolithic | 10,000 to present | Complex tools, agriculture |

**figure 15.4**

***Homo erectus.*** This artist's rendition is based on many fossils.

The distribution of *H. erectus* fossils suggests that they lived in families of male-female pairs (most primates have harems). The male hunted and the female nurtured young. Figure 15.4 is an artist's rendition of *Homo erectus.*

By 70,000 years ago, humans left evidence of what is considered modern behavior, in the form of intricately carved tools made of animal bones, and highly symmetrical hatchmarks made in red rock. Some anthropologists think that the markings indicate early language or counting. These fossils, discovered in a cave in South Africa, indicate human origin in Africa, a point we will return to.

*H. erectus* may have lived as recently as 35,000 years ago and probably coexisted with the very first *Homo sapiens.* Fossilized skulls from Java and China may be from individuals intermediate between *H. erectus* and *Homo sapiens.* These intermediate primates had big brains and robust builds and lived from 30,000 to 50,000 years ago. Several pockets of ancient peoples may have been dispersed throughout the world at that time; fossils have been found in Swanscombe, England; Steinheim, Germany; and the Middle East.

The Neanderthals, contemporaries of *H. erectus* and also members of genus *Homo,* appeared in Europe about 150,000 years ago. By 70,000 years ago, the Neanderthals had spread to western Asia. They had slightly larger brains than we do, prominent brow ridges, gaps between certain teeth, very muscular jaws, and large, barrel-shaped chests.

The Neanderthals take their name from Neander Valley, Germany, where quarry workers blasting in a limestone cave on a summer day in 1856 discovered the first preserved bones of this hominid that may or may not have been a member of our species. A fossilized, deformed skeleton buried with flowers in Shanidar Cave, Iraq, reveals that the Neanderthals may have been religious hunter-gatherers that were either clever enough or lucky enough to survive a brutal ice age. Anthropologists once thought the Neanderthals were hunchbacked brutes because the first fossil discovered was from an individual stooped from arthritis.

Fossil evidence indicates that from 30,000 to 40,000 years ago, the Neanderthals coexisted with the lighter-weight, finer-boned Cro-Magnons. The newcomers also had high foreheads and well-developed frontal brain regions, and signs of culture that the older Neanderthals lacked. The first Cro-Magnon fossils were found in a French cave. Five adults and a baby were arranged in what appeared to be a communal grave. Nearby were seashells, pierced in a way that suggested they may have been used as jewelry. Intricate art decorated the cave walls. In contrast, the few Neanderthal graves show no evidence of ritual, just quick burial.

Anthropological and fossil evidence indicate that the Neanderthals and Cro-Magnons coexisted for a time, but there is no evidence that they interacted, blending their genomes. It is more likely that Cro-Magnons replaced Neanderthals.

We don't know how it happened, but by about 28,000 years ago, the Neanderthals no longer existed, and the Cro-Magnons presumably continued on the path to humanity. We return to Neanderthals later in the chapter to examine DNA evidence of their relationship to us.

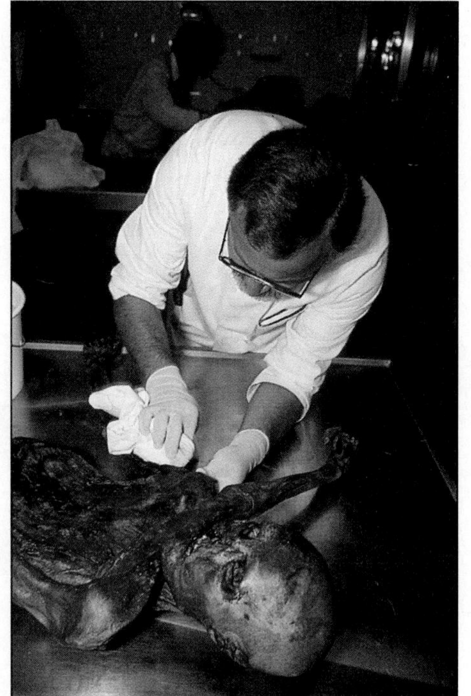

a.

b.

**figure 15.5**

**A 5,300-year-old man.** (*a*) Hikers discovered Ötzi, the Ice Man, in the Austrian/Italian Alps in 1991. He lived 5,300 years ago, recently enough to belong to the same gene pool as people living in the area today. (*b*) Ötzi wore well-made clothing, including a hat, used intricate arrows that demonstrate familiarity with ballistics and engineering; and carried mushrooms with antibiotic properties. He had tattoos, indentations in his ears that suggest he wore earrings, and evidence of a haircut. This depiction is derived from the evidence found on and near Ötzi's preserved body.

## Modern Humans

Cave art from about 14,000 years ago indicates that by that time, our ancestors had developed fine hand coordination and could use symbols—a milestone in cultural evolution. By 10,000 years ago, people had migrated from the Middle East across Europe, bringing agricultural practices.

In 1991, hikers in the Ötztaler Alps of northern Italy discovered an ancient man frozen in the ice (figure 15.5). After amateurs hacked away at him, causing much damage, the Ice Man, named Ötzi, ended up in the hands of several research groups. Unlike other remains that show evidence of intentional burial, the Ice Man was apparently tending sheep high on a mountaintop some 5,300 years ago when he perished. He was dressed for the weather. Berries found with him place the season as late summer or early fall. A wound to his shoulder detected in a CAT scan a decade after the discovery suggests that he may have died after being attacked, perhaps from infection of the wound. He probably fell into a ditch, where he froze to death and was soon covered by snow. After this safe burial, which preserved his body intact, a glacier sealed the natural tomb.

DNA analyses on Ötzi's tissues suggest that he belonged to the same gene pool as modern people living in the area. DNA from a sample of lung tissue held a surprise—it came from a fungus! A lung infection might have contributed to Ötzi's death. He was apparently in weakened condition, working hard in a low-oxygen environment.

Anthropologists try to fill in the gaps in our knowledge of what humans were like a few thousand years ago by studying vanishing indigenous peoples, such as the San (bushmen) and pygmies of Africa, the Basques of Spain, the Etas of Japan, the Hill People of New Guinea, the Yanomami of Brazil, and another Brazilian tribe, the Arawete, who number only 130 individuals. Studying DNA sequences within these populations provides information on their origins, as we'll see later in the chapter.

### Key Concepts

Monkeylike *Aegyptopithecus* lived about 30 to 40 million years ago and was ancestral to gibbons, apes, and humans. The first hominoid (ape and human ancestor), *Dryopithecus,* lived 22 to 32 million years ago and may have walked onto grasslands. Hominids (human ancestors) appeared about 19 million years ago. • About 4 million years ago, bipedalism opened up new habitats for *Australopithecus,* who walked upright and used tools. There were several types of australopithecines, and one, *A. garhi,* may have been a direct forebear of *Homo*. By 2 million years ago, *Australopithecus* coexisted with the more humanlike *Homo habilis*. • Later, *H. habilis* coexisted with *H. erectus,* who used tools in more complex societies. *H. sapiens* either coexisted with or arose from *H. erectus* 30,000 to 50,000 years ago. The Neanderthals preceded the Cro-Magnons. • Modern humans appeared about 40,000 years ago. A preserved man from 5,300 years ago is genetically like us.

## 15.2 Molecular Evolution

Fossils paint an incomplete picture of the past because only certain parts of certain organisms were preserved, and very few of these have been discovered. Additional information on the past comes from within the cell, where the informational molecules of life change over time. They evolve.

Determining and comparing genomes, chromosome banding patterns, or the building block sequences of DNA or proteins, is the field of **molecular evolution.** This approach is based on the fact that DNA sequences in nucleic acids and amino acid sequences in proteins change over time as mutations occur. The fewer differences between a gene or protein sequence in two species, the more closely related the two species are presumed to be—that is, the more recently they diverged from a common ancestor. Molecular evolution analyses are based on the assumption that it is highly unlikely that two unrelated species would evolve precisely the same sequence of DNA nucleotides or amino acids simply by chance.

### Comparing Genomes

On a whole genome level, humans, chimpanzees, and bonobos (pygmy chimpanzees) share 99.5 percent of the genes that encode proteins. This means that many of the differences between humans and chimpanzees are probably dictated by the 0.5 percent of the protein-encoding genome we don't share with them.

Uniquely human traits include spoken language, abstract reasoning ability, highly opposable thumbs, and larger frontal lobes of the brain. It might seem impossible that distinctions as great as those between human and chimp could reflect the actions of just a few genes, but actually individual genes can have great effects on appearance, physiology, and development.

One stark difference between chimp and human that could stem from a single gene is hairiness. Chimpanzees and gorillas express a keratin gene whose counterpart in humans has been silenced into pseudogene status by a nonsense mutation. That gene could, by itself, account for the great difference in hairiness between these other primates and us.

Another single gene that accounts for great differences among primates controls the switch from embryonic to fetal hemoglobin, which makes a drastic difference in the duration of fetal development. More primitive primates lack or have very little fetal hemoglobin. The presence of fetal hemoglobin in more recently evolved and more complex primates lengthened the fetal period, which in turn maximized brain growth—and larger brains make all sorts of capabilities possible. Single genes can also account for the lengthier childhood and adolescence in humans compared to chimpanzees.

It will be interesting to catalog the exact distinctions that the human and chimpanzee genome projects reveal. Meanwhile, comparisons of the human genome sequence to those of other species provide glimpses into how we are more complex—and how we are not (see figure 22.10). Overall, the human genome differs from those of the fruit fly and roundworm by having a more complex organization of the same basic parts. For example, the human genome harbors 30 copies of the gene that encodes fibroblast growth factor, compared to two copies in the fly and worm genomes. This growth factor is important for the development of highly complex organs.

Genome studies indicate that over deep time, genes and gene pieces provided animals that have backbones (vertebrates), including humans, with defining characteristics of complex neural networks, blood clotting pathways, and acquired immunity. In addition, the genomes of vertebrates make possible refined apoptosis, greater control of transcription, complex development, and more intricate signaling both within and between cells. Our close relationship to the other vertebrates is revealed by comparing the human genome sequence to that of the pufferfish *Tetraodon nigroviridis.* Its genome is like ours, minus many of the repeats and introns. So similar are the two genomes that researchers used the pufferfish genome to round down the estimate of the number of human genes to 28,000 to 34,000. On the other hand, it is somewhat sobering to think that the protein-encoding portion of our genome is nearly the same as that of a fish!

Evolutionary information can also be discerned within our genomes. Consider the 906 genes and pseudogenes that enable us to detect odors. The sense of smell derives from a one-inch-square patch of tissue high in the nasal cavity that harbors 12 million cells that bear odorant receptor (OR) proteins. (In contrast, a bloodhound has 4 billion such cells!) Molecules given off by something smelly—a wedge of pungent cheese or a freshly cut melon, for example—bind to distinct combinations of these receptors, which then signal the brain in a way that creates the perception of an associated odor.

Our odorant receptor genes include more than 30 million bases, or about 1 percent of the genome, and they occur in clusters. Yet the fact that 60 percent of the 906 genes are pseudogenes—with sequences similar to those of functional "smell" genes but riddled with mutations that render them unexpressed—suggests that today's gene family has shrunken, but retains remnants of times past. Our ancestors thousands and millions of years ago probably depended upon the sense of smell for survival more than we do today. Natural selection may have eliminated certain odorant receptor genes that were no longer crucial to survival. Yet other genetic evidence indicates that natural selection also acted in a positive direction. Specifically, the odorant receptor pseudogenes harbor many diverse SNPs (single nucleotide polymorphisms), whereas the OR genes that still function today are remarkably like one another. The nucleotide differences that do persist among the functional OR genes tend to be in codons that change the corresponding amino acids, rather than in the wobble position that is in effect a "silent" mutation. Such differences among similar genes in meaningful codon positions is very unusual. Geneticists interpret the finding to suggest that natural selection favored these particular sequences.

**table 15.2**

**Percent of Common Chromosome Bands Between Humans and Other Species**

| | |
|---|---|
| Chimpanzees | $99^{+}$% |
| Gorillas | $99^{+}$% |
| Orangutans | $99^{+}$% |
| African green monkeys | 95% |
| Domestic cats | 35% |
| Mice | 7% |

## Comparing Chromosomes

Similarities of band patterns and gene orders on stained and FISH-probed chromosomes provides measures of species relatedness (table 15.2). Chromosomes, genes, or proteins that are identical or very similar in different species are said to be **highly conserved.**

Human chromosome banding patterns match most closely those of chimpanzees, then gorillas, and then orangutans. The karyotypes of humans, chimpanzees, and apes differ from each other mostly by inversions, which are changes that occur within chromosomes. Karyotype differences between these three primates and more primitive primates are predominantly translocations, which are events that occur between chromosome types.

If both copies of human chromosome 2 were broken in half, we would have 48 chromosomes, as the three species of apes do, instead of 46. The banding pattern of chromosome 1 in humans, chimps, gorillas, and orangutans matches that of two small chromosomes in the African green monkey, suggesting that this monkey was ancestral to the other primates. Reading 15.1 describes the evolution of a gene that duplicated, traveled, and then duplicated again, in the genomes of various primates.

We can also compare chromosome patterns between species that are not as closely related. All mammals, for example, have identically banded X chromosomes. One section of human chromosome 1 that is alike in humans, apes, and monkeys is also remarkably similar to parts of chromosomes in cats and mice. A human even shares several chromosomal segments with a cat (figure 15.6).

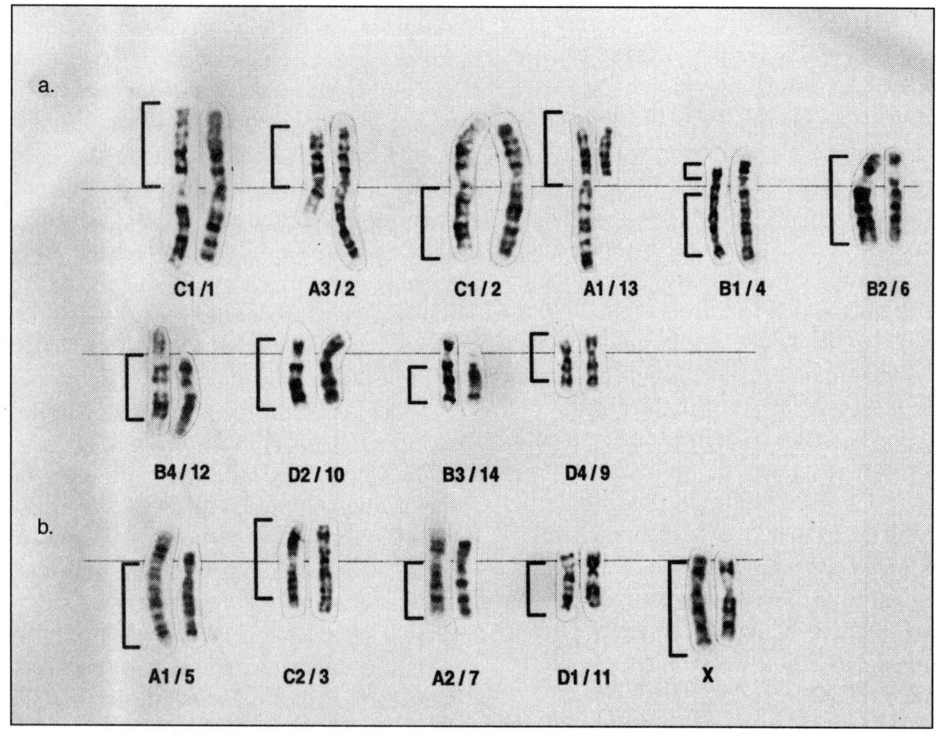

**figure 15.6**

**Conserved regions of human and cat chromosomes.** In each pair, the chromosome on the left is from a cat, and the chromosome on the right is from a human. The brackets indicate areas that appear to correspond. The chromosomes in (*a*) have similar banding patterns generated from traditional stains that target chromosome regions with generally similar DNA base content. The chromosomes in (*b*) do not look alike, but DNA probes indicate that specific gene pairs do indeed share DNA sequences. (*c*) The author and one of her cats, Nirvana.

Chromosome band pattern similarities, obtained with stains, although striking, are not ideal measures of species relatedness because a band can contain many genes that may not be the same as those within a band at a corresponding locus in another species. DNA probes used as part of a FISH analysis are more precise because they mark particular genes (see figure 12.8*b*). Direct correspondence of known gene order, or **synteny,** between species is better evidence of close evolutionary relationships. For example, 11 genes are closely linked on the long arm of human chromosome 21, mouse chromosome 16, and on a chromosome called U10 in cows. However, several genes on human chromosome 3 are found near the human chromosome 21 counterpart in mice and cows. One interpretation of this finding is that a mammal ancestral to these three species had all of these genes together and the genes dispersed to an additional chromosome in humans.

## Comparing Protein Sequences

The fact that all species utilize the same genetic code to synthesize proteins argues for a common ancestry to all life on earth. In addition, many different types of organisms use the same proteins, with only slight variations in amino acid sequence. The keratin genes that encode a sheep's wool protein, for example, have counterparts on chromosome 11 in humans. The similarities in amino acid sequences in human and chimpanzee proteins are astounding—many proteins are alike in 99 percent of their amino acids. Several are virtually identical. When analyzing a gene's function, researchers routinely consult databases of known genes in many other organisms. Two of the most highly conserved proteins are cytochrome *c* and homeobox proteins.

### *Cytochrome* c

One of the most ancient and well-studied proteins is cytochrome *c,* which helps to extract energy from nutrients in the mitochondria in the reactions of cellular respiration. Twenty of 104 amino acids occupy identical positions in the cytochrome *c* of all eukaryotes. The more closely related two species are, the more alike their cytochrome *c* amino acid sequence is (figure 15.7). Human cytochrome *c,* for example, differs from horse cytochrome *c* by 12 amino acids, and from kangaroo cytochrome *c* by eight amino acids. The human protein is identical to chimpanzee cytochrome *c.*

### *Homeobox Proteins*

Another gene that has changed little across evolutionary time is called a **homeobox** or HOX gene. It encodes a transcription factor that controls the order in which an embryo turns on genes. This cascade of gene action ultimately ensures that anatomical parts—whether a leg, petal, or segment of a larva—develop in the appropriate places. The

**Cytochrome *c* Evolution**

| Organism | Number of amino acid differences from humans |
|---|---|
| Chimpanzee | 0 |
| Rhesus monkey | 1 |
| Rabbit | 9 |
| Cow | 10 |
| Pigeon | 12 |
| Bullfrog | 20 |
| Fruit fly | 24 |
| Wheat germ | 37 |
| Yeast | 42 |

a.

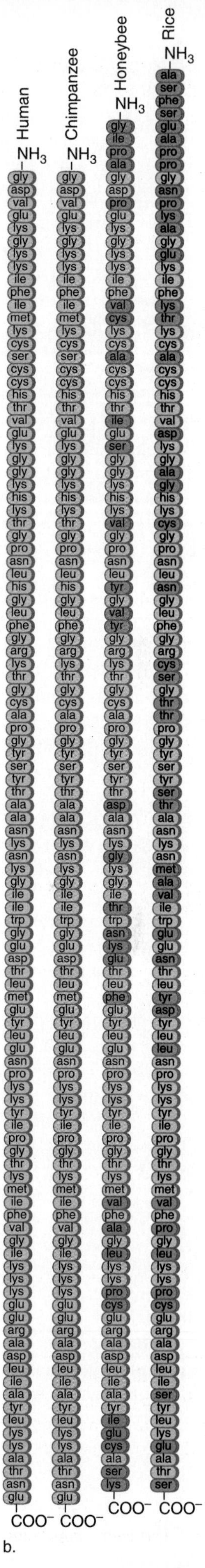

b.

**figure 15.7**

**Amino acid sequence similarities are a measure of evolutionary relatedness.** Similarities in amino acid sequence for the respiratory protein cytochrome *c* in humans and other species parallel the degree of relatedness among them. (*a*) compares the sequence differences of nine species for this highly conserved protein. (*b*) shows the differences in cytochrome *c* sequence among four species. Amino acids that differ from those in the human sequence are highlighted purple.

*Reading 15.1*

# Tracing the Evolution of a Gene in Primates

Knowing the DNA sequence of a protein-encoding gene provides a powerful tool to reconstruct evolution by tracking the sequence in the genomes of our closest relatives—other primates. So it is for two genes called PMCHL1 and PMCHL2. The presence of these genes in the human genome is the culmination of several events in the past that moved, mutated, and duplicated an original DNA sequence. Along the way, the newly created genes acquired functions—still unknown—that natural selection retained.

French researchers Anouk Courseaux and Jean-Louis Nahon had not been planning an evolutionary study when they began to look at the PMCHL genes, which are on chromosome 5. The initials stand for "pro MCH-like" because the two genes include much of the sequence of a gene called MCH that resides on chromosome 12. MCH encodes a precursor to a neuropeptide, and it functions in the brain. The researchers hunted for MCH-like DNA sequences in the genomes of several primates on the road toward humanity, and superimposed the genetic changes along a timeline that indicates when each type of primate diverged from the ancestral lineage (figures 1 and 2).

The work was painstaking. When the pieces were assembled, a story began to emerge. Before about 35 million years ago, when primates were limited to the primitive prosimians (such as lorises and tarsiers) that looked more like rodents than monkeys, a DNA sequence very similar to the human MCH gene resided on a chromosome that corresponds to human chromosome 12. Evolution continued, and gradually the Old World monkeys, distinguished by their long, grasping tails and agility in the trees, flourished. Sometime around 35 million years ago, the ancestral MCH gene moved to the counterpart to the short arm of human chromosome 5. Evidence suggests that this was a retrotransposition event—that is, an mRNA transcribed from the original MCH gene was reverse transcribed into a DNA sequence that then inserted itself back into a chromosome—but chromosome 5, not 12. At this time, as evidenced by the present-day genomes of New World monkeys, the early version of the PMCHL gene lacked introns.

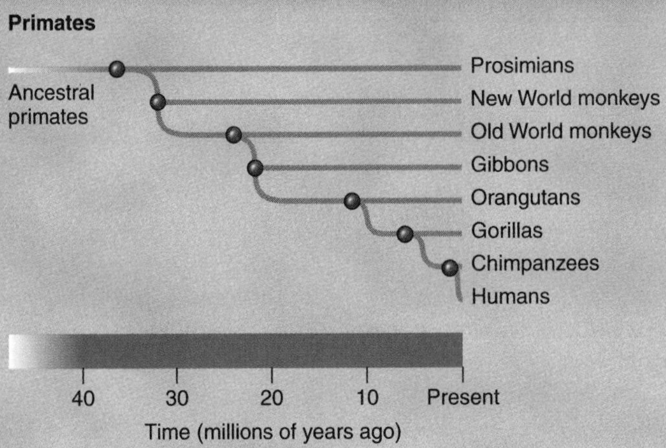


**figure 1**

**Evolutionary relationships of modern primates.**

By about 26 million years ago, the Old World monkeys, with their characteristic grasping thumb and specialized teeth, but shorter tails, had appeared. The PMCHL gene then underwent mutations that introduced splice sites, resulting in the appearance of introns. Perhaps other changes occurred that so damaged the health of the individual that natural selection weeded them out. But this first intron remained in the gene. By the time that gibbons had evolved, more introns were present, further mutations had occurred, and a part of the original gene that did not encode protein was included in the gene, and now did encode protein. The gene grew. Finally, sometime between the origin of the gibbon about 14 million years ago and the chimpanzee about 4 million years ago, the PMCHL gene underwent two profound events—it duplicated itself, and one copy jumped over the centromere to land in the long arm of chromosome 5. This is the gene organization that persists in humans today: an ancestral MCH gene on chromosome 12 and two PMCHL genes on either side of the centromere of chromosome 5. The functions of these two genes are still unknown. However, PMCHL1 is expressed in fetal, newborn, and adult brain, and PMCHL2 is expressed in the testis. These genes may illustrate exaptation, the arising of a new function from an existing DNA sequence that had a different original function.

The evolution of the PMCHL genes tells a broader tale, getting to the heart of what makes us human. We like to consider ourselves to be very different from the other primates, although to an observer from another world, we might just appear to be hairless, upright apes. Yet DNA sequences among the primates are remarkably, even disturbingly, alike. Moving and duplicating genes can explain how we can be simultaneously genetically alike yet genetically different.

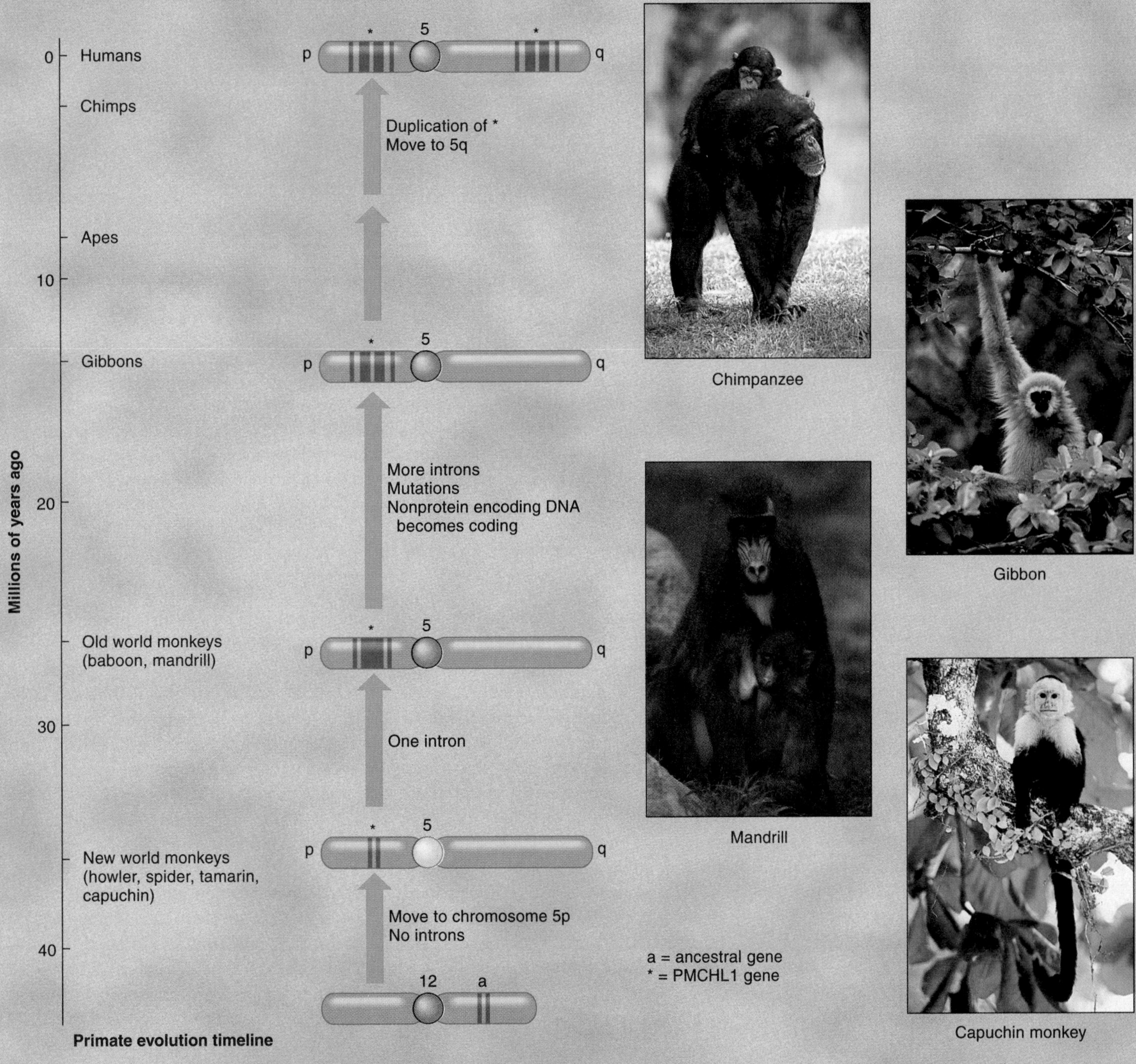

Chimpanzee

Gibbon

Mandrill

Capuchin monkey

## figure 2

**Gene evolution.** Comparing the same gene among primate species reveals how parts of the human genome arose from gene mutation, movement, and duplication.

highly conserved portion of a homeobox protein is a 60-amino-acid sequence, called the homeodomain, that is encoded by a 180-base DNA sequence called the homeobox. (Genes that include homeobox sequences are also termed homeotic.) The genes are known in organisms as simple as jellyfish and as complex as humans. In all the multicellular species in which homeobox sequences have been identified, they organize body parts. In humans and most other vertebrates, there are 39 HOX genes, and they are in four clusters called A, B, C, and D. The genes have very few introns. Another intriguing aspect of HOX gene clusters is that the individual genes are expressed in a sequence, in developmental time or anatomical position, that is the same as their order on the chromosome.

The terms homeobox and homeodomain derive from the homeotic mutants of the fruit fly *Drosophila melanogaster,* which have mixed-up body parts. Geneticists have studied them for half a century. In one mutant, for example, a leg grows where an antenna should be; in another, an antenna replaces mouthparts. Researchers sequenced the fruit fly homeobox gene in 1983 and then found it nearly everywhere—in frogs, mice, beetles, mosquitoes, slime molds, chickens, roundworms, corn, humans, petunias, and many other species. Because the homeobox protein is a transcription factor (see chapter 9), it controls the activities of other genes.

Mutations in homeobox genes cause certain human illnesses. In a form of leukemia, a homeobox mutation shifts certain immature white blood cells onto the wrong developmental pathway. The misguided cells retain the rapid cell division characteristic of immature cells, causing the cancer. DiGeorge syndrome is another condition that is caused by a homeobox gene. Although affected individuals hardly sprout legs from their heads as do *Antennapedia* flies, the signs and symptoms of DiGeorge syndrome are reminiscent of the flies. These include missing thymus and parathyroid glands and abnormal development of the ears, nose, mouth, and throat—structures corresponding to anatomical regions similar to the sites of abnormalities in the flies. Figure 15.8 shows another human disorder caused by a mutation in a HOX gene, synpolydactyly.

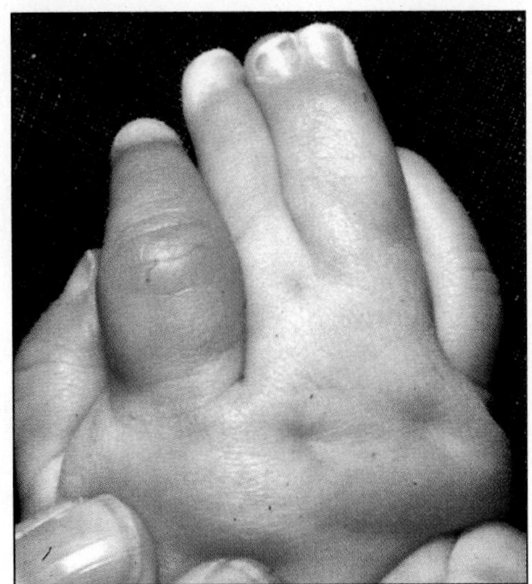

a.

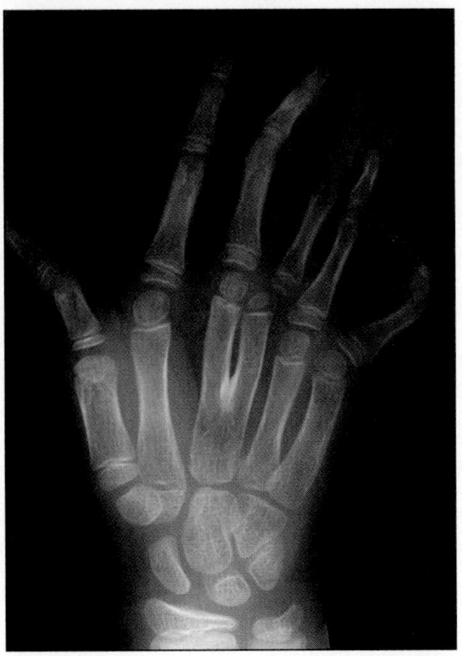

b.

**figure 15.8**

**A human HOX gene mutation causes synpolydactyly.** Mutation in the HOXD 13 gene disrupts development of fingers and toes, causing a very distinctive phenotype shown in this photograph (*a*) and X-ray image (*b*). The third and fourth fingers are partially fused, as are the fourth and fifth toes, with an extra digit within the webbed material. Other digits may be affected to lesser degrees. Researchers discovered the mutation in 182 members of an isolated family living in Turkey. The gene is on chromosome 2q. It has a type of mutation not seen before. The normal protein has a section of 15 alanines. In affected individuals, this polyalanine stretch is expanded by as many as 14 additional alanines.

Experiments that implant genes of one species into another reveal how alike the homeobox is—implying it is essential and ancient. If a mouse version of the *Antennapedia* gene is placed into the fertilized egg of a normal fly, the adult fly grows legs on its head, expressing the mouse gene as if it were the fly counterpart. Similarly, the human version of the gene, placed into a mouse's fertilized egg, disrupts the adult mouse's head development. Homeotic genes and the proteins they encode, therefore, are probably genetic instructions basic to the development of all complex organisms.

## Comparing DNA Sequences

We can assess similarities in DNA sequences between two species for a single gene or piece of DNA, or for the total genome. A comparison technique called DNA hybridization uses complementary base pairing to estimate how similar the genomes of two species are. The researcher unwinds DNA double helices from two species and mixes them. The rate at which hybrid DNA double helices re-form—that is, become DNA molecules containing one helix from each species—is a direct measure of how similar they are in sequence. The faster the DNA from two species forms hybrids, the more of the sequence they share and the more closely related they are (figure 15.9). DNA hybridization of entire genomes, including non-protein-encoding sequences, shows that human DNA differs in 1.0 percent of its base pairs from chimpanzee DNA; in 2.3 percent from gorilla DNA; and in 3.7 percent from orangutan DNA. The conclusion is that we are most closely related to chimps, then gorillas, and then orangutans. The sequencing of genomes will make this technique obsolete, but it remains a quick way to estimate evolutionary relatedness.

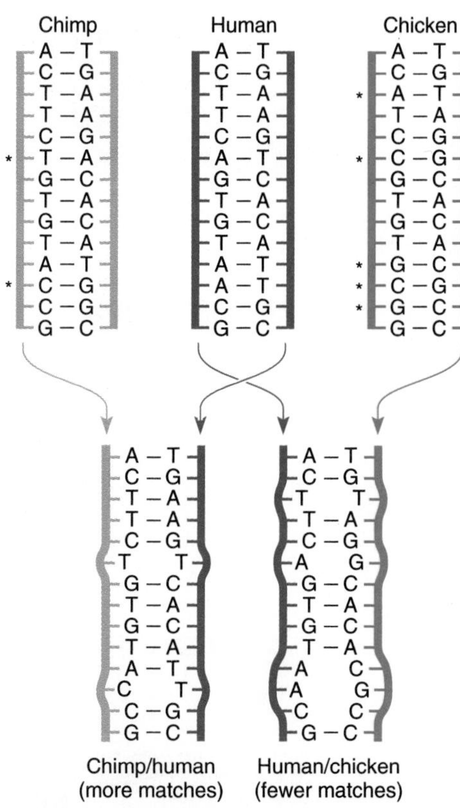


**figure 15.9**

**The rate of DNA hybridization reflects the degree of evolutionary relatedness.** This highly schematic diagram shows why DNA from a human hybridizes more rapidly with chimpanzee DNA than it does with chicken DNA. The * refers to sites where chimp or chicken DNA is different from human DNA.

## *Animal Models*

Identifying corresponding genes in different species is very important in medical research; animal models of human diseases are used to test experimental treatments. It is important that an animal model of a human disease have the same signs and symptoms. Because of differences in physiology, development, and lifespan, another mammal might not have the same phenotype as a human with a mutation in the same gene. However, corresponding genes can reveal basic abnormalities at the cell or molecular level. For example, two thirds of the genes known to cause Mendelian disorders in humans have counterparts in fruit flies!

For some genes, a close correspondence in phenotype can be seen among species. People with Waardenburg syndrome, for example, have a characteristic white forelock of hair; wide-spaced, light-colored eyes; and hearing impairment (figure 15.10). The responsible gene is very similar in sequence to one in cats who have white coats and blue eyes and who are deaf. Horses, mice, and minks also have this combination of traits, which is thought to stem from abnormal movements of pigment cells in the embryo's outermost layer.

## *Ancient DNA*

When comparing DNA of modern species, a researcher can easily repeat an experiment—ample samples of chimp or human DNA are available directly from the source. This isn't the case with ancient DNA, such as genetic material from insects preserved in amber, which is hardened resin from certain pine trees. The mix of chemicals in amber entombed whatever happened to fall into it when it was the consistency of maple syrup. Alcohols and sugars in the resin dried out the specimen, and other organic molecules acted as fixatives, keeping cellular contents in place. The resin itself sealed out oxygen and bacteria, which would otherwise have decomposed tissue before it could be preserved. Finally, organic molecules called tarpenes link together, hardening the resin over a period of 4 to 5 million years. Then, DNA must be extracted, and enough of it must be intact to amplify it using PCR (see figure 9.15).

Probing ancient, preserved DNA for clues to past life is an exciting field, but one that is subject to romanticization by the media. The novel and film *Jurassic Park* described the cloning of dinosaurs from blood in mosquitoes trapped in amber—not a very likely scenario. This is true too for the idea of bringing back a mammoth from its preserved DNA. Researchers are considering how to investigate mammoths that were flash-frozen high in the cliffs of Siberia. These elephant ancestors roamed the grasslands here from 1.8 million years ago until about 11,000 years ago. Starvation following the last ice

a.

b.

c.

**figure 15.10**

**One gene can cause a similar spectrum of effects in different species.** A gene in mice (*a*), cats (*b*), humans (*c*), and other species causes light eye color, hearing or other neurological impairment, and a fair forelock in the center of the head.

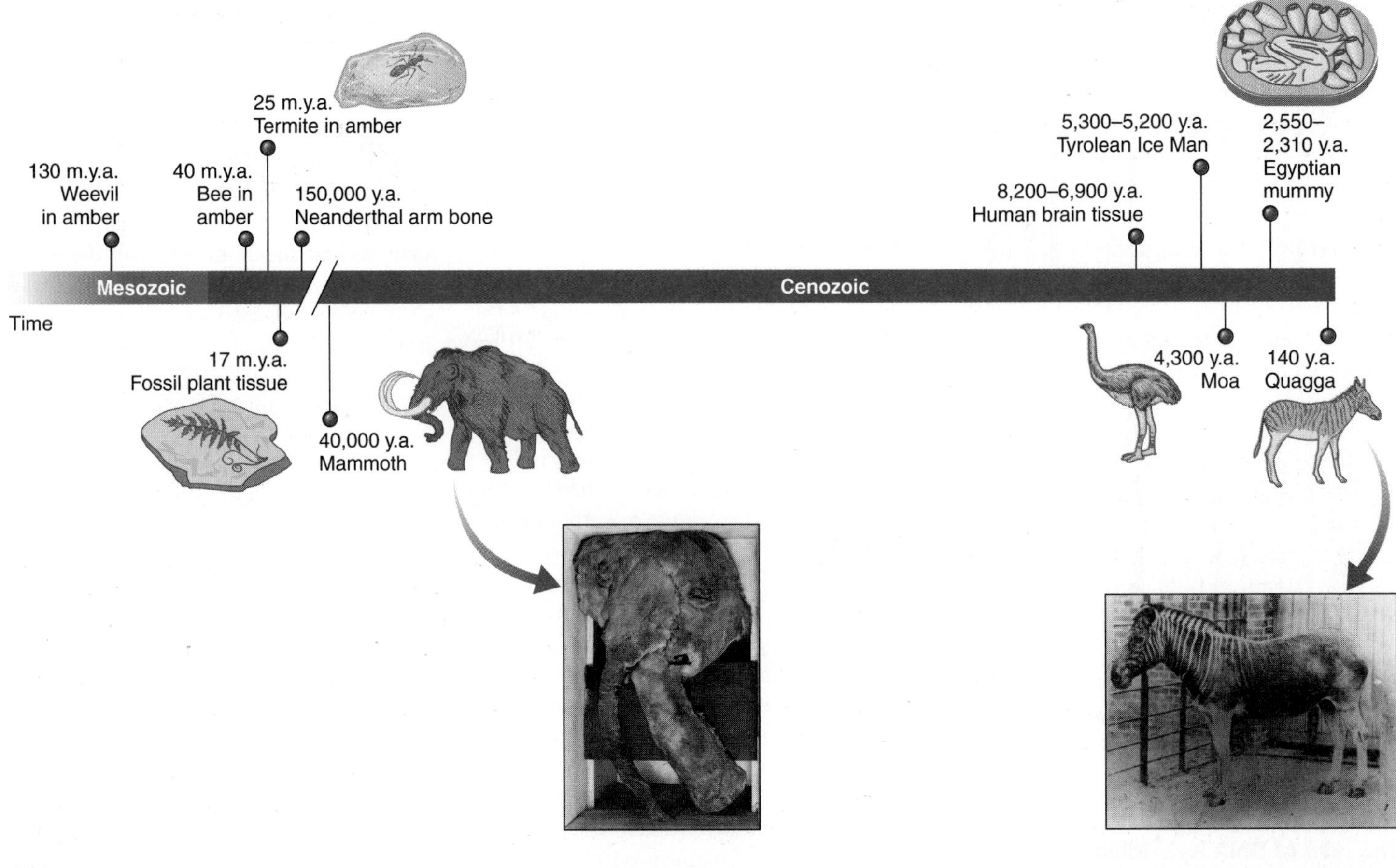


**figure 15.11**

**Ancient DNA.** Researchers have extracted DNA from these organisms.

Source: Data from Roger Lewis, Patterns in evolution, in *Scientific American,* 1997.

age drove their extinction, although a few isolated populations survived until about 3,800 years ago.

Biologists are searching frozen mammoth remains to see if cell nuclei might have been preserved, and can yield DNA. If so, researchers hope to take two approaches to recreating a mammoth. In the first strategy, sperm are found. Researchers plan to use those sperm bearing X chromosomes to artificially inseminate an elephant. If the 600-day pregnancy is successful, the elephant should give birth to a female that is half elephant, half mammoth. Some 13 or so years later, she can then be inseminated by more mammoth sperm, producing a baby that is three-quarters mammoth, and so on.

The second way to bring back a mammoth is to clone one. A nucleus from a somatic cell of a mammoth would be delivered into an elephant's oocyte whose nucleus has been removed. The offspring would be all mammoth, and likely fertile, in contrast to the half-breed that would result from artificial insemination. Males and females would need to be cloned to begin a population, and from 10 to 20 different genetic variants would be necessary to sustain a population—a tall order!

Mammoth cloning still lies in the realm of science fiction. In reality, the first successful extraction of bits of ancient DNA occurred in 1990, from a 17-million-year-old magnolia leaf entombed in amber (figure 15.11). The quest to probe ancient DNA has sent researchers into the back rooms of museums, dusting off specimens of pressed leaves, insects stuck with pins into Styrofoam, and old bones and pelts, in search of nucleic acid clues to life in the past.

### Key Concepts

Molecular evolution is based on the assumption that the more recently two species shared a common ancestor, the more alike their stained chromosome bands and protein and DNA sequences. • Whole genome comparisons indicate that humans differ from chimps by only 0.5 percent of their protein-encoding genes, but individual genes can exert great effects on phenotypes. Chromosome banding pattern similarities reflect species relationships. • Amino acid sequences of highly conserved proteins also reflect species relationships. • DNA hybridization experiments estimate how closely two species are related by how quickly their DNA forms hybrid double helices. Direct sequencing is also used to compare genes between species. Rarely, scientists can obtain DNA from preserved extinct organisms, amplify it, and compare it to sequences in modern species.

## 15.3 Molecular Clocks

A clock measures the passage of time by moving its hands through a certain degree of a circle in a specific and constant interval of time—a second, a minute, or an hour. Similarly, a polymeric molecule can be used as a **molecular clock** if its building blocks are replaced at a known and constant rate.

The similarity of nuclear DNA sequences in different species can help scientists estimate the time when the organisms diverged from a common ancestor, if they know the rate of base substitution mutation. For example, many nuclear genes studied in humans and chimpanzees differ in 5 percent of their bases, and substitutions occur at a rate of 1 percent per 1 million years. Therefore, 5 million years have presumably passed since the two species diverged. Mitochondrial DNA sequences are also tracked in molecular clock studies, as we will soon see.

Time scales based on fossil evidence and molecular clocks can be superimposed on evolutionary tree diagrams that are constructed from DNA or protein sequence data. However, evolutionary trees can become complex when a single set of data can be arranged into a large number of different tree configurations. A tree for 17 mammalian species, for example, can be constructed in 10,395 different ways! The sequence in which the data are entered into tree-building computer programs influences the tree's shape, which is vital to interpreting species relationships. With new sequence information, the tree possibilities change.

**Parsimony analysis** is a statistical method used to identify an evolutionary tree likely to represent what really happened. A computer connects all evolutionary tree sequence data using the fewest possible number of mutational events to account for observed DNA base sequence differences. For the 5-base sequence in figure 15.12, for example, the data can be arranged into two possible tree diagrams. Because mutations are rare events, the tree that requires the fewest mutations is more likely to reflect reality.

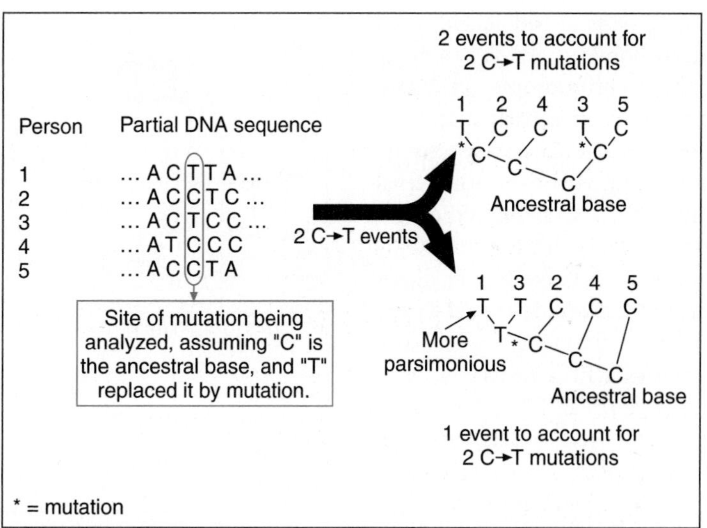


### figure 15.12

**Parsimony analysis.** Even a computer has trouble arranging DNA differences into an evolutionary tree showing species, population, or individual relationships. A parsimonious tree accounts for all data with the fewest number of mutations. Here, the two individuals who have a "T" in place of the ancestral "C" could have arisen in two mutational events or one, assuming that these individuals had a common ancestor. Since mutations are rare events, the more realistic scenario is one mutation.

### Neanderthals Revisited

Molecular clock data can provide clues to relationships among modern organisms, and also compensate for deficiencies in the fossil record. Consider our knowledge of Neanderthals.

Two technologies—analyzing ancient DNA and using mtDNA clocks—indicate that Neanderthals were a side branch on our family tree and diverged from us more than half a million years ago. This is much farther back than the fossil evidence indicates.

In 1997, a graduate student, Matthias Krings, ground up a bit of arm bone from the original French Neanderthal skeleton. He then performed PCR on several 100-base-pair-long pieces of a mitochondrial DNA that does not encode protein, and mutates very rapidly, perhaps because it does not affect the phenotype and is therefore not under selective pressure. The DNA pieces were sequenced, double-checked, and compared to corresponding sequences from 986 modern *Homo sapiens*. The Neanderthal DNA differed from corresponding DNA of modern people at 26 positions. Not only is this three times the number of differences seen between pairs of the most unrelated modern humans, but the locations of the base differences were completely different from the places where modern genes vary (SNPs). A more recent Neanderthal bone, from 29,000 years ago, was analyzed for a mitochondrial gene in early 2000 and the results supported the 1997 results. This genetic distinction suggests that it is highly unlikely that Neanderthals and modern humans ever interbred. Extrapolations from mutation rates of genes in modern humans and chimps indicate the last shared ancestor between Neanderthals and humans lived from 690,000 to 550,000 years ago. On a more philosophical level, the DNA evidence distancing us from Neanderthals suggests there may have been a time when two types of "people" roamed the planet.

### Choosing Clues

A complication of molecular clock studies is that genes change (mutate) at different rates. Another limitation is in deciding *which* genes to study. Different conclusions can arise from deciphering and comparing different DNA sequences. It can be difficult to know which genes to study to track human origins. But, as with any evidence, the more data, the more complete our picture of the past. Such reconstructions change when new evidence is added or existing evidence is questioned or discredited. Imagine how the Cinderella story might have turned out if the mean stepsisters or magic pumpkin didn't exist, or if a competing prince had arrived. So it is with the challenge of assembling clues to human origins into a coherent explanation of real events.

Tracking DNA sequences on autosomes and the X chromosome can provide information on groups of people. However, analyzing mitochondrial DNA and sequences on the Y chromosome can reveal which parent passed on a particular DNA sequence in females and males, respectively.

MtDNA is followed to trace maternal lineages because this DNA passes almost exclusively from mothers to offspring. Only about 100 mitochondria enter a fertilized egg, and usually these are destroyed by the 8-cell embryo stage. MtDNA is useful as a clock of recent events, because its DNA mutates faster than nuclear DNA because it lacks repair systems. Also, mtDNA is much more abundant in cells and therefore easier to extract to study—a cell has only one nucleus, but typically thousands of mitochondria and their small genomes (see section 5.2 in chapter 5). However, a drawback to mtDNA dating is that rarely paternal mtDNA enters an oocyte and recombines with the maternal mtDNA, which obscures conclusions.

Y chromosome sequences are used to trace paternal lineages. Much of the Y chromosome is identical in males from all population groups, but several sites in the part of the chromosome that does not recombine with the X have been used for molecular clock studies. In both mtDNA and Y chromosome investigations, researchers compare haplotypes as well as DNA sequences.

Researchers have been using mtDNA analysis longer than they have Y chromosome techniques. Interestingly, when Y chromosome data are added to existing mtDNA information, the interpretations sometimes change. This has been the case for two fascinating scenarios—pinpointing where modern humans arose, and the more recent origin of native Americans.

The logic behind mtDNA and Y chromosome analysis is that the more alike DNA sequences are between two individuals, the more recently they presumably shared a common ancestor. Put the reverse way, the greater the genetic differences between two individuals, the less closely related they are, and the more time has passed for those differences to have accrued. Researchers consider DNA sequence differences along with known mutation rates for those sequences to construct evolutionary tree diagrams that point to times and places of origin, or dispersal patterns, of groups of people.

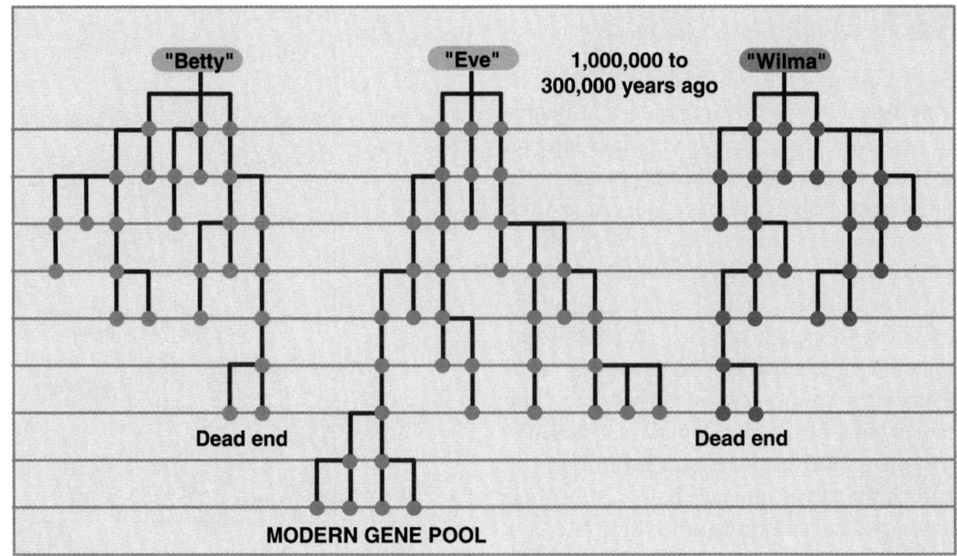


## figure 15.13

**Eve hypothesis.** According to the mitochondrial Eve hypothesis, modern mtDNA retains some sequences from a figurative first woman, "Eve," who lived in Africa 300,000 to 100,000 years ago. In this schematic illustration, the lines represent generations, and the circles, females. Lineages cease whenever a woman does not have a daughter to pass on the mtDNA.

### *Out of Africa: Mitochondrial Eve—or Eves?*

Theoretically, if a particular sequence of mtDNA could have mutated to yield the mtDNA sequences in modern humans, then that ancestral sequence may represent a very early human or humanlike female—a mitochondrial "Eve," or first woman. Figure 15.13 shows how one maternal line may have come to persist.

When might this theoretical first woman, the most recent female ancestor to us all, have lived? Most fossil and DNA evidence points to Africa, but some anthropologists maintain that modern peoples emerged from Asia, too. Berkeley researchers led by the late Allan Wilson compared mtDNA sequences for protein-encoding as well as noncoding DNA regions in a variety of people, including Africans, African Americans, Europeans, New Guineans, Australians, and others. They concluded from several methods that the hypothesized ancestral woman lived about 200,000 years ago, in Africa.

One way to reach this figure is by comparing how much the mtDNA sequence differs among modern humans to how much it differs between humans and chimpanzees. Specifically, the differences in mtDNA sequences among contemporary humans is 1/25 the difference between humans and chimpanzees. The two species diverged about 5 million years ago, according to extrapolation from fossil and molecular evidence. Multiplying 1/25 by 5 million gives a value of 200,000 years ago. This estimate assumes that the mtDNA mutation rate has remained constant over that time.

Where did Eve live? Mitochondrial DNA comparisons consistently find that African people have the most numerous and diverse collection of mtDNA mutations. The same is true for other regions of the genome. This indicates that Africans have existed longer than other modern peoples, because it takes time for mutations to accumulate. In many evolutionary trees constructed by parsimony analysis, the individuals whose DNA sequences form the bases are from Africa. The variants of other modern human populations are subsets of the genome of Africans.

The idea of mitochondrial Eve is part of the "out of Africa" view, or **replacement hypothesis** of human origins. It states that about 200,000 years ago, *H. sapiens* evolved from an *H. erectus* population in Africa. Descendants of these early *H. sapiens* migrated to the Middle East, Asia, and Europe, eventually replacing the people living there.

An alternate view is the **multiregional hypothesis,** which maintains that human

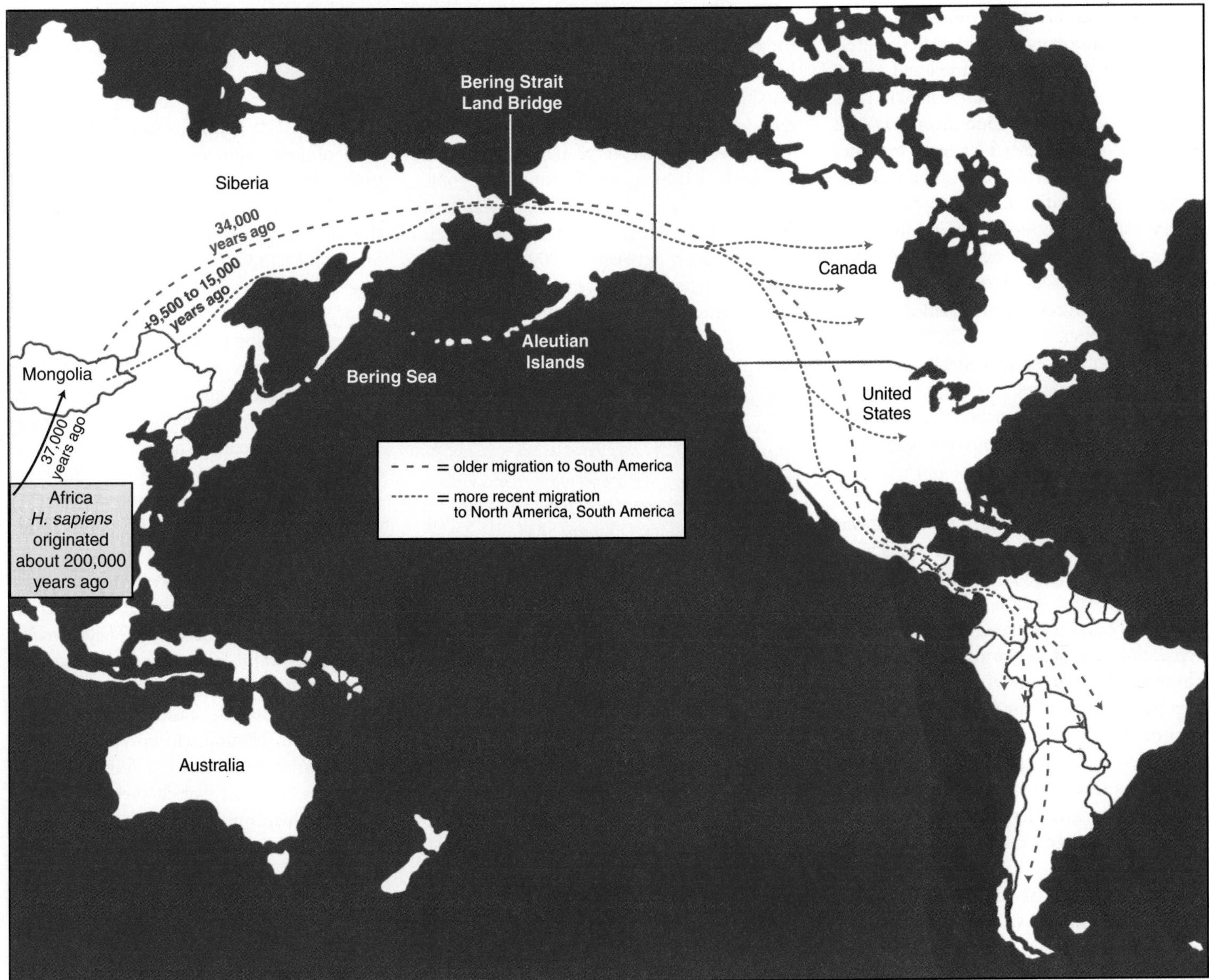


**figure 15.14**

**Tracing human origins.** Analyses of mitochondrial DNA and Y chromosome DNA sequences reveal that the ancestors of native Americans probably came from Mongolia in one migration.

traits originated in several places, and *H. erectus* migrated, mixing and sharing genes, gradually evolving into *H. sapiens.* While most of the mtDNA and Y chromosome data support the "out of Africa" scenario, conflicting evidence sometimes arises as more genes are compared. For example, one study traced the origin of a beta globin gene variant to Asia, and not Africa. However, gene variants seen in non-African populations that are not seen in Africans could have originated more recently than 200,000 years ago, after people had already migrated out of Africa.

The "out of Africa" and multiregionalism theories were once considered so opposed that scientists would actually fight over them at meetings and in scientific journals. Today, as more evidence fine-tunes the stories, it may be possible to reconcile the two views. That is, perhaps after the emergence of modern humans in Africa, one or more subgroups migrated. MtDNA, Y chromosome and chromosome 21 evidence suggests that just such a migration occurred, between 100,000 and 50,000 years ago, from eastern Africa. The people apparently migrated along the coast to southeast Asia, Australia, the Pacific Islands, and eventually throughout Asia. Yet this view is in conflict with fossil evidence that indicates migration northward out of Africa. Clearly, the unfolding story of human origins is not over!

## *A Native American Tale*

The ancestors of native Americans came to North America across the Bering Strait land bridge that formed between Siberia and Alaska during low glacial periods (figure 15.14). Anthropologists traditionally dated three waves of migration based on evidence of ancient human habitation and language differences among modern native populations—the Amerindians about 33,000 years ago, the Nadene 15,000 to 12,000 years ago, and the Eskimo-Aleuts from 7,000 to 5,000 years ago. But DNA evidence indicates one migration of people bringing in diverse genes—and they didn't originate in Siberia, as has been long suspected, but in Mongolia.

One of four major mtDNA haplotypes, B, is widespread in the New World but is not present in Siberians. Haplotype B is present, however, in people who live in Ulan Bator, the capital city of north central Mongolia. According to mtDNA evidence, they were more likely the people who trekked across the land bridge to populate the New World—not the Siberians.

Y chromosome analysis generally supports the Mongolian origin, but it also counters the anthropological evidence indicating three waves of migration. Instead, Y chromosome sequences reveal that all native Americans have descended from one founder population that came from the Mongolian/Chinese border in a nearly continuous migration between 37,000 and 23,000 years ago. Because the members of this population brought with them a variety of DNA sequences on their Y chromosomes, it appears that they included different groups.

Another study compared mtDNA from modern Mongolians, Tibetans, and Chinese to remains from a 700-year-old Amerindian graveyard in Illinois. All four of the mtDNA haplotypes common to native Americans were identified among all of these groups, although the Asian peoples had several others as well and the four native American haplotypes were unusual among the Asians. This suggests that a subset of people, carrying the four haplotypes, left Asia long ago to populate the Americas. The fact that the four haplotypes are very rare among Asians today argues for a single long-ago migration to America that brought in the subset. Since each haplotype is rare in Asia, the three-migration hypothesis requires a rare event to have happened three times. One event that introduced all four haplotypes seems more likely.

Like the competing "out of Africa" and multiregionalism hypotheses, whether native Americans came from Mongolia or Siberia still isn't clear. A study of different Y chromosome haplotypes indicates traces of an ancient Siberian group called the Kets among modern native Americans. The Kets today number fewer than 100, and they still have their own distinctive language. It is possible, then, that some Siberians also contributed to the native American gene pool. The emerging picture of native American origins is that many small groups of people wandered over the land bridge, in search of food, and introduced diverse genes. Their numbers grew between 37,000 and 23,000 years ago, when a rising sea blocked the land bridge and further input into the native American gene pool.

**KEY CONCEPTS**

Molecular clocks apply mutation rates to time scales to estimate when two individuals or types of organisms most recently shared ancestors. Different genes evolve at different rates. Parsimony analysis selects likely evolutionary trees from DNA data. Mitochondrial DNA clocks trace maternal lineages, and Y chromosome sequences trace paternal lineages. • Molecular clocks have been used to examine the relationship of Neanderthals to modern humans and the origin of modern humans. • Molecular clock data also explains the migration pattern of native Americans from Asia.

## 15.4 Eugenics

Fossil evidence, ancient DNA, and molecular clocks are useful in studying our past. But now people can control the future through reproductive choices, which affects the gene pool. You wouldn't think twice about pulling out the smaller, weaker individuals from a row of baby cucumber plants, or about mating your poodle only with another poodle. Attempting to control human reproduction is another matter.

Some people try to control the genes in their offspring by seeking mates with high intelligence or certain physical characteristics. This idea was taken to an absurd extreme in a sperm bank in California where the donors are all Nobel Prize winners, and in a website advertising eggs donated by supermodels or young women with high scores on college entrance exams, as if a perfect S.A.T. score is an inherited trait. The ability to control reproductive choices raises many bioethical issues.

**Eugenics** is the control of individual reproductive choices in humans to achieve a societal goal. Sir Francis Galton coined the term, meaning "good in birth," in 1883 (table 15.3). He defined eugenics as "the science of improvement of the human race germ-plasm through better breeding." The 2,500-year-old caste system in India, and the antimiscegenation laws in the United States from 1930 to 1967 that banned marriage between people of different races, were clearly eugenic because they sought to control reproduction to change society.

Galton's ideas were popular for a time. Eugenics societies formed in several nations and attempted to practice his ideas in various ways. Creating incentives for reproduction among those considered superior constitutes positive eugenics. Interfering with reproduction among those judged inferior is negative eugenics.

One vocal supporter of the eugenics movement was Sir Ronald Aylmer Fisher. In 1930, he published a book, *The Genetical Theory of Natural Selection,* which connected the concepts of Charles Darwin and Gregor Mendel and listed the basic tenets of population genetics. Natural selection and Mendelian inheritance provided a framework for eugenics. The final five chapters of Fisher's otherwise highly regarded work tried to apply the principles of population genetics to human society. Fisher maintained that those at the top of a society tend to be "genetically infertile," producing fewer children than the less-affluent classes. This, he claimed, was the reason why civilizations ultimately topple. He offered several practical suggestions, including state monetary gifts to high-income families for each child born to them.

Early in the 20th century, eugenics focused on maintaining purity. One prominent geneticist, Luther Burbank, realized the value of genetic diversity at the beginning of a eugenic effort. Known for selecting interesting plants and crossing them to breed plants with useful characteristics, such as less prickly cacti and a small-pitted plum, Burbank in 1970 applied his agricultural ideas to people. In a book called *The Training of the Human Plant,* he encouraged immigration so that advantageous combinations of traits would appear as the new Americans interbred. Burbank's plan ran into problems, however, at the selection stage, which allowed only those with "desirable" trait combinations to reproduce.

**table 15.3**

**A Chronology of Eugenics**

| | |
|---|---|
| **1883** | Sir Francis Galton coins the term *eugenics.* |
| **1889** | Sir Francis Galton's writings are published in the book *Natural Inheritance.* |
| **1896** | Connecticut enacts law forbidding sex with a person who has epilepsy or is "feebleminded" or an "imbecile." |
| **1904** | Galton establishes the Eugenics Record Office at the University of London to keep family records. |
| **1907** | First eugenic law in the United States orders sterilization of institutionalized mentally retarded males and criminal males, when experts recommend it. |
| **1910** | Eugenics Record Office founded in Cold Spring Harbor, New York, to collect family and institutional data. |
| **1924** | Immigration Act limits entry into the United States of "idiots, imbeciles, feebleminded, epileptics, insane persons," and restricts immigration to 7 percent of the U.S. population from a particular country according to the 1890 census—keeping out those from southern and eastern Europe. |
| **1927** | Supreme Court (*Buck vs. Bell*) upholds compulsory sterilization of the mentally retarded by a vote of 8 to 1, leading to many state laws. |
| **1934** | Eugenic sterilization law of Nazi Germany orders sterilization of individuals with conditions thought to be inherited, including epilepsy, schizophrenia, and blindness, depending upon rulings in Genetic Health Courts. |
| **1939** | Nazis begin killing 5,000 children with birth defects or mental retardation, then 70,000 "unfit" adults. |
| **1956** | U.S. state eugenic sterilization laws are repealed, but 58,000 people have already been sterilized. |
| **1965** | U.S. immigration laws reformed, lifting many restrictions. |
| **1980s** | California's Center for Germinal Choice is established, where Nobel Prize winners can deposit sperm to inseminate carefully chosen women. |
| **1990s** | Laws passed to prevent health insurance or employment discrimination based on genotype. |
| **2000+** | Human genome project completed. Many new genetic tests available. |

On the East Coast, Charles Davenport led the eugenics movement. In 1910, he established the Eugenics Record Office at Cold Spring Harbor, New York. There he headed a massive effort to compile data from institutions, prisons, and the general society. In the rather simplistic view of genetics at the time, he attributed nearly every trait to a single gene. "Feeblemindedness," he thought, was inherited as an autosomal recessive trait.

Eugenics was practiced in other nations, too. From 1934 until 1976, the Swedish government forced certain individuals to be sterilized as part of a "scientific and modern way of changing society for the better," according to one historian. At first, only mentally ill people were sterilized, but poor, single mothers were later included. Revelation of the Nazi atrocities did not halt eugenics in Sweden, but the women's movement in the 1970s pushed for an end to the forced sterilizations.

Seeking information on human genetics does not necessarily have anything to do with eugenics. Bioethics: Choices for the Future in chapter 1 describes how some nations are acquiring genetic information on citizens. These projects are not eugenic, however, because the information will not be used to make reproductive decisions. Eugenics, in contrast, uses such information to maximize the genetic contribution from those deemed desirable and minimize the contribution from those considered unacceptable. But a major fallacy of eugenics is its subjectivity. Who decides which traits are desirable or superior? Eugenic thinking arises from time to time, even today.

In 1994, for example, China passed the Maternal and Infant Health Care Law, which proposes "ensuring the quality of the newborn population" and forbids procreation between two people if physical exams show "genetic disease of a serious nature . . . that may totally or partially deprive the victim of the ability to live independently, that [is] highly possible to recur in generations to come, and that [is] medically considered inappropriate for reproduction." Such "genetic diseases" include mental retardation, mental illness, and seizures, conditions that are ill-defined in the law and are not necessarily inherited.

Because genetic technologies may affect reproductive choices and can influence which alleles pass to the next generation, the field of modern genetics has sometimes been compared to eugenics. Medical genetics and eugenics differ in their overall goals. Eugenics aims to skew allele frequencies in future generations by allowing only people with certain "valuable" genotypes to reproduce, for the supposed benefit of the population as a whole. The goal of medical genetics, in contrast, is usually to skew allele frequencies in order to prevent suffering on a family level. Bioethics: Choices for the Future on page 318 contrasts a population view of the distinction between genetic technology and eugenics with a very personal view.

One particularly frightening aspect of the eugenics movement early in the 20th century was the vague nature of the traits considered hereditary and undesirable, such as "feeblemindedness," "criminality," and "insanity."

# Two Views of Neural Tube Defects

Genetic technologies permit people to make reproductive choices that can alter allele frequencies in populations. Identifying carriers of a recessive illness, who then decide not to have children together, is one way to remove some disease-causing alleles from a population, by decreasing the number of homozygous recessive individuals. Screening pregnant women for fetal anomalies, then terminating affected pregnancies, also alters disease prevalence and, if the disorder has a genetic component, allele frequencies. This is the case for neural tube defects (NTDs), which are multifactorial.

An NTD forms at the end of the first month, when the embryo's neural tube does not completely close. If the opening is in the head, the condition is called anencephaly, and usually ends in miscarriage, stillbirth, or a newborn who dies within days. If the opening is in the spinal cord, the condition is called spina bifida. Usually the individual is paralyzed from the point of the lesion down, but can live into adulthood and have normal intelligence. Sometimes surgery can improve functioning in people with mild cases of spina bifida. People with spina bifida also often have hydrocephalus, or "water on the brain."

In 1992, the Centers for Disease Control and Prevention summarized studies indicating that taking the vitamin folic acid in pregnancy lowers the risk of NTD recurrence by 50 percent, from 3 to 4 percent to 1.5 to 2 percent. Women who had had an affected child began taking large doses of the vitamin in the months before conception. But when epidemiologists tried to monitor how well folic acid supplementation was working, they faced a problem—the prevalence values of NTDs were greatly underestimated. This happened because the statistics on NTD prevalence—vital to discovering whether folic acid was actually preventing the defect—included only newborns, stillborns, and older fetuses. Most reports did not account for pregnancies terminated following a prenatal diagnosis of an NTD, which accounts for prevalence of anencephaly being 60 to 70 percent underreported, and spina bifida, 20 to 30 percent underreported in some states. Screening for NTDs and ending affected pregnancies alters the allele frequencies by preventing causative genes from passing to new generations.

## A Personal View

Blaine Deatherage-Newsom has a different view of population screening for neural tube defects because he has one (figure 1). Blaine was born in 1979 with spina bifida. Paralyzed from the armpits down, he has endured much physical pain, but he has also achieved a great deal. He put the question, "If we had the technology to eliminate disabilities from the population, would that be good public policy?" on the Internet—initiating a global discussion. His view on NTD screening is one we do not often hear:

> **I was born with spina bifida and hydrocephalus. I hear that when parents have a test and find out that their unborn child has spina bifida, in more than 95 percent of the cases they choose to have an abortion. I also went to an exhibit at the Oregon Museum of Science and Industry several years ago where the exhibit described a child born with spina bifida and hydrocephalus, and . . . asked people to vote on whether the child should live or die. I voted that the child should live, but when I voted, the child was losing by quite a few votes.**
>
> **When these things happen, I get worried. I wonder if people are saying that they think the world would be a better place without me. I wonder if people just think the lives of people with disabilities are so full of misery and suffering that they think we would be better off dead. It's true that my life has suffering (especially when I'm having one of my 11 surgeries so far), but most of the time I am very happy and I like my life very much. My mom says she can't imagine the world without me, and she is convinced that everyone who has a chance to know me thinks that the world is a far better place because I'm in it.**

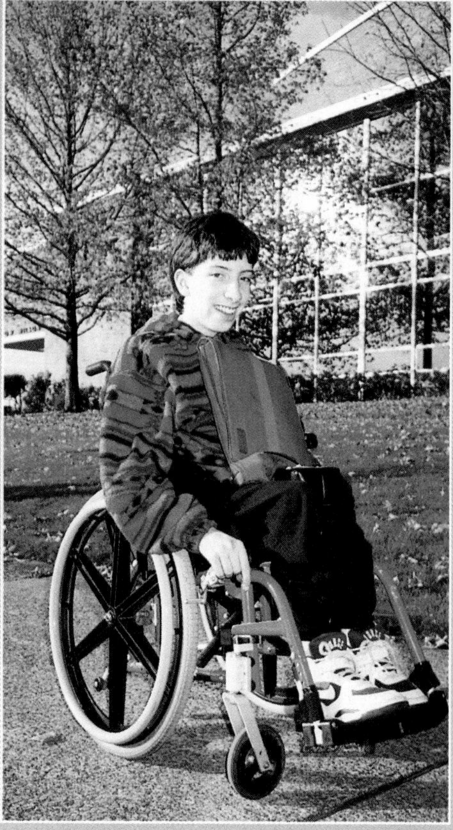

**figure 1**

**Blaine Deatherage-Newsom.**

Is eliminating disabilities good public policy? It depends on your point of view.

Excerpt by Blaine Deatherage-Newson, "If we could eliminate disabilities from the population, should we? Results of a survey on the Internet." Reprinted by permission.

Now, as a new century dawns and the human genome is revealed, will eugenics resurge? Will we use the abundance of new genetic information to pick and choose the combinations of traits that will make up the next generation?

Despite the ideological differences between eugenics and medical genetics, many people fear that tests to identify carriers or predisposition to disease will be used eugenically. Many such tests will be possible thanks to the sequencing of the human genome. We might not ever again enact forced sterilization laws, but more subtle denials, such as discrimination in insurance coverage or employment based on genetic status, might have the same effect on individuals. Laws are being proposed and implemented in many nations to prevent genetic discrimination. The American Society of Human Genetics issued a statement clearly discussing how genome information might be misused, and addressing ways to prevent this—guaranteeing reproductive freedom and opposing mandatory genetic testing. Let's hope that in addition to deciphering our genetic blueprints, we also learn how to apply that information wisely. Unlike other species, we have the ability to affect our own evolution.

**KEY CONCEPTS**

Eugenics is the control of individual human reproduction for societal goals, maximizing the genetic contribution of those deemed acceptable (positive eugenics) and minimizing the contribution from those considered unacceptable (negative eugenics). Some people consider modern genetic screening practices eugenic, but genetic testing usually serves a different goal—to prevent or alleviate human suffering.

# Summary

### 15.1 Human Origins

1. The first primates were rodentlike insectivores that lived about 60 million years ago. By 30 to 40 million years ago, monkeylike *Aegyptopithecus* lived. **Hominoids,** ancestral to apes and humans, lived 22 to 32 million years ago. They include *Dryopithecus* and other primates who began to walk upright.
2. **Hominids,** ancestral to humans only, appeared about 19 million years ago. These animals were more upright, dwelled on the plains, and had smaller brains than their forebears. The *Australopithecines* preceded and then coexisted with *Homo habilis,* who lived in caves and had strong family units and extensive tool use. *Homo erectus* was a contemporary who outsurvived *H. habilis,* lived in societies, and used fire. Evidence of tool use dates to 70,000 years ago. *H. erectus* overlapped in time with our own species.
3. Early *Homo sapiens* included possibly the Neanderthals and Cro-Magnons. Modern humans appeared about 40,000 years ago, and culture was apparent by 14,000 years ago.

### 15.2 Molecular Evolution

4. **Molecular evolution** considers differences at the genome, chromosome, protein, or DNA sequence level with mutation rates to estimate species relatedness.
5. Humans, chimps, and bonobos share 99.5 percent of their protein-encoding genes.
6. Closely related species have similar chromosome banding patterns. Genes in the same order on chromosomes in different species are **syntenic.**
7. Cytochrome *c* and homeobox proteins are **highly conserved.**
8. Genetic similarity can be estimated on a large scale by DNA hybridization or by sequencing and comparing individual genes.
9. Amplifying ancient DNA is difficult because contamination may occur.

### 15.3 Molecular Clocks

10. Gene sequence information in several species may be used to construct evolutionary tree diagrams, and a **molecular clock** based on the known mutation rate of the gene applied. Different genes mutate at different rates. Molecular trees indicate when species diverged from shared ancestors.
11. **Parsimony analysis** selects evolutionary trees requiring the fewest mutations, which are therefore the most likely.
12. Molecular clocks based on mitochondrial DNA are used to date recent events through the maternal line because this DNA mutates faster than nuclear DNA. Y chromosome genes are used to trace paternal lineages. Both types of evidence are used to study human origins in Africa and Asia, and the migration of native Americans from Asia.

### 15.4 Eugenics

13. **Eugenics** is the control of individual reproduction to serve a societal goal.
14. Positive eugenics encourages those deemed acceptable or superior to reproduce. Negative eugenics restricts reproduction of those considered inferior. Eugenics extends the concept of natural selection and Mendel's laws but does not translate well into practice.
15. Some aspects of genetic technology also affect reproductive choices and allele frequencies, but the goal is to alleviate or prevent suffering, rather than to change society.

# Review Questions

1. What is the difference between a hominoid and a hominid?
2. Which of the following animals is probably the closest relative to humans: *Australopithecus afarensis, Australopithecus garhi, Dryopithecus,* Neanderthals, or chimpanzees?
3. Some anthropologists classify chimpanzees along with humans in genus *Homo.* How does this conflict with fossil evidence of the *Australopithecus* species?
4. Give an example of how a single gene difference can have a profound effect on the phenotypes of two species.

5. What is the evidence that *Australopithecus garhi* may have been a direct forebear of *Homo?*

6. Give an example of molecular evidence that is consistent with fossil or other evidence, and an example of molecular evidence that conflicts with other information.

7. How does the information provided by Y chromosome and mitochondrial DNA sequences differ from the information obtained from nuclear DNA sequences?

8. List three aspects of development, anatomy, or physiology that were important in human evolution.

9. Why does comparing gene sequences offer more information for molecular evolution studies than comparing protein sequences?

10. Why can comparing the sequences of different genes or proteins lead to different conclusions about when two groups diverged from a common ancestor?

11. Why is comparing the DNA sequence of one gene a less accurate estimate of the evolutionary relationship between two species than a DNA hybridization experiment that compares large portions of the two genomes?

12. Cite a limitation of comparing chromosome banding patterns to estimate species' relationships.

13. What types of information are needed to construct an evolutionary tree diagram? What assumptions are necessary? What are the limitations of these diagrams?

14. Cite three examples of eugenic actions or policies.

15. How can the human and chimp genomes be 99 percent alike in DNA sequence, yet still be different?

# Applied Questions

1. A geneticist aboard a federation starship is given the task of determining how closely related Humans, Klingons, Romulans, and Betazoids are. Each organism walks on two legs, lives in complex societies, uses tools and technologies, looks similar, and reproduces in the same manner. Each can interbreed with any of the others. The geneticist finds the following data:
   - Klingons and Romulans each have 44 chromosomes. Humans and Betazoids have 46 chromosomes. Human chromosomes 15 and 17 resemble part of the same large chromosome in Klingons and Romulans.
   - Humans and Klingons have 97 percent of their chromosome bands in common. Humans and Romulans have 98 percent of their chromosome bands in common, and Humans and Betazoids show 100 percent correspondence. Humans and Betazoids differ only by an extra segment on chromosome 11, which appears to be a duplication.
   - The cytochrome *c* amino acid sequence is identical in Humans and Betazoids, and differs by one amino acid between Humans and Romulans and by two amino acids between Humans and Klingons.
   - The gene for collagen contains 50 introns in Humans, 50 introns in Betazoids, 62 introns in Romulans, and 74 introns in Klingons.
   - Mitochondrial DNA analysis reveals many more individual differences between Klingons and Romulans than between Humans or Betazoids.

   a. Hypothesize the chromosomal aberrations that might explain the karyotypic differences among these four types of organisms.

   b. Which are our closest relatives among the Klingons, Romulans, and Betazoids? What is the evidence for this?

   c. Are Klingons, Romulans, Humans, and Betazoids distinct species? What information reveals this?

   d. Which of the evolutionary tree diagrams is consistent with the data?

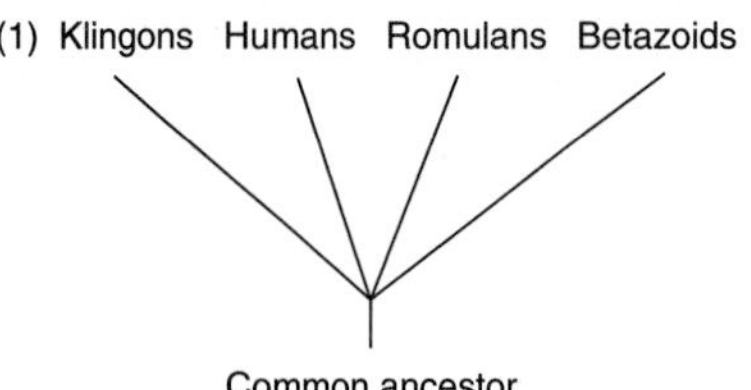


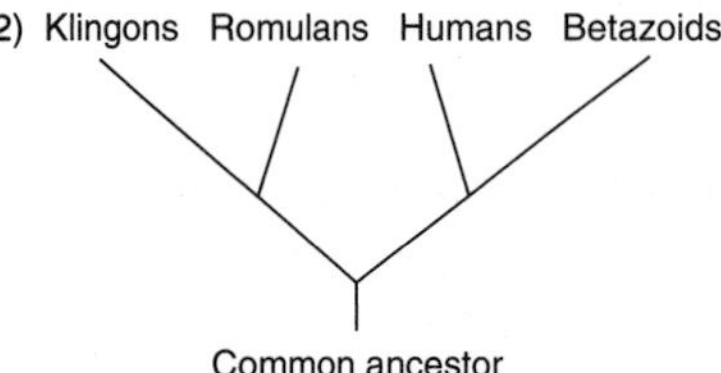


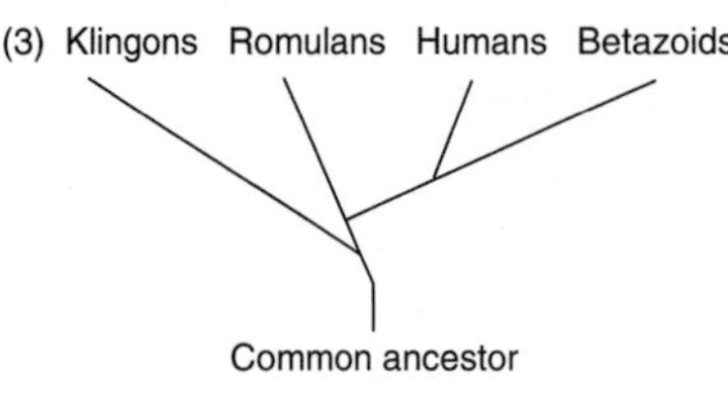


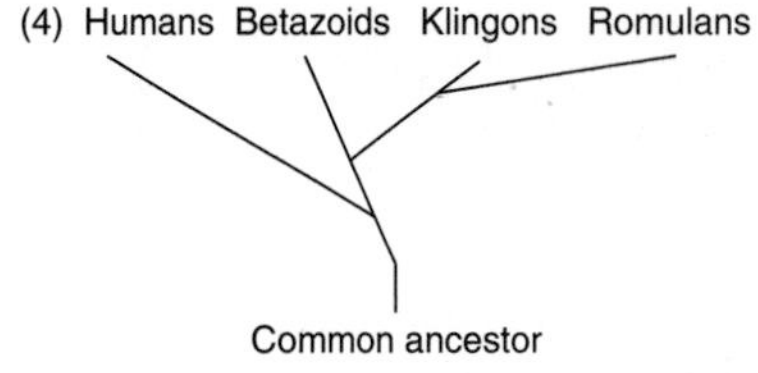


2. Give three examples of negative eugenic measures and three examples of positive eugenic measures.

3. A molecular anthropologist who is studying diabetes in native Americans feels that he can obtain information on why certain groups are prone to the disorder by analyzing genetic variants in isolated, small populations around the world. Do you think that the goal of understanding disease and alleviating suffering in one group of people justifies obtaining and studying the DNA of other people who have had little contact with cultures outside their own, even if they might be frightened by such attempts? Can you suggest a compromise intervention that might benefit everyone concerned?

4. In 1997, law schools in two states reversed their affirmative action policies and began evaluating all applicants on an equal basis—that is, the same admittance requirements applied to all. In fall 1997, classrooms of new law students had few, if any, nonwhite faces. How was this action eugenic, and how was it not?

5. Several women have offered to be artificially inseminated with sperm from the Ice Man, the human who died 5,300 years ago and was recently found in the Alps. However, he had been castrated. If sperm could have been recovered, and a woman inseminated, what do you think the child would be like?

6. The Human Genome Diversity Project will sample DNA from white blood cells collected from 1,000 people that represent 40 to 50 populations in order to learn about their differences. Suggest a plan for utilizing this information in ways to avoid eugenics.

7. Anthropologist Daniel E. Lieberman wrote in *Nature* magazine of the paper describing the discovery of the partial skull of *Kenyanthropus platyops*, "I suspect the chief role of *K. platyops* in the next few years will be to act as a sort of party spoiler, highlighting the confusion that confronts research into evolutionary relationships among hominins."
   a. What are the classification problems that *K. platyops* poses?
   b. What do you think are the limitations of attempting to classify pre-humans on the basis of skeletal remains?

# Suggested Readings


Balter, Michael. January 11, 2002. From a modern human's brow—or doodling? *Science* 295:247–48. Do hatchmarks on red rocks found in South Africa set the origin of modern human behavior back to 77,000 years ago?

Bear, Greg. 1999. *Darwin's radio.* New York: The Ballantine Publishing Group. A novel about modern humans evolving into a new species.

Butler, Declan. September 4, 1997. Eugenics scandal reveals silence of Swedish scientists. *Nature,* vol. 389. Until 1976, the Swedish government sterilized people deemed unfit.

Carlson, Elof Axel. 2001. *The Unfit: The History of a Bad Idea.* New York: Cold Spring Harbor Press. Within eugenics lay the seeds of Naziism.

Courseaux, Anouk, and Jean-Louis Nahon. February 16, 2001. Birth of two chimeric genes in the Hominidae lineage. *Science* 291:1293–97. Through evolutionary time, as the primates evolved, genes duplicated, diverged, and moved.

Crollius, Hugues Roest, et al. October 2000. Estimate of human gene number provided by genomewide analysis using *Tetraodon nigroviridis* DNA sequence. *Nature Genetics* 25:235–38. The pufferfish genome is similar to ours.

Fowler, Brenda. August 7, 2001. For 5,300-year-old ice man, extra autopsy tells the tale. *The New York Times,* p. D2. An arrow to the shoulder likely felled the Tyrolean ice man.

Gibbon, Ann. May 11, 2001. Modern men trace ancestry to African migrants. *Science* 292:1051–52. Most genetic evidence supports the replacement hypothesis of human origins.

Gillham, Nicholas Wright. 2001. *A Life of Sir Francis Galton: From African Exploration to the Birth of Eugenics.* New York: Oxford University Press. A history of eugenics, and other topics.

Goodman, F. R., and P. J. Scambler. January 2001. Human HOX gene mutations. *Clinical Genetics* 59:1–11. HOX genes help bodies to form in diverse species.

Ke, Yuehai, et al. May 11, 2001. African origin of modern humans in East Asia: A tale of 12,000 Y chromosomes. *Science* 292:1151–53. Three common mutations seen in many chromosomes sampled from an East Asian population descend from Africa, supporting the replacement hypothesis.

Laitman, Jeffrey T., and Ian Tattersall. April 2001. *Homo erectus Newyorkensis. Anatomical Record* 262:341–43. An Indonesian fossil rediscovered in Manhattan sheds light on the middle phase of human evolution.

Leakey, Meave G., et al. March 22, 2001. New hominin genus from eastern Africa shows diverse middle Pliocene lineages. *Nature* 410:433–40. A new partial skull discovery reveals a unique combination of traits that suggests that its owner lay somewhere on the path between *Australopithecus* and *Homo.*

Lewis, Ricki. March 4, 2002. Pufferfish genomes probe human genes, illuminate evolution. *The Scientist* 16(6):18. The genomes of two types of pufferfish are disturbingly similar to our own.

Lewis, Ricki. January 7, 2002. SNPs as windows on evolution. *The Scientist* 16(1):1. SNPs support an African origin.

Lewis, Ricki. June 25, 2001. From hair to eternity. *The Scientist* 15(13):21. A single gene difference helps distinguish ape from human.

Lewis, Ricki. October 16, 2000. Of SNPs and smells. *The Scientist* 14(2):8. Evolutionary clues lie in the olfactory receptor gene family.

Lewis, Ricki. March 15, 1999. Genetics society offers thoughts for future. *The Scientist* 13:1. A look at the anti-eugenics statement by the American Society of Human Genetics in light of the history of the eugenics movement.

Lieberman, Daniel E. March 22, 2001. Another face in our family tree. *Nature* 410:419–20. Classifying puzzling hominid remains is a challenge.

Ovchinnikov, Igor V., et al. March 30, 2000. Molecular analysis of Neanderthal DNA from the northern caucasus. *Nature* 404:490. MtDNA from a 29,000-year-old Neanderthal bone argues against the multiregional hypothesis.

Santos, Fabricio R., et al. February 1999. The central Siberian origin for native American Y chromosomes. *The American Journal of Human Genetics* 95:619. Nearly vanished Siberian populations may be remnants of groups that contributed to the native American gene pool.

Stone, Anne C., and Mark Stoneking. May 1998. MtDNA analysis of a prehistoric Oneota population: Implications for the people of the New World. *The American Journal of Human Genetics* 65:1153. Clues to native American origins come from a 700-year-old Illinois graveyard.

Tattersall, Ian. January 2000. Once upon a time, *H. sapiens* may not have been the only hominid. *Scientific American* 282:56. Modern humans may have emerged the sole survivor among several types of ancestral hominids.

Wade, Nicholas. May 2, 2000. The human family tree: 10 Adams and 18 Eves. *The New York Times,* p. F1. Mitochondrial Eve is a figurative woman.

Wilford, John Noble. February 20, 2002. When humans became humans. *The New York Times,* p. F1. Sophisticated tool use 70,000 years ago pushes back the date of modern humanity.

# On the Web

Check out the resources on our website at

**www.mhhe.com/lewisgenetics5**

On the web for this chapter, you will find additional study questions, vocabulary review, useful links to case studies, tutorials, popular press coverage, and much more. To investigate specific topics mentioned in this chapter, also try the links below:

Association for Spina Bifida and Hydrocephaly **www.asbah.demon.co.uk**

Eugenics Movement Website **www.vector.cshl.org/eugenics**

Image Archives on the American Eugenics Movement **vector.cshl.org/eugenics/**

Neural Tube Defect Information **www.cdc.gov/nceh/programs/cddh/folic/default.htm**

Online Mendelian Inheritance in Man **www.ncbi.nlm.nih.gov/entrez/query.fcgi?db=OMIM**
DiGeorge syndrome 188400
hereditary persistence of fetal hemoglobin 142470
synpolydactyly 186000
Waardenberg syndrome 193500

Paleo-DNA Laboratory **www.lakeheadu.ca/~lucas/pdnamain.htm**

CHAPTER

16

# Genetics of Immunity

## 16.1 The Importance of Cell Surfaces

Our cell surfaces are marked with molecules that indicate "self" as well as tissue type. The immune system is a vast army of cells, biochemicals, and associated vessels and organs that protects the body against "nonself" cells and molecules. It guards against infection and cancer—but can malfunction.

## 16.2 The Human Immune System

A pathogen faces a daunting task when it attempts to penetrate the human body's immune defenses. It must first breach physical barriers and torrents of body fluids, then overcome the broad defenses of innate immunity. Meanwhile, the antibodies and cytokines of the adaptive immune response are readying for attack. The adaptive part of the immune system "remembers" previous encounters and is prepared to rapidly combat future infections.

## 16.3 Abnormal Immunity

When the immune system malfunctions, the effects on health can be disastrous. Deficient immunity, whether inherited or acquired, opens the body to rampant infection and cancers. Autoimmunity sets the immune system against the body, and allergies represent misguided attacks against harmless substances.

## 16.4 Altering Immune Function

Understanding the immune response enables medical science to alter and direct it. Vaccines prevent infections by inducing a false first infection. Monoclonal antibodies and cytokines boost immunity to treat a variety of conditions. Conversely, immune function must be subdued for transplants to effectively replace body parts.

## 16.5 A Genomic View of Immunity—The Pathogen's Perspective

Infectious disease is a consequence of specific interactions between our genome and that of the pathogen. Sequencing the genomes of infectious organisms and viruses can suggest new ways to treat illness. The fact that pathogens can rapidly cause illness in unexposed populations is perverted to create bioweapons.

Leslie Hancock was born with cystic fibrosis. At the age of six, her liver began to harden as a complication of the CF, and the little girl desperately needed a new one. But because of a shortage of donor organs, a third of the children awaiting livers die before a transplant can be performed. Fortunately, Leslie's father, Jim, was able to donate part of his liver because his and Leslie's cell surfaces are so similar that her body was able to accept part of his. "When the subject of a living-relative transplant came up, I never had to think about it. It was just something I wanted to do," Jim recalls (figure 16.1).

Transplants between blood relatives succeed in about 90 percent of cases, compared to 80 percent for unrelated people whose cell surfaces are similar by chance. After the surgery, Leslie took drugs to suppress her immune system, further boosting the odds of success. The transplant worked. Two months after the operation, she was back in her kindergarten classroom in Dubuque, Iowa.

In the case of an organ transplant, the immune system is an obstacle. More often, this organ system protects the body against infection-causing viruses and microorganisms (collectively called pathogens) as well as cancer cells. Because the biochemicals that immune system cells produce are proteins, immunity is very much a matter of genetics. Yet environmental cues trigger many aspects of the immune response.

**figure 16.1**

**The importance of cell surfaces.** Jim Hancock, of Dubuque, Iowa, gave his six-year-old daughter, Leslie, part of his liver and a new chance at life. The transplant worked because the cell surfaces of this father and daughter are very similar. The little girl's immune system accepted the new liver as part of her body.

## 16.1 The Importance of Cell Surfaces

Because we share the planet with plants, microbes, fungi, and other animals, the human immune system has evolved in a way that keeps potentially harmful organisms out of our bodies. This system is a mobile army of about 2 trillion cells and the biochemicals they produce. Upon recognizing "foreign" or "nonself" surfaces of viruses, microbial cells, tumor cells, or transplanted cells, the system launches a highly coordinated, multipronged attack.

Early estimates from the human genome project suggest that 20,000 or more genes affect immunity, usually by contributing to polygenic, multifactorial traits. The effects of these genes add to confer susceptibilities or resistances to certain infectious diseases, or may increase the risk of developing an allergic or **autoimmune** condition (in which the immune system attacks an individual's own tissues or organs). A few single genes exert powerful effects on immunity. For example, a human variant of a mouse gene known to protect against certain infections and parasites confers resistance to three specific immune-related conditions: tuberculosis, sarcoidosis (build up of fibrous tissue in various organs), and juvenile rheumatoid arthritis. (The gene encodes natural-resistance-associated macrophage protein 1.)

Certain classes of genes and their encoded proteins oversee immunity. Genes encode **antibodies** and **cytokines** that directly attack foreign molecules, or antigens. An **antigen** is any molecule that elicits an immune response, and is usually a protein or carbohydrate. Genes specify the cell surface antigens that mark the body's cells as "self."

Because genes control immunity, mutations can impair immune function, causing immune deficiencies, autoimmune disorders, allergies, and cancer. But understanding how genes control immunity also makes it possible to develop technologies that enhance or redirect the system's ability to fight disease. We begin our look at normal immunity with some familiar examples of our personal cellular landscapes.

## Blood Groups

Transplanting an organ as complex as a liver is a major and risky medical procedure. A far simpler type of transplant, although still very dependent on matching cell surfaces that do not elicit "immune system" rejection, is a blood transfusion. Using one person's blood to restore another's health was an idea proposed centuries ago. To do so safely and successfully, however, it was necessary to understand the genetics of blood types.

### *ABO Blood Groups*

The first transfusions, performed in the late 1600s, used lamb blood. By the 1800s, physicians were trying to use human blood. The results seemed unpredictable—some recipients recovered, but others died. So poor was the success rate that, by the late 1800s, many nations banned transfusions.

Around this time, Austrian physician Karl Landsteiner began investigating why transfusions sometimes worked and sometimes didn't. In 1900, he determined that human blood was of differing types, and only certain combinations were compatible. In 1910, identification of the ABO blood antigen locus explained the observed blood type incompatibilities (figure 16.2). Today, we know of more than 20 different genes whose protein products are part of the surface topography of red blood cells.

Recall from chapter 5 that the *I* gene alleles encode enzymes that place antigens A, B, both A and B, or neither antigen on sugar chains on red blood cells (see table 5.1). Blood type incompatibility occurs when a person's immune system manufactures antibodies that attack the antigens his or her cells do not carry. A person with blood type A, for example, has antibodies against type B antigen. If he or she is transfused with type B blood, the anti-B antibodies cause the transfused red blood cells to clump, blocking circulation and depriving tissues of oxygen. A person with type AB blood doesn't manufacture antibodies against either antigen A or B, because if he or she did, the person's own blood would clump. Therefore, someone with type AB blood can

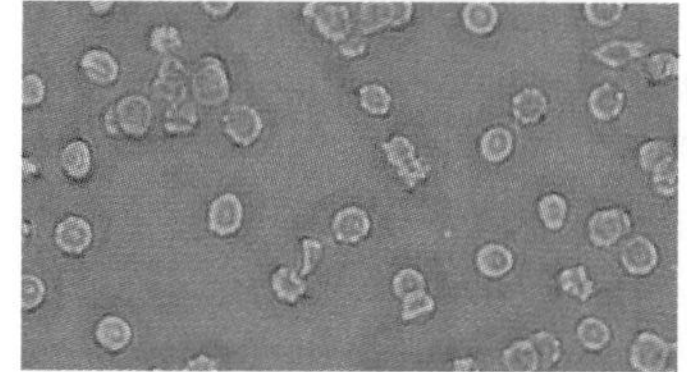

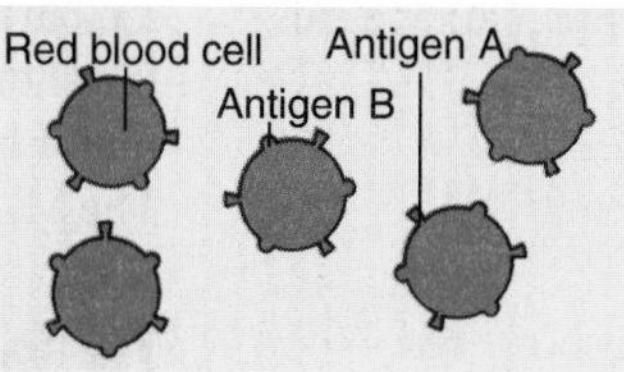

Red blood cell
Antigen A
Antigen B

Compatible Blood Types (no clumping)

| *Donor* | *Recipient* |
| --- | --- |
| O | O, A, B, AB |
| A | A, AB |
| B | B, AB |
| AB | AB |

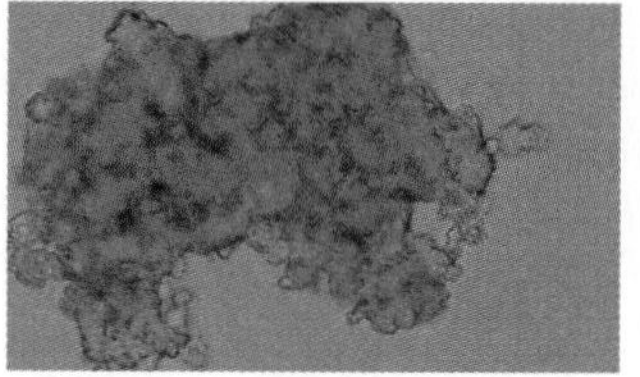

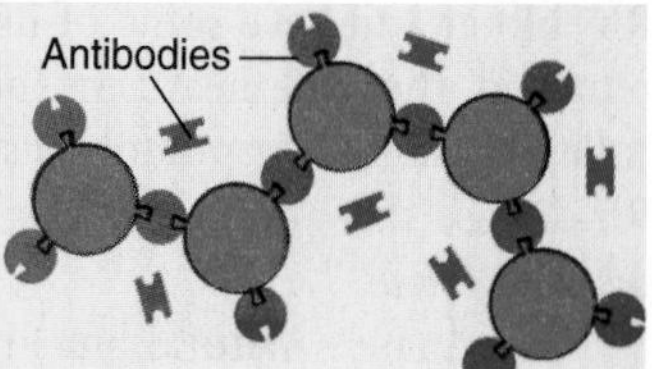

Antibodies

Incompatible Blood Types (clumping)

| *Donor* | *Recipient* |
| --- | --- |
| A | B, O |
| B | A, O |
| AB | A, B, O |

**figure 16.2**

**ABO blood types.** Genetics explains blood incompatibilities.

receive any ABO blood type. Type O blood has neither A nor B antigens, so it cannot stimulate an immune response in a transfusion recipient; people with type O blood can thus donate to anyone. However, the idea that a person with AB blood is a "universal recipient" and one with type O blood is a "universal donor" is more theoretical than practical because antibodies to other donor blood antigens can cause slight incompatibilities (for example, the Rh factor, discussed next). For this reason, blood is as closely matched as possible. An experimental "universal blood type" consists of red blood cells coated with a chemical that masks their surface antigens.

A person who receives mismatched blood quickly feels the effects—anxiety, difficulty breathing, facial flushing, headache, and severe pain in the neck, chest, and lower back. Red blood cells burst, releasing free hemoglobin that can damage the kidneys.

## *The Rh Factor*

ABO blood type is often further differentiated by a $^+$ or $^-$, which refers to another blood group antigen called the Rh factor. Whether a person has the Rh factor ($Rh^+$) or not ($Rh^-$) is determined by a combination of alleles of three genes. The antigens were originally identified in rhesus monkeys, hence the name.

Rh type is important when an $Rh^+$ man and an $Rh^-$ woman conceive a child who is $Rh^+$. The pregnant woman's immune system reacts to the few fetal cells that leak into her bloodstream by manufacturing antibodies against the fetal cells (figure 16.3). Not enough antibodies form to harm the first fetus, but if she carries a second $Rh^+$ fetus, the woman's now plentiful antibodies attack the fetal blood supply. In the fetus, bilirubin, a breakdown product of red blood cells, accumulates, damaging the brain and turning the skin and whites of the eyes yellow. The fetal liver and spleen swell as they rapidly produce new red blood cells. If the fetus or newborn does not receive a transfusion of

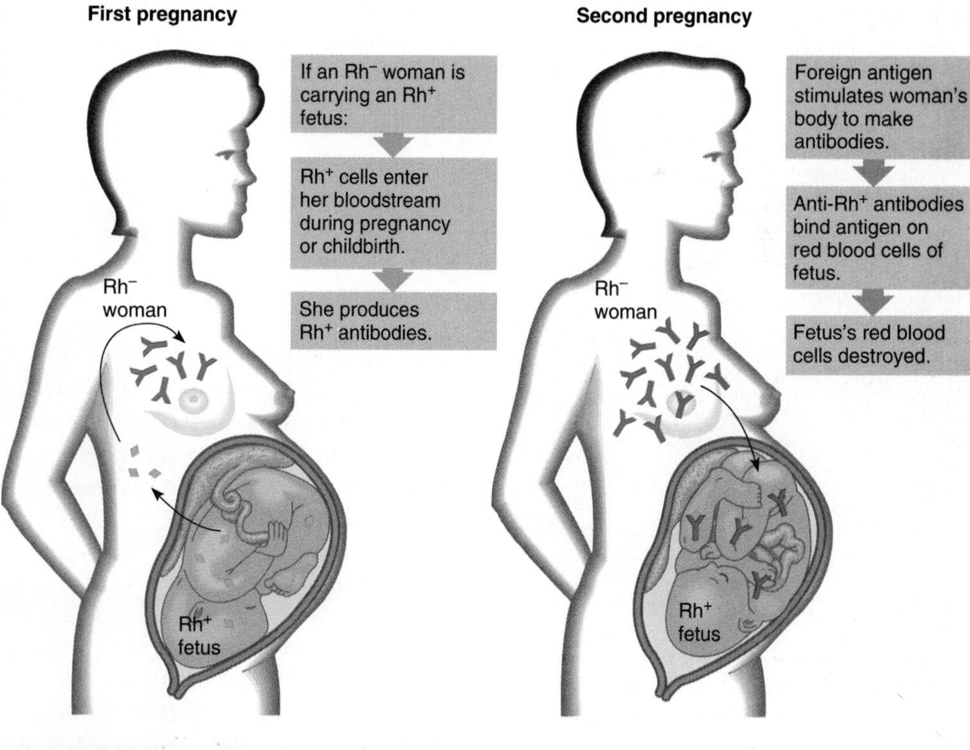

First pregnancy
If an Rh⁻ woman is carrying an Rh⁺ fetus:
Rh⁺ cells enter her bloodstream during pregnancy or childbirth.
She produces Rh⁺ antibodies.
Rh⁻ woman
Rh⁺ fetus
Second pregnancy
Foreign antigen stimulates woman's body to make antibodies.
Anti-Rh⁺ antibodies bind antigen on red blood cells of fetus.
Fetus's red blood cells destroyed.
Rh⁻ woman
Rh⁺ fetus

**figure 16.3**

**Rh incompatibility.** Fetal cells entering the pregnant woman's bloodstream can stimulate her immune system to make anti-Rh antibodies, if the fetus is $Rh^+$ and she is $Rh^-$. A drug called RhoGAM prevents attacks on subsequent fetuses.

$Rh^-$ blood and have some of its $Rh^+$ blood removed, then the heart and blood vessels collapse and fatal respiratory distress sets in. Rh disease that progresses this far is called hydrops fetalis.

Fortunately, natural and medical protections make hydrops fetalis rare today, although exchange blood transfusions were once common. Determining ABO blood types indicates whether an immune reaction against the fetus of an Rh-incompatible couple will take place. If the woman has type O blood and the fetus is A or B, then her anti-A or anti-B antibodies attack the fetal blood cells in her circulation before her system has a chance to manufacture the anti-Rh antibodies. This blocks the anti-Rh reaction.

Obstetricians routinely determine a pregnant woman's blood type. If she and her partner are Rh incompatible, doctors inject a drug (called RhoGAM) during pregnancy and after the birth. The drug covers antigens on fetal blood cells in the woman's circulation so that she does not manufacture anti-Rh antibodies. However, events other than pregnancy and childbirth can expose an $Rh^-$ woman's system to $Rh^+$ cells, placing even her first child at risk. These include amniocentesis, a blood transfusion, an ectopic (tubal) pregnancy, a miscarriage, or an abortion.

### *Other Blood Groups*

Another way of distinguishing red blood cells is by the *L* gene, whose codominant alleles *M, N,* and *S* combine to form six different genotypes and phenotypes (glycoprotein cell surface patterns). Another blood-type determining gene is called Lewis. It encodes an enzyme that adds an antigen to the sugar fucose, which the product of the *H* gene then places on red blood cells. (Recall that the *H* gene is necessary for ABO expression.) Individuals with genotype *LeLe* or *Lele* have the Lewis antigen on red blood cell membranes and in saliva, whereas *lele* people do not produce the antigen.

Another interesting gene that affects the blood is the secretor gene. People who have the dominant allele *Se* secrete the A, B, and H antigens in body fluids, including semen, saliva, tears, and mucus.

The blood group antigens, as cell surface markers, serve as cellular nametags, clearly delineating "self" to the immune system. But cell surfaces are dotted with many other molecules, too. Many of these protein surface features are encoded by genes that are part of a 6-million-base-long cluster on the short arm of chromosome 6, called the **major histocompatibility complex** (MHC). The MHC includes about 70 genes.

## The Human Leukocyte Antigens

The genes of the MHC are classified into three groups based on their functions. Class III MHC genes encode proteins that are in blood plasma and that carry out some of the innate immune functions discussed here. The class I and II genes of the MHC encode the **human leukocyte antigens** (HLA). The name comes from the fact that these antigens were first studied in leukocytes, a broad term for white blood cells. The HLA proteins link to sugars to form branchlike glycoproteins that emanate from cell surfaces. Some of these HLA glycoproteins latch onto bacterial and viral proteins, displaying them like badges in a way that alerts other immune system cells. This action, called **antigen processing,** is often the first step in an immune response. The cell that displays the foreign antigen held by an HLA protein is an **antigen-presenting cell.** Figure 16.4 shows how a large cell called a **macrophage** displays bacterial antigens. Certain white blood cells called T cells (or T lymphocytes) also

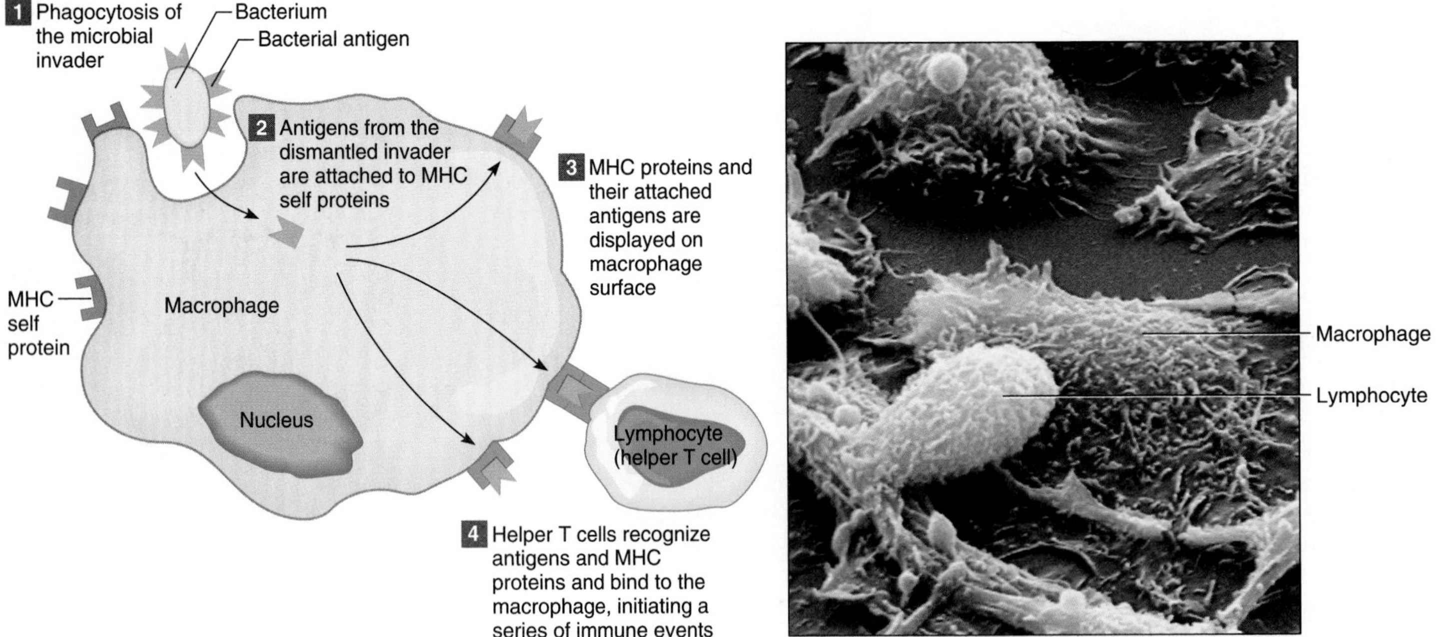


**figure 16.4**

**Macrophages are antigen-presenting cells.** A macrophage engulfs a bacterium, then displays foreign antigens on its surface, held in place by major histocompatibility complex (MHC) self proteins. This event sets into motion many immune reactions.

function as antigen-presenting cells. Class I and II HLA proteins differ in the types of immune system cells they alert.

All cells with nuclei (that is, all cells except red blood cells) have some HLA antigens, which identify them as "self," or belonging to the same individual. Beyond these common HLA cell surface markers are other, more specific markers that designate particular tissue types. Class I includes three genes, called *A*, *B*, and *C*, that are very polymorphic and are found on all cell types, and three other genes, *E*, *F*, and *G*, that are more restricted in their distribution. Class II includes three major genes whose encoded proteins are found mostly on antigen-presenting cells.

Because the HLA classes consist of several genes that have many alleles, individuals can be distinguished by an overall HLA "type." Only two in every 20,000 unrelated people match for the six major HLA genes by chance. When transplant physicians attempt to match donor tissue to a potential recipient, they determine how alike the two individuals are at these six loci. Usually at least four of the genes must match for a transplant to have a reasonable chance of success. Before the advent of DNA fingerprinting, HLA typing was the predominant type of blood test used in forensic and paternity cases to rule out involvement of certain individuals. However, HLA genotyping has become very complex because hundreds of alleles are now known. To make matters even more complicated, HLA genotype and disease associations differ in different populations.

For many years, researchers thought that the HLA genes exerted great control over immunity, but human genome sequence information has revealed that the effect is only modest—about 50 percent of the genetic influence on immunity stems from these genes. However, a handful of disorders are very strongly associated with particular HLA genotypes.

Consider ankylosing spondylitis, which inflames and deforms vertebrae. A person with either of two particular subtypes of an HLA antigen called B27 is 100 times as likely to develop the condition as someone who lacks either form of the antigen. HLA-associated risks are not absolute. More than 90 percent of people who suffer from ankylosing spondylitis have the B27 antigen, which occurs in only 5 percent of the general population. However, 10 percent of people who have ankylosing spondylitis do *not* have the B27 antigen, and some people who have the antigen never develop the disease.

**KEY CONCEPTS**

The immune system consists of cells and biochemicals that distinguish self from nonself antigens. • Blood types result from self antigen patterns on red blood cells. • The HLA complex is a highly diverse group of cell surface proteins, some of which display foreign antigens to other parts of the immune system, triggering a response.

## 16.2 The Human Immune System

On a macroscopic level, the immune system includes a network of vessels called lymphatics, which transport lymph, a watery fluid, to bean-shaped structures called lymph nodes. The spleen and thymus gland are also part of the immune system (figure 16.5). On a microscopic level, the immune system consists of white blood cells called **lymphocytes** and the wandering, scavenging macrophages that capture and degrade bacteria, viruses, and cellular debris.

The immune response consists of two lines of defense—an immediate generalized **innate immunity,** and a more specific, slower **adaptive immunity.** (The term "adaptive immunity" has replaced "acquired immunity.") Before these protective systems come into play, various physical barriers keep pathogens out. Table 16.1 and figure 16.6 summarize these basic components of immunity.

### Physical Barriers and the Innate Immune Response

Several familiar structures and fluids keep pathogens from entering the body in the innate immune response. Unbroken skin and mucous membranes such as the lining inside the mouth are part of this first line of

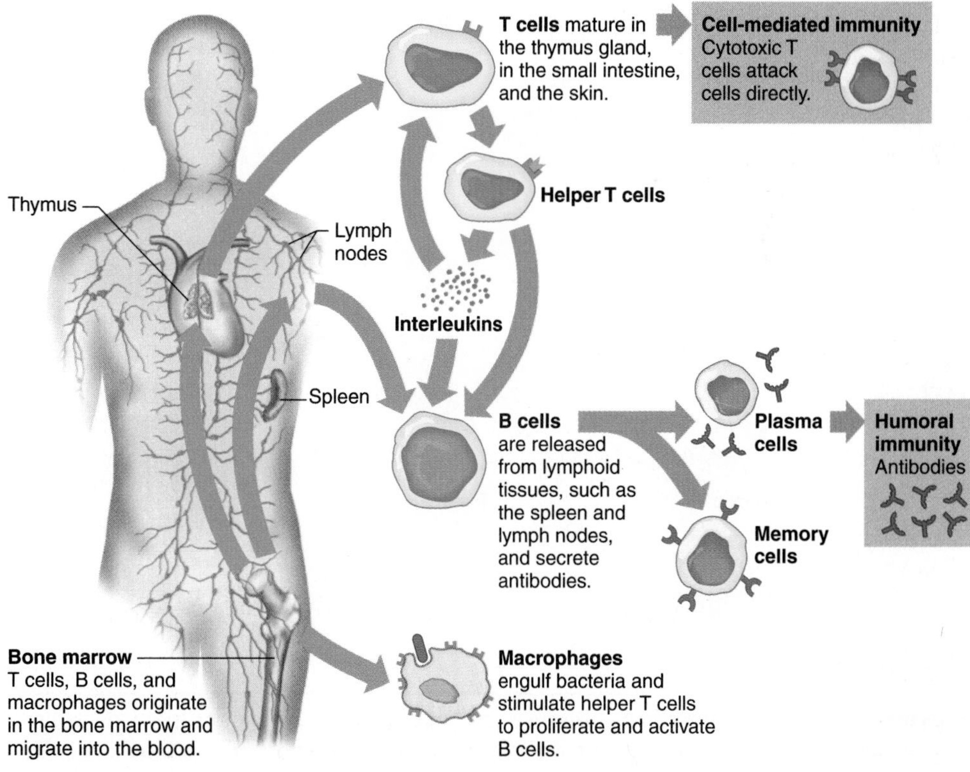


**figure 16.5**

**Immune cells are diverse.** T cells, B cells, and macrophages each contribute to an overall immune response. All three types of cells originate in the bone marrow and circulate in the blood.

## table 16.1

**Major Components of the Immune Response**

| First Line of Defense: Physical Barriers | Second Line of Defense: Nonspecific, Innate Defenses | Third Line of Defense: Specific, Adaptive Immunity |
|---|---|---|
| Unbroken skin | Phagocytosis (engulfing cells) | Humoral immune response |
| Mucous membranes and their secretions | Antimicrobial proteins | B cells, antibodies, memory cells |
| Infection-fighting chemicals in tears, saliva and other body fluids | Complement system | Cellular immune response |
| Flushing effect of tears, saliva, urination, and diarrhea | Collectins | T cells, cytokines, memory cells |
| | Cytokines | |
| | Inflammatory response | |
| | Fever | |

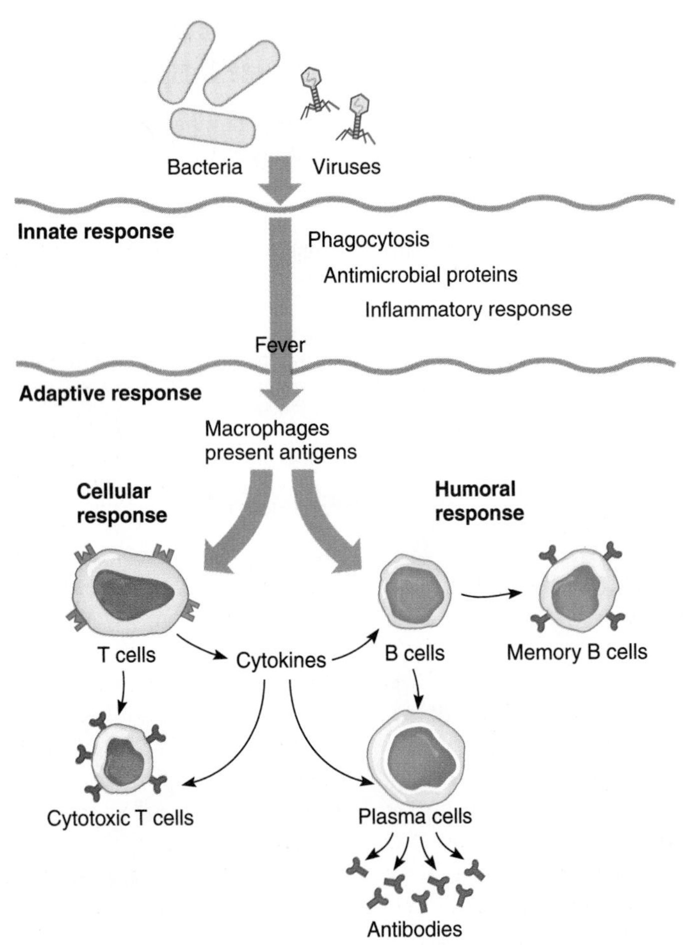


**figure 16.6**

**Levels of immune protection.** Disease-causing organisms and viruses (pathogens) first encounter barriers that prevent their entry into the body. If these barriers are breached, an array of non-specific cells and molecules attack the pathogen, in the innate immune response. If this is ineffective, the adaptive immune response begins: antigen-presenting cells stimulate T cells to produce cytokines, which activate B cells to differentiate into plasma cells, which secrete antibodies. Once activated, these specific cells retain a memory of the pathogen, allowing faster responses to subsequent attacks.

defense, as are earwax and the waving cilia that push debris and pathogens up and out of the respiratory tract. Most microbes that make it to the stomach perish in a vat of churning acid—though a notable exception is the bacterium that causes peptic ulcers. Other microbes are flushed out in diarrhea. These barriers are nonspecific—they keep out anything foreign, not particular pathogens.

If a pathogen breaches these physical barriers, innate immunity provides a rapid, broad defense. The term *innate* refers to the fact that these general defenses are in the body, ready to function should infection begin. A process called **inflammation** is a central part of the innate immune response (figure 16.7). Inflammation creates a hostile environment for microbes and viruses at an injury site, sending in cells that engulf and destroy the invaders. Such cells are called **phagocytes,** and their engulfing action is termed phagocytosis (figure 16.8) Certain blood cells, such as neutrophils, are phagocytes, as are the large, wandering macrophages. At the same time, plasma (the liquid portion of blood) accumulates, diluting toxins and bringing in antimicrobial chemicals. Increased blood flow warms the area, turning it swollen and red. The person may not be very comfortable, but often, the pathogen does not survive. Inflammation at the site of an injury can prevent infection.

At least three classes of proteins participate in the innate immune response—the complement system, collectins, and cytokines. Mutations in the genes that encode them can produce disorders that increase

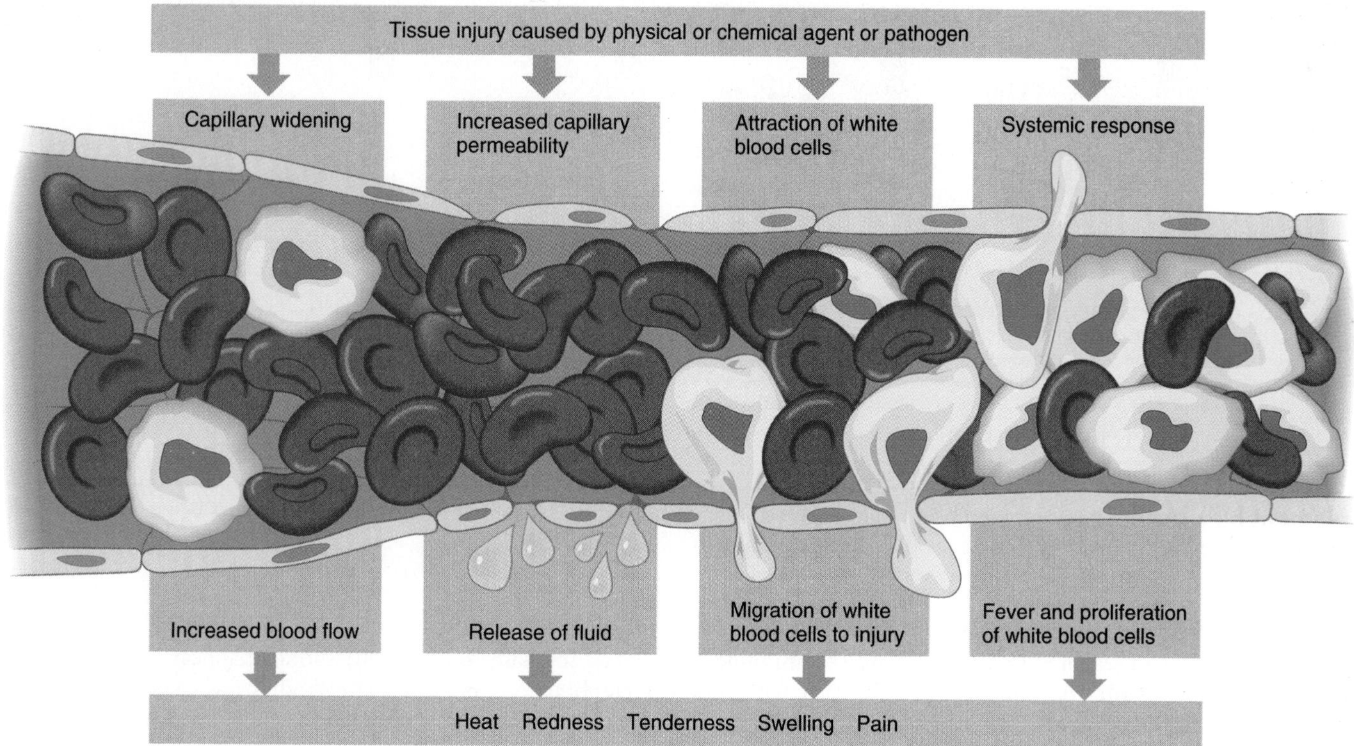


**figure 16.7**

**Inflammation.** Chemicals released at the site of an injury set into motion the several steps of the inflammatory response.

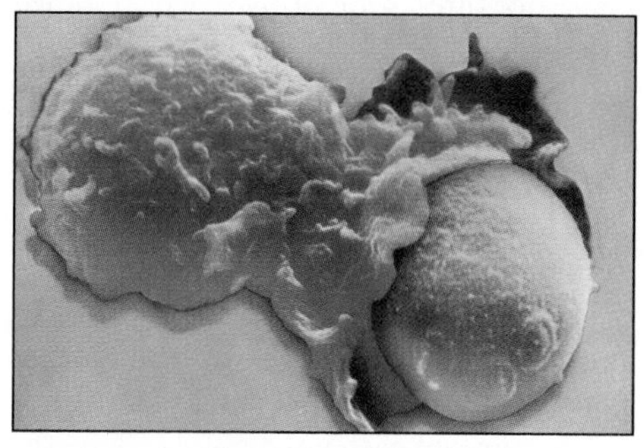

**figure 16.8**

**Nature's garbage collectors.** A human phagocyte engulfs a yeast cell.

susceptibility to infection. However, some mutations have no effect. This may indicate that other proteins can provide similar or the same functions.

The **complement system** consists of plasma proteins that assist, or complement, several of the body's other defense mechanisms. Some complement proteins trigger a chain reaction that punctures bacterial cell membranes, bursting the cells (figure 16.9). Complement proteins can also dismantle viral envelopes, destroying these pathogens. Other complement proteins assist inflammation by triggering the release of **histamine** from **mast cells,** another type of immune system cell type that is involved in allergies. Histamine dilates blood vessels, which enables fluid to rush to the infected or injured area. Still other complement proteins attract phagocytes to an injury site.

**Collectins** are proteins that provide broad protection against bacteria, yeasts, and some viruses. These proteins detect slight differences from human cells in the patterns of sugars that protrude from the surfaces of these pathogens. The human body's collection of collectins is diverse enough that groups of them correspond to the surfaces of different types of infectious agents. Different collectins recognize the distinctive sugars on infecting yeast, the linked sugars and lipids of certain bacteria, and the surface features of some RNA viruses.

Cytokines are proteins that play many roles in immunity. As part of the innate immune response, cytokines called **interferons** alert other components of the immune system to the presence of cells infected with viruses. These cells are then destroyed, which limits the spread of infection. **Interleukins** are cytokines that cause fever, temporarily maintaining a higher body temperature that directly kills some infecting bacteria and viruses. Fever also counters microbial growth indirectly, because higher body temperature reduces the iron level in the blood. Bacteria and fungi require more iron as the body temperature rises. Therefore, a fever-ridden body stops their growth. Phagocytes also attack more vigorously when the temperature rises. **Tumor necrosis factor** is another type of cytokine that activates other types of protective biochemicals, destroys certain bacterial toxins, and also attacks cancer cells. Many of the more unpleasant aspects of suffering from an infection are actually due to the immune response, rather than to the actions of the pathogens.

## The Adaptive (Acquired) Immune Response

Adaptive immunity must be stimulated into action, taking days to respond as compared to minutes for innate immunity. It is highly specific and directed.

Adaptive immunity has two lines of attack, based on the activities of two types of

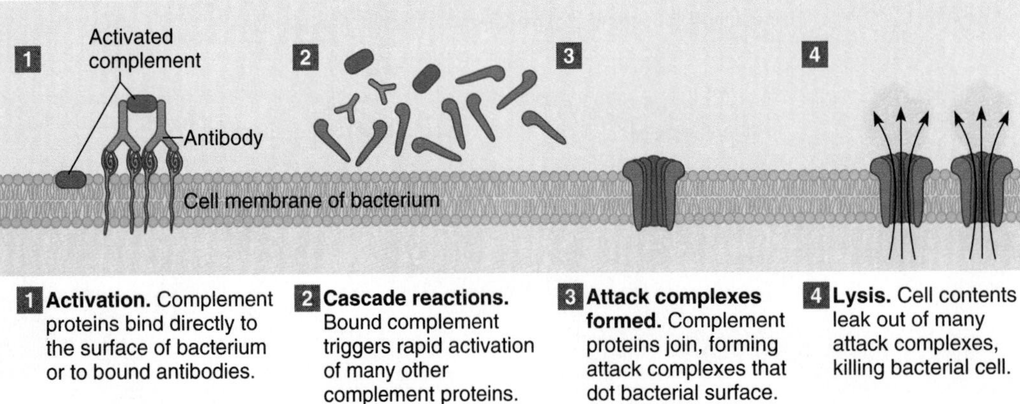


1 **Activation.** Complement proteins bind directly to the surface of bacterium or to bound antibodies.

2 **Cascade reactions.** Bound complement triggers rapid activation of many other complement proteins.

3 **Attack complexes formed.** Complement proteins join, forming attack complexes that dot bacterial surface.

4 **Lysis.** Cell contents leak out of many attack complexes, killing bacterial cell.

**figure 16.9**

**Complement kills bacteria.** Triggered by a bound antibody, complement proteins combine to riddle a bacterium's cell membrane with holes, shattering its physical integrity. The bacterial cell quickly dies.

lymphocytes, **B cells** and **T cells.** B cells produce antibody proteins in response to activation by T cells in the **humoral immune response.** T cells produce cytokines and activate other cells in the **cellular immune response.** Lymphocytes differentiate in the bone marrow and migrate to the lymph nodes, spleen, and thymus gland, as well as circulate in the blood and tissue fluid.

The adaptive arm of the immune system has three basic characteristics. It is *diverse,* vanquishing many types of pathogens. It is *specific,* distinguishing the cells and molecules that cause disease from those that are harmless. The immune system also *remembers,* responding faster to a subsequent encounter with a foreign antigen than it did the first time. The first assault initiates a **primary immune response.** The second assault, based on the system's "memory," is a **secondary immune response.** This is why we get some infections, such as chicken pox, only once. However, upper respiratory infections and influenza recur because the causative viruses mutate, presenting a different face to our immune systems each season. In addition, different viruses can cause "cold and flu" symptoms.

### *The Humoral Immune Response*

B cells secreting antibodies into the bloodstream constitute the humoral immune response. (*Humor* means fluid.) An antibody response begins when an antigen-presenting macrophage activates a T cell, which in turn contacts a B cell that has surface receptors that can bind the same type of foreign antigen. The immune system has so many B cells, each with different combinations of surface antigens, that there is almost always one or more available that corresponds to a particular foreign antigen. Each day, millions of B cells perish in the lymph nodes and spleen, while millions more form in the bone marrow, each with a unique combination of surface molecules.

Although antibodies have traditionally been considered part of adaptive immunity, some evidence indicates that the circulation of a fetus might contain some antibodies that originated in the fetus (that is, were not passed from the pregnant woman). This suggests that antibody production can also be innate. In Their Own Words on page 343 describes this evidence and what it might reveal about evolution.

Once the activated T cell finds a B cell match, it releases cytokines that stimulate the B cell to divide. Soon the B cell gives rise to two types of cells (figure 16.10). The first type of B cell descendants, **plasma cells,** are antibody factories, secreting up to 2,000 identical antibodies each per second at the height of their few-day life span. These cells provide the primary immune response. Plasma cells derived from different B cells secrete different antibodies, each type corresponding to a specific portion of the pathogen. This is called a polyclonal antibody response (figure 16.11). The second type of B cell descendant, **memory cells,** are far fewer and remain dormant. They respond to the foreign antigen faster and with more force in a secondary immune response should it appear again.

Antibodies are constructed of several polypeptides and are therefore encoded by several genes. The simplest antibody molecule is made up of four polypeptide chains connected by disulfide (sulfur-sulfur) bonds, forming a shape like the letter Y (figure 16.12). A large antibody molecule might consist of three, four, or five such Ys joined together.

In a Y-shaped antibody subunit, the two longer polypeptides are called **heavy chains,** and the other two **light chains.** The lower portion of each chain is an amino acid sequence that is very similar in all antibody molecules, even in different species. These areas are called **constant regions.** The amino acid sequence of the upper portions of each polypeptide chain, the **variable regions,** can differ greatly among antibodies.

Antibodies can bind certain antigens because of the three-dimensional shapes of the tips of the variable regions. These specialized ends of the antibody molecule are called **antigen binding sites,** and the specific parts that actually bind the antigen are called **idiotypes.** The parts of the antigens that idiotypes bind are called **epitopes.** An antibody contorts to form a pocket around the antigen.

Antibodies have several functions. Antibody-antigen binding inactivates a pathogen or neutralizes the toxin it produces. Antibodies can cause pathogens to clump, making them more visible to macrophages, which then destroy them. Antibodies also ac-

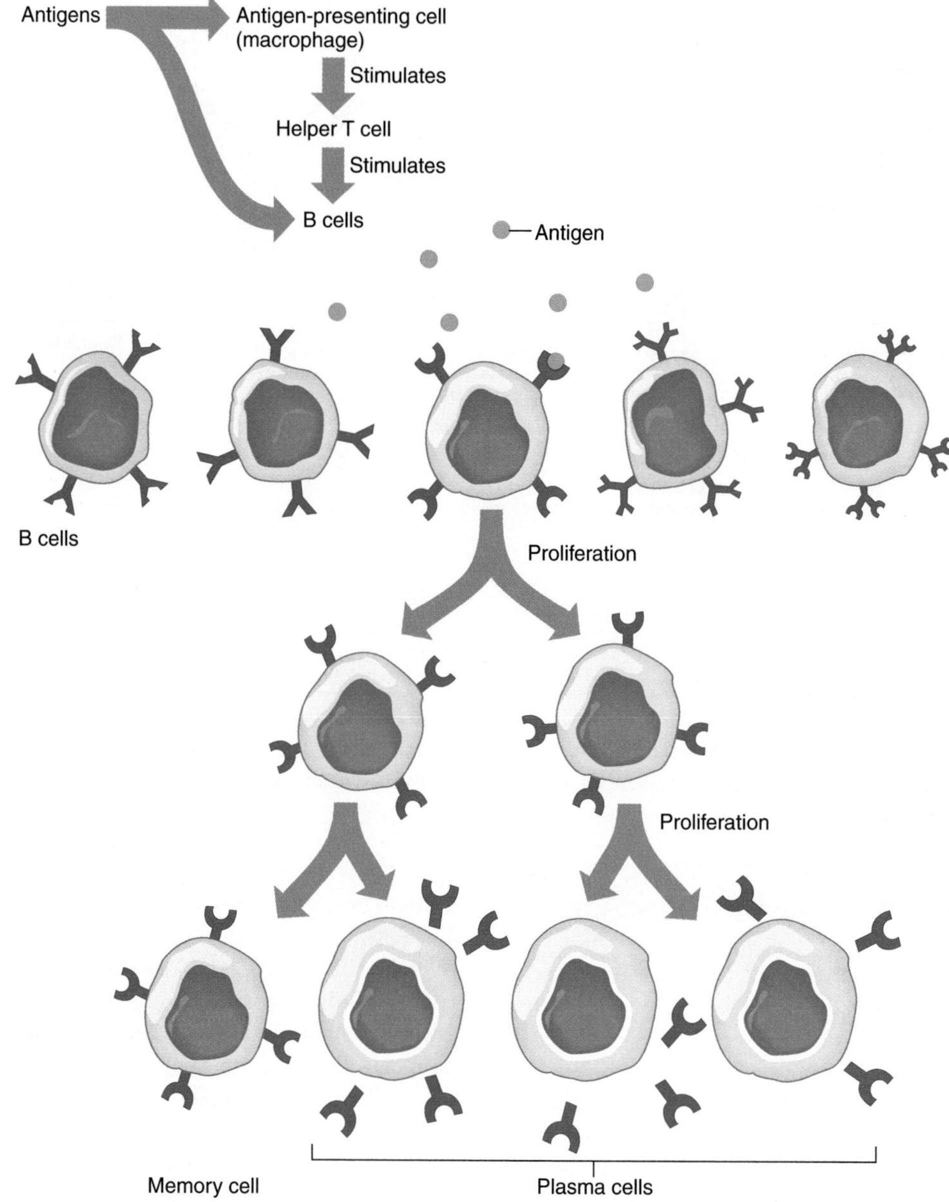


## figure 16.10

**Production of antibodies.** B cell proliferation and maturation into antibody-secreting plasma cells constitute the humoral immune response. Note that only the B cell that binds the antigen proliferates and develops into memory cells or plasma cells.

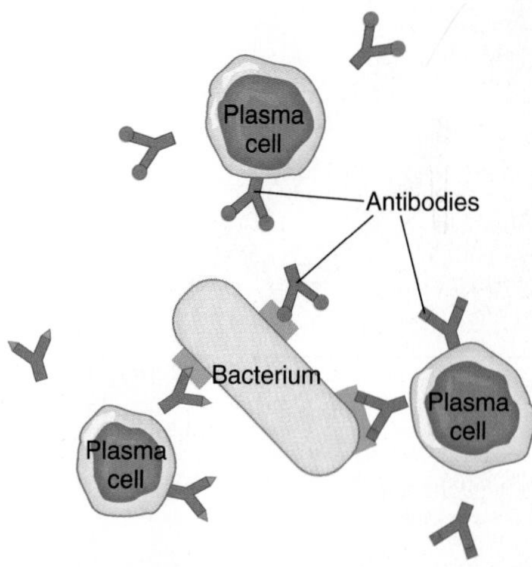


## figure 16.11

**An immune response recognizes many targets.** A humoral immune response is polyclonal, which means that different plasma cells produce antibody proteins that recognize and bind to different features of a foreign cell's surface.

tivate complement, extending the innate immune response.

Antibodies come in several types. Five classes of antibodies are distinguished by their locations in the body and by their functions (table 16.2). Antibodies are also called immunoglobulins, abbreviated *Ig.* Different antibody types predominate in different stages of an infection.

Perhaps the most astounding characteristic of the antibody response is that the human body can manufacture an apparently limitless number of different antibodies, though our genomes have a limited number of antibody genes. This great diversity is possible because different antibody gene products combine. During the early development of B cells, sections of their antibody genes move to other chromosomal locations, creating new genetic instructions for antibodies. Shuffling the polypeptide products of 200 genes generates 100 trillion different antibodies! Because of the tremendous diversity of antibodies, a human body can respond to nearly any infection. A single stimulated B cell is said to give rise to a clone of plasma and memory cells because they all express the same antibody gene combinations.

### *The Cellular Immune Response*

T cells provide the cellular immune response, so-called because the cells themselves travel to where they are needed, unlike B cells, which secrete antibodies into the bloodstream. T cells begin as stem cells in the bone marrow, then travel to the thymus gland (the "T" refers to thymus). As the immature T cells, called thymocytes, migrate towards the interior of the thymus, they display diverse cell surface receptors. An extensive selection process unfolds. As the wandering thymocytes touch lining cells in the gland that are studded with "self" antigens, thymocytes that do not attack the lining cells begin maturing into T cells, whereas those that do harm the lining cells die by apoptosis—in great numbers. Gradually, T cells-to-be that recognize self are selected and persist. The process of retaining only some thymocytes is termed "positive selection," while the process of killing others is called "clonal deletion."

Several types of T cells are distinguished by the types and patterns of receptors on their

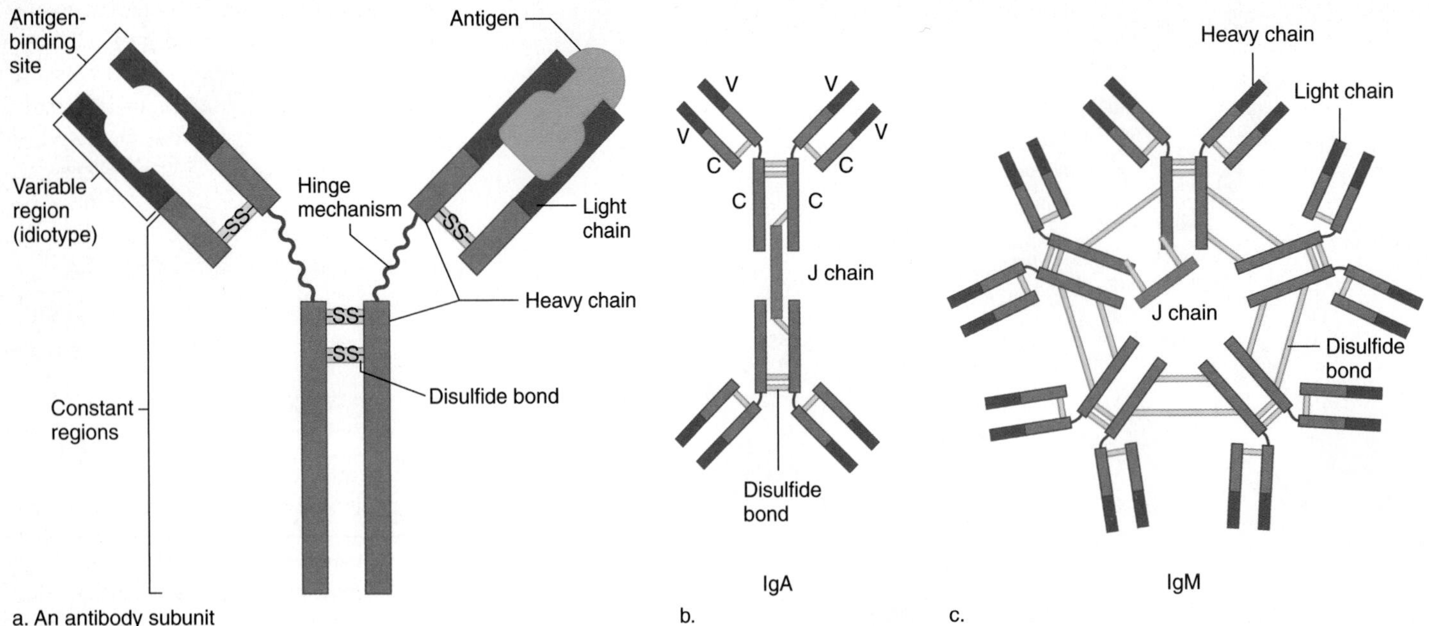


## figure 16.12

**Antibody structure.** The simplest antibody molecule (*a*) consists of four polypeptide chains, two heavy and two light, joined by two sulfur atoms that join, forming a disulfide bond. Part of each polypeptide chain has a constant sequence of amino acids, and the remainder of the sequence is variable. The tops of the Y-shaped molecules form antigen binding sites. (*b*) IgA consists of two Y-shaped subunits, and IgM (*c*) consists of five subunits. J chain proteins join the units.

## table 16.2

**Types of Antibodies**

| Type* | Location | Functions |
|---|---|---|
| IgA | Milk, saliva, urine, and tears; respiratory and digestive secretions | Protects against pathogens at points of entry into body |
| IgD | On B cells in blood | Stimulates B cells to make other types of antibodies, particularly in infants |
| IgE | In secretions with IgA and in mast cells in tissues | Acts as receptor for antigens that cause mast cells to secrete allergy mediators |
| IgG | Blood plasma and tissue fluid; passes to fetus | Protects against bacteria, viruses, and toxins, especially in secondary immune response |
| IgM | Blood plasma | Fights bacteria in primary immune response; includes anti-A and anti-B antibodies of ABO blood groups |

*The letters A, D, E, G, and M refer to the specific conformation of heavy chains characteristic of each class of antibody.

surfaces, and by their functions. **Helper T cells** recognize foreign antigens presented on macrophages, stimulate B cells to produce antibodies, secrete cytokines, and activate another type of T cell called a **cytotoxic T cell,** sometimes called a killer T cell. Certain T cells may help to suppress an immune response when it is no longer required. The cytokines that helper T cells secrete include interleukins, interferons, tumor necrosis factor, and **colony stimulating factors,** which stimulate white blood cells to mature in bone marrow (table 16.3). Cytokines interact and signal each other in complex cascades of gene action.

Distinctive surfaces distinguish subsets of helper T cells. Certain antigens called cluster-of-differentiation antigens, or CD antigens, enable T cells to recognize foreign antigens displayed on macrophages. One such cell type, called a CD4 helper T cell, is an early target of HIV. Considering the critical role of helper T cells in coordinating immunity, it is little wonder that HIV infection ultimately topples the entire system, a point we will return to soon.

Cytotoxic T cells are distinguished by a lack of CD4 receptors and the presence of CD8 receptors. These cells attack virally infected and cancerous cells by attaching to them and releasing chemicals. They do this by joining two surface peptides to form T cell receptors that bind foreign antigens. When a cytotoxic T cell encounters a nonself cell—a cancer cell, for example—the T cell receptors draw the two cells into physical contact. The T cell then releases a protein called perforin, which drills holes in the foreign cell's membrane. This disrupts the flow of chemicals in and out of the foreign cell and kills it (figure 16.13). Cytotoxic T cell receptors also attract body cells that are covered with certain viruses, destroying the

**table 16.3**

**Types of Cytokines**

| Cytokine | Function |
|---|---|
| Colony stimulating factors | Stimulate bone marrow to produce lymphocytes |
| Interferons | Block viral replication, stimulate macrophages to engulf viruses, stimulate B cells to produce antibodies, attack cancer cells |
| Interleukins | Control lymphocyte differentiation and growth, cause fever that accompanies bacterial infection |
| Tumor necrosis factor | Stops tumor growth, releases growth factors, stimulates lymphocyte differentiation, dismantles bacterial toxins |

Cancer cell
Foreign antigen
T cell receptor
Cytotoxic T cell
Perforin
Hole

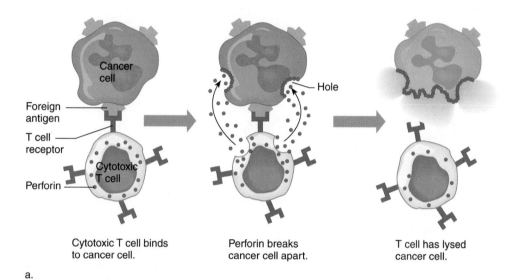

Cytotoxic T cell binds to cancer cell.

Perforin breaks cancer cell apart.

T cell has lysed cancer cell.

a.

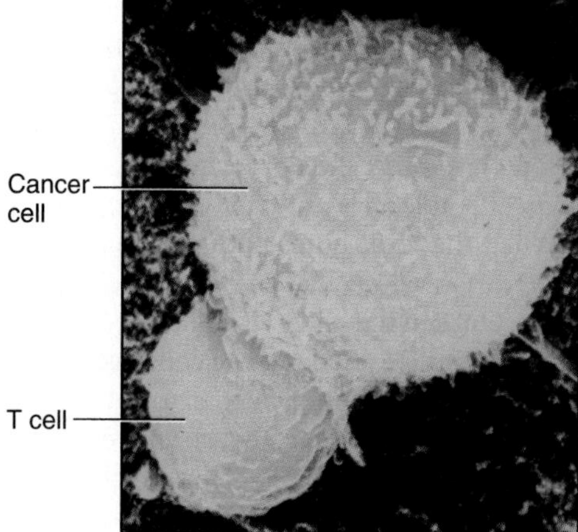


b.

## figure 16.13

**Death of a cancer cell.** *(a)* A cytotoxic T cell binds to a cancer cell and injects perforin, a protein that pokes holes in the cancer cell's membrane. As the holes form, the cancer cell dies, leaving behind debris that macrophages clear away. *(b)* The smaller cell is a cytotoxic T cell, which homes in on the surface of the large cancer cell above it. The cytotoxic T cell will shatter the cancer cell, leaving behind scattered fibers.

cells before the viruses on them can enter, replicate, and spread the infection. In this way, cytotoxic T cells continually monitor cells, recognizing and eliminating virally infected and tumor cells.

### *Integrating Immune Responses*

Yet another type of T cell, called a **gamma-delta T cell,** oversees the interactions of the innate and adaptive immune responses. These cells detect stress and injury, then activate other parts of the immune system to counter possible infection. Gamma-delta T cells begin as stem cells in the bone marrow, then they wander to patches of lymphocytes in the small intestine, where they specialize. Damaged epithelium (lining tissue) activates gamma-delta T cells. Only discovered recently because they make up a tiny fraction of the total population of lymphocytes, gamma-delta T cells:

- recognize nonprotein molecules that cells release when they are stressed.
- release epithelial growth factors that stimulate cell division in epithelium—an important step in wound healing.
- produce biochemicals that contribute to the inflammatory response.
- secrete cytokines that coordinate the functions of the innate and adaptive immune responses.

Gamma-delta T cells essentially link the two major branches of the immune system.

Table 16.4 summarizes immune system cell types.

**KEY CONCEPTS**

The immune system consists of physical barriers; an innate immune response of inflammation, phagocytosis, complement, collectins, and cytokines; and a more directed adaptive immune response. Adaptive immunity is diverse, specific, and it remembers. It has two components. • In the humoral immune response, stimulated B cells divide and differentiate into plasma cells and memory cells. A plasma cell secretes abundant antibodies of a single type. Antibodies are made of Y-shaped polypeptides, each consisting of two light and two heavy chains. Each chain consists of a constant and a variable region, and the tips of the Y form an antigen binding site with a specific idiotype. Antibodies make foreign antigens more visible to macrophages and stimulate complement. Shuffling gene pieces generates astounding antibody diversity. • In the cellular immune response, helper T cells stimulate B cells to manufacture antibodies and cytotoxic T cells to secrete cytokines. Using T cell receptors, cytotoxic T cells bind to nonself cells and virus-covered cells and burst them. Gamma-delta T cells coordinate the immune response.

**table 16.4**

**Types of Immune System Cells**

| Cell Type | Function |
|---|---|
| Macrophage | Presents antigens |
| | Performs phagocytosis |
| Mast cell | Releases histamine in inflammation |
| | Releases allergy mediators |
| B cell | Matures into plasma cell, which produces antibodies, or memory cell |
| T cells | |
| Helper | Recognizes nonself antigens presented on macrophages |
| | Stimulates B cells to produce antibodies |
| | Secretes cytokines |
| | Activates cytotoxic T cells |
| Cytotoxic | Attacks cancer cells and cells infected with viruses |
| Gamma-delta | Coordinates functioning of adaptive and innate immunity |

## 16.3 Abnormal Immunity

The immune system continually adapts to environmental change. Because the immune response is so diverse, its breakdown affects health in many ways. Immune system malfunction may be inherited or acquired, and immunity may be too strong, too weak, or misdirected.

### Inherited Immune Deficiencies

More than 20 types of inherited immune deficiencies are recognized, affecting both innate and adaptive immunity. In chronic granulomatous disease, neutrophils can engulf bacteria, but, due to deficiency of an enzyme called oxidase, cannot produce the activated oxygen compounds that kill bacteria. Because this enzyme is built of four polypeptide chains, four genes encode it, and there are four ways to inherit the disease, all X-linked. A very rare autosomal recessive form is caused by a defect in the vacuole that encloses bacteria. Antibiotics and interferon gamma are used as drugs in these patients to prevent bacterial infections, and the disease can be cured with a bone marrow or an umbilical cord stem cell transplant. However, these procedures have a high rate of failure and may cause death.

Mutations in genes that encode cytokines or T cell receptors impair cellular immunity, which primarily targets viruses and cancer cells. But because T cells activate the B cells that manufacture antibodies, abnormal cellular immunity (T cell function) causes some degree of abnormal humoral immunity (B cell function). Mutations in the genes that encode antibody segments, control how the segments join, or direct maturation of B cells impair immunity mostly against bacterial infection. Inherited immune deficiency can also result from defective B cells. In one form of B cell immune deficiency, B cells lack a protein, called B cell linker protein, that normally signals the cells to mature into plasma cells. A person with this type of immune deficiency is highly vulnerable to certain bacterial infections, particularly ear infections and sinusitis.

Severe combined immune deficiencies (SCID) affect both the humoral and cellular branches of the immune system. About half of SCID cases are X-linked. In a less severe form, the individual lacks B cells but has T cells. Before antibiotic drugs became available, these individuals died before the age of 10 years of overwhelming bacterial infection. In a more severe form of X-linked SCID, lack of T cells causes death by 18 months of age. Severe thrush (a fungal infection), chronic diarrhea, and recurrent lung infections usually kill these children.

A young man named David Vetter taught the world about the difficulty of life without immunity years before AIDS appeared. David had an autosomal form of SCID that caused him to be born without a thymus gland. His T cells could not mature and activate B cells, leaving him defenseless in a germ-filled world. Born in Texas in 1971, David spent his short life in a vinyl bubble, awaiting a treatment that never came. As he reached adolescence, David wanted to leave his bubble. An experimental bone marrow transplant was unsuccessful—soon afterward, David began vomiting and developed diarrhea, both signs of infection. David left the bubble but died within days of a massive infection. A commercial film recently poked fun at a "bubble boy." David Vetter's mother led many families with immune deficiencies in protesting the movie, which quickly vanished from theaters.

Another form of SCID, called adenosine deaminase deficiency, became the first illness to be successfully treated with gene therapy. It is discussed further in chapter 19.

### Acquired Immune Deficiency Syndrome

AIDS is acquired, which means that it is not an inherited disease. The human immunodeficiency virus (HIV) causes AIDS. HIV has RNA as its genetic material. It is a type of RNA virus called a retrovirus, because it converts RNA into DNA, the reverse of the usual direction of genetic information flow.

HIV infection gradually shuts down the immune system. First, HIV enters macrophages, impairing this first line of defense. In these cells and later in helper T cells, the virus adheres with its surface protein, called gp120, to two co-receptors on the host cell surface, CD4 and CCR5 (figure 16.14). Another glycoprotein, gp41, anchors gp120 molecules into the viral envelope. When the virus binds both co-receptors, it contorts in a way that enables it to enter the cell. Once in the cell, a viral enzyme, **reverse transcriptase,** catalyzes the construction of a DNA strand complementary to the viral RNA. The initial viral DNA strand replicates to form a DNA double helix, which enters the cell's nucleus and inserts into a chromosome. The viral DNA sequences are transcribed and translated, and the cell fills with viral pieces, which are assembled into complete new viral particles that eventually burst from the cell (figure 16.15).

Once helper T cells begin to die at a high rate, bacterial infections begin, because B cells aren't activated to produce antibodies. Much

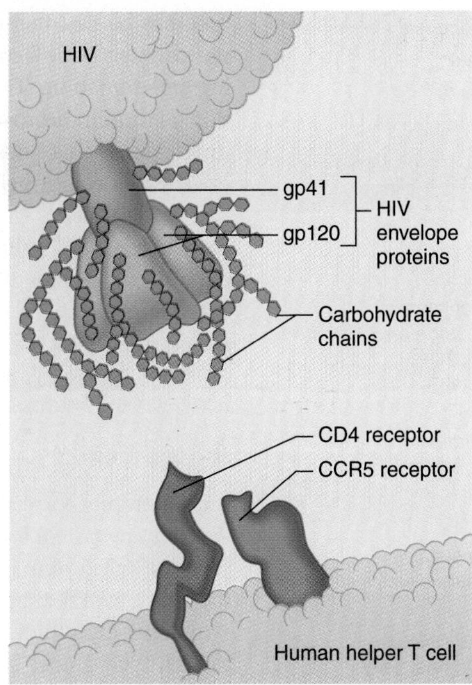

a.

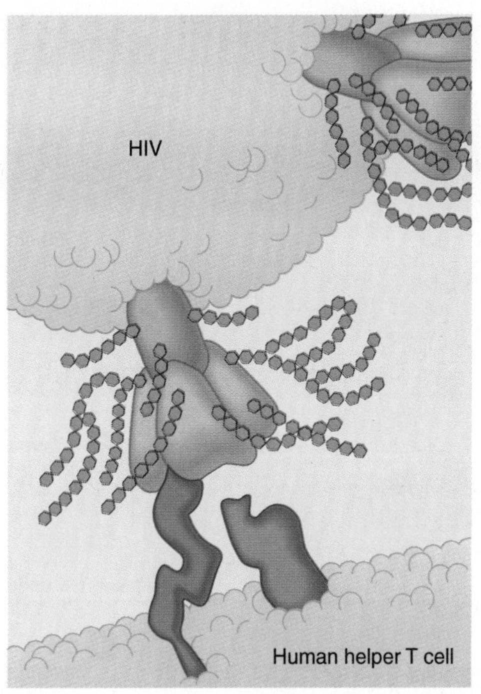

b.

## figure 16.14

**HIV binds to a helper T cell.** (*a*) The part of HIV that binds to helper T cells is called gp120 (gp stands for glycoprotein). (*b*) When the carbohydrate chains that shield the protein portion of gp120 move aside as they approach the cell surface, the viral molecule can bind to a CD4 receptor. Binding to a receptor called CCR5 is also necessary for HIV to dock at a helper T cell. Once bound to the cell surface, the viral envelope fuses with the cell membrane, enabling the virus to enter. A few lucky individuals lack CCR5, and cannot be infected by HIV. New types of drugs to fight HIV infection block the steps of viral entry. (The size of HIV here is greatly exaggerated.)

later in infection, HIV variants arise that can bind to a receptor called CXCR4 that is found on cytotoxic T cells. This explains the longstanding mystery of how the virus kills these cells, which lack CD4 receptors. When HIV binds to the cytotoxic T cells, it triggers apoptosis. Loss of these cells renders the body very vulnerable to viral infections and cancer.

HIV has an advantage over the human immune system because it replicates quickly, changes quickly, and can hide. The virus is especially prone to mutation, both because it cannot repair replication errors, and because those errors happen frequently—at a rate of 1 per every 5,000 or so bases—because of the "sloppiness" of reverse transcriptase. The immune system simply cannot keep up; antibodies against one viral variant are useless against another. For several years, the bone marrow produces 2 billion new T and B cells a day to counter the million to a billion new HIV particles that burst daily from shattered cells.

So genetically diverse is the population of HIV in a human host that within days of the initial infection, variants arise that resist the drugs used to treat AIDS. HIV's changeable nature has important clinical implications. Combining drugs that act in different ways provides the greatest chance of slowing the disease process.

Three types of drugs have cut the death rate from AIDS in half since their debut in 1996—two reverse transcriptase inhibitors, and protease inhibitors. The reverse transcriptase inhibitors block the conversion of viral RNA into DNA, which is necessary for it to insert into a human chromosome, where it would be replicated. The protease inhibitors block the trimming of certain viral proteins, which is required for new viral particles to assemble. The first of a fourth type of drug, called a fusion inhibitor. It is a 32 amino-acid-long peptide that corresponds to a segment of gp41 that is crucial for the virus to bind to a cell and enter it. The drug displaces the virus, and cell entry cannot occur. Fusion inhibitors belong to a more general class of HIV drugs called entry inhibitors. The goal is to keep coming up with new combinations of drugs that work by different mechanisms, to stay ahead of the natural selection of drug-resistant viral variants.

A very promising approach to conquering AIDS is to discover how certain people resist infection, and then attempt to mimic their conditions with drugs. So far, researchers have identified four receptors or the molecules that bind to them that are altered by mutation to keep HIV out in resistant individuals. To find these receptors, epidemiologists searched the DNA of people at high risk of HIV infection—people who had unprotected sex with many partners, and people with hemophilia who had received HIV-tainted blood in the 1980s—but who were not infected. Then, molecular biologists discovered the CCR5 receptor, and epidemiologists found that people who were homozygous recessive for a 32-base deletion in the CCR5 gene were among those who had, seemingly miraculously, avoided AIDS. Their CCR5 co-receptors were too stunted, thanks to a premature "stop" codon, to reach the cell's surface (figure 16.14). Like a ferry arriv-

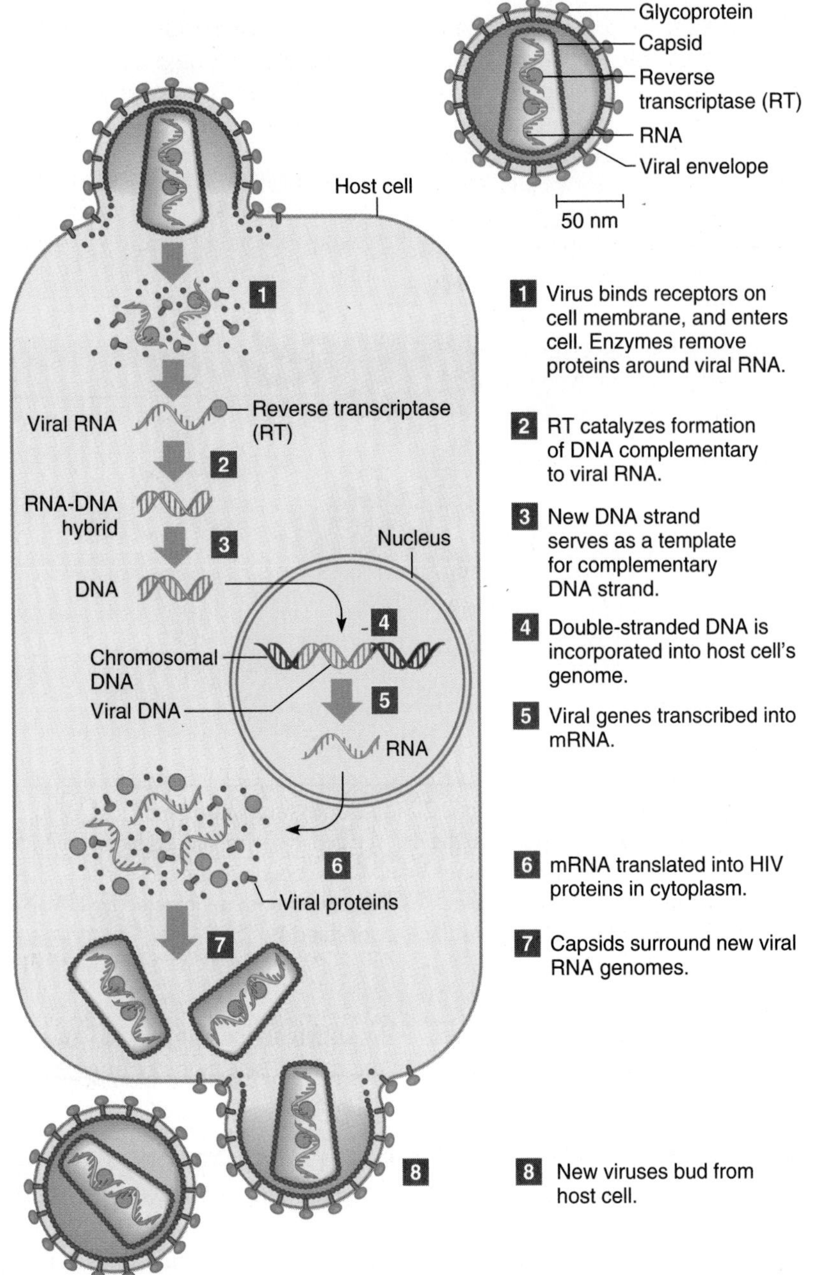
Glycoprotein
Capsid
Reverse transcriptase (RT)
RNA
Viral envelope
50 nm
Host cell
1 Virus binds receptors on cell membrane, and enters cell. Enzymes remove proteins around viral RNA.
Viral RNA
Reverse transcriptase (RT)
2 RT catalyzes formation of DNA complementary to viral RNA.
RNA-DNA hybrid
3 New DNA strand serves as a template for complementary DNA strand.
Nucleus
DNA
4 Double-stranded DNA is incorporated into host cell's genome.
Chromosomal DNA
Viral DNA
5 Viral genes transcribed into mRNA.
RNA
6 mRNA translated into HIV proteins in cytoplasm.
Viral proteins
7 Capsids surround new viral RNA genomes.
8 New viruses bud from host cell.

**figure 16.15**

**Replication of HIV.** HIV infects a cell by integrating first into the host chromosome, then producing more virus particles.

ing at shore to find no dock, HIV has nowhere to bind on the cells of these fortunate individuals. Heterozygotes, with one copy of the deletion, can become infected with HIV, but they remain healthy for several years longer than people who do not have the deletion. In the United States, 1 to 2 percent of Caucasians are homozygous recessive for the CCR5 mutation. Very few African Americans have it, and no Asians or native Americans are known to have it. The Japanese have their own form of HIV protection—a few individuals have a mutation in the gene that encodes a protein to which CCR5 binds. A type of entry inhibitor drug disables the CCR5 receptor, mimicking the situation in homozygotes for a mutation in this gene.

Another group of HIV-infected individuals whose conditions have suggested new treatment approaches are the "long term nonprogressors." These people are infected, but healthy. The secret to their success may be that their immune systems do not counter the infection with overwhelming inflammation. The huge overproduction of tumor necrosis factor and certain interleukins that normally accompanies infection does not help at all—in fact, inflammation speeds viral replication and the destruction of T cells. One company has developed a version of an adrenal hormone that dampens the inflammatory response in AIDS patients. Its effects on symptoms and disease progression are still being evaluated.

## Autoimmunity

Autoimmunity is a reaction that occurs when the immune system backfires, producing antibodies, called **autoantibodies,** that attack the body's own healthy tissues. The signs and symptoms of autoimmune disorders reflect the cell types under attack. For example, in autoimmune ulcerative colitis, colon cells are the target, and severe abdominal pain results.

A mutation in a single gene can cause the collection of symptoms that characterize an autoimmune disorder. For example, a mutation in a gene on chromosome 21q causes "autoimmune polyendocrinopathy syndrome type I." Malfunction of various endocrine glands occurs in a similar sequence in patients. Children under five develop candidiasis, a fungal infection (not an autoimmune disorder). By age 10, the individual's parathyroid glands begin to fail, affecting calcium metabolism. By age 15, most affected individuals also develop Addison disease, which reflects a deficiency in adrenal gland hormones. Other associated conditions include thyroid deficiency, diabetes mellitus, vitiligo (skin whitening), and alopecia (hair loss).

Many autoimmune conditions also have counterparts caused by other mechanisms. Hemolytic anemia, for example, may also be inherited, or provoked by exposure to a toxin.

Autoimmunity may arise in several ways:

- A virus replicating within a cell incorporates proteins from the cell's surface onto its own. When the immune system "learns" the surface of the virus to destroy it, it also learns to attack human cells that normally bear the protein.
- Some thymocytes that should have died in the thymus somehow escape apoptosis, persisting to attack "self" tissue later on.

- A nonself antigen coincidentally resembles a self antigen, and the immune system attacks both. In rheumatic fever, for example, antigens on heart valve cells resemble those on *Streptococcus* bacteria; antibodies produced to fight a strep throat also attack the heart valve cells. Some cases of insulin-dependent diabetes mellitus may also be due to this type of autoimmunity. Part of a protein on insulin-producing cells in the pancreas resembles part of bovine serum albumin (BSA), a protein in cow's milk. Children who are allergic to cow's milk may develop antibodies against BSA, which later attack the similar-appearing pancreas cells, causing diabetes.

Some disorders traditionally thought to be autoimmune in origin may in fact have a more bizarre cause—fetal cells persisting in a woman's circulation, even decades after the fetus has grown up! (See the discussion of fetal cell sorting in chapter 12.) In response to an as yet unknown trigger, the lurking fetal cells, perhaps "hiding" in a tissue such as skin, emerge, stimulating antibody production. If we didn't know the fetal cells were there, the resulting antibodies and symptoms would appear to be an autoimmune disorder. This mechanism, called microchimerism ("small mosaic"), may explain the higher prevalence of autoimmune disorders among women. It was discovered in a disorder called scleroderma, which means "hard skin" (figure 16.16).

Patients describe scleroderma, which typically begins between ages 45 and 55, as "the body turning to stone." Symptoms include fatigue, swollen joints, stiff fingers, and a masklike face. The hardening may affect blood vessels, the lungs, and the esophagus, too. Clues that scleroderma is a delayed response to persisting fetal cells include the following observations:

- It is much more common among women.
- Symptoms resemble those of graft-versus-host disease (GVHD), in which transplanted tissue produces chemicals that destroy the host's body. Antigens on cells in scleroderma lesions match those that cause GVHD.
- Mothers who have scleroderma and their sons have cell surfaces that are more similar than those of unaffected mothers and their sons. Perhaps the similarity of cell surfaces enabled the fetal cells to escape surveillance and destruction by the woman's immune system.
- Skin lesions from affected mothers of sons include cells that have Y chromosomes. Mothers can develop scleroderma from daughters too, but the fetal cells cannot be as easily distinguished because they are XX, like the mothers' cells.

It's possible that other disorders traditionally considered to be autoimmune may actually reflect an immune system response to lingering fetal cells. Table 16.5 describes some autoimmune disorders.

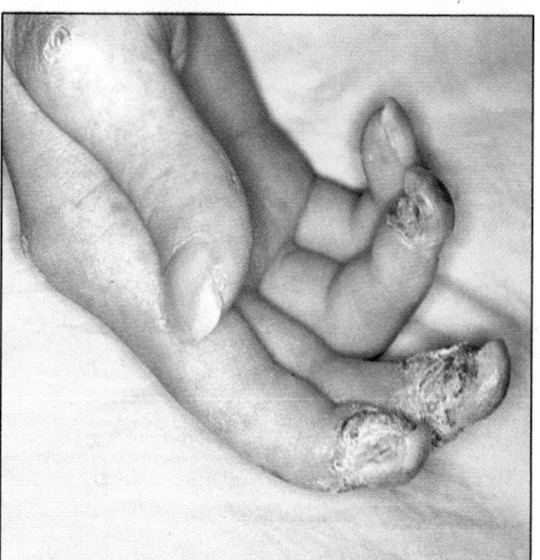

**figure 16.16**

**An autoimmune disorder—maybe.** Scleroderma hardens the skin. Some cases appear to be caused by a long-delayed reaction of the immune system to cells retained from a fetus—decades earlier.

**table 16.5**

**Autoimmune Disorders**

| Disorder | Symptoms | Targets of Antibody Attack |
|---|---|---|
| Glomerulonephritis | Lower back pain | Kidney cell antigens that resemble *Streptococcus* antigens |
| Graves disease | Restlessness, weight loss, irritability, increased heart rate and blood pressure | Thyroid gland antigens |
| Hemolytic anemia | Fatigue and weakness | Red blood cells |
| Myasthenia gravis | Muscle weakness | Nerve message receptors on skeletal muscle cells |
| Pernicious anemia | Fatigue and weakness | Binding site for vitamin B on cells lining stomach |
| Rheumatic heart disease | Weakness, shortness of breath | Heart valve cell antigens that resemble *Streptococcus* antigens |
| Rheumatoid arthritis | Joint pain and deformity | Cells lining joints |
| Scleroderma | Thick, hard, pigmented skin patches | Connective tissue cells |
| Systemic lupus erythematosus | Red rash on face, prolonged fever, weakness, kidney damage | DNA, neurons, blood cells |
| Type I diabetes mellitus | Thirst, hunger, weakness, emaciation | Pancreatic beta cells |
| Ulcerative colitis | Lower abdominal pain | Colon cells |

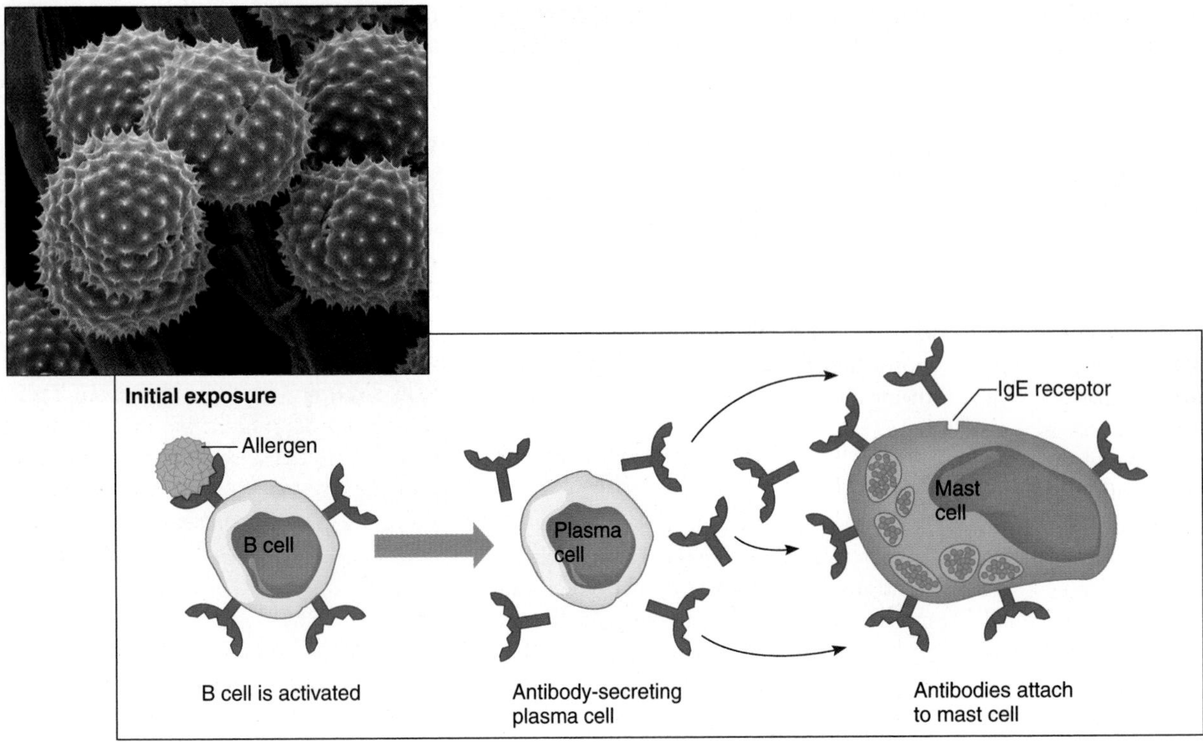


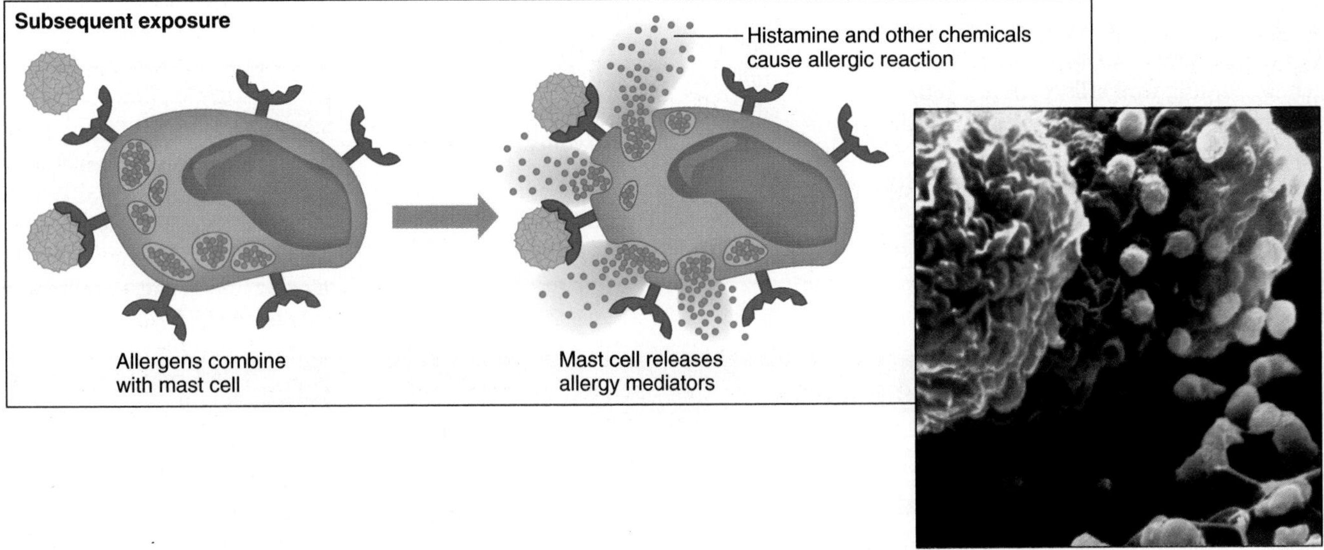


**figure 16.17**

**Allergy.** In an allergic reaction, B cells are activated by an allergen such as pollen (upper inset), and differentiate into antibody-secreting plasma cells. The antibodies attach to mast cells. When allergens are encountered again, they combine with the antibodies on the mast cells. The mast cells burst (lower inset), releasing the chemicals that cause itchy eyes and a runny nose.

## Allergies

An allergy is an immune system response to a substance, called an **allergen,** that does not actually present a threat. Many allergens are particles small enough to be carried in the air and into a person's respiratory tract. The size of the allergen may determine the type of allergy. For example, grass pollen is large and remains in the upper respiratory tract, where it causes hay fever. But allergens from house dust mites, cat dander, and cockroaches are small enough to infiltrate the lungs, triggering asthma (figure 16.17). Asthma is a chronic disease in which contractions of the airways, inflammation, and accumulation of mucus block air flow. One in five individuals in the United States has asthma.

Both the humoral and cellular arms of the adaptive immune system take part in an allergic response. Antibodies of class IgE are produced, and these bind to mast cells, sending signals that cause them to open and release allergy mediators such as histamine and heparin. Allergy mediators cause inflammation, with symptoms that may include the runny eyes of hay fever, the narrowed airways of asthma, rashes, or the overwhelming bodywide allergic reaction called anaphylactic shock. Allergens also activate a class of helper T cells that produce a particular mix of cytokines whose genes are clustered on chromosome 5q. Regions of chromosomes 12q and 17q have genes that control production of IgE.

The fact that allergies have become very common only during the past century suggests a much stronger environmental than genetic component. Still, people inherit susceptibilities to allergy. Twin studies of various allergies reveal about a 75 percent concordance, and isolated populations with a great deal of inbreeding tend to have high prevalence of certain allergies.

The allergies that people suffer today may be a holdover of an immune function that was important in the past. Evidence for this theory is that people with allergies have higher levels of white blood cells called eosinophils than do others, and these cells fight parasitic infections that are no longer common. In a more general sense, because allergies are more common in developed nations and have become more prevalent since the introduction of antibiotic drugs, some researchers hypothesize that some allergies may result from a childhood relatively free of infection, compared to times past—almost as if the immune system is reacting to being underutilized. Although we still have much to learn about how immune system genes respond to environmental stimuli to cause allergies, we can use what we know to develop treatments (table 16.6).

**Key Concepts**

Inherited immune deficiencies affect innate and adaptive immunity. • AIDS is caused by HIV. HIV replicates very rapidly, and T cell production matches it until the immune response is overwhelmed and AIDS begins. HIV is a retrovirus that injects its RNA into host cells by binding coreceptors. Reverse transcriptase then copies viral RNA into DNA. HIV uses the cell's protein synthesis machinery to mass produce itself; then the cell bursts, releasing virus. HIV continually mutates, becoming resistant to drugs. • In autoimmune disorders, autoantibodies attack healthy tissue. These conditions may be caused by a virus that borrows a self antigen, T cells that never learn to recognize self, or healthy cells bearing antigens that resemble nonself antigens. Some conditions thought to be autoimmune may actually reflect an immune system response to retained fetal cells. • An overly sensitive immune system causes allergies. In an allergic reaction, allergens bind to IgE antibodies on mast cells, which release allergy mediators. A subset of helper T cells secretes cytokines that contribute to allergy symptoms.

## 16.4 Altering Immune Function

Medical technology can alter or augment immune system functions in various ways. Vaccines trick the immune system into acting early. Antibiotic drugs, which are substances derived from organisms such as fungi and soil bacteria, have been used for decades to assist an immune response. Cytokines and altered antibodies are used as drugs to treat a variety of conditions. Transplants require suppression of the immune system so that the body will accept a nonself body part.

### Vaccines

A **vaccine** is an inactive or partial form of a pathogen that the immune system responds to by alerting B cells to produce antibodies. Upon encountering the pathogen in its natural state later, a secondary immune response ensues, even before symptoms arise. Vaccines consisting of entire viruses or bacteria can, rarely, cause illness if they mutate to a pathogenic form. A safer type of vaccine uses only the part of the pathogen's surface that elicits an immune response. Vaccines against different illnesses can be combined into one injection, or the genes encoding antigens from several pathogens can be inserted into a harmless virus and delivered as a "super vaccine."

Vaccine technology dates back to the 11th century in China. Because people observed that those who recovered from smallpox never got it again, they collected the scabs of infected individuals and crushed them into a powder, which they inhaled or rubbed into pricked skin. In 1796, the wife of a British ambassador to Turkey witnessed the Chinese method of vaccination and mentioned it to an English country physician, Edward Jenner. Intrigued, Jenner had himself vaccinated the Chinese way, and then thought of a different approach.

It was widely known that people who milked cows contracted a mild illness called cowpox, but did not get smallpox. The cows became ill from infected horses. Since the virus seemed to jump species, Jenner wondered, would exposing a healthy person to cowpox lesions protect against smallpox? A slightly different virus causes cowpox rather than smallpox, but Jenner's approach worked, leading to development of the first vaccine (in fact, the word comes from the Latin *vacca,* for "cow"). Unable to experiment on himself because he'd already taken the Chinese vaccine, Jenner instead tried his first vaccine on a volunteer, 8-year-old James Phipps. He dipped a needle in pus oozing from a small sore on a milkmaid named Sarah Nelmes, then scratched the

**table 16.6**

**Approaches to Treating Allergies**

| Agent | Mechanism |
|---|---|
| Corticosteroids | These drugs enter the nuclei of affected cells and suppress the activity of genes that encode cytokines and allergy mediator receptors. |
| Antihistamines | Antihistamines block receptors on mast cells that bind IgE antibodies, preventing the release of histamine. |
| Antibodies | Laboratory-produced antibodies bind to IgE, blocking contact with mast cells. |
| Sensitization | Gradual exposure to small amounts of the allergen can prevent allergy attacks. The mechanism is unknown. |

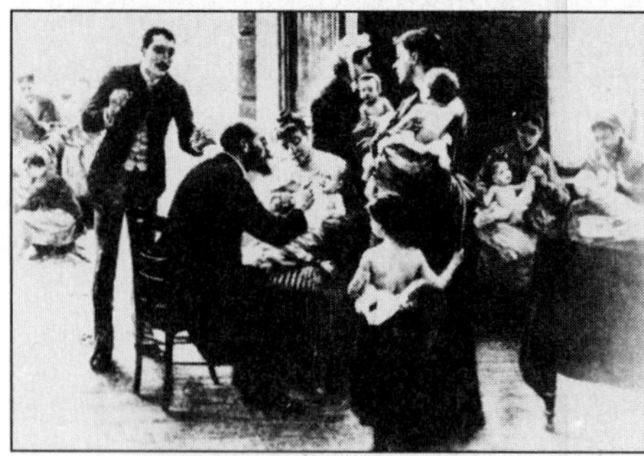

## figure 16.18

**Smallpox vaccine.** Edward Jenner inoculated children with smallpox vaccine in 1798.

boy's arm with it. Young James never became ill. Eventually, improved versions of Jenner's smallpox vaccine would eradicate a disease that once killed millions (figure 16.18). However, the United States government has ordered smallpox vaccine to again be mass produced, because of the threat of the virus being used as a bioweapon (discussed in section 16.5).

Polio was another vaccine success story. Many adults recall lining up in the 1950s to receive injections, and a few years later, oral vaccine squirted onto sugar cubes. These were the first polio vaccines. In the 1960s came vaccines to protect against diseases that were once a normal part of childhood—measles, mumps, and rubella. Several other vaccines are on the list today, including chicken pox and hepatitis B.

People still receive most vaccines as injections, but several new delivery methods are in development, including nasal sprays and genetically engineered fruits and vegetables (table 16.7). It may seem strange to eat a banana as a vaccine against an infection, but the idea makes sense. Edible plants are modified to have genes from pathogens, and these genes encode the antigens that evoke an immune response in the human body. The foreign antigens, when in the small intestine, travel through highly specialized cells called M cells to reach immune system cells (figure 16.19). Phagocytic cells called dendritic cells capture antigens from food and present them to T cells. From here, the antigens are passed to the bloodstream, where they stimulate B cells to differentiate into plasma cells that produce IgA. These anti-

## table 16.7

**Types of Vaccines**

| Type | Disease Prevented |
|---|---|
| Entire weakened (attenuated) pathogen | Polio |
| Inactivated toxin | Tetanus |
| Part of pathogen surface | Hepatitis B |
| Recombinant vaccine (pathogen's gene placed in harmless bacteria or yeast) | Lyme disease |
| "Naked" DNA from pathogen | Influenza, hepatitis B |
| Plant-based vaccine | |
| potato | Traveler's diarrhea |
| tomato | Hepatitis B, rabies |
| black-eyed peas | Infectious disease in mink |
| tobacco | Dental caries |

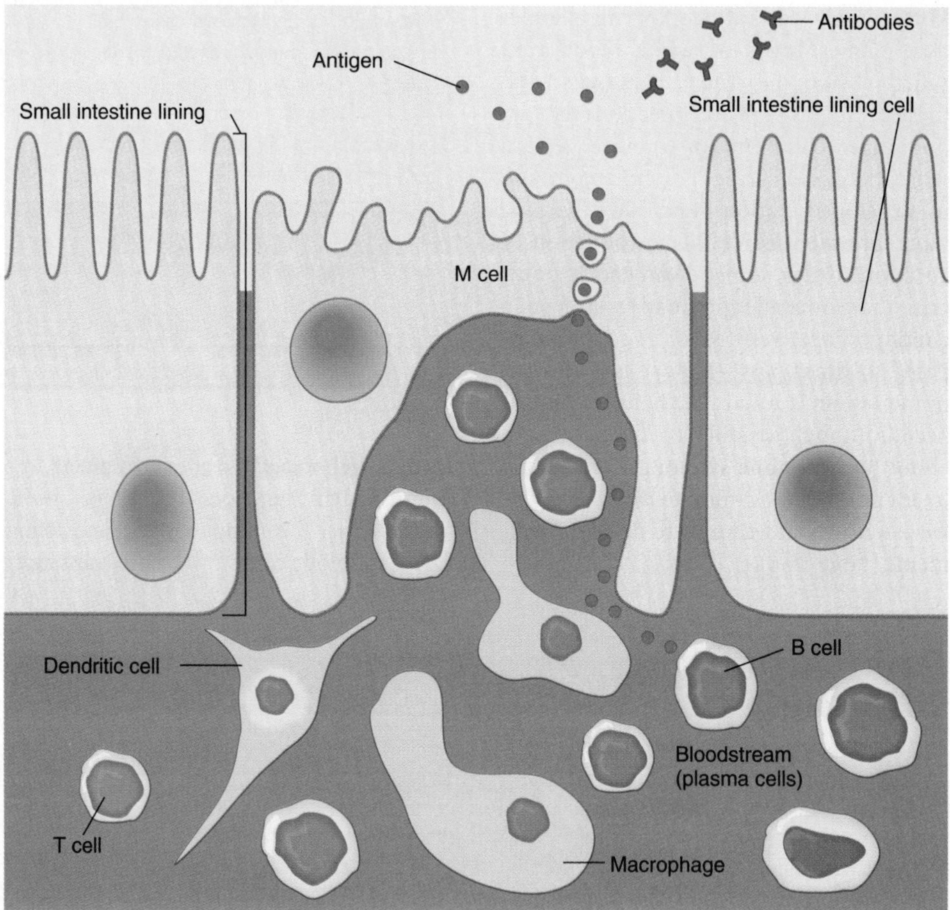


## figure 16.19

**M cells.** M cells set up immunity in the digestive tract. M cells located within the small intestinal lining sample antigens from food, and pass them to immune system cells that cluster below. These cells stimulate B cells to produce antibodies, which return to the small intestinal lining via the bloodstream.

bodies coat the small intestinal lining, protecting against food-borne pathogens. Infectious prions may reach the central nervous system through this route. Recall from chapter 1 that the prion disease kuru was linked to cannibalism, and both BSE in cattle and vCJD in humans are associated with consuming tainted animal flesh.

One of the first experimental plant-based vaccines is a potato given an *E. coli* gene that encodes a toxin that causes a type of infant and traveler's diarrhea. A week after mice eat the raw, shredded potatoes, antibodies against the toxin appear in their circulations. This vaccine is being tested on humans. A plant-based vaccine must be eaten raw, and because raw potato isn't very appetizing, future plant-based vaccines may be delivered in bananas. This would certainly be an easy way to vaccinate babies all over the world.

## Immunotherapy

Immunotherapy amplifies or redirects the immune response. It originated in the 19th century (see Technology Timeline: Immunotherapy). Today, many immunotherapies are in clinical trials, and a few are already part of medical practice.

### *Monoclonal Antibodies Boost Humoral Immunity*

When a single B cell recognizes a single foreign antigen, it manufactures a single, or monoclonal, type of antibody in response. A large amount of a single antibody type would be useful in targeting a particular pathogen or cancer cell because of the antibody's great specificity.

In 1975, British researchers Cesar Milstein and George Köhler devised **monoclonal antibody** (MAb) technology, which amplifies a single B cell, preserving its specificity. First, they injected a mouse with a sheep's red blood cells (figure 16.20). They then isolated a single B cell from the mouse's spleen and fused it with a cancerous white blood cell from a mouse. The fused cell, called a hybridoma, had a valuable pair of talents. Like the B cell, it produced large amounts of a single antibody type. Like the cancer cell, it divided continuously.

Today, MAbs are used in basic research, veterinary and human health care, agriculture, forestry, and forensics. Along the way, researchers have developed ways to make them more like human antibodies—the original mouse versions caused allergic reactions in many people.

Immunotherapy

**1890s** New York surgeon William Coley, after noting patients cured of cancer following bacterial infections, intentionally gives other cancer patients killed bacteria. Sometimes it works; the recipients' bodies make tumor necrosis factor in response to the bacteria.

**1890s** German bacteriologist Paul Ehrlich develops the concept of the "magic bullet," a substance that destroys diseased cells yet spares healthy cells.

**1950s** Attempts begin to immunize mice against future cancer by implanting their own tumors back into them. Later, scientists realize these tumors are not accurate models of human cancer because they are induced and small.

**1975** Monoclonal antibody technology harnesses the specificity of a single antibody type.

**1980s** Testing begins using cytokines (interleukins and interferons) to treat a variety of disorders. Researchers work out dosages, delivery protocols, and ways to minimize side effects.

**1990s–2000s** Several cytokines are approved for varied uses; others in clinical trials to treat cancers.

MAbs that detect tiny amounts of a molecule are used to diagnose everything from strep throat to turf grass disease. One common use is a home pregnancy test. A woman places drops of her urine onto a paper strip impregnated with a MAb that binds to hCG, the hormone present only during pregnancy. A color change ensues if the MAb binds its target.

MAbs can detect cancer earlier than other methods can. The MAb is attached to a fluorescent dye and injected into a patient or applied to a sample of tissue or body fluid. If the MAb binds its target—an antigen found mostly or only on cancer cells—the fluorescence is detected with a scanning technology or fluorescence microscope. MAbs linked to radioactive isotopes or to drugs can be used in a similar fashion to deliver treatment to cancer cells. A MAb-based drug called Herceptin is used to treat some breast cancers. "In Their Own Words" on page 343 describes a use of MAbs in basic research that may have important clinical implications.

### *Cytokines Boost Cellular Immunity*

As coordinators of immunity, cytokines are used to treat a variety of conditions. However, it has been difficult to develop these body chemicals into drugs for several reasons: They cause side effects, they remain active only for short periods, and they must be delivered precisely where they are needed.

Interferon (IF) was the first cytokine to be tested on a large scale. When researchers discovered it in the 1950s, they hailed it as a cure-all wonder drug. Although it did not live up to early expectations, various interferons are used today to treat a dozen or so conditions, including a few types of cancer, genital warts, and multiple sclerosis.

Interleukin-2 (IL-2) is administered intravenously to treat kidney cancer recurrence. Colony stimulating factors, which cause immature white blood cells to mature and differentiate, are used routinely to boost white blood cell supplies in people with suppressed immune systems, such as those with AIDS or individuals receiving cancer chemotherapy. This allows a patient to withstand higher doses of a conventional drug. Cancer treatment is increasingly combining immune system cells and biochemicals with standard therapies.

**figure 16.20**

**Monoclonal antibody technology.** Monoclonal antibodies are pure preparations of a single antibody type that recognize a single antigen. They are useful in diagnosing and treating disease because of their specificity.

Antigens injected into mouse
B cells with antibodies, extracted from mouse spleen
Cancer cells
Cancer cells are fused with spleen cells, so that newly formed cells (hybridomas) live longer
Hybridomas are separated and cloned to produce large populations
Monoclonal antibodies produced

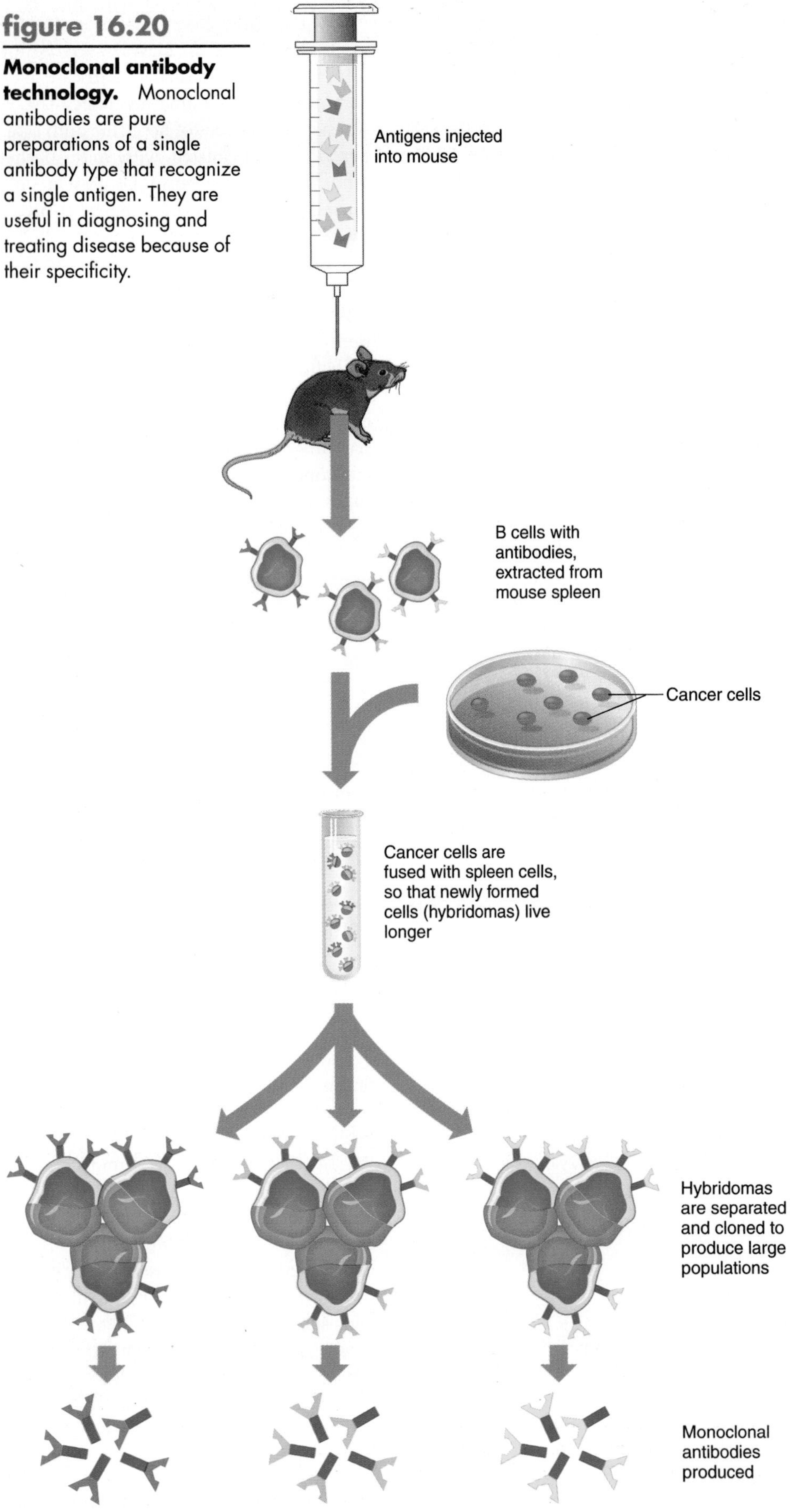

## Transplantation

When a car breaks down, replacing the damaged part often fixes the trouble. The same is sometimes true for the human body. Hearts, kidneys, livers, lungs, corneas, pancreases, skin, and bone marrow are routinely transplanted, sometimes several organs at a time. Although transplant medicine had a shaky start (see the Technology Timeline: Transplantation), many of the problems have been worked out. Today, thousands of transplants are performed annually and recipients gain years of life.

### *Transplant Types*

Transplants are classified by the relationship of donor to recipient (figure 16.21):

1. An **autograft** transfers tissue from one part of a person's body to another. A skin graft taken from the thigh to replace burned skin on the chest, or a leg vein that replaces a coronary artery, are autografts. The immune system does not reject the graft because the tissue is self. (Technically, an autograft is not a transplant because it involves only one person.)
2. An **isograft** is tissue from an identical twin. Because such twins are genetically identical, the recipient's immune system does not reject the transplant. Ovary isografts have been performed.
3. An **allograft** comes from an individual who is not genetically identical to the recipient, but is a member of the same species. A kidney transplant from a relative or other suitable donor is an allograft.
4. A **xenograft** transplants tissue from one species to another. (See the Bioethics Box on page 347.)

### *Rejection Reactions—Or Acceptance*

The immune system recognizes most donor tissue as nonself and may attempt to destroy it in a tissue rejection reaction that involves T cells, antibodies, and activation of complement. The greater the difference between recipient and donor cell surfaces, the more rapid and severe the rejection reaction. An extreme example is the **hyperacute rejection reaction** against tissue transplanted

*In Their Own Words*

# Innate Antibodies: An Ancient Protection Against Sperm or HIV?

Toby Rodman, professor emeritus at Cornell University Medical College, has long been interested in the components of umbilical cord blood, particularly the B cells. To study what those cells can do, she fuses them with Epstein-Barr virus, which makes them "immortal" (able to divide continuously) and able to produce single antibody types, much like monoclonal antibodies. A few years ago, she was startled to find that two such immortalized B cells attack an HIV protein called *tat.* Why and how would a fetus manufacture an antibody against an HIV protein? It might be a case of mistaken identity, an antibody directed against one antigen that can also target a chemically similar one, from the virus. More practically, Dr. Rodman hopes to direct these antibodies against HIV. She was born before World War I began and is still at work. Here, she explains how this apparent built-in protection against HIV infection may have arisen.

Several studies showed that with progression of HIV infection, these anti-*tat* antibodies decline. They control some of the bad effects associated with *tat,* which is able to cross the blood-brain barrier and therefore is probably directly concerned with progression to AIDS-related dementia. As long as antibodies are there, progression does not occur.

The predominant antibody that is reactive with *tat* has an arginine-rich sequence. Why is it important for the human genome to react to an arginine-rich protein? Now the story turns to protamines. These are proteins that are active as sperm form. During spermatogenesis, histones become depleted, and protamines substitute for them. This allows tighter packing of DNA into the sperm head. At fertilization, protamines are released, histones are restored, and it's back to the conventional DNA structure. Protamines are arginine-rich.

Years ago I began studying antibodies to protamines, and was systematically searching databases. I was struck by the fact that the *tat* database matched the database for sperm protamine! Why is it important to remove protamines from a fertilized egg? Apparently protamines are arginine-rich and can traverse the blood-brain barrier. Evolution wise, it is important to get rid of the protamines because they presented an innate hazard by getting into the brain.

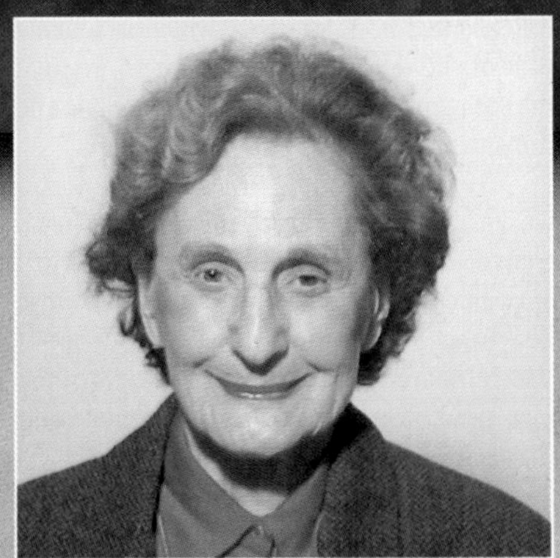

**figure 1**

**Toby Rodman.**

My work started out by looking at fertilization: I backed into the study of innate antibodies against HIV's *tat* protein. There are innate antibodies that can be captured out of cord blood cell-derived hybridomas, and now show a specific protective effect. The need to protect against a flood of protamines released at fertilization also prevents the arginine-rich *tat* protein from penetrating the blood-brain barrier to cause AIDS-related dementia. If I last long enough, I'll get it written up!

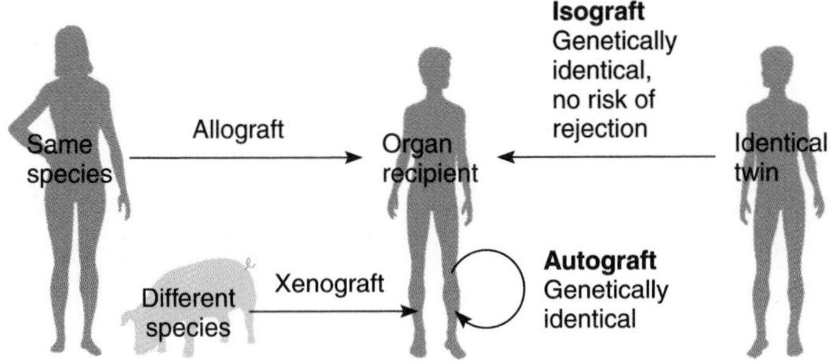


**figure 16.21**

**Transplant types.** An autograft is within an individual. An isograft is between identical twins. An allograft is between members of the same species, and a xenograft is between members of different species.

from another species—the donor tissue is usually destroyed in minutes as blood vessels blacken and cut off the blood supply.

Physicians use several approaches to dampen rejection so that a transplant can survive. These include closely matching HLA types between donor and recipient, and stripping donor tissue of antigens. Immunosuppressive drugs are given to dampen production of the antibodies and T cell subtypes that specifically attack transplanted tissue, while sparing other components of the immune system.

Graft-versus-host disease is a different type of immune problem that arise sometimes in bone marrow transplants used to correct certain blood deficiencies and cancers. The transplanted bone marrow, which is actually part of the donor's immune system, regards the recipient—its new body—as foreign. The transplant actually attacks the recipient.

Fortunately, many times transplants do work. A study of eight sex mismatched heart transplants vividly revealed that the recipient's body sent stem cells to help make the new heart part of the body. The evidence—Y chromosomes in unspecialized cells of the new heart, when the donor was female and the recipient male.

### Key Concepts

Vaccines are disabled pathogens or their parts that elicit an immune response, protecting against infection by the active pathogen. • Immunotherapy uses immune system components to fight disease. Hybridomas, artificial cells that consist of a B cell fused with a cancer cell, produce monoclonal antibodies (MAbs) that can target specific antigens. Cytokines boost immune function and destroy cancer cells. • Autografts transfer tissue from one part of a person's body to another; isografts are between identical twins; allografts are between members of the same species; and a xenograft is a cross-species transplant. Allografts can cause tissue rejection reactions, and xenografts can set off hyperacute rejection. In graft-versus-host disease, transplanted bone marrow rejects tissues of the recipient. Immunosuppressive drugs, stripping antigens from donor tissue, and matching donor to recipient improve the success rate of transplants. Sometimes, stem cells from the recipient colonize a donor organ.

### Technology Timeline

#### Transplantation

| | |
|---|---|
| 1899 | First allograft—a kidney from dog to dog. |
| 1902 | Pig kidney is attached to blood vessels of woman dying of kidney failure. |
| 1905 | First successful corneal transplant, from a boy who lost an eye in an accident to a man whose cornea was chemically damaged. Works because cornea cells lack antigens. |
| 1940s | First kidney transplants on young people with end-stage kidney failure. |
| 1950s | Blood typing predicts success of donor-recipient pairs for organ transplants. Invention of heart-lung bypass machine makes heart transplants feasible. |
| 1960s | First effective immunosuppressant drugs revive interest in human allografts. Kidney xenografts between baboons and chimpanzees. Heart transplants performed in dogs with mixed success. |
| 1967 | First human heart transplant. Patient lives 18 days. |
| 1968 | Uniform Anatomical Gift Act passes. Requires informed consent from next of kin before organs or tissues can be donated. |
| 1970s | Transplant problems: they extend life only briefly and do not correct underlying disease; surgical complications; rejection reaction. Many hospitals ban transplants. |
| 1980s | Improved immunosuppressant drugs, surgical techniques, and tissue matching, plus ability to strip antigens from donor tissue, reawaken interest in transplants. |
| 1984 | Doctors transplant a baboon's heart into "Baby Fae," who was born with half a heart. She lives 20 days before rejecting the xenograft. |
| 1992 | Surgeons transplant a baboon's liver into a 35-year-old man with hepatitis. The man lives for 71 days, dying of an unrelated cause. |
| 1995 | An AIDS patient receives bone marrow from an HIV-resistant baboon. |
| 1997 | Pig cell implants used to treat pancreatic failure and Parkinson disease. Pig liver used to maintain liver function for six hours as young man awaited a human liver. |
| 1998 | While researchers develop ways to make nonhuman cell surfaces more closely resemble those of humans, concern rises over risk of introducing retroviral disease through xenotransplants. |
| 2000 | Cloning of pigs brings xenotransplantation closer to reality. |

## 16.5 A Genomic View of Immunity—The Pathogen's Perspective

Immunity against infectious disease involves interactions of two genomes—ours and the pathogen's. At the same time that human genome information is revealing how the immune system functions to halt infectious disease, information from sequencing the genomes of pathogens also is useful. Table 16.8 lists some of the pathogens whose genomes have been sequenced.

Knowing the DNA sequence of a pathogen's genome, or the sequences of key genes, can reveal exactly how that organism causes illness in a human, and this information can suggest new treatment strategies. The sequence for *Streptococcus pneumoniae,* for example, revealed instructions for a huge protein that enables the bacterium to adhere to human cells. Pharmaceutical researchers can now search for compounds that dismantle this previously unknown adhesion protein. Another example is *Campylobacter jejuni,* which is the most common cause of food poisoning. In addition to causing severe diarrhea, infection by this bacterium also is sometimes accompanied by Guillain-Barré syndrome, a partial and temporary paralysis of facial muscles. The genome sequence revealed genes that control the placement of sialic acid molecules onto the bacterium's cell membrane, in patterns that mimic patterns on human cells. This tricks the immune system into attacking its own cells, causing the autoimmune Guillain-Barré syndrome.

**table 16.8**

**Some Human Pathogens with Sequenced Genomes**

| Pathogen | Associated Diseases |
|---|---|
| *Bacillus anthracis* | Cutaneous, inhalation, and gastrointestinal anthrax |
| *Borrelia burgdorferi* | Lyme disease |
| *Campylobacter jejuni* | Food poisoning |
| *Chlamydia pneumoniae* | Sexually transmitted disease |
| *Cryptococcus neoformans* | Cryptococcosis (fungal infection of central nervous system) |
| *Enterococcus faecalis* | Infections of urinary tract, abdomen, wounds, blood, and heart lining |
| *Haemophilus influenzae* | Otitis media (middle ear infection), respiratory infection, meningitis |
| *Helicobacter pylori* | Duodenal and gastric ulcers |
| *Listeria* | Food poisoning |
| *Mycobacterium tuberculosis* | Tuberculosis |
| *Neisseria meningitides* | Meningitis and septicemia (blood poisoning) |
| *Porphyromonas gingivalis* | Periodontitis (gum disease) |
| *Rickettsia prowazekii* | Typhus |
| *Staphylococcus aureus* | Toxic shock syndrome |
| *Streptococcus pneumoniae* | Otitis media, pneumonia, meningitis |
| *Treponema pallidum* | Syphilis |
| *Vibrio cholerae* | Cholera |

## Crowd Diseases

Because the adaptive arm of the immune system is a response to a stimulus in the environment—a pathogen—epidemics have often reflected the introduction of an infection into a population that has not encountered it before. History is riddled with such events.

When Europeans first explored the New World, they inadvertently brought lethal weapons—bacteria and viruses to which their immune systems had adapted. The immune systems of native Americans, however, had never encountered these pathogens before and were thus unprepared. Many people died.

Smallpox, a viral infection (figure 16.22), decimated the Aztec population in Mexico from 20 million in 1519, when conquistador Hernán Cortés arrived from Spain, to 10 million by 1521, when Cortés returned. By 1618, the Aztec nation had fallen to 1.6 million. The Incas in Peru and northern populations were also dying of smallpox. When explorers visited what is now the southeast United States, they found abandoned towns where natives had died from smallpox, measles, pertussis, typhus, and influenza.

The diseases that so easily killed native Americans are known as "crowd" diseases, because they arise with the spread of agriculture and urbanization and affect many people. Crowd diseases swept Europe and Asia as expanding trade routes spread bacteria and viruses along with silk and spices. More recently, air travel has spread crowd diseases. Returning soldiers introduced penicillin-resistant gonorrhea from southeast Asia to the United States during the Vietnam war. In 1998, travelers transported cholera by jet from Peru to Los Angeles.

Crowd diseases tend to pass from conquerors who live in large, intercommunicating societies to smaller, more isolated and more susceptible populations, and not vice versa. When Columbus arrived in the New World, the large populations of Europe and Asia had existed far longer than American settlements. In Europe and Asia, infectious diseases had time to become established and for human populations to adapt to them. In contrast, an unfamiliar infectious disease can quickly wipe out an isolated tribe, leaving no one behind to give the illness to invaders.

Fortunately, most crowd diseases vanish quickly. Vaccines or treatments may stop transmission. People may alter their behaviors to avoid contracting the infection, or the disease may kill before individuals can pass it on. Sometimes, we don't know why a disease vanishes or becomes less severe. We may be able to treat and control newly evolving infectious diseases one at a time, with new drugs and vaccines. But the mutation process that continually spawns new genetic variants in microbe populations—resulting in evolution—means that the battle against infectious disease will continue.

## Bioweapons

It may seem incomprehensible that pathogens would ever be used to intentionally harm people, but it is a sad fact of history—and the present—that such bioweapons exist. Biological weapons have been around since medieval warriors catapulted plague-ridden corpses over city walls to kill the inhabitants. During the French and Indian War, the British gave Indians blankets intentionally contaminated with secretions from smallpox victims. Although germ warfare was banned by international law in 1925, from 1932 until 1942 Japan field-tested bacterial bioweapons in rural China, killing thousands, including British and American prisoners of war.

In 1973, the Soviet Union established an organization called Biopreparat. Thousands of workers in 50 facilities prepared anthrax bombs and other bioweapons under the guise of manufacturing legitimate drugs, vaccines, and veterinary products. Soviet bioweapons were even more lethal than their natural counterparts. Plague bacteria, for example, were genetically modified to be resistant to 16 antibiotics and to manufacture a protein that strips nerve cells of their fatty coats, adding paralysis to the natural symptoms.

In 1979, an accident occurred in a Soviet city then called Sverdlovsk. At Military Compound Number 19, a miscommunication among shift workers in charge

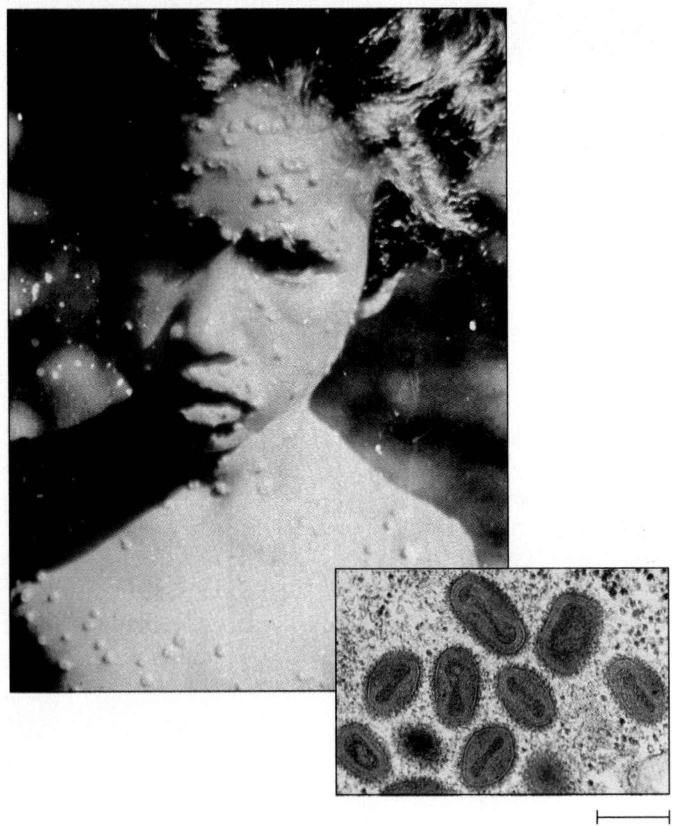


**figure 16.22**

**Death of a disease.** This boy is one of the last victims of smallpox, which has not naturally infected a human since 1977. Because many doctors are unfamiliar with smallpox, and people are no longer vaccinated against it, an outbreak would be a major health disaster. But because at least 16 nations have the smallpox virus stored in freezers and in light of the anthrax in the mail in October, 2001, the United States government is reactivating a vaccine program. The inset shows the smallpox virus.

of changing safety air filters resulted in the release of a cloud of dried anthrax spores over the city. Within weeks, more than 100 people died of anthrax, mostly young healthy men who were outside on that Friday night and breathed in enough anthrax spores to give them the inhalation form of the associated illness. Unfortunately, the government, which officially announced that the deaths were due to eating infected meat, worsened matters by spraying jets of water everywhere, which reaerosolized the spores and caused more infections.

The symptoms of inhalation anthrax result from a toxin that consists of three proteins. One protein forms a barrel-like structure that binds to macrophages and admits the other two components of the toxin. One of these overloads signal transduction and impairs the cell's ability to function as a phagocyte. The other toxin component breaks open macrophages, which release tumor necrosis factor and interleukins.

Early symptoms of inhalation anthrax resemble influenza, but the victim rapidly suffers respiratory collapse. In September 1992, Boris Yeltsin officially halted bioweapon research. On April 10, 1972, political leaders in London, Moscow, and Washington had signed the Biological Weapons Convention, an effort to prevent bioterrorism whose protocols are being strengthened today.

In the United States, a small-scale bioweapons effort began in 1942. A facility at Fort Detrick in Frederick, Maryland, stored 5,000 bombs loaded with anthrax spores; a production facility for the bombs was located in Terre Haute, Indiana; and Mississippi and Utah had test sites. President Richard Nixon halted the program in 1969, because he thought that conventional and nuclear weapons were sufficient. Today, the United States modest bioweapons effort at Fort Detrick may come back to haunt us, if it turns out to be the source of the anthrax contamination of mail in the weeks after the September 11, 2001 terrorist attacks.

Bioterrorism has come far since smallpox-infested blankets were catapulted over ancient city walls, and the Japanese dropped porcelain containers of plague-ridden fleas over China. Today's bioterrorists not only know how to grow and dry pathogens, but how to control particle size so that they can more easily infect a human body. In addition, genetic modification can alter characteristics of a virus or bacterium intended for use as a weapon.

### Key Concepts

Knowing the genome sequence of a pathogen can reveal how it evades the human immune system. • Crowd diseases happen when infectious agents are introduced into a population that hasn't encountered them before. • Bioterrorism is the use of pathogens—either in their natural state or genetically manipulated—to kill people.

*Bioethics: Choices for the Future*

# Pig Parts

In 1902, a German medical journal reported an astonishing experiment. A physician, Emmerich Ullman, had attached the blood vessels of a patient dying of kidney failure to a pig's kidney set up by her bedside. The experiment failed when the patient's immune system rejected the attachment almost immediately.

Nearly a century later, in 1997, an eerily similar experiment took place. Robert Pennington, a 19-year-old suffering from acute liver failure and desperately needing a transplant, survived for six and a half hours with his blood circulating outside of his body through a living liver removed from a 15-week-old, 118-pound pig named Sweetie Pie. The pig liver served as a bridge until a human liver became available. But Sweetie Pie was no ordinary pig. She had been genetically modified and bred so that her cells displayed a human protein that controlled the complement-mediated hyperacute rejection reaction against tissue transplanted from an animal of another species. Because of this slight but key bit of added humanity, plus immunosuppressant drugs, Pennington's body was able to tolerate the pig liver's help for the few crucial hours. Baboons have also been used as sources of organs for transplant, but they have not yet been genetically modified (figure 1).

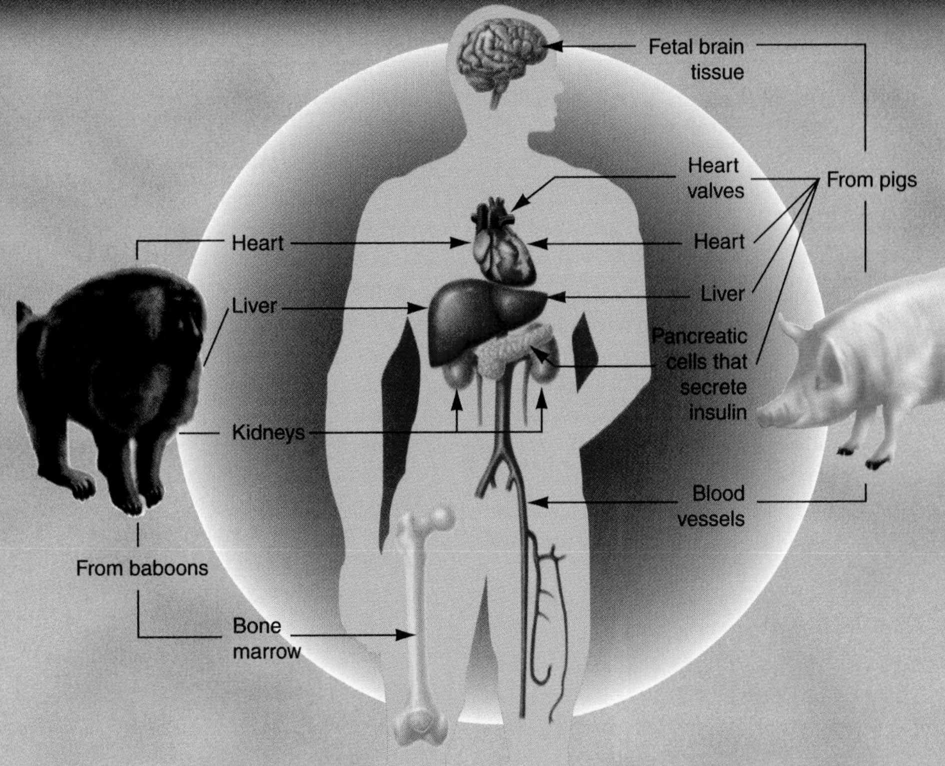


**figure 1**

Baboons and pigs can provide tissues and organs for transplant.

Successful xenotransplants would help alleviate the organ shortage. However, some people object to the idea of intentionally raising animals to use their organs as transplants because it requires killing the donors. One researcher counters such protests by comparing the use of animal organs to eating them. Those who eat ham or bacon can hardly justify objecting to a pig-to-human transplant.

A possible danger of xenotransplants is that people may acquire viruses from the organ donors. Viruses can "jump" species, and the outcome in the new host is unpredictable. So far, it is known that a virus called PERV—for "porcine endogenous retrovirus"—can infect human cells in culture. However, a study of several dozen patients who had received implants of pig tissue, for a variety of reasons, revealed that none showed evidence of PERV years later. That study, though, looked only at blood. We still do not know what effect pig viruses can have on a human body. Because many viral infections take years to cause symptoms, a new infectious disease in the future could be the trade-off for using xenotransplants to solve the current organ shortage.

# Summary

### 16.1 The Importance of Cell Surfaces

1. The cells and biochemicals of the immune system distinguish self from nonself, protecting the body against infections and cancer.
2. Most of the 20,000 or so genes whose products are involved in immunity function in a polygenic fashion.
3. Genetically encoded cell surface molecules determine blood types. A blood incompatibility occurs if a blood recipient manufactures **antibodies** against antigens in donor blood. Blood type systems include ABO and Rh.

4. The highly polymorphic HLA genes of the **major histocompatibility complex** are closely linked on chromosome 6. They encode cell surface antigens that present foreign antigens to the immune system.

### 16.2 The Human Immune System

5. If a pathogen breaches physical barriers, the **innate immune response** produces the redness and swelling of inflammation, and **complement, collectins,** and **cytokines.** The response is broad and general.
6. The **adaptive immune response** is slower and more specific. It also has memory. This response is both humoral and cellular.
7. The **humoral immune response** begins when macrophages display foreign antigens near HLA antigens. This activates **helper T cells,** which activate **B cells.** The B cells, in turn, mature into **plasma cells** and secrete specific antibodies. Some B cells give rise to **memory cells.**
8. An antibody is Y-shaped and made up of four polypeptide chains, two **heavy** and two **light.** Each antibody molecule has regions of **constant** amino acid sequence and regions of **variable** sequence.
9. The tips of the Y of each subunit form **antigen binding sites,** which include the more specific **idiotypes** that bind foreign antigens at their **epitopes.**
10. Antibodies bind antigens to form immune complexes large enough for other immune system components to detect and destroy. Antibody genes are rearranged during early B cell development, providing instructions to produce a great variety of antibodies.
11. T cells carry out the **cellular immune response.** Their precursors, called thymocytes, are selected in the thymus to recognize self. Helper T cells secrete cytokines that activate other T cells and B cells. A helper T cell's CD4 antigen binds macrophages that present foreign antigens. **Cytotoxic T cells** release biochemicals that bore into bacteria and kill them and also destroy cells covered with viruses.
12. **Gamma-delta T cells** coordinate innate and adaptive immune responses.

### 16.3 Abnormal Immunity

13. Mutations in antibody or cytokine genes, or in genes encoding T cell receptors, cause inherited immune deficiencies. Severe combined immune deficiencies affect both branches of the immune system.
14. HIV binds to the co-receptors CD4 and CCR5 on macrophages and helper T cells, and, later in infection, triggers apoptosis of cytotoxic T cells. As HIV replicates, it mutates, evading immune attack. Falling CD4 helper T cell numbers allow opportunistic infections and cancers to occur. People who cannot produce a complete CCR5 protein resist HIV infection.
15. In an **autoimmune disease,** the body manufactures **autoantibodies** against its own cells. Autoimmunity may result from a virus that incorporates and displays a self antigen, from bacteria or cancer cells that have antigens that resemble self antigens, from unselected T cells, or from lingering fetal cells.
16. In susceptible individuals, allergens stimulate IgE antibodies to bind to **mast cells,** which causes the cells to release allergy mediators. Certain helper T cells release certain cytokines. Allergies may be a holdover of past immune function.

### 16.4 Altering Immune Function

17. A **vaccine** presents a disabled pathogen, or part of one, in order to elicit a primary immune response.
18. **Immunotherapy** enhances or redirects immune function. **Monoclonal antibodies** are useful in diagnosing and treating some diseases because of their abundance and specificity. To create MAbs, individual activated B cells are fused with cancer cells to form hybridomas. Cytokines are used to treat various conditions.
19. Transplant types include **autografts** (within oneself), **isografts** (between identical twins), **allografts** (within a species), and **xenografts** (between species). A tissue rejection reaction occurs if donor tissue is too unlike recipient tissue.

### 16.5 A Genomic View of Immunity—The Pathogen's Perspective

20. Infectious disease involves interactions between the host and the pathogen's genomes. Learning the genome sequences of pathogens can reveal how they infect, which provides clues to developing new treatments.
21. Crowd diseases are infectious illnesses that spread rapidly through a population that has had no prior exposure, passed from members of a population that has had time to adapt to the pathogen.
22. Throughout history, people have used bacteria and viruses as weapons.

## Review Questions

1. Match the cell type to the type of biochemical it produces.

| | |
|---|---|
| 1. mast cell | a. antibodies |
| 2. T cell | b. HLA class II genes |
| 3. B cell | c. interleukin |
| 4. macrophage | d. histamine |
| 5. all cells with nuclei | e. interferon |
| 6. antigen presenting cell | f. heparin |
| | g. tumor necrosis factor |
| | h. HLA class I *A, B,* and *C* genes |

2. What is the physical basis of a blood type? of blood incompatibility?
3. What would be the consequences of lacking:
   a. helper T cells
   b. cytotoxic T cells
   c. B cells
   d. gamma-delta T cells
   e. macrophages
   f. M cells
4. State the function of each of the following immune system biochemicals:
   a. complement proteins
   b. collectins
   c. antibodies
   d. cytokines
5. Cite three reasons why developing a vaccine against HIV infection has been challenging.
6. It was once said that thymocytes are "educated" in the thymus, meaning that immature T cells are somehow "taught" to recognize and respect self cell surfaces. This is not exactly what happens. Why?
7. What part do antibodies play in allergic reactions and in autoimmune disorders?

8. How do each of the following illnesses disturb immunity?
   a. graft-versus-host disease
   b. SCID
   c. scleroderma
   d. AIDS
   e. hayfever
   f. systemic lupus erythematosus
9. Why is a deficiency of T cells more dangerous than a deficiency of B cells?
10. What do a plasma cell and a memory cell descended from the same B cell have in common? How do they differ?
11. Why is a polyclonal antibody response valuable in the body but a monoclonal antibody valuable as a diagnostic tool?
12. A person exposed for the first time to Coxsackie virus develops a painful sore throat. How is the immune system alerted to the exposure to the virus? When the person encounters the virus again, why doesn't she develop symptoms?

# Applied Questions

1. A man is flown to an emergency room of a major medical center, near death after massive blood loss in a car accident. There isn't time to match blood types, so the physician orders type O negative blood. Why did she order this type of blood?
2. Rasmussen's encephalitis is a rare and severe form of epilepsy that causes children to have 100 or more seizures a day. Affected children have antibodies that attack brain cell receptors that normally receive nervous system biochemicals. Is this condition an inherited immune deficiency, an adaptive immune deficiency, an autoimmune disorder, or an allergy?
3. Allergy to a protein in peanuts can cause anaphylactic shock. An experimental vaccine consists of the gene encoding this protein, wrapped in an edible carbohydrate, so it can be eaten and stimulate production of protective antibodies in the small intestine. When this vaccine was fed to rats who have a peanut allergy, their blood showed lowered levels of IgE, but increased levels of IgG. Also, their bowel movements contained higher than usual levels of IgA. Is the vaccine working? How can you tell?
4. Glomerulonephritis is a condition that causes dark urine, lower back pain, and other symptoms. It happens because antigens on certain kidney cells resemble antigens on *Streptococcus* bacteria. Explain the type of disorder glomerulonephritis is, and how the symptoms arise.
5. In people with a certain HLA genotype, a protein in their joints resembles an antigen on the bacterium that causes Lyme disease, an infection transmitted in a tick bite that causes flulike symptoms followed by arthritis (joint inflammation). When these individuals become infected, their immune systems attack not only the bacteria, but also their joints. Explain why antibiotic therapy helps treat the early phase of the disease, but not the arthritis.
6. Even in overwhelmingly deadly infectious diseases, such as bubonic plague and Ebola hemorrhagic fever, a small percentage of the human population survives. Suggest two mechanisms based on immune system functioning that can account for their survival.
7. State whether each of the following is an autograft, an isograft, an allograft, or a xenograft.
   a. Two parents each donate part of a lung to increase the respiratory capacity of their child, who has cystic fibrosis.
   b. A woman with infertility receives an ovary transplant from her sister, who is an identical twin. (This is the only way this particular transplant works.)
   c. A man receives a heart valve from a pig.
   d. A woman who has had a breast removed has a new breast reconstructed using fatty tissue from her thigh.
8. A man and woman are planning to have their first child, but they are concerned because they think that they have an Rh incompatibility. He is $Rh^-$ and she is $Rh^+$. Will there be a problem? Why or why not?
9. A young woman who has aplastic anemia will soon die as her lymphocyte levels drop sharply. What type of cytokine might help her?
10. In Robin Cook's novel *Chromosome Six,* a geneticist places a portion of human chromosome 6 into fertilized ova from bonobos (pygmy chimps). The bonobos that result are used to provide organs for transplant into specific individuals. Explain how this technique would work.
11. Suggest ways that local, state, and federal governments can prepare to handle a bioterrorism attack.

# Suggested Readings

Alibek, Ken. 1999. *Biohazard.* New York: Random House, Inc. The chilling tale of bioweaponry in the Soviet Union, written by a scientist who headed part of the program before defecting to the U.S.

Enserink, Martin. February 16, 2001. Finding the talismans that protect against infections. *Science* 291:1183. At least 20,000 genes provide immunity.

Ezekowitz, R. Alan. March 22, 2001. What is the best way to treat inherited disorders? *The New England Journal of Medicine* 344, no. 12:926–27. Stem cells from the blood of an HLA-matched sibling can cure chronic granulomatous disease, but the death rate is high.

Fahrer, Aude M. et al. February 15, 2001. A genomic view of immunology. *Nature* 409:836. The human genome sequence is revealing genes that provide immunity.

Hill, Adrian V. S. June 23, 2001. Immunogenetics and genomics. *The Lancet* 357:2037–40. The human genome sequence provides insight into immune function.

Lewis, Ricki. July 24, 2001. Portals for prions. *The Scientist* 15:1. M cells sample antigens in food so that we cannot be infected by what we eat.

Lewis, Ricki. October 1, 2001. New weapons against HIV. *The Scientist* 15(19):1. Entry inhibitors and immune modulators are new tools against AIDS.

Lewis, Ricki. October 29, 2001. Plague genome: The evolution of a pathogen. *The Scientist* 15(21):1. Will the publication of pathogen genomes give ideas to bioterrorists?

Lewis, Ricki. November 12, 2001. Attack of the anthrax 'virus.' *The Scientist* 15(22):42. Bioterrorism took many people in the U.S. offguard.

Lewis, Ricki. August 21, 2000. An eclectic look at infectious diseases: Participants in Atlanta conference discuss bioterrorism, AIDS, and new viral threats. *The Scientist* 14(16):1.

Lewis, Ricki. August 21, 2000. TIGR introduces *Vibrio cholerae* genome. *The Scientist* 14(16):8. The genome sequence of this pathogen revealed how it persists in aquatic habitats.

Lewis, Ricki. October 16, 2000. Porcine possibilities. *The Scientist* 14:1. The pros and cons of pig parts as transplants.

Quaini, Federico et al. January 3, 2002. Chimerism of the transplanted heart. *The New England Journal of Medicine* 346(1):5–15. Stem cells from a heart transplant recipient colonize the new organ and help it fit in.

Ridley, Matt. March 2000. Asthma, environment, and the genome. *Natural History,* vol. 109. The increase in asthma may reflect technologies that replace our immune systems.

Savoie, Keeley. April 2000. Edible vaccine success. *Nature Biotechnology* 18:367. Vaccines may soon be delivered in bananas.

# On the Web

Check out the resources on our website at

**www.mhhe.com/lewisgenetics5**

On the web for this chapter, you will find additional study questions, vocabulary review, useful links to case studies, tutorials, popular press coverage, and much more. To investigate specific topics mentioned in this chapter, also try the links below:

AIDS Information **www.cdc.gov/hchstp/hiv_aids/pubs/facts.htm**

American Autoimmune Related Diseases Assoc. Inc. **www.aarda.org**

Asthma and Allergy Information **www.hopkins-allergy.com**

Center for Civilian Biodefense Strategies **www.hopkins-biodefense.org/**

Centers for Disease Control and Prevention **www.cdc.gov/**

Immune Deficiency Foundation **www.primaryimmune.org**

Online Mendelian Inheritance in Man **www.ncbi.nlm.nih.gov/entrez/query.fcgi?db=OMIM**

ABO blood group 110300
autoimmune polyendocrinopathy syndrome 240300
chronic granulomatous disease 306400
complement deficiencies 120790, 217070
cystic fibrosis 219700
HLA class I 142800
HLA class II 142860, 143110
Lewis blood group (fucosyltransferase 3) 111100
Rh blood group 111700
scleroderma (familial) 181750
severe combined immune deficiency due to adenosine deaminase deficiency 102700
systemic lupus erythematosus 152700
X-linked severe combined immune deficiency 202500

Scleroderma Foundation **www.scleroderma.org**

CHAPTER

17

# The Genetics of Cancer

## 17.1 Cancer as a Genetic Disorder

Cancer starts with one cell that changes in a way that enables it to escape or ignore controls on the rate or progression of its cell cycle. That first errant cell may divide too fast or too often, or fail to die according to the developmental schedule. Whatever the initial error, the cancer cell proliferates at the expense of surrounding tissue, forming an abnormal growth. If it is not stopped, it spreads.

## 17.2 Characteristics of Cancer Cells

Compared to normal cells, cancer cells are rounder, oilier, less specialized, and pile atop one another. These aggressive cells squeeze into any space, invade tissue boundaries, travel in the bloodstream, then nestle into new areas, establishing their own blood supplies.

## 17.3 Genes That Cause Cancer

Oncogenes are genes that normally promote cell division, but are turned on at the wrong time or place. Tumor suppressor genes normally shut cell division off, but when mutant, they allow it to continue when or where it shouldn't. DNA repair genes, when mutant, allow too many mutations to slip through unchecked, some of which may cause cancer.

## 17.4 A Series of Genetic Changes Causes Some Cancers

Cancer can take many years to develop. Often a series of genetic changes must occur. Chromosome sections are lost, oncogenes are activated, and tumor suppressor genes are deleted or inhibited, leading to the cancerous state.

## 17.5 Cancer Prevention, Diagnosis, and Treatment

Environmental factors can cause the somatic mutations that often underlie cancer, but links between specific influences and cancers are usually correlations on a population level. Thanks to increased knowledge from genetics and genomics, cancer diagnosis has been refined and several new, more targeted treatments developed.

I had cancer. It seemingly came out of the blue in 1993, although it had been there, probably, for years. I'd been feeling fine, eating lots of vegetables, and exercising regularly. I had never smoked. But then a physician-friend noticed a swelling in my neck, which began a two-week period that today seems a blur—a time of biopsies, waiting, a frightening phone call, and, finally, surgery to remove my thyroid gland and its tumors. Two months later, I had radiation treatment to destroy any remaining cancerous cells. I was lucky. Many people with cancer undergo more extensive surgery than I did, receive more intense radiation, or take chemotherapy drugs that destroy healthy cells along with cancerous ones, making the experience most unpleasant. I never really had any pain.

Forty percent of us will develop cancer in our lifetimes—some of us more than once. A million new cases of cancer will be diagnosed in the United States this year. Ten million people are being treated for some form of cancer right now, and many of them will recover and live for many years. The experience of having cancer certainly brings a new appreciation of life!

## 17.1 Cancer as a Genetic Disorder

Cancer is a complication of being a many-celled organism. All of our different specialized cell types must stick to a schedule of mitosis—the cell cycle—so that organs and other body parts either grow appropriately during childhood, or stay a particular size and shape in an adult. If a cell escapes normal controls on its division rate, it forms a growth called a tumor. In the blood, such a cell divides to take over the population of blood cells. A tumor is benign if it grows in place but does not spread into surrounding tissue; it is cancerous, or malignant, if it infiltrates nearby tissue. The word *cancer* is Latin for "crab"; the appearance of a cancerous tumor as it extends into the healthy tissue around it resembles the shape of a crab's body.

Cancer is a group of disorders that arise from alterations in genes, 90 percent of which are environmentally induced. Unlike Mendelian conditions in which the faulty instructions are present in every cell at birth, mutations in cancer-causing genes usually occur in somatic cells over a lifetime. That is, cancer is a genetic disease at the cellular level, rather than at the whole-body level. Cancer often takes years to develop, as a sequence of genes mutate in the affected tissue. Then, the cells whose mutations enable them to divide more often than others gradually take over the tissue. Cancer is, in a sense, a form of natural selection within a body.

It took many years for scientists to view cancer as a genetic phenomenon. When President Richard Nixon declared a "war on cancer" in 1971, the targets were radiation, viruses, and chemicals. Today we know that these agents cause cancer by interfering with the precise genetic controls of cell division. One noted cancer researcher wrote, "A few years ago the question was whether there are genetic changes in cancer cells. Now you find a huge number of DNA changes. In cancer, the genome is shot to hell."

Researchers first discovered genes that could cause cancer in 1976, but there were indirect genetic hints earlier. For example, most substances known to be **carcinogens** (causing cancer) also proved to be mutagens (damaging DNA) when placed on cells growing in culture (the Ames test). Was the genetic change these chemicals promoted responsible for the cancer phenotype? A second line of evidence came from families in which colon or breast cancer was so prevalent that it could have been inherited as a Mendelian trait.

### Cancer in Families

In the 1980s and 1990s, searches for cancer-causing genes began with families that had many young members who had the same type of cancer. Researchers then identified parts of the genome that the affected individuals shared, such as a chromosomal aberration or a unique DNA sequence. Next, the search focused on specific genes in the identified region that could affect cell cycle control. This approach led to the discovery of more than 100 **oncogenes,** which cause cancer when they are inappropriately activated, and of more than 30 **tumor suppressor genes,** whose deletion or inactivation causes cancer.

The cancer family approach to identifying genes must have been quite thorough, because the first draft sequence of the human genome did not reveal any new cancer-causing genes—at least none that were recognized as such. However, searches for cancer-causing genes so far have been based on known genes whose function, if perturbed, could theoretically cause cancer, such as genes that encode growth factors or transcription factors. Researchers may have missed classes of cancer-causing genes because they did not know what to look for. The fact that two-thirds of families with more than four members who have breast cancer do not harbor known cancer-causing genes suggests that we do not know them all. Cancer might also result from the additive effects of several genes, each of which increases the risk slightly. That is, some cancers can be polygenic traits.

While many researchers continue to study the single genes that can cause or contribute to the development of cancer, a genomics-based approach to understanding cancer uses DNA microarrays, which highlight the differences in expression of thousands of genes between a tumor cell and the normal cell type from which it descended. Section 1.5 describes a "lymphochip" used to diagnose blood cancers and a breast cancer chip that can predict response to treatment.

Another DNA microarray that includes 6,000 cDNAs (DNA reverse transcribed from mRNA) enables oncologists (cancer specialists) to distinguish among four types of solid tumors in children. These cancers look so alike, when the cells are stained and viewed under a microscope, that they are collectively called "small round blue-cell tumors." It is not a very useful categorization for four very different cancers! The gene expression profiles revealed in the DNA microarrays, however, clearly distinguish Ewing sarcoma from neuroblastoma from rhabdomyosarcoma from non-Hodgkins lymphoma. This improvement in diagnostic power will lead to more targeted and therefore more effective treatment. For example, both Ewing sarcoma and rhabdomyosarcoma respond to the drug adriamycin, but a serious side effect is heart damage. Ewing sarcoma is more aggressive and responds to fewer drugs than does rhabdomyosarcoma. Therefore, the test enables doctors to choose adriamycin for Ewing sarcoma, and gentler treatment for the less aggressive rhabdomyosarcoma. Such DNA microarray expression panels will likely become widely used in cancer diagnosis. Summarized one researcher, "thousands of individual genes define the molecular portraits of each tumor."

## Loss of Cell Cycle Control

Cancer begins when a single abnormal cell divides to produce others like itself. It divides more frequently, or more times, than the normal cell type it descended from.

Cancer is a consequence of disruption of the cell cycle. The timing, rate, and number of mitoses depend on protein growth factors and signaling molecules from outside the cell, and on transcription factors from within. Because these biochemicals are under genetic control, so is the cell cycle abnormality that is cancer. Cancer cells probably arise in everyone, because mitosis occurs so frequently that an occasional cell escapes controls. However, the immune system destroys most cancer cells by recognizing tumor-specific antigens.

The discovery of the checkpoints that control the cell cycle revealed how cancer can begin (see figure 2.17). A mutation in a gene that normally halts or slows the cell cycle can lift the constraint, leading to inappropriate mitosis. Another cell cycle-related cause of cancer is failure to pause long enough to repair DNA (see table 11.9). Loss of control over telomere length may also contribute to causing cancer. Recall from figure 2.18 that telomeres, or chromosome tips, protect chromosomes from breaking. Human telomeres consist of the DNA sequence TTAGGG repeated thousands of times. The repeats are normally lost as a cell matures, at the rate of about 15 to 40 nucleotides per cell division. The more specialized a cell, the shorter its telomeres. The chromosomes in skin, nerve, and muscle cells, for example, have short telomeres. Chromosomes in a sperm cell or oocyte, however, have long telomeres. This makes sense—as the precursors of a new life, gametes must retain the capacity to divide many times.

Gametes keep their telomeres long, thanks to an enzyme, **telomerase,** that is a complex of RNA and protein. Part of the RNA—the sequence AAUCCC—serves as a template for the 6-DNA-base repeat that builds telomeres (figure 17.1). Telomerase moves down a chromosome tip like a zipper, adding six "teeth" at a time. In this way, telomerase repeatedly adds telomere material to the chromosomes of gametes.

In normal, specialized cells, telomerase is turned off, and telomeres shrink, signaling a halt to cell division when they reach a certain size. In cancer cells, telomerase is turned back on. Telomeres extend, and this releases the normal brake on rapid cell division. As daughter cells of the original abnormal cell continue to divide uncontrollably, a tumor forms, grows, and may spread. The longer the telomeres in cancer cells, usually the more advanced the disease. However, turning on telomerase production in a cell is not sufficient in itself to cause cancer.

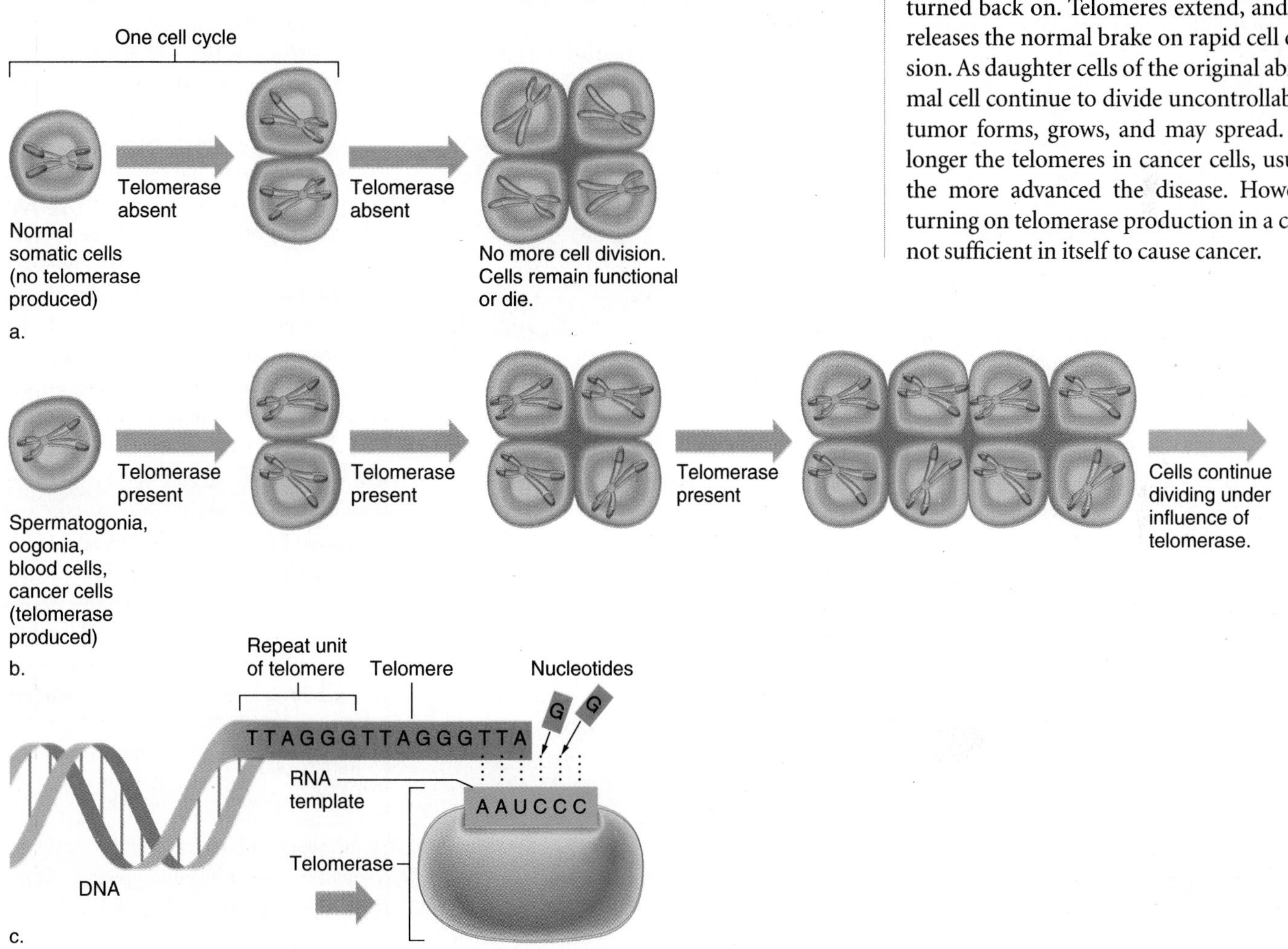


**figure 17.1**

**Telomeres.** (*a*) In normal somatic (nonsex) cells, telomeres shorten with each cell division because the cells do not produce telomerase. When the telomeres shrink to a certain point, the cell no longer divides. (*b*) Sperm-generating cells, blood cells, and cancer cells produce telomerase and continually extend their telomeres, resetting the cell division clock. (*c*) Telomerase contains RNA, which includes a portion (the nucleotide sequence AAUCCC) that acts as a template for the repeated DNA sequence (TTAGGG), which forms the telomere.

## Inherited Versus Sporadic Cancer

Most cancers are isolated, or sporadic, which means that the causative mutation occurs only in cells of the affected tissue. This is a **somatic mutation,** because it occurs in somatic (nonsex) cells. A sporadic cancer may result from a single, dominant mutation, or from two recessive mutations in the same gene. The cell harboring the mutation loses control of its cell cycle, divides continuously, and forms a tumor. Susceptibility to developing a sporadic cancer is *not* passed on to future generations because the gametes do not carry the mutant allele or alleles.

In contrast, a **germline,** or inherited, cancer susceptibility *is* passed to future generations because it is present in every cell, including gametes. Cancer develops when a second mutation occurs in a somatic cell. The cancer site in the body is the site of this second mutation (figure 17.2). Germline mutations may explain why some heavy smokers develop lung cancer, but many do not; the unlucky ones may have inherited a susceptibility allele present in every cell. Years of exposing lung tissue to the carcinogens in cigarette smoke eventually causes a mutation in the DNA of a lung cell, giving it a proliferative advantage. Without the susceptibility gene, it would take two such somatic mutations to trigger the cancer. The idea that two recessive mutations can cause a sporadic cancer is called the two-hit hypothesis of cancer causation, a point revisited later in the chapter. Germline cancers account for only about 5 percent of all cancers. However, their penetrance is high and they tend to strike earlier in life than sporadic cancers.

### Key Concepts

Cancer is a genetic disease, but it is not usually inherited as a Mendelian trait. • Single genes (oncogenes and tumor suppressors) when mutant can cause cancer, but new diagnostic approaches monitor expression of many genes. Cancer is caused by a loss of cell division control. Implicated genes include those encoding growth factors, transcription factors, and telomerase. Several mutations may contribute to development of a cancer. The genetic changes usually occur in somatic cells. Cancer may develop when an environmental trigger causes mutations in a somatic cell or when a somatic mutation compounds an inherited susceptibility.

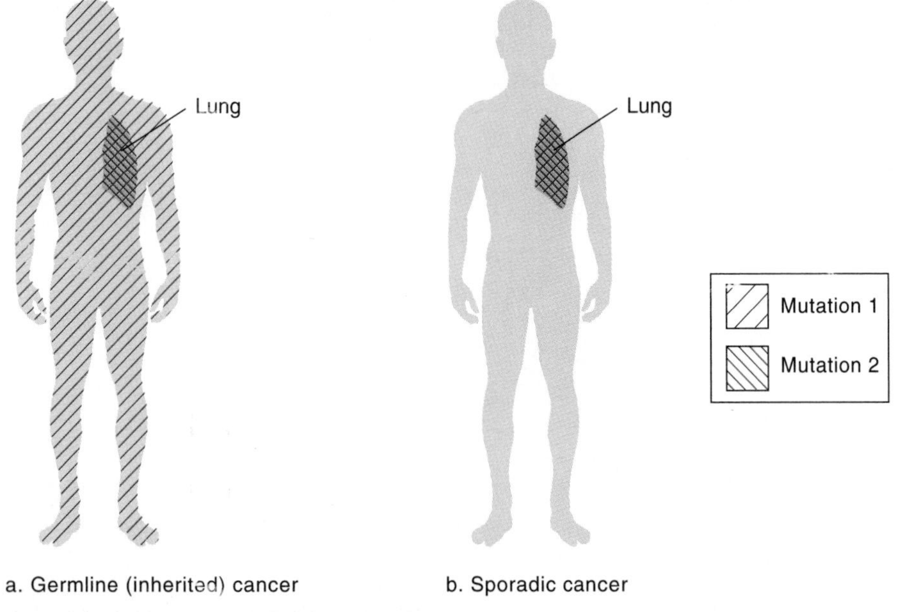


**figure 17.2**

**Germline versus sporadic cancer.** (*a*) A germline cancer (also called inherited or familial cancer) occurs when every cell has one cancer-susceptibility gene, and a second mutation occurs in the affected somatic tissue. This type of predisposition to cancer is transmitted as a Mendelian trait. (*b*) A sporadic cancer forms when two mutations occur in the same somatic cells.

## 17.2 Characteristics of Cancer Cells

Cell division is a rigorously controlled process. Whether a cell divides or stops dividing and whether it expresses the sets of genes that make it specialize as a particular cell type depend upon biochemical signals from surrounding cells. A cancer cell simply stops "listening" to those signals. Mutations that affect any stage of signal transduction or gene expression can catapult a cell toward unrestrained division.

Cancer cells can divide continuously if given sufficient nutrients and space. This is vividly illustrated by the cervical cancer cells of a woman named Henrietta Lacks, who died in 1951. Her cells persist today as standard cultures in many research laboratories. These "HeLa" cells divide so vigorously that when they contaminate cultures of other cells growing in the laboratory because of human error, they soon comprise most of the cells in the culture.

Cancer cells divide frequently or more times compared to the cell types from which they arose. Some cells normally divide frequently, and others rarely (table 17.1). Even the fastest-dividing cancer cells, which complete mitosis every 18 to 24 hours, do not divide as often as some normal human embryo cells do. A cancerous tumor eventually grows faster than surrounding tissue because a greater proportion of its cells is dividing.

**table 17.1**

**Some Normal Cells Divide More Often than Cancer Cells**

| Cell Type | Hours Between Divisions |
|---|---|
| **Normal Cells** | |
| Bone marrow precursor cells | 18 |
| Lining cells of large intestine | 39 |
| Lining cells of rectum | 48 |
| Fertilized ovum | 36–60 |
| **Cancer Cells** | |
| Stomach | 72 |
| Acute myeloblastic leukemia | 80–84 |
| Chronic myeloid leukemia | 120 |
| Lung (bronchus carcinoma) | 196–260 |

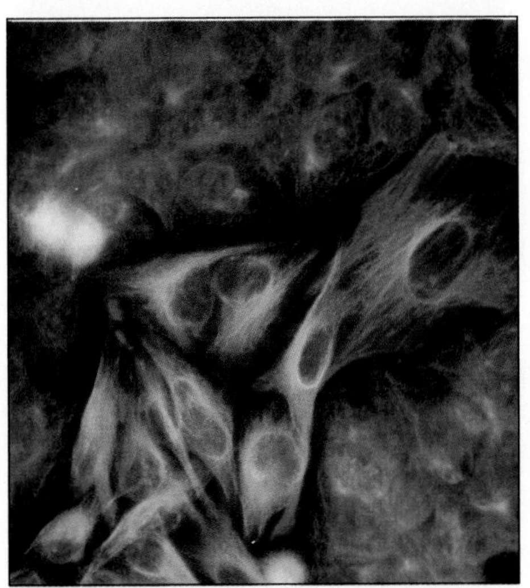

**figure 17.3**

**Staining highlights cancer cells.** The orange cells are a melanoma (skin cancer) that is invading normal skin. Cancer cells, when stained for the presence of gene variants characteristic of cancer cells only, look very different from surrounding healthy tissue.

Some cancers grow at an alarmingly fast rate. The smallest detectable fast-growing tumor is half a centimeter in diameter and can contain a billion cells. These cells divide at a rate that produces a million or so new cells in an hour. If 99 percent of the tumor's cells are destroyed, a million would still be left to proliferate. Other cancers develop very slowly and may not be noticed for several years. The rate of a tumor's growth is slower at first, because fewer cells divide. By the time the tumor is the size of a pea—when it is usually detectable—billions of cells are actively dividing.

A cancer cell looks different from a normal cell (figure 17.3). It is rounder because it does not adhere to surrounding normal cells as strongly as other cells do. Because the cell membrane is more fluid, different substances cross it.

When a cancer cell divides, both daughter cells are also cancerous, since they inherit the altered cell cycle control. Therefore, cancer is said to be **heritable** because it is passed from parent cell to daughter cell. A cancer is also **transplantable.** If a cancer cell is injected into a healthy animal of the same species, the disease begins there.

A cancer cell is **dedifferentiated,** which means that it is less specialized than the normal cell type it arose from. A skin cancer cell, for example, is rounder and softer than the flattened, scaly healthy skin cells above it in the epidermis. Cancer cell growth also differs from growth in normal cells. Normal cells placed in a container divide to form a single layer; cancer cells pile up on one another. In an organism, this pileup would produce a tumor. Cancer cells that grow all over one another lack **contact inhibition**—they do not stop dividing when they crowd other cells.

Cancer cells have surface structures that enable them to squeeze into any space, a property called **invasiveness.** They anchor themselves to tissue boundaries, called basement membranes, where they secrete chemicals that cut paths through healthy tissue. Unlike a benign tumor, an invasive malignant tumor grows irregularly, sending tentacles in all directions (figure 17.4).

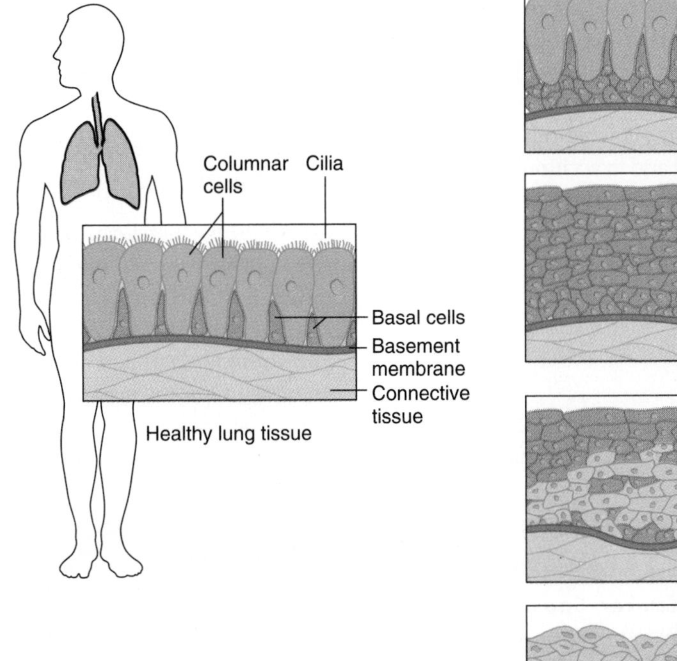


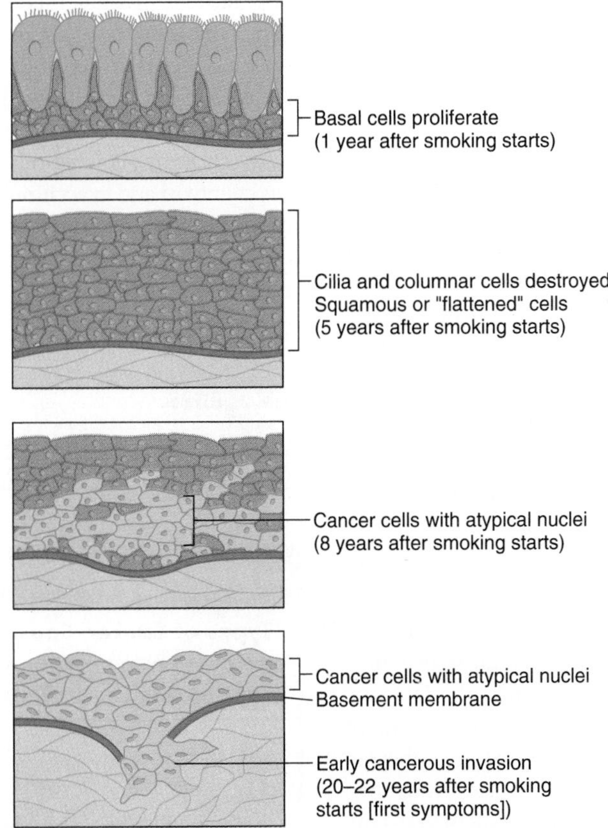


**figure 17.4**

**Cancers take many years to spread.** Lung cancer due to smoking begins with irritation of respiratory tubes. Ciliated cells die (but can be restored if smoking ceases), basal cells divide, and then, if the irritation continues, cancerous changes may appear.

**table 17.2**

| Characteristics of Cancer Cells |
|---|
| Different appearance |
| Loss of cell cycle control |
| Heritability |
| Transplantability |
| Dedifferentiation |
| Loss of contact inhibition |
| Ability to induce local blood vessel formation (angiogenesis) |
| Invasiveness |
| Increased mutation rate |
| Ability to spread (metastasis) |

Eventually, unless treatment (drugs, surgery, radiation, or immunotherapy) stops them, malignant cells reach the bloodstream or lymphatic vessels, which are conduits to other parts of the body. The traveling cancer cells settle into new sites and, once they've grown to the size of a pinhead, secrete factors that stimulate nearby capillaries (the tiniest blood vessels) to sprout new branches that extend toward it, bringing in oxygen and nutrients and removing wastes. This growth of new capillaries is called **angiogenesis.** Capillaries may snake out of the tumor. Cancer cells wrap around the blood vessels and creep out upon this scaffolding into nearby tissue in their characteristic crablike shape.

Cancer cells may also secrete hormones that encourage their own growth. For example, pancreatic cancer cells secrete the hormone gastrin, which stimulates the tumor's growth. Pancreas cells normally do not produce gastrin.

Once cancer cells move to a new body part, the disease has spread, or **metastasized** (from the Greek for "beyond standing still"). After a cancer spreads, it becomes very difficult to treat, because the DNA of secondary tumor cells often mutates, many times causing new chromosome aberrations. The metastasized cancer thus becomes a new genetic entity, often resistant to treatments that were effective against most cells of the original tumor.

Table 17.2 summarizes the characteristics of cancer cells.

**KEY CONCEPTS**

Cancer occurs when cells divide faster or more times than normal. Cancer cells are heritable, transplantable, and dedifferentiated. They lack contact inhibition, cutting through basement membranes and metastasizing.

## 17.3 Genes That Cause Cancer

Cancer is a normal process—mitosis—that is mistimed or misplaced. Genes are intimately involved in a cancer's genesis.

Mutations in three types of genes can cause cancer (table 17.3). Oncogenes are inappropriately activated genes that normally promote cell division. The oncogene functions in the wrong place or at the wrong time during development. Tumor suppressor genes normally constrain cell division, and mutations in these genes lift that control, leading to inappropriate cell division that passes a checkpoint. The third category of cancer-causing genes includes DNA repair genes, which were discussed in detail in chapter 10, section 10.6. Mutations in DNA repair genes indirectly cause cancer by allowing other mutations to persist unfixed and to accumulate. When such mutations activate oncogenes or inactivate tumor suppressor genes, cancer results. DNA repair disorders are often inherited in a Mendelian fashion, and are quite rare.

### Oncogenes

Genes that normally trigger cell division are called **proto-oncogenes.** They are active where and when high rates of cell division are necessary, such as in a wound or in a rapidly growing embryo. When proto-oncogenes are turned on at the wrong time or place, they function as oncogenes and cause cancer ("onco" means cancer). This abnormal activation may be the result of a mutation. A single base change in a proto-oncogene causes bladder cancer, for example. Alternatively, a proto-oncogene may be moved near a gene that is highly expressed; then it, too, is rapidly or frequently transcribed.

Consider a human proto-oncogene that is normally activated in cells at the site of a wound, where it stimulates production of growth factors that cause mitosis to fill the damaged area in with new cells. When that proto-oncogene is activated at a site other than a wound—as an oncogene—it still hikes growth factor production and stimulates mitosis. However, because the site of the action is not damaged tissue, the new cells accumulate into a tumor.

Some proto-oncogenes encode transcription factors that, as oncogenes, are abnormally highly expressed. Recall from chapter 10 that transcription factors bind to specific genes and activate transcription. The products of these activated genes then contribute to the cancer cell's characteristics.

#### *Increased Expression in a New Location*

A proto-oncogene can be transformed into its out-of-control oncogene counterpart when it is placed next to a different gene that boosts its expression. A virus infecting a cell, for example, may insert its genetic material next to a proto-oncogene. When the viral DNA begins to be rapidly transcribed, the adjacent proto-oncogene (now an oncogene)

**table 17.3**

**Types of Cancer-Causing Genes**

| Type of Gene | Mechanism of Carcinogenesis |
|---|---|
| Oncogene | Actively promotes cancer. Normal version, a proto-oncogene, controls cell cycle. Oncogene activates cell division at inappropriate time or place. |
| Tumor suppressor gene mutation | Mutation removes normal suppression of cell division. |
| DNA repair gene mutation | Indirect. Faulty DNA repair gene allows many mutations to accumulate, including those in proto-oncogenes and tumor suppressor genes. |

is also rapidly transcribed. The heightened activity of the oncogene increases production of its protein product, which switches on the genes that promote inappropriate mitosis. This triggers the cascade of cellular changes that leads to cancer. A few types of cancer in humans are caused by viral infection, including Kaposi's sarcoma and acute T cell leukemia. The first virus-cancer link discovered was the Rous sarcoma virus, found in chickens in 1911 and revealed to activate an oncogene in 1976.

A proto-oncogene can also be activated when it is moved next to a very active gene that is a normal part of the genome. This can happen when a chromosome is inverted or translocated. For example, a cancer of the parathyroid glands in the neck is associated with an inversion on chromosome 11, which places a proto-oncogene next to a DNA sequence that controls transcription of the parathyroid hormone gene. When the hormone is synthesized in this gland, the oncogene is expressed. The cells divide, forming a tumor.

Ironically, the immune system contributes to cancer when a translocation or inversion places a proto-oncogene next to an antibody gene. Recall from chapter 16 that antibody genes normally move into novel combinations when a B cell is stimulated and they are very actively transcribed. Cancers associated with viral infections, such as liver cancer following hepatitis, may be caused when proto-oncogenes are mistakenly activated along with antibody genes.

In Burkitt lymphoma, a cancer common in Africa, a large tumor develops from lymph glands near the jaw. People with Burkitt lymphoma are infected with the Epstein-Barr virus, which stimulates specific chromosome movements in maturing B cells to assemble antibodies against the virus. A translocation places a proto-oncogene on chromosome 8 next to an antibody gene on chromosome 14. The oncogene is overexpressed, and the cell division rate increases. Tumor cells of Burkitt lymphoma patients reveal the characteristic chromosome 8 to 14 translocation (figure 17.5).

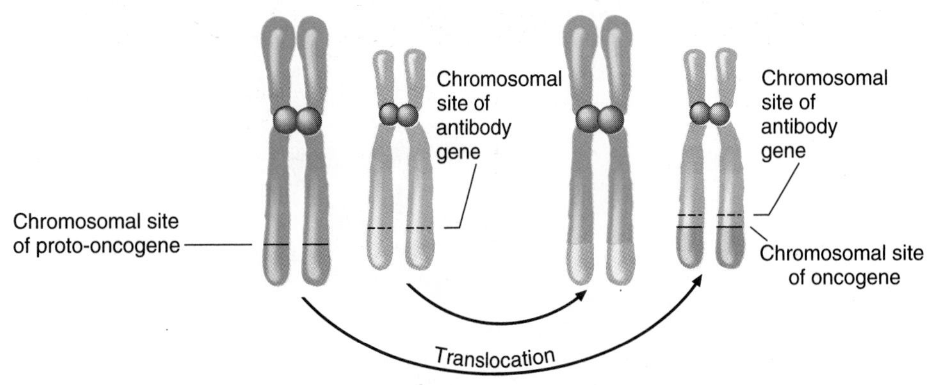


## figure 17.5

**A translocation that causes cancer.** The cause of Burkitt's lymphoma is translocation of a proto-oncogene on chromosome 8 to chromosome 14, next to a highly expressed antibody gene. Overexpression of the moved proto-oncogene, now an oncogene, triggers the molecular and cellular changes of cancer.

### *Fusion Proteins with New Functions*

Oncogenes are also activated when a proto-oncogene moves next to another gene, and the gene pair is transcribed and translated together, as if they form one gene. The double gene product, called a **fusion protein,** somehow activates or lifts control of cell division.

The first cancer-causing fusion protein was found in patients with chronic myeloid leukemia (CML). Most of these patients have a small, unusual chromosome called the Philadelphia chromosome, which consists of the tip of chromosome 9 translocated to chromosome 22. Discovering this cancer-chromosome link was quite a feat.

On August 13, 1958, two men entered hospitals in Philadelphia and reported weeks of unexplained fatigue. Each had very high white blood cell counts and were diagnosed with CML. Their blood had too many immature white blood cells, which were crowding out the healthy cells.

The men's blood samples eventually fell into the hands of University of Pennsylvania assistant professor Peter Nowell and graduate student David Hungerford. Nowell and Hungerford checked the cancer cells for unusual chromosomes. They arrested the white blood cells in mitotic metaphase, then stained and photographed them. But in 1960, cytogenetics was in its infancy—geneticists couldn't even tell similar-sized chromosomes apart. Nevertheless, Nowell and Hungerford studied many of the men's cells and found that each had a "minute chromosome," which they at first thought was a small Y chromosome. The telltale tiny chromosome was found in other people with this cancer, and later became known as the Philadelphia ($Ph^1$) chromosome (figure 17.6).

Many scientists and physicians at first doubted that the Philadelphia chromosome could cause leukemia because the idea of cancer as a genetic illness was so new. But as cases accumulated, the association strengthened. Then, with refinements in chromosome banding, important details emerged.

In 1972, Janet Rowley at the University of Chicago used new stains that distinguished AT-rich from GC-rich chromosome regions to tell that $Ph^1$ is a translocated chromosome. In 1983 and 1984, Dutch and U.S. researchers homed in on the two genes juxtaposed in the 9 to 22 translocation.

One gene from chromosome 9 is called the Abelson oncogene (*abl*), and the other gene, from chromosome 22, is called the breakpoint cluster region (*bcr*). Because the translocation is reciprocal, swapping parts of two chromosomes, two different fusion genes form. The *bcr-abl* fusion gene is part of the Philadelphia chromosome, and this is the one that causes CML. The encoded fusion protein, called the *bcr-abl* oncoprotein, is a form of the enzyme tyrosine kinase, which is the normal product of the *abl* gene. The cancer-causing form of tyrosine kinase is active for too long, which sends signals into the cell stimulating it to divide for too long. (The other translocated chromosome, which is mostly chromosome 9 material, includes the *abl-bcr* fusion gene. It is not known to affect health.)

The discovery that a fusion oncoprotein sets into motion the cellular changes

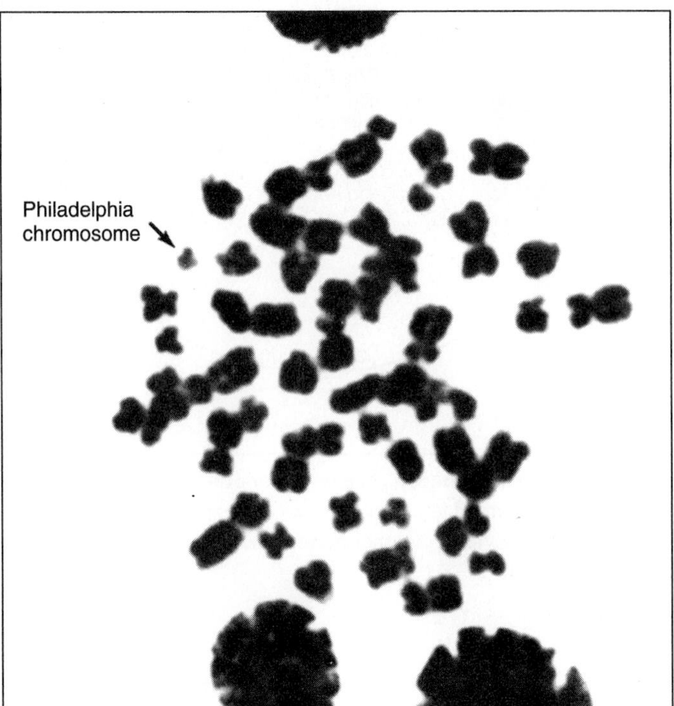


## figure 17.6

**The Philadelphia chromosome.** A tiny chromosome that appears consistently in the white blood cells of patients with chronic myeloid leukemia was named the Philadelphia chromosome in honor of the city where researchers identified it in 1960. Note the poor quality of the chromosome preparation, compared to the intricately banded and DNA-probed chromosomes we view today.

that cause CML has led directly to development of a very effective new drug called Gleevec. An older treatment, alpha interferon, helps only a few patients. Gleevec is a small molecule that nestles into the pocket on the tyrosine kinase that must bind ATP to stimulate cell division. With ATP binding blocked, cancer cells cease dividing—shown first in cell culture, then in mice, and, by the late 1990s, in astoundingly successful clinical trials. Of 54 participants who had not responded to interferon and had received the highest dose of the drug in the trial, 53 had blood counts return to normal, and 29 of them had significantly fewer cells that had the telltale Philadelphia chromosome. In seven participants, the unusual chromosome disappeared completely. The drug is being evaluated for treating other types of cancer.

A fusion oncoprotein also causes acute promyelocytic leukemia. (Leukemias differ by the type of white blood cell affected.) A translocation between chromosomes 15 and 17 brings together a gene coding for the retinoic acid cell surface receptor and an oncogene called *myl.* The fusion protein functions as a transcription factor, which when overexpressed causes cancer. The nature of this fusion protein explained an interesting clinical observation—some patients who receive retinoid (vitamin A-based) drugs enjoy spectacular recoveries. Their typically immature, dedifferentiated cancer cells, apparently stuck in an early stage of development where they divide frequently, suddenly differentiate, mature, and then die! Perhaps the cancer-causing fusion protein prevents affected white blood cells from getting enough retinoids to specialize, locking them in an embryoniclike, rapidly dividing state. Supplying extra retinoids allows the cells to continue along their normal developmental pathway.

### *Receiving a Too-Strong Signal to Divide*

In about 25 percent of women with breast cancer, the affected cells have 1 to 2 million copies of a cell surface protein called Her-2/neu that is the product of an oncogene. In other women, the number is only 20,000 to 100,000.

The Her-2/neu proteins are receptors for epidermal growth factor. The receptors traverse the cell membrane, extending outside the cell into the extracellular matrix and also dipping into the cytoplasm. They also have enzymatic activity, functioning as a tyrosine kinase. Tyrosine is an amino acid, and the kinase activity binds a phosphate group to the tyrosine. When the growth factor binds to the tyrosine of the receptor, it becomes phosphorylated, and this sends a signal into the cell that activates transcription of genes that stimulate cell division. The cause of the cancer is straightforward: too many tyrosine kinase receptors mean too many signals to divide.

At the human level, Her-2/neu breast cancer is particularly threatening. It usually strikes early in life and spreads quickly. Until recently, it was untreatable. However, a monoclonal antibody-based drug called Herceptin binds to the receptors, blocking the signal to divide. In some women, it has led to spectacular recoveries. Interestingly, Herceptin works when the extra receptors arise from multiple copies of the gene, rather than from extra transcription of a single *Her-2/neu* gene.

## Tumor Suppressors

Unlike cancers caused by oncogene activation, some cancers result from loss of a gene that normally suppresses tumor formation. The normal state—tumor suppression—results when cells respond to growth-inhibiting signals. Whereas oncogene activation is usually associated with a point mutation, chromosomal translocation or inversion, and a gain of function, a tumor-suppressor gene mutation that causes cancer is typically a deletion, which removes a function. More rarely, a point mutation can cause loss of function in a tumor-suppressing gene. Certain DNA viruses, such as SV40, adenovirus, and human papilloma virus, are also associated with a high risk of developing particular cancers. These viruses interact with the normal products of tumor suppressor genes, but their relationship to cancer isn't clear.

The childhood kidney cancer Wilms' tumor is an example of a form of cancer that develops from loss of tumor suppression. When a gene that normally halts mitosis in the rapidly developing kidney tubules in the fetus is absent, a tumor results because

*Reading 17.1*

# Retinoblastoma—The Two-Hit Hypothesis

Our current understanding of tumor-suppressing genes began with observations by Alfred Knudson, who developed the two-mutation hypothesis of cancer causation.

Knudson was interested in retinoblastoma (RB), a rare childhood eye cancer. Distinct tumors, representing individual original cancerous cells, develop in the eye (figure 1). Sometimes RB affects one eye, and sometimes both. Knudson examined the medical records of 48 children with RB admitted to M. D. Anderson Hospital in Houston between 1944 and 1969. He recorded the following information for each child:

1. Whether one eye or two were affected
2. How old the child was at the time of diagnosis
3. Whether any other relatives had RB
4. Sex
5. Number of tumors per eye

The fact that RB occurred in boys and girls told Knudson that any genetic control was autosomal. Pooling data from families with more than one case of RB revealed that approximately 50 percent of the children of an affected parent were also affected, suggesting dominant inheritance. Knudson also noted that in some families, a child with two affected eyes would have an affected grandparent, but both parents had healthy eyes. A picture of autosomal dominant inheritance with incomplete penetrance began to emerge.

Knudson, however, proposed a different explanation. He hypothesized that an initial, inherited recessive mutation had to be followed by a second, somatic mutation in the eye to trigger tumor formation. This idea became known as the "two-hit hypothesis" of cancer causation. Occasionally the second mutation would not occur, and this would explain the unaffected parents between two affected generations.

Knudson's two-mutation hypothesis explained another observation gleaned from his search of the medical records. Children with tumors in both eyes become affected much earlier than children with tumors in only one eye—generally before the age of five. This would make sense if a hereditary, bilateral (two-eye) form of RB requires a germline mutation followed by a somatic mutation, but a nonhereditary, unilateral (one-eye) form results from two somatic mutations in the same gene in the same cell. In other words, in the inherited form of RB, a child is already born halfway on the road to tumor development—just one somatic mutation in the eye is needed. The unilateral, noninherited form appears later in childhood because it takes longer for the required two somatic mutations to occur in the same cell.

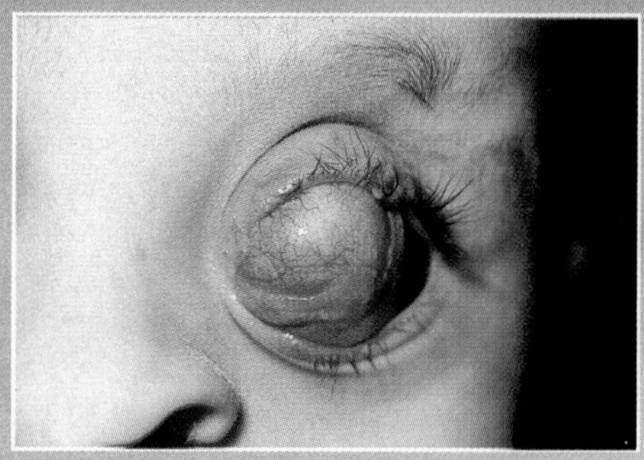

**figure 1**

In inherited retinoblastoma, all of the person's cells are heterozygous for a mutation in the *RB* gene. A second mutation, occurring in the original unmutated allele in cone cells in the retina, releases controls on mitosis, and a tumor develops.

Next, Knudson used a mathematical expression called a Poisson distribution to estimate the number of events required to account for a certain pattern of observations. He found that the average number of tumors per eye—three—was consistent with a two-hit mechanism, according to the equation.

Although it would be another 15 years before researchers identified the *RB* gene on chromosome 13, and longer still before its role in controlling the cell cycle was identified, Knudson's insights paved the way for that particular discovery and for recognition of the widespread action of tumor suppressors in general.

mitosis does not cease on schedule. The child's kidney retains pockets of cells dividing as frequently as if they were still in the fetus, forming a tumor. Among the best-studied tumor suppressor genes are the retinoblastoma gene, the *p53* gene, and *BRCA1*.

### *Retinoblastoma (RB) and the Two-Hit Hypothesis*

Alfred Knudson proposed the role of tumor-suppressing genes in causing cancer in 1971, as the intriguing story of a rare childhood eye tumor called **retinoblastoma** unfolded (Reading 17.1). Retinoblastoma has a long history.

A 2000 B.C. Mayan stone depicts a child with retinoblastoma who has an eye bulging out. In 1597, a Dutch anatomist provided the first clinical description of the eye cancer as a growth "the size of two fists." In 1886, researchers identified inherited cases. At that time, a child would survive the cancer only if the affected eye was removed—a drastic measure. Today, thanks to our understanding of how genes cause RB, children with an affected parent or sibling can be monitored from birth, so that noninvasive treatment can begin as soon as cancer cells appear. Full recovery is common, especially when the condition is detected early. Often the first indication of a problem is an unusual gray area that appears in an eye in a photograph—the tumor reflects light in a different way than unaffected parts of the eye.

About half of the 1 in 20,000 infants who develop RB inherit the susceptibility to the disorder. They harbor one germline mutant allele for the *RB* gene in each of their cells. Cancer develops in any somatic cell where the second copy of the *RB* gene mutates, illustrating the "two-hit hypothesis" of cancer causation. Therefore, inherited retinoblastoma requires two mutations or deletions, one germline and one somatic. In sporadic (noninherited) cases, cells undergo two somatic mutations in the *RB* gene. Whether inherited or sporadic, RB usually starts in a cone cell of the retina, which provides color vision.

Many children with RB have deletions in the same region of the long arm of chromosome 13, which led researchers to the cancer-causing gene. In 1987, they found the gene altered in RB and identified its protein product, linking the cancer to control of the cell cycle. The 928-amino-acid-long protein normally binds transcription factors so that they cannot activate genes that carry out mitosis. It normally halts the cell cycle at $G_1$. When the *RB* gene is mutant or missing, the hold on the transcription factor is released, and cell division ensues.

The *RB* gene may be implicated in other cancers. For example, children who are successfully treated for retinoblastoma often develop bone cancer in their teens or bladder cancer in adulthood. Mutant *RB* genes have been found in the cells of patients with breast, lung, or prostate cancers, or acute myeloid leukemia, who never had the eye tumors. These other cancers may be caused by expressions of the same genetic defect in different tissues.

## *p53* Causes Many Cancers

Another single gene that causes a variety of cancers is ***p53***. The unimaginative name comes from the fact that the gene product was initially known as "protein with molecular weight 53,000." Recall from chapter 11, section 11.6, that the p53 protein transcription factor "decides" whether a cell repairs DNA replication errors or dies by apoptosis. If a cell loses a *p53* gene, or if the gene mutates and no longer functions properly, the cell may lose cell cycle control and become cancerous.

More than half of human cancers involve an abnormal or deleted *p53* gene. The precise locations and types of mutations—a transition (purine to purine or pyrimidine to pyrimidine) or transversion (purine to pyrimidine or vice versa)—are different in different cancers. So far, cancers of the colon, breast, bladder, lung, liver, blood, brain, esophagus, and skin show distinct types of *p53* mutations.

Unlike the cancer-causing genes discussed so far, which are usually associated with chromosomal rearrangements, the *p53* gene seems especially prone to point mutation. The Technology Timeline on cancer and *p53* (see page 361) shows how researchers used the polymerase chain reaction (PCR) to posthumously identify a point mutation in *p53* as the cause of former U.S. vice president Hubert Humphrey's bladder cancer.

Mutational analysis and epidemiological observations reveal that *p53* may be a genetic mediator between environmental insults and development of cancer (figure 17.7). Consider a type of liver cancer prevalent in populations in southern Africa and Qidong, China. These two groups have in common exposure to the hepatitis B virus and to a food contaminant called aflatoxin B1. Most of the people with the liver cancer have a mutation in the *p53* gene, substituting a T for a G in the same codon. Could the food toxin, hepatitis virus, or both cause the mutation?

Similarly, in a type of lung cancer that develops in 1 in 8 heavy smokers, the same T to G base substitution occurs in the *p53* gene in some patients. It also occurs in cells growing in culture exposed to benzo(a)pyrene, a component of cigarette smoke. The carcinogen 4-aminobiphenyl, found in black tobacco, mutates the *p53* gene in a different way, causing a type of bladder cancer.

In the majority of *p53*-related cancers, mutations occur only in somatic cells. However, about 100 families worldwide suffer from a germline condition called the Li-Fraumeni family cancer syndrome. Family members who inherit a mutation in the *p53* gene have a very high risk of developing cancer—50 percent by age 30, and 90 percent by age 70. The risk of breast cancer is near 100 percent for women who inherit the syndrome. A somatic mutation in the affected tissue is necessary for cancer to develop, as is true for inherited retinoblastoma. Li-Fraumeni patients also develop cancers of the brain, blood, bone, adrenal glands, or soft solid tissues, such as muscle or connective tissue. The cancers tend to arise at an earlier age than they do in people who do not have the syndrome. Often the first sign is multiple tumors in one or several organs. National Cancer Institute researchers Fred Li and Joseph Fraumeni first described the syndrome in 1969, but it wasn't linked to the *p53* gene until 1990. Now that the

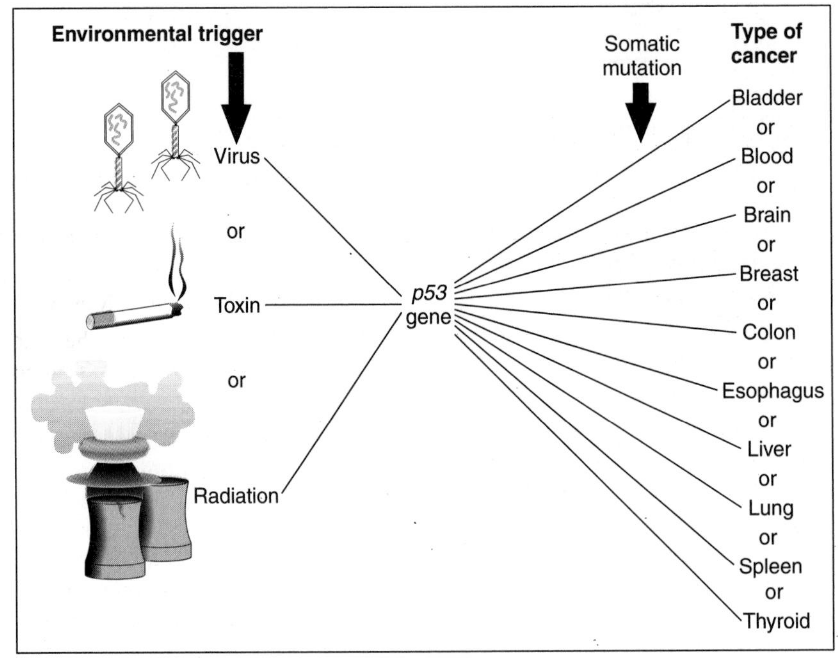


**figure 17.7**

***p53* cancers reflect environmental insults.** The environment triggers genetic changes that lead to cancer. The *p53* gene may act as the mediator.

genetic cause of the syndrome is clear, family members can be tested before cancer symptoms appear, when treatment is more likely to succeed.

### *BRCA1*—A Genetic Counseling Challenge

Breast cancer that recurs in families can reflect the inheritance of a germline mutation, or it can be due to two-hit somatic mutations that happen more than once in a family due to chance. Familial breast cancer touches on many of the complications of Mendel's laws discussed in chapter 5—multiple alleles, incomplete penetrance, variable expressivity, environmental influences, genetic heterogeneity, and polygenic inheritance and epistasis. This can make genetic counseling very challenging (table 17.4).

The *BRCA1* gene, which stands for "breast cancer predisposition gene 1," greatly increases the lifetime risk of inheriting certain forms of breast and ovarian cancer. *BRCA1* was linked to a marker on chromosome 17 in 1990, and then mapped and sequenced in 1994. It is a tumor suppressor gene, because the phenotype results from a loss of function. In the most common mutation, two adjacent bases are deleted, altering the reading frame in a way that shortens the protein. The mutation is inherited as an autosomal dominant trait, but with late onset of symptoms and incomplete penetrance.

The *BRCA1* gene encodes a large protein that normally resides in the nucleus, where it activates transcription of the genes that respond to p53 protein. Therefore, BRCA1 protein is necessary for DNA repair—specifically, mending double-stranded breaks that could threaten the stability of the chromosome.

The inherited breast cancer story also involves population genetics. BRCA1 breast cancer is especially common among the Ashkenazi Jewish population, where slightly more than 2 percent of all individuals have one mutant *BRCA1* allele. Three founder mutations appear in this population, which means that they were brought in with original settlers. The most common mutation is the allele with the two-base deletion, and the other two are a one-base insertion and a one-base deletion. In comparison to the 1 in 50 Ashkenazim, only 1 in 833 people in the general U.S. population has a mutant *BRCA1* allele. This very unequal distribution of disease-causing *BRCA1* alleles means that individual risks depend upon population group. The great difference in incidence among populations may mean that other genes or environmental factors affect the expression of *BRCA1*.

The *BRCA1* gene was discovered in a group of Jewish families with very distinct patterns of breast cancer recurrence—several members were affected, and they became ill at very young ages. Among this select group, a woman who inherits a founder mutation faces an 85 percent risk of developing breast cancer over her lifetime and a 50 percent risk of developing ovarian cancer—figures high enough to encourage some still-healthy relatives with the mutant gene to have the affected organs removed. This approach works, but is highly controversial because it is extreme—although women who have had it report that they are happy that their anxiety

**Technology TIMELINE**

### Cancer and *p53*

The purpose of using gene amplification to detect a cancer-causing mutation is not just to clarify recent medical history, but to diagnose these cancers at an earlier, more treatable stage. The Hubert Humphrey cancer timeline provides a peek at how cancer diagnosis has become increasingly molecular—and less invasive.

**1967** Hubert Humphrey goes to physician, reporting blood in urine. Biopsy of bladder tissue is inconclusive; doctors disagree on whether tissue is malignant.

**1973** Another biopsy diagnosed as "borderline malignancy."

**1976** Infiltrating cancer diagnosed; bulk of tumor removed surgically.

**1978** Humphrey dies of bladder cancer.

**1994** PCR amplification of *p53* gene in 1967 biopsy and 1976 surgical specimens reveal same single-base *p53* mutation, not present in noncancerous tissue sampled at the same time.

Hubert H. Humphrey

**table 17.4**

**The Complexities of Providing Genetic Counseling for Familial Breast Cancer**

1. Many mutations and polymorphisms are known in breast cancer genes. Not all are associated with disease.
2. Breast cancer can occur in other ways. When more than one case occurs in a family, it can be either familial or sporadic. A woman who does not have a *BRCA1* or *BRCA2* mutation can still develop breast cancer. Genetic testing may lead to false assurance.
3. *BRCA1* and *BRCA2* are incompletely penetrant—that is, inheriting a disease-causing allele does not always mean developing cancer.
4. The risk associated with inheriting disease-causing *BRCA1* or *BRCA2* alleles varies depending upon a woman's ethnic/population group, reflecting interactions with other genes.

has lessened. And it appears to work. In one study, 76 of 139 healthy women with *BRCA1* or *BRCA2* (a different gene) mutations had their breasts removed. Six years later, none of the 76 had developed cancer. But among the 63 women who did not undergo the surgery, 8 had developed cancer.

*BRCA1* mutations account for only 5 percent of all breast cancers, and are extremely rare in most population groups. Because the risk of developing breast cancer from a *BRCA1* mutation differs in different population groups (table 17.5), a genetic counselor offers *BRCA1* or *BRCA2* testing based upon whether the patient is part of a high-risk group.

BRCA2 breast cancer is also common among the Ashkenazim. This gene encodes a nuclear protein that is even larger than the BRCA1 protein. Ashkenazi women who inherit a mutation in *BRCA2* face a 60 to 85 percent lifetime risk of developing breast cancer and a 10 to 20 percent risk of developing ovarian cancer. Interestingly, men who inherit a *BRCA2* mutation have a 6 percent lifetime risk of developing breast cancer, which is 100 times the risk for men in the general population. Inheriting a *BRCA2* mutation also increases the risk of developing cancers of the colon, prostate, pancreas, gallbladder or stomach, or malignant melanoma. The fact that p53, BRCA1 and BRCA2 proteins all bind to each other in the nucleus suggests that they interact to enable a cell to repair double stranded DNA breaks.

Table 17.6 lists some oncogenes and tumor suppressor genes.

## Key Concepts

Proto-oncogenes normally control the cell cycle. They can become oncogenes when they mutate, when they move next to a gene that is highly expressed, or when they are transcribed and translated along with another gene, resulting in a fusion protein that triggers cancer. • Mutations in tumor suppressing genes that cause cancer usually are deletions causing loss of gene function, enabling a cell to ignore extracellular constraints on cell division. The *RB*, *p53*, and *BRCA1* genes encode tumor suppressors.

### table 17.5

**Lifetime Risk That Inheriting a *BRCA1* Mutation Will Lead to Breast Cancer**

| Group | Risk (%) |
|---|---|
| Ashkenazim with confirmed strong family history of early-onset cases | 87 |
| Ashkenazim with family history of breast cancer, but not early onset or many affected members | 56 |
| Ashkenazim with no family history of breast cancer | 36 |
| General population | 8–10 |

### table 17.6

**Cancer Genes**

| Oncogenes | Cancer Location/Type | Mechanism |
|---|---|---|
| *myc* | Blood, breast, lung, neurons, stomach | Alters transcription factor |
| *PDGF* | Brain | Alters growth factors or growth factor receptors |
| *RET* | Thyroid | Alters growth factors or growth factor receptors |
| *erb-B* | Brain, breast | Alters growth factors or growth factor receptors |
| *Her-2/neu* | Breast, ovarian, salivary glands | Alters growth factors or growth factor receptors |
| *ras* | Blood, lung, colon, ovary, pancreas | Affects signal transduction |
| *bcl-2* | Blood | Releases brake on apoptosis |
| *PRAD1* | Breast, head and neck | Disrupts cell cycle protein (cyclin) |
| *abl* | White blood cells | Translocation alters proto-oncogene and stimulates cell division |
| **Tumor Suppressors** | | |
| *MTS1* | Many sites | Releases brake on cell cycle |
| *RB* | Eye, bone, breast, lung, bladder | Releases brake on cell cycle |
| *WT1* | Kidney | Releases brake on cell cycle |
| *p53* | Many sites | Disrupts p53 protein, which normally determines whether DNA is repaired or cell dies |
| *DPC4* | Pancreas | Affects signal transduction |
| *NF1* | Peripheral nerves | Disrupts inhibition of normal *ras*, which stimulates cell division |
| *APC* | Colon, stomach | Makes nearby DNA more susceptible to replication errors |
| *BRCA1, BRCA2* | Breast, ovary, prostate | Faulty repair of double-stranded DNA breaks |
| *hMSH2, hMLH1, hPMS1, hPMS2* | Colon, uterus, ovary | Disrupts DNA mismatch repair |
| *Lkb1* | Peutz-Jeghers syndrome (many sites) | After birth, fails to block expression of vascular endothelial growth factor, normally active in embryo |

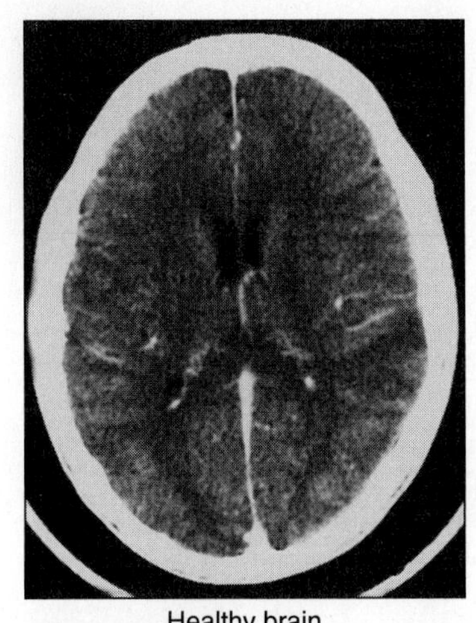

Healthy brain

Loss or mutation of both p53 tumor suppressor alleles
Loss of chromosome 9 genes encoding interferons and tumor suppressors
Oncogene activation increases growth signals
Loss of chromosome 10, role unknown

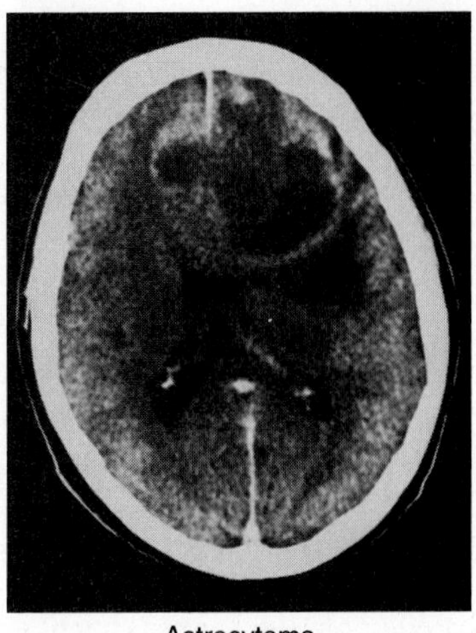

Astrocytoma

**figure 17.8**

**Several genes can contribute to a cancer.** A series of genetic changes transforms normal astrocytes, which support nerve cells in the brain, into a rapidly growing cancer.

## 17.4 A Series of Genetic Changes Causes Some Cancers

A two-step cancer, such as retinoblastoma, is simple. In contrast, some cancers are the culmination of a series of changes in several genes, which may explain why most cancers do not follow Mendel's laws, particularly when some of the genes are linked on the same chromosome.

Figuring out the sequence of genetic changes that leads to a particular cancer is like solving a mystery. To decipher the steps, researchers examine the genetic material of tumor cells from people in various stages of the same type of cancer. This approach is based on logic—the longer the tumor has existed, the more genetic changes have accumulated. A mutation present in all stages acts early in carcinogenesis, whereas a mutation seen only in the tumor cells of people near the end stages of cancer functions late in the process. By hypothesizing how each mutation contributes to the cancer, researchers can identify potential new points of treatment intervention. Following is a closer look at two types of cancer that reflect a series of genetic changes.

### A Rapidly Growing Brain Tumor

Astrocytomas, the most common types of brain tumors, occur when supportive cells called astrocytes divide uncontrollably. These tumors grow—and kill—quickly, unless they are removed at an early stage. The man whose brain is shown in figure 17.8 died just three months after noticing twitching in an eye. During those three months, a series of single-gene and chromosomal changes occurred. Loss of both *p53* alleles was early because it appears in many early-stage tumor cells, as well as in later ones.

By the time an astrocytoma has grown into a small tumor, another genetic change is apparent—loss of both alleles of several genes on chromosome 9. Some of the missing genes encode interferons, so the loss probably disrupts immune protection against the developing cancer. Two other deleted genes are tumor suppressors.

At least two additional mutations speed the tumor's growth. First, an oncogene on chromosome 7 is activated, overexpressing a gene that encodes a cell surface receptor for a growth factor. The cancer cells bear too many growth factor receptors, and receive too many messages to divide. Finally, the cancer cells lose one or even both copies of chromosome 10. This is a final change, because it is seen in all end-stage tumors, but not in early ones.

### Colon Cancer

Colon (large intestine) cancer does not usually occur in families with the frequency or pattern expected of a single-gene disorder. However, when family members with noncancerous growths (polyps) in the colon are considered with those who have colon cancer, a Mendelian pattern emerges. Five percent of cases are inherited. One in 5,000 people in the United States has precancerous colon polyps, a condition called familial adenomatous polyposis (FAP).

FAP begins in early childhood with tiny polyps, often hundreds, that progress over many years to colon cancer. Connecting FAP to the development of colon cancer enabled researchers to view the stepwise progression of a cancer. Several genes, including both oncogenes and tumor suppressors, take part.

On a cellular level, FAP may not be so much an excess of mitosis as it is a deficit of apoptosis. Colon lining cells typically live three days. In FAP, they fail to die on schedule, and instead build up, forming polyps.

The study of the hereditary nature of some colon cancers began in a genetics classroom at the University of Utah in Salt Lake City in the fall of 1947, when young professor Eldon Gardner stated that he thought cancer might be inherited. A student, Eugene Robertson, excitedly told the class that he knew of a family in which a grandmother, her three children, and three grandchildren had colon cancer.

Intrigued, Gardner delved into the family's records and began interviewing relatives. He eventually found 51 family members and arranged for each to be examined

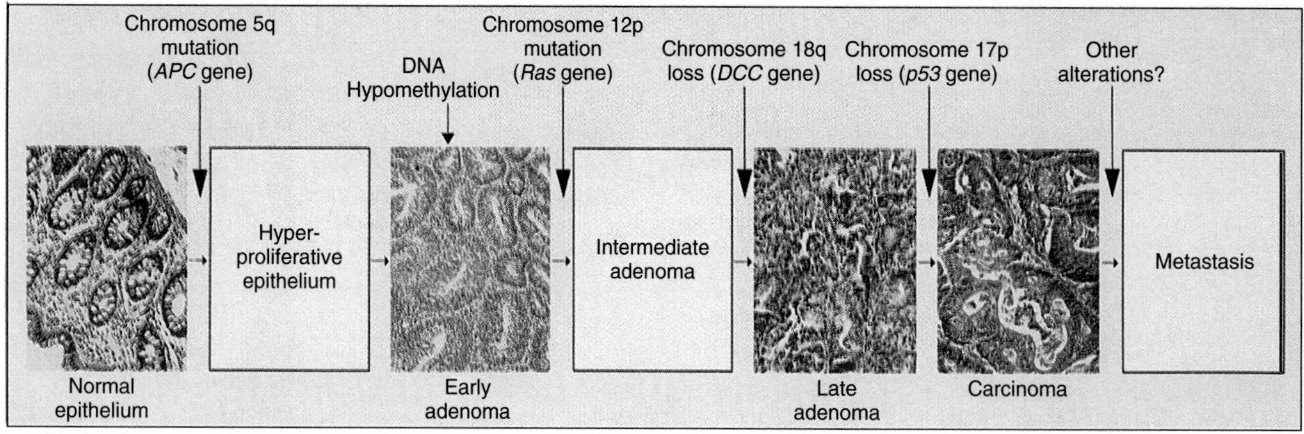


**figure 17.9**

**Several genes contribute to FAP colon cancer.** Cells lining the colon begin to divide more frequently when the *APC* gene on chromosome 5q undergoes a point mutation. This causes replication errors that disrupt the reading frame, shortening the protein product. The affected cell proliferates, forming a growth that becomes precancerous when DNA loses protective methyl groups. Next, the *Ras* oncogene is activated. Loss of the *p53* tumor suppressor gene produces cancer, and other genes may contribute to the cancer's spread. Researchers continue to fill in the gaps in the genetic orchestration of this cancer.

with a colonoscope, a lit instrument passed into the rectum that views the wall of the colon. The colons of 6 of the 51 people were riddled with the gobletlike precancerous growths, although they had no symptoms. Removal of their colons probably saved their lives.

In the years that followed, researchers identified other families with more than one case of colon polyps. Individuals with only polyps were diagnosed with FAP. If a person with colon polyps had cancer elsewhere, extra teeth, and pigment patches in the eye, the condition was called Gardner syndrome, named for the professor. The chromosomal defect that causes Gardner syndrome was identified in 1985, with the help of a 42-year-old man at the Roswell Park Cancer Institute in Buffalo, New York. He had several problems—no gallbladder, an incomplete liver, an abnormal kidney, mental retardation, and Gardner syndrome. To a geneticist, a seemingly unrelated combination of symptoms suggests a chromosomal abnormality affecting several genes. Sure enough, the man's karyotype revealed a small deletion in the long arm of chromosome 5. This was the first piece to the puzzle of colon cancer.

Since 1985, researchers have discovered other genes that contribute to colon cancer, including *p53*. The genes act at different points, culminating in cancer. A gene on chromosome 5q, called *APC*, may start the process. A point mutation that changes a T to an A in the *APC* gene results in a stretch of eight consecutive A's, which destabilizes replication enzymes. The result is a shift in the reading frame and a shortened protein. A plausible sequence of genetic events that causes FAP appears in figure 17.9, although other sequences are possible.

**Key Concepts**

Some cancers may be the culmination of a series of mutations in several genes. Determining which mutations are present in particular stages of a cancer can reveal the sequence of gene actions.

## 17.5 Cancer Prevention, Diagnosis, and Treatment

Because environmental factors can induce cancer-causing mutations, controlling those factors can, theoretically, lower the risk of developing cancer. Once cancer is diagnosed, medical science can usually offer many treatment options.

### Diet-Cancer Associations

Chemical carcinogens were recognized as long ago as 1775, when British physician Sir Percival Potts suggested that the high rate of skin cancer in the scrotums of chimney sweeps in London was due to their exposure to a chemical in soot. Since then, epidemiological studies have identified many chemicals as possibly causing cancer in certain populations (table 17.7). However, most studies reveal correlations rather than cause-and-effect relationships. In the best cases, genetic or biochemical evidence refines the observed environmental connection.

Epidemiologists use different statistical tools to establish links between environmental exposures and cancer. Links are strengthened when different types of investigations yield consistent results. This is true, for example, of the association between eating whole grain cereals and reduced incidence of colorectal cancer.

A **population study** compares the incidence of a type of cancer among very different groups of people. If the incidence differs, then some difference between the populations may be responsible. For example, an oft-mentioned study from 1922 found that primitive societies have much lower rates of many cancers than more developed groups. The study attributed the lack of cancer to the high level of physical activity among the

**table 17.7**

**Increase in Death Rates for Certain Cancers in Particular Geographical Areas**

| Cancer Type | Region | Possible Explanation |
|---|---|---|
| Breast | Northeast | *BRCA1* mutations, greater lifetime exposure to estrogens (early menstruation, late menopause, older age of first birth, exposure to pesticides) |
| Colon | Northeast | Dietary factors, medical screening |
| Lung | White men in south, white women in west, blacks in northern cities | Changes in regional trends in cigarette smoking |
| Lung | Men in southern coastal areas | Asbestos exposure while working in shipyards during World War II |
| Mouth, throat | Women in rural south | Smokeless tobacco |
| Esophagus | Washington, D.C., coastal South Carolina | Alcohol and tobacco, deficiencies of fruits and vegetables in diet |

primitive peoples—but diet might also explain the difference.

Population studies often have too many variables to enable a researcher to pinpoint a cause and effect. Consider the very high incidence of breast cancer on Long Island, New York. One hypothesis attributes the mini-epidemic to pesticide exposure. But the fact that this population is 16 percent Ashkenazim indicates a high frequency of *BRCA1* mutant alleles. Other studies have shown that this population is very careful about medical care, with women having frequent mammograms starting at a young age. As a result, the percentage of the population with recognized early stages of the disease may be higher than in other populations where women are less likely to have regular mammograms. All of these factors may contribute to the overall high reported breast cancer incidence in this area.

In a **case-control study,** people with a type of cancer are matched for age, sex, and other characteristics with individuals in a group of healthy people. Then researchers look for differences between the pairs. If, for example, the cancer patients had extensive dental X-rays at a young age but the control group didn't, that may be a causal factor. The problem with this type of study is that much of the information is often based on recall, people make mistakes, and not all relevant factors are recognized and taken into account.

The most informative type of epidemiological investigation is a **prospective study,** in which two or more groups of people follow a specified dietary regimen and are checked periodically for cancer. In contrast to population and case-control studies, a prospective study looks ahead, so the investigator has more control over the activities and can verify information. A limitation of this type of cancer study is that cancer usually takes many years to appear and progress. The study described in Reading 17.2 on page 366 circumvents this problem by assessing other measures of protection.

On the basis of all of these types of studies, the National Cancer Institute advises consuming whole grains, fruits and vegetables, and limiting consumption of animal fat to lower cancer risk. However, the meaning of these epidemiological studies is far from certain. For a correlation to be elevated to the status of possible cause or preventative, a biological explanation is necessary. For example, certain vegetables contain antioxidant compounds, which deactivate the free radicals that can damage DNA, thereby preventing mutations. Similarly, a high-fiber diet sweeps food through the digestive system faster, giving carcinogens less time to cause mutations. Until researchers know more about environmental influences on cancer causation, it is wise to eat a balanced diet, exercise, and not smoke.

## Diagnosing and Treating Cancer

Tests can identify members of families that have certain germline cancers who have inherited the mutation and are very likely to develop cancer. This information can be important, even lifesaving, if the tumor or affected body part can be removed, as is the case for breast cancer. Another example of a cancer that can be avoided with prophylactic (preventive) surgery is hereditary diffuse gastric cancer, which accounts for 3 percent of all stomach cancers.

In one family, 7 of 12 siblings died from stomach cancer, ranging in age from 14 to 40 years old. Before the gene was identified in 1998, one woman in the family chose to have her stomach removed to avoid her siblings' fate. After a gene test became available, she learned that she had *not* inherited the mutation! In another family, genetic tests showed that 5 of 14 siblings inherited the mutation, although all were still healthy. Three chose to have their stomachs removed, and cancer was found in all three of them, the tumors still too small to have caused symptoms or been detected with a biopsy. In these individuals, reconstructive surgery reattached the esophagus directly to the small intestine, so that they can eat and digest small amounts of certain foods.

Predicting that cancer will occur in a particular individual is only possible for a few disorders that are inherited in a Mendelian fashion and whose genes are known. More often, discovery of cancer follows a screening test such as mammography or high levels of prostate specific antigen in blood, or after symptoms occur or a person feels a lump. Then treatments begin—and there are usually many options.

Cancer treatments focus on particular characteristics of cancer cells. The oldest treatment, surgery, is straightforward—it prevents invasiveness by removing the tumor. Radiation and chemotherapy use a different approach, killing all cells that divide rapidly. Unfortunately, this also affects healthy cells in the digestive tract, hair follicles, and bone marrow, causing nausea, hair loss, great fatigue, and susceptibility to infection. Patients receive several other drugs to help them tolerate the side effects, including colony stimulating factors to replenish bone marrow. These other drugs enable patients to withstand higher, and more effective, doses of chemotherapy.

Reading 17.2

# Vegetables and Cancer Risk

A diet rich in cruciferous vegetables such as broccoli and Brussels sprouts correlates to lowered incidence of colon cancer, according to many population observations. Experiments conducted at the Imperial College of Medicine and the Institute of Food Research in the U.K. cleverly connected the epidemiological evidence with known pieces of the biochemical puzzle:

- Cooking meat induces formation of heterocyclic aromatic amines (HAs). These compounds are absorbed into the lining of the digestive tract and then metabolized by a liver enzyme into mutagens, some of which cause colon cancer.
- Cruciferous vegetables release glucosinolates, which activate "xenobiotic metabolizing enzymes," which metabolize the HAs into less dangerous forms (figure 1).

In the experiments, 20 healthy men followed three 14-day dietary regimens. In the first and third periods, food intake was closely monitored for 12 days to avoid foods that contain enzymes that could interfere with the HA metabolic pathways. On day 13, the men ate a steak in which levels of the two major HAs had been measured. Over the next two days, urine, saliva, and blood were collected at various times. During the middle 14-day period, the men had Brussels sprouts or broccoli soup for breakfast and either vegetable with dinner for the first 12 days, then ate the steak on day 13 and followed up with the body fluid samplings.

The researchers measured effects on metabolism in two ways: they assessed the levels of the two types of HAs in the body fluids, and they also measured the metabolism of a dose of caffeine given 12 hours after the steak meal. Because the enzymes from the vegetables affect liver enzymes that metabolize caffeine in much the same way that they affect metabolism of HAs, a change in caffeine metabolism would signal alteration of HA metabolism too. The results of the experiments strongly supported the epidemiological evidence: the vegetables lowered the concentration of the HAs in urine by 22 percent. This approach was much faster and safer than conducting dietary experiments over years and waiting to see who developed cancer! The only reported side effects of the dietary regimen were caffeine withdrawal and mild flatulence.

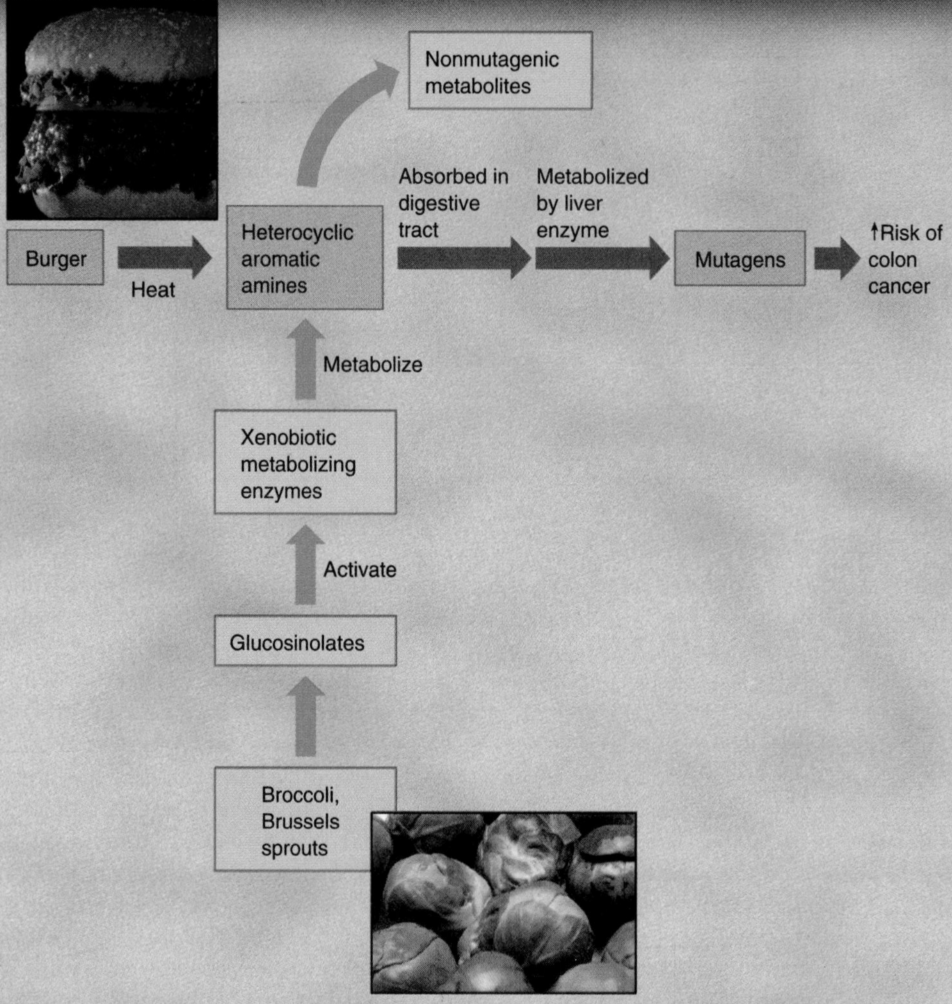


**figure 1**

**One way that cruciferous vegetables lower cancer risk.** Compounds called heterocyclic aromatic amines (HAs) form in cooking meat, are absorbed into the digestive tract, and are metabolized by a liver enzyme into mutagens, which may cause colon cancer. Broccoli and Brussels sprouts produce glucosinolates, which activate xenobiotic metabolizing enzymes that block part of the pathway that leads to production of the mutagens.

Newer treatments target properties of cancer cells other than their high division rate, which minimizes effects on rapidly dividing healthy cells. For example, chapter 16 (section 16.4) discussed the use of immune system biochemicals to treat cancer and chapter 19 (section 19.2) describes cancer vaccines that aim to prevent recurrences using the immune response.

Several new types of cancer drugs are available or in development. Experimental cancer treatments include factors that stimulate cells to regain specialized characteristics and an accompanying slower division rate. Retinoic acid is one such treatment. Another approach is to inhibit telomerase, which prevents cancer cells from elongating their telomeres and continually dividing. Drugs that are angiogenesis inhibitors, such as angiostatin and endostatin, rob a cancer of its blood supply. Other treatment approaches induce apoptosis, which counters the runaway cell division of cancer.

The evolution of diagnosis and treatments for breast cancer illustrates how genetic and genomic information will increasingly refine how physicians manage these diseases. The earliest treatments simply removed or destroyed the affected tissue (table 17.8). Then physicians began to determine whether tumor cells have receptors for estrogen or progesterone, two hormones. Women with estrogen receptor positive tumors typically begin a several-year course of a drug called Tamoxifen, which blocks these receptors from receiving signals to divide. Determining estrogen receptor status is subtyping by phenotype. With the discovery of single genes that cause cancer, diagnosis began to include genotyping too. A woman might have BRCA1 or Her-2/neu breast cancer. In the years to come, cancer diagnosis will be based on DNA microarrays that scan thousands of genes for their expression in tumor cells as described in section 1.5. This approach will reveal subtypes of subtypes that are not detectable in other ways, but that might respond to different combinations of treatments. A genomic view of cancer will enable physicians to match a particular patient to the treatments most likely to work—right from the start.

The limitation of any cancer treatment, old or new, is defined by the strength of the enemy. Cancer cells are incredibly abundant and ever-changing. Surgery followed by a barrage of drugs and radiation can slow the course of the disease, but all it takes is a few escaped cancer cells—called micrometastases—to sow the seeds of a future tumor. Cancer cells mutate in ways that enable them to pump out any drug sent into them. Although cancer treatments can completely cure the illness, it is more likely that they often kill enough of the cancer cells, and sufficiently slow the spread, that it takes the remainder of a lifetime for the tumors to grow back.

An ideal cancer treatment would selectively target only cancer cells—and accomplishing this long-sought goal requires continued research into how these strangely aggressive cells arise and persist. Already research has paid off in many new treatment approaches based on discoveries at the cellular and molecular levels, and population statistics are beginning to reflect those successes. Between 1990 and 1995, cancer incidence and death rates dropped for all cancers for the first time in many decades. Each year in the United States, there have been at least 27,000 fewer cancer deaths and 70,000 fewer new cases. This trend reflects new and improved treatments, earlier diagnosis and treatment, and perhaps healthier lifestyles.

**table 17.8**

**Evolution of Treatments for Breast Cancer**

| Strategy | Examples |
|---|---|
| Remove or destroy cancerous tissue | Surgery, radiation, chemotherapy |
| Use phenotype to select drug | Estrogen-receptor-positive women take Tamoxifen |
| Use genotype to select drug | *Her-2/neu*-positive cancers take Herceptin (monoclonal antibody) |
| Genomic level | Gene expression profile on DNA microarray used to guide drug choice |

**Key Concepts**

Lower cancer risk is associated with eating more fruits, vegetables, and whole grain cereals, and avoiding fats. • Treatments for cancer target the characteristics of cancer cells. Surgery removes tumors. Chemotherapy and radiation nonselectively destroy rapidly dividing cells. Newer treatments target receptors on cancer cells, block telomerase, stimulate differentiation, or attack a tumor's blood supply. Diagnosis and treatment of cancer will increasingly consider genetic information

# Summary

### 17.1 Cancer as a Genetic Disorder

1. Cancer is a genetically dictated loss of cell cycle control, creating a population of highly proliferative cells that outgrows and overwhelms surrounding tissue.
2. Sporadic cancers result from mutations in somatic cells only and are more common than germline cancers, which are caused by a **germline** mutation plus a somatic mutation in affected tissue. Cancer may also be a polygenic trait.
3. Mutations in genes that encode or control transcription factors, cell cycle checkpoint proteins, growth factors, repair proteins or telomerase may disrupt the cell cycle sufficiently to cause cancer.

### 17.2 Characteristics of Cancer Cells

4. A tumor cell divides more frequently or more times than cells surrounding it, has altered surface properties, loses the specializations of the cell type it arose from, and produces daughter cells like itself.
5. A malignant tumor infiltrates nearby tissues and can **metastasize** by attaching to basement membranes and secreting enzymes that penetrate tissues and open a route to the bloodstream. From there, a cancer cell can travel, establishing secondary tumors.

### 17.3 Genes That Cause Cancer

6. Cancer is often the result of a series of genetic changes involving the activation of **proto-oncogenes** to **oncogenes,** and the inactivation of **tumor suppressing genes.** Mutations in DNA repair genes can cause cancer by increasing the mutation rate.
7. Oncogenes normally promote controlled cell growth, but are overexpressed because of a point mutation, placement next to a highly expressed gene, or because they are transcribed and translated with another gene, producing a **fusion protein.** Oncogenes may also be overexpressed growth factor receptors.
8. A tumor suppressor is a gene that normally enables a cell to respond to factors that limit its division. Tumor suppressor genes include *RB,* ***p53,*** and *BRCA1.*

### 17.4 A Series of Genetic Changes Causes Some Cancers

9. Many cancers result from "two hits" or mutations, but some entail a longer series of genetic changes.
10. To decipher the gene action sequences that result in cancer, researchers examine the mutations in cells from patients at various stages of the same type of cancer. Those mutations present at all stages of the cancer are the first to occur.
11. Astrocytoma and FAP are two cancers that require several mutations to develop.

### 17.5 Cancer Prevention and Treatment

12. **Population, case-control,** and **prospective studies** can reveal correlations between environmental exposures and the development of certain cancers, but usually cannot establish cause and effect. Biochemical and/or genetic evidence is important to explain epidemiological observations.
13. Traditional cancer treatments consist of surgery, radiation, and chemotherapy. Newer approaches based on molecular biology include blocking hormone receptors, stimulating cell specialization, blocking telomerase, and inhibiting angiogenesis. A genomic approach identifies differences in gene expression that define cancer subtypes and may correlate to the success of certain treatments.

# Review Questions

1. How would mutations in genes that encode the following proteins lead to cancer?
   a. a transcription factor
   b. the p53 protein
   c. the retinoblastoma protein
   d. the *myl* oncogene's protein
   e. a repair enzyme
   f. the APC protein
2. How can the same cancer be associated with deletions as well as translocations of genetic material?
3. What would be the value of knowing whether a person's cancer is sporadic or inherited?
4. List four characteristics of cancer cells.
5. Cite three reasons why cancer may not follow a Mendelian pattern, but nevertheless involves abnormal gene function.
6. What is inaccurate about the statement that "cancer cells are the fastest dividing cells in the body?"
7. Distinguish among the following types of studies:
   a. population
   b. case-control
   c. prospective
8. Three percent of all cancer cells have chromosome rearrangements. What other type of genetic change might be present in a cancer cell?
9. List four new strategies for treating cancer, and explain how they work.

# Applied Questions

1. An individual can develop breast cancer by inheriting a germline mutation, then undergoing a second mutation in a breast cell; or by undergoing two mutations in a breast cell, one in each copy of a tumor suppressor gene. Cite another type of cancer, discussed in the chapter, that can arise in these two ways.
2. Several biotechnology companies offer tests to detect *p53* alleles that confer cancer susceptibility. Cite a limitation of such a test.
3. Humans missing both *p53* alleles in all cells are unknown. People with p53-related cancers either have a germline mutation and a somatic mutation in affected tissue, or two somatic mutations in the tissue. Experiments show that mice missing both copies of their *p53* genes die as embryos, with massive brain abnormalities.
   a. Why don't we see people with two missing or mutant *p53* alleles in all cells?
   b. Under what circumstances might a human with two mutant *p53* alleles be conceived?
4. Von Hippel-Lindau disease is an inherited cancer syndrome. The responsible gene lifts control over the transcription of certain genes, which, when overexpressed, cause tumors to form in the kidneys, adrenal glands, or blood vessels. Would the von Hippel-Lindau gene be an oncogene or a tumor suppressor? Cite a reason for your answer.
5. A tumor is removed from a mouse and broken up into cells. Each cell is injected into a different mouse. Although all the mice used in the experiment are genetically identical and raised in the same environment, the animals develop cancers with different rates of metastasis. Some mice die quickly, some linger, and others recover. What do these results indicate about the characteristics of the cells of the original tumor?
6. Colon, breast, and stomach cancers can be prevented by removing the affected organ. Why is this approach not possible for chronic myeloid leukemia?
7. A woman goes to see a genetic counselor, terribly distressed. Her sister developed breast cancer at age 52, and her husband's sister at age 34. The woman's mother was diagnosed with breast cancer at age 68. With these three cases in her family, she is convinced she has a mutant *BRCA1* gene and will suffer the same fate.
   a. What questions should the counselor ask her?
   b. Do you think she would benefit from a *BRCA1* gene test?
   c. What complications might arise from such testing?
8. A vegetarian develops pancreatic cancer and wants to sue the nutritionist who suggested she follow a vegetarian diet. Is her complaint justified? Why or why not?
9. Iron foundry workers in Finland and coke oven workers in Poland have high exposures to polycyclic aromatic hydrocarbons, and they tend to develop cancers caused by mutations in the *p53* gene. What information would help determine whether the chemical exposure causes the mutation and whether the mutation causes the cancers?
10. A woman who finds a small lump in her breast goes to her physician, who takes a medical and family history. The woman mentions that her father died of brain cancer, a cousin had leukemia, and her older sister was just diagnosed with a tumor of connective tissue. The doctor assures her that the family cancer history doesn't raise the risk that her breast lump is cancerous, because the other cancers were not in the breast. Is the doctor correct?
11. A project at the Sanger Centre in the United Kingdom is screening 1,500 types of cancer cells growing in culture to detect homozygous deletions. Will this approach identify oncogenes or tumor suppressor genes? Cite a reason for your answer.

# Suggested Readings


Culliton, Barbara J. May 5, 1994. Hubert Humphrey's bladder cancer. *Nature,* vol. 369. Years after the vice president's death, PCR revealed a single base change in a single gene as the cause of the cancer.

DeVilee, P. April 1999. *BRCA1* and *BRCA2* testing: Weighing the demand against the benefits. *The American Journal of Human Genetics* 64:943. The extent to which inheriting a *BRCA1* or *BRCA2* mutation raises cancer risk depends upon one's population group.

Druker, Brian J. et al. Efficacy and safety of a specific inhibitor of the BCR-ABL tyrosine kinase in chronic myeloid leukemia. *The New England Journal of Medicine* 344(14):1031–37. Understanding the nature of a fusion protein led to development of a successful cancer drug.

Eisen, Andrea, and Barbara L. Weber. July 19, 2001. Prophylactic mastectomy for women with *BRCA1* and *BRCA2* mutations—facts and controversy. *The New England Journal of Medicine* 345(3):207–8. Gene tests can predict which women in a family face extremely high risk of developing breast cancer.

Futreal, P. Andrew, et al. February 15, 2001. Cancer and genomics. *Nature* 409:850–52. The first draft human genome sequence did not reveal any new cancer-causing genes—but researchers might not have known what to look for.

Lewis, Ricki. April 30, 2001. Herceptin earns recognition in breast cancer arsenal. *The Scientist* 15(9):10. A monoclonal antibody blocks extra growth factor receptors on the cells of some people with breast cancer, halting it.

Meiser, Bettina, et al. July 2001. Cultural aspects of cancer genetics: Setting a research agenda. *The Journal of Medical Genetics* 38:425–26. In cultures where fate, karma, and astrology are thought to cause cancer, attempts by genetic counselors to discuss oncogenes and tumor suppressors may be futile.

Nathanson, Katherine N., et al. May 2001. Breast cancer genetics: What we know and what we need. *Nature Medicine* 7(5):552–56. Breast cancer may be Mendelian or polygenic.

Nowell, P., J. Rowley, and A. Knudson. October 1998. Cancer genetics, cytogenetics—defining the enemy within. *Nature*

*Medicine* 4:1107. A series of articles on the discoveries of cancer-causing genes, by the researchers who made the discoveries.

Page, Janice. August 7, 2001. Sometimes, it's the good news that makes the patient feel so bad. *The New York Times,* p. D7. A gene test that reveals absence of the *BRCA1* mutation causes another condition—survivor guilt.

Perou, Charles M., et al. August 17, 2000. Molecular portraits of human breast tumors. *Nature* 406:747–52. DNA microarrays reveal at a glance how tumor cells veer from normalcy.

Sorlie, Therese, et al. September 11, 2001. Gene expression patterns of breast carcinomas distinguish tumor subclasses with clinical implications. *Proceedings of the National Academy of Sciences* 98(19):10869–74. Neural network algorithms are used to match patterns of gene expression to cancer subtypes.

Weitzel, Jeffrey N., and Laurence E. McCahill. June 21, 2001. The power of genetics to target surgical prevention. *The New England Journal of Medicine* 344(25):1942–43. Removing the stomach is a drastic step, but it can prevent cancer.

# On the Web

Check out the resources on our website at:

**www.mhhe.com/lewisgenetics5**

On the web for this chapter, you will find additional study questions, vocabulary review, useful links to case studies, tutorials, popular press coverage, and much more. To investigate specific topics mentioned in this chapter, also try the links below:

American Cancer Society **www.cancer.org**

*Atlas of Cancer Mortality in the United States, 1950–1994* **www.nci.nih.gov/atlas**

Cancer Genome Anatomy Project **www.ncbi.nlm.nih.gov/CGAP/**

Cancer Information **www.cdc.gov/health/cancer.htm**

Ferguson Inherited Colorectal Cancer Registry **www.spectrum-health.org**

Hereditary Colon Cancer Newsletter **www.mdanderson.org/depts/hcc**

National Breast Cancer Coalition **www.natlbcc.org**

National Cancer Institute **nci.nih.gov**

National Ovarian Cancer Coalition **www.ovarian.org**

Oncolink **www.cancer.org**

Online Mendelian Inheritance in Man **www.ncbi.nlm.nih.gov/entrez/query.fcgi?db=OMIM**
adenomatous polyposis of the colon 175100
astrocytoma 155755
Burkitt lymphoma 151410
*BRCA1* 113705
*BRCA2* 600185
chronic myelogenous leukemia 151410
Gardner syndrome 175100
hereditary diffuse gastric cancer 137215
Li-Fraumeni familial cancer syndrome 151623
*p53* 191170
retinoblastoma 180200

CHAPTER

# 18

# Genetically Modified Organisms

### 18.1 Of Pigs and Patents

The techniques of modern biotechnology have produced a variety of organisms whose unique combinations of traits would probably not have arisen in nature. DNA sequences that form the basis of biotechnologies can receive patent protection if they are, like other inventions, new, useful, and not obvious. Patent law has had to evolve to keep pace with biotechnology.

### 18.2 Recombinant DNA Technology

So-called "genetic engineering" started in 1975 with fear and precautions, and has blossomed into an incredibly powerful tool to produce pharmaceuticals and to isolate and study individual genes. Recombinant DNA technology, the first technique to combine genes of different species, has spawned other biotechnologies.

### 18.3 Transgenic Organisms

Recombinant DNA on a multicellular scale is called transgenesis. The genetic manipulation is done on gametes, fertilized ova, or denuded plant cells. Transgenic organisms can produce proteins encoded by genes from other species, making it possible to "pharm" protein-based drugs.

### 18.4 Gene Targeting

Gene targeting takes advantage of genes' natural tendency to swap places with other versions of themselves, improving on the precision of transgenesis. The results of gene targeting experiments are called "knockouts" or "knockins." This technique can also be used to manufacture protein-based pharmaceuticals.

Pig manure presents a serious environmental problem. The animals do not have an enzyme that would enable them to extract the mineral nutrient phosphorus from a compound called phytate in grain, and so they must receive dietary phosphorus supplements. As a result, their manure is full of phosphorus, which, when it washes into natural waters, contributes to fish kills, oxygen depletion in aquatic ecosystems, algal blooms, and even the greenhouse effect (figure 18.1). But biotechnology may have solved the "pig poop" problem.

## 18.1 Of Pigs and Patents

In the past, pig raisers have tried various approaches to keep their animals healthy and the environment cleaner. Efforts include feeding animal by-products from which the pigs can extract more phosphorus, and giving supplements of the enzyme that liberates phosphorus from phytate, called phytase. But consuming animal by-products can introduce prion diseases, and giving phytase before each meal is costly. A "phytase transgenic pig," however, is genetically modified to secrete phytase in its saliva, which enables it to excrete low-phosphorus manure.

A **transgenic** animal has a genetic change in each of its cells. This particular type of pig has been given a phytase gene from the bacterium *E. coli*. Its manure has 75 percent less phosphorus than normal pig excrement. Says one researcher, "These pigs offer a unique biological approach to the management of phosphorus nutrition and environmental pollution in the pork industry."

The genetically modified pig, its genome manipulated so that its excrement is less polluting, is a product of **biotechnology**, which is the use or alteration of cells or biological molecules for specific applications. Biotechnology is, in a broad sense, hardly new. The ancient art of fermenting fruit with yeast to produce wine is a biotechnology, as is using yeast to make bread dough rise. Extracting biochemicals directly from organisms for various applications is also biotechnology.

The popular terms genetic engineering and genetic modification refer to a biotechnology that manipulates genetic material. This includes altering the DNA of an organism to suppress or enhance the activities of its own genes as well as combining the genetic material of different species. The latter is possible because all life uses the same genetic code.

It is this mixing of DNA from different species that some people object to as being unnatural. In fact, DNA does move—that is why we have viral DNA sequences nestled in our chromosomes and may even share some sequences with bacteria. But genetic modification usually endows organisms with traits that they would probably not acquire naturally, such as pigs with low-phosphorus manure, tomatoes that grow in salt water, and bacteria that synthesize human insulin, discussed later in this chapter. Figure 18.2 shows a dramatic example of a genetically modified organism—a tobacco plant that expresses a gene from a firefly that enables it to "glow."

Adding foreign genes to bacteria or other single cells is called **recombinant DNA technology,** and it was the first of the modern biotechnologies. When recombinant bacteria divide, they yield many copies, or clones, of the foreign DNA and produce

**figure 18.1**

**Genetically modified pig manure.** Transgenic pigs given a bacterial digestive enzyme excrete less-polluting manure. The author poses amidst a pile of the nonmodified material, at the University of Georgia.

a.

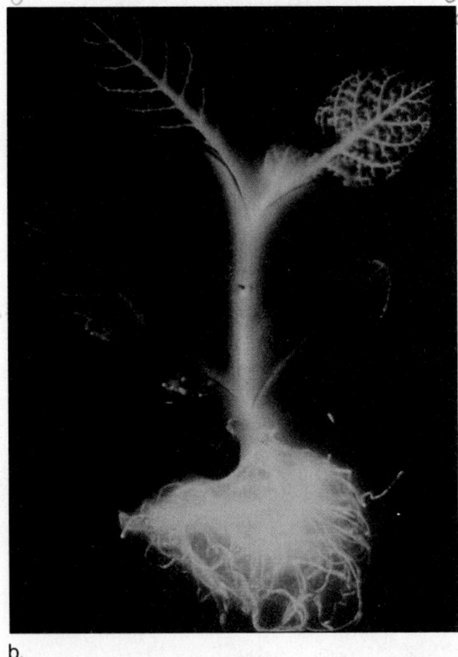

b.

**figure 18.2**

**The universality of the genetic code makes biotechnology possible.** Recombinant DNA and transgenic technologies are based on the fact that all organisms utilize the same DNA codons to specify the same amino acids. A striking illustration of the universality of the code appears in this transgenic tobacco plant (*a*) that contains firefly genes that specify the "glow" enzyme luciferase. When bathed in a chemical that allows the enzyme to be expressed, the plant glows (*b*).

many copies of the protein the foreign DNA specifies. In the 1980s, researchers began to apply recombinant DNA technology to multicellular organisms, producing transgenic plants and animals. Researchers add foreign DNA at the one-cell stage (a gamete or fertilized ovum). The transgenic organism that develops from the original altered cell carries the genetic change in every cell. Yet another biotechnology, called **gene targeting,** adds precision to transgenic technology. Gene targeting "knocks out" or "knocks in" the gene of interest at a particular chromosomal locus, where it trades places with an existing gene.

The ability to combine genes from different types of organisms has raised legal questions—is a recombinant or transgenic organism an invention, deserving of patent protection? By definition, to earn a patent an invention must be new, useful, and not obvious (see Technology Timeline).

Patent law has had to evolve in parallel to the rise of biotechnology. Early on, DNA sequences could be patented. In the mid-1990s, however, when the U.S. National Institutes of Health and biotech companies began seeking patent protection for thousands of pieces of protein-encoding DNA sequences, called expressed sequence tags (ESTs), the U.S. government's Patent and Trademark Office began to tighten the requirement for utility. Today, a DNA sequence alone is not patentable. It must be useful as a tool for research or as a novel and improved diagnostic test.

Despite the increasing stringency of patent requirements, problems still arise. A biotechnology company in the United States, for example, holds a patent on the *BRCA1* gene that includes any diagnostic tests based on the gene sequence. That company's tests, however, do not cover all mutations in the gene. A French physician working with a family that has a large deletion not seen anywhere else is challenging the patent, because her patients must pay a high licensing fee to the U.S. company that essentially "owns" the gene sequence to have their DNA screened. Bioethics: Choices for the Future on page 404 explores another case where a gene patent adversely affected families with inherited disease.

Patenting genes may become even more complex as genome information floods the Patent and Trademark Office. One problem is redundancy. For the same gene, it is possible to patent:

### Patenting Life and Genes

| | |
|---|---|
| **1790** | U.S. patent act is enacted. An invention must be new, useful, and not obvious to earn a patent. |
| **1873** | Louis Pasteur is awarded first patent on a life form for yeast used in industrial processes. |
| **1930** | New plant variants can be patented. |
| **1980** | First patent is awarded on a genetically engineered organism, a bacterium given four plasmids (DNA rings) that enable it to metabolize components of crude oil. The plasmids are naturally occurring, but do not all occur naturally in the manipulated bacteria. |
| **1988** | First patent is awarded for a transgenic organism, a mouse that manufactures human protein in its milk. Harvard University granted patent for "OncoMouse" transgenic for cancer. |
| **1992** | Biotechnology company is awarded a broad patent covering all forms of transgenic cotton. Groups concerned that this will limit the rights of subsistence farmers contest the patent several times. |
| **1996–1999** | Companies patent partial gene sequences and certain disease-causing genes as the basis for developing specific medical tests. |
| **2000** | With gene and genome discoveries pouring into the Patent and Trademark Office, requirements for showing utility of a DNA sequence are tightened. |

- Genomic DNA (the protein-encoding sequence as well as noncoding regions)
- expressed sequence tags
- cDNA (only the protein-encoding part of a gene)
- mutations
- SNPs

A researcher or company wanting to develop a tool or test based on a protein might infringe upon five different patents, based on essentially the same information. Now, as genetics begins to shift from a single-gene focus to analyzing expression patterns of suites of interacting genes, and even entire genomes, the Patent and Trademark Office will have to anticipate unprecedented change in the types of inventions submitted for protection from competition.

The origin of modern biotechnology dates to the 1970s, when researchers first began to ponder the potential uses and risks of mixing DNA from different species. The resulting recombinant DNA technology has led to transgenic technology and gene targeting. This chapter considers all three.

**Key Concepts**

Biotechnology is the use or modification of cells or biological molecules for a specific use, and includes recombinant DNA, transgenesis, and gene targeting. Patent law regarding DNA has evolved with the technology since the 1970s.

## 18.2 Recombinant DNA Technology

In February 1975, 140 molecular biologists convened at Asilomar, a seaside conference center on California's Monterey Peninsula, to discuss the safety and implications of a new type of experiment. Investigators had found a simple way to combine the genes of two species and were concerned about experiments requiring the use of a cancer-causing virus. Researchers were also concerned about where the field was headed.

The scientists discussed restricting the types of organisms used in recombinant DNA research and explored ways to prevent escape of a recombinant organism from the laboratory. The guidelines drawn up at Asilomar outlined measures of "physical containment," such as using specialized hoods and airflow systems that would keep the recombinant organisms inside the laboratory, and "biological containment," ways to weaken organisms so that they could not survive outside the laboratory.

A decade after the Asilomar meeting, many members of the original group reconvened at the meeting site to assess progress in the field. Nearly all agreed on two points: recombinant DNA technology was safer than expected, and the technology had spread to industry more swiftly and in more diverse ways than anyone had imagined. At a meeting 25 years after the event, attendees concluded that biotechnology had become so commercialized that the same open atmosphere would not be possible.

Recombinant DNA-based products have been slow to reach the marketplace because of the high cost of research and the long time it takes to develop any new drug. Today, several dozen such drugs are available, and many more are in the pipeline. Recombinant DNA research initially focused on direct gene products such as peptides and proteins with therapeutic actions, such as insulin, growth hormone, and clotting factors. However, the technology can target other biochemicals by affecting the genes that encode enzymes required to synthesize other substances, such as carbohydrates and lipids.

## Constructing Recombinant DNA Molecules

Manufacturing recombinant DNA molecules requires several components:

- enzymes that cut the donor and recipient DNA (**restriction enzymes**)
- DNA to carry the donor DNA (**cloning vectors**)
- recipient cells (bacteria or other cultured cells)

After inserting the donor DNA into the vectors, the procedure requires several steps:

- selecting cells that harbor DNA that, in turn, harbors foreign genes
- selecting those recombinant cells that contain the specific gene of interest
- stimulating expression of the foreign gene, so that its protein product can be collected and purified

The natural function of restriction enzymes is to protect bacteria by cutting and thereby inactivating the DNA of infecting viruses. Protective methyl ($CH_3$) groups shield the bacterium's own DNA from its restriction enzymes. Bacteria have hundreds of types of restriction enzymes. Each cuts DNA at a particular base sequence that typically consists of four, five, or six bases. These targets are symmetrical in a particular way—the recognized sequence reads the same, from the 5′ to 3′ direction, on both strands of the DNA. For example, the restriction enzyme EcoR1, shown in figure 18.3, cuts at the sequence GAATTC. The complementary sequence on the other strand is CTTAAG, which, read backwards, is GAATTC. (You can try this with other sequences to see that it rarely works this way!) This type of symmetry is called a palindrome in the English language, referring to a sequence of letters that reads the same in both directions, such as "Madam, I'm Adam." However, palindromic sequences in DNA reflect the sequences on two strands.

The cutting action of a restriction enzyme on double-stranded DNA creates single-stranded extensions of DNA called "sticky ends," because they are complementary to each other and therefore form hydrogen bonds as bases pair. The reason restriction enzymes work as molecular scissors in creating recombinant DNA molecules is that they cut at the same sequence in any DNA source. In other words, the same sticky ends result from the same restriction enzyme, whether the DNA is from a mockingbird or a maple. Any pieces of DNA bearing complementary sticky ends can join.

Another natural "tool" used in recombinant DNA technology is a cloning vector. This structure carries DNA from the cells of one species into the cells of another. The term *cloning* refers to the action of making many

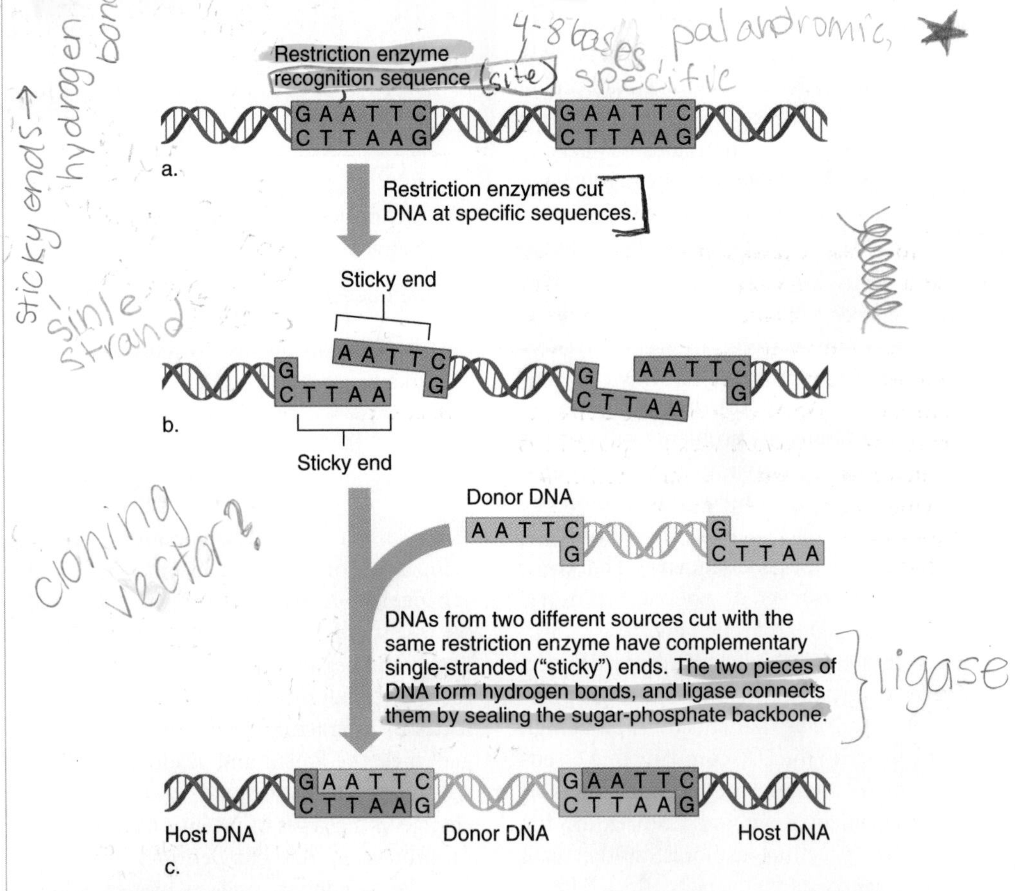


**figure 18.3**

**Recombining DNA.** A restriction enzyme makes "sticky ends" in DNA by cutting it at specific sequences. (*a*) The enzyme EcoR1 cuts the sequence GAATTC between G and the A. (*b*) This staggered cutting pattern produces "sticky ends" of sequence AATT. The ends attract through complementary base pairing. (*c*) DNA from two sources is cut with the same restriction enzyme. Pieces join, forming recombinant DNA molecules.

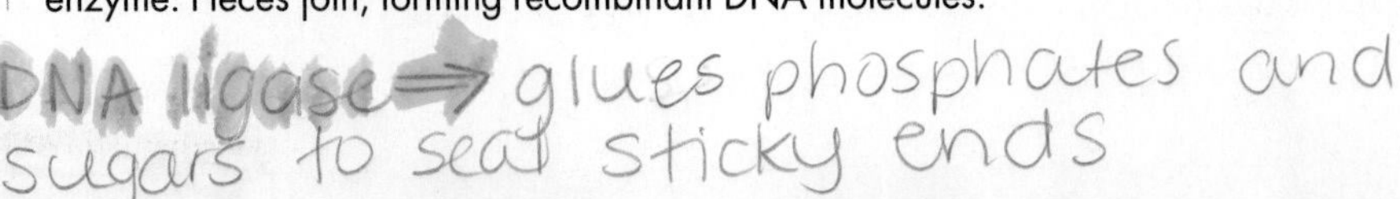

copies of a selected DNA sequence. (The use of the word in molecular biology predated its use as applied to animals by many years.)

A vector can be any piece of DNA that an organism's DNA can attach to for transfer into the cell of another organism. A commonly used type of vector is a **plasmid,** which is a small circle of double-stranded DNA that is naturally in some bacteria, yeasts, plant cells, and other types of organisms (figure 18.4).

Viruses that infect bacteria, called bacteriophages, provide another type of vector. Bacteriophages are manipulated so that they transport genetic material but do not cause disease. Disabled retroviruses are used as vectors too, as are artificially constructed chromosomes from bacteria and yeast. Bacterial artificial chromosomes, for example, were used by the international consortium of researchers that sequenced the human genome. Each "BAC" held about 150,000 DNA bases. A researcher chooses a cloning vector according to its capacity—that is, the desired gene must be short enough to insert into the vector. Gene size is typically measured in kilobases (kb), which are thousands of bases. Table 18.1 lists the capacities of a few types of cloning vectors.

To create a recombinant DNA molecule, a restriction enzyme cuts DNA from a donor cell at sequences known to bracket the gene of interest (figure 18.5). The enzyme leaves single-stranded ends dangling from the cut DNA, each bearing a characteristic base sequence. Next, a plasmid is isolated and cut with the same restriction enzyme used to cut the donor DNA. Because the same restriction enzyme cuts both the donor DNA and the plasmid DNA,

**table 18.1**

**Cloning Vectors**

| Vector | Size of Insert Accepted (kb) |
|---|---|
| Plasmid | up to 15 |
| Bacteriophage | up to 90 |
| Bacterial artificial chromosome (BAC) | 100–500 |
| Yeast artificial chromosome (YAC) | 250–1,000 |

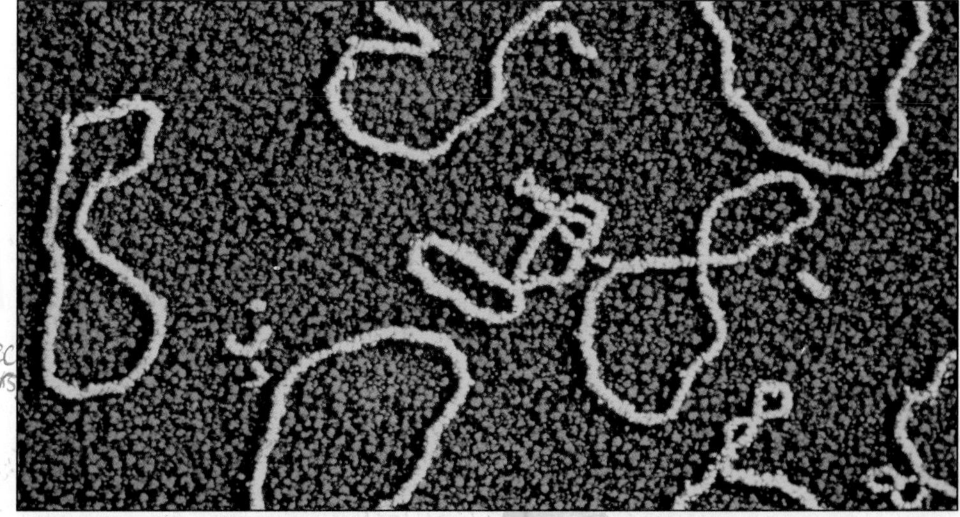

**figure 18.4**

**Plasmids.** Plasmids are small circles of DNA found naturally in the cells of some organisms. A plasmid can replicate itself as well as any other DNA inserted into it. For this reason, plasmids make excellent cloning vectors—structures that carry DNA from the cells of one species into the cells of another.

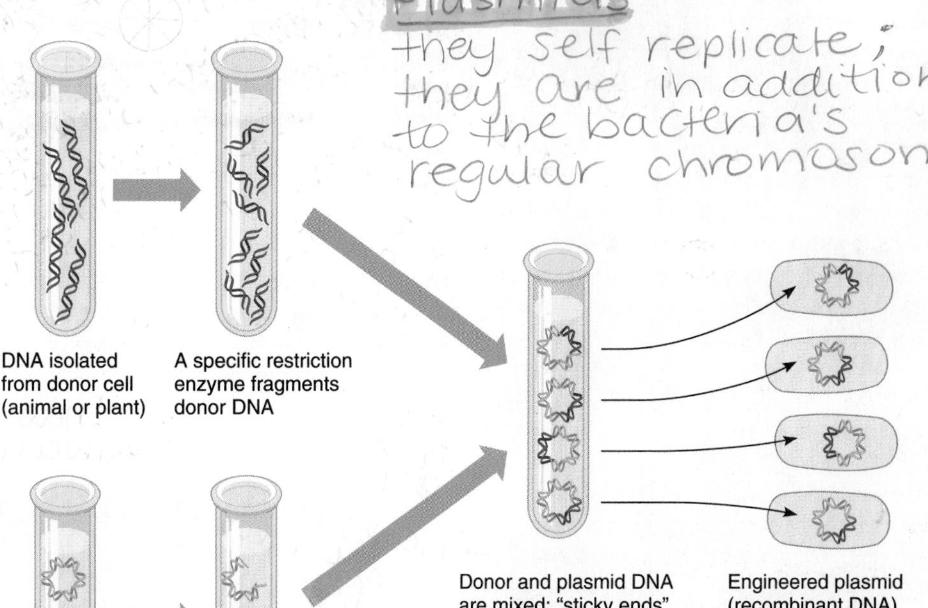
DNA isolated from donor cell (animal or plant)
A specific restriction enzyme fragments donor DNA
Plasmid isolated from bacterium
The same restriction enzyme that fragmented donor DNA is also used to open plasmid DNA
Donor and plasmid DNA are mixed; "sticky ends" of donor DNA form hydrogen bonds with sticky ends of plasmid DNA fragment; recombinant molecule is sealed with another specific enzyme
Engineered plasmid (recombinant DNA) is introduced into a bacterium, which reproduces and clones the gene from the donor cell that was spliced into the plasmid

**figure 18.5**

**Recombinant DNA.** To construct a recombinant DNA molecule, DNA isolated from a donor cell and a plasmid are cut with the same restriction enzyme and mixed. Some of the sticky ends from the donor DNA hydrogen bond with the sticky ends of the plasmid DNA, forming recombinant DNA molecules. When such an engineered plasmid is introduced into a bacterium, it is mass produced as the bacterium divides.

the same complementary single-stranded base sequences extend from the cut ends of each. When the cut plasmid and the donor DNA are mixed, the single-stranded "sticky ends" of some plasmids base pair with the sticky ends of the donor DNA. The result is a recombinant DNA molecule, such as a plasmid carrying the human insulin gene. The plasmid and its stowaway human gene can now be transferred into a cell from an individual of another species.

## Selecting Recombinant DNA Molecules

Much of the effort in recombinant DNA technology entails identifying and separating cells that contain the gene of interest, once the insertion of foreign DNA into the vector is accomplished. Three types of recipient cells can result:

1. Cells that lack plasmids
2. Cells that contain plasmids that do not contain a foreign gene
3. Cells that contain plasmids that have picked up a foreign gene

The procedure is cleverly set up so that recombinant plasmids are easily distinguished from plasmids that do not take up foreign DNA. In one strategy, a section of the foreign gene in the plasmid is detected by a color test. The plasmid includes a gene conferring resistance to an antibiotic drug, as well as a gene called *lac Z* that produces blue colonies in the presence of a compound called X-Gal. When the foreign DNA inserts into the plasmid, it does so at a restriction site that disables the *lac Z* gene. When bacteria are grown in the presence of the antibiotic, only those cells that have incorporated plasmids can grow, because the plasmids include the antibiotic resistance gene. Secondly, bacterial cells containing plasmids that have taken up the foreign DNA do *not* turn the cell blue when X-Gal is put into the medium. In contrast, cells with empty plasmids—and, therefore, a functional *lac Z* gene—turn blue.

When cells containing the recombinant plasmid divide, so does the plasmid. Within hours, the original cell gives rise to many cells harboring the recombinant plasmid. The enzymes, ribosomes, energy molecules, and factors necessary for protein synthesis transcribe and translate the plasmid DNA and its stowaway foreign gene, producing the desired protein.

## Isolating the Gene of Interest

The recombinant DNA process usually begins by cutting all of the DNA of the donor cell, termed genomic DNA. Researchers construct collections of recombinant bacteria (or other single cells) that harbor pieces of a genome. Such a collection is called a **genomic library.** For each specific application, such as using a human protein as a drug, a particular piece of DNA must be identified and isolated from a genomic library. There are several ways to do this "needle in a haystack" type of search.

If the sequence of the gene in question is known, a synthetic piece of DNA can be made and linked to a label, such as a radioactive or fluorescent molecule. This labeled fragment of the gene is called a **DNA probe.** It emits a signal when it binds to its complement in a bacterial cell that contains a recombinant plasmid. DNA probes can also be made using similar genes from other species. Many genes from mice and rats, for example, are so similar in sequence to those from humans that the genes can be used as probes to find their human counterparts. Using such a probe is a little like mistakenly using *hipropotamus* to search for *hippopotamus* on the Internet. You'd probably still come up with a hippo.

Genomic libraries can be enormous. The 3 billion base pairs of the human genome, for example, can occupy millions of plasmids, each harboring a small piece of DNA from one of our chromosomes, with many overlaps. A genomic library contains too much information for a researcher seeking a particular protein-encoding gene—it also contains introns, repeated sequences, the genes that encode rRNAs and tRNAs, and many repeated sequences. Another type of library, called a complementary DNA, or **cDNA library,** represents only protein-encoding genes. The difference between using a genomic library and a cDNA library is a little like a person looking for a particular book on yeast genetics searching in a huge university library that includes all possible subjects, versus a science library with more restricted choices.

A cDNA library is often made from the mRNAs in a differentiated cell, which reflect the proteins manufactured there. For example, a reticulocyte (an immature red blood cell, before the nucleus is extruded) is packed with the mRNAs that encode globin proteins, whereas a fibroblast has many mRNAs that encode the connective tissue proteins collagen and elastin. Researchers first extract the mRNAs from cells. These RNAs are used to construct complementary or "c" DNA strands using reverse transcriptase, DNA nucleotide triphosphates, and DNA polymerase (figure 18.6). (Reverse transcriptase synthesizes DNA complementary to RNA.) DNA

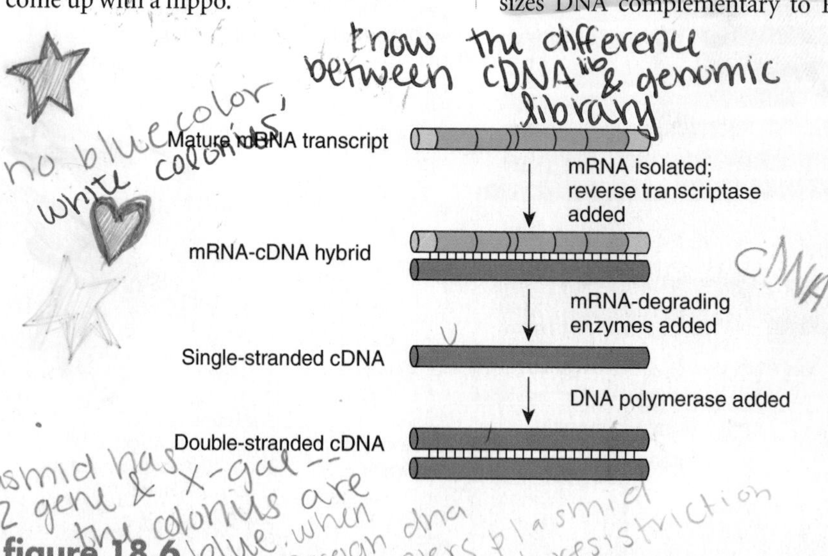


**figure 18.6**

**Copying DNA from RNA.** Researchers make cDNA from mRNA using reverse transcriptase, an enzyme from a retrovirus. A cDNA version of a gene includes the codons for a mature mRNA, but not sequences corresponding to promoters and introns. Labeled cDNAs are used as probes to locate genes in genomic libraries.

polymerase and bases then can synthesize the complementary strand to the single-stranded cDNA to form a double-stranded DNA. Different cell types yield different cDNA collections, or libraries, that reflect which genes are expressed.

A specific cDNA can be taken from a cDNA library and used to isolate the original gene of interest from the genomic library. If the goal is to harness the gene and eventually collect its protein product, then the genomic version, and not the cDNA version, is required, because it includes control regions such as promoters. Once a gene of interest is transferred to a cell where it can be transcribed into mRNA and that RNA translated, the protein is collected. Such cells are typically grown in devices called bioreactors, with nutrients sent in and wastes removed. A researcher collects the desired product from the medium the cells are growing in.

## Applications of Recombinant DNA Technology

Recombinant DNA technology provides a way to isolate individual genes from complex organisms and observe their functions on the molecular level. The technology was also developed to mass produce protein-based drugs. The first drug produced from recombinant DNA technology was human insulin.

Before 1982, 2 million people with type I diabetes mellitus in the United States got the insulin that they had to inject daily from pancreases removed from cattle in slaughterhouses. Cattle insulin is so similar to the human peptide, different in only 2 of its 51 amino acids, that most people with diabetes can use it. However, about 1 in 20 patients is allergic to cow insulin because of the slight chemical difference. Until recombinant DNA technology was possible, the allergic patients had to use expensive combinations of insulin from a variety of other animals or from human cadavers. Today, *E. coli* cells that manufacture human insulin grow in vats at a major pharmaceutical company. A person with diabetes can now purchase "Humulin" at a local drugstore.

Human growth hormone is another drug produced with recombinant DNA technology. It is used to treat pituitary dwarfism in children; these children are very short due to absence or deficiency of growth hormone. The hormone was formerly collected from the pituitary glands of cadavers, which can transmit diseases of the central nervous system, including the prion disorder Creutzfeldt-Jakob disease. In contrast, the hormone produced in genetically modified *E. coli* is pure and plentiful. Still, the drug undergoes many tests to assure its potency, purity, and biological activity before it is marketed. Table 18.2 lists some other drugs produced using recombinant DNA technology.

**table 18.2**

**Drugs Produced Using Recombinant DNA Technology**

| Drug | Use |
|---|---|
| Atrial natriuretic peptide | Dilates blood vessels, promotes urination |
| Colony stimulating factors | Help restore bone marrow after marrow transplant; restore blood cells following cancer chemotherapy |
| Deoxyribonuclease (DNase) | Thins pus in lungs of people with cystic fibrosis |
| Epidermal growth factor | Accelerates healing of wounds and burns; treats gastric ulcers |
| Erythropoietin (EPO) | Stimulates production of red blood cells in cancer patients |
| Factor VIII | Promotes blood clotting in treatment of hemophilia |
| Fertility hormones (follicle stimulating hormone, luteinizing hormone, human chorionic gonadotropin) | Treats infertility |
| Glucocerebrosidase | Treats Gaucher disease |
| Human growth hormone | Promotes growth of muscle and bone in people with very short stature due to hormone deficiency |
| Insulin | Allows cells to take up glucose in treatment of type I diabetes mellitus |
| Interferons | |
| Alpha | Treats genital warts, hairy cell leukemia, hepatitis C and B, Kaposi sarcoma |
| Beta | Treats multiple sclerosis |
| Gamma | Treats chronic granulomatous disease (a blood disorder) |
| Interleukin-2 | Treats kidney cancer |
| Lung surfactant protein | Helps alveoli in lungs to inflate in infants with respiratory distress syndrome |
| Renin inhibitor | Lowers blood pressure |
| Somatostatin | Decreases growth in muscle and bone in pituitary giants |
| Superoxide dismutase | Prevents further damage to heart muscle after heart attack |
| Tissue plasminogen activator | Dissolves blood clots in treatment of heart attacks, stroke, and pulmonary embolism |

Products of recombinant DNA technology are also used in the food industry. The enzyme rennin, for example, normally produced in calves' stomachs, is used in cheese making. The gene that encodes the enzyme is inserted into plasmids and transferred to bacteria, which are mass cultured to produce large quantities of pure rennin.

In yet another application, recombinant DNA technology is providing a new source of indigo, the dye used to make blue jeans blue (figure 18.7). The dye originally came from mollusks and fermented

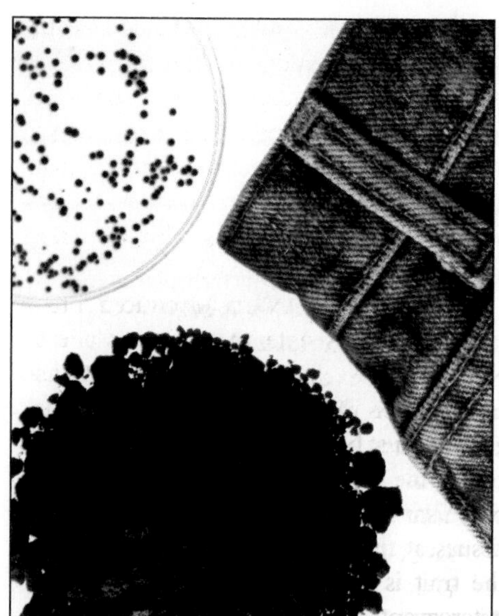

**figure 18.7**

**Genes for jeans.** *E. coli* are genetically modified to produce indigo, the dye used in denim.

leaves of the European woad plant or Asian indigo plant. The 1883 discovery of indigo's chemical structure led to the invention of a synthetic process to produce the dye using coal-tar. That method has dominated the industry, but it releases toxic by-products. In 1983, microbiologists discovered that *E. coli,* with a little help, can produce indigo. The bacterium converts glucose to the amino acid tryptophan, which then forms indole, a precursor to indigo. By learning the steps and offshoots of this biochemical pathway, researchers at a biotechnology company altered bacteria to suppress the alternative pathways for glucose, allowing the cells to synthesize much more tryptophan than they normally would. When the *E. coli* cells incorporate genes from another bacterial species, they extend the biochemical pathway all the way to producing indigo. The result: common bacteria that manufacture the blue dye of denim jeans from glucose, a simple sugar.

Products developed using recombinant DNA technology may be novel and interesting, but they still must compete in the marketplace with conventional products. This is especially apparent in the area of pharmaceuticals. Recombinant DNA-based drugs are expensive and the research to develop a new drug is very complex and time-consuming. Deciding whether a drug produced this way is preferable to an existing, similar drug is often a matter of economics, marketing, and plain common sense. For example, interferon beta-1b helps some people with a certain type of multiple sclerosis, but it costs more than $20,000 per year per patient. British researchers calculated what it would cost to treat the nation's patients who would benefit from the drug, and then determined how else the funds could be spent. They concluded that more people would be served if the money were spent on improved supportive care for many rather than on a costly new treatment for a few.

A classic case of determining the value of a recombinant DNA-derived drug is tissue plasminogen activator (tPA), a clot-busting drug developed in the mid-1980s. If injected within four hours of a heart attack, tPA dramatically limits damage to the heart muscle by restoring blood flow. Human tPA is also used to treat stroke, at a price of $2,200 a shot. However, several studies revealed that an older drug extracted from bacteria by conventional methods, streptokinase, is nearly as effective at a cost of $300 per injection. The recombinant drug is still very valuable for patients who have already had streptokinase and could have an allergic reaction to it if they were to use it again. Bioethics: Choices for the Future on page 384 considers another drug derived from recombinant DNA technology, erythropoietin (EPO).

**KEY CONCEPTS**

In recombinant DNA technology, a cell receives a cloning vector that contains foreign DNA encoding a protein of interest. The universality of the genetic code and restriction enzymes, which cut DNA at specific sequences and create sticky ends, make recombinant DNA technology possible. • After a cell containing the gene of interest is selected, it transcribes and translates the foreign gene. • The products of recombinant DNA technology are used in health care, food technology, agriculture, fabrics, and forensics.

## 18.3 Transgenic Organisms

A transgenic animal develops from a genetically altered gamete or fertilized ovum. A transgenic plant can be derived from these sources or from somatic cells. Different vectors and gene transfer techniques are sometimes used in plants because their cell walls, which are not present in animal cells, are difficult to penetrate. Some manipulations are done on plant cells that have had their cell walls removed, called **protoplasts.**

Transgenic technology permits the rapid introduction of new traits. For example, a gene that confers an agriculturally useful characteristic—such as the ability to withstand a particular pesticide—is isolated from one species and inserted into a vector; then the recombinant vector is placed into protoplasts. A whole plant regenerated from the genetically altered cell has the gene for the transferred trait in all of its cells. It is typically tested in the laboratory to see if the gene is expressed. Then the transgenic plant is grown in a greenhouse, and finally in experimental fields to see if the desired trait is present, and to assess effects of the transgenic plant on the ecosystem and beyond.

### Delivering DNA

Frequently used plant vectors include the ***Ti* plasmid** (for "tumor-inducing"), which occurs naturally in the bacterium *Agrobacterium tumefaciens,* and viruses found in plant cells (figure 18.8). Because a *Ti* plasmid normally causes a tumorlike growth to form, the genes controlling this process are removed. For example, a gene from the bacterium *Bacillus thuringiensis* specifies a protein that destroys the stomach linings of certain insect larvae. The "*bt*" gene is introduced into corn cells, and the cells are regenerated into plants that produce their own insecticide. More than two-thirds of the corn grown in the United States and more than half of the soybean crop are transgenic for the *bt* insecticide gene, whose protein product organic farmers have been using for years (see figure 20.9). Consumers have been eating *bt* corn and soybeans since 1996, with no reported ill effects.

Researchers use several methods to insert DNA into animal cells (table 18.3 and figure 18.9). Chemicals such as polyethylene

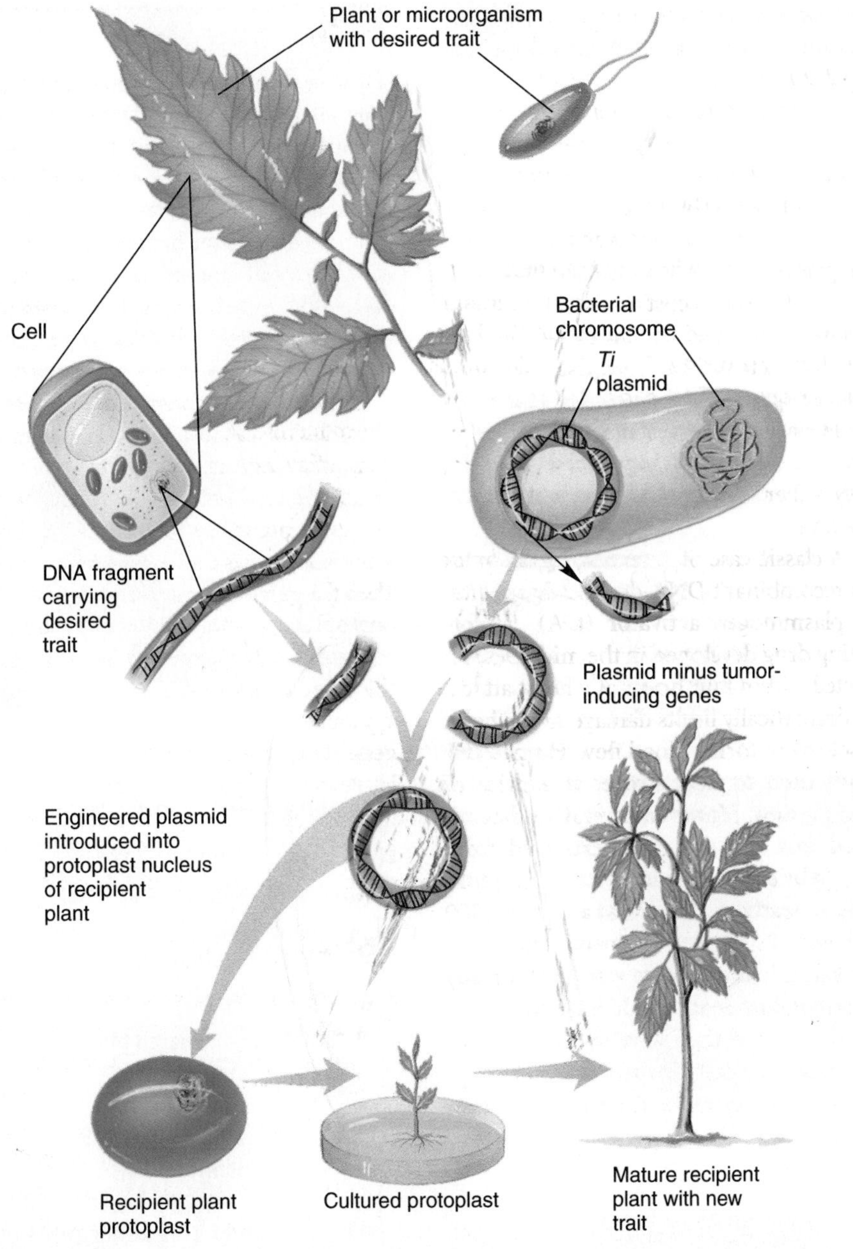


**figure 18.8**

**Producing a transgenic plant.** A fragment of DNA carrying the desired gene—conferring resistance to an herbicide, for example—is isolated from its natural source and spliced into a *Ti* plasmid with the tumor-inducing genes removed. The plasmid incorporating the foreign DNA is then allowed to invade a cell of the recipient plant, where it enters the nucleus and integrates into the plant's DNA. Finally, by means of cell culture, the cell is regenerated into a mature, transgenic plant that expresses the desired trait and passes it on to its progeny. A breeding step may be necessary to obtain plants homozygous for a recessive trait.

glycol and calcium phosphate are used to open transient holes in cell membranes, allowing DNA to enter. **Liposomes** are fatty bubbles that can carry DNA into cells as cell membranes envelop them. In **electroporation,** a brief jolt of electricity opens transient holes in cell membranes that may permit entry of foreign DNA. DNA is also injected into cells using microscopic needles (figure 18.10). This is called **microinjection.**

Another way to introduce DNA into cells is **particle bombardment.** A gunlike device shoots tiny metal particles, usually gold or tungsten, coated with foreign DNA. When aimed at target cells, usually in plants, some of the projectiles enter. For example, gene guns shoot dividing cells in soybean seeds with an *E. coli* gene that stains cells expressing it a vibrant blue, allowing detection of the gene transfer.

Once foreign DNA is introduced into a target cell, it must enter the nucleus, replicate along with the cell's own DNA, and be transmitted when the cell divides. Finally, an organism must be regenerated from the altered cell. If the trait is dominant, the transgenic organism must express it in the appropriate tissues at the right time in development. If the trait is recessive, then crosses between heterozygotes may be necessary to yield homozygotes that express the trait. Then the organisms must pass the characteristic on to the next generation. If cloning at the whole-animal level is perfected, it could be used to make additional homozygous individuals without repeating the complex breeding steps. Figure 18.11 illustrates the steps for an application of transgenesis called "pharming"—the secretion of human proteins that are useful as drugs into easily collected secretions, such as milk.

## Transgenic Pharming from Milk and Semen

Nancy is a sheep that produces human alpha-1-antitrypsin (AAT) in her milk. AAT is a glycoprotein normally present in blood serum that helps microscopic air sacs (alveoli) in the lungs inflate and function properly. Lack of AAT greatly increases the risk of developing emphysema. The alveoli coalesce, impairing breathing. Donated blood doesn't yield enough AAT to treat the 20,000 people who have this ailment. But a herd of transgenic sheep, genetically altered to secrete human AAT, can supply thousands of kilograms of the valuable substance in a single milking session.

The world's first transgenic "pharming" herds were developed at the Roslin Institute in Scotland, home of Dolly, the famed sheep clone. The transgenic sheep preceded her by several years, but they failed to capture much media attention.

To create a transgenic sheep, researchers give a ewe a drug to make her

**table 18.3**

**Gene Transfer Techniques**

| Approach | How It Works |
|---|---|
| Virus | A human gene is inserted into a virus, which infects a human cell, where it is expressed. |
| Retrovirus | An RNA virus carrying an RNA version of a human gene infects a somatic cell. The gene is reverse transcribed to DNA and inserts into a human chromosome. Here, it may produce a missing or abnormal protein. |
| Liposome transfer | A fatty bubble called a liposome carries a gene into a somatic cell. Here, the delivered gene may replace an abnormal one. |
| Chemical | Calcium phosphate or dextran sulfate opens transient holes in a cell membrane, admitting replacement DNA. |
| Electroporation | An electrical current opens transient holes in a cell's membrane, admitting replacement DNA. |
| Microinjection | A tiny needle injects DNA into a cell lacking that DNA sequence. |
| Particle bombardment | Metal pellets coated with DNA are shot with explosive force or air pressure into recipient cells. |

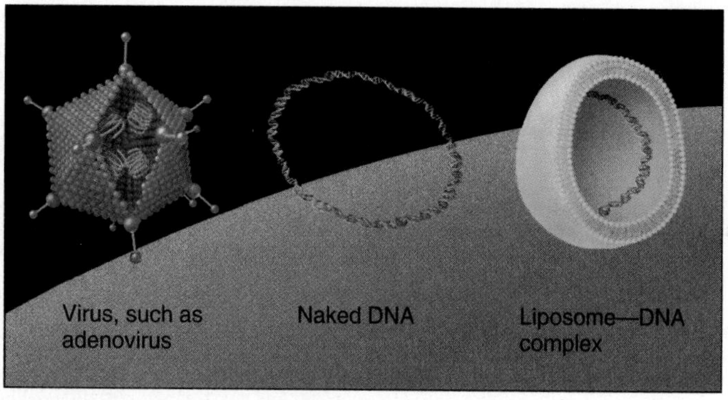


**figure 18.9**

**Adding DNA to a cell.** DNA can be sent into cells in viruses, alone (naked), or in liposomes.

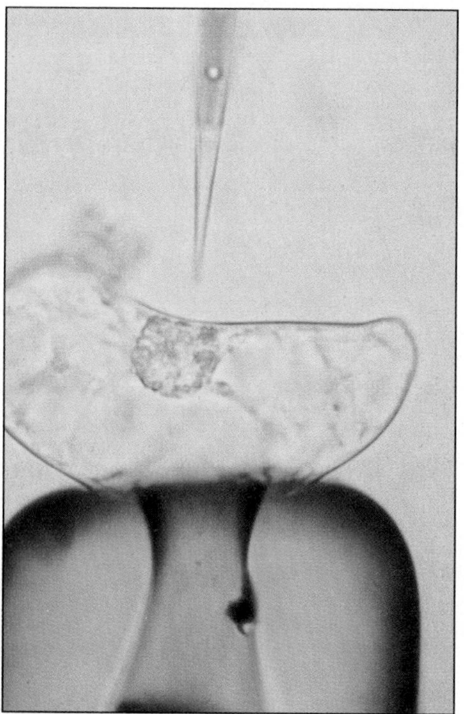

**figure 18.10**

**Injecting DNA.** One way to move "naked" DNA into a plant cell nucleus is by injecting it with a microscopic glass needle—an approach called microinjection. The DNA is "naked" because it is isolated from the cell and free of protein.

produce several mature oocytes at once, then artificially inseminate her with ram sperm. Fertilized ova are washed out of her body and microinjected with copies of the human AAT gene attached to a sheep promoter (gene activating) sequence. The manipulated fertilized ova are then implanted into sheep surrogate mothers. Sheep that carry the human transgene in their germline cells (sperm or eggs) are then mated to derive offspring that are homozygous for the human gene.

Transgenic pharming works, but it is difficult to carry out. A major limitation is that DNA typically inserts randomly into the host genome. Very few transgenics insert and are expressed in the tissue and at the point in development that the researcher intends. In the experiment that led to Nancy's birth, 152 surrogates implanted with the human AAT gene gave birth to 112 lambs. Of those, only one male and four females had the human transgene. Only one, a female, gave birth to a transgenic offspring. This was Nancy, whose milk contains pure, active, human alpha-1-antitrypsin—at a whopping yield of 35 grams per liter! Figure 18.11 outlines the steps in a similar process—transgenic pharming of a goat that secretes human tPA.

Ethel, the matriarch of a special flock in Scotland, is another pioneering transgenic sheep. She has a human gene for factor VIII, the protein required to clot the blood of people with hemophilia A, such as Don Miller, who writes of his life with hemophilia in chapter 1. Another piece of DNA activates the gene in the sheep's milk, and the valuable protein can then be separated from the milk. A herd of transgenic sheep can provide enough clotting factor to supply the world's hemophilia A patients. Table 18.4 lists some other products of transgenic pharming.

A new variation on transgenic technology harvests valuable proteins from semen. The ideal animals for this application are boars, which produce up to a pint of semen at a time. Unlike milk, which is a seasonal body fluid, boar semen can be obtained anytime. The technique was perfected in mice first, with an unexpected side effect. The human growth hormone gene was inserted into the mouse genome, linked to a

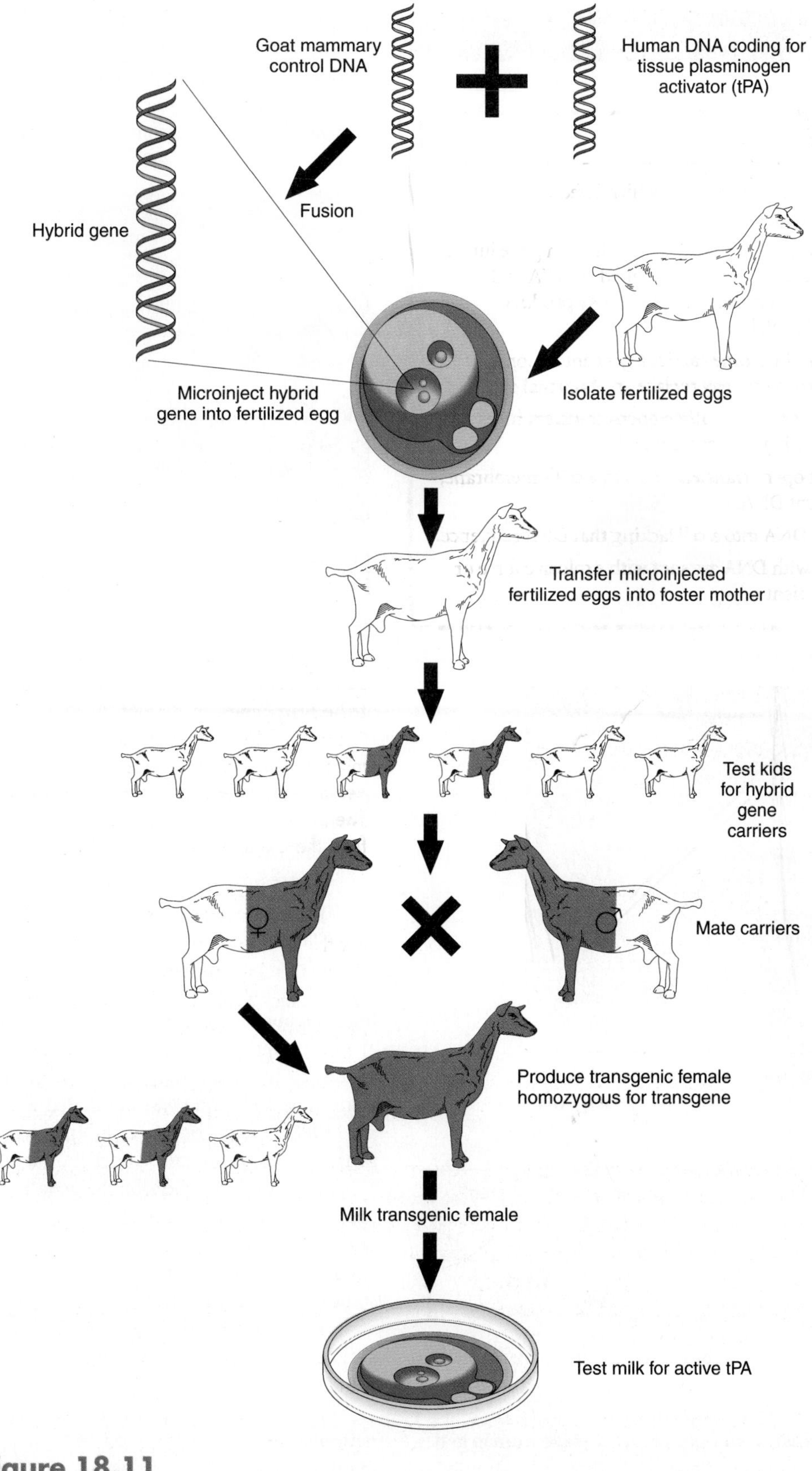

Goat mammary control DNA
Human DNA coding for tissue plasminogen activator (tPA)
Fusion
Hybrid gene
Microinject hybrid gene into fertilized egg
Isolate fertilized eggs
Transfer microinjected fertilized eggs into foster mother
Test kids for hybrid gene carriers
Mate carriers
Produce transgenic female homozygous for transgene
Milk transgenic female
Test milk for active tPA

**figure 18.11**

**Drugs from transgenic goats.** Transgenic technology enables goats to secrete human drugs, such as the clot-buster tPA, in their milk. Genetic manipulations are followed by breeding schemes that result in females homozygous for this recessive gene.

promoter that targeted its synthesis to the prostate gland, which contributes secretions to semen. The experiment actually worked too well. The male mice produced human growth hormone in their semen, but their kidneys unexpectedly secreted the hormone, too, into their bloodstreams. Flooded with human growth hormone, the genetically altered male mice grew so gigantic—three times normal mouse size—that they could not mate. For this reason, subsequent experiments to test seminal transgene pharming have not used growth hormone! No products are yet available from this twist on transgenic technology. Figure 18.12 is an example of how the genes of several organisms can be harnessed to produce a drug.

Transgenic animals are also valuable research tools as models of human disease. Mice are the most popular, because their genomes are remarkably similar to our own, and they are easy to raise, with a short lifespan. The Jackson Laboratory in Bar Harbor, Maine, has been supplying transgenic mice to researchers since 1993.

One strain of transgenic mouse harbors the human *BRCA1* gene that causes breast cancer, enabling researchers to study the very early development of the disease—which is not possible in humans. A transgenic mouse model for Huntington disease revealed the basis of the uncontrollable movements and personality changes of the disorder. Recall from chapter 11 that HD is a triplet repeat disorder that extends the encoded protein, huntingtin. A transgenic HD mouse, developed by Gillian Bates at Guy's Hospital in London, has symptoms very similar to those of human HD sufferers, including the movements and decreased brain size. Bates's team saw unusual clumps in parts of the mouse brains, and used a monoclonal antibody (MAb) to identify the material as human huntingtin. This experiment gave researchers at Columbia University in New York the idea to use the MAb to identify mysterious protein clumps they had discovered in the brain cells of people who had died of HD back in 1979. The clumps in the human cells turned out to be huntingtin protein.

Mouse models with human transgenes can provide practical information, too. Mice that have human growth hormone genes develop large bones early in life, but lose strength when they pass puberty. Physicians are tracking these symptoms in

## table 18.4

**Transgenic Pharming Products**

| Host | Product | Potential Use |
|---|---|---|
| Cows | Lactoferrin | Added to infant formula to bind iron and prevent bacterial infection |
| Goats | tPA | Breaks up blood clots |
| Pigs | Hemoglobin | Acts as a blood substitute |
| Rabbit | Erythropoietin | Treats anemia from dialysis |
| Rat | Human growth hormone | Treats pituitary dwarfism |
| Sheep | Alpha-1-antitrypsin | Treats hereditary emphysema |

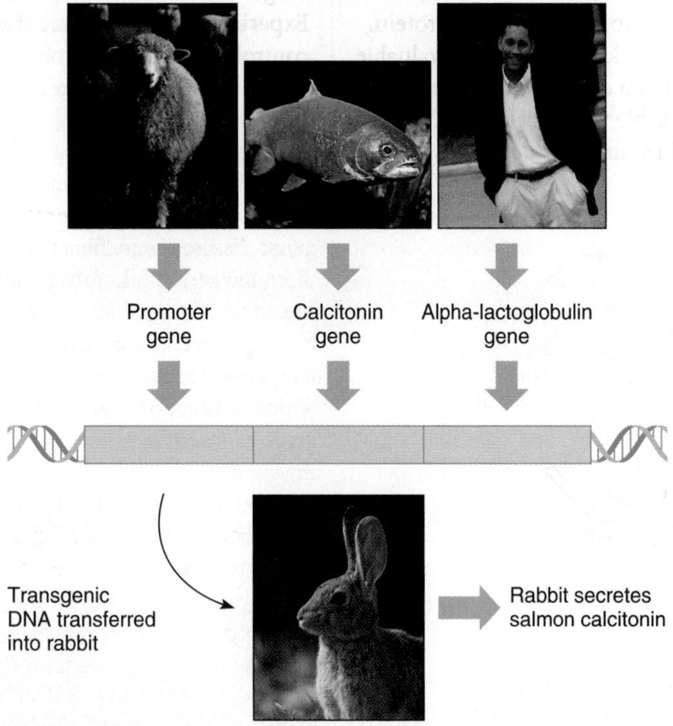


**figure 18.12**

**Combining genes.** Transgenic technology can combine gene segments from different organisms to produce a gene that does exactly what researchers want it to do. DNA regions from a sheep, a salmon, and a human are combined to make a transgenic rabbit that produces calcitonin for treating bone disorders.

mice to better follow their young human patients who received recombinant human growth hormone to treat short stature. Not enough time has passed to know all the effects of the treatment on people; the mice are providing clues.

Not all transgenic mice perfectly model human inherited disease. Differences in anatomy and physiology, developmental rate, age at breeding, and lifespan can lead to differences in gene expression and make it difficult to extrapolate from mice to humans. For example, mice given human genes for cystic fibrosis or Duchenne muscular dystrophy do not exhibit the same spectrum of symptoms as do affected children. Mice transgenic for human retinoblastoma genes develop tumors, as humans do, but in different parts of the body. Mice also develop cancer, and recover from it, much more easily than do humans. Because a mouse's lifespan is short, diseases that start in adulthood in humans might not have counterparts in mice.

**Key Concepts**

A transgenic organism is created by introducing foreign DNA into a single germline cell. The foreign DNA is then transmitted to each cell as the organism develops. • Animals are transgenically altered to "pharm" useful human proteins in their milk or semen and to provide models of human disease.

## 18.4 Gene Targeting

Transgenic technology is not very precise because it does not direct the introduced DNA to a particular chromosomal locus. The entry of the transgene can disrupt another gene's function, or the transgene can come under another gene's control sequence. Even if a transgene does insert into a chromosome and is expressed, the host's version of the same gene may overshadow the transgene's effect.

A more precise method of genetic modification is gene targeting, in which the introduced gene exchanges places with its counterpart on a host cell's chromosome. Gene targeting uses a natural process called **homologous recombination.** In this process, a DNA sequence locates and displaces a similar or identical sequence in a chromosome. The technique was developed in the late 1980s by introducing an inactivated gene into a mouse cell's nucleus, thereby "knocking out" function of the gene it replaced. By observing the effects of a gene's lack of function, researchers could deduce the gene's normal function as well as understand more about inherited disease. In a variation on gene targeting, genes that have an altered function are swapped in. These genes are called "knockins."

Gene targeting in mammals entails genetic alteration plus complex developmental manipulations because it does not work on fertilized ova. Instead, it is done in embryonic stem (ES) cells. The intervention must occur later in development. Most gene targeting is done on mice because their embryo cells are easiest to manipulate. (It is also possible to use the technique in the rhesus macaque, a type of monkey.) Gene targeting is theoretically possible in humans, now that ES cell lines exist.

The first gene targeting attempts followed transmission of coat color genes, so that results would be easy to see. In one scheme, the knockout mice were white. To begin, researchers used electroporation or microinjection to deliver an inactivated pigment gene into a pigmented mouse embryonic stem (ES) cell. Next, the engineered ES cells were injected into early embryos from colored mice. The embryos were implanted into surrogate mothers, where they continued development into individuals that had some cells bearing the targeted gene. The newborn mice were chimeras (mosaics), with patches of tissue whose cells contained the introduced inactivated pigment gene. When the chimeric mice were mated to each other, some of the pups developed from a sperm and egg that each had the knocked-out gene. These rare homozygotes were easily distinguished because they were white, while their siblings were pigmented (figure 18.13).

## Gene-Targeted Mice as Models

Gene targeting is very useful in developing animal models of human genetic diseases. First, researchers identify the animal version of a human disease-causing allele. Then they transfer a corresponding human mutant allele to mouse ES cells and follow the steps previously outlined to breed a homozygous animal. The mouse is a knockout if gene function is gone, and a knockin if it makes some version of the human protein.

Knockout or knockin mice are valuable models of human disease because they provide a controllable test population. Consider severe combined immune deficiency (SCID) due to adenosine deaminase (ADA) deficiency, an immune disorder (see figure 19.18). The phenotypes in humans with this illness differ, depending upon a person's environment. A child raised in a protective bubble might be relatively healthy; a child out among others would suffer frequent infections. This disorder is so rare, and people who have it so ill, that it is difficult to learn much about the condition. This is where knockout SCID mice come in.

Mice with knocked-out ADA genes can be raised under the same environmental conditions and bred to be genetically identical. Knockout SCID mice can then be exposed to particular infectious agents or undergo treatments such as gene therapies. Experiments using them, then, are highly controlled compared to observations of human patients. Transgenic mice are good models for these reasons, too, but the genetic alteration is not as precise.

Figure 18.14 compares a knockout mouse representing another human genetic disease, neurofibromatosis type I, to a heterozygous sibling. In people, this disorder causes benign tumors beneath the skin and distinctive pigmented areas of skin. It is autosomal dominant, and affected individuals have one mutant allele and one

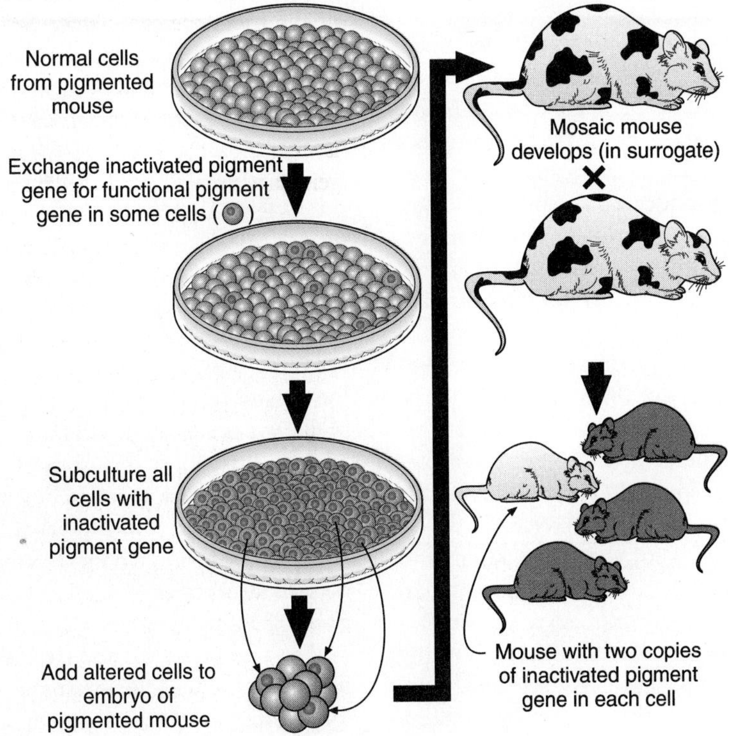

Normal cells from pigmented mouse
Exchange inactivated pigment gene for functional pigment gene in some cells ( )
Subculture all cells with inactivated pigment gene
Add altered cells to embryo of pigmented mouse
Mosaic mouse develops (in surrogate)
×
Mouse with two copies of inactivated pigment gene in each cell

**figure 18.13**

**Gene targeting.** Inactivated pigment-encoding genes are inserted into mouse embryonic stem (ES) cells, where they trade places with functional pigment-encoding alleles. The engineered ES cells are cultured and injected into early mouse embryos. Mosaic mice develop, with some cells heterozygous for the inactivated allele. These mice are bred to each other, and, if all goes well, yield some offspring homozygous for the knocked-out allele. It's easy to tell which mouse this is in the pigment gene example, but gene targeting is particularly valuable for revealing unknown gene functions by inactivating targeted alleles.

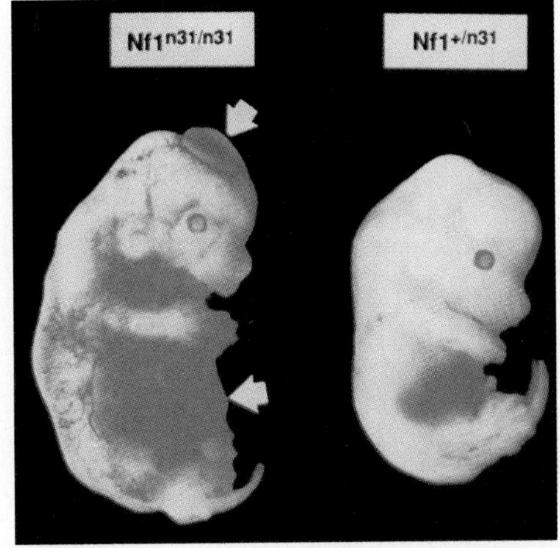

$Nf1^{n31/n31}$
$Nf1^{+/n31}$

**figure 18.14**

**Mouse embryos.** Gene targeting can produce mice that are homozygous for disease-causing dominant alleles, such as those that cause neurofibromatosis type I (mouse embryo on the left). These effects are difficult to study in humans because they disrupt development very early. The arrows point to tumors.

*Bioethics: Choices for the Future*

# The Ethics of Using a Recombinant Drug: EPO

EPO is a hormone produced in the kidneys that consists of a 165-amino-acid protein portion plus four carbohydrate chains. When the oxygen level in the blood dips too low, cells in the kidneys produce EPO, which travels to the bone marrow and binds to receptors on cells that give rise to red blood cell precursors. Soon, more red blood cells enter the circulation, and they carry more oxygen to the tissues (figure 1).

The value of EPO as a drug became evident in the aftermath of the invention of hemodialysis to treat kidney failure in 1961. This otherwise highly successful treatment also causes severe anemia, because dialysis removes EPO from the blood. To counteract dialysis-induced anemia, it was necessary to boost the patients' EPO levels. In 1970, the U.S. government sought ways to produce large amounts of the pure substance. But how?

Levels of EPO in human plasma are too low to make pooling from donors feasible. A more likely potential source was people suffering from disorders such as aplastic anemia and hookworm infection, which cause them to secrete large amounts of EPO. The National Institute of Health set up a program to extract EPO from the urine of South American farmers with hookworm infections. Government planes transported the EPO in diplomatic pouches! Then, in 1976, the National Heart, Lung and Blood Institute began a grant program in search of ways to purify EPO. In 1977, supplies came in the form of 2,550 liters of urine from Japanese aplastic anemia patients.

Problems loomed for those trying to purify EPO from these sources. Was it ethical to obtain a scarce substance from the urine of sick, usually poor, people from one country to treat comparatively wealthy people from the United States? Then AIDS arose. Extracting any biochemical from human body fluids was no longer safe.

Recombinant DNA technology solved the EPO problem. The hormone is produced in hamster kidney cells, which can attach EPO's four carbohydrate groups. Today EPO is sold under various names, including Epogen, Repotin, and Procrit. The Food and Drug Administration approved it in 1989 to treat anemia in dialysis patients, and in 1991 to fight anemia in AIDS patients. Today it is also widely used by people receiving chemotherapy, enabling them to avoid the need for transfusions.

Recombinant EPO found an unexpected market among Jehovah's Witnesses, whose religion forbids the use of blood, which they believe destroys the soul. They can, however, use EPO, because it has been declared a product of recent technology, rather than blood. Some Jehovah's Witnesses have used it before surgery to boost red blood cell supplies, or afterward to compensate for blood loss.

EPO's ability to increase the oxygen-carrying capacity of the blood, and thereby increase physical endurance, has attracted the attention of competitive athletes. Training at high altitudes increases endurance, and the reason is EPO. The scarcer oxygen in the air at high altitudes stimulates the kidneys to produce EPO, which stimulates production of more red blood cells. Athletes have attempted to reproduce this effect by abusing EPO. A few Dutch bicyclists developed dangerous blood clots in their legs from taking too high, and medically unnecessary, doses. It is now routine to screen Olympic athletes for EPO abuse.

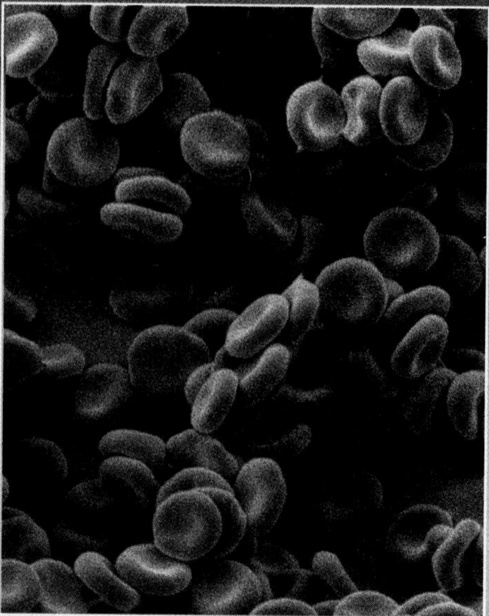

**figure 1**

These red blood cells are mass produced in a patient treated with erythropoietin (EPO).

One large Scandinavian family, however, gets its extra EPO quite naturally. They have an inherited condition called benign erythrocytosis, in which they overproduce the hormone. One of them won a gold medal for skiing in the Olympics. However, athletes who abuse EPO raise the issue of how to control the use of an otherwise valuable drug derived from biotechnology.

EPO is apparently not the only biotech-derived drug to be abused. Body builders abuse a recombinant version of the hormone somatropin, used to treat AIDS wasting syndrome, and the Internet is full of ads for recombinant human growth hormone.

normal one. For this condition and for most dominant genetic diseases, homozygous mutants are so severely affected that the individual does not develop beyond an embryo. Gene targeting can provide knockout mouse models of such homozygotes. The mouse embryo on the left in figure 18.14 has two neurofibromin genes knocked out. The embryo has a severely abnormal heart and stops developing in the middle of the embryo period. The mouse embryo on the right has one knocked-out gene. It has some tumors, but is not nearly as severely affected as its homozygous dominant sibling. Neurofibromin in mice and humans share 98 percent of their amino acid sequences. In humans, a double dose of the mutant gene would cause a miscarriage.

Animals with knocked-out genes are also useful in studying polygenic disorders. For example, researchers are studying atherosclerosis by inactivating combinations of genes whose products oversee lipid metabolism. Similarly, scientists can study multiple genetic changes responsible for some cancers by targeting the genes in various combinations.

In a series of gene targeting experi-
have de-
hemo-
e cell
cur n two
A re-
t Alabama
genes,
y N tional
nes—
abc the
all the
ed in the
sickle
am symp-
e, and
res rch to
pies.

on has
y w en ani-
nocked
ch ealthier
ce lack-
of collagen.
agen, in
the normal
bones.
Mutations in collagen genes cause a variety of syndromes in humans. Yet mice with knocked-out type X collagen genes have normal skeletons! How can this be?

A superhealthy knockout mouse forces researchers to rethink their assumptions about a gene's importance. Often these assumptions are based on knowing that the gene product has a vital function—such as contracting muscle or clotting blood. However, gene targeting experiments suggest that the importance of a gene's product must be considered in the context of the entire organism. Such a broader view of interacting genes presents several possible explanations for healthy knockout mice:

1. Other genes that encode the same or similar proteins as the knocked-out gene replace its function, so that disabling one gene does not affect the phenotype.
2. An absent protein in a knockout does not alter the phenotype, though an abnormal protein might.
3. The knocked-out gene does not do what we thought, and it may even have no function at all.
4. The knocked-out gene functions under different circumstances than those of the experiment. Various environmental challenges may be required to reveal the gene's function. For example, type X collagen, rather than being necessary for growth and development of a newborn's skeleton, may be called into action to repair fractures. It would therefore be unnecessary in embryonic and newborn mice.

Finally, the results of some gene targeting experiments are just weird. Consider one in which researchers took the fibroin gene from a silkworm—the gene that produces the major silk protein—and attached it to a gene from a jellyfish that encodes "green fluorescent protein" (GFP). GFP is used in many experiments to reveal where genes attached to it function. The researchers knocked out a silkworm's normal fibroin gene and replaced it with the GFP-tagged version. The result: larvae whose silk glands and silk glow green! In a related experiment, the spider silk gene spidroin was hooked up to GFP, producing spider silk that is used to manufacture bulletproof vests and parachutes—this silk, too, is a glowing green color.

Recombinant DNA, transgenic, and gene targeting technologies make possible new types of medicines. Chapter 19 explores how genes can be used to heal.

### Key Concepts

In gene targeting in mice, a gene of interest inserted into an embryonic stem cell recombines at the chromosomal site where it normally resides. The ES cell is then incorporated into a developing embryo from another individual, which is implanted into a surrogate. Animals with phenotypes indicating that they harbor cells with the targeted gene are bred to each other to yield homozygotes for the targeted gene. Swapping an inactivated allele for a gene of interest produces a knockout mouse, and replacing a gene with another that has an altered function creates a knockin mouse. These animals can model human disease, but they sometimes reveal that a gene does not function as we thought.

# Summary

### 18.1 Of Pigs and Patents

1. **Biotechnology** is the alteration of cells or biochemicals to provide a useful product. It includes extracting natural products, altering an organism's genetic material, and combining DNA from different species. A **transgenic** organism is multicellular and includes a gene from a different species in every cell.
2. DNA can be patented in several guises, but it must be useful.

### 18.2 Recombinant DNA Technology

3. **Recombinant DNA technology** is possible because of the universality of the genetic code. It is used to mass produce proteins in bacteria or other single cells. Begun hesitantly in 1975, the technology has matured into a valuable method to produce useful proteins.
4. Constructing a recombinant DNA molecule begins when **restriction enzymes** cut both the gene of interest and a **cloning vector** at a short palindromic sequence, creating "sticky ends" that are complementary. The cut foreign DNA and vector DNA are mixed, and vectors that pick up foreign DNA are selected.
5. **Genomic libraries** consist of recombinant cells containing fragments of a foreign genome.
6. **DNA probes** are used to select genes of interest from genomic libraries. DNA probes may be synthetic, taken from another species, or a **cDNA,** which is reverse transcribed from mRNA.

### 18.3 Transgenic Organisms

7. Recombinant DNA technology on a multicellular level produces a transgenic organism. A single cell—a gamete in an animal or plant, or a somatic cell in a plant—is genetically altered. The organism develops, including the change in each cell and passing it to the next generation. Heterozygotes for the transgene are then bred to yield homozygotes.
8. DNA is introduced into cells through **liposomes, electroporation, microinjection,** and **particle bombardment.**
9. Transgenic pharming produces valuable human proteins in the milk or semen of farm animals.

### 18.4 Gene Targeting

10. **Gene targeting** uses the natural attraction of a DNA sequence for its complementary sequence, called **homologous recombination,** to swap one gene for another. It is more precise than transgenic technology, which inserts a foreign gene but does not direct it to a specific chromosomal site.
11. Because homologous recombination will not occur in a gamete or fertilized ovum, the manipulation is done on an ES cell, which is then inserted into another embryo and transferred to a surrogate mother. Heterozygotes are bred to yield homozygotes.
12. Knockouts have the gene of interest inactivated. Knockins replace one gene with another allele with altered function.
13. Knockout mice with inactivated genes can model human disease. Sometimes, knockout mice reveal that a gene product is not vital to survival.

# Review Questions

1. Define each of the following terms:
   a. biotechnology
   b. recombinant DNA technology
   c. transgenic technology
   d. gene targeting
   e. homologous recombination
2. Describe the roles of each of the following tools in a biotechnology:
   a. restriction enzymes
   b. embryonic stem cells
   c. cloning vectors
3. How do researchers use antibiotics to select cells containing recombinant DNA?
4. List the components of an experiment to produce recombinant human insulin in *E. coli* cells.
5. Why would recombinant DNA technology be impossible if the genetic code was not universal?
6. Why must manipulations to create a transgenic organism take place at the single-cell stage?
7. Describe three ways to insert foreign DNA into cells.
8. Why isn't transgenic technology as precise as gene targeting?
9. How does Mendel's law of segregation for a monohybrid cross apply to carrying out transgenesis and gene targeting experiments?

# Applied Questions

1. Researchers have engineered a promoter that stimulates the expression of a particular gene in the nectary, or nectar-making organ of a flowering plant. By attaching the promoter to a gene of interest, they can produce the desired protein in the nectar, which bees collect and concentrate into honey. Then, the drug is extracted from the honey. What information is required to ensure that this is a safe new way to manufacture drugs?
2. Genetic engineering can creatively combine parts of organisms. From the following three lists, devise an experiment to produce a particular protein (choose one item from each list), and suggest what it might be used for.

| Organism | Biological Fluid | Protein Product |
|---|---|---|
| pig | milk | human beta globin chains |
| cow | semen | human collagen |
| goat | silk | human EPO |
| chicken | egg white | human tPA |
| aspen tree | sap | human interferon |
| silkworm | blood plasma | jellyfish GFP |
| rabbit | honey | human clotting factor |
| mouse | saliva | alpha-1-antitrypsin |

3. Collagen is a connective tissue protein that is used in skincare products, shampoo, desserts, and in artificial skin. For many years it was obtained from the hooves and hides of cows collected from slaughterhouses. Human collagen can be manufactured in transgenic mice. Describe the advantages of the mouse system for obtaining collagen.
4. How might cloning be used to speed transgenesis?
5. Tobacco plants given a transgene from bacteria can dismantle certain buried explosives and remove these organic pollutants from soil. What information is necessary to determine whether growing such plants is safe?
6. A human oncogene called *ras* is inserted into mice, creating transgenic animals that develop a variety of tumors. Why are mouse cells able to transcribe and translate human genes?

7. A healthy knockout mouse cannot manufacture what was thought to be a vital enzyme. Suggest three possible explanations for this surprising finding.
8. In a mouse model of a human condition called "urge syndrome," in which the feeling of impending urination occurs frequently, researchers inactivate a gene encoding nitric oxide synthase, which produces nitric oxide (NO). NO is the neurotransmitter that controls muscle contraction in the bladder. What type of biotechnology does this describe?
9. Mouse models for cystic fibrosis have been developed by inserting a human transgene, and by gene targeting to inactivate the mouse counterpart of the alleles that cause the disorder. How do these methods differ? Which method do you think produces a more accurate model of human cystic fibrosis, and why?

# Suggested Readings


Bobrow, Martin, and Sandy Thomas. February 15, 2001. Patents in a genetic age. *Nature* 409:763–64. The Patent and Trademark Office can hardly keep up with single-gene applications. What will happen in this new age of genomics?

Cibelli, J. B., et al. May 22, 1998. Cloned transgenic calves produced from nonquiescent fetal fibroblasts. *Science* 280:1256–58. Cloning can speed transgenesis.

Fox, Jeffrey L. July 2001. Fake biotech drugs raise concerns. *Nature Biotechnology* 19:603. Drug counterfeiting hasn't hurt anyone yet, but is potentially very dangerous.

Golovan, Serguei P., et al. August 2001. Pigs expressing salivary phytase produce low-phosphorus manure. *Nature Biotechnology* 19:741–42.

Lewis, Ricki. April 3, 2000. Clinton, Blair stoke debate on gene data. *The Scientist* 14:1. The public is very concerned about patenting genes.

Lewis, Ricki. November 8, 1999. Semen pharming. *The Scientist* 13:31. Boar semen may be a rich source of biopharmaceuticals.

Lewis, Ricki. October 26, 1998. How well do mice model humans? *The Scientist* 12:1. Many patient support groups for inherited diseases sponsor development of transgenic or knockout/knockin mice corresponding to the condition.

Marshall, Eliot. August 22, 1997. A bitter battle over insulin gene. *Science,* vol. 277. A legal dispute over experiments conducted during the early days of recombinant DNA technology continues.

Russo, Eugene. April 3, 2000. Reconsidering Asilomar. *The Scientist* 14:15. On Asilomar's 25th anniversary, those who were there agree that it couldn't happen again, due to the influences of the business community and consumer activism.

Sagar, Ambuj, et al. January 2000. The tragedy of the commoners: Biotechnology and its publics. *Nature Biotechnology* 18:2. Biotechnology affects politics and economics, and vice versa.

Schnieke, A. E., et al. December 19, 1997. Human factor IX transgenic sheep produced by transfer of nuclei from transfected fetal fibroblasts. *Science* 278:2130–33. Genetically modified sheep supply human clotting factors.


# On the Web

Check out the resources on our website at

**www.mhhe.com/lewisgenetics5**

On the web for this chapter, you will find additional study questions, vocabulary review, useful links to case studies, tutorials, popular press coverage, and much more. To investigate specific topics mentioned in this chapter, also try the links below:

FDA-approved recombinant DNA-derived drugs
**www.accessexcellence.org/AB/BA/The_Biopharmaceuticals.html**

Food and Drug Administration
**www.fda.gov/**

Genentech (recombinant DNA-derived drugs)
**www.gene.com/Medicine/index.html**

"Genetically engineered foods: Safety issues associated with antibiotic resistance genes," by A. Salyers.
**www.healthsci.tufts.edu/apua/salyersreport.htm**

The Jackson Laboratory **www.jax.org**

Online Mendelian Inheritance in Man
**www.ncbi.nlm.nih.gov/entrez/query.fcgi?db=OMIM**
alpha-1-antitrypsin (AAT) deficiency 107400
benign erythrocytosis 263400
*BRCA1* 113705
factor VIII deficiency (hemophilia A) 306700
growth hormone deficiency 139250
Huntington disease 143100
type I diabetes mellitus 222100
neurofibromatosis type I 162200
sickle cell disease 603903

U.S. Patent and Trademark Office
**www.uspto.gov/**

CHAPTER

19

# Gene Therapy and Genetic Counseling

### 19.1 Gene Therapy Successes and Setbacks

The 1990s dawned with the first gene therapy experiments, which were partially successful and paved the way for improved methods. The decade closed, however, with the death of a young participant in a gene therapy trial, leading to a reexamination of the tools and approaches used in this still very promising technology. Gene therapy continues to yield some successes.

### 19.2 The Mechanics of Gene Therapy

Designing a gene therapy requires the creative combining of genetic material. The gene of interest must be isolated and delivered to the tissue implicated in a particular illness, then coaxed to produce its encoded protein at the right time, in sufficient amounts, and for long enough to improve symptoms. Gene therapies are being developed for a variety of illnesses.

### 19.3 A Closer Look: Treating Sickle Cell Disease

We know more about sickle cell disease than perhaps any other inherited illness. Several "traditional" gene therapies are being applied to this disorder, as well as the unique approach of reawakening fetal genes.

### 19.4 Genetic Screening and Genetic Counseling

Genetic counselors have a special combination of scientific, medical, and psychological skills that enable them to educate and comfort people facing the possibility of inherited illness. These health care professionals guide individuals, couples, and families through the maze of decisions that accompany genetic testing.

Treatment for genetic diseases has evolved in three phases: (1) replacing missing proteins with material from donors, (2) obtaining pure proteins using recombinant DNA technology, and (3) delivering replacement genes to correct the problem at its source (**gene therapy**). The Technology Timeline on page 391 illustrates this evolution for hemophilia A, a disorder of blood clotting.

Preclinical testing and clinical trials of various gene therapies began in the 1990s. Tools developed in this first decade of experimental gene therapy will be wed to human genome information during the second decade. This chapter introduces some of the patients who have volunteered for the first gene therapies—then discusses the mechanics of the process.

## 19.1 Gene Therapy Successes and Setbacks

Any new medical treatment or technology begins with creative minds and courageous volunteers. The first individuals to take new vaccines or to try new treatments know that they may give their lives, either directly or indirectly, in the process. Gene therapy, however, is unlike conventional drug therapy. It attempts to alter an individual's genotype in a part of the body that has malfunctioned. Because the potentially therapeutic gene must usually be delivered along with other DNA, and it may be taken up by cell types other than those that are affected in the disease, the body's reactions are unpredictable.

### Adenosine Deaminase Deficiency—Early Success

For the first few years of her life, Laura Cay Boren couldn't recall what it was like to feel well (figure 19.1*a*). From her birth in July 1982, she fought infection after infection. Colds rapidly became pneumonia, and routine vaccines caused severe abscesses. In February 1983, doctors identified Laura's problem—severe combined immune deficiency (SCID) due to adenosine deaminase (ADA) deficiency. She had inherited the autosomal recessive inborn error of metabolism from two carrier parents.

Lack of ADA blocks a biochemical pathway that normally breaks down a metabolic toxin into uric acid, which is then excreted (figure 19.1*b*). Without ADA, the substance that ADA normally acts upon builds up and destroys T cells. Without helper T cells to stimulate them, B cells cannot mature into the plasma cells that produce antibodies. Both branches of the immune system therefore fail. The child becomes extremely prone to infections and cancer, and despite medical treatment, usually does not live beyond a year in the outside environment.

To Laura, the Duke University Medical Center, where she celebrated her first and second birthdays, became a second home. In 1983 and again in 1984, she received bone marrow transplants from her father, which temporarily bolstered her immunity. Red blood cell transfusions also helped for a time. Still, Laura was spending more time in the hospital than out. By the end of 1985, she was gravely ill. She had to be fed through a tube, and repeated infection had severely damaged her lungs. Laura's mother began to feel guilty for wishing that her child would die rather than suffer. Then, a medical miracle happened.

Laura was chosen to be the first recipient of a new treatment. She had been second

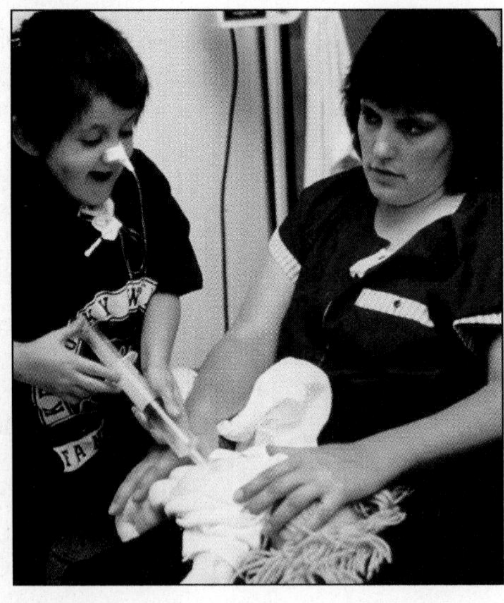

a.

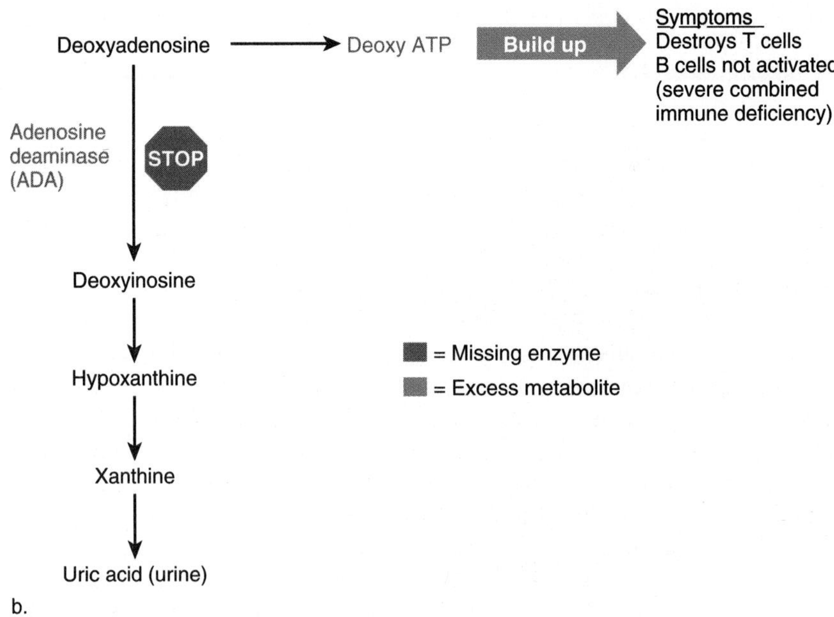


b.

**figure 19.1**

**Evolution of treatment for ADA deficiency.** (*a*) Laura Cay Boren spent much of her life in hospitals until she received the enzyme that her body lacks, adenosine deaminase (ADA). Here, she pretends to inject her doll as her mother looks on. Today, gene therapy is possible using cord blood stem cells. (*b*) ADA deficiency causes deoxy ATP to build up, which destroys T cells, which therefore cannot stimulate B cells to secrete antibodies. The result is severe combined immune deficiency.

### Hemophilia A (Factor VIII Deficiency)

Hemophilia A is a bleeding disorder caused by a mutation in the gene on the X chromosome that encodes clotting factor VIII. It affects 1 in 10,000 newborns. Hemophilia A is an excellent candidate for gene therapy because it is a single-gene disorder, many cell types can secrete the clotting factor, it is well studied in animal models, and just modest increases in clotting factor production can improve health. The treatment of hemophilia A parallels the evolution of biotechnology, from enzyme replacement to recombinant DNA technology to gene therapy. Even though clinical trials are well under way, researchers are seeking improvements in gene delivery in mice using a variety of viruses—including disabled HIV.

**1970s** Hemophilia A is treated with factor VIII pooled from donated plasma (cryoprecipitate). Recipients contract hepatitis.

**Early 1980s** Up to 70 percent of patients receiving cryoprecipitate contract HIV infection.

**1984** Factor VIII gene cloned, making prenatal diagnosis and carrier detection possible for some individuals.

**1985** Invention of PCR allows detection of many more hemophilia A mutations, previously unknown because they occur in an intron.

**1990** Recombinant factor VIII eliminates risk of infection from donated plasma while supplying missing gene product.

**1999** First gene therapy experiment, using viral vector given intravenously, works (see In Their Own Words, chapter 1)

**2001** In another gene therapy protocol, patients' skin fibroblasts are removed and cultured with cDNA for factor VIII, and reimplanted into abdominal fat. Four of six patients improve.

**2002** Another gene therapy protocol injects cDNA for factor VIII carried in disabled mouse leukemia retrovirus.

in line to a boy who was even more ill, but he died just before beginning treatment. In the spring of 1986, Laura received her first injection of PEG-ADA. This is the missing enzyme, ADA, from a cow and stabilized by adding polyethylene glycol (PEG) chains to it. PEG is the major ingredient in antifreeze.

Previous enzyme replacement therapy hadn't worked, because what remained of the immune system rapidly destroyed the injected, unaltered enzyme. Patients needed frequent doses, which provoked the immune system further, causing allergic reactions too severe to continue treatment. Laura's physicians hoped that adding PEG would keep ADA in her blood long enough to work.

Laura began responding to PEG-ADA almost immediately. Within hours her ADA level increased twentyfold. After three months, toxins were no longer in her blood, but her immunity was still suppressed. After six months, though, Laura's immune function neared normal for the first time ever—and stayed that way, with weekly doses of PEG-ADA. Her life changed drastically as she ventured beyond the hospital's germ-free rooms. By summer 1988, she could finally play with other children without fear of infection. She began first grade in fall 1989, but had to repeat the year—she had spent her time socializing!

PEG-ADA revolutionized treatment of this form of SCID, targeting the source of the disorder rather than trying to overcome the infections. But PEG-ADA was only the opening chapter of an ongoing story.

The second chapter began on September 14, 1990, at 12:52 P.M. Four-year-old Ashanti DaSilva sat up in bed at the National Institutes of Health in Bethesda, Maryland, and began receiving her own white blood cells intravenously. Earlier, doctors had removed the cells and patched them with normal ADA genes. Soon after, an eight-year-old, Cynthia Cutshall, received the same treatment. In the years following, both girls stayed relatively healthy. But this first gene therapy did not "heal" a sufficient percentage of the girls' cells. Both needed repeated treatments. One girl required continued PEG-ADA to remain healthy; the other, however, eventually showed the normal ADA gene in 25 percent of her T cells.

Even as Ashanti and Cynthia were being treated, researchers were onto the next step. Wouldn't the effect last longer if they could treat immature blood cells? The type of stem cell that produces T cells accounts for one in several billion bone marrow cells—not a very promising ratio. Umbilical cord blood was a more plentiful source. If fetuses who had inherited ADA deficiency could be identified and their parents agreed, then the appropriate stem cells could be extracted from the cord blood at birth, given ADA genes, and reinfused into the newborn. The third chapter in the ADA deficiency tale was about to begin.

Crystal and Leonard Gobea had already lost a five-month-old baby to ADA deficiency when amniocentesis revealed that their second fetus was affected. They and two other couples were asked to participate in the experiment. The May 31, 1993, issue of *Time* magazine featured newborn Andrew Gobea on the cover. He and the other two babies received their own bolstered blood cells on the fourth day after birth, and PEG-ADA to prevent symptoms in case the gene therapy did not work right away. The plan was to monitor the babies frequently to see if T cells carrying normal ADA genes would appear in their blood.

The experiment was a success, although the altered T cells accumulated slowly. After a few months, in each child, about 1 in 10,000 T cells had the genetic change. But after a year, that number rose to 1 in 100! By the time each child was 18 months of age, with the genetically altered T cell population still rising, researchers halved the PEG-ADA dose. The children remained healthy. By the summer of 1995, the three toddling two-year-olds each had about 3 in 100 T cells carrying the ADA gene. Gradually, the healthier, bolstered cells are replacing the ADA-deficient ones. Researchers hope that one day the children's ADA-producing T cells will be plentiful enough to discontinue the PEG-ADA treatment.

## Ornithine Transcarbamylase Deficiency—A Setback

The sad saga of Jesse Gelsinger stands in sharp contrast to the success stories of the children with ADA deficiency. In September 1999, the 18-year-old died, just days after receiving gene therapy. The cause of death was an overwhelming and unanticipated immune system reaction against the DNA used to introduce the healing gene.

Jesse had an inborn error of metabolism called ornithine transcarbamylase deficiency (OTC). In this X-linked recessive disorder, one of five enzymes required to break down amino acids liberated from dietary proteins is absent (figure 19.2). The nitrogen from the amino acids combines with hydrogens to form ammonia ($NH_3$), which rapidly accumulates in the bloodstream and travels to the brain, with devastating effects. The condition usually causes irreversible coma within 72 hours of birth. Half of affected babies die within a month, and another quarter by age five. The survivors can control their symptoms by following a special low-protein diet and taking drugs that bind ammonia.

Jesse wasn't diagnosed until he was two, because he was a mosaic—some of his cells could produce the enzyme, so his symptoms were milder. Still, when he went into a coma in December 1998 after missing a few days of his medications, he and his father began to consider Jesse's volunteering for a gene therapy trial they had read about. The researchers would not accept Jesse until he turned 18, so the next summer, four days after his birthday, Jesse underwent testing at the University of Pennsylvania, where the gene therapy center is located, and was admitted to the trial. He was jubilant. He knew he would not directly benefit, at least not for awhile, but he had wanted to volunteer to try to help the babies who die of the condition. A bioethics committee had advised that the experimental treatment could not be tried on newborns because the parents would be too distraught to give informed consent to an untried medical procedure. Instead, affected males and carrier females had volunteered. Said Jesse at the time, "What's the worst that can happen to me? I die, and it's for the babies."

The gene therapy consisted of an adenovirus—a type of virus that causes the common cold—with a functional human OTC gene stitched into it. This virus had already been used, apparently safely, in about a quarter of the 330 gene therapy experiments done on more than 4,000 patients since 1990. It is a disabled virus, with the genes that enable it to replicate and cause disease removed. Three groups of six patients each were to receive three different doses, with the trial designed to identify the lowest dose that would fight the genetic disease without causing dangerous side effects.

Jesse entered the hospital on Monday, September 13, after the 17 others in the trial had already been treated and suffered nothing worse than a fever and aches and pains. Several billion altered viruses were introduced into an artery leading into his liver. That night, Jesse developed a high fever—still not unusual. But by the morning, the whites of Jesse's eyes were yellow, indicating a high bilirubin level as his liver struggled to dismantle the hemoglobin released from burst red blood cells. A flood of hemoglobin meant a flood of protein, so the ammonia level in his liver soon skyrocketed, reaching 10 times normal levels by mid-afternoon. At the same time, his blood wasn't clotting well. Jesse became disoriented, then comatose. By Wednesday, doctors had controlled the ammonia buildup, but his lungs began to fail, and Jesse was placed on a ventilator. Thursday, Jesse's vital organs began to fail, and by Friday he was brain dead. His dedicated and devastated medical team

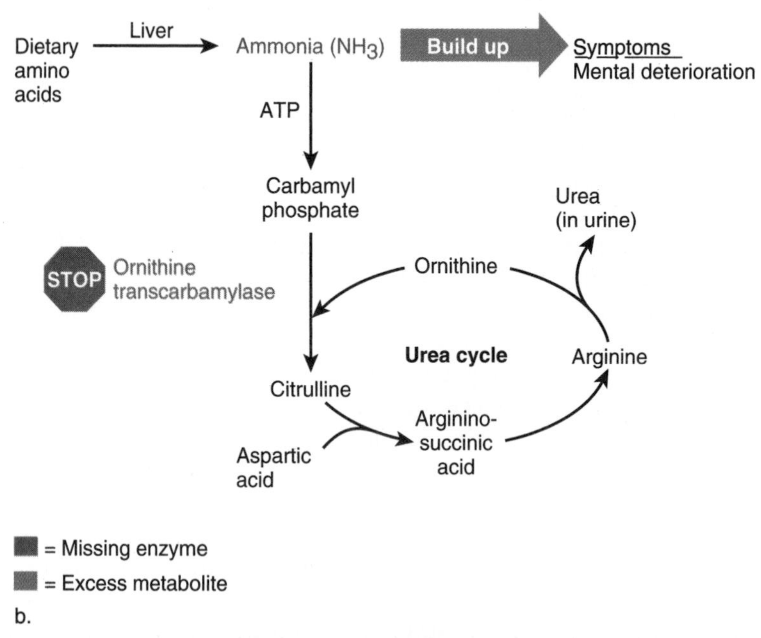


**figure 19.2**

**A brave example.** (*a*) Jesse Gelsinger received gene therapy for an inborn error of metabolism in September 1999. He died four days later from an overwhelming immune response. (*b*) Lack of ornithine transcarbamylase causes ammonia to accumulate, which is toxic to the brain.

stood by as his father turned off the life support, and Jesse died.

At a public hearing in December, doctors explained that the autopsy showed that Jesse had had an infection with parvovirus, which may have led his immune system to attack the adenovirus. In the liver, the adenovirus had traveled not to the targeted hepatocytes, but to a different cell type, the macrophages that function as sentries for the immune system. In response, interleukins flooded his body, and inflammation raged. Although parents of children with OTC who spoke at the hearing implored government officials to allow the research to continue, the death of Jesse Gelsinger led to suspension of several gene therapy trials. The death drew particular attention to safety because unlike most other volunteers, Jesse was not very ill. In the months following his death, representatives of the Food and Drug Administration and the National Institutes of Health (NIH) identified several possible contributing procedural factors to the tragedy: failure to report all side effects in a timely fashion, failure to fill out the proper eligibility forms, and inadequately documented informed consent. More careful attention should also have been paid to screening potential participants for underlying medical conditions. When the NIH examined other gene therapy trials, the agency discovered extreme underreporting of adverse side effects. However, this apparent oversight might reflect the fact that researchers considered many adverse effects to have been due to the underlying disease, and not the experimental treatment. Clearly the safety of the technology is a continuing story, but in the meantime, the NIH is inspecting facilities sponsoring gene therapy trials and requiring weekly reports of adverse effects.

## A Success in the Making—Canavan Disease

Jesse Gelsinger's death led to the halting or reevaluation of many gene therapy trials. However, efforts begun in 1995 to treat Canavan disease, which causes brain degeneration in children, continued. Canavan disease is an ideal candidate for gene therapy for several reasons: (1) the gene and protein are well known; (2) there is a window of time when affected children are healthy enough to be treated; (3) only the brain is affected; (4) brain scans can monitor response to treatment; and (5) there is no existing treatment.

Canavan disease disrupts a complex biochemical interaction between neurons and neighboring cells called oligodendrocytes, which produce the fatty myelin that coats the neurons and makes nerve transmission fast enough for the brain to function (figure 19.3*b*). Specifically, neurons normally release N-acetylaspartate (NAA), which is broken down into harmless compounds by an enzyme, aspartoacylase, that the oligodendrocytes produce. In Canavan disease, the enzyme is missing, and the resulting NAA buildup eventually destroys

a.

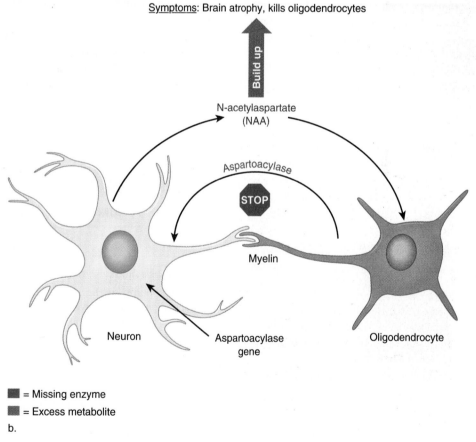

Symptoms: Brain atrophy, kills oligodendrocytes
Build up
N-acetylaspartate (NAA)
Aspartoacylase
STOP
Myelin
Neuron
Aspartoacylase gene
Oligodendrocyte

■ = Missing enzyme
■ = Excess metabolite

b.

**figure 19.3**

**Canavan disease.** (*a*) Max Randell, at two-and-a-half years old, is battling Canavan disease. (*b*) In Canavan disease, demyelination of brain neurons occurs because oligodendrocytes lack an enzyme that enables them to produce aspartoacylase. Gene therapy enables the neurons to secrete the enzyme, ultimately restoring the fatty covering that makes nerve transmission in the brain possible.

the oligodendrocytes. Without sufficient myelin, the neurons cease to function, and symptoms begin. The parents may first notice developmental delay—inability to sit and stand at ages when other children can. The children have poor vision, do not react much to their surroundings, may have seizures, require tube feeding, and may have muscle control so poor that they cannot even hold up their heads. In Their Own Words on page 395 and Bioethics: Choices for the Future on page 404 describe children with Canavan disease. Due to a powerful founder effect, Canavan disease is seen almost exclusively in the Ashkenazi Jewish population.

The first attempts at gene therapy for Canavan disease introduced the needed gene in a liposome, through holes bored into the skull. The first recipient, 18-month-old Lindsay Karlin, gained some skills for awhile. Previously, Lindsay could barely open her eyes and did not interact with anyone. But three months after the therapy, she looked around, moved, and vocalized. "It was as though she had awakened," wrote her mother. A magnetic resonance image of Lindsay's brain showed that myelination of neurons had begun in regions where it had vanished. Lindsay was not treated again until June 2001, when a viral vector replaced the liposomes. In the interim, while regulatory agencies argued about the safety of this therapy for a disease that had no other treatment, Lindsay lost some of the gains from the first round, such as being able to hold her head up. But the gene therapy does appear to be working, in Lindsay and several of 14 other children participating in current clinical trials.

**KEY CONCEPTS**

ADA deficiency was the first disorder researchers treated with gene therapy. They began by replacing the missing enzyme, progressed to genetically altering mature white blood cells in ill children, and finally moved to infusing genetically modified umbilical cord stem cells into newborns. • Gene therapy for OTC deficiency led to a tragic death due to an unexpected severe immune reaction to the virus used to deliver the genes. • For Canavan disease, gene therapy may provide the only treatment.

## 19.2 The Mechanics of Gene Therapy

The idea of fixing the symptoms of an inherited problem is not new. Altering genes to treat an inherited disorder, however, can provide a longer-lasting effect. The first gene therapy efforts have focused on inherited disorders that researchers know the most about, even though the conditions may be very rare. With the increasing understanding of human genes and their functions made possible by the human genome project, gene therapy efforts will be targeting more common illnesses, such as heart disease and cancer. Tables 19.1 and 19.2 list some general requirements and concerns related to gene therapy.

**table 19.1**

**Requirements for Approval of Clinical Trials for Gene Therapy**

1. Knowledge of defect and how it causes symptoms
2. An animal model
3. Success in human cells growing *in vitro*
4. Either no alternate therapies, or a group of patients for whom existing therapies are not possible or have not worked
5. Safe experiments

Gene therapy trials have progressively become more invasive. The first gene therapy, performed on Ashanti and Cynthia, altered cells outside their bodies. The corrected cells were then infused into the girls. This approach is called ***ex vivo* gene therapy.** In ***in situ* gene therapy,** the healthy gene plus the DNA that delivers it (the vector) is injected into a very localized and accessible body part, such as a single melanoma skin cancer. In the most invasive approach, ***in vivo* gene therapy,** the vector is introduced directly into the body, as it was into Jesse Gelsinger's liver.

### Treating the Phenotype

The phenotypes of some genetic disorders can be treated, often by replacing a missing protein. A child with cystic fibrosis sprinkles powdered cow digestive enzymes onto applesauce, which she eats before each meal to replace the enzymes her clogged pancreas cannot secrete. A boy with hemophilia receives a clotting factor. Even wearing eyeglasses is a way of altering the expression of one's inheritance. Today, newborns are routinely screened for certain inborn errors of metabolism whose symptoms can be prevented or alleviated by correcting the phenotype, such as by following a restrictive diet. Newborn screening is revisited at the chapter's end.

**table 19.2**

**Gene Therapy Concerns**

| Scientific | Bioethical |
|---|---|
| 1. Which cells should be treated? | 1. Does the participant in a gene therapy trial truly understand the risks? |
| 2. What proportion of the targeted cell population must be corrected to alleviate or halt progression of symptoms? | 2. If a gene therapy is effective, how will recipients be selected, assuming it is expensive at first? |
| 3. Is overexpression of the therapeutic gene dangerous? | 3. Should rare or more common disorders be the focus of gene therapy research and clinical trials? |
| 4. Is it dangerous if the altered gene infiltrates other than the intended tissues? | 4. What effect should deaths among volunteers have on research efforts? |
| 5. How long will the affected cells function? | |
| 6. Will the immune system attack the introduced cells? | |

*In Their Own Words*

## Gene Therapy for Canavan Disease

On September 8, 1998, at 11 months old, little Max Randell became the youngest person in the world to receive gene therapy for a degenerative brain disease. He was one of four children, out of 14 total, in the safety trial to show an increase of myelin in his brain. His progress was amazing after the surgery, but unfortunately, was short-lived.

Finally, on June 19, 2001, Max received another injection of corrected genes using a new improved gene transfer system. Dr. Paola Leone, and her wonderful team at Thomas Jefferson University, worked tirelessly to develop this new system. They are truly lifesavers! This time, 90 billion viral particles carrying a corrected copy of the new gene were injected directly into Max's brain at six different sites. He was the second person in the world to receive a gene transfer using this experimental method.

The surgery was a success, and Max started to show signs of improvement after only 10 days. After one month the changes were indisputable and were being recorded by doctors, family, and all of Max's therapists. Previous data show that this time the new gene should stay active in Max's brain for at least two years, possibly even up to five years.

To see Max do things he hasn't done in the past one to two years is simply amazing! He is a very happy and well-adjusted child. His social, emotional, and cognitive skills are those of any normal three-and-a-half-year old. I cannot even begin to describe the joy we feel just seeing him regain even the slightest bit of functional mobility; he just beams with pride when his body does what he wants it to.

By Ilyce and Mike Randell

An inherited illness with an unusual phenotypic treatment is hereditary hemochromatosis (HH). Because HH results in "iron overload," the treatment is to periodically remove blood, because this action also removes iron, which is part of the hemoglobin molecule (see figure 11.3).

In the United States, 1.5 million people have this autosomal recessive condition, and 32 million people—1 in 8—carry a mutant allele for the HH gene. It is most common among those of Irish, Scottish, or British descent. In HH, cells in the small intestine absorb too much iron from food. Over many years, the excess iron is deposited throughout the body, causing various symptoms and secondary conditions. The liver develops cirrhosis (scarring) and then sometimes cancer; the heart may fail or beat irregularly; an iron-loaded pancreas may cause diabetes; joints become arthritic; and the skin darkens. Early signs and symptoms include chronic fatigue, infection, hair loss, infertility, muscle pain, and feeling cold.

Diagnosis requires a blood test to detect the telltale blood-level increase in ferritin, a protein that carries iron, and a liver biopsy. Determining the genotype alone is not sufficient for diagnosis because the penetrance is very low. That is, although most people with iron overload have mutations in the HH gene, only a small percentage of people with a homozygous recessive genotype actually have symptoms. More men than women develop symptoms of HH, because a woman loses some blood each month when she menstruates. Once women pass the age of menopause, the sex ratio equalizes. Giving blood every few months, however, is a simple way to keep the body's iron levels down.

## Germline Versus Somatic Gene Therapy

Researchers distinguish two general types of gene therapy, depending upon whether it affects gametes or fertilized ova, or somatic tissue.

**Germline gene therapy** (also known as heritable gene therapy) alters the DNA of a gamete or fertilized ovum. As a result, all cells of the individual that develop have the change. Germline gene therapy is heritable—it passes to offspring. It is an application of the transgenic technology discussed in chapters 18 and 20, although such an alteration on a person evokes more bioethical concerns than a similar procedure performed on a plant or farm animal. Germline gene therapy is not being done in humans.

Correcting only the somatic cells that an illness affects is **somatic gene therapy.** This form of the technology is nonheritable, which means that a recipient does not pass the genetic correction to offspring. An example is clearing lungs congested from cystic fibrosis with a nasal spray containing functional CFTR genes. It doesn't alter the gametes (sperm or oocytes), and so a treated person could not pass a normal CFTR allele to offspring.

Many questions must be answered before somatic gene therapy can be applied. Which cells, in which tissues and organs, fail, and when in development do they do so? Which biochemical is abnormal or missing? What DNA sequence must be added to correct the defect? What percentage of cells in a tissue must be altered to alleviate symptoms? How long will the effects of the correction persist? Will it affect other cells in ways we can anticipate? Even when we have much of this information, designing a gene therapy is challenging. Some early gene therapy experiments have been disappointing because the correction isn't sufficient to overcome symptoms or because the immune system attacks the altered cells.

## Sites of Somatic Gene Therapy

Current somatic gene therapy clinical trials, which are summarized in figure 19.4 and in this section, target several different tissues. However, the discovery that stem cells in

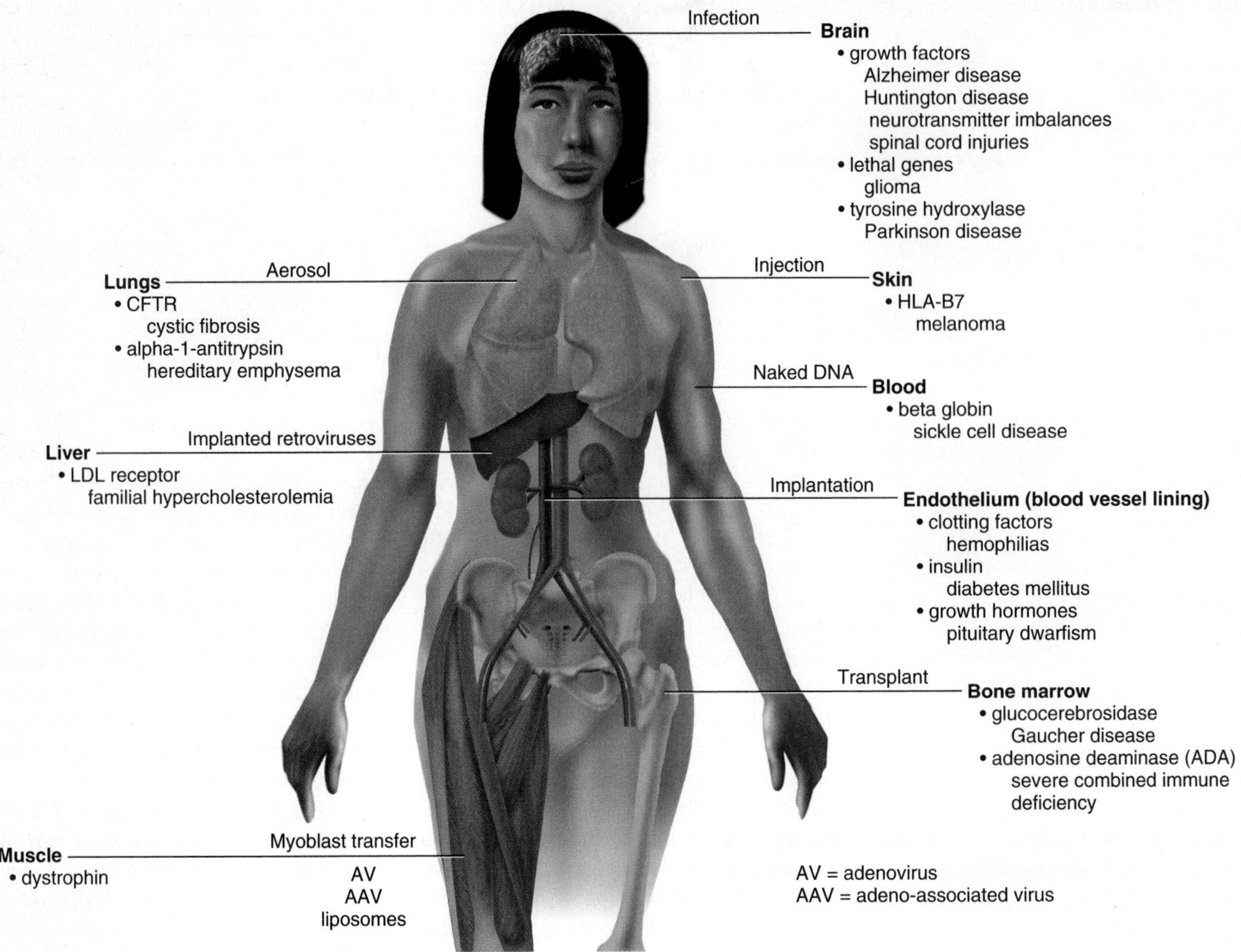


**figure 19.4**

**Gene therapy sites.** Beneath the label for each site are listed the targeted protein (•) and then the disease.

bone marrow can, under certain conditions, specialize as tissues other than blood—such as liver or muscle—suggests that gene therapy may someday consist of an injection of carefully selected stem cells, rather than more invasive implants.

### Endothelium

A tissue that is very amenable to gene therapy is endothelium, which forms capillaries. Genetically altered endothelium can secrete a needed protein directly into the bloodstream. A person with diabetes, for example, might receive capillaries that secrete insulin; someone with hemophilia might receive an implant that manufactures a clotting factor. Endothelium is implanted along with collagen to provide support and with angiogenesis factors to stimulate the growth of capillaries.

### Skin

Like endothelium, skin cells also grow well in the laboratory. A person can donate a patch of skin the size of a letter on this page; after a genetic manipulation, the sample can grow to the size of a bathmat within just three weeks. The skin can then be grafted back onto the person. Skin grafts genetically modified to secrete therapeutic proteins may provide a new drug delivery route.

### Muscle

Muscle tissue is a good target for gene therapy for several reasons. It comprises about half of the body's mass, is easily accessible, and is near a blood supply. However, a challenge is to correct enough muscle cells to alleviate symptoms. Consider gene therapy for Duchenne muscular dystrophy (DMD) (see figure 2.1*a*). An early challenge was to cut the dystrophin gene—about 3 million bases—down to a size small enough to deliver to cells. This was eventually accomplished, and the gene sent into immature muscle cells (myoblasts). But this approach worked only on small sections of muscle. An alternative strategy is to direct stem cells from bone marrow that can naturally migrate to muscle, where they differentiate and produce dystrophin.

### Liver

This largest organ in the body is an important candidate for gene therapy because it has many functions and can regenerate. An implant of corrected cells can take over liver function. For the inborn error of metabolism tyrosinemia, for example, effective gene therapy would have to "fix" only about 5 percent of the liver's 10 trillion cells.

Gene therapy to deliver the gene that encodes the LDL receptor can treat familial hypercholesterolemia, discussed in figure 5.2. Recall that when liver cells lack LDL receptors, cholesterol accumulates on artery interiors. Heterozygotes have half the normal number of LDL receptors and suffer heart attacks in early or mid-adulthood. Homozygotes die in childhood. Genetically altering liver cells to produce more LDL receptors can reverse the effects of FH. One young woman who is heterozygous for FH had 15 percent of her liver removed, the cells isolated and given functional LDL receptor genes, and redelivered into the body through a major liver vein. Eighteen months later, the grafted liver cells bore more LDL receptors, and the woman's serum cholesterol levels had improved.

### *Lungs*

The respiratory tract is easily accessed with an aerosol spray, eliminating the need to remove, treat, and reimplant cells. Several aerosols to treat cystic fibrosis attempt to replace the defective gene, but so far the correction is short-lived and localized. Another lung disorder that may be treatable with gene therapy is alpha-1-antitrypsin (AAT) deficiency. Absence of the enzyme AAT enables levels of another enzyme, called elastase, to accumulate and destroy lung tissue. White blood cells in the lungs normally produce elastase to destroy infecting bacteria, but in AAT deficiency, the levels of elastase rise too high. Delivering the AAT gene may normalize levels of both enzymes.

### *Nervous Tissue*

Many common illnesses and injuries affect the nervous system, including seizures, strokes, spinal cord injuries, and degenerative disorders such as Alzheimer disease, Parkinson disease, Huntington disease, and amyotrophic lateral sclerosis. Gene therapy on nervous tissue therefore isn't restricted to correcting inherited diseases. If a protein is known that could be used to correct an abnormal situation, then cell implants or gene delivery can possibly heal. However, neurons are difficult targets for gene therapy because they do not divide. Gene therapy efforts can alter other cell types, such as fibroblasts (connective tissue cells), to secrete nerve growth factors or manufacture the enzymes necessary to produce certain neurotransmitters. Then the altered cells can be implanted.

### *Cancer*

About half of current gene therapy trials target cancer. Two promising strategies are **suicide gene therapy** and manipulation of the immune response to create **cancer vaccines.**

Viruses are used to treat a type of brain tumor, called a glioma, which affects the glial cells that support and interact with neurons. Unlike neurons, glia can divide. Cancerous glia divide very fast, usually causing death within a year. Researchers reasoned that an agent directed against only the dividing cells might halt the cancer. One candidate was a "suicide" gene from the herpes simplex virus. In the presence of a certain drug, activation of the gene kills the cell that contains it. Would cancer cells infected with a virus carrying this gene self-destruct?

Figure 19.5 illustrates the herpes suicide gene therapy system. First, mouse fibroblasts are infected with a retrovirus or adenovirus vector that contains the herpes gene that encodes an enzyme called thymidine kinase. Any cell that produces thymidine kinase is susceptible to the anti-herpes drug ganciclovir. Because a retrovirus can only infect dividing cells, it does not harm nondividing, healthy brain neurons, but enters cancerous glia. The modified mouse fibroblasts are injected into a person's brain tumor through a hole drilled in the skull. The implanted cells release viruses, which infect neighboring tumor cells, which then produce thymidine kinase. When the patient takes ganciclovir, the drug is changed into a toxin that kills the cancer cells as well as nearby cells in what is called a "bystander effect." A few people have improved with this treatment, but studies on animal cells reveal a chronic immune response to the vectors.

Cancer vaccines enable tumor cells to produce immune system biochemicals or mark tumor cells so that the immune system recognizes them more easily. In one approach, a patient's cancer cells are removed, altered in a way that attracts an immune response, then reimplanted. The tumor cells are altered to overproduce cytokines or HLA cell surface molecules.

Melanoma, a skin cancer, is amenable to cancer vaccine treatment because it is on the body's surface and therefore accessible. In

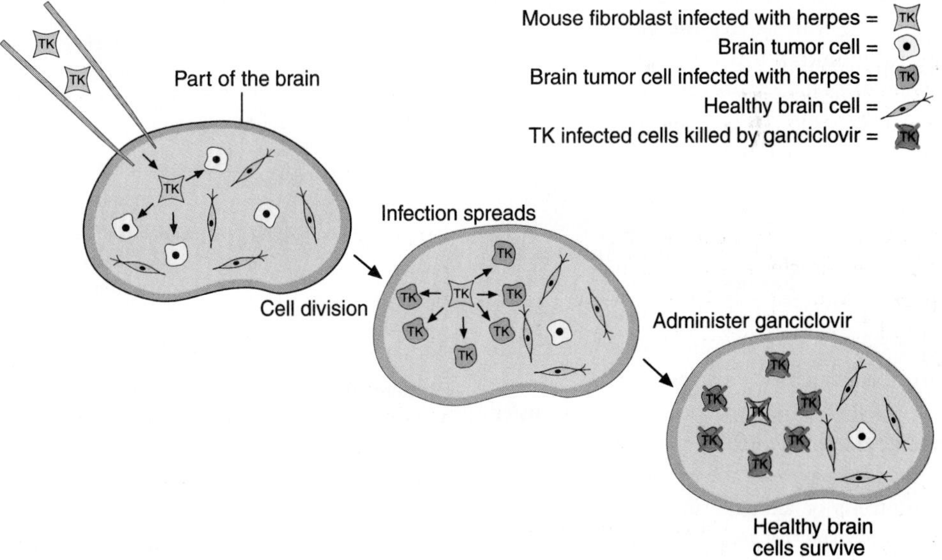


**figure 19.5**

**A herpes virus attacks cancer.** When mouse fibroblasts harboring a thymidine kinase gene from the herpes simplex virus, in a retrovirus vector, are implanted near the site of a brain tumor, the engineered viruses infect the rapidly dividing tumor cells. When the patient takes the antiviral drug ganciclovir, the cells producing thymidine kinase are selectively killed, providing a gene-based cancer treatment from within.

one group of experiments, melanoma cells were removed, given genes encoding interleukins, then reimplanted. The genes were then expressed and evoked an immune response. The tumors shrank. In another strategy, researchers inject liposomes bearing genes encoding HLA proteins directly into tumors. Figure 19.6 describes this approach for an HLA protein called HLA-B7. This protein, when displayed on tumor cell surfaces, stimulates the immune system to respond to the tumor as if it were foreign tissue.

## Gene Delivery

The idea behind gene therapy is straightforward—to introduce a piece of DNA into cells that do not function normally because they lack this DNA or have a mutant version of it. Researchers send foreign DNA into cells in several ways (see table 18.3). Physical methods include electroporation, microinjection, and particle bombardment. Chemical methods include liposomes that enclose the gene cargo and lipid molecules that carry DNA across the cell membrane. The lipid carrier can penetrate the cell membrane that DNA alone cannot cross. However, lipid-based methods often fail to deliver a sufficient payload, and gene expression is transient.

One way to improve upon lipid-mediated gene delivery is to link the lipid to a peptide that carries the gene of interest and also binds it to a specific integrin on the target cell. (Recall from chapter 2 that an integrin is a cellular adhesion molecule.) So far this approach has restored enzyme production in cells of patients with either of two types of lysosomal storage diseases—Fabry disease and fucosidosis. Another approach is to alter liposome surfaces so that they resemble viruses that can enter particular cell types. These liposomes are called artificial viruslike particles.

Biological approaches to gene transfer utilize a vector, which often is a viral genome. Researchers remove the viral genes that cause symptoms or alert the immune system and add the corrective gene. Different viral vectors are useful for different types of experiments. A certain virus may transfer its cargo with great efficiency but carry only a short DNA sequence. Another virus might carry a large piece of DNA but send it to many cell types, causing side effects. Still another virus may not infect enough cells to alleviate symptoms. Some retroviruses have limited use because they infect only dividing cells. Adenovirus as a vector has been replaced in many clinical trials because of Jesse Gelsinger's apparent allergic reaction to it. Table 19.3 lists the characteristics of viral vectors used in gene therapy experiments.

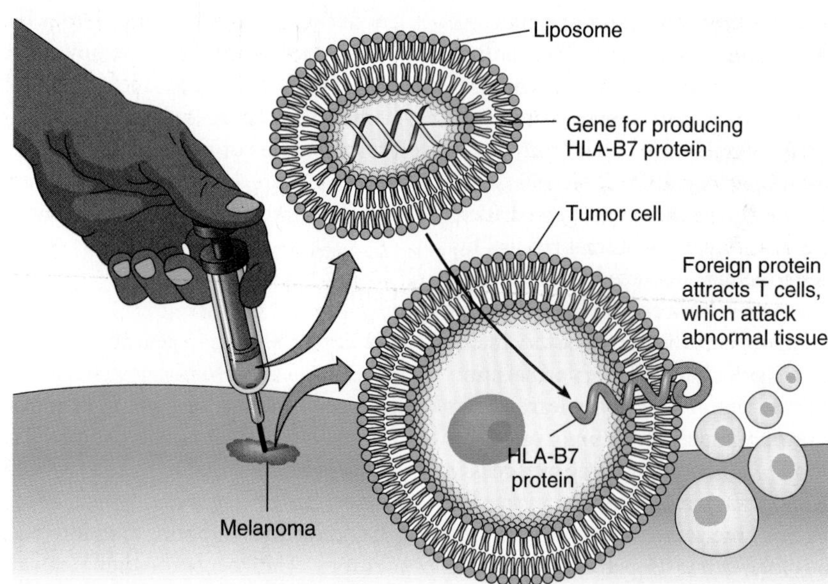


**figure 19.6**

**Gene therapy for skin cancer.** A gene that encodes a cell surface protein that attracts the immune system's tumor-killing T cells is injected directly into a melanoma tumor.

**table 19.3**

**Vectors Used in Gene Therapy**

| Vector | Characteristics | Applications |
|---|---|---|
| Adeno-associated virus (AAV) | Integrates into specific chromosomal site<br>Long-term expression<br>Nontoxic<br>Infects dividing and nondividing cells<br>Carries small genes | Cystic fibrosis<br>Sickle cell disease<br>Thalassemias |
| Adenovirus (AV) | Large virus, carries large genes<br>Transient expression<br>Evokes immune response<br>Infects dividing and nondividing cells, particularly in respiratory system | Cystic fibrosis<br>Hereditary emphysema |
| Herpes | Long-term expression<br>Infects nerve cells | Brain tumors |
| Retrovirus | Stable but imprecise integration<br>Long-term expression<br>Most types infect only dividing cells<br>Nontoxic<br>Most established in clinical experience | Gaucher disease<br>HIV infection<br>Several cancers<br>ADA deficiency |

Sometimes researchers attempt to send a virus where it would normally go. For example, the adenoviruses used to transport CFTR genes to the airway passages of people with cystic fibrosis normally infect lung tissue. By adding portions of other viruses, researchers can redirect a virus to infect a certain cell type. For example, adeno-associated virus (AAV) infects many cell types. Researchers add to the AAV genome a promoter from a parvovirus B19 gene that restricts AAV's infectivity to erythroid progenitor cells in bone marrow. These cells give rise to immature red blood cells, which contain nuclei and produce proteins. After a short time in the circulation, the immature cells extrude their nuclei, becoming doughnut-shaped red blood cells. If the AAV parvovirus B19 DNA is also linked to a human gene that encodes a protein normally found in red blood cells, then the entire vector can treat an inherited disorder of blood. Sickle cell disease is a good example. A gene therapy for sickle cell disease carries a normal human beta globin gene in an AAV genome to erythroid progenitor cells (figure 19.7).

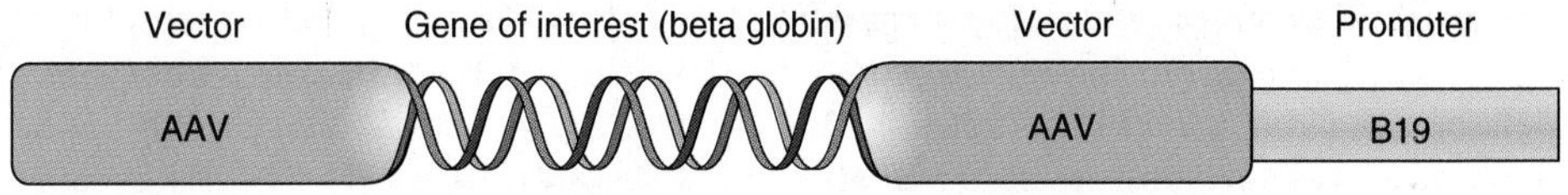

**figure 19.7**

**Correcting defects in red blood cells.** A promoter from a parvovirus B19 gene directs the adeno-associated virus (AAV) genome harboring a human beta globin gene to erythroid progenitor cells, which give rise to cells that mature into red blood cells.

An entirely different approach to gene therapy, called **chimeraplasty,** does not replace an abnormal gene, but attempts to repair it in place. This avoids using foreign genetic material and having to control the expression of the newly introduced gene. The chimeraplasty technique introduces a short, synthetic piece of DNA interspersed with some RNA, called a chimeraplast, into affected cells. A *chimera* is a Greek mythological creature with body parts from several types of animals. A chimeraplast is so-named because it consists of DNA and RNA.

The chimeraplast folds into a small double helix. The RNA portion provides stability and leads the chimeraplast to attach to its counterpart in a cell's genome. The chimeraplast is large enough, at 25 to 40 nucleotides, to bind to a specific target at the site of the mutation; a smaller molecule might bind to several sites in the genome. The chimeraplast introduces a mismatched base pair, which attracts the cell's natural DNA repair systems. Using the DNA portion of the chimeraplast as a template, the repair enzymes synthesize a normal sequence of nucleotides. The technique has worked on human cells in culture, correcting mutations in the beta globin gene.

Not all approaches to gene therapy utilize liposomes or disabled viruses. Researchers from the University of Alabama at Birmingham and Women's and Children's Hospital in South Australia have turned a commonly used antibiotic drug, gentamicin, into a promising gene therapy. Gentamicin kills bacteria by causing their ribosomes to read through the stop codons that normally signal the end of a protein-encoding gene. The drug doesn't harm humans because it doesn't bind as strongly to our ribosomes as it does to bacterial ribosomes.

The researchers reasoned that if gentamicin reads through just a few stop codons that cause genetic disease by shortening the protein product, enough of the gene's product might be made to improve health. They chose Hurler syndrome because two-thirds of known mutations in the gene are nonsense, in which premature stop codons shorten the product, an enzyme called alpha-L-iduronidase. Symptoms include hearing loss, dwarfism, mental retardation, heart and breathing problems, and an enlarged liver and spleen. Hurler syndrome is lethal in early childhood. So far, the antibiotic-based gene therapy works in cells growing in culture. A biotechnology company is exploring the approach to treat cystic fibrosis and Duchenne muscular dystrophy, too.

### Key Concepts

Gene therapy may be *in situ, ex vivo,* or *in vivo.* • Protein-based therapies replace gene products and treat the phenotype. Gene therapies replace malfunctioning or absent genes. • Germline gene therapy targets gametes or fertilized ova and is heritable. Somatic gene therapy targets various types of somatic tissue as well as cancer cells and is not heritable. • Vectors for delivering genes include liposomes and viruses. Chimeraplasty uses DNA repair to restore a normal gene sequence. An antibiotic may override nonsense mutations, restoring protein production.

## 19.3 A Closer Look: Treating Sickle Cell Disease

Sickle cell disease is perhaps the best-studied inherited illness, because the molecular defect was identified in 1949. We also understand the precise mutation that causes the disease, how the phenotype arises, and how the globin genes are regulated in the embryo, fetus, and newborn. Knowledge of this developmental regulation has led to a unique type of treatment.

Recall from chapter 11 that sickle cell disease is caused by a missense mutation in the beta globin gene that replaces a glutamic acid with valine at the sixth amino acid position (see figure 11.2). The valine protrudes from the otherwise globular molecule, causing it to latch onto other beta globin molecules and form sheets that bend the red blood cell. Then, the red blood cell's membrane changes and exposes receptors that glue it to blood vessel linings. The sickle-shaped, sticky cells lodge in the small passageways of the circulatory system—but only when the blood is low in oxygen. Globin molecules have slightly different shapes, depending upon the presence of oxygen. The blocked circulation causes the acute pain of a sickle cell crisis. *Crisis* is an apt term—one boy described the pain as similar to having your hand crushed in a car door.

Sickled red blood cells live only 20 days, compared to the normal 120-day lifespan, and as they die, anemia develops. The spleen works overtime to handle the onslaught of dying cells, abandoning its immune functions. Infections grow more troublesome. Newborn screening tests detect babies who have inherited sickle cell disease, and doctors administer prophylactic antibiotics from then on to prevent infections.

Both time-tested and experimental treatments are available for sickle cell disease. Blood transfusions and pain medications help patients through crises. Bone marrow transplants are risky but have led to complete cures, as the youngster in figure 19.8 experienced. Researchers can test new treatments using a knockout/transgenic mouse, described in chapter 18, that manufactures completely human sickled hemoglobin. Experimental gene therapy approaches use AAV and chimeraplasty.

A novel way to prevent sickling, using a cancer drug called hydroxyurea, is based on regulation of the globin protein chains during prenatal development. The story began with the observation that people who have a blood abnormality called "hereditary persistence of fetal hemoglobin" (HPFH) are healthy or have very mild anemia. Fetal hemoglobin, adapted to the different oxygen requirements before birth, consists of two alpha chains and two gamma chains. Normally, after six months of age, beta chains gradually replace the gamma chains. In people who have HPFH, the switch from gamma to beta chains doesn't happen. But the fact that people with HPFH are healthy shows that fetal hemoglobin does no harm in an adult. Could "turning on" fetal hemoglobin cure sickle cell disease by replacing the mutant beta chains with normal gamma chains?

In adults, the inactive gamma genes are normally silenced with methyl groups. A drug that removes the methyl groups might expose the gamma genes, enabling them to be expressed. In 1982, researchers found that a drug called 5-azacytidine indeed removed the methyl groups and raised the proportion of fetal hemoglobin in the blood. But this drug causes cancer, and therefore couldn't be used for the many years necessary to treat sickle cell disease.

The next piece in the puzzle fell into place when researchers discovered that 5-azacytidine not only removes methyl groups, but is also toxic to (or "stresses") red blood cells (erythrocytes). Which effect actually stimulates production of fetal hemoglobin? Researchers found that an existing, safer drug, hydroxyurea, also stresses the cells that give rise to red blood cells. Could it also turn on fetal globin genes?

From 1984 until 1992, experiments in monkeys, healthy humans, and sickle cell disease patients showed that hydroxyurea raises fetal hemoglobin levels. A trial began on 299 adults with sickle cell disease to see if this effect improved symptoms—half of the group received the drug, half received a placebo. By 1995, researchers called off the study. The participants receiving the drug were doing so much better than the others that it was unethical to deny the placebo group treatment. The patients receiving hydroxyurea had half the number of crises and needed fewer transfusions than the others. However, researchers are still studying the risks this drug may present if taken for many years.

The way hydroxyurea works involves biology, chemistry, and physics. First, the drug increases the proportion of gamma globin molecules in the blood. Some of the gamma globins bind to mutant beta globins, preventing them from forming the circulation-strangling sheets. This prolongs the time it takes for the mutant beta globin chains to join, simply because gamma chains are now in the mix (figure 19.9). The delay is long enough for a red blood cell to return to the lungs. Once the cell picks up oxygen, sickling cannot occur. By slowing

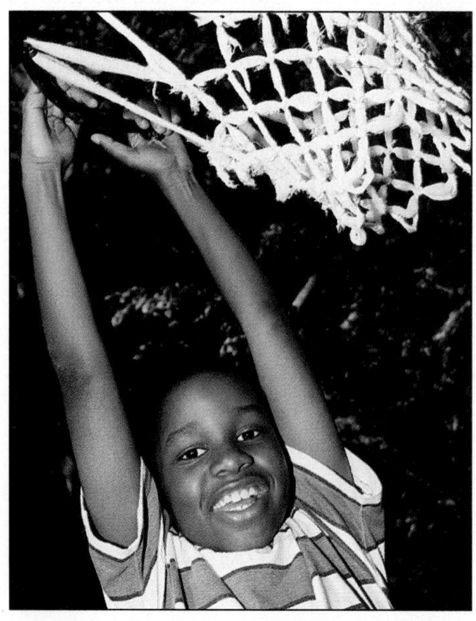

## figure 19.8

**Curing sickle cell disease.** Seye Arise was born with sickle cell disease. At a year of age, his hands and feet swelled due to blocked circulation, and he needed frequent transfusions to dilute his sickled red blood cells with normal ones. At age four, he began limping from a stroke. Because strokes can be lethal, Seye's parents allowed him to undergo an experimental bone marrow transplant, with brother Mayo donating the marrow. Four months later, Seye ran and played for the first time. He and 16 more of 22 children in the study were completely cured.

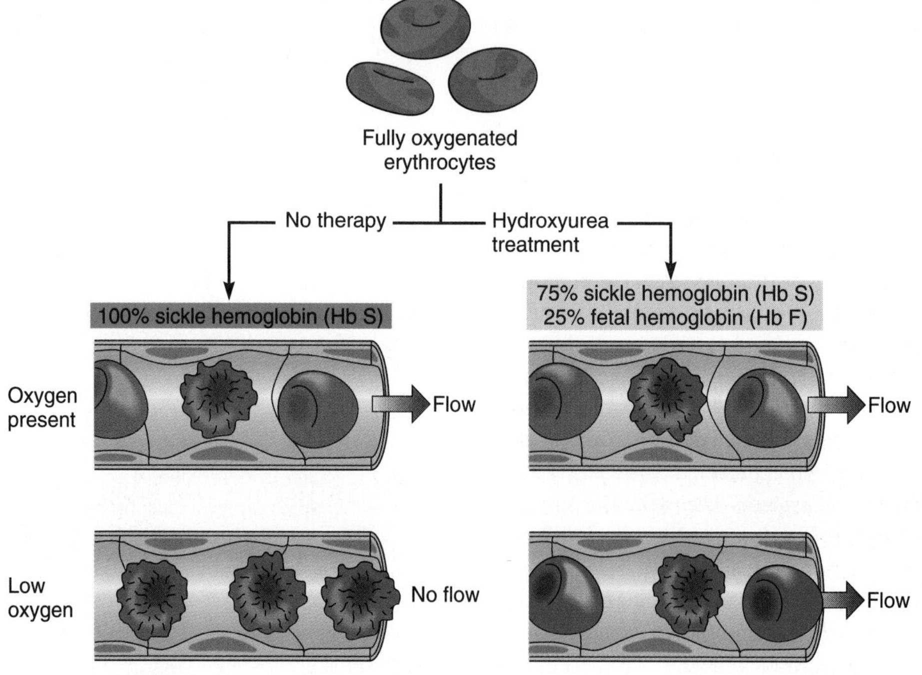


## figure 19.9

**Reactivating globin genes.** Hydroxyurea stimulates production of fetal hemoglobin (Hb F), which dilutes sickled hemoglobin (Hb S). With fewer polymerized hemoglobin molecules, the red cells do not bend out of shape as much and can reach the lungs. Oxygen restores the cells' shapes, averting a sickling crisis.

the sickling process, hydroxyurea corrects a devastating phenotype.

Teaming hydroxyurea with the kidney hormone erythropoietin (EPO) might work even better because EPO stimulates bone marrow to produce more red blood cells (see chapter 18, Bioethics: Choices for the Future). Other compounds are also being tested for their ability to reactivate dormant fetal globin genes.

**Key Concepts**

The more we know about a genetic disease, the more diverse treatment options we can develop. Conventional gene therapy protocols, as well as a unique approach that reactivates fetal hemoglobin genes, are being applied to treat sickle cell disease.

## 19.4 Genetic Screening and Genetic Counseling

Despite setbacks in gene therapy experiments, thousands of people have already participated in clinical trials, and, given the wealth of information the genome project is providing, gene therapies are likely to be part of medical practice in the future. Until very recently, however, most physicians and other medical professionals received limited training in genetics. Bridging the gap as human genetics becomes increasingly clinical has been a small group of healthcare workers, **genetic counselors.** About 99 percent of the 2,200 or so genetic counselors in the United States have masters degrees in the field. The other 1 percent of the members of the National Society of Genetic Counselors, including the author of this book, have PhDs. Genetic counselors often assist clinical geneticists, who usually have medical degrees and/or doctorates.

Since the 1970s, genetic counselors have led the way in explaining genetic disease inheritance patterns and recurrence risks to patients. Genetic counselors work in medical centers, hospitals, biotechnology companies, and medical practices. Recently, other healthcare workers have been incorporating explanations of genetics into their practices. One study surveyed hospitals, clinics, geriatric facilities and schools, and found that 70 percent of dietitians, social workers, physical therapists, psychologists, occupational therapists, audiologists, and speech-language pathologists regularly discuss genetic principles with their patients and clients.

### Genetic Counselors Provide Diverse Services

The knowledge that genetic counselors share with their patients reflects what you have read so far in this book, but it is presented in a personalized manner. A genetic counselor might explain Mendel's laws, but substitute the particular family's situation for the pea plant experiments.

A genetic counseling session usually begins with a discussion of the family's health history. Using a computer program or pencil and paper, the counselor derives a pedigree, then explains the risks of recurrence for particular family members other than the ones who are obviously already affected by the illness (figure 19.10). If the family has an inherited illness, the counselor provides detailed information and may refer the family to support groups. If a couple wants to have a biological child but to avoid the genome of the parent who is a carrier of an inherited illness, a discussion of assisted reproductive technologies (see chapter 21) might be in order.

A large part of the genetic counselor's job is to determine for whom specific biochemical, gene, or chromosome tests are appropriate, and arrange for people to take the tests (table 19.4). Certain tests for newborns are mandated by law, conducted on a small blood sample taken from the heels of newborns (table 19.5). The counselor interprets test results and helps the patient or family choose among medical options. Genetic counselors are also often asked to provide information on drugs that can cause birth defects, although this really is related to development, not genetics.

Counseling prospective parents on genetic issues is different from working with families dealing with a specific inherited disease. Prenatal genetic counseling is typically a straightforward analysis of population (empiric) and family-based risks, an explanation of tests, and a determination of whether or not the benefits of testing outweigh the risks. The family decides how to proceed, and, happily, most of the time, amniocentesis, chorionic villus sampling, or maternal serum screening indicates no detectable problem. A genetic counselor must explain that tests do not guarantee a healthy baby. For example, many people assume that amniocentesis checks every gene, when it actually is limited to detecting large-scale chromosome aberrations. Single-gene tests are separate. If a test reveals that the fetus

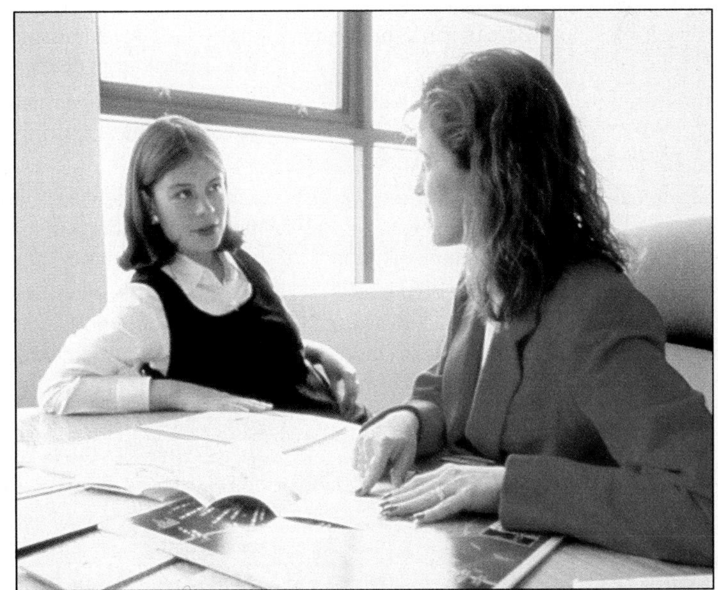

**figure 19.10**

**Genetic counselors provide genetic information to families.**

**table 19.4**

**Types of Genetic Screening Tests**

| Type | Examples |
|---|---|
| Prenatal | Amniocentesis, chorionic villus sampling, maternal serum markers, ultrasound for detection of increased risk for trisomy 21 and neural tube defects |
| Newborn screening | Sickle cell disease, PKU, maple syrup urine disease |
| Carrier screening | Heterozygote tests for cystic fibrosis, sickle cell disease, Tay-Sachs disease, thalassemia |
| Inherited predisposition | Breast and ovarian cancer genes *BRCA1* and *BRCA2*, Alzheimer disease |
| Characterizing tumor type | Breast cancer variant *Her-2/neu* indicates patient may respond to monoclonal antibody-based drug Herceptin |
| Diagnosis of genetic disease to confirm physical exam | Inborn errors of metabolism, muscular dystrophies, DNA repair disorders, inherited anemias, neurofibromatosis, Prader-Willi syndrome, Tay-Sachs disease, movement disorders |
| Presymptomatic testing for adult-onset Mendelian disorder | Huntington disease |
| Identity | Paternity testing, forensics to identify bodies |

has a serious medical condition, the counselor might discuss treatment plans and the option of ending the pregnancy.

Counseling when there is an inherited disease in a family is another matter. For recessive disorders, the affected individuals are usually children, whose condition often came as quite a surprise. Counseling for subsequent pregnancies requires great sensitivity and sometimes a little bit of mind reading. Many people become angry at the idea of terminating a pregnancy when the fetus has a condition that already affects their living child, and genetic counselors must recognize and respect these feelings, tailoring the options discussed accordingly. Counseling for adult-onset disorders may include preparing a patient to decide whether to take a presymptomatic test that will predict whether he or she will develop the family's disease. This is the case for Huntington disease, for example.

What a genetic counselor does *not* generally do is offer an opinion or suggest a course of action—instead he or she provides choices and tries to educate the family to make informed decisions. This objective approach is called "nondirective" genetic counseling.

**table 19.5**

**Newborn Screening Tests**

| Disease | Incidence | Symptoms | Treatment |
|---|---|---|---|
| Biotinidase deficiency | 1/70,000<br>Rare in blacks or Asians | Convulsions, hair loss, hearing loss, vision loss, developmental abnormalities, coma, sometimes death | Most physical symptoms reversed by oral biotin |
| Maple syrup urine disease | 1/250,000–300,000<br>More common in blacks and Asians | Lethargy, mental retardation, urine smells sweet, irritable, vomiting, coma, death by age one month | Diet very low in overproduced amino acids |
| Congenital adrenal hyperplasia | 1/12,000 whites<br>1/15,000 Jews<br>1/680 Yupik eskimos | Masculinized female genitalia, dehydration, precocious puberty in males, accelerated growth, short stature, ambiguous sex characteristics | Hormone replacement, surgery |
| Congenital hypothyroidism | 1/3,600–5,000 whites<br>Rare in blacks, more common in Hispanics | Mental retardation, growth failure, hearing loss, underactive thyroid, neurological impairment | Hormone replacement |
| Galactosemia | 1/60,000–80,000 | Muscle weakness, cerebral palsy, seizures, mental retardation, cataracts, liver disease | Galactose-free diet |
| Homocystinuria | 1/50,000–150,000 | Blood clots, thin bones, mental retardation, seizures, muscle weakness, mental disturbances | Low-methionine, high-cysteine diet, drugs |
| Phenylketonuria (PKU) | 1/10,000–25,000 | Mental retardation | Low phenylalanine diet |
| Sickle cell and other hemoglobinopathies | 1/400 U.S. blacks | Joint pain, severe infection, leg ulcers, delayed maturation | Prophylactic antibiotics |

## Scene from a Sickle Cell Disease Clinic

To get an idea of the diversity of skills a genetic counselor must develop, imagine a family with a newborn child who has sickle cell disease.

The family might first meet the counselor a few days after the child's birth, when the state-mandated newborn screening test detects the mutation. The counselor talks about the switch from fetal to adult hemoglobin, which explains why the newborn will be healthy for about six months. This might help the family to understand why the child will be started on antibiotics to prevent infection several weeks before the six-month mark. The counselor refers the family to support groups and provides as much information as the family seeks. Explaining how two carrier parents could conceive a child who inherits two mutant forms of the gene requires the counselor to reassure parents who may struggle with feelings of responsibility or guilt.

When the parents understand the autosomal recessive mode of inheritance, then the counselor might show them a pedigree for their family, indicating which other individuals might be carriers and suggesting that the parent's siblings be tested. Often the hardest information to communicate is the risk to subsequent children. Many people do not understand how probability works, and they assume that if one child has sickle cell disease, the next three will be healthy. Of course, genetic disease doesn't work this way. Each conception is a separate event.

The genetic counselor will probably meet with the family during future pregnancies to review the mode of inheritance and recurrence risk. Parents may want to test the fetus, to see if it has inherited the disease. Perhaps the already affected child will be a candidate for a bone marrow or cord blood stem cell transplant if a sibling is healthy and compatible—the counselor can explain these risky procedures and help make plans. The counselor becomes a lifeline of sorts, a source of communication and a liaison between the growing family and the staff of the medical center.

## Genetic Counseling Quandaries and Challenges

Because genetic counseling is a field based on communicating information, misunderstandings and tough situations involving confidentiality can arise. And because genetic counselors are human, they risk presenting information that reveals their opinions or feelings. Consider the following true examples:

- A couple has suffered several miscarriages, then have a child with multiple problems. A chromosome check finds that the father carries a translocation. The man's siblings may also carry the translocated chromosome, but the couple does not want to tell anyone the test result.
- A man and woman have the same autosomal recessive form of blindness but have children, even though they know that they will be blind because they too will be homozygous recessive. A genetic counselor suggests adoption or artificial insemination by donor. The couple considers this suggestion a value judgment on their choice to have a child.
- Mr. and Mrs. Gold know that their mildly retarded son has an extra Y chromosome. A questionnaire from the special education department in their school district asks if he's ever had a chromosome test. If the parents answer yes, their child may be stigmatized. If they answer no, he may not get appropriate support. They do not know what to do.
- A couple has a child with PKU. It costs \$8,000 a year to supply the special diet that prevents her from becoming mentally retarded. The woman becomes pregnant again. A test to detect PKU prenatally costs \$1,200. Insurance will not cover the cost of the test, nor will it pay to feed a second child with PKU. The couple elect not to have more children.
- A subway driver has familial hypercholesterolemia. Although he has chest pains and high blood pressure and serum cholesterol, he has never had a heart attack. He knows that he may suddenly die, but he won't tell the transportation department because he is retiring in a few months. The genetic counselor knows the diagnosis.

## Perspective: A Slow Start, But Great Promise

Medical genetics is no longer just the realm of a few physician specialists, genetic counselors, and the families that have very rare, so-called orphan inherited diseases. As we begin to understand precisely how genes control the functioning of the human body, genetic testing, counseling, and therapy will become a routine part of medical care, applied to common as well as rare disorders. But it has been a long time in coming.

When the age of gene therapy dawned in the 1990s, expectations were high—and for good reason. Work in the 1980s had clearly shown abundant, pure, human biochemicals, useful as drugs, could come from recombinant single cells and from transgenic organisms. It was merely a matter of time, researchers speculated, until genetic altering of somatic tissue would routinely treat a variety of ills.

In reality, gene therapy progress has been slow. Boys with Duchenne muscular dystrophy who receive myoblasts with healthy dystrophin genes do not walk again. People with cystic fibrosis who sniff viruses bearing the CFTR gene do not enjoy vastly improved lung function. Genetic correction for babies with SCID given normal ADA genes in umbilical cord stem cells takes years, with only partial correction achieved.

Perhaps we expected too much too soon, but at a molecular and cellular level, gene therapy *is* working. The patients with muscular dystrophy, cystic fibrosis, and SCID have cells that have accepted and expressed therapeutic genes. The challenge now is to find just the right vector to deliver a sustained, targeted, and safe genetic correction without alerting the immune system.

**KEY CONCEPTS**

A genetic counselor provides information to individuals, couples, and families on modes of inheritance, recurrence risks, genetic tests, and treatments. The counselor encourages informed decision making, but must be respectfully sensitive to individual choices.

*Bioethics: Choices for the Future*

## Canavan Disease: Patients Versus Patents

When Debbie Greenberg gave birth to Jonathan in 1981, she and her husband Dan had no idea that they would one day be leading the first effort to challenge how a researcher and a hospital obtained a patent on a gene. Debbie and Dan first suspected all was not right with Jonathan when he was only three months old. The baby could not control his wobbly head, didn't seem able to maintain eye contact, and could not maneuver his fist into his mouth, a trademark of young babies. By six months of age, when he still was not progressing in attaining motor skills, the Greenbergs finally received a diagnosis for their son—Canavan disease. They were each carriers of the autosomal recessive condition. Although Jonathan would live 11 years, his brain never developed past infancy. The couple had an affected daughter, Amy, a few years after Jonathan was born, and three healthy children.

Shortly after Jonathan's diagnosis, the Greenbergs joined the National Tay-Sachs and Allied Diseases Association, Inc. They also started a patient support group, the Canavan Foundation, which included establishment of a tissue bank that stored blood, urine, and autopsy tissue from affected children. In 1987, the Greenbergs met Dr. Reuben Matalon at a Tay-Sachs screening event in Chicago, and convinced him to begin a search for the Canavan gene. The Greenbergs helped to collect tissue from families from all over the world, which was critical to Dr. Matalon's success in identifying the gene and the causative mutation in 1993, when he was working at Miami Children's Hospital.

Finding the gene made it possible to detect the mutation, which could be used to confirm diagnoses and detect carriers, and in prenatal diagnosis. By 1996, the Canavan Foundation was offering free testing. But unknown to the members of the organization who had donated their childrens' tissues for the gene search, Dr. Matalon and Miami Children's Hospital had filed for a patent on their discovery. The U.S. Patent and Trademark office granted invention number 5,679,635—the Canavan gene—in 1997. A year later, the American College of Obstetricians and Gynecologists advised their physician members to offer carrier testing for Canavan disease to Ashkenazi Jewish patients, because 1 in 40 such women is a carrier. Identifying couples in which both people are carriers would give them the option of avoiding giving birth to affected children, a strategy that has reduced the number of children born with the similar Tay-Sachs disease to nearly zero. That same year, Miami Children's Hospital began to exercise its patent rights by requiring that doctors and diagnostic laboratories charge for a Canavan test. All of a sudden, families whose donations—both monetary and biological—had made the discovery of the gene possible found themselves having to pay for carrier and prenatal tests. They were outraged.

On November 30, 2000, a group of parents and three nonprofit organizations filed suit in Chicago against Dr. Reuben Matalon and Miami Children's Hospital. The suit does not challenge the patent, but does challenge the way in which it was obtained—in secret, they claim. They wish to recover earnings from the gene test to be turned over to the families who had to pay to offset licensing fees.

The Greenbergs' fight against a disease gene patent not only is a legal precedent, but also sounded a warning bell to other patient groups. As a direct result of the Canavan case, Sharon and Patrick Terry, of Sharon, Massachusetts, started a support group and tissue bank for the disorder that their child has—pseudoxanthoma elasticum (PXE), which causes connective tissue to calcify. Like the Greenbergs, the PXE parents supplied their children's tissue, but they stipulated that their group be listed as a coinventor on any gene patents. By doing this, the families hope that they will be able to retain control over the fate of the gene that they helped to discover.

# Summary

### 19.1 Gene Therapy Successes and Setbacks

1. Protein supplementation, from donors and then from recombinant DNA technology, preceded gene therapy.
2. **Gene therapy** replaces malfunctioning genes.
3. Gene therapy for ADA deficiency evolved in stages, beginning with enzyme replacement, then gene therapy in white blood cells, then gene therapy in stem cells that could more effectively replace the affected cells.
4. A death in a gene therapy trial for OTC deficiency appears to have been due to an immune system response to the viral genome used to introduce the gene.
5. Gene therapy for Canavan disease enables brain neurons to produce a missing enzyme.

### 19.2 The Mechanics of Gene Therapy

6. ***Ex vivo* gene therapy** is applied to cells outside the body that are then reimplanted or reinfused into the patient. ***In situ* gene therapy** occurs directly on accessible body parts. ***In vivo* gene therapy** is applied in the body.
7. Hereditary hemochromatosis is treated by having blood removed.
8. **Germline gene therapy** affects gametes or fertilized ova, affects all cells of an individual, and is transmitted to future generations. It is a form of transgenesis, and is not performed in humans.

9. **Somatic gene therapy** affects somatic tissue and is not passed to offspring.
10. Gene therapy delivers new genes, and encourages production of a needed substance at appropriate times and in therapeutic (not toxic) amounts.
11. Several types of vectors are used in gene therapy, including liposomes and viral genomes.
12. **Chimeraplasty** uses DNA repair of an introduced synthetic DNA/RNA hybrid to correct a mutation.
13. An antibiotic that causes a ribosome to read through a nonsense mutation may treat some diseases.

### 19.3 A Closer Look: Treating Sickle Cell Disease

14. We know the mutation that causes sickle cell disease, how the phenotype arises, and how the globin genes are developmentally regulated.
15. In addition to genes delivered on vectors, drugs to reactivate fetal globin genes may be useful in treating sickle cell disease.

### 19.4 Genetic Screening and Genetic Counseling

16. **Genetic counselors** are medical professionals who, since the 1970s, have provided information on inheritance patterns, disease risks and symptoms, and available tests and treatments.
17. Prenatal counseling and counseling a family coping with a particular disease pose different challenges.

# Review Questions

1. What are the three stages of the evolution of treatments for single-gene disorders?
2. Describe how a gene therapy works to treat
   a. ADA deficiency.
   b. ornithine transcarbamylase deficiency.
   c. Canavan disease.
   d. sickle cell disease.
3. What are two challenges in providing gene therapy for Duchenne muscular dystrophy?
4. Explain the differences among *ex vivo, in situ,* and *in vivo* gene therapies. Give an example of each.
5. Would somatic gene therapy or germline gene therapy have the potential to affect evolution? Cite a reason for your answer.
6. What factors would a researcher consider in selecting a viral vector for gene therapy?
7. Explain how the antibiotic gentamicin might be used for gene therapy. Why wouldn't it work for sickle cell disease?
8. Why is a bone marrow transplant from a healthy child to a sibling with sickle cell disease technically not gene therapy, while removing an affected child's bone marrow and replacing the mutant gene with a normal one is gene therapy?
9. Why is it easier to "fix" a liver with gene therapy than to treat a muscle disease?
10. Gene therapies for Duchenne muscular dystrophy have used AV, AAV, liposomes, and chimeraplasty. Explain how each approach works.

# Applied Questions

1. Inherited emphysema due to alpha-1 antitrypsin (AAT) deficiency is rare, but emphysema caused by cigarette smoking is fairly common. The same mechanism underlies both forms of the illness—chemical irritants in smoke attract white blood cells to the lungs, which produce an altered form of AAT so that elastase builds up and destroys lung tissue. Do you think that if gene therapy becomes available and is rationed, people with inherited AAT deficiency should have preference over those with smoking-induced emphysema? Cite a reason for your answer.
2. A lentivirus is a rare type of retrovirus that can infect nondividing cells, therefore widening its applicability as a gene therapy vector. HIV is a lentivirus that is being evaluated in a disabled form as a vector for gene therapy. What would have to be done to it to make this feasible?
3. Researchers have discovered that red blood cells from people with sickle cell disease hold oxygen longer when exposed to nitric oxide in a test tube. How can this observation be used to develop a new treatment for sickle cell disease?
4. Parkinson disease is a movement disorder in which neurons in a part of the brain called the substantia nigra can no longer produce the neurotransmitter dopamine. This neurotransmitter is not a protein. What are two difficulties in developing gene therapy for Parkinson disease?
5. Create a gene therapy by combining items from the three lists below. Describe the condition to be treated, and how a gene therapy might correct the symptoms.

| Cell Type | Vector | Disease Target |
|---|---|---|
| fibroblast | AV | Duchenne muscular dystrophy |
| skin cell | AAV | Alzheimer disease |
| glial cell | retrovirus | sickle cell disease |
| immature red blood cell | liposome | cystic fibrosis |
| immature muscle cell | herpes virus<br>naked DNA | glioma<br>melanoma |

6. How would you, as a genetic counselor, handle the following situations (all real)? (You might have to consult past chapters for specific information.)
   a. A couple in their early forties, who do not have any children, are expecting a child. Amniocentesis indicates that the fetus has XXX, a chromosomal syndrome that might never have been noticed without the test. When they learn of the abnormality, the couple asks to terminate the pregnancy.
   b. A couple's two sons have Duchenne muscular dystrophy. They want to have a third child, but would like it to be a girl, so that she could not inherit this illness.
   c. A 25-year-old woman gives birth to a baby with trisomy 21 Down syndrome. She and her husband are shocked—they thought that this could only happen to a woman over the age of 35.
   d. Two people of normal height have a child with achondroplastic dwarfism, an

autosomal dominant trait. They are concerned that subsequent children will also have the condition.

e. A newborn has a medical condition not associated with any known gene mutation or chromosomal aberration. The parents want to sue the genetics department of the medical center because the amniocentesis did not indicate a problem.

7. Genes can be transferred into the cells that form hair follicles. Would gene therapy to treat baldness most likely be *ex vivo, in situ,* or *in vivo?* Cite a reason for your answer.

8. Suggest three specific ways to use protein or gene therapy to treat sickle cell disease.

9. Why might a gene therapy for Canavan disease be more likely to pass requirements of a bioethics review board than the trial that Jesse Gelsinger took part in?

## Suggested Readings


Bird, Thomas D. May 1999. Outrageous fortune: The risk of suicide in genetic testing for Huntington disease. *The American Journal of Human Genetics* 64:1289. George Huntington called the disease named after him "that form of insanity that leads to suicide." The despair associated with a diagnosis makes genetic testing very difficult.

Estruch, Elaine, et al. September/October 2001. Nonviral, integrin-mediated gene transfer into fibroblasts from patients with lysosomal storage diseases. *Journal of Gene Medicine* 3, issue 5:488–97. Peptides bearing healing genes are used to attract carrier lipids to integrins on target cells.

Ferber, Dan. November 23, 2001. Gene therapy: Safer and virus-free? *Science* 294:1638–42. The death of Jesse Gelsinger focused attention on nonviral vectors for gene therapy.

Kling, James. July 19, 1999. Genetic counseling: the human side of science. *The Scientist* 13:10. Genetic counseling requires a special mix of skills and talents.

Lewis, Ricki. April 15, 2002. Hereditary hemochromatosis: Too soon for genetic testing? *The Scientist* 16, no. 8:22. Many people who inherit the genotype for this disorder do not develop the phenotype.

Lewis, Ricki. April 16, 2001. Antibiotic corrects genetic glitch. *The Scientist* 15, no. 8:16. A commonly used antibiotic can correct genetic disease by overriding nonsense mutations.

Lewis, Ricki. March 6, 2000. Stem cell research predicts a paradigm shift. *The Scientist* 14:1. The ability of stem cells to switch fates promises to impact design of gene therapy strategies.

Lewis, Ricki. December 11, 2000. Progress in treating DMD. *The Scientist* 14, no. 22:22. Gene therapy can treat muscular dystrophy, in mice.

Mannucci, Pier M., and Edward G. D. Tuddenham. June 7, 2001. The hemophilias—from royal genes to gene therapy. *The New England Journal of Medicine* 344, no. 23:1773–79.

Smaglik, Paul. January 10, 2000. Chimeraplasty potential. *The Scientist* 14:13. Researchers usurp DNA repair to create a new type of gene therapy.

Steinberg, Douglas. September 17, 2001. Gene therapy targets Canavan disease. *The Scientist* 15, no. 18:20–21. Sending genes into the brain aboard a virus may seem like an extreme treatment, but for children with this disease, it is the only option.

Stolberg, Sheryl Gay. February 2, 2000. Agency failed to monitor patients in gene research. *The New York Times Magazine,* p. F1. The death of Jesse Gelsinger forced a reevaluation of the safety of gene therapy trials.


## On the Web

Check out the resources on our website at

**www.mhhe.com/lewisgenetics5**

On the web for this chapter, you will find additional study questions, vocabulary review, useful links to case studies, tutorials, popular press coverage, and much more. To investigate specific topics mentioned in this chapter, also try the links below:

Access Excellence **www.accessexcellence.org/AB/BA/Gene_Therapy_Overview.html**

Alpha-1-Antitrypsin Deficiency Foundation **222.alpha1.org**

American Hemochromatosis Society **www.americanhs.org/**

Children Living with Inherited Metabolic Diseases (CLIMB) **www.CLIMB.org.uk**

Computer-based Genetic Counseling **www.jama.ama-assn.org/issues/v282n/8/full/jlt1110-2.html**

Iron Overload Disease Association **www.emi.net/-iron.iod/**

Kimeragen Inc. **www.kimerageninc.com**

MPS (Mucopolysaccharide and Mucolipidoses Society) **www.mpssociety.org**

National Society of Genetic Counselors **www.nsgc.org**

National Urea Cycle Disorder Foundation **www.nucdf.org**

Online Mendelian Inheritance in Man (OMIM) **www.ncbi.nlm.nih.gov/entrez/query.fcgi?db=OMIM**

adenosine deaminase (ADA) deficiency 102700
alpha-1-antitrypsin (AAT) deficiency 107280
Canavan disease 271900
Fabry disease 301500
familial hypercholesterolemia (FH) 143890
fucosidosis 230000
glioma 137800
hemophilia A 306700
hereditary hemochromatosis (HH) 235200
Huntington disease (HD) 143100
Hurler syndrome 252800
ornithine transcarbamylase (OTC) deficiency 311250
pseudoxanthoma elasticum 264800
sickle cell disease 141900

CHAPTER

# 20

# Agricultural Biotechnology

### 20.1 Traditional Breeding Compared to Biotechnology

Humans have been altering ecosystems to feed, shelter, and clothe themselves for at least 10,000 years—this is the essence of agriculture. Biotechnology adds speed and precision to the process, but raises concerns when it combines genes from different species and releases the resulting genetically modified (GM) organisms into the environment.

### 20.2 Types of Plant Manipulations

New plant varieties, courtesy of biotechnology, have been around since the 1980s. The first new types added traits that eased cultivation and harvesting. Future applications will be more targeted toward consumers. Transgenic plants can make drugs, vaccines, and plastic, and resist herbicides and pesticides. Some plants have their own genes over- or underexpressed.

### 20.3 Release of Genetically Modified Organisms into the Environment

Genetically modifying an organism is only the first step in agricultural biotechnology. The organism must also be bred to see if the trait is expressed, then tested in microcosms and in the field to assess survival, fertility, and interactions with other organisms. Bioremediation taps or transfers an organism's metabolic ability to detect or dismantle toxins or explosives.

### 20.4 Economic, Ecological, and Evolutionary Concerns

Genetically modified organisms have impacts beyond biotechnology. They affect ecology and evolution, as well as economics, politics, and the media.

### 20.5 The Impact of Genomics

Sequencing the genomes of plants lags behind similar efforts for humans and microorganisms. Instead, researchers are focusing on groups of linked genes that provide varied disease resistances in crop plants—information that will help both biotechnology and traditional breeding practices.

"Genetically modified" (GM) foods seem to have burst into the public eye over the last few years, but the technologies that genetically alter the organisms we eat, wear, or otherwise use were developed in the 1980s. Many people have been consuming GM crops for several years, with no apparent short-term harm. However, the long-term effects of the cultivation of GM crops on the environment, economics, and even on the course of evolution are not yet known.

Most GM plants and animals are transgenic; they have received genes from other species, as discussed in chapters 18 and 19. Figure 1.6 shows transgenic "golden rice," which has genes from petunia and a bacterium that enable it to produce beta-carotene—a precursor to vitamin A.

The creation of golden rice took a decade of genetic manipulations, and two decades of breeding research prior to that. The first golden rice plants were a "proof of principle"—that is, they were created just to show that the manipulation could work. The plant varieties were not edible, and the amounts of beta-carotene produced were small. Researchers at the International Rice Research Institute in the Philippines are using conventional breeding to transfer the qualities of golden rice into palatable varieties. It will still be a few years before the rice, which will be available for free, reaches the children who have vitamin A deficiency. Figure 20.1 and tables 20.1 and 20.2 describe other genetically modified crops.

"Genetic modification" can also mean that expression of an organism's own gene is suppressed or enhanced. **Antisense technology** can silence a gene by stamping a complementary nucleic acid onto the mRNA transcribed from it. Adding or altering promoters can increase the rate of transcription, boosting gene function. Golden rice, for example, overexpresses a rice gene whose protein product enables the plant to store more iron.

GM foods do not look any different from their natural counterparts, unless a difference is intended, such as fewer or smaller seeds. So far, GM foods appear to be safe, but there are objections to their creation. One concern is that transferring genes from one species to another could create a "superweed" or an animal with an unusual ability that will enable it to drive out natural populations. Other concerns pertain to food distribution, economics, and politics. Despite their ban in some nations, GM crops accounted for 16 percent of all crops grown worldwide in 2000, and slightly more in 2001.

A starting point to evaluate the potential risks and benefits of GM foods is to understand exactly how the organisms are altered, and how the techniques differ from those of conventional agriculture. In some ways biotechnology is the safer approach in the short term, but its long-term effects are unpredictable.

**table 20.1**

**Genetically Modified Traits in Crop Plants**

| Genetically Modified Traits in Crop Plants |
|---|
| Resistance to insects, insecticides, and herbicides |
| Slowed ripening |
| Larger fruits |
| Improved sweetness |
| Faster or more uniform growth |
| Additional nutrients |
| Easier processing |
| Pharmaceutical production |
| Insecticide production |
| Ability to fix nitrogen, decreasing fertilizer requirements |
| Frost resistance |

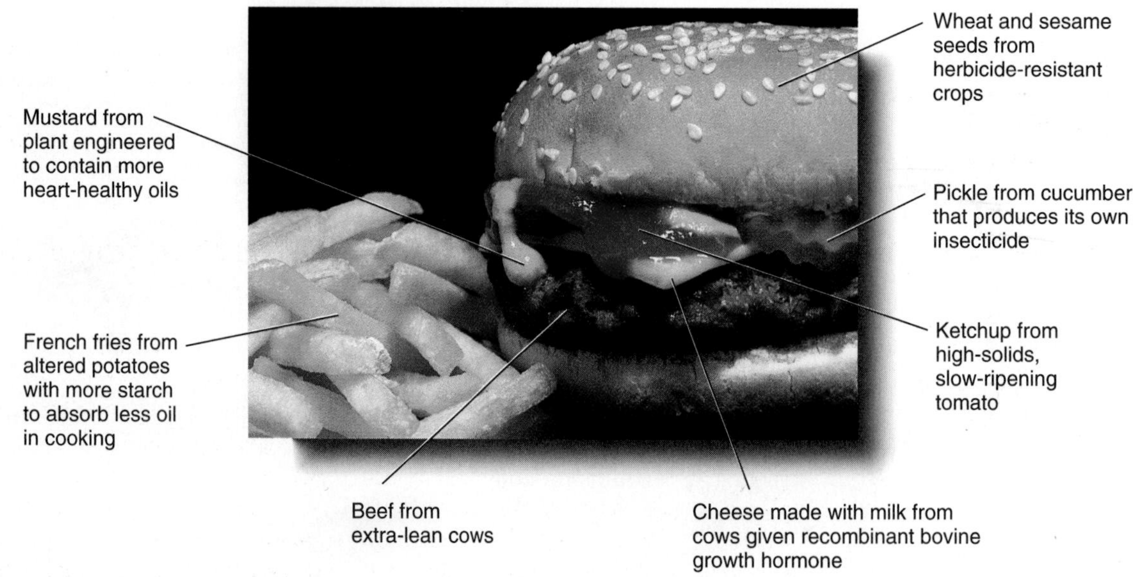
Wheat and sesame seeds from herbicide-resistant crops
Mustard from plant engineered to contain more heart-healthy oils
Pickle from cucumber that produces its own insecticide
French fries from altered potatoes with more starch to absorb less oil in cooking
Ketchup from high-solids, slow-ripening tomato
Beef from extra-lean cows
Cheese made with milk from cows given recombinant bovine growth hormone

**figure 20.1**

**GM junk food.** Crop plants are "genetically modified" to provide food with novel characteristics. Altering organisms in a gene-by-gene fashion is more precise, and controllable, than traditional breeding.

**table 20.2**

**Agricultural Biotechnology Products**

| Altered Plant | Effect |
|---|---|
| Rice with beta-carotene and extra iron | Added nutritional value |
| Canola with high-laurate oil | Can be grown domestically; less costly than importing palm and coconut oils |
| Delayed ripening tomato | Extended shelf life |
| Herbicide-resistant cotton | Kills weeds without harming crop |
| Minipeppers | Improved flavor, fewer seeds |
| Bananas resistant to fungal infection | Extended shelf life |
| Delayed-ripening bananas and pineapples | Extended shelf life |
| Elongated sweet pepper | Improved flavor, easier to slice |
| Engineered cotton fiber | Easier fabric manufacturing |
| Engineered paper pulp trees | Paper component (lignin) easier to process |
| High-starch potatoes | Absorb less oil on frying |
| Pest-resistant corn | Can resist European corn borer |
| Seedless minimelons | Single serving size |
| Sweet peas and peppers | Retain sweetness longer |

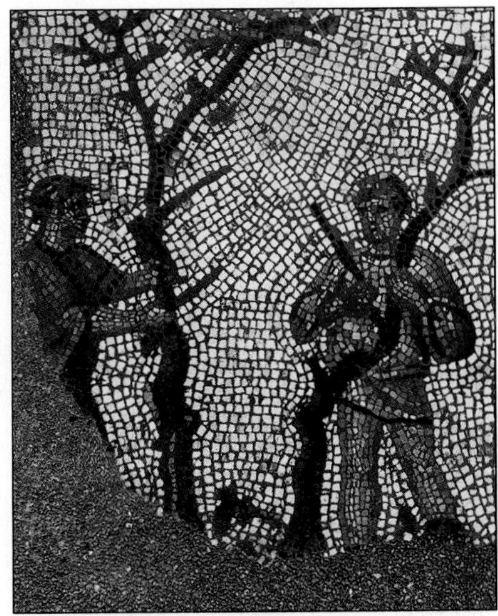

**figure 20.2**

**Ancient agricultural biotechnology.** A thousand years ago, Romans grafted branches of one variety of apple tree onto another to propagate their favorite fruits. It is commonplace today to purchase apple trees grafted to sport several different varieties of apple. Agricultural biotechnology in the 21st century manipulates plants and animals at the molecular and cellular levels.

## 20.1 Traditional Breeding Compared to Biotechnology

People have been sowing and saving seeds to control the characteristics of traits in crops and farm animals for 10,000 years. Many traditional crops are cross-species hybrids. Combining genes of different varieties of organisms is not a modern technology, although transgenic manipulations may combine genes that would never interact in nature. Nor is cloning new—taking "cuttings" to propagate plants is a form of cloning.

Figure 20.2 illustrates a technique of combining two plant varieties into one organism practiced in ancient Rome and in many societies since then—grafting part of one apple tree onto another, to increase the yield of tasty apples, but using the superior root system of a different variety. Apple grafting was popular in Europe, but in the United States Johnny Appleseed traveled the land distributing seeds but not grafts, claiming that the practice was ill-advised because it interferes with nature. After his death, however, apple grafting came into fashion in the United States.

In contrast to traditional agriculture, biotechnology manipulates organisms at the molecular, cellular, organismal, and, eventually, ecological levels. After introducing transgenes and perhaps crossing heterozygotes to obtain homozygotes, researchers observe the resulting organisms in settings similar to the natural environment. These include greenhouses and experimental fields for plants, soil microenvironments or bioreactors for bacteria, and tanks or pens for animals.

Although biotechnology is generally faster than conventional breeding, it can still be time-consuming to select the desired plant variant and then test it. Consider a chardonnay grape being developed at the New York State Agricultural Experimental Station in Geneva, New York, since 1994. The grape has been modified to overexpress a gene that encodes chitinase, an enzyme that kills the pest that causes powdery mildew. Only 1 in 50 transgenic cell lines showed resistance to the pest, and it took researcher Bruce Reisch, a professor at nearby Cornell University, four years to breed the resistant line so that it is hardy enough to withstand outside conditions. The plant is still so experimental that "taste" tests consist of chemical analyses of acid and sugar content—government regulations do not permit anyone to actually taste it. And when that becomes possible, there is no guarantee that wine experts will regard the modified plant as a true chardonnay. Until then, the location of the experimental fields is secret, for fear that environmental activists who consider the grape vines threatening will destroy them.

Figure 20.3 outlines the steps for creating GM plants, discussed throughout the chapter.

### Similar Steps, Different Degree of Precision

The steps in traditional plant breeding and biotechnology are similar. In both, the first step is for researchers to identify and breed, or engineer, an interesting trait into a plant whose other characteristics comprise a valuable package. For example, larger fruit size might be desirable in a plant that can already withstand temperature extremes and resist pests. The new variety is then tested in various habitats and seasons to determine the best growth conditions. Finally, seeds are distributed to growers.

Transgenesis

or

Gene inactivation (gene targeting, antisense)

STOP

or

Cell-based technologies (protoplast fusion, callus culture)

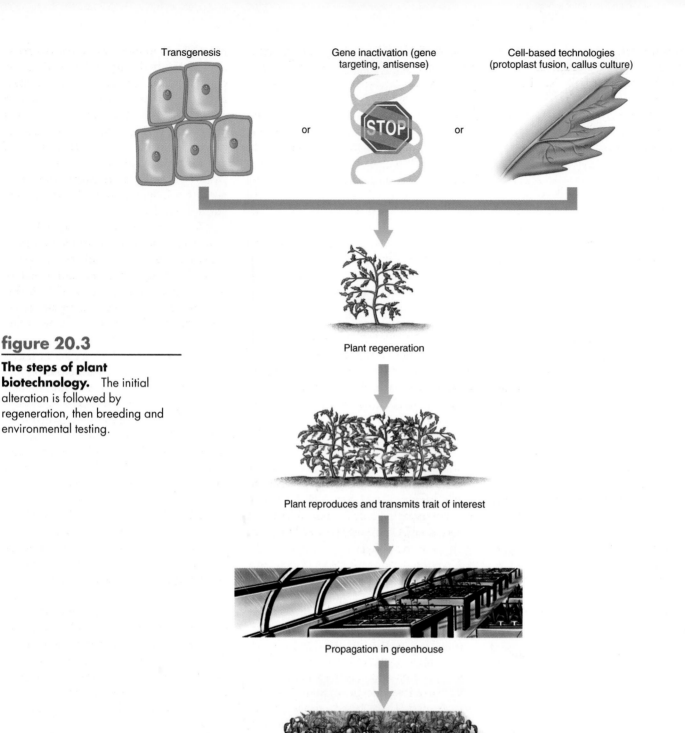

Plant regeneration

Plant reproduces and transmits trait of interest

Propagation in greenhouse

Growth from seed in field under normal conditions

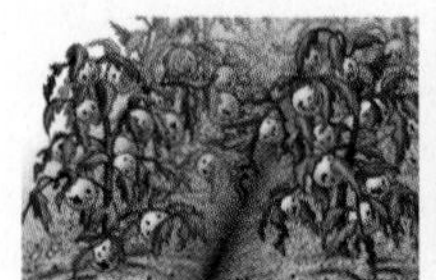

Pests

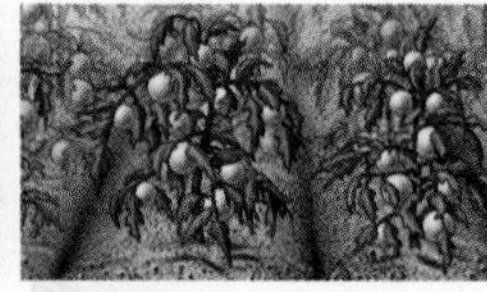

Pesticides Herbicides

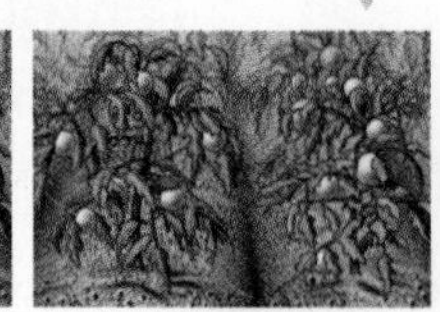

Temperature extremes

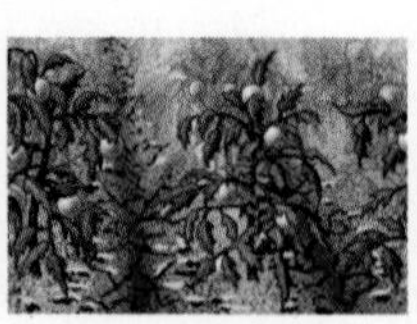

Salinity extremes

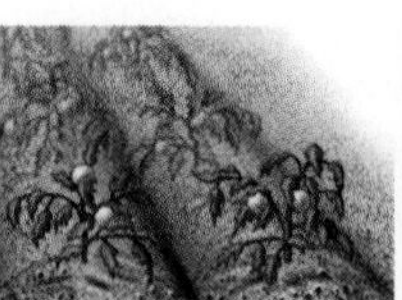

pH extremes

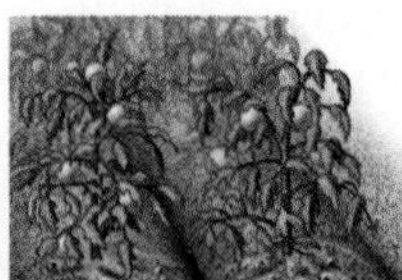

Drought

Plants tested under stressful conditions

## figure 20.3

**The steps of plant biotechnology.** The initial alteration is followed by regeneration, then breeding and environmental testing.

Traditional plant breeding introduces new varieties by a sexual route. The male gamete (sperm nucleus) in pollen carries a set of genes that may differ from the set in the female gamete (egg cell) of the ovule, even if both types of sex cells derive from the same plant. Because each gamete contributes a different combination of the parent plant's traits, offspring from a single cross are not necessarily genetically uniform—some plants may be taller, more robust, or may produce smaller seeds than others. Such genetic diversity is a direct consequence of meiosis and illustrates Mendel's laws. Traditional plant breeding is also time consuming. Breeding a polygenic trait into apples, with a generation time of four years, could take two decades or longer!

Conventional plant breeding uses gametes to derive new plant varieties, tapping into the preexisting genetic diversity of the species and selecting variants of interest. GM plants, in contrast, usually begin with somatic cells, which come from nonsexual parts of the plant such as leaves and stems. Cells taken from plant embryos are another source of somatic tissue. Somatic cells contain a complete set of genetic instructions. Plants regenerated from the somatic cells of the same plant do not have the unpredictable mixture of traits that plants derived from sexual reproduction do, because the somatic cells are usually genetic replicas, or clones, of each other. Biotechnology, then, offers a degree of precision and can ensure consistency in crop quality from season to season. It can also introduce several new traits at once.

## Government Regulation of Crops

The U.S. government regulates GM crops the same way it does conventionally bred crops. The plants are analyzed chemically, after breakdown in animals' stomachs, to see if they contain anything novel, toxic, or that could provoke an allergic reaction. The Food and Drug Administration (FDA), based on evaluations from many scientific organizations, follows a policy of "substantial equivalence." That is, if a GM food is very similar to the natural food, and does not contain a new allergen, a toxin, or changes in nutrient levels, then it is not subject to premarketing approval by the agency—just as other foods are not. In other words, a food is evaluated for its potential danger to the consumer, not for how that danger arose—a little like judging someone's behavior rather than their background. Evaluated in this manner, some products of traditional agriculture are actually more dangerous than GM crops. For example, development of a potato variety by conventional breeding was recently halted because it contained high levels of a naturally occurring toxin. But a transgenic potato with higher starch content, which absorbs less oil when fried, has safe toxin levels. The substantial equivalence requirement has been beneficial to the production and distribution of GM foods. However, in response to the concerns of some consumers, GM foods may bear labels stating their origins. This would give people the choice to avoid GM foods. About two-thirds of prepared foods in the U.S. contain GM plants, and consumers have no way of identifying them.

The FDA and U.S. Department of Agriculture approved the marketing of GM foods in 1994, then de-regulated them two years later after they continued to appear to be safe to eat. The Environmental Protection Agency agreed.

Government regulations, however, do not guarantee safety of GM crops. Humans make mistakes. For example, trouble arose over an environmental group's discovery that a variety of GM corn called Starlink that was approved only for animal feed was in taco shells. Consumers returned the taco shells, some reporting illness from the product to public health officials. But blood tests did not reveal the antibodies that would have indicated the allergic reactions that they claim to have suffered. The reason for restricting the corn to animal feed was that it contains a protein that stays longer in the digestive tract than most proteins, and some other proteins that linger in the gut are known to trigger allergic reactions. The risk was theoretical, not based on any animal tests of the specific protein. The media did not report the ending of the story—that the corn in the tacos was safe—nearly as prominently as they did the initial overreaction. But the incident did point out that government regulations can fail—a specific GM product meant to be excluded from the human food supply nonetheless got there.

## Biotechnology Provides Different Routes to Solving a Problem

An herbicide is not useful if it is so toxic that it kills crops as well as weeds. The traditional way to render a crop herbicide-resistant was to find a weed that was resistant to the herbicide and related to the crop plant. The hardy weed and its domesticated relative were then crossbred until a variant arose that retained the desired qualities of each parent plant—resistance to the herbicide, plus the characteristics that made the plant a valuable crop. In the past, if an herbicide-resistant crop could not be bred, then the herbicide was simply not used for that plant.

Today, instead of changing herbicides to fit crops, biotechnologists alter crops to withstand exposure to herbicides. In a cellular level technique called **mutant selection,** cells that grow in the presence of the herbicide are isolated and used to regenerate resistant plants. This technique uncovers naturally occurring mutations.

In a molecular level approach, transgenic technology is used to design an herbicide-resistant crop. Whereas mutant selection takes advantage of a naturally-occurring variant, transgenesis is directed—the researcher picks the transgene. The first step is to identify the biochemical reaction that the herbicide affects in the crop plants. Then, a gene is identified, from any organism, that might enable the plant to prevent or undo whatever damage the herbicide causes. Consider cotton. Since a commonly used herbicide, bromoxynil, kills cotton, researchers developed transgenic cotton containing a gene from a soil bacterium that encodes an enzyme called nitrilase, which enables cotton to grow in the presence of bromoxynil. A problem with genetically modified herbicide resistance, however, is that it ties a farmer to using a particular herbicide. Often the same company that provides GM seeds also manufactures the herbicide to which the plants are resistant.

### Key Concepts

Genetically modified organisms are transgenic or have one of their own genes over or underexpressed. GM plants have a variety of types of added traits. Traditional plant breeding and plant biotechnologies have similar steps, but traditional methods are less precise and start with gametes rather than somatic cells and take more time. • The FDA evaluates a crop's qualities, irrespective of its origin.

## 20.2 Types of Plant Manipulations

Biotechnology can alter plant structure or function at the level of the gene, the organelle, or the cell.

### Altering Plants at the Gene Level

Plants can be given genes from other species, or have their own genes silenced or overexpressed to affect traits in particular ways. Transgenesis in plants is technically challenging because the transgene must penetrate the tough cell walls, not present in animal cells, and purifying the protein product can be difficult. An advantage of plants in transgenesis is that the cells can add sugars and other side groups to proteins that the bacterial cells typically used in recombinant DNA technology cannot add.

#### *Transgenesis Revisited*

Bacteria are often the source of genes transferred to plants because they can confer useful traits such as built-in resistance to disease, insecticides, herbicides, and environmental extremes. Researchers also know more about the genomes of certain bacteria, and microorganisms are easier to grow and manipulate than other types of organisms.

In transgenesis, the donor DNA, as well as the vector DNA that transports it into the plant cell, are cut with the same restriction enzyme so they can attach to each other at the ends to form a recombinant molecule, as chapter 18 describes. The vector and its cargo gene are then sent into the plant cell.

Different types of plants take up different types of vectors. A *Ti* plasmid (see figure 18.8) easily enters the cells and nuclei of dicots, which are plants that produce two small "seed leaves" or first leaves when they germinate. Dicots include cucumbers, squashes, beans, beets, tomatoes, potatoes, soybeans, cassava, and sunflower. Examples of dicots with new traits abound, thanks to *Ti* plasmids that have carried genes into their cell nuclei. Sugar beets are rendered herbicide-resistant. This is an important trait because 37 percent of the world's sugar supply comes from this difficult-to-grow crop. Tomatoes are altered to display a protein on their cell surfaces that acts somewhat like a vaccine, "fooling" the tomato spotted wilt virus so that it does not infect the altered crop.

A dicot forms a lump of unspecialized tissue at the site of *Ti* plasmid infection and survives. In contrast, a monocot, which has one seed leaf, dies from *Ti* plasmid infection. Researchers must transfer genes into monocots in other ways, such as attached to viral genomes. Transgenesis of monocots is especially important because they include the cereals (rice, corn, wheat, barley, oat, millet, and sorghum), which feed more than half the world's people. Gene transfer techniques such as electroporation, microinjection, and particle bombardment (table 18.3) made genetic manipulation of monocots feasible for rice in 1988, for corn in 1990, for wheat in 1992, and for sorghum in 1994.

Sometimes monocots can borrow traits from each other, thanks to transgenesis. For example, sugarcane can be given a gene encoding a corn protein that dismantles a bacterial toxin. The altered sugarcane resists the corn pathogen and a fungus known to decimate sugarcane crops. Some transgenic combinations are stranger. Macintosh apples given an antimicrobial protein from a cecropia moth, for example, are resistant to a bacterial infection called fire blight.

Transgenic plants can be used in a variety of ways. The first applications provided traits that growers deemed valuable, such as pest and pesticide resistance (table 20.3). Chapters 1 and 18 mention the insertion of an insecticide-producing gene from the bacterium *Bacillus thuringiensis* into a variety of crops, including corn, soybeans, cotton, alfalfa, canola, wheat, and sorghum, which then secrete "*bt* toxin." Farmers have used the protein as an insecticide for more than 40 years by drying out bacterial spores and spreading them over crops. Different varieties and combinations of *bt* toxins tar-

**table 20.3**

**Transgenic Approaches to Agricultural Challenges**

| Challenge | Possible Solution |
|---|---|
| Frost damages crops | Spray crops with bacteria genetically altered to lack surface proteins that promote ice crystallization. Bacteria can also be manipulated to stimulate ice crystallization, then used to increase snow buildup in winter sports facilities. |
| Herbicides and pesticides damage crops | Isolate genes from an organism not affected by the chemical and insert it into the genome of a crop plant. |
| Crops need costly nitrogen fertilizer because atmospheric nitrogen is not biologically usable | Short-term: Genetically manipulate nitrogen-fixing *Rhizobium* bacteria to overproduce enzymes that convert atmospheric nitrogen to a biologically usable form in root nodules of legumes. Alter *Rhizobium* to colonize a wider variety of plants.<br>Long-term: Transfer *Rhizobium* nitrogen-fixation genes into plant cells and regenerate transgenic plants. |
| A plant food is low in a particular amino acid | Transfer gene from another species that controls production of a protein rich in the amino acid normally lacking in the crop plant. |
| A virus destroys a crop | Genetically alter crop plant to manufacture a protein on its cell surface normally found on the virus's surface. Plant becomes immune to virus. |
| Public concern about the safety of synthetic pesticides | Stimulate *Bacillus thuringiensis* to overproduce its natural pesticide, which destroys insects' stomach linings. Transfer *B. thuringiensis* bioinsecticide gene to crop plant. |

get flies, beetles, and moths. When applied this way, sunlight degrades the insecticide within days. But when expressed by GM plants, *bt* toxins are ever-present, which hastens the evolution of resistant insects.

Growers who raise *bt* transgenic crops do not have to repeatedly supply the protein. Although this genetic manipulation has resulted in greater crop yields, there may be effects on the environment and on other species, as discussed later in the chapter. Plus, growers may be forced to abandon these crops due to pressure from nations and companies that refuse to buy GM foods.

*Bt* transgenic crops decrease reliance on chemical pesticides. But another biotechnological approach makes plants resistant to chemical herbicides, which allows increased use of the herbicides. That is, a powerful herbicide can be used without the danger that it will kill crops along with weeds. Figure 20.4 shows the dramatic effects of placing a gene conferring herbicide-resistance into soybeans.

## figure 20.4

**GM herbicide resistance.** The soybeans on the left contain a viral gene that enables them to grow in the presence of glyphosate, a commercial herbicide. Exposure to the herbicide severely stunts growth of the unaltered plants on the right.

Most efforts to create transgenic plants have used genes in the nucleus. An approach called transplastomics instead targets genes in the chloroplast, a type of organelle called a plastid. Plant cells, especially leaf cells, contain thousands of chloroplasts, which are the organelles that house the biochemical reactions of photosynthesis. Like mitochondria, chloroplasts contain their own DNA.

Transplastomics has two advantages as well as two drawbacks. On the positive side, theoretically it can give high yields of protein products, because cells can have many copies of the chloroplast genome. A second advantage is that modified genes in the chloroplast are not released in pollen, and so do not present a risk of fertilizing non-GM plants. On the negative side, finding vectors to deliver genes into chloroplasts has been difficult. Also, transgene expression in this technique tends to be high in the greener parts of the plant, where cells are packed with chloroplasts, but not in the fruits or underground parts (tubers), which are usually eaten. However, researchers have modified tomato plants to produce a large amount of a protein that confers antibiotic resistance in the chloroplasts of the fruits. Antibiotic resistance was used so that the gene transfer could be easily detected by growing the plants in the presence of the drug. Someday, transplastomics may be used to create "medicinal fruits" that may include edible vaccines and antibodies.

Future applications of transgenic technology are likely to target traits of more interest to consumers.

**Nutrition** By altering genes that control the production or composition of lipids or carbohydrates, researchers can tailor the nutritional value of a food. For example, changing the number and types of genes for gluten proteins alters the quality of dough that can be made from a particular strain of wheat. Vegetable protein with a better diversity of amino acid types can be made by transferring genes from one species to another.

**Pharmaceutical Production** Transgenic plants, particularly their seeds, can serve as drug factories by harboring genes that encode proteins with therapeutic value, such as cytokines and clotting factors. Seeds work well because they are abundant, easy to store, resistant to environmental extremes, and are alive. Soybeans are especially efficient miniprotein factories, because they are 40 to 45 percent protein. Corn is also a candidate for pharmaceutical production. Protein comprises only 10 to 12 percent of a corn kernel, but a single plant yields many kernels.

Whole transgenic vegetables can serve as vaccines. The transgenic plant produces part of an antigen that triggers an immune response from T cells associated with the human digestive tract. Plant vaccines are discussed in chapter 16 (see figure 16.19).

A plant's roots can become drug factories in a process called **rhizosecretion.** By growing plants that have a transgene attached to a piece of DNA that stimulates secretion from roots in liquid medium, researchers can collect a protein of interest (figure 20.5). Roots offer several advantages over other transgenic pharming systems. Because they can be coaxed to secrete proteins directly into the medium, their cell walls do not have to be penetrated to collect the protein. Secondly, since roots naturally make few proteins, the transgene's product is easy to isolate and purify—it is abundant in the fluid surrounding the roots. Finally, plant cells add sugars and other chemical groups to proteins that recombinant bacteria cannot,

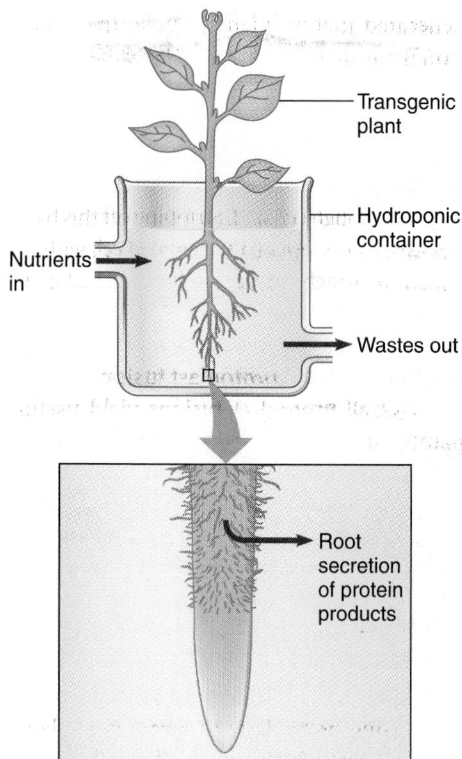


**figure 20.5**

**Rhizosecretion.** Transgenic plant technology has been problematic because of the difficulty of extracting protein products from plant cells. Enlisting roots to secrete—rhizosecretion—may solve the problem. A transgenic plant that expresses the gene for a desired protein in its roots can secrete the products directly into the surrounding medium.

thereby producing more complex molecules than genetically recombinant bacteria can.

The first rhizosecretion experiments used an interesting transgene designed to demonstrate the diversity of gene sources that roots can recognize. Researchers combined a bacterial gene that encodes an enzyme, the jellyfish gene for green fluorescent protein, a protein produced in the human placenta, and a DNA sequence to direct secretion. Then, the researchers introduced this multitransgene into tobacco cells using a cauliflower mosaic virus genome. It worked! The roots of the transgenic plants secreted green foreign protein—easily extracted, abundant, and continually produced.

**Textiles** Cotton plants transgenic for *bt* toxins resist bollworm and budworm infections. The plants require little or no chemical insecticide, and yields are usually higher than with traditional crops.

Cotton can be improved in other ways. For example, two genes from a bacterium, *Alcaligenes eutrophus,* interact with a cotton gene to produce poly-hydroxybutyrate (PHB), a biodegradable plastic. A gene gun sends bacterial genes into cotton plant embryos. Plastic granules form within the cellulose strands of the fibers, so they are not noticeable in cotton fabric. This naturally plasticized cotton retains heat longer than unaltered cotton fibers and may be very useful in making outdoor clothing.

**Paper and Wood Products** Perhaps nowhere is the speed of transgenic technology better appreciated than in the wood, pulp, and paper industries, where the generation time of a tree can far exceed the lifespan of the researcher. For this reason, and because regenerating trees from altered cells is difficult, the development of transgenic trees has lagged behind other biotechnological applications. However, several poplar species are emerging as valuable model systems for experimenting with transgenic trees. Researchers can evaluate new phenotypes on trees too young to reproduce, or incorporate sterility genes into experimental trees to control or contain the spread of novel traits.

Transgenic technology can protect trees from pests. For example, *B. thuringiensis* toxin renders larch, white spruce, pioneer elm, and American sweetgum trees resistant to attack by gypsy moth and forest tent caterpillars. Poplar trees transgenic for a gene from rice resist attack by the beetle *Chrysomella tremulae,* a major pest. The rice gene enables the trees to manufacture a polypeptide that blocks production of the insects' digestive enzymes.

### *Altering Levels of Gene Products*

Antisense technology selectively blocks expression of a gene—that is, protein synthesis. It does not combine genetic material from different species, as transgenesis does. The goal of antisense technology is to introduce a gene that already exists in the organism, but in reverse orientation. The mRNA transcribed from the inserted gene has a sequence that is complementary to the normal mRNA for that gene. The normal mRNA is called the "sense" sequence, and the complementary mRNA the "antisense" sequence. After transcription, the antisense sequence complementary base pairs to the sense sequence, preventing it from being translated into protein.

The creative application of antisense technology can lead to interesting new crops. Best known was the doomed antisense "Flavrsavr" tomato, which squelched production of a ripening enzyme that normally breaks down pectin in cell walls, softening the fruit. The Flavrsavr tomatoes could stay on the vine longer, accumulating sugars and becoming redder (figure 20.6). However, it failed to compete against tastier tomatoes already on the market, and once its origins were publicized, consumers' fear of genetically altered crops contributed to its failure.

Another application of antisense technology is to reduce the amount of lignin, a component of the cell walls in wood that must be removed in the manufacture of pulp and paper. Lignin removal is costly and damaging to the environment. Researchers have used antisense technology to block production of a key enzyme in the lignin biosynthetic pathway of trembling aspen. The result is less lignin to remove from the wood, but there were other unexpected benefits. Blocking the enzyme leads to buildup of its substrate, which is then routed onto the biosynthetic pathway for cellulose. The extra cellulose, combined with less lignin, gives the trees a growth spurt. A similar approach could be used on other tree species valuable in the pulp and paper industry.

Giving a biotech boost to a gene's expression can be powerful when the gene controls the activities of other genes. In *Arabidopsis thaliana,* a small flowering plant related to the mustard family and used widely in experiments, researchers altered the promoter of a gene encoding a transcription factor. Recall from chapter 10 that a transcription factor controls the expression of other genes. In this case, overexpressing the transcription factor causes overexpression of several genes that carry out "stress responses." As a result, the plants have increased tolerance to stressful conditions, including freezing, saltiness, and dehydration. Using traditional breeding to create crops with multiple stress resistances requires many generations of complex breeding schemes to funnel all of the desired traits into one plant, and rid it of undesirable traits.

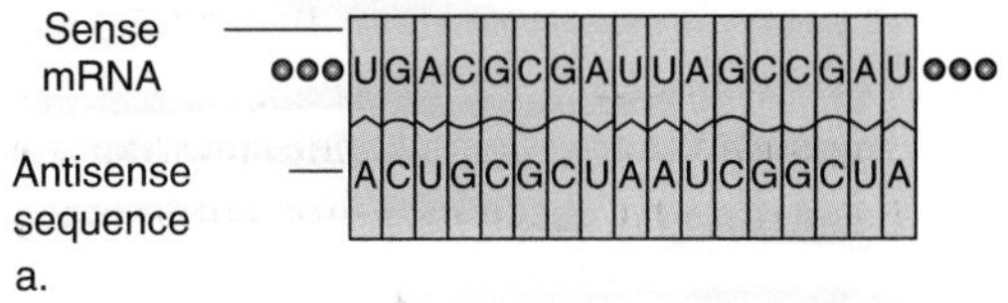

Sense mRNA ooo U G A C G C G A U U A G C C G A U ooo

Antisense sequence A C U G C G C U A A U C G G C U A

a.

b.

**figure 20.6**

**Antisense RNA blocks protein synthesis and produces new crops.** (*a*) The sense mRNA depicted on top cannot be translated into protein because an antisense sequence blocks it. (*b*) These tomatoes harbor an antisense gene that silences a ripening enzyme, allowing the fruit to remain longer on the vine, turning red but gaining a better flavor and texture, rather than becoming mushy. However, consumers rejected the tomatoes, which just didn't taste as good as older varieties.

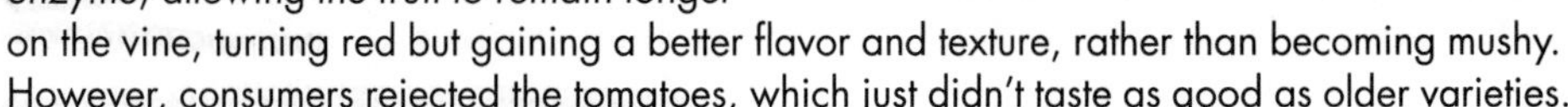

Like the experiment with the fast-growing antisense aspens, the multistress-resistant *Arabidopsis* experiments had an unanticipated side effect—in the absence of stress, the plants do not fare very well. These two experiments illustrate the importance of extensively testing new varieties of plants, whether they are obtained by transgenesis, by altering expression of a species' own genes, or by conventional breeding techniques. Genes interact with each other and with the environment in so many ways that we may not be able to predict the consequences of seemingly simple manipulations.

## Altering Plants at the Cellular Level

Several techniques that alter, transfer, or combine organelles and cells produce interesting new plant variants (table 20.4). These technologies are possible because somatic cells from plants can give rise to re-

generated mature plants, unlike the situation in animals.

### *Protoplast Fusion*

A natural impediment to manipulating plant cells is the tough cell wall. Stripping off this barrier leaves a protoplast, the denuded cell within, which is much more easily manipulated. Protoplasts fused from two different types of plants can yield interesting new hybrids. This technique is called **protoplast fusion.**

Not all protoplast fusions yield useful plants. Consider the "pomato" plant, which grows from a tomato protoplast fused with a potato protoplast. The regenerated plant produces both types of vegetables in the proper part of the plant, but they are stunted, with tiny seeds. It cannot be propagated. Even less useful is a fusion of radish and cabbage protoplasts. The not-very-tasty plant grows radish leaves and cabbage roots!

Fusing protoplasts to create new plant hybrids works better when the cells come from related species. For example, when a protoplast from a potato plant normally killed by the herbicide triazine is fused with a protoplast from the wild black nightshade, a poisonous relative that is naturally resistant to the herbicide, the resulting potato hybrid grows well in soil treated with triazine to control weeds.

**table 20.4**

**Cell-Based Plant Biotechnologies**

| Technique | How It Works | Advantage |
|---|---|---|
| Artificial seeds | An embryo grown in cell culture from a somatic cell is combined with nutrients and hormones and encased in a gel-like substance | Gives a crop uniform growth and other physical characteristics |
| Chlybridization | A chloroplast is introduced into a cell in a liposome | Endows cell with increased ability to harness solar energy |
| Clonal propagation | Cells are cultured and grown into genetically identical plants | Creates uniform crops that are easier to harvest<br>Speeds growth |
| Cybridization | Two cells are fused, and the nucleus of one is destroyed by radiation | Yields a cell with a desired combination of organelles |
| Mibridization | A mitochondrion is introduced into a cell in a liposome | Enhances energy-producing capacity of cell and introduces mitochondrial genes |
| Mutant selection | Cells in culture are exposed to a toxin; resistant surviving cells are used to start a crop | Creates crops resistant to herbicides, frost, disease |
| Protoplast fusion | Two cells from different species have their cell walls removed and are fused; a hybrid plant is regenerated | Introduces a plant with a new combination of traits |
| Somaclonal variation | A mutant somatic cell that is part of a callus grows into a plant with a new trait | Provides an abundant source of new variants |

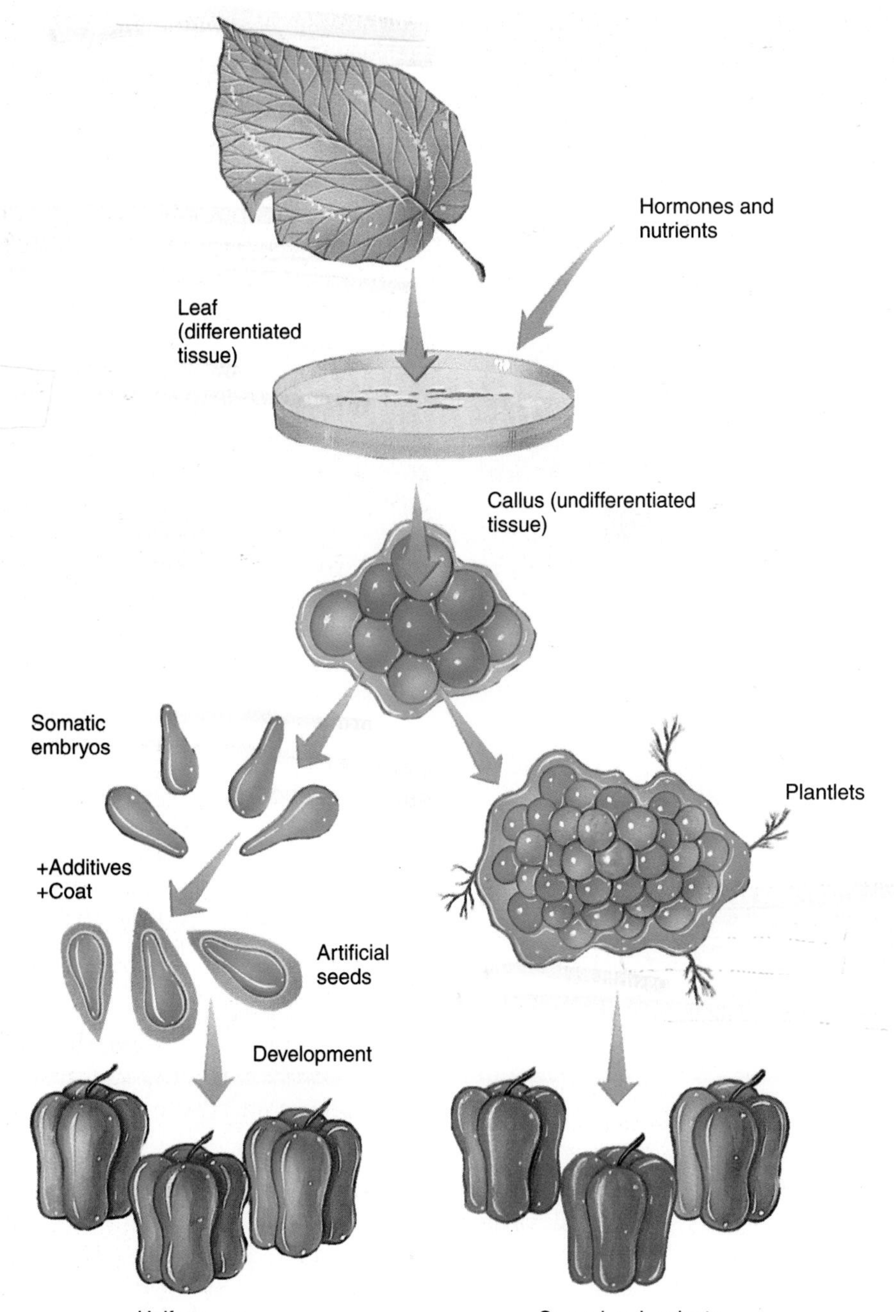


**figure 20.7**

**A callus yields embryos or plantlets.** Whether they are genetically identical or somaclonal variants depends upon the culture conditions. Wrapping identical somatic embryos in a gelatinous coat, with nutrients, hormones, and maybe even insecticide, forms artificial seeds, but is very costly. Plantlets that differ from each other provide interesting new variants, such as stringless celery, crunchier carrots, and buttery popcorn.

### *Callus Culture*

Some plant manipulations use undifferentiated (unspecialized) plant cells growing in culture. When protoplasts or tiny pieces of plant tissue, called **explants,** are nurtured in a dish with nutrients and plant hormones for a few days, the cells lose their special characteristics and form a white lump called a **callus.** The lump grows for a few weeks as its cells divide. Then certain cells grow into either a tiny plantlet with shoots and roots, or a tiny embryo. An embryo grown from callus is called a **somatic embryo** because it derives from somatic, rather than sexual, tissue. Callus growth is apparently unique to plants. In humans, it would be the equivalent of taking a cultured skin cell, multiplying it into a blob of unspecialized tissue, and then sprouting tiny humans or human embryos!

Most of the time, embryos or plantlets grown from a single callus are genetically identical, or clones. Sometimes, however, the embryos or plantlets differ because genes in certain of the callus cells mutate. This is an example of somatic mutation, and the plants with novel phenotypes are called **somaclonal variants.**

Biotechnologists can, to some extent, control whether growths from a callus are identical by altering the nutrients and hormones in the culture (figure 20.7). Controlling callus growth is important because agriculture sometimes benefits from a uniform crop, and at other times, seeks new varieties. Corn of the same height, for example, is easier to pick, and tomatoes of uniform water content are easier to boil into commercially prepared sauce than highly variable crops.

### Key Concepts

Biotechnology adds, subtracts, or modifies traits in organisms that provide products. Methods of gene transfer include the *Ti* plasmid in dicots and other methods in monocots. Transplastomics is transgenesis applied to chloroplast genomes. Transgenic traits provide resistance to toxins and the ability to produce novel substances with diverse applications. Antisense technology blocks translation. Altering promoters affects transcription. • Fusing protoplasts from different types of plants can produce novel variants. An explant grows into an undifferentiated callus, which may yield plantlets or somatic embryos. Culture conditions can be altered to produce genetically identical plantlets or somatic embryos, or somaclonal variants.

## 20.3 Release of Genetically Modified Organisms into the Environment

The first recombinant and transgenic organisms were not meant to exist outside the laboratory. They were manipulated so that the activation of "suicide" genes outside the laboratory would kill them, or were modified to make them unable to reproduce. Agricultural biotechnology has different goals. Since GM plants will ultimately grow outside of labs, it is important that they be able to persist in an ecosystem without harming its residents, or spreading to other ecosystems. To assess this, researchers test GM organisms in artificial models of ecosystems or under natural conditions.

**figure 20.8**

**Field tests.** Researchers conducting the first field tests of genetically modified bacteria in the 1980s wore spacesuits and protective gear. Today, after many successful field tests, scientists use far less protective equipment.

### Microcosm Experiments

A microcosm is a recreation of a portion of an ecosystem in a laboratory. To study genetically altered soil-borne bacteria, for example, researchers work with a cylinder of soil about the dimensions of a small drum. Such parameters as pH (acidity or alkalinity), moisture, antibiotics present, temperature, oxygen and nutrient content, and other organisms present are adjusted to mimic conditions in a natural ecosystem. The test bacteria are added to the microcosm and resulting changes monitored.

If the presence of the bacterium adversely affects any component of the microcosm, researchers can avoid a potentially dangerous field test. For example, a microcosm experiment prompted researchers to halt field testing of a bacterium genetically altered to convert rotting vegetation to ethanol, an alcohol that is a raw material for many industrial processes. In a microcosm, the altered bacteria not only made ethanol, but they also destroyed fungi that grow on the roots of several types of plants, including many crops. Researchers noticed plants in the microcosm dying. Had these bacteria gone directly from the laboratory to the field, results on native plants would have been disastrous.

Genetically altered bacteria or viruses that pass microcosm tests can be very useful. A genetically altered virus proved particularly helpful in a 1994 rabies epidemic. Researchers transferred a gene from the rabies virus into *Vaccinia* virus, which is used as a smallpox vaccine. The rabies gene converts *Vaccinia* into a rabies vaccine. After microcosm experiments showed that the vaccine could be produced safely, wildlife biologists added it to fish meal left in forests. Raccoons that ate the bait vaccinated themselves against rabies, preventing spread of the deadly infection to dogs and cats.

### Field Tests

Products of agricultural biotechnology must be tested in the field. Field tests are designed to answer the following questions:

- Can the GM organism survive outside the laboratory?
- Can the organism reproduce with non-GM members of its species?
- Does the organism have an advantage over species that are already part of a field environment or ecosystem?
- Can the transferred gene move to other species, or to nontransgenic members of the same species?
- Does the GM organism move beyond the test site? If so, does it move on its own, or by attaching to another organism?

The first field tests of genetically altered organisms sparked public panic (figure 20.8). In 1987, people living near fields where potato and strawberry seedlings were sprayed with "ice-minus" bacteria repeatedly sabotaged the experiments. The bacteria were actually *missing* a gene they normally carried.

In the years since, GM plants have *not* proven to be, as one researcher puts it, the "giant rutabagas that ate New Jersey." Thousands of field tests have shown that transgenic plants are safe to grow. Still, in 1999, a group of environmental activists destroyed an experimental forest of poplars near London. The goal of the experiment was to better understand wood chemistry so that less chemical herbicide would be necessary—partially in response to activists' requests for more such research. But the protest prompted the International Union of Forestry Research Organizations to publish guidelines to better educate the public about the nature of GM trees. The guidelines state that:

- GM trees are grown for short times (3 to 25 years).
- GM trees could ultimately increase wood production, reducing the need to destroy native forests.
- Precautions are taken to prevent the spread of genes until they are well studied. These precautions include isolating the trees from others, using only nonharmful genes, and controlling flowering time to avoid unwanted breeding.

Intentional destruction of GM crops continues. Non-GM crops have been destroyed by mistake. The controversy continues.

### Bioremediation

In **bioremediation,** bacteria or plants with abilities to detoxify certain pollutants are released or grown in a particular area to cleanse the environment. Natural selection has sculpted such organisms, perhaps as adaptations that render them unpalatable to predators. Bioremediation uses genes that give an organism an appetite for some substance that, to another species, is a toxin. The technology uses unaltered organisms, and also transfers "detox" genes to other species so that the protein products can more easily penetrate a polluted area.

Nature offers many organisms with (to us) strange tastes. One type of tree that grows in a tropical rain forest on an island near Australia, for example, accumulates so much nickel from soil that slashing its bark releases a bright green latex ooze. Up to 20 percent of the tree's dry weight may be nickel. This tree can be used to clean up nickel-contaminated soil. Similarly, a microbe called *Citrobacter* absorbs the nuclear wastes plutonium and uranium from soils.

Bioremediation is helpful in environmental disasters. Ten weeks after a tanker drenched Alaska's Prince William Sound in 11 million gallons of oil, U.S. Environmental Protection Agency workers spread nitrogen and phosphorus fertilizer on 750 areas along the 74 miles of oil-slicked shore. The fertilizer stimulated the growth of natural bacteria that consume organic toxins in the oil. The pollutants disappeared, and the beaches whitened, five times faster in the treated areas. Similarly, a series of bacteria with a taste for chlorine detoxify polychlorinated biphenyl (PCB) compounds in Hudson River sediments.

Transgenic bioremediation usually taps the metabolisms of microbes, sending them into plants whose roots then distribute the detox proteins in the soil. For example, transgenic yellow poplar trees can thrive in mercury-tainted soil because they have a bacterial gene that encodes an enzyme, mercuric reductase, that converts a highly toxic form of mercury in soil to a less toxic gas. The tree's leaves release the gas.

Bioremediation helps to clean up munitions dumps, the leftovers of wars past. One application uses bacteria that normally break down dinitrotoluene—better known as TNT, the major ingredient in dynamite and land mines. The enzyme that provides this capability is linked to a gene whose product makes the protein luminesce or fluoresce. When the bacteria are spread in a contaminated area, they glow where there are land mines, revealing the locations much more specifically than a metal detector could. Once the land mines are removed, the bacteria die, because their food is gone. Other experiments add the bacterial TNT-detecting gene plus a glowing marker gene to various plant species, whose roots reveal the locations of buried explosives by glowing.

Ultimately, transgenic plants may harbor a variety of microbial genes that bioremediate several types of explosives or toxins. Such bacteria have been identified in the lab under controlled conditions. But when they have been tested at a contamination site, they typically have not worked well, because toxins other than the ones that they can process harm them. Transferring several genes from various microbial sources into the same plant is a way to get them all to work at once.

**KEY CONCEPTS**

After organisms are genetically altered, they must be tested in microcosms and then in the field to see how they affect natural ecosystems. • Bioremediation is the intentional release of organisms that detect or detoxify heavy metals, organic wastes, or explosives in the environment. Enzyme-catalyzed biochemical reactions carry out the detoxification. These reactions may be a natural part of an organism's biochemistry or may be added transgenically or by using recombinant DNA technology.

## 20.4 Economic, Ecological, and Evolutionary Concerns

Agricultural biotechnology—altering DNA, organelles, or cells of organisms that we use for a variety of purposes—is controversial, because researchers can't predict all the consequences of manipulations. Consequences may not be obvious, because they go beyond science. For example, producing foods that are more healthful and easier to market may seem beneficial, particularly in the potential to feed populations suffering from food shortages if the foods can be fairly distributed. However, the opposite may also occur.

Consider canola oil, a product of the rapeseed plant. A transgenic rapeseed plant given a gene from the California bay tree gains an enzyme that alters the lengths of some of its fatty acids, producing substances called lauric oils that are used to manufacture soap and shampoos. We traditionally extract lauric oils from coconuts and palms imported from southeast Asia. Genetically rerouting domestic rapeseed plants to produce the valuable oils diverts funds that would otherwise go to poor southeastern Asian farmers.

Ecological concerns compound economic considerations. It takes many studies of pollen characteristics to determine whether genes that give crop plants the characteristics we want can be transferred to wild and weedy relatives. The two types of plants must use the same type of pollinator and must be close enough for pollination to occur, but an unwanted transfer of traits may occur. For example, herbicide-resistant rapeseed could pass the resistance gene to birdscrape mustard or wild mustard, which are troublesome weeds. However, so far many studies reveal that transgenes are no more likely to jump from one species to another than are any other genes.

Even if transgenes do not jump from crops to weeds, GM plants may nevertheless grow where they were not intended to grow. This has already happened. In September 2001, *bt* corn was found growing on a remote mountaintop in Mexico, and subsequent investigations repeated the finding. In India, a field of *bt* cotton was discovered. The farmer had unknowingly bred *bt* seeds from the United States with a local variant, then raised the hardy crop without realizing that he was evading regulations.

Questionable business practices have also added to rejection of GM crops. In the 1990s, certain large companies sold seeds resistant to the companies' herbicides, but also bearing mutations that made the seeds sterile. Farmers were forced to purchase not only the companies' herbicides, but the seeds anew each year, rather than saving them from the previous year's harvest. This action, which ran counter to the ideas of developing renewable resources and sustainability, may have sown the seeds for public distrust of GM products. Protest led to the demise of the "suicide seeds," but GM crops are still sold that are resistant to the same herbicide that the seed manufacturer markets.

Today, five large, multinational companies have acquired many small seed and biotechnology companies and hold patents on key gene transfer techniques and marker systems. Some people fear that this concentration of power in industry will restrict most research to developing crops with better disease resistance, improved nutritional qualities, and ease of cultivation, while making it harder for academic researchers to explore more basic goals, such as conservation and genetics.

On a broader scale, altering the genetic makeup of a species for human benefit affects evolution. Traditional farming, pet breeding, gardening, and the practice of medicine all do this. Whatever the method or goal, introducing an organism with a selective advantage can lower natural genetic diversity if the altered individuals come to predominate in a population under certain environmental conditions. For example, transgenic carp that grow 40 percent faster than normal because they have a growth hormone gene from rainbow trout will outcompete slower-growing fish and lead the population toward genetic uniformity. Should an environmental change render the alteration a liability, populations could vanish.

The first GM fish to appear on dinner tables will probably be Atlantic salmon given a growth hormone gene from Chinook salmon, with a promoter from a third fish, ocean pout. Atlantic salmon normally produce growth hormone only in the summer, but ocean pout, which lives in a different environment, produces it year round. The GM Atlantic salmon manufacture the hormone continually, and as a result enlarge at more than twice the normal rate.

Using biotechnology to create crops that are resistant to and thereby increase the use of insecticides and herbicides can encourage the persistence of pests that are resistant to the chemicals. This has happened with some insect pests that are resistant to *bt* toxin, as natural selection clashes with technology. Preexisting genetic variants of weed or pest species that can survive in the presence of a particular herbicide or pesticide will come to predominate in the population as those with sensitive phenotypes die off. To counter this relentless evolutionary force, farmers must grow nontransgenic sensitive crops in fields where the pesticides aren't used. This maintains some nonresistant insects, which breed with the resistant ones and slows the rise in resistance in pest populations.

Some nations ban GM foods. In the United States, however, the percentage of crops that are genetically modified is large and growing, as the years pass and no one has been harmed (figure 20.9). Evidence has even settled the issue of whether *bt* corn threatens monarch butterflies (see Bioethics: Choices for the Future on page 421). However, long-term ecological effects remain to be seen—as is the case for any form of agriculture.

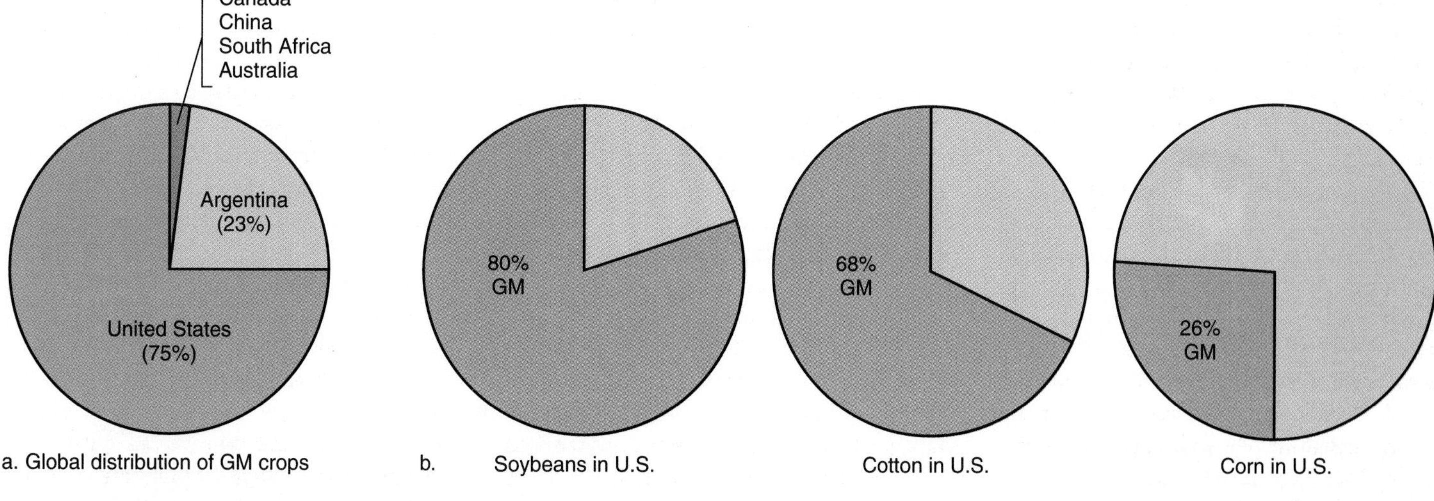


**figure 20.9**

**The global GM foods picture.** (*a*) Most GM crops are grown in the United States, due largely to the perception in other parts of the world that these crops are dangerous. (*b*) Percentage of selected crops in the United States that are genetically modified.

## Key Concepts

Economic impacts of GM foods are difficult to predict. These products may displace existing ones, or may not be equitably distributed. Ecological effects can be modeled in greenhouses and field tests, but GM organisms can escape their containments. Industry controls many aspects of agricultural biotechnology, which has raised concerns.

## 20.5 The Impact of Genomics

The benefits and risks of GM crops are still being hotly debated (table 20.5). If GM organisms can survive their negative image, genomics will provide researchers with many more traits to work with. However, plant genomics lags behind similar efforts in animals and microorganisms—only the model experimental plant *Arabidopsis thaliana* has had nearly all of its genome sequenced. A draft genome sequence for rice is about 90 percent complete. Sequencing of the 400 million or more bases of the rice genome, ironically, is expected to boost traditional breeding efforts by helping breeders identify genes of interest in existing varieties. In addition, the rice genome has many clusters of linked, or syntenic, genes, and many of these genes are similarly clustered in the genomes of related grasses, such as wheat and corn. This means that interesting new rice genes that researchers discover are likely to have homologs in other species. Although a large pharmaceutical firm published the first draft of the rice genome sequence, it has cooperated with an international consortium with the same goal and has made the information freely available to researchers.

Investigators need not have an entire genome in hand, however, to apply the many-gene approach of genomics. Consider another popular crop, the potato. The National Science Foundation supports a "potato functional genomics" project, in which a nonprofit organization, the Institute for Genomic Research (TIGR), offers DNA microarrays that hold 5,000 expressed sequence tags (ESTs) each. ESTs are pieces of protein-encoding genes that correspond to cDNAs reverse transcribed from the mRNAs present in a particular cell type. Different potato DNA microarrays correspond to the different tissues—root or shoot, stem or tuber—just as human DNA microarrays represent differential gene expression in nerve or muscle, skin or bone.

### table 20.5

**Potential Benefits and Risks of GM Crops**

**Benefits**

1. Enhanced nutritional qualities.
2. Resistance to pests, diseases, herbicides, pesticides, and environmental extremes.
3. Delayed fruit ripening to extend shelf life.
4. Delayed potato development to reduce need for chemicals to suppress sprouting.
5. Elimination of allergens.
6. Production of pharmaceuticals and edible vaccines.
7. Bioremediation of toxins and explosives.
8. Production of biodegradable plastics.
9. Changes in proteins, fats, and carbohydrates that serve as raw materials for paper, lubricants, detergents, and other products.
10. Control of plant height, flowering time, seed size and number, solids content and other traits important for harvesting and processing.
11. Ease of separation of transgenic plant seed from weed seeds.
12. Produce more of the edible portion of a plant.
13. Contribute less phosphorus to animal feces, making groundwater contamination less likely to cause algal blooms.

**Risks**

1. Movement of transgene to other species, harming them or making them toxic or allergenic.
2. Economic and political repercussions as existing crops are displaced.
3. Harm to farmers and U.S. companies as other nations boycott GM products.
4. Unexpected results from combinations of genes from different species.
5. Inability to predict long-term consequences of introducing GM organisms into the environment.
6. Reduction of biodiversity.
7. Possible ill effects on honey production.
8. Pollination of organically grown crops by pollen from GM plants may result in loss of "organic" qualities and status.
9. Growth of GM plants outside of intended areas.
10. Rise of pesticide-resistant insect variants, making conventional application of *bt* toxin ineffective.

# The Butterfly That Roared

A three-quarter page "Scientific Correspondence" in the May 20, 1999 issue of the journal *Nature,* entitled "Transgenic Pollen Harms Monarch Larvae," galvanized a multinational campaign against genetically modified (GM) organisms. The paper reported a study by Cornell University entomologist John Losey and colleagues. Their hypothesis was straightforward: Since milkweed plants grow near cornfields, and monarch butterfly larvae eat milkweed leaves, what would happen if monarch caterpillars ingested *bt* corn pollen, containing the toxin, that had blown onto milkweed leaves (figure 1)? To find out, the researchers placed lab-raised caterpillars in plastic containers and fed them *bt*-tainted milkweed leaves, leaves brushed with pollen from non-GM corn, or pollen-free leaves. The results: 44 percent of the larvae consuming *bt* pollen died, compared to none of those in the control groups. Surviving caterpillars that ate *bt* pollen were stunted.

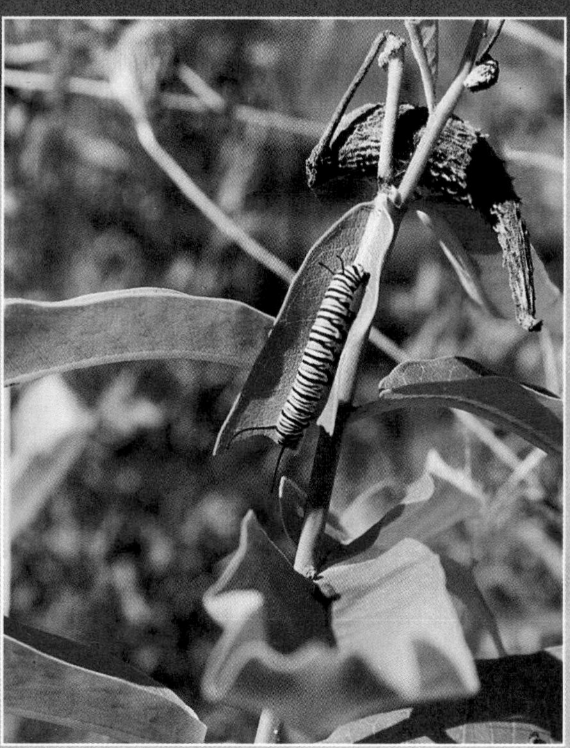

**figure 1**

**GM corn does not harm monarch butterfly larvae under natural conditions.** A monarch butterfly larva in its natural habitat, on a milkweed plant.

Conditions in the 1999 study did not resemble reality. Anthony Shelton, another Cornell entomologist, put the situation into perspective. "If I went to a movie and bought 100 pounds of salted popcorn, because I liked salted popcorn, and then I ate the salted popcorn all at once, I'd probably die. Eating that much salted popcorn simply is not a real-world situation, but if I died, it may be reported that salted popcorn was lethal." Plus, *Nature* published the short monarch paper, but rejected the paper reporting the transgenic "golden rice" with boosted iron and beta-carotene, mentioned earlier in the chapter, without sending it out for review. Some researchers suggested that the journal was bowing to public pressure to publicize evidence against GM crops.

In response to the Losey paper, and its implications, activists destroyed experiments, burglarized research facilities, called for special labeling, pushed for trade restrictions that have affected U.S. farmers, and spread fear that *bt* corn will produce what one researcher jokingly called a "toxic cloud of pollen saturating the Corn Belt." As activists dressed as butterflies led the antitechnology charge, the public joined in. Lost in the excitement was understanding the science behind the hype.

Logic argues against danger of *bt* corn. Most milkweed plants are not near cornfields. Monarchs are not endangered, and, as one scientist said, more are killed by lawnmowers, cars, and frost than will ever be killed by *bt* corn pollen. In addition, corn can be modified not to secrete the toxin in pollen.

In late 1999, researchers at a symposium in Chicago presented data that showed:

- The strength of the *bt* toxin varies.
- The concentration of corn pollen on milkweed plants growing in the field drops off very quickly with distance from the corn.
- Corn pollen that does land on milkweed leaves is blown or washed off quickly.
- About 95 percent of the Nebraska corn crop releases its pollen long before monarchs lay their eggs.

The results of the meeting were largely disregarded because of conflicts of interest—industry had funded many of the studies, and the work was presented to the press before it was published in peer-reviewed scientific journals. But then John Losey discovered that in the field, butterflies do not lay eggs on leaves coated with pollen! Just as a person is unlikely to eat 100 pounds of popcorn, a caterpillar, given a choice, probably wouldn't eat *bt* pollen.

In late 2001, six studies published in *The Proceedings of the National Academy of Sciences* presented even more compelling evidence that *bt* corn produces very little toxin in its pollen, and not enough of it lands on milkweed leaves to threaten monarch butterflies. But the image of the beautiful monarch butterfly as a victim of biotechnology persists.

Traditional breeding of potatoes has been tricky, because the leaves produce toxic alkaloid compounds that must be bred out of edible varieties. Cultivars (cultivated varieties) represent many years of breeding to maximize taste and texture, and minimize toxin levels. But genome analysis is also proceeding slowly, because plant genomes are large and very complex. So far, the potato genome has been cut into pieces that fit into bacterial artificial chromosomes (BACs), but these have not been sequenced and overlapped yet to reveal the sequence. For this reason, the potato genome project, being conducted in several university laboratories, is focusing first on several disease-resistance genes that cluster on a region of one chromosome. The genes provide resistances to various insects, nematode worms, viruses, and the most threatening pathogen, the water mold *Phytophthora infestans,* which causes a disease called late blight. It was late blight that caused the Irish potato famine, and today, no potato variety is completely resistant to it. Being able to manipulate and transfer these genes will help researchers quickly breed potatoes and perhaps also their relatives, such as tomatoes, peppers, and eggplants, that are resistant to several diseases.

Traditional agriculture has been with us for thousands of years. While it has provided human societies with food, clothing, and shelter, it has also profoundly altered ecosystems, sharply decreasing biodiversity as fields and farms replaced and displaced natural vegetation and the animals that feed and live among plants. The results of traditional agriculture were and are hard to predict, as growers attempted to identify interesting traits and then set up crosses in search of organisms with particular gene combinations.

Agricultural biotechnology extends the fundamental idea of traditional agriculture, also combining genes in novel ways, but with greater precision and control, yet also stranger possibilities. Some say this is interfering too much with nature. Perhaps in no other research field do economics, politics, and science clash so fiercely. It will be interesting to see how, in the years ahead, science and society deal with the applications of genetic technology to the age-old practice of agriculture.

**KEY CONCEPTS**

Progress in plant genomics has been slow but will guide researchers in identifying useful traits across species. The potato functional genomics project is beginning by analyzing a cluster of genes that confers a variety of disease resistances.

# Summary

### 20.1 Traditional Breeding Compared to Biotechnology

1. Biotechnology can produce plants with agriculturally useful traits. These plants must then be evaluated outside the laboratory.
2. Biotechnology differs from traditional plant breeding in that it is asexual, more precise, faster, and considers one trait at a time.
3. The FDA evaluates the characteristics of a food, not its origin.
4. Different biotechnologies can address the same problem. **Mutant selection** and transgenic technology can both generate herbicide-resistant plants.

### 20.2 Types of Plant Manipulations

5. Transgenic dicots are created using *Ti* plasmids as gene vectors. Physical techniques are necessary to add genes to monocots. Transgenes can be introduced into a chloroplast genome.
6. The first transgenic crops benefited growers; future engineered traits will target consumers. *Bt* crops have a bacterial transgene that harms certain caterpillar pests. Other transgenic traits include herbicide and pesticide resistance, altered flavor, increased uniformity, additional nutrients, and the ability to fix nitrogen.
7. Transgenic seeds and roots produce drugs; transgenic cotton makes plastic; and transgenic trees may ease paper and pulp production.
8. **Antisense technology** blocks expression of specific genes, such as those that encode ripening enzymes.
9. Altering promoters can increase production of a particular protein.
10. Fusing **protoplasts** from two species can create new plant variants.
11. Cultured protoplasts or explants can yield a **callus,** which produces **plantlets** or **somatic embryos.** These may be clones or **somaclonal variants,** depending upon culture conditions.

### 20.3 Release of Genetically Modified Organisms into the Environment

12. GM organisms are tested in microcosms, then in the field, to determine their survival, fertility, and interactions with other organisms.
13. **Bioremediation** uses an organism's natural or engineered metabolic reactions to detect or disarm pollutants, toxins, or explosives.

### 20.4 Economic, Ecological, and Evolutionary Concerns

14. The introduction of GM foods can have economic repercussions because they replace natural sources of substances.
15. Measures must be taken to prevent pollen from transferring transgenes to other species, and to prevent GM crops from growing where they weren't intended.
16. Herbicide- or pesticide-resistant crops can encourage persistence of this trait in weeds and pests.

### 20.5 The Impact of Genomics

17. Plant genomics lags behind efforts for animals and microorganisms.
18. The potato functional genomics project is focusing first on linked resistance genes.

# Review Questions

1. Explain how the following technologies work:
   a. antisense technology
   b. transgenesis
   c. rhizosecretion
   d. mutant selection
   e. bioremediation
   f. transplastomics
2. How are modern biotechnologies such as cell culture and transgenesis similar to and different from traditional plant breeding?
3. In most cases biotechnological approaches to creating a new plant variety are faster than traditional agricultural approaches.
   a. Why is this so?
   b. Cite hurdles that can make a biotechnological approach slow, too.
4. Why do plants that are derived sexually from the same parent plant differ from each other, whereas plants derived from somatic tissue from the same plant are identical?
5. Is a fruit or vegetable derived from mutant selection genetically modified?
6. Cite one reason why plants are easier to alter by biotechnology than animals are, and one reason why plants are harder to manipulate.
7. Give an example of how agricultural biotechnology can affect
   a. economics.
   b. ecology.
   c. evolution.
8. How does a somatic embryo differ from a normal embryo?
9. Under what circumstances might it be dangerous to genetically alter a crop plant to resist a herbicide?
10. How can releasing a genetically altered organism to a natural ecosystem influence natural selection?
11. Give two reasons why creating transgenic trees is challenging.

# Applied Questions

1. If many of the plants in the food supply of the United States are transgenic and have been for years, and there is no evidence that people are becoming ill from eating them, do you think that the public should be concerned about safety? What are other reasons that the widespread use of transgenic crops might be cause for concern?
2. Suggest three ways to develop tomatoes that are resistant to a particular herbicide.
3. How would you use biotechnology to create
   a. popcorn that glows in the dark
   b. an orangish-yellow potato
   c. a healthier potato chip by altering both the potato and canola
   d. a plant that has characteristics of zucchini and eggplant
   e. lawn grass that discourages growth of dandelions
   f. soybeans that cannot induce allergies in people allergic to ordinary soybeans
4. What are the potential benefits and risks of labeling food derived from GM organisms?
5. Do you agree or disagree with the FDA's standard of "substantial equivalence" in deciding whether new food varieties require further evaluation?
6. Does a transgenic plant pose the same potential threat to the environment as a new crop variant altered by antisense technology, such as the FlavrSavr tomato?
7. Growing identical plants is essential in the ornamental flower industry. Suggest a biotechnology that could meet this requirement.
8. There was no public outcry over the development of Humulin, the human insulin produced in bacterial cells and used to treat diabetes. Yet many people object to mixing DNA from different species in agricultural biotechnology. Why do you think that the same general technique is perceived as beneficial in one situation, yet a threat in another?

# Suggested Readings


Borisjuk, Nikolai V. May 1999. Production of recombinant proteins in plant root exudates. *Nature Biotechnology* 17:466. Plant roots make excellent bioreactors.

Chapela, Ignacio, and David Quist. November 21, 2001. Transgenic DNA introgressed into traditional maize landraces in Oaxaca, Mexico. *Nature,* 414:541–543. The original paper describing discovery of GM corn in Mexico.

Dale, Philip J. January 2000. Public concerns over transgenic crops. *Genome Research* 10:1. The benefits of GM crops tend to be scientific; the risks are more political and economic.

Fletcher, Liz. September 2001. GM crops are no panacea for poverty. *Nature Biotechnology* 19, no. 9:797–98. Use of GM crops depends as much on politics and economics as it does on biology.

Hall, Stephen S. September 1987. One potato patch that is making genetic history. *Smithsonian.* This story of the first release of recombinant bacteria onto plants reveals that history is repeating itself.

Hileman, Bette. September 17, 2001. Engineered corn poses small risk. *Chemical and Engineering News* 79, no. 38:11. Six studies vindicate *bt* corn.

Hooker, Brian S., and Rodney S. Skeen. May 1999. Transgenic phytoremediation blasts onto the scene. *Nature Biotechnology* 17:428. Plants given microbial genes that dismantle TNT can use their roots to disarm buried weapons.

Hubbell, Sue. 2001. Shrinking the cat: genetic engineering before we knew about genes. Houghton Mifflin, Boston. Humans dramatically altered the gene pools of cats, apples, corn, and silkworms.

Lewis, Ricki. March 6, 2002. Creating salt-tolerant plants. *The Scientist* 16:1. Salt tolerant GM crops can be grown more widely.

Lewis, Ricki, and Barry A. Palevitz. October 11, 1999. Science vs. PR: GM crops face heat of debate. *The Scientist* 13:1. Was Greenpeace's targeting of a major baby food manufacturer in their anti-GM campaign fair fighting?

Losey, John, et al. May 20, 1999. Transgenic pollen harms monarch larvae. *Nature* 399:214. This is the paper that made the monarch butterfly a symbol for dangers of GM organisms.

Maliga, Pat. September 2001. Plastid engineering bears fruit. *Nature Biotechnology* 19, no. 9:826–927. Genetic manipulation of plastid genomes prevents spread of foreign genes in the environment.

Potrykus, I. March 2001. Golden rice and beyond. *Plant Physiology* 123:1157–61. Ingo Potrykus describes his invention of golden rice—and how both environmental activists and the media have misunderstood its status.

Quist, David, and Ignacio H. Chapela. November 29, 2001. Transgenic DNA introgressed into traditional maize landraces in Oaxaca, Mexico. *Nature* 414:541–43. How did GM corn grow on a remote Mexican mountaintop?

Ruf, Stephanie, et al. September 2001. Stable genetic transformation of tomato plastids and expression of a foreign protein in fruit. *Nature Biotechnology* 19, no. 9:870–75.

Salt, David E. October 1998. Arboreal alchemy. *Nature Biotechnology* 16:905. Poplars given a bacterial gene can convert mercury in soil to a less toxic, gaseous form.

Sederoff, Ron. August 1999. Building better trees with antisense. *Nature Biotechnology* 17:750. Low-lignin trees can be produced by blocking an mRNA.

Strauss, Steven, et al. December 1999. Forest biotechnology makes its position known. *Nature Biotechnology* 17:1145. Protesters destroyed an experimental forest, countering their own demand for more research.

Thayer, Ann. September 17, 2001. Owning agbiotech. *Chemical and Engineering News* 79, no. 38:25–32. In the United States, five large companies control most of agricultural biotechnology.

Yoon, C. K. May 1, 2000. Altered salmon leading to dinner plates, but rules lag. *The New York Times*, p. A1. GM salmon may harm natural populations.

## On the Web

Check out the resources on our website at

**www.mhhe.com/lewisgenetics5**

On the web for this chapter, you will find additional study questions, vocabulary review, useful links to case studies, tutorials, popular press coverage, and much more. To investigate specific topics mentioned in this chapter, also try the links below:

The AgBio World Community **www.agbioworld.org**

Agricultural Biotechnology, U.S. Dept. of Agriculture **www.aphis.usda_gov/biotechnology/**

The Institute for Genomic Research **www.tigr.org**

National Science Foundation Potato Genome Homepage **www.bakerlab.usda.gov/NSFPotatogenome/project/background.html**

Rice Genome Research Program **http://rgp.dna.affrc.go.jp/**

CHAPTER

21

# Reproductive Technologies

### 21.1 New Ways to Make Babies

Male + female = baby is not the only route to parenthood. Third parties and technologies can assist in the process, leading to some unusual family situations.

### 21.2 Infertility and Subfertility

The male and female reproductive systems are each a series of tubes and associated structures. Blockages, abnormalities, and incompatibilities can disrupt the precise interactions necessary for sperm to approach, meet, and merge with an oocyte, kickstarting development. For 1 in 6 couples, something goes wrong, but most problems can be treated.

### 21.3 Assisted Reproductive Technologies

Technology and generous people can help others to have children. Gametes and embryos can be donated, uteruses borrowed, and early embryos screened for chromosomal problems.

### 21.4 On the Subject of "Spares"

*In vitro* fertilization usually generates extra fertilized ova or early embryos that couples must elect to implant, donate to infertile couples, store, or designate for research. Although studying such material is revealing new facts about early human development, many people object to its use in research.

A couple in search of an oocyte donor places an ad in the student newspaper of Stanford University, searching for an attractive young woman under 30 from an athletic family. A cancer patient has her oocytes stored before undergoing treatment. Two years later, she has several of them fertilized in a laboratory dish with her partner's sperm, and has a fertilized ovum implanted in her uterus. She survives the cancer and becomes a mother. A man paralyzed from the waist down has sperm removed and injected into his partner's oocyte. He, too, becomes a parent when he never thought he would.

Lisa and Jack Nash sought to have a child for a reason different from the people just mentioned. Their daughter Molly had been born on July 4, 1994, and was soon diagnosed with Fanconi anemia, an autosomal recessive condition that would destroy her bone marrow, severely impairing her immunity. An umbilical cord stem cell transplant from a sibling would offer up to an 85 percent chance of cure, but Molly had no siblings. Nor did her parents wish to conceive a child who would face the 1 in 4 chance of also inheriting the disorder, as Mendel's first law dictates. Technology offered another solution.

Shortly before Christmas in 1999, researchers at the Reproductive Genetics Institute at Illinois Medical Center mixed Jack's sperm with Lisa's oocytes in a laboratory dish. After allowing 15 of the fertilized ova to develop to the 8-cell stage, researchers then separated and applied DNA probes to one cell from each embryo. The genome of a cell that was free of the Fanconi anemia gene and also a match for Molly's human leukocyte antigen (HLA) cell surface pattern was identified and its 7-celled counterpart implanted into Lisa's uterus. Adam was born in late summer at Fairview University Hospital in Minnesota. A month later, physicians infused his umbilical cord stem cells into his sister, saving her life (figure 21.1). The Nash's journey wasn't easy—they began treatment in 1995, suffered several miscarriages, then faced media scrutiny for their action.

Increased knowledge of how the genomes of two individuals come together and interact has spawned several **assisted reproductive technologies** that help people have children. Assisted reproductive technologies replace the source of a male or female gamete, aid fertilization, or provide a uterus. Most of the time these procedures, usually invasive and expensive, are used to treat infertility. But as the Nash family illustrates, the techniques are also becoming a part of genetic screening, a role that will likely increase with the influx of human genome information.

**figure 21.1**

**Special siblings.** Adam Nash was conceived and selected to save his sister's life. But he is also a much-loved sibling and son.

## 21.1 New Ways to Make Babies

While the majority of new parents have children naturally, technology has made possible several new twists on the reproductive theme. Following are two unusual experiences, and a common one.

### Grandmother and Mother at the Same Time

When 42-year-old Arlette Schweitzer gave birth to twins Chelsea and Chad at St. Luke's Midland Hospital in Aberdeen, South Dakota, the event made national headlines. A woman in her forties having twins isn't odd, but Arlette's twins were rather unusual—they were her genetic grandchildren, but also her gestational children.

Arlette had given birth to Christa two decades earlier. Christa had healthy ovaries and oocytes, but she lacked a uterus, and therefore could not carry children. After Christa was married, she desperately wanted children; her mother stepped in to help. Christa's oocytes and her husband Kevin's sperm were mixed in a laboratory dish. Fertilization occurred, and after a few cell divisions, embryos were placed into Arlette's uterus. Arlette's hormonal cycle had been manipulated with drugs to make the uterine lining receptive to pregnancy. Christa and Kevin's route to parenthood worked because the technique worked, and because the participants shared the same goal.

### Midlife Motherhood

Rosanna Della Corte became a mother at age 62, after her teenage son died in an accident. An oocyte that a younger woman had donated was fertilized with sperm from Rosanna's husband Mauro, in a laboratory dish. The fertilized ovum divided twice. Meanwhile, hormone treatments thickened Rosanna's uterine lining. Her body nurtured the embryo that would develop into baby Riccardo.

The Cortes's success showed that it is not the condition of the uterine lining in an older woman that makes conceiving, carrying, and delivering a healthy baby difficult. Rather, the age of the oocyte is what limits pregnancy in older women. Soon after Riccardo's birth, a 63-year-old woman had a child!

### A Five-Year Wait

The experience of Pamela and Jonathon Loew is more typical than that of Arlette Schweitzer or the Cortes. For five years, the Loews underwent a long list of techniques trying to conceive, culminating in Alexandra's birth.

The Loews' quest began with ruling out various causes of the inability to conceive. Pamela received hormone therapy, then a physician placed Jonathon's sperm near her cervix eight times. One time, this resulted in pregnancy, but the embryo lodged in one of Pamela's fallopian tubes and had to be removed to save her life.

Next, antigens were removed from the surfaces of Jonathon's sperm to prevent Pamela's immune system from attacking. The "washed" sperm were placed in her fallopian tube, near the uterus, along with oocytes from Pamela's ovary. If all went well, the gametes would meet and merge. The first time the Loews used this procedure, it didn't work. The second time, it resulted in Alexandra. The Loew family was lucky. Another couple spent 10 years, and more than $100,000, and still did not have a child.

**Key Concepts**

Assisted reproductive technologies provide several ways for people to have children who might not otherwise be able to do so.

**table 21.1**

**Causes of Subfertility and Infertility**

**Men**

| Problem | Possible Causes | Treatment |
|---|---|---|
| Low sperm count | Hormone imbalance, varicose vein in scrotum, possibly environmental pollutants | Hormone therapy, surgery, avoiding excessive heat |
| | Drugs (cocaine, marijuana, lead, arsenic, some steroids and antibiotics), chemotherapy | |
| | Oxidative damage | |
| | Y chromosome gene deletions | |
| Immobile sperm | Abnormal sperm shape | Intracytoplasmic sperm injection |
| | Infection | |
| | Malfunctioning prostate | Antibiotics |
| | Deficient apoptosis | Hormones |
| Antibodies against sperm | Problem in immune system | Drugs |

**Women**

| Problem | Possible Causes | Treatment |
|---|---|---|
| Ovulation problems | Pituitary or ovarian tumor | Surgery |
| | Underactive thyroid | Drugs |
| | Polycystic ovary syndrome | |
| Antisperm secretions | Unknown | Acid or alkaline douche, estrogen therapy |
| Blocked fallopian tubes | Infection caused by IUD or abortion or by sexually transmitted disease | Laparotomy, oocyte removed from ovary and placed in uterus |
| Endometriosis | Delayed parenthood until the thirties | Hormones, laparotomy |

## 21.2 Infertility and Subfertility

**Infertility** is the inability to conceive a child after a year of frequent intercourse without the use of contraceptives. Some specialists use the term *subfertility* to distinguish those individuals and couples who can conceive unaided, but for whom this may take longer than is usual. On a more personal level, infertility is a seemingly endless monthly cycle of raised hopes and crushing despair. In addition, as a woman ages, the incidence of pregnancy-related problems rises, including chromosomal anomalies, fetal deaths, premature births, and low-birthweight babies. For most conditions, the man's age does not raise the risk of pregnancy complications.

There is also a societal gender bias concerning the age of new parents, which is perhaps a reflection of biology—that is, the indisputable fact that men can father children into old age, but women cannot. The 63-year-old woman who gave birth was sharply criticized in the media, yet a 77-year-old actor who became a father at about the same time was congratulated. But sometimes nature surprises us. In January 2000, a 55-year-old woman gave birth to naturally conceived triplets!

Physicians who specialize in infertility treatment can identify a physical cause in 90 percent of cases. Of these, 30 percent of the time the problem is primarily in the male, and 60 percent of the time it is primarily in the female. However, for cases where a physical problem is not obvious, the cause is usually a mutation or chromosomal aberration that impairs fertility in the male. The statistics are somewhat unclear, because in 20 percent of the 90 percent, both partners have a medical condition that could contribute to infertility or subfertility. A common combination, for example, is a woman with an irregular menstrual cycle and a man with a low sperm count.

One in six couples has difficulty in conceiving or giving birth to children. That's about 2.4 million couples in the United States. Table 21.1 summarizes causes of subfertility and infertility.

### Male Infertility

Infertility in the male is easier to detect but sometimes harder to treat than female infertility. One in 25 men is infertile. Some men have difficulty fathering a child because they produce fewer than the average 120 million sperm cells per milliliter of ejaculate, a condition called oligospermia. It has several causes. If a low sperm count is due to a hormonal imbalance, administering the appropriate hormones may boost sperm output. Sometimes a man's immune system produces IgA antibodies that cover the sperm and prevent them from binding to oocytes. Male infertility can also be due to a varicose vein in the scrotum. This enlarged vein produces too much heat near developing sperm, so that they cannot mature. Surgery can remove a scrotal varicose vein.

Until recently, no cause could be identified for 30 to 40 percent of cases of male infertility. Researchers then discovered that many of these men have small deletions of the Y chromosome that remove the only copies of key genes whose products control spermatogenesis. Clues to this cause of male infertility came from men who cannot make any sperm, and who also have very large Y chromosome deletions. If a deletion is very small, the man may make some sperm.

Other genetic causes of male infertility include mutations in genes that encode androgen receptors or protein fertility hormones, or that regulate sperm development or motility. Overall, most cases of male infertility are genetic.

For many men with low sperm counts, fatherhood is just a matter of time. They are subfertile, not infertile. If an ejaculate contains at least 60 million sperm cells, fertilization is likely eventually. To speed conception, a man with a low sperm count can donate several semen samples over a period of weeks at a fertility clinic. The samples are kept in cold storage, then pooled. Some of the seminal fluid is withdrawn to leave a sperm cell concentrate, which is then placed in the woman's reproductive tract. It isn't very romantic, but it is highly effective at achieving pregnancy.

Sperm quality is even more important than quantity. Sperm cells that are unable to move—a common problem—cannot reach an oocyte. If the lack of motility is due to a physical defect, such as sperm tails that are misshapen or missing, or bumps near the sperm head, there is no treatment (figure 21.2). If the cause is hormonal, however, replacing the absent hormones can sometimes restore sperm motility.

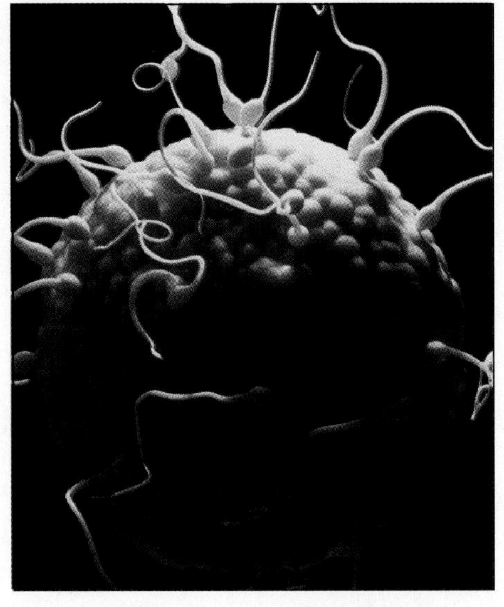

a.

b.

**figure 21.2**

**Sperm shape and motility are important.** (*a*) Healthy sperm in action. (*b*) A misshapen sperm cannot fertilize an oocyte.

Hampered sperm motility is also associated with the presence of white blood cells in semen. The blood cells produce toxic compounds called reactive oxygen species, which bind sperm cell membranes and destroy enzymes that are essential for the reactions within the sperm cell that generate ATP. With too little ATP to supply energy, sperm cannot move effectively. Fertility declines. Clinical trials are underway to test antioxidants to treat this form of male infertility.

Faulty apoptosis (programmed cell death) can also cause male infertility. Apoptosis selectively kills abnormally shaped sperm. Studies show that men with high percentages of abnormally shaped sperm often have cell surface molecules that indicate impaired apoptosis.

## Female Infertility

Female infertility can be caused by abnormalities in any part of the reproductive system. Many women with subfertility or infertility have irregular menstrual cycles, making it difficult to pinpoint when conception is most likely. In an average menstrual cycle of 28 days, ovulation usually occurs around the 14th day after menstruation begins, and this is when a woman is most likely to conceive.

For a woman with regular menstrual cycles who is under 30 years old and not using birth control, pregnancy is likely to happen within three or four months. A woman with irregular menstrual periods can use an ovulation predictor test, which detects a peak in the level of a certain hormone that precedes ovulation by a few hours, or keep a record of her body temperature each morning using a special thermometer with very fine subdivisions. She can then time intercourse for when she is most likely to conceive. Sperm can fertilize oocytes if they have been in the woman's body for up to five days before ovulation, but can fertilize for only a short time after ovulation.

The hormonal imbalance that usually underlies irregular ovulation has various causes—a tumor in the ovary or in the pituitary gland in the brain that hormonally controls the reproductive system, an underactive thyroid gland, or use of steroid-based drugs such as cortisone. Sometimes a woman produces too much prolactin, the hormone that normally promotes milk production and suppresses ovulation in new mothers. If prolactin is abundant in a nonpregnant woman, she will not ovulate and therefore cannot conceive.

Fertility drugs can stimulate ovulation, but they can also cause women to "superovulate," producing more than one oocyte each month. A commonly used drug, clomiphene, raises the chance of having twins from 1 to 2 percent to 4 to 6 percent. If a woman's ovaries

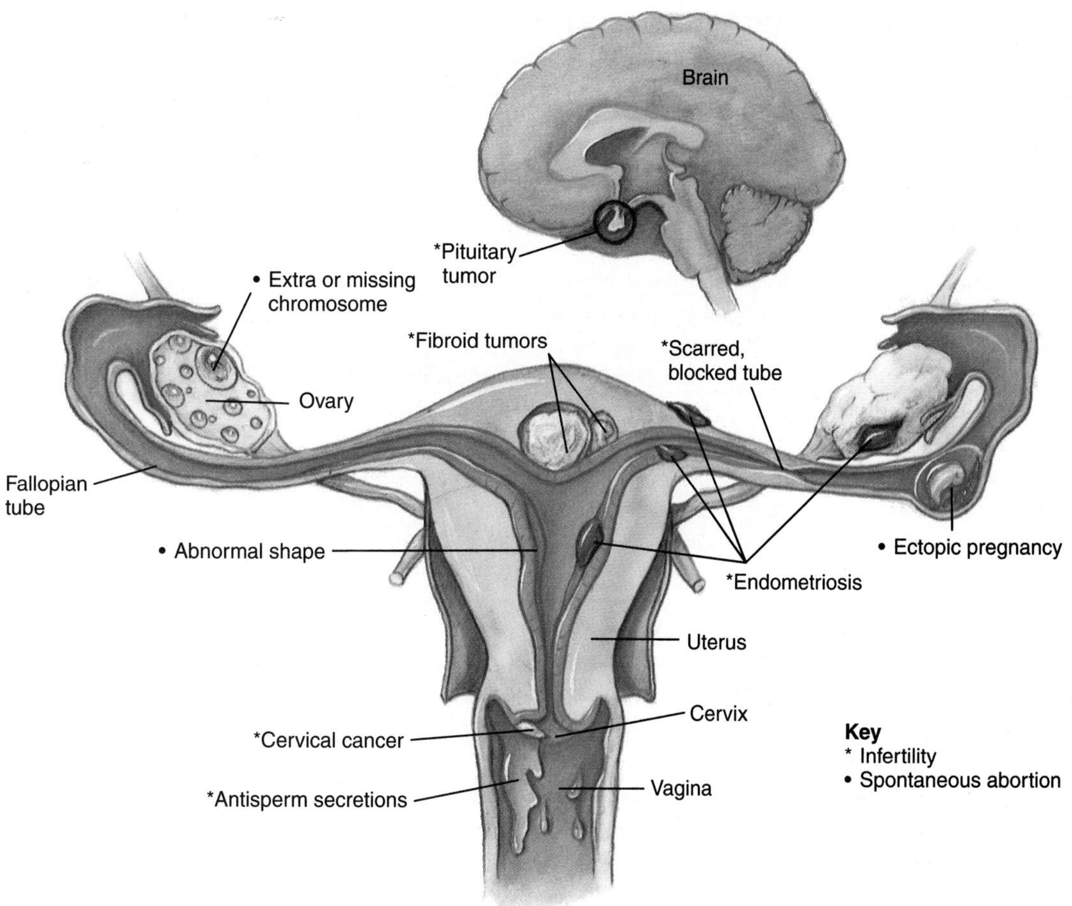


**figure 21.3**

**Sites of reproductive problems in the human female.**

are completely inactive or absent (due to a birth defect or surgery), she can become pregnant only if she uses a donor oocyte. Ovary transplants have succeeded only between identical twins, or in the same woman if she has ovarian tissue removed before cancer treatment and implanted after.

A common cause of female infertility is blocked fallopian tubes. Fertilization usually occurs in open tubes. Blockage can prevent sperm from reaching the oocyte, or entrap a fertilized ovum, keeping it from descending into the uterus (figure 21.3). The embryo begins developing in the tube and if it is not removed and continues to enlarge, the woman can die. This "tubal pregnancy" is called an **ectopic pregnancy.**

Fallopian tubes can also be blocked due to a birth defect or, more likely, from an infection such as pelvic inflammatory disease. A woman may not know she has blocked fallopian tubes until she has difficulty conceiving and medical tests uncover the problem. Surgery can sometimes open blocked fallopian tubes.

Excess tissue growing in the uterine lining may make it inhospitable to an embryo. This tissue can include benign tumors, called **fibroids,** or a condition called **endometriosis,** in which tissue builds up in the uterus and sometimes outside of it. In response to the hormonal cues to menstruate, the tissue bleeds, causing cramps. Endometriosis can make conception difficult, but curiously, once a woman with endometriosis has been pregnant, the cramps and bleeding usually subside.

Sometimes secretions in the vagina and cervix are hostile to sperm. If cervical mucus is unusually thick or sticky, as can happen during infection, sperm become entrapped and cannot move far enough to encounter an oocyte. Vaginal secretions may be so acidic or alkaline that they weaken or kill sperm. Douching daily with an acidic solution such as acetic acid (vinegar) or an alkaline solution, such as bicarbonate, alters the pH of the vagina so that it is more receptive to sperm cells. Too little mucus is treated with low daily doses of oral estrogen. Sometimes mucus in a woman's body harbors antibodies that attack sperm. Infertility may also result if the oocyte fails to release sperm-attracting biochemicals.

One reason that female infertility increases with age is that older women are more likely to produce oocytes with an abnormal chromosome number, which often causes spontaneous abortion. Losing very early embryos may appear to be infertility because the bleeding accompanying the aborted embryo resembles a heavy menstrual flow. The higher incidence of meiotic errors in older women may occur because their oocytes have been exposed longer to harmful chemicals, viruses, and radiation. However, some women are prone to producing oocytes with abnormal numbers of chromosomes.

## Infertility Tests

A number of medical tests can identify the cause or causes of infertility. The man is checked first, because it is easier, less costly, and certainly less painful to obtain sperm than oocytes. Sperm are checked for number (sperm count), motility, and morphology (shape). An ejaculate containing up to 40 percent unusual forms is still considered normal, but many more than this can impair fertility. A urologist performs sperm tests. A genetic counselor may also be of help in identifying the cause of male infertility. He or she can interpret the results of a PCR analysis of the Y chromosome to detect deletions that are associated with lack of sperm.

If a male cause of infertility is not apparent, the next step is for the woman to consult a gynecologist. This medical specialist checks to see that the structures of the woman's reproductive system are present and functioning.

Some cases of subfertility or infertility have no clear explanation. Psychological factors may be at play, or it may be that inability to conceive results from consistently poor timing. Sometimes a subfertile couple adopts a child, only to conceive one of their own shortly thereafter; at other times, the couple's infertility remains a lifelong mystery.

**KEY CONCEPTS**

Male infertility is due to a low sperm count or sperm that cannot swim or are abnormal in structure. • Female infertility can be due to an irregular menstrual cycle or blocked fallopian tubes. Fibroid tumors, endometriosis, or a misshapen uterus may prevent implantation of a fertilized ovum, and secretions in the vagina and cervix may inactivate or immobilize sperm. Oocytes may fail to release a sperm-attracting biochemical. Early pregnancy losses due to abnormal chromosome number may be mistaken for infertility; this is more common among older women. • A variety of medical tests can pinpoint some causes of infertility.

# 21.3 Assisted Reproductive Technologies

A growing number of couples with fertility problems are turning to alternative ways to achieve pregnancy, many of which were perfected in nonhuman animals. The Technology Timeline on reproductive technologies (see page 431) depicts the chronology for these assisted reproductive technologies.

## Donated Sperm—Artificial Insemination

The oldest assisted reproductive technology is **artificial insemination,** in which a doctor places donated sperm into a woman's reproductive tract. Her partner may be infertile or carry a gene for an inherited illness that the couple wishes to avoid passing to their child, or a woman may undergo artificial insemination if she desires to be a single parent without having sex.

The first artificial insemination in humans was done in 1790. For many years, physicians donated sperm, and this became a way for male medical students to earn a few extra dollars. By 1953, sperm could be frozen and stored and artificial insemination became much more commonplace. Today, donated sperm is frozen and stored in sperm banks, which provide the cells to obstetricians who perform artificial inseminations.

A couple who chooses artificial insemination can select sperm from a catalog that lists the personal characteristics of donors, such as blood type, hair and eye color, skin color, build, and even educational level and interests. Of course, not all of these traits are inherited. If a couple desires a child of one sex—such as a daughter to avoid passing on an X-linked disorder—the man's sperm can be separated by weight into fractions enriched for X-bearing or Y-bearing sperm.

Problems can arise in artificial insemination, as they can in any pregnancy. For example, a man who donated sperm years ago developed a late-onset genetic disease, cerebellar ataxia. Eighteen children conceived using his sperm now face a 1 in 2 risk of having inherited the mutant gene. In 1983, the Sperm Bank of California became the first to ask donors if they wished to be contacted by their children years later. In 2002, the first such meeting occurred, quite successfully.

A male's role in reproductive technologies is simpler than a woman's. A man can be a genetic parent, contributing half of his genetic self in his sperm, but a woman can be both a genetic parent (donating the oocyte) and a gestational parent (donating the uterus). Problems can result when a second female assists in either the conception and/or gestation of a child. Table 21.2 highlights some cases of assisted reproductive disasters.

**table 21.2**

**Assisted Reproductive Disasters**

1. A physician in California used his own sperm to artificially inseminate 15 patients and told them that he had used sperm from anonymous donors.
2. A plane crash killed the wealthy parents of two early embryos stored at –320°F (–195°C) in a hospital in Melbourne, Australia. Adult children of the couple were asked to share their estate with two 8-celled siblings-to-be.
3. Several couples in Chicago planning to marry discovered that they were half-siblings. Their mothers had been artificially inseminated with sperm from the same donor.
4. Two Rhode Island couples sued a fertility clinic for misplacing several embryos.
5. Several couples in California sued a fertility clinic for implanting their oocytes or embryos in other women without donor consent. One woman requested partial custody of the resulting children if her oocytes were taken, and full custody if her embryos were used, even though the children were of school age and she had never met them.
6. A man sued his ex-wife for possession of their frozen fertilized ova. He won, and donated them for research. She had wanted to be pregnant.
7. Jaycee Buzzanca once had five parents and now has none. In 1995, she was conceived using a sperm donor, an oocyte donor, and a surrogate mother, to be turned over at birth to John and Luanne Buzzanca. But John left his wife and refused to pay child support, and a judge agreed with him, calling Luanne a "temporary custodial person," rather than a mother. The other three parents did not want the child. Luanne eventually adopted Jaycee.

Landmarks in Reproductive Technology

| | In Animals | In Humans |
|---|---|---|
| **1782** | Artificial insemination in dogs | |
| **1790** | | Pregnancy reported from artificial insemination |
| **1890s** | Birth from embryo transplantation in rabbits | Artificial insemination by donor |
| **1949** | Cryoprotectant successfully freezes animal sperm | |
| **1951** | First calf born after embryo transplantation | |
| **1952** | Live calf born after insemination with frozen sperm | |
| **1953** | | First reported pregnancy after insemination with frozen sperm |
| **1959** | Live rabbit offspring produced from *in vitro* ("test tube") fertilization (IVF) | |
| **1972** | Live offspring from frozen mouse embryos | |
| **1976** | | First reported commercial surrogate motherhood arrangement in the United States |
| **1978** | Transplantation of ovaries from one cow to another | Baby born after IVF in United Kingdom |
| **1980** | | Baby born after IVF in Australia |
| **1981** | Calf born after IVF | Baby born after IVF in United States |
| **1982** | Sexing of embryos in rabbits | |
| | Cattle embryos split to produce genetically identical twins | |
| **1983** | | Embryo transfer after uterine lavage |
| **1984** | | Baby born in Australia from frozen and thawed embryo |
| **1985** | | Baby born after gamete intrafallopian transfer (GIFT) |
| | | First reported gestational surrogacy arrangement in the United States |
| **1986** | | Baby born in the United States from frozen and thawed embryo |
| **1989** | | First preimplantation genetic diagnosis (PGD) |
| **1992** | | First pregnancies from sperm injected into oocytes |
| **1994** | Intracytoplasmic sperm injection (ICSI) in mouse and rabbit | 62-year-old woman gives birth from donated oocyte |
| **1995** | Sheep cloned from embryo cell nuclei | Babies born following ICSI |
| **1996** | Sheep cloned from adult cell nucleus | |
| **1998** | Mice cloned from adult cell nuclei | Baby born 7 years after his twin |
| **1999** | Cattle cloned from adult cell nuclei | |
| **2000** | Pigs cloned from adult cell nuclei | |
| **2001** | | Sibling born following PGD to treat sister for genetic disease |
| | | Human preimplantation embryo cloned |

## A Donated Uterus—Surrogate Motherhood

If a man produces healthy sperm but his partner's uterus is absent or cannot maintain a pregnancy, a **surrogate mother** may help by being artificially inseminated with the man's sperm. When the child is born, the surrogate mother gives the baby to the couple. In this variation of the technology, the surrogate is both the genetic and the gestational mother. Attorneys usually arrange surrogate relationships. The surrogate mother signs a statement signifying her intent to give up the baby, and she is paid for her nine-month job.

The problem with surrogate motherhood is that a woman may not be able to predict her responses to pregnancy and childbirth in the cold setting of a lawyer's office. When a surrogate mother changes her mind about giving up the baby, the results are wrenching for all. A prominent early case involved Mary Beth Whitehead, who carried the child of a married man for a fee and then changed her mind about relinquishing the baby. Whitehead's ties to "Baby M" were perhaps stronger because she was both the genetic and the gestational mother.

Another type of surrogate mother lends only her uterus, receiving a fertilized ovum conceived from a man and a woman who has healthy ovaries but lacks a functional uterus. The gestational-only surrogate mother turns the child over to the donors of the genetic material.

## *In Vitro* Fertilization

In ***in vitro*** **fertilization** (IVF), which means "fertilization in glass," sperm and oocyte join in a laboratory dish. Then, the embryo is placed in the oocyte donor's uterus (or another woman's uterus), and, if all goes well, implants into the uterine lining.

Louise Joy Brown, the first "test-tube baby," was born in 1978. Initial media attention was great, with cartoons depicting a newborn with the word "Pyrex," a test-tube and glassware manufacturer, branded on her thigh. Yet Louise was, despite her unusual beginnings, a rather ordinary child. The technology has since mushroomed. More than half a million children have been born following the technique.

A woman might undergo IVF if her ovaries and uterus work but her fallopian tubes are blocked. To begin, the woman takes a superovulation drug that causes her ovaries to release more than one "ripe" oocyte at a time. Using a laparoscope, a physician removes several of the largest oocytes and transfers them to a dish. Chemicals that mimic those in the female reproductive tract are added, and sperm are applied to the oocytes.

If the sperm cannot readily penetrate the oocyte, they may be sucked up into a tiny syringe and microinjected into the female cell. This is called **intracytoplasmic sperm injection** (ICSI), and it is more effective than IVF alone (figure 21.4). ICSI is very helpful for men who have low sperm counts or high percentages of abnormal sperm. The procedure even works with immature sperm, making fatherhood possible for men who cannot ejaculate, such as those who have suffered spinal cord injuries. ICSI has been very successful, performed on thousands of men with about a 30 percent success rate. The Bioethics: Choices for the Future on page 436 considers an unexpected problem with ICSI—transmitting infertility.

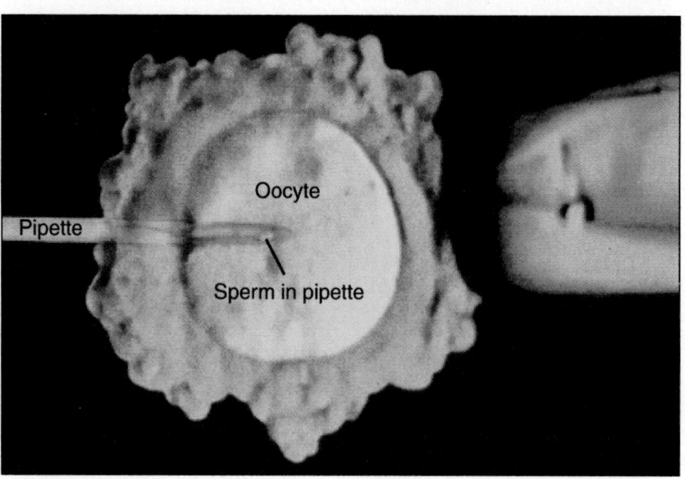


**figure 21.4**

**ICSI.** Intracytoplasmic sperm injection (ICSI) enables some infertile men, men with spinal cord injuries, or men with certain illnesses to become fathers. A single sperm cell is injected into the cytoplasm of an oocyte.

A day or so after sperm wash over the oocytes in the dish, or are injected into them, some of the embryos—balls of 8 or 16 cells—are transferred to the woman's uterus. If the hormone human chorionic gonadotropin appears in her blood a few days later, and its level rises, she is pregnant.

IVF costs from $8,000 to $15,000 per attempt, and the success rate is 15 to 20 percent. By contrast, two-thirds of embryos conceived through sexual intercourse implant. The low success rate of IVF is due both to the difficulty of the procedure, and the fact that couples who choose it have a higher incidence of subfertility or infertility than the general population. The children have twice the rate of birth defects (about 9 percent) compared to children conceived naturally, which may also reflect the underlying medical problem of a parent. In the past, several embryos were implanted, to increase the success rate, but this led to many multiple births. In many cases, embryos had to be removed to make room for others to survive. Section 21.4 considers what to do with extra embryos.

Measures to improve the chance that IVF will culminate in a birth include:

1. Blocking certain hormones during superovulation, which produces oocytes that are more mature and therefore more likely to be fertilized and develop.
2. Transferring embryos slightly later in development, at the blastocyst stage.
3. Culturing fertilized ova and early embryos with other cells that normally surround the oocyte in the ovary. These "helper" cells provide extra growth factors.
4. Screening early embryos for chromosome abnormalities, and implanting only those with apparently normal karyotypes.

## Gamete Intrafallopian Transfer

As the world marveled at the miracle of "test-tube" babies, disillusioned couples were learning that IVF is costly, time-consuming, and for years rarely worked. IVF may fail because of the artificial environment for fertilization. A procedure called GIFT, which stands for **gamete intrafallopian transfer,** improves the setting. Fertilization is assisted in GIFT, but it occurs in the woman's body rather than in glassware. Certain religions approve of GIFT, but not IVF, because GIFT preserves more natural reproductive function.

In GIFT, a woman takes a superovulation drug for a week, then several of her largest oocytes are removed. The man submits a sperm sample, and the most active cells are separated from it. The collected oocytes and sperm are deposited together in the woman's fallopian tube, at a site past any obstruction that might otherwise block fertilization. GIFT is about 26 percent successful, and costs about half as much as IVF.

A variation of GIFT is ZIFT, which stands for **zygote intrafallopian transfer.** In this procedure, an IVF ovum is introduced into the woman's fallopian tube. Allowing the fertilized ovum to make its own way to the uterus seems to increase the chance that it will implant. ZIFT is 23 percent successful.

## Oocyte Banking and Donation

Oocytes can be stored, as sperm are, but the procedure may introduce problems. Candidates for preserving oocytes for later use include:

- Women wishing to have children later in life
- Women undergoing chemotherapy or other toxic or teratogenic treatments
- Women who work with toxins or teratogens

Oocytes are frozen in liquid nitrogen at –30 to –40 degrees Celsius, when they are at metaphase of the second meiotic division. At this time, the chromosomes are aligned along the spindle, which is sensitive to temperature extremes. If the spindle comes apart as the cell freezes, the oocyte may lose a chromosome, which would devastate development. Another problem with freezing oocytes is retention of a polar body, leading to a diploid oocyte.

Alternative approaches try to overcome the difficulty of freezing oocytes. Researchers are developing ways to nurture in the laboratory 1-millimeter-square ovary sections that are packed with oocytes. Laboratory culture of oocytes appears to require a high level of oxygen and a complex combination of biochemicals. An approach to help women undergoing radiation treatments to the abdomen for cancer is to implant a strip of ovarian tissue beneath the skin of her arm. The oocytes there are easy to retrieve, and are not exposed to the radiation.

Women can also obtain oocytes from donors, who are typically younger women. Often these women are undergoing IVF and have "extra" harvested oocytes. The potential father's sperm and donor's oocytes are placed in the recipient's uterus or fallopian tube, or fertilization occurs in the laboratory, and an 8- or 16-celled embryo is transferred to the woman's uterus.

The first baby to result from oocyte donation was born in 1984 (figure 21.5). The success rate ranges from 20 to 50 percent, and the procedure costs at least $10,000. The technique is useful for the reasons cited for freezing oocytes, as well as to avoid transmitting a disease-causing gene.

**Embryo adoption** is a variation on oocyte donation. A woman with malfunctioning ovaries but a healthy uterus carries an embryo that results when her partner's sperm artificially inseminates a woman who produces healthy oocytes. If the woman conceives, the embryo is gently flushed out of her uterus a week later and inserted through the cervix and into the uterus of the woman with malfunctioning ovaries. The child is genetically that of the man and the woman who carries it for the first week, but is born from the woman who cannot produce healthy oocytes. "Embryo adoption" is also the term used to describe use of IVF "leftovers."

Another oocyte technology is **cytoplasmic donation.** Older women have their oocytes injected with cytoplasm from the oocytes of younger women to "rejuvenate" them. Although resulting children appear to be healthy, they are being monitored for a potential problem—heteroplasmy, or two sources of mitochondria in one cell. Researchers do not yet know the health consequences of having mitochondria from the donor cytoplasm plus mitochondria from the recipient's oocyte. After the birth of the first child from this technique in 2001, the media made much of the unnaturalness of having three parents—the father, the mother, and the ooplasm donor.

Because oocytes are harder to obtain than sperm, oocyte donation technology has lagged behind that of sperm banks, but is catching up. One IVF facility that has run a donor oocyte program since 1988 has a patient brochure that describes 120 oocyte donors of various ethnic backgrounds, like a catalog of sperm donors. The oocyte donors are young and have undergone extensive medical and genetic tests. Recipients may be up to 55 years of age.

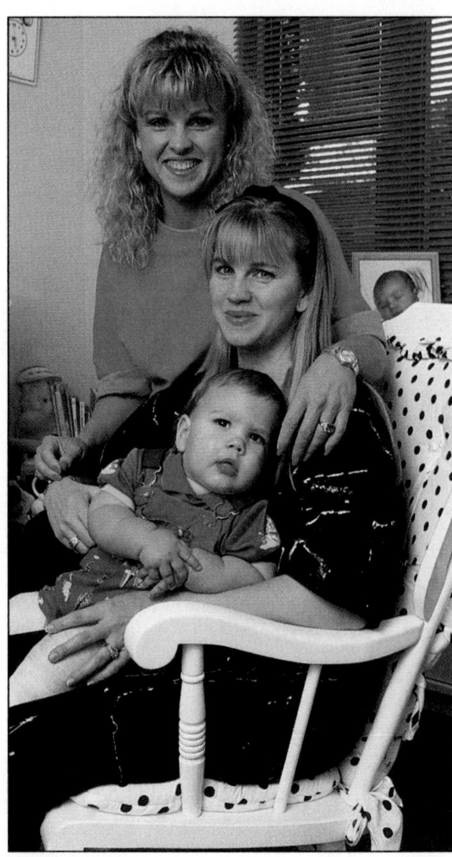

**figure 21.5**

**Oocyte donation.** Anthony Miceli was born from Vicki Miceli's uterus, but he was conceived when Jerry Miceli's sperm fertilized an oocyte donated by Bonny De Irueste, Vicki's sister (in pink). Says Bonny, "I'm his aunt, that's it." Not all reproductive technologies have such joyous outcomes.

## Preimplantation Genetic Screening and Diagnosis

Prenatal diagnostic tests such as amniocentesis, chorionic villus sampling, and fetal cell sorting can be used in pregnancies achieved with assisted reproductive technologies. A technique called **preimplantation genetic diagnosis** (PGD) detects genetic and chromosomal abnormalities *before* pregnancy starts. The couple can select a very early embryo—termed a "preimplantation embryo" because it would not normally have yet arrived at the uterus for implantation—that is free of a certain detectable genetic condition. PGD has about a two-thirds success rate.

PGD is possible because one cell, or blastomere, can be removed for testing from an 8-celled embryo, and the remaining 7 cells can complete development normally in a uterus.

Before the remaining embryo is implanted into the woman, the single cell is karyotyped, or its DNA amplified and probed for particular genes that the parents carry. Healthy ones are selected. At first, researchers implanted the remaining seven cells, but have found that letting the embryo continue developing in the laboratory until day five, when it is 80 to 120 cells, more often leads to success. Obtaining the cell to be tested is called "blastomere biopsy" (figure 21.6). This is the technique that the Nash's used to conceive Adam.

The first children who had PGD were born in 1989. In these cases, probes for Y chromosome-specific DNA sequences were used to select female preimplantation embryos, which developed into girls not affected by the X-linked conditions their mothers carried. The conditions avoided included Lesch-Nyhan syndrome (profound mental retardation with self-mutilative behavior) and adrenoleukodystrophy, in which seizures and nervous system deterioration are followed by sudden death in early childhood. The alternative would have been to risk Mendel's ratios and face the 25 percent chance of conceiving an affected male.

In March 1992, the first child was born who underwent PGD to avoid a specific inherited disease in the family. Chloe O'Brien was checked as an 8-celled preimplantation embryo to see if she had escaped inheriting the cystic fibrosis that affected her brother. Since then, PGD has helped to select hundreds of children free of several dozen types of inherited illnesses.

Preimplantation genetic diagnosis is increasingly being used to screen fertilized ova and early embryos derived from IVF for chromosome abnormalities before implanting them into women. This quality control of sorts ensures that the implanted embryos will not be spontaneously aborted due to chromosomal abnormalities, increasing the likelihood of success. PGD with IVF is primarily used for older women who have suffered spontaneous abortions, because they are at higher risk of having oocytes with abnormal numbers of chromosomes. One woman, for example, had undergone several IVF attempts without becoming pregnant. Finally, she and her partner had five preimplantation embryos karyotyped. Four were abnormal, yet in different ways. One embryo lacked chromosome 14; one lacked chromosome 4 but had an extra chromosome 16; a third embryo lacked a chromosome 8 and 9 but had an extra chromosome 1; another had three chromosomes missing and three extra! The couple's remaining embryo became their daughter. Obviously, one parent had very defective meiosis.

Like many technologies, preimplantation genetic diagnosis can introduce a bioethical "slippery slope" when it is used for reasons other than ensuring that a child is free of a certain disease, such as gender selection. A couple with five sons might, for example, use PGD to select a daughter. But this use of technology might just be a new expression of ages-old human nature, according to one physician who performs PGD. "From the dawn of time, people have tried to control the sex of offspring, whether that means making love with one partner wearing army boots, or using a fluorescence-activated cell sorter to separate X and Y bearing sperm. PGD represents a

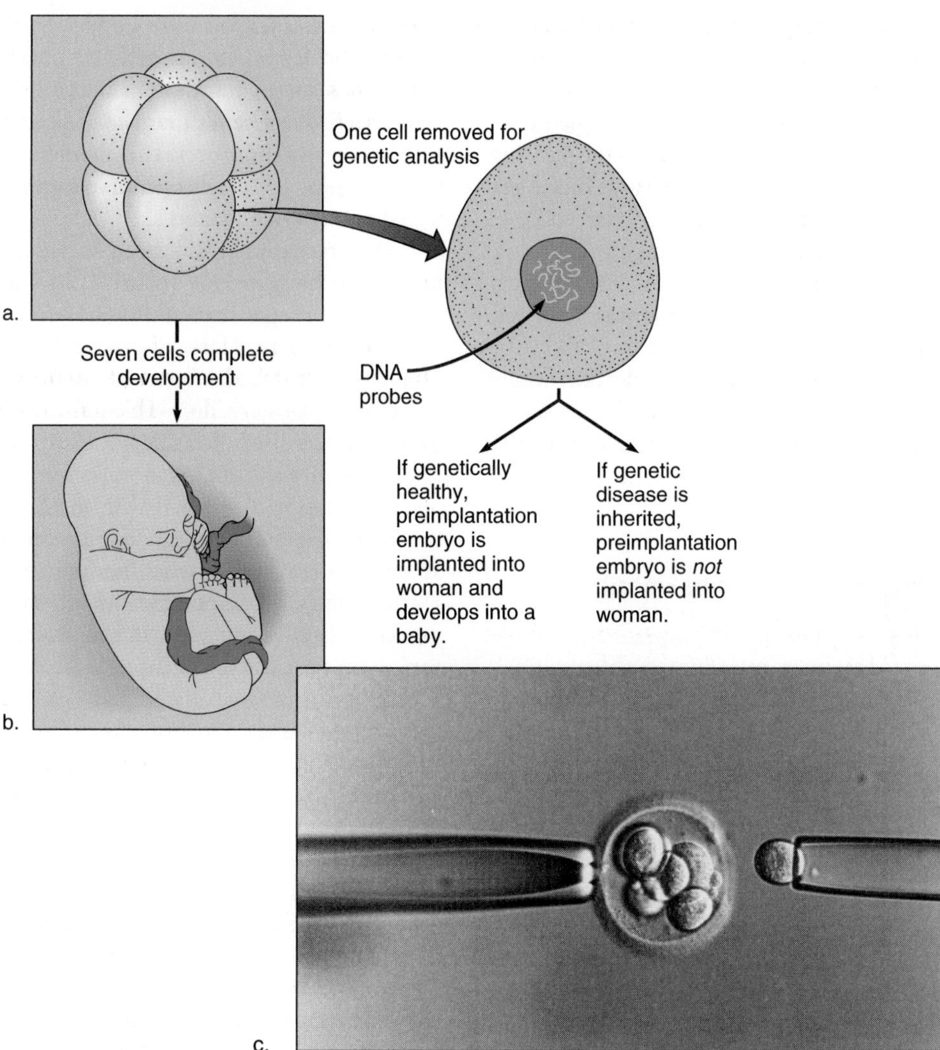


## figure 21.6

**Preimplantation genetic diagnosis.** A blastomere biopsy provides material for assessing the health of an embryo. Preimplantation genetic diagnosis probes disease-causing genes or chromosome aberrations in an 8-celled preimplantation embryo. (*a*) A single cell is separated from the ball of cells and tested to see if it contains a disease-causing gene combination or chromosome imbalance. (*b*) If it doesn't, the remaining seven cells or the embryo after a few more cell divisions, are implanted into the oocyte donor to complete development. (*c*) This preimplantation embryo is held still by suction applied on the left. On the right, a pipette draws up a single blastomere. Fertilization took place 45 hours previously, *in vitro*.

quantum leap in that ability—all you have to do is read the X and Y chromosome paints," he says.

While PGD used solely for family planning is certainly more civilized than placing baby girls outside the gates of ancient cities to perish, the American Society for Reproductive Medicine endorses the use of PGD for sex selection only to avoid passing on an X-linked disease. Yet even PGD to avoid disease can be controversial. A woman with early-onset familial Alzheimer disease used PGD to select a daughter free of the dominant mutant gene—but that child will have to experience her mother's illness.

**Key Concepts**

In artificial insemination, donor sperm is placed in a woman's reproductive tract. • A genetic and gestational surrogate mother is artificially inseminated, becomes pregnant, then gives the baby to the father and his partner. A gestational surrogate mother gestates a baby conceived *in vitro* with gametes from a man and a woman who cannot carry a fetus. • In IVF, sperm and oocyte unite outside the body, and the resulting embryo is transferred to the uterus. Early embryos can also be frozen and used later. • In GIFT, sperm and oocytes are placed in a fallopian tube at a site past a blockage. • In ZIFT, an IVF embryo is placed in a fallopian tube. • Oocytes can be donated. In embryo adoption, an artificially inseminated woman has an early embryo washed out of her uterus and transferred to a woman who lacks oocytes. • PGD removes cells from early embryos and screens them for genetic or chromosomal abnormalities.

## 21.4 On the Subject of "Spares"

In the United States, federal funds cannot be used for research that uses human embryos, and efforts are underway to criminalize such work. As a consequence, most strides in human reproductive research have been funded by patients or companies, or have occurred outside the United States. Although thousands of frozen embryos derived from IVF have been in United States deep freezers for years, this fact has recently surfaced in the media because these embryos are an obvious source of material for basic research on stem cells. If not used for research they might be discarded.

As government bioethics committees debate manipulating human embryos, privately funded researchers are gleaning information from "spares," the fertilized ova and early embryos that couples choose to discard, donate, or freeze. These glimpses into early human prenatal development sometimes challenge long-held ideas. This was the case for a study from Royal Victoria Hospital at McGill University in Montreal. Researchers examined the chromosomes of sperm from a man with Klinefelter (XXY) syndrome. Many of the sperm would be expected to have an extra X chromosome, due to nondisjunction (see figure 12.12), which could lead to a preponderance of XXX and XXY offspring. Surprisingly, only 3.9 percent of the man's sperm had extra chromosomes, but of 10 of his spare embryos, five had an abnormal X, Y, or chromosome 18. That is, even though most of the man's sperm were normal, his embryos weren't. The source of reproductive problems in Klinefelter disease, therefore, might not be in sperm, but in early embryos—a finding that was previously unknown and not expected.

In another study, Australian researchers followed the fates of single blastomeres that had too many or too few chromosomes. They wanted to see whether the abnormal cells preferentially ended up in the inner cell mass, which develops into the embryo, or the trophectoderm, which becomes extraembryonic membranes. The study showed cells with extra or missing chromosomes become part of the inner cell mass much more frequently than if distribution to either that or the trophectoderm occurs by chance. This finding indicates that a prediction of a healthy offspring based on preimplantation genetic diagnosis can go awry if a mutation occurs in a cell of the inner cell mass, a case of chromosomal mosaicism that arises after PGD.

Not everyone agrees that fertilized ova or embryos designated for discard should be used in research. Another assisted reproductive technology will avoid the problem by making spare embryos unnecessary—**polar body biopsy.** It is based on Mendel's first law, the segregation of alleles. In the technique, if a polar body of a woman known to be a carrier of an X-linked disorder has the mutant allele, then it is inferred that the oocyte to which it clings is free of that allele. Oocytes that pass this test can then be fertilized *in vitro* and the resulting embryo implanted in the woman. Polar body biopsy is possible because the polar body is attached to the much larger oocyte. A large pipette is used to hold the two cells in place, and a smaller pipette is used to pluck off the polar body. Then, DNA probes and FISH are used to assess the status of the polar body, and infer the genotype of the oocyte (figure 21.7). Polar body biopsy followed by PGD so far is quite effective in avoiding conceptions

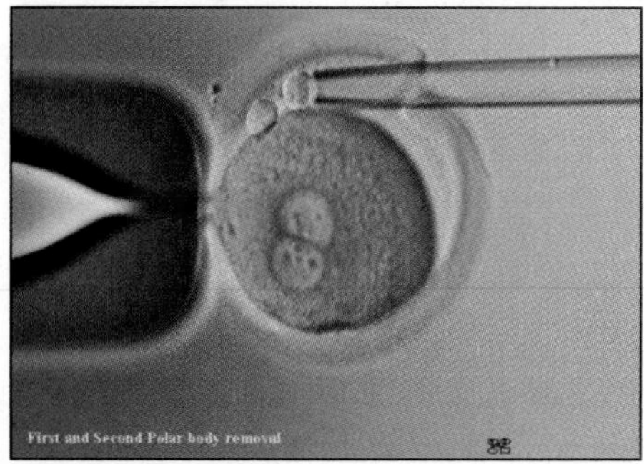


**figure 21.7**

**Polar body biopsy.** The fact that an oocyte shares a woman's divided genetic material with a much smaller companion, the polar body, is the basis for a polar body biopsy to screen oocytes for use in IVF. If a woman is known to be a carrier of a genetic disorder and the polar body contains the disease-causing gene variant, it can be inferred that the oocyte has received the "healthy" version of the gene. A polar body biopsy is possible because the polar body remains attached to the oocyte. In the procedure, the oocyte is held in place by a large pipette, and the attached polar body is captured by drawing it into a smaller pipette. Genetic tests may then be performed on the polar body.

*Bioethics: Choices for the Future*

# Technology Too Soon? The Case of ICSI

Intracytoplasmic sperm injection (ICSI), available since 1995, has been extremely successful in enabling men with AIDS, paralysis, very low sperm counts, or abnormal sperm to become fathers. The birth defect rate is the same as for normal conceptions, although abnormal embryos and sex chromosome aberrations are slightly more common. But as more ICSI procedures are performed and tests on nonhuman animals continue, potential problems are emerging, based on the fact that ICSI bypasses what one researcher calls "natural sperm selection barriers."

ICSI is now commonly used on men who have azoospermia—lack of sperm—or oligospermia—very few sperm. The rare sperm are sampled from an ejaculate or taken from the testes with a needle, then injected into an oocyte. Sometimes these men produce spermatids but not mature sperm, and spermatids that have already elongated can also successfully fertilize oocytes using ICSI. However, a complication has developed that involves not science, but logic. About 10 percent of infertile men have microdeletions in the Y chromosome. When their sperm is used in ICSI, they are passing on the source of their infertility to their sons. This is true for other causes of infertility and subfertility, too. For example, if a man's above-average proportion of abnormally shaped sperm is due to abnormal apoptosis, he could pass a susceptibility to cancer to his offspring.

Bioethicists are debating whether it is right to intentionally conceive a male who is genetically destined to be infertile. On the positive side, however, is the opportunity that ICSI is providing to study that infertility. It is possible that adolescents with microdeleted Y chromosomes can produce viable sperm or spermatids, and if so, these cells can be sampled and stored for later use. In the past, these men would not have suspected that they had a Y chromosome abnormality until they had difficulty fathering a child. Alternatives to transmitting deletions and other mutations include selecting and using only X-bearing sperm in ICSI, or selecting and implanting only female fertilized ova. Because of the transmission of Y-linked infertility with ICSI, men undergoing fertility testing and considering the procedure now have their Y chromosomes screened for deletions, which affect a region called AZFc. Genetic counseling is provided.

Meanwhile, work on rhesus monkeys is pinpointing the sources of damage to those ICSI embryos that do not continue to develop:

- Injecting sperm at the site of the polar body on the oocyte can disrupt the meiotic spindle, leading to nondisjunction (an extra or missing chromosome).
- Injected sperm DNA does not always condense properly, also leading to nondisjunction.
- Spermatids that have not yet elongated are often unable to fertilize an oocyte. Culturing them in the laboratory until they mature may help improve the odds of success.
- Spermatids may not be completely imprinted, leading to problems in gene expression.
- Mitotic cell cycle checkpoints are altered at the first division of the zygote following ICSI.
- Injected sperm sometimes lack a protein that normally associates with the sex chromosomes. This may explain the elevation in sex chromosome anomalies.
- Injected sperm can include surface proteins that are normally left outside the oocyte, producing unanticipated effects. They may also include mitochondria from the male.

Despite these largely theoretical concerns, parents of children conceived with ICSI do not have any cause to worry, researchers insist. Tests on these children so far have not revealed any problems. Says one researcher, "In spite of its potential risks, ICSI still seems to be remarkably safe." Still, it would be comforting to some to have more research to back up this new way to start development. Ongoing studies are following the children of ICSI for longer times, and investigating any correlations between health problems in offspring and the cause of subfertility or infertility in the fathers.

with chromosome anomalies or single gene disorders (table 21.3).

The area of assisted reproductive technology introduces a hotbed of bioethical issues, and is certain to intensify as human genome information provides more traits to track and perhaps control in coming generations. Already preimplantation genetic

**table 21.3**

**Results of the First Clinical Trial of Polar Body Biopsy Used with Preimplantation Genetic Diagnosis**

| | **Number** |
|---|---|
| Pregnancies | 102 |
| Births | 109 (80 single, 9 twin, 7 triplets reduced to twins) |
| Birth defects | 6 |
| Spontaneous abortions | 5 (quintuplets), plus 5 singletons |

diagnosis is beginning to be used to screen out embryos destined to face a high risk of developing a particular disorder later in life, such as Huntington disease or cancer. It is easy to envision, instead of a quick Y chromosome FISH test or a probe for a single gene that is mutant in a particular family, DNA microarrays widely applied to gametes, fertilized ova, or preimplantation embryos to screen for large numbers of genetic variants. This coming genomic fortune-telling could spiral out of control. Assisted reproductive technology is a field that operates on molecules and cells, but with repercussions on individuals and families. Ultimately, if widespread enough, it will affect population genetics. Let us hope that regulations will evolve with the technology to assure that it is applied sensibly and humanely.

**Key Concepts**

IVF produces extra fertilized ova and early embryos. They may be used, frozen, or donated. Used in basic research, such embryos are enabling investigators to make discoveries about early human development. Polar body biopsy enables physicians to identify and select out defective oocytes.

# Summary

### 21.1 New Ways to Make Babies

1. Assisted reproductive technologies help people to have children.
2. These technologies can enable postmenopausal women to have children, or help couples who could not otherwise have children to do so.

### 21.2 Infertility and Subfertility

3. **Infertility** is the inability to conceive a child after a year of unprotected intercourse. Subfertility refers to individuals or couples who manufacture gametes, but may take longer than usual to conceive.
4. Causes of infertility in the male include a low sperm count, a malfunctioning immune system, a varicose vein in the scrotum, structural sperm defects, drug exposure, and abnormal hormone levels. Of cases not associated with an obvious physical problem, a mutation is often the cause of the subfertility or infertility.
5. Female infertility can be caused by absent or irregular ovulation, blocked fallopian tubes, an inhospitable or malshaped uterus, antisperm secretions, or lack of sperm-attracting biochemicals. Early pregnancy loss due to abnormal chromosome number is more common in older women and may appear to be infertility.

### 21.3 Assisted Reproductive Technologies

6. In **artificial insemination,** sperm are obtained from a donor and introduced into a woman's reproductive tract in a clinical setting.
7. A gestational and genetic **surrogate mother** provides her oocyte to be artificially inseminated by a man whose partner cannot conceive or carry a fetus. The surrogate also provides her uterus for nine months. A gestational surrogate mother receives an ***in vitro* fertilized** ovum, which belongs genetically to the couple who ask her to carry it.
8. In IVF, oocytes and sperm meet in a dish, fertilized ova divide a few times, and the resulting embryos are placed in the woman's body, circumventing blocked tubes or the inability of the sperm to penetrate the oocyte.
9. Embryos can be frozen and thawed and then complete development when placed in a woman's uterus.
10. **GIFT** introduces oocytes and sperm into a fallopian tube past a blockage; fertilization occurs in the woman's body. **ZIFT** places an early embryo in a fallopian tube.
11. Oocytes can be frozen and stored, donated, or supplemented with cytoplasm from a younger cell. In **embryo adoption,** a woman is artificially inseminated. A week later, the embryo is washed out of her uterus and introduced into the reproductive tract of the woman whose partner donated the sperm.
12. Seven-celled embryos can develop normally if a blastomere is removed at the 8-cell stage and examined for abnormal chromosomes or genes.

### 21.4 On the Subject of "Spares"

13. In the United States, funding for assisted reproductive technologies has come from private sources. The same will be true for stem cell research.
14. Extra fertilized ova and early embryos generated in IVF must be used, donated to couples, stored, or donated for use in research. This material enables researchers to study aspects of early human development that they could not investigate in other ways.
15. Polar body biopsy enables physicians to perform genetic tests on polar bodies and to infer the genotype of the accompanying oocyte, eliminating the problem of having to decide the fate of extra fertilized ova or early embryos.

# Review Questions

1. Which assisted reproductive technologies might help the following couples? (More than one answer may fit some situations.)
   a. A woman who is born without a uterus, but who manufactures healthy oocytes.
   b. A man whose cancer treatments greatly damage his sperm.
   c. A woman undergoes a genetic test that reveals she will develop Huntington disease. She wants to have a child, but she does not want to pass on this presently untreatable illness.
   d. Two women wish to have and raise a child together.
   e. A man and woman are each carriers of sickle cell disease. They do not want to have an affected child, but they also do not want to terminate a pregnancy.
   f. A woman's fallopian tubes are scarred and blocked, so an oocyte cannot reach the uterus.

g. A young woman must undergo abdominal radiation to treat ovarian cancer, but wishes to have a child in the future.

2. Why are men typically tested for infertility before women?

3. A man reads his medical chart and discovers that the results of his sperm analysis indicate that 22 percent of his sperm are shaped abnormally. He wonders why the physician said he had normal fertility if so many sperm are abnormally shaped. Has the doctor made an error?

4. Cite a situation in which both man and woman contribute to subfertility.

5. How does ZIFT differ from GIFT? How does it differ from IVF?

6. A Tennessee lower court, in ruling on the fate of seven frozen embryos in a divorce case, called them "children *in vitro.*" In what sense is this label incorrect?

7. Explain how preimplantation genetic diagnosis is similar to and different from CVS and amniocentesis.

8. What are some of the causes of infertility among older women?

9. How do each of the following assisted reproductive technologies deviate from the normal biological process?

   a. *in vitro* fertilization

   b. GIFT

   c. embryo adoption

   d. gestational surrogacy

   e. artificial insemination

   f. cytoplasmic donation

# Applied Questions

1. At the same time that 62- and 63-year-old women gave birth, actors Tony Randall and Anthony Quinn became fathers at ages 77 and 78—and didn't receive nearly as much criticism as the women. Do you think this is an unfair double standard, or a fair criticism based on valid biological information?

2. Many people spend thousands of dollars pursuing pregnancy. What might be an alternative solution to their quest for parenthood?

3. An Oregon man anonymously donated sperm that were used to conceive a child. The man later claimed, and won, rights to visit his child. Is this situation for the man more analogous to a genetic and gestational surrogate mother, or an oocyte donor who wishes to see the child she helped to bring into existence?

4. Big Tom is a bull with valuable genetic traits. His sperm are used to conceive 1,000 calves. Mist, a dairy cow with exceptional milk output, is given a superovulation drug, and many oocytes are removed from her ovaries. Once fertilized, the oocytes are implanted into surrogate mothers, and with their help, Mist becomes the genetic mother of 100 calves—far more than she could give birth to naturally.

   Which two reproductive technologies performed on humans are based on these two agricultural examples?

5. State who the genetic parents are and who the gestational mother is in each of the following cases:

   a. A man was exposed to herbicides during the Vietnam War and abused drugs for several years before and after that. Now he wants to become a father, but he is concerned that his past exposures to toxins have damaged his sperm. His wife is artificially inseminated with sperm from the husband's brother, who has led a calmer life.

   b. A 26-year-old woman has her uterus removed because of cancer. However, her ovaries are intact and her oocytes are healthy. She takes drugs to superovulate, has oocytes removed, and has them fertilized *in vitro* with her husband's sperm. Two fertilized ova are implanted into the uterus of the woman's best friend.

   c. Max and Tina had a child by IVF in 1986. At that time, they had frozen three extra embryos. Two are thawed years later and implanted into the uterus of Tina's sister, Karen. Karen's uterus is healthy, but she has ovarian cysts that often prevent her from ovulating.

   d. Forty-year-old Christensen von Wormer wanted children, but not a mate. He donated sperm, and an Indiana mother of one was artificially inseminated with them for a fee. On September 5, 1990, von Wormer held his newborn daughter, Kelsey, for the first time.

   e. Two men who live together want to raise a child. They go to a fertility clinic, have their sperm collected and mixed, and used to artificially inseminate a friend, who turns the baby over to them.

6. Delaying childbirth until a woman is over age 35 is associated with certain physical risks, yet an older woman is often more mature and financially secure. Many women delay childbirth so that they can establish careers. Can you suggest societal changes, perhaps using a reproductive technology, that would allow women to more easily have both children and careers?

7. An IVF attempt yields 12 more embryos than the couple who conceived them can use. What could they do with the extras?

8. What do you think children born of an assisted reproductive technology should be told about their origins?

9. Wealthy couples could hire poor women as surrogates or oocyte donors simply because the adoptive mother does not want to be pregnant. Would you object to this practice? Why or why not?

10. Cloning could potentially help people who cannot make gametes. Madeline is fertile, but her partner Cliff had his testicles removed to treat cancer when he was a teenager. He did not bank testicular tissue. To have a child that is genetically theirs, the couple wishes to use a nucleus from one of Cliff's somatic cells, which would be transferred to an oocyte of Madeline's that has had its nucleus removed. The resulting fertilized ovum would be cultured in the laboratory and then transferred to Madeline's uterus via standard IVF. Is the couple correct in assuming that the child would be genetically theirs? Cite a reason for your answer.

# Suggested Readings


Boyle, Robert J., and Julian Savulescu. November 22, 2001. Ethics of using preimplantation genetic diagnosis to select a stem cell donor for an existing person. *British Medical Journal* 323:1240–43. An analysis of the Nash case.

Ford, W. C. L. April 21, 2001. Biological mechanisms of male infertility. *The Lancet* 357:1223–24. Deranged apoptosis can lead to sperm with an excess of abnormal forms.

Gottlieb, Scott. June 23, 2001. Scientists screen embryo for genetic predisposition to cancer. *British Medical Journal* 322:1505. A couple used PGD to ensure that their child did not inherit Li-Fraumeni cancer syndrome. The father was diagnosed at age two.

Hall, Judith G. July 2001. Neonatal outcome after preimplantation genetic diagnosis by analysis of the polar bodies. *Growth, Genetics, and Hormones* 17, no. 2:30–31. A mutant allele in a polar body may mean that the oocyte has been spared.

Hansen, Michéle. March 7, 2002. The risk of major birth defects after intracytoplasmic sperm injection and *in vitro* fertilization. *The New England Journal of Medicine* 346:725–730. Children born following IVF or ICSI face twice the risk of birth defects—but this may reflect the reason for a parent's infertility.

Hughes, Edward G., and Mita Giacomini. September 2001. Funding in *in vitro* fertilization treatment for persistent subfertility: The pain and the politics. *Fertility and Sterility* 76, no. 3:431–42. Assisted reproductive technologies remain prohibitively expensive.

Josefson, D. October 14, 2000. Couple select healthy embryo to provide stem cells for sister. *British Medical Journal* 321:917. The Nash family used PGD to save their daughter from an inherited disease.

Kutluk, Oktay, et al. September 26, 2001. Endocrine function and oocyte retrieval after autologous transplantation of ovarian cortical strips to the forearm. *Journal of the American Medical Association* 286, no. 12:1490–93. A woman about to undergo abdominal radiation or removal of her ovaries can preserve ovarian tissue in her arm.

Lewis, Ricki. November 13, 2000. Preimplantation genetic diagnosis—the next big thing? *The Scientist* 14, no. 22:16. PGD "leftovers" reveal details of the basic mechanisms of early human development.

Magli, M. C., et al. August 2000. Chromosome mosaicism in day three aneuploid embryos that develop to morphologically normal blastocysts *in vitro*. *Human Reproduction* 15:1781–86. Mutations that occur in the early embryo can confound the predictions of preimplantation genetic diagnosis.

Oehninger, Sergio. June 30, 2001. Place of intracytoplasmic sperm injection in management of male infertility. *The Lancet* 357:2068–69. ICSI is invasive and expensive, and so should be restricted to cases of male infertility—not used as a standard part of IVF.

Sheldon, Tony. March 16, 2002. Children at risk after sperm donor develops late onset genetic disease. *The British Medical Journal* 324:631. Eighteen English children conceived by artificial insemination face a 1 in 2 chance of developing cerebellar ataxia, because their biological father did not know he would develop it.


# On the Web

Check out the resources on our website at

**www.mhhe.com/lewisgenetics5**

On the web for this chapter, you will find additional study questions, vocabulary review, useful links to case studies, tutorials, popular press coverage, and much more. To investigate specific topics mentioned in this chapter, also try the links below:

American Society for Reproductive Medicine
**www.asrm.com/**

Assisted Reproductive Technologies
**www.phppo.cdc.gov/dls/art/art.asp**

National Council for Adoption
**www.ncfa-usa.org**

The Organization of Parents Through Surrogacy
**www.opts.com/**

Polycystic Ovarian Syndrome Association
**www.pcosupport.org/**

Pregnancy and Infant Loss Support Inc.
**www.nationalshareoffice.com**

Resolve: The National Infertility Association
**www.resolve.org**

CHAPTER

# 22

# The Human Genome Project and Genomics

### 22.1 Genome Sequencing: A Continuation of Genetics

The human genome project in the 1990s grew out of extensive linkage and cytogenetic maps amassed over the previous decades. Obtaining the first draft sequence turned into a race between an international public consortium and a private company.

### 22.2 The Origin of the Idea

Automated DNA sequencing and computer capabilities were crucial to the ability to sequence the human genome. Many people had the idea to sequence the human genome. The project began in 1990.

### 22.3 Technology Drives the Sequencing Effort

Expressed sequence tags provide a shortcut to protein-encoding parts of the genome. DNA microarrays enable researchers to study gene expression. The project switched about halfway through from mapping to sequencing.

### 22.4 Genome Information Answers and Raises Questions

The complexity of gene function that genomics is revealing indicates that we will have to reevaluate the definition of the gene. Comparisons of genomes among species reveal evolutionary leaps, as well as the humbling news that there is still much about genetics that we do not fathom.

On June 26, 2000, J. Craig Venter and Francis Collins flanked President Clinton in the White House garden to announce the shared accomplishment of obtaining a "first draft" of the human genome sequence. To those unfamiliar with genetics, the news must have seemed to have come out of nowhere, and perhaps to have been the work of just these two men. In actuality, the milestone capped a decade-long project, involving thousands of researchers, which in turn was the culmination of a century of discovery. The two men shaking hands and grinning amidst the presidential atmosphere had in fact been bitter rivals, and the historic date of June 26 came about because it was the only opening in the White House calendar! In other words, the work was monumental; its announcement was staged.

The completion of the first draft human genome sequence was, in fact, a huge accomplishment (figure 22.1). By the time that each competing group published a preliminary analysis of its results in February 2001, their discoveries had already greatly impacted the field of genetics, now evolving into genomics. Human genome information will affect us in many ways. This chapter is the story of how the human genome project came to be.

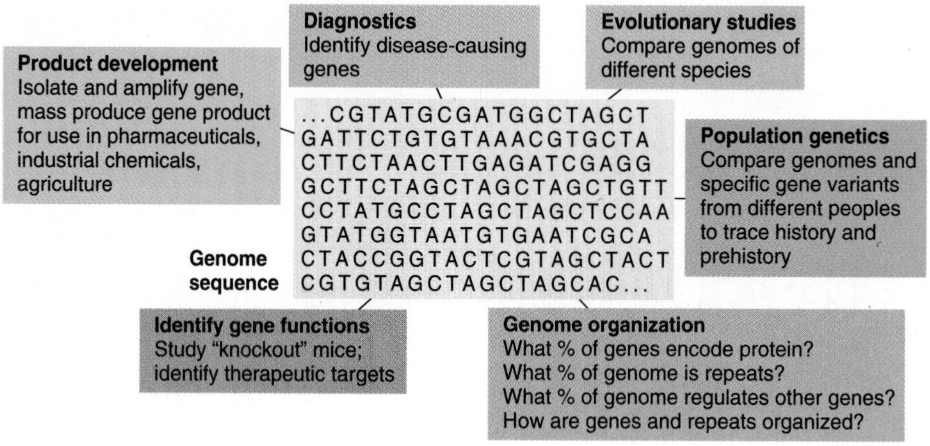

## 22.1 Genome Sequencing: A Continuation of Genetics

Francis Crick, writing in his 1966 book *Of Molecules and Men,* described the size of a human genome in terms of the amount of DNA in a single sperm cell. "This comes to about 500 large books, all different—a fair-sized private library." Just as you wouldn't read every word on every page of every book to locate a particular volume, but would scan the book titles, the human genome project began with deciphering signposts and discovering shortcuts. Many of the initial steps and tools grew from existing technology.

The linkage maps from as long ago as the 1950s and the many studies of families that associated chromosomal aberrations with syndromes provided a wealth of mapping data, assigning a handful of genes to their chromosomes. A technological revolution, in the form of automated DNA sequencing capability, enabled researchers to go from mapping genes to sequencing them, determining the actual order of A, C, T, and G that make up the chromosomes. Figure 22.2 illustrates the refinement and increasing resolu-

**Product development**
Isolate and amplify gene, mass produce gene product for use in pharmaceuticals, industrial chemicals, agriculture

**Diagnostics**
Identify disease-causing genes

**Evolutionary studies**
Compare genomes of different species

**Population genetics**
Compare genomes and specific gene variants from different peoples to trace history and prehistory

**Genome sequence**
...CGTATGCGATGGCTAGCT
GATTCTGTGTAAACGTGCTA
CTTCTAACTTGAGATCGAGG
GCTTCTAGCTAGCTAGCTGTT
CCTATGCCTAGCTAGCTCCAA
GTATGGTAATGTGAATCGCA
CTACCGGTACTCGTAGCTACT
CGTGTAGCTAGCTAGCAC...

**Identify gene functions**
Study "knockout" mice; identify therapeutic targets

**Genome organization**
What % of genes encode protein?
What % of genome is repeats?
What % of genome regulates other genes?
How are genes and repeats organized?

**figure 22.1**

**Uses for genome information.**

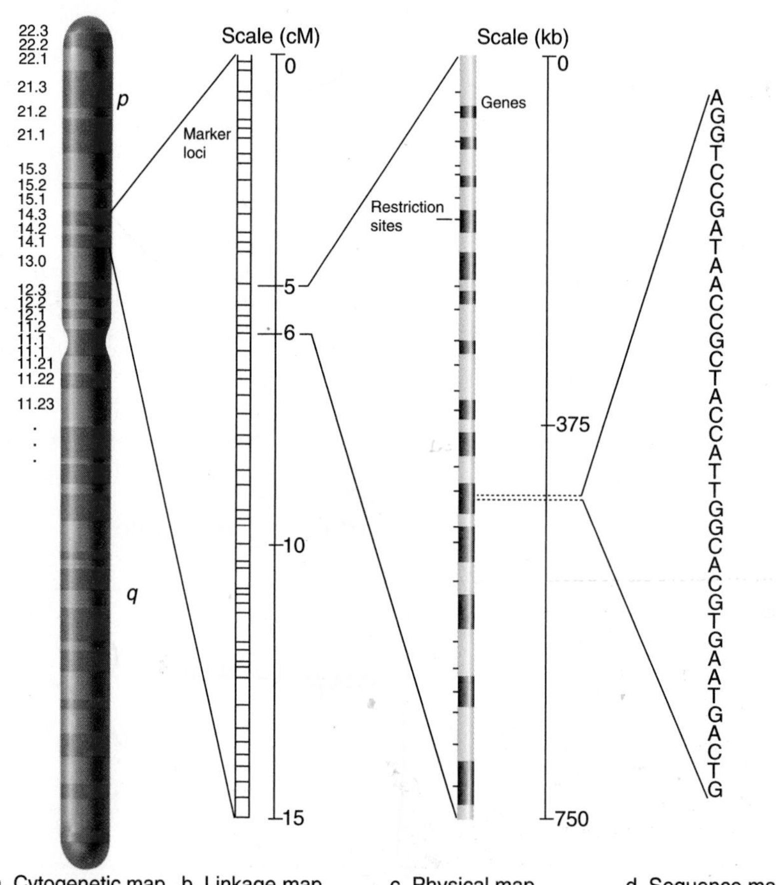


**figure 22.2**

**Different levels of genetic maps.** (*a*) A cytogenetic map, based on associations between specific chromosome aberrations and syndromes, can distinguish DNA sequences that are at least 5,000 kilobases (kb) apart. This is somewhat crude. (*b*) A linkage map is derived from recombination data and can distinguish genes hundreds of kb apart. (*c*) A physical map is constructed from DNA pieces cloned in vectors and then overlapped. It distinguishes genes tens of kb apart. (*d*) A sequence map is the ultimate genetic map, consisting of the nucleotide base sequence that makes up each gene.

tion of different types of genetic maps, from the chromosomal (cytogenetic) and linkage maps generated by family data, to the physical maps derived from restriction fragment length polymorphisms and overlapping large pieces of DNA cloned into vectors.

The human genome project examined hundreds or thousands of pieces of DNA at a time, their sequences catalogued in a public database called GenBank maintained by the National Library of Medicine, and at some companies. Figure 22.3 depicts schematically the two approaches used to determine the first draft sequence. The International Consortium, unofficially headed by Francis Collins, divided the genome into "BAC clones." Recall from table 18.1 that a BAC is a bacterial artificial chromosome. The ones used in the human genome project each housed about 100,000 bases of inserted human DNA. The BAC contents were known to correspond to specific places among the chromosomes called sequence tagged sites (STSs). Several copies of each BAC were cut, or "shotgunned," into up to 80 overlapping pieces, and the pieces sequenced. A powerful computer program then assembled the overlaps to derive the overall sequence for each chromosome.

Celera Genomics, Venter's company, skipped the BAC stage, instead shotgunning multiple copies of the genome into small pieces, and using a computer program to assemble the overlaps into larger pieces called scaffolds. They then also used STSs to anchor the resulting 119,000 scaffolds to the chromosomes. Celera used public database information to derive its sequence and then restricted access to it, causing quite a stir.

In parallel to the human genome project, many individual laboratories continued to investigate specific genes that cause specific diseases. This approach, called **positional cloning,** began with examining a particular phenotype corresponding to a Mendelian disorder, then gradually following clues to find a chromosomal locus or position for the responsible gene.

In contrast to the genome project's "sequence now—interpret later" strategy, positional cloning experiments pinpointed particular genes. It was a time-consuming approach, yet yielded a steady stream of discoveries of disease-causing genes throughout the 1980s and 1990s. The timeline inside the back cover highlights some of these discoveries. Each October, *Science* magazine would publish the human genome map, growing denser with entries each year. Reading 22.1 details how this approach identified the gene that causes Huntington disease. Many of the techniques are already obsolete.

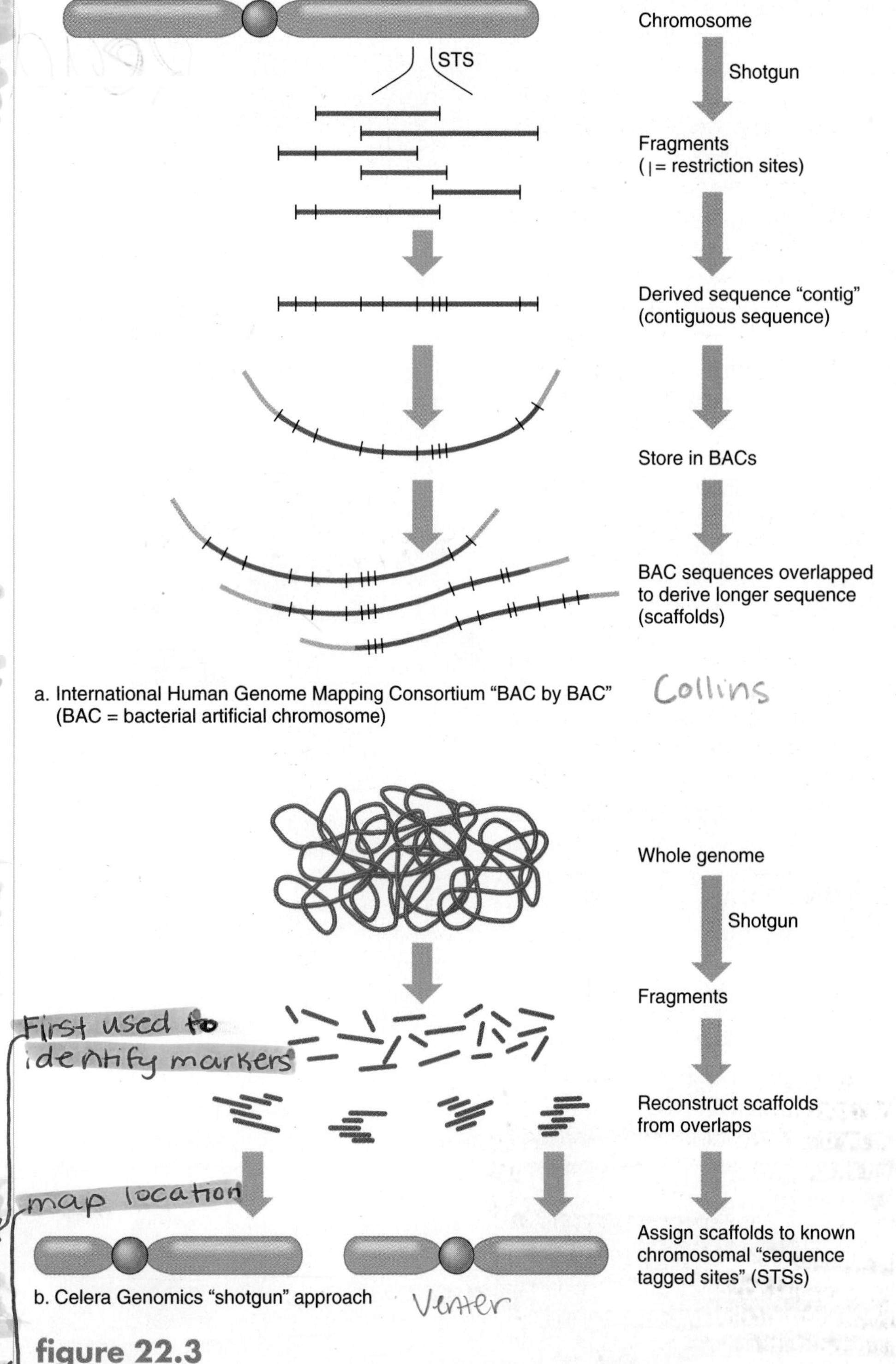


**figure 22.3**

**Two routes to the human genome sequence.** (*a*) The International Consortium began with known chromosomal sites and overlapped large pieces, called contigs, that in turn were reconstructed from many small, overlapping pieces. (*b*) Celera Genomics shotgunned several copies of a genome into small pieces, overlapped them to form scaffolds, and then assigned scaffolds to known chromosomal sites. They used some consortium data.

Reading 22.1

# The Search for the Huntington Disease Gene

Gene discovery occurs at a nonlinear pace. That is, the more gene sequences and functions are known, the easier it becomes to identify new genes, simply because there are more known sequences to compare them to.

Today, gene discovery is largely an informational science, mining databases of known genomes for sequence similarities. It was not always so. Consider the decade-long quest to identify the gene that causes Huntington disease (HD). The search was well underway when people first discussed sequencing the human genome. HD is an adult-onset, autosomal dominant condition that causes uncontrollable movements and personality changes.

The search for the HD gene began in a remote village on the shores of Lake Maracaibo, Venezuela. Seven generations ago, in the 1800s, a local woman married a visiting Portuguese sailor who, as the folklore goes, habitually walked as if intoxicated. Like most couples in the poor fishing village, the woman and her sailor had many children. Some grew up to walk in the same peculiar way as their father. Of their nearly 5,000 descendants, 250 living today have HD.

This extended family presented a natural experiment to geneticists, and they began by drafting a huge pedigree. Molecular studies followed. If a DNA sequence could be identified that was only in affected family members, then that sequence might contain the mutant gene that causes HD. This sequence would serve as a "marker." The family was theoretically large enough to detect a marker-disease gene association.

In 1981, Columbia University psychologist Nancy Wexler, daughter of an HD patient, began making yearly visits to Lake Maracaibo. The people lived in huts perched on stilts, as their ancestors did. Wexler traded candy and blue jeans for blood samples and skin biopsies to bring back to a team of geneticists. Meanwhile, investigators at Massachusetts General Hospital were sampling tissue from an Iowa family of 41, 22 of whom had HD. They extracted DNA from the samples, cut it with restriction enzymes, and tested the fragments with a set of labeled DNA probes. They were looking to see if any piece always bound to the DNA of sick people, but never to healthy ones. This would reveal a restriction fragment length polymorphism (RFLP), part of the DNA sequence unique to people with the disease (see figure 13.3). The team added the Venezuelan DNA to the protocol. A third research group, at Indiana University, matched the probe data to pedigrees, looking for a pattern.

With several hundred DNA probes and samples to sort through, the researchers expected the testing to take a long time. But one warm May night in 1983, the Indianapolis computer found a match on only the twelfth probe studied, called G8. In both families, G8 bound only the DNA of the family members with HD. Therefore, in these families, the G8 probe could serve as a genetic marker. It was apparently closely linked, and therefore usually inherited with, the mutant gene. Until the gene was discovered in 1993, the marker was the basis for presymptomatic testing.

The next step in finding the HD gene was to locate G8. A technique called somatic cell hybridization accomplished this. It uses the fact that when human and rodent cells are fused, a hybrid cell forms that loses only human chromosomes as the cell divides in culture (figure 1). Eventually, researchers had a collection of these cells that

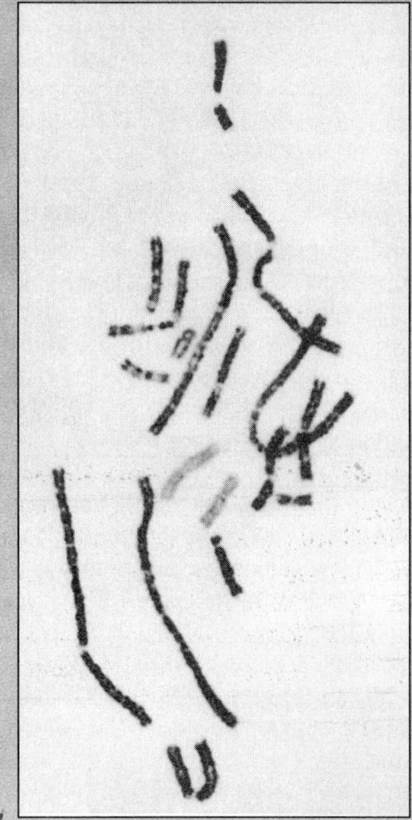

**figure 1**

**Isolating human chromosomes.** Somatic cell hybridization generates cells with only one or a few human chromosomes. When rodent and human cells are fused, human chromosomes are lost as the cell divides. This cell has a human X chromosome and chromosome 3, which appear lighter than the other, rodent chromosomes. If a DNA probe corresponding to a specific human gene binds only to somatic cell hybrids that contain chromosome 3, then researchers know that the probe identifies a gene on that chromosome. This technique was used to map the HD gene to chromosome 4.

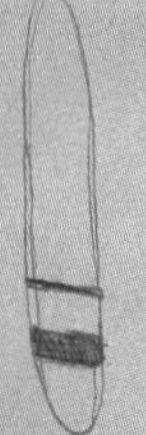

carried just one or a few different human chromosomes each. Only hybrid cells that harbored human chromosome 4 bound the G8 probe. Therefore, the HD gene is on chromosome 4. Researchers localized the gene to the tip of the short arm of chromosome 4 (figure 2). Then, to extend the DNA sequence from the marker to the gene of interest, the researchers located another probe that overlaps the first, then another probe that overlapped the second probe, and so on. Finally, a computer aligned the probes according to their sequence overlaps, creating a contiguous, or "contig" map of extended sequence.

The area surrounding a probe can include a lot of DNA, so researchers looked for clues to the locations of protein-encoding genes. One such clue is a stretch of CG repeats called "CpG islands." CpG island analysis indicated that the half-million-base-long contig map for HD could encode about 100 genes! How would researchers figure out which of these "candidate genes" causes HD? They needed another clue, something related to the phenotype. It came from David Housman at the Massachusetts Institute of Technology in 1992. He had recently discovered that myotonic dystrophy is caused by a new type of mutation, an expanding triplet repeat (see figure 11.11). Might this type of aberration cause HD, too? Both disorders affect the brain's control over nerves and muscles.

Looking for a triplet repeat was a long shot. But on February 24, 1993, the researchers found it. A stretch of DNA that was about 210,000 base pairs long in people without HD was longer in people with HD. The researchers named the long-sought gene a rather unremarkable IT15; IT stands for "interesting transcript." It was also an ancient and important transcript, because counterparts were known in diverse species. Next, the gene was tested against cDNA libraries made from the mRNAs in various tissue types. The gene hybridized to a brain library, meaning that the gene product is made in the brain. IT15 was the HD gene. The mutant form adds glutamines to the protein product, huntingtin, and the extra amino acids alter protein folding, disrupting the function of huntingtin and the proteins with which it interacts. Marker tests became instantly obsolete. Any person can be tested for the telltale expanded triplet repeat mutation that causes Huntington disease.

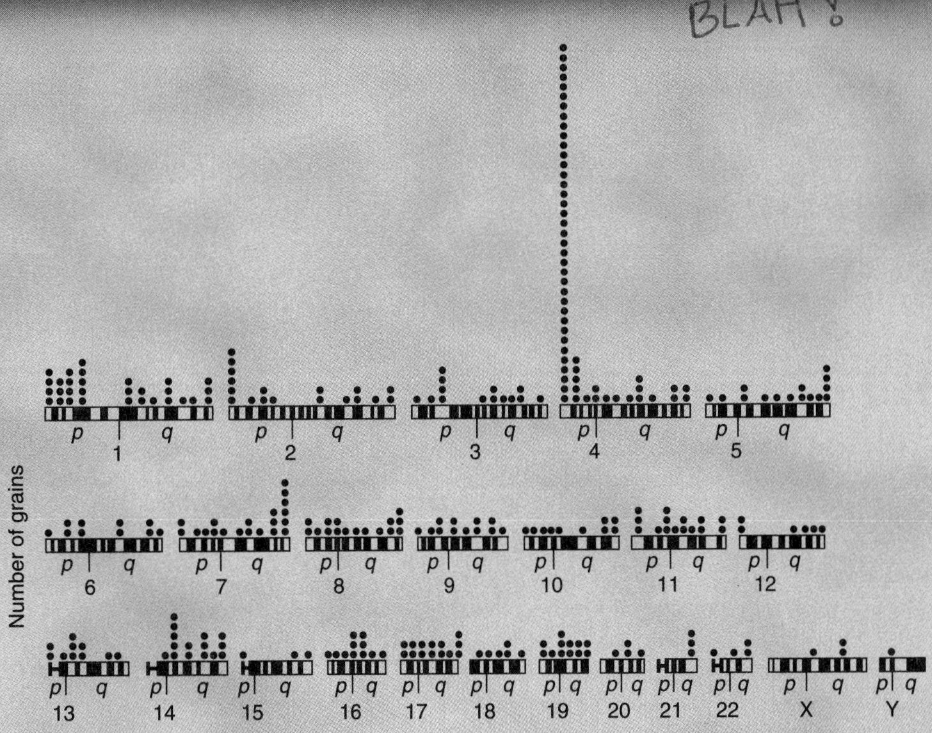


## figure 2

**Identifying the chromosomal locus of the HD gene.** In autoradiography, radioactively labeled probes bind to their chromosomal loci. A piece of photographic film is placed over a probed chromosome preparation, and the area where the most silver grains appear indicates where the probe bound. This diagram depicts the densest probe concentration, G8, on the short arm of chromosome 4.

**KEY CONCEPTS**

The human genome project built upon linkage and cytogenetic information gained from decades of work. Genome sequencing handles many pieces of DNA at once, with analysis later. The International Consortium pinned sequences in BACs to known chromosomal sites; Celera Genomics used a whole genome shotgunning approach. In contrast, positional cloning, which contributed much genetic information in the 1980s and 1990s, sought specific genes based on phenotypes behind particular Mendelian disorders.

## 22.2 The Origin of the Idea

The idea of sequencing an entire genome probably occurred to many researchers as soon as Watson and Crick determined the structure of DNA. In 1966, Francis Crick proposed sequencing all genes in *E. coli,* which would reveal how the organism works, writing that "this particular problem will keep very many scientists busy for a long time to come." Jacques Monod, who described the first genetic control system in bacteria, wrote at about the same time, "What is true for *Escherichia coli* is true for the elephant," meaning that to know genomes is to understand all life. Perhaps a quote from the Grateful Dead's Jerry Garcia is the most apt: "What a long, strange trip it's been." For although the number of sequenced genomes now exceeds 100, we are far from understanding life. The genomes sequenced to date only reveal how little we know—each has yielded genes whose sequences and functions are unknown.

For sequencing genomes to evolve from science fiction to science required the development of key tools and technologies (see Technology Timeline). First was the need to obtain the nucleotide base sequence of pieces of DNA. Then, computer programs would be required to rapidly detect overlaps in sequence of pieces cut from multiple copies of a genome, and assemble them to reconstruct each chromosome.

### The Sanger Method of DNA Sequencing

Modern DNA sequencing instruments utilize a basic technique developed by Frederick Sanger in 1977. The goal is to generate a series of DNA fragments of identical sequence that are complementary to the sequence of interest. These fragments differ in length from each other by one end base:

```
Sequence of interest:   T A C G C A G T A C
Complementary
sequence:               A T G C G T C A T G
Series of fragments:      T G C G T C A T G
                            G C G T C A T G
                              C G T C A T G
                                G T C A T G
                                  T C A T G
                                    C A T G
                                      A T G
                                        T G
                                          G
```

Note that the entire complementary sequence appears in the sequence of end bases of each fragment. If the complement of the gene of interest can be cut into a collection of such pieces, and the end bases distinguished with a radioactive or fluorescent label, then polyacrylamide gel electrophoresis (see figure 13.5) can be used to separate the fragments by size. Then once the areas of overlap are aligned, reading the labeled end bases in size order reveals the sequence of the complement. Replacing A with T, G with C, T with A, and C with G establishes the sequence of the gene in question.

Sanger invented a way to generate the gene pieces. In a test tube, he included the unknown sequence and all of the biochemicals needed to replicate it, including supplies of the four nucleotide bases. Some of each of the four types of bases were chemically modified at a specific location on the base sugar to contain no oxygen atoms instead of one—in the language of chemistry, they were *di*-deoxyribonucleotides rather than deoxyri-

**Technology TIMELINE**

Evolution of the Human Genome Project

| | |
|---|---|
| **1985–1988** | Idea to sequence human genome is suggested at several scientific meetings. |
| **1988** | Congress authorizes the Department of Energy and the National Institutes of Health to fund the human genome project. |
| **1989** | DNA chip (microarray) technology is explored at Stanford and Duke Universities. |
| **1990** | Human genome project officially begins. |
| **1991** | Expressed sequence tag (EST) technology is invented, which identifies protein-encoding sequences. |
| **1992** | First DNA microarrays became available. |
| **1993** | Need to automate DNA sequencing is recognized. |
| **1994** | U.S. and French researchers publish preliminary map of 6,000 genetic markers, one every 1 million bases along the chromosomes. |
| **1995** | Emphasis shifts from gene mapping to sequencing. |
| **1996** | Resolution passed at Second International Strategy Meeting on Human Genome Sequencing to make all data public and update it daily at GenBank (www.ncbi.nlm.nih.gov). |
| **1998** | Public consortium releases preliminary map of pieces covering 98 percent of human genome. Millions of sequences are listed in GenBank. Directions for developing DNA microarrays are posted on Internet. |
| **1999** | Rate of filing of new sequences in GenBank triples. |
| | A full 30 percent of human genome is sequenced by year's end. |
| | Public consortium and two private companies race to complete sequencing. |
| **2000** | Ninety-nine percent of genome sequenced privately. |
| **2000 and beyond** | Microarray technology explodes as researchers investigate gene functions and interactions. |

bonucleotides. A radioactive nucleotide, usually C, was also included in each reaction. DNA synthesis halted when DNA polymerase encountered a "dideoxy" base, leaving only a piece of the newly replicated strand.

Sanger repeated the experiment four times, each time using a dideoxy version of A, T, C, then G. The four experiments were run in four lanes of a gel (figure 22.4). Today, fluorescent labels are used, one for each of the four base types, allowing a single experiment to reveal the sequence. The data appear as a sequential readout of the wavelengths of the fluorescence from the labels (figure 22.5). In the mid-1980s, Leroy Hood automated Sanger's method of DNA sequencing, which was essential to the ability to analyze entire genomes.

DNA sequencing occurs on a vast scale today. Many individual laboratories or academic departments have one or more automated DNA sequencers, while companies devoted to genome sequencing may have hundreds. By the time you read this, it may be possible to sequence a person's genome in just a few days.

Four solutions contain unknown DNA sequence, primers (starting sequences), normal nucleotides A,T, C, and G, a radioactive nucleotide, and replication enzymes.

Replication

Fragments present in solution

Electrophoresis sorts fragments by size

Power source

⊖ ⊕

Terminator A added (A*). Replication continues until A* binds to unknown DNA sequence.

A*

*primer* A*
*primer* A T G C G C A*

A

Terminator C added (C*). Replication continues until C* binds to unknown DNA sequence.

C*

*primer* A T G C*
*primer* A T G C G C*

C

Terminator T added (T*). Replication continues until T* binds to unknown DNA sequence.

T*

*primer* A T*
*primer* A T G C G C A T*

T

Terminator G added (G*). Replication continues until G* binds to unknown DNA sequence.

G*

*primer* A T G*
*primer* A T G C G*
*primer* A T G C G C A T G*

G

Read off A T G C G C A T G

Deduce original unknown sequence T A C G C G T A C

**figure 22.4**

**Determining the sequence of DNA.** In the Sanger method of DNA sequencing, complementary copies of an unknown DNA sequence are terminated early because of the incorporation of dideoxynucleotide "terminators." A computer deduces the sequence by placing the fragments in size order. Radioactive labels are used to visualize the sparse quantities of each fragment.

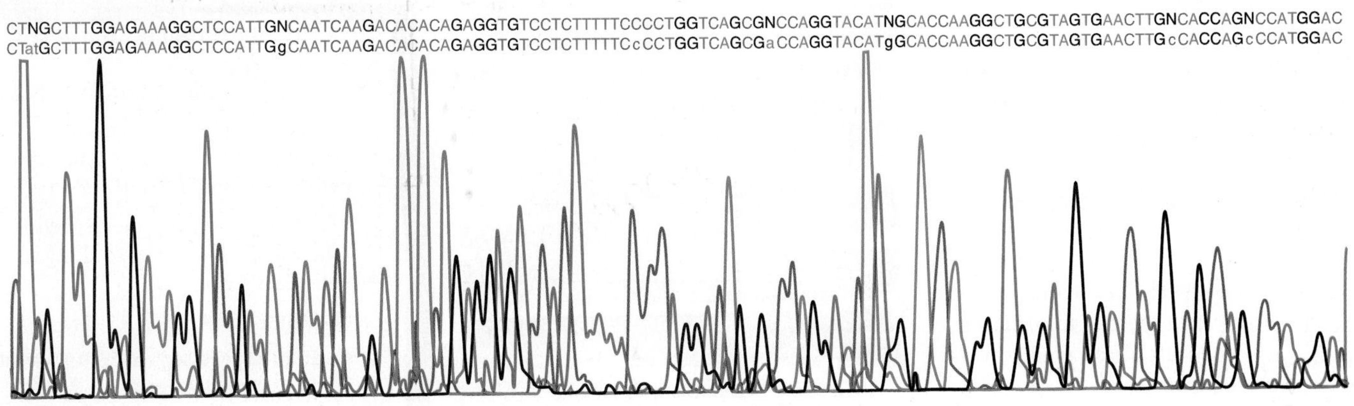


**figure 22.5**

**DNA sequence data.** A readout of sequenced DNA is a series of wavelengths that represent the terminally labeled DNA base.

## The Project Starts

During the 1980s, the idea to sequence the human genome evolved for several reasons. It surfaced at a meeting held by the Department of Energy in 1984 to discuss the long-term population genetic effects of exposure to low-level radiation. In 1985, at another gathering, Robert Sinsheimer, chancellor of the University of California, Santa Cruz, called for an institute to sequence the human genome, more or less because it could be done. The next year, virologist Renato Dulbecco proposed that the key to understanding the origin of cancer lay in knowing the human genome sequence. In the summer of 1986, many of the major players in genetics and molecular biology convened at the Cold Spring Harbor Laboratory on New York's Long Island to discuss the feasibility of a human genome project. Worldwide planning soon began.

The idea to systematically sequence the human genome introduced a fundamental shift in the goals of life science research. A furious debate ensued, with detractors claiming that the project would be more gruntwork than a creative intellectual endeavor, comparing it to conquering Mt. Everest just because it is there. Some researchers feared that such an unprecedented "big science" project would divert government funds from basic research and the AIDS budget. Finally, the National Academy of Sciences convened a committee composed of both detractors and supporters to debate the feasibility, risks, and benefits of the project. The naysayers were swayed to the other side. So in 1988, Congress authorized the National Institutes of Health (NIH) and the Department of Energy (DOE) to fund the $3 billion, 15-year human genome project. For the first few years, it appeared that the NIH was still unconvinced—DOE was the more active participant.

The government-sponsored human genome project officially began in 1990 with James Watson at the helm. Recognizing the profound effect that genetic information would have on public policy and on peoples' lives, the project set aside 3 percent of its budget for the ELSI program, which stands for ethical, legal, and social issues. ELSI helps ensure that genetic information is not misused to discriminate against people with particular genotypes.

The human genome project expanded both in scope and in number of participants. In addition to 10 major sequencing centers, the largest in England, France, Germany, and Japan, many hundreds of other basic research laboratories and biotechnology companies contributed DNA sequence data. The project also, from the start, sequenced the genomes of several different species to investigate how all life on earth is related.

### Key Concepts

DNA sequencing and computer software to align DNA pieces were essential technological developments necessary for genome projects to proceed. In the Sanger method of DNA sequencing, complementary copies of an unknown DNA sequence are terminated early by incorporating dideoxynucleotides. A researcher or automated DNA sequencer deduces the sequence by labeling the end bases and placing fragments in size order. • The idea to sequence the human genome emerged in the mid-1980s, with several goals. The project officially began in 1990.

## 22.3 Technology Drives the Sequencing Effort

The human genome project progressed in stages as sequencing technology and the computer capability to align and derive continuous sequence improved. At first the project focused on developing tools and technologies to divide the genome into pieces small enough to sequence, and to improve the efficiency of sequencing techniques. Table 22.1 summarizes the overall steps in genome sequencing.

In 1991, two key inventions entered the picture. Venter, a government researcher at the NIH, introduced a very powerful shortcut called **expressed sequence tag** (EST) technology. These are cDNAs that are pieces of the genes expressed in a particular cell type. Identifying ESTs enabled researchers to quickly pick out genes most likely to be implicated in disease, because they encode protein, and are expressed in the cell types affected in a particular illness. Venter left the NIH to develop ESTs at The Institute for Genomic Research (TIGR).

Also in 1991, Patrick Brown at Stanford University developed a way to embed DNA pieces of known sequence into tiny glass squares, and use this array to pull out complementary sequences from a sample of unknown DNA pieces. These devices were first called DNA chips, then **DNA microarrays.** The idea grew out of a whimsical suggestion in 1989 that genes be placed on chips, much as transistors are. Details of the technique were published in 1995, and a year later, a company, Affymetrix, marketed the first DNA chips. Today, a typical DNA microarray smaller than a postage stamp might hold up to 400,000 specific sequences of 25 DNA bases, some 30 to 40 of them corresponding to any one gene.

Sequencing and mapping techniques improved greatly from 1993 to 1998. At the start of the project, researchers cut the genome into overlapping pieces of about 40,000 bases (40 kilobases), then randomly cut the pieces into small fragments. Researchers would cut up several genomes in a single experiment, so that many of the resulting fragments overlapped. By finding the overlaps, the pieces could be assembled to reveal the overall sequence. The greater the number of overlaps, the more likely that the final assembled sequence would not omit anything. Today, computers recognize the overlaps and align and derive the sequences, either beginning with larger sections or using many small pieces. It is important that the sites of overlap be unique sequences, found in only one place in the genome. Overlaps of repeated sequences present in several places could lead to more than one derived overall sequence. Another problem is to determine which side of the double helix to sequence for each piece. Figure 22.6 illustrates the strategy of over-

### table 22.1

**Steps in Genome Sequencing and Analysis**

1. Obtain chromosome maps with landmarks from classical linkage or cytogenetic studies, or RFLP sites.
2. Obtain chromosome pieces, maintained in gene libraries or shotgun entire sequence.
3. Sequence the pieces.
4. Overlap aligned sequences to extend the known sequence.
5. Compare the sequence to those in other species.

```
                    FEELING WE'RE NOT
TOTO, I HAVE A FEE
                      ELING WE'RE NOT IN KANSAS ANYMORE
                A FEELING
            AVE A FEELING WE'RE NOT IN KANSAS A
                         ING WE'R
                                'RE NOT IN KANSAS ANYMORE
                      ELING WE'RE NOT IN KANS
        I HAVE A FEELING WE
                    FEELING WE'R
                          NG WE'RE NOT IN KANSAS ANYMO
TOTO, I HAVE A FEELING WE'RE NOT IN KANSAS ANYMORE
```

## figure 22.6

**Overlapping sequences.** Deriving a DNA sequence requires overlapping fragments according to where their sequences match. This English sentence provides an analogy.

```
Random fragments:      AGTCCT   CTAG       AGCTA
                           CTACT    TAGAGT     CCTAGC

Alignment:             CTAG
                        TAGAGT
                           AGTCCT
                              CCTAGC
                                 AGCTA
                                   CTACT

Sequence:              CTAGAGTCCTAGCTACT
```

## figure 22.7

**Deriving a DNA sequence.** Automated DNA sequencers first determine the sequences of short pieces of DNA, or sometimes of just the ends of short pieces, then algorithms search for overlaps. By overlapping the pieces, the software derives the overall DNA sequences.

lapping sequences using a sentence from the English language, and figure 22.7 depicts how a DNA sequence might be reconstructed from overlapping pieces.

By many people's accounts, 1995 was a turning point in the human genome project, as "proteomics" entered the growing vocabulary of genome jargon, referring to all the proteins that an organism can synthesize, as well as how the proteins are distributed among different cell types. Researchers began to see that proteomics would present an even greater challenge than sequencing the genome.

The year 1995 was also when efforts shifted from mapping genes to chromosomes, to DNA sequencing. For the first time, the number of gene sequences deposited in GenBank outnumbered the number of papers in the scientific literature introducing disease genes identified by positional cloning. The reason: the automation of DNA sequencing. Although similar to their 1980s prototypes in concept, devices could now produce lines of A, T, C, and G 10 times as fast as in the early days. The deadline for completion inched forward as optimism soared. Also in 1995, Venter's company, TIGR, developed computer software that could rapidly locate the unique sequence overlaps among many small pieces of DNA and assemble them into a continuous sequence, eliminating the preliminary step of gathering large guidepost pieces. Using this software, TIGR obtained the first complete genome sequence for an organism, in under a year! Many more would follow.

The distinction of being the first organism to have its genome sequenced went to *Haemophilus influenzae,* a bacterium that causes meningitis and ear infections. The researchers cut the bacterium's 1,830,137 bases into 24,000 fragments and sequenced the fragments, and the software assembled them. The human genome is 1,500 times the size of that of *H. influenzae.*

Meanwhile, as researchers at TIGR and elsewhere continued to sequence smaller genomes, the partners of the International Consortium met in Bermuda, which culminated with the "Bermuda rules." These guidelines bound researchers to release new DNA sequence to the public database within 24 hours of its determination. By 1998, Venter had founded Celera Genomics, with the goal of sequencing the human genome faster and at less cost than the consortium, a feat possible because of the company's powerful assembler software. Unlike the consortium's public posting of data, Celera planned to sell access to its sequence, even though they had to use some of the public data to get that sequence. And so began a fierce bioethical battle.

In 1999, the human genome project became intensely competitive, with Venter and Collins, director of the U.S. arm of the global public consortium since 1993, trying to beat each other to completion, with lower final cost projections by sacrificing a degree of accuracy for speed. The first map of an entire human chromosome, number 22, was published near year's end. In March 2000, Celera Genomics completed the sequence of the genome of the fruit fly, *Drosophila melanogaster,* in collaboration with a public consortium.

Venter claimed privately to have reached the first draft goal for the human genome by March 2000, while the consortium's final rendering of sequence was completed June 22, following a month of frantic work by a graduate student at the University of California at Santa Cruz, James Kent. His invention of the GigAssembler program immediately catapulted him to cult hero status in the genome project. The task that GigAssembler tackles has been compared to putting together a million-piece puzzle using thousands of preassembled pieces. The program looks for the connections between the pieces to present the final view of the human genome. Figure 22.8 is an overview of genome sequencing and the aftermath.

**KEY CONCEPTS**

The human genome project switched from mapping to sequencing in 1995. EST technology enabled researchers to focus on protein-encoding genes, and microarray technology considered many expressed genes at a time. In the second half of the 1990s, the race intensified.

## 22.4 Genome Information Answers and Raises Questions

Deciphering the human genome base by base has shown that the classical concept of gene structure as "beads on a string" is an oversimplification, as is the idea that each gene has just one clear function. Gene expression is, instead, a systemwide, multipart, dynamic phenomenon that we are only beginning to glimpse.

### The Definition of a Gene

It may seem odd to pose the question, "What is a gene?" at the end of a genetics textbook—odder still that researchers should be asking this age-old question yet again. But this is happening.

To Mendel, a gene was an ill-defined character that determined an easily observed trait. Later, biologists defined a gene as a unit of genetic function that maps to a particular chromosome and, when mutated, alters the phenotype. With knowledge of the nature of the genetic material, the definition of a gene evolved further, to a DNA sequence that is transcribed into RNA and encodes protein. We now know that none of these definitions suffices.

For many years, a gene's normal function was inferred and defined by what went wrong in a mutant. But DNA sequencing revealed that many polymorphisms—variations in a gene's sequence—have no apparent effect on phenotype. It also showed that a sequence that is an intron on one strand of the double helix may be an exon on the other strand, and that some stretches of DNA are parts of different genes. Knockout technology brought further surprises, obliterating a gene's function with no effect on phenotype and revealing unanticipated redundancy in genomes. Then microarray technology began to show that a gene can have different functions in different tissues at different times in development. To quote Alice in Wonderland, the definition of a gene is becoming curiouser and curiouser.

### Nonhuman Genome Projects

In addition to the changing definition of gene and gene function, genome projects reveal that there is still much we do not know. It isn't uncommon for a large percentage of an organism's genes to have no counterparts in GenBank. Yet genome researchers have

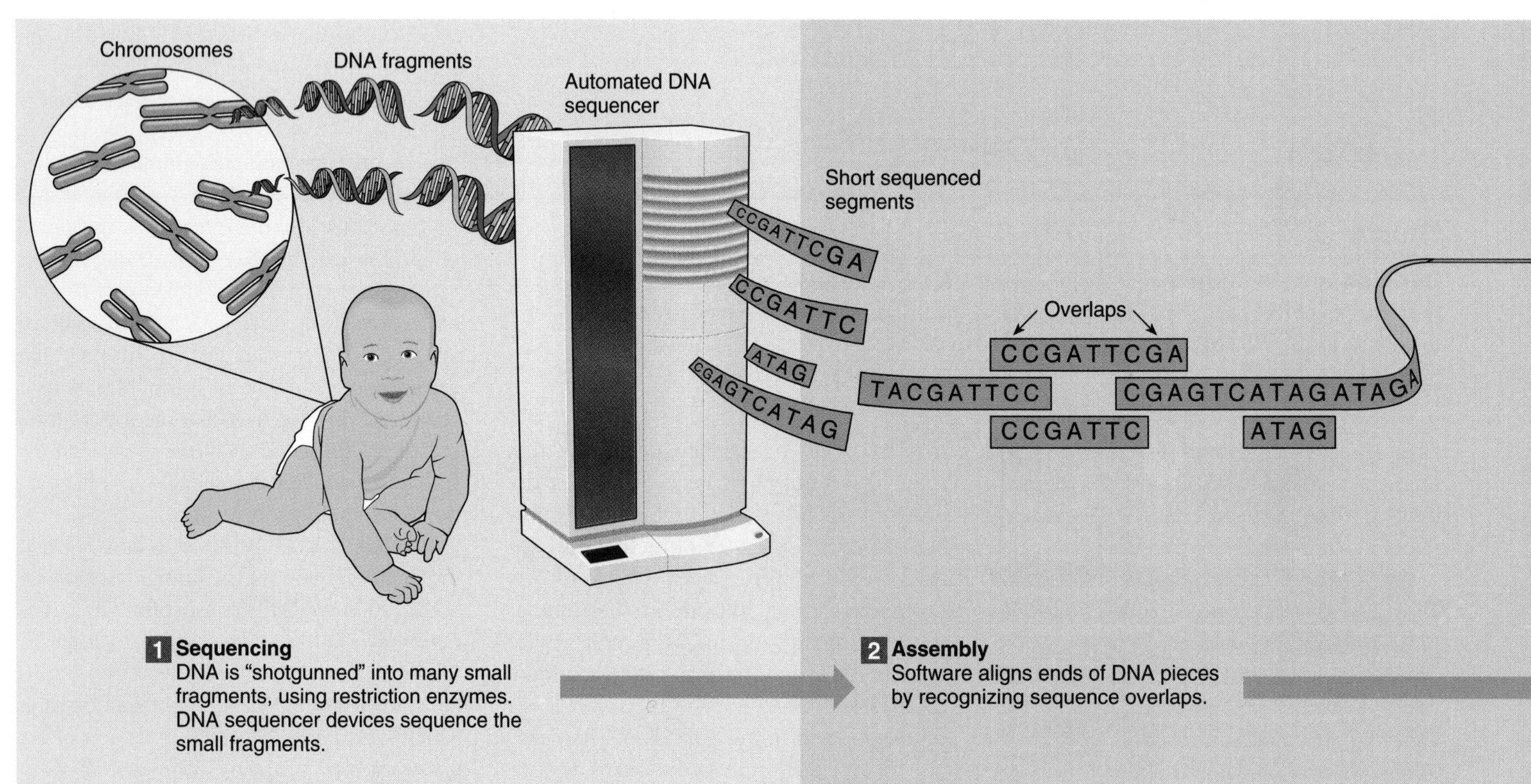


**figure 22.8**

**Sequencing genomes.** Determining the DNA sequence of a genome is just a first step—albeit a huge one. After pieces are assembled into a whole, protein-encoding genes are identified and then patterns of gene expression in different tissues assessed.

also glimpsed astounding similarities among earth's diverse residents, gene sequences that have remained relatively unchanged over the ages among startlingly different species, indicating a shared, if distant, ancestry.

It is clear from comparative genomics of primates that differences in genomes among some closely-related species may lie in their organization more than in DNA sequence. We also do not know the function of many repeats in the human genome. Their organization and number may encode a type of information that we do not yet understand. Why, for example, is the genome of a pufferfish very similar to that of a human, but with fewer repeats and much smaller introns?

Perhaps the greatest lessons to be learned from comparative genomics lie at the boundaries of great evolutionary leaps, when life changed drastically. Some of the lessons may reveal:

**The minimum set of genes required for life.** The smallest cell known to be able to reproduce is *Mycoplasma genitalium.* It infects cabbage, citrus fruit, corn, broccoli, honeybees, and spiders, and causes respiratory illness in chickens, pigs, cows, and humans. Researchers call its tiny genome the "near-minimal set of genes for independent life." Of the 480 protein-encoding genes, researchers think that 265 to 350 are essential for life. Considering how *Mycoplasma* uses its genes reveals the fundamental challenges of being alive (table 22.2). When the organism's genome was sequenced, nearly a quarter of the genes had no known counterpart in other organisms.

**Fundamental distinctions among the three domains of life.** *Methanococcus jannaschii* is a microorganism that lives at the bottom of a 2,600-meter-tall "white smoker" chimney in the Pacific Ocean, at high temperature and pressure and without oxygen. It is a member of the Archaea, which are cells that lack nuclei, yet that replicate DNA and synthesize proteins in ways similar to those of multicellular organisms. The genome sequence confirms that this organism represents a third form of life. Fewer than half of its 1,738 genes have known counterparts among the bacteria, other archaea, or eukaryotes. Even the genome of *E. coli,* the organism that geneticists supposedly knew the best, held surprises. The functions of more than a third of *E. coli*'s 4,288 genes remain a mystery.

**The simplest organism with a nucleus.** The paper in *Science* magazine that introduced the genome of the yeast *Saccharomyces cerevisiae* was entitled "Life with 6,000 Genes." But the unicellular yeast is more complex than the title implies. About a third of its 5,885 genes have counterparts among mammals, including those implicated in more than 70 human diseases. Understanding what a gene (such as one that controls the cell cycle) does in yeast can provide clues to how it affects human health.

**The basic blueprints of an animal.** The genome of the tiny, transparent, 959-celled nematode worm *Caenorhabditis elegans* is packed with information on what it takes to be an animal. Thanks to researchers who, in the 1960s, meticulously tracked the movements of each cell as the animal developed, much of the biology of this organism was already known before its 97 million DNA bases were revealed late in 1998. The worm's signal transduction pathways, cytoskeleton,

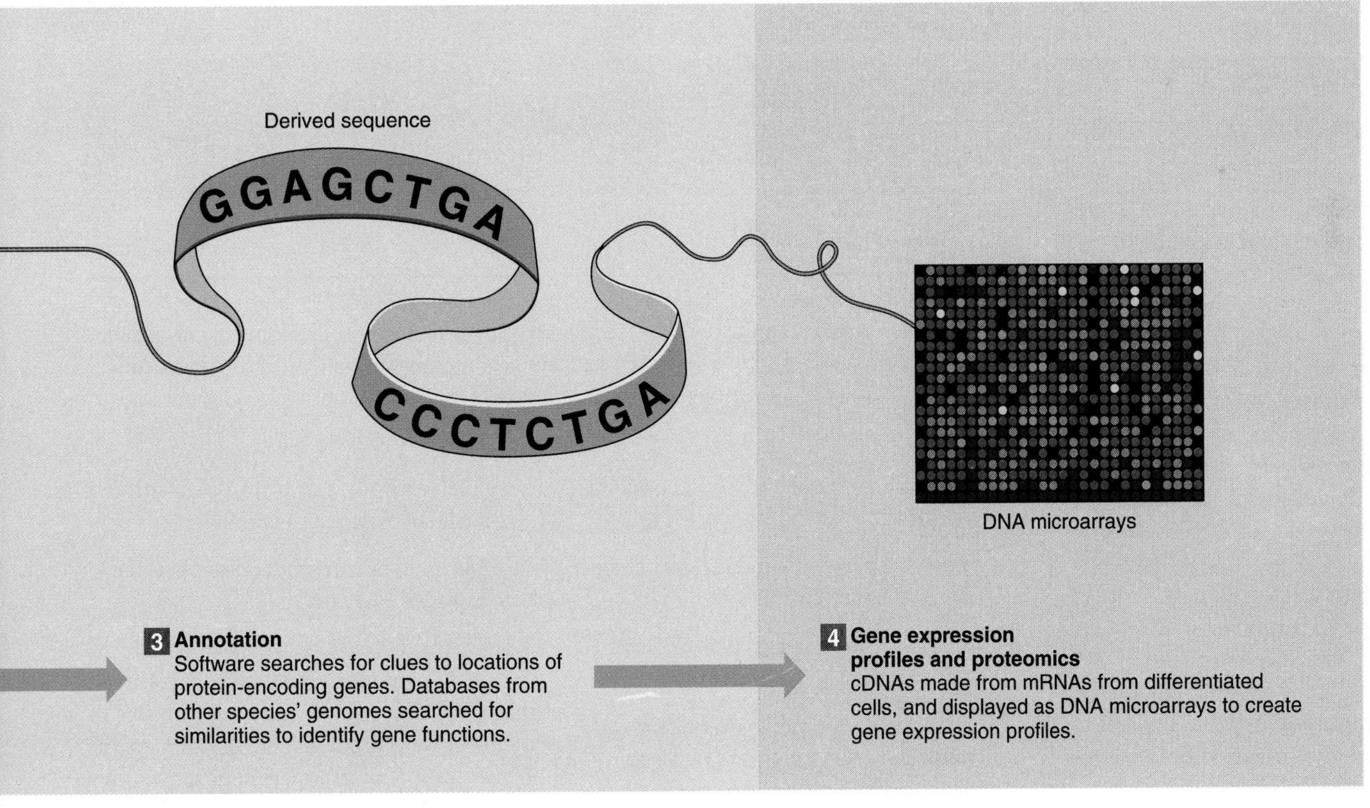
Derived sequence
GGAGCTGA
CCCTCTGA
DNA microarrays
3 Annotation
Software searches for clues to locations of protein-encoding genes. Databases from other species' genomes searched for similarities to identify gene functions.
4 Gene expression profiles and proteomics
cDNAs made from mRNAs from differentiated cells, and displayed as DNA microarrays to create gene expression profiles.

immune system, and even brain proteins are very similar to our own.

The sequencing of the fruit fly genome held a big surprise—it has 13,601 genes, less than the 18,425 in the much simpler *C. elegans* worm (figure 22.9). Apparently the way that genes control the development and functioning of an organism is more complex than studying single genes suggested. Also, of 289 human disease-causing genes, the fly sequencers identified 177—about 60 percent—in *Drosophila*. The fly might therefore serve as a model for humans in testing new treatments.

**table 22.2**

**The Minimal Functional Gene Set for Life (Distribution of Gene Function in *Mycoplasma genitalium*)**

| Percentage of Genome | Function |
|---|---|
| 30.0% | Maintaining cell membrane |
| 24.0% | Protein synthesis |
| 22.0% | Unknown (doesn't match any known gene) |
| 9.0% | DNA replication |
| 8.0% | Acquiring energy |
| 4.5% | Evading immune attack by host cell |
| 4.0% | Synthesizing and recycling nucleic acids |

### Key Concepts

Genome information is forcing yet another redefinition of a gene, based on function. Gene functions are diverse, overlapping, multiple, and redundant. • Comparing genomes of different species reveals the genetic changes that fostered evolutionary leaps, and that at the genetic level, we share a great deal with other species.

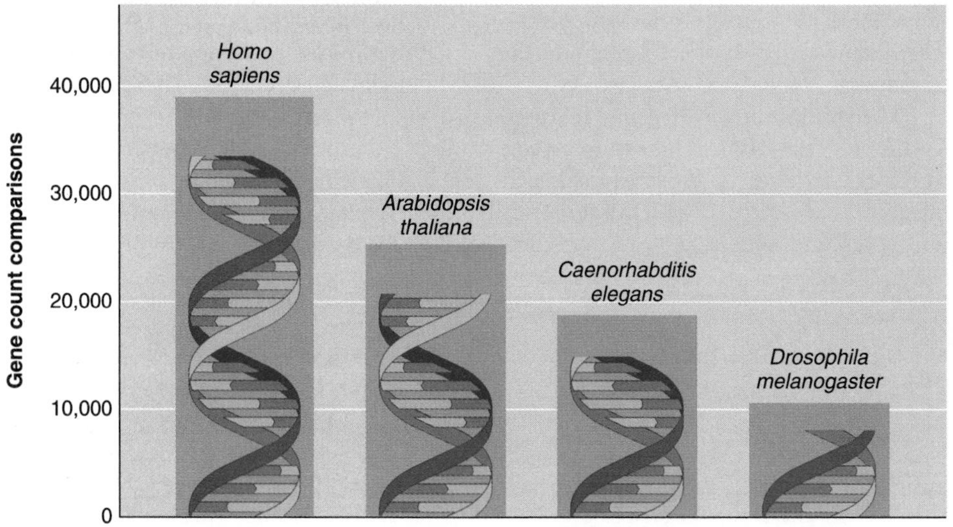


**figure 22.9**

**Comparative genomics.** The human genome is not much larger than that of other organisms and shares much of the DNA sequence. At a genomic level, we are actually similar to pufferfish and Brussels sprouts! Clearly, genome organization as well as DNA sequence is important in establishing the many differences among Earth's inhabitants.

## Epilogue: Genome Information Will Affect You

Since the dawn of humanity, people have probably noted inherited traits, from height and body build, to hair and eye color, to talents, to behavioral quirks, to illnesses. Genetics provides the variety that makes life interesting.

The science of genetics grew out of questions surrounding plant and animal breeding, then became human-oriented in the mid-20th century with the recognition that certain characteristics and medical problems tended to run in families, sometimes recurring with predictable frequencies. Today, genetics is beginning to impact many areas of clinical medicine, and appears in headlines with increasing frequency.

You will likely encounter many of the genetic technologies discussed in this book, as college roommates Mackenzie and Laurel did in chapter 1. You might

- be asked to take a DNA microarray test, matched to your ethnic group, to better diagnose or treat a medical condition.
- serve on a jury and be asked to evaluate DNA fingerprinting evidence.
- undergo a panel of genetic tests before trying to have a child.
- receive a chromosome report or DNA microarray analysis on your child-to-be.
- help a parent or other loved one through chemotherapy, with some assurance, thanks to DNA testing, that the most effective drugs will be tried first.
- eat a genetically modified fruit or vegetable.
- receive a body part from a pig, with cell surfaces matched to your own.
- take medicine manufactured in a transgenic organism.

The list of applications of genetic technology is long and ever-expanding. I hope that this book has offered you glimpses of the future and prepared you to deal personally with the choices that genetic technology will present to you. Let me know your thoughts!

Ricki Lewis
rickilewis@nasw.org

# Summary

## 22.1 Genome Sequencing: A Continuation of Genetics

1. Genetic maps have increased in detail and resolution, from cytogenetic and linkage maps to physical and sequence maps.
2. The International Consortium used BAC clones pinned to sequence tagged sites on chromosomes. Celera Genomics used a whole genome shotgun strategy.
3. **Positional cloning** was an approach to individual gene discovery that began with a phenotype and gradually identified a causative gene, localizing it to a particular part of a chromosome.

## 22.2 The Origin of the Idea

4. The two technologies that were key to the genome project were automated DNA sequencing and the ability of computers to align and derive long base sequences.
5. In the Sanger method of DNA sequencing, DNA fragments differing in size and with one labeled end base are aligned, and the sequence read off from the end bases. Computers can compare DNA sequences among species and spot certain parts of protein structure.
6. Many people thought of sequencing the human genome, for different reasons. The project officially began in 1990 under the direction of the DOE and NIH.

## 22.3 Technology Drives the Sequencing Effort

7. **Expressed sequenced tags** enable researchers to find protein-encoding genes. **DNA microarrays** reveal gene expression. Both were developed in the early 1990s.
8. In 1995, with better technology, the focus shifted from mapping to sequencing. Interest in proteomics began.
9. By 1999, competition had intensified. The first draft of the genome sequence was announced in 2000.

## 22.4 Genome Information Answers and Raises Questions

10. Genome sequences reveal that genes are more complex than just instructions for a protein. They may have different functions in different cell types at different times.
11. Comparisons of genomes from different species indicate shared ancestry. They also tell us that we do not know very much about gene function.

# Review Questions

1. How did the following technologies contribute to the success of the human genome project, or the start of genomics?
   a. expressed sequence tags
   b. positional cloning
   c. DNA microarray technology
   d. automated DNA sequencing
   e. assembler computer programs
2. Why is a cytogenetic map less precise than a sequence map?
3. In 1966, Francis Crick suggested that knowing the genome of a simple bacterial cell would reveal how life works. Did it?
4. Two difficulties in sequencing the human genome were the large number of repeat sequences, and the fact that DNA is double-stranded. Explain how these characteristics complicated sequencing efforts.
5. If a researcher wanted to attempt to create a genome, and therefore possibly life, which organism could serve as a model? Cite a reason for your answer.
6. What has comparative genomics revealed about our relationships to yeast, worms, and flies?

# Applied Questions

1. Why must several copies of a genome be cut up to sequence it?
2. Celera Genomics actually sequenced several different genomes (one of which was probably J. Craig Venter's). Why would the sequences not be identical?
3. How could DNA microarray technology grow before human genome sequencing was completed?
4. As a researcher scans new gene sequences into GenBank, she sees a sequence that encodes a protein whose shape suggests that it could, when abnormal, cause the disease she is investigating. Explain how this approach differs from positional cloning.
5. Axenfeld-Rieger anomaly is an autosomal dominant condition whose symptoms include absent eye muscles, flattened leg bones and associated abnormal gait, mild hearing loss, a flattened midface, prominent forehead, large nose, and bulging eyes. How might a researcher use DNA microarray technology and human genome information to identify the gene that causes this condition?
6. Which of the species listed below has the greatest density of protein-encoding genes?

| Species | Number of Protein-Encoding Genes | Genome Size (millions of bases) |
|---|---|---|
| *Haemophilus influenzae* (bacterium) | 1,743 | 1.8 |
| *Saccharomyces cerevisiae* (yeast) | 6,000 | 12.1 |
| *Caenorhabditis elegans* (roundworm) | 15,000 | 100.0 |
| *Homo sapiens* (human) | 70,000 | 3,000.0 |

7. Restriction enzymes break a sequence of bases into the following pieces:

T T A A T A T C G

C G T T A A T A T C G C T A G

G C T T C G T T

A A T A T C G C T A G C T G C A

C T T C G T

T A G C T G C A

G T T A A T A T C G C T A G C T G C A

How long is the original sequence? Reconstruct it.

8. One newly identified human gene has counterparts (homologs) in bacteria, yeast, roundworms, mustard weed, fruit flies, mice, and chimpanzees. A second gene has homologs in fruit flies, mice, and chimpanzees only. What does this information reveal about the functions of these two human genes with respect to each other?

# Suggested Readings


Baltimore, David. February 15, 2001. Our genome unveiled. *Nature* 409:814–16. A summary of the human genome project, including a great glossary of jargon and acronyms.

International Human Genome Mapping Consortium. February 15, 2001. A physical map of the human genome. *Nature* 409:934–41. The public first draft human genome sequence. Other articles in this issue discuss specific chromosomes.

Lewis, Ricki. November 26, 2001. A personal view of genomics. *The Scientist* 15:10. An interview with J. Craig Venter.

Lewis, Ricki. July 24, 2000. Keeping up: Genetics to genomics in four editions. *The Scientist* 14:46. A look at the co-evolution of this textbook and genomics.

Lewis, Ricki. May 1, 2000. Confessions of an ex fly pusher. *The Scientist* 14:1. Many human genes have counterparts in the fly genome.

Lewis, Ricki. January 5, 1998. Comparative genomics reveals the interrelatedness of life. *The Scientist* 12:5. Probing genomes shows that species are more alike than different.

Lewis, Ricki. March 18, 2002. Pufferfish genomes probe human genes. *The Scientist* 16(6):22. The pufferfish genome is one-eighth the size of ours, but disturbingly similar.

Maher, Brendan. February 4, 2002. The human genome—one year later. *The Scientist* 16:29. A look back at the human genome annotation.

Olson, Maynard. February 15, 2001. Clone by clone by clone. *Nature* 490:816–18. How the International Consortium sequenced the human genome.

Pevzner, Pavel A. September 2001. Assembling puzzles from preassembled blocks. *Genome Research* 11, no. 9:1–2. A clear explanation of how a DNA genome sequence assembler program works.

Roberts, Leslie. February 16, 2001. Controversial from the start. *Science* 291:1182–88. The first draft human genome project got off to a rocky start, and ended with a fight.

Singer, Peter A., and Abdallah S. Daar. October 5, 2001. Harnessing genomics and biotechnology to improve global health equity. *Science* 294:87–89. Researchers hope that genomic technologies will improve health care in all nations.

Venter, J. C., et al. February 16, 2001. The sequence of the human genome. *Science* 291:1304–51. Celera Genomics's first draft human genome sequence and annotation.

Wade, Nicholas. February 13, 2001. Grad student becomes gene effort's unlikely hero. *The New York Times*, p. F1. How a graduate student compressed the work of six months into four weeks to save the public human genome project.


# On the Web

Check out the resources on our website at

**www.mhhe.com/lewisgenetics5**

On the web for this chapter, you will find additional study questions, vocabulary review, useful links to case studies, tutorials, popular press coverage, and much more. To investigate specific topics mentioned in this chapter, also try the links below:

Annotated Human Genomic Database **www.DoubleTwist.com/genome**

Baylor College of Medicine **www.hgsc.bcm.tmc.edu**

Celera Genomics Corp. **www.celera.com**

Cold Spring Harbor Laboratory Learning Center **vector.cshl.org/resources/resources.html**

European Bioinformatics Institute **www.ensembl.org/genome/central/**

Functional Genomics Website **www.sciencegenomics.org**

GeneWise **www.sanger.ac.uk./software/wise2/**

Genome jokes and cartoons **cagle.slate.msn.com/news/gene/**

Genome News Network **www.celera.com/genomics/genomic.cfm**

How to Build a Microarray **cmgm.stanford.edu/pbrown/mguide/index.html**

The Institute for Genomic Research **www.tigr.org**

Microarray **http://ihome.cuhk.edu.hk/~6400559/array.html**

National Center for Biotechnology Information Splash Page **www.ncbi.nlm.nih.gov/genome/guide/human**

Nature Genome Gateway **www.nature.com/genomics/**

Washington University **http://genome.wustl.edu/gsc/human/mapping/**

# Answers to End-of-Chapter Questions

## Chapter 1 Overview of Genetics

### Answers to Review Questions

1. Gene pool, genome, chromosome, gene, DNA

2. Genomics is shifting the emphasis of genetic research from identifying and describing one gene at a time, to considering the co-expression and interactions of many genes.

3. a. An autosome does not carry genes that determine sex. A sex chromosome does.
   b. Genotype is the allele constitution in an individual for a particular gene. Phenotype is the physical expression of an allele combination.
   c. DNA is a double-stranded nucleic acid that includes deoxyribose and the nitrogenous bases adenine, guanine, cytosine, and thymine. DNA carries the genetic information. RNA is a single-stranded nucleic acid that includes ribose and the nitrogenous bases adenine, guanine, cytosine, and uracil. RNA carries out gene expression.
   d. A recessive allele determines phenotype in two copies. A dominant allele determines phenotype in one copy.
   e. Absolute risk refers to an individual's personal risk. Relative risk is in comparison to another group, and is less precise.
   f. A pedigree is a chart of family relationships and traits. A karyotype is a chart of chromosomes.

4. Inherited disease differs from other types of diseases in that recurrence risk is predictable for particular individuals in families; presymptomatic detection is possible; and different populations have different characteristic frequencies of traits or disorders.

5. CF is caused by one malfunctioning gene. Height reflects the actions of several genes and environmental influences, such as diet.

6. Empiric risk is based on observations of populations, whereas risk of a Mendelian disorder is based on family information.

7. A mutation is merely a genetic change, not necessarily a change for the worse.

### Answers to Applied Questions

1. 16

2. 46.66

3. The wife

4. Expression of the Crohn disease gene that increases susceptibility is affected by other genes and by environmental factors, such as diet. That is, certain conditions are necessary for the susceptibility to actually lead to disease.

5. More likely than general population: coronary artery disease, kidney cancer, lung cancer, depression.

   Less likely than general population: addictive behaviors, diabetes.

6. Multiple sclerosis is multifactorial, and because of interacting, multiple causes, precise recurrence risks are not possible.

7. If a woman picked up the crab lice during the rape, then the lice may contain the rapist's DNA in its gut contents. The forensic entomologist compares the DNA to DNA from the three suspects. It should match one.

8. Continued testing over the long term can supplement the existing information that GM foods are safe, with perhaps increased attention to reports of harm to health and spread of modified foods beyond their intended area.

9. a. It would help to see the scientific evidence that a DNA test can detect susceptibility to carpal tunnel syndrome.
   b. A predisposition is not the same as a diagnosis.
   c. Perhaps companies alerted to certain sensitivities among some workers can alter the workplace to remove the condition that can lead to illness. This could be done without identifying specific individuals.

## Chapter 2 Cells

### Answers to Review Questions

1. a. In signal transduction, an outside stimulus triggers a chain reaction among membrane proteins that causes production or activation of a second messenger, which activates the genes that directly provide the cell's response.
   b. In cell adhesion, cell surface molecules direct how cells attach and move.
   c. In the cell cycle, a cell proceeds through an interphase consisting of two gap phases, when proteins and lipids are synthesized, and a synthesis phase when DNA is replicated, then mitosis, when the DNA divides, followed by cytokinesis, when other cellular constituents are distributed between the resulting two daughter cells.
   d. In apoptosis, a death receptor receives a death signal, then activates caspases that destroy the cell in a series of steps. Membranes surround the pieces, preventing inflammation.
   e. In mitosis, the replicated DNA condenses, the nucleolus temporarily breaks down, and the centromeres part and two sets of chromosomes move to opposite ends of the cell. Then the cell physically divides in two.
   f. In secretion, proteins produced in the rough ER join carbohydrates and lipids in the Golgi apparatus, and exit the cell encased in lipid vesicles.

2. a. Tubulin forms microtubules and actin forms microfilaments, which comprise the cytoskeleton.
   b. Caspases carry out apoptosis.
   c. Changing levels of cyclins and kinases regulate the cell cycle.
   d. Checkpoint proteins provide choices during the cell cycle.
   e. Cellular adhesion molecules allow certain cell types to stick to each other.

3. Hormones, growth factors, cyclins, and kinases.

4. Specialized cells express different subsets of all the genes that are present in all cell types, except for red blood cells.

5. a. A bacterial cell is usually small and lacks a nucleus and other organelles. A eukaryotic cell contains membrane-bounded organelles, including a nucleus, that compartmentalize biochemical reactions.
   b. During interphase, cellular components are replicated. During mitosis, the cell divides, distributing its contents into two daughter cells.
   c. Mitosis increases cell number. Apoptosis eliminates cells.
   d. Rough ER is a labyrinth of membranous tubules, studded with ribosomes that synthesize proteins. Smooth ER is the site of lipid synthesis and lacks ribosomes.
   e. Microtubules are tubules of tubulin and microfilaments are rods of actin. Both form the cytoskeleton.

6. a. Mitochondria extract energy from nutrients to fuel cellular activities.
   b. Lysosomes are sacs of enzymes that break down debris.
   c. Peroxisomes break down certain lipids and rare biochemicals, synthesize bile acids and detoxify compounds that result from excess oxygen exposure.
   d. Smooth ER is the site of lipid synthesis.
   e. Rough ER is the site of protein synthesis.
   f. The Golgi apparatus is the site of carbohydrate addition to proteins to form secretions.
   g. The nucleus houses the genetic material.

7. Compartmentalization separates biochemicals that could harm certain cell constituents. It also organizes the cell so it can function more efficiently.

8. The cell membrane is the backdrop that holds many of the molecules that intercept incoming signals. These molecules contort in ways that amplify and spread the message.

9. A totipotent cell can become any cell type. A pluripotent cell can become any of several cell types. A progenitor cell can specialize as varieties of a cell type, such as blood cells.

10. Cells are complex structures plus a cell membrane, including genetic material and the means to synthesize proteins and other biomolecules. Viruses are not cells, but nucleic acids in protein coats. Prions are proteins that can become infectious if they fold in a certain way.

## Answers to Applied Questions

1. a. Lack of cell adhesion can speed the migration of cancer cells.
   b. Impaired signal transduction can block a message to cease dividing.
   c. Blocking apoptosis can cause excess mitosis, and an abnormal growth.
   d. Lack of cell cycle control can lead to too many mitoses.
   e. If telomerase is abnormal, a cell might not cease to divide when it normally would.

2. Because enzymes are proteins, genes encode them.

3. Nucleus and lysosomes

4. Stem cells maintain their populations because each mitosis produces a daughter cell that differentiates, as well as one that remains a stem cell.

5. Too frequent mitosis leads to an abnormal growth. Too little mitosis can limit growth or repair of damaged tissues. Too much apoptosis can kill healthy tissue. Too little apoptosis can lead to abnormal growths.

6. A cell in an embryo would not be in $G_0$ because it has to divide frequently to support the high growth rate.

7. Mitochondria

8. Stem cells can be obtained from adult tissues and organs.

9. Disrupting microtubules can prevent the spindle apparatus from either forming or breaking down, either of which would halt mitosis. A drug that disables telomerase might enable chromosomes to shrink, which might stop mitosis. A drug can intercept an extracellular signal to divide.

10. Signals from outside the cell interact with receptors embedded in the cell membrane, and the cell membrane's interior face contacts the cytoskeleton.

11. a. Abnormal chloride channels in cell membranes of lung lining cells and pancreas.
    b. Lack of a transport protein in peroxisomes leads to build up of long-chain fatty acids.
    c. Abnormal ankyrin collapses the cytoskeletal attachment to the cell membrane, causing red blood cells to balloon out.
    d. Abnormal growth factors and signal transduction cause nerve overgrowth under skin.
    e. Lack of CAMs impairs wound healing.
    f. Kuru is caused by stabilization and spread of an unusual conformation of prion protein.
    g. Syndactyly is a failure of apoptosis to fully separate digits.

# Chapter 3 Development

## Answers to Review Questions

1. **a.** 2 **b.** 2 **c.** 1 **d.** 2 **e.** 1 **f.** 1 **g.** 1

2. Male: Sperm are manufactured in seminiferous tubules packed into the testes, and mature and are stored in the epididymis. The epididymis is continuous with the vas deferens, which carries sperm to the urethra, which sends them out through the penis. The prostate gland and the seminal vesicles secrete into the vas deferens, and the bulbourethral glands secrete into the urethra, forming seminal fluid.

   Female: Oocytes develop within the ovaries and are released into the fallopian tubes, which carry them to the uterus, which narrows at its base to form the cervix, which opens to the vagina.

3. $2^{39}$. This is an underestimate because it does not account for crossing over.

4. Mitosis divides somatic cells into two daughter cells with the same number of chromosomes as the diploid parent cell. Meiosis forms gametes, in which the four daughter cells have half the number of chromosomes of the parent cell. Genetic recombination occurs in meiosis.

5. Both produce gametes, but oogenesis takes years and spermatogenesis takes months.

6. In female gamete maturation, most of the cytoplasm concentrates in one huge cell. In male gamete maturation, four same-size sperm derive from an original cell undergoing meiosis.

7. Hundreds of millions of sperm are deposited in the vagina during intercourse. There, the sperm are chemically activated and the oocyte secretes a sperm attractant. Sperm tail movement and contraction of the woman's muscles aid sperm movement. When sperm contact follicle cells, their acrosomes release enzymes that penetrate the oocyte. When the membranes of sperm and oocyte meet, a wave of electricity spreads physical and chemical changes over the oocyte surface, blocking other sperm. The chromosomes of the two cells meet. The fertilized ovum is a zygote.

8. Teratogens are more dangerous to an embryo than they are to a fetus because structures form in the first eight weeks.

9. Teratogens include thalidomide, alcohol, excess nutrients, cocaine, cigarettes, and infections. Exposure to teratogens during critical periods can have drastic effects.

10. Accelerated aging disorders and studies on adopted individuals indicate an inherited component to longevity.

11. Part of F.D. developed parthenogenetically. A clone develops from a somatic cell, rather than from a fertilized ovum.

## Answers to Applied Questions

1. Answers may vary. A commonly used cutoff point for experimentation is after the second week, when primary germ layers begin to form.

2. An inner cell mass is just a group of identical-appearing cells that are bunched along the inside of the blastocyst. A seven-week embryo has the rudiments of many organs.

3. Stem cells

4. A polar body does not have enough cytoplasm and organelles to support an embryo.

5. Some people believe that a pregnant woman should be held legally responsible for knowingly exposing an embryo or fetus to a teratogen.

6. The research subjects might outlive the researchers. Studying longevity using humans would take too long.

7. The man may have inherited a mutation that affects expression of telomerase, keeping it activated in all cell types. But this is science fiction.

8. Prenatal environmental stress, such as lack of nutrients, can alter gene expression in ways that enhance utilization of nutrients, which is adaptive for the fetus but may set the stage for future diabetes or heart disease. Evidence is epidemiological and based on animal studies.

# Chapter 4 Mendelian Inheritance

## Answers to Review Questions

1. The law of segregation derives from the fact that during meiosis, alleles separate into different gametes. The law of independent assortment is based on the fact that distribution of alleles of two unlinked genes into gametes occurs at random.

2. The two laws of inheritance were derived from pea crosses. Without knowing about chromosomes, Mendel observed offspring phenotypes and predicted results of crosses.

3. a. An autosomal recessive trait is inherited from carriers and affects both sexes. An autosomal dominant trait can be inherited from one parent, who is affected. Autosomal recessive inheritance can skip generations; autosomal dominant inheritance cannot.
   b. Mendel's first law concerns inheritance of one trait. The second law follows inheritance of two genes on different chromosomes.
   c. A homozygote has identical alleles for a particular gene, and a heterozygote has different alleles.
   d. The parents of a monohybrid cross are heterozygotes for a single gene. Parents of a dihybrid cross are heterozygous for a pair of genes.
   e. A Punnett square tracks the distribution of alleles of genes on different chromosomes from parents to offspring. A pedigree depicts family members and their inherited traits.

4. If Mendel had chosen two traits on the same chromosome, he would have observed a higher percentage of offspring inheriting two alleles together rather than independently.

5. Blood relatives share alleles inherited from common ancestors.

6. The Egyptian pedigree only traces genealogy and not inheritance of traits. The pedigree a genetic counselor would use today would trace inherited traits too.

7. The parents of a person who is homozygous dominant for HD must be heterozygotes.

8. Organismal: 4.3, 4.4, 4.7, 4.8, 4.10; chromosomal: 4.5, 4.9; gene: 4.4, 4.10, 4.11, 4.12; figures 4.14, 4.15, and 4.16 depict pedigrees, which indicate organisms and genes.

9. 100 percent

## Answers to Applied Questions

1. a. Autosomal recessive
   b. Both sexes are affected, and it skips generations
   c. Pedigree

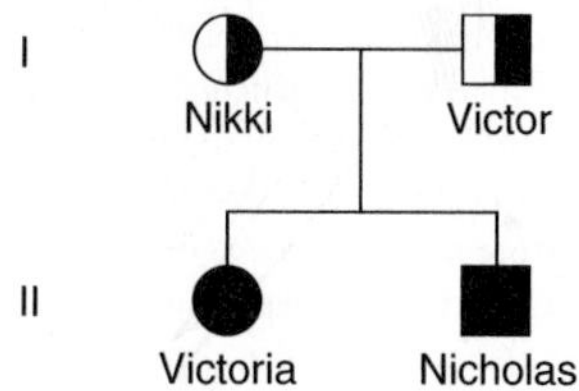


2. Autosomal dominant—it affects both sexes, and occurs every generation.

3. Pedigree

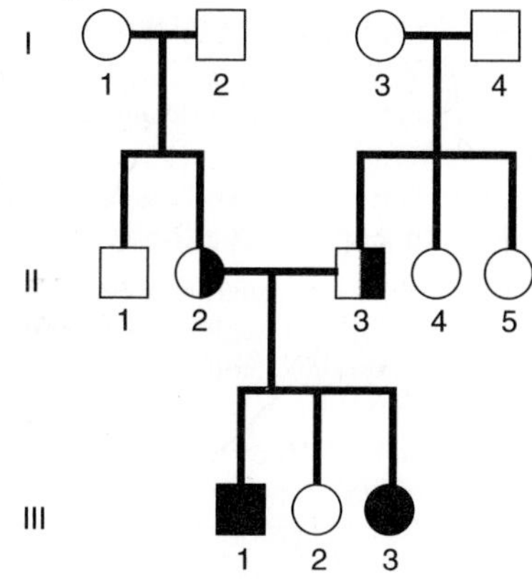


4. Carriers are I-3, I-4, II-4, III-3, III-4.

5. One in 2 chance the child has double eyelashes.

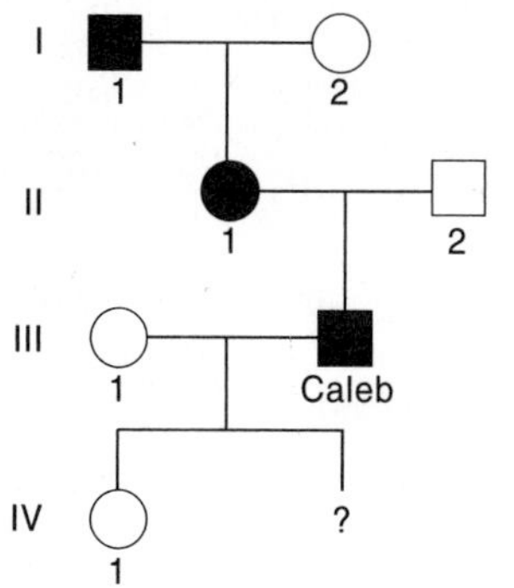


6. Consanguinity

7. Genotypic ratios:

| | |
|---|---|
| 4/32 = *BbHhEe* | 2/32 = *bbHhee* |
| 4/32 = *bbHhEe* | 1/32 = *BbHHEE* |
| 2/32 = *BbHhEE* | 1/32 = *BbhhEE* |
| 2/32 = *BbJJEe* | 1/32 = *BbHHee* |
| 2/32 = *BbhhEe* | 1/32 = *Bbhhee* |
| 2/32 = *BbHhee* | 1/32 = *bbHHEE* |
| 2/32 = *bbHhEE* | 1/32 = *bbhhEE* |
| 2/32 = *bbHHEe* | 1/32 = *bbHHee* |
| 2/32 = *bbhhEe* | 1/32 = *bbhhee* |

Phenotypic ratios:

9/32 = yellow urine, colored eyelids, short fingers
9/32 = red urine, colored eyelids, short fingers
3/32 = yellow urine, normal eyelids, short fingers
3/32 = red urine, normal eyelids, short fingers
3/32 = yellow urine, colored eyelids, long fingers
3/32 = red urine, colored eyelids, long fingers
1/32 = yellow urine, normal eyelids, long fingers
1/32 = red urine, colored eyelids, long fingers

8. II-1 and II-2 must be carriers. All of the individuals in generation 1 could be carriers.

9. Cloned mice bred to each other would produce offspring with different combinations of alleles because of meiosis, unless they are homozygous for all genes.

10. $.10 \times .002 = .0002 = 2$ in 10,000

11. a. Uncles and nieces having children together.
    b. Carriers = III-5, III-6, III-9, IV-1 and niece, IV-6, IV-11, V-4, V-5, V-6, V-9, V-10

12. No

13. First cousins

14. His normal allele enables enough of the enzyme to be produced so that his nervous system can function normally.

# Chapter 5 Extensions and Exceptions to Mendel's Laws

## Answers to Review Questions

1. a. Lethal alleles eliminate a progeny class that Mendel's laws predict should exist.
   b. Multiple alleles create the possibility of more than two phenotypic classes.
   c. Incomplete dominance introduces a third phenotype for a gene with two alleles.
   d. Codominance introduces a third phenotype for a gene with two alleles.
   e. Epistasis eliminates a progeny class when a gene masks another's expression.
   f. Incomplete penetrance produces a phenotype that does not reveal the genotype.
   g. Variable expressivity can make the same genotype appear to different degrees.
   h. Pleiotropy can make the same genotype appear as more than one phenotype because subsets of effects are expressed.
   i. A phenocopy mimics inheritance, but is an environmental effect.
   j. Conditions with the same symptoms but caused by different genes (genetic heterogeneity) will not recur with the frequency that they would if there was only one causative gene.

2. Dominant and recessive alleles are of the same gene, whereas epistasis is an interaction of alleles of different genes.

3. It can skip generations in terms of phenotype.

4. The $I^A$ allele is codominant with the $I^B$ allele; both are completely dominant to *i*.

5. Variably expressive means that some people have more abnormal digits than others. Incompletely penetrant means that some people who inherit the disease-causing genotype do not have symptoms.

6. Maternal inheritance describes transmission of mitochondrial genes, which sperm do not usually contribute to oocytes and therefore these traits are always passed from mothers only. Linked genes are transmitted on the same chromosome. Mendel's second law applies to genes transmitted on different chromosomes.

7. The *H* gene affects the *I* gene. If the *H* allele is present, the ABO genotype determines blood type. If *H* is not present, the ABO blood type is O.

8. Only females transmit maternally inherited traits. All of a woman's children inherit a mitochondrial trait, but a male does not pass the trait to his children.

9. Additional genes may affect expression of a particular gene, and therefore the same genotype for that gene may be associated with a different phenotype in different individuals.

10. Twenty-four linkage groups, including 22 pairs of autosomes, and the X and the Y.

## Answers to Applied Questions

1. a. D b. A c. E d. C e. H f. F g. G h. G i. E

2. Haplotypes can reveal if people without symptoms have the haplotype associated with the condition. These people are non-penetrant. If the genotype is lethal, individuals with the haplotype containing the lethal allele should not exist.

3. a. 1/2 b. 1/2 c. 0 d. 0

4. 1/4

5. The Bombay phenotype accounts for the young woman's type O blood. She is *hh*.

6. a. This alters the phenotype.
   b. Bombay phenotype

7. It would be Mendelian because the gene that causes the condition when mutated is in the nucleus.

8. $.45 \times .05 = .0225$

9. The sperm are all of genotype *hh se se*. The oocytes are of genotype *Hh Se se*, with the alleles in coupling:

*H* ____ *Se*
____
*h* *se*

All sperm are *h se*.

Oocytes: *H Se* 49.5% *h se* 49.5%
*H se* 0.5% *h Se* 0.5%

Chance offspring like father =
1 (*h se*) × 49.5 (*h se*) = 49.5

10. a. Male: *RrHhTt*
    Female: *rrhhtt*

b. Parental progeny classes: Round eyeballs, hairy tail, 9 toes; square eyeballs, smooth tail, 11 toes

Recombinant progeny classes: Round eyeballs, hairy tail, 11 toes; square eyeballs, smooth tail, 9 toes; round eyeballs, smooth tail, 11 toes; square eyeballs, hairy tail, 9 toes

c. Crossover frequency between eyeball shape (*R*) and toe number (*T*): Round eyeballs, 11 toes = 6 + 4; square eyeballs, 9 toes = 6 + 4; crossover frequency = 20/100 = 20%

# Chapter 6 Matters of Sex

## Answers to Review Questions

1. Sex is expressed at the chromosomal level as inheriting XX or XY; at the gonadal level by developing ovaries or testes; at the phenotypic level by developing male or female internal and external structures; and at the gender identity level by feelings.

2. Genes in the pseudoautosomal region are the same as certain X-linked genes, but X-Y genes share only sequence similarities.

3. **a.** Female **b.** Female **c.** Female

4. Sustentacular cells secrete anti-Müllerian hormone, which stops development of female reproductive structures. Interstitial cells secrete testosterone, which promotes development of male reproductive structures.

5. Absence of the *SRY* gene product causes the Müllerian ducts to develop into ovaries. The ovaries produce female hormones, which influence the development of external and internal reproductive structures.

6. Feelings of very young children; twin studies in which identical twins are more likely to both be homosexual than are fraternal twins; fruit fly behavior; genes on the X chromosome that segregate with homosexuality.

7. A female homozygous dominant for an X-linked dominant allele would probably be so severely affected that she would not be alive, or healthy enough, to reproduce.

8. Coat color in cats is X-linked. In females, one X chromosome in each cell is inactivated, and the pattern of a calico cat's coat depends on which cells express which coat color allele. A male cat, with only one coat color allele, would have to inherit an extra X chromosome to be tortoiseshell or calico.

9. Inactivation of the gene in some cells but not others results in a patchy phenotype.

10. Each cell in a female's body contains only one active X chromosome, which makes females genetically equivalent (in terms of X-linked genes) to males.

11. An X-linked trait appears usually in males and may affect structures or functions not distinct to one sex. A sex-limited trait affects a structure or function distinct to one sex. A sex-influenced trait is inherited as a recessive in one sex and dominant in the other.

12. Mouse zygotes with two female pronuclei or two male pronuclei are abnormal. In humans, two male and one female genome in the same embryo yields a placenta, while two female and one male genome yields a normal embryo with an abnormal placenta. Hydatidiform moles (placental tissue with no embryo) consist of cells with two male genomes; a teratoma contains cells with two female genomes.

## Answers to Applied Questions

1. **a.** 1/2 **b.** 1/2 **c.** 1/2 **d.** A carrier might have symptoms if the mutated gene is expressed in tissues affected by the condition.

2. Girls may have milder cases because X inactivation might inactivate a more severe allele in some cells, and because hormonal differences from males may affect the phenotype.

3. The unevenness of the teeth of affected females may reflect expression of the defect in only some cells as the result of X inactivation.

4. 1/2

5. Autosomal dominant

6. The female is "using" the gene conservatively, limiting the rate of weight gain in the embryo.

7. 1/4

8.

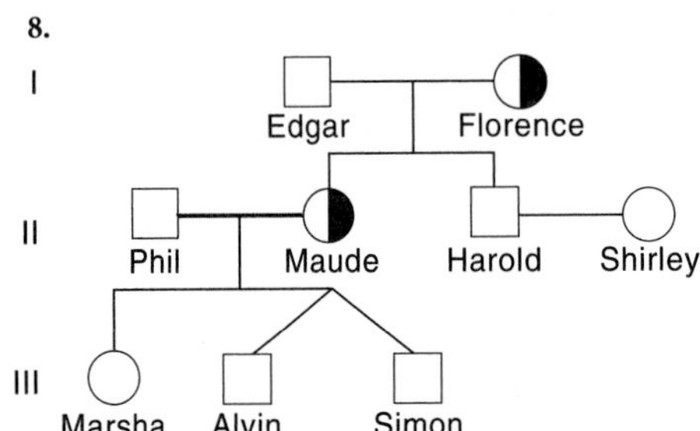


9. **a.** Because the phenotype is not severe.
**b.** 1 in 2, because she is a carrier.
**c.** 1 in 1

10. 1/2

11. **a.** Carriers = I-4, II-3, III-2, III-5 **b.** 1/2 **c.** 1/2 **d.** Women would have to inherit kinky hair disease from an affected father and a carrier mother; affected males would not live long enough to have children.

# Chapter 7 Multifactorial Traits

## Answers to Review Questions

1. A Mendelian multifactorial trait is caused by one gene and environmental influences, whereas a polygenic multifactorial trait is caused by more than one gene and environmental influences.

2. Eye color has a greater heritability because the environment has a greater influence on height than on eye color.

3. Heritability for skin color changes because the duration and intensity of sun exposure contribute to skin color.

4. The environment determines which infections a person contracts.

5. **a.** Empiric risk is based on observations of trait prevalence in a particular population or group of individuals.
**b.** Twin studies approximate the degree of heritability by comparing trait prevalence among pairs of MZ twins to DZ twins. The greater the difference, the higher the heritability.
**c.** Inherited traits may stand out in an adoptee's family where each member lives in the same environment, but the adopted individual has different genes.
**d.** Association studies establish correlations between particular gene variants or sections of chromosomes (haplotypes) and inheritance of certain traits or susceptibilities.

6. Many people must be surveyed to determine whether a certain haplotype (and its SNPs) is exclusively associated with a particular phenotype. Enough DNA must be considered so that several SNPs are analyzed, and a case-control strategy must be employed to ensure that the same SNPs are not found among affected individuals as well as unaffected ones. Finally, if many SNPs are considered, many combinations must be correlated to phenotypes.

7. Cardiovascular: apolipoproteins, lipoprotein lipase and proteins that regulate blood pressure or homocysteine metabolism

Body weight: neuropeptide Y, genes that control leptin, melanocortin-4 receptor

8. More genotypes specify a medium brown skin color than other colors.

## Answers to Applied Questions

1. A particular combination of alleles must be inherited for a polygenic disorder to recur. If many combinations are possible, the probability that a particular phenotype recurs will be low.

2. An association study correlates genotype (haplotype) and phenotype. Heritability estimates the relative contribution of genetics to a trait.

3. A preparatory course would more likely help a student to improve math skills because math ability has a lower heritability than verbal skills.

4. The heritability, 0.54, suggests that the influence of genes and the environment is about equal.

5. A case-control design is necessary to demonstrate that a particular haplotype is exclusively associated with the phenotype being studied—that is, that the haplotype does not also occur in nonaffected individuals.

6. An association study can identify individuals who are at high risk, and this information is valuable if they can then alter an environmental factor to lower the overall risk.

7. Lipoprotein A, HDL, BMI, triglycerides, diastolic blood pressure

8. **a.** .5 **b.** .5 **c.** Selfishness is difficult to assess, and is a normal part of childhood.

9. There is a large environmental component to obesity in this population.

10. Decrease in neuropeptide Y and increase in melanocortin-4 receptor expression could combat obesity.

# Chapter 8 The Genetics of Behavior

## Answers to Review Questions

1. Proteins whose products are involved in neurotransmission or signal transduction can affect behavior.

2. The older techniques of empiric risk and adoptee and twin studies estimate the proportion of the contribution of genetics to a trait. Linkage, SNP association studies, and genome-wide scans seek to identify particular genes that contribute to a particular behavior.

3. ADHD has a very distinctive phenotype; it is common; and linkage studies point to a candidate gene. In contrast, autism is rare, and linkage points to several candidate genes, suggesting genetic heterogeneity (different ways to inherit the same condition).

4. **a.** sleep disorders **b.** addiction **c.** addiction

5. We must understand how a gene variant affects behavior, and how other genes and the environment affect its expression.

6. **a.** N-CAM
 **b.** dopamine D(2) receptor
 **c.** ADH
 **d.** serotonin receptor or transporter
 **e.** dopamine or glutamine transporter

7. Genetic heterogeneity; when a behavior is part of several disorders; behaviors that fall within the range of normal, although they are extreme; ability to imitate a behavior.

8. A trait that is polygenic has varying degrees of expression. A trait that is genetically heterogeneic is the same even though caused by mutations in different genes.

9. With age, the environment has had longer to affect gene expression.

## Answers to Applied Questions

1. Association studies with negative results are important because they help researchers identify and focus on the genes that contribute to a trait. They also help to reveal the mechanism of how a gene causes a trait.

2. **a.** The same neurotransmitter controls different behaviors. The differences arise from which neurons are involved.
 **b.** Dopamine

3. Instead of demonstrating the existence and extent of inherited influences, DNA microarrays implicate specific genes, revealing clues to the biological basis of a behavior or behavioral disorder.

4. For all choices—drug addiction, eating disorder, and depression—a SNP profile indicating increased risk could be helpful if the person can take action to lower the risk of developing the condition, but can be harmful if the person just gives up, thinking that fate cannot be avoided. SNP profiling might also be used to discriminate.

5. This is an opinion question. If drugs are outlawed based on their addiction potential, then alcohol and nicotine should be included with other drugs that are illegal. An alternate approach would be to outlaw drugs that impair one's ability to function. Such an approach might outlaw alcohol and cocaine, yet permit tobacco or possibly marijuana use.

6. Yes—a criminal cannot alter his or her inherited behavior. No—the environment can override inherited criminal tendencies.

7. Depression and bipolar disorder each might actually be several conditions that have similar symptoms.

8. Identify what is abnormal in the Utah family, develop a drug to treat the condition, and then try the drug on elderly people who are having difficulty sleeping.

9. The IQ test could be flawed or measure intelligence in a way that does not compensate for differences in educational opportunities.

10. Better off back then: individuals with minor manifestations of behavioral disorders were treated as extremes of normal—not labeled or stigmatized as having a disorder.

Better off now: Person might be held responsible for behavior, or may be helped by drug treatment.

11. Research can isolate a chromosomal region where the DNA sequence is common to people with Wolfram syndrome, and eventually discover and describe the causative gene. The specific nature of the corresponding protein product may explain how a half normal gene dose causes the behavioral problems of relatives.

12. These data suggest a large inherited component to alcoholism.

# Chapter 9 DNA Structure and Replication

## Answers to Review Questions

1. DNA is replicated so that it is not used up in directing the cell to manufacture protein.

2. 1E 2C 3D 4B 5A

3. The nucleotide base sequence encodes information.

4. One end of a strand of nucleotides has a phosphate group attached to the 5′ carbon of deoxyribose. The other end has a hydroxyl group attached to the 3′ carbon. The opposite (complementary) strand has the reverse orientation.

5. Helicases, primase, DNA polymerase, ligase, photolyases

6. **a.** A G C T C T T A G A G C T A A
 **b.** G G C A T A T C G G C C A T G
 **c.** T A G C C T A G C G A T G A C

7. Histone protein, nucleosome, chromatin

8. **a.** Opposite orientations of the two DNA chains of a double helix.
 **b.** The fact that a new DNA double helix retains a parental strand.
 **c.** Hydrogen bonding of A with T, T with A, G with C, and C with G in the DNA double helix.

9. Hershey and Chase showed that DNA is the genetic material and protein is not. Meselson and Stahl showed that DNA replication is semiconservative, but not conservative or dispersive.

10. Strands separate and are held apart. Primase makes RNA primer. DNA polymerase adds DNA bases to RNA primer. Proofreading, repair. Continuous on one strand only. RNA primers removed. Ligase seals sugar-phosphate backbone.

## Answers to Applied Questions

1. The sugar-phosphate backbone of replicating DNA cannot attach.

2. Primase, DNA polymerase, helicase, binding proteins, and ligase are required for DNA replication.

3. 20 cycles × 2 minutes/cycle = 40 minutes

4. Such a molecule would bulge where purines paired with each other, yet be narrow where pyrimidines pair.

5. Lacking DNA polymerase makes life impossible, because cells cannot divide.

6. PCR can directly detect HIV RNA, rather than a sign that the human body is responding to the presence of the virus.

7. Determining the structure of DNA led to discoveries of the mechanism of heredity, whereas sequencing the human genome provided information.

# Chapter 10 Gene Action and Expression

## Answers to Review Questions

1. **a.** H bonds between A and T and G and C join the strands of the double helix.
 **b.** In DNA replication, a new strand is synthesized semiconservatively, with new bases inserted opposite their complementary bases to form a new strand.
 **c.** An mRNA is transcribed by aligning RNA nucleotides against their complements in one strand of the DNA.
 **d.** The sequence preceding the protein-encoding sequence of the mRNA base pairs with rRNA in the ribosome.
 **e.** A tRNA binds to mRNA by base pairing between the three bases of its anticodon and the three mRNA bases of a codon.
 **f.** The characteristic cloverleaf of tRNA is a consequence of H bonding between complementary bases.

2. The direction of the flow of genetic information in retroviruses is opposite that of the central dogma.

3. **a.** The start of a gene
 **b.** tRNA
 **c.** A pseudogene
 **d.** rRNA
 **e.** mRNA
 **f.** Transposon

4. **a.** Proteins with an incorrect sequence of amino acids may not function.
 **b.** If the initial amino acid is released, additional amino acids cannot add on.

c. If rRNA cannot bind to the ribosome, then mRNAs cannot be translated into protein.

d. If ribosomes cannot move, then a protein would not exceed two amino acids in length.

e. If a tRNA picks up the wrong amino acid, the protein's amino acid sequence will be abnormal.

**5.** Transcription is the DNA-directed synthesis of a strand of mRNA by the enzyme RNA polymerase. Translation is the synthesis of proteins by ribosomes, which read the triplet code in the mRNA and translate the code into a chain of amino acids.

**6.** RNA contains ribose and uracil and is single-stranded. DNA contains deoxyribose and thymine and is double-stranded. DNA preserves and transmits genetic information; RNA expresses genetic information.

**7.** DNA replication occurs in the nucleus of eukaryotes. Prokaryotes have no nucleus so replication occurs in the cytoplasm. The same is true of transcription. In eukaryotes, translation occurs in the nucleus and cytoplasm, and some ribosomes are attached to the ER.

**8.** Transcription controls cell specialization by turning different sets of genes on and off in different cell types.

**9.** The same mRNA codon can be at the A site and the P site because the ribosome moves.

**10.** In transcription initiation, the DNA double helix unwinds locally, transcription factors bind near the promoter, and RNA polymerase binds to the promoter.

**11.** mRNA is the intermediate between DNA and protein, carrying the genetic information to ribosomes. tRNA connects mRNA and amino acids, and transfers amino acids to ribosomes for incorporation into protein. rRNA associates with proteins to form ribosomes.

**12.** Post-transcriptional changes to RNA include adding a cap and poly A tail, and removing introns.

**13.** Ribosomes consist of several types of proteins and rRNAs in two subunits of unequal size.

**14.** Discovery of introns was surprising because genes were thought to be contiguous, and discovery of ribozymes was surprising because only proteins were thought to function as enzymes.

**15.** An overlapping code constrains protein structure because certain amino acids would always be followed by the same amino acids in every protein.

**16.** Transcription and translation recycle tRNAs and ribosomes.

**17.** Proinsulin is shortened to insulin after translation. RNA editing shortens apolipoprotein B post-transcriptionally.

**18.** The amino acid sequence determines a protein's conformation by causing attractions and repulsions between different parts of the molecule.

**19.** A two-nucleotide genetic code would only specify 17 types of amino acids. There are 20 amino acids in biological proteins.

**20.** To create a mature mRNA for translation, snurps (small RNAs and proteins) excise introns in pre-mRNAs. Ribozymes assist in peptide bond formation. Proteins make up part of the structure of ribosomes, and enzymes are involved in protein synthesis.

**21.** A gene need not be continuous. A gene can move. DNA does not have to be transcribed and translated into protein.

**22.** Prion protein

**23.** One product is broken down faster than the other.

**24.** Same tRNA recognizes CUU and CUC.

## Answers to Applied Questions

**1.** a. 46

b. 44

**2.** Exon shuffling; pseudogenes; DNA sequences where exons and introns are parts of different genes.

**3.** a. AAUGUGAACGAACUCUCAG

b. UGAACCCGAUACGAGUAAT

c. CCGACGUUAUCGGCAUCUA

d. CCUUAUGCAGAUCGAUCGU

**4.** a. CGATAGACAGTATTTTCTCCT

b. CACCGCATAAGAAAAGGCCCATCC

c. CTCCCTTAAGAAAGAGTTCGTTCA

d. TCCTTTTGGGGAGAATAATATCTA

**5.** Many answers are possible, using combinations of *his* (CAU or CAC), *ala* (CGU, GCC, GCA, GCG), *arg* (CGU, CGC, CGA, CGG), *ser* (AGU, AGC, AGA, AGG), *leu* (CUU, CUC, CUA, CUG), *val* (GUC, GUG), and *cys* (UGU, UGC).

**6.** Several answers are possible because of the redundancy of the genetic code. One answer is: CATACCTTTGGGAAATGG.

**7.** There is only one genetic code.

**8.** Use ACA with any triplet other than CAA, and see whether threonine or histidine occurs. Whichever occurs, ACA encodes.

**9.** The CFTR protein, if improperly folded, won't reach the cell membrane, where it functions as an ion channel.

**10.** Transcription factors are most needed before birth, when structures form.

**11.** 26,927

**12.** 125, but this doesn't account for degenerate codons.

**13.** 1

**14.** 3,777 DNA bases (plus promoter sequence)

**15.** a. Smaller

b. Smaller

c. Larger

d. Larger

e. Same

**16.** The number of repeats may have some meaning (as yet unknown).

# Chapter 11 Gene Mutation

## Answers to Review Questions

**1.** A germinal mutation occurs in a gamete or a fertilized ovum, and therefore affects all the cells of an individual and is more serious than a somatic mutation, which affects only some tissues and is not transmitted to future generations.

**2.** The gene for collagen is prone to mutation because it is very symmetrical.

**3.** A spontaneous mutation can arise if a DNA base is in its rare tautomeric form at the instant when the replication fork arrives. A wrong base inserts opposite the rare one.

**4.** Mutational hot spots are direct repeats or symmetrical regions of DNA.

**5.** Gaucher disease is caused by an insertion, a missense mutation, or a crossover with a pseudogene.

**6.** Mutations in the third codon position can be silent. Mutations in the second position may replace an amino acid with a similarly shaped one. 61 codons specify 20 amino acids.

**7.** (1) A degenerate codon, (2) a mutation that replaces an amino acid with a structurally similar one, (3) a mutation that replaces an amino acid in a nonessential part of the protein, and (4) a mutation in an intron.

**8.** A conditional mutation is only expressed under certain conditions, such as increased temperature or exposure to particular drugs or chemicals.

**9.** Frameshift, deletion, duplication, insertion, transposable element.

**10.** Retention of an intron and expanding triplet repeats may provide a new function for a gene, which may cause disease.

**11.** A jumping gene can disrupt gene function by altering the reading frame or shutting off transcription.

**12.** A new recessive mutation will not become obvious until two heterozygotes produce a homozygous recessive individual with a phenotype.

**13.** The gene is expanding.

**14.** Short repeats can cause mispairing during meiosis. Long triplet repeats add amino acids, which can disrupt the encoded protein's function, often adding a function. Repeated genes can cause mispairing in meiosis and have dosage-related effects.

**15.** Mutations in the globin genes result in different conditions.

**16.** Excision repair corrects ultraviolet-induced pyrimidine dimers. Mismatch repair corrects replication errors. They use different enzymes.

## Answers to Applied Questions

**1.** Glycine to arginine: GGU to CGU, GGC to CGC, GGA to AGA, GGG to AGG

**2.** The frameshift mutation can create a stop codon, leading to a shortened polypeptide.

**3.** Any change that produces UAA, UAG, or UGA.

**4.** A transcription factor controls the expression of several genes, in a time and tissue-specific manner. Therefore a mutation in it affects several genes, producing multiple symptoms.

**5.** GAU to AAU or GAC to AAC

**6.** The second boy's second mutation, further in the gene, restores the reading frame so that part of the dystrophin protein has a normal structure, providing some function.

**7.** *asn* to *lys:* AAU to AAA AAC to AAG
*ile* to *thr:* AUU to ACU AUC to ACC AUA to ACA

8. Nonsense

9. *Arg* to *his:* CGU to CAU  CGC to CAC

10. GAU to GUU or GAC to GUU

11. Nonsense

12. They shouldn't be concerned, because they have mutations in two different genes.

13. The repair enzymes could correct UV-induced pyrimidine dimers.

# Chapter 12 Chromosomes

## Answers to Review Questions

1. Essential parts of a chromosome are telomeres, the centromere, and origin of replication sites. A centromere includes repeats of alpha satellites; centromere-associated proteins; and centromere protein A.

2. Protein-encoding genes become denser from the telomeres inward toward the centromere.

3. Centromeres and telomeres contribute to chromosome stability and have many repeats.

4. **a.** Homologs do not separate in meiosis I or II, leading to a gamete with an extra or missing chromosome.
   **b.** DNA replicates, but is not apportioned into daughter cells, forming a diploid gamete.
   **c.** Increased tendency for nondisjunction in the chromosome 21 pair.
   **d.** Crossing over in the male yields unbalanced gametes, which can fertilize oocytes, but too much or too little genetic material halts development.
   **e.** A gamete including just one translocated chromosome will have too much of part of the chromosome, and too little of other parts. Excess chromosome 21 material causes Down syndrome.

5. Patches of octaploid cells in liver tissue may arise as a result of abnormal mitosis in a few liver cells early in development.

6. **a.** A XXX individual has no symptoms, but she may conceive sons with Klinefelter syndrome by producing XX oocytes.
   **b.** A female with XO Turner syndrome has wide-set nipples, flaps of skin on the neck, and no secondary sexual development.
   **c.** A female with trisomy 2 Down syndrome. Phenotype includes short, sparse, straight hair, wide-set eyes with epicanthal folds, a broad nose, protruding tongue, mental retardation, and increased risk of a heart defect, suppressed immunity, and leukemia.

7. Basketball players may have an extra Y chromosome that makes them tall.

8. Triploids are very severely abnormal. Trisomy 21 is the least severe trisomy, and involves the smallest chromosome. Klinefelter syndrome symptoms are worse if there is more than one extra X chromosome.

9. A balanced translocation causes duplications or deletions when a gamete contains one translocated chromosome, plus has extra or is missing genes from one of the chromosomes involved in the translocation. A paracentric or pericentric inversion can cause duplications or deletions if a crossover occurs between the inverted chromosome and its homolog. Isochromosomes result from centromere splitting in the wrong plane, duplicating one chromosome arm but deleting the other.

10. Chromosomes would not contort during meiosis because their genes are aligned.

11. **a.** High resolution chromosome banding: Cells in culture are synchronized in early mitosis, then the chromosomes are spread and stained.
   **b.** FISH: Fluorescently-labeled DNA probes bind homologous regions on chromosomes.
   **c.** Amniocentesis: Fetal cells and fluid are removed from around a fetus. Cells are cultured and their chromosomes stained or exposed to DNA probes, and karyotyped.
   **d.** Chorionic villus sampling: Chromosomes in chorionic villus cells are directly karyotyped.
   **e.** Fetal cell sorting: A fluorescence activated cell sorter separates fetal from maternal cells, and fetal chromosomes are karyotyped.
   **f.** Maternal serum markers: Abnormal levels of proteins such as AFP and hCG in a pregnant woman's blood indicate increased risk of certain disorders.
   **g.** Quantitative PCR amplifies sequences that are unique to specific chromosomes that are most often part of trisomies.
   **h.** Human artificial chromosomes are created by synthesizing and attaching the minimal requirements of chromosome components.

12. Trisomies 13 and 18 are highly represented in spontaneous abortions, therefore they are often lethal before birth. Chromosomes 5 and 16 are less often part of trisomies because they have high gene densities.

13. 48

14. A female cannot have Klinefelter syndrome because she does not have a Y chromosome, and a male cannot have Turner syndrome because he has a Y chromosome.

15. Nondisjunction in oocyte. Nondisjunction in sperm. Large deletion in X chromosome.

## Answers to Applied Questions

1. **a.** 35
   **b.** Karen Martini
   **c.** The risks in the right hand column are higher because they include many conditions.
   **d.** One in 66 is greater than 1 percent
   **e.** Age 40 trisomy 21 risk = 1/106. Age 45 risk = 1/30. Her risk approximately triples in the five years.

2. The person is a girl missing part of the short arm of chromosome 5. This is cri du chat syndrome, and she will be mentally retarded with a catlike cry.

3. **a.** Uniparental disomy
   **b.** The first condition might arise from an oocyte that has two copies of the long arm of chromosome 14 being fertilized by a sperm that lacks this segment. The situation would be the reverse for the second condition.
   **c.** A deletion mutation could remove the copy of the gene that is expressed.

4. FISH

5. One of the Watkins probably has a balanced translocation, because there is more than one Down syndrome case. The two spontaneous abortions were the result of unbalanced gametes. Their problems are likely to repeat with a predictable and high frequency, because the translocated chromosome is in half of the carrier parent's gametes. In contrast, the Phelps' child with Down syndrome is more likely the result of nondisjunction, which is unlikely to repeat. The Phelps child has trisomy Down syndrome; the Watkins' child may only have a partial extra copy of chromosome 21.

6. **a.** Reciprocal translocation
   **b.** She doesn't have extra or missing genes.
   **c.** She might have a child with translocation Down syndrome.

7. At the second mitotic division, replicated chromosomes failed to separate, yielding one of four cells with an extra two sets of chromosomes.

8. Down syndrome caused by aneuploidy produces an extra chromosome 21 in each cell. In mosaic Down syndrome, the extra chromosome is only in some cells. In translocation Down syndrome, unbalanced gametes lead to an individual with extra chromosome 21 material in each cell.

9. The XXY son could have gotten two X chromosomes from his mother, or an XY-bearing sperm from his father.

10. **a.** A translocation carrier can produce an unbalanced gamete that lacks chromosome 22 material.
   **b.** The microdeletion may be more extensive than the deleted region in the translocation individuals.
   **c.** Translocation family members might be infertile or have offspring with birth defects.

# Chapter 13 When Allele Frequencies Stay Constant

## Answers to Review Questions

1. A gene pool refers to a population.

2. Evolution is not occurring.

3. Knowing the incidence of the homozygous recessive class makes it possible to derive the "*q*" part of the Hardy-Weinberg equation.

4. The possibility of two unrelated Caucasians without a family history of CF having an affected child is $1/4 \times 1/23 \times 1/23 = 1/2{,}116$. If one person knows he or she is a carrier, the risk is $1/4 \times 1/23 = 1/92$.

5. Natural selection does not act on STRs.

6. Population databases are necessary to interpret DNA fingerprints because alleles occur with different frequencies in different populations.

7. For females use the standard formula. For males, gene frequency equals phenotypic frequency.

## Answers to Applied Questions

1. Autosomal recessive class = $900/10{,}000 = .09 = q^2$
   $q = .3 \quad p = .7$

Normal lashes = $p^2 + 2pq = .49 + .42$

Homozygote with normal lashes = $.49/.49 + .42 = .538$

2. $q^2 = .25$ $q = .5$ Carrier percentage = $2pq = (2)(.5)(.5) = .5$

3. a. Genes 2 and 4 have 2 alleles because they generate 2 bands on the gel

   b. $(.2)^2 + (2)(.3)(.7) + (.10)^2 + (2)(.4)(.2) = .04 \times .42 \times .01 \times .16 = .0000268$

   c. The man might have close male relatives who cannot be ruled out as suspects.

   d. It makes his guilt even more likely.

4. $q^2 = .001$ $q = .0316$ $p = .968$

   $2pq$ = carrier frequency = $(2)(.0316)(.968) = .061$

5. a. $q^2 = 1/190 = 0.005$. Square root of $0.005 = 0.071$ = frequency of mutant allele $q$

   b. Frequency of wild type allele = $p = 1 - 0.071 = 0.929$

   c. Carriers = $2pq = 2 \times 0.071 \times 0.929 = 0.132$

   d. Nonrandom mating

6. $0.1 \times 0.1 = 0.01$ = chance both are carriers. If both are carriers, $0.25 \times 0.01 = 0.0025 = 0.25\%$ chance child will be affected.

7. $q^2 = 1/8{,}000 = 0.000125$ $q = 0.011$ $p = 1 - 0.011 = 0.989$

   Carrier frequency = $2pq = 2 \times 0.011 \times 0.989 = 0.022$

8. $4/177 = 0.0225 = 2.25\%$ are carriers

9. Students genotypes: 6 *TT* 4 *Tt* 10 *tt*

   Parents genotypes:

   of 6 *TT* students = 9 *TT* 2 *Tt* 1 *tt*

   of 4 *Tt* students = 2 *TT* 4 *Tt* 2 *tt*

   of 10 *tt* students = 8 *Tt* 12 *tt*

   Students allele frequencies: $T = p = 1/2Tt + TT = 1/2\,(4) + 6 = 8/20$ students $= .4$

   $t = q = .6$

   Parents allele frequencies: $T = p = 1/2Tt + TT = 1/2\,(14) + 11 = 18/40$ parents $= .45$

   $t = q = .55$

   The gene is evolving because the allele frequencies change between generations.

# Chapter 14 Changing Allele Frequencies

## Answers to Review Questions

1. a. Agriculture

   b. Cities, as groups of immigrants arrive and mix in

   c. Endangered species, survivors of massacres or natural disasters and their descendants

   d. Introducing an inherited disease into a population

2. a. Highly virulent TB bacteria are selected against because they kill hosts quickly. Resistant hosts are selected for because they survive infection and live to reproduce. In this way, TB evolved from an acute systemic infection into a chronic lung infection. In the 1980s, antibiotic-resistant TB strains led to re-emergence of the disease.

   b. Bacteria that become resistant to antibiotics by mutation or by acquiring resistance factors from other bacteria selectively survive in the presence of antibiotics.

   c. Viral diversity is low at the start of infection because the immune system is vulnerable—viral variants aren't necessary. As symptoms ebb, viral diversity increases, then it decreases again as the immune system becomes overwhelmed.

   d. CF may be maintained in populations where heterozygosity protects against diarrheal disease.

3. Increasing homozygosity increases the chance that homozygous recessives will arise, who may be too unhealthy to reproduce. The population may decline.

4. Misuse of antibiotics provides a selective pressure that benefits and maintains populations of drug-resistant bacteria.

5. Recessive alleles persist in heterozygotes.

6. Sickle cell disease protects against malaria.

7. The incidence of the phenotype is less than that predicted by the genotype and mode of inheritance.

8. Historical records, geographical information, and linguistics

9. Decreased variation in mitochondrial and Y chromosome DNA sequences indicate the genetic uniformity of a population bottleneck.

10. In the Dunker population, the frequencies of blood type genotypes are quite different from those of the surrounding U.S. population and the ancestral population in Europe.

11. The most common CF allele is more common in France (70 percent) than in Finland (45 percent). Therefore the test would be more likely to detect a carrier in France than in Finland.

12. The same mutation in a population group and linkage disequilibrium (shared haplotypes) suggest a founder effect.

13. The causes of founder effects and population bottlenecks differ. A founder effect reflects a small group moving to start a new population, whereas a population bottleneck results from removal of individuals with certain genotypes from the population.

14. A gradual cline might reflect migration over many years. An abrupt cline could be due to a cataclysmic geological event that separates two populations, such as an earthquake or flood.

15. Genetic drift is the chance sampling of some genotypes from a population, and this may lead to nonrandom mating, in which the most fit individuals reproduce more successfully. Environmental conditions help determine which genotypes reproduce and are thus selected for.

16. a. A founder effect is a type of genetic drift that occurs when a few individuals found a settlement and their alleles constitute a new gene pool, amplifying the alleles they introduce and eliminating others.

    b. Linkage disequilibrium is the inheritance of certain combinations of alleles of different genes at a higher frequency than can be accounted for by their individual frequencies.

    c. Balanced polymorphism is the persistence of a disease-causing allele in a population because the heterozygote enjoys a health or reproductive advantage.

    d. Genetic load is the proportion of deleterious alleles in a population.

17. Shifts in allele frequency should parallel people's movements, which are often described in historical records, and explained by sociology and anthropology.

## Answers to Applied Questions

1. a. All modern Afrikaners with porphyria variegata descend from the same person in whom the disorder originated.

   b. One person, who had the dental disorder, contributed disproportionately to future generations.

   c. Heterozygotes for cystic fibrosis and sickle cell disease resist certain infectious diseases, maintaining the disease-causing allele in populations.

   d. The Pima Indian population has a high incidence of an allele that predisposes to develop type I diabetes mellitus, but it wasn't expressed until certain dietary and lifestyle changes became popular.

   e. The Amish and Pakistani groups have high incidences of certain inherited diseases because of consanguinity.

   f. Migration patterns are responsible for the different frequencies of the galactokinase deficiency allele across Europe.

   g. The same haplotype for CJD in different populations reflects descent from a shared ancestor followed by migration.

   h. Varying mtDNA sequences along the Nile river valley are due to migration.

   i. The alleles responsible for BRCA1 breast cancer originated in different ancestors for Ashkenazim and African-Americans.

2. a. Founder effect

   b. Geographical barriers and natural selection, acting over time, make two populations have different variants of inherited characteristics.

   c. Nonrandom mating

   d. Natural selection

3. Treating PKU has increased the proportion of mutant alleles in the population because without treatment, affected individuals would not have been able to reproduce.

4. The mutations arose independently.

5. The high incidence is due to extreme consanguinity—nearly everyone is related to nearly everyone else. Tracking the incidence of this condition is difficult because the symptoms exist independently, and may be caused by environmental factors rather than genes.

6. Natural selection

7. Balanced polymorphism

8. a. Balanced polymorphism b. Founder effect c. Migration

9. European males settled in India. Later, Asian immigrants arrived. The male European Indians had children with Asian Indian women.

# Chapter 15 Human Origins and Evolution

## Answers to Review Questions

1. Hominoids are ancestral to apes and humans, whereas hominids are ancestral to humans only. Therefore, hominoids are more ancient.

2. Neanderthals

3. Physically, chimpanzees are not as similar to us as were the australopithecines, yet the australopithecines are in a different genus from us.

4. A single gene can control the rates of development of specific structures, causing enormous differences in the relative sizes of organs in two species.

5. The hunting behavior of *A. garhi* and the time and place where it lived suggests that it could have been a bridge between *Australopithecus* and *Homo.*

6. Fossil evidence and mtDNA evidence support an "out of Africa" emergence of modern humans. Anthropological evidence indicates three waves of migration of native Americans from Siberia, but molecular evidence suggests one migration.

7. Y chromosome and mitochondrial DNA sequences enable researchers to trace the genetic contributions of fathers and mothers, respectively.

8. Bipedalism, larger brain, improved fine coordination

9. Gene sequences are more specific because different codons can encode the same amino acids.

10. Mutation rates are not the same across genomes. Genes mutate at different rates.

11. One gene encodes one polypeptide, and so comparing the evolution of a gene tracks a tiny part of the biology of the organism. DNA hybridization assesses relationships among many genes, and thus means more. Also, much of the genome does not encode protein.

12. A chromosome band may contain many genes, and so comparing them is not specific.

13. Knowledge of a DNA, RNA, or polypeptide sequence and the mutation rate is necessary to construct an evolutionary tree diagram. An assumption is that mutation rate is constant. A limitation is that only one biochemical is considered, and not large scale characteristics such as behavior and anatomy.

14. Sterilizing people with mental retardation.

Encouraging poor people to limit family size.

Avoiding marriage to a person who carries a disease-causing allele.

15. European men were among the first settlers, and contributed to the higher castes. Asian women arrived later and contributed to lower castes.

## Answers to Applied Questions

1. a. A small duplication occurred in human chromosome 11 to give rise to the Betazoid karyotype. The Klingon and Romulan karyotypes could have arisen from fusion of human chromosomes 15 and 17.
   b. The Betazoids are our closest relatives because of greater similarity in chromosome bands and chromosome arrangement. Cytochrome C sequences and intron pattern in the collagen gene are identical between humans and Betazoids.
   c. They are not distinct species because they can interbreed.
   d. (3)

2. Negative eugenics:
   —sterilizing people who are mentally retarded
   —restricting certain groups from immigrating
   —encouraging people with inherited disease or carriers not to reproduce

Positive eugenics:
   —a sperm bank where Nobel prizewinners make deposits
   —people seeking very smart people as mates
   —governments paying people with the best jobs and education to have larger families

3. People in third world nations might be alarmed by white-coated strangers wielding hypodermic needles seeking blood samples. A compromise would be to collect hair instead of blood, and use PCR to amplify the genes. Another compromise is to offer vaccines in exchange for tissue samples.

4. The action was eugenic because it indirectly selected against nonwhite students. It was not eugenic in that, superficially, all applicants were evaluated using the same criteria.

5. The child would be a modern human.

6. A law can be passed stating that no laws can be passed or restrictions imposed based upon genetic data.

7. a. *K. platyops* had facial characteristics of more than one type of australopithecine.
   b. Skeletal remains are incomplete.

# Chapter 16 Genetics of Immunity

## Answers to Review Questions

1. 1. d 2. e, g 3. a 4. b 5. h 6. b

2. Blood type is determined by specific glycoprotein antigens on red blood cell surfaces. Incompatibility results when a person's immune system manufactures antibodies that attack red blood cells bearing antigens of other blood types.

3. a. Collapse of entire immune system
   b. Increase in viral infections and cancer
   c. Increase in bacterial infections
   d. Collapse of coordination between innate and acquired immunity
   e. Total collapse of immune system
   f. Infection through the small intestinal lining

4. a. Destruction of bacteria and viruses, stimulation of inflammation
   b. Destruction of bacteria, yeasts, some viruses
   c. Bind to and stimulate destruction of bacteria
   d. Cause changes in body that are inhospitable to pathogens

5. HIV mutates rapidly, replicates rapidly, the site where it binds to T cells is shielded, and resistance alleles are very rare.

6. Thymocytes are selected as others die, rather than undergoing an "education."

7. Allergens stimulate production of IgE antibodies that bind mast cells, causing them to release allergy mediators. In an autoimmune disorder, antibodies attack the body's cells and tissues.

8. a. Transplanted bone marrow cells (the graft) recognize cells in the recipient (the host) as foreign. The donor's cells attack the host's cells.
   b. ADA deficiency may result in severe combined immune deficiency, in which a lack of ADA poisons T cells, which then cannot activate B cells.
   c. Autoantibodies attack cells that line joints.
   d. In AIDS, HIV infects helper T cells and reproduces, eventually killing enough T cells to overcome cellular immunity. Opportunistic infections occur.
   e. Grass antigens (allergens) induce production of IgE that causes mast cells to release histamines, causing allergy symptoms.
   f. IgE bursts mast cells in response to pollen.

9. T cells control B cell function.

10. Memory and plasma B cells respond specifically to one antigen following a cytokine cue from a T cell. Plasma B cells secrete antigen-specific antibodies and are in the circulation for only a few days. Memory B cells remain, providing a fast response the next time the antigen is encountered.

11. A polyclonal antibody response attacks an invader at several points simultaneously, hastening recovery. MAbs are useful as a diagnostic tool because of their specificity.

12. Memory cells alert the immune system of the first exposure, and ensure that a secondary immune response occurs on subsequent exposure to the Coxsackie virus.

## Answers to Applied Questions

1. O negative blood lacks ABO cell surface antigens and the Rh antigens, so it is less likely to evoke a rejection reaction than other types of blood.

2. Autoimmune

3. It is working. Lowered IgE signifies less of an allergic reaction, and increased levels of IgG and IgA indicate protection.

4. It is an autoimmune disorder that occurs when the immune system "mistakes" kidney cells for pathogenic bacterial cells.

5. Antibiotics treat the bacterial infection, but not the inflammation in the joints, which is an autoimmune response.

6. Innate antibodies

A mutation that prevents the pathogen from infecting.

7. **a.** Allograft **b.** Isograft **c.** Xenograft **d.** Autograft

8. There will not be an Rh incompatibility because the female would have to be $Rh^-$.

9. Colony stimulating factor

10. The bonobo's cells would bear antigens specific to a particular human.

11. Stockpile antibiotics locally, but control their use. Better vaccine coverage. Publicize what people should look out for, and how to disinfect themselves or suspicious materials.

# Chapter 17 The Genetics of Cancer

## Answers to Review Questions

1. **a.** An overexpressed transcription factor could function as an oncogene, causing too frequent cell division.
   **b.** Mutations in the *p53* gene lift tumor suppression, allowing too many cell divisions.
   **c.** Mutations in the retinoblastoma gene lift tumor suppression.
   **d.** The *myc* oncogene allows too frequent cell division.
   **e.** DNA repair fixes errors that would otherwise lead to cancer (turning on an oncogene or turning off a tumor suppressor).
   **f.** Mutations in the *APC* gene lift tumor suppression by making the DNA more prone to replication errors.

2. Cancer is a consequence of disruption of the cell cycle. Deletion of a tumor suppressor gene and translocation of an oncogene next to a highly active gene would have the same effect of uncontrolled division.

3. Only an inherited cancer susceptibility can pass to future generations.

4. Cancer cells divide continuously and indefinitely; they are heritable, transplantable, dedifferentiated, and lack contact inhibition.

5. One inherits a susceptibility to cancer, not the cancer itself. Cancer may affect only somatic tissue. Many mutations may be necessary for cancer to develop.

6. Cancer cells divide faster than cells from which they derive. Some normal cells divide faster than cancer cells.

7. **a.** A population study examines disease incidence in different populations, at any time.
   **b.** A case-control study analyzes pairs of people who differ only in the characteristic of interest.
   **c.** A prospective study is any study that evaluates results as they occur, rather than relying on recall.

8. A cancer cell could have a point mutation.

9. Inducing apoptosis halts runaway cell division. Inducing differentiation enables cancer cells to specialize, which would slow down their division rate. Blocking hormone receptors prevents cancer cells from receiving signals to divide. Inhibiting angiogenesis prevents tumors from building a blood supply. Blocking telomerase stops cell division.

## Answers to Applied Questions

1. Retinoblastoma

2. Before a p53 test can be developed, we need to identify other factors, including effects of other genes and the environment, that contribute to developing cancer.

3. **a.** Missing both *p53* alleles is lethal.
   **b.** A person with two mutant *p53* alleles could be conceived if both parents have one mutant allele.

4. It is an oncogene because it caused overexpression of a transcription factor.

5. The cells from the original tumor are not genetically identical.

6. Treatment for leukemia is more complex than removal of a solid organ because bone marrow must be replaced.

7. **a.** The counselor should ask about ethnic background, and other types of cancer in the family.
   **b.** This woman would probably not benefit from a BRCA1 test, because her two affected blood relatives were older when affected.
   **c.** A complication of BRCA1 testing is that if no mutation is found, a person might assume that she cannot develop breast cancer. We do not know all of the ways that this disease can develop, and in fact it may be several different disorders.

8. No diet can guarantee that an individual will not develop cancer.

9. Are certain *p53* mutations more prevalent among exposed workers, compared to people not exposed to PAHs? If so, the PAHs may cause the mutation.

10. The woman could have a familial cancer syndrome.

11. Tumor suppressors. These genes cause cancer when inactivated or removed.

# Chapter 18 Genetically Modified Organisms

## Answers to Review Questions

1. **a.** Biotechnology—Specific uses of altered cells or biochemicals.
   **b.** Recombinant DNA technology—Inserting foreign genes into bacteria or other single cells, which express them.
   **c.** Transgenic technology—A genetic alteration of a gamete or fertilized ovum, perpetuating the change in every cell of the individual that develops.
   **d.** Gene targeting—An introduced gene exchanges places with its counterpart in a chromosome.
   **e.** Homologous recombination—A gene introduced into a cell exchanging places with a gene on a chromosome.

2. **a.** Restriction enzymes cut DNA at specific sequences. They can be used to create DNA fragments for constructing recombinant DNA molecules.
   **b.** In gene targeting, genes of interest are added to embryonic stem cells where they exchange places with a homologous gene. After construction of chimeric embryos and crosses of mosaic animals, heterozygotes are bred to yield homozygous transgenic animals with a particular gene knocked out.
   **c.** Cloning vectors carry DNA molecules into cells.

3. Antibiotics are used to set up a system where only cells that have taken up foreign DNA can survive.

4. Human insulin DNA cut with a restriction enzyme, vector DNA cut with the same restriction enzyme, *E. coli*, DNA ligase, selection mechanism (such as antibiotic).

5. Bacteria couldn't manufacture human proteins if the genetic code was not universal.

6. For all cells of a transgenic animal to express the transgene, it must be introduced at the single-cell stage so that it is present in every cell.

7. Foreign DNA can be inserted on a virus, carried across cell membranes in liposomes, microinjected, electroporated, or sent in with particle bombardment.

8. Transgenic technology is not precise because introduced DNA is not directed to a particular chromosomal locus, as it is in gene targeting.

9. When a transgene or knocked out gene is present in one copy in animals, the animals must be bred to obtain homozygous individuals that express the phenotype. This is a monohybrid cross.

## Answers to Applied Questions

1. Ecologists study whether pollinators can transfer the transgene to other species—and what the effects are.

2. A goat produces human EPO in its semen.
   A mouse produces jellyfish GFP in its plasma.
   A chicken produces human clotting factor in its egg whites.

3. Human collagen produced in transgenic mice is less likely to include infectious agents than collagen obtained from hooves and hides. It is also the human protein, which is less likely to stimulate an immune response than the cow type.

4. Cloning eliminates the need to cross heterozygotes.

5. Does the transgene give the plant some other function that affects the ecosystem?

6. Mice can express human genes because they use the same genetic code.

7. Another gene specifies the same enzyme; another gene specifies a different enzyme with the same or a similar function; the enzyme isn't vital.

8. Gene targeting

9. Gene targeting shows the result of no CFTR protein. A transgenic model shows effects of an abnormal CFTR protein. Gene targeting models the most extreme expression of CF.

# Chapter 19 Gene Therapy and Genetic Counseling

## Answers to Review Questions

**1.** (a) Replace protein; (b) use recombinant DNA technology to obtain pure, human protein; (c) gene therapy.

**2.** **a.** ADA gene in cord blood cells.
**b.** OTC gene delivered to liver.
**c.** Implant gene in brain.
**d.** Activate fetal hemoglobin production.

**3.** Fitting the huge dystrophin gene into a vector, and getting gene expression in enough muscle cells to affect symptoms.

**4.** *Ex vivo* gene therapy alters cells outside the body, then injects or implants them. SCID is treated this way. *In situ* gene therapy is a localized procedure on accessible tissue, such as treating melanoma. *In vivo* gene therapy occurs inside the body, such as a nasal spray to deliver the CFTR gene to a person with cystic fibrosis.

**5.** Germline gene therapy can affect evolution because the changes are heritable.

**6.** Researchers should consider the amount of DNA the virus can carry, the types of cells the virus normally infects, how stable the incorporated vector is in the human genome, if toxic effects are associated with use of the virus, and whether or not the virus stimulates a strong immune response.

**7.** Gentamycin overrides nonsense mutations. Sickle cell disease results from a missense mutation.

**8.** A bone marrow transplant is a cell implant; genes are not altered. Removing bone marrow and adding a functional gene is gene therapy because the DNA is altered.

**9.** A liver is in only one place in the body. Muscle is more difficult to treat because it comprises much of the body's bulk. It is nearly everywhere.

**10.** AV and AAV are viral vectors, and the dystrophin gene is stitched into them using recombinant DNA technology. A liposome is a fatty bubble that can encase the gene and introduce it into a cell by traversing the lipid-rich cell membrane. Chimeraplasty swaps in a functional gene to replace the abnormal one, using DNA repair mechanisms.

## Answers to Applied Questions

**1.** A reason that some people might cite for offering gene therapy for emphysema preferentially to those who inherit it is that they did not behave in a way that caused their disease, whereas those with smoking-related emphysema did.

**2.** HIV would have to have the genes that make it destroy the immune system deleted.

**3.** Small amounts of nitric oxide could be used to prevent sickling. The cells sickle only in a low-oxygen environment.

**4.** There is no dopamine gene to manipulate—a gene therapy would be applied to an enzyme necessary to synthesize dopamine. It is difficult to deliver a gene therapy to the brain, because of the blood-brain barrier.

**5.** To treat Duchenne muscular dystrophy, the gene for dystrophin is delivered via a retrovirus into immature muscle cells.

To treat sickle cell disease, the beta globin gene is delivered as naked DNA into immature red blood cells.

To treat glioma, mouse fibroblasts are given retroviruses with a herpes gene that encodes thymidine kinase. The fibroblasts are implanted into the tumor, and an anti-herpes drug given, which selectively destroys dividing cells, including the cancer cells.

**6.** **a.** The counselor should explain to the couple that even though the fetus has an extra X chromosome, the individual will most likely not have any related symptoms other than perhaps great height.
**b.** The counselor might suggest a technique to separate out X-bearing sperm, to increase the chances of conceiving a female. Or, the couple could test the fetus for the muscular dystrophy gene and elect to terminate the pregnancy.
**c.** Explain that risk increases with age, but a young woman can still conceive a child with trisomy 21—it is just rare.
**d.** If the parents are of normal height, then their child with achondroplasia is a new mutation, and there should be no elevated risk to other children.
**e.** Genetic counselors always tell patients that amniocentesis can rule out certain chromosomal and biochemical disorders, but it cannot guarantee a healthy child.

**7.** *In situ,* because only the scalp need be treated.

**8.** (a) A viral vector (AAV) to add globin genes to red blood cell precursor cells, (b) a bone marrow transplant, (c) hydroxyurea to stimulate production of fetal hemoglobin.

**9.** Jesse Gelsinger was relatively healthy. Children with Canavan disease have no other treatment options.

# Chapter 20 Agricultural Biotechnology

## Answers to Review Questions

**1.** **a.** Antisense technology—Silence a gene by blocking it with a complementary RNA.
**b.** Transgenesis—Introduce a foreign gene into a gamete or fertilized ovum and regenerate an individual that bears the transgene in every cell. Traditional breeding can be used to obtain homozygotes.
**c.** Rhizosecretion—Genetically alter root cells to secrete a substance into liquid.
**d.** Mutant selection—Add a toxic chemical and breed a stock from the naturally resistant survivors.
**e.** Bioremediation—Identify a gene that dismantles a toxin, and deliver it to an organism that contacts the toxin.
**f.** Expression of genes from chloroplasts.

**2.** Identify an interesting trait. In breeding, new varieties are introduced by mating plants. The offspring are not genetically uniform. A biotechnological approach would begin with somatic cells, into which the gene of interest is introduced. The resulting somatic cells are clones, unless they undergo somaclonal variation.

**3.** **a.** Biotechnology for plants does not use sexual reproduction.
**b.** It can be difficult to get expression of the gene of interest in a part of the plant that is useful. A GM plant must be able to survive in the field.

**4.** Sexually derived plants have different combinations of parental genes. Plants derived from somatic tissue carry identical genes and are clones.

**5.** No

**6.** Plants are easier to manipulate because they can be regenerated from somatic tissue. They are harder to manipulate because cell walls protect them.

**7.** **a.** A new agricultural variant from biotechnology can replace a native species in the marketplace, adding new competition.
**b.** New variants from biotechnology can invade niches, depleting native populations.
**c.** Artificial selection (breeding) can change gene frequencies by setting up matings to perpetuate certain traits or select against others.

**8.** A somatic embryo is grown from a callus rather than from sexual tissues.

**9.** Nearby weeds could also become herbicide resistant.

**10.** If the genetic alteration offers a selective advantage to the altered organism, it will replace natural species in the environment.

**11.** Genetic alteration of trees is challenging because of the long generation time and the difficulty of regenerating plants from altered cells.

## Answers to Applied Questions

**1.** Concern still exists over GM crops because not enough time has elapsed to assess effects on ecosystems, and delayed health effects, such as cancer.

**2.** Spray many tomato plants with the herbicide and breed stock from those that are naturally resistant (mutant selection). Create a transgenic tomato that has a foreign gene that enables it to resist the herbicide. Select a resistant somaclonal variant.

**3.** **a.** Popping corn transgenic for the jellyfish green fluorescent protein.
**b.** Potato transgenic for genes enabling it to produce large amounts of beta-carotene.
**c.** Potato with a healthier balance of amino acids, fried in canola oil with lower saturated fat content.
**d.** Protoplast fusion of zucchini and eggplant.
**e.** Grass transgenic for gene that resists herbicide that kills dandelions.
**f.** Antisense technology to silence soybean antigens that trigger allergic response.

**4.** Labeling of GM foods can alert consumers to allergens that have been introduced into plants in which they do not normally occur. However, labeling implies that GM foods are hazardous—and so far there is no scientific evidence that this is so.

**5.** The FDA's policy evaluates whether or not a food is a health hazard—not its origin. It makes sense to many scientists.

**6.** A crop variant that is the result of antisense technology lacks a function, which would probably make it less of a threat than a crop that has been given a new function.

**7.** Transgenic ornamental plants could be cloned.

**8.** Public acceptance or rejection of the same basic technology can be opposite, depending upon the context and media spin. Humulin was regarded as supplying a missing natural biochemical to people with diabetes; yet the same technology used to create new plant variants raises great concern.

# Chapter 21 Reproductive Technologies

## Answers to Review Questions

**1.** **a.** Surrogate mother
- **b.** Artificial insemination
- **c.** Oocyte donation, preimplantation genetic diagnosis
- **d.** Artificial insemination
- **e.** Preimplantation genetic diagnosis or artificial insemination
- **f.** IVF, ZIFT, or GIFT
- **g.** Preservation of her ovarian tissue in her arm.

**2.** It is easier, less costly, and less painful to detect infertility in men than in women.

**3.** A man can have up to 40 percent abnormally shaped sperm and still be considered fertile.

**4.** A man with a low sperm count and a woman with an irregular menstrual cycle.

**5.** ZIFT and GIFT occur in the fallopian tube, whereas IVF takes place in the uterus. ZIFT and IVF transfer a zygote, whereas GIFT transfers gametes. In IVF, fertilization occurs outside the body.

**6.** They are "embryos *in vitro*" because a uterus is required for the embryos to develop—and there is no guarantee that this will happen.

**7.** Preimplantation genetic diagnosis is similar to amniocentesis and CVS in that it allows prenatal detection of disease-causing genes. It is different in that it takes place much earlier in gestation.

**8.** Endometriosis, scarred fallopian tubes, irregular ovulation, nondisjunction

**9.** **a.** Fertilization occurs outside of the body.
- **b.** Oocytes and sperm are collected and placed in the fallopian tubes.
- **c.** Conception occurs in a woman other than the one who gives birth.
- **d.** Conception occurs outside the body, and a woman other than the genetic mother carries the fetus.
- **e.** Conception does not occur as a result of sexual intercourse, but in a Petri dish.
- **f.** The nucleus and cytoplasm in a cell come from different individuals.

## Answers to Applied Questions

**1.** An older man fathering a child does not have to alter his physiology the way a postmenopausal woman must to conceive.

**2.** Adoption

**3.** Oocyte donor

**4.** Big Tom illustrates artificial insemination. Mist illustrates surrogate motherhood.

**5.** **a.** The genetic parents are the sperm donor and the woman, who is also the gestational mother.
- **b.** The genetic parents are the woman whose uterus is gone, and her husband. The gestational mother is the woman's friend.
- **c.** The genetic parents are Max and Tina; the gestational mother is Karen.
- **d.** The genetic parents are von Wormer and the Indiana woman, who was also the gestational mother.
- **e.** The genetic and gestational mother is the woman who is the friend of the men. The genetic father can be determined if DNA fingerprints of the child are compared to those of the sperm donors.

**6.** Younger women can freeze oocytes or early embryos, to be fertilized or implanted years later.

**7.** Extra preimplantation embryos can be donated to infertile couples.

**8.** People will vary in when they think children born from assisted reproductive technologies should be told of their origins.

**9.** Paying for reproductive services because one is lazy is not the same as seeking assistance because one has a fertility problem.

**10.** The child would be a clone of Cliff.

# Chapter 22 The Human Genome Project and Genomics

## Answers to Review Questions

**1.** **a.** Expressed sequence tags sped the discovery of protein-encoding genes by working with cDNAs reverse transcribed from mRNAs.
- **b.** Positional cloning identified many important disease-causing genes.
- **c.** DNA microarray technology enables researchers to analyze gene expression.
- **d.** Automated DNA sequencing sped determination of the base sequence of the genome.
- **e.** Assembler computer programs overlapped genome pieces.

**2.** A cytogenetic map provides only a few landmarks per chromosome, whereas a sequence map consists of the DNA bases.

**3.** No

**4.** Repeated sequences could be mapped to several chromosomal sites. Not knowing which strand a sequence comes from could lead to duplicating some parts of the genome sequence and missing others.

**5.** *Mycoplasma genitalium,* because it is the smallest genome sequenced.

**6.** We share a great deal, genetically speaking, with yeast and worms.

## Answers to Applied Questions

**1.** The pieces must be overlapped to derive the sequence, so several genome copies must be used.

**2.** SNPs differ among individuals.

**3.** DNA microarrays detect expression of selected genes. Many genes had been well-studied before the draft human genome sequence was unveiled.

**4.** Positional cloning begins with families and a disease. Scanning GenBank begins with looking at DNA sequences.

**5.** DNA microarrays could be used to identify a particular gene variant that is expressed in the affected tissues of affected individuals, but not in other tissues.

**6.** Bacterium = 1,743/1.8 million = .000968

Yeast = 6,000/12.1 million = .000496

Roundworm = 15,000/100 million = .00015

Human = 70,000/3 billion = .0023

The greatest density of protein-encoding genes is in the human genome.

(This comparison may not be valid, however, because of differences in genome organization and the sizes and structures of individual genes, such as introns.)

**7.** G C T T C G T T A A T A T C G C T A G C T G C A

**8.** The first gene is more ancient than the second.

# Glossary

## A

**A site** Part of ribosome that holds incoming amino acid in growing peptide chain.
**absolute risk** The probability that an individual will develop a particular condition, based on family history and/or test results.
**acrocentric chromosome** A chromosome in which the centromere is located close to one end, pinching off a very short piece.
**acrosome** A protrusion at the front end of a sperm cell containing enzymes that help cut through the oocyte's membrane.
**adaptive immunity** A slower, more specific immune response that develops after exposure to a foreign antigen.
**adenine** One of two purine nitrogenous bases in DNA and RNA.
**adhesion receptor proteins** Cellular adhesion molecules that extend from capillary walls.
**allantois** A membrane surrounding the fetus that gives rise to umbilical blood vessels.
**allele** An alternate form of a gene.
**allergen** A substance that elicits an allergic response.
**allograft** A transplant in which donor and recipient are members of the same species.
**alpha satellite** A repeated 171-base sequence that is an essential part of a centromere.
**amino acid** A small organic molecule that is a protein building block. Contiguous triplets of DNA nucleotide bases encode the 20 types of amino acids that polymerize to form biological proteins.
**amniocentesis** A prenatal diagnostic procedure in which a physician inserts a needle into the uterus to remove a small sample of amniotic fluid, which contains fetal cells and biochemicals. A chromosome chart is constructed from cultured fetal cells, and tests for certain inborn errors of metabolism are conducted on fetal biochemicals.
**amniotic cavity** A space between the inner cell mass and the outer cells anchored to the uterine lining.
**anaphase** The stage of mitosis when the centromeres of replicated chromosomes part.
**aneuploid** A cell with one or more extra or missing chromosomes.
**angiogenesis** Growth of new blood vessels.
**antibody** A multisubunit protein, produced by B cells, that binds a specific foreign antigen at one end, alerting other components of the immune system or directly destroying the antigen.
**anticodon** A three-base sequence on one loop of a transfer RNA molecule that is complementary to an mRNA codon, and therefore brings together the appropriate amino acid and its mRNA instructions.
**antigen** A molecule that elicits an immune response.
**antigen binding site** The region of an antibody molecule that includes the idiotype, where foreign antigens bind.
**antigen-presenting cell** A cell displaying a foreign antigen.
**antigen processing** A macrophage's display of a foreign antigen on its surface, next to an HLA self antigen. This alerts the immune system.
**anti-Müllerian hormone** A hormone that sustentacular cells in the fetal testes secrete, preventing female reproductive structures from developing.
**antiparallel** The head-to-tail arrangement of the two entwined chains of the DNA double helix.
**antisense technology** Using a piece of RNA that is complementary in sequence to a sense RNA to stop expression of a particular gene.
**apoptosis** A form of cell death that is a normal part of growth and development.
**artificial insemination** Placing a donor's sperm into a woman's reproductive tract. This is done in a medical setting to assist a couple to conceive when the man is infertile.
**assisted reproductive technologies** Procedures that replace a gamete or the uterus to help people with fertility problems have children.
**association study** A case-control study in which genetic variation, often measured as SNPs that form haplotypes, is compared between people with a particular condition and unaffected individuals.
**autoantibodies** Antibodies that attack the body's own cells.
**autograft** A transplant of tissue from one part of a person's body to another.
**autoimmunity** An immune attack against one's own body.
**autosomal dominant** The inheritance pattern of a dominant allele on an autosome. The phenotype can affect males and females and does not skip generations.
**autosomal recessive** The inheritance pattern of a recessive allele on an autosome. The phenotype can affect males and females and can skip generations.
**autosome** A non-sex-determining chromosome. A human has 22 pairs of autosomes.

## B

**bacteriophage** A virus that infects bacterial cells. Bacteriophages are used as vectors to introduce foreign DNA into cells.
**balanced polymorphism** Maintenance of a harmful recessive allele in a population because the heterozygote has a survival or reproductive advantage.
**Barr body** A dark-staining, inactivated X chromosome in a cell.
**base excision repair** Removal of five or fewer bases to correct damage due to reactive oxygen species.
**B cell** A type of lymphocyte that secretes antibody proteins in response to nonself antigens displayed on other immune system cells.
**bioethics** A field of study that analyzes personal issues that arise from the application of biological information.
**bioremediation** Use of an organism's natural or engineered metabolic abilities to remove toxins from the environment.
**biotechnology** The alteration of cells or biochemicals with a specific application, including monoclonal antibody technology, recombinant DNA technology, transgenic technology, and knockout and knockin technologies.
**bipedalism** Walking upright.
**bipolar disorder** A mood disorder in which long periods of depression alternate with periods of mania.
**blastocyst** A hollow ball of cells descended from a fertilized ovum.
**blastomere** A cell in a blastocyst.
**bulbourethral glands** Glands joined to the male urethra that contribute mucus to the seminal fluid, easing sperm release.

## C

**callus** A lump of undifferentiated plant somatic tissue growing in culture.
**cancer** A group of disorders resulting from loss of cell cycle control.
**capacitation** Activation of sperm in a woman's body.
**carbohydrate** A type of macromolecule; sugars and starches.
**carcinogen** A substance that induces cancerous changes in a cell.
**case-control study** An epidemiological method in which people with a particular condition are considered when paired with individuals as much like them as possible, but without the disease.

**cDNA library** A collection of cDNAs that represent the mRNAs in a particular cell type and therefore define gene expression in that cell type.
**cell** The fundamental unit of life.
**cell cycle** A cycle of events describing a cell's preparation for division and division itself.
**cell membrane** A structure consisting of a lipid bilayer with embedded proteins, glycoproteins, and glycolipids that forms a selective barrier around a cell.
**cellular adhesion molecules (CAMs)** Proteins that carry out cell-cell interactions by enabling cells to physically contact each other.
**cellular immune response** T cells release cytokines to stimulate and coordinate an immune response.
**centrioles** Structures consisting of microtubules oriented at right angles to each other near the nucleus that begin to form the spindle during mitosis.
**centromere** The largest constriction in a chromosome, located at a specific site in each chromosome type.
**centromere-associated proteins** Various types of proteins that appear at the centromeres in interphase, such as the cohesins, or during mitosis, such as those that form the kinetochore.
**cervix** The opening between the vagina and the uterus.
**checkpoint** A part of the cell cycle where a protein functions to control the process.
**chimeraplasty** A biotechnology that uses DNA repair to correct a mutation.
**chorionic villi** Fingerlike growths that extend from an embryo where it implants in the uterine wall.
**chorionic villus sampling** A prenatal diagnosis technique that analyzes chromosomes in chorionic villus cells, which, like the fetus, descend from the fertilized ovum.
**chromatid** A single, very long DNA molecule and its associated proteins forming half of a replicated chromosome.
**chromatin** DNA and its associated histone proteins.
**chromosomal mosaic** An individual in whom some cells have a particular chromosomal anomaly, and others do not.
**chromosome** A structure within a cell's nucleus that carries genes. A chromosome consists of a continuous molecule of DNA and proteins wrapped around it.
**cleavage** A series of rapid mitotic cell divisions after fertilization.
**clines** Allele frequencies that change from one area to another.
**cloning vector** A piece of DNA used to transfer DNA from a cell of one organism into that of another.
**coding strand** The side of the double helix for a particular gene from which RNA is not transcribed.
**codominant** A heterozygote in which both alleles are fully expressed.
**codon** A continuous triplet of mRNA that specifies a particular amino acid.
**collectins** Proteins that protect against bacteria, yeasts, and some viruses.
**colony stimulating factors** A class of cytokines that stimulate bone marrow to produce lymphocytes.
**complement** A set of biochemicals that destroy microbes and attack transplanted tissue.
**complementary base pairs** The pairs of DNA bases that bond together; adenine hydrogen bonds to thymine and guanine to cytosine in the DNA double helix.
**complementary DNA (cDNA)** A DNA molecule that is the complement of an mRNA, copied using reverse transcriptase.
**completely penetrant** A disorder or trait in which every individual inheriting the genotype displays symptoms or characteristics.
**concordance** A measure indicating the degree to which a trait is inherited, calculated by determining the percentage of twin pairs in which both members express a particular trait. High concordance among monozygotic (identical) twins indicates a considerable genetic component.
**conditional mutation** A genotype that is expressed only under certain environmental conditions.
**conformation** The three-dimensional shape of a molecule.
**consanguinity** "Same blood"; term for blood relatives having children together.
**constant regions** The lower portions of an antibody amino acid chain, which are similar in different species.
**contact inhibition** Tendency of non-cancer cells to cease dividing once they touch each other.
**correlation coefficient** Comparison of the actual incidence of a trait among related individuals and the expected incidence. Used to calculate heritability.
**critical period** The time during prenatal development when a structure is sensitive to damage from an abnormal gene or an environmental intervention.
**crossing over** An event during prophase I when homologs exchange parts, adding to genetic variability.
**cytogenetics** A discipline that matches phenotypes to detectable chromosomal abnormalities.
**cytokine** A biochemical that a T cell secretes that controls immune function.
**cytokinesis** Division of cellular parts other than DNA at the end of mitosis.
**cytoplasm** Cellular contents other than organelles.
**cytoplasmic donation** Adding cytoplasm from a young woman's oocyte to an older woman's oocyte which is then fertilized *in vitro.*
**cytosine** One of the two pyrimidine nitrogenous bases in DNA and RNA.
**cytoskeleton** A framework composed of protein tubules and rods that supports the cell and gives it a distinctive form.
**cytotoxic T cells** Lymphocytes that attack nonself cells by binding them and releasing chemicals that attack the cell.

## D

**dedifferentiated** A cell less specialized than the cell it descends from, such as a cancer cell.
**degenerate codons** Different codons specifying the same amino acid.
**deletion mutation** A missing sequence of DNA or part of a chromosome.
**density shift experiment** Experiment in which bacterial cultures labeled with radioactive isotopes are centrifuged to separate those that have incorporated the "heavier" isotopes into their DNA. This allows researchers to study DNA replication.
**deoxyribonucleic acid (DNA)** The genetic material. The biochemical that forms genes.
**deoxyribose** The 5-carbon sugar in a DNA nucleotide.
**dicentric** A chromosome that is abnormal because it has two centromeres.
**differentiation** The process by which cells develop distinctive characteristics, reflecting the expression of particular subsets of genes.
**dihybrid cross** A cross of individuals who are heterozygous for two traits.
**diploid cell** A cell containing two sets of chromosomes.
**dizygotic (DZ) twins** Twins that originate as two different fertilized ova and that are thus not identical. (Commonly known as fraternal twins.)
**DNA** *See* **deoxyribonucleic acid.**
**DNA microarray** Also called a DNA chip. A set of target genes embedded in a glass chip, to which labeled cDNAs from a sample bind and fluoresce. Microarrays are used to catalog gene expression.
**DNA polymerase** An enzyme that participates in DNA replication by inserting new DNA bases and correcting mismatched base pairs.
**DNA probe** A labeled short sequence of DNA that corresponds to a specific gene. When applied to a biological sample, the probe base pairs with its complementary sequence, and the label reveals its locus.
**DNA replication** Construction of a new DNA double helix using the information in parental strands as a template.
**domain** A taxonomic designation above kingdom.
**dominant** A gene variant expressed when present in even one copy.
**duplication** An extra copy of a gene or DNA sequence, usually caused by misaligned pairing in meiosis; a chromosome containing repeats of part of its genetic material.

**dynein** A protein necessary for microtubule movement.

## E

**ectoderm** The outermost primary germ layer.
**ectopic pregnancy** An embryo that grows outside of the uterus, usually in a fallopian tube.
**electroporation** Using a brief jolt of electricity to open transient holes in a cell membrane, allowing foreign DNA to enter.
**elongation** The stage of protein synthesis in which ribosomes bind to the initiation complex and amino acids join.
**embryo** A prenatal human until the end of the eighth week of development. The cells in an embryo can be distinguished from each other, but all basic structures are not yet present.
**embryonic adoption** A woman carries an embryo conceived with her partner's sperm and a donor oocyte.
**embryonic stem (ES) cell** A cell from a preimplantation embryo that is manipulated in gene targeting, and may be useful in regenerative medicine.
**empiric risk** Probability that a trait will recur based upon its incidence in a particular population.
**endoderm** The innermost primary germ layer.
**endometriosis** Abnormal buildup of uterine tissue in and on the uterus, causing cramps and impairing fertility.
**endoplasmic reticulum (ER)** A labyrinth of membranous tubules on which proteins, lipids, and sugars are synthesized.
**enzyme** A type of protein that speeds the rate of a specific biochemical reaction so that it is fast enough to be compatible with life.
**epididymis** A tightly coiled tube in the male reproductive tract where sperm cells mature and are stored.
**epigenetic** A layer of information placed on a gene that may be a chemical modification other than a change in DNA sequence.
**epistasis** One gene masking expression of another.
**epitope** Parts of an antibody that antigen binds.
**equational division** The second meiotic division, producing four cells from two.
**euchromatin** Parts of chromosomes that do not stain. It contains active genes.
**eugenics** The control of individual reproductive choices to achieve a societal goal.
**eukaryotic cell** A complex cell containing organelles, including a nucleus.
**euploid (cell)** A somatic cell with the normal number of chromosomes for that species. The human euploid chromosome number is 23 pairs.
**excision repair** Enzyme-catalyzed removal of pyrimidine dimers in DNA, which corrects errors in DNA replication.
**exon** The DNA base sequences of a gene that encode amino acids. Exons are interspersed with noncoding regions called introns.
**exon shuffling** Combining exons from different genes.
**expanding triplet repeat** A type of mutation in which a gene grows with each generation.
**explant** A small piece of plant tissue used to start a culture.
**expressed sequence tags (ESTs)** Short pieces of cDNAs that genome researchers use to locate and isolate protein-encoding genes.
***ex vivo* gene therapy** Genetic alteration of cells removed from a patient, then reinfused or implanted back into the patient.

## F

**fallopian tubes** Tubes leading from the ovaries to the uterus.
**fertilized ovum** An oocyte that a sperm has penetrated.
**fibroids** Noncancerous tumors in the uterus.
**first messenger** A hormone or growth factor.
**fluorescence *in situ* hybridization (FISH)** A technique that binds a DNA probe and an attached fluorescent molecule to its complementary sequence on a chromosome.
**follicle cells** Nourishing cells surrounding a developing oocyte.
**foreign antigen** A molecule that stimulates an immune system response.
**founder effect** A type of genetic drift in human populations in which a few members leave to found a new settlement, perpetuating a subset of the alleles from the original population.
**frameshift mutation** A mutation that alters a gene's reading frame.
**fusion protein** A protein that forms from transcription of two genes as a unit and translation. Can cause cancer.

## G

**gamete** A sex cell.
**gamete intrafallopian transfer (GIFT)** An infertility treatment in which sperm and oocytes are placed in a woman's fallopian tube, assisting fertilization in a natural setting.
**gamma-delta T cell** A type of T cell that controls the interactions of the innate and acquired immune responses.
**gap 1($G_1$) phase** The stage of interphase when proteins, carbohydrates, and lipids are synthesized in preparation for impending mitosis.
**gap 2($G_2$) phase** The stage of interphase when additional proteins are synthesized in preparation for impending mitosis.
**gastrula** A three-layered embryo.
**gene** A sequence of DNA that instructs a cell to produce a particular protein.
**gene expression** Transcription of a gene and its translation into protein.
**gene flow** Movement of alleles between populations.
**gene pool** All the genes in a population.
**gene targeting** A form of genetic engineering in which an introduced gene exchanges places with its counterpart on a host cell's chromosome by homologous recombination.
**gene therapy** Replacing a malfunctioning gene to alleviate symptoms.
**genetic code** The correspondence between specific DNA base sequences and the amino acids they specify.
**genetic counseling** A medical specialty in which a counselor calculates the risk of recurrence of inherited disorders in families, using pedigree charts and applying the laws of inheritance.
**genetic determination** The idea that the expression of an inherited trait cannot be modified by the environment.
**genetic drift** Changes in gene frequencies that occur when small groups of individuals are separated from or leave a larger population.
**genetic heterogeneity** One phenotype that can be caused by any of several genes.
**genetic load** The collection of deleterious recessive alleles in a population.
**genetics** The study of inherited variation.
**genome** All the genetic material in the cells of a particular type of organism.
**genomic imprinting** A process in which the phenotype differs depending upon which parent transmits a particular allele.
**genomic library** A collection of bacteria that harbor pieces of an entire other genome.
**genomics** The study of the functions and interactions of many genes at a time.
**genotype** The allele combinations in an individual that cause a particular trait or disorder.
**genotypic ratio** The ratio of genotype classes expected in the progeny of a particular cross.
**germline gene therapy** Genetic alterations of gametes or fertilized ova, which perpetuate the change throughout the organism and transmit it to future generations.
**germline mutation** A mutation that occurs in every cell in an individual and that was therefore inherited from a parent.
**glycolipid** A molecule that consists of a sugar bonded to a lipid.
**glycoprotein** A molecule that consists of a sugar bonded to a protein. A cell membrane component.
**Golgi apparatus** An organelle, consisting of flattened, membranous sacs, where secretion components are packaged.
**gonads** Paired structures in the reproductive system where sperm or oocytes are manufactured.
**growth factor** A protein that stimulates mitosis.
**guanine** One of the two purine nitrogenous bases in DNA and RNA.

## H

**haploid (cell)** A cell containing one set of chromosomes (half the number of chromosomes of a somatic cell).

**haplotype** A series of known DNA sequences linked on a chromosome.
**Hardy-Weinberg equilibrium** An idealized state in which gene frequencies in a population do not change from generation to generation.
**heavy chain** Either of the two longer amino acid chains of the four that comprise an antibody subunit.
**helicase** A type of enzyme that unwinds and holds apart strands of replicating DNA.
**helper T cells** Lymphocytes that recognize foreign antigens on macrophages, activate B cells and cytotoxic T cells, and secrete cytokines.
**hemizygous** The sex that has half as many X-linked genes as the other sex; a human male.
**heritability** An estimate of the proportion of phenotypic variation in a group due to genes.
**heterochromatin** Dark-staining genetic material that is inactive but that maintains the chromosome's structural integrity.
**heterogametic sex** The sex with two different sex chromosomes; a human male.
**heteroplasmy** The phenomenon of mitochondria within the same cell having different alleles of a particular gene.
**heterozygous** Having two different alleles of a gene.
**highly conserved** Genes or proteins whose sequences are very similar in different species.
**histamine** A biochemical that mast cells release that causes allergy symptoms.
**histone** A type of protein around which DNA entwines.
**hominids** Animals ancestral to humans only, such as australopithecines.
**hominoid** An animal ancestral to apes and humans only.
**homogametic sex** The sex with identical types of sex chromosomes; the human female.
**homologous pairs** Chromosomes with the same gene sequence.
**homologous recombination** A naturally occurring process in which a piece of DNA exchanges places with its counterpart on a chromosome.
**homozygous** Having two identical alleles of a gene.
**hormone** A biochemical secreted in one part of the body that travels in the bloodstream to another part, where it exerts an effect.
**human chorionic gonadotropin (hCG)** The hormone that prevents menstruation and indicates that a woman is pregnant.
**human leukocyte antigen (HLA) complex** Polymorphic genes closely linked on the short arm of chromosome 6 that encode cell surface proteins important in immune system function.
**humoral immune response** Process in which B cells secrete antibodies into the bloodstream.
**hyperacute rejection reaction** The immune system's very rapid destruction of a transplant from another species.

## I

**ideogram** A diagram of a chromosome showing bands and locations of known genes. An ideogram combines cytogenetic and molecular information.
**idiotype** The part of an antibody's antigen binding site that fits around a particular foreign antigen.
**inborn error of metabolism** An inherited disorder resulting from a malfunctioning or absent enzyme.
**incomplete dominance** A heterozygote intermediate in phenotype between either homozygote.
**incompletely penetrant** A disorder or trait in which not every individual inheriting the genotype displays symptoms or characteristics.
**independent assortment** The random arrangement of homologous chromosome pairs, in terms of maternal or paternal origin, down the center of a cell in metaphase I. The consequence is that inheritance of a gene on one chromosome does not influence inheritance of a gene on a different chromosome.
**infertility** The inability to conceive a child after a year of unprotected intercourse.
**inflammation** Part of the innate immune response that causes an infected or injured area to swell with fluid, turn red, and attract phagocytes.
**informatics** Comparison and analysis of DNA sequences.
**initiation complex** Aggregation of the components of the protein synthetic apparatus formed before mRNA is translated.
**initiation site** The site where DNA replication begins on a chromosome.
**innate immunity** Components of immune response that are present at birth, and do not require exposure to an environmental stimulus.
**inner cell mass** A clump of cells on the inside of the blastocyst that will continue developing into an embryo.
**insertion mutation** A mutation that adds DNA bases.
***in situ* gene therapy** Localized gene therapy in an easily accessible body part.
**integrins** Cellular adhesion molecules that bind to white blood cells and certain other cell types.
**interferons** A class of cytokines that fight viral infections and cancers.
**interleukins** A class of cytokines that control lymphocyte differentiation and growth.
**intermediate filament** A type of cytoskeletal component made of different proteins in different cell types.
**interphase** The stage of the cell cycle during which a cell is not dividing.
**interstitial cells** Cells in the fetal testes that secrete testosterone, causing the development of male internal reproductive structures.
**intracytoplasmic sperm injection** Injection of a sperm cell nucleus into an oocyte, to overcome lack of sperm motility.
**introns** Base sequences within a gene that are transcribed but are excised from the mRNA before translation into protein. Introns are interspersed with protein-encoding exons.
**invasiveness** The ability of cancer cells to squeeze into tight places.
***in vitro* fertilization (IVF)** Placing oocytes and sperm in a laboratory dish with appropriate biochemicals so that fertilization occurs, then after a few cell divisions, transferring the embryos to a woman's uterus.
***in vivo* gene therapy** Direct genetic manipulation of cells in the body.
**isochromosome** A chromosome with identical arms, forming when the centromere splits in the wrong plane.
**isograft** A transplant in which the donor and recipient are identical twins.

## K

**karyotype** A chart that displays chromosome pairs in size order.
**kinetochore** A structure built of centromere-associated proteins that extends from the centromere during mitosis and meiosis and contacts spindle fibers.

## L

**law of segregation** The distribution of alleles of a gene into separate gametes during meiosis. This is Mendel's first law.
**lethal allele** An allele that causes early death.
**ligand** A molecule that binds to a receptor.
**ligase** An enzyme that catalyzes the formation of covalent bonds in the sugar phosphate backbone of DNA.
**light chain** Either of the two shorter polypeptide chains of the four that comprise an antibody subunit.
**linkage** The relationship between genes on the same chromosome.
**linkage disequilibrium** Extremely tight linkage between DNA sequences.
**linkage maps** Maps that show how genes are ordered on chromosomes, determined from crossover frequencies between pairs of genes.
**lipoprotein lipase** An enzyme that breaks down fat along the linings of small-diameter blood vessels.
**liposomes** Fatty bubbles that can enclose and transport DNA into cells.
**LOD score** A statistical measurement that indicates whether DNA sequences are usually inherited together due to linkage or chance.
**lymphocytes** Types of white blood cells that provide immunity; they include B cells and T cells.
**lysosomal storage disease** An inborn error of metabolism that results from absence or malfunction of a lysosomal enzyme. A substrate builds up.
**lysosome** A saclike organelle containing enzymes that degrade debris.

## M

**macroevolution** Genetic change sufficient to form a new species.

**macrophage** A large, wandering scavenger cell that alerts the immune system by binding foreign antigens.

**major depressive disorder** A mood disorder characterized by prolonged, inexplicable sadness.

**major histocompatibility complex** A gene cluster, on chromosome 6 in humans, that includes many genes that encode components of the immune system.

**manifesting heterozygote** A female carrier of an X-linked recessive gene who expresses the phenotype because the normal allele is inactivated in some affected tissues.

**map unit** The relative distance between linked genes, determined from crossover frequencies.

**mast cells** Circulating cells that have IgE receptors. They release allergy mediators when IgE binds to receptors on their surfaces. This causes allergy symptoms.

**maternal serum marker test** Measurement of certain hormones in a pregnant woman's bloodstream that provides clues to fetal health.

**meiosis** A type of cell division that halves the usual number of chromosomes to form haploid gametes.

**memory cells** Descendants of activated B cells that participate in a secondary immune response.

**Mendelian trait** A trait that a single gene specifies.

**mesoderm** The middle primary germ layer.

**messenger RNA (mRNA)** A molecule of RNA complementary in sequence to the coding strand of a gene. Messenger RNA carries the information that specifies a particular protein product.

**metacentric chromosome** A chromosome with the centromere located approximately in the center.

**metaphase** The stage of mitosis when chromosomes align along the center of the cell.

**metastasis** Spread of cancer from its site of origin to other parts of the body.

**microevolution** Change of allele frequency in a population.

**microfilament** A solid rod of actin protein that forms part of the cytoskeleton.

**microtubule** A hollow structure built of tubulin protein that forms part of the cytoskeleton.

**mismatch repair** Proofreading of DNA for misalignment of short, repeated segments (microsatellites).

**missense** A single base change mutation that alters an amino acid in the gene product.

**mitochondrion** An organelle consisting of a double membrane that houses enzymes that catalyze reactions that extract energy from nutrients.

**mitosis** Division of somatic (nonsex) cells.

**mode of inheritance** The pattern in which a gene variant passes from generation to generation, determined by whether it is dominant or recessive and is part of an autosome or a sex chromosome.

**molecular clock** A tool for estimating the time elapsed since two species diverged from a shared ancestor, based on DNA or protein sequence differences and mutation rate.

**molecular evolution** Changes in protein and DNA sequences over time. We use this information to estimate how recently species diverged from a common ancestor.

**monoclonal antibody (MAb)** A single antibody type, produced from a B cell fused to a cancer cell (a hybridoma).

**monohybrid cross** A cross of two individuals who are heterozygous for a single trait.

**monosomy** A human cell with 45 (one missing) chromosomes.

**monozygotic (MZ) twins** Twins that originate as a single fertilized ovum and are thus identical.

**morula** The forming body in the prenatal stage preceding the embryo and resembling a mulberry.

**multifactorial trait** A trait or illness determined by several genes and the environment.

**multiregional hypothesis** The theory that *Homo erectus* gave rise to various geographically widespread human populations.

**mutagen** A substance that changes a DNA base.

**mutant** An allele that differs from the normal or most common allele, altering the phenotype.

**mutant selection** Growing cells in the presence of a toxin, such as an herbicide, so that only resistant cells survive.

**mutation** A change in a protein encoding gene that has an effect on the phenotype.

## N

**natural selection** Differential survival and reproduction of individuals with particular phenotypes in particular environments, which may alter allele frequencies in subsequent generations.

**neural tube** An embryonic structure that develops into the brain and spinal cord.

**nitrogenous base** A nitrogen-containing base that is part of a nucleotide.

**nondisjunction** The unequal partitioning of chromosomes into gametes during meiosis.

**nonsense mutation** A point mutation that changes an amino acid coding codon into a stop codon, prematurely terminating synthesis of the encoded protein.

**notochord** An embryonic structure that forms the framework of the skeleton and induces formation of the neural tube.

**nucleic acid** DNA or RNA.

**nucleolus** A structure within the nucleus where ribosomes are assembled from ribosomal RNA and protein.

**nucleosome** A unit of chromatin structure.

**nucleotide** The building block of a nucleic acid, consisting of a phosphate group, a nitrogenous base, and a 5-carbon sugar.

**nucleotide excision repair** Replacement of up to 30 nucleotides to correct damage of several types.

**nucleus** A large, membrane-bounded region of a eukaryotic cell that houses the genetic material.

## O

**oncogene** A dominant gene that promotes cell division. An oncogene normally controls the cell cycle but leads to cancer when overexpressed.

**oocyte** The female gamete (sex cell).

**oogenesis** Oocyte development.

**oogonium** The diploid cell that begins oogenesis.

**organelle** A specialized structure in a eukaryotic cell that carries out a specific function.

**organogenesis** Development of organs from a three-layered embryo.

**orgasm** Pleasurable sensations associated with sperm release or rubbing of the clitoris.

**ovaries** Paired structures in the female reproductive tract where oocytes mature.

## P

**P site** Part of a ribosome that holds a growing amino acid chain.

***p53*** A tumor suppressor gene whose loss of function is implicated in a number of different types of cancer. The functional allele enables a cell to repair damaged DNA or cease dividing.

**paracentric inversion** An inverted chromosome that does not include the centromere.

**parsimony analysis** A statistical method used to identify the most realistic evolutionary tree possible from a given data set.

**particle bombardment** Shooting DNA from a gunlike device into cells.

**PCR** ***See* polymerase chain reaction.**

**pedigree** A chart consisting of symbols for individuals connected by lines that depict genetic relationships and transmission of inherited traits.

**pericentric inversion** An inverted chromosome that includes the centromere.

**peroxisome** An organelle consisting of a double membrane that houses enzymes with various functions.

**phagocytes** Cells that surround a smaller cell or particle and destroy it.

**phenocopy** An environmentally caused trait that occurs in a familial pattern mimicking inheritance.

**phenotype** The expression of a gene in traits or symptoms.

**phenotypic ratio** The ratio of phenotypic classes expected in the progeny of a particular cross.

**phospholipid bilayer** The structural framework of the cell membrane formed by sheets of aligned fatty acid molecules.

**plasma cells** Descendants of activated B cells that produce large amounts of a single antibody type.
**plasmid** A small circle of double-stranded DNA found in some bacteria in addition to their DNA. Plasmids are commonly used as vectors for recombinant DNA technology.
**pleiotropic** A Mendelian disorder with several symptoms, different subsets of which may occur in different individuals.
**point mutation** A single base change in DNA.
**polar body** A product of female meiosis that contains little cytoplasm and does not continue development into an oocyte.
**polar body biopsy** Performing a genetic test on a polar body to infer the genotype of the attached oocyte.
**polygenic traits** Traits determined by more than one gene.
**polymerase chain reaction (PCR)** A technique in which a specific sequence of DNA from a gene of interest is replicated in a test tube to rapidly produce many copies.
**polymorphism** A DNA sequence at a certain chromosomal locus that varies in at least 1 percent of individuals in a population.
**polyploid (cell)** A cell with one or more extra sets of chromosomes.
**population** A group of interbreeding individuals.
**population bottleneck** Decrease in allele diversity resulting from an event that kills many members of a population.
**population genetics** The study of allele frequencies in different groups of individuals.
**population study** Comparison of disease incidence in different groups of people.
**positional cloning** Identifying a gene by beginning with a phenotype within a large family and gradually narrowing down the segment of a particular chromosome that includes the causative gene.
**preimplantation genetic diagnosis** Removing a cell from an 8-celled embryo and testing it for a disease-causing gene or chromosomal imbalance, to decide if the remaining embryo should be implanted in the uterus to continue development.
**primary germ layers** The three basic layers of an embryo.
**primary immune response** The immune system's response to a first encounter with a foreign antigen.
**primary oocyte** A cell in the female that undergoes reduction division.
**primary spermatocyte** A cell in the male that undergoes reduction division.
**primary (1°) structure** The amino acid sequence of a protein.
**primase** The enzyme that builds a short RNA primer at the start of a replicated DNA segment.
**primitive streak** A band along the back of a three-week human embryo that forms an axis other structures develop around and that eventually gives rise to the nervous system.
**prion** "Proteinaceous infectious particle"—a form of an infectious protein that can cause brain degeneration. Can be inherited or acquired.
**progeria** An inherited rapid-aging disorder.
**promoter** A control sequence near the start of a gene.
**pronuclei** Packets of DNA in the fertilized ovum.
**prophase** The first-stage of mitosis or meiosis when chromatin condenses.
**prospective study** A study in which two or more groups are followed into the future.
**protein** A type of macromolecule that is the direct product of genetic information; a chain of amino acids.
**proteomics** Study of the set of proteins that a cell produces.
**proto-oncogene** A gene that normally controls the cell cycle. When overexpressed, it functions as an oncogene, causing cancer.
**protoplast** A plant cell without its cell wall.
**pseudoautosomal region** Genes on the tips of the Y chromosome that have counterparts on the X chromosome.
**pseudogene** A gene that does not encode protein, but whose sequence very closely resembles that of a coding gene.
**Punnett square** A diagram used to follow parental gene contributions to offspring.
**purine** A type of organic molecule with a two-ring structure, including the nitrogenous bases adenine and guanine.
**pyrimidine** A type of organic molecule with a single-ring structure, including the nitrogenous bases cytosine, thymine, and uracil.

## Q

**quantitative trait loci** Genes that determine polygenic traits.
**quaternary (4°) structure** A protein that has more than one polypeptide subunit.

## R

**reading frame** The grouping of DNA base triplets encoding an amino acid sequence.
**receptor** A structure on a cell, usually a protein, that binds a specific molecule.
**recessive** An allele whose expression is masked by another allele.
**reciprocal translocation** A chromosome aberration in which two nonhomologous chromosomes exchange parts, conserving genetic balance but rearranging genes.
**recombinant** A series of alleles on a chromosome that differs from the series of either parent.
**recombinant DNA technology** Transferring genes into bacteria or other single cells from other species.
**reduction division** The first meiotic division, which halves the chromosome number.
**relative risk** Probability that an individual from a population will develop a particular condition in comparison to another group, usually the general population.
**replacement hypothesis** The theory that Africans replaced Eurasian descendants of *Homo erectus* 200,000 years ago.
**replication fork** Locally opened portion of a replicating DNA double helix.
**restriction enzyme** An enzyme, derived from bacteria, that cuts DNA at certain sequences.
**restriction fragment length polymorphism (RFLP)** Differences in restriction enzyme cutting sites among individuals at the same site among the chromosomes, resulting in different patterns of DNA fragment sizes.
**retrovirus** An RNA virus that uses reverse transcriptase to produce DNA from viral RNA. This DNA integrates into the host's genome, where it directs reproduction of the virus.
**reverse transcriptase** An enzyme that synthesizes a DNA molecule from an RNA molecule.
**rhizosecretion** A biotechnology that collects proteins secreted by a plant's roots into a liquid medium.
**ribonucleic acid (RNA)** A nucleic acid whose sequence of building blocks represents a gene's sequence (mRNA), or that assists protein synthesis (tRNA and rRNA).
**ribose** A 5-carbon sugar in RNA.
**ribosomal RNA (rRNA)** RNA that, with proteins, comprises ribosomes.
**ribosome** An organelle consisting of RNA and protein that is a scaffold for protein synthesis.
**ribozyme** RNA component of an RNA-protein complex that has enzymatic function.
**risk factor** A characteristic that is associated with increased likelihood of developing a particular medical condition.
**RNA** ***See* ribonucleic acid.**
**RNA polymerase (RNAP)** An enzyme that adds RNA nucleotides to a growing RNA chain.
**RNA primer** A short sequence of RNA that initiates DNA replication.
**Robertsonian translocation** A chromosome aberration in which two short arms of nonhomologous chromosomes break and the long arms fuse, forming one unusual, large chromosome.

## S

**schizophrenia** Loss of the ability to organize thoughts and perceptions, causing hallucinations and inappropriate behavior.
**secondary immune response** The immune system's response to a second or subsequent encounter with a foreign antigen.
**secondary oocyte** A cell resulting from meiosis I in the female.
**secondary spermatocyte** A cell resulting from meiosis I in the male.

**secondary (2°) structure** Folds in a polypeptide caused by attractions between amino acids close together in the primary structure.

**second messenger** A protein produced at the end of a signal transduction pathway that triggers the cell's response.

**selectins** Cellular adhesion molecules that bind white blood cells and to capillary walls.

**semiconservative replication** The synthesis of new DNA in which half of each double helix comes from a preexisting double helix.

**seminal fluid** Secretions in which sperm travel.

**seminal vesicles** Structures that secrete fructose and prostaglandins into semen.

**seminiferous tubules** A network of tubules in the testes where sperm are manufactured.

**sex chromosome** A chromosome containing genes that specify sex. A human male has one X and one Y chromosome; a female has two X chromosomes.

**sex-influenced trait** Phenotype caused when an allele is recessive in one sex but dominant in the other.

**sex-limited trait** A trait that affects a structure or function present in only one sex.

**signal transduction** A series of biochemical reactions and interactions that pass incoming information from outside a cell to inside, triggering a cellular response.

**single nucleotide polymorphism (SNP)** Single base sites that differ among individuals. A SNP is present in at least 1 percent of a population.

**site-directed mutagenesis** Introduction of a base change into a PCR protocol, so that the change is perpetuated.

**SNP map** A section of a chromosome (haplotype) with single nucleotide polymorphisms indicated.

**somaclonal variant** A plant variant that arises from a somatic mutation in a callus.

**somatic cell** A nonsex cell, with 23 pairs of chromosomes in humans.

**somatic embryo** An embryo that develops from a somatic cell, possible only in plants.

**somatic gene therapy** Genetic alteration of a specific somatic tissue, not transmitted to future generations.

**somatic mutation** A mutation occurring only in a subset of somatic (nonsex) cells.

**spermatid** The product of meiosis II in the male.

**spermatogenesis** Sperm cell development.

**spermatogonium** A diploid cell that gives rise to a cell that undergoes meiosis, developing into a sperm.

**spermatozoon (sperm)** A mature male reproductive cell (meiotic product).

**S phase** The stage of interphase when DNA replicates.

**spindle** A structure composed of microtubules that pulls sets of chromosomes apart in a dividing cell.

**spontaneous mutation** A genetic change that is a result of mispairing when the replication machinery encounters a base in its rare tautomeric form.

***SRY* gene** The sex-determining region of the Y, a gene that controls whether the unspecialized embryonic gonad will develop as testis or ovary. If the gene is activated, it becomes a testis; if not, an ovary forms.

**stem cells** Cells that retain the ability to differentiate (specialize) further.

**submetacentric** A chromosome in which the centromere establishes a long arm and a short arm.

**subtelomeres** The 10,000 or so DNA bases next to the telomeres. They contain short repeats but also protein-encoding genes.

**sustentacular cells** Cells in the fetal testes that secrete anti-Müllerian hormone, which halts development of female reproductive structures.

**synapsis** The gene-by-gene alignment of homologous chromosomes during prophase I of meiosis.

**synteny** The correspondence of genes located on the same chromosome in several species.

## T

**tandem duplication** A duplicated sequence of DNA located right next to the original sequence on a chromosome.

**T cell** A type of lymphocyte that produces cytokines and coordinates the immune response.

**telomerase** An enzyme, including a sequence of RNA, that adds DNA to chromosome tips.

**telomere** A chromosome tip.

**telophase** The stage of mitosis or meiosis when daughter cells separate.

**template strand** The DNA strand carrying the information that is transcribed.

**teratogen** A substance that causes a birth defect.

**tertiary structure** Folds in a polypeptide caused by interactions between amino acids and water. This draws together amino acids that are far apart in the primary structure.

**test cross** Crossing an individual of unknown genotype to an individual who is homozygous recessive for the trait being studied.

**testes** Paired sacs that hang outside the male's body that contain the seminiferous tubules, in which sperm develop.

**testosterone** A hormone that controls the development of male internal reproductive structures in the fetus and of secondary sexual characteristics at puberty.

**thymine** One of the two pyrimidine bases in DNA.

***Ti* plasmid** A virus that causes tumors in certain plants, used as a vector in transgenic technology in plants.

**transcription** Manufacturing RNA from DNA.

**transcription factor** A protein that activates the transcription of other genes.

**transfer RNA (tRNA)** A type of RNA that connects mRNA to amino acids during protein synthesis.

**transgenic organism** An individual with a genetic modification in every cell.

**transition** A point mutation altering a purine to a purine or a pyrimidine to a pyrimidine.

**translation** Assembly of an amino acid chain according to the sequence of base triplets in a molecule of mRNA.

**translocation** Exchange of genetic material between nonhomologous chromosomes.

**translocation carrier** An individual with exchanged chromosomes, but no signs or symptoms. The person has the usual amount of genetic material, but it is rearranged.

**transplantable** The ability of a cancer cell to grow into a tumor if it is transplanted into another individual.

**transposon** A gene or DNA segment that moves to another chromosome.

**transversion** A point mutation altering a purine to a pyrimidine or vice versa.

**trisomy** A human cell with 47 (one extra) chromosomes.

**trophoblast** The outermost cells of the preimplantation embryo.

**tumor necrosis factor** A cytokine that attacks cancer cells.

**tumor suppressor** A recessive gene whose normal function is to limit the number of divisions a cell undergoes.

## U

**uniparental disomy** Inheriting two copies of the same gene from one parent.

**uracil** One of the four types of bases in RNA. A pyrimidine.

**urethra** A tube leading from the bladder to the outside of the body.

**uterus** A saclike organ in a woman's reproductive tract where an embryo and fetus develops.

## V

**vaccine** An inactivated or partial form of a pathogen that alerts the immune system to produce antibodies when the native pathogen is encountered.

**variable regions** The upper parts of antibodies, which differ in amino acid sequence among individuals.

**variably expressive** A genotype producing a phenotype that varies among individuals.

**vas deferens** Paired tubes leading from the epididymis to the urethra that deliver sperm.

**vesicle** A membrane-bounded bubble-like structure in a cell.

**virus** An infectious particle built of nucleic acid in a protein coat.

## W

**wild type** The most common phenotype in a population for a particular gene.

**Wolffian ducts** Unspecialized tissue in an early embryo that can develop into a male reproductive tract.

## X

**xenograft** A transplant in which donor and recipient are different species.

**X inactivation** The inactivation of one X chromosome in each cell of a female mammal, occurring early in embryonic development.

**X inactivation center** A part of the X chromosome that inactivates the chromosome.

**X-linked** Genes on an X chromosome.

**X-Y homologs** Y-linked genes that are similar to genes on the X chromosome.

## Y

**Y-linked** Genes on a Y chromosome.

**yolk sac** A structure external to the embryo that manufactures blood cells and nourishes the developing embryo.

## Z

**zygote** A prenatal human from the fertilized ovum stage until formation of the primordial embryo, at about two weeks.

# Credits

## Line Art

The following have been reprinted from Lewis, et al. *Life,* 4th ed., © 2002. The McGraw-Hill Companies. Reprinted by permission. All Rights Reserved. 1.3*b,* 2.3, 2.4, 2.5, 2.6, 2.7, 2.8, 2.9, 2.10, figure 1, Reading 2.3, 2.11, 2.12, 2.13, 2.15, 2.19, 2.20, 2.21, 2.24, 3.1, 3.2, 3.3, 3.4, 3.5, 3.6, 3.7, 3.8, 3.9, 3.11, 3.12, figure 1, Reading 3.1, 3.13a, 3.14, 3.20, figure 1, Chapter 3 Bioethics Box, 4.2, 4.3, 4.5, 4.6, 4.8, 4.9, 4.10, 5.3, 5.4, 5.5, 5.6, 5.11, 5.12, 5.13, 5.14, 6.6, figure 1, Reading 6.1, 7.9 8.6, 8.7, 9.1, 9.2, 9.3, 9.4, 9.6, 9.7, 9.8, 9.9, 9.10, 9.11, 9.12, 9.13, 9.14, 9.15, 10.1, 10.2, 10.3, 10.4, 10.5, 10.6, 10.7, 10.8, 10.9, 10.11, 10.12, 10.14, 10.15, 10.16, 11.2, 11.4, 11.7, 11.9, figure 1, Reading 11.1, 11.13, 12.2, 12.5, 12.6, 12.18, figure 1, Reading 15.1, 15.7, 15.9 15.11, 16.3, 16.4, 16.5, 16.6, 16.7, 16.8, 16.9, 16.10, 16.11, 16.12, 16.13, 16.14, 16.15, 16.17, 16.20, 16.21, 17.1, 17.4, 18.3, 18.5, 18.9, 18.12, 20.5, 22.4.

### Chapter 1

Figure 1.2 from David Shier, et al., *Hole's Human Anatomy & Physiology,* 8th ed., © 1999 The McGraw-Hill Companies. Reprinted by permission. All Rights Reserved. Figure 1.6*b* reprinted with permission from Seymour Chwast, The Pushpin Group, Inc.

### Chapter 2

Figure 2.1*a* from David Shier, et al., *Hole's Essentials of Human Anatomy & Physiology,* 7th ed., © 2000 The McGraw-Hill Companies. Reprinted by permission. All Rights Reserved. Figure 2.1*b* from Sylvia Mader, *Inquiry Into Life,* 9th ed., © 2000 The McGraw-Hill Companies. Reprinted by permission. All Rights Reserved.

### Chapter 6

Figure 2, Reading 6.1 reproduced from *Ishihara's Tests for Colour Blindness,* published by Kanehara & Co., Ltd, Tokyo, Japan. Figure 6.15 HD Box (Ch. 21) from Leland Hartwell, et al., *Genetics: From Genes to Genomes,* © 2000 The McGraw-Hill Companies. Reprinted by permission. All Rights Reserved.

### Chapter 7

Figure 7.2 Data and print from Gordon Mendenhall, Thomas Mertens, and Jon Hendrix, "Fingerprint Ridge Court," in *The American Biology Teacher,* vol. 51, no. 4, April 1989, pp. 204–6. Figure 7.9 from Robert Plomin, et al., "The genetic basis of complex human behaviors." *Science,* vol. 264, 17 June 1994, pp. 1733–39, copyright 1994 American Association for the Advancement of Science. Figure 7.10 copyright The New Yorker Collection 1981 Charles Addams, from cartoonbank.com. All Rights Reserved.

### Chapter 8

Figure 8.2 from D. Shier, J. Butler, R. Lewis, *Hole's Human Anatomy and Physiology,* 9th ed. © 2002. The McGraw-Hill Companies. Reprinted by permission. All Rights Reserved. Figure 8.4 from *Scientific American,* November 1998, The general intelligence factor, by L. Gottfredson. Figure 8.9 by Robert E. Gilliam. Figure 8.22*c* from Leland Hartwell et al., *Genetics: From Genes to Genomes,* © 2000 The McGraw-Hill Companies. Reprinted by permission. All Rights Reserved.

### Chapter 10

Figure 10.18 adapted from *The New England Journal of Medicine,* vol. 344, no. 20, p. 1517, figure 1 by Stanley Prusiner.

### Chapter 12

Figure 2, Reading 12.1 modified from H. Willard (1998) *Current Opin Genet Dev* 8:219–25. Figure 12.14 Data reprinted from M. Eisheikh, et al. *Annals of Medicine* 1999, 31:99–105.

### Chapter 14

Figure 1, Reading 14.1 from CDC, *Emerging Infectious Disease,* vol. 4, no. 4, October–December, 1998. Figure 14.6 HD Box (Ch. 21) from Leland Hartwell, et al., *Genetics: From Genes to Genomes,* © 2000 The McGraw-Hill Companies. Reprinted by permission. All Rights Reserved. Figure 14.9 Data from Smadar Avigad et al., A single origin of phenylketonuria in Yemenite Jews. *Nature* 344:170, March 8, 1990. Figure 14.10 from European Working Group on Cystic Fibrosis Genetics, Gradient of distribution in Europe of the major CF mutation and of its associated haplotype. *Human Genetics* 85:436–45, 1990.

### Chapter 22

Figure 22.2 HD Box (Ch. 21) from Leland Hartwell, et al., *Genetics: From Genes to Genomes,* © 2000 The McGraw-Hill Companies. Reprinted by permission. All Rights Reserved. Figure 22.9 adapted from illustration by Marlene Viola. From Ricki Lewis and Barry A. Palevitz, "Genome Economy," *The Scientist* 15(12):21, June 2001. Reprinted by permission.

## Photos

### About the Author

Photo by Barry Palevitz

### Chapter 1

Figure 1.2(4): ©Courtesy David Ward, Department of Genetics; Yale University School of Medicine; Fig. 1.2(5-6): ©The McGraw-Hill Companies, Inc./Elite Images; Fig. 1.2(7): ©Richard Laird/Getty Images; Fig. 1.3a: ©Sunstar/Photo Researchers; Fig. 1.3b: ©Lester Bergman/ProjectMasters; Fig. 1.5a: Courtesy, Thierry LaCombe & Jean Pierre Bruno, I.N.R.A., France; Fig. 1.5b: ©Alexander Lowry/Photo Researchers; Fig. 1.6a: Courtesy Ingo Potrykus, Swiss Federal Institute of Technology

### Chapter 2

Figure 2.1a: ©Muscular Dystrophy Association; Fig. 2.1b: Courtesy, Cystic Fibrosis Foundation; Fig. 2.1c: ©Ann States/Corbis SABA; Fig. 2.2: ©Manfred Kage/Peter Arnold; Fig. 2.3 (top): ©David M. Phillips/The Population Council/Science Source/Photo Researchers ; Fig. 2.3 (left): ©K.R. Porter/Photo Researchers; Fig. 2.3 (right): ©Biophoto Associates/Science Source/Photo Researchers; Fig. 2.6: ©Prof. P. Motta & T. Naguro/SPL/Photo Researchers; Fig. 2.7: ©D. Friend-D. Fawcett/Visuals Unlimited; Fig. 2.8: ©Bill Longcore/Photo Researchers; Fig. 2.10: ©K.G. Murti/Visuals Unlimited; Reading 2.3, Fig. 1b: Courtesy Dr. Brett Casey, Baylor College of Medicine; Fig. 2.11b: ©AP/Wide World Photos; Fig. 2.12a: ©CNRI/SPL/Photo Researchers; Fig. 2.13: Photo Lennart Nilsson/Albert Bonniers Forlag, A CHILD IS BORN, Dell Publishing Company.; Fig. 2.15b: ©From Dr. A.T. Sumner, "Mammalian Chromosomes from Prophase to Telophase," Chromosoma, 100:410–418, 1991. @Springer-Verlag; Fig. 2.16b-e: ©Ed Reschke; Fig. 2.18: From L. Chong, et al. 1995. "A Human Telomeric Protein." *Science,* 270:1663–1667. @1995 American Association for the Advancement of Science. Photo courtesy, Dr. Titia DeLange; Fig. 2.19: Micrograph provided by Dr. John Heuser of Washington University of Medicine, St. Louis; Fig. 2.23a: ©Barry Dowsett/SPL/Photo Researchers; Fig. 2.25: ©The Nobel Foundation, 1976.

### Chapter 3

Figure 3.8: ©Larry Johnson, Dept. of Veterinary Anatomy and Public Health, Texas A&M University. Originally appeared in Biology of Reproduction, 47:1091–1098, 1992.; Fig. 3.9b: ©David M. Phillips/Visuals Unlimited; Fig. 3.9c: ©Granger Collection; Fig. 3.10: ©R.J. Blandau; Fig. 3.13b: ©Francis LeRoy/BioCosmos/SPL/Photo Researchers; Fig. 3.14(left, right): ©Petit Format/Nestle/Science Source/Photo Researchers; Fig. 3.14(center): ©P.M. Motta & J. Van Blerkom/SPL/Photo Researchers; Fig. 3.17: ©Steve Wernka/Impact Visuals; Fig. 3.18a: ©Petit Format/Nestle/Photo Researchers; Fig. 3.18b: ©Carolina Biological Supply Company/Phototake; Fig. 3.18c: ©Donald Yaeger/Camera M.D. Studios; Fig. 3.19: ©Richard Nowitz/Phototake; Fig. 3.21b-d: From Streissguth, A.P., Landesman-Dwyer, S., Martin, J.C., & Smith, D.W. July 1980. "Teratogenic effects of alcohol in human and laboratory animals." *Science,* 209(18):353–361. @1980 American Association for the Advancement of Science.; Fig. 3.22a: Courtesy Dr. Francis Collins; Fig. 3.22b: ©Reuters/Corbis Images; Fig. 3.23: ©J.L. Bulcao/Getty Images; p.70, fig. 1: ©AP/Wide World Photos; p.71, fig. 2: Courtesy UMass Amherst News Office; p.71, fig. 3: Ricki Lewis

### Chapter 4

Figure 4.1: ©Barry King/Getty Images; Fig. 4.15: ©Nancy Hamilton/Photo Researchers; Fig. 4.16: ©Courtesy James R. Poush

### Chapter 5

Figure 5.1a: ©Porterfield-Chickering/Photo Researchers; Fig. 5.2: From Genest, Jacques, Jr., Lavoie, Marc-Andre. August 12, 1999. "Images in Clinical

Medicine." *New England Journal of Medicine,* pp 490. ©1999, Massachusetts Medical Society. All Rights Reserved.; Fig. 5.6a: ©North Wind Picture Archives; p. 101, fig. 1a-c: From G. Pierard, A. Nikkels. April 5, 2001. "A Medical Mystery." *New England Journal of Medicine,* 344: p. 1057. ©2001 Massachusetts Medical Society. All rights reserved.

**Chapter 6**
Figure 6.2: ©Biophoto Associates/Photo Researchers; Fig. 6.5: ©Ward Odenwald, National Institute of Neurological Disease and Stroke; Fig. 6.7: Courtesy, Dr. Mark A. Crowe; Fig. 6.8: ©Historical Pictures Service/Stock Montage; Fig. 6.9b: Courtesy, Richard Alan Lewis M.D., M.S., Baylor College of Medicine; Fig. 6.10a: From J.M. Cantu et al. 1984. Human Genetics, 66:66–70. ©Springer-Verlag, Gmbh & Co. KG. Photo courtesy of Pragna I. Patel, Ph.D/Baylor College of Medicine.; Fig. 6.12a-b: From Wilson and Foster. 1985. *Williams Textbook of Endocrinol,* 7/e. © W.B. Saunders; Fig. 6.12c-d: Courtesy National Jewish Hospital & Research Center; Fig. 6.13a: ©William E. Ferguson; Fig. 6.13b: ©Horst Schafer/Peter Arnold; Fig. 6.14a-d: ©Bettmann/Corbis; Fig. 6.16a: ©The McGraw-Hill Companies, Inc./photo by Carla D. Kipper; Fig. 6.16b: Courtesy Roxanne De Leon and Angelman Syndrome Foundation

**Chapter 7**
Figure 7.3a: From Albert & Blakeslee, Corn and Man, *Journal of Heredity,* 1914, Vol. 5, pg. 51. By permission of Oxford University Press.; Fig. 7.3b: ©Library of Congress; Fig. 7.6: ©AP/Wide World Photos; Fig. 7.7: ©Dr. P. Marazzi/Science Photo Library/Photo Researchers; Fig. 7.12: ©Roche Molecular Systems

**Chapter 8**
Figure 8.1: ©AP/Pressefoto Baumann/Wide World Photos; Fig. 8.2: ©AP/Wide World Photos; Fig. 8.5: ©PhotoDisc/Vol.#94; p. 167, fig. 1: ©Redneck/Getty Images

**Chapter 9**
Figure 9.1: ©Dr. Gopal Murti/Photo Researchers; Fig. 9.5a: ©Science Source/Photo Researchers; Fig. 9.5b: ©Bettmann/Corbis; Fig. 9.9c: ©1948 M.C. Escher Foundation/Baarn-Holland, All Rights Reserved; Fig. 9.11 (top): ©1979 Olins and Olins/BPS; Fig. 9.11 (bottom): ©Science VU/Visuals Unlimited; p. 181, fig. 1: ©Stock Montage

**Chapter 10**
Figure 10.5c: ©Tripos Associates/Peter Arnold; Fig. 10.13: Courtesy, Alexander Rich; Fig. 10.16b: ©Kiseleva-Fawcett/Visuals Unlimited

**Chapter 11**
Figure 11.1: ©Susan McCartney/Photo Researchers; Fig. 11.2a-b: ©Bill Longcore/Photo Researchers; Fig. 11.5: ©SPL/Photo Researchers; p. 224, fig. 1a: ©David M. Phillips/Visuals Unlimited; p. 224, fig. 1b: From R. Simensen, R. Curtis Rogers, "Fragile X Syndrome," *American Family Physician,* 39:186 May 1989. @ American Academy of Family Physicians; Fig. 11.15a: Courtesy Dr. Alan Lehmann and Dr. David Atherton; Fig. 11.15b: ©Kenneth Greer/Visuals Unlimited; p. 230, fig. 1: Courtesy, Brush Wellman; p. 230, fig. 1 (inset): Reproduced with permission from Dr. Milton D. Rossman, Hospital of the University of Pennsylvania, Philadelphia, PA

**Chapter 12**
Figure 12.1a: Courtesy, Colleen Weisz; Fig. 12.1b: Courtesy Genzyme Corporationq; Fig. 12.2: ©Science VU/Visuals Unlimited; Reading 12.1, fig. 1: Courtesy of Dr. H.F. Willard, Case Western Reserve University; Fig. 12.6d: Courtesy Genzyme Corporation; Fig. 12.7: Courtesy of Jason C. Birnholz, M.D.; Fig. 12.8b: ©Courtesy David Ward, Department of Genetics; Yale University School of Medicine; Fig. 12.9: ©Courtesy Genzyme Corporation; Fig. 12.11: Courtesy Dr. Frederick Elder, Dept. of Pediatrics, University of Texas Medical School, Houston.; p. 252: : Photo taken at the 1999 National Klinefelter Syndrome Conference. Courtesy, Stefan Schwarz; p. 253, fig. 1: Photo courtesy of Kathy Naylor; Fig. 12.16b: Courtesy of Donna Bennett/IDEAS; Fig. 12.18b: Courtesy Lawrence Livermore National Laboratory; Fig. 12.19(all): From N.B. Spinner, et al. 1994. American Journal of Human Genetics, 55:239, fig. 1, published by the University of Chicago Press, 1994 by The American Society of Human Genetics. All rights reserved.

**Chapter 13**
Figure 13.2: ©Denise Grady/NYT Pictures; Fig. 13.3a: ©Paul A. Souders/Corbis Images; Fig. 13.7: Courtesy Wendy Josephs Colman

**Chapter 14**
Figure 14.2: Dr. Victor McKusick/Johns Hopkins University School of Medicine; Fig. 14.4: ©Bettmann/Corbis; Reading 14.2, fig. 1: ©Scott Camazine/Photo Researchers; Reading 14.2, fig. 2: ©Barb Zurawski

**Chapter 15**
Figure 15.3a: ©John Reader/SPL/Photo Researchers; Fig. 15.3b: Michael Hagelberg/Arizona State University Research Publications; Fig. 15.4: ©Volker Steger/Nordstar-4 Million Years of Man/SPL/Photo Researchers; Fig. 15.5a: ©G. Hinter Leitner/Getty Images; Fig. 15.5b: ©Burt Silverman/Silverman Studios; Fig. 15.6a-b: Courtesy, Dr. H. Hameister; Fig. 15.6c: Ricki Lewis; p. 309, fig. 2 (top left): ©John Glustina/GIUST/Bruce Coleman; p. 309, fig. 2 (top right): ©Roland Seitre/Peter Arnold; p. 309, fig. 2 (bottom left): ©G. C. Kelley /Photo Researchers; p. 309, fig. 2 (bottom right): ©Tom Ulrich/Visuals Unlimited; Fig. 15.8a-b: From F.R. Goodman and P.J. Scambler. Human Hox Gene Mutations, Clinical Genetics, Jan. 2001., page 2, Fig.s A and E.; Fig. 15.10a, c: Courtesy, James H. Asher, Jr.; Fig. 15.10b: ©Vickie Jackson; Fig. 15.11(left): ©Transparency # 320496, Courtesy Department of Library Services, American Museum of Natural History; Fig. 15.11(right): ©Zoological Society of London; P. 318, fig. 1: Courtesy of Marie Deatherage

**Chapter 16**
Figure 16.1: Courtesy, The Hancock Family; Fig. 16.2(both): ©Martin Rotker/Phototake; Fig. 16.4: ©Manfred Kage/Peter Arnold; Fig. 16.8: ©Biology Media/Photo Researchers; Fig. 16.13b: ©Dr. A. Liepins/SPL/Photo Researchers; Fig. 16.16: ©Courtesy, Dr. Maureen Mayes; Fig. 16.17(top): ©David Scharf/Peter Arnold; Fig. 16.17 (bottom): ©Phil Harrington/Peter Arnold; Fig. 16.18: ©Science VU/Visuals Unlimited; p. 343, fig. 1: Courtesy, Dr. Toby C. Rodman; Fig. 16.22: ©Science VU/Visuals Unlimited; Fig. 16.22 inset: ©Hans Gelderblom/Visuals Unlimited

**Chapter 17**
Figure 17.3: ©Nancy Kedersha/Immunogen/Photo Researchers; Fig. 17.6: From P.C. Nowell and P.A. Hungerford. 1960. "Chromosome studies normal and leukemic leukocytes." *Journal of the National Cancer Institute,* 25:1960, pp.85-109.; Reading 17.1, fig. 1: : ©Custom Medical Stock Photo; p. 361: ©Ron Bennett/Bettmann/Corbis; Fig. 17.8(both): Courtesy, Dr. Tom Mikkelsen; Fig. 17.9(all): From B. Vogelstein. Sept. 1990. "The Genes That Contribute To Cancer." *Journal of NIH Research,* 2(8):66. Reprinted with permission from Medical Economics Co., Montvale, N.J.; Reading 17.2, fig. 1(top): ©PhotoDisc/Vol.#30; Reading 17.2, fig. 1(bottom): ©PhotoDisc/Vol.#19

**Chapter 18**
Figure 18.1: Photo by Barry Paleritz; Fig. 18.2a-b: From D.W. Ow, V. Wood, M. Deluca, J.R. Dewet, D.R. Helsinki, S.H. Howell. November 1986. "Transient and Stable Expression of the Firefly Luciferase Gene in Plant Cells and Transgenic Plants." *Science,* 234, ©1986 American Association for the Advancement of Science; Fig. 18.4: ©SPL/Photo Researchers; Fig. 18.7: Courtesy of Genencor International, Inc.; Fig. 18.10: Courtesy, Calgene Fresh; Fig. 18.12(top left, top right): ©Corbis/R-F Website; Fig. 18.12(top center): ©Norbert Wu/Peter Arnold; Fig. 18.12(bottom): ©PhotoDisc/Vol.#44; Fig. 18.14: From Jacks, Tyler et al. July 1994. Nature Genetics, 7:357, fig. 6; p. 384, fig. 1: ©David Scharf

**Chapter 19**
Figure 19.1a: Reprinted with permission from The Courier-Journal.; Fig. 19.2a: ©Courtesy Paul and Migdalia Gelsinger, Photo: Arizona Daily Star; Fig. 19.3a: Courtesy Ilyce Randell ; Fig. 19.8: ©Ann States/Corbis Saba; Fig. 19.10: Courtesy of Genzyme Genetics

**Chapter 20**
Figure 20.1: ©The McGraw-Hill Companies, Inc./Bob Coyle; Fig. 20.2: ©Erich Lessing/Art Resource, NY; Fig. 20.4: Courtesy, Monsanto; Fig. 20.6b: Courtesy of Calgene Fresh, Inc.; Fig. 20.8: ©Michael Greenlar/The Image Works; p. 421, fig. 1: Ricki Lewis

**Chapter 21**
Figure 21.1: ©Keri Pickett/Timepix; Fig. 21.2a: ©Bob Schuchman/Phototake; Fig. 21.2b: ©Tony Brain/SPL/Photo Researchers; Fig. 21.4: ©CNRI/Phototake; Fig. 21.5: ©Steve Goldstein; Fig. 21.6c: Integra. Photo courtesy of Ronald Carson, The Reproductive Science Center of Boston; Fig. 21.7: Courtesy, Dr. Anver Kuliev

# Index

Page numbers followed by an "*f*" indicate figures; numbers followed by a "*t*" indicate tables.

## A

# B

# C

# D

# E

# F



# G

# H

## I

## K

# M

# N

# P